To Chakravarty,

my friend.

[signature]

4/15/93

Graph Theory, Combinatorics, and Applications

Graph Theory, Combinatorics, and Applications

Volume 1

PROCEEDINGS OF THE SIXTH QUADRENNIAL
INTERNATIONAL CONFERENCE ON THE THEORY
AND APPLICATIONS OF GRAPHS

Western Michigan University

Edited by

Y. Alavi
G. Chartrand
O.R. Oellermann
A.J. Schwenk

A Wiley-Interscience Publication
JOHN WILEY & SONS, INC.
New York / Chichester / Brisbane / Toronto / Singapore

Library of Congress Cataloging in Publication Data:

International Conference on the Theory and Applications of Graphs (6th
: 1988 : Western Michigan University)

Graph theory, combinatorics, and applications : proceedings of the
Sixth Quadrennial International Conference on the Theory and
Applications of Graphs / edited by Y. Alavi . . . [et al.].
    p.   cm.
"A Wiley-Interscience publication."
Includes bibliographical references.

ISBN 0-471-53245-2 (set). —ISBN 0-471-60917-X (v. 1)
                —ISBN 0-471-53219 (v. 2)
1. Graph theory—Congresses. 2. Combinatorial analysis—
Congresses. I. Alavi, Y. II. Title.
QA166.I55   1988
511'.5—dc20   90-28082
                                                          CIP

Printed in the United States of America

10 9 8 7 6 5 4 3 2 1

The Sixth Conference and these Proceedings are dedicated to
Lowell W. Beineke
and
Carsten Thomassen
and the editors herewith recognize and laud their outstanding contributions to
Graph Theory and its promotion around the world.

Preface

These volumes constitute the Proceedings of the Sixth Quadrennial International Conference on the Theory and Applications of Graphs, held at Western Michigan University in Kalamazoo, Michigan, 30 May–3 June 1988. Conference participants included research mathematicians from colleges, universities, and industry, as well as graduate and undergraduate students. Altogether 25 states and 18 countries were represented. The contributions to these volumes include many topics in current research in both the theory and applications in the areas of graph theory and combinatorics.

Contents

Volume 1

Acknowledgments   xiii

Y. Alavi, D.R. Lick, and H.B. Zou*
Second Order Degree Regular Graphs   1

T. Asano
Difficulty of the Maximum Independent Set Problem on Intersection Graphs of Geometric Objects   9

K.S. Bagga and F.W. Owens*
Some Graph Theoretic Properties of Deadlocks and Traps in Petri Nets   19

*K.S. Bagga, L.W. Beineke, M.J. Lipman, and R.E. Pippert**
The Concept of Leverage in Network Vulnerability   29

K.S. Bagga, L.W. Beineke, M.J. Lipman, and R.E. Pippert*
Extensions of an Algorithm for Computing the Edge-Integrity of Trees   41

*H.-J. Bandelt, K.S. Bagga, and H.M. Mulder**
Cartesian Factorization of Interval–Regular Graphs Having no Long Isometric Odd Cycles   55

J. Bang-Jensen
A Note on a Special Case of the 2-Path Problem for Semicomplete Digraphs   77

C.A. Barefoot, L.H. Clark, J. Douthett, R.C. Entringer, and M.R. Fellows*
Cycles of Length 0 Modulo 3 in Graphs   87

L.I. Basenpiler
Contractions of OP Graphs   103

D. Bauer, E. Schmeichel, and H.J. Veldman*
Some Recent Results on Long Cycles in Tough Graphs   113

M. Behzad and E.S. Mahmoodian
Graphs Versus Designs—A Quasisurvey   125

---

* Indicates the speaker at the Conference.

L.W. Beineke, K.S. Bagga, M.J. Lipman, and R.E. Pippert*
Explorations into Graph Vulnerability   143

*F. Boesch, X. Li, C. Stivaros, and C. Suffel**
Reliable Graphs with Unreliable Nodes   159

*I. Broere and M. Frick**
**A Characterization of the Sequence of Generalized Chromatic
Numbers of a Graph**   179

S.A. Burr
On the Automorphism Group of a Power of a Hypercube   187

S.A. Burr, P. Erdös, P. Frankl, R.L. Graham, and V.T. Sós*
Further Results on Maximal Anti-Ramsey Graphs   193

P.A. Catlin and H.-J. Lai*
Spanning Trails Joining Two Given Edges   207

J. Chen, J. Wang, and Y. Li
The Isomorphic Factorizations of a Class of 2-Circulant Graphs   233

A.G. Chetwynd
Total Colourings of Graphs—A Progress Report   237

H.-A. Choi, A.-H. Esfahanian, and B.C. Houck*
**Optimal Communication Trees with Application to Hypercube
Multicomputers**   245

F.R.K. Chung
Improved Separators for Planar Graphs   265

D.G. Collins
Canonical Forms for the Digraph Isomorphism Problem   271

A.M. Dean and J.P. Hutchinson*
Relations among Embedding Parameters for Graphs   287

A.M. Dean, C.J. Knickerbocker, P. Frazer Lock, and M. Sheard*
A Survey of Graphs Hamitonian-Connected from a Vertex   297

N. Dean
Which Graphs Are Pancyclic Modulo k?   315

I.J. Dejter and J. Quintana*
On an Extension of a Conjecture of I. Hável   327

*L.L. Deneen, G.M. Shute, and C.D. Thomborson**
A Dualizable Representation for Imbedded Graphs   343

*T. Deretsky, S.-M. Lee, and J.A. Mitchem**
On Vertex Prime Labelings of Graphs   359

*G.S. Domke, G. Fricke, S.T. Hedetniemi, and R.C. Laskar**
Relationships Between Integer and Fractional Parameters of Graphs   371

H. Enomoto and Y. Usami*
The Star Arboricity of Complete Bipartite Graphs   389

P. Erdös
Problems and Results in Combinatorial Analysis and Combinatorial Number Theory   397

*P. Erdös, R.J. Faudree, A. Gyárfás, and R.H. Schelp**
Odd Cycles in Graphs of Given Minimum Degree   407

P. Erdös, R.J. Faudree, C.C. Rousseau, and R.H. Schelp*
Edge Conditions for the Existence of Minimal Degree Subgraphs   419

*P. Erdös and J.G. Gimbel**
A Note on the Largest H-Free Subgraph in a Random Graph   435

P. Erdös, M.S. Jacobson, and J. Lehel*
Graphs Realizing the Same Degree Sequences and their Respective Clique Numbers   439

M. Escudero, J. Fábrega, M.A. Fiol, and N. Homobono*
On Surviving Route Graphs of Iterated Line Digraphs   451

R. Faudree, R.J. Gould, and T. Lindquester,*
Hamiltonian Properties and Adjacency Conditions in K(1,3)-Free Graphs   467

J. Feigenbaum
Lexicographically Factorable Extensions of Irreducible Graphs   481

M.A. Fiol
A Boolean Algebra Approach to the Construction of Snarks   493

*W.D. Goddard and O.R. Oellermann**
On the Cycle Structure of Multipartite Tournaments   525

*W.D. Goodard and H.C. Swart**
On Some Extremal Problems in Connectivity   535

R.J. Gould and V. Rödl*
Bounds on the Number of Isolated Vertices in Sum Graphs   553

D.L. Grinstead and P.J. Slater*
On the Minimum Intersection of Minimum Dominating Sets in Series-Parallel Graphs   563

Acknowledgments

The editors take pleasure in thanking the many people who contributed to the success of the Sixth Conference as well as the preparation of these Proceedings. In addition to the speakers, participants, and the contributors to these volumes, we gratefully acknowledge

The outstanding support of Western Michigan University,
Dr. Diether H. Haenicke, President
Dr. George M. Dennison, Provost

The excellent and overall support of the College of Arts and Sciences,
Dr. A. Bruce Clarke, Dean

The fine support of the Division of Research and Sponsored Programs,
Dr. Donald E. Thompson, Director and Chief Research Officer

The enthusiastic and overall support of the Department of Mathematics and Statistics,
Dr. Joseph T. Buckley, Chair and Conference Co-Director

The financial support of the Army Research Office (Durham), Air Force Office of Scientific Research, and *principally* the Office of Naval Research

The financial support of the Upjohn Company, in particular, the Computational Chemistry Unit

The very special assistance of a very special friend and Co-Director,
Doctor Ronald L. Graham

The fine assistance of our esteemed Graph Theory colleagues,
Professors S.F. Kapoor and Arthur T. White

The extraordinary work of our graph theory colleagues from around the world who in timely fashion, assisted with the refereeing of the manuscripts

The valuable and outstanding help of our Conference Assistant Directors,
Terry A. McKee, Farrokh Saba, and Curtiss E. Wall

The special administrative and overall assistance of Darlene Lard, and the general secretarial assistance and preparation of the book of abstracts by Margo Chapman and Lisa Morrison

The excellent and extensive work of Michelle Schultz, Junior Mathematics major, as Conference Secretary covering a span of two years

The dedicated work of our Conference Assistants: Ghidewon Abay Asmerom, Héctor Hevia, Sue Hollar, Ewa Kubicka, Grzegorz Kubicki, Jiuquang Liu, Paresh J. Malde, Bruce Mull, Jamal Nouh, Stavros Pouloukas, Reza Rashidi, Songlin Tian, Ignatios Vakalis

The superb work of our fine students Lisa Hansen, Karen Holbert, Cheryl Joan, Heather Jordon, Kris Parteka, Michelle Schultz, and our outstanding secretary Lisa Morrison for their technical typing.

Finally, we wish to thank Maria Taylor, Editor, and Rose Leo, Editorial Assistant, and Rose Ann Campise, Associate Managing Editor, Wiley Interscience, for their outstanding assistance with these Proceedings.

The editors apologize for any oversight in the acknowledgments or any errors in the manuscript.

Y.A.
G.C.
O.O.
A.S.

Second Order Degree Regular Graphs

Yousef Alavi
Western Michigan University

Don R. Lick
Eastern Michigan University

Hung Bin Zou
Intermagnetics General Corporation

Let G be a connected graph of order p. Let $d(u, v)$ denote the distance between the vertices u and v of G. For any integer k with $1 \leq k \leq p - 1$, we use the notation

$$N_k(v) = \{u \mid u \in V(G) \text{ and } d(u, v) = k\}$$

and

$$N_k[v] = N_k(v) \cup \{v\}.$$

Example

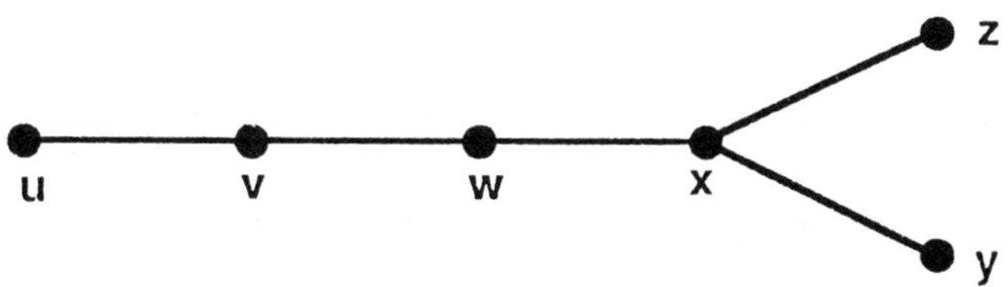

$$N_2(u) = \{w\}, \quad N_3(v) = \{y, z\} = N_4(u), \quad N_2(z) = \{w, y\}$$

The *kth order degree of a vertex* v, denoted by $\deg_k v$, is the cardinality of the set $N_k(v)$. In the above example, $\deg_4 u = 2$, $\deg_5 u = 0$, and $\deg_3 z = 1$.

It follows that for any connected graph G that

$$\sum_{k=1}^{p-1} \deg_k v = p - 1,$$

and

$$\sum_{k=1}^{i} \deg_k v \leq p - 1,$$

where $1 < i < p - 1$. We can now make the following observation. If G is a connected graph of order $p \geq 2$ and k is an integer with $1 \leq k \leq p - 1$, then $\deg_k v \leq p - k$ for every $v \in V(G)$.

It is well–known that for any graph G, the sum of the degrees of the vertices of G is even. We now show an analogous result for the kth order degree of a graph. We use the following notation throughout the remainder of this paper.

$$D_k = D_k(G) = \sum_{v \in V(G)} \deg_k v.$$

Example   (a)   From the example above, $D_2(G) = 10$, $D_3(G) = 6$, $D_4(G) = 4$, and $D_5(G) = 0$.

(b) $D_k(K_p) = 0$ for $k > 1$.

(c) $D_2(K(1, p - 1)) = (p - 1)(p - 2)$ and $D_k(K(1, p - 1)) = 0$ for $k > 2$.

Proposition 1   *Let k and p be integers with $1 \leq k \leq p - 1$. If G is a graph of order p, then $D_k(G)$ is even.*

Proof   Let k and p be integers with $1 \leq k \leq p - 1$ and let G be a graph of order p. For each pair of vertices u and v with $d(u, v) = k$, let P_{uv} be a path of length k joining u and v. Then in determining $D_k(G)$, P_{uv} is used twice, once for $\deg_k u$ and once for $\deg_k v$, and adds two the sum. Thus $D_k(G)$ is even. $\square$

Corollary 1   *In any graph there is an even number of vertices whose kth order degrees are odd.*

We can now look at the structure of some graphs with special kth order degrees.

Proposition 2   *Let k and p be positive integers with $p \geq k + 1$. Let G be a graph of order p. If there is a vertex v with $\deg_k v = p - k$, then there exists $k - 1$ vertices $u_1, u_2, ..., u_{k-1}$ of G with $\deg_k u_i = 0$ for $1 \leq i \leq k - 1$ and the subgraph S of G induced by the vertex set $V(G) - \{v, u_1, u_2, ..., u_{k-2}\}$ contains a star isomorphic to $K(1, p - k)$ with the center of S at u_{k-1}.*

Proof   Let $N_k(v) = \{x_1, x_2, ..., x_{p-k}\}$ and let $v, u_1, u_2, ..., u_{k-1}, x_1$ be a path of length k in G. Since $d(v, x_1) = k$, there is no shorter $v - x_1$ path in G. Then $V(G) = \{v, u_1, u_2, ..., u_{k-1}, x_1, x_2, ..., x_{p-k}\}$. Since $d(v, u_i) = i$, for $1 \leq i \leq k - 1$, and $d(v, x_j) = k$, for $1 \leq j \leq p - k$, the graph induced by $\{v, u_1, u_2, ..., u_{k-1}\}$ is a path of length $k - 1$ and the graph induced by $\{u_{k-1}, x_1, x_2, ..., x_{p-k}\}$ contains a subgraph S isomorphic to the star $K(1, p - k)$ and S has center u_{k-1}. Thus $\deg_k u_i = 0$ for $1 \le i \le k - 1$. $\square$

Remark Under the hypothesis of Proposition 2, G is isomorphic to a subgraph of the graph H of order p where H is composed of a complete graph of order $p - k + 1$ with a path of length $k - 1$ attached to one of the vertices.

Corollary 2a *Let G be a connected graph of order p. Then $D_2(G) \le (p - 1)(p - 2)$ and this inequality is the best possible.*

Proof Let q denote the size of G. Since, for every vertex v of G, $\deg_2 v = |N_2(v)| \le (p - 1) - \deg v$, $D_2(G) \le p(p - 1) - 2q \le p(p - 1) - 2(p - 1) = (p - 1)(p - 2)$. This is the best possible since $D_2(K(1, p{-}1)) = (p - 1)(p - 2)$. $\square$

Corollary 2b *Let G be a connected graph order $p \ge 3$. Then $D_2(G) = (p - 1)(p - 2)$ if and only if G is isomorphic to the star $K(1, p - 1)$.*

Proof It is clear that $D_2(K(1, p - 1)) = (p - 1)(p - 2)$. So we assume that G is a connected graph of order $p \ge 3$ with $D_2(G) = (p - 1)(p - 2)$. For every vertex v of G, $\deg v + \deg_2 v \le p - 1$, and so $D_1(G) + D_2(G) \le p(p - 1)$. Thus $D_1 \le p(p - 1) - D_2(G) = p(p - 1) - (p - 1)(p - 2) = 2(p - 1)$, but $2q = D_1(G)$, where q denotes the size of G. We have that $q = p - 1$, and so G is a tree. If $\deg_2 v \le p - 3$ for every vertex v of G, then $D_2(G) \le p(p - 3) = p^2 - 3p < p^2 - 3p + 2 = (p - 1)(p - 2)$. So at least one vertex of G, say v, has $\deg_2 v = p - 2$. But then there is a vertex u with $\deg_2 u = 0$. Thus the other $p - 2$ vertices also must have second order degrees $p - 2$. Hence G is isomorphic to $K(1, p - 1)$. $\square$

The graph G is said to be *kth order regular of degree d* if for every vertex v of G, $\deg_k v = d$. Then, being first order regular of degree d is equivalent to being regular of degree d. Before giving some examples of kth order regular graphs of degree d, we provide some notation and a definition.

(a) Let $K_{n(m)}$ denote the complete n–partitie graph with each partitie set containing exactly m vertices.

(b) Let G_1 and G_2 be two non–empty graphs. The *cartesian product* $G = G_1 \times G_2$ of G_1 and G_2 has $V(G) = V(G_1) \times V(G_2)$, and two vertices (u_1, u_2) and (v_1, v_2) are adjacent if and only if either

$$u_1 = v_1 \text{ and } u_2v_2 \in E(G_2)$$

or

$$u_2 = v_2 \text{ and } u_1v_1 \in E(G_1).$$

For $n \ge 3$, the graphs $C_n \times K_2$ are called *double cycles*.

(c) Let G_1 and G_2 be two non–empty graphs. The *lexicographic product* $G = G_1[G_2]$ of G_1 and G_2 has $V(G) = V(G_1) \times V(G_2)$, and two vertices (u_1, u_2) and (v_1, v_2) are adjacent if and only if $u_1v_1 \in E(G_1)$ or $u_1 = v_1$ and $u_2v_2 \in E(G_2)$. The *generalized cycle* $C(m, n)$, for integers $m \geq 3$ and $n \geq 1$, is the lexicographic product $C_m[\overline{K}_n]$.

Example

(a) The complete graphs K_n are second order regular of degree 0. In fact these are the only connected graphs that are second order regular of degree 0.

(b) The path P_4 of length 3 and the four cycle C_4 are second order regular of degree 1. The graphs $K_{n(2)}$ are second order regular of degree 1. More generally, the path P_{2k} of length $2k - 1$ and the 2k cycle C_{2k} are kth order regular of degree 1.

(c) The cycles C_n, $n \geq 5$, and the double star $S(2, 2)$ are second order regular of degree 2. The double cycle $C_3 \times K_2$ is second order regular of degree 2.

(d) The double cycle $C_4 \times K_2$ is second order regular of degree 3.

(e) The double cycles $C_n \times K_2$, $n \geq 5$, are second order regular of degree 4.

(f) Let m be any positive integer. Then $K_{n(m+1)}$ is second order regular of degree m and the double star $S(m, m)$ is second order regular of degree m.

(g) For $n \geq 2$, the complete bipartite graph $K(n + 1, n + 1)$ with a one factor removed is third order regular of degree one and regular of degree n.

(h) For the n–cube Q_n we have for any k, $1 \leq k \leq n$, that Q_n is kth order regular of degree $\binom{n}{k}$, the combinatorial coefficient.

Proposition 2 provides the following.

Corollary 2c *Let k and p be integers with $p \geq k + 1 \geq 3$. There exist no connected graphs G of order p that are kth order regular of degree $p - k$.*

Proposition 3 *Let k and d be positive integers. There exists graph $G(k, d)$ which is kth order regular of degree d.*

Proof 1. If $k = 1$, then $G(1, d)$ can be taken to be any regular graph of degree d.

2. If $k = 2$ and n is an integer at least 2, then $K_{n(d+1)}$ is second order regular of degree d.

3. Assume $k \geq 3$. The generalized cycle $C(2k, d)$ is a kth order regular graph of degree d. $\square$

We now look at the special case of second order regular graphs.

Proposition 4 *Let G be a connected graph of order $p \geq 4$. Then G is second order regular of degree $p - 3$ if and only if G is isomorphic to P_4, C_4, or C_5.*

Proof If G is isomorphic to P_4, C_4, or C_5, then G is second order regular of degree $p - 3$. so we assume that G is second order regular of degree $p - 3$. Then $\deg v + \deg_2 v \leq p - 1$ and so $\deg v \leq (p - 1) - \deg_2 v = (p - 1) - (p - 3) = 2$. Thus, since G is connected, $1 \leq \deg v \leq 2$. Hence G is a path or G is a cycle. If G is a path, then G is isomorphic to P_4 and if G is a cycle, then G is isomorphic to C_4 or C_5. ❑

Theorem 1 *The connected graph G is second order regular of degree 1 if and only if G is either a path of length 3 or G is $K_{n(2)}$.*

Proof Suppose that G is second order regular of degree 1 and that G is not a path of length 3. For each vertex v of G there is exactly one vertex v' of G whose distance from v is 2. Since the vertices of G can be paired this way, the order of G must be even, say $|V(G)| = 2n$. Suppose there is a vertex w of G with $d(v, w) = 3$. The path joining v and w must be of the form $vw'v'w$, where w' is the unique vertex of G whose distance from w is 2. If $V(G) = \{v, v', w, w'\}$, then, necessarily, G is a path of length 3, contradicting our choice of G. Hence, since the order of G is even, G has order at least six. Since G is connected, there is a vertex u in $G - \{v, v', w, w'\}$ which is adjacent to one of the vertices v, v', w, w'. If u is adjacent to v (or w), then u must also be adjacent to w', for otherwise, both u and w have distance 2 from w'. Now u must be adjacent to v', for otherwise, both u and v have distance 2 from v'. Also, u must be adjacent to w, for otherwise, both u and w' have distance 2 from w. Thus u is adjacent to each of the vertices v, v', w, w', and $d(v, w) \leq 2$. This contradicts the assumption that $d(v, w) = 3$. On the other hand, if u is adjacent to w' (or v'), then u must be adjacent to v, for otherwise, both u and v' has distance 2 from v. Thus u and v are adjacent and we have the previous case.

The above consideration shows that the diameter of G is 2. Let v be any vertex of G and let z be any vertex of G other than v and v'. Since the distance between v and z is at most 2 and cannot be 2, since $d(v, v') = 2$, v and z must be adjacent. That is, v is adjacent to every vertex of G except v'. Since v was an arbitrary vertex of G, G must be the complete graph K_{2n} with a one factor removed and so G is $K_{n(2)}$. ❑

This result indicates that if G is second order regular of degree one and G is not the path of length three, then G is $K_{n(2)}$. In all of the examples above, except for the paths, the graphs which are kth order regular are also regular. One might be led to the conclusion that if a graph is kth order regular and not a path, then it is regular. Figure 2 provides an example of a graph which is third order regular of degree one which is not regular. The graph in Figure 2 is an example of a third order regular graph of degree one which is not $K(n, n)$ with a one factor removed.

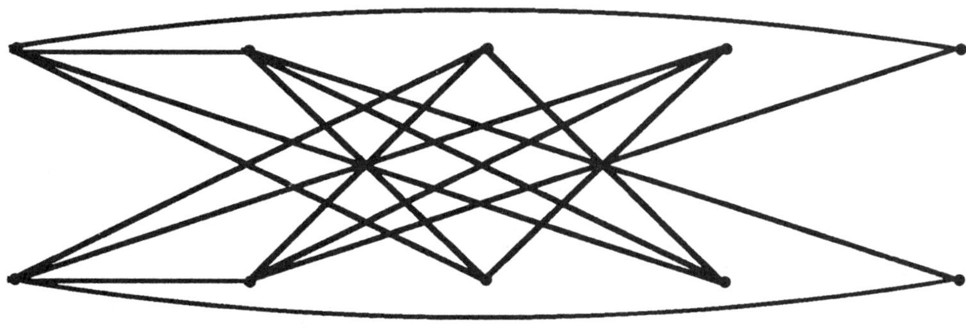

Figure 2

Theorem 2 *If G is a connected third order regular graph of degree 1, then G is either a path of length 5 or G has diameter 3.*

Proof Suppose that G is a connected third order regular graph of degree 1 and that G is not a path of length 5. Necessarily, the diameter of G is at least 3. First we show that the diameter is at most 4. If this is not the case, then there are vertices v and w of G with $d(v, w) = 5$; let $P : v, x, w', v', x', w$ be a shortest v–w path in G. Since G is connected and third order regular of degree 1, it follows that every vertex in $V(G) - V(P)$ is at distance at most 2 from some vertex in $V(P)$, i.e., $d(u, V(P)) \leq 2$, for all vertices $u \in V(G) - V(P)$.

If there is a vertex u belonging to $V(G) - V(P)$ with $d(u, V(P)) = 2$, then let u, u_1, u_2 be a shortest path from u to a vertex in $V(P)$. Thus u is not adjacent to any vertex of $V(P)$, since otherwise there exists neighboring vertices y_1 and y_2 in $V(P)$ such that $d(u, y_1) = 2$, $d(u, y_2) > 2$, and hence $\deg_3 y_2 > 1$, contradicting the hypothesis. This implies, however, that $d(v, w) \leq d(v, u) + d(u, w) \leq 4$, contrary to our assumption. Hence every vertex in $V(G) - V(P)$ is adjacent to some vertex in $V(P)$.

Assume that $u \in V(G) - V(P)$ and $uw' \in E(G)$. Since $\deg_3 v = 1$, it follows that $d(u, v) \leq 2$. Hence $d(u, w) \geq 3$. However, $\deg_3 w = 1$ and so $d(u, w) \geq 4$. Hence

$d(u, w) = 4$ $(w, x', v', w', u$ is a u–w path of length 4) and there is no u–x' path in G of length at most 2. This implies that both x and u have distance 3 from x', contradicting the hypothesis that $\deg_3 x' = 1$. Hence there can be no vertex of $V(G) - V(P)$ adjacent to w' and so $\deg w' = 2$ in G. In a similar manner, it follows that $\deg v' = 2$ in G.

Suppose that $u \in V(G) - V(P)$ and $ux \in E(G)$. Since $\deg_3 v' = 1$, we have that $d(u, v') \leq 2$. Since $\deg w' = 2 = \deg v'$, we must have that $ux' \in E(G)$. (The only u–v' path of length 2 is u, x', v'.) Then w, x', u, x, v is a v–w path of length 4 in G, which produces a contradiction. Hence $\deg x = 2$ and in a similar manner it follows that $\deg x' = 2$.

If $u \in v(G) - V(P)$ and $uv \in E(G)$, since $\deg_3 w' = 1$, $d(u, w') \leq 2$. This produces a contradiction since $\deg x = \deg w' = \deg v' = 2$. Hence $\deg v = 1$, and similarly, $\deg w = 1$. Thus G is a path of length 5 which is contrary to the hypothesis. We conclude that G has diameter at most 4.

We next show that the diameter of G is at most 3. If the diameter of G is 4, then there are vertices v and w of G such that $d(v, w) = 4$; let $P : v, w', x, v', w$ be a shortest v–w path in G. Let x' be the unique vertex of G whose distance from x is 3 and let $Q : x, x_1, x_2, x'$ be a shortest x–x' path in G. If $x_2 = v$, then, since the diameter of G is 4, it follows that x', v, w', x, v', w is not a shortest x'–w path of G. Let $Q' : x', z_1, z_2, ..., z_r = w$ be a shortest x'–w path in G. Since $d(v, w) = 4$, $r \geq 3$. However, since $\deg_3 x' = 1$, $z_3 = x$ and $r = 5$ (since $d(x, w) = 2$), contrary to the fact that r is at most the diamter of G. Hence $x_2 \neq v$, and in a similar manner, $x_2 \neq w$.

If $x_1 = w'$, since the diameter of G is 4, it follows that x', x_2, w', x, v', w is not a shortest x'–w path in G. Let $Q' : x', z_1, z_2, ..., z_r = w$ be a shortest x'–w path in G. If $r \geq 3$, then, since $\deg_3 x' = 1$, $z_3 = x$ and so $r = 5$, producing a contradiction. Thus $d(x', w) \leq 2$. However, since $\deg_3 x' = 1$, and x', x_2, w', v is an x'–v path of length 3, it follows that $d(x', v) = 2$. Let x', u, v be a shortest x'–v path. Since $d(v, w) = 4$, any u–w path must have length at least 3. In view of the fact that w' is the unique vertex of distance 3 from w, $d(u, w) \geq 4$. However, $d(u, w) \leq d(u, x') + d(x', w) \leq 3$, producing a contradiction. Hence, $x_1 \neq w'$; and by a similar argument, $x_1 \neq v'$.

We deduce that $V(P) \cap V(Q) = \{x\}$. This implies that some vertex of $V(Q) - \{x\}$ is at a distance 3 from v or w', producing a contradiction. Hence, if G is not a path of length 5, its diameter is 3. ∎

The above results lead us to the following conjecture.

Conjecture *For $n \geq 2$, if G is a connected nth order regular graph of degree 1, then G is either a path of length $2n - 1$ or G has diameter n.*

REFERENCES

[1]   G. Chartrand and L. Lesniak, *Graphs & Digraphs*, 2nd Edition, Monterey, CA, 1986.

[2]   R. Frucht, "A one regular graph of degree three", *Canad. J. Math.*, 4 (1952), 240 – 247.

Difficulty of the Maximum
Independent Set Problem on Intersection
Graphs of Geometric Objects

Tetsuo Asano

Osaka Electro–Communication University

ABSTRACT

Given a set S of geometric objects, we can define an intersection graph associated with S in which vertices correspond to those geometric objects and there is an edge between every pair of objects intersecting each other. When a geometric property of those objects is specified, the class of intersection graphs is somewhat limited. As is well known, the maximum independent set problem is NP–complete for general graphs. The purpose of this paper is to reveal the difficulty of the problem for the various classes of intersection graphs.

1. Introduction

The maximum independent set problem (MIS problem, in short) is known to be NP–complete for general graphs, but it is solvable in polynomial time for many restricted classes of graphs [GJ79]. In this paper we investigate the difficulty of the MIS problem on various kinds of intersection graphs.

Various kinds of intersection graphs have been considered such as interval graphs, circular arc graphs, straight–line–in–the–plane graphs, and so on. This paper starts by giving a precise definition of (geometric) intersection graphs. Let S be a set of geometric objects, such as line segments, rectangles, etc. Then, we can define the intersection graph G(S) where vertices correspond to those geometric objects and there is an edge between two vertices if and only if the corresponding objects intersect. For example, when S is a set of horizontal line segments on a line, the intersection graph G(S) is known as an interval graph. On the other hand, if S is a set of horizontal line segments such that any two of them do not overlap although they may touch each other, then the intersection graph G(S) consists of elementary paths. Note that intersection graphs to be considered in this paper are different from (more exactly, are restrictions of) those defined for arbirary sets (see [Ha72]). In our definition a base set S must be a set of geometric objects.

The MIS problem has been investigated for several intersection graphs. It is solvable in polynomial time for interval graphs and circular–arc graphs (if their representations are given) and NP–complete for intersection graphs of straight lines in the plane [Ka83] and those of unit circles [WK87]. Wang and Kuo [WK87] gave a very elegant proof of the NP–completeness for the MIS problem on unit–circle intersection graphs. This paper borrows a key idea for their NP–completeness proof although their proof is the reduction from the 3–satisfiability problem while ours is that from the planar 3–satisfiability. In the following we wish to find for what geometric objects the MIS problem on a class of intersection graphs defined by them is NP–complete.

2. Reduction from the Planar 3–Satisfiability Problem

The planar 3–satisfiability problem is stated as follows:
The instance is given by a boolean expression
$$E = E_1 \wedge E_2 \wedge ... \wedge E_m$$
where each clause E_j contains at most three literals which are variables or their negations and a planar graph is defined by the vertex set $\{E_1, E_2, ..., E_m, u_1, u_2, ..., u_q\}$ and the edge set $\{(E_i, u_j) \mid E_i$ contains u_j or $\bar{u}_j\}$. Then, the problem is to decide whether there exists an assignment $A \subseteq \{u_1, \bar{u}_1, u_2, \bar{u}_2, ..., u_q, \bar{u}_q\}$ such that
$$A \cap E_j \neq \phi \quad (j = 1, 2, ..., m)$$
and
$$|A \cap \{u_i, \bar{u}_i\}| = 1 \quad (i = 1, 2, ..., q),$$
where each clause E_j is regarded as the set of literals contained.

The reduction from the planar 3–satisfiability problem to the MIS problem will be established as follows. We begin with a planar representation of the bipartite graph G which has two kinds of vertices. At every vertex we draw a circle and replace every edge with two parallel edges, as shown in Fig. 1. For each vertex corresponding to a variable we define a non–self–intersecting cycle called a "circuit" which is the boundary of a collection of edges incident to the variable (see Fig. 2 for illustration). On the other hand each vertex corresponding to a clause E_j is represented by structures called "clause configurations" shown in Fig. 3 depending on the size of E_j, which determine how the different circuits meet each other. Since a given graph is drawn in the plane, by an appropriate plane embedding we can insure that any two distinct circuits do not intersect.

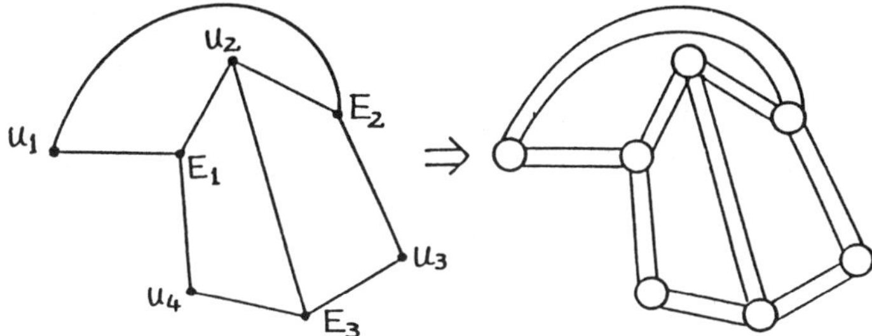

Fig. 1 The first step of transformation of a planar graph by circles and double edges.

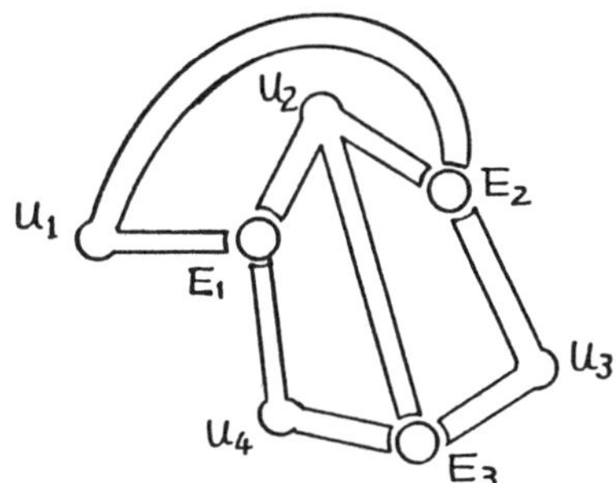

Fig.2 Cycles associated with vertices corresponding to variables.

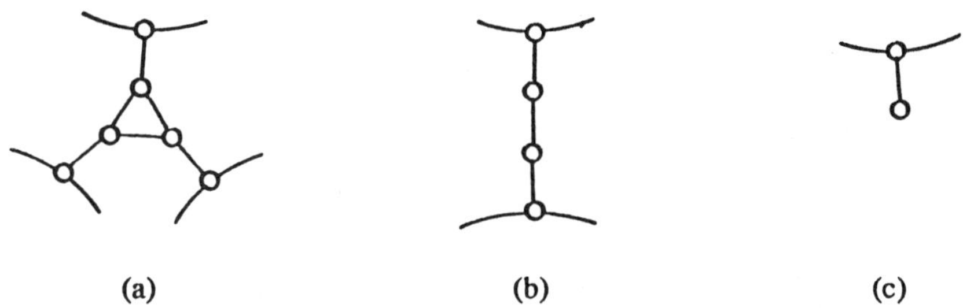

Fig. 3 Clause configurations
(a) for three–literal clause, (b) two–literal clause, and (c) one–literal clause.

We now describe the circuits and clause configurations in detail. A circuit associated with a variable u_i is a circuit consisting of an even number of vertices. A k–th vertex on the circuit is labeled as "true" or "false" depending on whether k is even or odd.

It may be easily observed that the number of vertices needed to form such a circuit is bounded by N^2, where N is the total number of variables and clauses if we imagine a embedding of a given planar graph into a grid plane, Also, it is easy to see that the coordinates of the vertices in all the circuits can be computed in polynomial time.

A clause configuration for a clause E_j is defined as follows. When a clause E_j includes three variables $u_1{}^j$, $u_2{}^j$ and $u_3{}^j$ (or their negations), we place a triangle consisting of three vertices $u_1{}^j$, $u_2{}^j$ and $u_3{}^j$ and then connect each of them with circuits corresponding to those variables. The important thing here is that if the clause E_j contains a variable $u_k{}^j$ (its negation $\bar{u}_k{}^j$, resp.) then the corresponding vertex $u_k{}^j$ is adjacent to a false (true, resp.) vertex.

When a clause consists of two literals, we remove one vertex from the corresponding triangle. The way of connection with circuits is the same as above. For a clause including only one literal u_i or $\bar{u}_i$, the clause is represented by a single vertex and connected to a false vertex or a true vertex on the circuit associated with the variable u_i, respectively.

Letting

$$p = \sum_{i=1}^{q} r_i/2 + m,$$

where r_i is the number of vertices included in the circuit corresponding to a variable u_i and m is the number of clauses, we claim that a given boolean expression E is satisfiable if and only if the constructed graph has an independent set W of size p or more.

Assume that E is satisfied by a truth assignment A. For $i = 1, 2, ..., q$, if A contains u_i, then include in W the true vertices of the circuit for u_i; if A contains $\bar{u}_i$, then include in W the false vertices of the circuit for u_i. We can add an additional vertex to W for each clause if at least one of their touching vertices is not in W. Note that if a clause contains a variable u_j (its negation $\bar{u}j$, resp.) then all true (false, resp.) vertices of the circuit for u_i are added to W and no false (true, resp.) vertices are included in W, and thus the vertex of the triangle for the clause which is adjacent to the circuit is independent from any vertex of W on the circuits associated with the clause. It is just the same for two–literal clauses and one–literal clauses.

To prove the converse, let W be an independent set of size p. Since each clause can contribute at most one vertex in W, each circuit C_i, $i = 1, 2, ..., q$, must have $r_i/2$ vertices in W. We have shown that each circuit C_i for u_i has either the true vertices or the false vertices in W. Let us include u_i or $\bar{u}_i$ in an assignment A depending on whether C_i has the true vertices or the false vertices in W. E is satisfied by this assignment since each clause configuration has contributed a vertex in W. ❑

An example of the transformation from a planar 3–satisfiability problem is illustrated in Fig. 4.

3. Difficulty of Maximum Independent Set Problem on Various Intersection Graphs

In the previous section we have established a reduction from the planar 3–satisfiability problem which is known to be NP–complete. Thus, given a class M of geometric objects (e.g., rectangles, line segments, etc.), what is required to show that the maximum independent set problem on the class of intersection graphs defined by M (referred to as MIS on M) is NP–complete is

(A) to show that the graph described in the previous section can be constructed as an intersection graph for some set of objects from M, and

(B) to guarantee that such a graph can be constructed in a polynomial time.

The main result in this paper is summarized as follows.

Theorem   *MIS on M is NP–complete if M is*

     (1)   *a class of unit squares,*

     (2)   *a class of unit circles [WK87],*

     (3)   *a class of line segments in the plane,*

     (4)   *a class of horizontal and vertical line segments any two of which may overlap, or*

     (5)   *a class of horizontal, vertical and 45–degree line segments with no two of the same type touching. (Thus, MIS problem on three–partite graphs is NP–complete.)*

Proof   The proof is somewhat straightforward. What we have to do is stated in (A) and (B) above. In order to guarantee that such a graph corresponding to a given instance of planar 3–satisfiability problem can be constructed as an intersection graph of a specified geometric object in polynomial time, we embed a planar graph into a grid of size n × n. First, since a vertex corresponding to a variable may have degree greater than 3, we replace such a vertex with vertices of degree at most 3 so that a subgraph defined by the vertex itself and a set of vertices adjacent to the vertex remains a tree. Note that the degree of a vertex corresponding to a clause is at most 3.

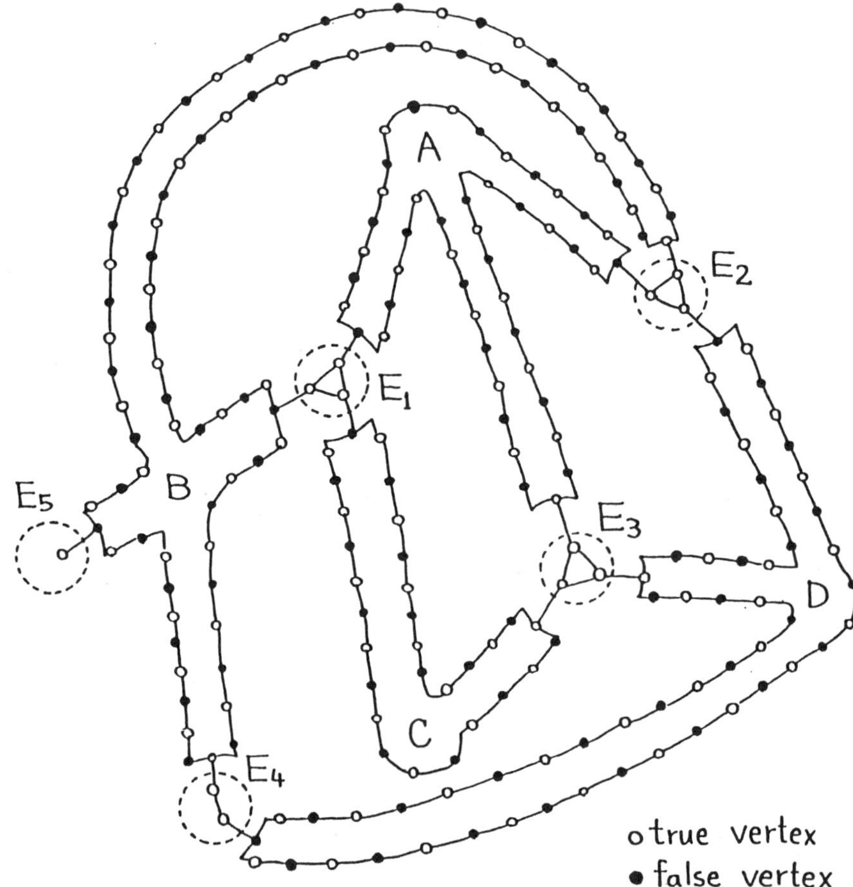

Fig. 4   An example of a graph corresponding to a boolean expression

$E = E_1 \wedge E_2 \wedge E_3 \wedge E_4 \wedge E_5$, where $E_1 = \{A, B, C\}$, $E_2 = \{A, \overline{B}, D\}$,

$E_3 = \{\overline{A}, \overline{C}, \overline{D}\}$, $E_4 = \{\overline{B}, D\}$, and $E_5 = \{B\}$.

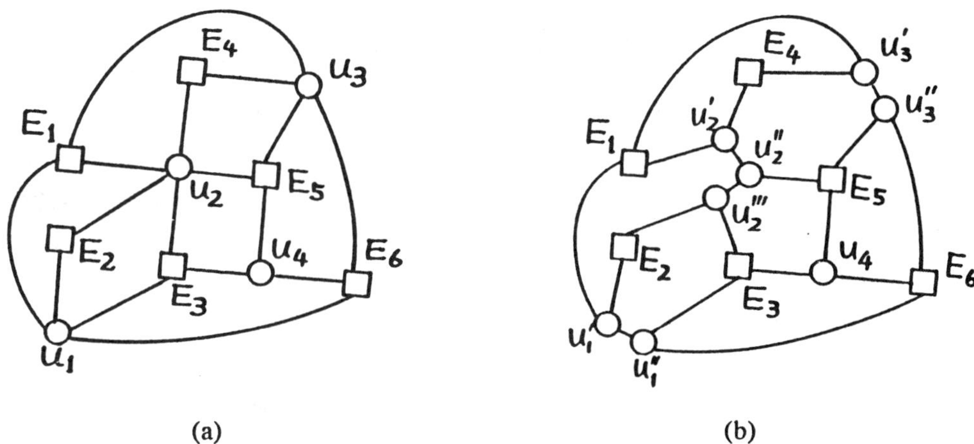

(a)   (b)

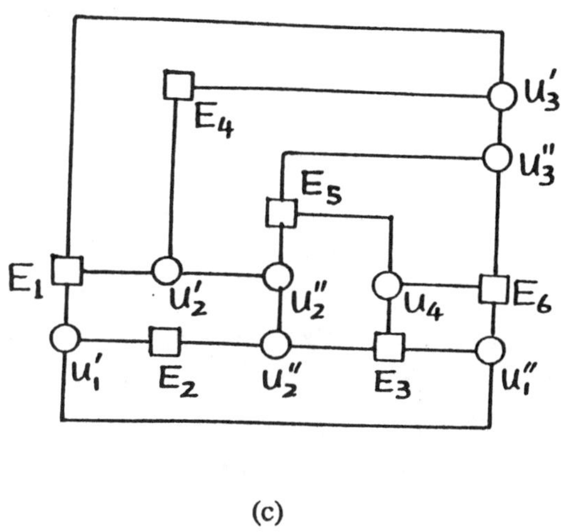

(c)

Fig. 5 Embedding of a graph into a grid graph.

(a) A given planar graph. (b) Decompostion of vertices of degree greater than 3 into those of degree at most 3. (c) Grid–graph embedding.

Now we have a planar graph such that every vertex has degree at most 3. Then, we can embed such a graph on the grid of size at most $n \times n$ as follows. We begin with some face F_1 of a planar graph. We draw a rectangle and put the vertices of F_1 on the upper side of the rectangle in the same order as in F_1. Then, we choose another face F_2 which is adjacent to the previous face F_1 so that $F_1 \cup F_2$ does not enclose any other face. Finding two vertices of degree 3 which belong to both F_1 and F_2, we draw another rectangle above the previous rectangle with the two vertices at their lower corners and then put vertices on the upper side of the rectangle. In this way we can place all the vertices of a given graph on upper sides of those rectangles. Now, it is not hard to make the resulting graph fit in the grid of size at most $n \times n$. An example of grid graph embedding is shown in Fig. 5.

If we imagine an embedding of an intersection graph into a grid plane, it is not so hard to show that the transformation is carried out in polynomial time.

As for the criterion (A), examples will be sufficient (see Fig. 6).

The MIS problem is known to be solvable in polynomial time on some intersection graphs. One such example is an interval graph, which is an intersection graph defined by a set of horizontal line segments. In this case a clause configuration for a three–literal clause cannot be constructed. If M is a class of horizontal and vertical line segments with no two of the same type touching, the resulting graph is always bipartite and thus cannot contain a triangle. (In this case the MIS problem can be solved using matching.)

More sophisticated is the case of a circle graph which is an intersection graph for a set of chords of a circle. (A polynomial time algorithm is known for circle graphs [Ga73].) In this case we can show that a simple graph defined for two three–literal clauses cannot be constructed (see Fig. 7).

4. Conclusion

In this paper we have presented a simple reduction from the planar 3–satisfiability problem to the maximum independent set problem. We showed that the maximum independent set problem is still NP–complete for intersection graphs defined by several sets of simple geometric objects. The reduction method presented in this paper is simpler than that proposed by Wang and Kuo who proved the NP–completeness of the maximum independent set problem on unit–circle intersection graphs. What is left is to find a necessary and sufficient condition on a class M of geometric objects that the MIS problem on intersection graphs for M is NP–complete.

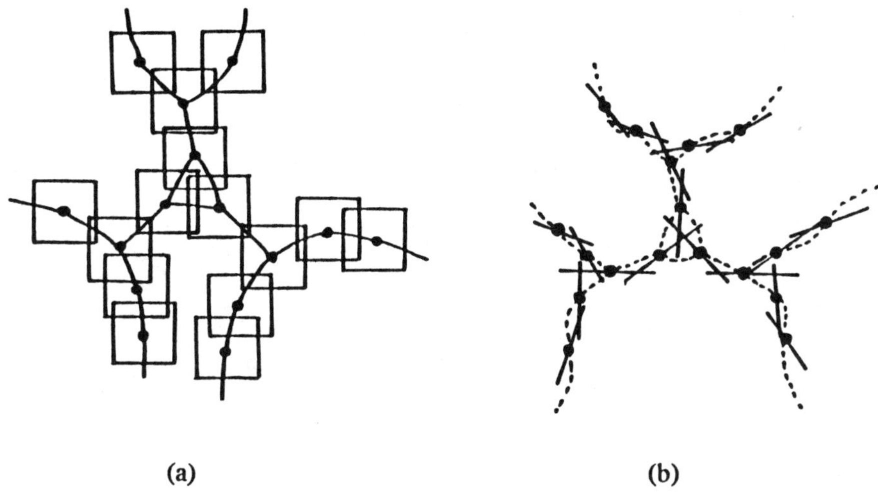

(a)　　　　　　　　　　　　　　(b)

Fig. 6 Construction of clause configurations for three–literal clauses.
(a) Construction by unit squares. (b) Construction by line segments.

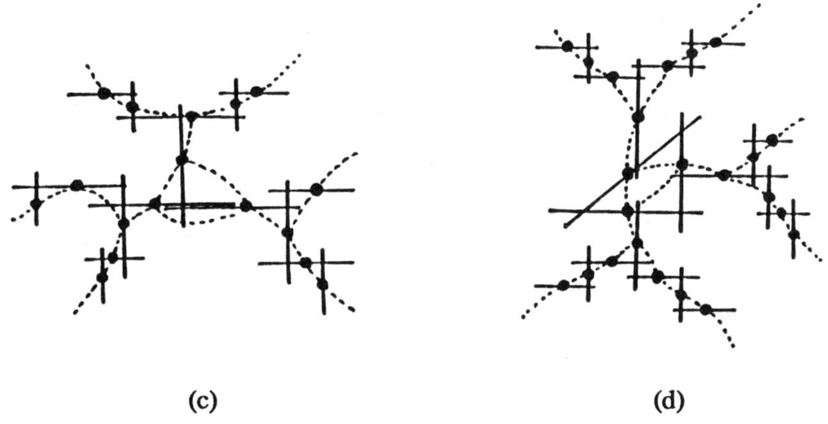

(c)　　　　　　　　　　　　　　(d)

(c) Construciton by horizontal and vertical line segments any two
of which may overlap. (d) Construction by horizontal, vertical and 45–degree line
segments no two of which overlap.

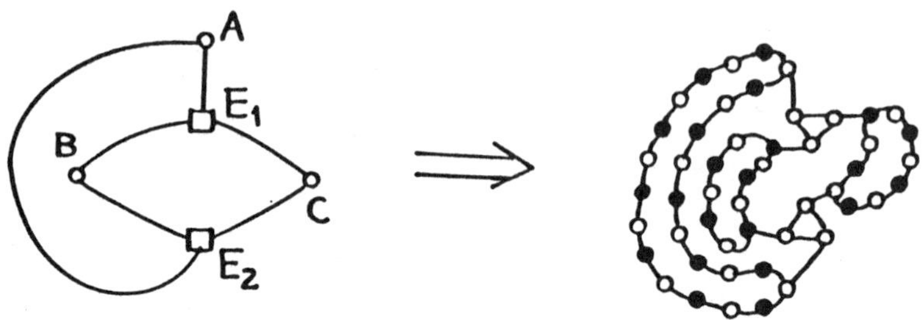

Fig. 7   An example of a graph which cannot be any circle graph.

REFERENCES

[Ga73]   F. Gavril, Algorithms for a Maximum Clique and a Maximum Independent Set of a Circle Graphs, Networks, 3, pp. 261 − 273, 1973.

[GJ79]   M.R. Garey and D.S. Johnson, Computers and Intractability: A Guide to the Theory of NP–completeness, W.H. Freeman, San Francisco, 1979.

[Ha72]   F. Harary, Graph Theory, Addison–Wesley, Reading Massachusetts, 1972.

[Ka83]   T. Kashiwabara, private communication.

[MS84]   N. Meggido and K.J. Supowit, On the Complexity of Some Common Geometric Location Problems, SIAM J. Comput., 13, 1, pp. 182 − 196, 1984.

[WK87]   D.W. Wang and Y.S. Kuo, A Study on Two Geometric Location Problems, Manuscript, 1987.

Some Graph Theoretic Properties of Deadlocks and Traps in Petri Nets

Kunwarjit S. Bagga*
Indiana University–Purdue University at Fort Wayne

Frank W. Owens
Ball State University

ABSTRACT

We consider the notions of deadlocks, traps, and liveness in Petri nets from a graph theoretic viewpoint. We characterize deadlocks and traps for a special class of Petri nets, namely bipartite tournaments, and then determine necessary and sufficient conditions for bipartite tournaments to be deadlock free. Similar conditions are determined for trap free bipartite tournaments. We also present a simpler version of a characterization of minimal deadlocks and traps in Petri nets obtained by J.–C. Bermond and G. Memmi.

Keywords: Petri net, deadlock, trap, liveness, bipartite tournament, strongly connected

Presented to the Sixth International Conference on the Theory and Applications of Graphs held at Western Michigan University, Kalamazoo, as an invited talk on May 30, 1988.

1. Introduction

Since their introduction by C.A. Petri in [5], Petri nets have been widely used as models in the study of networks involving information flows. Petri nets have also turned out to be useful models in the study of many asynchronous concurrent systems. These include, for example, distributed computer systems, operating systems of computers, and industrial process control systems.

* Research supported in part by the Office of Naval Research under contract N00014–86–K–0412.

The structure of a Petri net consists of a bipartite directed graph in which the nodes are distinguished into two types: places and transitions. Places may be thought of as conditions which when satisfied enable certain events or transitions. The flow of information is regulated by certain firing rules for transitions. Precise definitions are given in the following section.

Our interest in Petri nets mainly lies in their underlying graph structure. Many important properties of Petri nets as models of asynchronous concurrent systems have been studied. These include the notions of deadlocks, traps, boundedness, and liveness. We consider some of these in terms of graphs and show how graph theoretic tools can be applied in the study of Petri nets.

The references at the end of this paper provide more information on Petri nets and their applications.

2. Definitions

A *Petri net* is a quintuple (P, T, A, M_0, W), where P and T are the partite sets of a bipartite directed graph. The nodes in P are called *places*, and those in T are called *transitions*. A is the arc set of the bipartite directed graph. Thus $A \subseteq P \times T \cup T \times P$.

A *marking* M of a Petri net is a function $M: P \to N$, where $N = \{0, 1, 2, ...\}$ is the set of natural numbers. M_0 is a marking called the *initial marking*.

$W: P \times T \cup T \times P \to N$ is called the *weight* function. W satisfies $W(x, y) > 0$ if and only if $(x, y) \in A$.

In the figures, places are represented by circles and transitions by bars. If $p \in P$, the value of $M(p)$ is indicated by placing $M(p)$ *tokens* represented by dots inside the circle corresponding to the place p. The weight of an arc is indicated by a number next to the arc. The number is omitted for arcs having a unit weight. Figure 1 shows an example of a Petri net which models the flow of control of a simple computer program.

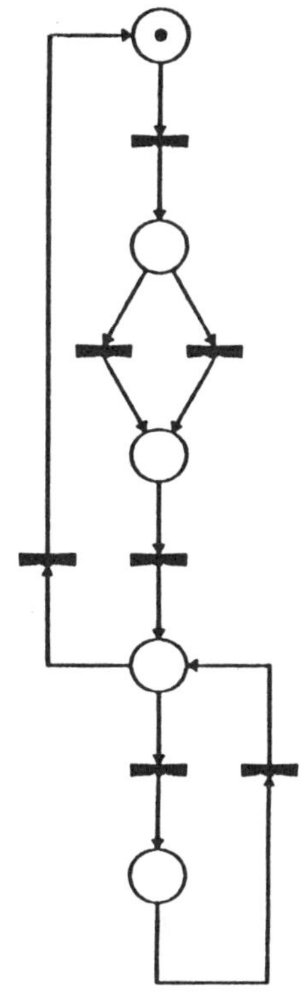

Figure 1

A transition $t \in T$ is *fireable at a marking* M if and only if $M(p) \geq W(p, t)$ for all $p \in P$. A transition t frieable at a marking M may *fire* to *reach* a new marking M_1 given by

$$M_1(p) = M(p) - W(p, t) + W(t, p)$$

for all $p \in P$. We then write $M \xrightarrow{t} M_1$. A marking M' is *reachable* from a marking M if and only if $M = M'$ or there exists sequence $t_1, t_2, ..., t_n$ of not necessarily distinct transitions such that

$$M \xrightarrow{t_1} M_1 \xrightarrow{t_2} ... \xrightarrow{t_n} M_n = M'.$$

A transition $t \in T$ is *live at a marking* M if and only if for every marking M' reachable from M there exists a marking M'' reachable from M' such that t is fireable at M''. A Petri net is *live at a marking* M if and only if every transition in T is live at M.

Several conditions for the liveness of a Petri net have been studied ([1] and [2]). We consider a particular aspect of the liveness property below.

Let x be a node of a directed graph. $N^+(x) = \{y \mid (x,y) \text{ is an arc}\}$ is called the set of *out–neighbors* of x and $N^-(x) = \{y \mid (y,x) \text{ is an arc}\}$ the set of *in–neighbors* of x.

Let X be a set of nodes of a directed graph. Then $N^+(X) = \bigcup_{x \in X} N^+(x)$ is called the set of out–neighbors of X and $N^-(X) = \bigcup_{x \in X} N^-(x)$ the set of in–neighbors of X.

A place $p \in P$ is *deficient for a marking* M if and only if $N^+(p) \neq \varnothing$, i.e., p has at least one out–neighbor, and $M(p) < \min_{t \in N^+(p)} \{W(p, t)\}$. A subset Q of P is *deficient for a marking* M if and only if every $p \in Q$ is deficient for M. In this case no transition in $N^+(Q)$ is fireable at M.

Commoner and Holt considered the following question in [2] and [3]. Which subsets of P deficient for a marking M remain deficient for any marking reachable from M? The answer is contained in the following theorem whose proof may be found in [1].

Theorem 2.1 *Let D be a subset of places such that $N^+(p) \neq \varnothing$ for each $p \in D$. Then the following conditions are equivalent:*

> *1) If D is deficient for a marking M, then D remains deficient for any marking reachable from M.*
>
> *2) $N^-(D) \subseteq N^+(D)$.*

A subset D of P is a *deadlock* if and only if $N^-(D) \subseteq N^+(D)$. Thus, if a deadlock D is deficient for a marking M, then no transition in $N^+(D)$ or $N^-(D)$ is fireable at any marking reachable from M. This means that no new tokens can enter D.

A subset R of P is a *trap* if and only if for every $t \in N^+(R)$ there exists $p \in N^+(t) \cap R$ such that either $N^+(p) = \varnothing$ or $W(t, p) \geq \min_{s \in N^+(p)} \{W(p, s)\}$. Clearly, if R is a trap, then $N^+(R) \subseteq N^-(R)$.

W is *nonblocking* if and only if for every $t \in T$ and every $p \in N^+(t)$ either $N^+(p) = \varnothing$ or $W(t, p) \geq \min_{s \in N^+(p)} \{W(p, s)\}$. If W is nonblocking, then a subset R of P is a trap if and only if $N^+(R) \subseteq N^-(R)$.

Note that W is nonblocking in the important case when $W(x, y) = 1$ for all $(x, y) \in A$. In this case the notion of a trap is dual to that of a deadlock in the sense that if all arcs of the Petri net are reversed, then deadlocks become traps, and traps become deadlocks. The result for traps corresponding to Theorem 2.1 is:

Theorem 2.2 *Let R be a subset of places such that $N^+(p) \neq \varnothing$ for each $p \in R$. Then the following conditions are equivalent:*

> *1) If R is not deficient for a marking M, then R is not deficient for any marking reachable from M.*
>
> *2) R is a trap.*

Thus, if a trap R is not deficient for a marking M, then for any marking M' reachable from M there exists $p \in R$ such that either $N^+(p) = \varnothing$ or $M'(p) \geq \min_{t \in N^+(p)} \{W(p, t)\} > 0$. This means that R cannot lose all of its tokens.

In the following sections we study several graph theoretic properties of deadlocks and traps. We first discuss a special class of Petri nets whose underlying graph structure is of interest in its own right. This is the class of bipartite tournaments.

3. Deadlocks and Traps in Bipartite Tournaments

A *bipartite tournament* is a directed bipartite graph B on partite sets P and T such that for every $p \in P$ and every $t \in T$ one and only one of (p, t) and (t, p) is an arc. Thus, B has $|P| \cdot |T|$ arcs. In what follows B is the underlying graph structure of a Petri net (P, T, A, M_0, W).

We first give some alternative characterization of deadlocks in B.

Proposition 3.1 *Let B be a bipartite tournament. Let D be a nonempty subset of places. Then the following conditions are equivalent:*

> *1) D is a deadlock.*
> *2) $D \not\subseteq N^+(t)$ for all $t \in T$.*
> *3) $N^+(D) = T$.*

Proof Assume that condition 1 is true. Then $N^-(D) \subseteq N^+(D)$. Let $t \in T$ and $p \in D$. If $p \rightarrow t$, i.e., if $(p, t) \in A$, then $D \not\subseteq N^+(t)$. If $t \rightarrow p$, then $t \in N^-(D) \subseteq N^+(D)$. Hence, $D \not\subseteq N^+(t)$.

Assume that condition 2 is true. Then for every $t \in T$ there exists $p \in D$ such that $p \rightarrow t$; thus, $t \in N^+(D)$. Hence, $N^+(D) = T$.

Assume that condition 3 is true. Then $N^-(D) \subseteq T = N^+(D)$, and D is a deadlock. $\square$

A Petri net is *deadlock free* if and only if no nonempty proper subset of places is a deadlock. We will use the above result to determine all deadlock free bipartite tournaments. Let B_m denote the bipartite tournament on partite sets P and T, where $|P| = |T| = m$, each $p \in P$ has unit outdegree, and $N^-(t)$ is nonempty for all $t \in T$. Then each $t \in T$ has outdegree $m - 1$. B_m is shown in Figure 2 with only the arcs from P to T being indicated.

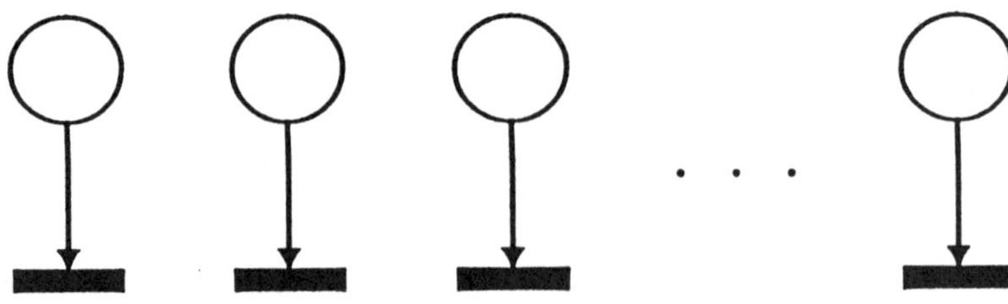

Figure 2

Theorem 3.2 *Let B be a bipartite tournament on partite sets P and T. Then B is deadlock free if and only if at least one of the following two conditions is satisfied:*

1) There exists $t \in T$ such that $N^+(t) = P$, equivalently, $N^-(t) = \emptyset$.

2) B_m is an induced subtournament of B, where $m = |P|$.

Proof   Assume that B is deadlock free and that condition 1 is false. If $|P| = 1$, then B_1 is an induced subtournament of B. Assume the $m = |P| > 1$, and let $P = \{p_1, p_2, ..., p_m\}$. Define $P_i = P - \{p_i\}$ for $1 \le i \le m$. By proposition 3.1 there exists $t_i \in T$ such that $P_i \subseteq N^+(t_i)$. Hence, $p_i \to t_i$ and $P_i = N^+(t_i)$. Clearly, $t_i \ne t_j$ if $i \ne j$. Thus, $m \le |T|$ and the subtournament induced by P and $\{t_1, t_2, ..., t_m\}$ is B_m.

Conversely, assume that condition 1 is true. Then $\emptyset$ is the only deadlock. Assume that condition 1 is false and that condition 2 is true. Let B_m be induced by $P = \{p_1, p_2, ..., p_m\}$ and $\{t_1, t_2, ..., t_m\} \subseteq T$ such that $p_i \to t_i$ for $1 \le i \le m$. Let X be a nonempty proper subset of P. If $p_i \notin X$, then $X \subseteq N^+(t_i)$. Hence, X is not a deadlock by proposition 3.1. $\square$

Next we consider traps in bipartite tournaments with a unit weight function W, i.e., $W(x, y) = 1$ if and only if $(x, y) \in A$. As observed in the previous section the notion of traps is dual to that of deadlocks in this case. The next two results are dual to the corresponding ones on deadlocks and are stated without proof. $\bar{B}_m$ denotes the dual of B_m. B_m and $\bar{B}_m$ are isomorphic as directed graphs.

Proposition 3.3 *Let B be a bipartite tournament with a unit weight function. Let R be a nonempty subset of places. Then the following conditions are equivalent:*

1) R is a trap.

2) $R \not\subseteq N^-(t)$ for all $t \in T$.

3) $N^-(R) = T$.

Theorem 3.4 *Let B be a bipartite tournament on partite sets P and T with a unit weight function. Then B is trap free if and only if at least one of the following two conditions is satisfied:*

      1) There exists $t \in T$ such that $N^+(t) = \varnothing$, equivalently, $N^-(t) = P$.

      2) $\overline{B}_m$ is an induced subtournament of B, where $m = |P|$.

We conclude this section by giving a characterization of bipartite tournaments with a unit weight function which are live at every nontrivial marking. A marking M is *nontrivial* if and only if $M(p) > 0$ for some $p \in P$. If $N^-(t) \neq \varnothing$ for all $t \in T$, then a Petri net can be live only at a nontrivial marking. This result has the following simple corollary. A bipartite tournament with a unit weight function is live at every marking containing exactly one token if and only if it is live at every nontrivial marking.

Theorem 3.5 *Let B be a bipartite tournament on partite sets P and T with a unit weight function. Then B is live at every nontrivial marking only if at least one of the following two conditions is satisfied:*

      1) There exists $t \in T$ such that $N^+(t) = P$, equivalently, $N^-(t) = \varnothing$.

      2) B_m is an induced subtournament of B, where $m = |P|$.

         If $N^+(t) \neq \varnothing$ for all $t \in T$, then the converse is also true.

Proof Assume that B is live at every nontrivial marking and that condition 1 is false. Then $m > 1$. Let $P = \{p_1, p_2, ..., p_m\}$. Choose any i such that $1 \le i \le m$. Define a nontrivial marking M_i of B by $M_i(p_i) = 1$ and $M_i(p_j) = 0$ for all $j \neq i$. There exists $t_i \in T$ such that t_i is fireable at M_i since B is live at M_i; hence, $p_i \to t_i$ and $t_i \to p_j$ for all $j \neq i$. Thus, B_m is an induced subtournament of B.

Conversely, assume that condition 1 is true. If $t \in T$ and $N^-(t) = \varnothing$, then t is fireable at any marking; hence, every transition is eventually fireable. Assume that condition 1 is false and that condition 2 is true with $N^+(t) \neq \varnothing$ for all $t \in T$. Then $m > 1$. Let M_0 be a nontrivial marking of B. By symmetry we may assume that $M_0(p_1) > 0$. Then t_1 is fireable at M_0, where $p_1 \to t_1$ and $t_1 \to p_j$ for all $j \neq 1$. Now let $M_0 \xrightarrow{t_1} M_1$. Then $M_1(p_i) > 0$ for all $i \neq 1$; hence, the transitions $t_2, ..., t_m$ are all fireable at M_1. If $m = 2$, then t_1 and t_2 are fireable infinitely often beginning at M_0. If $m > 2$, then let $M_1 \xrightarrow{t_2} M_2 \xrightarrow{t_3} M_3$. Then $M_3(p_i) > 0$ for all i so that any $t \in T$ may be fired. Finally, if there exists $t \in T$ such that $t \neq t_i$ for all i, $1 \le i \le m$, then $N^+(t) \neq \varnothing$. Thus, if t is fired, there exists $p \in P$ such that p contains at least one token, and the argument may be repeated. $\square$

Remark   The condition $N^+(t) \neq \emptyset$ for all $t \in T$ is necessary for the converse to be true. Consider the Petri net in Figure 3 which has B_2 as an induced subtournament. Let M be the marking defined by $M(p_1) = 1$ and $M(p_2) = 0$. Then t_3 is not fireable at any marking reachable from M.

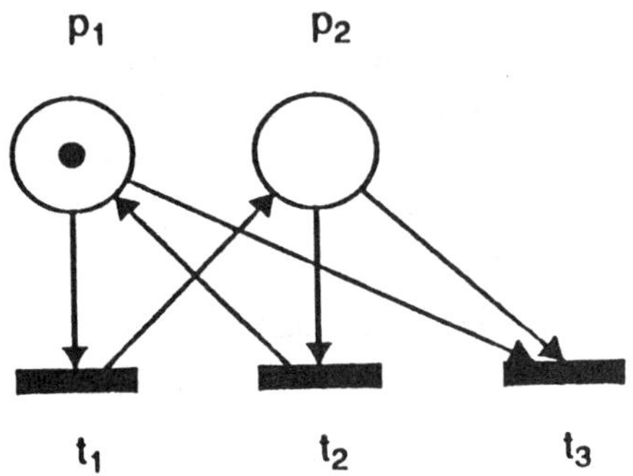

Figure 3

4. Minimal Deadlocks and Traps

A *minimal deadlock* is a deadlock containing no nonempty proper deadlocks. A *minimal trap* is a trap containing no nonempty proper traps. A bipartite graph is *strong* or *strongly connected* if and only if for every pair of nodes x and y there exists a directed path from x to y and a directed path from y to x.

We observed in Theorem 2.1 that a deadlock D which is deficient for a marking M remains deficient for any marking reachable from M; hence, no transitions in $N^+(D)$ are live at M. We also observed in Theorem 2.2 that a trap R which is not deficient for a marking M remains nondeficient for any marking reachable from M. Thus, a necessary condition for a Petri net to be live is that no nonempty deadlock becomes deficient, and a sufficient condition for this is that every nonempty deadlock contains a nondeficient trap. This latter condition is known as Commoner's *deadlock–trap property*. It is sufficient to check the deadlock–trap property for minimal deadlocks.

We now give a graph theoretic characterization of minimal deadlocks simpler than the one given in [1]. We also give a similar characterization of minimal traps in Petri nets with a nonblocking weight function.

Theorem 4.1 *Let D be a subset of places. Then D is a minimal deadlock if and only if D is minimal with respect to the property that the subgraph induced by $D \cup N^-(D)$ is strongly connected, i.e., if and only if the following two conditions are satisfied:*

> *1) The subgraph induced by $D \cup N^-(D)$ is strongly connected.*
>
> *2) The subgraph induced by $X \cup N^-(X)$ is not strongly connected for each nonempty proper subset X of D.*

Proof   Assume that D is a minimal deadlock. Let $X \cup Y$ be a nontrivial partition of the nodes of the subgraph induced by $D \cup N^-(D)$. To prove that this subgraph is strongly connected it is sufficient to show that there exist arcs in both directions between X and Y. First consider the case when $X \cap D = \varnothing$ or $Y \cap D = \varnothing$. If $X \cap D = \varnothing$, then $D \subseteq Y$ and $X \subseteq N^-(D) \subseteq N^+(D)$. Hence, if $t \in X$, there exist arcs in both directions between t and D. The subcase $Y \cap D = \varnothing$ is similar. Now consider the case when $X \cap D \neq \varnothing$ and $Y \cap D \neq \varnothing$. If $N^-(Y \cap D) = \varnothing$, then $Y \cap D$ is a nonempty proper subset of D and a deadlock in contradiction to the minimality of D. On the other hand, for every $t \in N^-(Y \cap D) \subseteq N^-(D) \subseteq N^+(D)$ there exists $p \in D$ such that $p \to t$. If there are no arcs from X to Y, then $p \in Y \cap D$ and we again have the same contradiction. The subcase for arcs from Y to X is similar. Thus, the subgraph induced by $D \cup N^-(D)$ is strongly connected.

Let X be a nonempty proper subset of D such that the subgraph induced by $X \cup N^-(X)$ is strongly connected. If $t \in N^-(X)$, then there exists $p \in X$ such that $t \to p$. There exists a directed path in $X \cup N^-(X)$ from p to t. Hence, $N^-(X) \subseteq N^+(X)$ and X is a deadlock in contradiction to the assumption that D is a minimal deadlock.

Conversely, assume that condition 1 and condition 2 are true. That D is a deadlock follows from condition 1 by the argument in the preceding paragraph. If D contains a nonempty proper minimal deadlock X, then the subgraph induced by $X \cup N^-(X)$ is strongly connected by the first paragraph in contradiciton to condition 2. Hence, D is minimal deadlock. ❑

Theorem 4.2 *Let R be a subset of places, and let W be a nonblocking weight function. Then R is a minimal trap if and only if R is minimal with respect to the property that the subgraph induced by $R \cup N^+(R)$ is strongly connected, i.e., if and only if the following two conditions are satisfied:*

> *1) The subgraph induced by $R \cup N^+(R)$ is strongly connected.*
>
> *2) The subgraph induced by $X \cup N^+(X)$ is not strongly connected for each nonempty proper subset X of R.*

Proof Assume that R is a minimal trap. Let $X \cup Y$ be a nontrivial partition of the nodes of the subgraph induced by $R \cup N^+(R)$. To prove that this subgraph is strongly connected it is sufficient to show that there exist arcs in both directions between X and Y. First consider the case when $X \cap R = \varnothing$ or $Y \cap R = \varnothing$. If $X \cap R = \varnothing$, then $R \subseteq Y$ and $X \subseteq N^+(R) \subseteq N^-(R)$. Hence, if $t \in X$, there exist arcs in both directions between t and R. The subcase $Y \cap R = \varnothing$ is similar. Now consider the case when $X \cap R \neq \varnothing$ and $Y \cap R \neq \varnothing$. If $N^+(X \cap R) = \varnothing$, then $X \cap R$ is a nonempty proper subset of R and a trap in the contradiction to the minimality of R. On the other hand, for every $t \in N^+(X \cap R) \subseteq N^+(R) \subseteq N^-(R)$ there exists $p \in R$ such that $t \to p$. If there are no arcs from X to Y, then $p \in X \cap R$ and we again have the same contradiction. The subcase for arcs from Y to X is similar. Thus, the subgraph induced by $R \cup N^+(R)$ is strongly connected.

Let X be a nonempty proper subset of R such that the subgraph induced by $X \cup N^+(X)$ is strongly connected. If $t \in N^+(X)$, then there exists $p \in X$ such that $p \to t$. There exists a directed path in $X \cup N^+(X)$ from t to p. Hence, $N^+(X) \subseteq N^-(X)$ and X is a trap in contradiction to the assumption that R is a minimal trap.

Conversely, assume that condition 1 and condition 2 are true. That R is a trap follows from condition 1 by the argument in the preceding paragraph. If R contains a nonempty proper minimal trap X, then the subgraph induced by $X \cup N^+(X)$ is strongly connected by the first paragraph in contradiction to condition 2. Hence, R is a minimal trap. $\square$

REFERENCES

[1] Bermond, J.–C. and G. Memmi, "A Graph Theoretical Characterization of Minimal Deadlocks in Petri Nets, *"Graph Theory with Applications to Algorithms and Computer Science*, Wiley–Interscience, New York, 1985. Edited by Y. Alavi, G. Chartrand, L. Lesniak, D.R. Lick, and C.E. Wall.

[2] Commoner, F.G., *Deadlocks in Petri Nets*, Applied Data Research Inc., Wakefield, Massachusetts, 1972. Report CA–7206–2311.

[3] Holt, A.W. and F. Commoner, *Events and Conditions,* Applied Data Research, New York, 1970.

[4] Peterson, J.L., *Petri Net Theory and the Modeling of Systems*, Prentice–Hall, Englewood Cliffs, New Jersey, 1981.

[5] Petri, C.A., "Kommunikation mit Automaten," *Schriften des Rheinisch–Westfalischen Institutes für Instrumentelle Mathematik an der Universitat Bonn,* 1962.

[6] Resig, W., *Petri Nets: An Introduction* (EATCS Monographs on Theoretical Computer Science, vol. 4), Springer–Verlag, Berlin, 1985.

The Concept of Leverage in Network Vulnerability[1]

K. S. Bagga

L. W. Beineke

M. J. Lipman

R. E. Pippert

Indiana University--Purdue University

at Fort Wayne

ABSTRACT

Measurement of the change in a specified graphical parameter induced by the removal of a set of edges or vertices from the graph has been the underlying theme in much recent work on network vulnerability and reliability. We formalize this concept as the "leverage" of the set with respect to the parameter under consideration. We then survey some results in the case where the parameter is the diameter, and add some specific results on the hypercube.

1. Introduction

The classification of parameters used in the measurement of vulnerability of networks leads to the observation that some parameters have a "second-order" nature in that they are based on changes in other parameters. One example is the persistence of a graph, introduced by Boesch, Harary, and Kabell [6] and defined to be the minimum number of vertices whose removal results in a graph with greater diameter. (There is, of course, an edge analogue.) Another example, introduced by Fink, Jacobson, Kinch, and Roberts [11], is called the bondage number of a graph, the minimum number of edges whose removal changes the domination number.

These examples are closely related to the changing and unchanging parameters defined by Harary [14]. For instance, if π is the domination number of G, then the

---

1 Research supported in part by the Office of Naval Research under contract N00014–86–K–0412.

changing parameter for deletion of edges (the "1" indicates edges, a "0" stands for vertices, and "+" or "–" refers to addition or deletion of elements),

$$ch(G, \pi, -, 1) := \min_{S \subseteq E(G)} \{ |S| \mid \pi(G-S) \neq \pi(G) \},$$

is the bondage number of G.

A limitation of the changing and unchanging parameters is that they do not quantify the changes made in the original parameter. For example, knowing that $ch(C_9, diam, -, 1) = 1$ tells us that the diameter of a 9–cycle can be changed by the removal of some edge, but it doesn't specify by how much (4 in this case).

We introduce leverage to give explicitly the change in a parameter produced by the deletion or addition of graph elements.

2. Definitions

A. Description

Several entities are involved in the concept of leverage--the graph, the parameter, the type of elements and the number of them, and whether they are to be added or deleted-- so notation is of necessity a bit cumbersome. Let π denote an arbitrary graph parameter and, as usual, $G = G(V, E)$ denotes a graph with vertex set V and edge set E, and we let $\overline{E}$ denote the edge set of the complement of G. Further, we adopt these conventions:

the subscript k denotes the size of the set being added or deleted;

absence of a superscript means that vertices are being deleted;
a prime means that edges are being deleted;
a superscript + means that edges are being added;

L represents the maximum change in π over all appropriate sets;
ℓ represents the minimum change.

In computing maximum or minimum values, we refer to the maximum or minimum absolute change, although we shall carry the sign of the change for computational purposes. The reason for this is that some parameters increase while others decrease upon removal of the same set of graph elements. This will be discussed in more detail in part B of this section.

Examples of this notation are:

$$L_k^+(\pi, G) = \pm \max_{|S|=k} | \pi (G \cup S) - \pi(G)|$$

where $S \subseteq \overline{E}$ and the sign is chosen to agree with the expression inside the absolute value, and

$$L_k' = \pm \min_{|S|=k} |\pi (G - S) - \pi(G)|$$

where $S \subseteq E$ and the sign is chosen as above.

In many cases the parameter is specified and perhaps also the graph, so that the notation can be simplified considerably through suppression of some of the variables in the expression.

Although further extensions of the concept of leverage are possible, we have chosen not to introduce more general notation due to its complexity.

Before considering leverage sequences, we make the observation that the removal of vertices from a graph can result in strange behavior with respect to certain parameters. Consider, for example, the graph in Figure 1, with π being diameter.

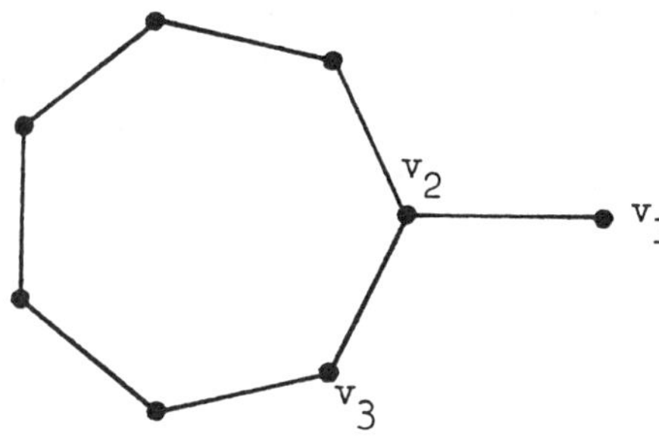

Figure 1

Here we observe that

$$\pi(G - v_i) - \pi(G) = \begin{cases} -1 & \text{if } i = 1, \\ +2 & \text{if } i = 2, \\ +\infty & \text{if } i = 3. \end{cases}$$

B. Leverage Sequences

When the leverage of sets of different sizes (with respect to a specified parameter π) is of interest, it is useful to consider the corresponding *leverage sequence* for π, for example

$$L(\pi, G) = L_1(\pi, G), L_2(\pi, G), \ldots$$

The length of the sequence will depend on π as well as G, since it may cease to be meaningful beyond a certain point; in particular, considering the diameter or average distance may require connectedness.

The following are some elementary observations about leverage sequences:

1. The changing and unchanging parameters are determined by the location of the first nonzero entry. For example, if a_i is the first nonzero entry in $L(\pi, G)$, then $ch(G, \pi, -, 0) = i$; if b_j is the first such entry in $\ell^+(\pi, G)$, then $un(G, \pi, +, 1) = j - 1$. More specifically, the location of the first nonzero entry in $L(\pi, G)$ is the persistence of G when π is the diameter and the bondage number when π is the domination number.

2. The mixed connectivity function introduced by Beineke and Harary [1] is in effect the leverage sequence $\ell(\pi, G)$ for π the edge–connectivity.

3. There is a dual relationship between the leverage sequence $L'(\pi, G)$ for π the order of a largest component, and the separation sequence $\eta(G)$ in which the ith entry is the least number of edges whose removal leaves no component of order greater than $|V| - i$. Consequently, the edge–integrity of a connected graph can be computed as

$$I'(G) = \min_k \{|V| + L'_k(\pi, G) + k\},$$

with a similar computation for the integrity.

4. Unusual behavior occurs in some leverage sequences when vertices are the elements removed, as can be seen if π represents the number of components:

$$L(\pi, P_7) = 1, 2, 3, 2, 1, 0.$$

On the other hand, removal of edges usually results in montonic behavior.

For purposes of comparing networks, good candidates for leverage include total distance or average distance, radius, and diameter. Similar parameters can be defined in other areas of graph theory, for example, the slimming number (also called the skewness) of a graph is the minimum number of edges whose removal yields a planar graph, and thus one can more generally ask for a reduction in the genus, thickness, or crossing number. However, because here we are interested in the vulnerability and reliability of graphs, we shall not pursue this further in this paper. The most widely studied leverage parameter is probably diameter, and that is the basis for the next section.

3. Leverage with Respect to Diameter

A. The General Case

For further information on leverage and diameter, the reader is referred to two good surveys: Bermond, Bond, Paoli, and Peyrat [3], and Chung [8]. Throughout this section, D will represent diameter.

Plesník [17] showed that if $L_1'(D, G)$ is finite, then it lies between 0 and D. He also defined the *elongation* of G, which is equal to $\ell_k'(D, G)$, and showed that it equals D if and only if G is a Moore graph.

More general results have been obtained by Chung and Garey [9]:

$$L_k'(D, G) \leq kD + o(k)$$

$$L_k(D, G) \leq \frac{1}{2}\left(\frac{n-k-2}{\kappa-k} - 1\right), \text{ where } k < \kappa \text{ (the connectivity of G)}.$$

The case of the removal of $\lambda - 1$ edges (λ = edge connectivity) was considered by Kerjouan [15]:

$$L_{\lambda-1}'(D, G) \leq \begin{cases} D & \text{if } \lambda = 2, \\ 2D - 1 & \text{if } \lambda = 3, \\ 3D - 2 & \text{if } \lambda = 4, \\ (\lambda-1)(D-1) + 3 & \text{if } D \geq 4. \end{cases}$$

Further, for $D \geq 4$, there are graphs of even diameter having $L_{\lambda-1}'(D, G) = (\lambda-1)(D-1)+3$ (which he conjectures to be the best upper bound), and there are graphs with d odd having $L_{\lambda-1}'(D, G) = (\lambda-1)(D-2)+4$.

Finally, in the case of adding edges, Chung and Garey [9] showed that if $d(G, t)$ is the minimum diameter of a graph obtained by adding t edges to a graph G of diameter $d(G)$, then $d(G, t) \geq \frac{d(G) - t}{t + 1}$. This, of course, gives a bound on $L_t^+(D, G)$. The extremal cases are achieved by paths, and Kerjouan [15] showed that

$$\frac{n+t-4}{t+1} \leq d(P_n, t) \leq \frac{n+t-1}{t+1} \qquad (t \text{ even}); \text{ and}$$

$$\frac{n+t-4}{t+1} \leq d(P_n, t) \leq \frac{n+2t-5}{t} \qquad (t \text{ odd}).$$

It follows that if t is even, then $\frac{(n-2)t}{t+1} \leq L_t'(D, G) \leq \frac{(n-2)t+3}{t+1}$ for any graph G of order n. There is a similar result for t odd.

B. Special Graphs

A number of special graphs have been proposed as an interconnection model for distributed–processing networks, including de Bruijn graphs, Kautz graphs, circulant graphs, and hypercubes. In this section, we examine some known results for the first graphs mentioned and compare them to corresponding results for the hypercube.

Those who have studied the Kautz and de Bruijn graphs include Bermond, Bond, Homobono, and Peyrat [4, 7, 16], who obtained results which can be expressed in the following form:

If B is a de Bruijn graph of maximum degree $\Delta > 4$ and diameter $D \geq 5$, then

   i) $L_1(D, B) = 0$, and

   ii) $L_k(D, B) \leq 1$ for $k \leq \Delta - 2$.

If K is a Kautz graph of maximum degree $\Delta > 4$ and diameter $D \geq 5$, then

   i) $L_1(D, K) = 0$,

   ii) $L_k(D, K) \leq 1$ for $k < \Delta - 6$, and

   iii) $L_k(D, K) \leq 2$ for $k < \Delta - 2$.

Furthermore, if G is any de Bruijn or Kautz graph of diameter at least 3, then $L_1'(D, G) = 0$.

Certain circulant graphs, classified as distributed loop computer networks by Bermond, Illiades, and Peyrat [5], have been shown to have both $L_1(D, G) \leq 1$ and $L_1'(D, G) \leq 1$.

Other results of this type can be obtained from work done on (Δ, M, M', s)–graphs. These are graphs of maximum degree Δ, diameter at most M, whose diameter is at most M' upon removal of any set of s vertices. For example, Fiol, Yebra, and Fabrega [12] have shown that sequence graphs yield $(\Delta, M, M+2, 1)$–graphs.

One of the most widely used computer networks is the n–dimensional cube, Q_n. We make some observations pertinent to the cube, which we list simply as propositions.

Proposition 1 *If G is an r–regular graph with edge–connectivity r, then $L'_{r-1}(D, G) > 0$.*

Proof Let u and v be vertices at distance D and let x be a vertex adjacent to v. Delete the $r-1$ edges at x other than xv. Since the only remaining path from u to x passes through v, the distance from u to x is greater than D. $\square$

Proposition 2 *If u and v are vertices at distance k in Q_n, then there is a collection of n vertex–disjoint u–v paths, of which k have length k and the other $n-k$ have length $k + 2$.*

Proof   Consider Q_n with the usual binary labeling so that vertices having k 1's in their labels are situated at level k, and are ordered from left to right in numerically decreasing order at each level. By symmetry, u and v may be taken to be the antipodal vertices $(0,0, ..., 0)$ and $\underbrace{(1,1, ..., 1,}_{k} \underbrace{0, ..., 0)}_{n-k}$ of the k–dimensional subcube Q_k spanned by them. This subcube Q_k contains k vertex–disjoint paths of length k between u and v.

We construct the remaining n − k paths as follows. Begin at $(0, 0, ..., 0)$ and move to a vertex $(0, 0, ..., 0, 1, 0, ..., 0)$ at level 1, having a 1 in one of the last n − k positions, say the jth. Then proceed "parallel" to the left–most path in Q_k, that is, simply add 1's successively in the left–most available position until reaching vertex

$$\underbrace{(1, 1, ..., 1,}_{k} 0, ..., 0, \underset{\underset{j^{th} \text{ position}}{\uparrow}}{1}, 0, ..., 0)$$

which is at level k + 1 and is obviously adjacent to v. $\square$

As a consequence of these observations, we see that $L'(D, Q_n) = 0, 0, ..., 0, 1, \infty$. This means that among all graphs in the class of n–regular graphs with $\lambda = n$, the n–cube has the maximum possible resistance to change of diameter through removal of edges.

This contrasts with the other networks mentioned above, for which the diameter may be increased by the removal of just two edges (or even one edge in some cases).

4.   Observations and Questions

1. The result of Plesník mentioned earlier shows that there are very "few" graphs which have the property that the removal of any edge results in maximum diameter change (the Moore graphs). On the other hand there are many graphs for which the removal of any edge or vertex increases the diameter, as seen from the following result of Gliviak and Plesník [13]:

Let $d \geq 2$. Any graph H is an induced subgraph of some graph G of diameter d such that both $\ell_1(D, G)$ and $\ell'_1(D, G)$ are positive.

We have not yet investigated constraints which may exist in regard to the sequences $\ell(D, G)$ and $\ell'(D, G)$.

2. An interesting question is which sequences are diameter–leverage sequences of regular graphs? At the extremes, the hypercubes are regular graphs having $L'(D, G) = 0, 0, ..., 0, 1, \infty$; on the other hand, there are families of graphs for which $L'_1(D, G)$ is the diameter of G.

In Figure 2, we show a regular graph G with $L'(D, G) = 1, 2, 2, 3, 3, \ldots, n, n$ for a specified n. Our construction requires $O(n^2)$ vertices; can it be done with fewer? Can $1, 2, 3, 4, \ldots, n$ be part of such a sequence, and if so, what is the minimum order of a graph which achieves it?

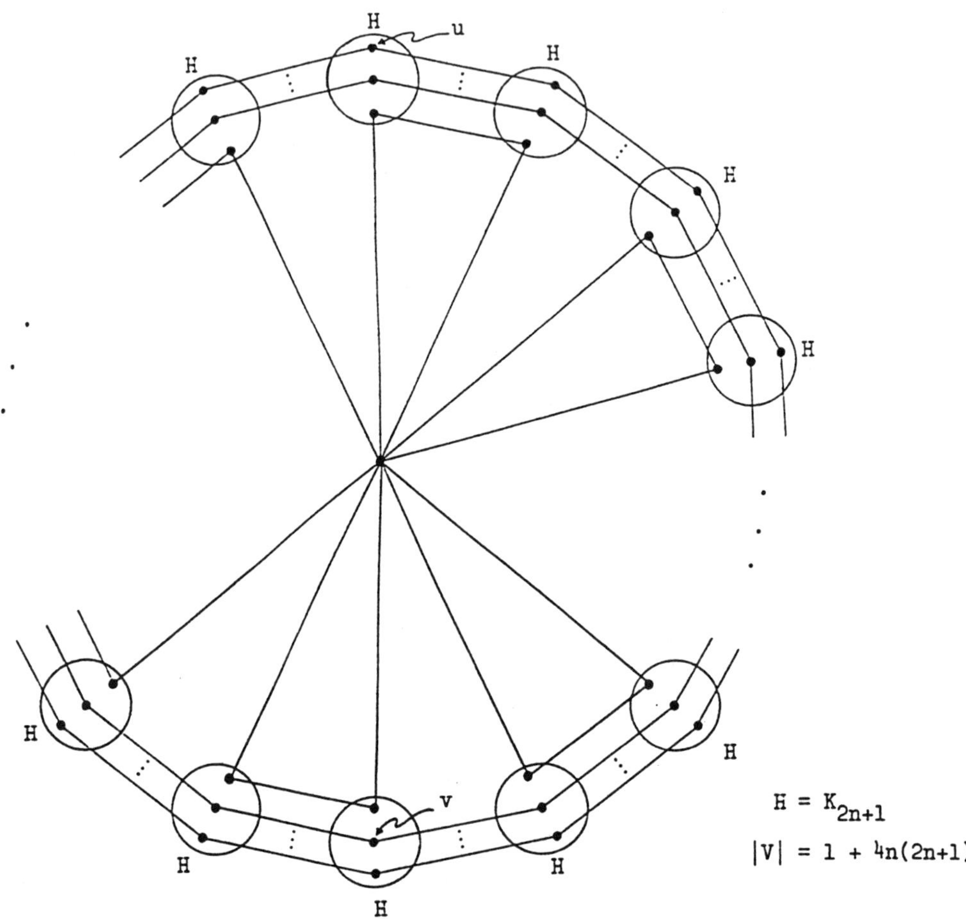

Figure 2

3. We conclude with a problem of leverage with respect to a different parameter. Let π denote the sum of the distances between all pairs of vertices of a graph. Favaron, Kouider, and Maheo [10] showed that $L_1'(\pi, G) \le \dfrac{\sqrt{2} - 1}{3} n^3 + O(n^2)$, where n is the number of vertices of G. We are interested in the minimum change that must be induced in π by the removal of k edges from Q_n. In particular, what is the largest value of k for which $\ell_k'(\pi, Q_n) = 2k$?

Clearly $\ell_k'(\pi, Q_n) \geq 2k$ since the removal of any edge leaves its incident vertices at distance 3 instead of 1. On the other hand, the removal of a single edge affects no other distances, since Proposition 2 shows that vertices at distance 2 or more have at least two disjoint paths of minimum length.

Now, to achieve the desired result, no two edges can be removed from a 4–cycle, since this is easily seen to produce an increase of more than 4 in the distance sum. It is interesting that this is the only restriction which must be met.

Lemma *The removal of k edges from Q_n, no two of which lie on a 4–cycle, yields $\ell_k'(\pi, Q_n) = 2k$.*

Proof Consider two vertices u and v at distance d in Q_n, $d > 1$. If d is even, there is a series of 2–cubes (4–cycles) connecting u and v as in Figure 3(a). If d is odd, the connection can be made using one 3–cube and a series of 2–cubes as in Figure 3(b).

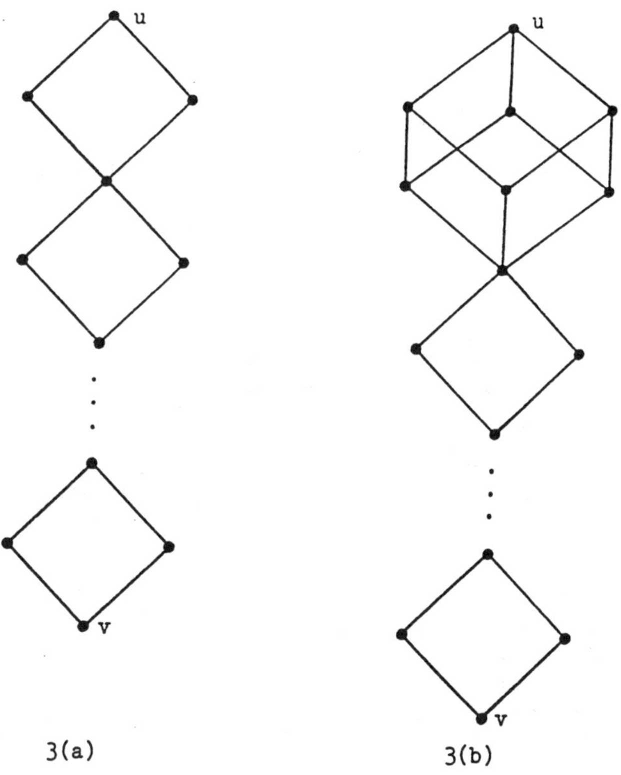

Figure 3

In the cases of the 2– and 3–cubes, it is straightforward to check that the restriction of removing no two edges of a 4–cycle is sufficient to achieve the desired result. Consequently, vertices at distance 2 in the 2–cube and at distance 3 in the 3– cube remain at those distances. From Figure 3, we see that the same restriction thus leaves u and v at distance d. ❑

Since any pair of adjacent edges in Q_n lies in a 4–cycle, an upper bound for the number of edges which can be removed is the size of a perfect matching, which is 2^{n-1}. Clearly a full 1–factor cannot be achieved in Q_2 or Q_3. However, for large values of n, the limit can be approached asymptotically. This can be readily accomplished by removing sets of non–adjacent edges from alternate edge–levels of Q_n, choosing a matching from each smaller vertex set to a subset of the larger. For small values of n, this is only moderately efficient, yielding somewhat more than half the upper bound. However, Stirling's Formula can be used to show that this method achieves the value asymptotically.

Theorem *Let π be the distance sum, and let k be maximum so that $\ell_k(\pi, Q_n) = 2k$. Then $\lim\limits_{n \to \infty} \dfrac{k}{2^{n-1}} = 1$.*

REFERENCES

[1] L. W. Beineke and F. Harary. The connectivity function of a graph. *Mathematika* **14** (1967) 197–202.

[2] L. W. Beineke, L. Lesniak and R. E. Pippert. On the edge separation sequence of a graph. In process.

[3] J.–C. Bermond, J. Bond, M. Paoli and C. Peyrat. Graphs and interconnection networks: diameter and vulnerability. Proc. of the Tenth British Comb. Conf. (1983), Cambridge University Press. *London Math. Soc. Lect. Notes Series* **82** (1983) 1–30.

[4] J.–C. Bermond, N. Homobono, and C. Peyrat. Large fault–tolerant interconnection networks. *Proc. of the First Japan International Conf. on Graph Theory and Applications*, Hakone, Japan (1986), to appear in *Graphs and Combinatorics*.

[5] J.–C. Bermond, G. Illiades and C. Peyrat. An optimization problem in distributed loop computer networks. *Laboratoire de Recherche en Informatique, Université de Paris–Sud, Rapport de Recherche* **419** (1988).

[6] F. T. Boesch, F. Harary and J. A. Kabell. Graphs as models of communication network vulnerability: connectivity and persistence. *Networks,* **11** (1981) 57–63.

[7]   J. Bond and C. Peyrat. Diameter vulnerability of some large interconnection networks. *Laboratoire de Recherche en Informatique, Université de Paris–Sud, Rapport de Recherche* **171** (1984).

[8]   F. R. K. Chung. Diameters of graphs: old problems and new results. *Congressus Numerantium* **60** (1987) 295–317.

[9]   F. R. K. Chung and M. R. Garey. Diameter bounds for altered graphs. *Journal of Graph Theory* **8** (1984) 511–534.

[10] O. Favaron, M. Kouider and M. Maheo. Edge–vulnerability and mean distance. *Laboratoire de Recherche en Informatique, Université de Paris–Sud, Rapport de Recherche* **248** (1985). To appear in *Networks*.

[11] J. R. Fink, M. S. Jacobsen, L. F. Kinch and J. Roberts. The bondage number of a graph. Preprint.

[12] M. A. Fiol, J.L.A. Yebra, and J. Fabrega. Sequence graphs and interconnection networks. *Ars Combinatoria* **16–A** (1983) 7–14.

[13] F. Gliviak and J. Plesník. On the existence of certain overgraphs of given graphs. *Acta Fac. R. N. Univ. Comen. Math.* **23-1969** (1970) 113–119.

[14] F. Harary. Changing and unchanging invariants for graphs. *Bull. Malaysian Math. Soc.* (2) **5** (1982) 73–78.

[15] R. Kerjouan. Arête–vulnerabilite du diamètre dans les réseaux d'interconnexion. *Laboratoire de Recherche ên Informatique, Université de Paris–Sud, Rapport de Recherche* **261** (1986).

[16] C. Peyrat. Diameter vulnerability of graphs. *Discrete Applied Math.* **9** (1984) 245–250.

[17] J. Plesník. Note on diametrically critical graphs. *Recent Advances in Graph Theory (Proc. Second Czechoslovak Sympos.,* Prague, 1974) Academia, Prague, 455–465.

EXTENSIONS OF AN ALGORITHM FOR COMPUTING THE EDGE-INTEGRITY OF TREES

Kunwarjit S. Bagga

Lowell W. Beineke

Marc J. Lipman

Raymond E. Pippert

Indiana University - - Purdue University

at Fort Wayne

ABSTRACT

We consider algorithms for computing the edge-integrity of trees and tree-like graphs. In an earlier paper we presented an algorithm of complexity no worse that $0(p^3)$ for the computation of the edge-integrity of trees. Our algorithm can easily be extended to a similar algorithm for the computation of the edge-separation sequence of trees. It can also be extended, not so easily, to an algorithm for cacti with cycles of bounded order. In this paper we present those extensions, paying particular attention to the simplest examples, $k = 3$ and 4, and this includes all $(2,0)$ - trees.

1.   Introduction

Edge-integrity, a measure of the vulnerability of a graph, was defined by Barefoot, Entringer, and Swart [2]. Given a graph G and set S of its edges, we let $m(G - S)$ denote the largest order of a component of $G - S$; the edge integrity is then defined as

Research supported in part by the Office of Naval Research under Contract N00014-86-0412.

$$I'(G) = \min_{S} \{ \ m(G - S) + |S| \ \}.$$

The interplay between the order of components in $G - S$ and the size of S itself suggested examining the separation sequence, defined by Beineke, Lesniak, and Pippert [3] to be

$$\eta(G) = (\eta_1, \eta_2, \cdots, \eta_{p-1}),$$

where $|G| = p$, and η_i is the minimum size of a set S of edges for which $m(G - S \leq p - i$. It is more convenient, and in some ways more natural, to consider the reverse of η, which we call ζ. Notice that the edge-integrity is related to ζ in that

$$I'(G) = \min_{i} \{ \ i + \zeta_i \ \}.$$

In [1], we presented an algorithm of complexity at most $0(p^3)$ for computing $I'(T)$, where T is a tree. The algorithm actually computes, for each value i in a range of values guaranteed to produce the minimum sum, a minimal S such that $m(T - S) \leq$. That is, the algorithm computes portions of $\zeta(T)$.

A non-trivial graph is a cactus if each block is an edge or a cycle. In this paper we extend the algorithm to one which computes $\zeta(C)$, and hence $I'(C)$, where C is a cactus with a specified bound on the cycle length, and show that this extension is also of complexity at most $0(p^3)$. In fact, the algorithm works if at most one end-block exceeds the bound. We present a specific example of the algorithm where the bound on the cycle length is 4.

2. The Labeling

For a given C we wish to compute $\zeta(C)$ and $I'(C)$. We will compute each entry of ζ by selecting an upper bound M for $m(C - S)$ and determining the size of a smallest set S of edges for which $m(C - S) \leq M$. We say that such an S achieves M. That is, $|S| = \zeta_M$. Choose an end-block B of C with cut-vertex $v = v_B$ and a value for M. Suppose we know the following about the subcactus $C' = C - (B - v)$ obtained by deleting all of B except the connecting vertex v:

(1) The order s of a smallest set of edges achieving M in C';

(2) among all the sets of s edges achieving M, the smallest number a of vertices in the component containing v in C' when such a set is deleted from C'; and

(3) among all the sets of $s + 1$ edges achieving M, the smallest number b of vertices in the component containing v in C' when such a set is deleted from C'.

From that information, it is easy to comput ζ_M. Suppose that B has $t + 1$ vertices. If $t + a \leq M$ then s edges can be selected from C' to achieve M in C, so $\zeta_M = s$. If $t + a > M$, but $t + b \leq M$ then $s + 1$ edges can be selected from C' to achieve M in C, so $\zeta_M = s + 1$. Otherwise, since $a \leq M$ and B is a cycle (or an edge), we can select an additional $q = |(t + a)/M|$ edges in B to achieve M in C, so $\zeta_M = s + q$. That is, it is easy to select edges in a cycle whose deletion leaves components of exactly the orders we need. Two things should be noted about this process: First, the set of s edges achieving M in C' does not have to known, only its size. Second, ζ_M depends only on $M, t, s, a,$ and b.

The algorithm we present here computes this kind of information for a chosen block B with cut- vertex $v_B = v$. The information is stored in a label for that vertex. In order to do so, the algorithm must similarly label every vertex not in B and every block except B. Below we give the specific details of these labelings and describe procedures that use vertex labels to label blocks and block labels to label vertices. From those two types of procedures a complete algorithm can be constructed.

For a given cactus C, fix an end-block B. For each vertex v, we define the subcactus rooted at v, denoted C_v, as follows: If v is a cut-vertex of C, then exactly one component of $C - v$ contains vertices from B. In this case, C_v is the subcactus of C which consists of C minus that component. If v is not a cut-vertex of C then C_v is the trivial graph consisting of v.

Select a value for M, the upper bound on $m(G - S)$. Excepting only B and its non-cut-vertices, for this value of M the algorithm assigns to each vertex and block in C a label of the form (s, a, b).

A vertex v is assigned the label (s, a, b) as follows:

s is the smallest number of edges achieving M in C_v;

among all the sets of s edges achieving M, a is the smallest number of vertices in the component containing v in C' when such a set is deleted from C'; and

among all the sets of $s + 1$ edges achieving M, b is the smallest number of vertices in the component containing v in C' when such a set is deleted from C'.

Notice that if v is not a cut-vertex then it has label $(0, 1, 1)$ regardless of the value of M.

For a block A other than the end-block B, let v_A be the cut-vertex in A separating A from B and let A_v be the cactus consisting of v_A together with the component of $C - v_A$ that contains vertices of A. That is, A_v is C minus every component of $C - v_A$ not containing vertices of A. The label (s, a, b) of A is assigned so that:

s is the smallest number of edges achieving M in A_v;

among all the sets of s edges achieving M, $a + 1$ is the smallest number of vertices in the component containing v in A_v when such a set is deleted from A_v; and

among all the sets of $s + 1$ edges achieving M, $b + 1$ is the smallest number of vertices in the component containing v in A_v when such a set is deleted from A_v.

Figure 1 shows a modest-sized cactus. Figure 2 shows some parts of that cactus labeled for $M = 5$.

3. The Algorithm

We describe the complete algorithm for computing ζ, and hence I', of the cactus C. We require that C have no cycle of length greater than k'. The algorithm uses the labeling described above by means of two procedures, *Label-Block* and *Label-Vertex*, each of which uses the output of the other.

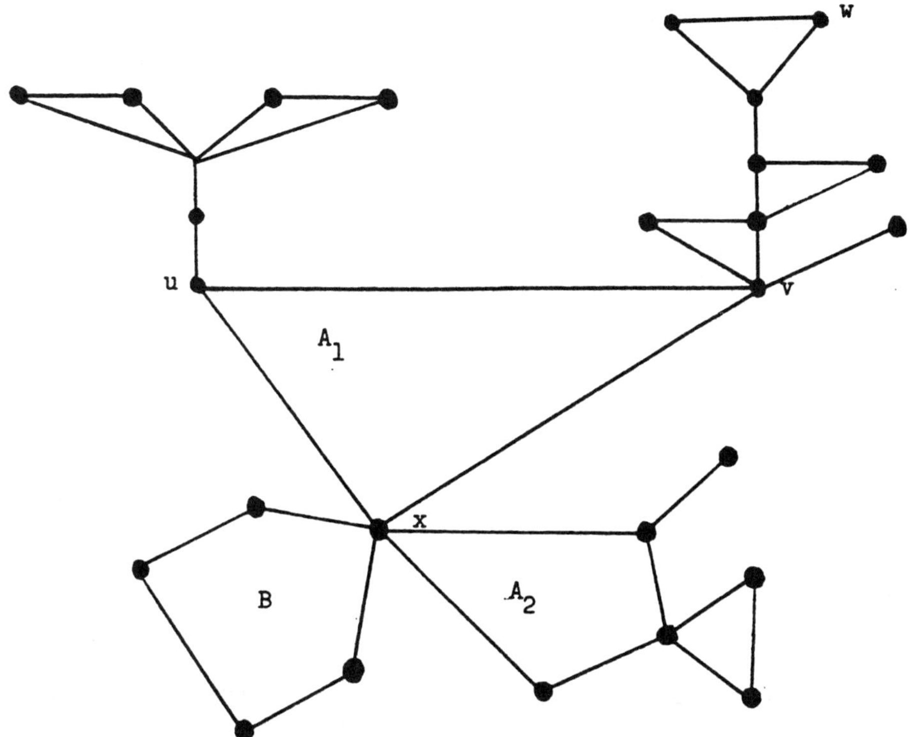

Figure 1. A modest-sized cactus

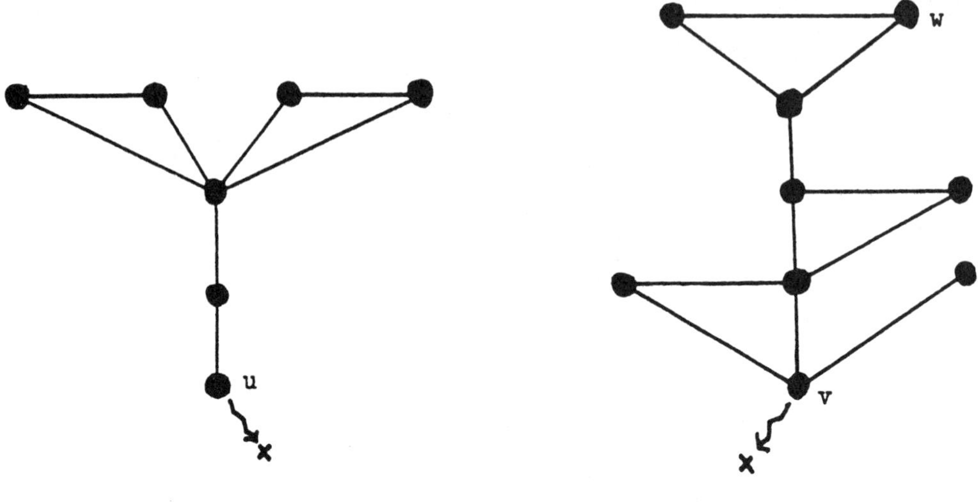

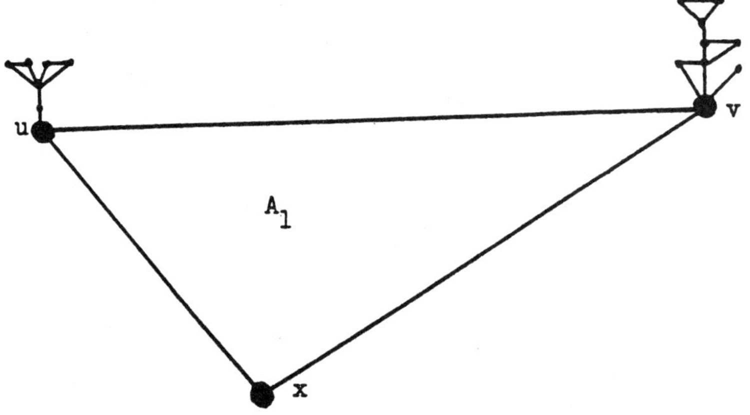

$$\text{label(w)} = (0,1,1) \quad \text{label(u)} - (1,2,1)$$

$$\text{label(v)} = (2,4,2) \quad \text{label(A}_1) = (4,4,2)$$

Figure 2: Some labels of parts of a cactus for $M = 5$

Algorithm ζ for k'-cactus :

Given: Cactus C as described above.

Choose an end-block B with cut-vertex v_B.

Assign permanent labels $(0, 1, 1)$ to every non-cut-vertex not in B.

$\zeta_1 =$ the number of edges in C.

For M between 2 and $p - 1$ *do*

$\{$compute $\zeta_M\}$

Until v_B has a label *do*

If every block in C_v is labeled, call Label-Vertex for v.

If every vertex in block A except v_A is labeled, call

Label-Block for A.

(Note: Label-Vertex is executed once per vertex, and Label-Block once per

block.)

Compute ζ_M as follows:

$$\text{If } t + a \leq M, \text{ then } \zeta_M = s;$$

$$\text{else if } t + b \leq M, \text{ then } \zeta_M = s + 1;$$

$$\text{else for } t + a = qM + r \text{ with } 0 < r \leq M,$$

$$\zeta_M = s + q + 1.$$

Procedure Label-Block assigns a label to a block A of order k provided that every vertex in A except v_A is labeled and the appropriate sub-procedure for k-cycles has already been constructed. In these sub-procedures, the *case value* is the smallest number of vertices that might remain adjacent to v when the *case number* of edges has been deleted from B. Therefore, the smallest case number for which the case value drops below M determines both s and a, and the next case value yields b.

Unfortunately, there does not appear to be a general method to determine these case values other than by exhaustively checking every reasonable set of edges. Reasonable here means that M is still achieved. Procedure Label-Vertex assigns a label to vertex x provided that every block in C_v containing x has a label. The *cut value* of label (A) = (s,a,b) is the maximum number of vertices which can be separated from x at the cost of deleting a single additional edge. By the definition of a and b, this is $a - b$. However, since A is a cycle, deleting the two edges in A incident with x separate all a vertices from x. This means that an average of $a/2$ vertices can be separated from x per edge. If $a/2 > a - b$, it is reasonable to consider making two edge deletions in A, and $a/2$ is the value of each cut. Label-Vertex selects cut values until the remaining (uncut) vertices form a legal component (order $\leq M$).

Next we present Procedure Label-Block with sub-procedures k-Cases for $k = 2, 3, 4$.

Procedure Label-Block:

> Given: M; k-block $A = \{v, u_1, \cdots u_{k-1}\}$ in cyclic order, where v separates A from B; and label $(u_i) = (s_i, a_i, b_i)$ for $i = 1, \cdots k - 1$.

> Call k-Cases.

> Label $(A) = (s, a, b)$, where

> $a =$ first case value less than M,

> $b =$ next case value, and

> $s = s_1 + s_2 + \cdots + s_{k-1} +$ case number which yields a.

Procedure 2-Cases:

Compute the cases below until a case has value $< M$. Return the case number and its value. Compute the next case and return its value.

> Case 0: a.

> Case 1: 0.

Procedure 3-Cases:

Compute the cases below until a case has value $< M$. Return the case number and its value. Compute the next case and return its value.

$$\text{Case 0: } a_1 + a_2.$$

$$\text{Case 1: } \min\{a_1 + b_2, a_2 + b_1\}.$$

$$\text{Case 2: If } a_1 + a_2 = M \text{ then } 0,$$

$$\text{else } \min\{b_1 + b_2, a_1, a_2\}.$$

$$\text{Case 3: } 0$$

Procedure 4-Cases:

Compute the cases below until a case has value $< M$. Return the case number and its value. Compute the next case and return its value.

$$\text{Case 0: } a_1 + a_2 + a_3.$$

$$\text{Case 1: } \min\{a_1 + a_j, b_k + b_k | i, j, k \text{ distinct}\}.$$

$$\text{Case 2: If } a_1 + a_2 + a_3 = M \text{ then } 0,$$

$$\text{else } \min\{a_1 + a_3, X, Y, Z\}, where$$

$$X = \min\{a_i + b_j + b_k | i, j, k \text{ distinct}\};$$

$$\text{If } a_1 + a_2 \leq M, \text{ then } Y = a_3,$$

$$\text{else } Y = a_2 + a_3;$$

$$\text{If } a_2 + a_3 \leq M, \text{ then } Z = a_1,$$

$$\text{else } Z = a_1 + a_2;$$

$$\text{Case 3: If } a_1 + a_2 \leq M \text{ or } a_2 + a_3 \leq M, \text{ then } 0;$$

$$\text{else } \min\{a_1, a_3, b_1 + b_2 + b_3, a_2 + b_1, a_2, +b_3\}.$$

$$\text{Case 4: } 0.$$

Next we give Procedure Label-Vertex, which includes details to guarantee that cuts are taken legally. Formally, we define the cut value of label(A) $= (s, a, b)$ to be the maximum of $a - b$ and $a/2$. Recall that $a - b$ is the number of vertices that a single additional edge from A or beyond it can separate from

v_A, while $a/2$ is the average number of vertices that each of two edges would separate from v_A by cutting off all of A. The cut value represents the maximum to be gained by deleting an edge from A or beyond it. Procedure Label-Vertex destroys some of the block labels. However, label(A) exists only to allow the labeling of v_A.

Procedure Label-Vertex:

> Given: $M;v$; and label $(A_i) = (s_i, a, i, b_i)$, for each block A_i incident with v in C_v, say $i = 1, \cdots, q$.

Assign $a(v) := 1 + a_1 + a_2 + \cdots + a_q$.

> While $a(v) > M$ do

> {Make a cut}

Find the largest cut value c' among the set of labels.

> Among all labels with cut value c' find the label with the largest a value, say label $(A) = (s, a, b)$.

Assign label$(A) := (s + 1, b, 0)$

Assign $a(v) : a(v) - (a - b)$.

Let $s(v) = s_1 + s_2 + \cdots + s_q$.

Suppose that the last cut made had order c and that the largest value of $a - b$ currently among the labels is d. (This means that it is possible to make an additional cut of value d.)

If $d \leq c$, then assign label(v) $:= (s(v), a(v), a(v) - d)$.

If $d > c$, then the final cut made was not the best possible.

> Therefore, find d', the second largest value of $a - b$ currently among the labels. $d > c$ implies that $d > d'$.

Then, if $a(v) + c - d' \leq M$ then assign label(v) $:= (s(v), a(v) + c - d', a(cv) - d)$.

Otherwise, assign

> label(v) $:= (s(v) + 1, a(v) - d, a(v) - d - d')$. For example, suppose that $M = 10, G - v$ has five components, and the labels of the four blocks incident with v in C_v yield:

block number	block label	cut value
1	(2, 4, 0)	4
2	(1, 6, 2)	4
3	(1, 5, 3)	5/2
4	(1, 3, 2)	3/2

Then cuts are made as follows:

cut value	a(v)	block #	and new label	new cut value
- -	19			
4	15	1	(3,0,0)	0
4	11	2	(2,2,0)	2
5/2	9	3	(2,3,0)	3 (3 > 2)

Hence label $(v) = (3 + 2 + 2 + 1, 9, 6) = (8, 9, 6)$.

We observe that the algorithm is of reasonable complexity.

Theorem. The algorithm has complexity $0(p^3)$.

Proof. We use as a measure of complexity the number of times either a vertex or a block is consulted. A vertex or block is consulted when it is labeled and when it is used to help label something else. Given a cactus C, denote by p, q and b, the number of vertices, edges, and blocks in C respectively, and denote the maximum degree of a vertex in C by Δ. Each of these numbers is $O(p)$. M varies from $p - 1$ to 1, so the number of M values (the number of entries in ζ) is also $0(p)$.

A non-cut-vertex is consulted exactly once to assign it a permanent label and once for each value of M as its block is being labeled. For each value of M, a cut-vertex is consulted once as it is labeled, and, after it is labeled, once more as the unlabeled block incident with it is labeled. Therefore, the total number of vertex consultations is at most $2p(p - 1)$, which is $0(p^2)$.

A block is consulted when it is labeled. This occurs $(p - 1 \cdot b$ times, which is at most $0(p^2)$. A specified block A is consulted during the labeling of A_v every time it is necessary to change a block-label during that labeling. Therefore,

M	label(u)	label(v)	label(A_1)	label(A_2)	label(x)	ς_M
>22	(0,7,1)	(0,9,6)	(0,16,10)	(0,6,5)	(0,23,17)	0
22-21	"	"	"	"	(1,17,7)	1
20-17	"	"	"	"	"	2
16	"	"	(1,10,0)	"	(2,7,6)	2
15-14	"	"	(1,10,7)	"	(2,14,7)	3
13-12	"	"	"		(3,7,6)	3
11	"	"	"	"	(3,7,6)	3
10	"	"	(2,7,0)	"	(3,7,6)	4
9	"	"	"	"	"	5
8	"	(1,6,2)	(2,7,3)	"	(4,7,4)	5
7	"	"	(3,3,0)	"	(4,7,6)	6
6*	(1,1,1)	(1,6,3)	(3,4,1)	(1,5,0)	(5,5,2)	6
5	(1,2,1)	(2,4,2)	(4,4,2)	(2,1,0)	(7,4,2)	9
4	(3,1,,1)	(3,2,1)	(6,3,2)	(2,2,0)	(9,4,3)	11
3	(3,2,1)	(3,3,2)	(8,2,0)	(3,1,0)	(12,2,1)	14
2	(5,1,1)	(7,2,1)	(14,1,0)	(5,0,0)	(19,2,1)	23
1						35

Table 1: Labels for various values of M for the cactus in Figure 1

the total number of block-label changes is at most $\Delta 1(p-1)$. This number is at most $0(p^3)$. The sum of both parts of the algorithm is $0(p^3)$ and the proof is complete.

As an example, we perform the algorithm for the cactus C in Figure 1, which has 27 vertices, 9 cycles, and hence 35 edges. Vertex w is assigned permanent label $(0,1,1)$. B is the 5-cycle and $B_v = x$. In Table 1 we use the ditto to indicate that the computation of the label proceeds exactly as above (a stronger statement than that the label is the same). The line corresponding to $M = 6$ yields $I'(C) = 12$.

REFERENCES

[1] K.S. Bagga, L. W. Beineke, M. J. Lipman, R. E. Pippert, and R. L. Sedlmeyer, *A Good Algorithm for the Computation of the Edge-Integrity of Trees*, submitted for publication.

[2] C. A. Barefoot, R. Entringer, and H. Swart, *Integrity of Trees and powers of cycles*, Congressus Numerantium 58, 1987, pp. 103-114.

[3] L. W. Beineke, L. Lesniak, and R. E. Pippert, *On the Edge-Separation Sequence of a Graph*, in process.

CARTESIAN FACTORIZATION OF INTERVAL-REGULAR GRAPHS HAVING NO LONG ISOMETRIC ODD CYCLES

Hans-Jürgen Bandelt

Kunwarjit S. Bagga

Universität Bielefeld

Henry Martyn Mulder

Vrije Universiteit

ABSTRACT

A graph is interval-regular if, for each pair u, v of vertices, the number of neighbours of u on all shortest paths from u to v equals their distance $d(u, v)$. Every (finite) interval- regular graph, in which the isometric odd cycles have length at most 5 and the vertices of each induced 5-cycle have a common neighbour, admits a Cartesian factorization, where each factor is the join of K_1 and a geodetic graph of diameter at most 2. In particular, the Hamming graphs are characterized as the interval-regular graphs without induced $K_{1,1,2}$ and isometric odd cycles other than triangles.

1. Introduction

Hypercubes constitute the simplest examples of Cartesian products: the n-*cube* Q_n is the product K_2^n of n copies of the two-vertex graph K_2. More generally, a *Hamming graph* is the Cartesian product $K_{j_1} \cdots K_{j_n}$ of complete graphs (so that two vertices are adjacent if and only if they differ in exactly one coordinate). Characterizations of these graphs abound: see Foldes (1977), Mulder (1979, 1980 a,b), Laborde and Rao Hebbare (1982), Bandelt and Mulder (1983), Ceccherini and Sappa (1986) for hypercubes, and Mulder (1980 b,

1982), Wilkeit (1986), Mollard (a) for arbitrary Hamming graphs. Typically, one aims at descriptions in terms of local properties, where triples of vertices, the geodesics between two vertices, the neighbourhoods of vertices, and the like are involved. Hamming graphs certainly constitute an important class of graphs, but they are perhaps a bit too special and thus not quite susceptible of exciting alternative descriptions. In searching for larger classes that still admit a nice Cartesian factorization, let us first reflect upon some salient characteristics of Hamming graphs. One of the properties used to characterize them is that of interval-regularity; see Mulder (1980 b, 1982), cf. Foldes (1977). Before we explain this notion we outline some items from metric graph theory.

The *distance* $d(u,v)$ between two vertices u and v of a graph G is just the length of any (u,v)-*geodesic*, i.e., shortest path from u to v. The *diameter* of G is the largest distance in G. The *interval* between u and v is the set

$$I(u,v) := \{w \mid w \text{ is a vertex on some } (u,v) - geodesic\}.$$

The *interval function* I is extensively studied in Mulder (1980 b). The set of vertices at distance k from u is denoted by $N_k(u)$. Then $N_1(u)$ is the usual neighbourhood $N(u)$ of u. A graph G is *interval-regular* if u has exactly $d(u,v)$ neighbours in $I(u,v)$, for any two vertices u and v of G. The main information about interval-regular graphs is provided by Theorem 4 of Mulder (1982): a graph G is interval-regular if and only if, for any two vertices u and v of G, the geodesics between u and v give a $d(u,v)$-cube. Loosely speaking, this means that, for $d(u,v) = k$, the interval $I(u,v)$ induces a k-cube with possibly some "horizontal" edges within levels $N_i(u) \cap I(u,v)$ of the interval. Recall that the *Cartesian product* of two arbitrary graphs G and H is the graph $G\,H$ with vertex-set $V(G) \times V(H)$, where two vertices (u,x) and (v,y) are adjacent whenever they are equal in one coordinate and adjacent in the other. Note that the distance between (u,x) (v,y) is just $d_G(u,v) + d_H(x,y)$, and that the interval between (u,x) and (v,y) is precisely the Cartesian product of the interval $I_G(u,v)$ and $I_H(x,y)$. So, interval-regularity is preserved under Cartesian multiplication. This feature is shared by another property of graphs, which for obvious reasons is denoted by ∇: if u, v, w are vertices such that vw is an edge and $d(u, v) = d(u, w)$, then there exists a common neighbour x of v and w with $d(u, x) = d(u, v) - 1$. We call such an edge vw with $d(u, v) = d(u, w)$ *horizontal* with respect to u. We call a triple u, v, w of vertices an I-*triangle* whenever we have

$$I(u, v) \cap I(u, w) = \{u\} ,$$
$$I(v, u) \cap I(v, w) = \{v\} ,$$
$$I(w, u) \cap I(w, v) = \{w\} ,$$

Such an I-triangle is *equilateral* if, in addition, we have $d(u, v) = d(u, w) = d(v, w)$. This distance is then the *size* of the equilateral I-triangle. A subgraph H of a graph G is *isometric* in G if all distances in H equal the corresponding distances in G. A subgraph H of a graph G is *convex* if all geodesics between vertices from H lie completely in H. Note that convexity of subgraphs is preserved under taking intersections. The *convex closure* of a subgraph H of G is the smallest convex subgraph containing H. A graph is *geodetic* whenever between any two vertices there is just one geodesic. Finally, the *join* $K_1 + H$ of the one-vertex graph K_1 and the graph H is obtained from H by adding a new vertex and making it adjacent to all vertices of H.

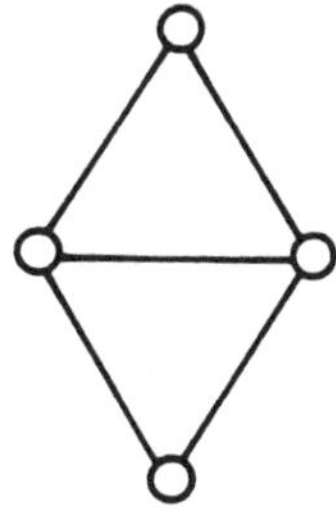
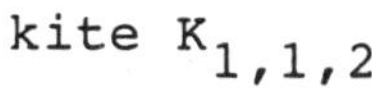
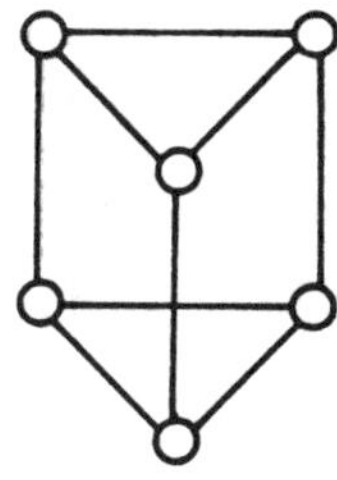
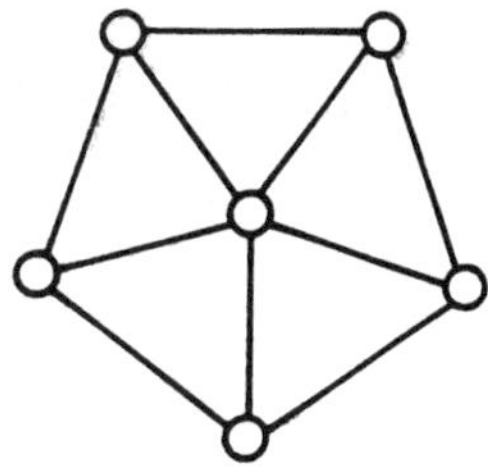

kite $K_{1,1,2}$ prism $K_2 \,\square\, K_3$ 5-wheel $W_5 = K_1 + C_5$

Figure 1. Three interval-regular graphs of diameter 2

The indecomposable graphs in our factorization theorem are interval regular graphs of diameter at most 2 which are constructed from geodetic graphs; see Bandelt and Mulder (1984). Namely, the join of K_1 and any geodetic graph of diameter at most 2 gives an interval-regular graph. A good deal of information is available for geodetic graphs of diameter 2; see Bosák (1978), Parthasarathy and Srinivasan (1986), Scapellato (1986). Then potential generalizations of the following theorem may involve other interval-regular graphs of diameter 2 (see also Bandelt and Wilkeit (1987)) or, more generally, particular F-geodetic graphs sensu Scapellato (a); but this, inevitably, would require considerably more efforts. Note that the *kite* and the *5-wheel* are given in Figure 1, whereas the *hammock* is depicted in Figure 2.

Theorem Let G be a finite, connected graph. Then the following conditions are equivalent:

(i) G is the Cartesian product of graph $G_1, \cdots, G_n$ ($n \geq 1$), where each G_i is the join $K_1 + H_i$ of K_1 and a geodetic graph H_i of diameter at most 2;

(ii) G is interval-regular and has property $\triangledown$;

(iii) G is interval-regular and each I-triangle in G is equilateral;

(iv) G contains no isometric hammock or isometric odd cycle of length greater than 5, the convex closure of an induced 5-cycle C_5 is the 5-wheel W_5, the convex closure of an isometric even cycle C_{2k} is the k-cube Q_k ($k \geq 3$), and the convex closure of an induced 2-path P_2 is either the 4-cycle C_4 or the kite $K_{1,1,2}$.

It is easy to see that condition (ii) of the Theorem can be tested in $0(m \cdot p^2)$ time, where m and p are the numbers of edges and vertices. Moreover, the factors can actually be determined in $0(m \cdot p^2)$ time, too. Indeed, for any vertex z of maximum degree, we can identify the factors as the components of the neighbourhood $N(z)$ of z joined with z. This, of course, is not really surprising since the general theory of Cartesian factorization guarantees that the decomposition of an arbitrary graph G can be performed in time $0(p^4)$ in the worst case; see Winkler (1987) for a survey on these matters.

All graphs in this paper are finite and connected. For convenience, we will not always distinguish between a set W of vertices in a graph G and the subgraph of G induced by W.

Some Lemmas We commence by investigating the interrelation of the properties listed in the Theorem. For this purpose, it is useful to consider a certain condition that is much weaker than interval-regularity. We say that a graph G has the *property* $\diamond$ if, for each triple u, v, w of vertices in G such that

$$d(v, w) = 2,$$

$$d(u, v) = d(u, w) = k \geq 2,$$

$$N_{k+1}(u) \cap N(v) \cap N(w) \neq \phi,$$

there exists also a common neighbour of v and w at distance $k - 1$ to u. This condition first came up in the study of median graphs and their generalizations; see Bandelt (a), Bandelt and Mulder (1986), and Chung, Graham, and Saks (a). Typically, even cycles of length greater than 4 violate $\diamond$ (although they can be extended to graphs fulfilling $\diamond$, e.g., hypercubes). Whether or not all I-triangles are equilateral essentially depends on $\diamond$ and $\triangledown$. Figure 2 includes some critical graphs that are used in the sequel.

Lemma 1 Let G be a graph with interval function I. Then G has properties $\triangledown$ and $\diamond$ if and only if every I-triangle in G is equilateral and any triple of pairwise non-adjacent vertices in an isometric 6-cycle or an isometric chorded 6-cycle has a common neighbour.

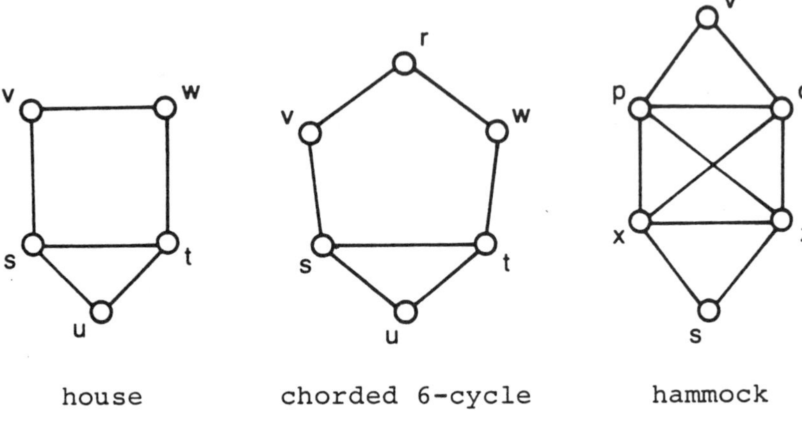

Figure 2. Three critical graphs

Proof   First, let G have properties $\triangledown$ and $\diamond$. The property guarantees the existence of a common neighbour for any triple of pairwise non-adjacent vertices in isometric 6-cycles or chorded 6-cycles. Let u, v, w be any I-triangle in G with $d(u,v) = k$, $d(u,w) = \ell$, and $d(v,w) = m$. Assume that the triangle is not equilateral, say $k < \ell$. Let P be a geodesic $v = v_0 \to \cdots \to v_m = w$ between v and w such that $\sum_{i=0}^{m} d(u,v_i)$ is as small as possible. Walking along P from v to w we can classify the edges of P as upward, horizontal, or downward edges according to the distance of their endpoints from u. Since $k < \ell$, we must traverse at least one upward edge of P. Since u, v, w is an I-triangle, we have $d(u, v_{m-1}) \geq d(u,w) = \ell$. So on P there must be an upward edge $v_{j-1}v_j$ such that the next edge $v_j v_{j+1}$ is either horizontal or downward. Note that $d(v_{j-1}, v_{j+1}) = 2$. If $v_j v_{j+1}$ is downward, then by $\diamond$, we find a common neighbour x of v_{j-1} and v_{j+1} with $d(u,x) = d(u,v_j) - 2$. Now we replace v_j by x in P and thus obtain a geodesic between v and w that is nearer to u than

P. This contradicts our minimality assumption on P. Hence $v_j v_{j+1}$ has to be horizontal. Then by $\diamond$, we have a common neighbour y of v_j and v_{j+1} with $d(u, y) = d(u, v_j) - 1 = d(u, v_{j-1})$. Because of the minimality condition on P, we know that y is not adjacent to $v_{(j-1)}$. Then by $\diamond$, we find a common neighbour z of y and v_{j-1} with $d(u, z) = d(u, y) - 1 = d(u, v_{j+1}) - 2$. Further by ∇, we find a common neighbour x of v_{j-1}, z, and v_{j+1}. As above, we replace v_j by x on P and get a geodesic between v and w nearer to u than P. This refutes our assumption that $k < \ell$, whence the I-triangle u, v, w is equilateral.

Conversely, we have to establish ∇ and $\diamond$. Let vw be a horizontal edge with respect to u. Let x in $I(u, v) \cap I(u, w)$ be such that $d(u, x)$ is maximal. then x, v, w is an I-triangle, whence x is adjacent to v and w. This settles ∇. Now, let u, v, w be vertices with $d(u, v) = d(u, w) = k \geq 2$ and $d(v, w) = 2$. Let z be a common neighbour of v and w with $d(u, z) = k + 1$. Let x be a vertex in $I(u, v) \cap I(u, w)$ with $d(u, x)$ as large as possible, and suppose that x is not a common neighbour of v and w. Then x, v, w must be an I-triangle of size 2. Let s be a common neighbour of x and v, and let t be a common neighbour of x and w. Then s cannot be adjacent to w, and t cannot be adjacent to v. In particular, s and t are distinct. If s and t were adjacent, then we would have an isometric chorded 6-cycle. The hypothesis in the lemma, however, then conflicts with the fact that x, v, w is an I-triangle. So, x and t are not adjacent. Suppose that $d(s, w) = 2$. Then by ∇, we find a common neighbour t' of s, x, and w. Note that t', being a common neighbour of x and w, cannot be adjacent to v. But then x, s, t', v, w, z form an isometric chorded 6- cycle. As above, this conflicts with the fact that x, v, w form an I-triangle. Therefore $d(s, w) = 3$, and similarly, $d(t, v) = 3$, whence x, s, t, v, w, z induce an isometric 6-cycle. Hence we find a common neighbour of x, v, and w, by the condition on isometric 6-cycles in G. And again we have a contradiction. So we better not assume that x was non-adjacent to v and w. This settles $\diamond$ and completes the proof.

In order to link the properties $\bigtriangledown$ and $\diamond$ to conditions involving isometric cycles, we investigate to which extent the definition of $\bigtriangledown$ and $\diamond$ is redundant. It turns out that it suffices to require $\bigtriangledown$ only for triples of vertices on isometric odd cycles and houses and require $\diamond$ only for those on isometric even cycles and chorded 6-cycles.

Lemma 2   Let G be a graph satisfying the following conditions: for every isometric cycle of length $2k+1$ or $2k+2$ and for each triple of distinct vertices u, v, w with $d(u, v) = d(u, w) = k \geq 2$, there exists a common neighbour x of v and w with $d(u, x) = k - 1$; for every isometric house or isometric chorded 6-cycle in G, the vertices u, v, w as indicated in Figure 2 have a common neighbour. Then G has properties $\bigtriangledown$ and $\diamond$.

Proof   We prove $\bigtriangledown$ and $\diamond$ for triples of vertices u, v, w with $d(u, v) = d(u, w) = k$ and $1 \leq d(v, w) \leq 2$ by induction on k. We distinguish two cases.

Case 1 $d(v, w) = 1$.

Assume that there is no common neighbour of v and w at distance $k - 1$ from u. Then it follows from the induction hypothesis that $I(u, v) \cap I(u, w) = \{u\}$. Let $P = u \to 1 \cdots \to v_{k-1} \to v$ be a geodesic in $I(u, v)$, and let $Q = w \to w_1 \to \cdots \to w_{k-1} \to u$ be a geodesic in $I(u, w)$. Then P and Q have no vertices in common except u. We have $d(v_1, v) = k - 1 = d(u, w_1), d(v_1, w) = k$, and $k-1 \leq d(v_1, w_1) \leq k$. Suppose that $d(v_1, w_1) = k-1$. Then, by $\bigtriangledown$ for $k - 1$, we find a common neighbour s of v and w_1 with $d(v_1, s) = k - 2$. So $d(u, s) = k - 1$. Hence, by $\bigtriangledown$ for $k - 1$, we find a common neighbour t of s and w_1 with $d(u, t) = k - 2$. Now s, t, v, w, w_1 induce a house, so that there is a common neighbour x of t, v, and w. But then we have $d(u, x) = d(u, t) + 1 = k - 1$. This contradicts our assumption that no such vertex exists for the triple u, v, w. We conclude that $d(v_1, w_1) = k$. Now we proceed along P, and simultaneously along Q, and prove by an analogous argument that $d(v_i, w_i) = k$, for $i = 2, \cdots, k - 1$. Therefore $P \to Q$ induces an isometric cycle of length $2k + 1$. By the hypothesis of the lemma, we arrive at a contradiction. This settles Case 1.

Case 2 $d(v, w) = 2$, and z is a common neighbour of v and w with $d(u, z) = k + 1$.

Again assume that there is no common neighbour of v and w at distance $k - 1$ from u. By the induction hypothesis, we get $i(u, v) \cap I(u, w) = \{u\}$. Let P be a geodesic $u \to v_1 \to \cdots \to v_k = v$ in $I(u, v)$, and let Q be a geodesic $w = w_1 \to \cdots \to w_k \to u$ in $I(u, w)$. Then P and Q have no vertices in common except u. We have $d(v_1, v) = k - 1$, $d(v_1, z) = k$, and $k \leq d(v_1, w) \leq k + 1$. Suppose that $d(v_1, w) = k$. By Case 1, we have a common neighbour r of z and w with $d(v_1, r) = k - 1$. Then $\diamond$ or $\triangledown$ for $k - 1$ provides us with a common neighbour s of r and v with $d(v_1, s) = k - 2$. Note that $d(u, s) = k - 1$ and s cannot be adjacent to w. Since $d(u, r) = d(v_1, r) + 1 = k$, Case 1 provides us with a vertex t adjacent to r and w with $d(u, t) = k - 1$. Then by $\diamond$ or $\triangledown$ for $k - 1$, we get a common neighbour y of s and t with $d(u, y) = k - 2$. If s and t were adjacent, then y, s, t, v, w, z would induce an isometric chorded 6-cycle. But any common neighbour of y, v, and w would be at distance $k - 1$ from u. Therefore s and t are not adjacent. Now s, y, r, t, w induce a house in G. Then s, y, and w have a common neighbour t'. Since t' is a common neighbour of y and w, it cannot be adjacent to v. Hence y, s, t', v, w, z induce an isometric chorded 6-cycle. Again we are in trouble, and we cannot avoid it. So we have to conclude that $d(v_1, w) = k + 1$. Similarly, as in Case 1, we now proceed along P and Q, and eventually prove that $d(v_i, w_i) = k + 1$, for $i = 1, 2, \cdots, k$. Then $P \to z \to Q$ induces an isometric cycle of length $2k + 2$. By the hypothesis on such cycles, we arrive at a final contradiction, thus concluding the proof.

An immediate consequence of this lemma is the following observation. Every isometric subgraph of a graph G has the properties $\triangledown$ and $\diamond$ if and only if G does not contain an isometric house or any isometric cycle of length greater then 4.

The next lemma clarifies how isometric cycles can arise in a Cartesian product of graphs $G_1, \cdots, G_n$. It turns out that no "skew" isometric odd cycles are produced and that the even ones are built up from smaller even cycles in the factors and copies of K_2.

Lemma 3 Let $G = G_1 \cdots G_n$ be the Cartesian product of $n \geq 1$ graphs G_i, and let C be an isometric cycle in G such that each projection π_i from G to G_i maps C onto a non-singleton graph. If C is odd, then $n = 1$ must hold. If C has length $2k \geq 4$, then each image $\pi(C_i)$ either has $2k_i = 2$ vertices or is an isometric cycle in G_i of length $2k_i \geq 4$ such that $k = k_1 + \cdots + k_n$.

Proof It suffices to prove the lemma for $n = 2$. We colour the edges of C as follows: an edge gets colour i whenever its endpoints are equal in the i-th coordinate, for $i = 1, 2$. Then projection of C to G_i amounts to contracting the edges of colour i to vertices, for $i = 1, 2$. The conditions on the projections of C imply that there is a "pivot" vertex (u, v) on C incident with an edge from each colour class, say one neighbour of (u, v) is of the form (t, v), while the other is of the form (u, w). Clearly, C cannot be a triangle. Suppose that C has length $2k + 1 \geq 5$. Then we may assume that the two vertices on C that are at distance k from (u, v) are of the form (x, y) and (x, z). So we have

$$d_2(v, y) = k - d_1(u, x) = d_2(v, z),$$

where d_i is the distance function of $G_i(i = 1, 2)$. Hence (t, v) has distance $d_1(t, x) + d_2(v, y)$ to (x, y) as well as (x, z), which, however, is impossible. Therefore C has even length $2k$. We assert that opposite edges on C have the same colour. Assume the contrary, and let $(u, v)(t, v)$ and $(x, y)(x, z)$ be opposite edges on C, where (u, v) and (x, y) are opposite vertices as well as (t, v) and (x, z). Then on the one hand we have

$$d_2(v, y) = k - d_1(x, u) = d_2(v, z) + 1,$$

and on the other hand we have

$$d_2(v, z) = k - d_1(t, x) = d_2(v, y) + 1.$$

This is impossible, so opposite edges of C have the same colour. In particular, there is an even number of edges of each colour, say $2k_i$ edges of colour i, for $i = 1, 2$. If say, $k_1 = 1$, then π_2 projects C onto an edge in G_1. Otherwise, if $k_1 > 1$, then π_2 projects C onto an isometric cycle of length $2k_1$. In any case we get $k = k_1 + k_2$.

Proof of the Theorem   We will establish the following equivalence and implications:

$$\text{(iii)} \longleftrightarrow \text{(ii)} \longrightarrow \text{(i)} \longrightarrow \text{(iv)} \longrightarrow \text{(ii)}.$$

Since interval-regularity implies $\diamond$, the equivalence of (ii) and (iii) is an immediate consequence of Lemma 1.

As to the implication (i) $\longrightarrow$ (iv), recall from Lemma 3 that every isometric odd cycle C in G is already contained in a copy

$$\{u_1\} \cdots \{_{i-1}\} \, G_i \, \{u_{i+1}\} \; \cdots \; \{u_n\}$$

of G_i, for some i. Then C must have length 3 or 5, since H_i has diameter at most 2. The convex closure of a 5-cycle in G_i is a 5-wheel, and hence so is the convex closure of a 5-cycle in G. Now consider an isometric cycle C in G of length $2k \geq 6$. No factor G_i contains an isometric even cycle. Therefore, by virtue of Lemma 3, there are exactly k distinct factors, say $G_1, \cdots, G_k$, such that each projection to $G_i(i = 1, \cdots, k)$ maps C onto some edge of G_i. Pick any two vertices u and v on C at distance k. The interval $I(u, v)$ is the Cartesian product of intervals in the factors. Since $I(u, v)$ includes C, each factor interval is a copy of K_1 or K_2, whence $I(u, v)$ induces a k-cube in G. Evidently, the convex closure of every 2-path in G is isomorphic to C_4 or the kite $K_{1,1,2}$. Finally, for vertices u and v of G with $d(u, v) = 3$, the interval $I(u, v)$ induces either $K_2^3 = Q_3$ or $K_2 \, K_{1,1,2}$. Hence the hammock cannot be isometric in G.

Next we prove that (iv) implies (ii). First we show that G has properties $\triangledown$ and $\diamond$. By Lemma 2, it suffices to check the triples u, v, w in some isometric copies of the first two graphs of Figure 2. If u, v, w are as in the house of Figure 2, then there must be a common neighbour x of u and v different from s. Certainly x is not adjacent to t (for otherwise, t and v have three common neighbours). Now, if x were not adjacent to w, then t, u, x, v, w induce a 5-cycle. The convex closure of this cycle contains s and so cannot be a wheel, yielding a contradiction. If u, v, w are as in the chorded 6-cycle of Figure 2, then let x be the vertex making the 5-cycle induced by s, t, v, w, r into a wheel. Since we assume that the chorded 6-cycle is isometric, x cannot be adjacent to u. We will arrive at a contradiction by producing an isometric 6-cycle having at least three more vertices in its convex closure. Let p be the second common neighbour of u and v. To avoid that p serves as some third common neighbour, p is not adjacent to x, t or w. So p, u, t, x, v induce a 5-cycle with s in its convex closure, whence s is adjacent to p. Then p cannot be adjacent to r. Similarly, we get a common neighbour q of u, t and w that is not adjacent to r, s, v or x. Suppose that p and w have a common neighbour y (necessarily distinct from all previous vertices). then y cannot be adjacent to s or t. Hence p, s, t, w, y induce a 5-cycle having u as well as x in its convex closure. So p and w are at distance 3. Similarly q and v are at distance 3. Thus we have produced an isometric 6-cycle $u \to p \to v \to r \to w \to q \to u$ with s, t and x in its convex closure. Since this is far from being a 3-cube, we conclude that no chorded 6-cycle is isometric in G. Hence G has properties $\triangledown$ and $\diamond$.

Next we prove by induction on $k = d(u, v)$ that the interval $I(u, v)$ contains exactly k neighbours of v. For $k \leq 2$ this is covered by the assumptions on G. So let $k \geq 3$. Pick any neighbour w of v in $I(u, v)$. Then, by the induction hypothesis, w has exactly $k - 1$ distinct neighbours $s_1, \cdots, s_{k-1}$ $I(u, w)$. Then, for $i = 1. \cdots, k - 1$, v and s_i have a common neighbour t_i besides w. If $t_i = t_j$, for some $i \neq j$, then w and t_i must be adjacent, as well as s_i and s_j, since they have a common neighbour y at distance 3 from v and $k-3$ from u (by property $\triangledown$ or $\diamondsuit$). But then v, w, t_i, s_i, s_j, y induce an isometric hammock. So all $t_i(i = 1, \cdots, k-1)$ are distinct. Suppose that there is yet another neighbour x of v in $I(u, v)$, besides w and $t_1, \cdots, t_{k-1}$. Since w and x have a neighbour at distance $k-2$ to u, some s_i and v have w, t_i, and x as common neighbours, giving a contradiction. This proves that G is interval-regular, and thus establishes (ii).

Finally we prove that (ii) implies (i). A number of steps is necessary: first the convex closures of houses and 5-cycles are checked, and then the neighbourhood of a vertex with maximum degree is investigated. This subgraph entails information about the potential factors of the Cartesian factorization we aim at.

(1) The convex closure of an induced 5-cycle is a 5-wheel.

Let $u \to p \to v \to w \to q \to u$ be an induced 5-cycle. By $\triangledown$, we have a common neighbour x of u, v, and w. Assume that x is not adjacent to p. Then by $\triangledown$, there is a vertex y distinct from x and adjacent to u, p and w. Certainly y is distinct from q. But now u and w have three distinct common neighbours x, y and q, which contradicts interval-regularity. Hence x is adjacent to p, and similarly to q. So x and the 5-cycle induce a 5-wheel. Because of interval-regularity, none of the pairs of non-adjacent vertices can have a common neighbour outside the wheel. Therefore the convex closure of the 5-cycle is the wheel.

(2) The convex closure of an induced house is a prism.

Let u, s, t, v, w induce a house, where u is the top of the house adjacent to s and t (see Fig. 2). By $\triangledown$, we get a common neighbour x of u, v and w. By interval-regularity, x cannot be adjacent to s or t, and moreover, none of the non-adjacent pairs among u, s, t, v, w, x can have a common neighbour outside this subgraph. Hence the convex closure of the house is the prism $K_2 \ K_3$ (see Figure 1).

(3) For any vertex z, the neighbourhood $N(z)$ is the disjoint union of

geodetic graphs of diameter at most 2.

Let $u \to v \to w$ be an induced 2-path in $N(z)$. Then z and v are the common neighbours of the non-adjacent vertices u and w. So u and w have v as the unique common neighbour in $N(z)$. Hence it suffices to prove that every component in $N(z)$ has diameter at most 2. Let $u \to v \to w \to x$ be an induced 3-path P in $N(z)$. Since u and x have z as a common neighbour, they must have a second common neighbour y. Then y cannot be adjacent to w, for otherwise, u and w have v, y, z as three distinct common neighbours. Similarly, y is not adjacent to v, whence the 3-path P together with y induce a 5-cycle. So, by (1), the vertices y and z must be adjacent, that is, y is in $N(z)$. Hence u and x have distance 2 in $N(z)$.

Next we need some elementary information on geodetic graphs of diameter 2; cf. also Bosák (1978).

(4)   A geodetic graph of diameter 2 either is a block or consists of complete graphs all attached at a single vertex.

Let H be a geodetic graph of diameter 2 that is not a block. Let x be a cut-vertex. Then any two vertices in different components of $H - x$, must have x as their unique common neighbour in H. In particular, x is adjacent to all other vertices of H. Assume that some component F of $H - x$ is not complete. Let $u \to v \to w$ be an induced 2-path in F. Then the non-adjacent vertices u and w have two distinct common neighbours v and x in H. This conflicts with H being geodetic. So all components of $H - x$ are complete.

(5)   Every edge of a geodetic block H of diameter 2 lies on an induced 5-cycle. To see this, consider an edge uv of H and a maximal complete subgraph K containing uv. Since H has diameter 2 but no cut-vertex, there exists a vertex s at distance 2 to each vertex in K (cf. Lemma 3 of Bandelt and Mulder (1984)). Suppose that there exists a common neighbour x of s, u, v. Then, by maximality of K, there is at least one vertex z in K which is not adjacent to x. Now, x and z have two common neighbours, which is

impossible. Therefore s, u, v together with any common neighbour t of s and u and with a common neighbour w of s and v induce a 5-cycle.

For the remainder of the proof, let z be a vertex of *maximum degree* in G. So, the next statements essentially rely on the finiteness assumption. In contrast, the above lemmas and statements (1) to (5) remain valid for infinite graphs, too.

(6)   For a vertex u in a component A of $N(z)$ the edges between $N(z) - u$ and $N(u) \cap N_2(z)$ form a perfect matching between $N(z) - A$ and $N(u) \cap N_2(z)$.

For every vertex x in $N(z) - A$ there is a unique common neighbour $\varphi(x)$ of u and x different from z by interval-regularity. Necessarily, $\varphi(x)$ belongs to $N(u) \cap N_2(z)$. Since every vertex in $N_2(z)$ has exactly two neighbours in $N(z)$, each vertex $\varphi(x)$ has no other neighbour in $N(z) - u$ than x. It remains to show that there is no edge between $A - u$ and $N(u) \cap N_2(z)$. Suppose by way of contradiction that there is a vertex v in $N_2(z)$ having two distinct neighbours p and q in A. Note that p and q must be adjacent, for otherwise, they have three distinct common neighbours, viz., v,z, and one in A. We distinguish two cases.

Case 1 A is not a block of diameter 2. Then A is complete or consists of complete graphs all attached at a single vertex. Let x be a vertex in A adjacent to all other vertices of A. If, say p equals x, then we get

$$|N(p)| = |A - p| + |N(p) \cap N_2(z)| + |\{z\}|$$
$$\geq |A - p| + |N(z) - A| + |\{v\}| + 1$$
$$= |N(z)| + 1.$$

This contradicts the maximality of the degree of z. So, x is distinct from p and, similarly, from q. Hence neither p nor q is adjacent to all vertices of A, that is, x is a cut-vertex of A.

Let s be a vertex in a component of $A - x$ which does not contain p, q. Then p, q, s, v, x, z induce a hammock (see Figure 2). This graph cannot belong to the interval $I(s,v)$ because G is interval-regular. Hence, as v and s cannot be adjacent, there exists a common neighbour t of v and s. Then t must be in $N_2(z)$. By interval-regularity, t is not adjacent to p, q, and x. Hence $s \to x \to q \to v \to t \to s$ is an induced 5-cycle with p in its convex closure. Since p is not adjacent to t, we get a conflict with (1). This settles Case 1.

Case 2 A is a block of diameter 2.

Then, according to (5), the edge pq is in some induced 5-cycle $p \to q \to r \to s \to t \to p$ in A. Then v is not adjacent to r, s, or t. Suppose there is a common neighbour x of s and v. Since x cannot be adjacent to p or z, the vertices x, s, z, p, v induce a 5-cycle. The convex closure of this cycle contains q and t, thus violating (1). Therefore s and v are at distance 3. By interval-regularity, v must have a third neighbour $y \neq p, q$ in $I(s, v)$. Necessarily, y is adjacent to r and t within the 3-cube in $I(s, v)$. Now, however, r and t have three distinct common neighbours, a contradiction. This completes the proof of (6).

(7)   For a vertex u in a component A of $N(z)$, the perfect matching between $N(z) - A$ and $N(u) \cap N_2(z)$ induces a graph isomorphism φ between the subgraphs induced by these sets.

Let p and q be distinct vertices of $N(z) - A$, and let p be matched to s and q to t in $N(u) \cap N_2(z)$, that is: $s = \varphi(p)$ and $t = \varphi(q)$. First, if p and q are adjacent, then $u \to s \to p \to q \to t \to u$ is a 5-cycle having z in its convex closure. Since this is not a 5-wheel, the cycle cannot be induced. The only possible chord is st, whence s and t are adjacent. Similarly, if p and q are not adjacent, then adjacency of s and t would establish an induced 5-cycle $s \to p \to z \to q \to t \to s$ with u in its convex closure, equally impossible. Thus we have shown that p and q are adjacent if and only if s and t are adjacent.

(8)   Let $A_1, \cdots, A_n$ be the components of $N(z)$. Then G can be embedded (as an induced subgraph) into the Cartesian product F of the subgraphs

$A_i + z(i = 1, \cdots, n)$.

Recall that the vertices of F are the n-vectors $(u_1, \cdots, u_n)$ with u_i in $A_i + z$, for $i = 1, \cdots, n$. Let $\bar{z}$ denote the vertex $(z, z, \cdots, z)$ of F. Then $N_k(\bar{z})$ in F consists of all n-vectors with exactly k coordinates different from z. Consider any vertex u in $N_k(z)$. By Theorem 4 of Mulder (1982), the interval $I(u, z)$ induces a k-cube with possibly some horizontal edges with respect to z. Since any two distinct vertices in $I(u, z) \cap N(z)$ have a common neighbour in $I(u, z) \cap N_2(z)$, we infer from (6) that the k vertices in $I(u, z) \cap N(z)$ come from k different components of $N(z)$. This enables us to represent u by a vector $\psi(u) = (u_1, \cdots, u_n)$, where u_i is the vertex in $I(u, z) \cap A_i$ whenever $I(u, z)$ intersects A_i, and $u_i = z$ whenever $I(u, z) \cap A_i = \Phi$. This gives a mapping Ψ from G to F, where z is mapped onto $\bar{z}$ in F, and the vertices in $N_k(z)$ are mapped onto vertices in $N_k(\bar{x})$, for $k \geq 1$. Statements (3) and (7) above amount to the fact that the subgraph of G induced by $N_0(z) \cup N_1(z) \cup N_2(z)$ is mapped isomorphically onto the subgraph of F induced by $N_0(\bar{z}) \cup N_1(\bar{z}) \cup N_2(\bar{z})$. Now we show by induction on k that the mapping Ψ embeds the subgraph of G induced by $\displaystyle\bigcup_{i=0}^{k} N_i(z)$ into the subgraph of F induced by $\displaystyle\bigcup_{i=0}^{k} N_i(\bar{z})$. So, for $k \geq 3$, we have to show that Ψ embeds the subgraph $N_k(z)$ into the subgraph $N_k(\bar{z})$ so that $t \in N_{k-1}(z)$ is adjacent to $u \in N_k(z)$ if and only if $\Psi(t)$ and $\Psi(u)$ are adjacent in F. Since, by interval-regularity, the geodesics from z to u give a k-cube, t is adjacent to u if and only if the n-vectors $\Psi(t)$ and $\Psi(u)$ differ in exactly one coordinate so that $t_i = z \neq u_i$ for some i. This, however, is equivalent to the adjacency of $\Psi(t)$ and $\Psi(u)$ in F. Then, in order to show that Ψ maps $N_k(z)$ into $N_k(\bar{z})$ injectively, it suffices to verify that any two distinct vertices u and v in $N_k(z)$ have at most one common neighbour in $N_{k-1}(z)$. Suppose the contrary, that is: let $s \neq t$ in $N_{k-1}(z)$ be neighbours of u and v. Then s and t have yet another common neighbour, viz. one in $N_{k-2}(z)$, so that they must be adjacent. By the induction hypothesis, $\Psi(s)$ and $\Psi(t)$ are adjacent vertices in $I(\bar{z}, \Psi(u)) \cap N_{k-1}(\bar{z})$. But this is impossible since each interval between $\bar{z}$ and any other vertex in F has no horizontal edges. This establishes injectivity. Finally, let u and v be distinct vertices in $N_k(z)$. We claim that u and v are adjacent if and only if their images under Ψ are. If uv is an edge, then there is a common neighbour t in $N_{k-1}(z)$ by $\bigtriangledown$. Certainly, there is another neighbour s of u in $N_{k-1}(z)$, which is not adjacent to v. Let r be the common neighbour of s and t in $N_{k-2}(z)$. Since r, s, t, u, v induce a house, there exists a common neighbour x of r, s, and v (which is not adjacent to t or u). By virtue of the induction hypothesis, the only edge of this house that might not be preserved by Ψ is the one between $\Psi(u)$ and $\Psi(v)$. But in that case, we would get an induced 5-cycle in F of which the convex closure is not a 5-wheel. Therefore $\Psi(u)$ and $\Psi(v)$ must be adjacent. Since we also know that F is interval-regular and satisfies $\bigtriangledown$ (being a Cartesian product of such graphs), we can reverse the preceding argument and conclude that u and v are adjacent whenever $\Psi(u)$ and $\Psi(v)$ are. This settles (8).

As to surjectivity of Ψ, we prove by induction on k that every vertex of F at distance k from $\bar{z}$ belongs to $\Psi(G)$. For $k \leq 2$ this follows from the construction of F according to (3) and (7). Now, for $k \geq 3$, every vertex in $N_k(\bar{z})$ has two neighbours (and more) in $N_{k-1}(\bar{z}) = \Psi(N_{k-1}(z))$. Since these two vertices are non-adjacent and thus already have two common neighbours in $\Psi(G)$, we must have $N_k(\bar{z}) = \Psi(N_k(z))$. Therefore $\Psi(G)$ is all of F, and so G is isomorphic to F. This concludes the proof of the Theorem.

Some corollaries and remarks   There a number of equivalent conditions that allow us to characterize the graphs of the Theorem. We abstain from listing all possible characterizations. they are generated by combining several of the properties investigated in this paper. For instance, under the presence of property $\diamond$ all I-triangles are equilateral if and only if property $\bigtriangledown$ holds (see Lemma 1). We have also seen in the proof of the Theorem that a graph having

∇ is interval-regular if and only if it has $\diamond$ but no isometric hammock and any two vertices at distance 2 have exactly two common neighbours. Now, in view of Lemma 2 we may substitute ∇ in the preceding statement by requiring that there are no isometric odd cycles of length greater than 3. Hence the interval-regular graphs without such cycles form a proper subclass of the graphs described in the Theorem, viz. no factor is the join of K_1 and a geodetic block of diameter 2. Indeed, a geodetic graph of diameter at most 2 does not contain an induced 5-cycle if and only if it consists of complete graphs all attached at a single vertex; see statement (4) in the proof above. Therefore we deduce from the Theorem the following result.

Corollary 1   A graph G is the Cartesian product of graphs $G_i (i = 1, \cdots, n)$, where each G_i consists of a family of complete graphs all attached at a single edge, if and only if G is interval-regular and does not contain any isometric odd cycle of length greater than 3.

The Hamming graphs are characterized among the graphs of Corollary 1 as those graphs that do not contain induced kites.

Corollary 2   A graph G is a Hamming graph if and only if G is interval-regular and contains neither the kite $K_{1,1,2}$ nor any odd cycle of length greater than 3 as an isometric subgraph.

Substituting in the preceding corollary the requirement that all odd cycles of length greater than 3 are forbidden by the condition that all I-triangles are equilateral, we obtain Theorem 9 of Mulder (1982), cf. Mulder (1980 b). From the Theorem it is clear that the Hamming graphs are precisely the interval-regular graphs that are regular and have property ∇. It would be interesting to characterize the interval-regular graphs that do not contain any induced kite. Are they necessarily Cartesian products of complete graphs and extended odd graphs (for the latter graphs, see Mulder (1982))?

REFERENCES

[1] H. -J. Bandelt (a), Characterizing median graphs, *European J. Combinatorics*, to appear.

[2] H. -J. Bandelt and H. M. Mulder (1983), Infinite median graphs, (0,2)-graphs, and hypercubes, *J. Graph Theory 7*, 487-497.

[3] H. -J. Bandelt and H. M. Mulder (1984), Interval-regular graphs of diameter two, *Discrete Math.* 50, 117-134.

[4] H. -J. Bandelt and H. M. Mulder (1986), Pseudo-modular graphs, *Discrete Math.* 62, 245-260.

[5] H. -J. Bandelt and E. Wilkeit (1987), Dwarf, brick, and triangulation of the torus, *Discrete Math.* 67, 1-14.

[6] J. Bosák (1978), Geodetic graphs, in: *Combinatorics, Colloq. Math. Soc. János Bolyai 18* (North-Holland, Amsterdam), 151-172.

[7] P. V. Ceccherini and A. Sappa (1986), A new characterization of hypercubes, *Annals Discrete Math. 30*, 137-142.

[8] F. R. K. Chung, R. Graham, and M. E. Saks (a), A dynamic location problem for graphs, to appear.

[9] S. Folds, (1977), A characterization of hypercubes, *Discrete Math. 17*, 155-159.

[10] J. M. Laborde and S. P. Rao Hebbare (1982), Another characterization of hypercubes, *Discrete Math. 39*, 161-166.

[11] M. Mollard (a), Two characterizations of generalized hypercube, to appear.

[12] H. M. Mulder (1979), $(0,\lambda)$-graphs and n-cubes, *Discrete Math. 28*, 179- 188.

[13] H. M. Mulder (1980 a), n-Cubes and median graphs, *J. Graph Theory 4*, 107-110.

[14] H. M. Mulder (1980 b), The interval function of a graph, *Math. Centre Tracts 132*, Amsterdam.

[15] H. M. Mulder (1982), Interval-regular graphs, *Discrete Math. 41*, 253-269.

[16] K. R. Parthasarathy and N. Srinivasan (196), On geodetic blocks of diameter 2: some structural properties, *Ars Combinatoria 20*, 49-60.

[17] R. Scapellato (1986), Geodetic graphs of diameter two and some related structures, *J. Combinatorial Theory Ser. B*, 218-229.

[18] R. Scapellato (a), On F-geodetic graphs, to appear.

[19] E. Wilkeit (1986), Isometrische Untegraphen von Hamming-Graphen, doctoral dissertation, Oldenburg.

[20] P. Winkler (1987), The metric structure of graphs: theory and applications, in: C. Whitehead ed., *Surveys in combinatorics 1987, London Math. Society Lecture Note Series 123* (Cambridge University Press, Cambridge) 197-221.

A Note on a Special Case of the 2–Path Problem for Semicomplete Digraphs

Jørgen Bang–Jensen

Department of Mathematics and Computer Science

Odense University, DK–5230 Odense M, DENMARK

ABSTRACT

In this note we treat a special case of the so called 2–path problem for semicomplete digraphs. We consider the following problem: Given a semicomplete digraph T and special vertices $x, y,$ and z. When does T have an (x, z)–path through y? The corresponding decision problem for digraphs is well known to be NP–complete [5]. It follows from a result by Thomassen and Bang–Jensen that the problem is in P for semicomplete digraphs [1]. In this note we give a complete characterization of those semicomplete digraphs T with special vertices $x, y,$ and z, which do not have an (x, z)–path through y. This characterization also leads to a simpler algorithm for the problem.

1. Introduction

The so called 2–path problem for digraphs is the following:

    (P1):   Given a digraph D and four different vertices x_1, x_2, y_1, y_2.

            Decide whether or not D has two disjoint paths which start in

            x_1 (respectively x_2) and end in y_1 (respectively y_2).

This problem is well known to be NP–complete for digraphs in general [5]. A polynomial algorithm for the problem (P1) when restricted to semicomplete digraphs (i.e. digraphs with no two nonadjacent vertices) is given in [1]. In [3] we give best possible conditions in terms of connectivity for a semicomplete digraph T with special vertices x_1, x_2, y_1, y_2 to have a pair of disjoint (x_1, y_1)–, (x_2, y_2)–paths. In [3] we give examples of 2–connected tournaments and of 4-connected semicomplete digraphs which contain special vertices x_1, x_2, y_1, y_2 without a pair of disjoint (x_1, y_1)–, (x_2, y_2)–

paths. Thus there seems to be no simple structural characterization of the semicomplete digraphs with no such pair of paths for given special vertices x_1, x_2, y_1, y_2. Consider the following special case of problem (P1):

(P2):   Given a digraph D and special vertices x, y and z all different.

Decide whether or not D has an (x, z)–path through y.

This problem is also NP–complete for digraphs in general [5].

It is easy to see how to reduce the problem (P1) to (P2) by a polynomial time reduction: Just add a new vertex y to the digraph D and join y to D by an edge $y_1 \rightarrow y$ and an edge $y \rightarrow x_2$. Now the new digraph has an (x_1, y_2)–path through y if and only if the original digraph has a pair of disjoint (x_1, y_1)–, (x_2, y_2)–paths. (P2) also reduces to (P1) by a polynomial time reduction: Let D with special vertices x, y and z be given. Partition the vertices of $V(D) - y$ into four sets I, O, C, E, such that

I = {u ∈ V(D): the only edge between u and y is the edge u → y},

O = {u ∈ V(D): the only edge between u and y is the edge y → u},

C = {u ∈ V(D): u and y induce a directed 2–cycle in D, i.e. u → y and y → u},

E = {u ∈ V(D): there are no edges between u and y in D}.

Construct a new digraph D' by deleting y and adding two new vertices y_1 and x_2 with edge $y_1 \rightarrow x_2$. Let the edges between {y_1, x_2} and the rest of V(D) be as between y and these in D. (compare Figure 1.1 below). Now it is easy to see that D has an (x, z)–path through y if and only if D' has a pair of disjoint (x, y_1)–, (x_2, z)– paths.

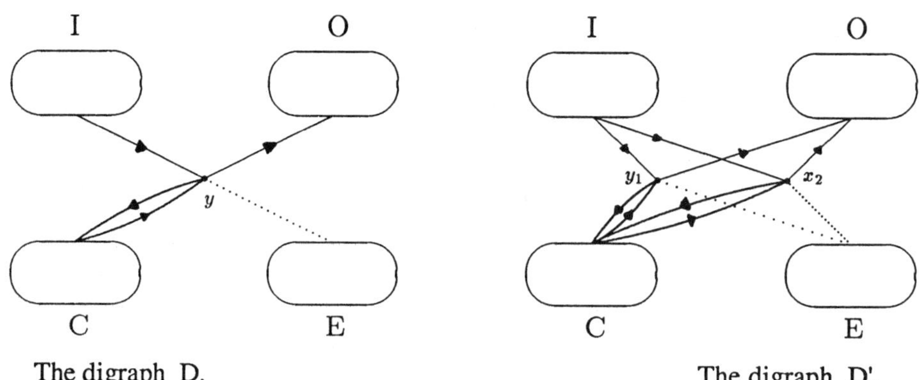

The digraph D.   The digraph D'.

Figure 1.1

In the case when D is a semicomplete digraph the digraph D' is also a semicomplete digraph. Thus the result in [1] implies the existence of a polynomial algorithm for problem (P2) for semicomplete digraphs. In this paper we give a complete characterization of those semicomplete digraphs T with special vertices x, y, z which do not have an (x, z)–path through y. The characterization is given in terms of two operations by which all semicomplete digraphs (T, x, y, z) without an (x, z)–path through y can be constructed from a class of quite simple semicomplete digraphs.

The analogous problems (P1'), (P2') dealing with edge–disjoint paths are also NP–complete [5]. In [2] both of these problems where completely solved in the case of tournaments.

2 . Terminology and Preliminaries

Most of the notion is the same as in [2], but for the sake of completeness we will include all the needed notation here, except for the very standard notation, for which we refer to [4].

A *digraph* D consists of a pair V(D), E(D), where V(D) is a finite set of *vertices* and E(D) is a set of ordered pairs xy of vertices called *edges.* In our definition of a digraph we do not allow multiple edges in the same direction between two vertices. A *double edge* is a pair of edges inducing a directed 2–cycle. An *oriented* graph is a digraph with no cycle of length 2. A *semicomplete* digraph is a digraph with no non adjacent vertices. A *tournament* is an oriented graph with no non adjacent vertices. Thus tournaments are a special subclass of the semicomplete digraphs.

If xy is an edge of the digraph D then we say that x *dominates* y and we will use the notation $x \rightarrow y$ to denote this. With slight abuse of notation $x \rightarrow y$ also denotes the edge xy. For any subset A of $V(D) \cup E(D)$, $D - A$ denotes the subgraph obtained by deleting all vertices of A and their incident edges and then deleting the edges of A still present. We write $D - x$ instead of $D - \{x\}$ when $x \in V(D) \cup E(D)$.

The subgraph *induced* by a vertex set A of D is defined as $D - (V(D) \setminus A)$ and is denoted by D(A). We will often write $x \in D$ instead of $x \in V(D)$ or $x \in E(D)$, but the meaning will always be clear. A *path* is a digraph with vertex set $\{x_1, x_2, ..., x_n\}$ and edge set $\{x_1 \rightarrow x_2, x_2 \rightarrow x_3, ..., x_{n-1} \rightarrow x_n\}$, such that all the vertices and edges shown are distinct. We call such a path an (x_1, x_n)–path and denote it by $x_1 \rightarrow x_2 \rightarrow ... \rightarrow x_n$. If P is a path containing vertices x, y in that order then we let P[x, y] denote the part of P from x to y. In this paper a path is always a directed path. A *component* D' of a digraph D is maximal subdigraph, such that for any two vertices x, $y \in D'$, D' contains an (x, y)–path and a (y, x)–path. A digraph D is *strong* or *strongly connected* if it has only one component. D is *k-connected* if for any set A of at most $k - 1$ vertices, $D - A$ is strong.

3. A Best Possible Condition in Terms of Local Connectivities

Lemma 3.1 *Let u, v, w be different vertices in a semicomplete digraph T. Suppose that there are at least two internally disjoint (u, v)–paths in T and that P_1, P_2 are internally disjoint (v, w)–paths, which together induce a semicomplete digraph with no (v, w)–path of length 2. Then T has a (u, w)–path through v.*

Proof Let Q be a (u, v)–path in $T - w$. Q intersects both P_1 and P_2. Let z be the first vertex on Q, which is on one of P_1, P_2 as we go along Q from u towards v. If $z = v$ then we are done. So suppose $z \neq v$. We may assume without loss of generality that $z \in V(P_1)$. Let y be the predecessor of w on P_1. Then the only edge between y and v is $y \to v$, since otherwise $v \to y \to w$ is a (v, w)–path of length 2 in $T(V(P_1) \cup V(P_2))$, contradiction the assumption on P_1 and P_2. Now let $Q' = Q[u, z] \cup P_1[z, y] \cup \{y \to v\}$. Then $Q' \cup P_2$ is a (u, w)–path through v. $\square$

Corollary 3.2 *Let x, y and z be different vertices of a semicomplete digraph T. Suppose that the number of internally disjoint (x, y)–paths in $T - z$ is at least 2 and that the number of internally disjoint (y, z)–paths in $T - x$ is at least 2. Then T has an (x, z)–path through y.*

Proof Let P_1, P_2 be internally disjoint (y, z)–paths in $T - x$. If $T' = T(V(P_1) \cup V(P_2))$ does not contain a (y, z)–path of length 2 then the result follows by Lemma 3.1. So suppose that $y \to a \to z$ is a (y, z)–path of length 2 in T'. Then $T - \{a, z\}$ contains an (x, y)–path, since the number of internally disjoint (x, y)–paths in $T - z$ is at least 2, and we are done. $\square$

This result is best possible in terms of local connectivities as shown by the infinite family of semicomplete digraphs T_n defined as follows:

$$V(T_n) = \{x = x_1, x_2, ..., x_{n-3}, u, y, z\},$$

$$z \to y \quad \text{is the only edge between } y \text{ and } z.$$

The edges between u and $\{x_{n-3}, y, z\}$, and between x and z are all double edges,

$$x_i \to x_{i+1} \quad \text{for } i = 1, 2, ..., n - 4,$$

$$x_j \to x_i \quad \text{for } j \geq i + 2, i = 1, 2, ..., n - 5,$$

$$x_i \to z \quad \text{for } i = 1, 2, ..., n - 3,$$

$$y \to x_i \quad \text{for } i = 1, 2, ..., n - 3,$$

$$u \to x_i \quad \text{for } i = 1, 2, ..., n - 4.$$

Compare figure 3.1 below. It is easy to see that T_n does not have an (x, z)–path through y. In fact $T_n - x$ has $n - 2$ internally disjoint (y, z)–paths, there is an (x, y)–path in $T_n - z$, but there are no two internally disjoint (x, y)-paths in $T_n - z$.

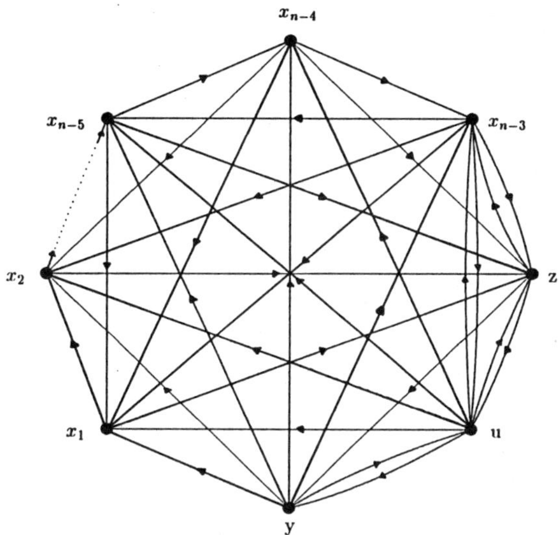

Figure 3.1 The semicomplete digraph T_n

In the next sections we shall show that if (T, x, y, z) (meaning the semicomplete digraph T with special vertices x, y, and z) has an (x, y)–path in $T - z$ and a (y,z)–path in T–x, but does not have an (x, z)–path through y, then T bears some resemblance to the semicomplete digraph T_n above.

4. Definitions and Operations

Lemma 4.1 *Let T be a semicomplete digraph with 4 vertices, $V(T) = \{x, y, z, u\}$. Suppose that $T - z$ has an (x, y)–path and that $T - x$ has a (y, z)–path. Then T does not have an (x, z)–path through y if and only if u lies on all (x, y)–paths in T − z and on all (y, z)–paths in $T - x$.*

Proof Suppose that $T - z$ has an (x, y)–path and $T - x$ has a (y, z)–path, but T does not have an (x, z)–path through y. Then the only edge between x and y is y → x and the only edge between y and z is z → y. Thus every (x, y)-path in $T - z$ must pass through u and every (y, z)–path in $T - x$ must pass through u. Therefore u has the desired property. The other half of the proof is trivial. ❑

Definition 4.1 *Let (T, x, y, z) denote a semicomplete digraph with special vertices x, y and z and $|V(T)| \geq 4$. We say that the tuple (T, x, y, z) is of type 1 if there is an (x, y)–path in $T - z$, there is a (y, z)–path in $T - x$, and there exists a vertex $u \in V(T) - \{x, y, z\}$, such that there is no (x, y)–path in $T - \{u, z\}$ and no (y, z)–path in T − \{u, x\}.*

Definition 4.2   *We say that the semicomplete digraph (T, x, y, z) is of type 2 if there is an (x, y)–path in T − z, there is a (y, z)–path in T − x, but T has no (x, z)– path through y.*

Clearly if (T, x, y, z) is of type 1 then (T, x, y, z) is also of type 2. The other implication does not hold as we shall see below.

We will define two operations, each of which will take a semicomplete digraph (T, x, y, z) of type 2 into a new semicomplete digraph (T*, x*, y*, z*), such that T is a subdigraph of T* and (T*, x*, y*, z*) is of type 2.

Operation 1   Let (T, x, y, z) be given. Let T' be a semicomplete digraph, with $V(T) \cap V(T') = \varnothing$, containing a vertex x' such that x' can reach all other vertices in T' by directed paths in T'. (Possibly V(T') = {x'}). Let (T*, x', y, z) denote the semicomplete digraph obtained from T' and T, by adding edges as follows:

      Let all edges between T' and T − {x, z} go from T to T',

      Let at least one vetex in T' dominate x,

      Let all edges between z and T' − x' go from z to T' − x', and finally

      choosing the edge between x' and z at random or making it a double

      edge.

Compare figure 4.1.

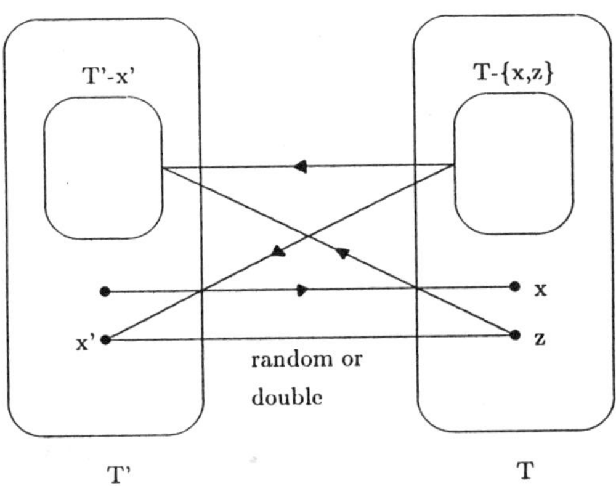

Figure 4.1: Operation 1.

Operation 2   Let (T, x, y, z) be given. Let T" be a semicomplete digraph with $V(T") \cap V(T) = \varnothing$, containing a vertex z" such that all vertices in T" − z" can reach z" by directed paths in T". (Possibly V(T") = {z"}). Let (T*, x, y, z") be the semicomplete digraph obtained from T and T" be adding edges as follows:

      Let all edges between T − {x, z} and T" go from T to T,

      Let at least one vertex in T" be dominated by z,

Let all edges between x and T" – z" go from T" to x, and finally choosing the edge between x and z" at random or making it a double edge.

Compare figure 4.2.

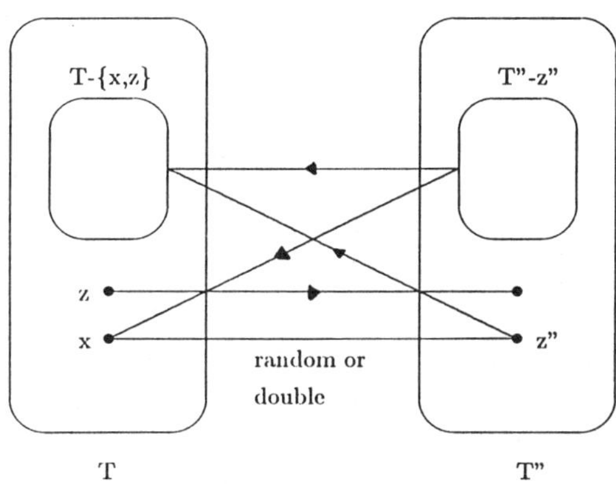

Figure 4.2 Operation 2.

It is not obvious that operations 1 and 2 produce semicomplete digraphs of type 2. The proof of that is a part of Theorem 5.1 in the next section.

Definition 4.3   Let (T, x, y, z) and (T^, x^, y^, z^) be semicomplete digraphs with special vertices x, y, and z, respectively x^, y^, and z^. We say that (T, x, y, z) is *(T^, x^, y^, z^)–obtainable* if (T, x, y, z) can be constructed from (T^, x^, y^, z^) by zero or more applications of operations 1 and 2 in any order.

5 . The Main Result

We are ready to formulate the main result.

Theorem 5.1 *Let the sets M and M′ consisting of semicomplete digraphs be defined as follows:*

$$M = \{(T, x, y, z): \ |V(T)| \geq 4, \text{ and } (T, x, y, z) \text{ is of }$$
type 2},

$$M' = \{(T^*, x^*, y^*, z^*): \ (T^*, x^*, y^*, z^*) \text{ is } (T^, x^,$$
y^, z^)–obtainable for some semicomplete digraph (T^,
x^, y^, z^) of type 1}.

Then M = M′.

Proof Suppose that (T, x, y, z) is $(T^\wedge, x^\wedge, y^\wedge, z^\wedge)$–obtainable for some $(T^\wedge, x^\wedge, y^\wedge, z^\wedge)$ of type 1. We want to prove that $(T, x, y, z) \in M$. It is sufficient to prove that M contains all semicomplete digraphs of type 1 and that operations 1 and 2, applied to a semicomplete digraph in M, results in a semicomplete digraph which is in M. The first assertion follows from the remark after definition 4.2. For the second assertion we may assume without loss of generality that $(T^\wedge, x^\wedge, y^\wedge, z^\wedge) \in M$ and that $(T, x', y^\wedge, z^\wedge)$ is obtained from $(T^\wedge, x^\wedge, y^\wedge, z^\wedge)$ by operation 1. (x' denotes the special vertex of the semicomplete digraph T' in operation 1). Now let us show that $(T, x', y^\wedge, z^\wedge) \in M$. Clearly there is an $(x', y^\wedge)$–path in $T - z^\wedge$ and there is a $(y^\wedge, z^\wedge)$–path in $T - x'$. It is easy to see that every $(x', y^\wedge)$–path in $T - z^\wedge$ must pass through $x^\wedge$. All edges between $z^\wedge$ and $T' - x'$ go from $z^\wedge$ to T', by the definition of operation 1. Thus the fact that all edges between $T^\wedge - x^\wedge$ and $T' - x'$ go from $T^\wedge$ to T' implies that no $(y^\wedge, z^\wedge)$–path in $T - \{x', x^\wedge\}$ passes through T'. Hence, the fact that $(T^\wedge, x^\wedge, y^\wedge, z^\wedge) \in M$ implies that there can be no $(x', z^\wedge)$–path in T which passes through $y^\wedge$, and therefore $(T, x', y^\wedge, z^\wedge)$ is of type 2, i.e. $(T, x', y^\wedge, z^\wedge) \in M$. The proof in the case of operation 2 is analogous.

Now let $(T, x, y, z) \in M$. We want to prove that (T, x, y, z) is $(T^\wedge, x^\wedge, y^\wedge, z^\wedge)$–obtainable for some $(T^\wedge, x^\wedge, y^\wedge, z^\wedge)$ of type 1. The proof is by induction on $n = |V(T)|$. If $n = 4$ then the result follows from Lemma 4.1. Thus we may proceed to the induction step and assume that $n \geq 5$. By Corollary 3.2 we may assume that either there are no two internally disjoint (x, y)–paths in $T - z$, or there are no two internally disjoint (y, z)–paths in $T - x$. We may assume without loss of generality that we are in the first case. The proof in the other case is anaolgous. By Menger's theorem there exists a vertex $a \in V(T) - z$, such that there is no (x, y)–path in $T - \{a, z\}$. Let x^* be chosen such that there is no (x, y)–path in $T - \{x^*, z\}$ and the set $A = \{u \in T:$ there exists an (x, u)–path in $T - \{x^*, z\}\}$ is minimal with respect to inclusion. Let $B = V(T) - A - \{x^*, z\}$. Then all edges between A and B go from B to A. Furthmore $y \in B$.

Claim: *There are no edges from $A - x$ to z.*

Proof Suppose that $a \in A - x$ and $a \to z$. Then $y \to a \to z$ is a (y, z)–path in $T - x$. Thus the fact that $(T, x, y, z) \in M$, implies that every (x, y)–path in $T - z$ contains a. Let $A = A' \cup A'' \cup \{a\}$, where

$$A' = \{u \in T(A): \text{ there exists an } (x, u)\text{-path in } T(A - a)\}.$$

Then there is no edge from A' to x^*, since otherwise we can find an (x, y)–path in $T - \{a, z\}$, contradicting $(T, x, y, z) \in M$. (There is an (x^*, y)–path in $T(B \cup \{x^*\})$ by the choice of x^*). Also all edges between A' and A'' go from A'' to A' (possibly $A'' = \varnothing$). Thus choosing a instead of x^* gives a contradiction to the choice of x*, since A' is a proper subset of A. This contradiction proves the claim. ❏

By the claim and the fact that all edges between A and B go from B to A, there can be no (y, z)–path in T − {x, x*} which passes through A. Let T* = T(B ∪ {x*, z}). If every (y, z)–path in T − x contains x* then (T, x, y, z) is of type 1 and thus belongs to M'. Therefore we can assume that T − {x, x*} contains a (y, z)–path. By the above this path lies entirely in T*. Thus in T* we have at least one (y, z)–path in T* − x* and also at least one (x*, y)–path in T* − z (there is at least one (x, y)–path in T − z and every (x, y)–path in T − z contains x*). Since (T, x, y, z) ∈ M we must have that T* does not contain an (x*, z)–path through y, as such a path can easily be extended to the desired path in T. Now it follows easily that $|V(T^*)| \geq 4$ and thus (T*, x*, y, z) ∈ M. By the induction hypothesis (T*, x*, y, z) ∈ M', i.e. there exists a semicomplete digraph (T^, x^, y^, z^) of type 1, such that (T*, x*, y, z) is (T^, x^, y^, z^)–obtainable. Thus (T, x, y, z) is (T^, x^, y^, z^)–obtainable too, since we can add one more application of operation 1 to the sequence of applications of operations 1 and 2 which produces (T*, x*, y, z) from (T^, x^, y^, z^). In this last application of operation 1 we take T' = T(A) and let the edges be those induced from T. This proves that (T, x, y, z) ∈ M' and the proof is complete. ❏

Corollary 5.2   *There exists a polynomial algorithm P which, given a semicomplete digraph (T, x, y, z), decides whether or not T has an (x, z)–path through y.*

Proof   We give a sketch of such an algorithm below:
1. If there is no (x, y)–path in T − z or no (y, z)–path in T − x then the algorithm reports the non−existence of the desired path and stops.
2. If for all a ∈ T − {x, y, z}: T − {a, z} contains an (x, y)–path, and T − {a, x} contains a (y, z)–path, then the algorithm reports the existence of the desired path and stops.
3. If there are at least 2 internally disjoint (x, y)–paths in T − z then turn all edges and interchange the names x and z. (* I.e. from now on we may assume that there are no two internally disjoint (x, y)–paths in T − z *)
4. Choose x* such that there is no (x, y)–path in T − {x*, z} and A = {u ∈ V(T): there exists an (x, u)–path in T − {x*, z}} is minimal with respect to inclusion. Then y ∈ T − A − {x*, z}.
5. If there is no (y, z)–path in T − {x, x*} then the algorithm reports the non− existence of the desired path and stops.
6. If there is an edge from A − x to z then the algorithm reports the existence of the desired path and stops.

(* Now we know that T contains an (x, z)–path through y if and only if $T^* = T - A$ contains an (x^*, z)–path through y. Also there is an (x^*, y)–path in $T^* - z$ and there is a (y, z)–path in $T^* - x^*$. *)

7. If $|V(T^*)| = 3$ then the algorithm reports the existence of the desired path and stops.

8. Let $x = x^*$, $T = T^*$ and go to step 2.

This algorithm, whose validity follows from the proof of Theorem 5.1, is clearly polynomial in the size of the semicomplete digraph T. ❏

The reader can verify for himself/herself that the algorithm can be modified to find such a path if it exists.

The algorithm presented above is considerably simpler than the one, implied by the more general algorithm in [1].

REFERENCES

[1]   Bang–Jensen, J. and C. Thomassen, A polynomial algorithm for the 2–path problem for semicomplete digraphs, submitted.

[2]   Bang–Jensen, J., Edge–disjoint In– and Out–branchings in tournaments and related path problems, Journal of Combinatoral Theory Ser. B, to appear.

[3]   Bang–Jensen, J., On the 2–linkage problem for semicomplete digraphs, to appear in the proceedings of the Dirac conference in Denmark 1985.

[4]   Bondy, J.A. and U.S.R. Murthy, Graph Theory with applications, Mac Millan Press (1976).

[5]   Fortune, S., J. Hopcroft and J. Wyllie, The directed subgraph homeomorphism problem. *Theoretical Computer Science* **10** (1980), 111 – 121.

CYCLES OF LENGTH 0 MODULO 3 IN GRAPHS

C. A Barefoot

New Mexico Institute of Mining and Technology

L. H. Clark

Jack Douthett

R. C. Entringer

University of New Mexico

M. R. Fellows

University of Idaho

ABSTRACT

It is shown that every graph in any of the following three classes contains a cycle of length 0 mod 3: (i) cubic graphs, (ii) subdivisions of 3-connected cubic graphs with order $n \geq 10$ and (iii) subdivisions of graphs with order n and $3n - 5$ or more edges. The bound in (iii) is sharp. The proof of (ii) is by induction on n, the base case being established by computer.

1.　Introduction

In 1975 Burr and Erdös (see [7]) asked if, for given integers a and b, a graph G must contain a cycle of length a mod b even though G is sparse. Clearly, restrictions must be put on a to accommodate the absence of odd cycles in bipartite graphs. Their question can then be formulated precisely as follows. Given integers $a \geq 0$ and $b > 0$, with a even if b is even, is there a constant $c_{a,b}$ depending only on a and b such that every graph with order n and at least $c_{a,b}n$ edges contains a cycle of length a mod b? Two years later Bollobás [2] provided an affirmative answer with the following result.

Theorem　　(Bollobás). Let $k \geq 3$ be an odd natural number and let G be a graph of order n. If

$$\delta(G) \geq \frac{(k+1)^k - 1}{k}$$

or

$$e(G) \geq \frac{(k+1)^k - k - 1}{k} n,$$

then for every natural number ℓ, G contains a cycle of length ℓ modulo k.

In 1983 Thomassen [9], in extending and refining the work of Bollobás, obtained the following result.

Theorem　　(Thomassen). If k and d are natural numbers, then any graph G of minimum degree at least $4d(k+1)$ contains a collection of paths $P_0, P_1, \cdots, P_{k+1}$ such that the paths $P_1, P_2, \cdots, P_{k+1}$ have length d and have exactly an end-vertex x_0 in common and P_0 and P_i have precisely the end of P_i distinct from x_0 in common for each $i = 1, 2, \cdots, k+1$. In particular, G contains a cycle of length $2d$ modulo k.

Although Thomassen's result is stronger in general, the result of Bollobás tells us that $\delta(G) \geq 21$ implies G contains a cycle of length 0 mod 3.

Our ultimate goal is the determination of all graphs that contain a cycle of length 0 mod 3. We will call such graphs *good*. A graph that is not good is *bad*. We note the obvious, but frequently used, consequence that graph G is bad iff every subgraph of G is bad. The next section will be devoted exclusively to good and bad graphs; while in the Conclusion, we briefly consider graphs with cycles of length a mod b for other values of a and b.

All results of the next section will depend crucially on several observations. The verification of certain of these observations is left to the Appendix. The following general rule was applied in determining the type of verification. If it is required to show that every graph in a given class is bad, this is accomplished by exhaustive computer search. On the other hand, if it is required to show that every graph G in a given class is good, this is done "manually" by exhibiting a good cycle in G.

We will generally follow the notation and terminology of Bondy and Murty [3] except that all graphs will be simple, i.e., they will have no loops or multiple edges. In particular, the *order* of a graph is $|V(G)|$ and the *size* is $|E(G)|$. Multigraphs will be considered in a paper in preparation in which circuits, i.e., closed trails, of a certain size are studied.

2. Three Classes of Good Graphs

We begin with a simple, but suggestive observation.

Observation A   Every cubic graph of order $n \le 14$ is good.

Verification   This can be confirmed using the tables of Bussemaker et al. [4]. We will show that the restriction $n \le 14$ is unnecessary, but do this only as a consequence of a more general result concerning subdivisions of cubic graphs. We will use the labeling system illustrated in Figure 1 to indicate subdivisions of graphs. There the labels assigned to an edge, sometimes called the *weight* of that edge, indicates the number of edges, mod 3, into which that edge is to be subdivided. The weight of a subgraph is the sum of the weights of its edges and we sometimes refer to the weighted length of a path or cycle. Associated with a subdivision G of a graph is the *companion* to G obtained by multiplying the edge labels of G by 2 mod 3. Clearly, a graph is good iff its companion is good.

An infinite class of bad graphs is depicted in Figure 2. These graphs are subdivisions of certain cubic graphs with connectivity 2 and containing m copies of $K_4 - e$ where m is any positive integer that is not a multiple of 3. We will show that the situation changes dramatically when we consider subdivisions of cubic graphs that are 3-connected.

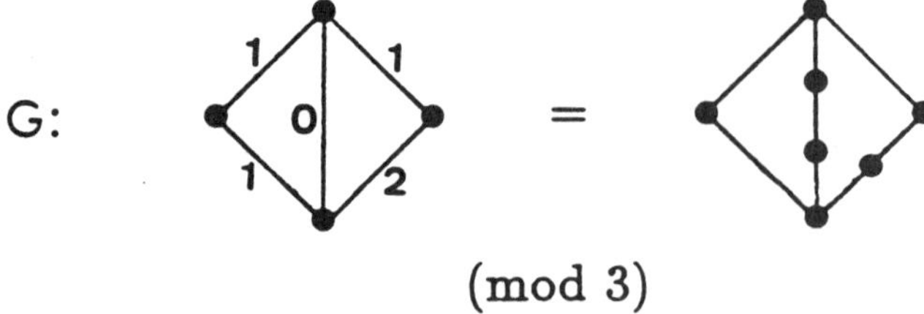

Figure 1. Subdivisions and companions

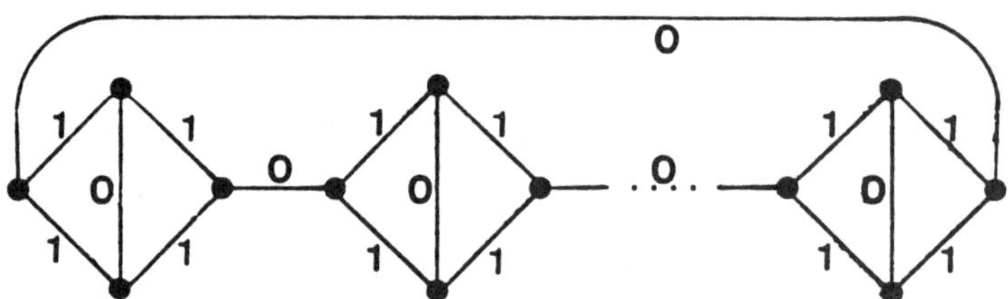

Figure 2. Bad subdivisions of 2-connected cubic graphs

The next result is essential to all that follows.

Observation B   Except for companions, Figure 3 is a list of all bad subdivisions of 3-connected cubic graphs of order $n \leq 8$.

Verification   By computer.

Observation C   Every subdivision of every 3-connected cubic graph of order 10 is good.

Verification   (see appendix)

Observation C leads directly to one of our main results.

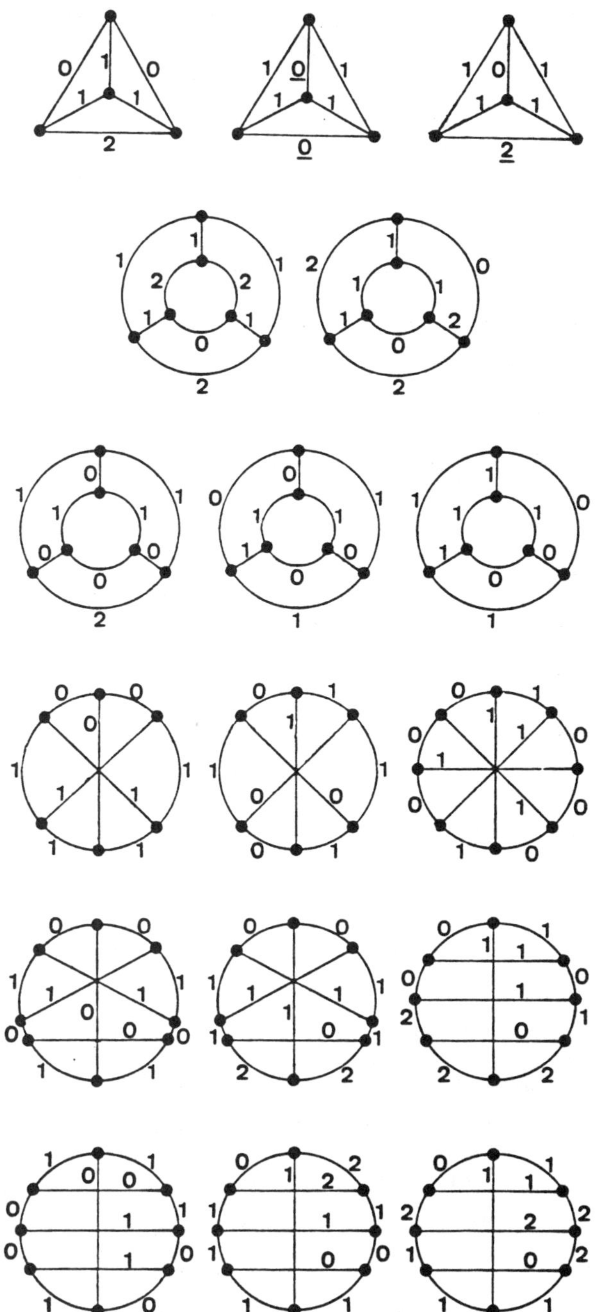

Figure 3. The bad subdivisions of the 3-connected cubic graphs

of order $n \leq 8$

Theorem 1   Every subdivision of every 3-connected cubic graph G of order $n \geq 10$ is good.

Proof (by induction on the order n of G)   Observation C provides the basis for the induction. Suppose, now, that G' is a bad subdivision of a 3-connected cubic graph G of order $n \geq 12$. Barnette and Grünbaum [1] have shown that every 3-connected graph H different from K_4 contains an edge e for which $H - e$ is a subdivision of a 3-connected graph. Let uv be such an edge of G and let P be the corresponding $u - v$ path in G'. Then the nontrivial component of $G' - E(P)$ is a bad subdivision of a 3-connected cubic graph of order $n - 2 \geq 10$ which is impossible by the inductive hypothesis.

We are now able to extend Observation A, but will require the following result.

Lemma   If G is a cubic graph with connectivity 2, then G contains subgraphs H_i, $i = 1, 2$ satisfying:

(i) $V(H_1) \cap V(H_2) = \phi$,

(ii) H_i contains two vertices u_i and v_i of degree 2, all others have degree 3,

(iii) $u_i v_i \notin E(G)$,

(iv) G contains disjoint $u_1 - u_2$ and $v_1 - v_2$ paths P_u and P_v, respectively, for which $(V(P_u) \cup V(P_v)) \cap V(H_i) = \{u_i, v_i\}$ and

(v) $\kappa(H_i + u_i v_i) = 3$.

Proof   We use the fact that connectivity and edge-connectivity are equal in any cubic graph. Let $\{e_1', e_2'\}$ be an edge cut of G and let $H_i', i = 1, 2$ be the components of $G - \{e_1', e_2'\}$. Of all subgraphs of H_i', that can be obtained as a component of $G - \{e_{i,1} e_{i,2}\}$ for some $e_{i,1}, e_{i,2} \in E(G)$, let H_i be one with minimal order. Clearly, H_i satisfies (i), (ii) and, since G is cubic, (iii) (see Figure 4). Because G has connectivity 2, there exist disjoint paths $P_u = (u_1 \cdots u_2)$ and $P_v = (v_1 \cdots v_2)$ for which (iv) holds. Finally, the minimality of H_i can be seen to guarantee $\kappa(H_i + u_i v_i) = 3$.

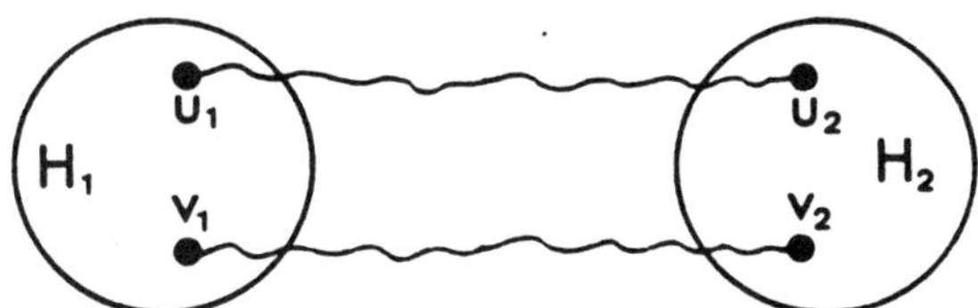

Figure 4. The subgraphs H_i

Theorem 2   Every cubic graph G is good.

Proof   (i) $\kappa(G) = 3$. In this case, the theorem follows immediately from Observation A and Theorem 1.

(ii) $\kappa(G) = 2$. If G is bad, then the subgraphs $H_i + u_i v_i, i = 1, 2$, described in the Lemma must both be bad when weighted as follows. Each edge except $u_i v_i$ has weight 1. The weight of edge $u_i v_i$ is the weight of an arbitrary $u_i - v_i$ path in G that also contains u_{3-i} and v_{3-i}. Thus, each H_i is a bad weighted 3-connected cubic graph in which at most one edge has weight different from 1 mod 3. In view of Observation B and Theorem 1, no such graphs exist and we conclude that G is good.

(iii) $\kappa(G) = 1$. Let H be an end block of G and let v be the cut vertex of G in $V(H)$. We may assume the vertices u and w, say, adjacent to v in H are not adjacent for otherwise G is good. We now apply the lemma to $H - v + uw$ argue as in (ii), and exhibit at least one bad weighted 3-connected cubic graph with at most one edge having a weight different from 1 mod 3. As before, we conclude that G is good.

Theorem 1 suggests the possibility that every subdivision of every 3-connected graph of sufficiently large order is good. The graphs $G_n, n \geq 4$, of Figure 5 show this is false. Here T is any tree of order $n - 2 \geq 2$. All edges of T have weight 0, the edge uv has weight 0 and both u and v are adjacent to all vertices of T and all of these edges have weight 1. It is easily verified that G_n is bad.

These graphs also are extremal graphs for our next result, a density theorem. First, we need certain properties of K_5 and the graph L constructed from K_4 with vertex set $\{w_1, w_2, w_3, w_4\}$ by adding two new vertices u and v along with the edges $uv, uw_i, i = 1, 2$ and $vw_j, j = 3, 4$.

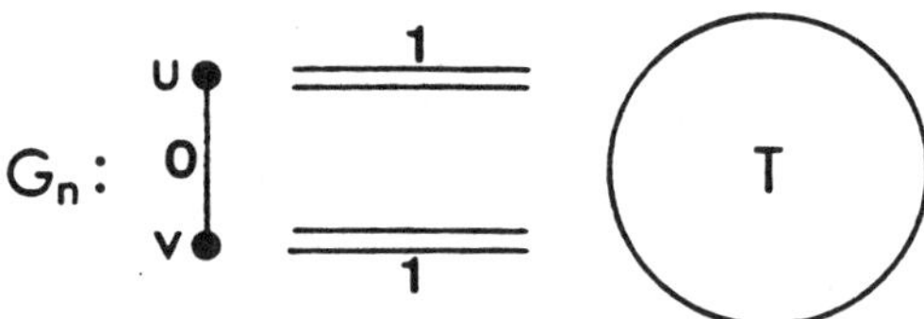

Figure 5. A bad subdivision of a 3-connected graph order n

and size $3n - 6$

Observation D   Every subdivision of K_5 or of L is good.

Verification   (see Appendix)

Theorem 3   If G has order $n \geq 3$ and size $\geq 3n - 5$, then every subdivision of G is good.

Proof   Dirac [6] has conjectured that every graph of order $n \geq 5$ and size at least $3n - 5$ contains a subdivision of K_5. Dirac's conjecture remains unsettled, but Thomassen [8] has shown that such a graph contains either a subdivision of K_5 or of the graph L. The theorem now follows immediately from Observation D; the graphs of Figure 5 show the bound is sharp.

We conclude this section with three conjectures, the first of which gives a lower bound for $c_{0,3}$ (see Introduction). Note that we are not necessarily concerned with subdivisions now.

Conjecture 1   If the graph G has order $n \geq 3$ and size at least $2n - 3$, then G is good. The graphs $K_{2,n-2}$ show that the bound on $|E(G)|$ cannot be improved.

Conjecture 2   (N. Dean[5]). If $\kappa(G) \geq m \geq 3$, then G contains a cycle of length 0 mod m.

Conjecture 2 implies Conjecture 1. This is obvious if $\kappa(G) = 3$. If $\kappa(G) \leq 2$, suppose G is a minimal counterexample and let S be a minimal cut set of $G(S = \phi$ if G is not connected). Let $G_i, i = 1, \cdots, m$ be the S-components of G and set $n_i = |V(G_i)|$. Clearly, $n_i \geq 3(i = 1, \cdots, m)$ and

$$\sum_{i=1}^{m} n_i = n + (m - 1)|S|$$

so that, since each of the G_i, $i = 1, \cdots, m$, is bad,

$$2n - 3 \leq |E(G)| \leq \sum_{i=1}^{m} |E(G_i)| \leq \sum_{i=1}^{m} (2n_i - 4)$$
$$= 2n + 2(m - 1)|S| - 4m \leq 2n - 4.$$

Thus, Conjecture 1 follows from Conjecture 2 by induction on n.

Perhaps even a stronger result holds.

Conjecture 3   If $\delta(G) \geq 3$, then G is good.

3. Conclusion

Theorem 1 suggests the generalization: for fixed $b \geq 3$, there is a constant n_b such that every subdivision of every 3-connected cubic graph of order $n \geq n_b$ contains a cycle of length 0 mod b.

The graphs of Figure 6 show that no such theorem is possible for even b, since the graph pictured has no cycle of length 0 mod b. To see this, we simply observe that the number of edges with weight 1 in any cycle is either 2 or odd.

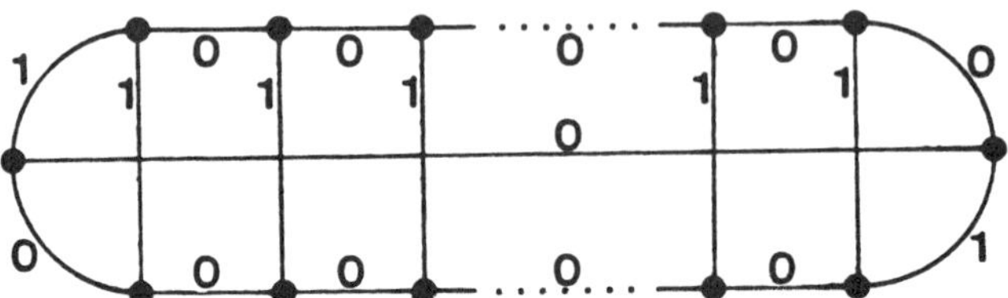

Figure 6. A subdivision of a cubic graph of order $n \geq 4$
with no cycle of length 0 mod b, $b \geq 4$ and even

Now suppose b is odd and at least 3. The graph pictured in Figure 7 has no cycle of length 0 mod b. To see this, note that the graph contains exactly $2b$ edges with weight 1, every cycle must contain at least one edge with weight 1 and that no cycle can contain all $2b$ edges of weight 1. Finally, one observes that no cycle can contain exactly b edges of weight 1, since b is odd. Consequently we have, for odd $b \geq 3$, a graph with no cycles of length 0 mod b and of order $n = 4b - 4$. It appears, then, that any attempt to prove the above generalization of Theorem 1 along the lines of our proof will require computational time that increases exponentially with b.

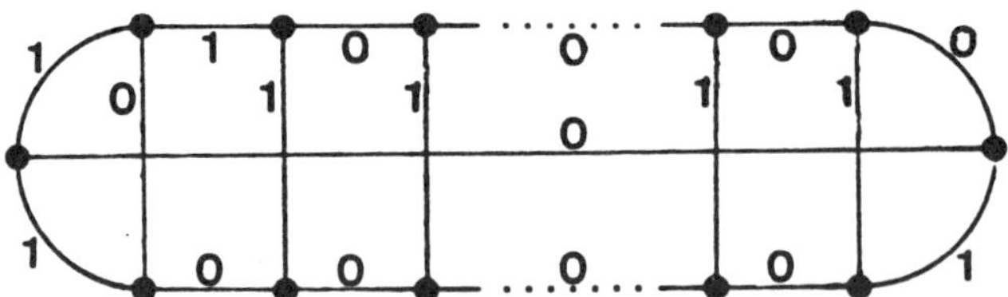

Figure 7. A subdivision of a cubic graph of order $n = 4b - 4, b \geq 3$
and odd, with no cycle of length 0 mod b

In connection with Theorem 2 and Conjecture 3, Erdös [7] has reported that H. Jochens of Hamburg has proved that every graph G with minimum degree 3 contains a cycle of length 1 mod 3 unless every block of G is isomorphic to the Petersen graph.

In connection with Theorem 3 and Conjecture 1, Burr and Erdös (see [7]) have shown that every graph of order n and size at least $(b+2)n$ contains a cycle of length 2 mod b.

REFERENCES

[1] D. Barnette and B. Grünbaum, On Steinitz's theorem concerning convex 3-polytopes and on some properties of planar graphs. "The Many Facets of Graph Theory". *Lecture Notes in Mathematics*, vol. 110, Springer, Berlin, 1969, 27-40.

[2] B. Bollobás, Cycles modulo k. *Bull. London Math. Soc. 9* (1977), 97-98.

[3] J. A. Bondy and U. S. R. Murty, "Graph Theory with Applications". Elsevier, New York, 1976.

[4] F. C. Bussemaker, S. Cobeljić, D. M. Cvetković, and J. J. Seidel, Computer investigation of cubic graphs. *T. H. Report 76-WSK-01, Department of Mathematics, Technological University Eindhoven*, Eindhoven, 1976.

[5] N. Dean, private communication.

[6] G. A. Dirac, Homomorphism theorems for graphs. *Math. Ann. 153* (1964), 69-80.

[7] P. Erdös, Some recent problems and results in graph theory, combinatorics and number theory. In *Proceedings of the Seventh Southeastern Conference on Combinatorics, Graph Theory, and Computing.* Congress. Numer. 17, Utilitas Mathmatica Publishing, Winnipeg (1976), 3-14.

[8] C. Thomassen, Some homeomorphism properties of graphs. *Math. Nachr. 64* (1974), 119-133.

[9] C. Thomassen, Graph decomposition with applications to subdivisions and path systems modulo k. *J. Graph Theory 7* (1983), 261-271.

4. Appendix

Verification of Observation C   The graph numbers referred to in the following are those assigned by Bussemaker et al. [4] to the cubic graphs of order 10. Our verification of Observation C consists of showing that

(i) nos. 12, 15, 17 and 18 are obtained by adding an edge to a subdivision of Q_3,

(ii) nos. 6,7,8,9,11,13, and 14 are obtained from certain 3-connected graphs of order 8 by expanding a vertex to a triangle,

(iii) nos. 16 and 19 are obtained from a subdivision of the Möbius ladder of order 8 by adding an edge, and

(iv) no. 10 is obtained by adding an edge to a subdivision of the cubic graph of order 8 and connectivity 2.

(i) We see, from Observation B, that every subdivision of the hypercube Q_3 is good. Since it is easily demonstrated that every subdivision of no. 12, 15, 17 or 18 contains some such subdivision as a subgraph, it follows that they are good.

(ii) Suppose G is a bad subdivision of a 3-connected cubic graph G_{10} of order 10 and that G_{10} contains a triangle T which, when contracted to a single vertex t, say, reduces G_{10} to a cubic graph G_8 of order 8. then G_8 is 3-connected and has a subdivision H which is bad. Such a subdivision may be constructed by weighting each edge of G_8 not incident with t with its weight in G_{10} and weighting each edge of G_8 incident with t as indicated in Figure 8. The graph H is bad since each cycle in H corresponds in an obvious manner to a cycle in G_{10} with the same weighted length. Thus, H is one of the bad subdivisions of cubics of order 8 listed in Observation B.

Now let a, b and c be the weights assigned to the edges of T in G_{10} in obtaining G. Since G is bad, $a + b + c \not\equiv c \bmod 3$ and it follows that, say, $a + b \not\equiv c \bmod 3$. Let uw be the edge of T with weight c and let v be the remaining vertex T. It is now easily argued that all $u - w$ paths in G that do not contain vertex v or edge uw must have weighted length $a + b + v \bmod$

3. Thus in H there are two edges incident with t having the property that all cycles in H containing these two edges have the same length. But examination of each vertex of every bad subdivision of 3-connected cubic graphs of order 8, as listed in Observation B, reveals that no vertex can play the role of t.

(iii) Cubic graph no. 16 of order 10, say G_{16}, consists of a Hamilton cycle $v_1 v_2 \cdots v_{10}$ together with edges $v_i v_{i+5}, 1 \le i \le 5$. This graph has edges of two types: those that lie in exactly one C_4 and those that lie in two C_4's. We remove one of the edges that lies in two C_4's, say $v_5 v_{10}$, and obtain a subdivision of the cubic graph, G_5 say, of order 8 that consists of a Hamilton cycle $u_1 u_2 \cdots u_8$ together with edges $u_i u_{i+4}, 1 \le i \le 4$. The edges of G_5 are of the same two types as those of G_{16}. Consequently, we may assume G_{16} was obtained from G_5 by replacing $u_1 u_2$ and $u_5 u_6$ with the paths $u_1 v u_2$ and $u_5 w u_6$, respectively, and adding edge vw. Now, it is easily verified from Observation B that every subdivision of G_{16} is good by verifying that $u - v$ paths, different from uv, of all-weighted lengths mod 3 exist.

Similar arguments show the Petersen graph, no. 19, is good since it is obtained from G_5 by replacing edges $u_1 u_5$ and $u_3 u_7$ by paths $u_1 v u_5$ and $u_3 w u_7$, respectively, and adding edge vw.

(iv) The remaining 3-connected cubic graph of order 10, no. 10, is pictured in Figure 9. It is easily shown that if, in some subdivision G of this graph, there are $u_1 - v_1$ paths of different weighted lengths mod 3 that use only vertices from $S_1 = \{u_1, v_1, w_1, x_1, y_1\}$ and also $u_1 - v_1$ paths of different weighted lengths mod 3 using only vertices from $S_2 = \{u_1, v_1, u_2, v_2, w_2, x_2, y_2\}$, then G is good. Consequently, we may assume all $u_1 - v_1$ paths using only vertices from S_1, have the same length and similarly for all $v_1 - w_1$ paths. Thus if $u_1 x_1, v_1 x_1, w_1 x_1, w_1 x_1, u_1 y_1, v_1 y_1$ and $w_1 y_1$ have weights a, b, c, d, e and f, respectively, we obtain

$$a + b \equiv d + e \equiv a + c + f + e \equiv d + f + c + b \bmod 3$$

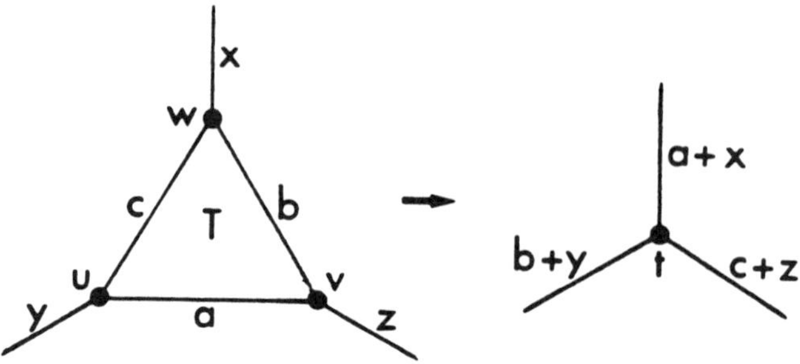

Figure 8: Contraction of a triangle to a vertex.

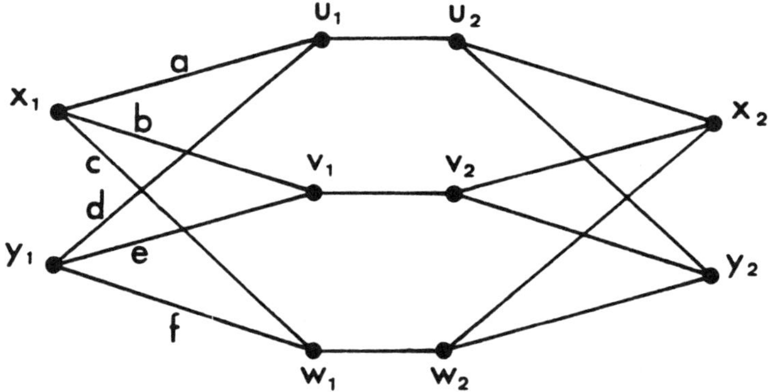

Figure 9: A special graph.

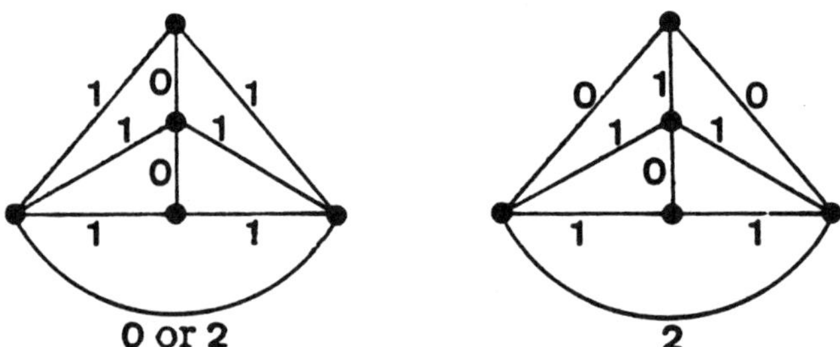

Figure 10: The bad subdivisions of $K_5 - e$.

and

$$b + c \equiv e + f \text{ mod } 3$$

so that

$$a \equiv d, b \equiv e \quad \text{and} \quad c \equiv f \text{ mod } 3.$$

If any two of a, b and c have weight sum 0 mod 3, then G is good. The remaining cases give $u_1 - v_1$ paths of different weighted lengths mod 3 and we conclude G is good.

Verification of Observation D Using the bad subdivisions of K_4 given in Observation B one easily shows that the graphs pictured in Figure 10 are the only bad subdivisions of $K_5 - e$. Now it is easily checked that every subdivision of K_5 is good.

The graph L is obtained from $K_5 - e$ by subdividing with a single vertex u, say, an edge incident with a vertex of degree 3 and adding the edge uv to the other vertex v, say, with degree 3 in $K_5 - e$. There are, altogether, five ways this can be done using the graphs of Figure 10. Just one of these is from the graph on the left. Now, using arguments similar to those in the verification of Observation C part (iii), it can be shown that all subdivisions of L are good.

CONTRACTIONS OF OP GRAPHS

Larry I. Basenpiler

ABSTRACT

In this paper, we show that such properties as outer planarity and maximal outerplanarity of a graph can be determined from a deck of its non-isomorphic 2-connected elementary contractions. Moreover, maximal outerplanar graphs are reconstructible, up to 2-isomorphism, from such a deck.

1.   Introduction

The graph theoretic notation is essentially that of Harary [4]. Some basic definitions and results are briefly reviewed below, and others are introduced as necessary. Let G denote a finite simple graph with vertex set $V(G) = \{v_1, v_2, \cdots, v_n\}$, and edge set $E(G) = \{e_1, e_2, \cdots, e_m\}$. A graph obtained from G by identifying a pair of adjacent vertices connected by edge e, with subsequent removal of a loop and possible multiple edges, is called an elementary contraction G/e of G. We note that the use of 'contraction' for 'elementary contraction' hereafter should not cause confusion since throughout the paper we consider exclusively elementary contractions. A reconstruction (edge-reconstruction) of a graph G is a graph H such that $V(H) = V(G)(E(H) = E(G))$ and $H - v \cong G - v(H - e \cong G - e)$ for $v \in V(G)$ (for all $e \in E(G)$). A C - reconstruction of a graph G is a graph H such that $E(H) = E(G)$ and $H/e \cong G/e$ for all $e \in E(G)$. G is reconstructible if every reconstruction of G is isomorphic to G.

The Reconstruction Conjecture states that a graph is reconstructible, uniquely up to isomorphism, from its proper subgraphs [4]. The C-Reconstruction

103

Conjecture states that a graph is reconstructible, uniquely up to isomorphism [1,3], from its contractions. In this paper we will deal with the set of non-isomorphic contractions of a graph G, called a C-deck of G. Since separable graphs are C-reconstructible [6] (and, therefore, their C-decks can be easily identified), one may assume that any C-deck under investigation is a C-deck of a non-separable graph. The subset of 2-connected contractions of the C-deck is called a δ-deck. We investigate properties of outerplanar (OP) and maximal outerplanar (MOP) graphs which can be deduced from these subsets.

2. Recognizability

Along with attempts to prove the Reconstruction Conjecture directly, there exists another approach to the problem which will be referred to as "recognizability", namely: a class of graphs is recognizable if, for each graph G in M, every reconstruction of G is also in M. A graph is C-recognizable if it is recognizable from a set of its contractions. In this section, we show that most of OP and MOP graphs are C-recognizable from their δ-decks. A graph is outerplanar if it can be embedded in the plane with all vertices lying on a single face. The edges of this face are called exterior; all other edges are called interior. Observe that contraction of any interior edge of an OP graph G produces a separable graph whereas contraction of an exterior edge of G produced an OP graph. That is, a δ-deck of an OP graph G consists of all non-isomorphic contractions of its exterior edges. We consider graphs on $n \geq 5$ vertices. Notice that an OP graph G on three vertices is MOP, and its C-deck consists of one element - an edge. G is not recognizable from its C-deck but is recognizable (and reconstructible) from the set of all its contractions. The same holds for a graph on four vertices.

Lemma 1   Non-separable OP and MOP graphs of order 5 are recognizable and reconstructible from their C-decks.

Proof   We use the following criterion of outerplanarity [4] (OP-criterion): a graph is outerplanar if and only if it does not have a subgraph contractible to

K_4 or to $K_{2,3}$. We also use the fact that contractions and deletions commute. A graph $G(V,E)$ of order 5 with a subgraph contractible to K_4, would obviously have a K_4 as one its contractions, and the assertion of the lemma concerning recognizability is trivial. However, all subgraphs of G isomorphic to $K_{2,3}$ would vanish in its C-deck containing graphs on 4 vertices. Consider all 2-connected graphs with $n = |V| = 5$ and $m = |E| > 4$ (their list can be found in [4]) and their C-decks. Direct observation shows that, for $m = 8, 9$ and 10, an appropriate C-deck contains K_4. Further analysis reveals that each graph, with $n = 5$ and $m < 8$, which is not contractible to K_4, has its own unique C-deck. For example, the C-deck of C_5 contains a single element C_4, and $K_{2,3}$ contains a single element $K_4 - e$, etc. Interestingly enough, δ-decks of the only two non-outerplanar graphs not contractible to K_4 (depicted in Figure 1), both consist of a single element, $K_4 - e$.

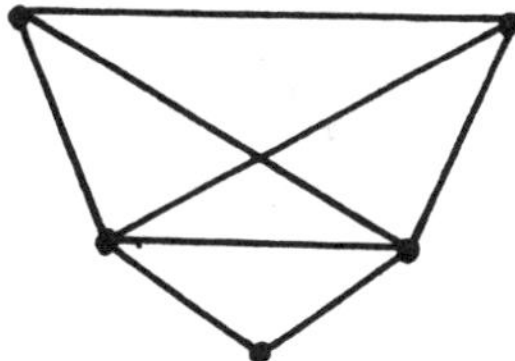

Figure 1

Proposition 1   A non-separable graph G of order at least 6 is outerplanar if and only if every element of its δ-deck is outerplanar.

Proof   Applying the same argument as in the proof of lemma 1, it is easy to deduce that an OP graph G with $|V| > 5$ is C-recognizable from its C-deck. We prove that it suffices to conduct the outerplanarity test only on the δ-deck of G (there must be at least one 2-connected graph in the C-deck of non-separable graph G.) Assume to the contrary that G is not an OP graph, and this can be determined by the OP-criterion only from a separable G/e. The edge e is then a cut edge of G. Let $G - H \cup H'$, and $H \cap H' = e$.

Since G is 2-connected, G/e consists of 2-connected blocks B and B', where $B = H/e$ and $B' = H'/e$. Let B be a non-OP block. Obviously, $|V(B)| \geq 4$. By OP - criterion, $H \subseteq G(|V(H) \geq 4)$ is not an OP graph also. Contract any edge of B' in G/e. If B' is a single edge, then H' has been a triangle, and contraction of any of its edges other than e would produce H as a 2-connected element of the C-deck of G, i.e. H is an element of δ-deck of G. If B' is 2-connected block, then contraction of at least one of its edges, say e_j, does not make B' separable. That is, both H'/e_j and G/e_j are 2-connected and contain subgraphs (B and H , respectively) contractible to K_4 or to $K_{2,3}$.

Proposition 2   A non-separable graph G of order at least 6 is a MOP graph if and only if every element of its δ-deck is a non-separable MOP graph.

Proof   Since OP graphs are C-recognizable from their δ-decks, we assume that a δ deck of a graph under consideration represents an OP graph. That is, it remains to prove that one can deduce from investigation of the δ-deck, if, in accordance with the definition of MOP graphs, an edge can be added to G without losing outerplanarity. Since contraction of an exterior edge of G produces a MOP graph, the necessity of the proposition follows. We prove sufficiency by contradiction. Assume that G is not a MOP graph but all its 2-connected contractions are MOP. If G is an OP but not a MOP graph, then there exists at least one interior face (an induced cycle) C in G of length greater than three. The case when C is the only cycle of G is easily recognizable. Suppose otherwise, and contract an exterior edge e which does not belong to C. Notice that in general, edges of C do not necessarily form an induced cycle in G/e: when, for example, there exist cycles C_1 and C_2 such that $C_1 \cap C_2 = \{e, e'\}$, and $C = C_1 \cup C_2 \backslash \{e, e'\}$. However, such a construction is impossible in an OP graph, since the common vertex of edges e and e' would be interior in the OP graph G. Therefore, the C-deck of the OP graph will contain at least one element with the cycle C, which contradicts the assumption.

The propositions and lemma 1 combined produce the following theorem.

Theorem 1   Non-separable outerplanar and maximal outerplanar graphs of order at least 5 are recognizable from their δ-decks.

3. C_2-Reconstructions

Two graphs G and G' are cyclic isomorphic if there is a 1-1 correspondence between the edges of G and G' so that cycles correspond to cycles [8]. In fact, graphs G and G' are cyclic isomorphic if induced cycles in G correspond to those in G' [2]. The cyclic isomorphism is equivalent to so-called 2-isomorphism [8], thus we say that two MOP graphs G and G' are 2-isomorphic if there is 1-1 correspondence between the edges of G and G' so that faces correspond to faces. In this section we prove that MOP graphs are reconstructible, up to 2-isomorphism, from their δ-decks. We call such reconstruction a C_2- Reconstruction.

Let G be a MOP graph, and $F = \{F_i\}$ the set of its interior faces. Place a vertex f_i in each interior face F_i of G (such that vertices will be referred as nodes) and, when two faces have an edge e in common, join the corresponding nodes with an edge e'' crossing only e. We call the edges $\{e''\} = E''$ internal. Cross every edge e lying on the exterior face of G with a pendant edge e' incident to the corresponding node. The resulting tree is called a semidual graph (or simply a semidual) G_S of G [5]. Figure 2 gives an example of two graphs, G and H, and their (dashed) semiduals. Notice the following: all nodes f_i are of degree 3, and the endpoints correspond to the edges of the exterior face. Obviously, every MOP graph G uniquely determines its semidual with the natural one-to-one mapping $\varphi : E(G) \to E'(G_S) \cup E''(G_S)(E''(G_S) \cap E''(G_S) = \oslash)$.

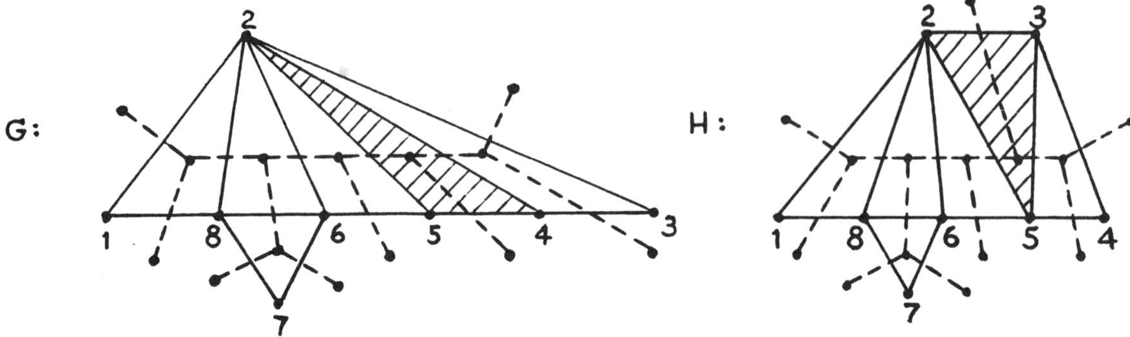

Figure 2

Let G_S be a semidual of G. Construct a graph H using the following recursive procedure called the f-procedure. If G_S is a star whose pendant edges are, say e_1', e_2', and e_3', then H is a triangle $\{e_1, e_2, e_3\}$. If the order of G_S is greater than 3, then there must be at least two nodes in G_S which are incident to two pendant edges each: a MOP graph has at least two vertices of degree 2 which produce such nodes. Choose one of the nodes, say f_i, incident to pendant edges e_i' and e_j', and to an internal edge e_k''. These three edges will constitute the face $F_i = \{e_i, e_j, e_k\}$ of H. Remove from G_S the used pendant edges e_i' and e_j', along with their endpoints, turning the node f_i into an endpoint and the edge e_k'' into a pendant edge e_k' (later on, e_k' will correspond to the edge e_k in another face, say F_j). Proceed to decompose the new tree, which is a semidual of some MOP graph [5], into triangular faces until the resulting tree becomes a star. At the end, coalesce all the triangles along their mutual edges $e_k \in F_i$ and $e_k \in F_j$ into a graph H called the f-reconstruction G.

Lemma 2　　A non-separable MOP graph G and its f-reconstruction H are 2-isomorphic.

Proof　　Let H be an f-reconstruction of G obtained as above from a semidual G_S. By the f-procedure, the edges of H form a face F if and only if their pre-images are incident to a node f of G_S. On the other hand, G_S has the node f incident to the appropriate edges if and only if their pre-images under φ form a face in G. So every single face of H consists of the edges corresponding to the edges of a face in G. Let $\Psi : E(G) \longrightarrow E(H)$ be such a correspondence. In order to prove that Ψ is a 2-isomorphism, one has to show that two faces F_i and F_j in H have a common edge e if and only if faces $\Psi^{-1}F_i$ and $\Psi^{-1}F_j$ in G share the edge $\Psi^{-1}e$. By the f- procedure, two faces, F_i and F_j, in H have a common edge e if and only if the nodes f_i and f_j in G_S are connected by the edge e''; the latter holds if and only if the corresponding faces of G have a mutual edge $e = \Psi^{-1}e''$.

The 2-isomorphism does not, however, necessarily induce an isomorphism: edges e_i and e_j, incident in G, may have nonincident images in H, and conversely. In Figure 2, for example, a pair (3,4) and (5,6) of nonincident external edges in G and their incident images, (4,5) and (5,6), respectively, in H illustrate this case.

Lemma 3   Semiduals G_S and H_S of a MOP graph G and its f-reconstruction H are isomorphic.

Proof   Let $\Psi : E(G) \longrightarrow E(H)$ be a 2-isomorphism, and φ_G and φ_H the mappings $E(G) \longrightarrow E(G_S)$ and $E(H) \longrightarrow E(H_S)$, respectively. We prove first, that G_S and H_S are edge isomorphic, i.e. that $\varphi_G e_i$ and $\varphi_G e_j$ are incident if and only if $\varphi_H \Psi e_i$ and $\varphi_H \Psi e_j$ are incident if and only if $\varphi_H \Psi e_i$ and $\varphi_H \Psi e_j$ are. This indeed holds due to the following considerations. $\varphi_G e_i$ and $\varphi_G e_j$ are incident if and only if e_i and e_j belong to the same face in G. Since Ψ is a 2-isomorphism, Ψe_i and Ψe_j then belong to one face in H too. That is, $\varphi_H \Psi e_i$ and $\varphi_H \Psi e_j$ are incident in H_S. The edge isomorphism of G_S and H_S induces their isomorphism except for the case of a 4-star and triangle. The latter, however, is impossible since the semiduals are tees.

Lemmas 2 and 3 combined render the following proposition.

Proposition 3   A non-separable MOP graph is C_2-reconstructible from its semidual.

Theorem 2   A non-separable MOP graph is C_2-reconstructible from its δ-deck.

Proof   Let G be a MOP graph, and G_S its semidual. Consider connections between G_S and $(G/e)_S$ where G/e is an element of the δ-deck of G. All exterior edges of G fall into one of two categories: (1) an exterior edge of a face with two exterior edges, and (2) an exterior edge of a face with only one exterior edge. We do not consider the trivial case when G is a triangle with three exterior edges. Let $f = \{e_1, e_2, e_3\}$ be a face in G. If e_1 is the only interior edge of f, then call the edges $\{\varphi e_1, \varphi e_2, \varphi e_3\}$ of G_S a 3-star and the edges $\{\varphi e_2, \varphi e_3\}$ a 2-fork. Contracting an exterior edge e_2 (or e_3) in G results in contracting the pendant edges φe_2 and φe_3 in G_S (see, for example, faces $\{1,2,8\}$ or $\{6,7,8\}$ of G in Figure 2). Observe that G/e_2 and G/e_3 are isomorphic and contribute one member to the δ-deck of G. A δ-dual T_G of G is a tree obtained from a semidual G_S of G by merging each 2-fork into one edge. Thus, each 3-star of G_S becomes a 2-path in T_G. A δ-dual, then, may have degrees 1,2, or 3. It is easy to see that a δ-dual is uniquely determined by the semidual from which it was obtained, and in fact a semidual is uniquely reconstructible from a δ-dual. To implement the latter, attach to each node of degree 2 a pendant edge. Now contracting either exterior edges e_2 or e_3 in G results in contracting one appropriate pendant edge in T_G. Let δ_1 denote the set of all elements of the δ-deck of G obtained by contracting an edge from the first category. Contracting an exterior edge e_1 from the second category in G results in contracting the pendant edge φe_1 and the internal edge φe_2 (or φe_3) in G_S (and in T_G) (see, for example, faces $\{2,4,5\}$ and $\{2,5,6\}$ of G in Figure 2). Denote the set of such contractions δ_2.

Given a δ-deck, create the set of semiduals for each element of the δ-deck, and then the set of δ-duals of the semiduals. The δ-duals are contractions of T_G. Our objective is to C-reconstruct a tree T_G from its contractions. The algorithms [1,3] implementing this task are based on the separation of contractions of internal edges (in this paper's terminology) from contractions of pendant edges from the first category of T_G. Applying this technique, find δ_1 (with exceptions which will be handled later), which is a set of non-isomorphic subtrees. δ_1 provides a basis for using the vertex reconstruction algorithm (M-algorithm) for trees by B. Manvel [7]. The M-algorithms cannot be applied, however, without noticing that δ_1 does not contain all the subtrees of T_G; it is missing the subtrees obtained by deletion of pendant edges from the second category. An analysis of the M-algorithm so applied to δ-duals shows that due to the primitivity of δ-duals many of the numerous ramifications of the M-algorithm are inapplicable. For example, a new endvertex must always be adjacent to an endvertex. The pendant edges from the second category are never used in the crucial stages of the M-algorithm which determine the diameter, the centricity, or the largest branch of an original tree, since there always is a 2-path beyond the internal node incident to a pendant edge. Once T_G is reconstructed, G_S can be reconstructed from T_G. By proposition 3, G then is f-reconstructible from G_S up to 2-isomorphism. Exceptions to the above discourse are an infinite family of MOP graphs whose δ-deck contains only one element. One such graph can be obtained by deleting vertex 7 in Figure 2. All other graphs can be obtained by adding triangles from the right and/or left to the graph. However, the semiduals of such graphs are unique and easily recognizable which implies their C_2-reconstructibility. Since we use the M-algorithm, it is necessary to mention that two pairs of trees are exceptions to the vertex reconstruction theorem for trees. One of these contains a graph which cannot be a δ-dual, the other - produces isomorphic semiduals. Paraphrasing theorem 3, non-separable MOP graphs are reconstructible, up to 2-isomorphism, from the set of their non-isomorphic MOP contractions.

Acknowledgments I am indebted to an anonymous referee for numerous helpful comments which enhanced the style and presentation of this paper.

Especial thanks to the referee for pointing out a flaw in the proof of Theorem 2.

REFERENCES

[1] Basenpiler, L., Choizonova K. L., On Reconstruction Problem for Graphs. *Proceedings in Applied Mathematics and Cybernetics*, Siberia Section of Academia of Science of USSR, State Research Institute of Technical and Scientific Information, 5285-72 DEP (1973), 49-55.

[2] Basenpiler, L., Keselman D. J., On Whitney Theorems Related to 2-isomorphic Graphs. *Proceedings in Applied Mathematics and Cybernetics*, Siberia Section of Academia of Science of USSR, State Research Institute of Technical and Scientific Information, 5285-72 DEP (1973), 43-48.

[3] Bhave V. N., Kundu S., and Sampathkumar, E., Reconstruction of a tree from its homomorphic images and other related transform. *J. Combin. Theory Ser. B. 20*, # 2, (1976), 117- 123.

[4] Harary, F., *Graph Theory*. Addison-Wesley, Reading, MA, 1969.

[5] Hedetniemi, S. M., Proskurowski A., and Syslo M. M., Interior Graphs of Maximal Outerplane Graphs. *J. Comb. Theory, B 38*, 156-167 (1985).

[6] Krishnamoorthy V. and Parthasarathy K. R., On Reconstruction of Separable Graphs from Elementary Contractions. *Discrete Math. 38* (1982) 197-205.

[7] Manvel B., Reconstruction of Trees. *Can. J. Math. 22* (1976), 55-60.

[8] Whitney H., 2-isomorphic Graphs. *Amer. J. Math. 55*, 1933, # 2, 245-254.

Some Recent Results
on Long Cycles in Tough Graphs

D. Bauer
Stevens Institute of Technology

E. Schmeichel
San Jose State University

H.J. Veldman
University of Twente

ABSTRACT

We survey a number of recent results concerning the relationship between the toughness of a graph and the maximum length of a cycle in a graph.

1. Introduction

In the last few years a number of results have been obtained concerning the relationship between the toughness of a graph and the maximum length of a cycle in a graph. While some of these results are discussed in [6], an equal number have been established since [6] was written. Consequently we feel it is desirable to survey here the current state of the art. We will not include any proofs, only statements of theorems and in some instances examples to show that the theorems are best possible. Our "survey" is undoubtedly not comprehensive. The only thing we are reasonably sure of is that we have not omitted any of our own results.

We consider only finite undirected graphs without loops or multiple edges. Our terminology is standard except as indicated below. A good reference for any undefined terms is [15]. We need a few definitions and some convenient notation. Let $\omega(G)$ denote the number of components of a graph G. A graph G is *t–tough* if $|S| \geq t\omega(G - S)$ for any subset S of the vertex set V of G with $\omega(G - S) > 1$. The *toughness* of G, denoted $t(G)$, is the maximum value of t for which G is t–tough ($t(K_n) = \infty$ for all $n \geq 1$). We will denote by κ and α the vertex connectivity and the cardinality of a maximum set of independent vertices of G, respectively. A cycle

C of G is a *dominating cycle* if every edge of G has at least one of its vertices on C. The length of a longest cycle in a graph G, called the *circumference* of G, will be denoted by c(G). A *k–factor* is a k–regular spanning subgraph. Finally, for $k \geq 2$ we let $\sigma_k = \min \{ \sum_{i=1}^{k} d(v_i) \mid \{v_1, ..., v_k\}$ is an independent set of vertices$\}$.

Chvátal introduced the notion of tough graphs in [17]. The interest in t–tough graphs stems primarily from their connection with the existence of hamiltonian cycles. It is easy to see that a necessary condition for a graph G to be hamiltonian is that G is 1-tough. Chvátal conjectured that there exists a finite number t_0 such that all t_0–tough graphs are hamiltonian. He showed in [17] that there exist 3/2–tough nonhamiltonian graphs and later Thomassen [12, p. 132] found t–tough nonhamiltonian graphs with $t > 3/2$. More recently Enomoto et al. [21] have found $(2 - \varepsilon)$–tough graphs having no 2–factor for arbitrary $\varepsilon > 0$. Nevertheless it is still unknown whether all 2–tough graphs are hamiltonian, although some progress has recently been made (see Section 4).

Another question, namely "How difficult is it to recongnize t–tough graphs?" was open for some time. The question was first raised by Chvátal [16] and subsequently appeared in [32] and [18, p. 429]. It was recently shown by Bauer, Hakimi and Schmeichel [4] that for any fixed positive rational number t, it is NP–hard to recognize t–tough graphs (see Section 6).

In Sections 2 through 5 we survey some recent results concerning the relationship between the toughness of a graph and the circumference of a graph.

2. Some Results

We begin our discussion with a well–known theorem of Dirac [19].

Theorem 1 *Let G be a graph on $n \geq 3$ vertices with $\delta \geq n/2$. Then G is hamiltonian.*

A "long cycle version" of Theorem 1 was also proved by Dirac.

Theorem 2 *Let G be a 2–connected graph on n vertices. Then $c(G) \geq \min (n, 2\delta)$.*

In 1960 Ore [30] generalized Theorem 1 as follows.

Theorem 3 *Let G be a graph on $n \geq 3$ vertices with $\sigma_2 \geq n$. Then G is hamiltonian.*

A long cycle version of Theorem 3 was later established independently by Bermond [11] and Linial [28].

Theorem 4 *Let G be a 2–connected graph on n vertices with $\sigma_2 \geq q$. Then $c(G) \geq \min(n, q)$.*

As remarked earlier, a necessary condition for a graph G to be hamiltonian is that G is 1–tough. A natural question, answered by Jung in 1978 [26], is how far can the lower bound of n in Ore's Theorem be reduced under the assumption that G is 1-tough.

Theorem 5 *Let G be a 1–tough graph on $n \geq 11$ vertices with $\sigma_2 \geq n - 4$. Then G is hamiltonian.*

The proof of Theorem 5 in [26] is lengthy and complicated. A much simpler proof for $n \geq 16$ appears in [5]. Using arguments of the latter proof, Skupień [31] recently obtained the following improvement of Theorem 5.

Theorem 6 *Let G be a 1–tough graph on $n \geq 11$ vertices such that $\sigma_2 \geq n - 4 - \varepsilon(n)$, where $\varepsilon(n) = 1$ if n is even and $n \neq 12$, and $\varepsilon(n) = 0$ otherwise. Then G is hamiltonian.*

To see that Theorem 6 is best possible we define, for $n \geq 11$, a nonhamiltonian 1-tough graph A_n on n vertices with $\sigma_2(A_n) = n - 5 - \varepsilon(n)$. If n is odd then A_n is obtained from $\overline{K}_{(n-1)/2} \cup K_{(n-5)/2} \cup K_3$ by joining every vertex in $K_{(n-5)/2}$ to all other vertices and adding a matching between the vertices of K_3 and three vertices of $\overline{K}_{(n-1)/2}$. If n is even, then A_n is obtained from A_{n-1} by adding a new vertex and joining it to a vertex of degree $(n-6)/2$ in A_{n-1} and the $(n-6)/2$ vertices of degree $n-2$ in A_{n-1}.

It is reasonable to consider a "long cycle version" of Jung's Theorem. The first step in this direction was taken by Ainouche and Christofides [1].

Theorem 7 *Let G be a 1–tough graph on $n \geq 3$ vertices with $\sigma_2 \geq q$. Then $c(G) \geq \min(n, q+1)$.*

They also conjectured that $q + 1$ in Theorem 7 could be replaced by $q + 2$, and presented the following graphs showing this would be best possible. For $n = 3r + 1 \geq 7$, construct the graph B_n from $3K_r + K_1$ by choosing one vertex from each copy of K_r, say u, v and w, and adding the edges uv, uw and vw. The graph

B_n is a 1-tough graph on $n = 3r + 1$ vertices with $\sigma_2 = 2r$ and $c(B_n) = 2r + 2$. Their conjecture was recently established by Bauer and Schmeichel [7].

Theorem 8 *Let G be a 1–tough graph on $n \geq 3$ vertices with $\sigma_2 \geq q$. Then* $c(G) \geq min \ (n, q + 2)$.

While Theorem 8 is "best possible", a stronger result may be obtained if $q \geq 2n/3$. A useful intermediate result concerns the existence of a longest cycle which is also a dominating cycle. Since dominating cycles have recently played a useful role in hamiltonian cycle theory we disgress to mention a few results about them.

3. Dominating Cycles

Dominating cycles were first introduced by Nash–Williams in [29] and were later studied in more detail by Veldman [33]. In [29], Nash–Williams proved the following.

Theorem 9 *Let G be a 2–connected graph on n vertices with $\delta \geq (n + 2) / 3$. Then every longest cycle in G is a dominating cycle.*

The next result follows easily [29].

Theorem 10 *Let G be a 2–connected graph with $\delta \geq max \ ((n + 2) / 3, \alpha)$. Then G is hamiltonian.*

In 1980 bondy [14] generalized Theorem 9.

Theorem 11 *Let G be a 2–connected graph on n vertices with $\sigma_3 \geq n + 2$. Then every longest cycle in G is a dominating cycle.*

An analogous generalization of Theorem 10 occurs in [2].

Theorem 12 *Let G be a 2–connected graph on n vertices with* $\sigma_3 \geq max \ (n + 2, 3\alpha)$. *Then G is hamiltonian.*

Theorem 12 is an immediate consequence of the followint long cycle version, established in [6].

Theorem 13 *Let G be a 2–connected graph on n vertices with $\sigma_3 \geq n + 2$. Then* $c(G) \geq min \ (n, n + \sigma_3/3 - \alpha)$.

In [6] Theorem 13 is proved by combining Theorem 11 with a technical lemma, which we state explicitly because of its central role in proofs of several results in this survey.

Lemma 14 *Let G be a graph on n vertices with $\delta \geq 2$ and $\sigma_3 \geq n$. Suppose G contains a longest cycle C which is a dominating cycle and v is a vertex in $V(G) - V(C)$. With respect to some orientation of C, let S be the set of immediate successors on C of the vertices adjacent to v. Then $(V(G) - V(C)) \cup S$ is an independent set of vertices.*

The previous results on dominating cycles have all assumed that G is 2–connected. If instead G is assumed to be 1–tough, the bounds in Theorems $9 - 11$ can be improved. The next two theorems are due to Bigalke and Jung [13].

Theorem 15 *Let G be a 1–tough graph on n vertices with $\delta \geq n/3$. Then every longest cycle in G is a dominating cycle.*

Theorem 16 *Let G be a 1–tough graph on $n \geq 3$ vertices with $\delta \geq max\,(n/3, \alpha - 1)$. Then G is hamiltonian.*

We close this section with the following generalization of Theorem 15 apearing in [6].

Theorem 17 *Let G be a 1–tough graph on n vertices with $\sigma_3 \geq n$. Then every longest cycle in G is a dominating cycle.*

4. Some More Results

By combining Lemma 14 with Theorem 17, a result similar to Theorem 13 was proved in [6].

Theorem 18 *Let G be a 1–tough graph on $n \geq 3$ vertices with $\sigma_3 \geq n$. Then $c(G) \geq min\,(n, n + \sigma_3/3 - \alpha)$.*

It is easy to see that if $t(G) = \tau$ then $\alpha(G) \leq n/(\tau + 1)$. In particular, $\alpha \leq n/2$ for 1-tough graphs. Thus if G is a 1–tough graph on $n \geq 3$ vertices with $\delta \geq n/3$, then $c(G) \geq 5n/6$. Note that from Theorem 8 we could only conclude that $c(G) \geq 2n/3 + 2$. Also note that Theorem 18 is only a partial generalization of Theorem 16. Theorem 18 allows us to draw conclusions concerning long, but not necessarily hamiltonian, cycles in G. However if $\delta = \alpha - 1 \geq n/3$ we can not conclude from Theorem 18 that G is hamiltonian. We have been able, however, to modify the proof of Theorem 16 in [13] to prove the following generalization of Theorem 16 [9].

Theorem 19 *Let G be a 1–tough graph on $n \geq 3$ vertices with $\delta \geq n/3$. Then $c(G) \geq min\,(n, n + \delta - \alpha + 1)$.*

The graphs B_n show that the condition $\delta \geq n/3$ can not be relaxed. The graphs A_n show that Theorem 19 is best possible when $\delta = \lfloor (n-5)/2 \rfloor$. Note that Theorem 19 allows us to conclude that $c(G) \geq 5n/6 + 1$. We do not believe, however, that this is best possible.

Conjecture 1 *Let G be a 1–tough graph on $n \geq 3$ vertices with $\sigma_3 \geq n$. Then $c(G) \geq min\ (n, (3n+1)/4 + \sigma_3/6)$.*

Conjecture 1, if true, is best possible. For odd $n \geq 15$ construct the graph $D_{n,\delta}$ from $\overline{K}_{(n-1)/2} \cup K_\delta \cup K_{(n+1)/2-\delta}$, where $n/3 \leq \delta \leq (n-5)/2$, by joining every vertex in K_δ to all other vertices and by adding a matching between all vertices in $K_{(n+1)/2-\delta}$ and $(n+1)/2-\delta$ vertices in $\overline{K}_{(n-1)/2}$. Note that $D_{n,(n-5)/2} = A_n$. It is easily seen that $D_{n,\delta}$ is 1–tough with $c(D_{n,\delta}) = (3n+1)/4 + \delta/2$ $((3n+3)/4 + \delta/2)$ if $(n+1)/2-\delta$ is odd (even). Note that the truth of Conjecture 1 would allow us to conclude that under its hypothesis $c(G) \geq (11n+3)/12$. It would also imply the following generalization of Jung's Theorem.

Conjecture 2 *Let G be a 1–tough graph on $n \geq 13$ vertices with $\sigma_3 \geq (3n-14)/2$. Then G is hamiltonian.*

We now consider graphs G with $t(G) = \tau \geq 1$. By combining Theorem 18 with the observation that $\alpha(G) \leq n/(\tau+1)$, the next result was obtained [6].

Theorem 20 *Let G be a graph on $n \geq 3$ vertices with $t(G) = \tau \geq 1$. If $\sigma_3 \geq n$ then $c(G) \geq min\ (n, n\tau/(\tau+1) + \sigma_3/3)$.*

Corollary 21 *Let G be a 2–tough graph on $n \geq 3$ vertices. If $\sigma_3 \geq n$ then G is hamiltonian.*

As mentioned earlier, Enomoto, et al. [21] have shown that there exist $(2-\varepsilon)$–tough graphs having no 2–factor for arbitrary $\varepsilon > 0$. More precisely, they showed that for any positive integer k and for any positive real number ε, there exists a $(k-\varepsilon)$–tough graph G on n vertices with kn even and $n \geq k+1$ which has no k–factor. While this might lead one to believe that there exist nonhamiltonian 2–tough graphs, they established another result which, together with Corollary 21, points in the opposite direction.

Theorem 22 *Let G be a k–tough graph on n vertices with $n \geq k+1$ and kn even. Then G has a k–factor.*

Other results have appeared relating toughness to the existence of k–factors. In particular, Enomoto [20] has strengthened Theorem 22 by weaking the assumption of "k–toughness" to the assumption that $|S| \geq k\omega(G-S) - 7k/8$ for all $S \subseteq V(G)$ with $\omega(G-S) > 1$. Also Katerinis [27] has shown that a 1–tough bipartite graph on $n \geq 3$ vertices has a 2–factor.

We conclude this section with a sufficient condition for a 1–tough graph G to be hamiltonian based on the vertex connectivity of G. The background for this result begins with a theorem of Häggkvist and Nicoghossian [25].

Theorem 23 *Let G be a 2-connected graph on n vertices with* $\delta \geq (n + \kappa)/3$. *Then G is hamiltonian.*

This obviously represents a great improvement over Dirac's Theorem for 2–connected graphs with small vertex connectivity. Theorem 23 was recently generalized in [2].

Theorem 24 *Let G be a 2-connected graph on n vertices with* $\sigma_3 \geq n + \kappa$. *Then G is hamiltonian.*

Both Theorems 23 and 24 are best possible. We illustrate this for the latter theorem; a similar class of graphs, described in [25], shows Theorem 23 is best possible. For $2 \leq \kappa < n/2 - 1$ let $m = \lfloor (n + \kappa + 1)/3 \rfloor$. Consider the graph G obtained from $K_{m,m}$ by joining each of κ vertices in one color class to every vertex in a complete graph on $n - 2m$ vertices. Clearly $\sigma_3 = n + \kappa - 1$ and G is not hamiltonian.

Recently Bauer and Schmeichel [8] ahve established a result analogous to Theorem 23 for 1–tough graphs.

Theorem 25 *Let G be a 1-tough graph on* $n \geq 3$ *vertices with* $\delta \geq (n + \kappa - 2)/3$. *Then G is hamiltonian.*

For $n \geq 11$, the graphs A_n show that Theorem 25 is best possible when $\kappa = (n - 5)/2$. If $\kappa = 2$, then graphs B_n show Theorem 25 is best possible.

5. On Generalizing Jung's Theorem

As mentioned in Section 2, Skupień [Theorem 6] recently generalized Jung's Theorem. Bauer, Broersma and Veldman [3] have further generalized Skupień's result by imposing two degree conditions, each of which is weaker than the degree condition of Theorem 6. One of them is of a type introduced by Fan [22].

Theorem 26 *Let G be a 2-connected graph on n vertices. If for all vertices x, y, $d(x, y) = 2$ implies* $max(d(x), d(y)) \geq n/2$, *then G is hamiltonian.*

Generalizations of Theorem 26 were obtained by Benhocine and Wojda [10]. A simple proof of Fan's Theorem was recently found by Veldman [34].

Suppose G satisfies the conditions of Theorem 6. If $d(x, y) = 2$, then max $(d(x), d(y)) \geq \lceil (n - 4 - \varepsilon(n)) / 2 \rceil = \lceil (n - 4) / 2 \rceil$. Futhermore, $\sigma_3 \geq \lceil 3(n - 4 - \varepsilon(n)) / 2 \rceil \geq n$, since $n \geq 11$. Hence the following result generalizes Theorem 6 [3].

Theorem 27 *Let G be a 1–tough graph on $n \geq 3$ vertices such that $\sigma_3 \geq n$ and, for all vertices $x, y, d(x, y) = 2$ implies max $(d(x), d(y)) \geq (n - 4) / 2$. Then G is hamiltonian.*

Theorem 27 is best possible in the sense that neither of the two degree conditions can be relaxed. For $n \geq 17$, the nonhamiltonian 1–tough graph A_n has $\sigma_3 \geq n$ and satisfies the second degree condition of Theorem 27 with $(n - 4) / 2$ replaced by $\lceil (n - 4) / 2 \rceil - 1$. For $t = 2, 3$, $n \geq t + 5$ and $n - t - 1 \equiv 0 \pmod 2$, obtain the graph $E_{n,t}$ from $K_1 + (K_t \cup 2K_{(n-t-1)/2})$ by choosing one vertex from K_t and one vertex from each copy of $K_{(n-t-1)/2}$ and adding the edges of a triangle between them. The graph $E_{n,t}$ is a nonhamiltonian 1–tough graph satisfying the second degree condition of Theorem 27, but with $\sigma_3 = n - 1$.

Note that $\kappa(E_{n,t}) = 2$ for all n and t. Another result in [3] shows that the condition on σ_3 is Theorem 27 can be dropped completely if G is required to be 3-connected and large enough.

Theorem 28 *Let G be a 3–connected 1–tough graph on $n \geq 35$ vertices such that, for all vertices $x, y, d(x, y) = 2$ implies max $(d(x), d(y)) \geq (n - 4) / 2$. Then G is hamiltonian.*

The 3–connected graphs A_n show that Theorem 28 is, in a sense, best possible.

Theorem 28 generalizes Theorem 6 within the class of 3–connected graphs that are large enough. Note that within the class of graphs on at least 17 vertices with $\kappa = 2$ Theorem 6 is generalized by Theorem 24. We do not believe that the requirement $n \geq 35$ in Theorem 28 is tight. It would be interesting to know the smallest value n_0 by which 35 could be replaced in Theorem 28. The graph A_{12} shows that $n_0 \geq 13$.

We close this section with a very recent improvement of Theorem 27 by Skupień [31].

Theorem 29 *Let G be a 1–tough graph on $n \geq 3$ vertices such that $\sigma_3 \geq n$ and, for all veritces $x, y, d(x, y) = 2$ implies max $(d(x), d(y)) \geq (n - 5) / 2$. Then either G is hamiltonian or else n is odd, $n \geq 15$, and G is a subgraphs of A_n.*

6. Complexity

As mentioned earlier, the question "How difficult is it to recognize t–tough graphs?" has remianed an interesting open problem for some time. The question was first raised by Chvátal [16] and subsequently appeared in [32] and [18, p. 429]. It has recently been shown by Bauer, Hakimi, and Schmeichel [4] that for any fixed positive rational number t, it is NP–hard to recognize t–tough graphs. The proof is accomplished in two stages. First, a well–known NP–hard variant of INDEPENDENT SET [23, p. 194] is reduced to the problem of recognizing 1–tough graphs. Then the latter problem is reduced to recognizing t–tough graphs, for any fixed positive rational number t.

In fact it is easy to use an argument analgous to that used in [4] to reduce INDEPENDENT SET to the problem of recognizing 1–tough graphs. Since the reduction is fairly simple we outline it here. Given a graph G with vertex set $\{v_1, ...,$ $v_n\}$ construct G' from G as follows. Add to G a set $A = \{w_1, w_2, ..., w_n\}$ of independent vertices, and join v_i and w_i by an edge for $i = 1, 2, ..., n$. Then add another set B of $k-1$ vertices which induce a complete graph, and join each vertex of B to every vertex of $V(G) \cup A$. It is easy to show that $\alpha(G) \geq k$ if and only if G' is not 1–tough. It is possible that 1–tough graphs can be recognized in polynomial time over a suitably restricted class of graphs. For example, Häggkvist [24] has shown that if $\delta(G) \geq n/2 - 2$ then there is a polynomial time algorithm to determine if G is hamiltonian. As a consequence of Jung's Theorem, a graph G on $n \geq 11$ vertices satisfying $\delta(G) \geq n/2 - 2$ is hamiltonian if and only if G is 1–tough. It follows that 1–tough graphs can be recognized in polynomial time when $\delta(G) \geq n/2 - 2$. Nevertheless we conjecture that it remains NP–hard to recognize 1-tough graphs even when $\delta(G) \geq rn$, for any fixed $r < 1/2$. This is shown to be true in [4] for any fixed $r < 1/3$.

REFERENCES

[1] A. Ainouche and N. Christofides, Conditions for the existence of hamiltonian circuits in graphs based on vertex degrees, J. London Math. Soc. (2) 32 (1985), 385 – 391.

[2] D. Bauer, H.J. Broersma, Li Rao and H.J. Veldman, A generalization of a result of Häggkvist and Nicoghossian, to appear in J. Combin. Theory Ser. B.

[3] D. Bauer, H.J. Broersma and H.J. Veldman, On generalizing a theorem of Jung, preprint 1988.

[4] D. Bauer, S.L. Hakimi, and E. Schmeichel, Recognizing tough graphs is NP-hard, preprint 1988.

[5] D. Bauer, A. Morgana and E. Schmeichel, A simple proof of a theorem of Jung, to appear in Discrete Math.

[6] D. Bauer, A. Morgana, E. Schmeichel and H.J. Veldmand, Long cycles in graphs with large degree sums, to appear in Discrete Math.

[7] D. Bauer and E. Schmeichel, Long cycles in tough graphs, Stevens Research Reports in Mathematics 8612, Stevens Institute of Technology, Hoboken, New Jersey 07030, 1986.

[8] D. Bauer and E. Schemichel, A sufficient condition for hamiltonian cycles in 1-tough graphs, preprint 1988.

[9] D. Bauer, E. Schmeichel and H.J. Veldman, A generaliztion of a theorem of Bigalke and Jung, to appear in ARS Combinatoria.

[10] A. Benhocine and A.P. Wojda, The Geng–Hua Fan conditions for pancyclic or Hamilton–connected graphs, J. Combin. Theory B Ser. 42 (1987), 167 – 180.

[11] J.C. Bermond, On Hamiltonian walks, in *Proc. Fifth British Combinatorial Conference, Aberdeen, 1975* Utilitas Math. (1976), 41 – 51.

[12] J.C. Bermond, Hamiltonian graphs, in *Selected Topics in Graph Theory*, L. Beineke and R.J. Wilson, ed., Academic Press, London and New York, (1978), 127 – 167.

[13] A. Bigalke and H.A. Jung, Über Hamiltonische Kreise und unabhängige Ecken in Graphen, Monatsh. Math. 88 (1979), 195 – 210.

[14] J.A. Bondy, Longest paths and cycles in graphs of high degree, Research Report CORR 80 – 16, University of Waterloo, Waterloo, Ontario, 1980.

[15] G. Chartrand and L. Lesniak, *Graphs and Digraphs,* Wadsworth Inc., Belmont,CA, 1986.

[16] V. Chvátal, private communication.

[17] V. Chavátal, Tough graphs and hamiltonian circuits, Discrete Math. 5 (1973), 215 – 228.

[18] V. Chvátal, Hamiltonian Cycles, in *The Traveling Salesman Problem,* edited by E.L. Lawler, J.K. Lenstra, A.H.G. Rinnooy Kan and D.B. Shmoys, 1985, J. Wiley.

[19] G.A. Dirac, Some theorems on abstract graphs. Proc. London Math. Soc. 2 (1952), 69 – 81.

[20] H. Enomoto, Toughness and the existence of k–factors II, Graphs and Combinatorics (1986), 37 – 42.

[21] H. Enomoto, B. Jackson, P. Katerinis and A. Saito, Toughness and the existence of k–facotrs, J. Graph Theory 9 (1985), 87 – 95.

[22] Fan Geng–Hua, New sufficient conditions for cycles in graphs, J. Combin. Theory Ser. B 37 (1984), 221 – 227.

[23] M.R. Garey and D.S. Johnson, *Computers amd Intractability: A Guide to the Theory of NP–Completeness*, Freeman, San Francisco, 1979.

[24] R. Häggkvist, On the structure of non–hamiltonian graphs I, preprint 1987.

[25] R. Häggkvist and G.G. Nicoghossian, A remark on hamiltonian cycles, J. Combin. Theory Ser. B 30 (1981), 118 – 120.

[26] H.A. Jung, On maximal circuits in finite graphs, Annals of Discrete Math. 3 (1978), 129 – 144.

[27] P. Katerinis, Two sufficient conditions for a 2–factor in a bipartite graph, J. Graph Theory 11 (1987), 1 – 6.

[28] N. Linial, a lower bound on the circumference of a graph, Discrete Math. 15 (1976), 297 – 300.

[29] C. St. J.A. Nash–Williams, Edge–disjoint hamiltonian circuits in graphs with vertices of large valency, *Studies in Pure Mathematics*, Academic Press, London, 1971, 157 – 183.

[30] O. Ore, Note on hamiltonian circuits, Amer. Math. Monthly 67 (1960), 55.

[31] Z. Skupień, Sharp sufficient conditions for hamiltonian cycles in tough graphs, preprint 1987.

[32] C. Thomassen, Long cycles in digraphs, Proc. London Math. Soc. (3) 42 (1981), 231 – 251.

[33] H.J. Veldman, Existence of dominating cycles and paths, Discrete Math. 43 (1983), 281 – 296.

[34] H.J. Veldman, Short proofs of some Fan–type results, preprint 1988.

Graphs Versus Designs – A Quasisurvey*

M. Behzad
Iran University Press

E.S. Mahmoodian
Sharif University of Technology

ABSTRACT

From some graphs one can produce designs. On the other hand, with a given design D numerous graphs can be associated; in fact, D by itself is a hypergraph. The literature is full of these interesting interactions, and reveals that graph theory is a source of tools for design theorists. A complete survey of the very specific topic, graphs associated with designs, is a subject for a book. So, in this article, among all such graphs, we consider several which are widely used and are, in our judgement, the most important ones. Based on their definitions, we present the selected graphs under three headings: V–graphs, B–graphs, and (V ∪ B)–graphs. Finally, we associate with a given design D new graphs, called the Δ–graphs of D, and present a few results without proofs. Throughout the paper we propose several unsolved problems.

1. Introduction

Graph theory is not only an interesting and dynamic field, but has numerous applications in other disciplines, especially in different branches of discrete mathematics. One such branch is design theory. To each design, in fact to each combinatorial structure such as a Hadamard matrix [L1, W2], an affine plane [H7], and a projective plane [H5] one can associate different graphs; as from some graphs one can produce other combinatorial structures. Designs, too, can arise in connection with Latin Squares, finite projective geometries, finite euclidean geometries, graphs, and in general hypergraphs [W1]. These interrelationships are both interesting and beneficial. In fact, in graph theory, which seems to be a source of tools for all such investigations, sometimes it is

* The authors are grateful to the Research Center of the Atomic Energy Organization of Iran.

125

possible to define a notion and obtain a theorem which generalizes several combinatorial concepts, and apparently unrelated theorems. See, e.g., [H5], [A2], and [H1, page 110].

Specifically, in the last several decades, much research in the theory of designs has been oriented toward the theory of graphs [R1]. Many graphs are associated directly with a given design, and some indirectly [C3, D1]; quite a few have no specific names, and some have several. Very many of them are used to study specific designs such as maximal partial Steiner triple systems of order ≤ 11; see [C7].

One main reason for all these studies is that many such graphs serve as design invariants; invariants which can be used effectively, as far as computational complexity is concerned, to obtain all nonisomorphic designs with given conditions [C10]; or even to decide whether two given designs are isomorphic. Another reason is that, in some context, the only important feature is the structure of some specific graphs associated with the given design; see, e.g. [C9].

Since a thorough investigation of all graphs associated with designs, within a limited scope, is not possible, we confine ourselves to several such graphs which are widely used and, based in our judgements, are most interesting and useful. We completely ignore digraphs and hypergraphs associated with designs [C11, B4], and concentrate on (multi)graphs which are directly associated with, mainly, 2–designs.

In order to avoid confusion and clarify statements, aside from replacing (multi)graphs by graphs, on several occasions we change terminologies and notations without even mentioning the originals. However, we provide appropriate references for further investigations.

To simplify matters, after the introduction of some notions and notations we categorize the associated graphs of a design $D = (V,B)$ into three main sections: V–graphs, graphs associated with the elements of D; B–graphs, graphs associated with the blocks of D; and $(V \cup B)$–graphs, those associated with the elements and with the blocks of D. In each section we define appropriate concepts, briefly mention a result or two, and provide a few references.

Next, we associate with a t–design $D = (V,B)$, new graphs, called Δ–graphs of D with respect to elements of V. In this regard, we state several new results. Furthermore, throughout the article we propose a few unsolved problems.

2. Main Definitions and Preliminaries

In this section we provide some of the basic terminology and notation, along with a few preliminaries. This is essential, since even design theorists often include such terminologies.

An (n, r, p, q)–*strongly regular graph* G, introduced by Bose [B5], is a graph of order n which is r–regular having the property that each pair of distinct adjacent (nonadjacent) vertices of G are simultaneously joined to p (q respectively) vertices of G. For $p > 0$, an (n, r, p)–graph is an (n, r, p, p)–strongly regular graph. In other words, an (n, r, p)–graph is an r–regular graph G of order n, such that if S_k is the set of vertices of G which are joined to a vertex v_k, then for $v_i \neq v_j$, $|S_i \cap S_j| = p$. Complete graphs and empty graphs of any order are trivially strongly regular – which are often excluded. The Petersen graph is a good example. It can be seen that if G is a strongly regular graph, so is its complement, $\overline{G}$. By using some simple results of design theory it can be shown that for a given value of p, there exist only finitely many (n, r, p)–graphs; see [C2].

Based on the definition of a graph, two distinct vertices are joined by n edges, where $n = 0$ or $n = 1$. If we relax this restriction, we obtain the definition of a *multigraph* [H2] which is of great value in design theory. In this article by a "graph" we mean a multigraph which might or might not have multiple edges. In the literature, this term is frequently written in the form "(multi)graph". For other graph theoretic notions and results see for example, [B1] or [H2].

Let t, λ, k, and v be fixed positive integers with $\lambda \geq 1$, and $1 \leq t \leq k$. A t–*design*, denoted by t–B $[k,\lambda;v]$ or by $S_\lambda(t,k,v)$, is an ordered pair $D = (V,B)$, where V is a v–set each of whose members is an *element* (*treatment* or *variety*) of D, and B is a collection, with possible repetition of elements, of k–subsets of the set V, called *blocks* of D, such that every t–subset of V is contained in precisely λ blocks of D.

In a t–design, if $k = v$, then it is *trivial* in the sense that its construction is trivial. Hence we consider t–designs with $k < v$. Among other trivial t–designs are those with $t = k$. Therefore, we assume that $t < k$. For a t–design the following assertions hold.

1. Each i–subset, $1 \leq i \leq t$, of V is contained in precisely λ_i blocks, where

$$\lambda_i = \lambda_{i+1} \, \frac{v - i}{k - i} \, , \text{ for } 1 \leq i \leq t - 1, \text{ and } \lambda_t = \lambda.$$

2. Each element of D belongs to precisely λ_1 blocks of D, where λ_1 is usually denoted by r. Hence

$$r = \lambda_2 \, \frac{v - 1}{k - 1} \, .$$

3. The total number of the blocks of D, denoted by b, is $b = \frac{vr}{k}$. The number of distinct blocks of D is denoted by b*.

4. **Fisher's Theorem**. In a 2–design we have $b \geq v$; hence in every 2–design $r \geq k$.

5. It can be shown that in a 2–design D the following statements are equivalent:

    i. $b = v$,

    ii. $r = k$

    iii. The cardinality of the intersection of any two distinct blocks of D is λ.

A 2–design is also called a *balanced incomplete block design*, (BIBD), a *block design*, or a *design*, and is usually denoted by $B[k,\lambda;v]$. Unless otherwise stated the word design will mean a BIBD. A *symmetric design* (SD), denoted by $SB[k,\lambda;v]$, is a design in which $b = v$. A symmetric design is sometimes called a λ-*plane*, or a *projective design*. In addition, 1–planes, i.e. $SB[k,1;v]$ designs, are referred to as *finite projective planes* (FPP) of order $k - 1$.

A design with $k = 3$, i.e., a $B[3,\lambda;v]$ design is called a *triple system* (TS). A *Steiner system* (SS) is a t–$[k,1;v]$ design, and is often denoted by $S(t,k,v)$. An $S(2,3,v)$, or equivalently a $B[3,1;v]$ design is called a *Steiner triple system of order* v $(STS(v))$, and a $B[3,2;v]$ design is known as a *twofold triple system of order* v $(TTS(v))$. It is well known that a $STS(v)$ exists if and only if $v \equiv 1$ or $3 \pmod 6$, and that a $TTS(v)$ exists if and only if $v \equiv 0$ or $1 \pmod 3$.

A *partial triple system of order* v $(PTS(\lambda;v))$ is a pair (V,B) where V is a v–set, and B is a collection of 3–subsets of V with the property that each 2–subset of V is contained in at most λ blocks. A *maximal partial triple system of order* v (MPTS $(\lambda;v)$) is a $PTS(\lambda;v)$, (V,B), such that there is no $PTS(\lambda;v)$, (V,B') with $B \subsetneq B'$.

If $\lambda = 1$, then these two systems are called *partial Steiner triple system* (PSTS(v)) and *maximal partial Steiner triple system* (MPSTS(v)), respectively.

Two designs $D = (V,B)$, and $D' = (V',B')$ are *isomorphic*, written $D \cong D'$, if there exists a bijection $f: V \rightarrow V'$ such that $a \in B$ if and only if $f(a) \in B'$.

A design $D = (V,B)$ is *irreducible* or *indecomposable* if there do not exist designs $D_1 = (V,B_1)$ and $D_2 = (V,B_2)$ such that $B = B_1 \cup B_2$. It is obvious that if D_1 is a t–$B[k,\lambda_1;v]$ design, and D_2 is a t–$[k,\lambda_2;v]$ design, then the union of these two designs is a t–$B[k,\lambda_1+\lambda_2;v]$ design. So, designs which cannot be expressed in this form are irreducible designs.

A design $D = (V,B)$ is called a *quasi–symmetric design* (QSD) if for any two distinct blocks c and d of D, $|c \cap d| \in \{x, y\}$, where x and y, $x \leq y$, are two fixed nonegative integers. Note that, based on Fisher's Theorem, every symmetric design is a QSD as well. One simple class of nonsymmetric quasi–symmetric designs is the class of designs with $\lambda = 1$ which are not finite projective planes; in this case $\{x, y\} = \{0, 1\}$.

There are several generalizations of designs. One such generalization is a *general design*, which is an ordered pair (V,B) where V is a set and B is a collection of subsets of V. For further reference the interested reader is referred to [S3, H1, and C10].

3. V–Graphs

There are several graphs associated with the elements of a design; some of which have no specific names, and a few different ones are named the same. As expected, in the literature there is no systematic study of these graphs. In this section we mention just a few of these graphs to demonstrate their importance.

3.1   Leaves

In [C7] the *leave* of a partial triple system D = (V,B) is defined to be a graph L(D) = (V,E), where E contains all pairs that appear less than λ times in B; the multiplicity of an edge is λ minus the number of its appearances in B. A partial triple system is maximal if and only if its leave contains no triangles. In [C7] leaves are used to obtain all MPSTS of order v, v ≤ 11.

In [C8] a PSTS of order v, D = (V,B), is considered, and with D a graph L(D) is associated. The vertex set of L(D) is V, and its edge set consists of 2–subsets of V which are not subsets of any block of D. Clearly, L(D) is the leave of D defined above, and if L(D) is an empty graph, then D is an STS. Some basic necessary conditions for a graph to be a leave can easily be obtained. In the same paper special classes of graphs which are leaves of maximal partial triple systems are determined.

In [H4] an extensive use is made of a special class of graphs which are, though not mentioned explicitly, leaves as defined above. It is mentioned there that for embedding purposes an important feature is the structure of the leaves; hence *leave problem*, i.e., characterization of graphs which arise as leaves of a PB[3,λ;v], is an important problem. In the aforenamed papers some basic necessary conditions for a graph to be a leave are mentioned, and some sufficient conditions are obtained. For other results on this subject, which needs a survey of its own, see [C4].

3.2   Single–Edge Graphs

It is known that for each v ≡ 0 or 1 (mod 3) there exists a TTS(v) without repeated blocks [B6]. In 1986 Rosa and Hoffman [R3] considered the following problem: Given v ≡ 0 or 1 (mod 3) and a nonnegative integer p, does there exist a TTS(v) with exactly p repeated blocks? By using an auxiliary notion, a graph, this question is answered completely. With a TTS(v), say D = (V,B), a graph S(D), called the *single–edge graph* of D, is associated as follows: V(G) = V, and for each distinct pair u and w of elements of D, uw $\in$ E(G) if and only if {u, w} is contained in two distinct blocks of D. In [R3] a few rather elementary properties of G are used to achieve the goal.

3.3 Adjacency Graphs

In [S2] a general design, D = (V,B), is considered. For each u, w $\in$ V, the multiplicity, $m_D(u,w)$, of u and w is the number of blocks containing u and w. The *adjacency graph* Ad(D) of D has vertex set V; two vertices u and w, u $\neq$ w, are joined by $m_D(u,w)$ edges. Here, adjacency graphs of partial geometric designs are characterized.

3.4 Neighbourhood Graphs

In [M3], a B[k,λ;v] design, D = (V,B), is considered. With an element x of V, a graph $N_x(D)$, called the *neighbourhood graph of* D *with respect to* x, is associated as follows: $N_x(D)$ has v − 1 vertices, each of which corresponds to a member of V − {x}. Two distinct vertices are joined with m edges, $0 \leq m \leq \lambda$, if and only if the two elements represented by these two vertices along with x appear in m blocks of D. Figure 1 shows all the neighbourhood graphs of a B[3,2;9] design whose blocks are:

$$
\begin{array}{ccccccccccccccccccccccc}
1 & 1 & 1 & 1 & 1 & 1 & 1 & 1 & 2 & 2 & 2 & 2 & 2 & 2 & 3 & 3 & 3 & 3 & 4 & 4 & 4 & 5 & 6 & 7 \\
2 & 2 & 3 & 3 & 4 & 5 & 6 & 7 & 3 & 3 & 4 & 4 & 5 & 5 & 4 & 5 & 5 & 6 & 5 & 6 & 6 & 7 & 8 & 8 \\
8 & 9 & 4 & 9 & 5 & 6 & 7 & 8 & 6 & 7 & 8 & 9 & 6 & 7 & 7 & 8 & 9 & 8 & 8 & 7 & 9 & 9 & 9 & 9
\end{array}
$$

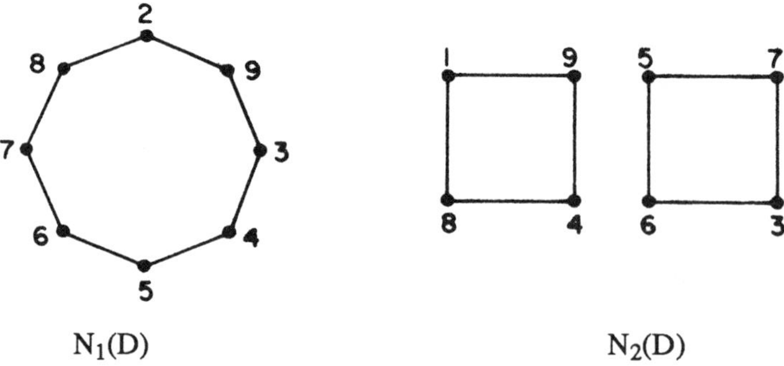

$N_1(D)$ $N_2(D)$

$$N_i(D) \cong N_1(D), \ 3 \leq i \leq 9$$

Figure 1

In [M3] this graph is used to obtain all nonisomorphic B[3,2;9] designs, and all nonisomorphic B[3,3;7] designs. In fact, as indicated in [M3], this method was first used by W.D. Wallis for finding the four nonisomorphic B[3,2;7] designs.

It is easy to see that a B[3,2;v] design D is irreducible, if for some x, $N_x(D)$ contains an odd cycle; however, for irreducibility this condition is not necessary. This can be seen by the example given in Figure 1. For details see [S3].

Let D = (V,B) be a TS, and x be a fixed element of D. Consider all blocks of D which contain x, and eliminate x from them to obtain a collection of 2–subsets of V. This collection, denoted by N(x), is called the *neighborhood of* x *in* D. The collection N(x) can be interpreted as the edges of a λ–regular graph of order v − 1, which is the neighbourhood graph defined above. N(x) is an effective tool for design theorists. This often has no specific name, and sometimes is referred to as the adjacency graph for the particular element x; see, e.g., [S3]. The term "figure around the variety (treatment) x of D" is used for this graph by Foody and Hedayat [F1].

In [H3] a B[3,3;9] design without repeated blocks is considered. For such a design D, $N_x(D)$ which has no multiple edges is again referred to as the neighbourhood graph of D with respect to x. Incidently by using $N_x(D)$ "apparently" all 332 nonisomorphic B[3,3;9] designs without repeated blocks are generated.

In [C5] the graph defined in this section is called the neighbourhood of the element. The *neighbourhood problem* is to determine which λ–regular graphs are produced from B[3,λ;v] designs. Here some basic necessary conditions for the problem are obtained. For sufficiency, emphasis is on designs with λ = 3 in which case every cubic graph is a candidate. By using graph theoretic arguments the problem is settled for this case.

In conclusion, systematic strudy of V–graphs and their interrelationships are proposed.

4. B–Graphs

In this section we consider three main classes of graphs whose vertices correspond to the blocks of a given design. Then we briefly mention, in chronological order, some special classes of designs and their related graphs which are B–graphs. A brief remark ends this section.

4.1 Intersection Graphs

In [S2] a general design D = (V,B) is considered. The *intersection graph* I(D) of D is a graph whose vertices correspond to the blocks of D; two distinct vertices u and w are joined by $|b_u \cap b_w|$ edges, where b_u and b_w are the blocks which correspond, to u and w, respectively. Then I(D) is used to obtain several results about some specific designs. It is seen that for a B[k,λ;v] design D, I(D) is a k(r − 1)–regular

graph. Figure 2 shows $\overline{I(D)}$, the complement of I(D), where D is a B[3,1;9] design.

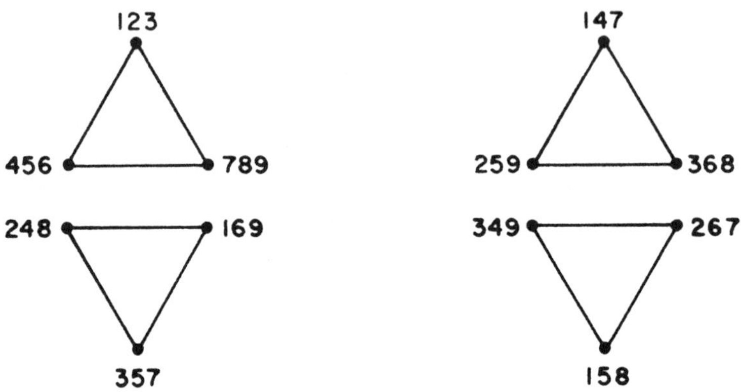

Figure 2 $\overline{I(D)}$ of the indicated B[3,1;9] design D

4.2   Block Intersection Graphs

It is known that, see for example [C10], there is no known polynomial time algorithm for design isomorphism. Hence, it is of value to look for "effective" isomorphism invariants.

Let D = (V,B) be a B[k,λ;v] design. For each i, $0 \le i \le k$, the *block intersection graph* $B_i(D)$ of D is the graph whose vertices corredspond to the blocks of D; two vertices are adjacent if and only if their corresponding blocks contain exactly i elements in common.

For the design of Figure 3, $B_0(D)$ is the union of $\overline{K}_2$ and the 1–regular graph of order 12; while $B_3(D)$ is $K_2 \cup \overline{K}_{12}$. If in Figure 3, we remove the 3 edges joining the two vertices 124, we obtain $B_2(D)$.

If D and D' are two isomorphic designs, then for each i, $0 \le i \le k$, $B_i(D) \cong B_i(D')$. Hence, for design isomorphism, invariants of these graphs are important. One such invariant is the number of K_c's in $B_i(D)$, $0 \le i \le k$. This and some other invariants were first employed by Gibbons [G1], and later by others, for distinguishing small designs; see, e.g., [M1] and [C9]. Note that for each i, $B_i(D)$ contains no multiple edges; and that $B_i(D)$, in general, is not regular.

4.3 Associated Graphs

The *associated graph* A(D) of a design D = (V,B) is a graph whose vertices correspond to the blocks of D; two vertices u and w are joined by an edge for each pair of elements common to their corresponding blocks.   Hence,   A(D)   is a $(k(k-1)(\lambda - 1)/2)$–regular graph.   Figure 3 shows A(D) for a reducible B[3,2;7] design D. Note that the associated graph of the B[3,1;9] design of Figure 2 is the empty graph $\overline{K}_{12}$. For some results concerning A(D) see, e.g., [S3] and [B3]. One such result is that a B[k,2;v] design D is irreducible if and only if A(D) contains an odd cycle.

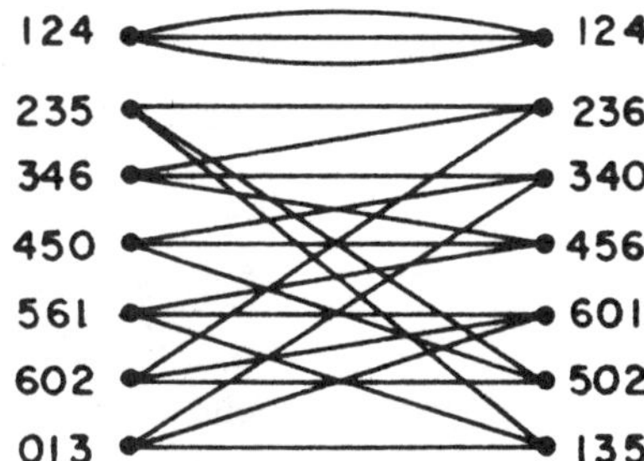

Figure 3   A(D) of the indicated B[3,2;7] design D

4.4 Other B–Graphs

In [B5] Bose investigates Steiner systems D which are not projective planes. Then for each block, referred to as a *line*, a vertex is considered, and two vertices are joined by an edge if and only if the intersection of the corresponding lines is nonempty. (This graph, called the *line graph* of D, is nothing but $B_1(D)$.)

In [A1], Aliev and Seiden characterize a special class of Steiner triple systems in terms of strongly regular graphs. For this purpose, for an arbitrary STS, D, they define a graph whose vertices correspond to the blocks of D; two vertices are joined by an edge if and only if their corresponding blocks have exactly one element in common. (This graph is again $B_1(D)$.)

Goethals and Seidel [G2] consider quasi–symmetric designs D with x < y, and define the *block graph* of D as a graph whose vertices correspond to the blocks of D; two vertices are joined by an edge if and only if the intersection of the corresponding blocks has cardinality y. (This graph is $B_y(D)$.) Among other results, they show that the block graph of such a design is strongly regular; and based on this, they construct several interesting classes of strongly regular graphs. See also [C1] and [S1].

The term "block intersection graph" was introduced by Kramer [K1] who constructed the block intersection graph G of a B[k,2;v] design D = (V,B) as follows: B is the vertex set of G and two vertices of G are adjacent if and only if their intersection contains a pair of elements. (This graph is simply $B_2(D)$.) Then Kramer showed that G is a bigraph if and only if the design is reducible.

The term block intersection graph also appears in [C6] with a different meaning. In this case a twofold triple system D is considered. The block intersection graph G of D has a vertex for each block of D; two vertices are joined by an edge for each 2–subset which is in the intersection of the corresponding blocks. (This time G is A(D); moreover G is a cubic graph all of whose components are 3–connected.)

The above considerations, which are by no means exhaustive, indicate that B–graphs associated with designs are of considerable importance. The ones which have been used most are I(D), A(D), and $B_i(D)$. Systematic study of these B–graphs and their interrelationships are proposed.

5. (V ∪ B)–Graphs

In this brief section we consider graphs associated with designs D = (V,B) where vertices correspond to the members of V ∪ B. As far as we know this line of work was originated with the work of A.J. Hoffman [H5], who was interested in characterizing some regular connected graphs by means of the eigenvalues of their adjacency matrices.

Let D = (V,B) be a finite porjective plane, SB[k + 1,1;v] design, known as a projective plane of order k. Each member of V is a *point* and each member of B is a *line* of D. Thus D has $k^2 + k + 1$ points and $k^2 + k + 1$ lines. To D a bigraph In(D) of order $2(k^2 + k + 1)$ (one vertex for each point and one for each line of D) is associated, in which two vertices are adjacent if and only if one corresponds to a point and the other to a line containing that point. The results obtained in [H5] are generalized to a SB[k,λ;v] design, D = (V,B), by Hoffman and Ray–Chaudhuri [H6]. Here In(D) is defined as above; it is seen that In(D) is a k–regular bigraph of order 2v with the property that any two vertices of the same part are adjacent to exactly λ vertices of the other part. Finally, in [H7], a similar graph is associated with a finite affine plane and its properties are obtained.

In [R2], the same graph is associated with a B[k,1;v] design and a characterization of the line graph of this graph is presented; see also [D2].

In order to compute the automorphism group of a design $D = (V,B)$ the same graph as above, denoted by $In(D)$, is used in [M2], and the following theorem is stated.

Theorem *The automorphism group of a design D is isomorphic with the automorphism group of $In(D)$. Moreover, two designs D_1 and D_2 are isomorphic if and only if $In(D_1) \cong In(D_2)$.*

In [S2], with a general design $D = (V,B)$ a graph, called the *incidence graph* $In(D)$ of D, is associated as above – the vertex set of $In(D)$ is $V \cup B$; two vertices are adjacent if and only if one is an element, the other is a block, and the block contains the element – and the resulting bigraph is used as an aid.

To the best of our knowledge in the literature there is only one main type of $(V \cup B)$–graph associated with a design which we mentioned above. In the following section with a given design $D = (V,B)$ we associate other $(V \cup B)$–graphs, called delta graphs of D.

6. Δ–Graphs

Let $D = (V,B)$ be a t–$B[k,\lambda;v]$ design, x be an arbitrary element of D, called the *root*, and i be an integer, where $0 < i \leq k$. Based on the observations mentioned in sections 3.4, 4.3, and 5, we associate with D a graph $\Delta_x^i (D)$, or simply Δ^i, called the *delta–graph* of D *with respect to* x, as follows. The vertex set of Δ^i is $V \cup B$, i.e., to each element and to each block of D we associate a vertex. We join two vertices u and w, $u \neq w$, of Δ^i by an edge if:

1. one is an element of D, and the other is a block of D which contains that element,
2. u and w are two elements of D which are different from x, and the set $\{u, w, x\}$ is contained in a block of D; the multiplicity of this edge is the number of blocks of D which contain the set $\{u, w, x\}$, and
3. u and w are two blocks of D; in this case u and w are joined by m edges, where m is the number of distinct i–sets which are subsets of both u and w.

These graphs can be used as invariants for design isomorphism. Several results on this line will appear elsewhere.

In what follows, we confine ourselves to 2–$B[k,\lambda;v]$ designs; to avoid trivial cases we assume that $3 \leq k < v$, $v \geq 6$, and $\lambda \geq 1$. If D is such a design, we will denote $\Delta_x^2 (D)$ by $\Delta_x(D)$ or simply by Δ. As an example we consider the graph $\Delta_6(D)$ of the $B[3,2;6]$ design whose blocks are indicated in Figure 4.

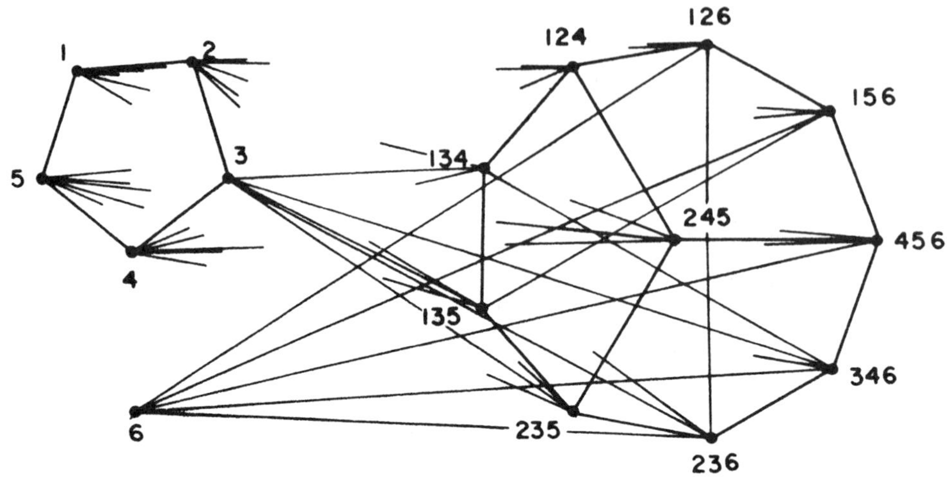

Figure 4 $\Delta_6(D)$

Regarding the graph $\Delta_x(D) = (V \cup B, E)$, where $D = (V,B)$ is a $B[k,\lambda;v]$ design, we state next several elementary results, observations and comments which are independent of the choice of the root x.

6.1 $p(\Delta) = v + b$, *where* $p(\Delta)$ *denotes the number of vertices of* Δ.

6.2 $deg_{\langle V \rangle} x = 0$, *and* $deg_\Delta x = r$, $r \geq 3$. *For further references we denote* $deg_\Delta x$ *by* ρ. *Here* $\langle V \rangle$ *denotes the subgraph of* Δ *induced by* V.

6.3 $deg_{\langle V \rangle} u = \lambda(k-2)$, *where* $u \neq x$ *is a vertex of* Δ *belonging to* V. *This follows from the fact that each of the* λ *blocks which contains* x *and* u, *contain exactly* $k-2$ *other elements of* D. *Thus* $\langle V \rangle$ *is a nonempty subgraph of* Δ.

6.4 $deg_\Delta u = \lambda(k-2) + r$, *where as before* $u \neq x$ *is a vertex of* Δ *belonging to* V. *We denote the number* $\lambda(k-2) + r$ *by* υ.

6.5 $deg_{\langle B \rangle} u = k(k-1)(\lambda-1)/2$, *where* u *is a vertex of* Δ *belonging to* B. *Note that* u *contains* $\binom{k}{2}$ *pairs of elements of* D *each of which appears in* λ *blocks. Also, note that* $\langle B \rangle$ *is an empty graph if and only if* $\lambda = 1$.

6.6 $deg_\Delta u = k(k\lambda - k - \lambda + 3)/2 \geq k$, *where again* u *is a vertex of* Δ *belonging to* B; *equality holds if and only if* $\lambda = 1$. *We denote the number* $k(k\lambda - k - \lambda + 3)/2$ *by* β.

6.7 $q(\Delta) = \lambda(v - 1)\{2(k - 1)(k - 2) + v(k\lambda - k - \lambda + 5)\}/4(k - 1)$, where $q(\Delta)$ *denotes the number of edges of* Δ.

6.8   *The neighbourhood graph* $N_x(D)$, *and the associated multigraph* $A(D)$, *defined in Sections 3.4, and 4.3, are both (vertex) induced subgraphs of* $\Delta_x(D)$ *which are isomorphic to* $\langle V - \{x\}\rangle$ *and* $\langle B\rangle$, *respectively.*

6.9   *The degree set of* Δ *has either 3 or 2 distinct members.*

   i) Assume that the degree set of Δ has 3 members. Then Δ contains exactly one vertex of degree ρ, namely the root x; has at least 5 vertices of degree υ, $\upsilon > \rho$, namely those belonging to $V - \{x\}$; and contains at least $p(\Delta)/2$ vertices of degree β, namely those belonging to B. Thus, for a given design D, if the cardinality of the degree set of Δ is 3, then one of the three possibilities: $\beta > \upsilon > \rho$, $\upsilon > \rho > \beta$, or $\upsilon > \beta > \rho$ holds.

Next, assume that the degree set of Δ has 2 (distinct) members. Then $\upsilon > \rho$ implies that either

   ii) $\upsilon > \rho = \beta$, or

   iii) $\beta = \upsilon > \rho$.

In case iii), Δ has exactly one vertex of degree ρ, namely the root x; any other vertex of Δ has degree $\beta = \upsilon > \rho$. In case ii) more than half of the vertices of Δ, namely those of $B \cup \{x\}$ have degree $\rho = \beta$, and the remaining ones (there are at least 5 of them, namely those belonging to $V - \{x\}$) have degree $\upsilon > \rho$.

Based on the above observations we define the following notion.

Definition   A design of type I, II, or III is a design whose Δ–graph is of type i, ii, or iii, respectively.

Designs of type I, II, and III exist. In fact, the design D_i, $1 \le i \le 3$, given below is of type I with degree set $\{5, 6, 7\}$, of type II with degree set $\{3, 4\}$, and of type III with degree set $\{6, 10\}$, respectively.

D_1: The B[3,2;6] design given in Figure 4
    $\beta = k + k(k - 1)(\lambda - 1)/2 = 6$, $\rho = r = 5$, $\upsilon = r + \lambda(k - 2) = 7$.

D_2: The B[3,1;7] design

    1 2 3 4 5 6 7
    2 3 4 5 6 7 1
    4 5 6 7 1 2 3
    $\rho = \beta = 3$, $\upsilon = 4$.

D_3: The B[4,2;10] design

$$
\begin{array}{ccccccccccccccc}
1 & 1 & 3 & 2 & 1 & 1 & 1 & 1 & 2 & 2 & 5 & 2 & 3 & 4 & 3 \\
2 & 2 & 4 & 4 & 6 & 3 & 4 & 5 & 3 & 8 & 6 & 5 & 4 & 5 & 7 \\
3 & 4 & 5 & 6 & 7 & 6 & 8 & 7 & 7 & 9 & 9 & 6 & 6 & 7 & 8 \\
5 & 0 & 0 & 7 & 0 & 8 & 9 & 9 & 9 & 0 & 0 & 8 & 9 & 8 & 0
\end{array}
$$

$\rho = 6, \ \beta = \upsilon = 10.$

6.10 *A B[k,λ;v] design D is of type II if and only if*
    (a) r = k + k(k – 1)(λ – 1)/2;
It is of type III if and only if
    (b) r = 2λ + k(k – 3)(λ – 1)/2;
and D is of type I if and only if neither (a) nor (b) hold.

6.11 *No symmetric design D is of type III. A symmetric design is of type II if and only if it is a finite projective plane.*

6.12 *Assume that a graph Δ is given, and that there exists a design D with a root x whose Δ–graph is isomorphic with Δ. Then from Δ all the parameters v, k, λ; hence b, and r, of D can be determined. Moreover, if Δ is of type I or II, then from Δ we can determine the support size b* of the design D; in fact, by labeling appropriate vertices of Δ we can determine all the blocks, hence the unique (up to isomorphism) design D.*

6.13 *If two designs D = (V,B), and D'= (V',B') are isomorphic, then for each element x of D there exists an element x' of D' such that $\Delta_x(D) \cong \Delta_{x'}(D')$.*

6.14 *If two graphs Δ and Δ' are isomorphic, and if Δ is a Δ–graph of a design D, then Δ' is also a Δ–graph of a design D'. Moreover, $D \cong D'$.*

6.15 *Assume that D and D' are two nonisomorphic designs of type I or II. Let x and x' be arbitrary elements of D and D', respectively. Then $\Delta_x(D) \not\cong \Delta_{x'}(D')$.*

6.16 *Note that if for a design D, and an arbitrary root x, the unlabeled graph $\Delta_x(D)$ is given, then, in all cases, based on the degrees of the vertices of $\Delta_x(D)$, one can distinquish x. Moreover, in the first two cases, vertices which are associated with the elements or with the blocks of D can be differentiated.*

6.17 *For a given design D, an important invariant is the collection of the eccentric sequences, and another one is the collection of the eccentric sets, of the graphs $\Delta_x^i\,(D)$, where x varies over V; see, e.g., [B2]. It can be shown that the eccentric set of D with* *respect to a root x, is {2}, {3}, {2, 3}, {3, 4}, or {2, 3, 4}. With the exception of {2, 3, 4} which is still open, all these sets are realizable. In fact, if D is symmetric, then its eccentric set is {2}, {3}, or {2,3}. Results on these lines, their generalizations, and computer algorithms, appear elsewhere.*

In conclusion, it is worth mentioning that with each new special class of graphs a host of open problems pop up. In general, characterizing such special gaphs, evaluating some of their graph–theoretic parameters, and obtaining some of their properties are of interest.

REFERENCES

[A1]   I.S.O. Aliev and E. Seiden, Steiner triple systems and strongly regular graphs, *J. Comb. Th.* **6**(1969) 33 – 39.

[A2]   D. Archdeacon, J. Dinitz and F. Harary, Orthogonal edge colorings of graphs. Presented to *The Sixteenth Southeastern Conference on Combinatorics, Graph Theory and Computing*, Boca Raton, Florida Atlantic Univ. (1985).

[B1]   M. Behzad, G. Chartrand and L. Lesniak–Foster, *Graphs and Digraphs*, Prindle. Weber and Schmidt (1979).

[B2]   M. Behzad and J. E. Simpson, Eccentric sequences and eccentric sets in graphs. *Discrete Math.* **16** (1976) 187 – 193.

[B3]   E.J. Billington, Further constructions of irreducible designs. *Congressus Numerantium* **35** (1982) 77 – 89.

[B4]   E.J. Billington, Balanced n–ary designs: A combinatorial survey and some new results. *Ars Combinatoria* **17A** (1984) 37 – 72.

[B5]   R.C. Bose, Strongly regular graphs, partial geometries and partially balanced designs. *Pacific J. Math.* **13** (1963) 389 – 419.

[B6]   J. van Buggenhaut, On the existence of 2–designs $S_2(2,3,v)$ without repeated blocks. *Discrete Math.* **8** (1974) 105 – 109.

[C1]   P.J. Cameron, Strongly regular graphs. *Selected Topics in Graph Theory* (eds. L.W. Bieneke and R. J. Wilson) Academic Press (1978).

[C2]   P. J. Cameron and J. H. van Lint, Graphs, Codes and Designs. *London Math. Soc. Lec. Note Series* **43**, Cambridge U. Press (1980).

[C3]   C.J. Colbourn, Disjoint cyclic Steiner triple systems. *Congressus Numerantium* **32** (1981) 205 – 212.

[C4]   C.J. Colbourn and R.A. Mathon, Leave graphs of small maximal partial triple systems. *Journal of Comb. Math. and Comb. Computing* **2** (1987) 13–28.

[C5]   D.J. Colbourn and B.D. McKay, Cubic neighbourhoods in triple systems. *Amm. Discrete Math.* (to appear).

[C6]   D.J. Colbourn, W.R. Pulleyblank, and A. Rosa, Hybrid triple systems and cubic feedback sets. *Graphs and Combinatorics* (to appear).

[C7]   C.J. Colbourn and A. Rosa, Maximal partial Steiner triple systems of order $v \leq 11$. *Ars Combinatoria* **20** (1985) 5–28.

[C8]   C.J. Colbourn and A. Rosa, Quadratic leaves of maximal partial triple systems. *Graphs and Combinatorics* **2** (1986) 317 – 337.

[C9]   M.J. Colbourn, An analysis technique for Steiner triple systems. *Proc. 10th Southeastern Conf. Combinatorics, Graph Theory and Computing* (1979) 289 – 303.

[C10]   M.J. Colbourn, Algorithmic aspects of combinatorial designs: A survey. *Annals of Discrete Math.* **26** (1985) 67 – 136.

[C11]   M.J.Colbourn, C.J. Colbourn, and W.L. Rosenbaum, Train: An invariant for Steiner triple systems. *Ars Combinatoria* **13** (1982) 149 – 162.

[D1]   J.W. Di Paola, Block designs and graph theory. *J. Comb. Th.* **1** (1966) 132 – 148.

[D2]   T.A. Dowling and R. Laskar, A geometric characterization of the line graph of a finite projective plane. *J. Comb. Th.* **3** (1967) 402 – 410.

[F1]   W. Foody and A. Hedayat, Supports of BIB designs: An algebraic and graphical study. *Tech. Report #86–02 Stat. Lab. Dept. of Math. Stat. & Comp. Sc., U. of Ill. at Chicago* (March 1986).

[G1]   P.B. Gibbons, *Computing Techniques for the Construction and Analysis of Block Designs*. Ph.D. Thesis, U. of Toronto (1976).

[G2] J.M. Goethals and J.J. Seidel, Strongly regular graphs derived from combinatorial designs. *Can. J. Math.* **22** (1970) 597 – 614.

[H1] M. Hall, Jr., *Combinatorial Theory*. Wiley Publishing Co., New York (1986).

[H2] F. Harary, *Graph Theory*. Addison–Wesley, Reading, Mass. (1969).

[H3] J.J. Harms, C.J. Colbourn and A.V. Ivanov, A census of (9,3,3) block designs without repeated blocks. *Congressus Numerantium* **57** (1987) 147 – 170.

[H4] A. Hartman, Partial triple system and edge colourings. *Discrete Math.* **62** (1986) 183 – 196.

[H5] A.J. Hoffman, On the line graph of a projective plane. *Proc. Amer. Math. Soc.* **16** (1965) 297 – 302.

[H6] A.J. Hoffman and D. K. Ray–Chaudhuri, On the line graph of a symmetric balanced incomplete block design. *Trans. Am. Math. Soc.* **116** (1965) 238 – 252.

[H7] A.J. Hoffman and D.K. Ray–Chaudhuri, On the line graph of a finite affine plane. *Can J. Math.* **17** (1965) 687 – 694.

[K1] E.S. Kramer, Indecomposable triple systems. *Discrete Math.* **8** (1974) 173 – 180.

[L1] C.H.C. Little and D.J. Thuente, The Hadamard conjecture and circuits of length four in a complete bipartite graph. *J. Austral. Math. Soc.* (A) **31** (1981) 252 – 256.

[M1] R.A. Mathon and A. Rosa, A census of Mendelsohn triple systems of order nine. *Ars Combinatoria* **4** (1977) 309 – 315.

[M2] B.D. McKay, Naughty user's guide. *Technical Report TR–CS–87–03 Dept. of Computer Sc. The Aust. Nat. U.* (September 1987).

[M3] E.J. Morgan, Some small quasi–multiple designs. *Ars Combinatoria* **3** (1977) 233 – 250.

[R1] D. Raghavarao, *Constructions and Combinatorial Problems in Design of Experiments*. Wiley, New York (1971).

[R2]   S.B. Rao and A. R. Rao, A characterization of the line graph of a BIBD with $\lambda = 1$. *Sankhya* (A) **31** (1969) 369 – 370.

[R3]   A. Rosa and D. Hoffman, The number of repeated blocks in twofold triple systems. *J. Comb. Th.* (A) **41** (1986) 61 – 88.

[S1]   J.J. Seidel, Strongly regular graphs. *Surveys in Combinatorics, Proc. 7th British Comb. Conf,* (ed. Bollabas) London Math. Soc. Lec. Notes Ser. **38** (1979) 157 – 180.

[S2]   S.S. Shrikhande and N.M. Singhi, Designs, adjacency multigraphs and embeddings: A survey. *Proc. Comb. and Graph Th. Calcutta, 1980,* (ed. S.B. Rao), Springer–Verlag (1981) 113 – 132.

[S3]   A.P. Street and D.J. Street, *Combinatorics of Experimental Design.* Oxford Soc. Pub. (1987).

[W1]   A.T. White, Block designs and graph imbeddings. *J. Comb. Th.* (B) **25** (1978) 166 – 183.

[W2]   W.D. Wallis, Certain graphs arising from Hadamard matrices. *Bull. Austral. Math. Soc.* **1** (1969) 325 – 331.

EXPLORATIONS INTO GRAPH VULNERABILITY*

Lowell W. Beineke

Kunwarjit S. Bagga

Marc J. Lipman

Raymond E. Pippert

Indiana University - - Purdue University

at Fort Wayne

ABSTRACT

A description is given of work on a comprehensive survey of graph vulnerability, including a discussion of some of the problems encountered. Some vulnerability parameters in each of three categories – cutting, covering, and closeness – are treated briefly. Finally, some recent work on edge- integrity is described.

1.   Introduction

This work is divided into three parts. The first is an overview of the topic of graph vulnerability in a very broad sense and describes our attempt at organizing and classifying results in this area. The next section includes a rather personal and eclectic look at some examples within each of the three areas in our classification scheme: covering, closeness, and cutting. In the last section, we take an extended look at one parameter, the edge-integrity, that we have been investigating.

*Work supported in part by the Office of Naval Research under Contract N00014-86-K-0412.

2. Vulnerability - - A Classification Scheme A project that we have been working on (see also [2]) is a survey of parameters related to graphs as communication networks. One of the problems which arises is deciding what terminology should be used. Not only have different names and notation been used for the same concept, but different concepts have been given the same name, so we must make some choices. It would also be convenient to have some general terms to describe broad areas of the subject. Frequently, the word "vulnerability" has been used to cover the deterministic aspects, while "reliability" has been used for the probabilistic portion. However, these two terms do not fit together well, since one has positive connotations and the other negative. In addition, if these terms were to be adopted, what word (if any) should be used to indicate both? We would appreciate suggestions from our audience on these matters.

There are various other questions which also arise. For example, what structures should be considered? What types of parameters qualify? Should probabilistic aspects be included? In addressing these questions, we have adopted the following guidelines:

1. For the most part, we choose to consider only graphs - - simple, undirected, and unweighted - - although on occasion digraphs will be included. This is not to dismiss networks with weighted edges (for example), but rather to keep things manageable.

2. Similarly, we focus our attention on deterministic aspects, even though the probabilistic area is also of great importance.

3. Occasionally, synthesis problems, constructing graphs with specific properties, will be considered, but we are primarily interested in the analysis of graphs.

4. We take a broad interpretation of vulnerability, not restricting ourselves to just the topic of disrupting communication.

5. We allow both vertices and edges to be the "active agents", the objects being deleted. At times,we will also allow for the addition of edges.

Partitioning the subject in various ways - - probabilistic and deterministic aspects, synthesis and analysis, vertex- and edge-measures, and so forth - - can be useful not only in examining what is already known, but in pointing out where some important gaps in our knowledge still exist. For example, for a given vertex-based measure of vulnerability of graphs, what of

its edge counterpart, or a combination?

synthesizing resistant graphs?

the corresponding probabilistic measure?

directed and weighted versions?

Investigation also continues into "second-order" parameters, such as measures of how a particular parameter changes when vertices or edges are removed. For an introduction to this topic, called "leverage", see the contribution of Pippert et al. [22] to these Proceedings.

If some parameters are better measures than others, then which ones are best for particular purposes? What properties are desirable in a vulnerability parameter? monotonicity? normalization? We would appreciate suggestions from the audience regarding parameters and results to be included in our survey as well as these various questions.

3. Some Examples of Vulnerability Measures

As we mentioned in the introduction, our interpretation of vulnerability is broad. Taking graphs as models of communication networks, we consider three types of measures of their vulnerability according to

(A) how efficiently communication can be monitored,

(B) how long communication takes,

(C) how many elements must be deleted for communication to be disrupted.

These three categories of parameters are called *covering, closeness,* and *cutting,* respectively.

A.   Covering

The literature on covering and domination in graphs is itself vast, with there being numerous interesting variations. Broadly speaking, we consider "domination" to be a vertex-to-vertex relationship, while " covering" is a relationship between vertices and edges. However, this is only a crude distinction, and to put these parameters into perspective, we look at other types also.

Let $G = G(V, E)$ be a graph and let X be a subset of its vertex set. Then X is called

(1) *a covering set* if every edge of G is incident with at least one vertex in X;

(2) an *independent* set if no two vertices in X are adjacent;

(3) a *dominating* set if every vertex in $V - X$ is adjacent to at least one in X;

(4) an *irredundant* set if, for each vertex v in X, v or one of its neighbors has no neighbor in $X - \{v\}$.

We illustrate these concepts with the graph in Figure 1, in which

{a,b,c,d,i} is an independent set;

{e,f,g,h} is a covering set;

{e,f,i} is a dominating set;

{a,b,f,i} is an irredundant set.

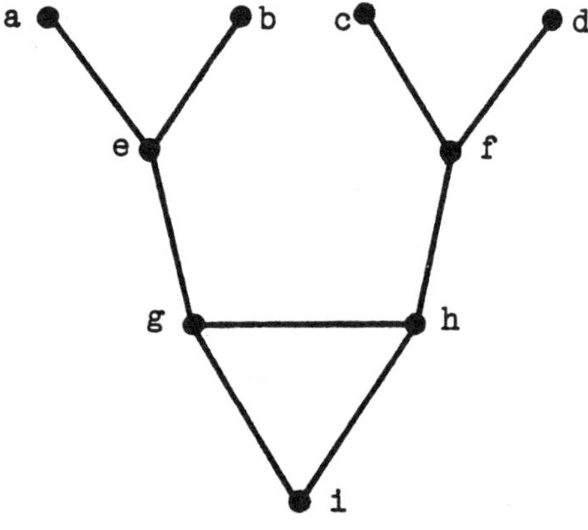

Figure 1

There are some relationships among these types of sets. For instance, a set is a covering set if and only if its complement is an independent set. Two parameters associated with these sets are well-known: the *covering number* is the minimum order of a covering set and the *independence number* is the maximum order of an independent set.

[We observe that notation for these parameters is far from standardized. Two sets of notation predominate, but they are opposites of each other. In Bondy and Murty [14], the independence number is denoted by α (as it is in Berge [9]) and the covering number by β, while Chartrand and Lesniak [16] use β for the independence number and α for the covering number (Bollobás [13] and Harary [20] add subscripts: β_0 and α_0)].

From the above observation on complements, we deduce the well-known result that the sum of the independence and covering numbers of a graph equals the order of the graph. At first hearing, it may sound as though a maximal irredundant set is a dominating set. However, this is not the case, as demonstrated by the set $\{g,h\}$ in Figure 1. This can be seen to be a maximal irredundant set, but it is certainly not a dominating set. In fact, Cockayne, Favaron, Payan, and Thomason [17] have proved that the complement of an irredundant set is dominating but the converse is not necessarily true, so all we can say is that the sum of the domination number (the minimum order of a dominating set) and the irredundance number (the maximum order of an irredundant set) is at most the order of the graph.

Other parameters, in a sense dual to these four, have also been studied. For purposes of clarity, we introduce our own notation:

Cov G: = the maximum order of a minimal covering set
cov G: = the minimum order of a covering set

Ind G: = the maximum order of an independent set
ind G:= the minimum order of a maximal independent set

Dom G: = the maximum order of a minimal dominating set
dom G: = the minimum order of a dominating set

Irr G: = the maximum order of an irredundant set
irr G: = the minimum order of a maximal irredundant set

(Thus, cov G, Ind G, dom G, and Irr G are the covering, independence, dominating, and irredundance numbers defined above.) In the example of Figure 1, Cov G = 6 and cov G = 4, Ind G = 5 and G = 3, Dom G = 5 and dom G = 3, and Irr G = 5 and irr G = 2.

Because of the basic complementary property of covering and independence, we do not consider the former any further in this discussion. Among the other six measures, there is a linear relationship, as shown by Cockayne, Hedetniemi, and Miller [18]:

$$\text{irr } G \leq \text{dom } G \leq \text{ind } G \leq \text{Ind } G \leq \text{Dom } G \leq \text{Irr } G.$$

In fact, they have shown that there exist graphs in which all six of these parameters have different values.

Many variations of these parameters have also been investigated; but because we cannot discuss them to any extent here, we only mention some of those that involve domination:

total domination: the vertices in X must also be dominated;

k-domination: every vertex must be dominated by at least k vertices in X;

distance domination: every vertex must be within a given distance of X;

connected domination: X must induce a connected subgraph.

B. Closeness

Distances between vertices are an important factor in the ability to maintain communication or supply lines within a network, especially when one is considering locations of facilities or the probability of disruption of links. (Of course, other factors, such as the number of paths between sites, are also important.) In this section, we take a brief look at some parameters that involve distance. (For convenience, we assume that the graphs being considered are connected.)

The *eccentricity* of a vertex v in a graph G is defined to be

$$e(v) := \max_{w} \mathrm{dist}(v, w);$$

The *radius* of G is

$$rad\ G := \min_{v} e(v);$$

and the *diameter* is

$$diam\ G := \max_{v} e(v).$$

It is readily observed that

$$rad\ G \leq \mathrm{diam}\ G \leq 2\ rad\ G.$$

My personal interest in this topic comes from working with Harary, Norman, and Cartwright on their pioneering monograph [21] on directed graphs as mathematical models in the social sciences. In a strongly connected digraph D, the out-radius, in-radius, and diameter can be defined analogously to the undirected concepts:

out-radius:

$$r + (D) := \min_{v}\ \max_{w}\ \mathrm{dist}(v, w)$$

in-radius:

$$r - (D) := \min_{w}\ \max_{v}\ \mathrm{dist}(v, w).$$

diameter: Again referred to the contribution [22] by Pippert et al. to these *Proceedings* and to other articles referred to there.

C. Cutting

When vulnerability of communication networks is mentioned, most of us think first of disruption because elements of the network break down or are destroyed. That is, we think of the deletion of vertices or edges from the underlying graph.

Some concepts of this type give rise in a natural way to sequences of parameters, and it is on some of these "cutting sequences" that we focus in this section. Most of those we consider utilize edge-removal, but many of them have natural vertex counterparts.

Let $G = G(V, E)$ be a connected graph with p vertices and q edges. In this discussion X and S will always denote subsets of V and E respectively.

Sequence 1 $\mu_i(G) := \min_{|X|=1} \lambda(G - X)$: that is, the minimum edge-connectivity among all induced subgraphs of G of order $p - i$, for $i = 0, 1, \cdots, \kappa(G)$.

This was introduced by Beineke and Harary [7] in 1967 under the name *connectivity function*. Clearly, this is a strictly decreasing sequence (it could be extended to length p by adding zeros) with $\mu_0 = \lambda(G)$ and $\mu_{\kappa(G)} = 0$.

The principal results on this sequence are the following:

Theorem 2 Every decreasing sequence $\{\mu_0, \mu_1, \cdots, \mu_{,-1}, 0\}$ is the connectivity function of some graph.

Theorem 3 If $\mu_i(G) = j$, then for each pair of vertices s and t, G contains a family of $i + j$ edge-disjoint $s - t$ paths, of which i are disjoint except for s and t.

Sequence 2 $\nu_i(G) := \min \{|S| : G - S \text{ has } i+1 \text{ components}\}$; that is, $\nu_i(G)$

is the fewest edges whose removal leaves $i+1$ components, for $i = 1, 2, \cdots, p-1$.

Note that $\nu_1 = \lambda$, that $\nu_{p-1} = q$ (the number of edges), and that this is a strictly increasing sequence. The origins of this concept are in a 1978 paper [10] by Boesch and Chen; the following is a result of Goldsmith [19]:

Theorem 4   If $\nu_i < \delta(G)$ (the minimum degree of G), and if S is a set of edges achieving ν_i, then each component of $G - S$ has at least two vertices not incident with S.

The vertex counterpart was studied by Chartrand, Kapoor, Lesniak, and Lick [15]. Among their results is a determination of which sequences can be so realized.

Sequence 3 $\delta_i := \min_{|X|=i} |$, the minimum size of an edge cut joining exactly i vertices to the other $p - i$. This sequence was introduced by Boesch and Thomas [12] in 1970 and was called the *minimum* i-*degree*. This sequence need not be monotonic (even within the first half).

An example of a result on this sequence is the following, due to Boesch and Felzer [11]:

Theorem 5   Let G be an r-regular graph of girth g. Then

$$\delta_i(G) = \begin{cases} i(r-2) + 2 & \text{for } i < g \\ i(r-2) & \text{for } g \leq i < \lfloor \frac{3g-1}{2} \rfloor \end{cases}$$

Sequence 4 $\lambda_i = \min_{i \leq |X| \leq p/2} |X, V - X)|$, the minimum size of an edge cut separating at least i vertices, but not more than half the total, from the remainder (for $1 \leq i \leq p/2$). This is called the *edge-connectivity sequence* of G; clearly it is nondecreasing and $\lambda = \lambda(G)$. It was first studied by Pippert and Lipman [23], and the following theorem is due to them. It can be extended, but even in this form, shows that this family of sequences is quite restricted.

Theorem 6   If $\lambda_1 = m$ and $\lambda_2 \leq 2(m-1)$, then $\lambda_2 = \lambda_3 = \cdot = \gamma_m$.

Sequence 5 $\varsigma_i := \min\{|S| : \text{no component of } G - S \text{ has order greater}$ than $i\}$, for $i = 1, 2, \cdots, p-1$. Clearly $\varsigma_1 = q, \varsigma_{p-1} = \lambda(G)$, and this is a nonincreasing sequence. Originally, this sequence was studied in its reverse order, as $\eta_i = \varsigma_{p-i}$, because of its connections with the edge-connectivity sequence λ_i. Introduced by Beineke, Lesniak, and Pippert [8], the sequence $\{\eta_i\}$ was called the *separation sequence*. The followint theorem develops the relationship between λ_i and η_i.

Theorem 7   Let G be a connected graph of order p. For $i \leq \frac{1}{2}p, \nu_i \leq \lambda_i$. If $\nu_j < \lambda_j$ but $\nu_i = \lambda_i$ for $i < j$, then $j \leq \frac{1}{3}p+1$ and $\nu_j = \nu_{j+1} = \cdots = \nu_{p-j+1}$.

4. The Integrity of Graphs

A. Introduction

In this last section, we discuss some recent work on two cutting parameters: the vertex-integrity and the edge-integrity. As my co-authors and I have been investigating the latter, let me begin there.

Consider a graph as a model of a communication network. One measure of vulnerability is the edge- connectivity λ, but it says little about how much disruption can be caused by removing λ edges - - perhaps only one vertex is isolated, perhaps half are split off. Our goal is broader, to cut few links and yet leave little communication possible between sites. Thus, we are interested in removing a set S of edges so that both of the following quantities are small:

$|S|$, the size of the cut, and

$m(G - S)$, the maximum order of a remaining component.

From among the many ways of measuring a combination of the two, we choose what seems most natural, their sum, which has been named the *edge-*

integrity of G:

$$I'(G) := \min_{S \subseteq E}\{|S| + m(G - S)\}.$$

Historically, this parameter was introduced after its vertex counterpart: the (*vertex-*) *integrity* of G is defined to be

$$I(G) := \min_{X \subseteq V}\{|X| + m(G - X)\}.$$

Both parameters were introduced by Barefoot, Entringer, and Swart [4]. Note that both are monotonic in that if H is a subgraph of G, then $I(H) \leq I(G)$ and $I'(H) \leq I'(G)$. After reviewing some results on both for trees, we turn to some of our work exclusively on edge-integrity. For convenience, we consider only connected graphs here.

B.   Trees

The integrities of stars are among the easiest to compute: The removal of the central vertex leaves only isolated vertices, while removing any edge reduces the order of the biggest component by only one. It follows that $I(K_{1,n}) = 2$ and $I'(K_{1,n}) = n + 1$. These simple graphs illustrate that the two parameters behave quite differently, since among all connected graphs of a given order, stars have the smallest possible vertex-integrity but the largest possible edge-integrity.

Among the trees, paths are at the opposite extremes, having the largest vertex-integrity and the smallest edge-integrity:

$$I(P_n) = \lceil 2\sqrt{n} + 1 \rceil - 2 \text{ and } I'(P_n) = \lceil 2\sqrt{n} \rceil - 1.$$

In each case, every integer between the extremes is achieved by some "star with a tail." The *comet* $C_{n,t}$ is obtained by attaching a path of length $n - t - 1$

to the center of the star $K_{1,t}$ see (Figure 2). It is not difficult to show that $I'(C_{n,k+1})$ is at most 1 greater than $I'(C_{n,k})$, while $I(C_{n,k})$ is at most 1 greater than $I(C_{n,k+1})$, and the assertion follows. We summarize these results in the following theorem (see [5]):

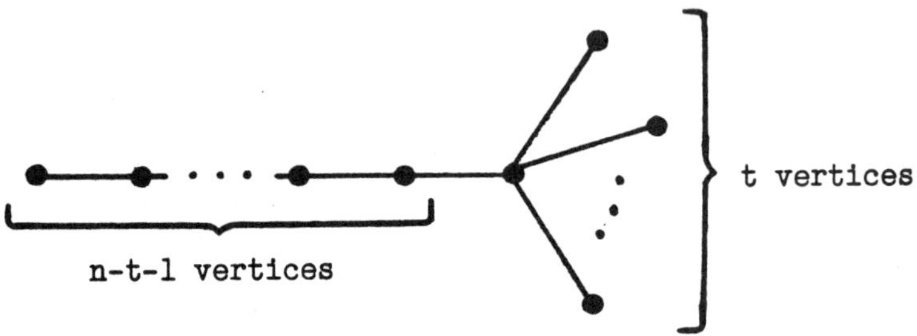

Figure 2

Theorem 8   (Integrity Interpolation for Trees)

(a) For any tree T of order n, $2 \leq I(T) \leq \lceil 2\sqrt{n} + 1 \rceil - 2$. Furthermore, if $2 \leq r \leq 2\lceil 2\sqrt{n} + 1 \rceil - 2$, then there exists a tree of order n and vertex-integrity r.

(b) For any tree T of order n, $\lceil 2\sqrt{n} \rceil - 1 \leq I'(T) \leq n$. Furthermore, if $\lceil 2\sqrt{n} \rceil - 1 \leq r' \leq n$, then there exists a tree of order n and edge-integrity r'.

We now turn to a theorem (see [3]) which explores the edge-integrity of trees in greater detail. The bound in (b) is in fact sharp.

Theorem 9   Let T be a tree of order n and maximum degree Δ.

(a) If $\Delta \geq \frac{1}{2}n$, then $I'(T) = \Delta + 1$.

(b) If $\Delta < \frac{1}{2}n$, then $I'(T) \leq \lfloor \frac{1}{2}(n + 3) \rfloor$.

C.   Cartesian Products. In this subsection, we look at some results on the edge-integrity of the Cartesian product $G \times H$ of two graphs G and H, beginning with one graph being complete (see [1]).

Theorem 10   For any graph G and any positive integer n,

$$I'(G \times K_n) = nI'(G).$$

The edge-integrity of hypercubes follows by induction:

Corollary $I'(Q_n) = 2^n.$

We note in passing that the vertex-integrity of Q_n has been conjectured to be $2^{n-1} + 1$, but this has not yet been proved.

Our next theorem gives bounds on $I'(G \times H)$ in general. Clearly, the roles of G and H can be interchanged so that we have two bounds on each side. The upper bound is simply a corollary of the previous theorem; the lower bound is due to Goddard (unpublished). (As usual, $\delta(G)$ denotes the minimum degree.)

Theorem 11   For any graphs G and H,

$$(\delta(G) + 1)I'(H) \leq I'(G \times H) \leq |V(G)|I'(H).$$

Next we turn to some results on grids (the products of paths), due to Beineke, Goddard, and Lipman [6]:

Theorem 12   For any paths P_r and P_s,

$$I'(P_r \times P_s) \geq \lceil 3(rs)^{2/3} \rceil - (r + s).$$

We note two families in which the bound is exact: (i) $r = s$ and (ii) $r = s = t^3$. In the first instance, cutting $P_r \times P_r$ into r blocks of $P_r \times P_r$ achieves the lower bound, while in the second, cutting the grid into t^2 square blocks each t^2-by-t^2 accomplishes the task.

We now consider products in which one of the factors is a short path or cycle and the other is arbitrary. Since $P_2 = K_2, C_3 = K_3$, and $C_4 = K_2 \times K_2$, the following results hold for all graphs G as consequences of Theorem 10:

$$I'(P_2 \times G) = 2I'(G)$$

$$I'(C_3 \times G) = 3I'(G)$$

$$I'(C_4 \times G) = 4I'(G).$$

However, similar results do not hold for longer paths or cycles. For example, in Figure 3 we show a cut of five edges in $P_3 \times K_{1,3}$ whose removal leaves two components of order 6, so that $I'(P_3 \times K_{1,3}) \leq 11$ (in fact, equality holds). On the other hand, we can show that for all $n \geq 3, I'(P_3 \times P_n) = 3I'(P_n)$ and $I'(C_5 \times C_n) = 5I'(C_n).$

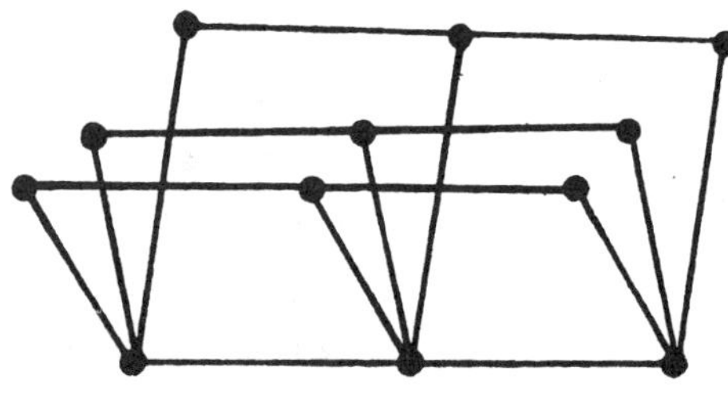

Figure 3

Although a great deal more could be said about each of the topics we have touched on here, we conclude with one result which combines two of them - - cutting and closeness - - in the following theorem [1].

Theorem 13   If a graph has diameter 2, then its edge-integrity equals its order.

Concluding Remarks

One aspect of vulnerability that we have not discussed here is the computability of the various parameters. Quite a bit of work has been done on that important topic, but a discussion of those endeavors is not possible within the limitations of this paper. Of course, there is a need for much more to be done. In addition to various questions that arose in our discussion of specific parameters, other broad areas of graph vulnerability remain largely uninvestigated. Some of these were suggested in Section 1, and include relationships between vulnerability measures, probabilistic aspects of vulnerability, and methods for strengthening vulnerable networks efficiently. We look forward to advances in the study of graph vulnerability in the near future.

REFERENCES

[1] K. S. Bagga, L. W. Beineke, M. J. Lipman, and R. E. Pippert, On the edge-integrity of graphs. Proceedings of the Eighteenth Southeastern Conference on Combinatorics, Graph Theory, and Computing, *Congressus Numerantium* 60(1987), 141-144.

[2] K. S. Bagga, L. W. Beineke, M. J. Lipman, and R. E. Pippert, A classification scheme for vulnerability and reliability parameters of graphs. *Computers and Mathematics with Applications* (To appear).

[3] K. S. Bagga, L. W. Beineke, M. J. Lipman, and R. E. Pippert, Some bounds and an algorithm for the edge-integrity of trees. (Submitted).

[4] C. A. Barefoot, R. Entringer, and H. Swart, Vulnerability in graphs - - a comparative survey. *J. Combin. Math. Combin. Comput.* 1(1987), 12-22.

[5] C. A. Barefoot, R. Entringer, and H. Swart, Integrity of trees and powers of cycles. *Congressus Numerantium* 58 (1987), 103-114.

[6] L. W. Beineke, W. Goddard, and M. J. Lipman, On the edge-integrity of graph products. (In process).

[7] L. W. Beineke and F. Harary, The connectivity function of a graph. *Mathematika* 14(1967), 197-202.

[8] L. W. Beineke, L. Lesniak, and R. E. Pippert, On the edge-separation sequence of a graph. (In process).

[9] C. Berge, *Graphs* (2nd rev. ed.), North Holland, 1985.

[10] F. T. Boesch and S. Chen, A generalization of line connectivity and optimally invulnerable graphs. *SIAM J. Appl. Math.* 34(1978), 657-665.

[11] F. T. Boesch and A. P. Felzer, On the minimum m-degree vulnerability criterion. *IEEE Trans. Circuit Theory*, CT-18 (1971), 224-228.

[12] F. T. Boesch and R. E. Thomas, On graphs of invulnerable communication nets. *IEEE Trans. Circuit Theory*, CT-17 (1970), 183-192.

[13] B. Bollobás, *Graph Theory.* Springer-Verlag, 1979.

[14] J. A. Bondy and U. S. R. Murty, *Graph Theory with Applications*, North-Holland, 1976.

[15] G. Chartrand, S. F. Kapoor, L. Lesniak, and D. R. Lick, Generalized connectivity in graphs. (Submitted).

[16] G. Chartrand and L. Lesniak, *Graphs & Digraphs.* Wadsworth, 1986.

[17] E. J. Cockayne, O. Favaron, C. Payan, and A. Thomason, Contributions to the theory of domination, independence and irredundance in graphs. *Discrete Math.* 33(1981), 249-258.

[18] E. J. Cockayne, S. T. Hedetniemi, and D. J. Miller, Properties of hereditary hypergraphs and middle graphs. *Canad. Math. Bull.* 21(1978), 461-468.

[19] D. L. Goldsmith, On the n-th order connectivity of a graphs. *Congressus Numerantium* 32(1981), 375-382.

[20] F. Harary. *Graph Theory.* Addison-Wesley, 1969.

[21] F. Harary, R. Z. Norman, and D. Cartwright, *Structural Models: An introduction to the Theory of Directed Graphs.* Wiley, 1965.

[22] R. E. Pippert, K. S. Bagga, L. W. Beineke, and M. J. Lipman, The concept of leverage in network vulnerability. These *Proceedings.*

[23] R. E. Pippert and M. J. Lipman, Toward a measure of vulnerability I. The edge- connectivity vector. *Graph Theory with Applications to Algorithms and Computer Science*, (Y. Alavi, et al., eds.) Wiley (1985), 651-657.

Reliable Graphs With Unreliable Nodes

F. Boesch[1]

X. Li

C. Stivaros

C. Suffel[2]

Stevens Institute of Technology

ABSTRACT

There is vast literature on graph reliability utilizing a model where edges fail with independent probabilities. In contrast, there are few results concerning the case where edges are perfectly reliable but nodes fail with equal but independent probabilities. Herein we consider this node failure case and introduce several problems regarding synthesis of optimally reliable graphs. One such problem, which arises when the failure probability of a node is large, is to synthesize graphs having the maximum number of induced 3−node connected subgraphs among all graphs having n nodes and e edges. We consider the cases $n - 1 \leq e \leq 2(n - 2)$ and $\frac{n(n-1)}{2} - 2n \leq e \leq \frac{n(n-1)}{2}$.

1. Introduction and Preliminaries

A large–scale parallel processor system of hardwired microcomputers located in a single room may be topologically modelled as an undirected simple graph with perfectly reliable edges and nodes subject to failure. If the computers are all of the same make, it is reasonable to assume that each node has the same probability of failure. Furthermore, in most instances, failures are independent of each other. Of course this is just one of the many situations which fit the model described above. In the remainder of this paper we are concerned with the reliability of such a network and, in preparation for the discussions to follow we present some preliminaries.

[1] Supported by NSF grant ECS–8600782.
[2] Supported by ONR contract N00014–87–K–0376.

For basic notation and terminology we refer the reader to the book of Harary [8]. Some ideas are repeated here for the sake of completeness. A *graph* G is a pair $(V(G), E(G))$, where $V(G)$ is a finite non–empty set called the *nodes* or *points* of G and $E(G)$ is a set of two element subsets of $V(G)$ referred to as the *edges* or *lines* of G. If $|V(G)| = n$ and $|E(G)| = e$, G is called an *(n, e) graph*.

The nodes of G are assumed to operate independently, each with probability p. The graph G is said to be *operational* at any given time if the surviving nodes induce a connected subgraph having at least two nodes. Hence the *reliability of G*, defined to be the probability that G is operational, may be written

$$R(G, p) = \sum_{i=2}^{n} s_i\, p^i\, (1-p)^{n-i} ,$$

where $s_i(G)$, or simply s_i, denotes the number of induced connected subgraphs of G having i nodes. Of course, $s_2 = e$ for any (n, e) graph. Furthermore, if G is connected and $i > n-\kappa$, where κ is the (node) connectivity of G, then $s_i = \binom{n}{i}$. Thus

$$R(G, p) = ep^2\, (1-p)^{n-2} + \sum_{i=3}^{n-\kappa} s_i\, p^i\, (1-p)^{n-i}$$

$$+ \sum_{i=n-\kappa+1}^{n} \binom{n}{i}\, p^i\, (1-p)^{n-i}$$

for any connected (n, e) graph G.

Now the problem of designing the kind of probabilistic network just described, given only n and e amounts to selecting an (n, e) graph with high reliability. In fact, the first thought that comes to mind is to search for an (n, e) graph having the largest value of reliability for all $p \in (0, 1)$. However, as we shall see, such a graph does not always exist. On the other hand, it has been shown by Frank [5] that if p is sufficiently close to 1 then $R(G, p)$ is maximized only if G has the maximum possible connectivity κ of all (n, e) graphs and the maximum value of $s_{n-\kappa}$ over all (n, e) graphs having connectivity κ. Such a graph is referred to as a *super–κ* graph. Although a general solution to the problem of constructing super–κ graphs has yet to be discovered, partial results are known. Boesch and Felzer [2] have shown that every regular complete κ–partite graph is super–κ. Hakimi and Amin [6] have given constructions of (n, e) graphs having the maximum possible connectivity κ and no more than n disconnecting sets of size κ. Smith [10] gave constructions of super–κ graphs for various cases where $2e/n \geq 3$. Smith and Doty [11] give constructions of super–κ graphs when $\frac{\kappa}{n}$ is "small". Finally, Boesch and Li [3] have shown that constructions of super–λ graphs given by Bauer, Boesch, Suffel and Tindell [1] for $n \leq e < 3n/2$ are super–κ with the exception of the cases $6 \leq n \leq 7$ and $e = n + 2$. (Recall that an (n, e) graph is *super–λ* if and only if it has the maximum possible edge connectivity $\lambda = \left\lfloor \frac{2e}{n} \right\rfloor$ over all (n, e) graphs and the smallest number of minimum size edge cuts over all (n, e) graphs with edge connectivity $\lambda = \left\lfloor \frac{2e}{n} \right\rfloor$.)

Now, using Frank's approach, it can be shown that when p is sufficiently small, $R(G, p)$ is maximized only when $s_3(G)$ is maximized over all (n, e) graphs. Indeed, if G and H are any two (n, e) graphs with $s_3(G) > s_3(H)$, then

$$R(G, p) - R(H, p) = (1 - p)^n \left(\tfrac{p}{1-p}\right)^3 \left[s_3(G) - s_3(H) + \sum_{k=4}^{n} (s_k(G) - s_k(H))\left(\tfrac{p}{1-p}\right)^{k-3} \right].$$

Thus, if p is small enough we may force

$$\left| \sum_{k=4}^{n} (s_k(G) - s_k(H))\left(\tfrac{p}{1-p}\right)^{k-3} \right| < s_3(G) - s_3(H),$$

from which the contention follows. In the next two sections we present constructions of *3–optimal graphs*, i.e. graphs which maximize $s_3(G)$ over all (n, e) graphs G. Section 2 is devoted to consideration of some sparse cases $(e \leq 2(n-2))$ while the dense cases $(e \geq \binom{n}{2} - 2n)$ are discussed in Section 3. In the final section we present some conjectures. Thus we end this section with the derivation of two expressions for $s_3(G)$ to be used later.

First observe that if d_v is the degree of node v in G then $\binom{d_v}{2}$ is the number of induced three–point connected subgraphs in which v has degree equal to two. Now consider the sum

$$\sum_{v \in V(G)} \binom{d_v}{2}.$$

This sum counts each induced three point path exactly once and each induced triangle three times. Thus

$$(1): \quad s_3(G) = \sum_{v \in V(G)} \binom{d_v}{2} - 2\tau(G),$$

where $\tau(G)$ denotes the number of triangles of G. We may also note that $\binom{e}{2}$ counts each induced path on three points once, each induced triangle three times and each induced matching with two edges once. Thus

$$(2): \quad s_3(G) = \binom{e}{2} - 2\tau(G) - m_2(G)$$

where $m_2(G)$ denotes the number of pairs of independent edges.

2. 3–Optimal Sparse Graphs

Since we consider only connected (n, e) graphs, the extreme sparse case occurs when $e = n - 1$, i.e. when the class of graphs under consideration is the collection of all trees T on n nodes. Of course $\tau(T) = 0$ for any tree T so

$$s_3(T) = \binom{n-1}{2} - m_2(T) \le \binom{n-1}{2}.$$

Now only the star $K_{1,n-1}$ has $m_2(K_{1,n-1}) = 0$, so that

$$s_3(T) < \binom{n-1}{2} = s_3(K_{1,n-1})$$

for each $T \ne K_{1,n-1}$. Summarizing this discussion we have our first theorem.

Theorem 1 *($s_3(K_{1,n-1})$ IS MAXIMUM) The unique 3–optimal graph having $e = n-1$ edges is the star $K_{1,n-1}$.*

It is natural to conjecture that $H := K_{1,n-1} + x$ is a 3–optimal (n, n) graph. To see that this is so first realize that by (1) of Section 1,

$$s_3(H) = \sum_{v \in V(H)} d_{v_2}(H) - 2\tau(H)$$

$$= \binom{n-1}{2} + 2 - 2(1) = \binom{n-1}{2}.$$

Now consider the relaxed combinatorial optimization problem:

$$\text{maximize} \quad \sum_{i=1}^{n} \binom{d_i}{2}$$

$$\text{such that} \quad \sum_{i=1}^{n} d_i = 2n,$$

and $1 \le d_i \le n - 1$ for $i = 1, \dots, n$. Suppose that $2 \le d_j \le d_i \le n - 2$. Then the sequence obtained by replacing d_i by $d_i + 1$ and d_j by $d_j - 1$ is feasible and has a larger objective value by virtue of the equation

$$(3): \quad \left[\binom{d_i+1}{2} + \binom{d_j-1}{2}\right] - \left[\binom{d_i}{2} + \binom{d_j}{2}\right] = d_i - d_j + 1.$$

Now arrange the elements of the sequence in increasing order:

$$d_1 \le d_2 \le \dots \le d_{n-1} \le d_n .$$

Of course $d_{n-1} \ge 2$, so if $d_n < n-1$, then $d_1, \dots, d_{n-2}, d_{n-1}-1, d_n+1$ is again feasible and has objective value strictly larger than that of the original sequence. If $d_{n-1} < d_n$, then by (3) the new sequence improves the objective value by at least 2. Thus if $d_n \le n - 3$, two steps of the above procedure yields a feasible solution which has objective value larger than that of the original sequence by at least 3. If we iterate this procedure, beginning with a graphicallly *realizable* sequence, we obtain either $1, \dots, 1, 3, n - 1$ or $1, \dots, 1, 2, 2, n - 1$. Next we observe that $1, \dots, 1, 3, n - 1$ is not possible since backtracking from $1, \dots, 1, 3, n - 1$ we obtain the sequences $1, \dots, 1, 4, n - 2$; $1, \dots, 1, 5, n - 3$; $\dots$; all of which are not realizable. Thus, if G is an (n, n) graph with $\Delta(G) \le n - 3$ then

$$s_3(G) \le \sum_{v \in V(G)} \binom{d_v(G)}{2}$$

$$\le \sum_{v \in V(H)} \binom{d_v(H)}{2} - 3$$

$$< s_3(H)$$

If $\Delta(G) = n - 2$, then G must be a 4–cycle with $n - 4$ pendant edges emanating from a single node of the 4–cycle (see Figure 1).

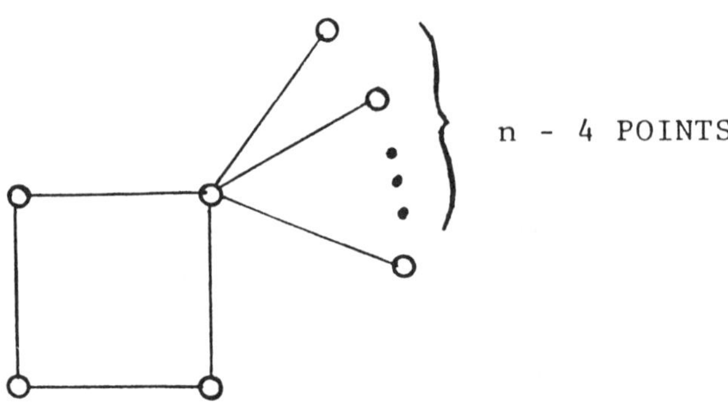

Figure 1.

Thus $s_3(G) = \binom{n-2}{2} + 3$. But $s_3(H) - s_3(G) = n - 5$.

Thus, if $n > 5$, $H := K_{1,n-1} + x$ is the unique 3–optimal connected (n, n) graph. If $n = 5$, then $K_{1,4} + x$ and the graph G shown in Figure 1 are both 3–optimal. If $n = 4$, a comparison of $K_{1,3} + x$ and C_4, the only two realizations, reveals C_4 to be the unique 3–optimal connected graph. We have thus proved our second theorem.

Theorem 2 *(3–OPTIMAL (n, n) GRAPHS) If $n = 4$, then the 4–cycle is 3–optimal among the two connected $(4, 4)$ graphs. If $n = 5$, $K_{1,4} + x$ and the graph shown in Figure 1 are the only 3–optimal $(5, 5)$ graphs. Finally, if $n > 5$, then $K_{1,n-1} + x$ is the unique 3–optimal (n, n) graph.*

Remark   For a graph to be the most reliable in its class for all possible $p \in (0, 1)$ it would have to be super–κ and 3–optimal. However if $n > 5$, the only 3–optimal (n, n) graph is $K_{1,n-1} + x$, while the unique super–κ graph is C_n — the cycle on n nodes. Thus "uniformly" optimal (n, n) graphs do not exist.

An examination of the previous discussion reveals some interesting observations. We see that for all n sufficiently large, the best construction is the one which maximizes

$$\sum_{v \in V(G)} d_{v_2}(G)$$

while suffering the existence of a maximum number of triangles. It is only for small values of n that it is better to minimize $\tau(G)$ first and then maximize

$$\sum_{v \in V(G)} d_{v_2}(G)$$

over the resulting sequences. This is the case in general, as our next and last result of this section shows. We shall not provide a proof here due to its many details; rather we refer the reader to [4].

In preparation for Theorem 3, we consider $e = n + a < 2(n-2)$, where $a \geq 0$, and define $G_1(n, a)$ and $G_2(n, a)$ as shown below in Figure 2:

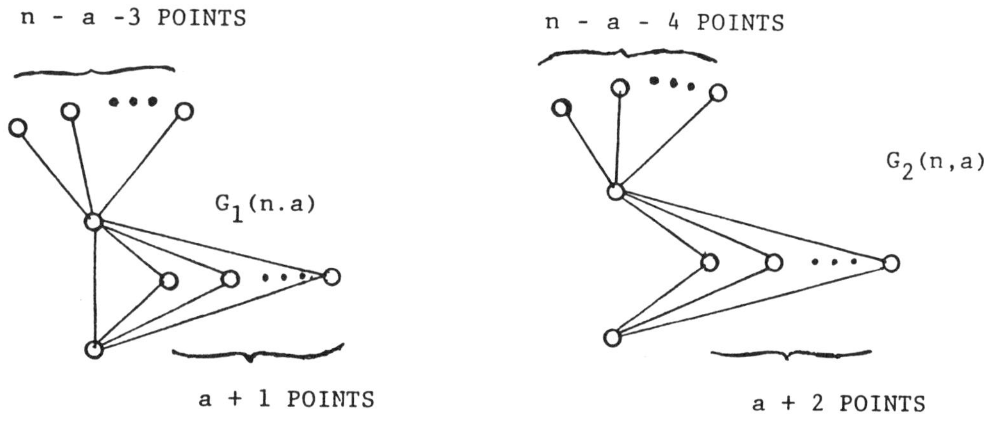

Figure 2.

Theorem 3 *(SPARSE 3–OPTIMAL GRAPHS) If $a \geq 0$ and $e = n + a < 2(n-2)$, then*

(1) $G_1(n, a)$ is the unique 3–optimal (n, e) graph for $n > 2a + 5$;

(2) $G_2(n, a)$ is the unique 3–optimal (n, e) graph for $n < 2a + 5$;

and (3) $G_1(n, a)$ and $G_2(n, a)$ are the only 3–optimal (n, e) graphs for $n = 2a + 5$.

Remark If $n \geq 4$, this result states that $K_{2,n-2}$ is the unique 3–optimal $(n, 2(n-2))$ graph.

3. 3–Optimal Dense Graphs

The reliability function of a network with n perfectly reliable nodes and e edges which fail independently and with the same probability may be written

$$R(G, p) = \sum_{k=n-1}^{e-\lambda} C_k(G)p^k(1-p)^{e-k} + \sum_{k=e-\lambda+1}^{e} \binom{e}{k} p^k(1-p)^{e-k}$$

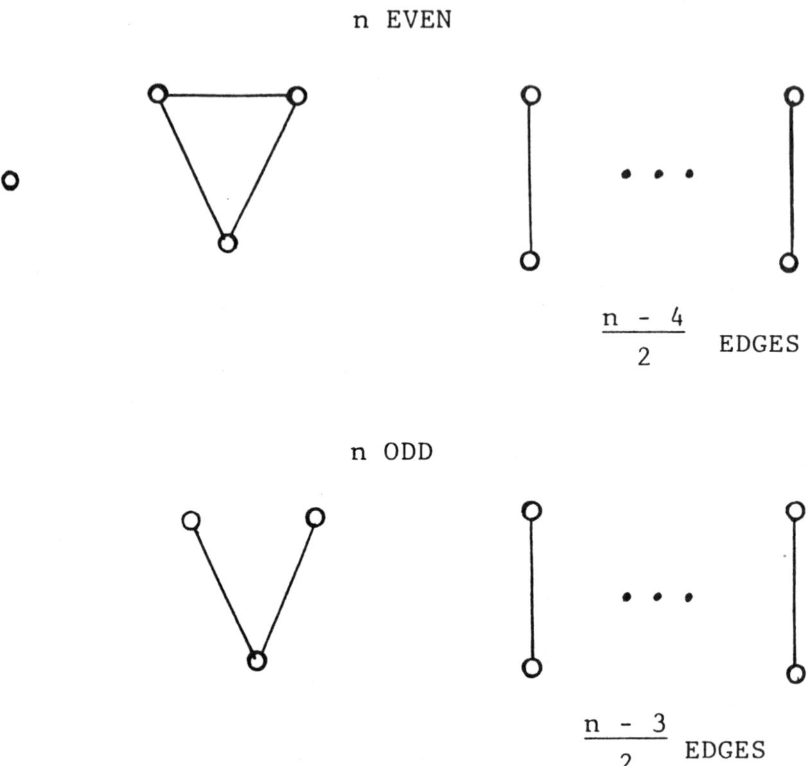

Figure 3.

Thus, by taking complements we obtain 3–optimal graphs.

Proceeding in this greedy fashion, we are led to the following results which we state without proof. Proofs may be found in [4]. For ease of notation we define

$$\left\lfloor \frac{2n}{2} \right\rfloor^* = \begin{cases} n \text{ if } n \text{ is a multiple of 3} \\ n-1 \text{ otherwise} \end{cases}$$

$$\left\lfloor \frac{3n}{2} \right\rfloor^* = \begin{cases} \left\lfloor \frac{3n}{2} \right\rfloor \text{ if } n \text{ is a multiple of 4} \\ \left\lfloor \frac{3n}{2} \right\rfloor - 1 \text{ if } n-1 \text{ or } n-3 \text{ is a multiple of 4} \\ \left\lfloor \frac{3n}{2} \right\rfloor - 2 \text{ if } n-2 \text{ is a multiple of 4} \end{cases}$$

where λ is the edge connectivity of the graph and $C_k(G)$ denotes the number of spanning connected subgraphs having k edges. Using an argument analogous to the one given in Section 1, it can be shown that for p sufficiently small $R(G, p)$ is maximized only when $C_{n-1}(G)$, the number of spanning trees of G, is maximized over all (n, e) graphs. It was first shown by Shier [9] that for $e \geq \binom{n}{2} - \lfloor \frac{n}{2} \rfloor$ the unique (n, e) graph with the maximum number of spanning trees is the graph obtained by removing a matching from K_n. The natural question arises as to whether this graph also plays the analogous role of maximizing $s_3(G)$ over all (n, e) graphs. Although the answer is yes, the proof, unlike that of the optimality of C_{n-1}, is quite easy. First we require the following useful lemma.

Lemma 1 *(MAX $s_3(G)$ = MIN $s_3(G)$)* *The graph G has the maximum number of induced three–point connected subgraphs over all (n, e) graphs if and only if the complement of G, $\overline{G}$, has the minimum number of induced three–point subgraphs over all $(n, \binom{n}{2}-e)$ graphs.*

Proof It is enough to note that for any (n, e) graph H,

$$s_3(H) + s_3(\overline{H}) = \binom{n}{3}. \quad \square$$

Now a matching is the only graph which has no induced three–point connected subgraphs, so our next theorem follows from Lemma 1.

Theorem 4 *(K_n–matching IS 3–OPTIMAL)* *If $e \geq \binom{n}{2} - \lfloor \frac{n}{2} \rfloor$, the unique 3–optimal (n, e) graph is obtained by removing $\binom{n}{2} - e$ independent edges from K_n.*

Again resorting to Lemma 1, we may solve the 3–optimal problem for $e = \binom{n}{2} - \left(\lfloor \frac{n}{2} \rfloor + 1 \right)$ by finding the $(n, \lfloor \frac{n}{2} \rfloor + 1)$ graph with the smallest number of induced three–point connected subgraphs. Of course it is obvious that the graphs shown in Figure 3 minimize s_3 for $e = \lfloor \frac{n}{2} \rfloor + 1$.

and

$$\left\lfloor \frac{4n}{2} \right\rfloor^* = \begin{cases} 2n-2 & \text{if } n-1 \text{ or } n-4 \text{ is a multiple of } 5 \\ 2n-3 & \text{if } n-2 \text{ or } n-3 \text{ is a multiple of } 5 \\ 2n & \text{if } n \text{ is a multiple of } 5 \end{cases}$$

Theorem 5 *(min s_3 GRAPHS) If $\left\lfloor \frac{n}{2} \right\rfloor + 1 \le e \le \left\lfloor \frac{4n}{2} \right\rfloor^*$ then the graphs given below are the (n, e) graphs with the minimum number of induced three-point connected subgraphs. Furthermore if we partition the range into the subsets $\frac{n}{2} + 1 \le e \le \left\lfloor \frac{2n}{2} \right\rfloor^*$, $\left\lfloor \frac{2n}{2} \right\rfloor^* + 1 \le e \le \left\lfloor \frac{3n}{2} \right\rfloor^*$, and $\left\lfloor \frac{3n}{2} \right\rfloor^* + 1 \le e \le \left\lfloor \frac{4n}{2} \right\rfloor^*$, then, as is depicted below, the plot of min s_3 versus e exhibits a partial periodicity in each of the intervals. The graphs and plots for the three subsets are given in Figures 4, 5 and 6.*

Corollary 5–1 *If $\binom{n}{2} - \left\lfloor \frac{4n}{2} \right\rfloor^* \le e \le \binom{n}{2} - \left(\left\lfloor \frac{n}{2} \right\rfloor + 1\right)$, then the 3–optimal graphs are the complements of those given in Theorem 5.*

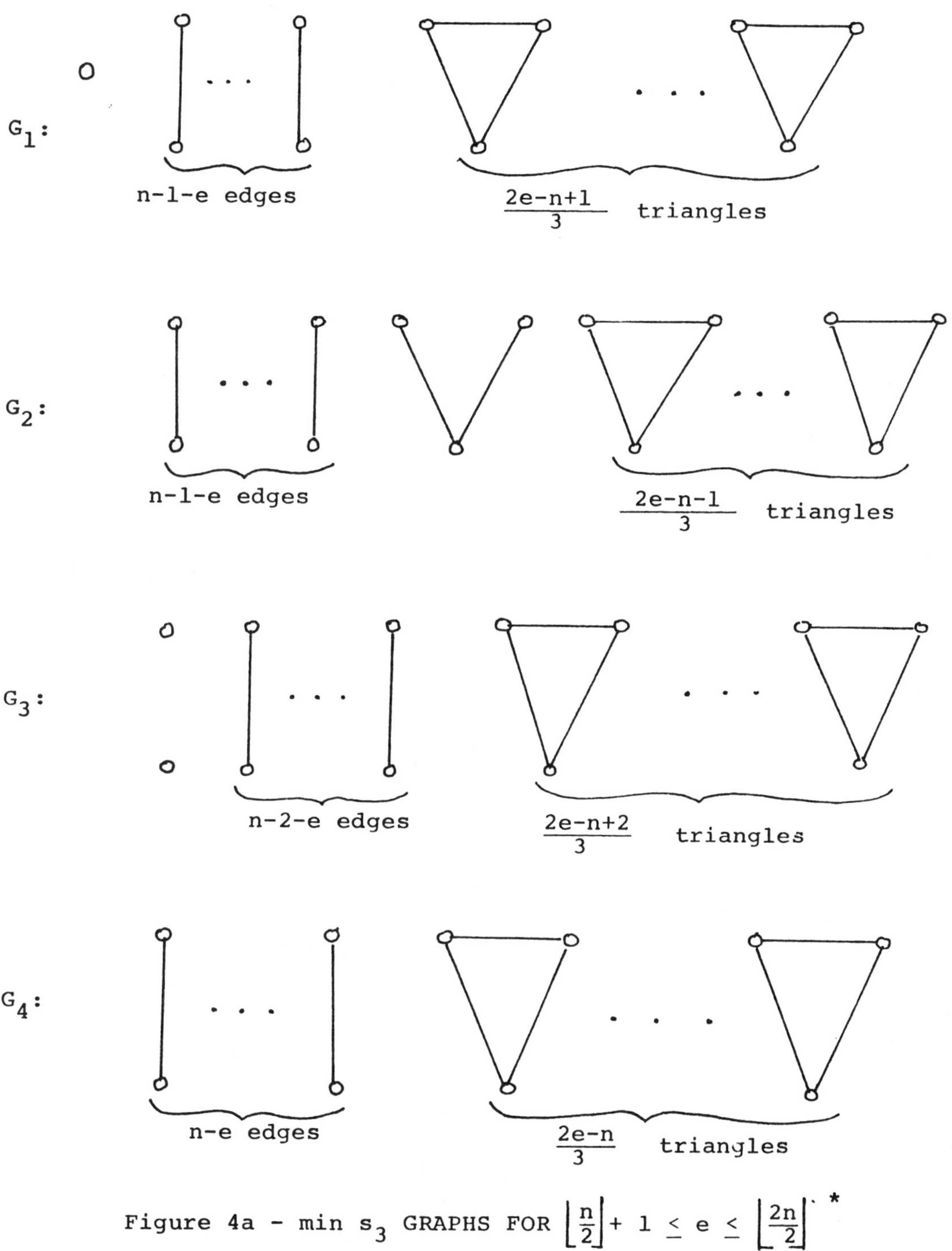

Figure 4a - min s_3 GRAPHS FOR $\left\lfloor \dfrac{n}{2} \right\rfloor + 1 \leq e \leq \left\lfloor \dfrac{2n}{2} \right\rfloor^*$

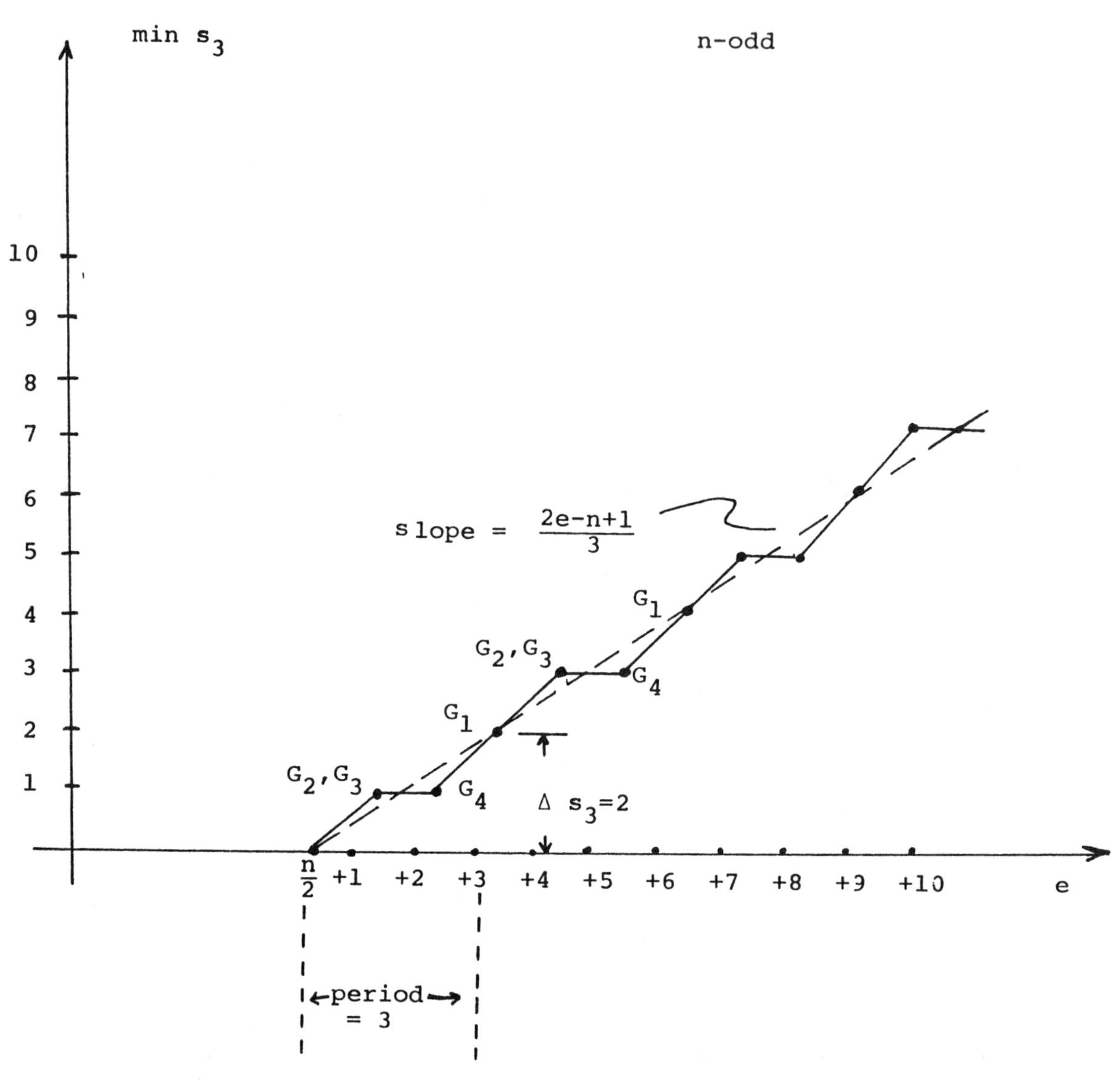

Figure 4b — min s_3 VERSUS e FOR $\left\lfloor \dfrac{n}{2} \right\rfloor + 1 \leq e \leq \left\lfloor \dfrac{2n}{2} \right\rfloor^{*}$

G_1:

G_2:

G_3:

G_4:

G_5:

G_6:

G_7:

G_8:

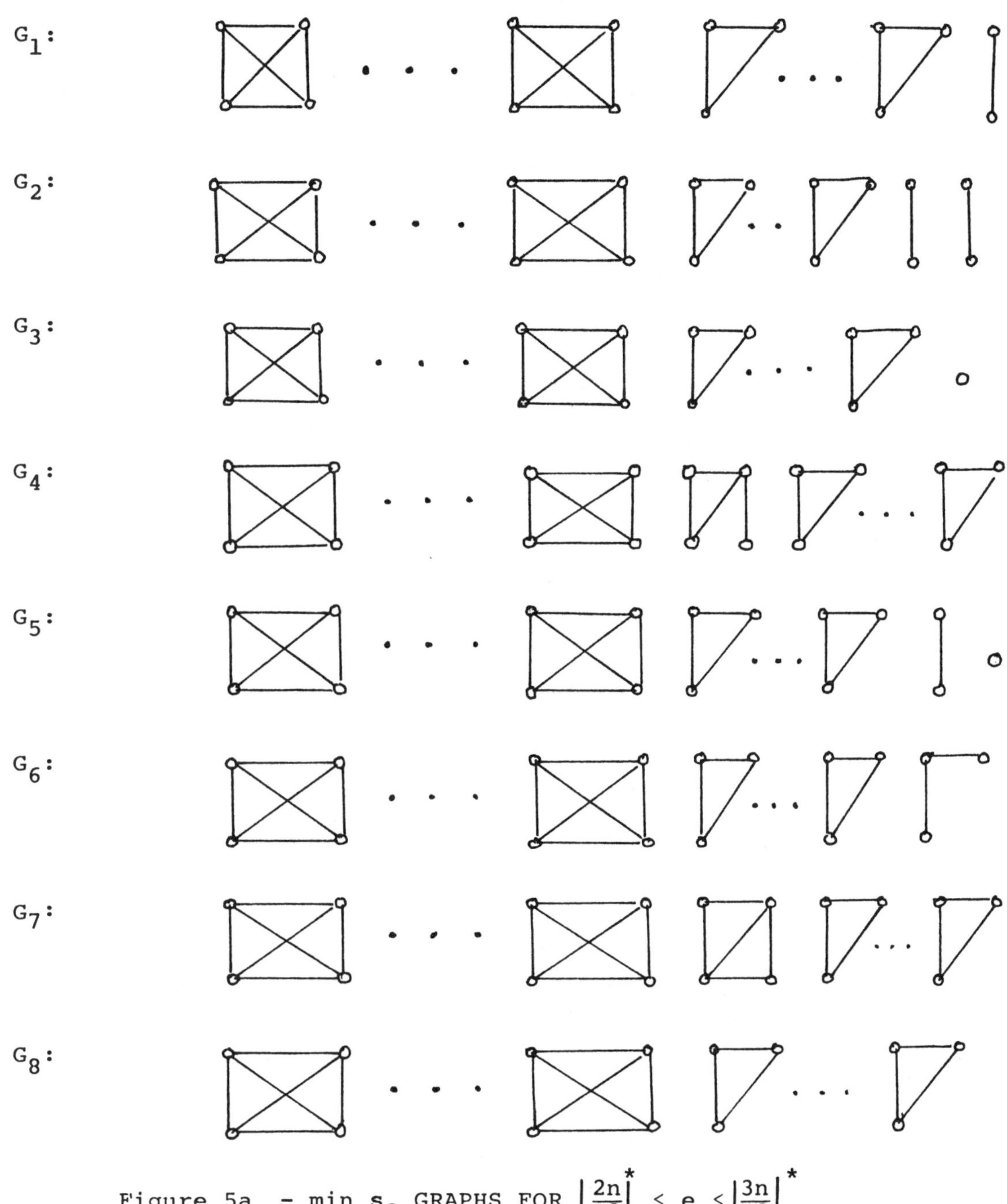

Figure 5a $-$ min s_3 GRAPHS FOR $\left\lfloor \frac{2n}{2} \right\rfloor^* < e \leq \left\lfloor \frac{3n}{2} \right\rfloor^*$

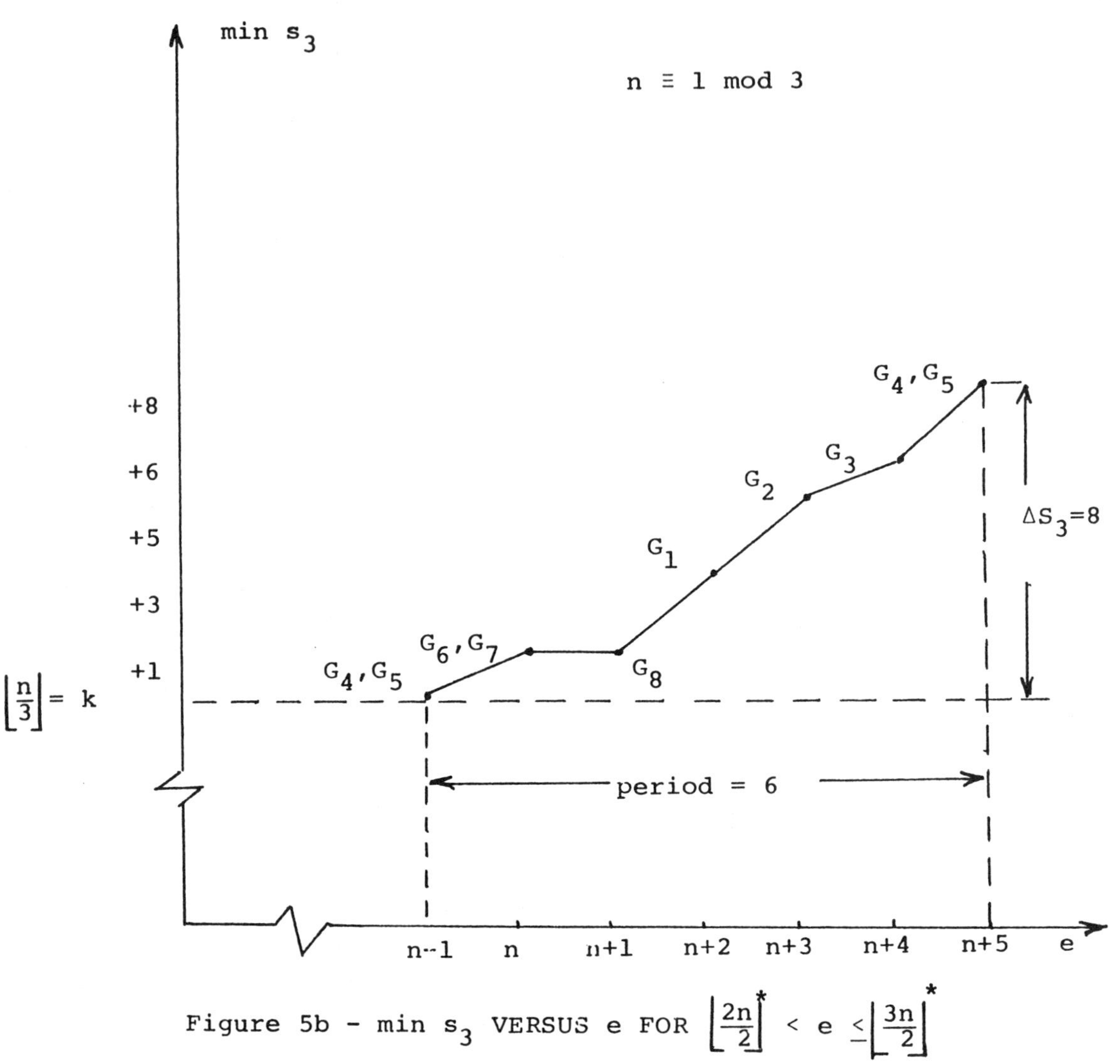

Figure 5b - min s_3 VERSUS e FOR $\left\lfloor \dfrac{2n}{2} \right\rfloor^{*} < e \leq \left\lfloor \dfrac{3n}{2} \right\rfloor^{*}$

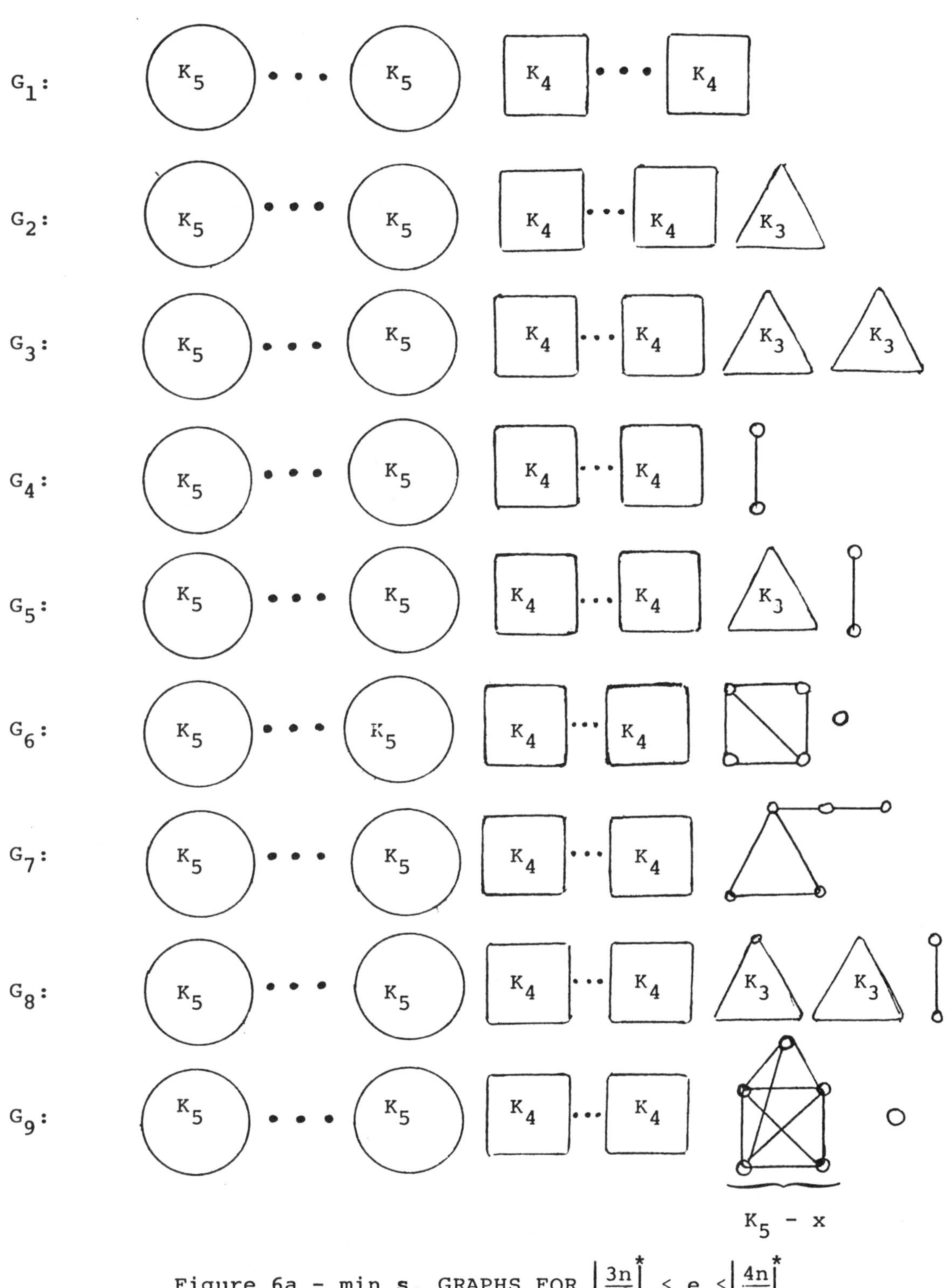

$$\text{Figure 6a} - \min s_3 \text{ GRAPHS FOR } \left\lfloor \frac{3n}{2} \right\rfloor^* < e \le \left\lfloor \frac{4n}{2} \right\rfloor^*$$

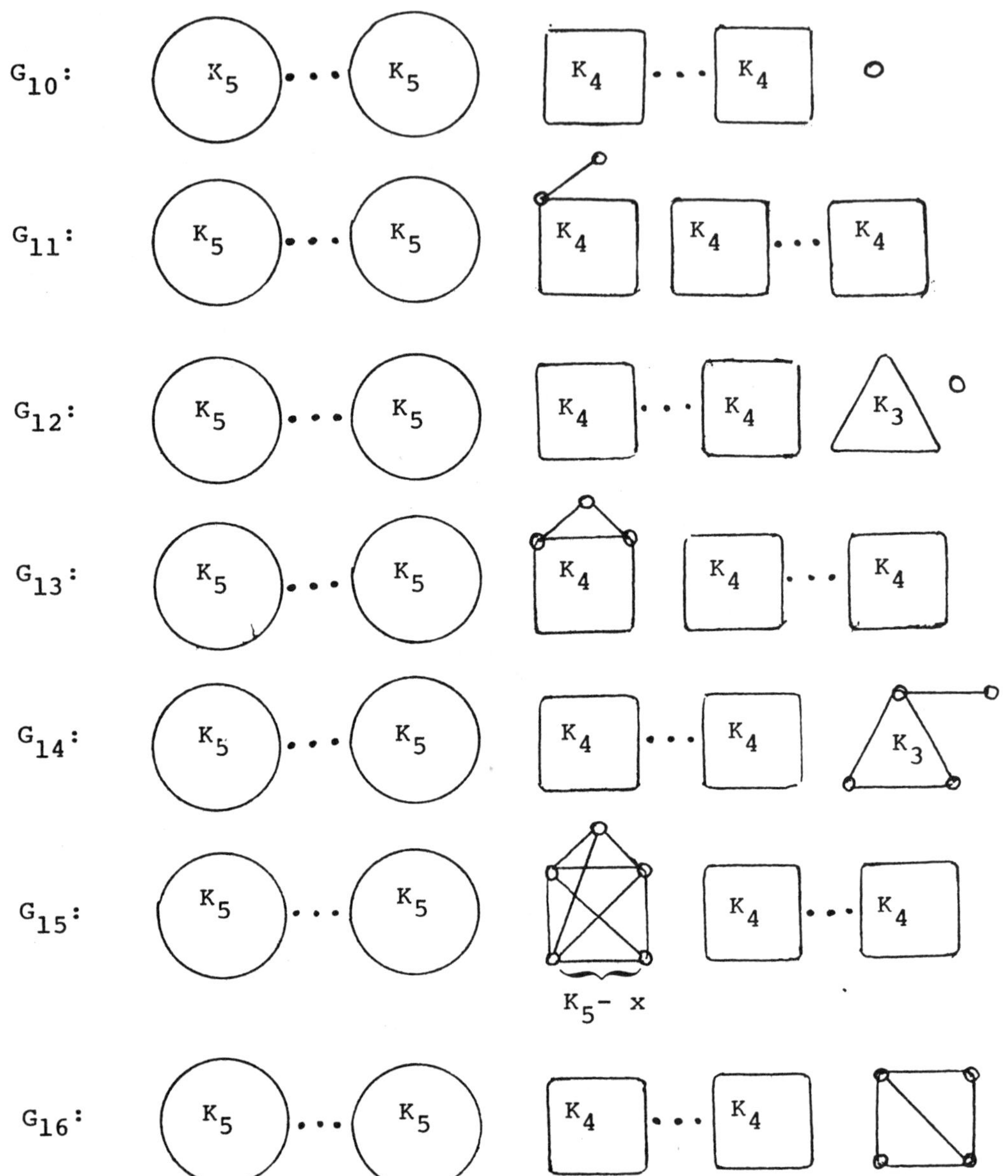

G_10:
K_5
K_5
K_4
K_4

G_11:
K_5
K_5
K_4
K_4
K_4

G_12:
K_5
K_5
K_4
K_4
K_3

G_13:
K_5
K_5
K_4
K_4
K_4

G_14:
K_5
K_5
K_4
K_4
K_3

G_15:
K_5
K_5
K_4
K_4
K_5 - x

G_16:
K_5
K_5
K_4
K_4

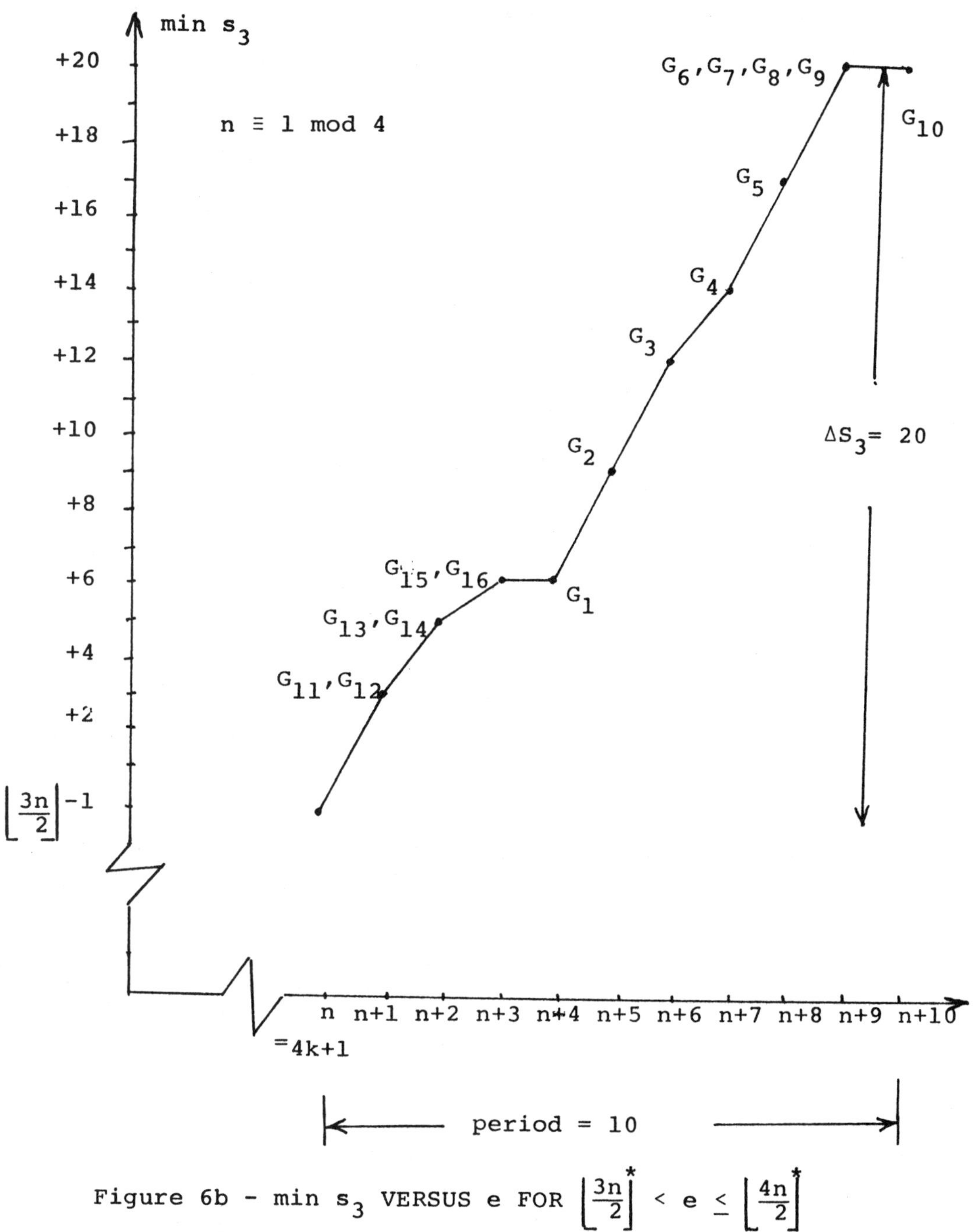

Figure 6b – min s_3 VERSUS e FOR $\left\lfloor \dfrac{3n}{2} \right\rfloor^{*} < e \leq \left\lfloor \dfrac{4n}{2} \right\rfloor^{*}$

4. Remarks and Conjectures

As a result of Theorem 3 and Corollary 5–1, we may conclude that $K_{2,n-2}$, $K_{3,4}$, $K_{3,5}$, $K_{4,5}$, and $K_{5,5}$ are the unique 3–optimal graphs in their classes.

Conjecture 1 *The unique 3–optimal $(n, m(n-m))$ graph is $K_{m,n}$.*

As a consequence of Theorem 4 and Theorem 5 note that $K_{p+2,p+2,\ldots,p+2}$ is 3-optimal for $p = 0, 1, 2$, and 3. In response to a conjecture made by us at the Sixth International Conference on the Theory and Applications of Graphs, J.C. Bermond proved that $K_{k,\ldots,k}$ and $K_{k,\ldots,k,k+1,\ldots,k+1}$ are the unique 3–optimal graphs in their class. A proof will appear in a future publication. An examination of the plots given in Theorem 5 yields the second conjecture.

Conjecture 2 The plot of min s_3 versus e inside the range $\frac{pn}{2} < e < \frac{(p+1)n}{2}$ exhibits a periodicity of period $\binom{p+2}{2}$ with a corresponding change of min s_3 over a period equal to $\frac{2p}{3}\binom{p+2}{2}$.

We close by remarking that no progress has been made to date on the general nature of the plot of min s_3 versus e.

REFERENCES

[1] D. Bauer, F. Boesch, C. Suffel, and R. Tindell, Combinatorial optimization problems in the analysis and design of probabilistic networks, *NETWORKS*, **15**, (1985) 257–271.

[2] Boesch and Felzer, On the invulnerability of the regular complete k-partite graphs, *SIAM J. Appl. Math.* **20**, (1971) 176-182.

[3] Boesch and Li, On the existence of uniformly optimally reliable networks, *Oper. Res.* (to appear).

[4] Boesch, Li, Stivaros and Suffel, On Graphs having the maximum number of three point induced connected subgraphs, *Stevens Inst. Res. Rep. in Computer Science* (1988) 8817.

[5] Frank, Maximally reliable node weighted graphs, *Proc. 3rd Annual Conf. on Inf. Sciences and Systems*, Princeton University (1969) 1–6.

[6] Hakimi and Amin, On the design of reliable networks, *NETWORKS* **3** (1973) 241–260.

[7] Harary, The maximum connectivity of a graph, *Proc. Nat. Acad. Sci. USA* (1962) 1142–1146.

[8] Harary, *Graph Theory*, Addison–Wesley, Reading (1964).

[9] Shier, Maximizing the number of spanning trees in a graph with n nodes and m edges, *J. Res. NBS,* **78B**, (1974) 193–196.

[10] Smith, Graphs with the smallest number of minimum cutsets, *NETWORKS* **14** (1984) 47–62.

[11] Smith and Doty, On the construction of optimally reliable graphs, submitted to *NETWORKS,* Dec. 1987.

A Characterization of the Sequence
of Generalized Chromatic
Numbers of a Graph

Izak Broere

Rand Afrikaans University

Marietjie Frick*

University of South Africa

ABSTRACT

The m th chromatic number $\chi_m(G)$ of a graph G is the least number of colours required to colour the vertices of G such that no m–clique of G is monocoloured. For every graph G one has the nonincreasing sequence $\chi_2(G)$, $\chi_3(G)$, ... of generalized chromatic numbers which has an infinite tail of ones. We characterize these sequences by giving necessary and sufficient conditions for a sequence (in terms of inequalities on the terms of the given sequence) for it to be the sequence of generalized chromatic numbers of a graph. We also determine the longest possible length of a constant subsequence of such a sequence. Furthermore, we show that the difference between any two consecutive terms of such a sequence can be arbitrarily large.

1. Introduction

All graphs considered in this paper are finite and simple. For undefined concepts, we refer the reader to [1]. Following Sachs (see [6]) we define, for an integer $m \geq 2$, an *m–admissible colouring* of a graph G as a colouring of the vertices of G such that no m–clique of G is monocoloured. The m th *chromatic number* of G, denoted by $\chi_m(G)$, is the least number of colours with which G can be m–admissibly coloured. Obviously, $\chi_2(G) = \chi(G)$, the chromatic number of G. Graphs which are *critical* with

---

* For part of this research this author was supported by the South African Council for Scientific and Industrial research.

respect to χ_m are studied in [2], and graphs which are *uniquely colourable* in this context are studied in [3].

These *generalized chromatic numbers* obviously satisfy

$$\chi_2(G) \geq \chi_3(G) \geq \ldots$$

and, for any given graph G, the sequence $\chi_2(G)$, $\chi_3(G)$, ... has an infinite tail of ones. (To be precise: $\chi_m(G) = 1$ if and only if the clique number of G satisfies $\omega(G) \leq m - 1$. The following result is due to Sachs [6]. (See also [4] and [5].)

Theorem 1 *For all integers $k \geq 2$ and $m \geq 2$ there exists a graph G with $\chi_m(G) = k$ and $\omega(G) = m$.*

One of the aims of this paper is to characterize these sequences of generalized chromatic numbers. Call a finite sequence of integers $k_2, k_3, \ldots, k_m$ with $k_2 \geq k_3 \geq \ldots \geq k_m \geq 2$ *chromatic* if there exists a graph G such that $\chi_i(G) = k_i$ for $i = 2, 3, \ldots, m$ and $\chi_{m+i}(G) = 1$ for all $i \geq 1$. Our main result is then: Such a sequence is chromatic if and only if $\lceil k_i/(k_j - 1) \rceil > (j - 1)/(i - 1)$ whenever i and j satisfy $2 \leq i < j \leq m$.

The sufficiency of the condition in our main result is proven by constructing a graph with suitable properties. The basic step of this construction is introduced and studied in Section 2. This construction is also employed in Section 4, where we determine the longest possible length of a constant subsequence of a chromatic sequence. We also prove that the difference between any two consecutive terms of a chromatic sequence can be arbitrarily large.

2. The Construction and its Properties

Let $k \geq 2$ and $m \geq 2$ be given integers and G a given graph of order of at least $m - 1$. We construct the graph $G(k,m)$ as follows: Take $k - 1$ pairwise disjoint copies of G, say $G_1, G_2, \ldots, G_{k-1}$ and let the elements of the set $X = \{(X_1, X_2, \ldots, X_{k-1}) \mid X_i \subseteq V(G_i)$ and $|X_i| = m - 1$ for $i = 1, 2, \ldots, k - 1\}$ also be vertices of $G(k,m)$. Each vertex $X = (X_1, X_2, \ldots, X_{k-1})$ of X is adjacent in $G(k,m)$ to each vertex of X_i in G_i, $i = 1, 2, \ldots, k - 1$. In order to describe the required properties of $G(k,m)$, we need some definitions. If G is a graph and c is a colouring of the vertices of G in $1, 2, \ldots, k$, then the *clique number of colour* i with respect to c, *denoted by* $\omega_c(i)$, is the maximum order of an i – coloured clique in G, $i = 1, 2, \ldots k$. Our first result describes bounds on the sum of these clique numbers.

Lemma 1 *If $\chi_m(G) \leq k$ and c is an m–admissible colouring of G in 1, 2, ..., k,*
then $\sum_{i=1}^{k} \omega_c(i) \leq k(m-1)$. Also, if $\chi_m(G) = k$ and c is an m–admissible colouring
of G in 1, 2, ..., k, then $\sum_{i=1}^{k} \omega_c(i) \geq km/2$.

Proof   Since c is an m–admissible colouring of G we have $\omega_c(i) \leq m - 1$ for
$i = 1, 2, ..., k$ and the upper bound follows immediately.

For the lower bound, note that $\omega_c(i) + \omega_c(j) \geq m$ if $i \neq j$, $i, j = 1, 2, ..., k$
(since if $\omega_c(i) + \omega_c(j) \leq m - 1$ for some $i \neq j$, one could recolour G using only one
colour for all the i–coloured and all the j–coloured vertices of G to obtain $\chi_m(G) \leq$
$k - 1$, a contradiction). Hence

$$\sum_{i=1}^{k} \omega_c(i) = (\omega_c(1) + \omega_c(2) + \omega_c(2) + \omega_c(3) + ... + \omega_c(k) + \omega_c(1)) / 2 \geq km/2. \quad \square$$

If G is a graph with $\chi_m(G) \leq k$, then we define $M_{k,m}(G) = \min_{c} \sum_{i=1}^{k} \omega_c(i)$,

where the minimum is taken over all the m–admissible colourings of G in 1, 2, ..., k.

Corollary 1 *If $\chi_m(G) \leq k$, then $M_{k,m}(G) \leq k(m-1)$.*

We now study $M_{k-1,m}(G(k,m))$.

Lemma 2 *If G is a graph of order at least m – 1 and $\chi_m(G(k,m)) \leq k - 1$, then*
$M_{k-1,m}(G(k,m)) > M_{k-1,m}(G)$.

Proof   Let c be any m–admissible colouring of G(k,m) in 1, 2, ..., k – 1 and let
$c(i)$ be the colouring of the copy G_i (of G in G(k,m)) induced by c, $i = 1, 2, ...,$
$k - 1$. Since $c(i)$ is then an m–admissible colouring of G_i, $\omega_{c(i)}(j) \leq m - 1$ for each
$j = 1, 2, ..., k - 1$. Hence each G_i contains a set of vertices X_i with $|X_i| = m - 1$
and with the vertices of some $\omega_{c(i)}(i)$ – clique contained in X_i, $i = 1, 2, ..., k - 1$.
Since the vertex $(X_1, X_2, ..., X_{k-1})$ is adjacent in G(k,m) to each vertex of each X_i,
it follows that $\omega_c(j) > \omega_{c(j)}(j)$ if c assigns j to $(X_1, X_2, ..., X_{k-1})$. Note also that
$\omega_c(i) \geq \omega_{c(j)}(i)$ for this j and $i = 1, 2, ..., k - 1$. Hence $\sum_{i=1}^{k-1} \omega_c(i) > \sum_{i=1}^{k-1} \omega_{c(j)}(i) \geq$
$M_{k-1,m}(G)$ from which it follows, since c is arbitrary, that $M_{k-1,m}(G(k,m)) >$
$M_{k-1,m}(G)$. $\square$

We also need to investigate $\chi_m(G(k',m'))$.

Lemma 3 *If* $\chi_m(G) \leq k$ *and* $\lfloor (m'-1)/(m-1) \rfloor < k/(k'-1)$, *then* $\chi_m(G(k',m'))$ $\leq k$.

Proof Any m–admissible colouring d of G in 1, 2, ..., k can be extended to an m–admissible colouring of G(k',m'): Colour each of the k' – 1 copies of G with d. For every vertex X of X, a colour is now available since the number of disjoint (m – 1)–cliques whose vertices are all adjacent to X is at most $(k'-1)\lfloor (m'-1)/(m-1) \rfloor < k$. ❑

Corollary 2 *If* $\chi_m(G) \leq k$ *then* $\chi_m(G(k,m)) \leq k$.

Our next result deals with the clique number $\omega(G(k,m))$.

Lemma 4 *If G is a graph with* $\omega(G) \leq m$, *then* $\omega(G(k,m')) \leq m$ *for all k and all* $m' \leq m$.

Proof Any clique of G(k,m') not contained in a copy G_i of G, i = 1, 2, ..., k – 1, contains one vertex of X and at most m' – 1 other vertices, all of them in some G_i. ❑

Our final result of this sections deals with a recursive construction which will be used in Theorems 2, 3 and 4.

Lemma 5 *Given integers* $k \geq 2$ *and* $m \geq 2$ *and a graph* H_0 *of order at least* $m-1$ *with* $\chi_m(H_0) \leq k$. *If we define* H_l *recursively by* $H_{l+1} = H_l(k,m)$ *for* $l = 0, 1,...,$ *then there is an integer* l_0 *such that* $\chi_m(H_{l_0}) = k$.

Proof Let $l_0 = (k-1)(m-1)$. It follows from Corollary 2 that $\chi_m(H_l) \leq k$ for $l = 0, 1, ...$ .If $\chi_m(H_{l_0}) \leq k-1$, then $\chi_m(H_l) \leq k-1$ for $l = 0, 1, ..., l_0$ and then Lemma 2 applies to each of $H_0, H_1,..., H_{l_0-1}$ and hence $M_{k-1,m}(H_{l_0-1}) > ... > M_{k-1,m}(H_0) \geq 1$. But then $M_{k-1,m}(H_{l_0}) \geq l_0 + 1 > (k-1)(m-1)$, contradicting Corollary 1. ❑

3. The Characterization

The first lemma of this section will indicate which sequences are not chromatic.

Lemma 6 *If i and j are integers with* $2 \leq i < j$, *and* k_i *and* k_j *are integers with* $k_i \geq k_j \geq 2$ *such that* $\lceil k_i/(k_j-1) \rceil \leq (j-1)/(i-1)$, *then there does not exist a graph G with* $\chi_i(G) = k_i$ *and* $\chi_j(G) = k_j$.

Proof   If G is such a graph, consider an i – admissible colouring of G in k_i colours.

Let $\lceil k_i/(k_j - 1) \rceil = r$ so that $k_i/(k_j - 1) \le r \le (j - 1)/(i - 1)$. The union of any r colour classes of this colouring of G induces a subgraph of G containing no j–clique since $r(i - 1) \le j - 1$. But then G has a j–admissible colouring using only $\lceil k_i/r \rceil \le k_j - 1$ colours, contradicting $\chi_j(G) = k_j$. $\square$

Corollary 3   *If $k_2 \ge k_3 \ge \ldots \ge k_m \ge 2$ are given integers and for some i and j with $2 \le i < j \le m$ we have $\lceil k_i/(k_j - 1) \rceil \le (j - 1)/(i - 1)$, then the sequence $k_2, k_3, \ldots, k_m$ is not chromatic.*

The next lemma will be useful to provide a starting point for the inductive proof of our theorem.

Lemma 7   *Let $k_2 \ge k_3 \ge \ldots \ge k_m \ge 2$ be given integers such that $\lceil k_i/(k_j - 1) \rceil > (j - 1)/(i - 1)$ whenever i and j satisfy $2 \le i < j \le m$ and let $k_r(r - 1) = \min\limits_{2 \le i \le m} k_i(i - 1)$. Then there exists a graph H with $\chi_i(H) \le k_i$ for $i = 2, 3, \ldots, m$ and $\chi_r(H) = k_r$ and $\chi_{m + 1}(H) = 1$.*

Proof   We define H_l recursively, starting with a suitable complete graph: Let $n = \min\{m, k_r(r - 1)\}$ and let $H_0 = K_n$ and $H_{l + 1} = H_l(k_r, r)$ for $l = 0, 1, \ldots$.

We start by proving inductively that (for any fixed j with $2 \le j \le m$) $\chi_j(H_l) \le k_j$ for all $l \ge 0$: Note that $\chi_j(H_0) = \chi_j(K_n) = \lceil n/(j - 1) \rceil \le \lceil k_j(j - 1)/(j - 1) \rceil = k_j$. Hence assume that $\chi_j(H_l) \le k_j$ for some $l \ge 0$. We investigate the applicability of Lemma 3 in three cases:

Case 1   If $r < j$, then $\lfloor (r - 1)/(j - 1) \rfloor = 0 < k_j/(k_r - 1)$.

Case 2   If $r = j$, then $\lfloor (r - 1)/(j - 1) \rfloor = 1 < k_j/(k_r - 1)$.

Case 3   If $j < r$, then $\lceil k_j/(k_r - 1) \rceil > (r - 1)/(j - 1)$ by hypothesis and hence $\lfloor (r - 1)/(j - 1) \rfloor < k_j/(k_r - 1)$.

Hence in each case Lemma 3 applies and $\chi_j(H_{l + 1}) \le k_j$ follows. This completes our induction.

By Lemma 5 there is an l_0 such that $\chi_r(H_{l_0}) = k_r$. We now take $H = H_{l_0}$ to obtain a graph with $\chi_i(H) < k_i$ for $i = 2, 3, \ldots, m$ and $\chi_r(H) = k_r$. Note, to conclude the proof, that $\omega(H_l) \le m$ for each $l \ge 0$ (by an easy inductive proof using Lemma 4). Hence $\omega(H) \le m$, that is, $\chi_{m + 1}(H) = 1$. $\square$

We are now ready to give a characterization of the sequence of generalized chromatic numbers.

Theorem 2 *Suppose $k_2' \geq k_3' \geq \ldots \geq k_m' \geq 2$ are given integers. Then the sequence k_2', k_3', ..., k_m' is chromatic if and only if the inequality $\lceil k_i'/(k_j' - 1) \rceil > (j-1)/(i-1)$ is satisfied for all i and j with $2 \leq i < j \leq m$.*

Proof   The necessity of the condition is given by Corollary 3.

We prove the sufficiency by constructing a graph G satisfying $\chi_i(G) = k_i'$ for $i = 2, 3, \ldots, m$ and $\chi_{m+1}(G) = 1$ for all $i \geq 1$.

Let $(i_2, i_3, \ldots, i_m)$ be a permutation of $\{2, 3, \ldots, m\}$ such that
$$k_{i_2}'(i_2 - 1) \leq k_{i_3}'(i_3 - 1) \leq \ldots \leq k_{i_m}'(i_m - 1).$$
Then, denoting each k_{i_j}' by k_j we have $k_2(i_2 - 1) \leq k_3(i_3 - 1) \leq \ldots \leq k_m(i_m - 1)$. In this notation, we need to construct a graph G with $\chi_{i_r}(G) = k_r$ for $r = 2, 3, \ldots, m$ and $\chi_{m+i}(G) = 1$ for all $i \geq 1$.

We start our proof (which is inductive by nature) with a graph H satisfying $\chi_{i_2}(H) = k_2$, $\chi_{i_r}(H) < k_r$ for $r = 3, 4, \ldots, m$ and $\chi_{m+1}(H) = 1$. The existence of this graph is guaranteed by Lemma 7. Hence we assume that, for some n with $2 \leq n < m$, we have a graph H with $\chi_{i_r}(H) = k_r$ for $r = 2, 3, \ldots, n$ and $\chi_{i_r}(H) \leq k_r$ for $r = n+1, n+2, \ldots, m$ and $\chi_{m+1}(H) = 1$. Construct a sequence of graphs by letting $H_0 = H$ and $H_{l+1} = H_l(k_{n+1}, i_{n+1})$ for $l = 0, 1, \ldots$. We prove that $\chi_{i_r}(H_l) \leq k_r$, for any given r and all $l \geq 0$, by induction on l:

Note that $\chi_{i_r}(H_0) \leq k_r$ is given. Assume, for some $l \geq 0$, that $\chi_{i_r}(H_l) \leq k_r$ and consider the applicability of Lemma 3 in three cases:

Case 1   If $i_{n+1} < i_r$, then $\lfloor (i_{n+1} - 1)/(i_r - 1) \rfloor = 0 < k_r/(k_{n+1} - 1)$.

Case 2   If $i_{n+1} = i_r$, then $\lfloor (i_{n+1} - 1)/(i_r - 1) \rfloor = 1 < k_r/(k_{n+1} - 1)$.

Case 3   If $i_r < i_{n+1}$, then $(i_{n+1} - 1)/(i_r - 1) < \lceil k_r/(k_{n+1} - 1) \rceil$ by hypothesis and hence $\lfloor (i_{n+1} - 1)/i_r - 1) \rfloor < k_r(k_{n+1} - 1)$. Hence $\chi_{i_r}(H_{l+1}) \leq k_r$ follows in each case from Lemma 3.

Hence $\chi_{i_r}(H_l) \leq k_r$ for $r = 2, 3, \ldots, m$ and all $l \geq 0$. Since H_0 is a subgraph of H_l for all $l \geq 0$, we also have that $\chi_{i_r}(H_l) = k_r$ for $r = 2, 3, \ldots, n$. By Lemma 5 there is an l_0 such that $\chi_{i_{n+1}}(H_{l_0}) = k_{n+1}$.

For such an l_0 we then have a graph H_{l_0} satisfying $\chi_{i_r}(H_{l_0}) = k_r$ for $r = 2, 3, \ldots,$ $n + 1$ and $\chi_{i_r}(H_{l_0}) \le k_r$ for $r = n + 2, n + 3, \ldots, m$. Note also that $\omega(H_0) = \omega(H) \le m$ since $\chi_{m+1}(H) = 1$. Hence, by repeated application of Lemma 4, we have that $\omega(H_{l_0}) \le m$, that is, $\chi_{m+1}(H_{l_0}) = 1$. $\square$

4. Further Results on the Sequence

Our first result of this section concerns the length of a constant subsequence of a chromatic sequence.

Theorem 3   *If $n \ge r \ge 2$ and $k \ge 2$ are given integers, then there exists a graph G with $\chi_r(G) = \chi_{r+1}(G) = \ldots = \chi_n(G) = k$ if and only if $n \le 2r - 2$.*

Proof   The necessity of the condition follows from Lemma 6 since $\lceil k/(k-1) \rceil = 2$, and $(n-1)/(r-1) < 2$ only if $n \le 2r - 2$.

To prove the sufficiency, take $H_0 = K_{k(r-1)}$ and $H_{l+1} = H_l(k,n)$ for $l = 0, 1,$ $\ldots$. Then $\chi_n(H_0) \le k$ and hence, by Lemma 5 there is an l_0 such that $\chi_n(H_{l_0}) = k$. Also $\chi_r(H_0) = k$, and, since $n - 1 < 2r - 2$, we have $\lfloor (n-1)/(r-1) \rfloor = 1 < k/(k-1)$. We can therefore apply Lemma 3 repeatedly, to prove that $\chi_r(H_{l_0}) \le k$, and hence $\chi_r(H_{l_0}) = k$.

We now take $G = H_{l_0}$, then $\chi_r(G) = \chi_{r+1}(G) = \ldots = \chi_n(G) = k$. $\square$

Our final result concerns consecutive terms of a chromatic sequence.

Theorem 4

    (a)   *If $m \ge 3$ and k_m and k_{m+1} are any positive integers with $k_m \ge k_{m+1}$ then there exists a graph G with $\chi_m(G) = k_m$ and $\chi_{m+1}(G) = k_{m+1}$.*

    (b)   *Given positive integers k_2 and k_3 there exists a graph G with $\chi_2(G) = k_2$ and $\chi_3(G) = k_3$ if and only if $k_3 \le \lceil k_2/2 \rceil$.*

Proof   (a) The case $k_m = k_{m+1} = 1$ is trivial, and the case $k_m > k_{m+1} = 1$ follows from Theorem 1 (because $\chi_{m+1}(G) = 1$ if $\omega(G) = m$). Now suppose $k_{m+1} > 1$. We distinguish two cases. In each case we construct a graph H_{l_0} satisfying $\chi_m(H_{l_0}) = k_m$ and $\chi_{m+1}(H_{l_0}) = k_{m+1}$. The graph H_{l_0} is then the desired graph G.

Case 1 $k_m(m-1) \le k_{m+1}m$. In this case take $H_0 = K_{k_m(m-1)}$ and $H_{l+1} = H_l(k_{m+1}, m+1)$ for $l = 0, 1, \ldots$. Then $\chi_{m+1}(H_0) = \lceil k_m(m-1)/m \rceil \le k_{m+1}$, and hence it follows from Lemma 5 that there is an integer l_0 such that

$\chi_{m+1}(H_{l_0}) = k_{m+1}$. Since $\chi_m(H_0) = k_m$ and $\lfloor m/(m-1) \rfloor = 1 < k_m/(k_{m+1} - 1)$, Lemma 3 can be applied repeatedly to prove that $\chi_m(H_{l_0}) \leq k_m$ and hence $\chi_m(H_{l_0}) = k_m$.

Case 2 $k_{m+1}m \leq k_m(m-1)$. In this case take $H_0 = K_{k_{m+1}m}$ and $H_{l+1} = H_l(k_m, m)$ for $l = 0, 1, \ldots$ .Then $\chi_m(H_0) = \lceil k_{m+1}m/m - 1 \rceil \leq k_m$, and hence it follows from Lemma 5 that there is an integer l_0 such that $\chi_m(H_{l_0}) = k_m$. Since $\chi_{m+1}(H_0) = k_{m+1}$, Lemma 3 can be applied repeatedly to prove that $\chi_{m+1}(H_{l_0}) \leq k_{m+1}$, and hence $\chi_{m+1}(H_{l_0}) = k_{m+1}$.

(b) The necessity follows from Lemma 6. We now prove the sufficiency. The case $k_2 = k_3 = 1$ is trivial, and the case $k_2 > k_3 = 1$ follows from Theorem 1. Now suppose $k_3 > 1$. If $k_3 = \lceil k_2/2 \rceil$, take $G = K_{k_2}$. Then $\chi_2(G) = k_2$ and $\chi_3(G) = \lceil k_2/2 \rceil = k_3$ as required. If $k_3 < \lceil k_2/2 \rceil$ then $2k_3 \leq k_2$. In this case we proceed as for Case 2 of (a), taking $m = 2$, to find the required graph $G = H_{l_0}$. □

REFERENCES

[1]   Behzad, M., Chartrand, G. and Lesniak–Foster, L. *Graphs and Digraphs*, Prindle, Weber and Schmidt, Boston 1979.

[2]   Broere, Izak and Frick, Marietjie. On the order of color critical graphs, *Congressus Numerantium* **47** (1986), 125 – 130.

[3]   Broere, Izak and Frick, Marietjie. On the order of uniquely (k,m) – colourable graphs (to appear).

[4]   Broere, Izak and Frick, Marietjie. Two results on generalized chromatic numbers (in preparation).

[5]   Hakimi, S. Louis and Stillman, Maureen J. A generalization of vertex – coloring in graphs (manuscript).

[6]   Sachs, Horst. Finite graphs in *Recent progress in combinatorics* (Proc. Third Waterloo Conf. on Combinatorics, 1968), pp. 175 – 184. Academic Press, New York 1969.

ON THE AUTOMORPHISM GROUP OF A POWER OF A HYPERCUBE

S. A. Burr

City College, C.U.N.Y.

ABSTRACT

Let Q_n be the n-dimensional hypercube graph, and let $Q_n^k(1 \leq k \leq n)$ be its k-th power, that is, the graph formed from Q_n^k by joining all vertices at (Hamming) distance $\leq k$. We study the automorphism groups of these graphs, determining them exactly when k is odd and when $k = n$ or $n-1$. When k is even and $\leq n - 2$, we show only that the automorphism group is a subgroup of that of Q_n^2 and a supergroup of Q_n.

1.   Introduction

Let Q_n denote the (graph-theoretic) n-dimensional hypercube. That is, the vertices of Q_n are n-bit binary strings, and two strings are adjacent if and only if they differ in exactly one coordinate, which is the same as having Hamming distance 1. Here, we will be concerned with Q_n^k, which is formed from Q_n by joining all pairs of vertices with Hamming distance $\leq k$. In particular, we are interested in its automorphism group $A(Q_n^k)$. The three cases $k = 1, n - 1, n$ are straightforward. If A and B are groups, let A[B]

This research was partially supported by ONR grant N00014-85-K-0704 and PSC-CUNY grant 6-68284.

This paper was submitted to the editors in May 1989 by the author, who was invited to the Conference but did not speak there.

denote their wreath product, which is also known as their composition (see [1, pg. 164], for instance). It is easy to see that $A(Q_n) \cong S_n[S_2]$, $A(Q_n^{n-1}) \cong S_2n-1[S_2]$, and $A(Q_n^n) \cong S_2^n$, with orders $n!2^n$, $(2^{n-1})!2^{2^{n-1}}$, and $(2^n)!$, respectively. Our purpose here is to prove the following result for the remaining values of k.

Theorem   If $k \leq n-2$ then $A(Q_n^k) \cong A(Q_n) \cong S_n[s_2]$ when k is odd, and $A(Q_n^k)$ is a subgroup of $A(Q_n^2)$ and a supergroup of $A(Q_n) \cong S_n[S_2]$ when k is even.

Z. Miller and M. Perkel [2] have also proved the above, at least when n is large relative to k, and have further shown that the automorphism group of Q_n^k is actually the same as that of Q_n^2 when k is even, at least when n is large. We mention here that the problem of determining the automorphism group of Q_n^k (or at least its order) arose while studying the hypercube bandwidth of random graphs [3].

In what follows, we will usually treat n and k as fixed positive integers. We will continue to consider the vertices of Q_n^k as n-bit binary strings.

Therefore, we may speak of the Hamming distance between vertices, even though their distance in Q_n^k will generally be smaller. When there is no danger of confusion, we will often omit the "Hamming". If u and v are adjacent vertices of Q_n^k with (Hamming) distance d, we denote the number of common neighbors in Q_n^k by N_d. By symmetry, this terminology is meaningful, since the number of such common neighbors depends only on d. We also observe that $N_d = 0$ if $d > k$. We will be using binomial coefficients, and we adopt the convention that $\binom{a}{b} = 0$ if $b > a$ or b is not an integer.

2. The Main Lemma

Lemma   Let $2 \leq d \leq k \leq n$ be given. Then we have

$$N_{d-1} - N_d = \binom{c-1}{(d-1)/2} \binom{n-d}{k-(d-1)/2}$$

Proof   Consider Q_n^k. Choose some vertex v_0. Since all vertices are equivalent, it is clear that we lose no generality by assuming that $v_0 = (0, 0, \cdots, 0)$. We wish to compute $N_{d-1} - N_d$. To this end, we will choose two vertices, one at Hamming distance $d - 1$ from v_0, and one at distance d from v_0, calling these v_{d-1} and v_d, respectively. Because of symmetry, we are free to choose any two such vertices that please us. We choose

$$v_{d-1} = (\underbrace{1, \cdots, 1}_{d-1}, \underbrace{0, \cdots, 0}_{n-d+1}),$$

$$v_d = (\underbrace{1, \cdots, 1}_{d}, \underbrace{0, \cdots, 0}_{n-d}).$$

The advantage of these two choices is that we can perform a great deal of cancellation while computing $N_{d-1} - N_d$. First, consider any vertex $v \neq v_0$ whose Hamming distance from one of $\{v_{d-1}, v_d\}$ is $\leq k - 1$. Then v is at distance $\leq k$ from the other. Therefore, if the distance from v_0 to v is at distance $\leq k$ from the other. Therefore, if the distance from v_0 to v is $\leq k - 1$, then v is in the common neighborhood of v_{d-1} and v_d, and hence contributes 0 to the difference $N_{d-1} - N_d$. Similarly, any v which is at Hamming distance at least $k + 2$ from one of $\{v_{d-1}, v_d\}$ is not in the common neighborhoods of either, so such a v also contributes 0 to the difference. Therefore, we can ignore all vertices except those whose Hamming distances from $\{v_{d-1}, v_d\}$ are $\{k, k+1\}$ (in some order). Such a vertex may be described as one whose Hamming distance from the string

$$(\underbrace{1, \cdots, 1}_{d-1}, *, \underbrace{0, \cdots, 0}_{n-d})$$

is k, where $*$ is a symbol which is considered to match either a 0 or a 1.

Let v be a vertex, considered as a string. Let a and b be the number of 1's in the first $d - 1$ bits and in the last $n - d$ bits of v, respectively. Then the condition at the end of the last paragraph becomes $d - a - 1 + b = k$. Such vertices come in pairs, one with a 0 in position d and one with a 1 in position d. The first has Hamming distance k from v_{d-1} and $k + 1$ from v_d, and the opposite is true for the second. Therefore, these pairs of vertices contribute

0 to the difference $N_{d-1} - N_d$, and hence can be ignored, *unless* one vertex of the pair is at Hamming distance k from v_0 and the other is at distance $k + 1$. The condition for this to happen is $a + b = k$. Any pair satisfying this condition and $d - a - 1 + b = k$ contributes 1 to N_{d-1} and 0 to N_a, and thus contributes 1 to the difference, while all other pairs contribute 0 to the difference.

If we solve the two equations $d - a - 1 + b = k$ and $a + b = k$, we get

$$a = (d - 1)/2 \text{ and } b = k - (d - 1)/2.$$

The number of ways of choosing a bits from among the first $d - 1$ is $\binom{d-1}{a}$ and the number of ways of choosing b from the last $n - d$ is $\binom{n-d}{b}$. (By convention, both binomial coefficients are 0 if d is even.) Hence, we have

$$\begin{aligned}
N_{d-1} - N_d &= \binom{d - 1}{a} \binom{n - d}{b} \\
&= \binom{d - 1}{(d - 1)/2} \binom{n - d}{k - (d - 1)/2}.
\end{aligned}$$

Proof of the Theorem We now derive four corollaries. The Theorem follows from the last two.

Corollary 1 If $2 \le l \le n$, where l is even, then $N_{l-1} = N_l$.

Proof By the Lemma, $N_{l-1} - N_l = 0$.

Corollary 2 Let $k \le n - 2$. Then N_1 and N_2 are strictly greater than N_d for every $d > 2$.

Proof By the Lemma, and the trivial fact that $N_d = 0$ whenever $d > k$, the difference $N_{d-1} - N_d$ is never negative. Furthermore, if we set $d = 3$ in the Lemma, we have

$$N_2 - N_3 = \binom{2}{1} \binom{n - 3}{k - 1} > 0,$$

since $n - 3 \ge k - 1$.

Corollary 3 $A(Q_n^k)$ is a subgroup of $A(Q_n^2)$.

Proof   This follows from Corollary 2, since that says that we can uniquely recover all pairs of vertices whose Hamming distance is ≤ 2, so that we can uniquely recover the corresponding copy of Q_n^2 that lies within our Q_n^k. Therefore, any automorphism in $A(Q_n^k)$ must also be an automorphism of that Q_n^2.

Corollary 4   If $k \leq n - 2$ and k is odd, then $A(Q_n^k)$ is a subgroup of $A(Q_n)$.

Proof   By Corollary 3, we can determine all pairs of vertices whose Hamming distance is ≤ 2. We therefore can determine all pairs with Hamming distance $\leq d$ whenever d is even. In particular, we can do this for $d = k - 1$ and $d = k + 1$. Furthermore, Q_n^k itself gives us all pairs whose Hamming distance is $\leq k$. Given this information, we can easily determine all pairs whose Hamming distance is exactly k and all whose distance is exactly $k + 1$. Now fix some vertex v, and consider all pairs of vertices v_1, v_2 for which the Hamming distances from v to v_1 and to v_2 are (exactly) k and $k + 1$, respectively. Among these pairs, some will be at Hamming distance ≤ 2, and these pairs can be determined from Q_n^2. But since Q_n is bipartite, any such pairs will be at distance exactly 1. Furthermore, it is clear that every pair of vertices has Hamming distances k and $k + 1$ from some vertex, so that the underlying Q_n is uniquely determined, and hence any automorphism of Q_n^k must also be one of Q_n.

Proof of the Theorem   In view of Corollaries 3 and 4, it suffices to observe that $A(Q_n)$ is a subgroup of $A(Q_n^k)$ for all k, since the operation of taking the kth power obviously destroys none of the automorphisms of a graph.

REFERENCES

[1] F. Harary. *Graph Theory*. Addison-Wesley, Reading, MA, 1969.

[2] Z. Miller and M. Perkel, "Automorphisms of Powers of the n-Cube", (to appear).

[3] S. A. Burr, M. Goldberg, and Z. Miller, "Hypercube Bandwidth in Random Graphs" (to appear).

FURTHER RESULTS ON MAXIMAL ANTI-RAMSEY GRAPHS

S. A. Burr

CUNY

P. Erdös

V. T. Sós

Hungarian Acad. of Sciences

P. Frankl

University of Paris

R. L. Graham

AT & Bell Laboratories

ABSTRACT

A well-studied question in graph theory asks the following: Given a graph L and an integer $r > 0$, which graphs G have the property that no matter how the edges of G are r-colored, a monochromatic copy of L must always occur in G? (More precisely, G has a subgraph isomorphic to L in which all edges have the same color. We will typically use this type of informal description when the meaning is clear**.) Indeed, the forthcoming book [BFRS] will list several hundred papers which deal with various aspects of this subject. In [BEGS], we recently initiated a study of a related problem which in a certain sense goes in the opposite direction.*

---

* for undefined terminology in graph theory, see [H].

** to us.

1. Introduction

Our motivation actually originated from a question of Berkowitz (see [BEGS]) concerning time-space tradeoffs for models of computation (which still is unresolved). Basically, we investigated the following. Given a graph L and integers n and e, what is the *smallest* number r so that for *some* graph $G = G(n, e)$ with n vertices and e edges, the edges of G can be r-colored so that in every copy of L in G, all edges have *different* colors. We say that such a copy of L is *totally multicolored* (abbreviated TMC), and we denote this least value r by the expression $\chi_S(n, e, L)$. This notation comes from the fact that the value we seek is also the strong chromatic number of the hypergraph which has as its set of points the edges of G, with (hyper)edges consisting of the sets of edges of G which form copies of L.

In [BEGS] a substantial body of results are given concerning $\chi_S(n, e, L)$, as well as an unexpectedly large number of open problems. Some of these we will mention at the end. Several of the nicest results, however, could not be included there because of space limitations. Our purpose in this note is to fulfill the promise made in [BEGS] and present the details of these results here.

2. Preliminaries

We will call two edges of a graph *strongly independent* if they are disjoint and their four endpoints span no other edges. Our two main results will contrast the different behavior of $\chi_S(n, e, L)$ depending on whether or not L has two strongly independent edges.

Lemma 1   If a bipartite graph contains at least $(m - 1)^2 + 1$ edges, then it either contains m independent edges or else a vertex of degree at least m. This result is sharp.

    Proof   This follows immediately from the well-known result of König: a bipartite graph with maximum degree Δ is the union of Δ matchings. If G is a bipartite graph with $q \geq (m - 1)^2 + 1$ edges and $\Delta(G) \leq m - 1$, then some matching contains at least $\lceil q/\Delta \rceil \geq m$ edges. The sharpness is shown by any bipartite graph in which one part has exactly $m - 1$ vertices, each of degree $m - 1$.

Lemma 2   Let $0 < \delta < 1$ be a fixed real number and let p be a fixed positive integer. There exists a positive number $\alpha = \alpha(\delta, p)$ such that for all sufficiently large n, every graph with n vertices and average degree at least δn contains the complete bipartite graph $K(p, \lceil \alpha n \rceil)$.

Proof   Just use the standard argument: if the average degree d of a graph satisfies

$$n \binom{d}{p} > (q - 1)\binom{n}{p}.$$

then the graph contains $K(p, q)$. With $\alpha = \frac{1}{2}\delta^p$ and $q = \lceil \alpha n \rceil$, the above inequality holds for all sufficiently large n.

Notation   If A and B are disjoint subsets of $V(G)$, the bipartite subgraph of G with vertex partition (A, B) and edge set $\{xy | x \in A, y \in B, xy \in E(G)\}$ will be denoted $< A, B >_G$ or, if G is understood, simply $< A, B >$.

Lemma 3   Let $0 < \epsilon < 1$ be fixed and let p be a fixed positive integer. There are then numbers $\alpha, \beta, \gamma \in (0, 1)$ so that for all sufficiently large n, every graph G with n vertices and at least $e(n) = \lceil \frac{1}{2}\epsilon n^2 \rceil$ edges contains a bipartite subgraph $< A, B >$ such that

(i) $|A| = \lceil \alpha n \rceil$ and $|B| = \lceil \beta n \rceil$,

(ii) every vertex in B is adjacent to at least $\lceil \gamma n \rceil$ vertices of A, and

(iii) there exists $X \subset B$ with $|X| = p$ such that $< A, X >$ is complete.

Proof   Since G has average degree at least ϵn, it contains a subgraph G' with $|V(G')| > \epsilon n$ in which every vertex has degree at least $\epsilon n/2$. [Just remove vertices of degree less than $\epsilon n/2$ until there are none left; for each such removal, the average degree does not decrease. Since the remaining graph has average degree at least ϵn, it has more than ϵn vertices.] By Lemma 2 (with $\delta = \epsilon/2$ and n replaced by ϵn)G' contains a bipartite subgraph

$< X, A > \cong K(p, \lceil \alpha n \rceil)$, where $\alpha = (\epsilon/2)^{p+1}$. Set $R = V(G')\backslash\{A \cup X\}$. For simplicity of notation, let $|A| = a, |R| = r$.

Suppose that each vertex of A is adjacent to at least λr vertices of R. Then there are at least

$$s = \lceil [\frac{\lambda r}{2 - \lambda} \rceil$$

vertices in R which are each adjacent to at least $\lambda a/2$ vertices of A. If not, the number of $A - R$ edges is less than $sa + (r - s)\lambda a/2$ and since

$$sa + (r - s)\lambda a/2 \leq a\lambda r,$$

this contradicts the fact that every vertex in A is adjacent to at least λr vertices of R. Since $a = \lceil \alpha n \rceil$ and each vertex in A adjacent to at least $(\epsilon/2 - p - a + 1)$ vertices of R, it follows that for all sufficiently large n, the above hypothesis is satisfied with $\lambda = \epsilon/2 - \alpha$. Since $r \geq (\epsilon - \alpha)n - 0(1)$ and $s/r > \lambda/2$, the stated conclusion follows by setting $\beta = \lambda(\epsilon - \alpha)/2$ and $\gamma = \lambda\alpha/2$.

The Main Results

Theorem 1   Let L be a bipartite graph satisfying $\Delta(L) \geq 2$ and having at least two strongly independent edges. Let ϵ be a real number in $(0,1)$ and set $e(n) = \lceil \frac{1}{2}\epsilon n^2 \rceil$. Then there is a positive number $\kappa(L, \epsilon)$ such that

$$\chi_S(n, e(n), L) > \kappa n^2$$

for all sufficiently large n.

Two proofs of this result will be given. The first is based on results from extremal graph theory while the second one relies heavily on Szemerédi's regularity lemma.

First Proof   Let G be an arbitrary graph with n vertices and $e(n)$ edges. We want to show that there is a positive number κ such that in every coloring of $E(G)$ using Kn^2 or fewer colors, there is a non-TMC copy of L. Choose p

and q so that L is a subgraph of $K(p,q)\backslash 2K_2$. Consider the bipartite subgraph of G which is guaranteed by Lemma 3 and let $C = B\backslash X$. Choose $m \geq q$ so that

$$m\binom{\lceil \gamma n \rceil}{q} > \binom{\lceil \alpha n \rceil}{q}.$$

(If n is large enough, $m > 2(\alpha/\gamma)^q$ will suffice to ensure this inequality.) Then for every set of at least m vertices in C, there will be two vertices which have a common neighbourhood in A of at least q vertices. Implicit in the proof of Lemma 3 is the fact that there is a number ζ such that every vertex in A is adjacent to at least $\lceil \zeta n \rceil$ vertices of C. Suppose that $\kappa < \alpha\zeta/m^2$ and G is colored using κn^2 or fewer colors. Then at least m^2 of the $A - C$ edges have the same color and by Lemma 1 the bipartite subgraph contains either m independent monochromatic edges or else a monochromatic star of degree at least m. We now distinguish three cases.

Case (i) - a matching of m independent monochromatic edges. In view of our choice of m, there are two vertices in C which are incident with edges of the monochromatic matching which have a common neighborhood in A of at least q vertices. Now it is clear that there is an embedding of L into G using the two matching edges and this copy of L is non-TMC.

Case (ii) - monochromatic star of degree m with center in A. Then there are two vertices in C which are end vertices of this star and which have a common neighborhood in A of at least q vertices. Thus we have a $K(p+2, q+1)$ subgraph of G which has two edges of the same color and so a non-TMC copy of L.

Case (iii) - monochromatic star of degree m with center in C. In this case, there is a $K(p+1, m)$ subgraph which has a monochromatic star of degree M and again a non-TMC copy of L.

Before giving the other proof let us recall the regularity lemma.

For a graph G with vertex set $V = V(G)$ and disjoint subsets $A, B \subset V$, define $e(A, B)$ as the number of edges of G with one endpoint in each of A and B. The *density* of the pair (a, B) is simply $e(A, B)/|A| \cdot |B|$ and it is denoted by $d(A, B)$.

For $0 < \epsilon < 1$, a pair (A, B) is called *ϵ-regular* if $|d(A_0, B_0) - d(A, B)| \leq \epsilon$ holds for all $A_0 \subset A, B_0 \subset B$ with $|A_0| > \epsilon|A|, |B_0| > \epsilon|B|$.

The regularity lemma (Szemerédi) [Sz]. For every $\epsilon > 0$ there exists an integer $M_0 = M_0(\epsilon)$ such that the vertices of every graph G can be partitioned into $m + 1$ classes $C_0, \cdots, C_m$ for some $m, 1/\epsilon < m < M_0$, so that the following hold: $|C_0| < \epsilon n, |C_1| = \cdots = |C_m|$ and all but $\epsilon\binom{m}{2}$ of the pairs $C_i, C_j)$ are ϵ-regular.

Because we shall often need it, let us state the following immediate consequence of the definition of ϵ-regularity.

Lemma 5 If (C_i, C_j) is a regular pair with density β and $A \subset C_i, |A| > \epsilon|C_i|$, then the number of vertices of C_j which are connected to fewer than $(\beta - \epsilon)|A|$ vertices in A is less than $\epsilon|C_j|$.

Second Proof of Theorem 1 Let $\alpha > 0$ be a fixed real number and suppose that G is a graph with n vertices and at least $\alpha\binom{n}{2}$ edges. Set $\epsilon = (\alpha/3)^l/25$, where $l = |L|$, and apply the regularity lemma.

Suppose without loss of generality that (C_1, C_2) is a ϵ-regular pair with density at least $\alpha/2$. From now on we shall only deal with the bipartite graph H spanned by this pair.

Set $u = |_1| = |C_2|$.

By Lemma 5 all but at most $2\epsilon u^2$ edges of H have both of their endpoints of degree at least $[\frac{\alpha}{2} - \epsilon]u > \frac{\alpha}{3}u$. Call these edges *good* and note that there are more than $\frac{\alpha}{3}u^2$ of them.

Set $k = \lceil 6/\alpha \rceil$ and let r denote the Ramsey number $r(k, k)$ (see [GRS]). Now let $\alpha' > 0$ be a real number satisfying

$$a' M_0^2 kr < \alpha/6.$$

Suppose that the edges of G, and consequently those of H, are colored by at most $\alpha'\binom{n}{2}$ colors. By the choice of a', some color, say red, will occur among the good edges in H with multiplicity at least kr. Therefore, either there are k good red edges forming a star or r good red edges forming a matching.

Lemma 6   Let X be a set of u elements, $0 < \alpha < 1, k = \lceil 6/\alpha \rceil$. Suppose that $B_1, \cdots, B_k \subset X$ satisfy $|B_i| \geq \alpha u/3$. Then there exist $1 \leq i < j \leq k$ with

$$|B_i \cap B_j| > \alpha^2 u/25.$$

Proof   Suppose the contrary and let us bound the size of the union of the sets. Thus,

$$u \geq |B_1 \cup \cdots \cup B_k| \geq \sum |B_i| - \sum |B_i \cap B_j| \geq k\alpha u/3 - \binom{k}{2}\alpha^2 u/25 > 2u - \frac{49}{50}u,$$

which is a contradiction.

If there is a red star of size k then Lemma 6 implies that we can select two of its edges so that the endpoints have at least $\alpha^2 u/25$ common neighbors.

If we have a red matching, say $e_1, \cdots, e_r$, then let us form an auxiliary graph on $\{1, \cdots, r\}$ in the following way.

If the endpoints of e_i and e_j have fewer than $\alpha^2 u/25$ common neighbors in $C_t, t = 1, 2$, then join i and j by an edge of color t. By Lemma 6 no monochromatic complete graph of size k occurs this way. Thus, by the choice of r there are vertices $1 \leq i \leq j \leq r$ which are not connected by an edge. Thus, e_i and e_j have at least $\alpha^2 u/25$ common neighbors in both C_1 and C_2.

From now on the case of the star and the matching are very similar - we just find a "large" complete bipartite graph in the common neighborhood of the two red edges, thereby constructing a copy of L containing two red edges, i.e., which is not TMC.

This is done by repeated applications of Lemma 5. We only deal with the case of the matching, the other being nearly identical.

Let $e_i = (x_1, y_1), e_j = (x_2, y_2)$ with $x_1, x_2 \in C_1$ and $y_1, y_2 \in C_2$. Let $A \subset C_1$ be the set of common neighbors of y_i and y_j. The set $B \subset C_2$ is defined analogously. Recall that l denotes the number of vertices of L.

By Lemma 5 we can choose $x_3 \in A - \{x_1, x_2\}$ such that the neighborhood $B^{(3)}$ of x_3 in B satisfies

$$|B^{(3)}| > \frac{\alpha}{3}|B|.$$

Continuing in this way we find $x_1, \cdots, x_l \in A$ such that they have at least $(\frac{\alpha}{3})^{l-2}|B|$ common neighbors in B. Let $y_1, y_2, \cdots, y_l$ be any l of them including y_1 and y_2.

Then these $2l$ vertices span a bipartite graph which is complete except possibly for the two edges (y_1, x_2) and (x_1, y_2). Therefore it contains a copy of L in which $(x_1, y_1), (x_2, y_2)$ are the two (red) strongly independent edges. This concludes the proof.

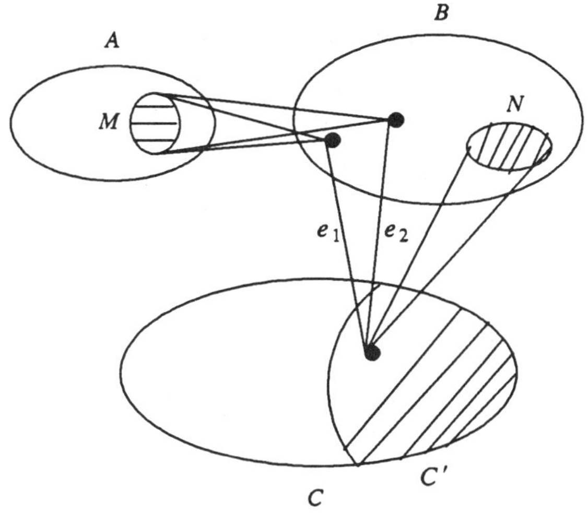

Figure 1

Using slightly more sophisticated arguments we will prove the following generalization in a later paper. Note that for $k = 2$ we get Theorem 1.

Theorem 2   Let L be a graph which is not the disjoint union of complete graphs, and which has two strongly independent edges e and e'. Suppose the vertices of L can be partitioned into k independent sets so that e and e' both have endpoints on the same two sets. Let α be a real number satisfying $\frac{k-2}{k-1} < \alpha < 1$. Then there is a $\beta > 0$ depending only on L and α, such that if $e > \alpha\binom{n}{2}$,

$$\chi_S(n, e, L) \geq \beta \binom{n}{2}.$$

The next result applies to graphs not having two strongly independent edges. We first state an auxiliary result.

Lemma 7   There is a function f such that if $y \leq f(x)$ then

$$\binom{a + b}{a} < n$$

for all a, b which satisfy $1 \le a \le x \log n$ and $1 \le b \le y \log n$.

Proof   First we note that

$$\binom{a+b}{a} < \frac{(a+b)^{a+b}}{a^a b^b}$$

for all $a, b \ge 1$. This follows immediately from Robbins' form of Stirling's formula:

$$n! = \sqrt{2\pi n}\left[\frac{n}{3}\right]^n e^{\alpha_n} \text{ where } \frac{1}{12n+1} < \alpha_n < \frac{1}{12n}.$$

Since $\binom{a+b}{a}$ increases with both a and b, we have by a simple calculation

$$\log \binom{a+b}{a} < \log \left[\frac{(x+y)^{x+y}}{x^x y^y}\right]\log n$$

for all a, b satisfying $1 \le a \le x \log n$ and $1 \le b \le y \log n$. Thus to obtain

$$\binom{a+b}{a} < n,$$

it suffices to have

$$\frac{(x+y)^{x+y}}{x^x y^y} < e. \tag{1}$$

For fixed $x > 0$, the left-hand side of (1) approaches 1 as $y \downarrow 0$ and thus the existence of the desired function f is assured. In fact, writing (1) as

$$x \log(1 + y/x) + y \log(1 + x/y) < 1$$

and using the fact that $\log(1 + t) < \min(t, \sqrt{t})$ for all $t > 0$, it follows that (1) is satisfied if $y + \sqrt{xy} \le 1$ and we may take $f(x) = 1/(x + 2)$.

Theorem 3   Let $0 < \epsilon < \frac{1}{2}$ be fixed and set $e(n) = \lfloor \binom{n}{2} - \epsilon n^2 \rfloor$. Assume that n is appropriately large. There exists a graph G with n vertices and $e(n)$ edges which can be colored using $C(\epsilon)n^2/\log n$ or fewer colors so that for every graph L with no two strongly independent edges, every copy of L in

G is TMC. In particular, for every $q \leq e(n)$ and every graph L with no two strongly independent edges,

$$\chi_S(n, q, L) \leq \frac{C(\epsilon)n^2}{\log n}$$

for all sufficiently large n.

Proof   Let L be an arbitrary graph having no two strongly independent edges. We shall show that there is a graph G with n vertices and at least $e(n)$ edges which can be colored using $C(\epsilon)n^2/\log n$ or fewer colors so that every copy of L in G is TMC. To prove this existence of such a graph, we use the probabilistic method. Consider the random graph $G_{n,p}$ on n vertices in which each edge is chosen with independent probability $p = 1 - \epsilon$. The expected number of edges of $G_{n,p}$ is $(1 - \epsilon)\binom{n}{2}$ and it has at least $\binom{n}{2} - \epsilon n^2$ edges almost surely. Extend the notion of strong independence by defining disjoint pairs of vertices $\{u_1, v_1\}$ and $\{u_2, v_2\}$ in G to be strongly independent if none of four pairs $u_1 u_2, u_1 v_2, v_1 u_2, v_1 v_2$ is an edge in G (whether or not $u_1 v_1$ and/or $u_2 v_2$ are edges). Let X be the number of m-tuples of pairs of vertices in $G_{n,p}$ no two of which are strongly independent. The probability that two disjoint pairs are not strongly independent is $1 - \epsilon^4$ and the expected value of X is

$$E(X)\frac{n!}{2^m m!(n - 2m)!}(1 - \epsilon^4)^{\binom{m}{2}}$$
$$< \left(\frac{en^2(1 - \epsilon^4)^{(m-1)/2}}{2m}\right)^m$$
$$\left(\frac{en^2 \mathrm{enp}(-(m - 1)\epsilon^4/2)}{2m}\right)^m.$$

Now if $m = \lceil 5\log n/\epsilon^4 \rceil$ then

$$E(X) = o(1) \quad (n \to \infty).$$

Thus, for all sufficiently large n, there exist graphs with n vertices and at least $\binom{n}{2} - \epsilon n^2$ edges such that every list of $m = \lceil 5\log n/\epsilon^4 \rceil$ disjoint pairs of vertices has two strongly independent pairs. Pick such a graph and simply refer to it as G.

Partition the edges of G of into n or fewer matching. We claim that the edges of an arbitrary matching in G can be colored using $C(\epsilon)n/\log n$ or fewer colors so that every pair of edges which receive the same color are strongly independent. The truth of this claim means that the edges of G can be colored with $C(\epsilon)n^2/\log n$ or fewer colors so that each copy of L is TMC. If a matching has $n/\log n$ or fewer edges then the claim is trivial; just give each edge a different color. Thus consider a matching M which contains between $n/\log n$ and $n/2$ edges.

There is a positive number $\kappa(\epsilon)$ so that M contains at least $\kappa \log n$ strongly independent edges. To see this, consider the graph with vertex set M in which two vertices are independent whenever the corresponding edges in M are strongly independent. By the previous conclusion, this graph has no clique of size $m = \lceil 5\log n/\epsilon^4 \rceil$. It must then have an independent set of k vertices (so M must contain a set of k strongly independent edges) if $|M| \geq r(k,m)$, where $r(k,m)$ denotes the classical Ramsey number. Since (see [GRS])

$$r(k,m) \leq \binom{k+m-2}{k-1},$$

we certainly get the desired result if $\kappa(\epsilon)$ can be chosen so that $k = \lceil \kappa(\epsilon)\log n \rceil$ satisfies

$$\binom{k+m}{k} < \frac{n}{\log n}.$$

This follows easily from Lemma 7 (by setting $x > 5/\epsilon^r$ and replacing n by $n/\log n$).

Set

$$C(\epsilon) = \frac{1}{2\kappa(\epsilon)} + 1.$$

Since we may remove sets of $\lceil \kappa(\epsilon)\log n \rceil$ or more strongly independent edges, giving each such set a distinct color, until there are at most $n/\log n$ edges of M left, it follows that the edges of M can be colored with $C(\epsilon)n/\log n$ or fewer colors so that every pair of edges receiving the same color are strongly independent. This completes the proof of the earlier claim and thus of the theorem.

Note that we have really done more than we needed to. The coloring of G that we have given actually makes every copy of every graph in the entire class of graphs without strongly independent edges TMC, not just copies of L.

4.   Concluding Remarks

As mentioned in [BEGS], our knowledge of the behavior of $\chi_S(n,e,L)$ is strikingly incomplete even for some very simple graphs L. For example, how does $\chi_S(n,e,C_5)$ behave? We only know:

$$\chi_S(n,e,C_5) > c_n N \text{ for } e = (\frac{1}{4} + \epsilon)n^2, \epsilon > 0,$$

where $c_n \to \infty$ (very slowly) as $n \to \infty$; and

$$\chi_S(n,e,C_5) = 0(n^2/\log n) \text{ for } e = (\frac{1}{2} - \epsilon)n^2, \epsilon > 0.$$

The gap between these two bounds is embarrassing.

For P_4, the path of length 3, the situation is only slightly better.

$$\chi_S(n,e,P_4) > c'_n n \text{ for } e = \epsilon n^2, \ \epsilon > 0$$

where $c'_n \to \infty$ (very slowly) as $n \to \infty$; and

$$\chi_S(n,e,P_4) \leq n \text{ for } e = n^2/\exp(c\sqrt{\log n})$$

for a suitable $c > 0$.

Clearly a great deal more remains to be done in this area.

5. Acknowledgement

The authors wish to express their gratitude to Cecil Rousseau, who essentially wrote the final version of this paper.

REFERENCES

[1] S. A. Burr, P. Erdös, R. L. Graham and V. T. Sós, [BEGS] "Maximal anti-Ramsey graphs and the strong chromatic number" (to appear in the *Journal of Graph Theory*, 1989).

[2] S. A. Burr, R. Faudree, C. C. Rousseau and R. Schelp, [BFRS] "Graph - ical Ramsey Theory (to appear; title tentative).

[3] R. L. Graham, B. L. Rothschild and J. H. Spencer, [GRS] "Ramsey Theory", John Wiley and Sons, New York, 1980.

[4] R. L.Graham, D. E. Knuth and O. Patashnik [GKP], "Concrete Mathematics" Addison- Wesley, Reading, MA, 1989.

[5] F. Harary [H], *Graph Theory*, Addison-Wesley, Reading, MA, 1969.

[6] E. Szemerédi, "Regular partitions of graphs", in *Problèmes Combinatories et Théorie des Graphes*, (J. C. Bermond et. al., eds.) Editions du CNRS, Paris, 1978, pp. 399-401.

Spanning Trails Joining Two Given Edges

Paul A. Catlin
Wayne State University

Hong–Jian Lai
University of Waterloo

ABSTRACT

Let G be a graph and let $e_1, e_2 \in E(G)$. If G has two edge–disjoint spanning trees, then either G has a spanning trail whose first edge is e_1 and last edge is e_2, or $\{e_1, e_2\}$ is an edge cut of G such that both components of $G - \{e_1, e_2\}$ contain at least one edge. This strengthens a result of S.–M. Zhan.

Keywords: Graph theory, spanning trail, spanning eulerian subgraph, hamiltonian line graph, edge–disjoint spanning trees

1. Notation

We shall use the notation of Bondy and Murty [1], except where noted otherwise. We forbid loops but allow multiple edges in graphs. An edge–cut X of a connected graph G is called *essential* if at least two components of $G - X$ contain at least one edge. The symmetric difference of sets R and S is denoted $R \triangle S$.

Let $e_1, e_2 \in E(G)$. A trail in G whose first edge is e_1 and whose last edge is e_2 is called an (e_1, e_2)–*trail.* An (e_1, e_2)–trail T is called a *spanning* (e_1, e_2)–trail if $V(T) = V(G)$ and if every edge of G is incident with an internal vertex of T. For $v_1, v_2 \in V(G)$, a trail in G whose origin is v_1 and whose terminus is v_2 is called a (v_1, v_2)–*trail*, and it is a *spanning* (v_1, v_2)–*trail* if it contains every vertex of G.

The *line graph* of a graph G is the graph $L(G)$ with $E(G)$ as its vertex set, where e_1 and e_2 are adjacent vertices in $L(G)$ whenever they are adjacent edges in G.

2. The Problem

S.–M. Zhan [11] proved the following result:

207

"

Theorem 1 *(Zhan [11]) If G is a 4–edge–connected graph, then for any edges $e_1, e_2 \in E(G)$ there is a spanning (e_1, e_2)–trail in G.*

A graph G is *Hamilton–connected* if for every pair of vertices v_1, v_2 of G, there is a Hamilton (v_1, v_2)–path in G. Combination of Theorems 1 and 2' gives this result:

Corollary 1A *If a graph G is 4–edge–connected, then $L(G)$ is Hamilton–connected.*

In this paper we shall improve on Theorem 1 by using a weaker hypothesis. An exceptional case arises.

Zhan [11] emphasized Hamilton paths in $L(G)$, as in Corollary 1A, and he did not state Theorem 1 in the form given above. However, Theorem 1 can be obtained as a case of Theorem 4 of [11].

Harary and Nash–Williams [5] demonstrated this relationship between trails in G and Hamilton cycles in $L(G)$:

Theorem 2 *(Harary and Nash–Williams [5]) Let G be a graph of order at least 4. Then $L(G)$ is hamiltonian if and only if G has a closed trail Γ such that each edge of $E(G)$ has at least one end in $V(\Gamma)$.*

(In Theorem 2, $V(\Gamma)$ need not equal $V(G)$, and so Γ may not be a spanning trail. Also, we regard a single vertex as a closed trail.) A slight change in the proof of Theorem 2 gives:

Theorem 2' *Let G be a graph and let $e_1, e_2 \in E(G)$. Then $L(G)$ has a Hamilton (e_1, e_2)–path if and only if G has an (e_1, e_2)–trail whose internal vertices contain at least one end of each edge of G.*

For any $k \in N$, it is a consequence of a theorem of Tutte [10] and Nash–Williams [8] that a 2k–edge–connected graph has k edge–disjoint spanning tress (see, e.g., [7] or [4]). For the case $k = 2$ (the case of interest for this paper), Zhan (in the proof of his Lemma 6 [11]) proved the "$\Rightarrow$" part of the next result:

Theorem 3 *(Catlin [3]) Let $k \in N$, let G be a graph with $|E(G)| \geq k$, and let ε_k be the family of all k–element subsets of $E(G)$. Then G is 2k–edge–connected if and only if for any $E \in \varepsilon_k$, the graph $G - E$ has k edge–disjoint spanning trees.*

We shall prove the following result which, by Theorem 3 with $k = 2$, is stronger than Theorem 1:

Theorem 4 *Let G be a graph and let $e_1, e_2 \in E(G)$. If G has two edge–disjoint spanning trees, then exactly one of the following holds:*

    (a) G has a spanning (e_1, e_2)–trail;

    (b) $\{e_1, e_2\}$ is an essential edge–cut of G.

Corollary 4A *Let G be a graph of order at least 3 containing two edge–disjoint spanning trees. Then $L(G)$ is Hamilton–connected if and only if $L(G)$ is 3-connected.*

The proof of Theorem 4 appears in subsequent section, and it requires Theorems 8 and 9 and an application of the following reduction method. Corollary 4A is proved by combining Theorems 2' and 4.

3. The Reduction Method

For any graph H, define

$$O(H) = \{\text{odd–degree vertices of } H\}.$$

Let G be a graph, and let S be an even subset of $V(G)$. An *S–subgraph* Γ of G is a subgraph $\Gamma \subseteq G$ such that both

$$O(\Gamma) = S$$

and

$$G - E(\Gamma) \text{ is connected.}$$

We call G *collapsible* if G has an S–subgraph for every even set $S \subseteq V(G)$. The family of collapsible graphs is denoted CL. If $G \in CL$, then we can set $S = O(G)$ in the definition and see that $G - E(\Gamma)$ is a spanning eulerian subgraph of G, and hence that G has a spanning closed trail, by Euler's Theorem ([1], p. 51). Of course, $K_1 \in CL$, and any nontrivial graph in CL must be 2–edge–connected.

For any graph G, define

$$F(G) = \max_{E \subseteq E(G)} 2[\omega(G - E) - 1] - |E|.$$

Thus, $F(G) = 0$ if and only if G has two edge–disjoint spanning trees (see [8], [9], or [10]). Let $F'(G)$ denote the minimum number of edges that must be added to $E(G)$ in order to create a graph with two edge–disjoint spanning trees.

Proposition *For any graph G, $F(G) = F'(G)$.*

The proof of this proposition appears later.

Theorem 5 *(Catlin [2]) If a graph G satisfies $F'(G) \leq 1$ (equivalently, $F(G) \leq 1$), then exactly one of the following holds:*

    (a) $G \in CL$;

    (b) G has a cut–edge.

Corollary 5A   *(Jaeger [6]) If $F(G) = 0$ then G has a spanning closed trail.*

Let H be a connected subgraph of G and let $S \subseteq V(G)$. Let G/H denote the graph obtained from G by contracting H to a vertex called v_H in G/H. Contractions are defined so that $E(G/H) = E(G) - E(H)$. Define

$$S/H = \begin{cases} S - V(H) & \text{if } |S \cap V(H)| \text{ is even;} \\ S - V(H) \cup \{v_H\} & \text{if } |S \cap V(H)| \text{ is odd.} \end{cases}$$

We shall need the following result:

Theorem 6   *(Catlin [2]) Let G be a graph, let H be a subgraph of G, and let $S \subseteq V(G)$. If $H \in CL$, then G has an S–subgraph if and only if G/H has an (S/H)-subgraph.*

Corollary 6A   *[2] If H is a collapsible subgraph of G, then*
$$G \in CL \Leftrightarrow G/H \in CL.$$

Corollary 6B   *[2] If H is a collapsible subgraph of G, then G has a spanning closed trail if and only if G/H has a spanning closed trail.*

For a graph G, let $H_1, H_2, \ldots, H_c$ be the maximal collapsible subgraphs of G. We proved in [2] that these H_i's are uniquely determined and pairwise vertex–disjoint. Each vertex of G is in some H_i ($1 \le i \le c$), because $K_1 \in CL$. Let G' denote the graph of order c obtained from G by contracting each H_i to a distinct vertex ($1 \le i \le c$). We call G' the *reduction* of G. If G has no nontrivial subgraph in CL, then we call G *reduced*. We also proved [2] that the reduction of G is reduced. Examples of reduced graphs include forests and $K_{2,t}$ ($t \ge 2$). By Corollary 6B, G has a spanning closed trail if and only if the reduction G' has a spanning closed trail.

4. Associated Results

Lemma 7   *([2], Lemma 1) Let H be a graph and let $S \subseteq V(H)$ have evenly many vertices in each component of H. Then there is a subgraph $\Gamma \subseteq H$ such that $O(\Gamma) = S$.*

Proof   Let $P_1, P_2, \ldots, P_m$ be $m = |S|/2$ paths in H that join the vertices of S in distinct pairs. Thus, each $x \in S$ is an end of exactly one of the m paths. Define Γ by the rule that $e \in E(\Gamma)$ if and only if e lies in an odd number of the paths P_i ($1 \le i \le m$). $\square$

We say that an edge $e \in E(G)$ is *subdivided* when it is replaced by a path of length 2 whose internal vertex, denoted $v(e)$, has degree 2 in the resulting graph. The process of taking an edge e and replacing it by that length 2 path is called *subdividing* e. For a graph G and edges $e_1, e_2 \in E(G)$, let $G(e_1)$ denote the graph obtained from G by subdividing e_1, and let $G(e_1, e_2)$ denote the graph obtained from G by subdividing both e_1 and e_2. Thus,

$$V(G(e_1, e_2)) - V(G) = \{v(e_1), v(e_2)\}.$$

Theorem 8 *Let G be a graph and let $e_1, e_2 \in E(G)$. If G has edge–disjoint spanning trees Γ_1 and Γ_2 such that $e_1, e_2 \notin E(\Gamma_1)$, then G has a spanning (e_1, e_2)– trail.*

Proof Suppose that G, e_1, e_2, Γ_1, and Γ_2 satisfy the hypothesis. Denote $H = G - E(\Gamma_1)$. Since $E(\Gamma_2) \subseteq E(H)$, H is connected, and so by Lemma 7, there is a subgraph Γ of $H(e_1, e_2)$ such that

$$O(\Gamma) = O(G) \cup \{v(e_1), v(e_2)\}.$$

It follows that

$$O(G(e_1, e_2) - E(\Gamma)) = \{v(e_1), v(e_2)\},$$

and hence $G(e_1, e_2) - E(\Gamma)$ has an Euler trail joining $v(e_1)$ and $v(e_2)$. This Euler trail induces a spanning (e_1, e_2)–trail in G. ❏

Note that Theorem 1 follows from Theorem 8 and the case $k = 2$ of Theorem 3. If G satisfies the hypothesis of Theorem 1, then by Theorem 3, edge–disjoint spanning trees Γ_1 and Γ_2 can be chosen so that $e_1, e_2 \notin E(\Gamma_1)$, and so the hypothesis of Theorem 8 is satisfied. This is essentially the method used by Zhan [11] to prove Theorem 1.

In Theorem 4, we consider a graph G having two edge–disjoint spanning trees, say Γ_1 and Γ_2. To apply Theorem 8, we would want to know whether Γ_1 and Γ_2 can be chosen so that $e_1, e_2 \notin E(\Gamma_1)$. This motivates the following definition and theorem.

Let G be a graph and let $e_1, e_2 \in E(G)$. An $\{e_1, e_2\}$–*forbidden subgraph* G_0 is any subgraph G_0 of G such that

    (i) $\{e_1, e_2\}$ is an edge–cut of G_0; and

    (ii) $F(G_0) = 0$.

An $\{e_1, e_2\}$–forbidden subgraph will also be called a *forbidden subgraph* if there is no confusion about the values of e_1 and e_2.

Theorem 9 *Let G be a graph, and let $e_1, e_2 \in E(G)$. If $F(G) = 0$ and if G has no $\{e_1, e_2\}$–forbidden subgraph, then G has 2 edge–disjoint spanning trees Γ_1 and Γ_2 such that $e_1, e_2 \notin E(\Gamma_1)$.*

Of course, if e_1 and e_2 are parallel edges in G, then $G_0 = G[\{e_1, e_2\}]$ is a forbidden subgraph. If $\{e_1, e_2\}$ is an edge–cut of G and if $F(G) = 0$, then $G_0 = G$ is a forbidden subgraph. In these two cases, it is obvious that the conclusion of Theorem 9 fails. There are other instances when forbidden subgraphs cause the conclusion of Theorem 9 to fail.

Theorem 10   *Let G be a graph and let $e_1, e_2 \in E(G)$. If $F(G) = 0$ and if G has no $\{e_1, e_2\}$–forbidden subgraph, then $G(e_1, e_2) \in CL$ (i.e., the reduction of $G(e_1, e_2)$ is K_1).*

5. Proof of Theorem 9

Lemma 11   *Let G be a graph and let H be a connected subgraph of G. If $F(H) = 0$ and $F(G/H) = 0$, then H has edge–disjoint spanning trees, say U_1 and U_2, and G/H has edge–disjoint spanning trees, say T_1 and T_2. The pair (Γ_1, Γ_2) with $\Gamma_i = G[E(U_i) \cup E(T_j)]$ $(i, j \in \{1, 2\})$ is a pair of edge–disjoint spanning trees of G.*

Proof   Suppose $F(H) = 0$ and $F(G/H) = 0$. By the theorem of Tutte [10] and Nash–Williams [8], H and G/H each have two edge–disjoint spanning trees. These trees can be combined as indicated to form the trees Γ_1 and Γ_2 that span G. $\square$

Lemma 12   *If G is a counterexample to Theorem 9 with*

(1) $|V(G)| + |E(G)|$ *minimized,*

then for any proper nontrivial subgraph H of G, $F(H) \geq 1$.

Proof   By way of contradiction, suppose that G is a counterexample to Theorem 9 that satisfies (1), and let H be a nontrivial proper subgraph of G with

(2) $F(H) = 0$.

It follows from (2) that H is connected.

Since G is a counterexample to Theorem 9,

(3) $F(G) = 0$.

By (3), G has two edge–disjoint spanning trees, and thus G/H does also. Therefore,

(4) $F(G/H) = 0$.

Case 1   Suppose that $|V(H)| < |V(G)|$ and $\{e_1, e_2\} \cap E(H) = \varnothing$.

Suppose, by way of contradiction, that G_0 is a subgraph of G/H with $F(G_0) = 0$ and with $\{e_1, e_2\}$ as an edge–cut, i.e., that G_0 is a forbidden subgraph of G/H. Since G satisfies the hypothesis of Theorem 9, G_0 is not a subgraph of G, and so the vertex v_H of G/H corresponding to H must be in G_0. By (2) and $F(G_0) = 0$, both

H and G_0 have two edge–disjoint spanning trees. But then $F(G[E(G_0) \cup E(H)]) = 0$ and $\{e_1, e_2\}$ is an edge–cut of $G[E(G_0) \cup E(H)]$. Thus, $G[E(G_0) \cup E(H)]$ is a forbidden subgraph of G, contrary to the assumption that G is a counterexample.

Therefore, G/H has no forbidden subgraph, and since G is a smallest counterexample to Theorem 9, G/H has edge–disjoint spanning trees Γ'_1 and Γ'_2 with $e_1, e_2 \notin E(\Gamma'_1)$. Since $F(H) = 0$, Lemma 11 implies that Γ'_1 and Γ'_2 induce edge–disjoint spanning trees Γ_1 and Γ_2 of G with $e_1, e_2 \notin E(\Gamma_1)$. Thus, G satisfies the conclusion of Theorem 9, a contradiction.

Case 2   Suppose the $|V(H)| < |V(G)|$ and $|\{e_1, e_2\} \cap E(H)| = 1$.

Without loss of generality, suppose that
$$e_1 \in E(H) \text{ and } e_2 \notin E(H).$$
Let v_H be the vertex of G/H onto which H is contracted. By (2), there are edge–disjoint spanning trees U_1 and U_2 of H, with $e_1 \notin E(U_1)$; and by (4) there are edge–disjoint spanning trees T_1 and T_2 of G/H with $e_2 \notin E(T_1)$. By Lemma 11, G has the edge–disjoint spanning trees
$$\Gamma_i = G[E(T_i) \cup E(U_i)] \quad (1 \le i \le 2).$$
Thus, Γ_1 and Γ_2 satisfy the conclusion of Theorem 9, a contradiction.

Case 3   Suppose that H is a proper subgraph of G (possibly a spanning subgraph), such that $e_1, e_2 \in E(H)$.

No forbidden subgraph H_0 exists in H, for otherwise H_0 would be a forbidden subgraph of G, and G would not be a counterexample.

Since H has no forbidden subgraph, (2) and the minimality of G imply that the proper subgraph H has edge–disjoint spanning trees, say U_1 and U_2, with $e_1, e_2 \notin E(U_1)$. If H is a spanning subgraph of G, then $\Gamma_1 = U_1$ and $\Gamma_2 = U_2$ satisfy the conclusion of Theorem 9. If H is not a spanning subgraph of G, then by Lemma 11 and since $F(G/H) = 0$, $E(U_1)$ and $E(U_2)$ are contained in edge–disjoint spanning trees Γ_1 and Γ_2, say, of G, where $e_1, e_2 \notin E(\Gamma_1)$. Again G satisfies the conclusion of Theorem 9, contrary to our assumption.

Case 4   Suppose that H is a spanning proper subgraph of G and that $e_i \notin E(H)$ for some $i \in \{1, 2\}$.

Define $G' = G - e_i$. Since H is a spanning subgraph of G', G' has two edge–disjoint spanning trees, say Γ_1 and Γ_2, and since neither contains e_i, there is no loss of generality in assuming that $e_1, e_2 \in E(\Gamma_1)$. Thus, G satisfies the conclusion of Theorem 9, a contradiction. ❑

Proof of Theorem 9   By way of contradiction, let G and $\{e_1, e_2\}$ be a counterexample of Theorem 9 satisfying (1), i.e., a smallest counterexample. By Lemma 12, if G'' is a proper nontrivial subgraph of G, then

$$(5) \qquad\qquad F(G'') \geq 1.$$

Since G is a counterexample to Theorem 9, G has no $\{e_1, e_2\}$–forbidden subgraph, and so

$$G - \{e_1, e_2\} \text{ is connected.}$$

Hence, G has a spanning tree T with

$$E(T) \subseteq E(G - \{e_1, e_2\}).$$

Let $F_1, F_2, ..., F_k$ be the $\omega(G - E(T)) = k$ components of $G - E(T)$.

If $G - E(T)$ is a forest, then G is exactly $k - 1$ edges short of having two edge–disjoint spanning trees, and so $0 = F(G) = k - 1$. Therefore, $k = 1$ and so $\Gamma_1 = T$ and $\Gamma_2 = F_1$ are spanning trees that satisfy the conclusion of Theorem 9.

Therefore, we suppose that $G - E(T)$ is not a forest, and hence that at least one component F_i has a cycle. If F_i has at least one cycle, then we call F_i a *cyclic component* $(1 \leq i \leq k)$. Define

$$\sigma(T) = \min_{1 \leq i \leq k} \{|E(F_i)| : F_i \text{ is a cyclic component of } G - E(T)\}.$$

Choose a spanning tree T of $G - \{e_1, e_2\}$ so that

$$(6) \qquad\qquad \omega(G - E(T)) \text{ is minimized}$$

and, subject to (6), so that

$$(7) \qquad\qquad \sigma(T) \text{ is minimized.}$$

Let H be a cyclic component of $G - E(T)$ with

$$|E(H)| = \sigma(T).$$

Let $T_1, T_2, ..., T_m$ denote the components of $T[V(H)]$, and denote

$$V_i = V(T_i) \ (1 \leq i \leq m).$$

Set $H^* = G[V(H)]$. Since H is cyclic component, $|E(H)| \geq |V(H)|$, and so

$$(8) \qquad |E(H^*)| = |E(H)| + \sum_{i=1}^{m} |E(T_i)|$$

$$\geq |V(H)| + (|V(H)| - m)$$

$$= 2|V(H)| - m.$$

Let E be a subset of $E(H^*)$ such that $F(H^*) = 2[\omega(H^* - E) - 1] - |E|$. If H' is any component of $H^* - E$, and E' a subset of $E(H')$ such that

$$F(H') = 2[\omega(H' - E') - 1] - |E'|,$$

then

$$F(H^*) \geq 2[\omega(H^* - (E \cup E')) - 1] - |E \cup E'|$$

$$= 2[\omega(H^* - E) + \omega(H' - E') - 2] - |E| - |E'|$$

$$= 2[\omega(H^* - E) - 1] - |E| + 2[\omega(H' - E') - 1] - |E'|$$

$$= F(H^*) + F(H'),$$

and hence $F(H') = 0$. Using (5), we conclude that every component of $H^* - E$ is trivial, implying that $E = E(H^*)$ and $F(H^*) = 2[|V(H^*)| - 1] - |E(H^*)|$. Again by (5),

$$(9) \qquad |E(H^*)| = 2|V(H^*)| - 2 - F(H^*) \leq 2|V(H^*)| - 3.$$

Combination of (8) and (9) gives

$$(10) \qquad m \geq 3.$$

For $i, j \in \{1, 2, ..., m\}$, denote

$$Y_{i,j} = \{uv \in E(H) : u \in V_i, v \in V_j\},$$

and denote

$$Y = \bigcup_{i \neq j} Y_{i,j}.$$

Since H is connected, $Y \neq \varnothing$.

Case 1 Suppose $Y - \{e_1, e_2\} \neq \varnothing$. Therefore, $Y - \{e_1, e_2\}$ has an edge z_1z_2, say, and without loss of generality, suppose that $z_1 \in V_1$ and $z_2 \in V_2$. Let C be the unique cycle of $T + z_1z_2$. There are edges $u_1v_1, u_2v_2 \in E(T)$ with $u_i \notin V(H)$ and $v_i \in V_i$ $(1 \leq i \leq 2)$.

1A Suppose that z_1z_2 is a cut–edge of H.

Since H is cyclic, one of the components of $H - z_1z_2$ has a cycle. Without loss of generality, we assume that the component of $H - z_1z_2$ containing z_1 has a cycle. Then

$$T' = T + z_1z_2 - u_2v_2$$

is a spanning tree of $G - \{e_1, e_2\}$ such that

$$\sigma(T') \leq \sigma(T) - 1 \text{ and } \omega(G - E(T)) = \omega(G - E(T')),$$

contrary to (6) and (7).

1B Suppose that z_1z_2 is not a cut–edge of H. Hence, $H - z_1z_2$ is connected. Define

$$T'' = T + z_1z_2 - u_2v_2$$

Then

$$\omega(G - E(T'')) = \omega(G - E(T)) - 1,$$

contrary to (6).

Case 2 Suppose that $Y \subseteq \{e_1, e_2\}$. Since H is connected, (10) forces $Y = \{e_1, e_2\}$, $m = 3$, and $H[V_1]$, $H[V_2]$, and $H[V_3]$ must all be connected. Therefore, each $G[V_i]$ is an edge–disjoint union of the spanning connected graphs T_i and $H[V_i]$ $(1 \leq i \leq 3)$, and hence $F(G[V_i]) = 0$. By (5) with $G'' = G[V_i]$, this forces $G[V_i] = K_1$ $(1 \leq i \leq 3)$.

Hence, H is acyclic, a contradiction. This completes Case 2 and the proof of Theorem 9. $\square$

6. Proof of Theorem 10

Let G, e_1, and e_2 satisfy the hypothesis of Theorem 10. The hypothesis of Theorem 9 holds, and so G has two edge–disjoint spanning trees, say T and U, such that

$$e_1, e_2 \notin E(T).$$

If $e_1 \notin E(U)$ or if $e_2 \notin E(U)$, then $F(G(e_1, e_2)) \leq 1$ and hence by Theorem 5, $G(e_1, e_2) \in \mathcal{CL}$. Thus, assume that $e_1, e_2 \in E(U)$. To prove $G(e_1, e_2) \in \mathcal{CL}$, we must prove that $G(e_1, e_2)$ has an S–subgraph Γ, for any even set $S \subseteq V(G(e_1, e_2))$. Let S be an even subset of $V(G(e_1, e_2))$.

Case 1 Suppose that $v(e_1), v(e_2) \notin S$. By Lemma 7, there is a subgraph Γ in T with $O(\Gamma) = S$. Since $E(\Gamma) \subseteq E(G(e_1, e_2))$ and since $U(e_1, e_2)$ is a spanning tree in $G(e_1, e_2) - E(\Gamma)$, Γ is an S–subgraph of $G(e_1, e_2)$.

Case 2 Suppose $v(e_1), v(e_2) \in S$. By Lemma 7, $U(e_1, e_2)$ has a subgraph Γ with $O(\Gamma) = S$. Then T is a spanning tree of $G(e_1, e_2) - E(\Gamma) - \{v(e_1), v(e_2)\}$, and since $d(v(e_i)) = 1$ $(1 \leq i \leq 2)$ in $G(e_1, e_2) - E(\Gamma)$, the subgraph $G(e_1, e_2) - E(\Gamma)$ is connected and spans $G(e_1, e_2)$. Therefore, Γ is an S–subgraph of $G(e_1, e_2)$.

Case 3 Suppose that $v(e_1) \in S$ and $v(e_2) \notin S$. Let C be the unique cycle of $T + e_2$ in G. Then $C - e_2$ contains an edge, say e_3, that joins the two components of $U - e_2$ in G. Define the edge–disjoint spanning trees

$$T' = T + e_2 - e_3, \quad U' = U - e_2 + e_3.$$

Thus, $e_1 \in E(U')$, $e_2 \in E(T')$, and $S \subseteq V(U'(e_1))$. By Lemma 7, $U'(e_1)$ has a subgraph Γ such that $O(\Gamma) = S$. Then $G(e_1) - E(\Gamma)$ is a spanning connected subgraph of $G(e_1)$ containing e_2, and so $G(e_1, e_2) - E(\Gamma)$ is a spanning connected subgraph of $G(e_1, e_2)$. Hence Γ is an S–subgraph of $G(e_1, e_2)$.

The case $v(e_1) \notin S, v(e_2) \in S$ is similar. This proves Theorem 10. $\square$

7. Proof of Theorem 4

Lemma 13 *Let G be a graph, let H be a subgraph of G, let S be a subset of $V(G)$, and let R be an even subset of $V(H)$. If $H \in \mathcal{CL}$, then G has an S–subgraph if and only if G has an $(S \triangle R)$–subgraph.*

Proof   Suppose G, H, R, and S satisfy the hypothesis. Since $H \in CL$, Theorem 6 implies these two equivalences:

$$G \text{ has an } S\text{–subgraph} \Leftrightarrow G/H \text{ has an } (S/H)\text{–subgraph};$$
$$G \text{ has an } (S \triangle R)\text{–subgraph} \Leftrightarrow G/H \text{ has an } ((S \triangle R)/H)\text{–subgraph}.$$

By the definition of S/H and since R is an even subset of V(H),

$$S/H = (S \triangle R)/H,$$

and Lemma 13 follows. $\square$

Proof of Theorem 4   Suppose that G and $\{e_1, e_2\}$ satisfy the hypothesis of Theorem 4. Thus, $F(G) = 0$.

Suppose that G has no $\{e_1, e_2\}$–forbidden subgraph. By Theorems 9 and 8, G has a spanning (e_1, e_2)–trail, and (a) of Theorem 4 holds.

Next, suppose that G has an $\{e_1, e_2\}$–forbidden subgraph, say G_0. If $\{e_1, e_2\}$ is an edge–cut of G, then either (b) of Theorem 4 holds, or one component of G – $\{e_1, e_2\}$ is a single vertex. In the latter case, Corollary 5A implies (a) of Theorem 4. Suppose henceforth that $\{e_1, e_2\}$ is not an edge–cut of G.

Let G_1 and G_2 be the two components of $G_0 - \{e_1, e_2\}$. Since G_0 is a forbidden subgraph, $F(G_0) = 0$, and so G_0 has two edge–disjoint spanning trees. It follows that each component G_1 and G_2 of $G_0 - \{e_1, e_2\}$ has two edge–disjoint spanning trees, and so

$$(11) \qquad\qquad\qquad F(G_1) = F(G_2) = 0.$$

By (11) and Theorem 5,

$$(12) \qquad\qquad\qquad G_1, G_2 \in CL.$$

Since $\{e_1, e_2\}$ is not an edge–cut of G, it follows that $G - e_2$ is 2–edge–connected. Also, $F(G) = 0$ gives $F(G - e_2) \le 1$, and so $G - e_2 \in CL$, by Theorem 5. By setting $S = O(G - e_2)$ in the definition of CL, we see that $G - e_2$ has a spanning eulerian subgraph H, say. Define $e_1 = x_1 x_2$ and $e_2 = y_1 y_2$, where

$$x_1, y_1 \in V(G_1) \text{ and } x_2, y_2 \in V(G_2).$$

Case 1   Suppose that $e_1 \in E(H)$.

Then $H - e_1$ has an eulerian (x_1, x_2)–trail that spans $G - \{e_1, e_2\}$. Define

$$\Gamma = G - \{e_1, e_2\} - E(H),$$

and set $S = O(G - \{e_1, e_2\}) \triangle \{x_1, x_2\}$, i.e., $S = O(G - \{e_1, e_2\}) \triangle O(H - e_1)$. Then Γ is an S–subgraph of $G - \{e_1, e_2\}$, and Γ is the complement of $H - e_1$ in $G - \{e_1, e_2\}$. Set

$$R = \begin{cases} \{e_1, e_2\} & \text{if } x_2 \ne y_2 \\ \varnothing & \text{if } x_2 = y_2 \end{cases}$$

Then by Lemma 13, by (12), and since R is an even subset of $V(G_2)$, it follows that $G - \{e_1, e_2\}$ has an $(S \triangle R)$–subgraph Γ', say. Note that Γ' is the complement in $G - \{e_1, e_2\}$ of a spanning connected subgraph H', say, where

$$O(H') = O(H - e_1) \triangle R = \{x_1, y_2\}.$$

By Euler's Theorem ([1], p. 52), H' contains an eulerian (x_1, y_2)–trail that spans $V(G - \{e_1, e_2\})$. By adding e_1 and e_2 to this trail, we extend it to a spanning (e_1, e_2)–trail of G.

Case 2   Suppose that $e_1 \notin E(H)$.

We imitate Case 1. Define

$$\Gamma = G - \{e_1, e_2\} - E(H),$$

and

$$S = O(G - \{e_1, e_2\}),$$

so that Γ is an S–subgraph of $G - \{e_1, e_2\}$ and Γ is the complement in $G - \{e_1, e_2\}$ of H. Set

$$R = \begin{cases} \{e_1, e_2\} & \text{if } x_2 \neq y_2 \\ \varnothing & \text{if } x_2 = y_2 \end{cases}$$

By Lemma 13, by (12), and since $R = \{x_2, y_2\}$ is an even subset of $V(G_2)$ (set $R = \varnothing$ if $x_2 = y_2$), it follows that $G - \{e_1, e_2\}$ has an $(S \triangle R)$–subgraph Γ' that is the complement in $G - \{e_1, e_2\}$ of a spanning (x_2, y_2)–trail. By adding e_1 at x_2 and e_2 at y_2, we extend this trail to form a spanning (e_1, e_2)–trail in G. This completes the proof of Theorem 4. ❑

8. Proof of Proposition

In [3], we used the terminology

$$\mathcal{S}_{2,t} = \{G \mid F'(G) \leq t\},$$

where $t \geq 0$, and we proved that if a graph G has a subgraph H with 2 edge–disjoint spanning trees, then

$$(13) \hspace{3cm} F'(G) = F'(G/H).$$

(In [3], this was expressed by saying that $\mathcal{S}_{2,0}$ is the "kernel" of $\mathcal{S}_{2,t}$, where $t = F'(G)$, and where "kernel" is defined in [3].) We shall now also show

$$(14) \hspace{3cm} F(G) = F(G/H),$$

when H is a subgraph of G having two edge–disjoint spanning trees. Let $E'' \subseteq E(G/H)$ be such that

$$F(G/H) = 2[\omega((G/H) - E'') - 1] - |E''|.$$

Since $E'' \subseteq E(G/H) \subseteq E(G)$,

$$\begin{aligned} F(G/H) &= 2[\omega((G/H) - E'') - 1] - |E''| \\ &= 2[\omega(G - E'') - 1] - |E''| \leq F(G) \end{aligned}$$

Hence,

(15) $$F(G/H) \le F(G)$$

Suppose next that the subset $E' \subseteq E(G)$ is minimized, subject to

(16) $$F(G) = 2[\omega(G - E') - 1] - |E'|.$$

Let

$$E_1 = \{e \in E' \mid \text{both ends of } e \text{ are in } V(H)\} \text{ and } E_2 = E' - E_1.$$

Since H has two edge–disjoint spanning trees,

(17) $$F(H) = 0,$$

by a theorem of Tutte [10] and Nash–Williams [8]. By (16) and (17),

$$\begin{aligned}
F(G) &= 2[\omega(G - E') - 1] - |E'| \\
&= 2[\omega(G - E_2) - 1] - |E_2| + 2[\omega(H - E_1) - 1] - |E_1| \\
&\le 2[\omega(G - E_2) - 1] - |E_2| + F(H) \le F(G)
\end{aligned}$$

By the minimality of E', $E_1 = \varnothing$ and $E' = E_2$. Hence $E' \subseteq E(G/H)$ and so

(18) $$\begin{aligned}
F(G) &= 2[\omega(G - E') - 1] - |E'| \\
&\le 2[\omega(G/H - E') - 1] - |E'| \le F(G/H).
\end{aligned}$$

Combination of (15) and (18) yields (14).

The *arboricity* $a(G)$ of a graph G is the minimum number of edge–disjoint spanning trees whose union is G. Nash–William [9] proved

$$a(G) = \max_{H \subseteq G} \left\lceil \frac{|E(H)|}{|V(H)| - 1} \right\rceil,$$

where the maximum is taken over all nontrivial subgraphs H of G. In [3] (Theorem 11), we used this to show that if no nontrivial subgraph H of G has two edge–disjoint spanning trees, then $a(G) \le 2$.

By way of contradiction, suppose that G is the smallest graph with $F(G) \ne F'(G)$. If G has a nontrivial subgraph H that contains two edge–disjoint spanning trees, then by (13) and (14),

$$F(G/H) = F(G) \ne F'(G) = F'(G/H),$$

contrary to the minimality of G. Thus, G has no nontrivial subgraph containing two edge–disjoint spanning trees, and it follows that $a(G) \le 2$, by prior remarks.

Since the arboricity of G is at most 2, the definition of $F'(G)$ yields

$$F'(G) = 2(|V(G)| - 1) - |E(G)|.$$

Let E be a subset of $E(G)$ that attains the maximum in the definition of $F(G)$:

$$F(G) = 2[\omega(G - E) - 1] - |E|,$$

and let $H_1, H_2, \ldots, H_c$ be the $c = \omega(G - E)$ components of $G - E$. To prove $F(G) = F'(G)$, it suffices to prove that each H_i $(1 \le i \le c)$ is a K_1, because this would imply $c = \omega(G - E) = |V(G)| - 1$. Since no nontrivial subgraph of G has two edge–disjoint spanning trees, it suffices to prove that H_i has two edge–disjoint spanning trees, for then H_i must be trivial. By way of contradiction, therefore, suppose that H_i

does not have two edge–disjoint spanning trees. By the theorem of Tutte [10] and Nash–Williams [8], $F(H_i) > 0$, and so there is a subset $X \subseteq E(H_i)$ such that

$$2[\omega(H_i - X) - 1] - |X| > 0.$$

Then

$$2[\omega(G - (E \cup X)) - 1] - |E \cup X|$$
$$= 2[\omega(G - E) - 1 + \omega(H_i - X) - 1] - |E| - |X|$$
$$= F(G) + 2[\omega(H_i - X) - 1] - |X| > F(G),$$

a contradiction. As already remarked, the Proposition follows. $\square$

9. Examples

The hypothesis of Theorem 4, that G has two edge–disjoint spanning trees, is equivalent to the statement $F(G) = 0$. The three connected graphs illustrated in Figure 1 have $F(G) = 1$. For the designated edges e_1 and e_2, each fails to satisfy the conclusion of Theorem 4, except when $\{e_1, e_2\}$ is an essential edge–cut in the first graph. Also, each has no Hamilton (e_1, e_2)–path in its line graph, except for the first one when the small circle represents a lone vertex.

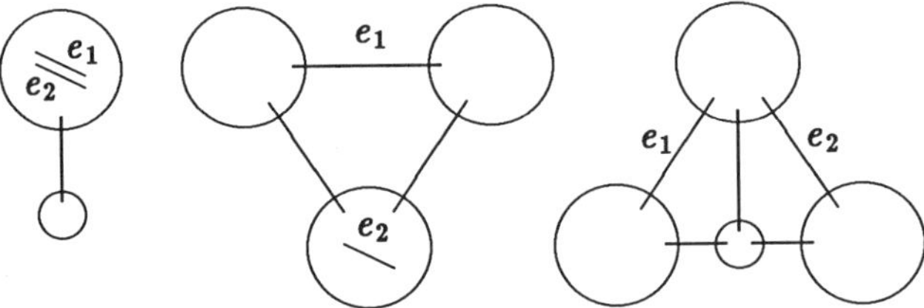

Figure 1: Three graphs with no spanning (e_1, e_2)–trail

In each graph of Figures 1 and 2, a circle denotes a subgraph having two edge–disjoint spanning trees, and if the circle is large, that subgraph is necessarily nontrivial.

In Figure 2, we give a typical example of a graph G with $F(G) = 0$ that has an $\{e_1, e_2\}$–forbidden subgraph G_0 such that $G_0 \neq G$ and $G_0 \neq G[\{e_1, e_2\}]$, where G has no edge–disjoint spanning trees Γ_1 and Γ_2 such that $e_1, e_2 \notin E(\Gamma_1)$. In this figure, $G_0 = G[V(G_1) \cup V(G_2)]$, and e_1 and e_2 are not parallel.

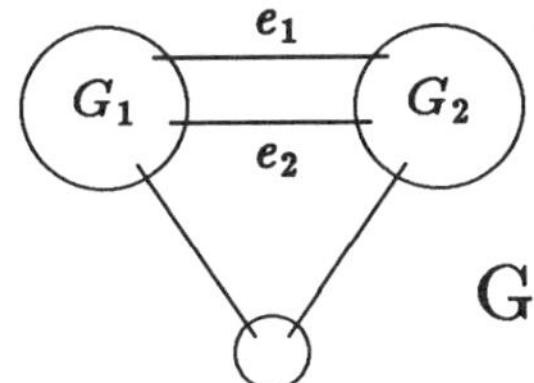

Figure 2

For the graph G of Figure 2, the reduction of $G(e_1, e_2)$ is $K_{2,3}$, which is not collapsible. Thus, the hypothesis in Theorem 10 that G has no $\{e_1, e_2\}$–forbidden subgraph is needed. We conjecture that if that hypothesis were omitted from Theorem 10, then the reduction of G (in the conclusion of Theorem 10) would be either K_1 or $K_{2,t}$ ($t \geq 2$). This would follow from a conjecture of Catlin, that if a connected graph G satisfies $F(G) = 2$, then the reduction of G is either K_1, K_2, or $K_{2,t}$ ($t \geq 1$).

Let $t \geq 3$ and let G_t denote the graph containing parallel edges e_1 and e_2 such that $G_t(e_1, e_2)$ is $K_{2,t}$. Then $F(G_t) = 0$ and G_t satisfies (a) of Theorem 4. However, every spanning (e_1, e_2)–trail in G is open (respectively, closed) if t is odd (respectively, even). Thus, even when e_1 and e_2 are adjacent and $F(G) = 0$ and (a) of Theorem 4 holds, we cannot guarantee that there is always a closed (resp., open) spanning (e_1, e_2)–trail in G.

Call the graph G *essentially 3–edge–connected* if for any $e, e' \in E(G)$, at most one component of $G - \{e, e'\}$ has an edge. A corollary of Theorem 4 is that if $F(G) = 0$ and if G is essentially 3–edge–connected, then G has a spanning (e_1, e_2)–trail, for any $e_1, e_2 \in E(G)$. However, this corollary does not appear to be sharp, because it may be possible to substitute $F(G) \leq 1$ for $F(G) = 0$ in the hypothesis. It would not be possible to substitute $F(G) \leq 2$, though, because $G = K_{2,t}$ ($t \geq 3$) satisfies $F(G) = 2$, and G is essentially 3–edge–connected but does not have a spanning (e_1, e_2)–trail when t is odd and e_1 and e_2 are incident with a common divalent vertex of G.

Theorem 1 and Corollary 1A are best–possible in the sense that 3–edge–connectedness would not suffice. Let G be obtained by attaching ten disjoint copies of K_4 to Petersen graph, with just one K_4 attached at each vertex of the Petersen graph. Since the Petersen graph is not hamiltonian, there are edges of G (in a common K_4) such that G has no spanning (e_1, e_2)–trail.

Acknowledgment: The authors are grateful to the referee for many useful suggestions.

REFERENCES

[1]   J.A. Bondy and U.S.R. Murty, "Graph Theory with Applications." American Elsevier, New York (1976).

[2]   P.A. Catlin, A reduction method to find spanning eulerian subgraphs. J. Graph Theory 12 (1988) 29 – 44.

[3]   P.A. Catlin, The reduction of graph families closed under contraction. Submitted.

[4]   D. Gusfield, Connectivity and edge–disjoint spanning trees. Information Processing Letters 16 (1983) 87 – 89.

[5]   F. Harary and C. St. J.A. Nash–Williams, On eulerian and hamiltonian graphs and line graphs. Canad. Math. Bull. 8 (1965) 701 – 709.

[6]   F. Jaeger, A note on subeulerian graphs, J. Graph Theory 3 (1979) 91 – 93.

[7]   S. Kundu, Bounds on the number of edge–disjoint spanning trees. J. Combinatorial Theory (B) 17 (1974) 199 – 203.

[8]   C. St. J.A. Nash–Williams, Edge disjoint spanning trees of finite graphs. J. London Math. Soc. 36 (1961) 445 – 450.

[9]   C. St. J.A. Nash–Williams, Decompositions of finite graphs into forests. J. London Math. Soc. 39 (1964) 12.

[10] W.T. Tutte, On the problem of decomposing a graph into n connected factors. J. London Math. Soc. 36 (1961) 221 – 230.

[11] S.–M. Zhan, Hamiltonian connectedness of line graphs. Ars Combinatoria 22 (1986) 89 – 95.

THE ISOMORPHIC FACTORIZATIONS OF A CLASS OF 2-CIRCULANT GRAPHS

Jingfa Chen

Jianfang Wang

Academia Sinica, Beijing

Yuqiang Li

Henan University, Kaifeng

ABSTRACT

G $(2n; s, g, f)$ *denotes the 2-circulant graph with symbol* $(S.G.F)$. $(2n; S, 1, S)$ *is simply written* $G(2n; S, S)$. *In this paper, we proved that* $G(2n; S, S)$ *is 1-factorizable and if* $|S| = p$ *is prime then* $G(2n; S, S)$ *is rational.*

1.   Introduction

Let $\subseteq \{1, 2, \cdots, n-1\}$ with the property that $i \in S$ implies $n - i \in S$. The circulant graph $G(n; S)$ is defined so that

$$V(G(n; S)) = \{0, 1, \cdots, n - 1\}$$

$$E(G(n; S)) = \{(i, j) | j - i \in S\}.$$

where we take $j - i$ modulo n.

We write Z_n for the ring of integers modulo n and Z_n^* for the multiplicate group of units in Z_n. If $q \in Z_n^*$, then the mapping $i \longrightarrow qi$ is an isomorphism of $G(n, S)$ onto $G(n, qS)$. If $S \subseteq \{1, 2, \cdots, n - 1\}, S = -S$ and $S \neq \phi$, then $E(S)$ denotes the largest-order subgroup H of Z_n^* such that S is a union of cosets of H. A set $S \subseteq \{1, 2, \cdots, n-1\}$ is called symmetric if $S = -S$ mod n.

Alspach and Sutcliffe [2] gave the definition of 2-circulant.

A graph constructed on $2n$ vertices according to the following rule is called a 2-circulant. It has parameters S, q, F and the triple (S, q, F) is called the symbol of the 2-circulant. We denote it by $G(2n; S, q, F)$. To construct $G = G(2n; S, q, F)$, first partition the set of $2n$ vertices into two sets of vertices each of cardinality n and label them $u_{1,0}, u_{1,1}, \cdots, u_{1,n-1}$ and $u_{2,0}, u_{2,1} \cdots, u_{2,n-1}$ respectively. On $\{u_{1,0}, u_{1,1} \cdots, u_{1,n,1}\}$ construct the circulant $G_1 = G(n; S)$ with $u_{1,i}$, adjacent to $u_{1,j}$ if and only if $j - i \in S$. Now there is a restriction on q, namely that q may be chosen if and only if $q^2 \in E(S)$. Notice that $q = n - 1$ always will work because $q^2 = 1 \in E(S)$. Once q has been chosen, we then construct the circulant $G_2 = G(n; q, S)$ on $\{u_{2,0}, u_{2,1}, \cdots, u_{2,n-1}\}$ with $u_{2,i}$ adjacent to $u_{2,j}$ if and only if $j - 1 \in qS$. By the principle theorem of Turner [3] $G_1 \cong G_2$ if n is a prime.

We have now defined all the edges that are internal to either $V(G_1)$ or $V(G_2)$ and the only remaining edges are the "cross-edges" with one vertex in $V(G_1)$ and the other vertex in $V(G_2)$. We first insist that the "rotation" property

$$(u_{1,i}, u_{2,j}) \in E(G) \text{ if and only if } (u_{1,i+1}, u_{2,j+1}) \in E(G) \tag{1}$$

hold. This guarantees that $(u_{1,0}, u_{1,1}, \cdots, u_{1,n-1})(u_{2,0}, u_{2,1}, \cdots, u_{2,n-1})$ is an element of Aut(G) and means that the cross-edges incident with $u_{1,0}$ completely determine all the cross-edges. We define $F = \{j \in Z_n | (u_{1,0}, u_{2,j}) \in E(G)\}$. Again there is a restriction because we want to guarantee that the resulting graph is vertex transitive. We do this by choosing F so that there exists $\sigma \in Aut(G)$ such that $\sigma(G_1) = G_2$ and $\sigma(G_2) = G_1$. There are essentially two ways to build F.

If $qS = S$ then F may be any subset of Z_n $\quad$ (2a)

If $qS \neq S$, then $j \in F$ implies $-qj \in F$ $\quad$ (2b)

Alspach and Sutcliffe showed that if n is a prime, then $G(2n; S, q, F)$ is vertex transitive.

Let $G(2n; S, S) = G(2n; S, 1, S)$.

An isomorphic factorization of a graph $G = (V, E)$ is a partition $\{E_1, E_2, \cdots, E_t\}$ of E such that $(V, E_1 \cong (V, E_2) \cong \cdots \cong (V, E_t)$. If there is an isomorphic factorization of G into t factors, we then say that G is divisible by t and write $t|G.t| \, |E(G)|$ is a necessary condition for $t|G$ and is called divisibility condition. In general, the divisibility condition may not be sufficient. For a given graph G, if for any integer $t > 0t| \, |E(G)|$ is sufficient for $t|G$, then G is called rational.

In this paper, we shall prove that the 2-circulant $G(2n; S, S)$ is 1-factoriz - able and rational.

Let $\phi_\alpha = (u_{\alpha,0}, u_{\alpha,1}, \cdots, u_{\alpha,n-1})$ be a permutation of the symmetric group S_n and ϕ'_α is the induced permutation of ϕ_α in the pair group of S_n defined by $\phi'_\alpha \{u_{\alpha,i}, u_{\alpha,j}\} = \{\phi_\alpha u_{\alpha,i}, \phi_\alpha u_{\alpha,j}\}$.

2. Theorems and Their Proofs

Theorem 1   For any symmetric set $S \subseteq \{1, 2, \cdots, n - 1\}$, the 2- circulant graph $G = G(2n; S, S)$ is 1-factorizable.

Proof   Assume $|S| = m$, and $S - \{a_1, a_2, \cdots, a_m\}$. Then G is a 2m-regular graph. G is 1-factorizable if and only if G is 2m-edge colourable. It is discussed in two cases.

i) n is even. By the corollary 2.3.1 in [4]. G_1 and G_2 is m-edge colourable. Since $E(G_1) \cap E(G_2) = \phi.G_1 \cup G_2$ is m-edge colourable. Let $\varphi_1 : E(G_1 \cup G_2) \longrightarrow \{1, 2, \cdots, m\}$ be an m-edge colouring of $(G_1 \cup G_2)$ such that

$$\varphi_1(u_{1,i}, u_{1,j}) = \varphi_1(u_{2,i}, u_{2,j}) \quad \text{for} \quad j - 1 \in S.$$

$$E_2 - \{(u_{1,i}, u_{2,i+a_i} | i = 0, 1, \cdots, n - 1, j = 1, 2, \cdots, m\}$$

Let φ_2 is a mapping from E_2 onto $\{m + 1, \cdots, 2m\}$ with $\varphi_2(u_{1,i}, u_{2,i+a_j}) =$

$m + j$.

$$\Psi(e) = \begin{cases} \varphi_1(e), & \text{if } e \in E(G_1 \cup G_2), \\ \varphi_2(e), & \text{if } e \in E_2, \end{cases}$$

It is easy to see that Ψ is a proper 2m-edge coloring of $G(2n; S, S)$.

ii) n is odd. Without loss of generality, we assume G_1 is connected. Since n is odd, G_1 is even regular graph. Let $m = 2k$ then G is 4k-regular graph. We prove the proposition by induction on k.

When $k = 1, G_1$ is an odd cycle. We may assume that

$$G_1 \text{ is } (u_{1,0}, u_{1,1}, u_{1,2}, \cdots, u_{1,n}, u_{1,0}).$$

The colour 1 is assigned to the following edges:

$(u_{1,0}, u_{1,1}), (u_{1,2}u_{1,3}), \cdots, (u_{1,n-3}, u_{1,n-2}), (u_{1,n-1}, u_{2,n-2}), (u_{2,n-3}, u_{2,n-4}),$

$(u_{2,n-5}, u_{2,n-6}), \cdots, (u_{2,0}, u_{2,n-1})$

The colour 2 is assigned to the following edges:

$(u_{,1,2}, u_{1,2}), (u_{1,3}, u_{1,4}), \cdots, (u_{1,n-2}, u_{1,n-1}), (u_{1,0}, u_{2,n-1}),$

$(u_{2,n-2}, u_{2,n-3}), (u_{2,n-4}, u_{2,n-5}), \cdots, (u_{2,1}, u_{2,0}).$

The colour 3 is assigned to the following edges:

$(u_{1,0}, u_{1,n-1}), (u_{1,1}, u_{2,0}), (u_{1,2}, u_{2,1}), \cdots, (u_{1,n-1}, u_{2,n-3}), (u_{2,n-1}, u_{2,n-2}).$

The colour 4 is assigned to the following edges:

$(u_{1,0}, u_{2,1}), (u_{1,1}, u_{2,2}), \cdots, (u_{1,n-1}, u_{2,0}).$

It is easy to see that this colouring is a proper edge colouring. Thus when $k = 1$ the theorem holds.

Suppose the proposition holds for $k - 1$, we shall prove it holds for k. Let

$$G' = G - \{(u_{\alpha,i}, u_{\alpha,i+\alpha_1}) | \alpha = 1, 2, i = 0, 1, \cdots, n - 1\}$$

$$- \{(u_{1,i}, u_{2,i+\alpha_1}), (u_{1,i}, u_{2,i-\alpha}) | i = 0, 1, \cdots, n - 1\}$$

Let $S' = S \setminus \{\alpha_1, -\alpha_1\}$, and $G' = G(2n; S', S')$. By the induction hypothesis, there exists a $4(k - 1)$-edge coloring of G'.

Let

$$G' = G[\{(u_{\alpha,i}, u_{\alpha,i+1}) | i = 0, 1, \cdots, n - 1, \alpha = 1, 2\} \cup$$

$$- \{(u_{1,i}, u_{2,i+\alpha_1}), (u_{1,i}, u_{2,1-\alpha_1}) i = 0, 1, \cdots, n - 1\}]$$

It is easy to see that G'' consists of $ged(n, \alpha_1)$ components, each of which is a 4-regular 2-circulant. Therefore there is a proper 4-edge colouring of G''. Hence there is a proper 4k-edge colouring of G. The proof is completed.

Theorem 2　$G(2n; s, s)$ can be decomposed into 4-polygons for any symmetric set $S \subseteq \{1, 2, \cdots, n-1\}$.

Proof　By the construction of $G = G(2n; S, S)$, $G[\{u_{1,j}, u_{1,j+a_i}),$ $(u_{2,j}, u_{2,j+\alpha_i}),\ (u_{1,j}, u_{2,j+\alpha_i}),\ (u_{1,j+\alpha_1}, u_{2,j})\}]$ is a 4-polygon for any $j \in \{0, 1, \cdots, n-1\}$ and $i \in \{1, 2, \cdots, m\}$. This 4-polygon is said to be determined by (j, i). It is easy to see that when $(j_1, i_1) \neq (j_2, i_2)$, the 4-polygon determined by (j_1, i_1) and one by (j_2, i_2) have no edges in common. The proof is completed.

Theorem 3　If $S \subseteq \{1, 2, \cdots, n-1\}$ is symmetric and $|S| = p$ is a prime, then 2-circulant $G(2n; S, S)$ is rational.

Proof　First suppose $p > 2$. Then p is an odd number. Thus $\frac{n}{2} \in S$. Also $|E(G(2n; S, S))| = 2pn$. We discuss 4 cases.

Case 1　$t|n$.

Subcase 1-1　$t|\frac{n}{2}$. Let

$$E_0^\alpha = \{(u_{\alpha,0}, u_{\alpha,a_1}), \cdots, (u_{\alpha,0}, u_{\alpha,a_{\frac{p-1}{2}}}), (u_{\alpha,0}, u_{\alpha,a_{\frac{n}{2}}})\}, \alpha = 1, 2,$$

$$E_{\frac{n}{2}}^\alpha = \{(u_{\alpha,\frac{n}{2}}, u_{\alpha,\frac{n}{2}+a_1}), \cdots, (u_{\alpha,\frac{n}{2}}, u_{\alpha,\frac{n}{2}+\alpha_{\frac{p-1}{2}}})\}\alpha = 1, 2,$$

$$E_0^3 = \{(u_{1,0}, u_{2,\pm\alpha_1}), \cdots, (u_{1,0}, u_{2,\pm a_{\frac{p-1}{2}}}), (u_{1,0}, u_{2,\frac{n}{2}})\},$$

$$E_{\frac{n}{2}}^3 = \{(u_{1,\frac{n}{2}}, u_{2,\frac{n}{2}\pm a_1}), \cdots, (u_{1,\frac{n}{2}}, u_{2,\pm\alpha,\frac{p-1}{2}}), (u_{1,\frac{n}{2}}, u_{2,0})\},$$

$$E_1^* = E_0^1 \cup E_{\frac{n}{2}}^1, E_2^* = E_0^2 \cup E_{\frac{n}{2}}^2, E_3^* = E_0^3 \cup E_{\frac{n}{2}}^3,$$

$$E = \cup_{k=0}^{\frac{n}{2t}-1} [(\phi_1')^k E_1^* \cup (\phi_2')^k E_2^* \cup (\phi_1'\phi_2')^k E_3^*]$$

Then $\{\phi_1'\phi_2')^{\frac{in}{2t}} E | j = 0, 1, \cdots, t-1\}$ forms an isomorphic factorization of $G(2n; S, S)$.

Subcase 1.2　$\frac{n}{2} \neq 0(mod\, t)$. Since $t|n$, there is t_1 such that $t = 2t_1$ and $t_1|\frac{n}{2}$.

Let E_β^α be defined as in subcase 1-1 for $\alpha = 1, 2$ and $\beta = 0, \frac{n}{2}$. Also define

$$E_1 = \{(u_{1,0}, u_{2,\alpha_1}), \cdots, (u_{1,0}, u_{2,\alpha,\frac{p-1}{2}}), (u_{1,0}, u_{2,\frac{n}{2}})\},$$

$$E_1' = \{(u_{2,0}, u_{1,\alpha_1}), \cdots, (u_{2,0}, u_{1,\alpha,\frac{p-1}{2}}), (u_{2,0}, u_{1,\frac{n}{2}})\},$$

$$E_{\frac{n}{2}} = \{(u_{1,\frac{n}{2}}, u_{2,\frac{n}{2}+\alpha_1}), \cdots, (u_{1,\frac{n}{2}}, u_{2,\frac{n}{2}+\alpha_{\frac{p-1}{2}}})\},$$

$$E_{\frac{n}{2}}' = \{(u_{2,\frac{n}{2}}, u_{1,\frac{n}{2}+a_1}), \cdots, (u_{2,\frac{n}{2}}, u_{1,\frac{n}{2}+\alpha_{\frac{p-1}{2}}})\}.$$

Let

$$E_1^* = E_0^1 \cup E_1 \cup E_{\frac{n}{2}}^1 \cup E_{\frac{n}{2}}, \quad E_2^* = E_0^2 \cup E_1' \cup E_{\frac{n}{2}}^2 \cup E_{\frac{n}{2}}',$$

$$E = \cup_{k=0}^{\frac{n}{2t_1}-1} (\phi_1' \phi_2')^k E_1^*, \quad E' = \cup_{k=0}^{\frac{n}{2t_1}-1} (\phi_1' \phi_2')^k E_2^*,$$

Then $(V, E) \cong (V, E')$.

Thus $\{\{(\phi_1' \phi_2')^{\frac{jn}{2t_1}} E | j = 0, 1, \cdots, t_1 - 1\}, \{(\phi_1' \phi_2')^{\frac{jn}{2t_1}} E' | j = 0, 1, \cdots, t_1 - 1\}\}$ forms an isomorphic factorization of $G(2n; S, S)$ into $2t_1 = t$ factors.

Case 2 $t = 2pt_1, t_1 | n.$

According to Theorem 1, $G(2n; S, S)$ is 1-factorizable, because $G(2n; S, S)$ is 2p-regular, therefore there is a factorization of $G(2n; S, S)$ into 2p 1-factors. Since each 1-factor has n independent edges and $t_1 | n$, each 1-factor can be decomposed into t_1 independent edge sets in which each set include $\frac{n}{t_1}$ edges. Hence the edge set of $G(2n; S, S)$ can be decomposed into $2pt_i = t$ independent edge sets, and so spanning subgraphs corresponding to them are isomorphic to each other.

Case 3 $t = 2t_1$ and $t_1 | n.$

If $t_1 | \frac{n}{2}$, then we can construct an isomorphic factorization of $G(2n; S, S)$ into $2t_1$ factors by the method used in the subcase 1.1. Now we assume

$$\frac{n}{2} \neq 0(mod\, t_1), t_1 = 2t_1', t_1' | \frac{n}{2}, \text{ Let}$$

$$\overline{E}_0^\alpha = E_0^\alpha \backslash \{(u_{\alpha,0}, u_{\alpha,\frac{n}{2}})\}, \text{for } \alpha = 1, 2,$$

$$\overline{E}_{\frac{n}{2}}^\alpha = E_{\frac{n}{2}}^\alpha \cup \{(u_{\alpha,\frac{n}{2}}, u_{\alpha,\frac{n}{2}+\frac{n}{2}})\}, \text{for } \alpha = 1, 2,$$

$$\overline{E}_1 = \cup_{k=0}^{\frac{n}{2t_1'}-1} (\phi_1' \phi_2')^k (\overline{E}_0^1 \cup E_1),$$

$$\overline{E}_2 = \cup_{k=0}^{\frac{n}{2t_1'}-1} (\phi_1' \phi_2')^k (\overline{E}_{\frac{n}{2}}^1 \cup E_{\frac{n}{2}}),$$

$$\overline{E}_1' = \cup_{k=0}^{\frac{n}{2t_1'}-1} (\phi_1' \phi_2')^k (\overline{E}_0^2 \cup E_1'),$$

$$\overline{E}_2' = \cup_{k=0}^{\frac{n}{2t_1'}-1} (\phi_1' \phi_2')^k (\overline{E}_{\frac{n}{2}}^2 \cup E_{\frac{n}{2}}').$$

It is easy to see that $E = \overline{E}_1 \cup \overline{E}_2, E' = \overline{E}_1' \cup \overline{E}_2'$ and $(V, \overline{E}_1) \cong (V, \overline{E}_1') \cong (V, \overline{E}_1') \cong (V\overline{E}_2')$.

By an argument similar to that used in subcase 1.2, there is an isomorphic factorization of $G(2n; S, S)$ into $4t_1' = 2t_1 = t$ factors.

Case 4 $t = pt_1, t_1 | n.$

If t_1 is even then this case is included in the Case 2.

Now we assume, t_1 is odd. We shall prove that the union of any two 1-factors of $G(2n; S, S)$ can be decomposed into t_1 independent edge sets. Let F_1, F_2 be any two given 1- factors. F_1 is decomposed into t_1 independent edge sets. Let F_1, F_2 be any two given 1-factors. F_1 is decomposed into $\frac{t_1-1}{2}$ subsets each one of which has $\frac{2n}{t_1}$ edges and one subset with $\frac{n}{t_1}$ edges denoted E_1. The edges of E_1 are adjacent to $\frac{2n}{t_1}$ at most edges in F_2. Thus since $t \geq 3$ there are $\frac{n}{t_1}$ edges in F_2 which are not adjacent to any edge of E_1. We write these $\frac{n}{t_1}$ edges as E_2. Then $E_1 \cup E_2$ consists of $\frac{2n}{t_1}$ independent edges. $F_2 \backslash E_2$ can be decomposed into $\frac{t_1-1}{2}$ subsets each one of them has $\frac{2n}{t_1}$ edges. Hence each two 1-factors can be partitioned into t_1 sets, each one of them consisting of $\frac{2n}{t_1}$ independent edges. Thus there exists an isomorphic factorization of $G(2n; S, S)$ into pt_1 factors.

Subcase 4.3 $t_1 = 1$

According to Theorem 2, $G(2n; S, S)$ can be decomposed into 4-polygons. Because $G(2n; S, S)$ is 2p-regular. There is a 2-factorization of $G(2n; S, S)$

into p factors, each component of which is 4-polygon. Hence these factors are isomorphic to each other.

Now assume $p = 2$. Then $|E(G(2n; S, S))| = 4n$.

(I) t is odd. Then $t|n$.

(I.1) $G(2n; S, S)$ is connected.

Let $S = \{a, -a\}$.

Let

$$E^\alpha = \{(u_{a,i}, u_{a,i+a}) | i = 0, 1, \cdots, \frac{n}{t} - 1\} \text{ for } \alpha = 1, 2.$$

$$E_0 = \{(u_{1,i}, u_{2,i+a}), (u_{1,i}, u_{2,i-a}) | i = 0, 1, \cdots, \frac{n}{2} - 1\}.$$

$$E = E^1 \cup E^2 \cup E_0$$

Then $\{(\phi_1' \phi_2')^{\frac{in}{t}} E | j = 0, 1, \cdots, t - 1\}$, forms an isomorphic factorization of $G(2n; S, S)$.

(I.2) $G(2n; S, S)$ is not connected. Assume $G(2n; S, S)$ consists of n_1 components, $H_1, H_2, \cdots, H_{n1}$ denote these components.

Obviously, $H_1 \cong H_2 \cong \cdots \cong H_{n1} \cong G(2n_1; S, S)$. To decomposition $t = t_1 t_2$ such that $t_1 | n_1$ and $t_2 | n_2$, where $n_2 = n/n_1$. We give an isomorphic factorization of H_i into t_2 factors by the method used in (I.1). $F_1^i, F_2^i, \cdots, f_{ti}^i$ denote these factors, $i = 1, 2, \cdots, n_1$.

Let

$$A_{j,k} = \cup_{i=1}^{n_1/t_i} F^{\frac{in_1}{t_1} + i}, k = 1, 2, \cdots, t_2$$

$$j = 0, 1, \cdots, t_1 - 1.$$

Then $\{A_{j,k} | j = 0, 1, \cdots, t_1 - 1, k = 1, 2, \cdots, t_2\}$ forms an isomorphic factorization of $G(2n; S.S)$ into t factors.

II t is even. We can give an argument for this case similar to that for $p > 2$, omitted here.

The proof is completed #.

Acknowledgements

The first author thanks K. C. Wang Education Foundation Limited, who jointly sponsored him to attend the conference in the United States.

REFERENCES

[1] Wang Jianfang and Zhou Yong Sheng. "Isomorphic factorizations of circulant graph with prime degree". *Chinese Quarterly Journal of Mathematics*, Vol. 3, No. 1, (1988), 66-70.

[2] B. Alspach and R. J. Sutcliffe. "Vertex-transitive graph of order $2p$", *Am. N.Y. Acad. Sci.* 319(1970), 18-27.

[3] J. Turner, "Point-Symmetric Graphs with a prime number of points". *J. Combinatorial Theory*, 3(1967) 136-145.

[4] R. A. Stong. "On 1-factorizability of Cayley graphs", *J. Combinatorial Theory Series B* 39 (1985) 298-307.

Total Colourings of Graphs – A Progress Report

A.G. Chetwynd

University of Lancaster

ABSTRACT

It is twenty years since Behzad wrote his survey paper, The Total Chromatic Number of a Graph. This paper outlines the progress on total colourings since its publication.

1. Introduction

There are three types of graph colouring. *Vertex colouring*, where we colour the vertices so that adjacent vertices have different colours and the minimum number of colours needed for a total colouring of a graph G is called the chromatic number $\chi(G)$. *Edge colouring*, where we colour the edges so that adjacent edges have different colours and the minimum number of colours needed for a graph G is called the chromatic index $\chi'(G)$. *Total colouring*, which combines the above colourings by colouring both the vertices and the edges so that adjacent vertices, adjacent edges and edges and their incident vertices all have different colours. The minimum number of colours needed for a total colouring of a graph G is called the total chromatic number $\chi''(G)$. Total colourings were introduced by Behzad [1] in 1965.

2. Bounds on the Total Chromatic Number

Let G be a connected simple graph with maximum degree Δ. Then it was shown by Brooks [6] that for vertex–colouring

$$\chi(G) \leq \Delta + 1$$

and by Vizing [25, 26] that for edge–colourings

$$\Delta \leq \chi'(G) \leq \Delta + 1.$$

This led Behzad to conjecture a similar result for the total chromatic number.

Total Chromatic Number Conjecture *Let G be a simple graph with maximum degree Δ then*

$$\Delta + 1 \leq \chi''(G) \leq \Delta + 2.$$

Behzad also made the following conjecture for loopless multigraphs G with the maximum number of edges between any two vertices m(G).

$$\Delta + 1 \leq \chi''(G) \leq \Delta + m(G) + 1$$

It is obvious that the lower bound given in the conjecture is exact since a vertex of maximum degree is incident with Δ edges which require Δ colours plus another colour for the vertex. We call graphs which use $\Delta + 1$ colours Type 1 and those which need at least $\Delta + 2$ colours Type 2.

An obvious upperbound is $\chi''(G) \leq 2\Delta + 1$. This can be obtained by colouring the vertices of G with at most $\Delta + 1$ colours and then colouring the edges. At each uncoloured edge vw there will be at most $\Delta - 1$ colours used on edges at v and similarly at w plus the 2 colours used on v and w making at most 2Δ colours already used. Thus there remains at least one colour to colour the edge.

The first published result on the upper bound was due to Behzad, Chartrand and Cooper [3] and they showed:

Theorem $\chi''(G) \leq \chi'(G) + \chi(G)$ *with equality only if G is bipartite.*

Using Konig's result [16] for edge colouring bipartite graphs B the above theorem then becomes $\chi''(B) \leq \Delta + 2$, thus they proved the total chromatic number conjecture for bipartite graphs.

In 1974, Cook [10] obtained the following result of the Nordhaus–Gaddum class.

Theorem *Let G be a graph with n vertices then*
$$n + 1 \leq \chi''(G) + \chi''(\overline{G}) \leq 4\left\lfloor \frac{n}{2} \right\rfloor + 2$$
and
$$2\left\lfloor \frac{n}{2} \right\rfloor + 1 \leq \chi''(G)\,\chi''(\overline{G}) \leq \left(2\left\lfloor \frac{n}{2} \right\rfloor + 1\right)^2$$
Further these bounds are attained whenever n is odd.

Using the list chromatic number of a graph other authors have deduced new upper bounds.

The list chromatic number of a graph G, $\chi'_{\ell}(G)$, is the minimum size of a set of colours available at each edge such that G has an edge colouring, where each edge is coloured with a colour from its own set. If we restrict this to the same set of colours at each edge then this is equivalent to the chromatic index. An example of a list colouring is shown in figure 1.

Bollobas and Harris [5] have shown that

$$\chi'_{\ell}(G) \;\leq\; 2\Delta - \left\lceil \frac{\Delta}{6} - \frac{1}{3}\left(\log \Delta\right)^{\frac{1}{2}} \Delta^{\frac{1}{2}} \right\rceil \quad \text{for } \Delta \geq 3917.$$

Chetwynd and Häggkvist [7] have slightly improved this

$$\chi'_{\ell}(G) \;\leq\; \frac{9}{5}\Delta \quad \text{for triangle–free graphs.}$$

We can use these upper bounds on the list chromatic number to obtain better upper bounds for the total chromatic number. We first colour the vertices of our graph, this reduces by at most two the number of colours available at each edge and provided this number is at least $\chi'_{\ell}(G)$ we can complete the colouring.

Theorem $\chi''(G) \leq \chi'_{\ell}(G) + 2.$

The list chromatic number of a graph is conjectured to be at most $\Delta + 1$.

In 1977 Kostochka [18] showed that for most multigraphs (with a few unsettled cases) of maximum degree at least 6 that

$$\chi''(G) \;\leq\; \frac{3}{2}\Delta(G).$$

In 1987 Hind [15] gave a simple proof that for all graphs the following is true.

Theorem $\chi''(G) \leq \frac{3}{2}\Delta(G) + 2.$

In the same paper Hind extended this argument to show that

Theorem $\chi''(G) \leq \chi'(G) + 2\sqrt{\chi(G)}$

This result is the best known upperbound.

Goldberg in [12] conjectured that multigraphs of high chromatic index have total chromatic number equal to the chromatic index and has given a partial result to this conjecture in the following theorem.

Theorem *Let G be a critical multigraph with $\chi'(G) > \dfrac{9\Delta + 6}{8}$, then $\chi''(G) = \chi'(G)$.*

If we restrict our graphs to regular simple graphs then Chetwynd and Hilton [9] have shown the following result for graphs of high degree.

Theorem *Let G be a regular graph with degree $d(G)$, at least $\frac{3}{4}\,|V(G)|$ then*

$$\chi''(G) \;\leq\; d(G) + 3.$$

3. Graphs Which Satisfy the Conjecture

In 1970 Rosenfeld [23] considered several classes of graphs.

Theorem *Let G be a complete balanced K–partite graph or be a complete 3–partite graph or have maximum degree 3 then G satisfies the total chromatic number conjecture.*

Vijayaditya [24] in 1971 independently showed the result for graphs with maximum degree 3. In 1977 Kostochka [17] increased the bound on the maximum degree to 4. Behzad [2] then extended the maximum degree 3 case to include multigraphs.

By extending results on edge colourings Chetwynd and Hilton [9] gave a new class of graphs which satisfy the conjecture and includes the complete graphs.

Theorem *If G is a regular graph with*

$$d(G) \geq \begin{cases} \frac{6}{7}|V(G)| \ for \ |V(G)| \ odd \\ \frac{3}{4}|V(G)| \ for \ |V(G)| \ even \end{cases}$$

then G satisfies the total chromatic number conjecture.

In [27] Yap proves that the total chromatic conjecture is true for complete r–partite graphs, and in [28] Yap, Wang Jian–Fang, and Zhang Zhongfu prove that it is true for any graph of order p having maximum degree at least $p - 4$.

4. Graphs of Known Total Chromatic Number

The first graphs to be classified as Type 1 or Type 2 were the complete graphs and the complete bipartite graphs by Behzad, Chartrand and Cooper in 1967 [3].

Theorem

$$\chi''(K_n) = \begin{cases} \Delta + 1 \ if \ n \ is \ odd \\ \Delta + 2 \ if \ n \ is \ even \end{cases}$$

$$\chi''(K_{m,n}) = \begin{cases} \Delta + 1 \ if \ m \neq n \\ \Delta + 2 \ if \ m = n. \end{cases}$$

The result for circuits which had previously been shown to be $\Delta + 1$ or $\Delta + 2$ totally colourable by Vijayaditya [24] was finished by Behzad [2] in 1971.

Theorem

$$\chi''(C_n) = \begin{cases} \Delta + 1 \ if \ n \equiv 0 \ mod \ 3 \\ \Delta + 2 \ otherwise \end{cases}$$

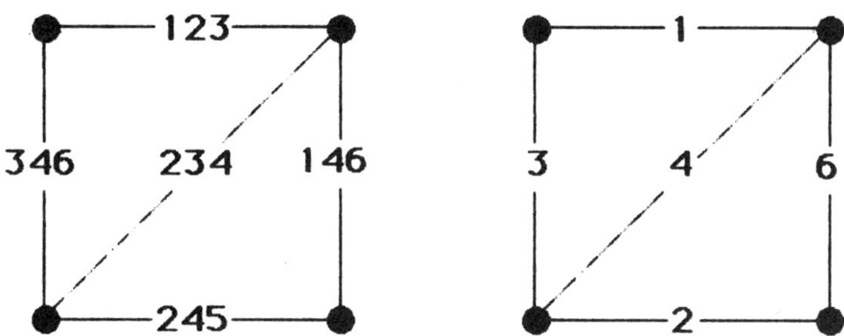

Figure 1: An example of a list colouring

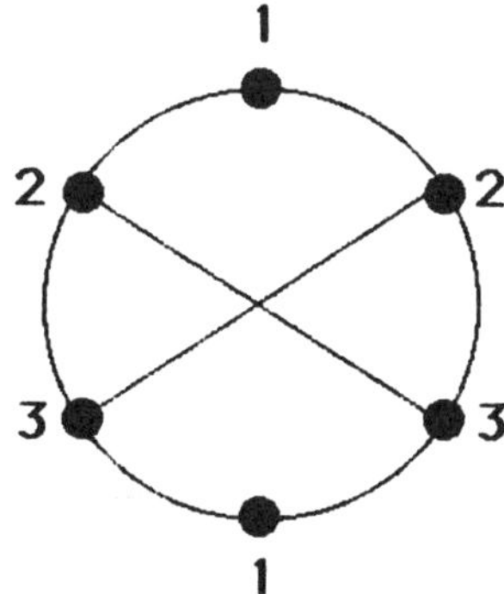

Figure 2: Conformable graph of Type 2

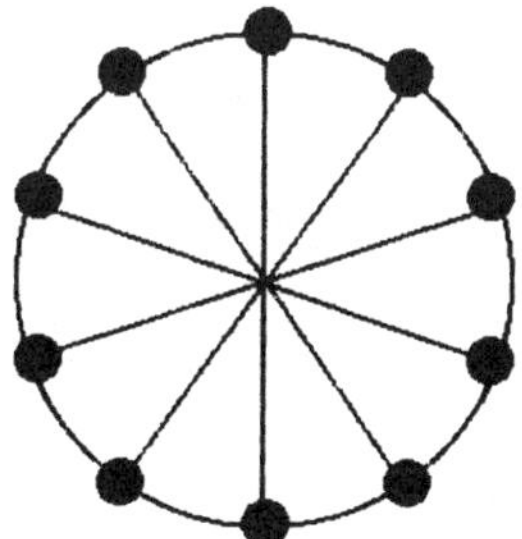

Figure 3: Mobius Ladder M_{10}

This result was generalized in 1976 by Meyer [20]. He considered the cartesian product of, G_r, a cycle of length r and the null graph on n vertices S_n, and completely characterised such graphs.

Theorem

$$\chi''(G_4 \otimes S_n) = 2n + 2 \ \text{for all} \ n \geq 2$$
$$\chi''(G_7 \otimes S_2) = 6$$
$$\chi''(G_r \otimes S_n) = 2n + 1 \ \text{for all} \ n \geq 2, r \neq 4 \ \text{and} \ (r, n) \neq (7, 2).$$

For $n = 1$ the graphs are just the ordinary cycle graphs which were classified above.

In 1972 Lasker and Hare [19] conjectured a result for balanced r–partite graphs. This conjecture was proved in a slightly modified form by Bermond [4] in 1974, and also by Mynhardt [21].

Theorem

$$\chi''(K_{r \times n}) = \begin{cases} \Delta + 1 & \text{if } r \text{ odd or both } r \text{ and } n \text{ even and } r \neq 2 \\ \Delta + 2 & \text{if } r = 2 \text{ or } r \text{ even and } n \text{ odd.} \end{cases}$$

Before discussing further results it will be useful to introduce two new definitions.

Conformable and Biconformable Graphs

In edge colouring graphs there is a nice numerical condition on the number of edges for a graph not to be colourable with Δ colours. It would be helpful to have a similar result to use when testing for Type 2 graphs.

A result of this kind uses the facts that if the graph is Type 1 then any colour can be absent from not more than a certain number of vertices and if a colour is on a number of vertices which is congruent to $|V(G)| + 1$ modulo 2 then it is certainly missing from at least one vertex.

The following two lemmas are due to Chetwynd and Hilton [8]. Let

$$\text{def}(G) = \sum_{v \in V(G)} (\Delta(G) - d_G(v)).$$

Lemma *Let G be a Type 1 graph and let G be totally coloured with colours $c_1, ..., c_s$ where $s = \Delta(G) + 1$. Suppose that for $1 \leq j \leq s$ there are i_j vertices of colour c_j and that for some $r \in \{1, ..., s\}$*

$$i_1 \equiv ... \equiv i_r \equiv |V(G)| \ mod \ 2 \qquad (1)$$

and

$$i_{r+1} \equiv ... \equiv i_s \equiv |V(G)| + 1 \ mod \ 2. \qquad (2)$$

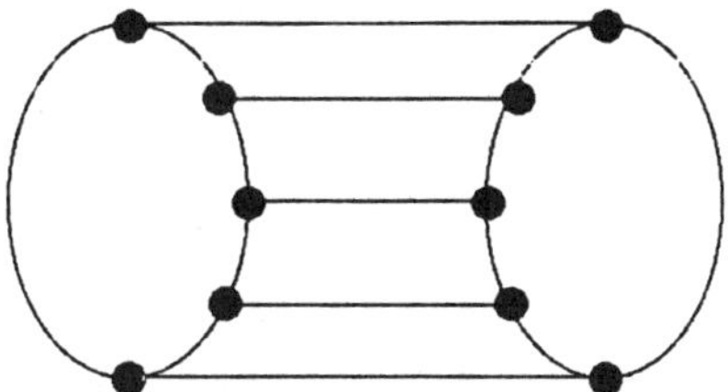

Figure 4: Ladder L_{10}

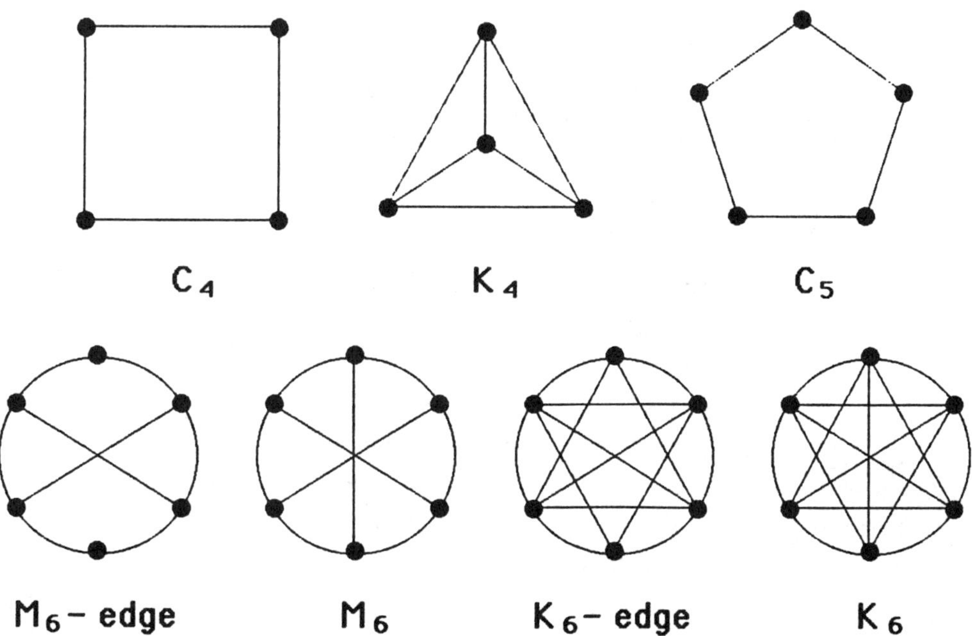

Figure 5: Type 2 graphs on at most 6 vertices

Then

$$def(G) \geq s - r. \tag{3}$$

Clearly for regular graphs we need to be able to find a vertex colouring with s colours where the number of vertices of each colour is congruent to $|V(G)|$ modulo 2.

If G has a vertex colouring with colours $c_1, \ldots, c_s$ where $s = \Delta(G) + 1$, such that, for some $r \in \{1, \ldots, s\}$, (1), (2) and (3) are satisfied, then we call the vertex colouring *conformable*, and we call G itself conformable.

The fact that a graph is conformable does not guarantee it is Type 1 as is shown by the example in figure 2. A conformable colouring is given but the graph can easily be shown to be of Type 2.

Let B be a bipartite graph with the same number of vertices in each partition then B is equibipartite.

Type 1 bipartite graphs also have a simple structural property.

Lemma *Let B be a Type 1 equibipartite graph with partition (A, B). Let G be totally coloured with colours $c_1, \ldots, c_s$ where $s = \Delta(G) + 1$ and for $1 \leq j \leq s$, let a_j vertices of A and b_j vertices of B be coloured c_j. Then*

$$def(G) \geq \sum_{k=1}^{s} |a_j - b_j|. \tag{4}$$

For regular equibipartite graphs this condition means that there must be the same number of vertices in each partition of each colour. If an equibipartite graph G with bipartition (A, B), where $|A| = |B| = n$, has a vertex–colouring with colours $c_1, \ldots, c_s$, where $s = \Delta(G) + 1$ such that (4) is satisfied, then we call the vertex colouring *biconformable*. Similarly we call G itself biconformable. Since complete graphs of even order are not conformable and complete bipartite graphs are not biconformable and hence they are of Type 2 it is interesting to see how many edges must be removed before the graphs become Type 1. Hilton [13], [14] recently proved the following result.

Theorem

(i) *Let G be a simple graph with $\Delta(G) = |V(G)| - 1$ then G is Type 2 if and only if G contains a non–conformable subgraph H with $\Delta(H) = \Delta(G)$.*

(ii) *Let G be a simple bipartite graph with bipartition (A, B) where $|A| \geq |B|$ and $\Delta(G) = |A|$ then G is Type 2 if and only if G contains a non–conformable equibipartite subgraph H with $\Delta(H) = \Delta(G)$.*

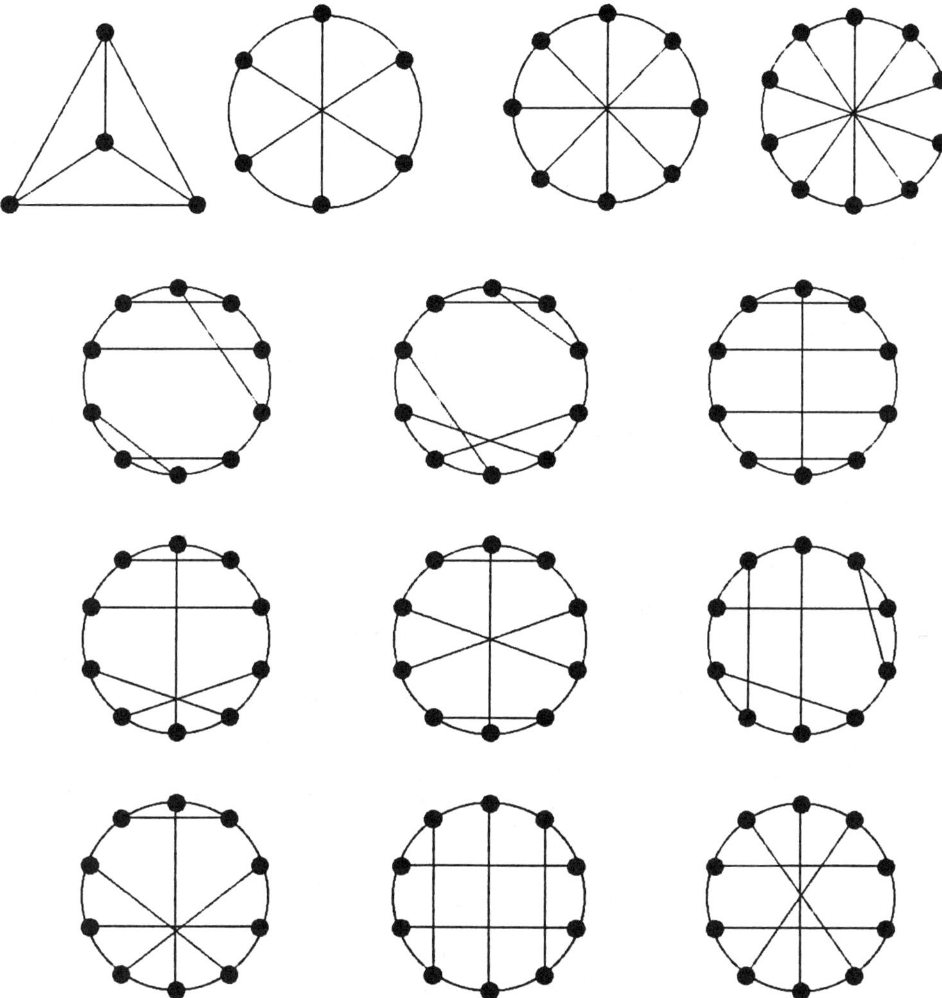

Figure 6: Type 2 graphs of degree three on at most 10 vertices

For regular graphs of odd order with high degree the classification is given by Chetwynd and Hilton [9] in the next result. However for graphs of even order no such result has as yet been found.

Theorem *Let G be a regular graph of odd order with $d(G) \geq \frac{19}{21} |V(G)| - \frac{3}{7}$ then G is of Type 2 if and only if G contains a non–conformable subgraph H with $\Delta(H) = \Delta(G)$.*

Chetwynd and Hilton [8] have classified Mobius ladders M_{2n} (which can be thought of as circuits C_{2n} with diagonally opposite vertices joined, see figure 3).

Theorem *The Mobius ladders M_{2n} are Type 2 for all $n \geq 1$.*

In fact these remain Type 2 even if multiple rungs are allowed (i.e. the diagonally opposite vertices are joined by r edges) that is $M_{2n}(r)$ is Type 2 for all natural numbers n and r. Apart from some small cases these Mobius Ladders are all conformable and moreover those that are equibipartite are biconformable. If instead of considering Mobius Ladders we look at Ladders L_{2n}, then these are all Type 1 except for the ladder L_{10}, shown in figure 4, which is Type 2.

Theorem *For $n \geq 2$ the ladders L_{2n} are Type 1 except for the ladder L_{10} which is Type 2.*

For ladders with multiple rungs $L_{2n}(r)$ with $r \geq 2$ these are all Type 1. Although L_{10} is Type 2 it is conformable.

5. Small Graphs of Type 2

An exhaustive search of all connected graphs on at most 6 vertices has produced just the following Type 2 graphs, shown in figure 5.

If we restrict our search to regular graphs of degree 3 on at most 10 vertices, we have the Type 2 graphs, shown in figure 6.

REFERENCES

[1]   M. Behzad. Graphs and their chromatic numbers. Doctoral thesis (Michigan State University), 1965.

[2]   M. Behzad. The total chromatic number of a graph, a survey, *Combinatorial Maths and its Applications*, (Acad. Press, New York 1971 ed. D.J.A. Welsh) 1 – 9.

[3]   M. Behzad, G. Chartrand, and J.K. Cooper Jr. The colour numbers of complete graphs. *J. London Math. Soc.* **42**, (1967) 225 – 228.

[4]   J.C. Bermond. Nombre chromatique total du graphe r–parti complet. *J. London Math. Soc.* **(2) 3** (1974) 279 – 285.

[5]   B. Bollobas and A.J. Harris. List colourings of graphs, *Graphs and combinatorics*, **1** (1985), 115 – 127.

[6]   R.L. Brooks. On colouring the nodes of a network, *Proc. Camb. Phil. Soc.* **13** (1941) 194 – 197.

[7]   A.G. Chetwynd and R. Häggkvist. A note on list–colourings, *J. Graph Theory* **13** (1989), 87 – 95.

[8]   A.G. Chetwynd and A.J.W. Hilton. Some refinements of the total chromatic number conjecture, *congressus numerantium* **66** (1988) 195 – 216.

[9]   A.G. Chetwynd and A.J.W. Hilton. The total chromatic number of regular graphs of high degree, submitted.

[10]   R.J. Cook. Complementary graphs and total chromatic numbers. *SIAM J. Appl. Math,* **27, 4** (1974) 626 – 628.

[11]   K. O'Donoghue. Total colourings and total graphs. M.Sc. Thesis, University of Natal (1985).

[12]   M. Goldberg. Edge colourings of multigraphs; Recolouring Technique. *J. Graph Theory,* **8** (1984) 123–137.

[13]   A.J.W. Hilton. A total chromatic number analogue of Plantholt's theorem, *Discrete Math.*, to appear.

[14]   A.J.W. Hilton. The total chromatic number of nearly complete bipartite graphs, submitted.

[15]   H.R. Hind. An upper bound for the total chromatic number, submitted.

[16]   D. Konig. Theorie der endlichen und unendlichen graphen, Chelsea, New York, 1950.

[17]   A.V. Kostochka. The total colouring of a multigraph with maximal degree 4. *Discrete Math.* 17 (1977) 161 – 163.

[18]   A.V. Kostochka. An analogue of Shannon's estimate for complete colourings (Russian). *Diskret. Analiz.* **30** (1977) 13 – 22.

[19]   R. Lasker and W. Hare. Chromatic numbers for certain graphs. *J. London Math. Soc.* **4**, (1972).

[20]   J.C. Meyer. Nombre Chromatique total du joint d'un ensemble stable par un cycle. *Discrete Math.* **15** (1976) 41 – 54.

[21]   C.M. Mynhardt. The total chromatic number of complete balanced m–partite graphs (preprint).

[22]   M. Rosenfeld. The total chromatic number of certain graphs. *Notices Amer. Math. Soc.* **15**, (1968) 360 – 365.

[23]   M. Rosenfeld. On the total colouring of certain graphs. *Israel J. Math.* **9** (1971), 396 – 402.

[24]   N. Vijayaditya. On total chromatic number of a graph. *J. London Math Soc.* **(2) 3** (1971) 405 – 408.

[25]   V.G. Vizing. On an estimate of the chromatic class of a p graph. *Diskret Analiz.* **3** (1964) 25 – 30.

[26]   V.G. Vizing. Some unsolved problems in graph theory (Russian). *Upsekhi Math. Nauk.*, **23** (1968) 117 – 134.

[27]   H.P. Yap. Total colouring of graphs, submitted.

[28]   H.P. Yap, Wang Jian–Fang, and Zhang Zhongfu. Total chromatic number of graphs of high degree, submitted.

OPTIMAL COMMUNICATION TREES
WITH APPLICATION TO HYPERCUBE MULTICOMPUTERS

Hyeong-Ah Choi

George Washington University

Abdol-Hossein Esfahanian
Bradley C. Houck

Michigan State University

ABSTRACT

A routing strategy for sending a message from a node (called source to any number of other nodes (called its destinations) in a multicomputer system is studied. The strategy requires that every destination receive the source message in a minimum number of time steps while generating a minimum amount of traffic. This strategy is formulated as finding a particular subtree (called an optimal communication tree) in a graph G, where G represents the underlying topology of the system. We prove that the problem of finding such a tree is NP-hard even if G is bipartite. The problem is then considered for the hypercube multicomputers. Optimal and suboptimal solutions are finally discussed.

1. Introduction

This work has been motivated by problems which arise in routing messages in *MultiComputer* (MC) systems. An MC is a collection of linked autonomous processors in which each processor has its own local memory. Processors may communicate by explicit exchange of messages [HwBr84]. A prime example of

245

an MC is the *hypercube multicomputer* which is now commercially available in several variations [Amet86, GrRe86, HMSC86, Seit85].

In an MC, each processor is directly connected to a small number of other processors (called *neighboring* processors). Neighboring processors can communicate directly, whereas non- neighboring processors have to communicate indirectly through other processors. Thus, in an MC, a message generally has to go through several intermediate processors before reaching its destination. As indicated in [Foxg83], interprocessor communication is an important factor affecting the efficiency of an MC. The message handling strategies and techniques, and how they are implemented, constitute the effectiveness of interprocessor communication.

A natural communication problem in an MC is that a processor (called the *source*) wants to send a message to $k, k \geq 1$, other processors (*destinations*). This problem will be referred to as 1-to-k communication. Clearly, from the source processor viewpoint, this is the most general type of communication, and the main issue here is that of determining which path(s) should be used to deliver the source message to all its destinations. This path selection process is commonly referred to as *routing*. Since, in an MC, there are generally many paths joining pairs of processors, routing should be done according to some criteria.

The two generic routing parameters considered in this paper are *time* and *traffic*. Parameter *time* is a measure of the actual time used to inform each destination, whereas parameter *traffic* quantifies the amount of system resources utilized to send the source message to its destinations. These parameters are then used to describe routing strategies.

A number of routing strategies have been proposed in the literature [BhJa83], Farl79]. These strategies generally involve only one of the two routing parameters. In [Farl79], the main parameter considered is time, whereas in [BhJa83] it is traffic. While these criteria are suitable for certain applications, they do not conform to the MC environment that we envision. We introduce a new routing strategy which minimizes both time and traffic. As it will be seen shortly, these parameters are not, in general, totally independent, and therefore some ordering must be used for their minimization.

The underlying topology of an MC will be modeled by a graph with nodes representing the processors and the edges representing the bidirectional communication links between processors. We will often use the terms node and processor interchangeably, and similarly for the terms edge and link. The rest of the paper is organized as follows. Our graph theoretical notation and terminology are given in the next section. In Section 3, our routing strategy is described in more detail. This strategy is then formulate as finding a particular subtree, called an *optimal communication tree* (OCT). We prove in Section 4 that the problem of finding an OCT is NP-hard even if the graph is bipartite. Section 5 discusses how our routing strategy can be implemented for a very popular MC, namely, the hypercube multicomputers.

2.   Notation and Terminology

Graph theoretical terms not defined here can be found in [Hara72]. Let $G(V, E)$ be an *undirected graph* (or simply graph) with the *node set* $V(G) = V$ and *edge set* $E(G) = E$. If an edge $e = (u, v) \in E$, then the nodes u and v are said to be *adjacent* (or *neighbors*) and the edge e is said to be *incident* to these nodes. A graph is said to be *connected* if every pair of its nodes are joined by a *path*. A *tree* is a connected graph which contains no *cycles*. A graph $H(V, E)$ is a *subgraph* of another graph $G(V, E)$, if $V(H) \subseteq V(G)$ and $E(H) \subseteq E(G)$. A subgraph which is a tree is referred to as a *subtree*. The *distance*, $d_G(u - v)$, between a pair of nodes u and v in G, is equal to the length (in number of edges) of a shortest path joining u and v.

3.   Routing Strategies

In this section, we first describe the routing parameters *time* and *traffic* in detail. The routing strategy is then formulated using these parameters.

We measure *time* in *time steps*, where a *time step* is the actual time needed to send a unit of information (message) from a node to one of its neighboring nodes. This time is assumed to be constant for all pairs of neighboring nodes. Furthermore, we will assume that any node can inform any number of its neighboring nodes during a time step. This is contrary to the assumption made in the traditional study of system level communication in networks, by which any node could inform no more than one neighboring node during a time step [Farl79, ScWu84]. Our assumption is, however, more compatible with the current technology [BrSc86, SaSc85].

Parameter traffic is quantified by the number of communication links which are employed to deliver the source message to its destination(s). Observe that if G is the underlying topology of an MC, then the edges which are used for message delivery *induce* a connected subgraph of G. Furthermore, minimizing traffic necessitates that this subgraph be a tree. Such a tree will be referred to as a *communication tree* (CT).

Clearly, in 1-to-k communication, the number of time steps required to inform a destination is at least equal to its distance from the source. Also, the number of destinations gives a lower bound for the amount of traffic required to complete communication. It is desirable to develop a routing scheme that completes communication while minimizing both time and traffic. The following example will, however, illustrate that these parameters, in general, are not totally independent and thus an ordering should be used for their minimizations.

Consider a 1-to-4 communication in an MC whose topology is given in Figure 1(a). Node v_0 is the source, and nodes v_2, v_9, u_{10}, and v_{11} are its destinations. Clearly, this communication will generate at least 4 units of traffic, and by using path $v_0 \longrightarrow v_2 \longrightarrow v_{10} \longrightarrow v_{11} \longrightarrow v_9$, the source node can inform all its destinations while generating that minimum amount of traffic, see Figure 1(b). However, node v_9 receives the source message after 4 time steps, whereas it could receive the message after 2 time steps if a shortest (v_0, v_9)-path were used. It can be easily verified that there is no routing by

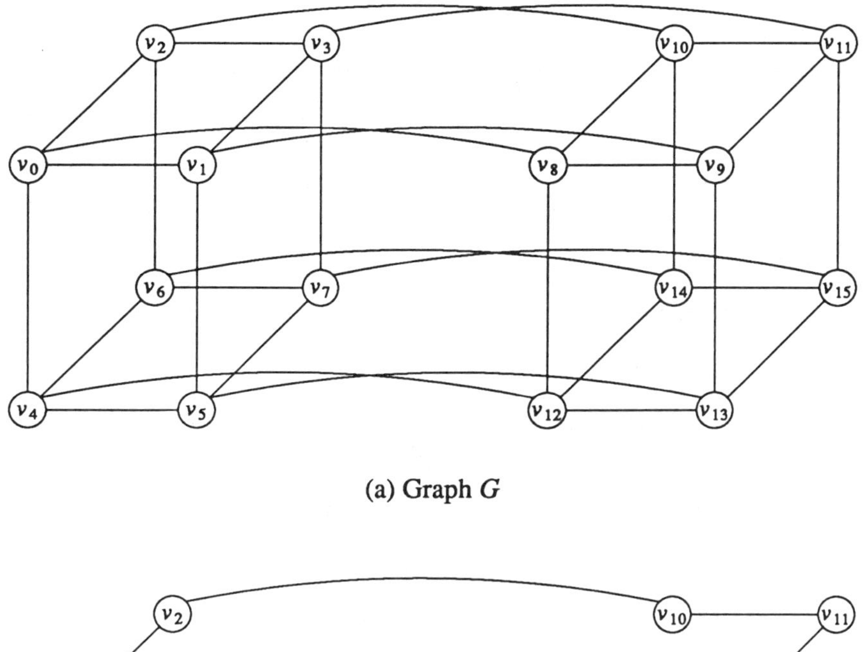

(a) Graph G

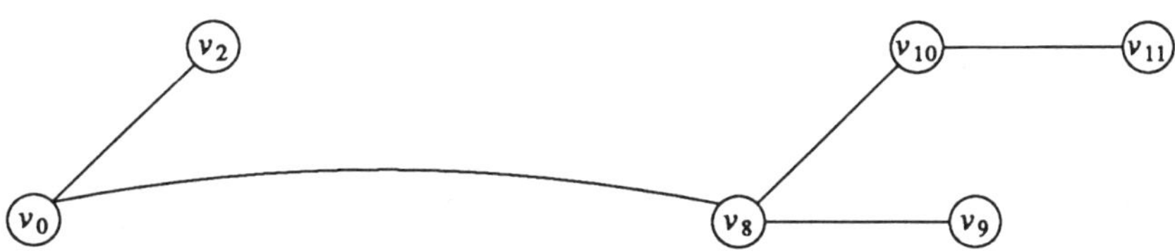

(b) A CT when source is v_0 and destinations are v_2, v_9, v_{10} and v_{11}, and traffic is minimized first.

(c) A CT when source is v_0 and destinations are v_2, v_9, v_{10} and v_{11}, and time is minimized first.

Figure 1.

which each of the destination nodes v_2, v_9, v_{10}, and v_{11}, can receive the source message in minimum number of time steps while generating 4 units of traffic. Thus, it will take at least 5 units of traffic in order to have each destination node receive the source message in minimum number of time steps. This can be accomplished by using the communication tree given in Figure 1(c). Using this CT, the source node sends its message to both of its neighboring node v_2 and v_8, in the first time step. Nodes v_{10} and v_9 are informed in the second time step and finally node v_{11} receives the source message in the third time step. Note that in this case node v_8 is an *intermediate* node.

The above example suggests that an ordering should be used in minimizing the parameters time and traffic. Considering an MC environment, our communication criterion will be to first minimize time and then traffic. By minimizing time we mean that each individual destination must receive the source message in a minimum number of time steps. This implies that, for each destination node, the message should be delivered through a shortest path, from the source, to that destination. Also, note that to minimize traffic it suffices to minimize the number of intermediate nodes used to deliver the source message to all destinations. Using the above criterion, we now formulate the above routing strategy as a graph theoretic problem. Let graph $G(V, E)$ represent the topology of a MC and let $M = \{s, v_1, \cdots, v_k\}$ be a subset of V. Node s corresponds to the source node, and nodes $\{v_1, \cdots, v_k\}$ correspond to k destination nodes in a 1-to-k communication. To implement the above routing strategy one needs to find a subtree $T(V, E)$ of $G(v, E)$, called *optimal communication tree* (OCT), such that

(a) $M \subseteq V(T)$,

(b) $d_T(s - v_i) = d_G(s - v_i)$, for $1 \leq i \leq k$, and

(c) $|V(T)|$ is as small as possible.

It should be noted that if we drop condition (b) above, the above problem becomes the well- known *Steiner tree* problem. Thus, *Steiner tree* problem can be used to model the situation in which the only concern is to minimize traffic [GaGJ77]. This is the criterion studied in [BhJe83]. From hereafter, we will exclusively refer to any subtree of G satisfying conditions (a) and (b) above as CT. Thus a CT with minimum number of nodes is an OCT.

4. Finding an Oct if NP-Hard

In this section we prove that finding an OCT is NP-hard by showing the corresponding decision problem is NP-complete. As usual, our proof will be based on a transformation from a known NP-complete problem, namely the 3SAT problem, which is described below (for further exposure to the theory of NP- completeness the reader is referred to [GaJo79].

3SAT Problem

Instance   Set U of variables, collection C of clauses over U such that each clause $c \epsilon C$ has $|c| = 3$.

Question   Is there a truth assignment for U such that the clause in C has at least one true literal?

Theorem   The following decision problem is NP-complete.

Instance   A bipartite graph $G(V, E)$, a subset $M = \{s, v_1, \cdots, v_k\}$ of V, and a constant l.

Question   Does G contain a subtree $T(V, E)$ such that

$$(a) M \subseteq V(T),$$

$$(b) d_T(s - v_i) = d_G(s - v_i)m \text{ for } 1 \leq i \leq k, \text{and}$$

$$(c) |V(T) = l ?$$

Proof   We will refer to the above decision problem as Optimal Communication Tree Decision Problem (OCTDP). It is not difficult to see that OCTDP belongs to the NP class. We next present a transformation from an instance of 3SAT problem to an instance of OCTDP.

Let $U = \{u_1, u_2, \cdots, u_n\}$ be the set of variables and $C = \{c_1, c_2, \cdots, c_m\}$ be the set of clauses as defined in the 3SAT problem. We first construct a bipartite graph $G(V, E)$ as follows. The node set of G is:

$$V(G) = \{u_1, u_2, \cdots, u_n\} \cup \{\overline{u}_1, \overline{u}_2, \cdots, \overline{u}_n\} \cup \{p_1, p_2, \cdots, p_n\}$$

$$\cup \{s\} \cup \{c_1, c_2, \cdots, c_m\}$$

$$\cup \{u_{ij} \mid \text{literal } u_i \text{ appears in clause } c_j\}$$

$$\cup \{\overline{u}_{ij} \mid \text{literal } \overline{u}_i \text{ appears in clause } c_j\},$$

and the edge set of G is:

$$E(G) = \{s, u_i) \mid 1 \le i \le n\} \cup \{(s, \overline{u}_i) \mid 1 \le i \le n\}$$

$$\cup \{(u_i, p_i) \mid 1 \le i \le n\} \cup \{(\overline{u}_i, p_i) \mid 1 \le i \le n\}$$

$$\cup \{(u_i, u_{ij}) \mid u_{ij} \epsilon V(G), 1 \le i \le n, 1 \le j \le m\}$$

$$\cup \{(\overline{u}_i, \overline{u}_{ij}) \mid \overline{u}_{ij} \epsilon V(G), 1 \le i \le n, 1 \le j \le m\}$$

$$\cup \{(u_{ij}, c_j) \mid u_{ij} \epsilon V(G), 1 \le i \le n, 1 \le j \le m\}$$

$$\cup \{\overline{u}_{ij}, c_j) \mid \overline{u}_{ij} \epsilon V(G), 1 \le i \le n, 1 \le j \le m\}.$$

It is not difficult to see that the above construction produces a bipartite graph. Note that $|V(G)| = 3n + 4m + 1$ and $|E(G)| = 4n + 6m$. Figure 2 shows $G(V, E)$ when $U = \{u_1, u_2, u_3, u_4\}$ and $C = \{c_1, c_2\}$ where $c_1 = \{u_1, u_2, \overline{u}_3\}$ and $c_2 = \{u_1, u_2, u_4\}$.

Now we define a 1-to-$(n + m)$ communication in G as follows. Let $M = \{s, c_1, c_2, \cdots, c_m, p_1, p_2, \cdots, p_n\}$, where s corresponds to the source node, and $c_1, c_2, \cdots, c_m, p_1, p_2, \cdots, p_n$ correspond to its destination nodes. It will be shown that there is a truth assignment for U such that each clause in C has at least one true literal (we will refer to such a truth assignment as *proper* truth assignment) if and only if there exists a subtree $T(V, E)$ in G such that

(a) $M \subseteq V(T)$,

(b) $d_T(s - c_j) = d_G(S - c_j), 1 \le j \le m$, and $d_T(s - p_i) = d_G(s - p_i)$,

for $1 \le i \le n$

(c) $|V(T)| = 2m + 2n + 1$.

Suppose there exists a truth assignment for U as described above. One can construct a subtree $T(V, E)$ of G as follows. The node set of T is:

$$V(T) = \{s\} \cup \{c_1, c_2, \cdots, c_m\} \cup \{p_1, p_2, \cdots, p_n\}$$

$\cup \{u_i \epsilon V(G) | u_i \text{ is true }\} \cup \{\overline{u}_i \epsilon V(G) | \overline{u}_i \text{ is true}\}$

$\cup \{u_{ij}, \overline{u}_{ij} \epsilon V(G) | \text{ for each } j, 1 \leq j \leq m, i \text{ is the least index such that either}$

$u_i \text{ or } \overline{u}_i \text{ is true }\}$

and the edge set of T is:

$$E(T) = \{(x, y) \epsilon E(G) | x \text{ and } y \epsilon V(T)\}.$$

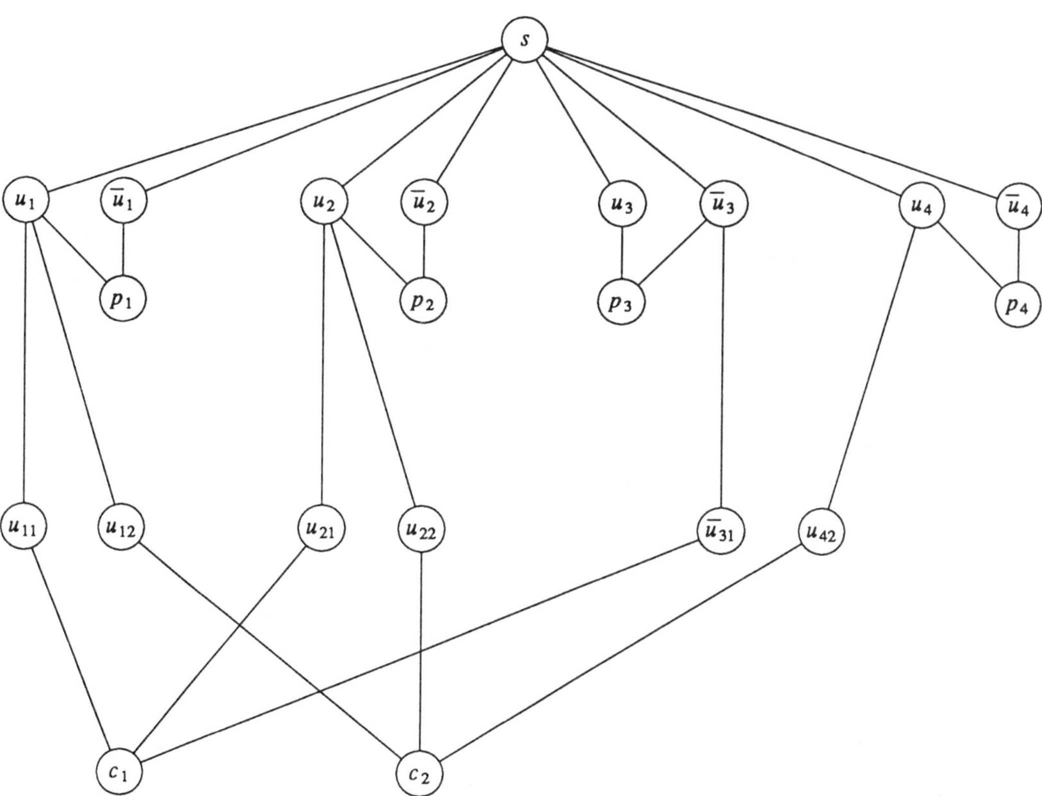

Figure 2. The bipartite graph G constructed for

$$U = \{u_1, u_2, u_3, u_4\} \text{ and } C = \{ \{u_1, u_2, \overline{u}_3\}, \{u_1, u_2, u_4\} \}.$$

It is not difficult to verify that $T(V, E)$ as constructed above satisfies conditions (a), (b), and (c). Thus it remains to show that if such a tree $T(V, E)$ exists then we can find a proper truth assignment for U.

Let $T(V, E)$ be such a tree. Note that from condition (a) we have $\{s, c_1, c_2, \cdots, c_m, p_1, p_2, \cdots, p_n\} \subseteq V(T)$. Further, condition (b) implies that each node in $\{c_1, c_2, \cdots, c_m, p_1, p_2, \cdots, p_n\}$ is a *leaf* in T. This means that for each $i, 1 \leq i \leq n$, either u_i or $\overline{u}_i$, but not both, is in T. Moreover, each $c_j, 1 \leq j \leq m$, being a destination node causes that a node of type u_{ij} or $\overline{u}_{ij}$ be included in T. Finally, if u_{ij} (or $\overline{u}_{ij}$) is in T, then u_i (respectively, $\overline{u}_i$) must be in T. These observations establish the existence of a proper truth assignment for U. In particular, assigning true value to literal u_i (or $\overline{u}_i$) if node $u_i \epsilon V(T)$ (respectively, if node $\overline{u}_i \epsilon V(T)$) gives a proper truth assignment. This completes the proof of the theorem.

It should be noted that the graph $G(V, E)$ constructed in the above proof has *diameter* six or less, and no two nodes in set $M - \{s\}$ are adjacent in G, and the distance from the source to each destination in G is no more than three. This implies a stronger NP-completeness result for OCTDP.

Finding an OCT in Hypercube Multicomputers In this section we discuss the problem of finding an OCT in hypercube multicomputer. These machines have gained much popularity due to their many applications in parallel processing. Hypercube MCs have been the subject of many research articles in the past few years. In particular, much work has been done on internode communication in a hypercube [DaSe87, HoJo86, LaEN87, LeHa88, MoSc86, ReBA88]. When we studied the routing strategy in the first commercially available hypercube, namely the Intel's first generating iPSC, we noticed that only 1-to-1 communication was supported. Any 1-to-$k, k > 1$, communication was implemented as k, 1-to-1 communications. Such as scheme has the potential of generating more traffic than required. This gave rise to our formulation of OCTs. A hypercube has the n-cube graph, Q_n, as its underlying topology. Although the n-cube is known to be bipartite, it cannot be deduced from the

NP-completeness result on OCTs, presented in the previous section, that the problem remains NP-hard for the n-cube. However, we conjecture the OCTDP will remain NP- complete for the n-cube.

Once a problem is shown to be NP-hard, finding *suboptimal* solutions become very attractive. In the absence of any polynomial-time algorithm for finding an OCT in the n-cube, we will consider, in this section, heuristics for generating suboptimal CTs in the n-cube. Previous work as well as our approach will be discussed.

We proceed by defining what we mean by a suboptimal CT. Let G be the topology of an MC and M be defined as in Section 3. A *suboptimal communication tree* (SOCT) is a CT which satisfied conditions (a) and (b) in the definition of an OCT. However, the total numer of nodes in a SOCT can be larger than that of an OCT. Further, let us remind the reader what the n-cube is. The n-cube, $Q_n(V, E)$, has $N = 2^n$ nodes, and is constructed as follows. Each node is labeled by a distinct n-bit binary number $(b_{n-1}b_{n-2} \cdots b_1 b_0)$. Then two nodes are joined by an edge if and only if their binary labels differ in exactly one bit position. If $v \epsilon V(Q_n)$ then $b(v)$ represents its n-bit binary label. We will often refer to a node by its binary label.

Now, consider a 1-to-k communication in Q_n, with set $M = \{s, v_1, v_2, \cdots, v_k\}$ as defined before. Without loss of generality, assume that source node has binary label $(00 \cdots 0)$. The first heuristic for generating a SOCT was proposed by Lan, Esfahanian, and Ni (we will refer to their heuristic as LEN heuristic) [LaEN87]. To describe this heuristic the following terminology is in order. Any subset of $D = \{v_1, v_2, \cdots, v_k\}$ is called a *destination sublist*. A node v in a communication there $T(V, E)$ is a *forwarding* node if it is either the source node or it is a nonleaf node in T. In Q_n, a node x is a *descendant* of another node y, with respect to the source node s, if $d(s - x) = d(s - y) + d(y - x)$. The *relative address* of a node $b(v_i)$ with respect to another node $b(v_j)$, denoted $b(v_i|v_j)$, is given by a bit-wise Exclusive-ORing of $b(v_i)$ with $b(v_j)$. For example, if $b(v_i) = 11010$ and $b(v_j) = 01111$ then $b(v_i|v_j) = 10101$. Note that

$b(v_i|v_j) = b(v_j|v_i)$. Finally, let X be a collection of m n-bit binary numbers $a_{n-1}^k, a_{n-2}^k \cdots, a_0^k$ where $1 \leq k \leq m$. Then $B(X) = (X_{n-1}, X_{n-2}, \cdots, X_o)$ where $X_i = \sum_{k=1}^{m} a_i^k$. For example, let $X = \{10110, 00111, 111000\}$. Then $B(X) = (2, 1, 3, 2, 1)$.

Len Heuristic   Len heuristic is based on the following strategy. The CT should start *growing* from the source. Each forward node should contribute to the overall growth of the CT. Moreover, each node, upon receiving a destination sublist, should decide which of its adjacent nodes in Q_n should get included in the CT. The decision criterion is to recursively select an adjacent node with a maximum number of descendants in the received destination sublist. Note that the distributed nature of the above strategy makes the LEN heuristic quite suitable for a parallel processing environment. A detailed description of the algorithm follows. Each node v in Q_n, upon receiving a destination sublist X, performs the following. Note that the first node is to execute this algorithm is the source node s.

1. For each node $b(x)\epsilon X$, replace $b(x)$ with $b(x|v)$. (This will produce the relative addresses for use by the algorithm).

2. Compute $B(X) = (X_{n-1}, X_{n-2}, \cdots, X_0)$, and then select the smallest i such that $X_i \geq X_j$ for any j.

3. If $X_i = 0$ STOP, otherwise, create a destination sublist Y containing all $b(x)\epsilon X$ such that $b(x)$ has a one in bit position i.

4. Send Y to the node u in Q_n, where u is adjacent to v and $b(u|v)$ has a one in bit position i.

5. Set $X \longleftarrow X - Y$, and go to step 2.

Perhaps the following example will clarify the LEN heuristic. Consider Q_4, and let node 0000 be the source of a 1-4 communication, with destination list $\{0100, 1010, 1010, 1011\}$. The source node will execute the LEN algorithm as follows. Step 1 produces $X = \{0100, 0101, 1010, 1011\}$. Step 2 given $B(X) = \{2, 2, 2, 2\}$, and $i = 0$. Execution of Step 3 produces $Y = \{0101, 1011\}$. The node selected by Step 4 is 0001, and Y is sent to this node. Step 5 resets X to{0100}. Finally, the third pass of the algorithm will select the node 0100, and the destination sublist sent to this node contains only 0100.

Now, each of the neighboring nodes of 0000 selected in Step 4 of the algorithm, will perform the same procedure with their corresponding destination sublist. The complete communication tree generated is depicted in Figure 3.

One problem with the LEN heuristic is that it often splits adjacent destination nodes among different destination sublists, which may result in unnecessary intermediate nodes. We propose a modification to the LEN algorithm, called the *Covered* Heuristic. The following definition is necessary to an explanation of this heuristic.

Let x and y be a pair of adjacent nodes in Q_n. If x and y are contained in a destination sublist, and x is a descendant of y, then x is said to be *covered*. Note that in the above example both destinations 0101 and 1011 are covered. The Covered Heuristic is described below.

Covered Heuristic   Each node v in Q_n, upon receiving a destination sublist X, performs the following. Note that the first node to execute this algorithm is the source node s.

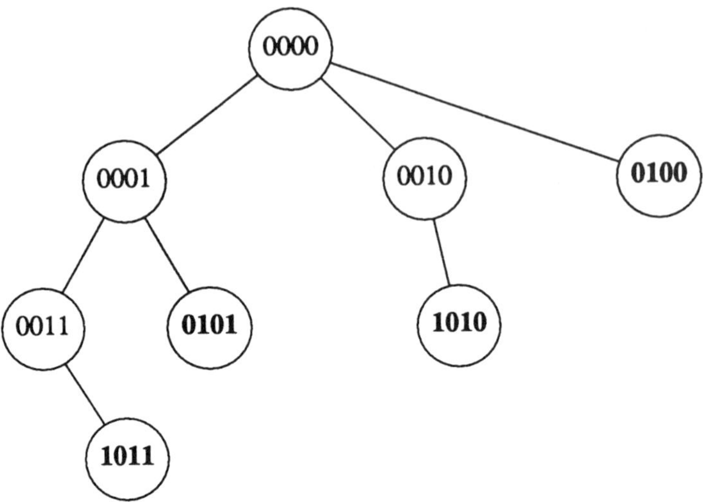

Figure 3. The communication tree generated by the LEN heuristic
for a 1-to-4 communication in Q_4 with source node 0000
and destination list {0100, 0101, 1010, 1011}.

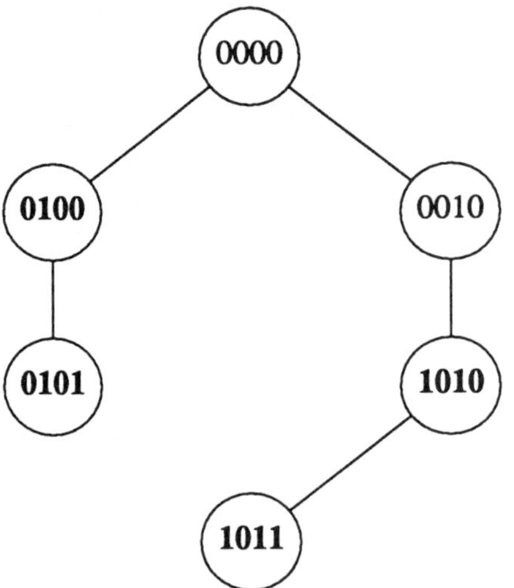

Figure 4. The communication tree generated by the Covered heuristic
for a 1-to-4 communication in Q_4 with source node 0000 and
destination list {0100,0101, 1010, 1011}.

1. For each node $b(x) \epsilon X$, replace $b(x)$ with $b(x|v)$. (This will produce

the relative addresses for use by the algorithm). Also, determine and then mark each node in X as "covered" or "not covered".

2. Compute $B(X) = (X_{n-1}, X_{n-2}, \cdots, X_0)$. Let I be the set of all integers i, such that $X_i \geq X_j$ for any j. If $|I| = 1$ then let i be that element in I and proceed to Step 3. Otherwise, let $W = \{b(x)\epsilon X | b(x)$ is not covered$\}$. Calculate $B(W)$, and then select the smallest i such that $W_i \geq W_j$ for any j.

3. If $X_i = 0$ STOP, otherwise, create a destination sublist Y containing all $b(x)\epsilon X$ such that $b(x)$ has a one in bit position i.

4. Send Y to the node u in Q_n, where u is adjacent to v and $b(u|v)$ has a one in bit position i.

5. Set $X \longleftarrow X - Y$, and go to step 2.

It is not difficult to see that the changes in Step 2 attempt to keep a node together with its covered descendants in the same destination sublist. The Covered Heuristic will generate the communication tree depicted in Figure 4 for the same 1-to-4 communication explained in the above example. Note that in this particular example, the number of intermediate nodes will be reduced by two, and in fact the CT in this case is also an OCT. However, it is not difficult to construct destination lists for which neither of the above heuristics will generate an optimal communication tree.

We have not yet determined the performance of the above heuristics in comparison to an optimal solution. In the absence of such theoretical results, we have resorted to simulation to obtain comparative performance of these heuristics.

Prior to addressing these results, it should be pointed out that an earlier attempt to solve the 1-to-k communication problem was made by researchers at Intel [MoSc86]. The resulting communication heuristic did not necessarily produce a tree and so the heuristic is not included here as a solution to the OCT problem. The results for this heuristic were evaluated and have been included for comparison.

To evaluate the three heuristics, random destination lists were produced and each heuristic was run. In addition, exact solutions were determined by testing all possible communication trees until the smallest was found for the destination set. Figure 5 shows the average curves obtained for different destination lists sizes in Q_6. Figure 6 shows similar results for Q_{10}, except that the exact solution curve was excluded. Note that in Q_6, the Covered heuristic results were at most 0.8 extra nodes beyond the exact solution, and show significant improvement over LEN and Intel's iPSC heuristics.

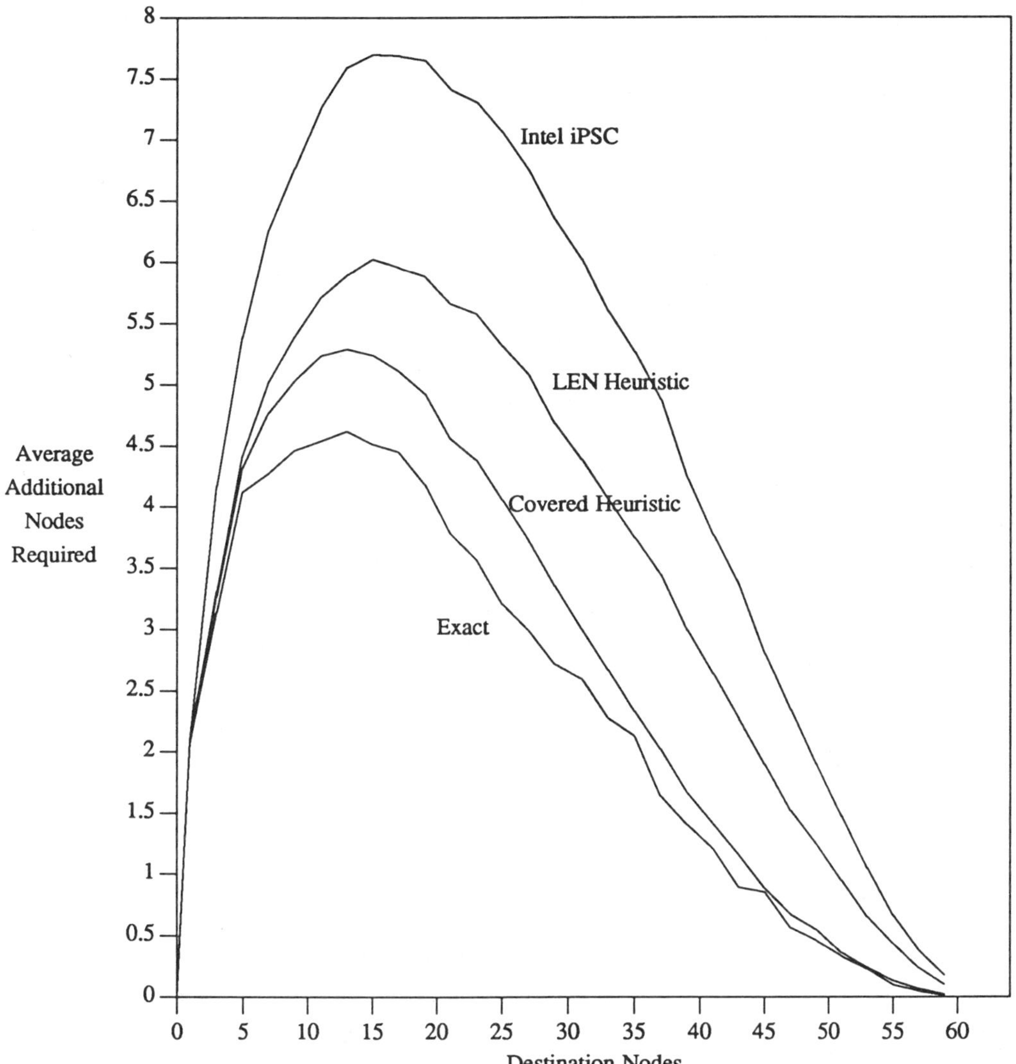

Figure 5. Performance of different heuristics for 1-to-k communication in Q_6

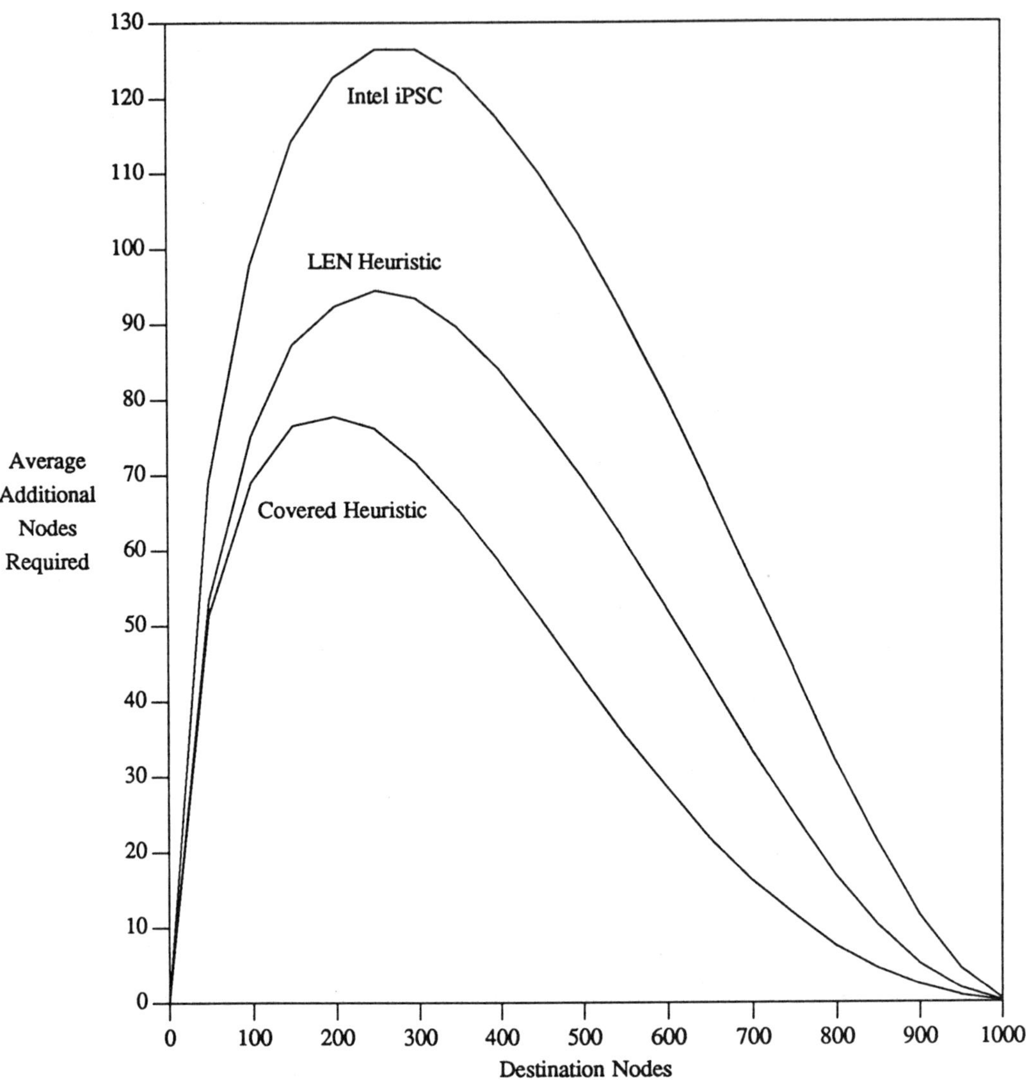

Figure 6. Performance of different heuristics for 1-to-k communication in Q_{10}

REFERENCES

[1] Ametek Computer Research Division, [Amet86] "Ametek system 14 user's guide: C edition," Version 2.0, May 1986.

[2] K. Bharath-Kumar and J. M. Jaffe, [BhJa83] "Routing to multiple destination in computer networks," *IEEE Transactions on Communications*, Vol. 31, No. 3, pp. 343-351., March 1983.

[3] J. E. Brandenburg and D. S. Scott, [BrSc85] "Embeddings of communication trees and grids into hypercubes," *Technical Report*, Intel. Scientific Computers, 1985.

[4] W. J. Dally and C. L. Seitz, [DaSe87] "Deadlock-free message routing in multiprocessor interconnection networks," *IEEE Transactions on Computers*, May 1987, pp. 547-553.

[5] Farley, A. M., [Far179] "Minimal Broadcast Networks," *NETWORKS*, Vol. 9, pp. 313-332, (1979).

[6] Fox, G. C., [Foxg83] "The Impact of Specialized Processors in Elementary Particles Physics," *Conference on Scientific Calculation with Ensemble Computers*, Pauda, Italy, 1983.

[7] Garey, M., Graham, R. and D. Johnson, [GaGJ77] "The complexity of computing Steiner minimal trees," *SIAM Journal of Applied Mathematics*, 32, pp. 835-859, 1977.

[8] Garey, M. R. and D. S. Johnson, [GaJo79] *Computer and Intractability*, W. H. Freeman and Company, San Francisco, 1979.

[9] D. C. Grunwald and D. A. Reed, [GrRe86] "Benchmarking Hypercube Hardware and Software," *Technical Report*, UIUCDCS-R-86-1303, Department of Computer Science, University of Illinois at Urbana - Champaign, 1986.

[10] F.Harary, [Hara72] *Graph Theory*, Addison Wesley, 1972.

[11] C. T. Ho and L. S. Johnson, [GaJo79] "Distributed routing algorithms for broadcasting and personalized communication in hypercubes, *Proceedings of the 1986 International Conference on Parallel Processing*, pp. 640-648, August 1986.

[12] K. Hwang and F. A. Briggs, [HwBr84] *Computer Architecture and Parallel Processing*, McGraw-Hill Book Co., 1984.

[13] J. P. Hayes, T. N. Mudge, Q. F. Stout, S. Colley, S. and J. Palmer, [HMSC86] "Architecture of a Hypercube Supercomputer," *Proceedings of 1986 International Conference on Parallel Processing*, August 1986, pp. 653-660.

[14] Lan, Y. A. Esfahanian, and L. M. Ni, [LaEN87] "Multicast in Hypercube Multiprocessors," to appear in *Journal of Parallel and Distributed Computing*.

[15] T. C. Lee and J. P. Hayes, [LeHa88] "Routing and Broadcasting in Faulty Hypercube Computers," *The 3rd Conference on Hypercube Concurrent Computers and Applications*, pp. 346- 354, January 1988.

[16] C. Moler and D. Scott, [MoSc86] "Communication utilities for the iPSC," *iPSC Technical Report*, No. 2, Intel Scientific Computers, 1986.

[17] A. L. N. Reddy, P. Banerjee and A. G. Abraham, [ReBA88] "I/O Embedding in Hypercubes," *Proceedings of the 1988 International Conference on Parallel Processing*, Vol. I, pp. 331-338, August 1988.

[18] Saad, Y. and Schultz, M., [SaSc85] "Data communication in hypercubes," *Research Report 428*, Department Computer Science, Yale University, 1985.

[19] Scheuermann P. and G. Wu, [ScWu84] "Heuristic Algorithms for Broadcasting in Point-to-Point Computer Networks," *IEEE Transactions on Computers*, Vol. 33, pp. 804-812, Sept. 1984.

[20] Seitz, C. [Seit85] "The COSMIC Cube," *Communications of the ACM*, Vol. 28, No. 1, Jan. 1985, pp. 22-23.

IMPROVED SEPARATORS FOR PLANAR GRAPHS

F. R. K. Chung

Bellcore

Morristown, NJ

ABSTRACT

The n vertices of a planar graph can be partitioned into three sets A, B, C, such that no edge connects a vertex in A with a vertex in B, A and $B each have at most $n/2$ vertices, and C contains no more than $3\sqrt{6}\sqrt{n}$ vertices. The constant $3\sqrt{6}$ is an improvement over previous bounds in several papers by Lipton and Tarjan, Djidjev, and Venkatesan. A version for planar graphs with vertex-costs is also given.

1.   Introduction

Lipton and Tarjan [1] first proved the planar separator Theorem:

Theorem [LT1]   The n vertices of a planar graph can be partitioned into three sets, A, B, C, such that no vertex in A is adjacent to a vertex in B, A and B each have at most $n/2$ vertices, and C contains no more than $\sqrt{8}/(1 - \sqrt{2/3})\sqrt{n}$ vertices.

Due to the great impact of this theorem in a wide range of areas, much attention has been focused upon improving the constants involved. The constant $\sqrt{8}(1 - \sqrt{2/3})$ (~ 15.413) was improved to $\sqrt{6}(1 - \sqrt{2/3})$ (~ 13.348) by Djidjev [D] and later to about 9.587 [D]. The best constant known so far is $7 + 1/\sqrt{3}(\sim 7.587)$ by Venkatesan [V]. In this paper, we will improve the constant to $3\sqrt{6}(\sim 7.348)$. Similar improvements can be made for planar graphs with vertex costs. Suppose G is an n-vertex planar graph with non-negative vertex costs summing to 1. Then G can be separated into two parts, each with total vertex cost not exceeding $\frac{1}{2}$, by removing $c\sqrt{n}$ vertices, where the best known constant for c is $8 + \frac{16}{81} \cdot (\sqrt{6}/(1 - \sqrt{2/3}))(\sim 10.636737)$ by Venkatesan [V]. In this note we will show that this constant can be improved to $\frac{108}{19}\sqrt{3}(\sim 9.845)$.

We remark that if we only require A and B each have at most $2n/3$ vertices (instead of $n/2$ in the Theorem stated above, the best bound due to Gazit [G] for C is $7\sqrt{n}/3$, improving the original bound of $2\sqrt{2}\sqrt{n}$ in [LT1].

Preliminaries

The method we use here is based on the results of Lipton and Tarjan [LT1] [LT2], Djidjev [D] and Venkatesan [V]. The underlying approaches are very similar. Only the optimization formulation in the next section is somewhat different. Here we state the facts in [LT1], [D], [V] that we need here.

Lemma 1: [LT1] Let G be a planar graph of radius s, with non- negative vertex-costs summing to 1. Then the vertices of $G]$] can be partitioned into three parts, A, B, C, such that there is no edge between A and B, neither A nor B has total vertex cost exceeding $\frac{2}{3}$, and C contains most $2s + 1$ vertices.

Lemma 2: [d] Let G be an n-vertex planar graph of radius s. For any real number r, $\frac{1}{2} \leq r \leq 1$, there exists a set $S \subseteq V(G)$ with at most $3s + 1$ vertices such that by removing vertices in S from G the remaining graph is separated into three parts A, B, C such that A, B each contain at most $(1 - r)n$ vertices, and C contains at most rn vertices.

Lemma 3: [V] For any integer s, an n-vertex planar graph G contains a subgraph H of at least $n - n/s$ vertices so that any subgraph of H can be embedded into another planar graph of radius $s - 1$.

3. Separation into two halves

Theorem 1   The n vertices of a planar graph can be partitioned into three parts A, B and C, such that no vertex in A is adjacent to a vertex in B, A and B each have at most $n/2$ vertices, and C contains at most $3\sqrt{6}\sqrt{n}$ vertices.

Proof   Let G denote a planar graph on n vertices. We will determine A, B and C iteratively as follows:

> **Step 0** Set $s = \lfloor \sqrt{\frac{n}{6}} \rfloor$ and use Lemma 3 to find a set S_0 with at most n/s vertices and embed $G - S_0$ into a graph G' of radius $s - 1$. Set $A = B = \varnothing, C = S_0$ and $j = 1$.

In general, for $i = 1, 2, \cdots$, the step i can be described as follows:

> **Step i** Set $r = (\frac{n}{2} - |A|)/n'$ where $n' = |V(G')|$. For $j \leq 2$, use Lemma 2 to find a separator S' containing at most $3(s - 1)$ vertices which separate G' into A', B' and C' with $|A'| \leq |B'| \leq (1 - r)n'$ and $|B'| \leq |C'| \leq rn'$. Set A to be the smaller one of $A \cup C'$ and $B \cup B'$; B to be the larger one of $A \cup C'$ and $B \cup B'$; and C to be $C \cup S'$. If $j < 2$, apply Lemma 3 to the induced subgraph of A' and form a new graph G' of radius $s - 1$, set $j = j + 1$ and repeat Step i. If $j = 2$, set i to be $i + 1, j$ to be 1. Then set s to $\sqrt{\frac{n'}{6}}$ when applying Lemma 3 to A' and form G' of radius $s - 1$, add n'/s vertice to C. Then go to the next step until G' is empty.

The correctness of this algorithm can be established by verifying the following facts in Step i. Suppose at the beginning of Step i A has p vertices, B has q vertices and G' has w vertices. The new A and B have no more than the maximum of $p + |C'|$ and $q + |B'|$. Since $p + |C'| \leq p + rw \leq p + (\frac{n}{2} - p) \leq \frac{n}{2}$ and $q + |B'| \leq q + (1 - r)w \leq q + w - \frac{n}{2} + p \leq \frac{n}{2}$, the new A and B have no more than $n/2$ vertices. Furthermore $|A'| \leq w/3$ since $|A'| \leq |B'| \leq |C'|$ and $|A'| + |B'| + |C'| \leq w$. This implies the new G' has at most $w/3$ vertices. Because of our choice of j, for each i, Step i is repeated twice for $j = 1$ and 2 (except for possibly the last step). So altogether $|G'|$ is reduced by a factor of 9.

We can bound the size of the separator $C = C(n)$ as follows:

$$C(n) \leq 2\sqrt{6}\sqrt{n} + \sum_{i \geq 1} 2\sqrt{6}\sqrt{\frac{n}{9^i}}$$

$$\leq 2\sqrt{6}\sqrt{n} + |C(\frac{n}{9})|$$

Therefore

$$|C(n)| \leq 3\sqrt{6}\sqrt{n}$$

This completes the proof for Theorem 1.

4. Separating planar graphs with vertex costs

Theorem 2   Let G be an n-vertex planar graph with non-negative vertex costs summing to 1. Then the vertices of G can be partitioned into three sets A, B, C such that no vertex in A is adjacent to a vertex in B, neither A nor B has total cost exceeding $\frac{1}{2}$, and C contains no more than $\frac{108\sqrt{3}}{19}\sqrt{n}$ vertices.

Proof   We will use the following algorithm, which is slightly different from that in Theorem 1.

Step 0 Set $s = \lfloor \sqrt{\frac{n}{12}} \rfloor$ and use Lemma 3 to find a set S_0 with at most n/s vertices and embed $G - S_0$ into a graph G' of radius $s-1$. Then use Lemma 1 to partition G' into A', B' and C' with $|c(A')| \leq |c(B')| \leq \frac{2}{3}$ and $|C'| \leq 2s + 1$ where $c(A')$ denotes the total cost of the vertices in A'. Apply Lemma 3 to find a set S'_0 with at most $|V(B')|/s$ vertices and set G' to be the graph of radius $s - 1$ into which $B' - S')_0$ can be embedded. Apply Lemma 1 to partition B' into A'', B'', and C'' with $|c(A'')| \leq |c(B'')| < \frac{2}{3}|c(B')|$. Now choose A to be the smaller (in vertex cost) of A' and B'', and B to be the larger one. Set G' to be A'', C to be

$$C' \cup C'' \cup S_0 \cup S''_0, \text{ and } j \text{ to be } 3.$$

In general for $i = 1, 2, \cdots$, Step i can be described as follows:

Step i For $j \leq 6$, using Lemma 1, the n' vertices of G' are partitioned into A', B' and C' with $|c(A')| \leq |c(B')| \leq \frac{2}{3}|c(G')|$, and $|C'| \leq 2s - 1$ where $c(A')$ denotes the total cost of the vertices in A'. Set A to be $A \cup A'$. Then set A to be the smaller of A and B, and B to be the larger one. If $j < 6$, apply Lemma 3 to the induced subgraph of B' and form G' of radius $s - 1$. Set $j = j + 1$ and repeat Step i. If $j = 6$, set i to be $i + 1$ and j to be 1. Then set s to be $\sqrt{\frac{V(G')}{12}}|$ when applying Lemma 3 on B' and form G' or radius $s - 1$. Add $|V(G')|/s$ vertices to C and repeat Step i until G' becomes empty.

To verify that A, B, C satisfy the required conditions we note that in Step i, the new A and B have vertex-cost no more than the sum of $c(A')$ and the vertex-cost of the old A. Since the old A have vertex-cost no more than old B and $c(A') \leq c(B')$, the new A and B have vertex-cost no more than $1/2$ and the new G' has vertex cost at most $2/3$ of the old G'. Because of our choice of j, Step i is repeated 6 times (except for Step 1, which is repeated 4 times). The total size of $C = C(n)$ can be bounded as follows:

$$|C(n)| \leq 2\sqrt{12n} + \sum_{i \geq 1} \sqrt{12n[\frac{2}{3}]^{6i}} \leq 2\sqrt{12n} + |C(n(\frac{2}{3})^6)|$$

Therefore,

$$|C(n)| \leq \frac{108\sqrt{3}}{19}\sqrt{n}$$

This completes the proof of Theorem 2.

5. Concluding remarks

There exists an n-vertex planar graph for which any separator bisecting the graph contains at least $\frac{2}{\sqrt{3}}\sqrt{n} \approx 1.155\sqrt{n}$. It is not hard to check that the graph with vertex set consisting of points $\{x(1,0) + y(0, \frac{\sqrt{3}}{2}) : x, y \epsilon Z\}$ in a regular hexagon of side $\sqrt{\frac{n}{3}}$ and edges between points of distance ≤ 1 satisfies the above property. So the best constant for planar separators lies between 1.155 and 7.348 for (unweighted) planar graphs; and between 1.155

and 9.845 for the weighted version. There is still considerable room for further improvement.

REFERENCES

[1] H. N. Djidjev, [D] "On the problem of partitioning planar graphs", *SIAM J. Algebra Discrete Methods* 3 (1982), 229-241.

[2] H. Gazit, [G] "An improved algorithm for separating a planar graph, preprint.

[3] R. J. Lipton and R. E. Tarjan, [LT1] "A separator theorem for planar graphs, *SIAM J. Appl. Math.* 36 (1979), 177-189.

[4] R. J. Lipton and R. E. Tarjan, [LT2] "Applications of a planar separator theorem", *SIAM J. Comput.* 9 (1980), 615-627.

[5] S. M. Venkatesan, [V] "Improved constants for some separator theorems", *J. of Algorithms* 8 (1987), 572-578.

Canonical Forms for the
Digraph Isomorphism Problem

Dennis Glenn Collins
University of Puerto Rico, Mayaguez

ABSTRACT

The paper shows that no canonical form may have the <u>sub–digraph property</u> that upon removing any set of vertices the remainder of the ordering still has the given canonical form, but that the MAX canonical form has the <u>truncation sub–digraph property</u> and presents other desirable properties of this canonical form.

I

Canonical forms, such as the unique row–reduced echelon form and the Jordan normal form, have wide application to matrix problems. Some, such as the Jordan normal form, or eigenvalue expansion, even have application to the study of the adjacency matrix of a graph, and give valuable information about the graph. However eigenvalues do not provide by any means a complete set of invariants for the graph or digraph isomorphism problem, nor does any other set of invariants seem to be known at present. An obvious reason for the insufficiency of the spectrum (i.e. the existence of co–spectral graphs) is that the eigenvalue expansion is not adapted to the symmetry group (the symmetric group) of the adjacency matrix (so that a similarity transformation may exist mapping one adjacency matrix onto another without there being a permutation of the vertices or symmetric group element that does so). The current view seems to be that the adjacency matrix is too cumbersome to study under the symmetric group [1, 4]. However this paper will try to show some potentialities of a canonical form for the adjacency matrix of a graph or digraph under the symmetric group.

First we may show that we cannot expect too much from any canonical form. The most desirable property of a canonical form for the digraph isomorphism may be termed the *sub–graph* or *sub–digraph property:* that if we have the adjacency matrix in its canonical form and remove certain vertices (and thus the rows and columns that correspond to them,) then the square matrix that forms from the remaining entries (as when calculating a minor of a determinant) will still have the given canonical form.

We show that it is impossible for a canonical form to have this property. By a canonical form we mean a way (or ways) of ordering the vertices, so that the adjacency matrix with respect to that ordering becomes uniquely determined, regardless of the original ordering of vertices.

First we prove the digraph case: Consider the folloing digraph with two vertices: $1 \to 2$. There are only two possible canonical forms for it:

$$
A = \begin{array}{c@{}c} & \begin{array}{cc} 1 & 2 \end{array} \\ \begin{array}{c} 1 \\ 2 \end{array} & \left[\begin{array}{cc} 0 & 1 \\ 0 & 0 \end{array} \right] \end{array}
\quad \text{or} \quad
B = \begin{array}{c@{}c} & \begin{array}{cc} 2 & 1 \end{array} \\ \begin{array}{c} 2 \\ 1 \end{array} & \left[\begin{array}{cc} 0 & 0 \\ 1 & 0 \end{array} \right] \end{array}
$$

Now consider the following digraph with three vertices:

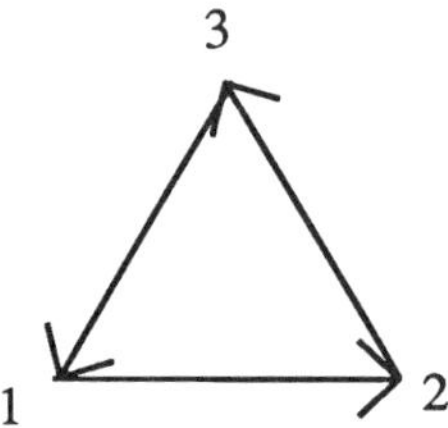

There are six (= #S$_3$) possible orderings for it, which reduce to only two different possible canonical matrices: C (representing orderings 123, 231, and 312) or D (representing orderings 132, 213, and 321), as illustrated below.

$$
C = \begin{array}{c@{}c} & \begin{array}{ccc} 1 & 2 & 3 \end{array} \\ \begin{array}{c} 1 \\ 2 \\ 3 \end{array} & \left[\begin{array}{ccc} 0 & 1 & 0 \\ 0 & 0 & 1 \\ 1 & 0 & 0 \end{array} \right] \end{array}
\quad \text{or} \quad
D = \begin{array}{c@{}c} & \begin{array}{ccc} 1 & 3 & 2 \end{array} \\ \begin{array}{c} 1 \\ 3 \\ 2 \end{array} & \left[\begin{array}{ccc} 0 & 0 & 1 \\ 1 & 0 & 0 \\ 0 & 1 & 0 \end{array} \right] \end{array}
$$

Now a canonical form must select one of A or B, and one of C or D. There are four cases: AC, BC, AD, BD. However it is possible to remove a vertex, so as to invalidate each possible selection. For example if AC is selected then removing vertex 2 from C yields canonical matrix B instead of A:

$$B = \begin{array}{c} 1 \\ \\ 3 \end{array} \left[\begin{array}{cc} 0 & 0 \\ \\ 1 & 0 \end{array} \right]$$

a contradiction if the sub–digraph property holds, since the sub–digraph for verices 1 and 3 should be equal to matrix A, given that the canonical form selects AC. The other three cases are proved similarly, removing vertex 3 from C, vertex 2 from D, and vertex 3 from D, respectively.

Since graphs are a proper subset of digraphs, the subgraph property might hold for a canonical form for graphs, even though the sub–digraph property does not hold for the more general digraph. The next example shows that the sub–graph property cannot hold for graphs.

Graph case: Consider the following graph with three vertices:

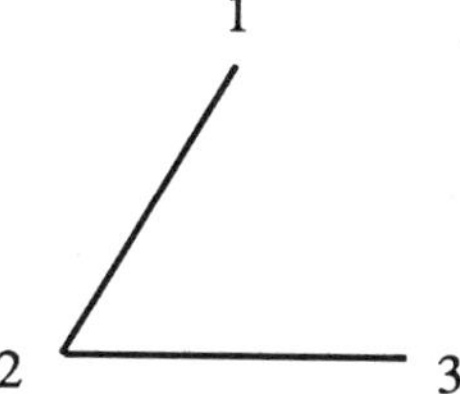

There are three possible canonical matrices for it, A (representing orderings 123, 321), B (representing 132, 312), and C (representing 213, 231).

$$A = \begin{array}{c} 1 \\ 2 \\ 3 \end{array} \begin{array}{ccc} 1 & 2 & 3 \end{array} \left[\begin{array}{ccc} 0 & 1 & 0 \\ 1 & 0 & 1 \\ 0 & 1 & 0 \end{array} \right] \qquad B = \begin{array}{c} 1 \\ 3 \\ 2 \end{array} \begin{array}{ccc} 1 & 3 & 2 \end{array} \left[\begin{array}{ccc} 0 & 0 & 1 \\ 0 & 0 & 1 \\ 1 & 1 & 0 \end{array} \right] \qquad C = \begin{array}{c} 2 \\ 1 \\ 3 \end{array} \begin{array}{ccc} 2 & 1 & 3 \end{array} \left[\begin{array}{ccc} 0 & 1 & 1 \\ 1 & 0 & 0 \\ 1 & 0 & 0 \end{array} \right]$$

Now consider the following graph, the cycle C_4:

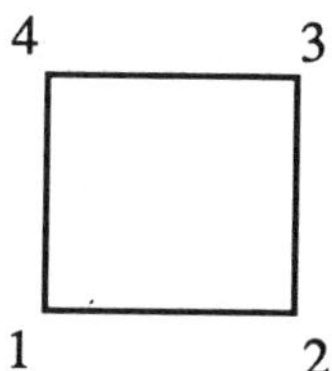

There are only three possible canonical forms for it, D (representing orderigns 1234, 1432, 3214, 3412, 2143, 4123, 2341, 4321), E the <u>min</u>imal form (representing orderings 1324, 1342, 3124, 3142, 2413, 4213, 2413, 4231), and F the <u>max</u>imal form (representing orderings 1234, 1423, 3214, 3421, 2134, 4132, 2314, 4312). However as the following diagram of matrices illustates, for any selection of a canonical form D, E, or F, it is possible to remove a vertex, so as to obtain at least *two* different of the canonical forms A, B, and C. This situation invalidates any of the nine combinations of a selction of A, B, or C with D, E, or F as a possible canonical form with the sub–graph property:

$$
D = \begin{array}{c|cccc} & 1 & 2 & 3 & 4 \\ \hline 1 & 0 & 1 & 0 & 1 \\ 2 & 1 & 0 & 1 & 0 \\ 3 & 0 & 1 & 0 & 1 \\ 4 & 1 & 0 & 1 & 0 \end{array}
\qquad
A = \begin{array}{c|ccc} 1 & 0 & 1 & 0 \\ 2 & 1 & 0 & 1 \\ 3 & 0 & 1 & 0 \end{array}
\qquad
C = \begin{array}{c|ccc} 1 & 0 & 1 & 1 \\ 2 & 1 & 0 & 0 \\ 4 & 1 & 0 & 0 \end{array}
$$

$$
E = \begin{array}{c|cccc} & 1 & 3 & 2 & 4 \\ \hline 1 & 0 & 0 & 1 & 1 \\ 3 & 0 & 0 & 1 & 1 \\ 2 & 1 & 1 & 0 & 0 \\ 4 & 1 & 1 & 0 & 0 \end{array}
\qquad
B = \begin{array}{c|ccc} 1 & 0 & 0 & 1 \\ 3 & 0 & 0 & 1 \\ 2 & 1 & 1 & 0 \end{array}
\qquad
C = \begin{array}{c|ccc} 3 & 0 & 0 & 1 \\ 2 & 1 & 0 & 0 \\ 4 & 1 & 0 & 0 \end{array}
$$

$$
F = \begin{array}{c c} & \begin{array}{cccc} 1 & 2 & 4 & 3 \end{array} \\ \begin{array}{c} 1 \\ 2 \\ 4 \\ 3 \end{array} & \left[\begin{array}{cccc} 0 & 1 & 1 & 0 \\ 1 & 0 & 0 & 1 \\ 1 & 0 & 0 & 1 \\ 0 & 1 & 1 & 0 \end{array} \right] \end{array}
$$

$$
B = \begin{array}{c} 2 \\ 4 \\ 3 \end{array} \left[\begin{array}{ccc} 0 & 0 & 1 \\ 0 & 0 & 1 \\ 1 & 1 & 0 \end{array} \right]
$$

$$
C = \begin{array}{c} 1 \\ 2 \\ 4 \end{array} \left[\begin{array}{ccc} 0 & 1 & 1 \\ 1 & 0 & 0 \\ 1 & 0 & 0 \end{array} \right]
$$

II

The above examples, besides showing that the sub–digraph and sub–graph properties cannot hold for any canonical form, also show that there may be only a limited number of possible canonical forms of a graph with n vertices. In fact if the graph has m symmetries, then $n! = \#S_n = m \times r$, where r is the number of different possible canonical forms for the graph. In the extreme case of the complete graph K_n there is only $r = 1$ possible canonical form, since all orderings yield the same adjacency matrix. In the above example C_4 has $m = 8$ symmetries (including the identity), so that there are only $\#S_4/m = 24/8 = 3$ different canonical forms from which to choose.

As can be observed from the above examples, two well–defined canonical forms are 1) the MAX canonical form [2, 3] and 2) the MIN canonical form. These are defined by forming the adjacency matrix into one large number by putting one row after another from the first row to the last row, and then obtaining the maximal (for the MAX canonical form) or minimal (for the MIN canonical form) value that can be obtained over all permutations of vertices. This number clearly exists, even for a weighted digraph, assuming the number of vertices is finite. Of course, the ordering of vertices that yields the given canonical form is not necessarily unique, but two graphs will be isomorphic if and only if they have the same MAX (or MIN) canonical form, since in that case the two sets of vertices can be matched up in at least one way, to give the same adjacency matrix.

III

Next we cover some of the properties of the MAX canonical form. In particular, we will try to show that the MAX form possesses a restricted from of the sub–graph property which may be termed the *truncation sub–graph property*. The following example will try to illustrate the type of "chaining" that the MAX form possesses, and also serve to illustrate some of the other propositions.

Example 1   The MAX canonical form of the following graph is shown below:

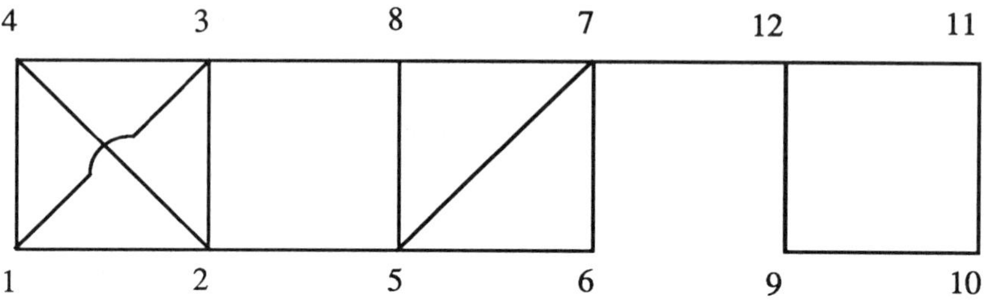

	2	3	1	4	5	8	7	6	12	9	11	10
2	0	1	1	1	1	0						
3	1	0	1	1	0	1						
1	1	1	0	1	0	0						
4	1	1	1	0	0	0						
5	1	0	0	0	0	1	1	1				
8	0	1	0	0	1	0	1	0				
7					1	1	0	1	1			
6					1	0	1	0	0			
12							1	0	0	1	1	0
9									1	0	0	1
11									1	0	0	1
10									0	1	1	0

Proposition 1   *(interchangeability of vertices)   The interchange of two vertices, corresponding to position i and j of an ordering, has no effect on the adjacency matrix, or any other adjacency matrix of the digraph or graph, if the two rows and columns that correspond to the two vertices 1) are equal, or 2) are equal except for the matrix positions*

$$\begin{bmatrix} a_{ii}\, a_{ij} \\ a_{ji}\, a_{jj} \end{bmatrix} = \begin{bmatrix} 0 & 1 \\ 1 & 0 \end{bmatrix}.$$

In case 1), the matrix for positions

$$\begin{bmatrix} a_{ii}\, a_{ij} \\ a_{ji}\, a_{jj} \end{bmatrix}$$

contains all zeros:

$$\begin{bmatrix} 0 & 0 \\ 0 & 0 \end{bmatrix}.$$

There is no point trying to decide which of such vertices should come first, since any ordering of them will yield the same adjacency matrix. In Example 1, vertices 9 and 11 satisfy case 1) of Proposition 1 and vertices 1 and 4 satisfy case 2). As far as the MAX ordering goes, any set of interchangeable vertices will come as a unit into the ordering, as Example 1 illustrates. The proof of Proposition 1 is merely to observe that interchanging two vertices changes the row ordering by interchanging the two rows and at the same time the column ordering by interchanging the two columns. Except for the above four matrix positions, an entry of the changing rows or columns makes only one move under the given interchange, and thus must be equal to the quantity it changes with. The above four matrix positions make two moves, and the two cases of Proposition 1 correspond to whether the interchanging vertices are adjacent (case 2) or not (case 1). The diagonal entries are always 0. If Proposition 1 holds, the two vertices have exactly the same adjacency structure and therefore will be interchangeable in any adjacency matrix. In the complete graph K_n on n vertices, any interchange satisfies case 2).

Proposition 2   *(initial vertex) In the MAX ordering, the first vertex will come from the set of vertices with maximal degree (or out–degree in the case of a digraph).*

This rule holds because the MAX possible value for the first row is a 0 (the 11– diagonal entry) followed by the greatest number of 1's; this greatest number of 1's is obtained by taking one of those vertices with the greatest number of adjacent edges, i.e. maximal degree. In Example 1, vertices 2, 3, 5 and 7 have a maximal degree of 4, and the first vertex of the ordering is one of these vertices, namely 2.

Proposition 3 *("chaining" rule for a graph) If a vertex v_i comes into the MAX ordering, then all remaining vertices adjacent to it must come into the ordering (in some order) as soon as possible, that is, as soon as all the vertices adjacent to the previous vertex v_{i-1} have entered the ordering.*

If the rules holds and all vertices adjacent to v_{i-1} have entered the ordering (as columns), then in the row of v_i, the next column of the ordering may have a 1 if a vertex adjacent to v_i enters, whereas otherwise it will have a 0. The maximal value for the row of v_i is then found by putting one of those vertices adjacent to vertex v_i, which puts a 1 in this column position. For Example 1, once vertex 3 enters in column (and row) 2, then vertex 8, which is adjacent to vertex 3, must enter as soon as all vertices (3, 1, 4, 5) adjacent to the first vertex of the ordering, vertex 2, have entered the ordering. The entry of vertex 8 thus puts a 1 in row 2, column 6, whereas the entry of any other vertex (excluding of course 2, 3, 1, 4, 5) would put a 0 in row 2, column 6.

Thus, after the starting vertex, it is possible to assign to each vertex a parent vertex, which causes the given vertex to enter the ordering. The set of all vertices which enter the ordering because of their adjacency to certain vertex may be termed the "children" of the vertex. In Example 1, the children of vertex 2 are vertices 3, 1, 4, 5; the child of vertex 3 is vertex 8; vertices 1 and 4 have no descendents; the children of vertex 5 are vertices 7 and 6, and so on. The entry of vertices into the ordering thus corresponds to a breadth–first spanning tree for the connected component of the first vertex to enter the ordering.

Corollary 1 to Proposition 3 *(invariant sub–space theorem) In the MAX canonical form, connected components of a graph enter as units, that is, the MAX adjacency matrix has a block–diagonal form, where the blocks correspond the the connected components of the graph.*

Corollary 2 to Proposition 3 *(zero rows) In the MAX canonical adjacency matrix of a graph, rows with all zeros (corresponding to isolated vertices) come at the bottom.*

Proposition 4 *(binary tree number of an ordering) The MAX ordering of a graph will be that ordering with the maximal binary tree number.*

What is the binary tree number of an ordering or part of an ordering--it is found by counting the number of vertices satisfying adjacency/non–adjacency conditions for each added vertex and forming all these numbers into one large number.

Stage 1: Count the first vertex of the ordering plus the number of vertices adjacent to the first vertex of the ordering. This sum will be the first integer of the binary tree number. Then count the number of vertices not adjacent to the first vertex; this sum will be the second number of the binary tree number of the given ordering. The sum of these two integers will be n, the total number of vertices of the graph. It is helpful to list the vertices in each group.

Stage 2: Divide each group (i.e. subset), denoted 1 and 0, of Stage 1 into two groups; one may contain the second vertex of the ordering and/or those vertices adjacent to the second vertex and one contains vertices not adjacent to the second vertex of the ordering. Now there are four groups, which can be denoted 11, 10, 01, 00, where for example group 10 contains vertices adjacent to the first vertex but not the second. The first vertex is omitted at Stage 2, so that there are only n − 1 elements in the four groups. The binary tree number at this stage is the number of vertices in groups 1, 0, 11, 10, 01, and 00 respectively. If an ordering has the largest binary tree number at any stage, it determines the MAX ordring up to that point, because it is able to muster the largest number of 1's toward the beginning of each row successively determined in the adjacency matrix.

Stage 3 and higher: At the next stage, eight groups are created by dividing each group of Stage 2 into two new groups, depending on whether each vertex is adjacent to the third vertex of the ordering, or not, and so on for higher stages, if necessary. Theoretically it may be necessary to check all n! orderings with this method if the graph is the complete graph K_n on n vertices. In the case of Example 1, there are four vertices, 2, 3, 5, and 7 to check with maximal index 4.

Sample constructions are illustrated next, where some possibilities are omitted but in general only vertices from the highest–ranking non–empty class at any stage should be selected for the next vertex of the ordering.

ordering	binary tree number
2	57
3	3116
1	20010106
4	1000000100010006
5	00000000000000100000001000000024

ordering	binary tree number
3	57
2	3116
1	20010106
4	1000000100010006
8	00000000000000100000001000000015

ordering	binary tree number
5	57
7	3116
8	11010115

As with the adjacency matrix itself in the MAX ordering, the integers are not actually put one after another, but listed row after row.

The ordering 23145 beats the ordering 32148 because the last digits 24 of the binary tree number of 23145 are greater than the last two digits 15 of the binary tree number 32148, all previous digits being the same. The larger number 24 > 15 at the end represents the extra edge from vertex 5 to vertex 7, since vertex 8 is adjacent to only vertex 7, but vertex 5 is adjacent to both vertices 6 and 7. The ordering 578... goes out of the running for MAX at stage three.

It remains to be seen if an efficient way of obtaining the binary tree numbers by computer can be worked out, whereby the exponential growth is offset by saving only essential information and the reduction of the number of remaining vertices by one at each stage.

In any case, to obtain the binary tree number of an ordering or the beginning of an ordering, such as 578, it is helpful to list the various subsets, as for example:

ordering	subsets
5	{2 5 6 7 8} {1 3 4 9 10 11 12}
7	{6 7 8} {2} {12} {1 3 4 9 10 11}
8	{8} {6} ∅ {2} ∅ {12} {3} {1 4 9 10 11}

To go from one stage to the next it is only necessary to ask if the given vertices are adjacent to the next vertex of the ordering, considering a vertex as adjacent to itself and removing previous vertices of the ordering.

Proposition 5   *(truncation sub–graph property)   If a given ordering yields the MAX canonical form and a subgraph is obtained by truncating the ordering at any point (i.e. eliminating all vertices after some vertex of the ordering), then the truncated ordering yields the MAX canonical form of the subgraph so obtained.*

Proof   It suffices to consider removing only one vertex at a time. If the vertex removed is an isolated vertex, then by the invariant subspace theorem, the MAX canonical form yields a block–diagonal adjacency matrix with a zero in the second one–dimensional block. Since the last row then contains all zeros, removing it cannot affect the ordering of the remaining vertices.

In the other case that the last vertex is adjacent to some other vertices, we may find the parent of the last vertex. Now consider the siblings of this parent. Apparently the parent is the last vertex in the ordering of those siblings that have offspring, since the offspring of any later sibling vertex must follow the offspring of the given parent vertex, but the offspring of the given parent vertex is by hypothesis last in the ordering. Now removing this last vertex may cause the given parent vertex to join the later siblings that have no offspring. However as these offspring–less vertices will be on an equal footing, it is permissible according to the MAX canonical form to have the given parent vertex listed as the first of them. As for the siblings ahead of the given parent in the ordering, they will stay ahead when the given parent vertex is weakened by removing the last vertex. Thus no change need be made in the order of the siblings of the given parent vertex to preserve MAX canonical form when the last vertex is removed. A similar argument applies to the parent of the siblings, and so on toward the beginning of the ordering. This type of argument applies if the given parent has no siblings. In short, the removal of the last vertex will weaken its ancestor vertices, but these vertices are already ordered last within competitive vertices because the vertex removed is last.

For *Example 1*, the parent of the last vertex 10 of the given ordering is vertex 9. The sibling of vertex 9 is vertex 11; removing vertex 10 causes vertices 9 and 11 to become on an equal offspring–less footing. However the truncated ordering is still a MAX canonical ordering.

To summarize, the MAX canonical form has the advantages that connected components of a graph enter as diagonal blocks of the adjacency matrix (invariant subspace theorem) and that truncating a MAX ordering yields a MAX ordering (truncation sub–graph property). In view of section I, we cannot expect any canonical form to have the (full) sub–graph property.

IV

To extend Propositions 3 to 5 to the digraph case, it is necessary mainly to replace edge by outward–directed edge, assuming the adjacency matrix is given in the form:

TO

FROM [   ]

as for section I.

With this substitution the "chaining rule" for graphs still applies to the outward–directed edges of digraphs, although it must be extended to include inwardly–directed edges:

Proposition 3'   *("chaining" rule for digraphs)*

> *1)   The "chaining" rule applies if we consider only directed edges (v_i, v) going outward from a vertex v_i, i.e. all such remaining vertices v must enter the ordering in some order as soon as all vertices w that have an outward edge (v_{i-1}, w) have entered, where v_{i-1} is the vertex before v_i in the ordering.*
>
> *2)   Once all vertices that can be reached from v_i or its ancestors (in the sense of the "chaining" rule for graphs, with edge replaced by outwardly–directed edge) by going outward (through any number of vertices) have entered the ordering (call these vertices the $\underline{v_i\text{–out}}$ $\underline{family}$), then the remaining vertices v with directed edges (v, v_i) coming into v_i must enter the ordering as soon as possible (that is, as soon as all vertices going into v_{i-1} have entered) in some order, where each such v brings together with it all remaining vertices that can be reached from v by going outward, i.e. the remainder of the v–out family.*

Thus we span all possible vertices going outward from a vertex with maximal out–index, and then gradually pick up vertices coming in, together with all additional vertices spanned by them going outward.

Proposition 4'   *(binary tree number for a digraph)   The binary tree number remains the same as the graph case, except that it is necessary to include all vertices at each stage, so that the number of vertices does not decrease. The binary tree number will be obtained by checking whether or not a given vertex of a group can be reached from the next vertex to enter the ordering (by an outwardly–directed edge), where all vertices must be considered at each stage, since an outward edge from vertex i to vertex j does not imply an outward edge from vertex j to vertex i (i.e. both in–and out–adjacency information is not obtained by one binary division).*

The reason that this proposition holds is basically that symmetry of the adjacency matrix is nowhere involved in deriving it (except in eliminating vertices in the binary tree number), and that row–by–row maximization corresponds to out–index maximization. To deal with in–degree it would be possible to consider column–by–column maximization, or take transposes of the adjacency matrices involved.

Example 2. The following digraph has the MAX canonical form below:

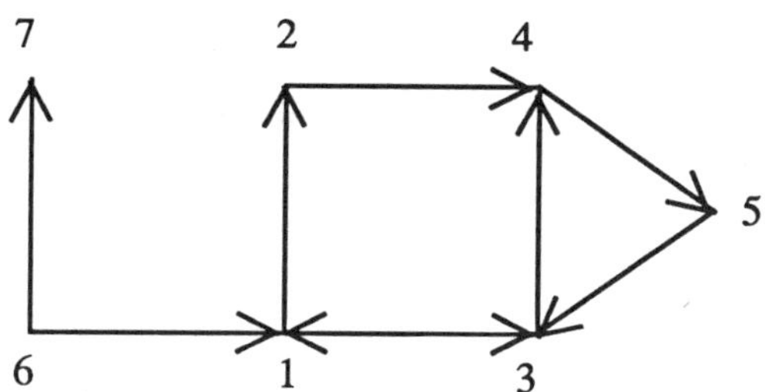

$$\begin{array}{c|ccccccc}
 & 1 & 3 & 2 & 4 & 5 & 6 & 7 \\
\hline
1 & 0 & 1 & 1 & 0 & 0 & 0 & 0 \\
3 & 1 & 0 & 0 & 1 & 0 & 0 & 0 \\
2 & 0 & 0 & 0 & 1 & 0 & 0 & 0 \\
4 & 0 & 0 & 0 & 0 & 1 & 0 & 0 \\
5 & 0 & 1 & 0 & 0 & 0 & 0 & 0 \\
6 & 1 & 0 & 0 & 0 & 0 & 0 & 1 \\
7 & 0 & 0 & 0 & 0 & 0 & 0 & 0 \\
\end{array}$$

Here the first vertex of the ordering will be one of those with maximal out–degree, i.e. vertex 1 or 3, which have out–degree 2. The two main candidates for the MAX ordering start with 132 or 314. The subsets involved in comparing these two orderings are as follows:

ordering	subsets
1	{1 2 3} {4 5 6 7}
3	{1 3} {2} {4} {5 6 7}
2	∅ {1 3} {2} ∅ {4} ∅ ∅ {5 6 7}

ordering	subsets
3	{1 3 4} {2 5 6 7}
1	{1 3} {4} {2} {5 6 7}
4	∅ {1 3} {4} ∅ ∅ {2} {5} {6 7}

In the second row of the 314 construction, we see that 1 has an outward edge to 3 (and assumed 1 for the construction) but not 4, and an outward edge to 2 but not 5,

6, or 7. These subsets correspond to digraph binary tree numbers (obtained by counting the number of elements in each subset):

ordering	digraph binary tree number
1	3 4
3	2 1 1 3
2	0 2 1 0 1 0 0 3

ordering	digraph binary tree number
3	3 4
1	2 1 1 3
4	0 2 1 0 0 1 1 2

Thus 132 beats 314 because the last four integers 1003 of the binary tree number of 132 are greater than the last four integers 0112 of the binary tree number of 314, all previous integers being the same. Then by the "chaining" proposition, vertex 3 brings in vertex 4 and vertex 4 brings in vertex 5, to yield 13245 as the MAX canonical ordering of the vertices spanned outward from vertex 1, i.e. the 1–out family. Next comes vertex 6 which maps into vertex 1, and vertex 7, which is the remaining part of the 6–out family. (Remark 1: Rows of zeros need not come at the end of the matrix for the digraph case. Remark 2: The v–out families for different v are not necessarily disjoint. For example the 6–out family contains the 1–out family in the above example. The v–out family may also contain vertices coming into v if they appear before v in the ordering.)

Corollary 1 to Proposition 3'   *(invariant subspace theorem) Isolated components of a digraph (i.e. subsets of vertices with no directed edges between them) correspond to diagonal blocks of a MAX canonical adjacency matrix.*

Corollary 2 to Proposition 3' and 4'   *(root of a tree) If a directed tree is defined by directed edges outward from the root node, then in the MAX canonical form of the adjacnecy matrix of the tree, the root node will either be the first vertex of the ordering, or be the vertex v_i such that row i is the lowest row with a 1 below the diagonal.*

Proof of Corollary 2   Directed edges coming into a given vertex of the ordering correspond to 1's below the diagonal, and directed edges coming out of a given vertex correspond to 1's above the diagonal. Once the root node of the directed tree enters the ordering, all directed edges go outward, and therefore are above the diagonal. ❑

Proposition 5'   *(truncation sub–digraph property) If a MAX canonical ordering for a digraph is truncated at any point, the sub–digraph that remains still has a MAX canonical ordering.*

Proof   As for the graph case, removing the last vertex may weaken some vertices (or systems of vertices) until they are interchangeable with later vertices (or systems of vertices), but such vertices (or systems of vertices) already enter later than stronger vertices (or systems of vertices); as a consequence no change is necessary to preserve the MAX canonical form if the last vertex (or a series of vertices at the end of the ordering) are removed. ❏

The form of the above results is such that the MAX canonical ordering may be amenable to calculation by some artificial intelligence type of programs, together with some number crunching.

V

One application of the MAX canonical form is that it gives a precise ordering of the power ranking of vertices, at least if we average over symmetries which yield the same MAX adjacency matrix. In fact this criterion was decisive in selecting the MAX canonical form over other possibilities, although the MAX canonical form also seems to possess several characteristics of the Jordan and unique row–reduced echelon form of matrices.

REFERENCES

[1]   A.Z. Berztiss, A Backtrack Procedure for Isomorphism of Directed Graphs, *J. of the ACM 20*, No. 3 (July 1973) 365 – 377.

[2]   J.F. Nagle, On Ordering and Identifying Undirected Linear Graphs, *J. Math Physics* (1968) 1588 – 1592.

[3]   M. Randic, On the Recognition of Identical Graphs Representing Molecular Topology, *J. Chem Phys 60* (1974) 3920– 3928.

[4]   Ronald Read & Corneil, Derek, The Graph Isomorphism Disease, *J. Graph Theory 1* (1977) 339 – 363.

The author wishes to thank seminar students D. Feliciano and A. Gonzalez for discussion of these results; of course they are not responsible for any errors.

The reuslts of Section I were presented in March 1988 by D. Feliciano at the annual meeting of Puerto Rico Junior Technical Society at Humacao, Puerto Rico; these results however were attributed to the author of the present paper.

Relations among Embedding Parameters for Graphs

Alice M. Dean
Skidmore College

Joan P. Hutchinson[1]
Smith College

ABSTRACT

Let G be a simple graph with n vertices, and let a_1, bt, γ, and θ denote, respectively, the arboricity, book thickness (also called pagenumber), genus, and thickness of G. We establish inequalities, each of which is best possible up to a constant, between pairs of these parameters. In particular, we show:

$$\theta \leq 6 + \sqrt{2\gamma - 2},$$
$$a_1 \leq bt + 1, \qquad and$$
$$\gamma \leq (\theta - 1)(n - 1).$$

1. Introduction

There are several different ways to characterize the embeddability of a graph G. Some of the numerical measures include the genus γ, the thickness θ, the book thickness (or pagenumber) bt, and the arboricity a_1. These parameters have many applications, including VLSI design and complexity theory.

In this paper, we study relationships between these parameters. We prove several inequalities, each of which is best possible up to a constant.

2. Definitions and Basic Properties

Throughout, **G** will denote a simple, *connected* graph, and **n** and **e** (or n_G and e_G, if more than one graph is being considered) will denote, respectively, the number of

---

[1] Research supported in part by N.S.F. grant # DCR–8411690.

vertices and edges of G. We will consider four parameters that measure the embeddability of G in different ways.

First, the (orientable) *genus of G*, denoted by γ, is the minimum number k such that G can be embedded on a sphere with k handles. Second, the *thickness of G*, denoted by θ, is the minimum number of edge–disjoint planar subgraphs whose union is G. Third, the *arboricity of G*, denoted by a_1, is the minimum number of edge–disjoint forests whose union is G. Finally, the *book thickness* or *pagenumber of G*, denoted by bt, is the minimum number of pages joined at a common spine such that G can be embedded with its vertices on the spine and edges on the pages.

All these parameters are known for the complete graphs, the complete bipartite graphs, and the generalized cubes. For example, $\gamma(K_n) = \lceil (n-3)(n-4)/12 \rceil$ for $n \geq 3$, $\theta(K_n) = \lfloor (n+7)/6 \rfloor$ (except that $\theta(K_9) = \theta(K_{10}) = 3$), $a_1(K_n) = \lceil n/2 \rceil$, and $bt(K_n) = \lceil n/2 \rceil$ for $n \geq 4$ (see [BCL, BK]).

There are two fundamental results concerning these parameters that will be used throughout the paper. The first is Euler's formula (and a corollary), and the second is Nash–Williams' formula for arboricity:

Theorem 1 *If G is embedded on S_g, the sphere with g handles, then, letting f denote the number of faces:*

(i) $n - e + f \geq 2 - 2g$, with equality if and only if the embedding is a 2–cell embedding,

(ii) $e \leq 3n - 6 + 6g$.

Theorem 2 [N] *For any graph G,*

$$a_1 = \max_{H < G} \left\lceil \frac{e_H}{n_H - 1} \right\rceil,$$

where H ranges over all nontrivial induced subgraphs of G.

It follows easily from Theorem 2 that a_1 is bounded in terms of the minimum degree δ and maximum degree Δ of G:

Corollary 2   *(i)* $\left\lceil \dfrac{\delta+1}{2} \right\rceil \leq a_1 \leq \left\lceil \dfrac{\Delta+1}{2} \right\rceil.$

(ii) [AH] If G is d–regular, then $a_1 = \left\lceil \dfrac{d+1}{2} \right\rceil = \left\lceil \dfrac{e}{n-1} \right\rceil.$

The arboricity and thickness of a graph G are closely related in size; namely, $\theta \leq a_1 \leq 3\theta$. The first inequality follows from the definitions, and the second follows from Theorems 1 and 2, which together imply that any planar graph has arboricity at most 3. Furthermore, these bounds are sharp, in the sense that for any θ, there are graphs that achieve each extreme; see [T, BH].

Another related parameter is the *vertex arboricity*, denoted by a. The vertex arboricity of a graph G is the minimum number of vertex–disjoint forests whose union is G. In [B] it is shown that $a \leq a_1$, and so any bound on arboricity will give the same bound for vertex arboricity.

3. Upper Bounds on Thickness and Arboricity

In [A], Asano showed that triangle–free graphs and toroidal graphs have thickness $\theta \leq \gamma + 1$. A recent result of Scheinerman [S] shows that for any G, the arboricity a_1 satisfies $a_1 \leq 2 + \sqrt{3\gamma}$. Since $\theta \leq a_1$, it follows that, for any G, $\theta \leq 2 + \sqrt{3\gamma}$. Archdeacon and Richter (unpublished manuscript) have obtained a similar bound on θ. Furthermore, the example of the complete graph K_n shows that this result is best possible up to a constant, since $\theta(K_n) \approx \sqrt{\gamma(K_n) / 3}$.

Using an idea due to M. Albertson and C. Thomassen (personal communication), we improve on Scheinerman's result for thickness, reducing the constant from $\sqrt{3}$ to $\sqrt{2}$.

Theorem 3 *If G is a simple graph, then*
$$\theta \leq 6 + \sqrt{2\gamma - 2}.$$

Proof   Let $k = 6 + \lfloor \sqrt{2\gamma - 2} \rfloor$ and let $G_1, G_2, ..., G_k$ each be graphs with the same vertex set as G, but initially no edges. For each vertex v with $\deg(v) \leq k$, remove its edges from G and place one edge in each G_i for $1 \leq i \leq \deg(v)$. Repeat this process for each vertex of degree at most k in the resulting graph until all the remaining vertices in the final graph G' have degree strictly greater than k.

Each of the graphs G_i generated by this process is acyclic: Suppose some G_i contained a cycle C. Consider the "oldest" vertex v on the cycle C, i.e., the vertex on C whose edges were removed from G before any other vertices of C. Then by the construction of the G_i's, the degree of v in C is 0 or 1, a contradiction.

Since all remaining vertices of G' have degree at least $k + 1 \geq 6 + \sqrt{2\gamma - 2}$, Thm. 1 (ii) yields
$$(6 + \sqrt{2\gamma(G) - 2}) \, n' \leq 2e' \leq 6n' + 12(\gamma(G') - 1),$$
and, since $\gamma(G') \leq \gamma(G)$,
$$(6 + \sqrt{2\gamma(G) - 2}) \, n' \leq 2e' \leq 6n' + 12(\gamma(G) - 1),$$
where n' and e' are the number of vertices and edges of G'. Thus,
$$n' \leq 6\sqrt{2\gamma(G) - 2}.$$

Now G' is a subgraph of the complete graph $K_{n'}$, which satisfies
$$\theta(K_{n'}) = \begin{cases} \lfloor (n' + 7) \rfloor / 6, & n' \neq 9, 10 \\ 3, & n' = 9, 10 \end{cases}$$

It follows that

$$\begin{aligned}
\theta(G') &\leq \theta(K_{n'}) \\
&\leq \lfloor (n' + 9) / 6 \rfloor \\
&\leq \lfloor (6\sqrt{2\gamma(G) - 2} + 9) / 6 \rfloor \\
&\leq \sqrt{2\gamma(G) - 2} + 6.
\end{aligned}$$

Hence, G' can be partitioned into at most $k = \lfloor \sqrt{2\gamma(G) - 2} \rfloor + 6$ planar subgraphs, say $P_1, \ldots, P_{k'}$, where $1 \leq k' \leq k$. Form the unions $P_i \cup G_i$ for $i = 1, \ldots, k'$.

We claim that each tree of G_i is incident with at most one vertex of P_i and hence $P_i \cup G_i$ is a planar graph. If not, find a tree T_i of G_i and two vertices x and y of $P_i \cap T_i$ such that the path P_{xy} from x to y in T_i contains no other vertex of P_i. Note that x and y are not adjacent since each edge of G_i is incident with at least one vertex not in P_i. As before, let z be the "oldest" internal vertex of P_{xy}. Its degree in P_{xy} is 0 or 1, a contradiction.

In summary, the original graph G has thickness at most k, where $k \leq 6 + \sqrt{2\gamma(G) - 2}$. $\square$

Since $a_1 \leq 3\theta$, we obtain as an immediate corollary that $a_1(G) \leq 3\sqrt{2\gamma(G) - 2} + 18$. However, we can improve the constant if, in the proof of Theorem 3, we replace "thickness" by "arboricity" and use the arboricity of $K_{n'}$ to estimate $a_1(G')$. The resulting inequality is weaker than Scheinerman's best possible result that $a_1(G) \leq 2 + \sqrt{3\gamma(G)}$ (cf. [S]):

Corollary 3 *If G is a simple graph, then*

$$a_1(G) \leq 6 + \sqrt{6\gamma(G) - 6}.$$

The examples of the complete graph, the complete bipartite graph, and the generalized cube all have $\theta \approx 2\gamma/n$, and this suggests a natural bound on θ in terms of γ and n. Since for all graphs $\gamma \leq n^2/12$, we have $\gamma/n \leq \sqrt{\gamma}$, suggesting a tighter upper bound: $\theta = O(\gamma/n)$. However, this bound does not hold in general: Let G be any graph with $\theta = O(\gamma/n)$, and embed G on its genus surface. Append an arbitrary number of vertices of degree one to G and note that the resulting graphs may no longer satisfy a $O(\gamma/n)$ bound, since n can become arbitrarily large. In contrast, in the next section we will show that $\gamma = O(\theta n)$ for all graphs.

However, for many graphs including the regular graphs it is true that $a_1 = O(\gamma/n)$ and hence $\theta = O(\gamma/n)$.

Theorem 4 *Let $0 < \varepsilon \leq 1$. If G is a graph with at least $1 + 1/\varepsilon$ vertices and such that $a_1(G) = \lceil e / (n - 1) \rceil$, then*

$$a_1(G) \leq \lceil (3 + 6\gamma/n)(1 + \varepsilon) \rceil.$$

Proof   By Theorem 1 (ii),

$$a_1(G) = \lceil e / (n - 1) \rceil$$
$$\leq \lceil (3n - 6 + 6\gamma) / (n - 1) \rceil$$
$$\leq \lceil (3 + 6\gamma /n) (1 + 1 / (n - 1)) \rceil$$
$$\leq \lceil (3 + 6\gamma /n) (1 + \varepsilon) \rceil. \quad \square$$

From Corollary 2 (ii) we have:

Corollary 4   *Let $0 < \varepsilon \leq 1$. If G is a regular graph with at least $1 + 1/\varepsilon$ vertices, then*

$$a_1(G) \leq \lceil (3 + 6\gamma /n) (1 + \varepsilon) \rceil = O(\gamma /n).$$

Since $\theta \leq a_1$, we get the same bound on θ.

In [H], Hutchinson shows that a graph G with a 2–cell embedding on the k–handled sphere S_k is 5-colorable if the embedding has "short" edges. There, the length of an edge is measured relative to the length of a side of the standard 4k–gon on which the embedding is represented. The following theorem shows that under those conditions, G has thickness at most 2 (and hence arboricity at most 6).

Theorem 5   *Suppose G has a 2–cell embedding on a surface of genus $k \geq 1$, and suppose that G has a representation G_k on the standard 4k–gon (i.e., a 4k–gon in which each side has unit length) such that every edge of G_k has length $< \varepsilon$, where ε is at most 1/2. Then $\theta(G) \leq 2$.*

Proof   We begin by extending G_k to a triangulation of the surface. Subdivide every nontriangular face by adding a new vertex adjacent to all vertices on the face boundary. If any new edge has length ε or more, subdivide it by adding vertices. Repeat this process until the resulting graph G_k' is a triangulation of the surface having all edges of length less than ε. G_k is a subgraph of G_k' which implies that $\theta(G) \leq \theta(G')$. Thus, we may assume that G_k was a triangulation to begin with.

For each $i = 1, ..., k$, consider the "handle" in the polygon P_k with sides labelled $a_i, b_i, a_i^{-1}, b_i^{-1}$. Note that one point S is common to all 4k sides of P_k. Let p_i (resp., q_i) be the set of all points of P_k at distance 2ε (resp., ε) from side b_i. Note that $\varepsilon \leq 1/2$ and so $2\varepsilon \leq 1$. Thus, p_i and q_i are paths going from a_i to a_i^{-1} representing a non–null–homologous cycle (henceforth abbreviated "nnh–cycle") on the surface (but not necessarily in the graph). Let L_i be the closed region bounded by p_i, q_i, a_i, and a_i^{-1}, i.e., L_i is the set of all points of P_k to the left of p_i (as we travel from a_i to a_i^{-1}) and within distance ε of p_i (see Fig. 1 below).

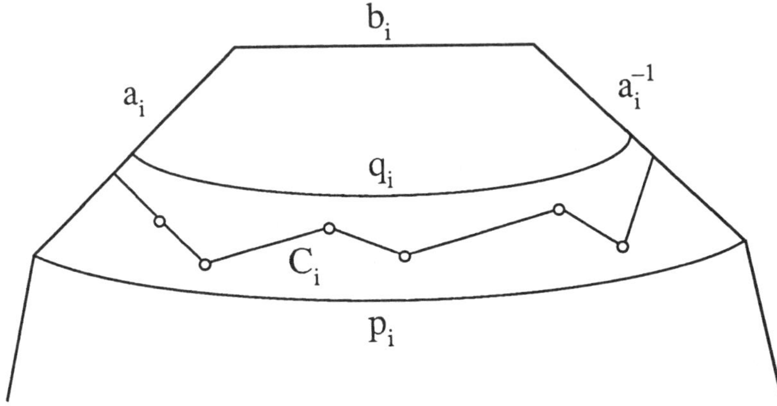

Figure 1

Within L_i, there is a path, starting at corresponding edges or vertices on a_i and a_i^{-1}, that represents a nnh–cycle in G_k: To find such a path, color a face or region of G_k blue if it meets L_i, but does not cross p_i. Color the remaining regions red. Because G_k is a 2–cell triangulation and the width of L_i is ε, we can only cross red faces as we walk along the path q_i. Similarly, we can only cross blue faces as we walk along p_i. Therefore, the boundary between the red and blue regions must lie within L_i and it must reach from a_i to a_i^{-1}, giving us a nnh–cycle in G.

For each $i = 1, ..., k$, choose such a cycle C_i, and let H_i be the subgraph of G induced by the set of all edges incident to a vertex of C_i and lying to the left of C_i. Since all vertices of C_i lie within L_i, no vertex of H_i can lie on b_i. Thus H_i is planar, since it is embedded on a cylinder. Since, by construction, the H_i's are vertex–disjoint, their union H is also planar.

Now let K be the subgraph of G induced by the edges of $G - H$. Then K is planar because it is embedded on the surface that remains when we cut P_k along all the nnh–cycles C_i, i.e., K is embedded on a sphere with open disks cut out and is therefore planar. Finally, $G = H \cup K$, and so $\theta(G) \le 2$, as claimed. $\quad\square$

It follows easily from the defintions that $\theta \leq \lceil bt/2 \rceil$, where bt is the book thickness of G (also cf. [BK]). This in turn implies that $a_1 \leq 3\lceil bt/2 \rceil$. In fact, $a_1 \leq bt + 1$, as shown in the next theorem.

Theorem 6 *If G is a simple graph, then*
$$a_1 \leq bt + 1.$$

Proof   The inequality clearly holds if $a_1 = 1$, so assume that $a_1 \geq 2$. By [BK Thm. 3.3], the following inequality holds:

$$bt \geq (e - n) / (n - 3), \text{ or equivalently,}$$
$$e \leq n(bt + 1) - 3bt.$$

Let H be a subgraph of G that gives the maximum value in the Nash–Williams arboricity formula. Then

$$\begin{aligned} a_1 &= \lceil e_H / (n_H - 1) \rceil \\ &\leq \lceil [n_H(bt_H + 1) - 3bt_H] / (n_H - 1) \rceil \\ &\leq \lceil bt_H + 1 + (1 - 2bt_H) / (n_H - 1) \rceil. \end{aligned}$$

Since $2 \leq a_1 = \lceil e_H / (n_H - 1) \rceil$, H is not a path and hence $bt_H \geq 1$. Thus, $1 - 2bt_H$ is negative, and therefore $a_1 \leq bt_H + 1 \leq bt + 1$. $\square$

Example 1   A variation on a construction in [BK] provides an infinite class of graphs whose arboricity achieves this upper bound: For $k \geq 3$, fix $n \geq 2k + 1$ and let T_k be the triangulated n–gon shown in Figure 2.

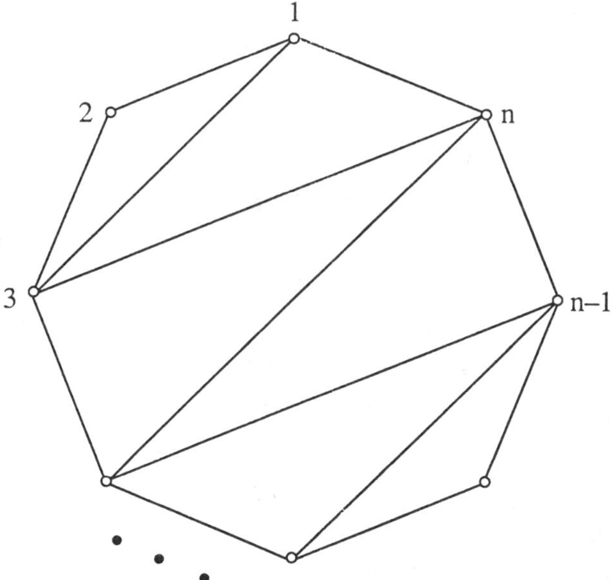

Figure 2

For $i = 0, \ldots, k - 1$, let $G_k{}^i$ be the graph obtained by rotating T_k through i successive positions (but leaving vertex labels fixed). Let G_k be the graph of order n obtained by taking the union of the $G_k{}^i$ for $i = 0, \ldots, k - 1$. Clearly, $\theta(G_k) \leq k$, and so $a_1(G_k) \leq k + 1$ by Theorem 5. Thus, it suffices to show that $a_1(G_k) \geq k + 1$.

By construction of G_k, we have

$$e = k(n - 3) + n$$
$$= (k + 1)n - 3k.$$

Next,

$$a_1 \geq \lceil e / (n - 1) \rceil$$
$$= \lceil [(k + 1)n - 3k] / (n - 1) \rceil$$
$$= \lceil k + 1 - (2k - 1) / (n - 1) \rceil.$$

Since $n \geq 2k + 1$, it follows that $a_1 \geq k + 1$, and thus $a_1 = \theta + 1$, as claimed.

4. Upper Bounds on Genus

It is impossible to find an upper bound for the genus γ in terms of thickness θ or arboricity a_1, as is demonstrated by the example of $K_{4,n}$, which has $\theta = 2$ and $a_1 = \theta = 4$ for $n \geq 4$, but which has unbounded genus $\gamma = \lceil (n - 2) / 2 \rceil$ (see [BCL]). The genus of a graph G can be bounded, however, in terms of both θ and the number of vertices n (or in terms of a_1 and n, or in terms of θ and n), as shown in the following theorems. As in Theorem 3 and its corollary, the complete graph K_n shows that these results are best possible up to a constant.

Theorem 7   *If G is a simple graph with n vertices, then*
$$\gamma(G) \leq (\theta(G) - 1)(n - 1).$$

Proof   Suppose G can be written as the union of $G_1, \ldots, G_\theta$, where the G_i's are planar and pairwise edge–disjoint. Embed each G_i on a sphere S_i; we will sew the spheres together in such a way that the resulting surface has $(\theta - 1)(n - 1)$ handles.

On sphere S_1, cut out $\theta - 1$ tangents disks at each vertex i. Call the disks $D_i{}^{1,2}$, $D_i{}^{1,3}, \ldots, D_i{}^{1,\theta}$. On each other sphere S_k, for $k = 2, \ldots, \theta$, cut out only one tangent disk at each vertex i and call the disk $D_i{}^k$.

Join the boundaries of the cut–out disks with cylinders as follows: For $k = 2, \ldots, \theta$, attach a cylinder from $D_i{}^{1,k}$ (the kth tangent disk to vertex i on sphere S_1) to $D_i{}^k$ (the disk tangent to vertex i on sphere S_k).

The number of cylinders from S_1 to S_k equals the total number of vertices n. This is equivalent to adding $n - 1$ handles to S_1. Since we do this for each value of k from 2 to θ, the resulting surface is a sphere with $(\theta - 1)(n - 1)$ handles. We now erase the vertices on all spheres except S_1 and extend incident edges on S_k, $k > 1$, along the incident cylinder to the corresponding vertex on S_1. (An equivalent way to view this pasting operation is that we are simply identifying all points of the boundaries of the two disks $D_i^{1,k}$ and D_i^k, for $i = 1, ..., n$ and $k = 2, ..., \theta$.)

This gives an embedding of G on a surface of genus $(\theta - 1)(n - 1)$, and therefore, $\gamma(G) \leq (\theta(G) - 1)(n - 1)$. ❑

Since $\theta \leq \lceil bt/2 \rceil$, we obtain the following bound on γ in terms of bt and n.

Corollary 7   *If G is a simple graph of order n, then*
$$\gamma(G) \leq (\lceil bt(G) / 2 \rceil - 1)(n - 1).$$

Since $\theta \leq a_1$, Theorem 7 also implies that $\gamma(G) \leq (a_1(G) - 1)(n - 1)$. Using Euler's formula, we can cut this bound in half:

Theorem 8   *If G is a simple graph with n vertices, then*

$$\gamma(G) \leq (a_1(G) - 1)(n - 1) / 2.$$

Proof   Suppose G can be written as a union of a_1 forests, each on n vertices. Then $e \leq a_1(n - 1)$. If G has a 2–cell embedding on a surface of genus γ, then Euler's formula implies
$$2 - 2\gamma = n - e + f \geq n - a_1(n - 1) + f, \text{ and so}$$
$$2\gamma \leq (a_1 - 1)(n - 1) - f + 1,$$
where f is the number of faces. Since $f \geq 1$, the result follows. ❑

Note that this result, together with the fact that $a_1 \leq bt + 1$, gives another bound on γ in terms of bt and n, almost identical to, but slightly weaker than, Corollary 7.

5. Conclusions

There are several places where the results in this work could be sharpened or improved. First, the examples known to the authors suggest that the constants can be lowered in Theorems 3, 7, and 8, and in Corollary 7. Second, it would be of interest to obtain a characterization of the graphs that satisfy $\theta = O(\gamma/n)$. Third, Theorem 5 would be improved by replacing the class of "short edge" graphs with another, more natural class of graphs of genus γ, for which a_1 and θ are bounded by a constant.

Finally, our work contains no upper bounds on book thickness. Yannakakis has proved a best possible result that $bt \leq 4$ for G planar (cf. [Y]). Heath and Istrail use that result in [HI] to show that $bt = O(\gamma)$; most recently, Malitz [M] has obtained a result of best possible order by showing that $bt = O(\sqrt{\gamma})$.

Acknowledgements: The authors would like to thank Michael Albertson, Carsten Thomassen, and Stan Wagon for helpful conversations and computer graphics.

REFERENCES

[AH] J. Akiyama and T. Hamada, The decompositions of line graphs, middle graphs and total graphs of complete graphs into forests, *Discrete Math.* 26 (1979) 203 – 208.

[A] K. Asano, On the genus and thickness of graphs, *J. Combinatorial Theory Ser. B* 43 (1987) 287 – 292.

[BCL] M. Behzad, G. Chartrand, and L. Lesniak–Foster, *Graphs & Digraphs*, Wadsworth, Belmont, CA (1979).

[BH] L. Beineke and F. Harary, The thickness of the complete graph, *Canad. J. Math.* 17 (1965) 850 – 859.

[BK] F. Bernhart and P. Kainen, The book thickness of a graph, *J. Combinatorial Theory Ser. B* 27 (1979) 320 – 331.

[B] S. Burr, An inequality involving the vertex arboricity and edge arboricity of a graph, *J. Graph Th.* 10 (1986) 403 – 404.

[HI] L. Heath and S. Istrail, The pagenumber of genus g graphs is $O(g)$, *Wesleyan Univ. Computer Science Technical Report* #86–2 (1986).

[H] J. Hutchinson, A five–color theorem for graphs on surfaces, *Proc. Amer. Math. Soc.* 90 (1984) 497 – 504.

[M] S. Malitz, Genus g graphs have pagenumber $O(\sqrt{g})$, *Proc. 29th Ann. Symp. FOCS* (1988), 458 – 468.

[N] C. Nash–Williams, Decomposition of finite graphs into forests, *J. London Math. Soc.* 39 (1964) 12.

[S] E. Scheinerman, The maximum interval number of graphs with given genus, *J. Graph Theory* 11 (1987) 441 – 446.

[T] W. Tutte, The thickness of a graph, *Nederl. Akad. Wetensch. Proc. Ser. A* 66 (1963) 567 – 577.

[Y] M. Yannakakis, Four pages are necessary and sufficient for planar graphs, *18th ACM Symposium of Theory of Computing* (1986) 104 – 108.

A Survey of Graphs Hamiltonian–Connected from a Vertex

Alice M. Dean

C.J. Knickerbocker*

Patti Frazer Lock*

Michael Sheard

ABSTRACT

A graph G is called hamiltonian–connected from a vertex v if a hamiltonian path exists from v to every other vertex w ≠ v. We present a survey of the main results known about such graphs, including a section on graphs uniquely hamiltonian–connected from a vertex and a section on the computational complexity of determining whether a given graph is hamiltonian–connected from a vertex or uniquely hamiltonian–connected from a vertex.

1. Introduction

In 1963, Ore [19] defined a graph to be *hamiltonian–connected* if there is a hamiltonian path between every pair of distinct vertices. Since that time, a great deal of work has been done on this subject (see, for example, [2], [23], [25], [4]). At the Fourth International Conference on the Theory and Applications of Graphs in 1980, Chartrand and Nordhaus [3] extended this concept by defining a graph to be *hamiltonian–connected from a vertex* v if there is a hamiltonian path from a distinguished vertex v to every other vertex. This generalization arises quite naturally in many situations. For example, one might consider the problem of a communications network in which all communications arise at a central office, are routed through all offices and end at any one of the other offices, or vice versa. (In addition, one of the authors is particularly fond of the "Halloween party" example, in which the participants start at their respective houses, trick–or–treat at all other houses, and end at the

297

Halloween party.) Examples of graphs hamiltonian–connected from a vertex v are given in Figure 1.

We use HCv to denote hamiltonian–connected from a vertex v (and, similarly, HC to denote hamiltonian–connected). We use p to denote the order of a graph, and we denote the set of all neighbors of a vertex u by N(u).

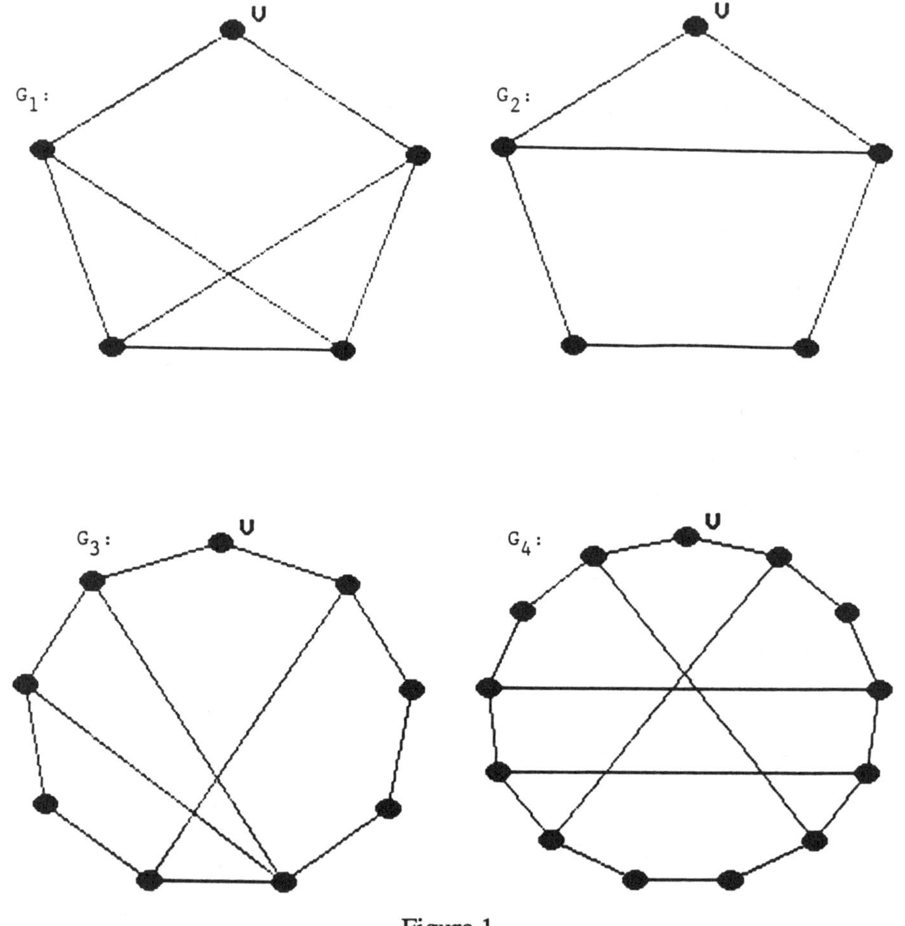

Figure 1

One of the most interesting problems in the study of graphs hamiltonian–connected from a vertex has been to determine the minimum number of edges for such a graph. G.R.T. Hendry originally raised the possibility that this minimum number of edges would be attained by graphs with the minimum number of hamiltonian paths originating at v. As we will see, Hendry went on to produce the surprising result that this is false. However, it was this idea that led him to make the following definition in 1984 ([8]): A graph is called *uniquely hamiltonian–connected from a vertex* v (UHCv) if there is a unique hamiltonian path from v to every other vertex. In Figure 1, graphs G_2 and G_3 are uniquely hamiltonian–connected from v, while G_1 and G_4 each have multiple paths from v to at least one other vertex.

The diagram in Figure 2 is taken from [8]. It is clear that the set of graphs UHCv and the set of graphs HC are both properly contained in the set of graphs HCv. The intersection of UHCv and HC was shown by Hendry in [8] to consist solely of the set $\{K_1, K_2, K_3\}$. (Similarly, the obvious definition of graphs "uniquely hamiltonian–connected" yields only these three graphs.) Finally, it is shown in [3] by Chartrand and Nordhaus that the set of graphs HCv is properly contained in the set of all hamiltonian graphs.

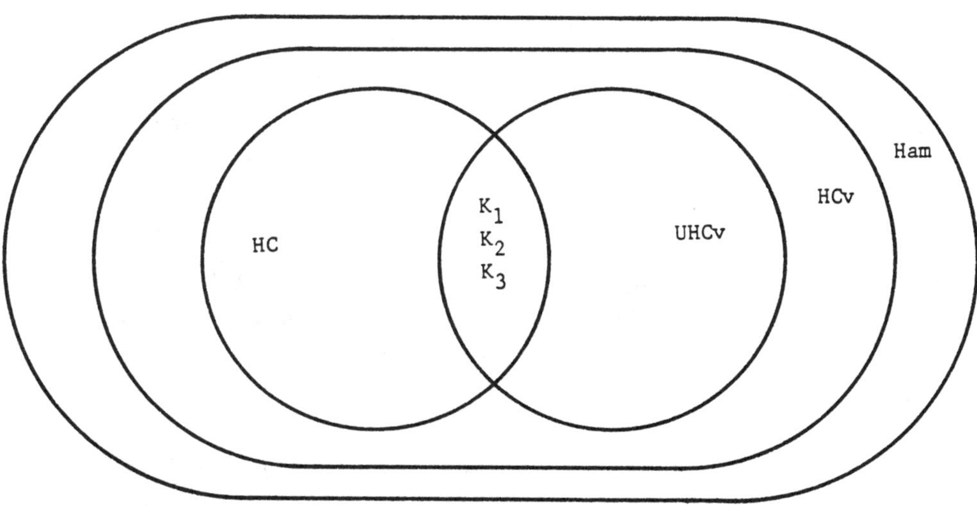

Figure 2

In the next section, we present some general results about graphs hamiltonian–connected from a vertex. In Section 3, a survey of graphs uniquely hamiltonian–connected from a vertex is given. We then return to the motivating problem of minimizing the number of edges in graphs HCv in Section 4. In Section 5, we discuss the computational complexity of HCv and of UHCv, and we conclude with a section discussing possible avenues for further exploration.

2. General Results on Graphs HCv

In the first paper on the subject of graphs hamiltonian–connected from a vertex, Chartrand and Nordhaus [3] presented the following results:

Theorem 1   *(Chartrand and Nordaus, [3]) Let G be a graph of order p with $p \geq 4$.*

   1. If G is HCv, then both G and $G - v$ are 2–connected.

   2. If G is HCv with $u \in N(v)$, then $deg(u) \geq 3$.

   3. If G contains a vertex adjacent to two vertices u and v each of degree 2, then G is hamiltonian–connected from at most two vertices, namely u and v.

The graph G_2 in Figure 1 is hamiltonian–connected from exactly one vertex, namely v. The complete graph K_p is clearly hamiltonian–connected from every vertex. It is easily seen that a graph cannot be hamiltonian–connected from exactly $p - 1$ vertices. Chartrand and Nordhaus gave a class of examples in [3] illustrating that all other possibilities are realized: a graph G of order p may be hamiltonian–connected from any of $0, 1, 2, ..., p - 2$, p vertices.

Ore proved in [18] that if a graph G with order $p \geq 2$ satisfies $deg\ x + deg\ y \geq p + 1$ for all pairs x, y of nonadjacent vertices, then G is hamiltonian–connected. Chartrand and Nordhaus proved a similar result in [3] to obtain the following sufficient condition for a graph to be hamiltonian–connected from a vertex. Examples were given in [3] to show that this result is best possible.

Theorem 2   *(Chartrand & Nordhaus, [3]) Let G be a hamiltonian graph of order p. If there exists a vertex u such that $deg\ u + deg\ w \geq p + 1$ for each vertex w not adjacent to u, then G is hamiltonian–connected from at least two vertices. (In particular, G is hamiltonian–connected from the two vertices adjacent to u on a hamiltonian cycle.)*

Theorem 3   *(Chartrand & Nordhaus, [3]) Let G be a hamiltonian graph of order p and let C: $v_1, v_2, ..., v_p, v_1$ be a hamiltonian cycle of G. If v_1 and v_j $(j \neq 1)$ are nonadjacent and $deg\ v_1 + deg\ v_j \geq p + 1$, then G is hamiltonian–connected from v_p and v_2 if and only if $G + v_1 v_j$ is hamiltonian–connected from v_p and v_2.*

Theorem 4   *(Chartrand & Nordhaus, [3]) Let G be a hamiltonian graph of order p. Suppose there exists a vertex u such that $deg\ u + deg\ v \geq p$ for each vertex v not adjacent with u. Define*

$$S = \{v:\ v \neq u,\ uv \notin E(G),\ deg\ u + deg\ v = p\}.$$

Then G contains at least two vertices, namely those consecutive with u on a hamiltonian cycle, from which hamiltonian paths exist to all other vertices with at most $|S|$ exceptions.

3. Graphs Uniquely Hamiltonian–Connected from a Vertex

The study of graphs *uniquely* hamiltonian–connected from a vertex has yielded some of the richest results in the field. It is hoped that the methods and ideas used in the study of these graphs will enable us to better understand the larger class of all graphs hamiltonian–connected from a vertex. As mentioned above, the complete graphs K_1, K_2, and K_3 are uniquely hamiltonian–connected from a vertex. These are trivial cases, however, and in all that follows we assume that the order of the graph is greater than 3.

The following theorem gives some of the major results from [8] and [10.

Theorem 5 *(Hendry, [8, 10]) Let G be a graph UHCv. Then:*

1. *$\deg v$ is even.*
2. *G has $(\deg v)/2$ hamiltonian cycles.*
3. *p is odd.*
4. *$|E| = 3(p-1)/2$*
5. *$G - v$ has a unique hamiltonian cycle.*
6. *Every vertex other than v has degree $2, 3,$ or 4.*
   If $u \in N(v)$, then $\deg u = 3$.
7. *Every edge of G lies on at least 2 hamiltonian paths from v.*
   Futhermore, every edge is traversed at least once in each direction by some hamiltonian path from v.

A *forced edge* is defined to be an edge which lies on every hamiltonian path from v. Hendry made many observations about forced edges in [9]; for example, he notes that no forced edge can lie on a triangle.

The definition of forced edge, given for graphs UHCv, applies equally well to graphs HCv. The complete graphs K_p give examples of graphs HCv which contain no forced edges. However, every graph UHCv must contain at least one forced edge. Indeed, the following theorem gives sharp upper and lower bounds for the number of forced edges in a graph UHCv.

Thoerem 6 *(Hendry, [8]) Let G he UHCv. Let n_2 denote the number of vertices of degree 2 in G. Then the number, f_G, of forced edges in G satisfies:*
$$n_2/2 \leq f_G \leq (p - 1 - \deg v)/2.$$

If G is a graph UHCv, and u is a vertex in $G - v$, we use the notation H_u to denote the hamiltonian path from v to u. We say vertex x is *penultimate* to vertex y if x is adjacent to y in H_y.

Theorem 7 *(Hendry, [10]) Let G be UHCv. Let x be a vertex with $x \neq v$. Then:*
   1. *if deg $x = 2$, then x is penultimate to both its neighbors,*
   2. *if deg $x = 4$ or if $x \in N(v)$, then x is not penulitmate to any neighbor,*
   3. *if deg $x = 3$ with $x \notin N(v)$, then x is penultimate to exactly one of its neighbors.*

The next result relates the number of vertices of different degrees in a graph UHCv. Part (1) follows from counting vertices, part (2) follows from the previous theorem by counting penultimate vertices in hamiltonian paths from v, and part (3) is obtained by combining the first two parts.

Proposition 8 *Let G be UHCv. Let $n_i =$ the number of vertices in $V(G) - \{v\}$ of degree i for $i = 2, 3, 4$. Then:*
   1. $n_2 + n_3 + n_4 = p - 1$,
   2. $2n_2 + n_3 - deg\ v = p - 1$,
   3. $n_2 = n_4 + deg\ v$.

The following theorem relating forced edges to penultimate vertices reinforces the belief that penultimate vertices are an important tool in understanding graphs UHCv.

Theorem 9 *(Knickerbocker, Lock, Sheard [16]) Let G be UHCv. Then xy is a forced edge if and only if x is penultimate to y and y is penultimate to x.*

The proofs of many of the previous theorems, and most of the results to follow, rely heavily on the concept of the "transform graph" of a graph UHCv. This idea originated with Pósa and Thomason (see [22] and [24]) and is one of the most important concepts in this field.

Definition 10   (Hendry, [9]) Let G be a graph UHCv. Define the *transform graph of G at* v, denoted T(G, v) or T, as follows:
$$V(T) = \{H_x : x \in V(G) - \{v\}\},$$
$$H_x H_y \in E(T) \text{ if and only if } H_y : v...wx...y \text{ with } w \in N(y).$$

In Definition 10, notice that, since $w \in N(y)$, we have a path to x in the form H_x : v...wy...x by simply reversing the last part of the path to y. In this case, we say that H_y can be *transformed* to H_x. It is easily seen that $deg_T(H_x) = deg_G(x) - 1$ for all vertices $x \neq v$. An example is given in Figure 3 of a graph G which is UHCv, together with its corresponding transform graph T.

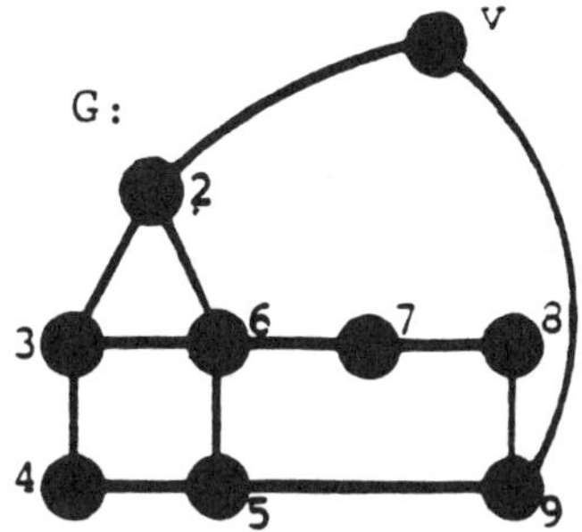

T:

$$H_7 - H_3 - \underset{|}{H_6} - H_9 - H_2 - H_5 - H_8$$
$$H_4$$

Figure 3

Theorem 11 *(Hendry, [8, 10]) Let G be UHCv with transform graph T. Then T is a forest. Furthermore, the number of components of T is (deg v) / 2.*

In [9], Hendry extends the definition of the transform graph to any graph G with a distinguished vertex v as follows: Each vertex of T corresponds to a hamiltonian path from v. An edge is drawn whenever the hamiltonian path corresponding to one vertex can be transformed to the hamiltonian path corresponding to the other vertex. Hendry has shown, however, that the converse of the preceding theorem is false. Even if we assume that G is HCv and that T(G, v) is a forest, G need not be UHCv.

Transform graphs have not yet been explored in depth for graphs HCv. One of the few results known is due to Hendry (unpublished): If G is HCv and T is its transform graph, then girth(T) $\geq$ 6 with equality if and only if G has a hamiltonian path from v in which the last four vertices induce a K_4. The following theorem describes more of the structure of the transform graphs for graphs which are UHCv.

Theorem 12 *(KLS, [16]) Let G be UHCv and let T be the transform graph of G. Then every component of T contains a path in the form:*
$$H_{x_1}H_{x_2} \ldots H_wH_u \ldots H_{y_2}H_{y_1}$$
where $u, w \in N(v)$, $x_1, x_2 \in N(u)$ with x_1 penultimate to u, and $y_1, y_2 \in N(w)$ with y_1 penultimate to w. Furthermore, in each component, these are the only vertices H_x in which x is a neighbor of v or a neighbor of a neighbor of v.

The path described in the above theorem is called the *central path* of a component of T. The importance of this path is readily apparent from the following observations. If x is a neighbor of a neighbor of v (say $x \in N(u)$ with $u \in N(v)$), then H_x has the form: $vu \ldots x$. Notice, then, that $H_x - vu + ux$ is the unique hamiltonian cycle of $G - v$. Thus we say this cycle is *represented* by vertices H_x in T for every $x \in N(N(v))$. Furthermore, if $u \in N(v)$, then $H_u + uv$ is a hamiltonian cycle in G. It is easily seen that two vertices corresponding to paths to neighbors of v are in the same component of T if and only if they represent the same hamiltonian cycle in G. Thus the components of T are in a natural one–to–one correspondence with the hamiltonian cycles of G, and the unique hamiltonian cycle of $G - v$ is represented in every component of T.

Knickerbocker, Lock, and Sheard [16] developed a classification system for all edges in $G - v$. Let ab be an edge in $G - v$. If edge ab is included in the path H_x, we say ab *covers* H_x. The induced subgraph of T generated by the set of all T-vertices H_x covered by ab will be denoted $C(ab)$ and will be called the *cover of ab in T*. An edge in T will be called a *border of $C(ab)$* if it is not in $C(ab)$ but it is incident with a vertex in $C(ab)$.

Theorem 13   *(KLS, [16]) Assume G is UHCv with deg $v = 2$, and let T be the transform graph of G. Then every edge ab in $G - v$ falls into one of the following four categories:*

    1. $C(ab)$ has no borders. This is the case if and only if ab is a forced edge.

    2. $C(ab)$ has exactly one border. This is the case if and only if a is penultimate to b and b is not penultimate to a, or vice versa. We will call such an edge an ultimate edge.

    3. $C(ab)$ has two borders and is connected. We will call such an edge a stick.

    4. $C(ab)$ has two borders and is disconnected. We will call such an edge an arrow.

We extend these definitions to any graph G UHCv (i.e. with deg $v > 2$) as follows: The definitions of forced edge and ultimate edge are still valid. We call edge ab a *local stick* if $C(ab)$ has two borders, both on the same component of T, and if the intersection of $C(ab)$ with this component is connected. We call edge ab a *local arrow* if $C(ab)$ has two borders, both on the same component of T, and if the intersection of $C(ab)$ with this component is disconnected.

Knickerbocker, Lock, and Sheard [16] described the interconnections between forced edges, ultimate edges, (local) sticks, and (local) arrows. The results obtained from this classification of edges in $G - v$ produce the following three theorems:

Theorem 14 *(KLS, [16]) Let G be UHCv and assume that deg v = 2. Then the unique hamiltonian cycle of G consists exactly of the sets of forced edges, ultimate edges, and sticks, together with the two edges incident with v.*

Theorem 15 *(KLS, [16]) Let G be UHCv, and let ab be an edge in G. If a is penultimate to b, then edge ab lies on the unique hamiltonian cycle of G – v and on every hamiltonian cycle of G.*

Theorem 16 *(KLS, [16]) A graph G can be uniquely hamiltonian–connected from at most three vertices. Furhtermore, if G is UHC from more than one vertex, then all distinguished vertices must have degree two.*

Whether a graph can be uniquely hamiltonian–connected from more than one vertex remains an open question. However, the authors do have examples of graphs which are uniquely hamiltonian–connected from one vertex, and hamiltonian–connected from additional vertices. The graph given in Figure 4, for example, is uniquely hamiltonian–connected from v and also hamiltonian–connected from a and b.

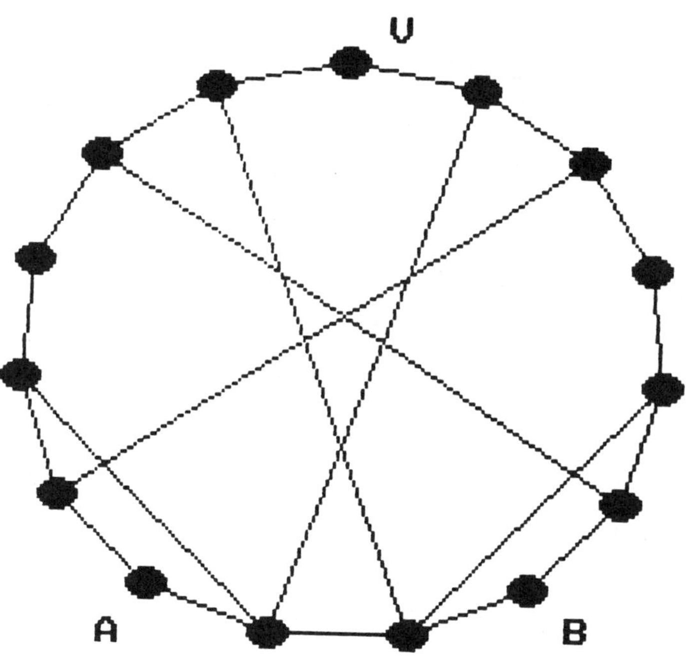

Figure 4

One of the broadest remaining open questions involving graphs UHCv is whether it is possible to completely characterize all such graphs. There are several ways one might approach this question, two of which are discussed here: considering forbidden subgraphs, and finding a class of constructions. We will discuss the progress made on each.

First, we would hope to answer the following question: Which graphs H may not be subgraphs of any graph which is UHCv? Hendry [9] has shown that if G is UHCv, then the following cannot be subgraphs of G: K_4, $K_{2,3}$, $K_1 + P_4$. Knickerbocker, Lock, and Sheard have extended this list to include $P_3 \times P_3$, the "triple bowtie", and the "fish". These six forbidden subgraphs are given in Figure 5. (Graphs a, b, c, e, f in Figure 5 are known to be best possible in that deleting even one edge yields a graph which is not forbidden. It is unknown whether graph d of Figure 5 is best possible.) Ideally, it may be possible to identify a class of graphs such that if H does not contain any member of this class as a subgraph, then H may be embedded in a graph which is UHCv.

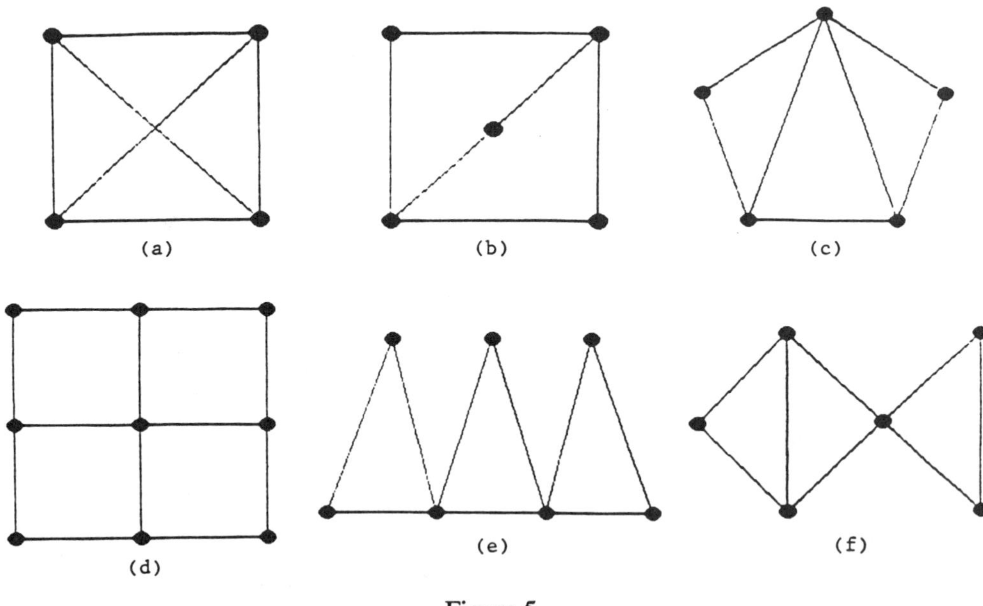

Figure 5

Second, consider the problem of constructing graphs UHCv. Hendry [8] described four constructions (that is, four methods of obtaining larger graphs UHCv from one or more smaller graphs UHCv), and a fifth construction is described by him in [9]. Kirchdorfer, Knickerbocker, Lock, and Sheard [14] added to and extended these constructions to obtain a current total of eight constructions. Hendry gave examples in [9] of four graphs UHCv which are not constructible in this manner, and the authors have found more. (Such graphs have been dubbed "rogues".) Still, a constructive characterization of these graphs may well be possible.

4. The Number of Edges in a Graph HCv

Recall that the original motivating question behind the study of graphs UHCv was to find the minimum number of edges in a graph HCv. The following lower bound for the number of edges in such a graph is given in [3]:

Theorem 17 *(Chartrand and Nordhaus, [3]) If G is HCv with $p \geq 4$, then the number of edges in G is greater than or equal to $\lceil (5p - 1) / 4 \rceil$.*

The proof of this result is based on the following observation: if G is HCv, then no vertex in $V(G) - \{v\}$ can be adjacent to two vertices in $V(G) - \{v\}$ both of degree 2. In [3], Chartrand and Nordhaus gave a class of graphs HCv with approximately $3p/2$ edges. Hendry improved on this when he developed a class of graphs HCv in [8] which have approximately $7p/5$ edges. Finally, in [15], Knickerbocker, Lock, and Sheard gave a class of examples of arbitrarily large order which exactly meet the lower bound given in Theorem 17. Given the "local" nature of the proof of this theorem, it is surprising that the bound is sharp. In addition, it is interesting to note that there are graphs HCv with fewer edges (but more paths!) than graphs UHCv.

Figure 6 is an example of a graph which exactly meets this lower bound. Knickerbocker, Lock, and Sheard have shown that if a graph which is HCv has the minimum number of edges as given in the bound of Theorem 14, then the distinguished vertex v is unique.

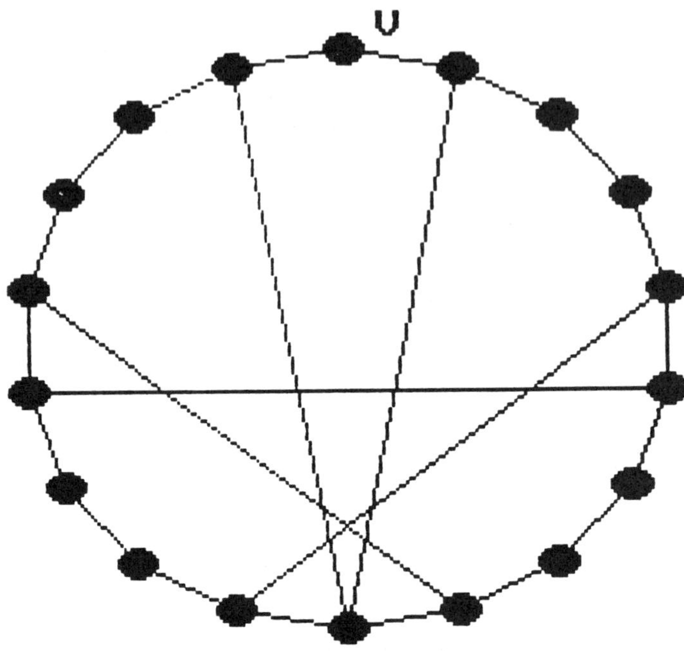

Figure 6

Approaching the problem of counting edges from the other direction, Hendry [7] determined the maximum number of edges in a graph which is *not* hamiltonian–connected from a vertex, as follows:

Theorem 18   *(Hendry, [7])   If G is a graph of order p > 2   which is not hamiltonian–connected from a vertex, then:*

1. $|E(G)| \leq (p^2 - 3p + 4)/2$, *and*

2. *if equality holds in 1, then G is C_4, $K_2 + \overline{K}_3$, or for p > 3, $K_2.K_{p-1}$ (i.e. the graph of order p having two blocks: K_2 and K_{p-1}).*

5. The Complexity of HCv and UHCv

One of the most famous computationally difficult problems in graph theory is determining whether a given graph G has a hamiltonian circuit; this problem, which we will call HAM, is NP–complete ([13]). Since determining whether G is HCv or UHCv are closely related problems, it is natural to ask whether these problems are also hard in this sense. Because of the many structural restrictions (odd number of vertices, $(3p - 3)/2$ edges, all vertices of degree 2, 3, or 4, etc.), the complexity of deciding whether G is UHCv is of particular interest. One might hope that all these restrictions might make the time complexity of this problem polynomial. On the other hand, very restricted forms of the hamiltonian cycle problem remain NP–complete (e.g., restricted to planar bipartite graphs with all vertex degrees either 2 or 3 ([12])). In this section, we discuss what is known about the complexity of HCv and UHCv.

For a review of the basic ideas and terminology of complexity theory, we refer the reader to [6], [17], [20], or [12]. Throughout this section, we will follow the notation and terminology of [12].

In Table 1, we list three decision problems that are relevant to our discussion:

HCv   *Instance:*   A graph G and distinguished vertex v.
         Question: Is G hamiltonian–connected from v?

2HCv   *Instance:*   A graph G and distinguished vertex v.
         Question: Is G hamiltonian–connected from v in more than
                       one way, i.e., is there a hamiltonian path from v to
                       each other vertex x, and is there at least one vertex x
                       for which there is more than one such path?

UHCv   *Instance:*   A graph G and distinguished vertex v.
         Question: Is G uniquely hamiltonian–connected from v?

Table 1

Using a transformation from HAM, Dean [5] has determined the complexity of the first two problems:

Theorem 19   *(Dean, [5]) HCv and 2HCv are NP–complete.*

The complexity of UHCv is apparently much more difficult to determine. To begin with, it is not at all obvious that UHCv is even an element of NP. To verify that a particular graph G is uniquely hamiltonian–connected from vertex v, one would not only have to demonstrate that hamiltonian paths exist from v to each other vertex x, but also that each such path was unique. On the other hand, if UHCv graphs could be characterized in terms of a list of structural restrictions that could be checked in polynomial time, then UHCv would actually be an element of P.

There is a class of decision problems for which UHCv is a natural member; it is the class D^P, which was defined by Papadimitriou and Yannakakis in 1984 ([21]). D^P is defined to be the class consisting of those decision problems which are the intersection of a problem in NP with one in coNP. In other words, each problem Q in D^P is defined in terms of two problems Q_1 and Q_2 in NP with the same set of instances, such that the answer for Q is yes if and only if the answer for Q_1 is yes and the answer for Q_2 is no.

Note that NP and coNP are subclasses of D^P, since any problem Q is the intersection of itself with the trivial problem having the same instances and a question whose answer is always yes. It is easy to see that UHCv is an element of D^P, since UHCv = HCv $\cap$ $\overline{2HCv}$.

The problem UHAM (Given G, does it have a unique hamiltonian cycle?) is another natural member of D^P. UHAM is NP–hard and has the additional property that if it is in NP, then NP = coNP. The proof follows from a transformation used in [12] which shows that 2HAM is NP–complete. However, an argument using the analogous transformation for 2HCv fails to show that UHCv has these properties.

A problem Q in D^P is D^P–complete if any problem in D^P can be polynomially transformed to Q. Note that any D^P–complete problem Q has the properties described above. It is not known whether UHAM is D^P–complete. Indeed, Blass and Gurevich [1] have shown that it may be impossible to answer this question using standard techniques, by constructing two oracles for which the relative D^P classes properly contain NP $\cup$ coNP, but such that the closely related problem USAT (Given a boolean formula, does it have a unique satifying truth assignment?) is D^P–complete relative to one oracle and not D^P–complete relative to the other. For more details on the complexity of unique problems and the class D^P, see [11], [12], or [21].

6. Conclusion

In this section we discuss open problems and current research directions, some of which have already been mentioned in passing. The concentration of previous research in the more restrictive area of graphs UHCv leads to a larger collection of specific questions there than for the more general case of HCv, but this should not be taken as a measure of relative importance. Indeed, in the realm of graphs HCv, we suspect that the most interesting questions have not yet been formulated.

The most promising area for research on graphs HCv appears at present to be an exploration of transform graphs. The hope is that a fuller understanding of transform graphs will provide the kind of insights into the structure of graphs HCv which has already been provided for graphs UHCv. Currently, very little is known. It is easy to generate natural conjectures and questions; for example, Hendry [private communication] has posed the following:

Problem   Determine necessary and sufficient conditions for a graph HCv to have a connected transform graph.

Problem   Determine necessary and sufficient conditions for two graphs HC from distinguished vertices to have isomorphic transform graphs.

If the class of graphs HCv appears too large to tackle all at once, a reasonable starting point is the class of graphs *minimally* hamiltonian–connected from v: G is minimally HCv if it is HCv, but the deletion of any edge yields a graph which is not HCv. Such graphs may have nice structural properties, similar to those of graphs UHCv (which are a subclass). Any results obtained may then lift in some fashion to the broader class of HCv, since clearly any graph HCv contains a minimally HCv subgraph. We add the following:

Problem   Determine max $\{ \, |E(G)| \, : \, |V(G)| = p \,$ and G is minimally HCv$\}$.

As suggested earlier, perhaps the outstanding open problem concerning graphs UHCv is to provide a constructive characterization: a small list of graphs with which to begin and a few methods of making larger graphs out of smaller ones, whereby the class of all graphs UHCv can be generated. The evidence is ambiguous as to whether such a characterization is possible. On one hand, at the largest orders for which a systematic study has been carried out (p = 11, 13, and 15), rogues appear which bear little resemblance to any smaller graphs UHCv. On the other hand, the huge majority of graphs UHCv of each of these orders are easily constructible using known constructions. The issue may well depend on whether more regularity or less emerges at higher orders.

While the problem of characterizing graphs UHCv remains mysterious, a characterization of their transform graphs may be within reach. In particular, the structure of the transform tree where deg v = 2 is already partially understood.

We turn now to forbidden subgraphs for graphs UHCv. Some organized understanding of this topic is sorely needed, if it is to be more than just a collection of amusing but largely unrelated puzzles. The ideal would be to give a simple characterization of a class Φ of graphs (the "atomic forbidden subgraphs") such that H is a subgraph of a graph UHCv if and only if H has no member of Φ as a subgraph. What such a Φ might look like, or exactly how simple its description might be, is currently most unclear. At the moment, the proofs for the few known forbidden subgraphs appear to be very much *ad hoc*, so nothing like an organized pattern has yet emerged.

The problem of showing that a graph can be UHCv from at most one vertex remains stubbornly open, although Theorem 16 and some more technical work of the authors have greatly restricted the terrain in which a counterexample might be found. We continue to believe that the distinguished vertex is unique. In any case counterexamples will not be easily found; we have shown that a graph which is UHC from three vertices (the maximum possible under Theorem 16) must have order at least 27 and size at least 39, which makes even a theory–driven search virtually impossible.

Finally, there are several open questions concerning the computational complexity of UHCv: Is UHCv in NP or does it have the property, like UHAM, that if it is in NP, then NP = coNP? If the answer to the latter question is yes, then does UHCv have the stronger property of being D^P–complete?

While a great deal of progress has been made in understanding graphs hamiltonian–connected from a vertex and graphs uniquely hamiltonian–connected from a vertex, much remains to be done. With open questions ranging from hands–on construction techniques to computational complexity to grand classification schemes, the field of graphs hamiltonian–connected from a vertex is ripe for further research.

REFERENCES

[1]   A. Blass and Y. Gurevich, On the unique satisfiability problem, *Information and Control*, 55 (1982), 80 – 88.

[2]   G. Chartrand, S.K. Kapoor, H.V. Kronk, A generalization of hamiltonian–connected graphs, *J. Math. Pures Appl.*, 49 (1969), 109 – 116.

[3]   G. Chartrand and E.A. Nordhaus, Graphs hamiltonian–connected from a vertex, *The Theory and Application of Graphs,* G. Chartrand et. al., eds., Wiley, New York, 1981, 189 – 201.

[4]   G. Chartrand and L. Lesniak, *Graphs and Digraphs, Second Edition*, Wadsworth and Brooks/Cole, Monterey, 1986.

[5]   A. Dean, The complexity of graphs hamiltonian–connected from a vertex, Technical Report, Skidmore College, 1987.

[6]   M.R. Garey and D.S. Johnson, *Computers and Intractability*, W.H. Freeman, 1979.

[7]   G.R.T. Hendry, Maximum graphs non–hamiltonian–connected from a vertex, *Glasgow Math. J.*, 25 (1984), 97 – 98.

[8]   G.R.T. Hendry, Graphs uniquely hamiltonian–connected from a vertex, *Discrete Math*, 49 (1984), 61 – 74.

[9]   G.R.T. Hendry, Ph.D. Thesis, Aberdeen University, 1985.

[10] G.R.T. Hendry, The size of graphs uniquely hamiltonian–connected from a vertex, *Discrete Math*, 61 (1986), 57 – 60.

[11] D.S. Johnson, The NP–completeness column: an ongoing guide, *J. Algorithms*, 6 (1985), 291 – 305.

[12] D.S. Johnson and C.H. Papadimitriou, Computational complexity, *The Travelling Salesman Problem*, E.L. Lawler and J.K. Lenstra, eds., Wiley, 1985, 37 – 85.

[13] R.M. Karp, Reducibility among combinatorial problems, *Complexity of Computer Computations*, R.E. Miller and J.W. Thatcher, eds., Plenum Press, New York, 1972, 85 – 103.

[14] D. Kirchdorfer, C.J. Knickerbocker, P.F. Lock, M. Sheard, Constructions of UHCv graphs, Technical Report, St. Lawrence University, 1987.

[15] C.J. Knickerbocker, P.F. Lock, M. Sheard, The minimum size of graphs hamiltonian–connected from a vertex, *Discrete Math*, to appear.

[16] C.J. Knickerbocker, P.F. Lock, M. Sheard, On the structure of graphs uniquely hamiltonian–connected from a vertex, *Discrete Math*, submitted.

[17] H.R. Lewis and C.H. Papadimitriou, *Elements of the Theory of Computation*, Prentice–Hall, Englewood Cliffs, NJ, 1981.

[18] O.Ore, Note on hamiltonian circuits, *Amer. Math Monthly*, 67 (1960), 55.

[19] O. Ore, Hamiltonian connected graphs, *J. Math Pures Appl.*, 42 (1963), 21 – 27.

[20] C.H. Papadimitriou and K. Steiglitz, *Combinatorial Optimization: Algorithms and Complexity*, Prentice–Hall, Englewood Cliffs, NJ, 1982.

[21] C.H. Papadimitriou and M. Yannakakis, The complexity of facets (and some facets of complexity), *J. Comput. System Sci.*, 28 (1984), 244 – 259.

[22] L. Pósa, Hamiltonian circuits in random graphs, *Discrete Math*, 14 (1976), 359 – 364.

[23] Z. Skupien, Homogeneously traceable and hamiltonian connected graphs, *Demonstratia Mathematica*, 17 (1984), 1051 – 1067.

[24] A.G. Thomason, Hamiltonian cycles and uniquely edge–colourable graphs, *Ann. Discrete Math*, 3 (1978), 259 – 268.

[25] I. Tomescu, On hamiltonian–connected regular graphs, *J. Graph Theory*, 7 (1983), 429 – 436.

Which Graphs Are Pancyclic Modulo k?

Nathaniel Dean

Bellcore

ABSTRACT

Several results are known about graphs which contains cycles of length 0 modulo some integer k. We investigate the structure of graphs which contain cycles of every length (i.e., pancyclic graphs) modulo k, and we obtain results which in general do not hold for nonplanar graphs. For example, we show that every 3–connected planar graph (except K_4) with minimum degree at least k is pancyclic modulo k. This is accomplished by applying Steinityz's characterization of 3–connected planar graphs and the theory of Euler contributions.

1. Introduction

One of the most conspicuous and still trivial observations about 2–connected graphs is that *every such graph which is not an odd cycle contains an even cycle.* We use [5] for our basic notation and terminology, and so our graphs have no loops or multiple edges and the length of a cycle C is $\varepsilon(C)$ (i.e., the number of edges in C). For any integer k, we define a graph to be pancyclic modulo k if it contains a cycle of every length modulo k. Hence, another interesting observation about 2–connected graphs is that *every such graph which is not an odd cycle is pancyclic modulo 2 if and only if it is not bipartite.* To some extent, both of these observations have been extended to results about cycles of various lengths modulo some given integer. This paper is mainly concerned with extensions of the second observation.

The following problem is alluded to in [1] but is motivated mainly by discussions with E.M. Arkin.

Open Problem 1 Characterize the 3–connected graphs which are pancyclic modulo 3.

Letting v represent the number of vertices in a graph, a graph is pancyclic if and only if it is pancyclic modulo v. Even when there is no pancyclic result, there may be a modular pancyclic result. For example, J.A. Bondy [4] proved that if a graph G contains more than $v^2/4$ edges, then it contains a cycle of every length p where $3 \leq p \leq (v + 3) / 2$. If G also has at least $2k + 1$ vertices, then G is pancyclic modulo k (B. Bollobas [3]). Several results on the cycle modularity of graphs (not necessarily planar) can be found in [7].

In this paper we restrict our attention to planar graphs, and we obtain results which do not hold for graphs in general. This is accomplished by using Steinitz's characterization of 3–connected planar graphs for global structure and the theory of Euler contributions for local structure. These ideas are introduced in section 2 and are applied with respect to conditions on minimum degree and connectivity in sections 3 and 4, respectively.

2. Some Useful Results About Planar Graphs

This section describes certain somewhat unfamiliar properties of planar graphs which will be used in sections 3 and 4, and the more popular properties will be used without mention. We begin with an extremely important but surprisingly neglected theorem of E. Steinitz [12]: A graph is planar and 3–connected if and only if it is the one–skeleton of a convex 3–dimensional polyhedron. The basic importance of this theorem stems from the fact that it transforms problems about a 3–dimensional object into 2–dimensional considerations. The main consequence of interest in this paper is that it restricts the manner in which faces can intersect.

Lemma 2.1 *In a 3–connected plane graph, the non–empty intersection of two distinct faces is either a vertex or an edge.*

The theory of Euler contributions began with Euler's polyhedron formula in 1750 which was later generalized by others to the now familiar formula $v - \varepsilon + \phi = 2$ for connected plane graphs. In 1940 H. Lebesgue [8] showed that certain corollaries of Euler's formula could be considerably strengthened by a new method. He associated certain numbers with the face angles of a polyhedron, defined certain sums, and investigated how certain faces "contribute" to these sums. This approach, referred to in [9] as the theory of Euler contributions, was further developed in [9] and applied in [10] and [11] to the study of special colorings of plane graphs. The remainder of this section summarizes the aspects of this theory which are of use in our investigation of cycle modularity. These results and others are essentially contained in [8], but a more refined explanation can be found in [10] and [11].

Let G be a 2–connected plane graph. Let $x_i(v)$ denote the size of the ith face $F_i(v)$ at vertex v. Often the vertex label is understood or unimportant, and so we usually write x_i for $x_i(v)$ and F_i for $F_i(v)$. Also, the faces are labeled so that the x_i's occur in nondecreasing order. The $d(v)$–tuple $(x_1, ..., x_{d(v)})$ is called the face configuration vector at v. The Euler contribution of v is given by the formula

$$\Phi(v) = 1 - \frac{d(v)}{2} + \sum_{i=1}^{d(v)} \frac{1}{x_i} \; .$$

Lemma 2.2 *If G is a 2–connected plane graph, then $\Sigma_v \, \Phi(v) = 2$.*

Lemma 2.3 *If v is a vertex in a 2–connected plane graph G, then $\Phi(v) \leq 1 - d(v) / 6$.*

These two lemmas have some interesting comsequences. First of all, there must be a vertex v with $\Phi(v) > 0$; such a vertex is called a control point, and Lemma 2.3 implies that every control point has degree at most 5. Furthermore,

$$\sum_{i=1}^{d(v)} \frac{1}{x_i} > \frac{d(v)}{2} - 1.$$

All solutions to this inequality for $d(v) \geq 3$ are given in Tables 1 and 2 where $x = |F_3|$, $|F_4|$, or $|F_5|$ depending on whether $d(v) = 3, 4,$ or 5, respectively. Since the face boundaries of a 2–connected graph are cycles, the numbers in Table 2 and the first three columns of Table 1 represent cycle lengths. For 3–connected graphs, it follows from Lemma 2.1 that every number in Table 1 is a cycle length.

Sometimes it is useful to know that a plane graph has more than one control point. In fact, every vertex of the triangle, the tetrahedron, the octahedron, and the icosahedron is a control point with face configuration vector $(3, 3), (3, 3, 3), (3, 3, 3, 3),$ and $(3, 3, 3, 3, 3)$, respectively. More generally, we make the following observation which follows immediately from Lemmas 2.2 and 2.3.

F_1	F_2	F_3	$F_1 \Delta F_2$	$F_1 \Delta F_3$	$F_2 \Delta F_3$	$F_1 \Delta F_2 \Delta F_3$
3	3	3, 4, ...	4	$1 + x$	$1 + x$	x
3	4	4, 5, ...	5	$1 + x$	$2 + x$	$1 + x$
3	5	5, 6, ...	6	$1 + x$	$3 + x$	$2 + x$
3	6	6, 7, ...	7	$1 + x$	$4 + x$	$3 + x$
3	7	7, ..., 41	8	$1 + x$	$5 + x$	$4 + x$
3	8	8, ..., 23	9	$1 + x$	$6 + x$	$5 + x$
3	9	9, ..., 17	10	$1 + x$	$7 + x$	$6 + x$
3	10	10, ..., 14	11	$1 + x$	$8 + x$	$7 + x$
3	11	11, 12, 13	12	$1 + x$	$9 + x$	$8 + x$
4	4	4, 5, ...	6	$2 + x$	$2 + x$	$2 + x$
4	5	5, ..., 19	7	$2 + x$	$3 + x$	$3 + x$
4	6	6, ..., 11	8	$2 + x$	$4 + x$	$4 + x$
4	7	7, 8, 9	9	$2 + x$	$5 + x$	$5 + x$
5	5	5, ..., 9	8	$3 + x$	$3 + x$	$4 + x$
5	6	6, 7	9	$3 + x$	$4 + x$	$5 + x$

Table 1: Cycle structure at a trihedral control point

Lemma 2.4   *Every 2–connected plane graph of minimum degree δ contains at least $\lceil 12 / (6 - \delta) \rceil$ control points.*

$d(v) = 4$	$(3, 3, 3, x)$	3, 4, ...
	$(3, 3, 4, x)$	4, ..., 11
	$(3, 3, 5, x)$	5, 6, 7
	$(3, 4, 4, x)$	4, 5
$d(v) = 5$	$(3, 3, 3, 3, x)$	3, 4, 5

Table 2: Configuration vectors at a control point of degree ≥ 4

3. Pancyclicity and Minimum Degree

In this section we show how the minimum degree of a graph can insure the existence of cycles of various lengths.

Lemma 3.1   *Every 2–connected plane graph G which contains a control point of degree at least 4 is pancyclic modulo 3.*

Proof   Refer to the tables, and recall that our graphs have no loops or multiple edges. The (0 mod 3) case is trivial because each configuration contains a triangle. For the (1 mod 3) case notice that, except for (3, 3, 5, x), each configuration which has no 4-face contains two adjacent triangles which contain a 4–cycle. We can assume that the faces containing v have the cyclic ordering indicated in Figure 1. Since u ≠ w, the 5-face cannot contain both u and w. If it contains neither of them, then the symmetric difference of these three faces is a 7–cycle. If it contains one of them, we get a 4–cycle as shown in Figure 2, and this completes the case for (1 mod 3)–cycles. In any cyclic ordering of the faces containing v notice that there is either a 5–face, a triangle adjacent to a 4–face, or three consecutive triangles. It follows that G contains a (2 mod 3)–cycle. ❑

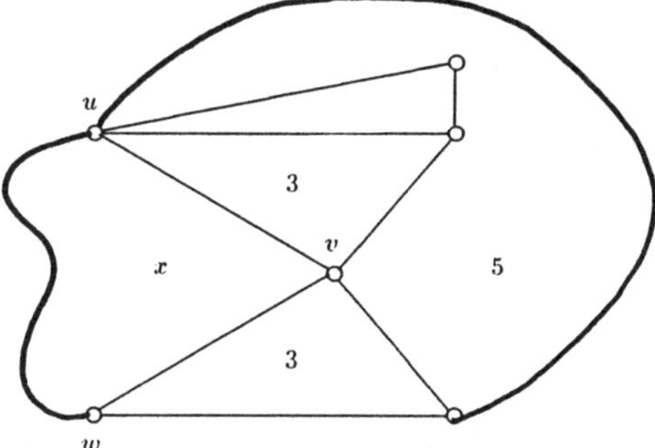

Figure 1: Construction used in the proof of Lemma 3.1

Even though a graph is planar and has minimum degree at least 3, it is possible that is does not contain a (2 mod 3)–cycle. For example, K_4 has no such cycle, and if two vertices, one from each of two copies of K_4, are respectively identified with the two vertices of degree 2 in $K_{3,3}^-$ (i.e., $K_{3,3}$ minus one edge), we have yet another example. Clearly infinitely many examples can be constructed in this way. However, if every vertex had degree at least 4, we have cycles of every length.

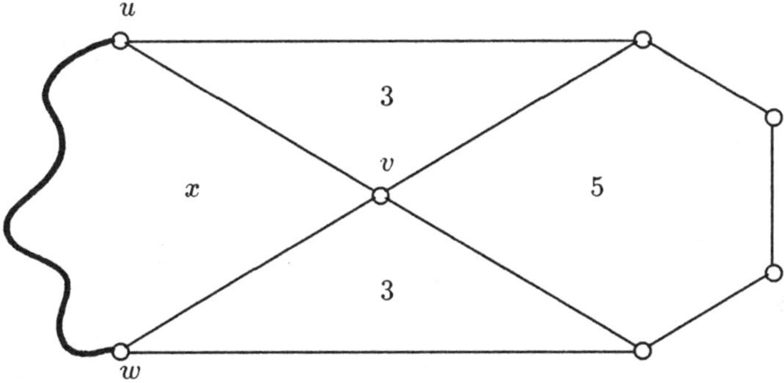

Figure 2: Construction used in the proof of Lemma 3.1

Theorem 1   *Every planar graph   G   with minimum degree at least   4   is pancyclic modulo   3.*

Proof   By inducting on   v   we can assume that   G   is connected. Since the smallest such graph is   2–connected, Lemma 3.1   settles the basis case and allows us to assume that   G   has a cut vertex   x.   Let   H   be a component of   G – x.   Choose   x   and   H   so that   v(H)   is as small as possible. Then $\hat{H}$ = G[V(H) $\cup$ x]   is   2–connected and every vertex of $\hat{H}$ (except possibly   x)   has degree at least   4.   Therefore, every planar embedding of $\hat{H}$ has a control point of degree at least   4   (Lemma 2.4) and, hence, is pancyclic modulo   3   (Lemma 3.1).   ❑

The following two results are even easier to prove.

Lemma 3.2   *Every   2–connected plane graph which contains a control point of degree at least   5   is pancyclic modulo   4.*

Theorem 2   *Every planar graph with minimum degree at least   5   is pancyclic modulo   4.*

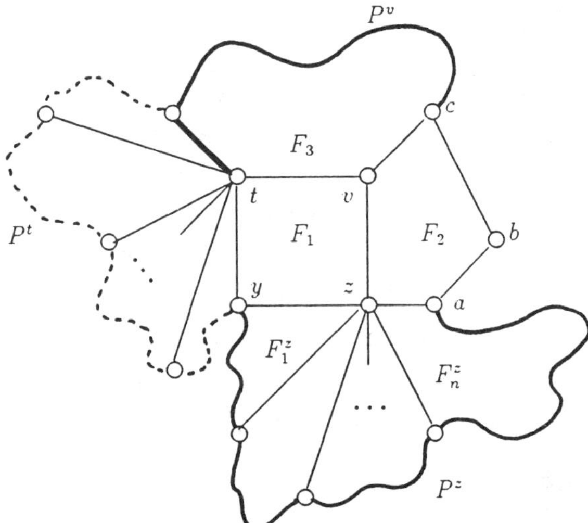

Figure 3: Case 1 of the proof of Proposition 4.1

4. Pancyclicity and Connectivity

In this section we examine how connectivity plays a role in producing cycles of every length in planar graphs. For this purpose, the most effective tools are not the most common, i.e., instead of Menger's theorem and Euler's formula, we use the consequences of Steinitz's theorem and the theory of Euler contributions outlined in section 2. To summarize this section, we prove that *every   3–connected planar graph* *with minimum degree at least k (except K_4) is pancyclic modulo k.* Only the cases $3 \le k \le 5$ deserve consideration.

Proposition 4.1 *Every 3–connected planar graph contains a (0 mod 3)–cycle.*

Proof Assume by contradiction that G has no such cycle. From Lemma 3.1 and Table 1 it follows that every control point v is cubic and has face configuration (4, 5, x) or (5, 5, x). We can assume the ordering is counterclockwise around v.

Case 1: (4, 5, x). From Table 1 we have $|F_1| = 4$, $|F_2| = 5$, and $|F_3| = x \equiv$ 2 mod 3. Clearly $P^v = bd(F_3) - v$ is a path. Using the labeling shown in Figure 3 we construct two more paths as follows. Let $F^z = F_1{}^z \Delta \dots \Delta F_n{}^z$ where $\{F_1{}^z, \dots, F_n{}^z\}$ is the set of faces $\notin \{F_1, F_2\}$ containing z, and let F^t be defined similarly except that only faces F_1 and F_3 are excluded. Clearly $P^t = bd(F^t) - t$ and $P^z = bd(F^z) - z$ are paths. Since $P^z + azy$ and $P^z + azvty$ are cycles, $\varepsilon(P^z) \equiv 0$ mod 3. Similarly, $\varepsilon(P^v) \equiv$ $\varepsilon(P^t) \equiv 0$ mod 3. Now P^v must intersect P^z for, otherwise, $P^v + P^z + abc + ty$ is a (0 mod 3)–cycle. Simiarly, P^t and abc must meet (otherwise, $P^t + yzabc + P^v - t$ is a (0 mod 3)–cycle) which is impossible.

Case 2: (5, 5, x). From Table 1 it follows that x = 7. Hence, $|F_1| = |F_2| = 5$, and $|F_3| = 7$. Refering to Figure 4 let $F^u = F_1{}^u \Delta \dots \Delta F_5{}^u$ where $\{F_1{}^u, \dots, F_5{}^u\}$ is the set of faces $\notin \{F_1, F_3\}$ containing u, and let F^w be defined similarly except that only the faces F_2 and F_3 are excluded. By assumption $|F^u|, |F^w|, |F^u \Delta F_3|,$ $|F^w \Delta F_3| \not\equiv 0$ mod 3. Hence, $|F^u| \equiv |F^w| \equiv 2$ mod 3. Now F^u intersects F_2 for, otherwise, $F^u \Delta F_1 \Delta F_2 \Delta F_3$ is a (0 mod 3)–cycle. Similarly, F^w must intersect F_1 which is impossible. $\square$

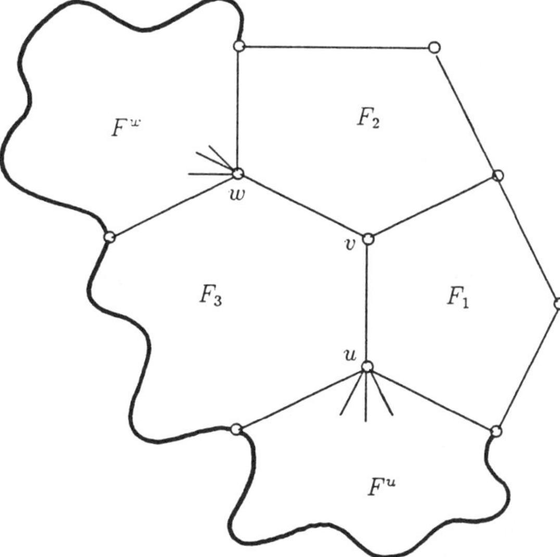

Figure 4: Construction used in the proof of Propositions 4.1 and 4.2

Proposition 4.2 *Every 3–connected planar graph contains a* (1 mod 3)–*cycle.*

Proof To obtain a contradiciton we assume that G is a counterexample. By Lemma 3.1 we have a cubic control point. Notice that in Table 1, except for configurations (5, 5, 5) and (5, 5, 8), at least one of the cycles F_1, F_2, F_3, $F_1 \triangle F_2$, $F_1 \triangle F_3$, $F_2 \triangle F_3$, $F_1 \triangle F_2 \triangle F_3$ is of length 1 mod 3. As in Case 2 of the proof of Proposition 4.1 we use the labeling shown in Figure 4 where $|F_3| = x \equiv 2$ mod 3. By assumption $|F^u|$, $|F^w|$, $|F^u \triangle F_1 \triangle F_3|$, $|F^w \triangle F_2 \triangle F_3| \not\equiv 0$ mod 3. Hence, $|F^u| \equiv |F^w| \equiv 2$ mod 3. Now F^u intersects F_2 for, otherwise, $F^u \triangle F_1 \triangle F_2 \triangle F_3$ is a (1 mod 3)–cycle. Similarly, F^w must intersect F_1 which is impossible. $\square$

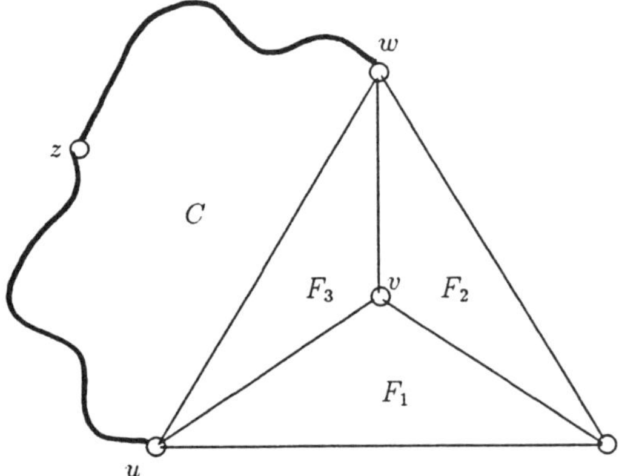

Figure 5: Construction used to prove Proposition 4.3

Proposition 4.3 *Every 3–connected planar graph except K_4 contains a* (2 mod 3)–*cycle.*

Proof Assume that there exist a 3–connected planar graph on at least 5 vertices which has no (2 mod 3)–cycle. Using Lemma 3.1 and Table 1 we are left with the following cases.

Case 1: (3, 3, x), (3, 6, x), (3, 9, x) where $x \equiv 0$ mod 3. Label the faces as shown in Figure 5 where, for the moment, we assume $uw \in E$. Since the graph $H = F_1 \cup F_2 \cup F_3 \cong K_4$, there are two internally disjoint paths from some vertex $z \in V - V(H)$ to H, and we can assume that these paths end at u and w to form a cycle C with uw. Clearly one of the cycles C, $C \triangle F_3$, $C \triangle F_1 \triangle F_3$ is of length 2 mod 3, a contradiction. Hence, $uw \notin E$. Let the cycles F^u and F^w be as defined in the proof of Propositions 4.1 and 4.2. Since $|F^u| \not\equiv 2$ mod 3 and $|F^u \triangle F_1| \not\equiv 2$ mod 3, $|F^u| \equiv 0$ mod 3. Furthermore, $w \in bd(F^u)$ for otherwise $F^u \triangle F_1 \triangle F_2$ is a (2 mod 3)–cycle. Similarly, $u \in bd(F^w)$ which is impossible.

Case 2: $(4, 4, x)$ and $(4, 7, x)$ where $x \equiv 1 \bmod 3$. At this point our approach should be obvious. We use the same labeling as before to eventually conclude that F^u intersects F_2 and F^w intersects F_1 which is impossible. $\square$

Thoerem 3 *Every 3–connected planar graph except K_4 is pancyclic modulo 3.*

This result is best possible in at least two ways. First of all, any cycle demonstrates that it is not sufficient for the graph to be planar and 2–connected. Also, planarity cannot be dropped because $K_{3,v-3}$ has no $(2 \bmod 3)$–cycle.

Theorem 4 *Every 3–connected planar graph with minimum degree at least 4 is pancyclic modulo 4.*

Proof Let G be a planar embedding of a 3–connected graph whose minimum degree is at least 4, and let v be a control point of G. Lemma 3.2 permits us to assume that $d(v) = 4$. In Table 2 notice that v lies in a triangle; also, v lies in either a 5–face, a triangle which is consecutive with a 4–face, or three consecutrive tirangles. Hence, G contains cycles of length 1 and 3 mod 4. On the other hand, if v is contained in neither a 4–face nor two consecutive triangles, then the cyclic ordering of faces (if not clockwise, then counterclockwise) containing v is given by (F_1, F_3, F_2, F_3) where $|F_4| = 5$, 6, or 7, and a $(0 \bmod 4)$–cycle is easily found in each case.

Assume that G contains no $(2 \bmod 4)$–cycle. Then the configuration vector for v is neither $(3, 3, 4, x)$ nor $(3, 3, 5, x)$ because these vectors yield the $(2 \bmod 4)$–cycles $F_1 \triangle F_2 \triangle F_3$ and $F_2 \triangle F_3$, respectively. Only two cases remain.

Case 1: The configuration vector at v is $(3, 3, 3, x)$. From Lemma 2.1 is easily follows that $x \equiv 3 \bmod 4$. We can assume that F_4 and F_1 share an edge vy. Thus, $bd(F_1)$ can be denoted by $tvyt$, and some face F_{ty} other than F_1 contains the edge ty. Since $F_1 \cup F_2 \cup F_3$ contains a ty–path of every length modulo 4, not all of these paths can be internally disjoint from the path $bd(F_{ty}) - ty$. Since $\delta(G) \geq 4$, this implies that $F_{ty} \cap F_i$ is neither a vertex nor an edge for some integer $i \in \{1, 2, 3\}$, a contradiction to Lemma 2.1.

Case 2: The configuration vector at v is $(3, 4, 4, x)$. Clearly no two 4–faces are adjacent. Hence, we can assume that the cyclic ordering of faces is given by (F_1, F_2, F_4, F_3) where $|F_4| = 5$, and we can use the labeling shown in Figure 6 where F_{ab} is the face $\neq F_3$ which contains ab. Notice that the paths ab, $auzwvcb$, $avcb$, and $auvcb$ are of length 1, 2, 3, and 0 mod 4 (respectively), and so from Lemma 2.1 and the fact that $\delta(G) \geq 4$ it follows that F_{ab} must intersect $\{w, z\}$. Similarly, uz must lie in a face F_{uz} which intersects $\{b, c\}$, a contradiction. $\square$

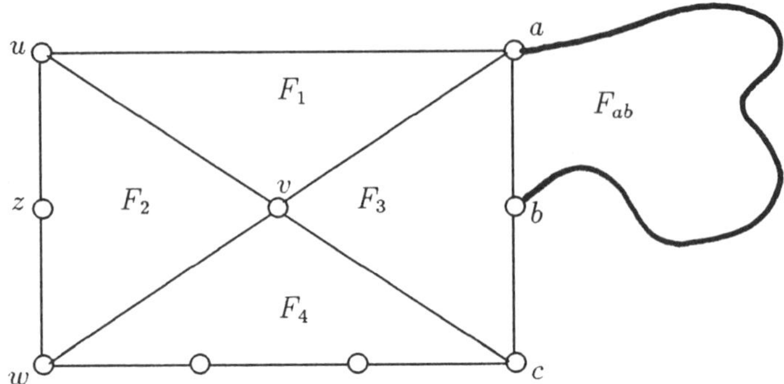

Figure 6: Construction used in the proof of Theorem 4

In a sense this theorem is best possible. The minimum degree condition cannot be dropped because no graph which is 3–connected, planar, and bipartite (for example, the cube) can contain cycles of length 1 or 3 mod 4. Furthermore, planarity is necessary (for example, consider K_5 and $K_{4,v-4}$). It is tempting to conjecture that *every 3-connected graph with minimum degree at least 4 is pancyclic modulo 3*. The Petersen graph, K_4 and $K_{3,v-3}$ show that the bound on the minimum degree cannot be dropped.

The only challenge in proving the next theorem arises when we try to obtain a (2 mod 5)–cycle with the configuration vector (3, 3, 3, 3, 3). Since this case is easily settled in several different ways, we leave it to the reader. Observe that the octahedron forces the minimum degree requirement and $K_{5,v-5}$ forces us to require planarity.

Thoerem 5 *Let G be a 3–connected planar graph with minimum degree at least 5. Then G is pancyclic modulo 5.*

5. More Unsolved Problems

One truely curious theorem in [2] states that every cubic graph contains a (0 mod 3)–cycle. Furthermore, it is shown that every subdivision of a cubic 3-connected graph on at least 10 vertices contains a (0 mod 3)–cycle. The following statement also seems to be true.

Conjecture 2 Every k–connected graph (k $\geq$ 3) contains a (0 mod k)–cycle.

For whatever the reason, questions about algorithms inevitably arise, and so the following recent results are of interest. In [13] it is shown that, for any fixed integer k, there is a polynomially bounded algorithm for deciding if an arbitrary graph contains a (0 mod k)–cycle. Also, [1] and [6] independently give a linear time algorithm for recognizing graphs in which every cycle is of length p mod k. Finally, the following question was raised by M. Fellows (personal communication), and, as far as I know, it remains open even for k = 2.

Open Problem 3   For a fixed integer k, what is the computational complexity of deciding whether or not an arbitrary graph contains an induced (0 mod k)–cycle?

REFERENCES

[1]   E.M. Arkin, C.H. Papadimitriou and M. Yannakakis, Modularity of cycles and paths in graphs. Technical Report, Department of Operations Research, Standford University, May 1986.

[2]   C.A. Barefoot, L.H. Clark, J. Douthett, R.C. Entringer and M.R. Fellows, Cycles of length 0 modulo 3 in graphs. Preprint for Sixth International Conference on the Theory and Applications of Graphs, Kalamazoo, Michigan (1988).

[3]   B. Bollobas, Cycles modulo k. *Bull. London Math. Soc.* **9** (1977) 97 − 98.

[4]   J.A. Bondy, Large cycles in graphs. *Discrete Math.* **1** (1971) 121 − 132.

[5]   J. A. Bondy and U.S.R. Murty, *Graph Theory with Applications.* American Elsevier, New York (1976).

[6]   G.J. Chang and C.–S. Wu, Graphs whose cycles are of length r modulo k. Report No. 86439–OR, Institut für ÖKonometrie und Operations Research, Bonn, West Germany, November 1986.

[7]   P. Erdős, Some recent problems and results in graph theory, combinatorics and number theory. *Proceedings of the Seventh Southeastern Conference on Combinatorics, Graph Theory, and Computing*, Congressus Numerantium XV, Utilitas, Winnepeg (1976) 3 − 14.

[8]   H. Lebesgue, Quelques conséquences simples de la formule d'Euler. *J. de Math.* **9** (1940) 27 − 43.

[9]   O. Ore, *The Four Color Problem*, Academic Press, New York (1966).

[10]   O. Ore and M.D. Plummer, Cyclic coloration of plane graphs. *Recent Progress in combinatorics*, Proceedings of the Third Waterloo conference on Combinatorics, Academic Press, New York (1969) 287 − 293.

[11]   M.D. Plummer and B. Toft, Cyclic coloration of 3–polytopes. *J. Graph Theory* **11** (1987) 507 − 515.

[12]   E. Steinitz, Polyheder und Raumeinteilungen. *Enzykl. math. Wiss.* **3** (Geometrie), Part 3AB **12** (1922) 1 − 139.

[13]   C. Thomassen, On the presence of disjoint subgraphs of a specified type. *J. Graph Theory* **12** (1988) 101 − 111.

ON AN EXTENSION OF A CONJECTURE OF I. HÁVEL

I. J. Dejter

J. Quintana

University of Puerto Rico

ABSTRACT

A conjecture of I. Hável asserts that there is a Hamilton cycle for the graph induced by the middle levels of any odd Boolean lattice. In this paper, we deal with an extension of that conjecture. In fact, we explicit some Hamilton cycles for the graphs induced by the levels contiguous to the middle levels of some even Boolean lattices. This uses some group actions on graphs and the notion of an F- diagram of a graph.

1.   Introduction

Let $i, j, \epsilon Z$ with $0 < i < j$ and $n = i + j$. Let RG_{ij} be the graph induced by the ith and the jth levels of the Boolean lattice of the set $I_n = \{0, 1, ..., n-1\}$, which are symmetrical levels. The Kneser graph O_{ij} and its bipartite covering G_{ij} will be considered in Section 3 so as to show their relation to RG_{ij}. A conjecture of Hável [11] (also Kelly [13] and attributed to Erdös in [4]) is that every $RG_{i,i+1}$ is hamiltonian. To treat this conjecture, we considered group actions on graphs and their quotients. This

*Partially supported by the Experimental Program to Stimulate Competitive Research (EPSCOR-NSF) of Puerto Rico, Computational Mathematics Component and by the University of Puerto Rico.

approach gave in [4] a sufficient condition for the conjecture in terms of Hamilton paths of a quotient. This, in turn, yielded Hamilton cycles of RGi_{i+1} in [4,5,6,7].

The aim of this paper is to show that some $RG_{i,i+2}$ and $G_{i,i+2}$ are hamiltonian. In doing so, we interpret some quotients of RG_{ij} under group actions as F-diagrams. These diagrams were studied by Frucht [10], Hevia [12] and Arlinghaus [2].

Another approach to Hável's conjecture was studied by Duffus, Hanlon and Roth [8], Duffus, Sands and Woodrow [9] and Kierstead and Trotter [14]. As far as we know,their ideas on canonical matchings on $RG_{i,i+1}$ do not extend to $RG_{i,j}$, for $j - 1 > 1$, and there are not other attack approaches than ours here to the immediate extension of Hável's conjecture to the graphs $RG_{i,i+2}$.

2. F-diagrams

Let Γ be a group and let $G = (V, E)$ be a graph. A Γ-action $\tau = (T, t)$ on G is given by a pair of actions $T : \Gamma \times V \longrightarrow V$ and $t : \Gamma \times E \longrightarrow E$ such that $t(k, e)$ has ends $T(k, u)$ and $T(k, v)$, for every $k\epsilon\Gamma$ and every $e\epsilon E$ with ends $u, v\epsilon V$. Let V/T be the quotient set of V under T. Let Γ be a finite cyclic group and let $\tau = (T, t)$ be a Γ-action on a finite graph G. We define the quotient graph G/τ as a graph $(V/T, E/t)$ such that there is a graph map $G \longrightarrow G/\tau$ with subjacent vertex-set map $V \longrightarrow V/T$ and edge-set map $E \longrightarrow E/t$ and satisfying conditions (a) and (b) below. Select a main representative b in each T-orbit $B\epsilon V/T$.

Let 1_Γ be a fixed generator of Γ. Let b' be a representative of B. Then $b' = T(z \cdot 1_\Gamma, b)$ for some $z\epsilon Z$. We set $b' = B.z$. We orient arbitrarily each $e_t\epsilon E/t$ into an arc $\hat{e}_t$.

(a) For each $e_t\epsilon E/t$, if the arc $\hat{e}_t$ joins $B\epsilon V/T$ to $C\epsilon V/T$, then there exists a unique representative $e\epsilon E$ of e_t with B.0 and C.z as its ends, for some integer z in the interval $[0, d)$, where $d = gcd\{|B|, |C|\}$. We label $\hat{e}_t$ with z.

(b) The structure imposed by (a) on the graph $(V/t, E/t)$ does not depend on the orientation selected for E/t in (a). If the orientation of an arc $\hat{e}_t$ with label z, as in (a), is reversed, then the label of the reversed arc is -z + d.

The graph G/τ together with a labelling of V/T by the T-orbit cardinalities, an orientation of E/t and a labelling of the arcs of this orientation as in (a) is called an F-diagram and was studied essentially in [10, 12, 2].

A Γ-action $\tau = (T, t)$ on a graph G is said to be free if both T and t are free. In [7], we compared F-diagrams for free actions with voltage graphs. Also in [7], we extended the notion of an F-diagram for the case of Γ being a solvable group and applied this to the determination of some Hamilton cycles of $RG_{i,i+1}$. We do the same here for $RG_{i,i+2}$, by means of a sequence of F-diagrams as above, avoiding using directly the mentioned extension.

3.   Boolean and odd graphs

Up to a graph isomorphism, we can take the graph $RG_{i,j}$ as the bipartite graph having vertex classes D_i and D_j formed by the n-tuples whose Hamming weights are i and j, respectively, and whose adjacency is given by changing exactly j-1 coordinate values 0 to 1, or vice versa. A graph isomorphism from the first definition of RJ_{ij} in the Introduction to the second one now is given by means of the function associating to each subset of I_n its Z_2-characteristic vector, with vector coordinate subindices taken in I_n.

Any n-tuple d will be denoted as $d = (d_0, \cdots, d_{n-1})$, with coordinate subindices $0, 1, \cdots, n-1$ taken mod n, as elements of Z_n. Let $0 < r \epsilon Z$ be fixed and such that $gcd(r, n) = 1$. Certain operations on vector coordinates or their values yield group actions $\tau_s = (T_s, t_s)$ on RG_{ij} $(s = 0, 1, 2, 3)$ which appeared in [4,5,6,7]. They are determined as follows.

(0) Let T_0 be *complementation*, i.e. a Z_2-action determined by $T_0(1, d) = (1 + d_0, \cdots, 1 + d_{n-1})$ $(d \epsilon D_i U D_j)$. Note that T_0 is a free action.

(1) Let T_1 be *reflection*, i.e. a Z_2-action determined by $T_1(1, d) = (d_{n-1}, \cdots, d_0)$ $(d \epsilon D_i U D_j)$.

(2) Let T_2 be *rotation* or *shifting*, i.e. a Z_n-action given by $T_2(k, d) = (d_k, d_{-k-1}, \cdots, d_{-k-1})$ $(k \epsilon Z_n, d \epsilon D_i U D_j)$.

(3) Let T_3 be r-*interleaving*, i.e. a Z_μ-action (μ = lowest positive $m \epsilon Z$ such that $r^m \equiv 1$ mod n) determined by $T_3(-1, d) = d_0, d_r, d_{2r} \cdot \cdot, d_{-r})$ $(d \epsilon D_i U D_j)$.

The graph G_{ij} is defined as having the same vertex subset of RG_{ij}, so that any two vertices u, v are adjacent in G_{ij} if and only if u and $T_1(v)$ are adjacent in RG_{ij}. The corresponding actions on G_{ij} to those above will be denoted, by abuse of notation, with the same symbols $\tau_i = (T_i, t_i)$ $(i = 0, 12, 3)$.

Let $R\tau_0 = (RT_0, Rt_0)$ be *reflected complementation*, ie. the Z_2-action obtained as the composition $\tau_0\tau_1$ of the actions τ_0 and τ_1. From the next Lemma and on we will indicate between parentheses "(" and")", respective alternatives in our exposition. In view of the Lemma, we will not use F-diagram labels explicitly for the actions $R\tau_0$ and τ_0.

Lemma 1   [4] If $a, b, \epsilon D_i$, then a and $RT_0(1, b)$ $(T_0(1, b))$ are adjacent in $RG_{ij}(G_{ij})$ and only if $RT_0(1, a)$ $(T_0(1, a))$ and b are adjacent in RG_{ij} (G_{ij}).

From Lemma 1 we see that if D_i and D_j are drawn in vertical columns so that each RT_0-orbit is given in a row of its own, then the *nonhorizontal links* appear always as *crossed pairs*, i.e. with their ends at the same corresponding rows. The underlined terms are to be used below.

According to Hável [11], Laborde observed essentially that, for $j - i = 1$, the quotient graph G_{ij}/τ_0, is isomorphic to the Kneser graph 0_{ij} [3,4]. This fact holds in general, as stated in the theorem below, a proof of which is contained for $j - i = 1$ in [7] and be extended immediately.

We define O_{ij} as having vertex set equal to the family of i-subsets of I_n, so that two vertices are adjacent in O_{ij} if they are disjoint in I_n. A proof of the following extension of Laborde's observation may be found in [7].

The graph G_{ij} as having the same vertex subset of RG_{ij}, so that any two vertices u, v are adjacent in G_{ij} if and only if u and $T_1(v)$ are adjacent in RG_{ij}. The corresponding actions on G_{ij} to those above will be denoted, by abuse of notation, with the same symbols $\tau_i = (T_i, t_i)(i = 0, 1, 2, 3)$.

The graph $RO_{ij} = RG_{ij}/R\tau_0$ is given as having the same vertex set of O_{ij} but its adjacency differs from the one of O_{ij} by the action of T_1, in the same way that the adjacency of RG_{ij} differs from the one of G_{ij}.

Every graph $G = (V, E)$ has associated a bipartite graph $G' = (V', E')$ with vertex set $V' = Vx\{0, 1\}$ formed by the union of vertex classes $Vx\{0\}$ and $Vx\{1\}$, such that there is a subjective correspondence $\theta : E' \longrightarrow E$, with $\theta^{-1}(e)$ composed by two edges with end sets $\{(u, 0), (v, 1)\}$ and $\{u, 1), (v, 0)\}$, for every edge with end set $\{u, v\}\epsilon E$ with $u, v\epsilon V$. It is easy to check that $RG_{ij} (G_{ij})$ is isomorphic to $RO'_{ij}(O'_{ij})$, or by means of a joint "()" notation that we will use subsequently: $(R)G_{ij}$ is isomorphic to $(R)O'_{ij}$.

Each vertex of $(R)OG_{ij}$ will be denoted by the vector expression of its representative in D_j. As a consequence, if $d\epsilon D_j$ then the vertices adjacent to d in $(R)O_{ij}$ are determined by reading (backwards) the vectors obtained from d by:

(1) selecting $j - 1$ coordinates of d with subindices $k(t)$ for which $d_{k(t)} = 1(t = 1, \cdots, j - 1)$;

(2) complementing the values of the coordinates of d subindexed in $z_n - \{k(t) : t = 1, \cdots, j - 1\}$.

Vectors in $D_i \cup D_j$ are represented as words, without commas or parentheses. Arrows "$\longrightarrow$" and "$\longleftarrow$" are used as follows. In $RG_{2,4}, d = 111100$ is adjacent to $d' = 110000$ by selecting $k(1) = 2$ and $k(2) = 3$. Then d^1 is adjacent to $RT_0(1, d^2)$ in $RO_{2,4}$, which can be checked, with arrows, as follows:

$$\text{forward columns} : 012345$$

$$d \qquad = \longrightarrow 111100$$

$$d' \qquad = \longleftarrow 001111$$

$$\text{backward columns} : \quad 543210$$

where forward columns 2 and 3 rest unchanged, the other columns are complemented from d to d' and arrows $\longrightarrow$ ($\longleftarrow$) mean that the vector reading is to be taken from left to right (right to left).

We represent each vertex of O_{ij} *by the complement* of its z_2- characteristic vector. Thus, if two vertices are adjacent in O_{ij}, then the $j-1$ common coordinates equal in their representations have their values equal to 1.

Theorem 2 $RG_{ij}/R\tau_0$ (G_{ij}/τ_0) is isomorphic to RO_{ij} (O_{ij}).

We identify $RG_{ij}/R\tau_0$ with RO_{ij}; and also G_{ij}/τ_0 with O_{ij}. We agree that notations G, O, τ_0, etc., respectively, stand either for $RG_{ij}, RO_{ij}, RG'_{ij}$, Retc. or for G_{ij}, O_{ij}, RG'_{ij}, etc. The given actions generate commutative square diagrams of projection maps between these graphs and their quotients. We adopt the same notation for graph maps represented in parallel in these commutative diagrams. With these conventions, we now denote

$$(^*) \qquad H = O/\tau_2 \text{ and } J = H/\tau_3.$$

We recover, from the notations G, O, O', H, H', J and J', the original notations, subindexing with i and prefixing R if needed.

Let A be a vertex of $(R)H'_{ij}$. Then A represents a T_2-orbit:

$$A = (a_0 a_1 \cdot \cdot a_{n-1} \quad a_1 a_2 \cdot \cdot a_0 \cdots a_{n-1} a_0 \cdot \cdot a_{n-2})$$

in $D_i \cup D_j$. The dot notation used in the definition of an F-diagram in Section 2 will be replaced, for the action τ_2, by colon notation. (We reserve the dot notation just for F-diagram edge labels arising from the action τ_3): We will select, for each $A \epsilon (R)H'_{ij}$, a distinguished representative $A : 0 \epsilon D_i \cup D_i \cup D_j$ in the corresponding T_2-orbit. Suppose that $A : 0 = a_0 a_1 \cdot \cdot a_{n-1}$. We denote

$$A : 1 = a_1 \cdot \cdot a_{n-1} a_0, \cdot \cdot, A : n - 1 = a_{n-1} a_0 \cdot \cdot a_{n-2},$$

so that $A = (A : 0 \ A : 1 \cdot \cdot A : n - 1)$. We also denote $A = (A : 0)$, to expose the particular selection of $A : 0$.

A vertex of RH_{ij} is said to be *palindromic* if and only if some representative of it in D_j is T_1-invariant. Let a^n be the vertex of RH_{ij} having a representative $(a_0 a_1 \cdots a_{n-1})$ with $a_k = 1$ if and only if $0 \leq k \leq j - 1$. Then a^n is palindromic.

From the adjacency of $RH_{i,i+1}$, the following facts can be proved [4]: Let $a \epsilon D_j$. Each coordinate $a_k = 1$ of a for which $a_{k-h} + a_{k+h} = 1$ for every $h = 1, \cdots, [n/2]$ determines a loop incident to a. In particular, a^n has two incident loops. In n is composite, say $n = m.q$, then given $a^m = (a_0, \cdots, a_{n-1})$ and a (q-1)- tuple b different from $RT_0(1, b)$, then the ends of a 2-link are obtained by interleaving b and $T_0(1, b)$, respectively, between each a_k and $a_{k+1}(k = 0, 1, .., n - 1$ taken mod n). This provides the two ends of a 2-link. For example in $RH_{7,8}$, ($\underline{11}00\underline{11}1\underline{1}00\underline{1}0\underline{1}00\underline{1}$) and (1 $\underline{0}\underline{1}10\underline{1}0\underline{1}1000\underline{11}0$) are two such ends for a 2-link, where $a^m = (110)$ and $b = (1001)$. Similar facts hold for RG_{ij} with $j - i > 1$. The following is a straight extension of a theorem of [4]. We say that a cycle C in RG_{ij} is obtained by cyclotomic lifting from a path P in RH_{ij} having loops 1_1 and 1_2 at its endvertices if the different liftings of P, 1_1 and 1_2 to RG_{ij} compose the cycle C.

Theorem 3   If RG_{ij} has T_2 free, then, in order to get a Hamilton cycle Q of RG_{ij} by cyclomatic lifting from a Hamilton path S in RH_{ij}, it is sufficient that S has as ends a^n and another vertex with an incident loop, and in case that n is composite, that S has two interior vertices adjacent by means of a 2-link.

Remark 4   Some palindromic vertices of RH_{ij} having two incident loops, like a^n, form a T_3-orbit. Palindromicity and the property of a vertex having a fixed number of incident loops in RH_{ij} are preserved by T_3. Combining these facts with Theorem 3 simplifies the Hamilton cycle search, see [4,5,6,7] and below. The dot notation given for representatives of a T-orbit in Section 2 will be used to distinguish representatives in RH_{ij} of vertices of RJ_{ij} considered as T_3-orbits. A Hamilton path S as in theorem 3 is said to be a *cyclable path* [5]. A path P in RH'_{ij} made up by a horizontal link joined to the union of two paths formed by crossed pairs of nonhorizontal links, with each of these pairs having one link in each path, or any of the liftings of P in RG_{ij}, is called a *spring*. If a path P in RH_{ij} with a loop in one of its endvertices lifts into such a spring, then P is called an *unfolder*. The proof of Theorem 3 consists on the construction of cyclable unfolder in RH_i.

4. Interleaving the coordinates

In this section, we make some observations about the action t_3 and present an algorithm to determine the orbits of RJ_{ij}. The vertices $(a_0, \cdots, a_{n-1})$ of RH_{ij} or H_{ij} will be called *necklaces* and represented with lowercase letters. We noticed that RT_0, T_0 and T_2 are free, so that their orbits have the same cardinality. On the other hand, T_3 may yield different orbit cardinalities.

We introduce some cyclotomic coset facts and notation about the action of T_3 into a necklace $A = (a_0 a_1 \cdots a_{n-1})$ RH_{ij}. By considering here the positions of the value a_1, as A:0 undergoes the successive action of T_3 at $1, 2, .., n-1, 0 \epsilon Z_n$, we get certain z_n-coordinate orbits which happen to be a cyclotomic coset mod n. We may set the values of such a coset in the form of an array, in such a way that, if possible, each nonzero column adds up to n. A cyclotomic coset arranged in two lines like these will be called a 2-*deck*.

Not necessarily 2-decks happen in the place of cyclotomic cosets: the z_7-coordinate orbits can be arranged for $r = 2$ with its elements forming two cyclotomic cosets of opposite signs, instead of a 2-deck. Cyclotomic cosets like these, contained in two contiguous rows, so that values in a column add up to n (-7, for the case) will be called 1-*decks*. We also say that 0 constitutes a 1-deck. These facts illustrate the following fact.

Lemma 5   The following conditions are equivalent:

(<u>a</u>) The deck that contains $1 \epsilon z_n$ is a 2-deck;

(<u>b</u>) τ_1 is a restriction of τ_3 for RG_{ij};

(<u>c</u>) $-1 \equiv 1 \bmod n$.

Any of these conditions holds for n equal to a prime power.

Let oc stand for T_3-orbit cardinality. The following algorithm determines every $\underline{oc}$ in $RJ_{ij} = RH_{ij}/\tau_3$ and the vertices of RJ_{ij} that, as T_3- orbits of RH_{ij}, have each $\underline{oc}$ in RJ_{ij}.

(1) Let $N = \{M_0, M_1, \cdots\}$ be the collection of cyclotomic cosets of Z_n. Let $N' \underline{C} N$ be such that $\sum\{|M_k| : k\epsilon N'\} = j$. Such a subset N' produces a vertex $a = (a_0, \cdots, a_{n-1})\epsilon D_j$ with $a_k = 1$ exactly for those coordinates k in some M_s, where $s\epsilon N'$. Moreover, the orbit of a in RH_{ij} has $oc= 1$.

(2) Let c be the cardinality of a largest cyclotomic coset mod n. For each $\delta = 2, \cdots, c$ with $gcd(r, \delta) > 1$, we subdivide each cyclotomic coset $M_s = \{sr^0, sr^1, \cdots, sr^{d-1})$ into subsets

$$M_{s,0} = \{sr^0, sr^\delta, \ldots\ldots, sr^{-\delta}\},$$

$$M_{s,1} = \{sr^1, sr^{\delta+1}, \ldots\ldots, sr^{-\delta+1}\},$$

$$\ldots\ldots\ldots\ldots\ldots\ldots\ldots\ldots\ldots\ldots$$

$$M_{s,\delta-1} = \{sr^{\delta-1}, sr^{2\delta-1}, \ldots\ldots, sr^{-1}\}.$$

We perform successively, for each value of δ as above and in a similar fashion to (1), the determination of necklaces associated to the new subdivision of N and not contained in T_3-orbits of previously determined necklaces.

For example, for $(n, r, \delta) = (9, 2, 2)$ we obtain $M_{0,0} = \{0\}$, $M_{1,0} = \{1, 4, 7\}, M_{1,1} = \{2, 8, 5\}$, $M_{3,0} = \{3\}$ and $M_{3,1} = \{6\}$. An easy calculation leads to the values to be given to the coordinates in each of these $M_{s,\sigma}$ to yield the representatives of T_3-orbits in RH_{ij} that have $oc= 2$.

5. Notation management for Hamilton cycles of RG_{ij}

We first remark that every Hamilton cycle C of RG_{ij} yields a Hamilton cycle of G_{ij} by changing alternative vertices of C in D_j via the reflection $T_1(1, -)$. From now on, capital letters will represent vertices of RJ_{ij}. If A is such a vertex and if A.0 $= (a)$ then we denote $A - [a]$.

Let PRJ_{ij} be the subset of palindromic vertices of RJ_{ij}. For the vertices of RJ_{ij} (PRJ_{ij}), there is a maximal oc. This is denoted $moc(\text{n})$ ($pmoc(\text{n})$). Either $pmoc(\text{n}) = \frac{1}{2} moc(\text{n})$ or $pmoc(\text{n}) = moc(\text{n})$, according to whether τ_1 is a restriction of τ_3 or not, respectively. We try to find a subgraph of RJ_{ij} consisting of:

(a) Either a collection $\{P\}$ of disjoint paths with the ends (interior vertices) of each P having $oc = pmoc(\text{n})$ ($oc = moc(\text{n})$);

(b) Or a collection $\{S\}$ of disjoint cycles on vertices having $oc = moc(\text{n})$.

If τ_1 is a restriction of τ_3 then: (i) each P as in (a) above can be blown up to $pmoc(\text{n})$ directed cycles $P.\mu$ in RH_{ij}; (ii) we may get a cyclable path by splitting these cycles $P.\mu$ into paths and plugging to their ends the remaining vertices. These cycles are taken with the orientations given in their descriptions below, from left to right, so that if we write

$$P.\mu = (.., P_s.\mu', P_{s+1} \cdot \mu', ..),$$

where $\mu' = \mu$ or $\mu + moc\,(n)$, then we adopt the notation $p_{s+1} \cdot \mu'(_sp \cdot \mu' = (P_{s+1} \cdot \mu')^{-1}))$ for $P.\mu - (P_s.\mu', P_{s+1} \cdot \mu')$ as an oriented path from $P_{s+1} \cdot \mu'$ to $P_s.\mu'$ ($P_s.\mu'$ to $P_{s+1} \cdot \mu'$), where lower-case letters, like p, replace capital letters, like P. The path obtained by deleting (saving exactly) from $q.\mu'$ its last y vertices is denoted by $q; y.\mu'(qy; .\mu')$, where $q = p_{s+1}$ or $_sp$.

For example, for $RJ_{5,6}$ with $r = 2$, (a) is applicable with the only path $P = (P_1, \cdots, P_5)$, where

$$P_1 = \longrightarrow [10111101000], P_2 = \longleftarrow [01000011111],$$

$$P_3 = \longrightarrow [10111100001], P_4 = \longleftarrow [01000111110] \text{ and}$$

$$P_5 = \longrightarrow [1111100000],$$

so that P blows up to five cycles

$$P.s = (P_1.s, P_2.s, \cdots, P_5 \cdot s, P_4 \cdot (s+5), \cdots, P_2 \cdot (s+5), P_1.s),$$

for $s = 0, \cdots, 4 \epsilon Z_{10}$. Then the representatives of the remaining vertex $A =$ [11011100010] in $RJ_{5,6}$ ($oc = 2$) can be integrated into a cyclable path $(_5p.0, A.0,_4 p.2,_2 p.1,_4 p.3, A.1, p_5.9)$.

A path P in $RH_{i,i+1}(RH_{i,i+2})$ may be given by its first vertex $\Phi(P)$ and the sequence $\pi(P) = \{\pi_s(P)\}$ of the z_n-coordinates (pairs of z_n-coordinates) that remain unchanged from the sth vertex of P to the $s+1^{th}$ one, where the first vertex is written from left to right using "$\longrightarrow$" and the remaining path vertices are alternatively written from right to left and from left to right using "$\longleftarrow$" and "$\longrightarrow$". For $RJ_{5,6} : (\Phi(P), \pi(P)) = (P_1, \{7, 10, 5, 1\})$.

6.   Hamilton cycles of RG_{ij} for $j - i = 2$

If $j = i+2$, then T_2 is free if and only if either $i \neq 2i'$ or $j \neq 2j'$, where $i', j' \epsilon Z$, in which case Theorem 3 applies. Otherwise, for $i = 2i'$ and $j = 2j'$, there is a copy of $RG_{i'j'}$ inside RG_{ij} by means of the graph monomorphism

$$(F, f) : RG_{i'j'} = (V'E') \longrightarrow RG_{ij} = (V, E),$$

given by $F(c) = (c, c)$ and accordingly for f. Moreover, (F, f) induces corresponding graph monomorphisms between the considered quotient graphs. These monomorphisms are denoted again (F, f).

Now, the cardinality of each necklace in $F(RG_{i'j'})/\tau_2(RH_{ij} - F(RG_{i'j'})/\tau_2))$ is $i' + j'$ $(i + j)$, constraining our strategy in the search of Hamilton cycles for these cases.

Example $i = 2$. Let $S = (S_1 =\!\longrightarrow 101011, S_2 =\!\longleftarrow 111100)$ be a path in $RH_{2,4} - (F, f)RH_{1,2}$ having $\Phi(S) = 101011$ and $\pi(S) = \{(0, 2)\}$. The path S is an unfolder with spring in $RH'_{2,4}$ given by

$$S' = (010100, 111100, 110000, 110101).$$

Then, a Hamilton cycle of $RG_{2,4}$ is:

$$(S', 100100, (\tau_2(3, S'))^{-1}, 101101, \tau_2(5, S'), 001001, (\tau_2(2, S'))^{-1},$$

$$011011, \tau_2(4, S'), 010010, (\tau_2(1, S'))^{-1}, 110110, 010100).$$

Example $i = 3$. A Hamilton unfolder S of $RH_{3,5}$ has $\Phi(S) = (11111000)$ and $\pi(S) = \{(2,3), (3,6), (1,6), (2,5), (0,4), (1,7)\}$. The path S unfolds in $RG_{3,5}$ into a spring S' that represents the pattern of Theorem 3 and Remark 4.

Example $i = 4$. In $RH_{4,6} - (F, f)RH_{2,3}$, we have unfolders S and U covering all vertices, with $\Phi(S) = (1101101001)$, $\pi(S) = \{(1,4), (5,8), (3,9), (4,9), (5,7), (2,5), (4,9), (0,4), (0,8), (3,8), (4,5), (0,5), (3,8), (1,3)\}$, $\Phi(U) = (1111001100)$ and $\pi(U) = \{(7,0), (9,4), (9,6), (7,0)\}$. Let $S' : 0$ $(U' : 0)$ be the spring

$$(\Phi(S) : 0, \cdots, 0100011010) \quad ((\Phi(U) : 0, \cdots, 1100110000)),$$

A tenth part W of a Hamilton cycle C of $RG_{4,6}$ is given by $(F(11010), S'^{-1} : 5, F(01001), S'^{-1} : 0, U' : 5, F(11001), U' : 0)$. The vertex of $RG_{4,6}$ starting the next tenth part of C is $F(01100)$.

Example $i = 5$. The graph $RJ_{5,7}/\tau_1$ for $r = 5$ has the following vertex types (where we set $\beta(-) = T_3(-1, -)$):

A: Four palindromic vertices A with $oc = 1$.

H: Three nonpalindromic vertices H with $oc = 1$.

Y: Three palindromic vertices Y with $oc = 2$.

X: Five nonpalindromic X with $oc = 2$ and $\beta(X) = \tau_1(1, X)$.

P: Ten nonpalindromic vertices P with $oc = 4$.

All the representatives of P-vertices in $RH_{5,7}$ form a cycle P, a fourth part $+P.1$ of it with $\Phi(+P.1) = +P_1.1 = (001110110101)$ and $\pi(+P.1) = \{(2,6), (2,10), (3,9), (3,5), (7,10), (1,8), (5,11), (5,6), (1,5)\}$. Continuing $\pi(+P.1)$ with $(1,10)$ means continuing $+P.1$ with $+P_{10}.0$. Then $P = (+P.1, (+P.0)^{-1}, -P.1, (-P.0)^{-1}, +P_1.1)$. Let $+H_1.0 = (\longleftarrow 010001101111)$,

$-H_1.0 = (101111010001)$, $A_3.0 = (110011100110)$, and Ω and Σ be paths with $\Phi(\Omega) = (1011011010110)$, $\pi(\Omega) = \{(0,2), (6,8), (5,9), (3,11), (4,8), (0,8), (6,10), (5,9), (0,8), (2,6), (3,6)\}$, $\Phi(\Sigma) = (011110001110)$ and $\pi(\Sigma) = \{(2,8), (6,7), (4,6), (0,2), (0,10), (4,6), (2,4), (0,10), (9,10), (2,8)\}$. Then a cyclable path in $RH_{5,7}$ is $+p^{-1}; 5.0, -P; 4.1, +H_1.0, -H_1.0, -P4; .1, -p^{-1}7; .0,$ $A_3.0, -p^{-1}; 7.0, +P.1, +P^{-1}4; 0, \Omega, +P_6.0, \Sigma$.

If each vertex is denoted with its type symbol, then we can set $\Omega = (Y_2.0, A_4.0, Y_2.1, Y_3.1, Y_1.1, -Y_2.0, Y_1.0, +H_2.0, -H_3.0, A_2.0, +H_3.0, Y_3.0)$ and $\Sigma = (-X_5.1, -X_4.1, -X_3.1, -X_2.1, -X_1.1, A_1.0, +X_1.0, +X_2.0, +X_3.0, +X_4.0, +X_5.0)$.

Example $i = 6$. In $RJ_{6,8}$ with $r = 3$, there are only two vertices of $(F, f)RJ_{3,4}$ whose oc's are 2 and 3, denoted respectively X and Y. We will come back to them after treating the remaining vertices. All nine vertices with $oc = pmoc14 = 3$ (thirty vertices with $oc = pmoc$ $(14 = 6)$ of $RJ_{6,8} - \{X,Y\}$ form a path $Q(P)$, such that $U = (Q,P)$ is a path in $RJ_{6,8}$ with $Q_1 = [01111100011100]$ and $\mu(U) = \{((1,9), (8,13), (2,5), (9,2), (3,4), (3,7), (7,13), (1,5), (8,12), (8,13), (1,13), (2,12), (2,9), (1,13), (10,11), (2,10), (1,10), (6,11), (3,5), (0,7), (7,12), (10,12), (8,10), (3,7), (7,13), (5,7), (0,3), (1,9), (9,12), (8,9), (1,6), (1,13), (5,13), (6,13), (8,13), (3,4), (3,7), (0,3), (2,11)\}$. U has three liftings $U.s = (Q_1.s, ..)$, for $s = 0, 1, 2, \epsilon Z_3$, respectively z_6. Two more vertices with $oc = 1$ in $RJ_{6,8}$ are the middle vertices of the following paths A' and C'.

$$Q_{10}.0 = \longleftarrow [10101011100011] \quad P_{29}.0 = \longleftarrow [01111100101001]$$

$$A\ .0 = \longrightarrow [01010101011101] \quad C\ .0 = \longrightarrow [11000011011110]$$

$$P_1.3 = \longleftarrow [10101010111100]. \quad P_{30}.0 = \longleftarrow [10111100100101].$$

Let Q' be Q modified by replacing the first integer pair $(1,9)$ in $\pi(U)$ by $(5,11)$, maintaining the other edges of U. Let $P_C.0$ be P modified by inserting $C.0$, as in C'. Then, in $RH_{6,8}$ we consider a cycle $\Omega = (Q.0, A.0, P_C.0, (P.3)^{-1}, Q'.2, P.2, (P.5)^{-1}, Q.1, P.1, (P.4)^{-1}$. Thus, $(\beta_2^1)^{-1}(\Omega)$

is the disjoint union of seven double covering paths $= (Q_1.0 : k, \cdots)$, for $k = 0, \cdots, 6$. If a:s is a vertex of $\Omega(k, k + 7)$, then also $a : (s + 7)\epsilon\Omega : (k + 7)$. Now, there is a path $W = (Y.1, X.1, P_4.3, P_5.3, Y.2, P_5.0, P_4.0, X.0, Y.0)$ in $RH_{6,8}$ with $Y.1 = (10011011001101)$ and $\pi(W) = \{(3,10),\ (9,12),\ (0,12),$ $(0,2),\ (10,12),\ (1,12),\ (1,4),\ (1,8)\}$, having seven liftings W:k in $RO_{6,8}$. They produce in $RO_{6,8}$ the path

$$\Sigma = (W : 0, (W : 1)^{-1}, W : 4, (W : 5)^{-1}, W : 3, (W : 6)^{-1}), W : 2).$$

By inserting into Σ the paths $\Omega - (P_4.s, P_5.s)$, for $s = 0, \cdots, 6$, we get a Hamilton path Σ' of $RO_{6,8}$. In fact, the ends of Σ' are looped vertices in $RO_{6,8}$ and Σ' lifts to two zigzag paths in $RG_{6,8}$ forming together with corresponding horizontal links a Hamilton cycle of $RG_{6,8}$.

REFERENCES

[1] W. C. Arlinghaus, "The Classification of Minimal Graphs with Given Abelian Automorphism Group", (*Memoirs Amer. Math. Soc.*, 330, 1985).

[2] N. Biggs, "Algebraic Graph Theory", (Cambridge University Press, England, 1972).

[3] N. Biggs, "Some odd graph theory, in: A. Gewritz and L. V. Quintas, eds., *II Conf. Combin. Math.*. (Ann. New York Ac. Sci. 319, 1979) 71-81.

[4] I. J. Dejter, "Hamilton cycles and quotients of bipartite graphs", in: Y. Alavi et al., eds. *Proc. V^0 Int. Conf. Theory Appl. Graphs* (Wiley, New York, 1985) 189-199.

[5] I. J. Dejter, "Stratification for hamiltonicity", *Congressus Numerantium* 47 (1985) 265-272.

[6] I. J. Dejter, J. Córdova and J. Quintana, "Two Hamilton cycles in bipartite reflective Kneser graphs", *Discrete Math.*, 72 (1988).

[7] I. J. Dejter, W. Cedeno and V. Jáuregui, "F-diagrams, Boolean graphs and Hamilton cycles", submitted.

[8] D. Duffus, P. Hanlon and R. Roth, "Matchings and Hamiltonian cycles in some families of symmetric graphs", preprint.

[9] D. Duffus, B. Sands and R. Woodrow, "Lexicographic matchings cannot form Hamiltonian cycles", preprint.

[10] R. Frucht, "How to describe a graph", in L. V. Quintas ed., *Int. Conf. Combin. Math.* (Ann. New York Ac. Sci. 175, 1970) 159-167.

[11] I. Hável, "Semipaths in directed cubes", in: Graphs and Other Combinatorial Topics, B. 59 (Teubner Texte zum Mathematik, Teubner, Leipzig) (3^0 Symposium Tchescoslovaque de Th. Graphes, Nov. 1982).

[12] H. Hevia, Representación orbital grafos y número mínimo de puntos para grafos n- cíclicos, n una potencia de primo", *Scientia* 148 (1977) 102-122; (Math. Rew. 81g : 05068).

[13] D. Kelly, Problem 2.5, in I. Rival, ed., "Graphs and Order: The Role of Graphs in the Theory of Ordered Sets and its Applications" (D. Reidel, Dordrecht, 1985).

[14] H. Kierstead and T. Trotter, "Explicit matchings in the middle levels of the Boolean lattice", Order (to appear).

A Dualizable Representation for Imbedded Graphs

Linda L. Deneen

Gary M. Shute

Clark D. Thomborson (a.k.a. Thompson)*

Computer Science Department

University of Minnesota

ABSTRACT

A new representation is described for graphs imbedded in pseudomanifolds. The graphs may have loops and parallel edges, and they may be disconnected. It is shown that our representation has a well–defined dual that naturally extends the definition of surface duality. This representation is easily transformed to a data structure with a set of operators for modifying a graph, a set of query operators, and a set of navigation operators. Three variants of the data structure are described. For a graph with n vertices, the first provides $O(1)$–time updates and $O(n)$–time queries. The second provides $O(n)$–time updates and $O(1)$–time queries. The third, a tradeoff between the first two, provides $O(\log n)$–time updates and queries. The space used in all variants is linear in the size of the graph.

1. Introduction

In this paper we extend several known representations for imbedded graphs [1, 4, 16, 18] to allow for disconnected components, loops, parallel edges, and isolated vertices. Our representation allows for natural imbeddings into pseudomanifolds. Most noteworthy is that our representation gives rise naturally to a well–defined dual graph. Our dual can be obtained directly from the representation, and it has a natural imbedding that generalizes the usual notion of a dual imbedding. Our presentation of a graph in terms of half–edges is succinct and more general than previous presentations of imbedded graphs.

---

* Supported by the National Science Foundation, through its Design, Tools and Test Program under grant number MIP 84–06139.

We also develop a set of primitive operators and three data structure implementations for use in representing and manipulating graphs. This work extends previous work as well. Isolated vertices are not allowed in a quadedge structure [7], and they are tolerated as special cases at best in a DCEL or winged–edge representation [11]. The quadedge structure has a second deficiency: it does not allow one to imbed a component of a disconnected graph inside a face of some other component of that graph.

Our primitive operators are divided into three groups: modification operators, navigation operators, and query operators. Six special modification operators can be viewed as extensions of Guibas and Stolfi's splice operator [7]; we need six instead of one because of the added flexibility of our representation.

Our paper is organized as follows. Section 2 describes the basic graph theory background and gives a history of previous work that our work generalizes. Section 3 gives the mathematical description of our graph representation for the orientable case. In section 4 we describe our abstract operators, and in section 5 we describe our three implementations, again for the orientable case. The representation is extended to the nonorientable case in section 6. Conclusions and open problems are discussed in section 7.

2 . Fundamentals

Graph theorists have expended much effort studying the imbeddings of graphs in geometric spaces. It is common to study imbeddings in spaces that are locally similar to $\mathcal{R}^2$. Many investigators [1, 4, 6, 12, 14, 16, 17, 18] have studied imbeddings in *2–manifolds,* which are connected compact topological spaces with the property that each point has a neighborhood homeomorphic to the open unit disk in $\mathcal{R}^2$. These 2–manifolds can be either orientable or nonorientable. In the following we will be concerned only with spaces that are locally 2–dimensional so that we will drop the dimensional prefix from our terminology.

We will investigate imbeddings into pseudomanifolds. Our definition of a pseudomanifold will diverge from White's pseudosurface in that our pseudomanifolds need not be connected.

Definition A pseudomanifold is a compact topological space M in which each point x has a neighborhood that is homeomorphic to a finite number of open disks identified at x.

The number of disks in the neighborhood of a point x is called the *degree* of x. If the degree of x is greater than one, then x is called a *singular point* of M.

Following Biggs [1], we define a *graph* as follows:

Definition (Biggs) A graph is an ordered quadruple (E, V, λ, τ) where E and V are finite sets, λ is a surjective function from E to V, and τ is an involution on E.

We call the elements of E *half–edges* and the elements of V *vertices* of G. If the half–edge e is not fixed by τ, then we may interpret e as a directed edge with $\lambda(e)$ as the initial vertex of e and $\tau(e)$ as the edge directed opposite to e. In keeping with this interpretation, we say that e is *incident* on v if $\lambda(e) = v$. We defer the interpretation of half–edges that are fixed by τ until later. We note that this definition of a graph admits loops, parallel edges, and disconnected graphs.

An imbedding of a graph in a topological space identifies the vertices of the graph with distinct points in the space and the edges with homeomorphic images of line segments. The image of an edge must have the images of its vertices as endpoints, and images of edges may not intersect except at the endpoints. If a graph G is imbedded in a pseudomanifold M, the components of $M \setminus G$ are the *faces* of the imbedding, and the imbedding is a *2–cell imbedding* if the faces are all homeomorphic to the open unit disc.

In classifying imbeddings of graphs most authors use variants of the *permutation technique* or the *rotation system technique*. White [17] attributes this technique to Edmonds [4] and calls it *Edmonds' permutation technique*. Gross and Tucker [6] call it the method of *rotation systems* and jointly credit Heffter (1891) and Edmonds (1960), hypothesizing that Edmonds rediscovered the technique without being aware of Heffter's work. Both White and Gross and Tucker credit Youngs [18] with making Edmonds' work accessible. The central result of this work is that there is a one–to–one correspondence between the orientable 2–cell imbeddings of G in manifolds and certain permutations of the directed edges of G called *rotations*. The rotations specify the rotational order of the imbedded edges around a vertex.

Given a 2–cell imbedding of a connected graph G in a pseudomanifold M, one can define the *dual graph G** like this: choose one vertex for G* for each face of G's imbedding, and for each edge e in G, add an edge e^* to G* connecting the vertices corresponding to the two faces on either side of e. The use of the term *dual* refers to the fact that the dual graph of the dual graph of G is G again. In studying representations of graphs and their imbeddings, we consider two questions: does a given imbedded graph have a well–defined dual graph, and if so, does the particular graph representation make it easy to find the dual graph? Edmonds [5] considers only 2–cell imbeddings of connected graphs in spheres; Biggs [1] extends this to manifolds with his representation, which gives rise to the dual graph in a natural way. Tutte [16] allows imbeddings in either orientable or nonorientable manifolds, and he allows disconnected components in his graphs, but his dual graph representation requires that each separate component of the graph be imbedded in a distinct manifold; his duals are found component–wise. In our presentation, we will extend all of the above results to include graphs that are not necessarily connected and to imbeddings on pseudomanifolds.

3. An Extended Graph Representation

We present a new graph representation that extends the previously discussed work in several ways. First of all, with this new representation, we can represent graphs that may be disconnected or that may contain isolated vertices. Secondly, these graphs may be imbedded on pseudomanifolds. Thirdly, in an imbedding of a disconnected graph, we can specify clearly whether or not a component lies inside a face of another component. Finally, all of our graphs have well–defined duals, and the dual graph of a graph in our representation is easily obtained. This is a significant advance in duality theory, since duals were previously defined only for connected graphs or in the case of Tutte, disconnected graphs with separate components on separate manifolds.

Edmonds [4, 18] uses the notion of a *rotation* on a graph G to characterize 2–cell imbeddings of G in compact manifolds. Our definition of a rotation follows Biggs [1].

Definition   (Biggs) A rotation on a graph $G = (E, V, \lambda, \tau)$ is a permutation O on E such that $\lambda O = \lambda$.

The definition insures that for any vertex v, O restricts to a permutation on the half–edges incident on V. We call each cycle of O an *origin* and loosely regard each vertex as a set of origins. If each vertex contains only one origin then O is said to be *smooth*. For the imbedding construction of Edmonds, rotations are required to be smooth with the cycles inducing a circular ordering on the half–edges incident on a vertex.

We eliminate the smoothness condition by considering imbeddings in pseudomanifolds. If a vertex v contains k origins, then in the imbedding v must be a point of degree k. The neighborhood of a singular point can be visualized as a set of cones sharing a common apex. When a vertex is located at a singular point, the half–edges incident on that vertex are not cyclically ordered. Instead, the ordering can be described as a not necessarily smooth rotation mapping a half–edge onto the next half–edge counterclockwise on the same cone. In the limiting case of a vertex with one half–edge per cone, the edges are completely unordered. Another interpretation of a vertex located at a singular point is that its incident edges are only partially imbedded. This notion of partial imbedding is useful when representing the intermediate results of a graph–imbedding algorithm, such as Hopcroft and Tarjan's linear–time planarity tester [8].

If a cone about v contains no edges that connect to v we say that the cone is a *blank cone*. If a half–edge e is fixed by τ then we will require that e is also fixed by

O. Then e is by itself an origin. We interpret the origin e as an *isolated* origin so that the cone corresponding to e is a blank cone.

Edmonds begins by defining a permutation P on E by $P = O\tau$. We may view P as dual to O, with cycles of P determining a cyclic ordering of half–edges around a boundary of a face in the imbedding of G. We call the cycles of P *panes*, so that panes are dual to origins. To construct a theory of imbeddings in which every imbedding has a dual, we must also dualize the notion of a vertex. Thus to specify an imbedding we must have a finite set W, which is dual to V, and a function μ from E to W, which is dual to λ. To preserve duality, we require that μ be surjective and that $\mu P = \mu$. We call the elements of W *windows*. In an imbedding of a graph G in a space M, windows correspond to the connected components of $M \setminus G$ and panes correspond to the connected components of boundaries of windows. We may loosely regard each window as a set of panes. Analogous to the terminology for vertices, we say that a half–edge e is *incident* on a window w if $\mu(e) = w$.

We are now ready to extend the concept of a rotation to allow imbeddings of graphs in orientable pseudomanifolds.

Definition   A generalized rotation system on a graph $G = (E, V, \lambda, \tau)$ consists of a rotation O that fixes each half–edge that is fixed by τ, a finite set W, and a surjective function μ from E to W satisfying $\mu O\tau = \mu$.

When discussing a generalized rotation system on a graph we will use the terminology of vertices, origins, windows, and panes as introduced above.

Given a graph G and a generalized rotation system for G, we define an imbedding of G in a pseudomanifold M as follows. As in Edmonds' construction, we begin by defining a permutation P on E by $P = O\tau$. Then $\mu P = \mu$ so for any window w, P restricts to a permutation of the half–edges incident on w. Thus the concept of a pane of w is meaningful. For each window w we form a surface M_w by removing open disks with disjoint boundaries from a sphere, with one disk removed for each pane in w. Since $P = O\tau$, and O fixes the fixed half–edges of τ, it follows that if a half–edge e that is fixed by τ occurs in a pane then e must in fact be the only half–edge in that pane. In this case we retract the boundary of the hole for that pane to a point and label the point by the vertex $\lambda(e)$. Otherwise, we divide the boundary of the hole for the pane into segments, with one segment labeled by each half–edge in the origin. The segments are ordered so that the segment labeled by $P(e)$ appears counterclockwise around the hole from the segment labeled by e, as seen from the outside of the sphere. For the segment labeled by e, label the clockwise endpoint by the vertex $\lambda(e)$.

Now, we paste the surfaces M_w together to form M' by identifying each boundary segment labeled e to the segment labeled $\tau(e)$, provided $e \neq \tau(e)$. Boundary segments are identified so that the segments labeled e and $\tau(e)$ are oppositely directed. Finally, we form M from M' by identifying all sets of points labeled by the same vertex.

It is straightforward to show that M is an orientable pseudomanifold and that the constructed imbedding satisfies the following definition:

Definition　　A simple imbedding of a grpah G in a pseudomanifold M is an imbedding such that

   (i)　Each component of M contains at least one vertex of G,

   (ii)　Each singular point of M is a vertex of G,

   (iii)　The number of blank cones about each vertex v of G is equal to the number of half–edges e that are incident on v and fixed by τ, and

   (iv)　Each connected component of $M \setminus G$ is a sphere with a finite number of holes whose boundaries are disjoint.

Moreover, the imbedding we have constructed is unique up to isomorphism. This follows from the fact that given a sphere S with a finite number of open holes and a permutation σ of holes, there is a homeomorphism of S onto itself that permutes the holes as does σ.

Finally, if we are given an imbedding of a graph G in an orientable pseudomanifold M, then guided by the interpretations given above, we can recover W, μ, and P. More precisely, let W be the set of components of $M \setminus G$. We associate each blank cone at a vertex v with a half–edge e that is incident on v and fixed by τ. We set $P(e) = e$ and $\mu(e) = w$, where w is the component that contains the cone. For other half–edges, e, $\mu(e)$ is the component on the right of e as a directed edge, and $P(e)$ is the next half–edge walking in the direction of e around $\mu(e)$. Defining W, μ, and P in this way, we have $\mu P(e) = \mu(e)$ and $\lambda P\tau(e) = \lambda(e)$, for all half–edges e. See Figure 1. We now recover O by computing $O = P\tau$. Since $\lambda O = \lambda P\tau = \lambda$, we are assured that O is a rotation. Since τ is an involution, $P = O\tau$, and it follows that $\mu O\tau = \mu$. We have thus shown

Theorem 1　*Given a graph G, there is a one–to–one correspondence between generalized rotation systems on G and simple imbeddings of G in orientable pseudomanifolds.*

We can present a graph, together with a simple imbedding in an orientable pseudomanifold, in a way that makes duality more apparent. We do this by specifying the set E of half–edges, together with two sets, V and W, of permutations. There is one permutation in V for each vertex v, obtained by restricting P to the half–edges incident on v. Similarly, W is obtained from O. When writing the description of a graph, the set E may be omitted if the permutations in V and W are given in cycle notation with 1–cycles for fixed half–edges. We will use this notation in our illustrations.

The dual of an imbedded graph is then obtained by interchanging V and W. Clearly, every simple imbedding of a graph in an orientable pseudomanifold has a uniquely defined dual. The duality operator is an involution and in the case of planar imbeddings, our dual is the usual planar graph dual except for a reversal of orientation.

For example, the graphs in Figures 2 and 3 are duals of one another. A more interesting pair of dual graphs is given in Figures 4 and 5. In Figure 4 the dotted line joins two origins that belong to the same vertex. This vertex is mapped onto the singular point of a sphere whose north and south poles are identified. In the dual graph this two–origin vertex becomes a window with two panes: as drawn in Figure 5, these are the innermost pane and the exterior pane. This window is thus isomorphic to a cylinder with circular panes on opposite ends. The cylinder forms the hole through the torus on which the dual graph is imbedded.

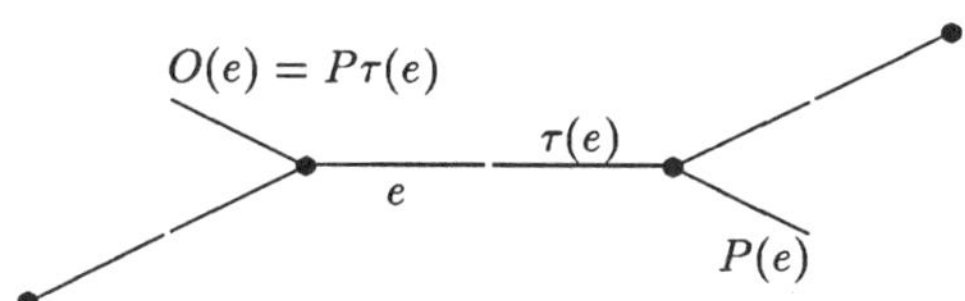

Figure 1: A portion of a graph, showing that $\lambda O(e) = \lambda(e)$.

4. Abstract Operators for the Extended Graph Representation

In this section we describe a set of abstract operators to be used to express the extended graph representation described in the previous section. These operators are part of a computer program to manipulate graphs; three implementations are described in the next section. The operators in this section can be grouped into two classes: navigation operators and modification operators.

There are five navigation operators. *Onext* maps a half–edge to the next half–edge in the cyclic order around an origin. *Onext* corresponds to the permutation O of the previous section. The operator *Oprev* corresponds to O^{-1}. *Pnext* corresponds to the permutation P in the previous section. The operator *Pprev* corresponds to P^{-1}. Our *Onext* is the same as Guibas and Stolfi's *Onext* [7], while *Pnext* is the same as their *Rprev*. *Mnext* maps a half–edge a to the opposing half–edge in its edge pair if it has one; otherwise a represents an isolated vertex and *Mnext*(a) = a. *Mnext* corresponds to the involution τ in the previous section.

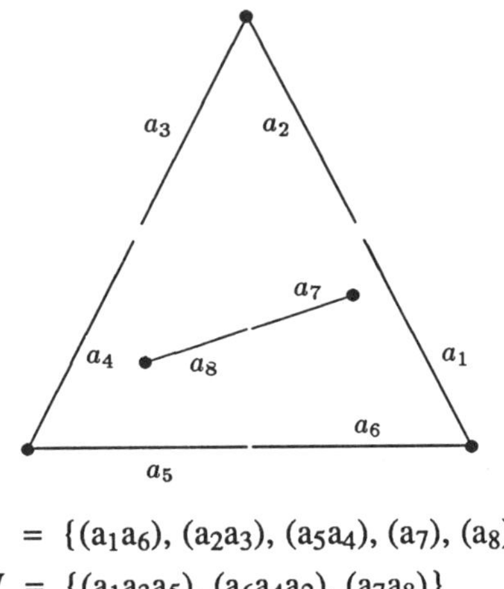

$$V \;=\; \{(a_1a_6),\,(a_2a_3),\,(a_5a_4),\,(a_7),\,(a_8)\}$$
$$W \;=\; \{(a_1a_3a_5),\,(a_6a_4a_2),\,(a_7a_8)\}$$

Figure 2: A graph with a 2–pane window.

There are nine modification operators. *MakeNullg* simply creates an empty graph. *Addh* returns a half–edge that is *Onext*, *Pnext*, and *Mnext* to itself. It is thus an isolated vertex imbedded in its own sphere. *Delh(a)* removes a from the graph. Before we can apply *Delh* to a, however, we must ensure that a is an isolated vertex; it must not be linked in any way to any other half–edge in the graph.

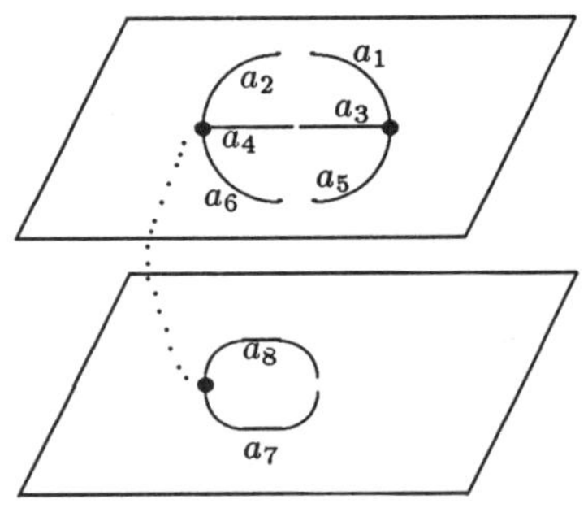

$$V \;=\; \{(a_1a_3a_5),\,(a_6a_4a_2),\,(a_7a_8)\}$$
$$W \;=\; \{(a_1a_6),\,(a_2a_3),\,(a_5a_4),\,(a_7),\,(a_8)\}$$

Figure 3: A graph with a 2–origin vertex.

The remaining six modification operators come in pairs, corresponding to the vertex–face duality of our presentation. The first pair of operators, *Osplice* and *Psplice* affect the origin and pane cycles, respectively. Like Guibas and Stolfi's quadedge splice operator [7], our *Osplice* and *Psplice* are their own inverses. Given two half–edges a_1 and a_2 lying in distinct origins, $Osplice(a_1, a_2)$ merges the two origins. This is done by writing a_1's origin as a cycle starting with a_1 and writing a_2's origin as a cycle starting with a_2 and then concatenating the two cycles. If the operands of *Osplice* are distinct half–edges with a common origin, this origin will be split in two. We require that the arguments of *Osplice* be in the same vertex. The *Psplice* operator behaves similarly, with its arguments required to be in the same window.

The remaining modification operators modify the vertex and window structures. Given two half–edges a_1 and a_2 in distinct vertices, *Vmerge* (a_1, a_2) merges a_1's vertex with a_2's vertex. *Vsplit* (a) splits the vertex containing the half–edge a into two vertices: a's origin becomes the only member of a new vertex, and the remaining origins stay with the original vertex. Similarly, *Wmerge* merges two windows and *Wsplit* splits a single window in two.

As an example, we show in Figure 6 how we can add the new edge with half–edges c and d to an already existing graph, using the operators described above. Note that at intermediate stages of the construction the medial function τ is not always an involution. We are currently developing higher level operators, built on top of our primitive operators, that will preserve the property of being a graph in the sense of Biggs' definition.

5. Three Implementations for the Orientable Extended Graph Representation

For these three implementations, we choose to store panes and origins in circular linked lists, thus fitting the order of a cycle making up a pane or origin into the data structure. The problem of storing windows and vertices is slightly more complex. A window is represented by linking together a distinguished half–edge, called a *wedged edge*, one from each pane in the window, into a circularly linked list. Similarly, a vertex is represnted by linking together a distinguished half–edge, called a *vedged edge*, one from each origin in the vertex, into a circularly linked list.

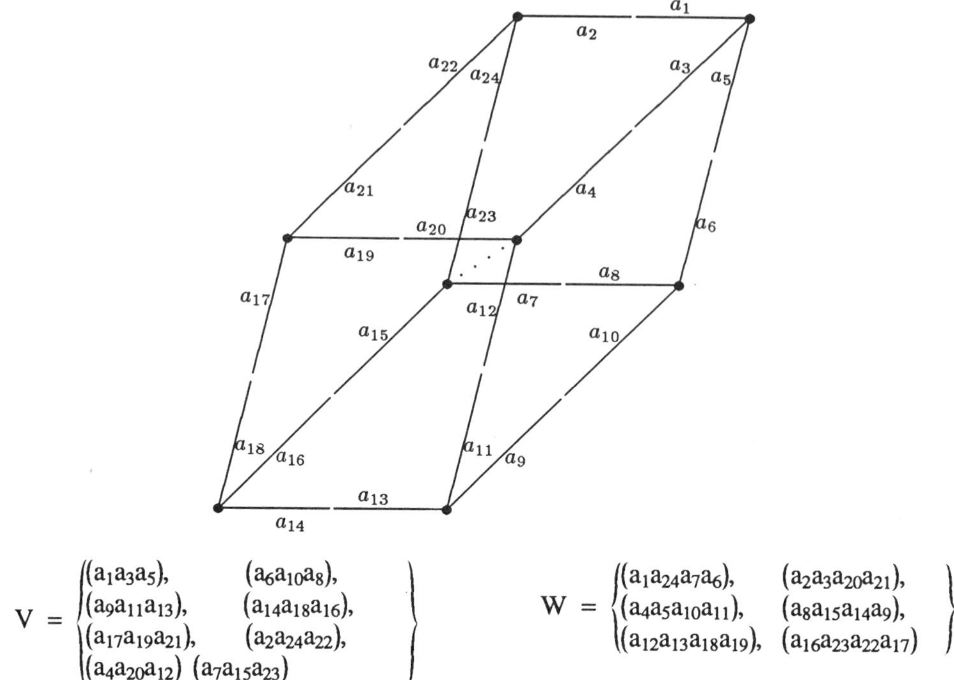

$$V = \begin{cases} (a_1a_3a_5), & (a_6a_{10}a_8), \\ (a_9a_{11}a_{13}), & (a_{14}a_{18}a_{16}), \\ (a_{17}a_{19}a_{21}), & (a_2a_{24}a_{22}), \\ (a_4a_{20}a_{12}) & (a_7a_{15}a_{23}) \end{cases} \qquad W = \begin{cases} (a_1a_{24}a_7a_6), & (a_2a_3a_{20}a_{21}), \\ (a_4a_5a_{10}a_{11}), & (a_8a_{15}a_{14}a_9), \\ (a_{12}a_{13}a_{18}a_{19}), & (a_{16}a_{23}a_{22}a_{17}) \end{cases}$$

Figure 4: A cube with 8 origins and 7 vertices on a pseudosphere.

In the data structure a record is allocated for each half–edge. In the first implementation we store values for *Onext, Pprev, Vnext, Vprev, Wnext,* and *Wprev*. As is common practice in pointer list implementations, we use the special value *nil* to denote the value of an undefined function such as *Wnext*(a) when a is not wedged. We need not store *Oprev, Pnext,* or *Mnext,* because these can be computed in O(1) time from the others: *Mnext*(a) = *Pprev*(*Onext*(a)), *Pnext*(a) = *Onext*(*Mnext*(a)), *Oprev*(a) = *Mnext*(*Pprev*(a)).

Finally, we add four query operators that allow the user to ask questions about the form of the graph. *Vfind* returns the vedged half–edge in th origin of its argument, and *Wfind* returns the wedged half–edge in the pane of its argument. The boolean function *Vsame* indicates whether its two arguments are in the same vertex, and *Wsame* indicates whether its two arguments are in the same window.

We now examine the run times for the first implementation and suggest two variants. The first implementation requires six pointers per half–edge, using O(n) space to store a graph of size n. The navigation and splicing operators take O(1) time. Unfortunately, queries are slow in this implementation. It takes O(n) time to execute a *Vfind, Wfind, Vsame,* or *Wsame*. This is comparable to the less general quadedge scheme [7], which uses O(n) space, O(1) time for navigation and splicing, and O(n) time to determine if two edges are in the same face or same vertex. A second implementation stores the *Vfind*(a) and *Wfind*(a) values with each half–edge a, in addition to the fields of the first implementation. This scheme allows O(1)–time queries, but now the splicing operations take O(n) time. This implementation is comparable to the DCEL representation [11]. A third scheme, of intermediate speed, maintains the *Vfind* and *Wfind* information by means of a self–adjusting binary tree on the half–edges in each vertex and window [13]. A self–adjusting tree supports O(log n)–time tree splices and root–finds, providing a graph representation with O(log n)–time queries and splices, and O(1)–time navigation operators.

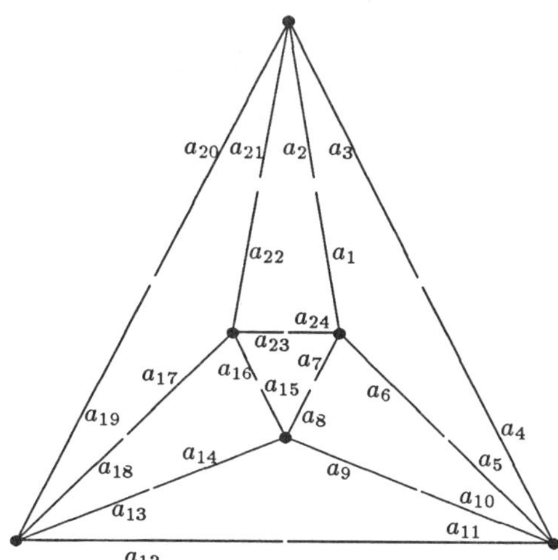

$$V = \left\{ \begin{array}{ll} (a_1 a_{24} a_7 a_6), & (a_2 a_3 a_{20} a_{21}), \\ (a_4 a_5 a_{10} a_{11}), & (a_8 a_{15} a_{14} a_9), \\ (a_{12} a_{13} a_{18} a_{19}), & (a_{16} a_{23} a_{22} a_{17}) \end{array} \right\}$$

$$W = \left\{ \begin{array}{ll} (a_1 a_3 a_5), & (a_6 a_{10} a_8), \\ (a_9 a_{11} a_{13}), & (a_{14} a_{18} a_{16}), \\ (a_{17} a_{19} a_{21}), & (a_2 a_{24} a_{22}), \\ (a_4 a_{20} a_{12}) & (a_7 a_{15} a_{23}) \end{array} \right\}$$

Figure 5: The dual of Figure 4, an octahedron on a torus.

6. An Extension to the Nonorientable Case

Ringel [12], Stahl [14] and Tutte [16] have investigated imbeddings of graphs on nonorientable manifolds as well as orientable manifolds. Tutte represents each edge as a set of four *crosses*, called a *cell*. If X is a cross, the other members of its cell are θX, φX, and θφX, where θ and φ are involutions on the finite set of all crosses making up the graph. The pair X and θX can be thought of as a *half–edge* with φX and θφX making up the opposing half–edge in the edge. The involution θ is a flip, or reversal of orientation, so that X and θX represent different sides of a half–edge. The involution ϕ is a facial rotation, so that X and ϕX are adjacent elements of the boundary of the face on one side of the edge. A permutation Q is then defined on the crosses to give the rotations of the half–edges around their vertices. Breaking an edge into four crosses and defining θ, ϕ, and Q on the crosses gives enough flexibility to represent imbeddings of these graphs in nonorientable manifolds. Guibas and Stolfi [7] have developed a graph representation reminiscent of Tutte's, and they have also designed a data structure for representing and manipulating graphs with a computer.

Following Tutte, we use four records (crosses or quarter–edges) to represent an edge. An origin is then a pair of cycles of quarter–edges, one for each local orientation. As before, a vertex is a set of origins. Thus, in our presentation, we tabulate a rotation O of quarter–edges around origins, as well as a set V of sets of origins. We also tabulate Tutte's involution θ, mapping quarter–edges onto quarter–edges of opposite orientation. The involution θ is restricted by Tutte's axiom: $O\theta = \theta O^{-1}$.

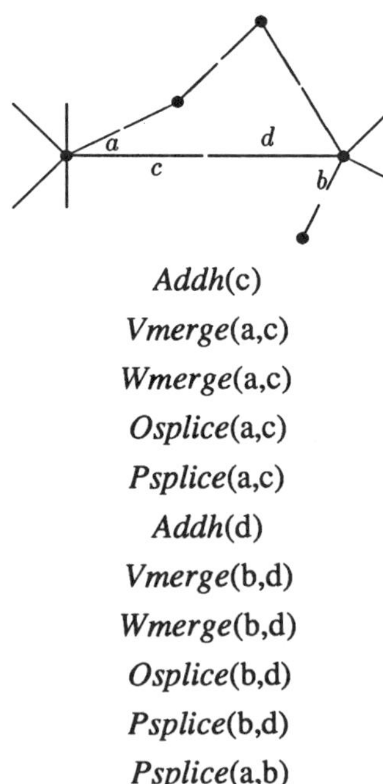

$Addh(c)$

$Vmerge(a,c)$

$Wmerge(a,c)$

$Osplice(a,c)$

$Psplice(a,c)$

$Addh(d)$

$Vmerge(b,d)$

$Wmerge(b,d)$

$Osplice(b,d)$

$Psplice(b,d)$

$Psplice(a,b)$

Figure 6: Adding edge with half-edges c and d to an existing graph.

The facial structure of a graph is treated analogously. A pane is a pair of cycles of quarter–edges, one for each orientation of its boundary. A window is a set of panes. We thus tabulate a rotation P of quarter–edges around panes as well as a set W of sets of panes. The rotation P is restricted by Tutte's axiom: for each quarter–edge q, the orbits of P through q and θq are distinct. Our medial rotation τ as well as Tutte's involution ϕ are implict in our presentation: $\tau = P^{-1}O$ and $\phi = \theta\tau$.

In a straightforward data structure based on the above, we could represent a quarter–edge with one seven–pointer record: *(Onext, Pprev, θ, Vnext, Vprev, Wnext, Wprev)*. Preliminary research indicates that a much more compact representation is possible. We believe we can represent each pair of oppositely oriented quarter–edges $(q, \theta q)$ with a single record containing four pointer fields and one boolean field.

7. Conclusions and Open Problems

This paper makes contributions in two areas: graph theory and data structures. We have given a succinct and general presentation for graphs imbedded on pseudomanifolds. We have also extended surface duality theory by showing how our representation gives rise to a natural dual for graph that may be disconnected or contain isolated vertices. This representation is useful in graph data structures: we have described three implementations with various query and update time preformances. Perhaps most importantly, we have defined a general set of operators for navigating, querying, and manipulating graphs.

At present, we are testing a fast–update and slow–query implementation of our data structure, in preparation for coding an algorithm for generating optimal rectilinear Steiner trees [15]. Dobkin and Laszlo have expressed interest in incorporating our ideas into their extension of the quadedge concept to three–dimensional subdivisions [3]. Relaxing the requirement that the medial function τ be involution gives rise to graphs whose edges have more than two sides and endpoints. It is possible to interpret imbeddings of such graphs. The k-sided edges may be useful in representing the output of an edge–detector in their computer vision application [2]. It may be possible to extend our results to graphs on manifolds with boundary, following Mohar [10]. We expect that many other graph–theoretic algorithms, from connectivity testing to Voronoi diagram computation, would gain clarity and conciseness if our data structures were used.

A host of open questions arise from this work. It remains to see whether our representation can be used to prove extensions of the major theorems in the area of surface imbeddings and duality. In particular, it will be interesting to see whether the notions of genus and characteristic of a graph can be extended to our generalized graphs and whether equations linking these two integers can be obtained.

Acknowledgements

The authors would like to thank T.A. McKee for a helpful discussion and a copy of a manuscript [9]. A. White sent us a number of instructive comments on an early version of this manuscript. We also appreciate critical suggestions by a referee, which have had a significant impact on the final version of this manuscript. Finally, we are grateful to Chris Jackman and Clyde Rogers for their patience in coding C language implementations of three preliminary versions of our splice primitives.

REFERENCES

[1]   N. Biggs. Spanning trees of dual graphs. *Journal of Combinatorial Theory*, 11: 127 – 131, 1971.

[2]   D.P. Dobkin. Private communication. September 1987.

[3]   D.P. Dobkin and M.J. Laszlo. *Primitives for the Manipulation of Three-Dimensional Subdivisions*. Technical Report CS–TR–089–87, Department of Computer Science, Princeton University, April 1987.

[4]   J.R. Edmonds. A combinatorial representation for polyhedral surfaces. *Notices of the American Math Society*, 7:646, 1960.

[5]   J.R. Edmonds. On the surface duality of linear graphs. *Journal of Research of the National Bureau of Standards*, 69B(1 – 2):121 – 123, January–June 1965.

[6]   J.L. Gross and T.W. Tucker. *Topological Graph Theory. Wiley Interscience Series in Discrete Mathematics and Optimization*, Wiley–Interscience, 1987.

[7]   L. Guibas and J. Stolfi. Primitives for the manipulation of general subdivisions and the computation of Voronoi diagrams. *ACM Transactions on Graphics*, 4(2):74–123, April 1985.

[8]   J. Hopcroft and R. Tarjan. Efficient planarity testing. *Journal of the ACM*, 21(4):549 – 568, October 1974.

[9]   T.A. Mckee. Transfer principles for graph semiduality. April 1988. Submitted to *Fundamenta Mathematica*.

[10] B. Mohar. Simplicial schemes. *Journal of Combinatorial Theory, Series B*, 42:68–86, 1987.

[11] F.P. Preparata and M.I. Shamos. *Computational Geometry: An Introduction*. Springer–Verlag, New York, 1985.

[12] G. Ringel. *Map Color Theorem*. Springer–Verlag, New York, 1974.

[13] D.D. Sleator and R.E. Tarjan. Self–adjusting binary search trees. *Journal of the ACM*, 32(3):652 – 686, July 1985.

[14] S. Stahl. Generalized embedding schemes. *J. Graph Theory*, 2:41–52, 1978.

[15] C.D. Thomborson, L.L. Deneen, and G.M. Shute. Computing a rectilinear Steiner minimal tree in $n^{O(\sqrt{n})}$ time. In A. Albrecht, et. al., editor, *Parallel Algorithms and Architectures*, pages 176 – 183, Akademie–Verlag, Berlin, June 1987.

[16] W.T. Tutte. *Graph Theory*. Volume 21 of *Encyclopedia of Mathematics and its Applications*, Addison–Wesley, 1984.

[17] A.T. White. *Graphs, Groups and Surfaces*. North–Holland, Amsterdam, 1973.

[18] J.W.T. Youngs. Minimal imbeddings and the genus of a graph. *Journal of Mathematics and Mechanics,* 12(2):303 – 315, 1963.

ON VERTEX PRIME LABELINGS OF GRAPHS

Tatiana Deretsky

Sin-Min Lee

John Mitchem

San Jose State University

ABSTRACT

It has been conjectured that every tree has a prime labeling. Although this conjecture remains unsettled, it is shown that every tree, in fact that every connected graph, has a vertex prime labeling.

1.   Introduction

A graph has a *vertex prime labeling*, denoted vpl, if its edges can be labeled $1, 2, \cdots, |E|$ such that for each vertex of degree at least 2 the g.c.d of the labels on its incident edges is 1.

A dual concept which has been considered by several people is a prime labeling. A graph G is said to have a prime labeling if one can label the vertices by $1, 2, \cdots, |V|$ such that for each edge (u, v) the GCD of the labels on its adjacent nodes is 1. (see [1], [2] and [3]).

The complete graph K_n, with $n \geq 4$ has a vpl but no prime labeling. Every cycle C_n has a vpl as well as a prime labeling.

Ten years ago, R. C. Entringer conjectured that every tree has a prime labeling. The problem is still unsettled. However, we show that any tree has a vpl in Section 2. In fact, we show that all connected graphs have vpl's.

In Section 3, we consider the conjecture that a 2-regular graph has a vpl if and only if it does not contain more than two components which are odd cycles.

We give some partial results for this conjecture. In Section 4, we exhibit some other disconnected graphs with vpl's.

2.   Forests and Connected Graphs Have vpl's

Before proving our first theorems we define a vertex as *labeled* if at least 2 of its incident edges are labeled and the GCD of all labels on incident edges is 1.

Theorem 1   Let G be a graph with components $G_1, G_2, \cdots, G_r, r \geq 1$. Suppose each G_i has a connected subgraph H_i where each endvertex of H_i is also an endvertex of G_i. If $H = H_1 \cup H_2 \cup ... \cup H_r$ has a vpl, then G has a vpl also.

Proof   We vertex prime label H and then extend that label consecutively to $G_i, i = 1, 2 \cdots, r$. Assume that the vertex prime labeling of H has been extended to $G_1 \cup G_2 \cup ... \cup G_{i-1} \cup H_i \cup H_{i+1} \cup ... \cup H_r$ and t is the largest label used.

Use $t+1, t+2, \cdots, t_1$ to label all unlabeled edges of G_i which are incident with 2 vertices of H_i. By hypothesis these vertices are not endvertices of H_i. Thus each vertex remains labeled.

Next choose a vertex v_1 not in H_1 but adjacent to some vertex v_0 in H_i. Let $P = v_0 v_1 \cdots v_r$ be a longest path with initial edge $v_0 v_1$ and with $v_i, 1 \leq i \leq r$, unlabeled. Consecutively label the edges of P with $t_1 + 1, \cdot t_1 + r$. Now v_r is either an endvertex of G_i or is adjacent to a labeled vertex v_{r+1}. In the latter case, label $v_r v_{r+1}$ with t_{r+1}. In either case, we have a labeled subgraph of G_i that has the properties of H_i. By repeating the above two operations we obtain a vpl of $G_1 \cup G_2 \cup \cdots \cup G_i \cup H_{i+1} \cup \cdots \cup H_r$.

Corollary 2   If graph G is connected with a cycle, then G has a vpl.

Proof   Consecutively label the edges of any cycle C of G with $1, 2, \cdots, n$. Thus C has a vpl and from Theorem 1 it follows that G has a vpl also.

Corollary 3   Any forest F has a vpl.

Proof   In each component G_i find a longest path H_i. A vpl of $H_1 \cup H_2 \cup \cdots \cup H_r$ is given by labeling the edges of the paths consecutively $1, 2, \cdots, k_1, k_1 + 1, k_1 + 2, \cdots, k_2 \cdots$ It follows then from Theorem 1 that F is vpl.

Corollary 4   All connected graphs have a vpl.

3. Vertex Prime Labelings of 2-Regular Graphs

A graph G is *2-regular* if all vertices have degree 2. It is clear that G is 2-regular iff it is disjoint union of cycles. Thus for 2-regular graphs prime labeling and vertex prime labeling are equivalent. We state the following theorem which was stated (for prime labelings) and proved in [3].

Theorem 5   If G is 2-regular with at least two odd cycles, then G does not have a vpl.

Also the following conjecture (stated for prime labelings) was first given in [3].

Conjecture 1   A 2-regular graph G has a vpl if and only if it does not have two odd cycles.

We now prove various special cases of the conjecture.

Theorem 6   Let G be a disjoint union of C_{2k} and C_n, where n can be odd or even, then G has a vpl (and by symmetry G also has a prime labeling).

Proof   Label the edges of C_{2k} consecutively with $2, 3, 4, \cdots, 2k + 1$, and the edges of C_n consecutively with $1, 2k + 2, 2k + 3, \cdots, 2k + n$.

Theorem 7   The 2-regular graph $C_{2n} \cup C_{2k} \cup C_{r\prime}$ where r is odd or even has a vpl.

Proof   Case 1: $n \not\equiv 2 \pmod 3$

Label the cycles as follows:

$$C_{2n} : 3, 4, 5, \cdots 2n + 2,$$

$$C_{2k} : 2n + 3, 2n + 4, \cdots, 2n + 2k + 1, 2,$$

$$C_r : 2n + 2k + 2, 2n + 2k + 3, \cdots, 2n + 2k + r, 1.$$

Case 2: $n = 2 \pmod 3$

Label the cycles as follows:

$$C_{2n} : 4, 5, 6, \cdots, 2n + 3,$$

$$C_{2k} : 3, 2n + 4, 2n + 5, \cdots, 2n + 2k + 1, 2,$$

$$C_r : 2n + 2k + 2, 2n + 2k + 3, \cdots, 2n + 2k + r, 1.$$

Theorem 8 $G = C_{2r+1} \cup C_{2n} \cup C_{2k} \cup C_{2s}$ has a vpl.

Proof We consider four different cases.

Case 1 : $r \equiv 1 \pmod 3$

$$C_{2r+1} : 3, 5, 6, 7, \cdots 2r + 3, 2r + 4,$$

$$C_{2n} : 2r + 5, \cdots, 2r + 2n + 3, 2$$

$$C_{2k} : 2r + 2n + 4, \cdots, 2r + 2n + 2k + 2, 1$$

$$C_{2s} : 2r + 2n + 2k + 3, \cdots, 2r + 2n + 2k + 2s + 1, 4$$

Case 2: $r \equiv 1 \pmod 3$ and 5 does not divide $2r + 5$.

$$C_{2r+1} : 2r + 5, 5, 6, \cdots 2r + 4,$$

$$C_{2n} : 3, 2r + 6, \cdots, 2r + 2n + 3, 2,$$

$$C_{2k} : 2r + 2n + 4, \cdots, 2r + 2n + 2k + 2, 1$$

$$C_{2s} : 2r + 2n + 2k + 3, \cdots, 2r + 2n + 2k + 2s + 1, 4$$

Case 3: $r \equiv 1 \pmod 3$, 5 divides $(2r + 5)$ and 7 does not divide $(2r + 5)$

$$C_{2r+1} : 5, 6, 2r + 5, 7, \cdots, 2r + 4,$$

$$C_{2n} : 3, 2r + 6, \cdots, 2r + 2n + 3, 2,$$

$$C_{2k} : 2r + 2n + 4, \cdots, 2r + 2n + 2k + 2, 1$$

$$C_{2s} : 2r + 2n + 2k + 3, \cdots, 2r + 2n + 2k + 2s + 1, 4$$

Case 4: $r \equiv 1 \pmod 3$, 5 and 7 both divide $(2r+5)$

$$C_{2r+1} : 3, 5, 6, 11, \cdots, 2r + 3, 2r + 4, 7, 8, 9, 10,$$

$$C_{2n} : 2r + 5, \cdots, 2r + 2n + 3, 2$$

$$C_{2k} : 2r + 2n + 4, \cdots, 2r + 2n + 2k + 2, 1$$

$$C_{2s} : 2r + 2n + 2k + 3, \cdots, 2r + 2n + 2k + 2s + 1, 4$$

Theorem 9   The 2-regular graph $C_{2q} \cup C_{2r} \cup C_{2s} \cup C_{2t}$ has vpl.

Proof   If each of q, r, s, t is congruent to 1 modulo 3, we label

$$C_{2q} : 5, 6, 7, \cdots, 2q + 3, 4,$$

$$C_{2r} : 2q + 4, \cdots, 2q + 2r + 2, 1,$$

$$C_{2s} : 2q + 2r + 3, \cdots, 2q + 2r + 2s + 1, 2,$$

$$C_{2t} : 2q + 2r + 2s + 2, \cdots, 2q + 2r + 2s + 2t, 3.$$

If only $1, 2$, or 3 of q, r, s, t are congruent to 1 (mod 3), let $q \equiv 1 \pmod 3$ and $r \equiv 1 \pmod 3$, then interchange 1 and 3 in the labeling above. If none of q, r, s, t is congruent to 1 modulo 3 then let $r \equiv q \pmod 3$ and interchange 1 and 3 in the labeling above.

Theorem 10   The graph $5C_{2n}$ has a vpl.

Proof   We consider a number of cases.

Case 1.   5 does not divide $2n + 4$ and 3 does not divide $n + 1$.

In this case label the five components consecutively:

$$5, 6, \cdots, 2n + 4$$

$$2n + 5, 2n + 6, \cdots, 4n + 3, 2$$

$$4n + 4, 4n + 5, \cdots, 6n + 2, 3$$

$$6n + 3, 6n + 4, \cdots, 8n + 1, 4$$

$$8n + 2, 8n + 3, \cdots, 10n, 1.$$

Case 2.   5 does not divide $2n + 4$ and 3 divides $n + 1$.

In this case change the labels on the second, third and fourth components of case 1 to :

$$2n + 5, \cdots, 4n + 2, 3, 2$$

$$4n + 3, \cdots, 6n + 1, 6n + 4$$

$$6n + 3, 6n + 2, 6n + 5, \cdots, 8n + 1, 4.$$

Case 3.   5 divides $2n + 4$ and 7 divides $2n + 4$. In this case label the five components:

$$6, 5, 8, 9, \cdots, 2n + 4, 1$$

$$2n + 5, \cdots, 4n + 3, 4$$

$$4n + 4, \cdots, 6n + 2, 7$$

$$6n + 3, \cdots, 8n + 1, 2$$

$$8n + 2, \cdots, 10n3$$

This is a vpl as long as $n = 3k + 1$ for some integer k.

If $n = 3k + 2$ we obtain a vpl by relabeling the last two components

$$6n + 3, \cdots, 8n, 3, 2$$

$$8n + 3, 8n + 2, 8n + 1, 8n + 4, \cdots, 10n$$

Note that $(10n, 8n + 3) = (8n + 3, 2n - 3) = (2n - 3, 15)$.

However 3 does not divide $2n - 3$ and since 5 divides $2n + 4$, 5 does not divide $2n - 3$. Thus $(2n - 3, 15) = 1$.

If $n = 3k$ we change the labels on the third and fifth components to

$$4n + 4, \cdots, 6n + 2, 3$$

$$8n + 4, 8n + 3, 8n + 2, 8n + 5, 8n + 6, \cdots, 10n, 7$$

The facts $(8n + 4, 7) = 1 = (10n, 7)$ follows from the hypothesis that 7 divides $2n + 4$. Thus we obtain a vpl in all subcases of case 3.

Case 4.   5 divides $2n + 4$, 7 does not divide $2n + 4$, and $n \neq 3k + 2$.

In this case the labels are:

$$7, 6, 5, 8, \cdots, 2n + 4$$

$$2n + 5, \cdots, 4n + 3, 4$$

$$4n + 4, \cdots, 6n + 2, 3$$

$$6n + 3, \cdots, 8n + 1, 2$$

$$8n + 2, \cdots, 10n, 1$$

Case 5.   5 divides $2n + 4$, 7 does not divide $2n + 4$ and $n = 3k + 2$.

In this case we label the components:

$$7, 6, 5, 8, 9, \cdots, 2n + 4$$

$$2n + 5, \cdots, 4n + 3, 4$$

$$4n + 4, \cdots, 6n + 2, 1$$

$$un + 3, \cdots, 8n, 3, 2$$

$$8n + 3, 8n + 2, 8n + 1, 8n + 4, 8n + 5, \cdots, 10n.$$

In the last component $(8n + 3, 10n) = (8n + 3, 2n - 3) = (2n - 3, 15) = 1$.

Thus in all cases we have a vpl of $5C_{2n}$.

Theorem 11   Let $p = 2n - 1$ be prime, then $rC_p t_{+1}$ has vpl for each positive integer $r < p^2$ and each positive integer t.

Proof   Let the components of $rC_p t_{+1}$ be $G_1, G_2, \cdots, G_r$.

Let $i = kp + j$ where $0 \leq k < p, 0 \leq j < p$, and if $j > 0$, we label G_i with $(i-1)(p^t + 1) + 1, (i-1)(p^t + 1) + 2, \cdots, (i-1)(p^t + 1) + (p^t + 1) = i(p^t + 1)$. Now the only vertex which is not obviously labeled is the one with edges labeled $i(p^t + 1)$ and $(i - 1)(p^t + 1) + 1$. The difference of these numbers is p^t. Thus their only possible prime divisor is p. However $i(p^t + 1) = (kp + j)(p^t + 1) = kp^{t+1} + jp^t + kp + j, 0 < j < p$, does not contain p as a factor. Thus each vertex is labeled.

In the case where $i = kp, k \geq 1$ we not only label G_i but change the labels on various other components. Specifically relabel $G_{kp-k+1}, G_{kp-k+2}, G_{kp-k+3}, \cdots, G_{kp-1}$

and label G_{kp} as follows:

$$G_{kp-k+1} : (kp-k)(p^t+1)+1, (kp-k)(p^t+1)+2, \cdots,$$

$$(kp-k)(p^t+1)+p^t, kp(p^t+1)$$

Simplifying, we obtain

$$kp^{t+1}-kp^t-k+kp+1, kp^{t+1}-kp^t-k+kp+2, \cdots, kp^{t+1}-k-kp^t+kp+p^t, kp(p^t+1).$$

$$G_{kp-k+2} : kp^{t+1}-kp^t+kp+p^t-k+1, \cdots, kp^{t+1}-kp^t+kp+2p^t-k+1.$$

$$G_{kp-k+3} : kp^{t+1}-kp^t+kp+2p^t-k+2, \cdots, kp^{t+1}-kp^t+kp+3p^t-k+2.$$

$$G_{kp}-k+4 : kp^{t+1}-kp^t+kp+3p^t-k+3, \cdots, kp^{t+1}-kp^t+kp+4p^t-k+3.$$

$$\vdots$$

$$\vdots$$

$$G_{kp} : kp^{t+1}-kp^t+kp+(k-1)p^t-k+(k-1), \cdots, kp^{t+1}-kp^t+kp+kp^t-k+(k-1)$$

which simplifies to $kp^{t+1}+kp-p^t-1, \cdots, kp^{t+1}+kp-1.$

In order to complete the proof that $rC_p t_{+1}$ has a vpl we only need to verify that the labels above for G_i, $kp-k+1 \leq i \leq kp$ are vpl. In order to do this we note that most consecutive labels differ by 1. Thus we need only consider the final label on each G_i. In $G_{kp-k+1'}$ any prime which divides the last label $kp(p^t+1)$ must be p or a divisor of k or a divisor of p^t+1. However, such a factor does not divide the first label $(kp-k)(p^t+1)+1 = kp^{t+1}-kp^t+kp+(1-k)$ nor the next to the last label $(kp-k)(p^t+1)+p^t = kp^{t+1}-kp^t+kp+(p^t-k)$ thus $G_{kp}-k+1$ has all vertices labeled. For $G_i, kp-k+2 \leq i \leq kp$, the difference between the first and last labels is p^t. Thus the only possible prime divisors of those labels is p, but neither is divisible by p. It follows that each G_i has all vertices labeled and $rG_p t_{+1}$ has a vpl.

Theorem 12   Let $p > 3$ be a prime. Then $kC_{2p} t$ has a vpl if $k \leq 14$.

Proof   Every cycle $C_{2p} t$ will be labeled by numbers $2np^t+1, 2np^t+2, \cdots, 2(n+1)p^t$, where $0 \leq n \leq k-1$ as follows:

If $n = 0, 1, 3, 7$ then $(2np^t + 1, 2(n+1)p^t) = 1$, and we label the edges consecutively $2np^t + 1, 2np^t + 2, \cdots, 2(n+1)p^t$.

In all other cases, $2(n+1) = 2^a q^b$, where q is an odd prime. We see that $q = 3$ $n = 2, 5, 8, 11; q = 5$ when $n = 4, 9; q = 7$ when $n = 6, 13; q = 11$ when $n = 10$; and finally, $q = 13$ when $n = 12$.

In each of these cases, start by consecutive labeling again.

If $(2np^t + 1, 2(n+1)p^t) = 1$, then the graph has a vpl.

If not, $(2np^t + 1, 2(n+1)p^t) = (2np^t + 1, 2^a q^b) = (2np^t + 1, q^b) \neq 1$, and hence $2np^t + 1 \equiv 0 \bmod q$. Now modify the consecutive labeling as follows. Take the first two numbers $2np^t + 1$ and $2np^t + 2$, and place them in between of $(2n+1)p^t + 1$ and $(2n+1)p^t + 2$, so that the edges are labeled

$$2np^t + 3, 2np^t + 4, \cdots, (2n+1)p^t + 1, 2np^t + 1, 2np^t + 2,$$

$$(2n+1)p^t + 2, \cdots, 2(n+1)p^t$$

Now, $(2np^t + 3, 2(n+1)p^t) = (2np^t + 3, q^b) = 1$, since $2np^t + 1 \equiv 0 \pmod{q}$. Also $((2n+1)p^t + 1, 2np^t + 1) = (2np^t + p^t + 1, 2np^t + 1) = (p^t, 2np^t + 1) = 1$, and $((2n+1)p^t + 2, 2np^t + 2) = (2np^t + p^t + 2, 2np^t + 2) = (p^t, 2np^t + 2) = 1$. Thus the graph has a vlp.

Theorem 13   Let G be disconnected with exactly 2 components, G_1 and G_2, one of which is not an odd cycle, then G has a vpl.

Proof   We consider three cases.

Case 1.   Both components are trees.

In this case G is a forest and the result follows from Corollary 3.

Case 2.   One component, say G_1, is cyclic and the other is a tree.

By Corollary 2, G_1 has a vpl, and thus we label its edges appropriately $1, 2, \cdots, n$. Now label the edges of a longest path of G_2 with $n+1, n+2, \cdots, t_1$. It follows from Theorem 1, that G has a vpl.

Case 3.   Both components are cyclic.

Subcase 1. At least one component contains an even cycle, C. Let C' be a cycle of the other component. Now by Theorem 6, CuC' has a vpl and thus, by Theorem 1, G has a vpl.

Subcase 2. Neither component contains an even cycle.

However, by hypothesis one of the components, say G_1 contains an edge e, not on the odd cycle C, but incident with a vertex v of C. Since G_2 is connected, we can vertex- prime label it $1, 2, ..., t$. Now cyclically label C from v with $t+1, t+2, \cdots, s$ and label e with $s+1$. Now if the vertex $u \neq v$ incident with e is an endvertex, the results follows from Theorem 1. If u is incident to other edges which go to C, label these edges with $s+2, \cdots, s+k$ and apply Theorem 1. Otherwise, there is a new vertex w and edge uw which we label $s+2$. By continuing in this way we eventually obtain a vpl of G.

We close with the following

Conjecture 2   Let G have components $G_1, G_2, \cdots, G_t$. Also let $q = |E(G)|, p_i =$ number of vertices of degree at least two in $G_i, m_i =$ size of a maximum matching in the subgraph H_i induced by the p_i vertices of degree at least two and $n_i = p_i - 2m_i$, then G has a vpl if and only if

$$\lceil q/2 \rceil \geq \sum_{i=1}^{t}(n_i + m_i).$$

We note that if G has a vpl, then each of the p_i vertices of degree at least two is incident with an edge labeled with an odd integer. These edges need at least $n_i + m_i$ odd integers. However there are exactly $\lceil q/2 \rceil$ odd integers available and thus the inequality follows. If we assume the inequality, it is apparently difficult to prove that G has a vpl. We observe that Conjecture 1 is included in Conjecture 2.

REFERENCES

[1] B. D. Acharya, "On vertex valuations of a graph which induce a constant GCD function on its line set", Research Report, Mehta Research Institute of Mathematics and Mathematical Physics, Allahabad, (1980).

[2] Sin-Min Lee, Indra Wui and Jeran Yeh, "On the amalgamation of prime graphs, to appear in *Bull. Malaysian Math. Society* 11(12), 1989.

[3] R. Tout, A. N. Daboucy and K. Howalla, *Nat. Acad. Sci. Letters*, 5(11) (1982), 365- 368.

Relationships Between Integer and Fractional Parameters of Graphs

Gayla S. Domke[*]

Stephen T. Hedetniemi[*]

Renu C. Laskar[*]
Clemson University

Gerd Fricke
Wright State University

ABSTRACT

For $G = (V, E)$ let $g : V \to \{0, 1, 2, ..., k\}$ and let $g(S) = \sum\limits_{v \in S} g(v)$ for any set $S \subseteq V$, where $|g| = g(V)$. Such a function g is a k–dominating function if for every $v \in V$, $g(N[v]) \geq k$. The k–domination number of G is given by $\gamma_k(G) = min \{ |g| : g$ is a k–dominating function of $G\}$. A function g is a k–packing function if for every $v \in V$, $g(N[v]) \leq k$. The k–packing number of G is given by $P_k(G) = max \{ |g| : g$ is a k–packing function of $G\}$. A function g is a k–irredundant function if for every $v \in V$ with $g(v) > 0$, there exists a vertex $u \in N[v]$ where $g(N[u]) = k$. This paper gives results for these parameters as well as for the other related min/max parameters. The results include relationships between the k–parameters in addition to relationships between k–parameters and fractional parameters.

1. Fractional Parameters

A dominating set of a graph $G = (V, E)$ is a set S of vertices where every vertex of G either belongs to S or is adjacent to a vertex in S. The domination number, $\gamma(G)$, is the minimum cardinality of a dominating set of vertices. The closed neighborhood of a vertex x is defined as follows: $N[x] = \{x\} \cup \{v \in V : xv \in E\}$. An irredundant set of

* Research was partially supported by the Office of Naval Research
contract #N00014-86-K-0693

vertices is a set S of vertices where every vertex v in S has a "private neighbor": a vertex in $N[v]$ which is not in any other $N[u]$ for $u \in S - \{v\}$. That is for every $x \in S$, $N[x] - \bigcup_{y \in S-x} N[y] \neq \varnothing$. The irredundance number, $ir(G)$, is the minimum cardinality of a maximal irredundant set of vertices of G.

An alternate way to define the above sets is to assign a value of 1 to every vertex in the set S and assign a value of 0 to the vertices not in S. For example, a set S would be a dominating set of vertices if for every $v \in V$, the sum of the values assigned to the vertices in the closed neighborhood of v is greater than or equal to one. This definition motivates the definitions of fractional parameters. The values assigned to the vertices would not be restricted to be 0 or 1 but any real value in the closed interval between 0 and 1.

This idea of fractional parameters has been studied before in different contexts. Berge [2] has studied fractional transversals of hypergraphs, Pulleyblank [19] has studied fractional matchings, and Aharoni [1] has studied fractional matchings and covers in infinite hypergraphs. Chung, Furedi, Garey, and Graham [6] have studied fractional covers of hypergraphs. The fractional chromatic number has been looked at by Larsen, Propp, and Ullman as well as by others (see [17]). It appears that Farber [12] introduced indirectly the concept of fractional domination, whereas Hedetniemi, Hedetniemi, and Wimer [16] were the first ones to study fractional domination. Grinstead and Slater [14] have also studied fractional domination as well as fractional packings, fractional redundance, and fractional efficient domination. Domke, Hedetniemi, and Laskar [10] also studied fractional domination and fractional packings as well as fractional irredundance.

If g is a function mapping the vertex set V of G into some set of real numbers, then let $g(S) = \sum_{v \in S} g(v)$, where $S \subseteq V$, and let $|g| = g(V) = \sum_{v \in V} g(v)$. A real-valued function $g : V \to [0, 1]$ is a dominating function of a graph G if for every $v \in V$, $g(N[v]) \geq 1$. Thus, if S is a dominating set of vertices of G and g is a function where $g(v) = 1$ if $v \in S$, $g(v) = 0$ if $v \notin S$, then this function g is a dominating function of G. A dominating function g is minimal if for every $v \in V$ with $g(v) > 0$, there exists a vertex $u \in N[v]$ such that $g(N[u]) = 1$. Then the fractional domination number of G, denoted $\gamma_f(G)$, and the upper fractional domination number of G, denoted $\Gamma_f(G)$, are defined as follows: $\gamma_f(G) = \min \{ |g| : g$ is a minimal dominating function$\}$, and $\Gamma_f(G) = \max \{ |g| : g$ is a minimal dominating function$\}$.

A real-valued function $g : V \to [0, 1]$ is a packing function of a graph G if for every $v \in V$, $g(N[v]) \leq 1$. A packing function is maximal if for every $v \in V$ with $g(v) < 1$, there exists a vertex $u \in N[v]$ where $g(N[u]) = 1$. Then the lower fractional packing number of G, denoted $p_f(G)$, and the (upper) fractional packing number of G, denoted $P_f(G)$, are defined as follows: $p_f(G) = \min\{\,|g| : g$ is a maximal packing function$\}$ and $P_f(G) = \max\{\,|g| : g$ is a maximal packing function$\}$.

An irredundant set of vertices is a set S of vertices where every vertex v in S has a "private neighbor": a vertex which is in the closed neighborhood of v but not in the closed neighborhood of any other vertex in S. That is, for every $x \in S$, $N[x] - \bigcup_{y \in S-x} N[y] \neq \varnothing$ The irredundance number of G, denoted $ir(G)$, and the upper irredundance number of G, denoted $IR(G)$, are the minimum and maximum cardinalities, respectively, of all maximal irredundant sets of vertices of G. We define fractional irredundance as follows: a real-valued function $g : V \to [0, 1]$ is an irredundant function if for every $v \in V$ with $g(v) > 0$, there exists a vertex $u \in N[v]$ such that $g(N[u]) = 1$. An irredundant function g is maximal if there does not exist an irredundant function h obtained from g by increasing at least one function value of g. The fractional irredundance number of G, denoted $ir_f(G)$, and the upper fractional irredundance number of G, denoted $IR_f(G)$, are respectively the minimum and maximum values of $|g|$ where g is a maximal irredundant function of G.

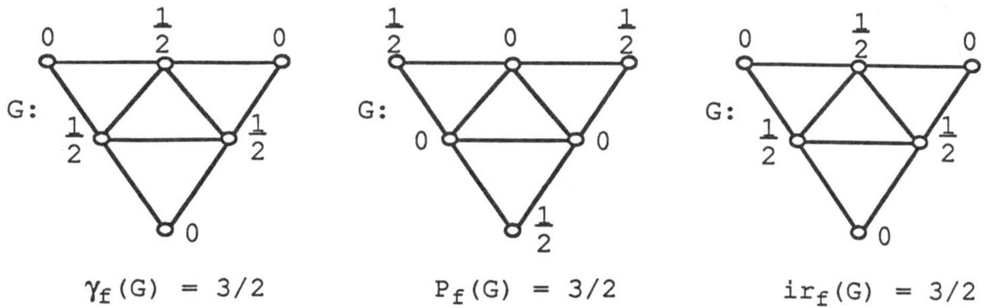

Figure 1. Examples of fractional domination, packings and irredundance.

The following results involving fractional parameters will be used in the following sections.

Theorem (Domke [9], see also Domke, Hedetniemi, and Laskar [10]) *For every graph G, $ir_f(G) \le \gamma_f(G) \le \Gamma_f(G) \le IR_f(G)$.*

Theorem (Chandrasekharan, Hedetniemi, Laskar, and Majumdar [3], see also Domke, Hedetniemi, and Laskar [10]) *For every graph G, $\gamma_f(G) = P_f(G)$.*

Theorem (Domke, Hedetniemi, and Laskar [10]) *For every graph G, $p_f(G) \le P_2(G) \le P_f(G)$, where $P_2(G)$ is the 2–packing number defined by Meir and Moon [18].*

Theorem (Domke, Hedetniemi, and Laskar [10]) *For every graph G with p vertices, $\gamma_f(G) = 1$ if and only if $\Delta(G) = p - 1$.*

2. k–Domination, k–Packings, and k-Irredundance

The idea of k–domination was presented by Hare [15] while trying to find values for the fractional domination number. Cheston, Fricke, Hedetniemi, and Pokrass [4] have studied upper k–domination. Also, other k–parameters have been studied. For example, Berge [2] has studied k–transversals of hypergraphs. In this paper we will introduce k–packings and k–irredundance and relate these k–parameters as well as the k–domination parameters to fractional parameters in addition to other packing and covering parameters.

Once again, if g is a function from the vertex set V of G into some set of real numbers, then for $S \subseteq V$, let $g(S) = \sum_{v \in S} g(v)$, and let $|g| = g(V) = \sum_{v \in V} g(v)$. For any fixed positive integer k, a function $g : V \to \{0, 1, 2, ..., k\}$ is a k–dominating function of G if for every $v \in V$, $g(N[v]) \geq k$. Thus, if S is a dominating set of vertices and $g(v) = 1$ if $v \in S$, $g(v) = 0$ if $v \notin S$, then this function is a 1-dominating function of G. A k–dominating function is minimal if for every $v \in V$ with $g(v) > 0$, there exists a vertex $u \in N[v]$ such that $g(N[u]) = k$. For each positive integer k, the k–domination number of G, denoted $\gamma_k(G)$, and the upper k–domination number of G, denoted $\Gamma_k(G)$, are the minimum and maximum values, respectively, of $|g|$ where g is a minimal k–domination function of G. Examples of k–domination are shown in Figure 2. The k–domination numbers can be verified with results following in this paper.

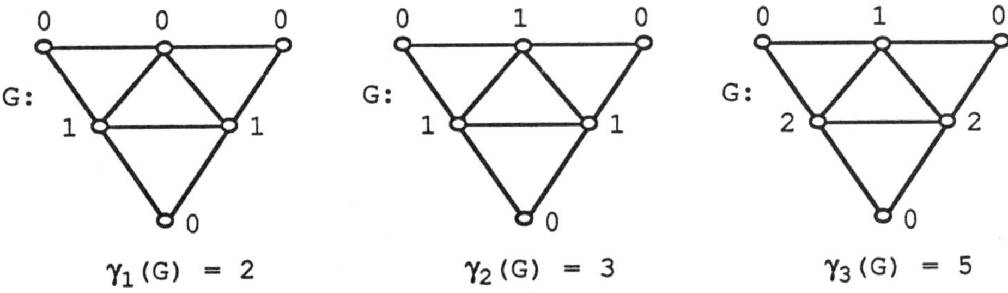

Figure 2. Examples of k–domination for k = 1, 2, 3.

It should be noted that this definition of k–domination differs from the definition of n–domination given by Fink and Jacobson in [13]. They defined a set $S \subseteq V$ to be an n–dominating set of vertices if each vertex $v \in V - S$ is adjacent to at least n vertices in S. Then the n–domination number, denoted $\gamma_n(G)$, is the minimum cardinality of an n–dominating set of vertices of G.

For a fixed positive integer k, a function $g : V \to \{0, 1, 2, ..., k\}$ is a k–packing function of G if for every $v \in V$, $g(N[v]) \leq k$. A k–packing function is maximal if for every $v \in V$ with $g(v) < k$, there exists a vertex $u \in N[v]$ such that $g(N[u]) = k$. Then for each positive integer k, the lower k–packing number of G, denoted $p_k(G)$, and the (upper) k–packing number of G, denoted $P_k(G)$, are defined respectively as the minimum and maximum values of $|g|$ among all maximal k–packing functions g. Examples of k–packing functions are given in Figure 3 for $k = 1, 2, 3$. The (upper) k–packing numbers can be verified with results found in this paper.

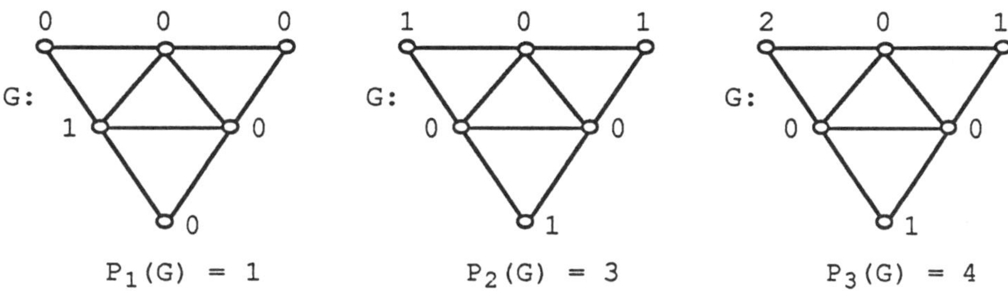

$$P_1(G) = 1 \qquad P_2(G) = 3 \qquad P_3(G) = 4$$

Figure 3. Examples of k–packings for $k = 1, 2, 3$.

Once again, for a fixed positive integer k, a function $g : V \to \{0, 1, 2, ..., k\}$ is a k–irredundant function of G if for every vertex $v \in V$, with $g(v) > 0$, there exists a vertex $u \in N[v]$ where $g(N[u]) = k$. A k–irredundant function is maximal if there does not exist a k–irredundant function h obtained from g by increasing at least one function value of g. For a fixed positive integer k, the k–irredundance number of G, denoted $ir_k(G)$, and the upper k–irredundance number of G, denoted $IR_k(G)$, are respectively the minimum and maximum values of $|g|$ where g is a maximal k–irredundant function (see Figure 4).

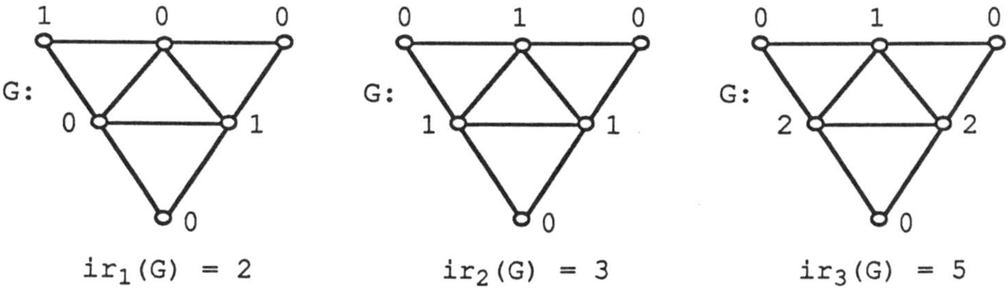

$$ir_1(G) = 2 \qquad ir_2(G) = 3 \qquad ir_3(G) = 5$$

Figure 4. Examples of k–irredundance for $k = 1, 2, 3$.

It should be noted here that when finding the minimum value of $|g|$ where g is a k–dominating function, it will necessarily be true that g is a minimal k–dominating function. Similarly, if $|g|$ is a maximum where g is a k–packing (k-irredundant) function, then g is a maximal k–packing (k–irredundant) function. Therefore, $\gamma_k(G) = \min \{|g| : g$ is a k–dominating function$\}$, $P_k(G) = \max \{|g| : g$ is a k–packing function$\}$, and $IR_k(G) = \max \{|g| : g$ is a k–irredundant function$\}$.

At this point little work has been done on these parameters other than that of Hare [15] who has computed the k–domination number a variety of grid graphs. For example, she has proven that $\gamma_k(P_2 \times P_{2n}) = 2n(n + 1)$ when $k = 2n + 1$, and this result gives $\gamma_f(P_2 \times P_{2n}) = \gamma_k(P_2 \times P_{2n})/k = 2n(n + 1)/(2n + 1)$. Figure 5 shows some examples of k–dominating functions on grid graphs which were found computationally [15].

It is important to notice that the definitions of k–domination and k–irredundance are closely related to those of domination and irredundance. In particular, if S is a dominating (irredundant) set of vertices of a graph, and if a function value of k is placed on those vertices in S while a fuction value of zero is placed on the others, this function

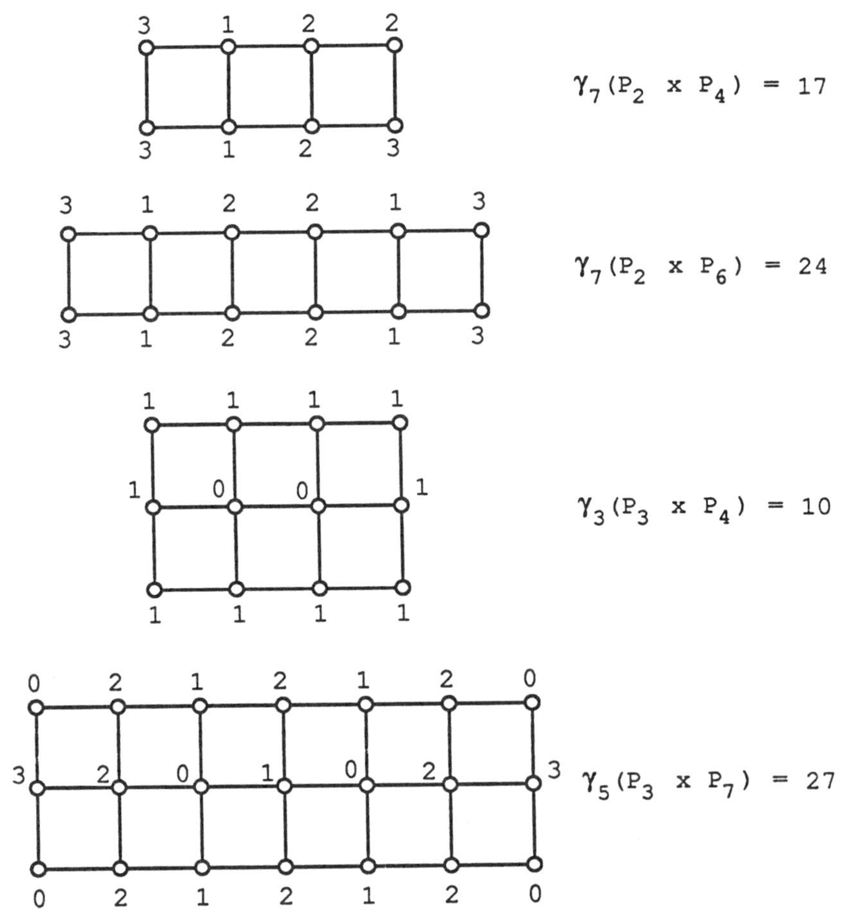

Figure 5. Examples of k–domination for grid graphs.

will be a k–dominating (k–irredundant) function. The next results relate the domination and irredundance numbers to their corresponding k–parameters.

Theorem 1   *For every graph G and positive integer k, $\gamma_k(G) \leq k\cdot\gamma(G) \leq k\cdot\Gamma(G) \leq \Gamma_k(G)$.*

Proof   Let S be any minimal dominating set of vertices of G, and let g : V $\to$ {0, 1, 2, ..., k} where $g(v) = \begin{cases} k & \text{if } v \in S \\ 0 & \text{if } v \notin S; \end{cases}$ that is g is a {0, k}–function. It is easily seen that g is a k–dominating function of G.

Now suppose there exists a vertex v with g(v) > 0 where for every u $\in$ N[v], g(N[u]) > k. Since every function value is either zero or k, g(N[u]) $\geq$ 2k for u $\in$ N[v], and g(v) = k. Thus, v $\in$ S. Hence, every vertex adjacent to v is adjacent to at least one other vertex other than v in S. Also, v is adjacent to at least one other vertex in S. So, S – {v} is a dominating set of vertices of G. But this contradicts the minimality of S. Therefore, every vertex v with g(v) > 0, has a vertex u $\in$ N[v] where g(N[u]) = k, and g is a minimal k–dominating function. Thus, $\gamma_k(G) \leq$ k$\cdot$|S| = k$\cdot\gamma$(G).

A similar argument, in which S is taken to be a minimal dominating set of maximum cardinality, shows that k$\cdot\Gamma$(G) $\leq \Gamma_k$(G). ❏

Theorem 2   *For every graph G and positive integer k, $ir_k(G) \leq k\cdot ir(G) \leq k \cdot IR(G) \leq IR_k(G)$.*

Proof   Let S be any maximal irredundant set of a graph G, and let g : V $\to$ {0, 1, 2, ..., k} where $g(v) = \begin{cases} k & \text{if } v \in S \\ 0 & \text{if } v \notin S; \end{cases}$ that is g is a {0, k}–function. If g(v) > 0, then g(v) = k, and v $\in$ S. Since S is an irredundant set, there exists a vertex u in N[v] such that u $\notin$ N[S – v]. If u = v, then N[v] $\cap$ S = {v}, and g(N[u]) = g(v) + 0 = k. If u $\neq$ v, then g(N[u]) = g(v) + 0 = k. Hence, g is a k– irredundant function.

Now suppose there exists a k–irredundant function h obtained from g by increasing at least one function value of g. Thus, there exists a vertex v $\notin$ S where h(v) = m > 0, and of course g(v) = 0. Since h is a k–irredundant function obtained from g, there exists a vertex u $\in$ N[v] and u $\notin$ S with h(N[u]) = k. Since v $\in$ N[u] and h(v) > 0, there cannot exist a w $\in$ S with w $\in$ N[u] (else h(N[u]) $\geq$ h(v) + h(w) $\geq$ m + k > k). i.e. S $\cap$ N[u] = $\varnothing$, and u is a private neighbor of v. We claim that S $\cup$ {v} is an irredundant set, which contradicts the maximality of S. Certainly, u is a private neighbor of v in S $\cup$ {v}, i.e. N[v] – N[S] $\neq \varnothing$. Suppose that there is a w $\in$ S such that N[w] – N[S $\cup$ {v} – w] = $\varnothing$. This would imply that for every u $\in$

N[w], h(N[u]) > k, contradicting our assumption that h is a k–irredundant function. Therefore, g is a maximal k–irredundant function.

If S is a maximal irredundant set where $|S|$ = ir(G), then $|g|$ = k·$|S|$ = k · ir(G). Since $ir_k(G)$ = min {h : h is a maximal k–irredundant function} and g is a maximal k–irredundant function, then $ir_k(G) \leq |g|$ = k·ir(G).

Similarly, since $IR_k(G)$ = max {h : h is a maximal k–irredundant function}, if $|g|$ = k·IR(G), then $IR_k(G) \geq |g|$ = k·IR(G). Thus, the result holds. ❑

There is always an important relationship between domination and irredundance. As a matter of fact, every minimal dominating set of vertices is a maximal irredundant set. Recall that Cockayne and Hedetniemi [7] showed that ir(G) $\leq \gamma$(G), for every graph G. It was shown by Domke [9] (see also [10]) that every minimal dominating function is a maximal irredundant function. As the following shows, an analagous statement is true for minimal k–dominating functions and maximal k–irredundant functions.

Lemma 3 *If k is a positive integer, every minimal k–dominating function is a maximal k–irredundant function.*

Proof Let g : V → {0, 1, 2, ..., k} be a minimal k–dominating function of G. Since g is minimal, for every vertex v ∈ V with g(v) > 0, there is a vertex u ∈ N[v] such that g(N[u]) = k. Therefore g is a k–irredundant function.

Now let h : V → {0, 1, 2, ..., k} be a function obtained from g by increasing at least one function value of g. That is, for some v ∈ V where g(v) < k, define h(v) = g(v) + m where m is a positive integer and h(v) ≤ k. Now, for any vertex u ∈ N[v], h(N[v]) ≥ g(N[v]) + m ≥ k + m > k. Thus, for every u ∈ N[v], h(N[u]) > k, and hence h is not a k–irredundant function. Therefore, g is a maximal k–irredundant function. ❑

Theorem 4 *For every graph G and positive integer k, $ir_k(G) \leq \gamma_k(G) \leq \Gamma_k(G) \leq IR_k(G)$.*

Proof Let g be a minimal k–dominating function of G where $|g|$ = γ_k(G). Since $ir_k(G)$ = min {h : h is a maximal k–irredundant function}, and g is a maximal k–irredundant function by Lemma 3, then $ir_k(G) \leq |g|$ = γ_k(G).

Similarly, let g be a minimal k–dominating function where $|g|$ = Γ_k(G). Since $IR_k(G)$ = max {h : h is a maximal k–irredundant function}, then $IR_k(G) \geq |g|$ = Γ_k(G), and the result holds. ❑

It is interesting to note that there is no such relationship between maximal k–packing functions and maximal k–irredundant functions. Figure 3 shows examples of maximal k–packing functions which are k–irredundant functions, but which are not maximal. Also, Figure 4 gives examples of maximal k–irredundant functions which are not k–packing functions.

These next results show that the sequences of k–domination numbers (similarly k–packing numbers) are strictly increasing. This is helpful to know since once $\gamma_k(G)$ (similarly $P_k(G)$) is known, it gives a bound on $\gamma_{k-1}(G)$ and $\gamma_{k+1}(G)$ ($P_{k-1}(G)$ and $P_{k+1}(G)$).

Theorem 5 *For every graph G and positive integer k, $\gamma_k(G) < \gamma_{k+1}(G)$.*

Proof Let $g : V \to \{0, 1, 2, ..., k + 1\}$ be a $(k + 1)$–dominating function of G where $|g| = \gamma_{k+1}(G)$.

Case 1. Suppose there exists at least one vertex $w \in V$ with $g(w) = k + 1$. Define a function $g' : V \to \{0, 1, 2, ..., k\}$ as follows:

$$g'(v) = \begin{cases} k & \text{if } g(v) = k + 1 \\ g(v) & \text{if } g(v) \le k. \end{cases}$$

Now let v be any vertex in V. If there exists a vertex $u \in N[v]$ where $g(u) = k + 1$, then $g'(u) = k$, and $g'(N[v]) \ge g'(u) = k$. If there is no vertex $u \in N[v]$ where $g(u) = k + 1$, then $g'(u) = g(u)$ for every $u \in N[v]$, and $g'(N[v]) = g(N[v]) \ge k + 1 > k$. Thus, $g'(N[v]) \ge k$, for every $v \in V$, and g' is a k–dominating function of G. Since $\gamma_k(G) = \min \{ |h| : h$ is a k–dominating function$\}$, $\gamma_k(G) \le |g'| < |g| = \gamma_{k+1}(G)$.

Case 2. If $g(v) \le k$ for every vertex $v \in V$, let $w \in V$ be any vertex where $g(w) > 0$. Define a new function $g'' : V \to \{0, 1, 2, ..., k\}$ by $g''(w) = g(w) - 1$, and $g''(v) = g(v)$ for $v \ne w$. Since g is a $(k + 1)$–dominating function, $g(N[v]) \ge k + 1 > k$. If $w \notin N[v]$, then $g''(N[v]) = g(N[v]) > k$. If $w \in N[v]$, then $g''(N[v]) = g(N[v]) - 1 \ge k + 1 - 1 = k$. In either case, $g''(N[v]) \ge k$, and g'' is a k–dominating function of G. Now, either g'' is a minimal k–dominating function or it is not. If not, then by defintion, there exists a vertex $v \in V$ with $g(v) > 0$, and for every $u \in N[v]$, $g(N[u]) > k$. We can then define a new k–dominating function $g''' : V \to \{0, 1, 2, ..., k\}$, where $g'''(v) = g''(v) - 1$ and $g'''(w) = g''(w)$ for $w \ne v$. By repeating this process at most a finite number of times we can obtain a minimal k–dominating function g^* with $\gamma_k(G) \le |g^*| < |g| = \gamma_{k+1}(G)$. $\square$

Theorem 6 *For every graph G and positive integer k, $P_k(G) < P_{k+1}(G)$.*

Proof Let $g : V \to \{0, 1, 2, ..., k\}$ be a k–packing function of G where $|g| = P_k(G)$. Thus, for every $v \in V$, $g(N[v]) \le k < k + 1$. Hence, g is a $(k + 1)$–packing function of G. Since $P_{k+1}(G) = \max \{ |h| : h$ is a $(k + 1)$–packing function$\}$, $P_{k+1}(G) \ge |g| = P_k(G)$.

Let w be any vertex of G. Define $g' : V \to \{0, 1, 2, ..., k + 1\}$ by $g'(w) = g(w) + 1$ and $g'(v) = g(v)$ for $v \neq w$. If $w \notin N[v]$, then $g'(N[v]) = g(N[v]) \leq k < k + 1$. If $w \in N[v]$, then $g'(N[v]) = g(N[v]) + 1 \leq k + 1$. Hence, g' is a $(k + 1)$–packing function of G and $P_{k+1}(G) \geq |g'| = |g| + 1 > |g| = P_k(G)$. Therefore, $P_k(G) < P_{k+1}(G)$. ◻

It would be interesting to know if similar results are true for the other k–parameters. For example, is it true that $\Gamma_k(G) < \Gamma_{k+1}(G)$? At this point that result is unknown.

3. Relating k–Parameters and Fractional Parameters

As we stated in the last section, there is a close relationship between the domination (irredundance) number and the k–domination (k–irredundance) number of a graph. The same is true between fractional parameters and their associated k–parameters. These first results show that any k–dominating (k–packing) function provides a dominating (packing) function when each of its function values is divided by k. If $g : V \to \{0, 1, 2, ..., k\}$ is a minimal k–dominating function, then the function $h : V \to [0, 1]$ defined by $h(v) = g(v)/k$ is a minimal dominating function.

Theorem 7   *For any graph G and positive integer k, $\gamma_f(G) \leq \gamma_k(G)/k \leq \Gamma_k(G)/k \leq \Gamma_f(G)$.*

If $g : V \to \{0, 1, 2, ..., k\}$ is a maximal k–packing function, then the function $h : V \to [0, 1]$ defined by $h(v) = g(v)/k$ is a maximal packing function.

Theorem 8   *For any graph G and positive integer k, $p_f(G) \leq p_k(G)/k \leq P_k(G)/k \leq P_f(G)$.*

Now we can use the above two theorems to show that $P_k(G) \leq \gamma_k(G)$ for every graph G and positive integer k.

Corollary 9   *For any graph G and positive integer k, $P_k(G) \leq k \cdot P_f(G) = k \cdot \gamma_f(G) \leq \gamma_k(G)$.*

Proof   By Theorems 7 and 8 and the fact that $P_f(G) = \gamma_f(G)$, $\dfrac{P_k(G)}{k} \leq P_f(G) = \gamma_f(G) \leq \dfrac{\gamma_k(G)}{k}$. Thus, the result holds. ◻

Theorem 10   *For any graph G and positive integer k, $IR_f(G) \geq IR_k(G)/k$.*

Proof   Let k be any positive integer, and let $g : V \to \{0, 1, 2, ..., k\}$ be a k–irredundant function where $|g| = IR_k(G)$. Now define a function h where $h(v) =

$g(v)/k$ for every $v \in V$. Since $0 \leq g(v) \leq k$, then $0 \leq h(v) \leq 1$ and $h : V \to [0, 1]$. Since g is a k–irredundant function, to each $v \in V$ with $g(v) > 0$, there corresponds a vertex $u \in N[v]$ where $g(N[u]) = k$. Thus, if $v \in V$, and $h(v) > 0$, then $g(v) > 0$, and there exists a vertex $u \in N[v]$ where $h(N[u]) = \sum_{w \in N[u]} h(w) = \sum_{w \in N[u]} \frac{g(w)}{k} = \frac{1}{k} g(N[u]) = \frac{1}{k} k = 1$. Therefore, h is an irredundant function of G. Since $IR_f(G) = \max \{ |g'| : g'$ is an irredundant function of $G\}$, then $IR_f(G) \geq |h| = \sum_{v \in V} h(v) = \sum_{v \in V} \frac{g(v)}{k} = \frac{1}{k} |g| = \frac{1}{k} IR_k(G)$, and the result holds. $\square$

Since $\gamma_f(G) \leq \gamma_k(G)/k$ for every positive integer k, we wondered if equality holds for any value of k (similarly for $P_f(G)$). By the next results, we can see that this is true. Recall that $\gamma_f(G) = P_f(G)$ can be found using linear programming (see [3] or [10]). This fact is crucial in proving the next two results.

Theorem 11 *For any graph G, $\gamma_f(G) = min \{\gamma_k(G)/k : k$ is a positive integer}.*

Proof By Theorem 7, $\gamma_f(G) \leq \gamma_k(G)/k$ for every positive integer k. Therefore, $\gamma_k(G) \leq min \{\gamma_k(G)/k : k$ is a positive integer}.

Since $\gamma_f(G)$ can be found using a linear program with rational coefficients, there exists an optimal solution with rational values. Let $g : V \to [0, 1]$ be a dominating function of G corresponding to those rational values found using linear programming. Note that $|g| = \gamma_f(G)$. Now let $k =$ least common denominator $\{g(v) : v \in V\}$. Define $h(v) = k \cdot g(v)$ for every $v \in V$. Since $0 \leq g(v) \leq 1$, then $0 \leq h(v) = k \cdot g(v) \leq k$. Also, since k is a common denominator of all of the $g(v)$, $h(v)$ is an integer. Thus, $h : V \to \{0, 1, 2, ..., k\}$. Since g is a dominating function, $g(N[v]) \geq 1$ for every $v \in V$. Thus, $h(N[v]) = \sum_{w \in N[v]} h(w) = \sum_{w \in N[v]} k \cdot g(w) = k \cdot g(N[v]) \geq k \cdot 1 = k$. Therefore h is a k–dominating function of G, and $\gamma_k(G)/k \leq |h|/k = k \cdot |g|/k = |g| = \gamma_f(G)$. Hence, $min \{\gamma_m(G)/m : m$ is a positive integer$\} \leq \gamma_k(G)/k \leq \gamma_f(G)$, and the equality holds. $\square$

Theorem 12 *For any graph G, $P_f(G) = max \{ P_k(G)/k : k$ is a positive integer}.*

Proof By Theorem 8, $P_f(G) \geq P_k(G)/k$ for every positive integer k. Therefore, $P_f(G) \geq max \{P_k(G)/k : k$ is a positive integer}.

Since $P_f(G)$ can be found using a linear program with rational coefficients, there exists an optimal solution with rational values. Let $g : V \to [0, 1]$ be a packing function of G corresponding to those rational values found using linear programming. Note that $|g| = P_f(G)$. Now let $k =$ least common denominator $\{g(v) : v \in V\}$. Define $h(v) =$

$k \cdot g(v)$ for every $v \in V$. Since $0 \le g(v) \le 1$, then $0 \le h(v) = k \cdot g(v) \le k$. Also, since k is a common denominator of all of the $g(v)$, $h(v)$ is an integer. Thus, $h : V \to \{0, 1, 2, ..., k\}$. Since g is a packing function, $g(N[v]) \le 1$ for every $v \in V$. Thus,

$$h(N[v]) = \sum_{w \in N[v]} h(w) = \sum_{w \in N[v]} k \cdot g(w) = k \cdot g(N[v]) \le k \cdot 1 = k.$$

Therefore h is a k-packing function of G and $P_k(G)/k \ge |h|/k = k \cdot |g|/k = |g| = P_f(G)$. Hence, $\max\{P_m(G)/m : m$ is a positive integer$\} > P_k(G)/k > P_f(G)$, and the equality holds. $\square$

Cheston, Hare, and Pokrass [5] have proved a similar result involving the upper fractional domination number. The result states that $\Gamma_f(G) = \max\{\Gamma_k(G)/k : k$ is a positive integer$\}$. Similar results should hold for the other parameters.

4. Expanding Fractional Parameter Results to k–Parameters

In [10] Domke, Hedetniemi, and Laskar show that for any graph G with p vertices, $\gamma_f(G) = 1$ if and only if $\Delta(G) = p - 1$. A similar result holds for $\gamma_k(G)$. Note that if g is any k-dominating function, then $|g| \ge g(N[v]) \ge k$, for any vertex $v \in V$. Therefore, $\gamma_k(G) \ge k$ for any graph G and any positive integer k.

Theorem 13 *For any graph G with p vertices and any positive integer k, $\gamma_k(G) = k$ if and only if $\Delta(G) = p - 1$.*

Proof ($\Leftarrow$) Let G be a graph with p vertices and maximum degree $\Delta(G) = p - 1$, and let $v \in V$ be a vertex of G with $\deg(v) = \Delta(G) = p - 1$. Also let k be any positive integer. Define a function $g : V \to \{0, 1, 2, ..., k\}$ where $g(v) = k$ and $g(u) = 0$ for every $u \in V$, $u \ne v$. Since v is adjacent to every other vertex of G, $v \in N[u]$ for every $u \in V$ and $g(N[u]) = k$ for every vertex u. Thus g is a k-dominating function of G and $|g| = k$. Since $\gamma_k(G) = \min\{|g| : g$ is a minimal k-dominating function$\}$, and as was observed above, $\gamma_k(G) \ge k$, then $\gamma_k(G) = k$.

($\Rightarrow$) Let $\gamma_k(G) = k$, and assume that $\Delta(G) \ne p - 1$. Also let g be a k-dominating function of G where $|g| = \gamma_k(G) = k$. Now let $v \in V$ be a vertex where $g(v) = m > 0$. Since $\deg(v) \le \Delta(G) < p - 1$, there is a vertex $u \in V$, $u \ne v$, such that u is not adjacent to v. Thus, $v \notin N[u]$ and $g(N[u]) \le |g| - g(v) = k - m < k$. But since g is a k-dominating function, $g(N[u]) \ge k$, which is a contradiction. Therefore, our assumption was wrong, and $\Delta(G) = p - 1$. $\square$

For any maximal k-packing function g, if $g(v) < k$, then there is a vertex $u \in N[v]$ where $g(N[u]) = k$. Hence, $|g| \ge g(N[u]) = k$. If $g(v) = k$, then $|g| \ge g(v) = k$. In either case, $|g| \ge k$. Therefore, it is easy to see that $p_k(G) \ge k$ for every positive integer k. Similarly, it is easy to see that $ir_k(G) \ge k$. Now, since $ir_k(G) \le \gamma_k(G)$ and $P_k(G) \le \gamma_k(G)$, the following result can be easily seen.

Corollary 14   *For any positive integer* k, $ir_k(K_p) = P_k(K_p) = \gamma_k(K_p) = k$.

The 2–packing number of G, denoted $\mathbf{P_2}(G)$ was defined by Meir and Moon [18] as the maximum cardinality of a set S of vertices where $d(x, y) > 2$ for every distinct pair of vertices x and y in S. Domke, Hedetniemi, and Laskar [10] showed that $p_f(G) \leq P_2(G) \leq P_f(G)$ for any graph G. A similar result holds for the k–packing numbers of G.

Theorem 15   *For any graph* G *and positive integer* k, $p_k(G) \leq k \cdot P_2(G) \leq P_k(G)$.

Proof   Let G be any graph, and let $S \subseteq V$ be a 2–packing set of maximum cardinality (i.e. $|S| = P_2(G)$). Define a function $g : V \to \{0, 1, 2, ..., k\}$ as follows:
$$g(v) = \begin{cases} k & \text{if } v \in S \\ 0 & \text{if } v \notin S; \end{cases} \quad \text{that is } g \text{ is a } \{0, k\}\text{-function. Thus, } |g| = \sum_{v \in S} k =$$
$|S| \cdot k = k \cdot P_2(G)$.

For every pair of distinct vertices $x, y \in S$, $d(x, y) > 2$ since S is a 2–packing set. Thus, $N[x] \cap N[y] = \varnothing$ for every pair of distinct vertices $x, y \in S$. So every $v \in V$ has at most one element of S in its closed neighborhood, and $g(N[v]) \leq k$ for every $v \in V$. Therefore, g is a k–packing function of G.

Claim: g is maximal.

Proof of claim: It must be shown that for every $v \in V$, $g(v)$ cannot be increased. There are three cases to consider.

Case 1.   Let $v \in S$. If $v \in S$, then $g(v) = k$ and cannot be increased.

Case 2.   Let $v \notin S$, with v adjacent to a vertex u in S. Then $g(v) = 0$ and $g(u) = k$, so $g(N[v]) = k$. Thus, $g(v)$ cannot be increased.

Case 3.   Let $v \notin S$ and v not be adjacent to any vertex in S. Then $g(v) = 0$, and for every $u \in S$, $d(u, v) \geq 2$. If $d(u, v) > 2$ for every $u \in S$, then $S \cup \{v\}$ is a 2–packing set of greater cardinality than S – a contradiction. So there exists $u \in S$ where $d(u, v) = 2$. Then there is a vertex $x \in V$ such that $ux, xv \in E$. Thus, $g(N[x]) = k$ since $u \in N[x]$ and $u \in S$. Thus, $g(v)$ cannot be increased.

Therefore, g is a k-maximal packing function.

Now, $p_k(G) = \min \{ |h| : h$ is a maximal k–packing function of $G\} \leq |g| = k \cdot P_2(G) \leq \max \{ |h| : h$ is a maximal k–packing function of $G\} = P_k(G)$.

Therefore, $p_k(G) \leq k \cdot P_2(G) \leq P_k(G)$. $\square$

Corollary 16   *For any graph* G *and positive integer* k, $k \leq p_k(G) \leq k \cdot P_2(G) \leq P_k(G) \leq \gamma_k(G) \leq k \cdot \gamma(G)$.

Recall that a block graph is a graph in which each of its blocks is complete. Domke, Hedetniemi, Laskar, and Allan [11] showed that if G is a connected block graph, then $P_2(G) = \gamma(G)$. This result combined with the above corollary gives the following.

Corollary 17 *For any positive integer* k, *if* G *is a connected block graph, then*
$$k \cdot P_2(G) = P_k(G) = \gamma_k(G) = k \cdot \gamma(G).$$

Domke, Hedetniemi, Laskar, and Allan [11] defined the degree-covering number of order k of G, denoted $C_k°(G)$, as the minimum cardinality of a set S, where for every vertex $v \in V$, $\deg_S v \geq k$ where $\deg_S v$ is the number of vertices in S adjacent to v. Also, $C_k°(G)$ is defined only for $k \leq \delta(G)$. The following result relates the degree-covering number of order k and the k–domination number of G.

Theorem 18 *For any graph* G *and positive integer* $k \leq \delta(G)$, $\gamma_k(G) \leq C_k°(G)$.

Proof Let S be a degree-covering set of order k where $|S| = C_k°(G)$. Define a function $g : V \rightarrow \{0, 1, 2, ..., k\}$ where $g(v) = \begin{cases} 1 & \text{if } v \in S \\ 0 & \text{if } v \notin S; \end{cases}$ that is g is a $\{0, 1\}$–function. Now, $|g| = C_k°(G)$. Since S is a degree-covering function of order k, $\deg_S v \geq k$ for every vertex $v \in V$. Thus, v is adjacent to at least k vertices in S and $g(N[v]) \geq k$ for every $v \in V$. Therefore, g is a k–dominating function. Since $\gamma_k(G) = \min \{ |h| : h$ is a k–dominating function$\}$, $\gamma_k(G) \leq |g| = C_k°(G)$. ❏

5. Inequalities Involving Domination Parameters

There is a string of inequalities relating the upper and lower irredundance numbers, domination numbers, and independence numbers. An independent set of vertices is a set S of vertices which are pairwise nonadjacent. The independence number of G, denoted $\beta_0(G)$, is the maximum cardinality of an independent set of vertices. The lower independence number, which is also known as the independent domination number of G, denoted $i(G)$, is the minimum cardinality of a maximal independent set of vertices. The following inequalities are given by Cockayne, Hedetniemi, and Miller in [8].

Theorem (Cockayne, Hedetniemi, and Miller [8]) *For any graph* G, $ir(G) \leq \gamma(G)$ $\leq i(G) \leq \beta_0(G) \leq \Gamma(G) \leq IR(G)$.

A similar inequality chain involving fractional parameters also holds, namely $ir_f(G) \leq \gamma_f(G) \leq \Gamma_f(G) \leq IR_f(G)$ (see [9]) as well as an inequality chain involving k–parameters, namely $ir_k(G) \leq \gamma_k(G) \leq \Gamma_k(G) \leq IR_k(G)$ (see Theorem 4).

Still another inequality holds involving the upper fractional domination number and the upper irredundance number of a graph.

Theorem 19 *For any graph G, $\Gamma_f(G) \le IR(G)$.*

Proof Let $g : V \to [0, 1]$ be a minimal dominating function of G where $|g| = \Gamma_f(G)$. Also let $S = \{v_1, v_2, ..., v_m\} \subseteq V$ be the set of vertices with $g(N[v_i]) = 1$ for $i = 1, 2, ..., m$. Note that since g is minimal, every vertex $u \in V$ with $g(u) > 0$ is adjacent to at least one vertex in S, i.e. S dominates the set P of vertices with positive function values. Let $D \subseteq S$ be a minimal subset of S which dominates P. Now, D is an irredundant set of $\langle D \cup P \rangle$ and hence of G. Thus, $IR(G) \ge |D|$. Since D is a dominating set of P, every $v \in V$ with $g(v) > 0$ is in $N[v_i]$ for some $v_i \in D$. Also, $g(N[v_i]) = 1$ for every $v_i \in D$. Hence, $|D| = \sum_{v_i \in D} 1 = \sum_{v_i \in D} g(N[v_i]) \ge \sum_{v \in V} g(v) = |g| = \Gamma_f(G)$. Therefore, $IR(G) \ge \Gamma_f(G)$. $\square$

It should be noted that a similar inequality is not true for the fractional domination number and the irredundance number. For a cycle on four vertices (see Figure 6a) $\gamma_f(C_4) = 4/3$ and $ir(C_4) = 2$. Hence, $\gamma_f(C_4) < ir(C_4)$. For the tree given in Figure 6b, $\gamma_f(T) = \gamma(T) = 5$ and $ir(T) = 4$. Hence, $\gamma_f(T) > ir(T)$. Therefore, in general there is no strict order relationship between $\gamma_f(G)$ and $ir(G)$.

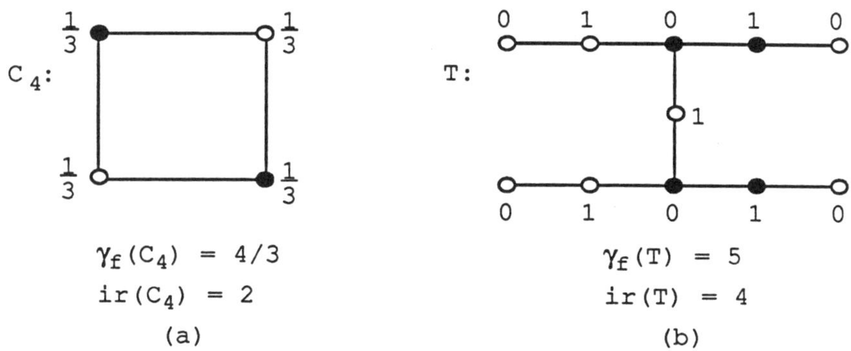

Figure 6. Examples where $\gamma_f(G) < ir(G)$ and $\gamma_f(G) > ir(G)$.

Combining Theorem 19 with Theorems 1, 2, 7, 8, 10, and 15 and other results stated in this paper, gives the following extended inequality chains.

Corollary 20 *For any graph G and positive integer k, $\dfrac{ir_k(G)}{k} \le ir(G) \le \gamma(G) \le$*

$i(G) \le \beta_0(G) \le \Gamma(G) \le \dfrac{\Gamma_k(G)}{k} \le \Gamma_f(G) \le IR(G) \le \dfrac{IR_k(G)}{k} \le IR_f(G)$.

Corollary 21 *For any graph G and positive integer k, $1 \le p_f(G) \le \dfrac{p_k(G)}{k} \le$*

$$P_2(G) \le \frac{P_k(G)}{k} \le P_f(G) = \gamma_f(G) \le \frac{\gamma_k(G)}{k} \le \gamma(G) \le i(G) \le \beta_0(G) \le \Gamma_f(G) \le$$

$$\frac{\Gamma_k(G)}{k} \le \Gamma(G) \le IR(G) \le \frac{IR_k(G)}{k} \le IR_f(G).$$

6. Conclusions and Open Problems

In this paper we have studied several k–parameters, namely k–domination, k–packings, and k–irredundance, with the hope that they will help us to better understand their associated fractional parameters. This idea of k–parameters involving packings and coverings is just beginning to surface. It is hoped that these results spur an interest by other researchers in this area.

There are several open problems in this area. For example, is it true that $\Gamma_k(G) < \Gamma_{k+1}(G)$ for every graph G? Also, it has been shown that $\gamma_f(G) = \min \{\gamma_k(G)/k : k$ is a positive integer} and $\Gamma_f(G) = \max \{\Gamma_k(G)/k : k$ is a positive integr} for every graph G, but no upper bound has been found on the value of k for which the minimum and maximum values are assumed. Is it possible to define fractional (or k–) independence so that the corresponding max/min parameters, i_f and β_f, fit into the string of inequalities (i.e. $\gamma_f \le \gamma \le i_f \le i \le \beta_0 \le \beta_f \le \Gamma$)? Some work along these lines is in progress and will appear elsewhere.

REFERENCES

[1]   R. Aharoni, Fractional matchings and covers in infinite hypergraphs, *Combinatoria* **5(3)** (1985), 181-184.

[2]   C. Berge, Packing problems and hypergraph theory: A survey, *Ann. Discrete Math.* **4** (1979), 3-37.

[3]   N. Chandrasekharan, S.T. Hedetniemi, R. Laskar, and A. Majumdar, Fractional domination in graphs, unpublished manuscript (1988).

[4]   G.A. Cheston, G. Fricke, S.T. Hedetniemi, D.J. Pokrass, On the computational complexity of upper fractional domination, *Discrete Appl. Math.*, to appear.

[5]   G.A. Cheston, E.O. Hare, and D.J Pokrass, private communication (1988).

[6]   F.R.K. Chung, Z. Furedi, M.R. Garey and R.L. Graham, On the fractional covering number of hypergraphs, *SIAM J.Discrete Math. Vol.1, No.1,* (1988), 45-49.

[7] E.J. Cockayne and S.T. Hedetniemi, Towards a theory of domination in graphs, *Networks* **7** (1977), 247-261.

[8] E.J. Cockayne, S.T. Hedetniemi, and D.J. Miller, Properties of hereditary hypergraphs and middle graphs, *Canad. Math. Bull.* **21**, No. 4 (1978), 461-468.

[9] G.S. Domke, Variations of colorings, coverings, and packings of graphs, Ph.D. dissertation, Department of Mathematical Sciences, Clemson University, Clemson, South Carolina, August 1988.

[10] G.S. Domke, S.T.Hedetniemi, and R.C. Laskar, Fractional packings, coverings and irredundance in graphs, *Congr. Numer.* **66** (1988), 227–238.

[11] G.S. Domke, S. Hedetniemi, R.Laskar, and R. Allan, Generalized packings and coverings of graphs, *Congr. Numer.* **62** (1988), 259-270.

[12] M. Farber, Characterizations of strongly chordal graphs, *Discrete Math.* **43** (1983), 173-189.

[13] J.F. Fink and M.S. Jacobson, n-domination in graphs, *Graph Theory with Applications to Algorithms and Computer Science*, (Kalamazoo, MI, 1984) 283-300, Wiley-Intersci. Publ., Wiley, New York, 1985.

[14] D. Grinstead and P.J. Slater, Fractional domination and fractional packings in graphs, *Congr. Numer.*, to appear.

[15] E.O. Hare, private communication (1988).

[16] S.M. Hedetniemi, S.T. Hedetniemi, and T.V. Wimer, Linear time resource allocation algorithms for trees, Dept. Mathematical Sciences, Clemson Univ. Tech. Rept. 428, July 1983.

[17] M. Larsen, J. Propp, and D. Ullman, The asymptotic chromatic number of a graph, to appear.

[18] A. Meir and J.W. Moon, Relations between packing and covering numbers of a tree, *Pacific J.Math.* **61,** No.1 (1975), 225-233.

[19] W.R. Pulleyblank, Fractional matchings and the Edmonds-Gallai Theorem, *Discrete Appl. Math.* **16** (1987), 51-58.

The Star Arboricity of Complete Bipartite Graphs

Hikoe Enomoto

Keio University

Yoko Usami

Ochanomizu University

ABSTRACT

A graph H is called a star forest if every component of H is a star. The minimum number of star forests that cover the edges of a graph G is called the star arboricity of G, and is denoted by $sa(G)$. In this paper, we study the star arboricity of complete bipartite graphs, and prove that

$$sa(K_{n,kn}) \leq min\left\{ n, \left\lceil \frac{kn + 2}{k + 1} \right\rceil + 1 \right\}$$

if $(k, n) \neq (1, 5)$.

In this paper, we consider finite undirected simple graphs without loops and multiple edges. For a graph G, $V(G)$ and $E(G)$ denotes the vertex set and the edge set of G, respectively. For a vertex v in $V(G)$,

$$N_G(v) := \{u \in V(G) \mid uv \in E(G)\}$$

is the neighborhood of v in G, and

$$d_G(v) := |N_G(v)|$$

is the degree of v in G. For a subset S of $V(G)$,

$$N_G(S) := \bigcup_{v \in S} N_G(v).$$

For a real number r, $\lceil r \rceil$ denotes the least integer not less than r. A graph H is called a *star forest* if every component of H is a star. The minimum number of star forests that cover the edges of a graph G is called the *star arboricity* of G, and is denoted by $sa(G)$.

The star arboricity is known for several families of regular graphs: For complete graphs, Akiyama and Kano [1] proved that

$$sa(K_n) = \left\lceil \frac{n}{2} \right\rceil + 1 \quad (n \geq 4).$$

For complete bipartite regular graphs, Egawa et al. [3] and Truszczynski [4] independently proved that

$$sa(K_{n,n}) = \left\lceil \frac{n}{2} \right\rceil + 2 \quad (n \geq 6).$$

Truszczynski [4] also proved some upper and lower bounds for the star arboricity of n–dimensional cube Q_n. For complete multipartite regular graphs, Aoki [2] proved that

$$sa(\overline{mK_n}) = \left\lceil \frac{(m-1)n}{2} \right\rceil + 2$$

if $(m-1)n$ is even or if $n \geq 2m + 5$.

In this paper, we shall determine the star arboricity of complete bipartite graphs $K_{n,kn}$.

Theorem 1 *If $(k, n) \neq (1, 5)$,*

$$sa(K_{n,kn}) = min\left\{n, \left\lceil \frac{kn + 2}{k + 1} \right\rceil + 1\right\}.$$

In the proof of Theorem 1, we use the following proposition.

Proposition 2 *Let k and m be integers greater than 1. Then*

$$sa(K_{(k+1)m,k(k+1)m-1)} \geq km + 2.$$

Proof of Theorem 1 By the results of [3] or [4], we may assume that $k \geq 2$. Let $G := K_{n,kn}$ with the partite sets A and B ($|A| = n$, $|B| = kn$), and $s := sa(K_{n,kn})$.

First, suppose $s < min\left\{n, \left\lceil \dfrac{kn + 2}{k + 1} \right\rceil + 1\right\}$, that is, $s < n$ and

$$s \leq \left\lceil \frac{kn + 2}{k + 1} \right\rceil \leq \frac{kn + k + 2}{k + 1}. \tag{1}$$

By the definition of the star arboricity, there exist subgraphs $H_1, H_2, ..., H_s$ of G satisfying

$$E(G) = E(H_1) \cup ... \cup E(H_s) \quad \text{(disjoint)}$$

and Hi are star forests ($1 \leq i \leq s$). Let C_i be the set of centers of components in H_i, that is, for each connected component L of H_i, $|L \cap C_i| = 1$ and $d_{H_i}(x) = 1$ for all

$x \in V(H_i) - C_i$. (From a component isomorphic to K_2, either one of the vertices may be chosen as a center.) Then

$$|E(H_i)| = |V(G)| - |C_i|. \tag{2}$$

Let
$$I_i := \{x \in V(H_i) \mid d_{H_i}(x) = 0\}$$

be the set of isolated vertices in H_i. Note that for each vertex $x \in V(G), d_{H_i}(x) \geq 2$ for some i, since we have assumed that $s < n \leq d_G(x)$.
Hence
$$\bigcup_{i=1}^{s} (C_i - I_i) = V(G),$$

and
$$\sum_{i=1}^{s} |C_i| \geq |V(G)| + \sum_{i=1}^{s} |I_i|. \tag{3}$$

By (2),
$$|E(G)| = kn^2 = \sum_{i=1}^{s} |E(H_i)| = s(k+1)n - \sum_{i=1}^{s} |C_i|,$$

and therefore
$$\sum_{i=1}^{s} |C_i| = s(k+1)n - kn^2 \geq |V(G)| = (k+1)\,n \tag{4}$$

by (3). Together with (1), we get
$$\frac{kn+k+2}{k+1} \geq s \geq \frac{kn+k+1}{k+1}.$$

Case 1 $s = \dfrac{kn+k+1}{k+1}$.

In this case, $n \equiv 0 \pmod{k+1}$. Let $n = (k+1)m$. Then $s = km+1 \leq n-1$ means that $m \geq 2$. By proposition 2,
$$sa(G) = km + 1 \geq sa(K_{n,kn-1}) \geq km + 2,$$
but this is a contradiction.

Case 2 $s = \dfrac{kn+k+2}{k+1}$.

Note that $n \equiv 1 \pmod{k+1}$ in this case. Let $n = (k+1)m + 1$. Then $s = km+2 \leq n-1$ implies that
$$n \geq 2k + 3, \tag{5}$$

or $m \geq 2$. Define
$$M(x) := \{i \mid x \in C_i - I_i\}$$
for $x \in V(G)$, and
$$U_i := \{x \mid M(x) = \{i\}\}.$$

Note that $U_i \cap U_j = \varnothing$ $(i \neq j)$ and that $d_{H_i}(x) \geq d_G(x) - (s - 1)$ for $x \in U_i$, since $d_{H_j}(x) \leq 1$ for $j \neq i$.

Claim 1   Suppose x and y are in U_i. Then x and y are nonadjacent in G.

Proof   In H_i, both x and y are centers of different components. Hence they are nonadjacent in H_i. Suppose $xy \in E(H_j)$ $(j \neq i)$. Then x or y is a center of a nontrivial component in H_j. This is impossible, because $M(x) = M(y) = \{i\}$. $\square$

Claim 2   Suppose $|U_i| \geq 2$. Then U_i is contained in B.

Proof   Suppose U_i is not contained in B. By Claim 1, U_i is contained in A. Suppose $x_1, x_2 \in U_i$ $(x_1 \neq x_2)$. Then $N_{H_i}(x_1)$ and $N_{H_i}(x_2)$ are disjoint subsets of B, and
$$d_{H_i}(x_j) \geq d_G(x_j) - (s - 1) = kn - s + 1 \quad (j = 1, 2).$$
Hence
$$2(kn - s + 1) \leq kn,$$
which is a contradiction. $\square$

Let
$$U := \bigcup_{i=1}^{s} U_i = \{x \in V(G) \mid |M(x)| = 1\},$$
and
$$u := |U| = \sum_{i=1}^{s} |U_i|.$$
Then
$$\sum_{i=1}^{s} |C_i| \geq u + 2(|V(G)| - u),$$
which implies that
$$u \geq 2|V(G)| - \sum_{i=1}^{s} |C_i| = kn > ks.$$

Hence $|U_i| > k$ for some i $(1 \leq i \leq s)$. By Claim 2, U_i is contained in B. Suppose $|U_i| \geq k + 2$. Then

$$|A| = n \geq |N_{H_i}(U_i)| = \sum_{x \in U_i} d_{H_i}(x) \geq (k + 2)(n - s + 1) = \frac{(k + 2)(n - 1)}{k + 1}.$$

This implies that $n \leq k + 2$, which contradicts (5). Hence $|U_i| = k + 1$. Then
$$|N_{H_i}(U_i)| \geq (k + 1)(n - (s - 1)) = n - 1.$$
Let A' be any subset of $N_{H_i}(U_i)$ with $|A'| = n - 1$, $B' := B - U_i$, and G' be the subgraph of G induced by $A' \cup B'$. Then there are no edges between A' and B' in H_i. This means that

$$E(G') \subseteq \bigcup_{\substack{1 \leq j \leq s \\ j \neq i}} E(H_j),$$

and therefore
$$sa(G') \geq s - 1 = km + 1.$$
This contradicts Proposition 2, since G' is isomorphic to $K_{(k+1)m,k(k+1)m-1}$. This completes the proof of

$$sa(K_{n,kn}) \leq \min\left\{ n, \left\lceil \frac{kn + 2}{k + 1} \right\rceil + 1 \right\}$$

if $(k, n) \neq (1, 5)$.

Next we shall prove that

$$sa(K_{n,kn}) \leq \min\left\{ n, \left\lceil \frac{kn + 2}{k + 1} \right\rceil + 1 \right\}.$$

It is obvious that $sa(K_{n,kn}) \leq n$. so, let
$$n = (k + 1)t + r \quad (2 \leq r \leq k + 1),$$
and we shall give a decomposition of $K_{n,kn}$ into $kt + r + 1$ star forests $F^{(1)}, F^{(2)}$ and $F_{i,j}$ $(1 \leq i \leq k, 1 \leq j \leq t_i)$, where $t_i := t + 1$ for $0 \leq i \leq r - 1$ and $t_i := t$ for $r \leq i \leq k$. (Note that if $n \equiv 1 \pmod{k + 1}$, $\left\lceil \frac{kn + 2}{k + 1} \right\rceil = \left\lceil \frac{k(n + 1) + 2}{k + 1} \right\rceil$. Hence we may assume that $2 \leq r \leq k + 1$.) Let

$$A := \{a_{i,j} \mid 0 \leq i \leq k, 1 \leq j \leq t_i\},$$
and
$$B := \{b_{i,j}^{(h)} \mid 1 \leq h \leq k, 0 \leq i \leq k, 1 \leq j \leq t_i\}.$$

In the following construction, the centers of $F^{(1)}$ are $a_{0,j}$ $(1 \leq j \leq t + 1)$, $a_{1,t+1}$ and $b_{0,t+1}^{(h)}$ $(r \leq h \leq k)$, the centers of $F^{(2)}$ are $a_{i,j}$ $(1 \leq i \leq k, 1 \leq j \leq t_i)$ and $a_{0,t+1}$, and the centers of $F_{i,j}$ are $a_{i,j}, a_{0,j}, b_{0,j}^{(i)}$ and $b_{i,j}^{(h)}$ $(1 \leq h \leq k)$. More precisely, the forest that contains the edge joining $a_{i,j}$ and $b_{p,q}^{(h)}$ is determined as follows:

(Case I) $i = 0$

   (Subcase I – a) $1 \leq h < r$

      (Subcase I – a – i) $p = 0$

         $F^{(1)}$ if $j = q$,

         $F_{h,q}$ otherwise.

      (Subcase I – a – ii) $p \neq 0$

         $F^{(1)}$ if $h = p$ and $j = q$,

         $F_{h,j}$ otherwise.

   (Subcase I – b) $r \leq h \leq k$

      (Subcase I – b – i) $p = 0$

         $F^{(1)}$ if $j = q \leq t$,

         $F^{(2)}$ if $j = q = t + 1$,

         $F_{h,j}$ if $j \neq q = t + 1$,

         $F_{h,q}$ otherwise.

      (Subcase I – b – ii) $1 \leq p < r$

         $F^{(1)}$ if $j = t + 1$,

         $F_{h,j}$ otherwise.

      (Subcase I – b – iii) $r \leq p \leq k$

         $F^{(1)}$ if $j = t + 1$ and $p \neq h$,

         $F^{(1)}$ if $j \leq t$ and $p = h$,

         $F_{1,t+1}$ if $j = t + 1$ and $p = h$,

         $F_{h,j}$ otherswise.

(Case II) $i \neq 0$

   (Subcase II – a) $p = 0$

         $F^{(1)}$ if $h \geq r$ and $q = t + 1$,

         $F^{(2)}$ if $h = i$ and $q = j$,

         $F_{i,j}$ otherwise.

   (Subcase II – b) $p > 0$

      (Subcase II – b – i) $i = 1$ and $j = t + 1$

         $F^{(1)}$ if $h = 1$ and $p \geq r$,

         $F^{(2)}$ if $p = 1$ and $q = t + 1$,

         $F_{p,q}$ if $h = p \geq r$,

         $F_{p,q}$ if $h = 1$, $p < r$ and $(p, q) \neq (1, t + 1)$,

         $F_{1,t+1}$ otherwise.

      (Subcase II – b – ii) $(i, j) \neq (1, t + 1)$

         $F^{(2)}$ if $p = i$ and $q = j$,

         $F_{p,q}$ if $h = i$ and $(p, q) \neq (i, j)$,

         $F_{i,j}$ otherwise. □

Proof of Proposition 2   Proposition 2 is proved by a similar argument as the first part of the proof of Theorem 1.   So, only the sketch is given.   Let $G := K_{(k+1)m,k(k+1)m-1}$ with the partite sets A and B ($|A| = (k + 1)m$, $|B| = k(k + 1)m - 1$), $s := sa(G)$, and suppose $s \leq km + 1$.   Then

$$E(G) = E(H_1) \cup ... \cup E(H_s) \qquad \text{(disjoint)}$$

with H_i star forests. Let C_i be the set of centers of components in H_i,

$$I_i := \{x \in V(H_i) \mid d_{H_i}(x) = 0\},$$
$$M(x) := \{i \mid x \in C_i - I_i\},$$
$$U_i := \{x \mid M(x) = \{i\}\}$$

and

$$u := \sum_{i=1}^{s} |U_i|.$$

Then

$$\sum_{i=1}^{s} |C_i| \geq |V(G)| + \sum_{i=1}^{s} |I_i| \tag{6}$$

and

$$\sum_{i=1}^{s} |C_i| = s \cdot |V(G)| - |E(G)| \geq u + 2(|V(G)| - u).$$

Hence

$$u \geq (k^2 + 2k)m - 1 > sk,$$

which implies that $|U_i| \geq k + 1$ for some i. We can easily show that U_i is contained in B, and then

$$|N_{H_i}(U_i)| \geq (k + 1)((k + 1)m - (s - 1)) \geq (k + 1)m = |A|.$$

This means that all the vertices in $B - U_i$ are isolated in H_i, or

$$|I_i| = k(k + 1)m - 1 - (k + 1).$$

This contradicts (6). $\qquad \Box$

It appears to be difficult to determine the exact value of the star arboricity of complete bipartite graphs in general. However, it is not difficult to prove that

$$sa(K_{l(k+1)n,k(k+1)n}) \geq kln + 2$$

if $k, l \geq 2$.

Conjecture 3   *Let k and l be integers greater than 1. Then*

$$sa(K_{l(k+l)n,k(k+l)n}) = kln + 2.$$

REFERENCES

[1]   J. Akiyama and M. Kano, Path factors of a graph, in Graphs and Applications (Proc. First Colorado Symp. on Graph Theory, ed. by F. Harary & J.S. Maybee), John Wiley & Sons (1985) 1 – 21.

[2]   Y. Aoki, The star–arboricity of the complete regular multipartite graphs, to appear.

[3]   Y. Egawa, M. Urabe, T. Fukuda and S. Nagoya, A decomposition of complete bipartite graphs into edge–disjoint subgraphs with star components, Discrete Mathematics 58 (1986) 93 – 95.

[4]   M. Truszczynski, Decomposing graphs into forests of star, Congressus Numerantium 54 (1986) 73 – 86.

Problems and Results
in Combinatorial Analysis
and Combinatorial Number Theory

P. Erdös

The Hungarian Academy of Sciences

ABSTRACT

In this short paper I discuss a few of my old and new problems and try to concentrate on the problems I discussed in my talk.

1. First an old problem of mine on the n dimensional cube

The graph $H(n)$ determined by the vertices of the n–dimensional cube has 2^n vertices and $n2^{n-1}$ edges. Let $F_k(n)$ be the smallest integer for which every subgraph of $H(n)$ having $f_k(n)$ edges contains a C_{2k}. I conjectured more than 20 years ago that for $n > n_0(\varepsilon)$ $f_2(n) < \left(\frac{1}{2} + \varepsilon\right)n2^{n-1}$, in other words slightly more than half the edges of $H(n)$ force a C_4. It is surprising that this simple and I think attractive conjecture is still open. Fan Chung proved (unpublished) that $f_2(n) < 0.6\, n2^{n-1}$. The best possible result which could hold would be

$$f_2(n) < n2^{n-2} + c2^n \tag{1}$$

for some sufficiently large c. It is perhaps not hopeless to determine $f_2(n)$ exactly.

I further conjectured

$$f_3(n) < \varepsilon n2^n$$

for every $\varepsilon > 0$ if $n > n_0(\varepsilon)$. Perhaps for $k \geq 3$

$$f_k(n) < cn^{a_k} 2^n$$

where $a_k < 1$ and $a_k \to 0$ as $k \to \infty$. It is easy to give lower bounds for a_k by the probability method but as far as I know $f_k(n) = o(n2^n)$ has not yet been proved – even if k is large.

Remarks to problem 1. Furedi just informed me that he and Jeff Kahn proved some time ago that any set of $2^n\left(\frac{2}{3} + \varepsilon\right)$ sets of vertices of the n–dimensional cube always contains a C_4. We thought that perhaps this can be improved to $2^n\frac{2}{3} + O(1)$. We also asked: is it true that every set of $(1 + O(1))2^{n-1}$ vertices induces a C_6. It is easy to see that one needs for this at least $\left(1 + \frac{1}{n^2}\right)2^{n-1}$ vertices.

2. Now some problems on coloration of graphs

S. Burr, R. L. Graham, V. T. Sós and I have several forthcoming papers on anti–Ramsey theorems, the first of which will soon appear in the Journal of Graph Theory. We state several unsolved problems in the paper, here I only mention two of them. Let L be a graph. Denote by

$$\chi_s(n;e,L)$$

the smallest integer r for which there is a graph of n vertices and e edges whose edges can be coloured by r colors so that every subgraph isomorphic to L is T.M.C. i.e. all its edges get different colors. We proved that for $k \geq 3$

$$\chi_s\left(n;\left[\tfrac{n^2}{4}\right] + 1,\ C_{2k+1}\right) > cn^2$$

and conjecture that

$$\chi_s\left(n;\left[\tfrac{n^2}{4}\right] + 1,\ C_{2k+1}\right) = (1 + o(1))\frac{n^2}{8} \tag{2}$$

(2) if true will perhaps not be very difficult to prove. Observe that the problem is meaningful only if e is greater than the Turán number $T(n;L)$ of L. ($T(n;L)$ is the largest integer for which there is a $G(n; T(n;L))$ which does not contain a subgraph isomorphic to L).

For $K=2$ i.e. for C_5 the situation is very much more complicated. We proved that if $e > \left(\frac{1}{4} + \varepsilon\right)n^2$ then

$$\chi_s(n,e,C_5)/n \to \infty$$

and if $e < \left(\frac{1}{2} - \varepsilon\right)n^2$ then

$$\chi_s(n,e,C_5) \le cn^2/\log n.$$

Simonovits and I have some sharper results for $\chi_s(n,e,C_5)$.

Is it true that

$$\chi_s(n,\varepsilon n^2,C_4) < n \qquad\qquad \text{for } n > n_0$$

if e is sufficiently small? For further problems and results I have to refer to our forthcoming papers.

3. Pyber, Tuza and I considered the following set of problems:

Let G_e be a graph of e edges and let $n > n_0(G_e)$ be sufficiently large. For which graphs G_e does there exist a $d(G_e)$ such that if we color the edges of $K(n)$ by e colors so that every color in every vertex has degree $\ge d(G)$ then there is a T.M.C. subgraph isomorphic to G_e? We must of course have $d(G) < \frac{n-1}{e}$. If such a $d(G)$ exists, determine if possible the best possible value of $d(G)$. A very special case of this problem was posed in 1986 at the Kürschák competition in Hungary: Color the edges of a K_{3n+1} by three colors so that every vertex in every color has degree n (n must of course be even). Then G has a T.M.C. triangle. Several of the competitors (freshmen) proved this. For the triangle Kostochka proved that $d = \frac{n}{4}$ is best possible and Lovász, Pyber and Tuza observed that this follows from a general result of Gallai. Tuza proved that $d(C_4) \le \left(\frac{1}{4} - \varepsilon\right)n$, but the best value of $d(C_4)$ is not known. Here I just state one of our problems (for further problems and positive and negative results I refer to a forthcoming paper). Let n be even, color the edges of a $K(6n+1)$ by 6 colors so that every vertex has in every color degree n. Is it true that there is a C_6 which is T.M.C.? Is it true that there is a $K(4)$ which is T.M.C.? (Brightwell and Trotter proved that $d(C_6) > (1 - \varepsilon)\frac{n}{6}$ for every $\varepsilon > 0$.)

4. Some extremal problems

I published many papers on this subject some of which are joint papers with Simonovits and others. Here I just state two conjectures with Simonovits. Let G be a bipartite graph which has no induced subgraph every vertex of which has degree > 2. Is it then true that the Turán number $T(n;G)$ satisfies the inequality

$$T(n;G) < cn^{3/2} \qquad\qquad (3)$$

and probably

$$\lim T(n;G)/n^{3/2} \to c \ ? \tag{4}$$

(3) and (4) if true are probably very deep. A slightly weaker conjecture states as follows: Let G be bipartite, assume $T(n;G) < cn^{3/2}$. Let x be a vertex not in G and join x to two vertices of G which belongs to the same color class. Denote this new bipartite graph by G_1. is it true that $T(n;G_1) < c_1 n^{3/2}$?

Is it true that if G is bipartite and

$$T(n;G)/n^{3/2} \to \infty,$$

Then $T(n;G)$ contains an induced subgraph each vertex of which has the degree ≥ 3 and then

$$T(n;G) > n^{(3/2)+\varepsilon}. \tag{5}$$

(3), (4) and (5) can perhaps be generalized as follows: Assume that G has no induced subgraph of degree $> r$, then

$$T(n;G) < cn^{2-1/r} \text{ and } T(n;G)/n^{2-1/r} \to \lambda \ ?$$

Further if

$$T(n;G)/n^{2-1/r} \to .\infty$$

then G has an induced subgraph each vertex of which has degree $> r$ and then

$$T(n;G) > n^{2-\frac{1}{r}+\varepsilon}.$$

Finally is it true that for every bipartite G there is a rational α, $1 \leq \alpha < 2$ for which

$$T(n;G)/n^{\alpha} = c_G. \tag{6}$$

(6) if true is probably deep. Perhaps c_G is always rational. See references [1] and [6] for further information.

5. On Turán–Ramsey theorems

Hajnal, V. T. Sós, Szemerédi and I recently published a paper on this subject and we have a fairly long forthcoming paper with Simonovits. Here I state only two unsolved problems. Let $G(n;e)$ be a graph which contains no $K(2,2,2)$ and the largest independent set of which is $o(n)$. Is it then true that $e = o(n^2)$? This problem seems very difficult, unless we overlooked a simple argument.

Let next $G(n;e)$ contain no $K(5)$ and assume that every set of εn vertices contains a triangle (for every $\varepsilon > 0$ if $n > n_0(\varepsilon)$). Do these conditions imply $e = o(n^2)$? We only could prove $e \leq \left(\frac{1}{12} + o(1)\right)n^2$. If we assume that our $G(n;e)$ contains no $K(6)$ then we proved that $e < \left(\frac{1}{6} + o(1)\right)n^2$ but we showed $e > \left(\frac{1}{8} - o(1)\right)n^2$ is possible. If G contains no $K(7)$ then $\max e = \left(\frac{1}{4} + o(1)\right)n^2$. for the many further problems in this subject I have to refer to the literature. See references [2] and [3].

I often stated problems on inequalities about the Ramsey numbers $r(u,v)$, since unfortunately as far as I know no new results have been obtained. I just state two old questions which are still open.

Is it true that for every $\varepsilon > 0$ and $n > n_0(\varepsilon)$

$$r(4,n) > n^{3-\varepsilon} \ ?$$

and in fact, probably

$$r(4,n) > cn^3/(\log n)^{\alpha} \ .$$

Is it true that

$$r(n+1,n+1) > (1+c)r(n,n) \ ?$$

In fact it is not even known that

$$r(n+1,n+1) - r(n,n) > cn^2 \ .$$

6. Finally I state some miscellaneous problems

Hajnal and I considered the following problem. First I state a special case. Is it true that for $n > n_0$ there is no graph $G(n)$ for which every set of $(\log n)^3$ vertices contains both a complete subgraph of $\log n$ vertices and an independent set of $\log n$ vertices? The answer is almost certainly affirmative. Probably if every subgraph of $(\log n)^3$ vertices contains a complete subgraph of $\log n$ vertices then $G(n)$ contains a very large click perhaps one of size $> (\log n)^3$ which of course would answer our question affirmatively. We also considered the following question: Suppose $G(n)$ is such that every subgraph of $(\log n)^2$ vertices contains a complete subgraph of size $(\log n)$. Then we thought that $G(n)$ must contain an enormous click probably one of size $> n^{1/2-\varepsilon}$ but in fact we could not even show that $G(n)$ contains a click of size $2\log n$. if $G(n)$ contains no independent set of size 3 then it is well known that every set of k^2 vertices contains a complete subgraph of size k but $G(n)$ does not have to contain a click of size $n^{\frac{1}{2}+\varepsilon}$. This was our reason that we did not ask for a click larger than $n^{\frac{1}{2}+\varepsilon}$.

Now our general problem. Let $f(n) \to \infty$ $g(n) \to \infty$, $g(n) < n$ but $g(n)$ tends to infinity much faster than $f(n)$. For which functions $f(n)$ and $g(n)$ are there graphs of n vertices every subgraph of $g(n)$ vertices of which contains both a complete graph and an independent set of $f(n)$ vertices? Trivially $g(n) > f(n)^2$ but probably much stronger results will hold. Also if we only assume that every induced subgraph of $g(n)$ vertices contains a complete graph of $f(n)$ vertices, what is the largest click our graph must contain? If we live we hope to investigate this question further.

We also looked at the following special question. Let $G(n)$ be a graph and assume that every set of seven vertices spans a triangle. Denote by $h(n)$ the size of the largest click our graph must contain. We expect that $h(n) > n^{\frac{1}{3}-c}$ but there is a $G(n)$ for which $h(n) < n^{\frac{1}{2}-c}$. The reason for our conjecture is that we can assume that $G(n)$ does not contain an independent set of four vertices, but that it can contain an independent set of three vertices.

In my paper in the fourth international conference on graph theory and its applications at Kalamazoo held in 1980 I stated that Faudree, Rousseau, Schelp and I can prove that every graph G of $\binom{2n+1}{2} - \binom{n}{2} - 1$ edges is the union of a bipartite graph and a graph each vertex of which has degree $< n$. Faudree has a nice proof if G has $2n + 1$ vertices but we could never prove it in general. The "theorem" must be relegated to a conjecture.

The following problems which we posed several years ago are still open: Let $G(n;2n-1)$ be a graph of n vertices and $2n-1$ edges. Is it true that it has an induced subgraph of $m < n(1-\varepsilon)$ vertices each vertex of which has degree ≥ 3. Faudree found a proof with $m < n-c\sqrt{n}$, several other authors independently proved this, but nobody managed to get a better result, perhaps after all it is best possible.

Is it true that if a $G(n;2n-2)$ does not have a proper induced subgraph each vertex of which has degree ≥ 3, then it contains all small cycles? We could only prove it for cycles of size ≤ 6.

One of our many problems which we posed during my visits to Western Michigan University states as follows: Is it true that every $G(2n+1; n^2+n+1)$ contains two vertices x_1 and x_2 of the same degree which are joined by a path of length three? The complete bipartite graph $K(n+1; n)$ shows that this fails for a $G(2n+1; n^2+n)$.

7. Now I would like to state a few old and new problems in combinatorial number theory.

Perhaps my oldest problem states as follows: Let $a_1 < a_2 < ... < a_k$ be a sequence of integers for which all the sums

$$\sum_{i=1}^{k} \varepsilon_i a_i, \quad \varepsilon_i = 0 \text{ or } 1$$

are all distinct.

Determine or estimate $\min a_k = f(k)$ as accurately as possible. $f(k) > c2^k/k$ is very easy and by the second moment method Leo Moser and I proved $a_k > c2^k/\sqrt{k}$ which is still the current record. I conjectured more than 50 years ago that

$$a_k > 2^{k-c} \tag{7}$$

for some absolute constant c. I offer 500 dollars for a proof of disproof of (7). Conway and Guy showed that $a_k < 2^{k-2}$ is possible for $k \geq 24$. It has been conjectured that $a_k > 2^{k-3}$ holds. Perhaps $f(k)$ would be determined accurately at least for small values of k. Recently Sárközy and I considered the following related problem: Let $b_1 < b_2 < ... < b_k$ be a sequence of integers and assume that the set $\sum_{i=1}^{k} \varepsilon_i b_i$, $\varepsilon_i = 0$ or 1 (ε_i not all 0) does not contain an arithmetic progression of r terms with non zero difference d. Put $\min a_k = g_r(k)$. Perhaps

$$g_3(k) = 3^{k-1} \tag{8}$$

we could only prove $g_3(k) > 3^{k-c \log k}$ and are convinced that $g_3(k) > 3^{k-c}$ holds. The behaviour of $g_r(k)$ for $r > 3$ seems more complicated. We hope to publish a paper on this subject soon.

Szemerédi and I considered the following problem. Let $a_1 < a_2 < ... < a_m$ be a set of m integers. Denote by $f_r(A,m)$ the number of distinct integers which can be written in the form

$$\sum_{i=1}^{m} \varepsilon_i a_i, \quad \text{or} \quad \prod_{i=1}^{m} a_i^{\varepsilon_i}, \quad \sum \varepsilon_i = r$$

where $\varepsilon_i = 0$ or 1. In other words $f_r(A,m)$ is the number of integers which can be written as the distinct sum or product of r a's. Put

$$\min f_r(A,m) = F_r(m)$$

where the minimum is to be taken over all sets of distinct integers $a_1 < a_2 < \ldots < a_m$.

We proved

$$m^{1+c} < F_2(m) < m^2 \exp(-c_2 \log m / \log\log m). \tag{9}$$

The upper bound is probably closer to the truth than the lower one. Probably for every r and $\varepsilon > 0$ if $n > n_0(m,\varepsilon)$.

$$f_r(m) > m^{r-\varepsilon}. \tag{10}$$

Put

$$f(A,m) = \sum_{r=1}^{m} f_r(A,m)$$

in other words $f(A,m)$ is the number of integers which can be written as the distinct sum or product of distinct a's. Put $g(m) = \min f(A,m)$ where the minimum is extended over all choices of distinct integers $1 \le a_1 < a_2 < \ldots < a_m$. We of course conjecture that for every r and $m > m_0(r)$

$$g(m) > m^r \tag{11}$$

(11) would of course follow from (10). we proved that

$$g(m) < \exp(c(\log m^2(\log\log m))). \tag{12}$$

Perhaps (12) is close to the truth.

These problems can also be posed if the a's are real or complex numbers or elements of a vector space. Since our paper with Szemerédi has nearly been forgotten I perhaps can be permitted to state two more problems from our paper: Let $G(m;e)$ be a graph of m vertices and e edges. Let the vertices of our graph be the a's and if a_i and a_j are joined we consider the $2e$ numbers $a_i + a_j$, $a_i a_j$. It seems likely that the number of distinct numbers of this form is greater than e/m^ε. This conjecture would of course imply (10) for $r=2$. But we could not even prove that the number of distinct integers of the form $a_i + a_j$, $a_i a_j$ is greater than m^{1+d} if $c > m^{1+\varepsilon}$.

Finally assume that the number of distinct integers of the form $a_i + a_j$ is $< cm$. Then we expect that the number of distinct integers of the form $a_i a_j$ is greater than $m^2/\log m)^c$, perhaps the deep results of Freiman will imply this conjecture.

The inexperienced reader might wonder why we did not ask for the explicit determination of $g(m)$ or $\min F_2(m)$. We considered this problem as hopeless and it would be a good joke on us if it would turn out that we were wrong. See references [4] and [5].

An old problem of Sidon states: Let $a_1 < a_2 < \ldots < a_t \leq m$ be a sequence of integers for which all the sums $a_i + a_j$ are distinct. Put $\max t = f_2(m)$. Determine $f_2(m)$ as accurately as possible. I conjectured

$$f_2(m) = m^{1/2} + 0(1). \tag{13}$$

Ruzsa and H. Taylor convinced me that (13) is perhaps too optimistic and should be replaced by

$$f_2(m) = m^{1/2} + 0(m^\varepsilon)$$

for every $\varepsilon > 0$. Turán and I proved

$$f_2(m) \leq m^{1/2} + cm^{1/4}$$

and this was improved by Lindstrom to

$$f_2(m) \leq m^{1/2} + m^{1/4} + 1$$

which is the current record. Chowla and I proved

$$f_2(m) > m^{1/2} - m^{1/2-c}$$

for some $c > 0$. Here I would like to state an old and forgotten conjecture of mine: Let $a_1 < a_2 < \ldots < a_m$ be a sequence of integers for which all the sums $a_i + a_j$ are distinct. Is it then true that our sequence can be continued to be a perfect difference set for some prime p (or perhaps for all sufficiently large primes p)? In other words is there a continuation of our sequence

$$a_1 < a_2 < \ldots < a_n < a_{n+i} < \ldots < a_{p+1} \leq p^2 + p \tag{14}$$

so that every non zero residue $(\mod(p^2 + p + 1))$ can be uniquely written in the form $a_i - a_j$? If (14) is too difficult or false, is it true that there is a continuation

$$a_1 < \ldots < a_n < a_{n+i} < \ldots < a_m, a_m = (1 + 0(1))m^2$$

for which the sums $a_i + a_j$ are all distinct?

Let now $a_1 < a_2 < ...$ be an infinite sequence of integers for which all the sums $a_i + a_j$ are distinct. I proved that for infinitely many n

$$a_n > cn^2 \log n \tag{15}$$

and Ajtai, Komlós and Szemerédi constructed such a sequence for which all n

$$a_n < cn^3/\log n . \tag{16}$$

Probably both (15) and (16) are very far from the truth. Finally, here is an attractive old conjecture of mine: Let $a_1 < a_2 < ...$ be an infinite sequence for which all the triple sums $a_i + a_j + a_l$ are distinct. Is it then true that

$$\overline{\lim} \ a_n/n^3 = \infty \ ? \tag{17}$$

I offer 500 dollars for a proof or disproof of (17). For older literature see the excellent book of Halberstam and Roth, Sequences, Springer Verlag 1983.

Finally let $1 \le a_1 < a_2 < ... < a_t \le n, t = [n^{1/2}]$. Denote by $f(n)$ the largest number of integers $m \le n$ which can be written in the form $a_i + a_j$. Determine or estimate $f(n)$ as accurately as possible. Determine $\lim f(n)/n = i$. Freud and I showed $\frac{1}{4} < i < \frac{1}{2}$.

REFERENCES

[1]   B. Bullobás, Extremal graph theory, Academic Press London (1978).

[2]   P. Erdös, A. Hajnal, V. T. Sós and E. Szemerédi, More results on Ramsey Turán type problems, *Combinatorics* (1983) 69-82.

[3]   P. Erdös, A. Hajnal, V. T. Sós and E. Szemerédi, Turán-Ramsey theorems and simple asymptotically extremal structures, (it is written but not yet submitted).

[4]   P. Erdös and E. Szemerédi, On sums and products of integers, *Studies in Pure Mathematics,* to the memory of Paul Turán, 213–218.

[5]   G.A. Freiman, Foundations of structural theory of the set addition, *Translations of Math.* Monographs, Amer. Math. Soc. **37**, Providence, RI, (1973).

[6]   M. Simonovits, Extremal graph theory, Selected topics in graph theory (ed. by Reineke and Wilson) Academic Press London, New York, San Francisco. pp. 109-200.

Odd Cycles in Graphs of Given Minimum Degree

P. Erdös[1]

R. J. Faudree[2]

A. Gyárfás[3]

R. H. Schelp
Memphis State University

ABSTRACT

The principal result of the paper is that any nonbipartite 2–connected graph on n vertices of minimum degree $\geq 2n/(k + 2)$ (k a fixed odd integer and n large) contains a k–cycle or is isomorphic to the following graph H. The graph H has n vertices (with n divisible by k + 2) and is obtained from the k + 2– cycle by replacing each of its k + 2 vertices by an independent set of order $n/(k + 2)$.

1. Results

Let G be a nonbipartite graph of order n and minimum degree δ. A natural extremal question is the following: What is the smallest value of δ such that G contains a cycle C_k of fixed, odd length k?

To address this question first consider two special n–vertex graphs H and L. Let $G_i = G_i(A_i, B_i)$, $1 \leq i \leq 3$, be three vertex disjoint copies of the complete bipartite graph $K_{\lceil (n-3)/6 \rceil, \lceil (n-3)/6 \rceil}$ where A_i and B_i denote the partite sets of G_i. Take a triangle C_3 with vertices a_1, a_2, a_3 (the vertices of C_3 disjoint from each G_i) and for each i join vertex a_i completely to the set of vertices A_i in G_i. Let L denote the graph that results. To ensure that L is an n–vertex graph one can assume that $n - 3$ is a multiple of 6. For n divisible by k + 2 let H be the n–vertex graph obtained from a C_{k+2} by replacing each of its k + 2 vertices by an independent set of order $n/(k + 2)$.

---

1   The Hungarian Academy of Sciences
2   Research partially supported under ONR grant no. N00014-88-K-0070
3   On leave from the Computer and Automation Institute of the Hungarian Academy of Sciences

Since L has C_3 as its only odd cycle and has minimum degree $\lceil (n-3)/6 \rceil >$ $2n/(k+2)$ for $k \geq 11$ and n large, minimum degree $\delta = 2n/(k+2)$ is not sufficient to guarantee the existence of a C_k in G. If one insists that G be 2–connected as well as nonbipartite, then H shows $\delta > 2n/(k+2)$. In fact under these conditions G contains a C_k when $\delta \geq 2n/(k+2)$ unless G is isomorphic to H. This is the content of the principal result in the paper and is stated as the first theorem.

Theorem 1   *Let $k \geq 3$ be a fixed odd positive integer. If G is a 2–connected nonbipartite graph on n vertices of minimum degree $\geq 2n/(k+2)$, then for n large $(n \geq f(k))$ either G contains the k–cycle C_k or is isomorphic to H.*

What happens if G is regular? This has particular meaning when n is odd, since then the graph G must be nonbipartite. Also in this case the 2–connected condition can be dropped as is seen in the next theorem.

Theorem 2   *Let $k \geq 3$ be a fixed odd positive integer. If G is $2n/(k+2)$–regular on n vertices with n odd, then for n large $(n \geq g(k))G$ contains a C_k or is isomorphic to H.*

As noted earlier the 2–connectedness of G assumed in Theorem 1 is essential, at least for $k \geq 11$. What happens if $3 \leq k \leq 9$? Theorem 3 shows for these cases that the 2–connectedness can be dropped.

Theorem 3   *Let G be a nonbipartite graph on n vertices of minimum degree $\geq 2n/(k+2)$, where k is a fixed odd integer. For n large $(n \geq l(k))G$ either is isomorphic to H or contains a C_k for $k \in \{3, 5, 7, 9\}$, but may fail to contain a C_k and also not be isomorphic to H when $k \geq 11$.*

Clearly if the conditions of Theorem 1 hold, then for n large G contains all C_{2t+1} for $k < 2t + 1 \leq d$, d any fixed number larger than k. Also if $G' = K_{2n/(k+2), n-(2n/(k+2))}$ and G is obtained from G' by adding an edge to its smaller part, then G contains C_{2t+1} for all $3 \leq 2t + 1 \leq 4n/(k+2) - 1$ and no larger odd cycle. Thus it is reasonable to inquire whether the conditions of Theorem 1 guarantee all odd cycles C_{2t+1} for $k < 2t + 1 \leq 4n/(k+2) - 1$. This is partially answered in Theorem 4.

Theorem 4   *Let G be a 2–connected nonbipartite graph of order n and minimum degree $\geq cn, 0 < c < \frac{1}{3}$. For each $\varepsilon, 1 > \varepsilon > 0$, there exist functions $h_1(c,\varepsilon)$ and $h_2(c,\varepsilon)$ such that for large $n(n \geq h_2(c,\varepsilon))G$ contains the cycle C_{2t+1} for $h_1(c,\varepsilon) \leq 2t + 1 \leq 4(1 - \varepsilon)cn/3.$*

From the discussion given above, letting $c = 2/(k + 2)$, it is clear that the upper bound in the last result can at most be improved to $2(1 - \varepsilon)cn$. Although this is most likely true the present proof does not seem to work beyond the bound given in the theorem.

It is easy to give examples which show that the degree condition of Theorem 1 cannot be replaced by a reasonable edge condition, even if the graph G contains a very small odd cycle. For example take a cycle C_{k+2}, and assume k is not too small, say k is fixed and odd with $k \geq 15$. Mark four consecutive vertices of the cycle x_1, x_2, x_3, x_4. Join a new vertex x to precisely x_1 and x_2, and replace each of x_3 and x_4 by an independent set of order $(n - k - 1)/2$. The resulting graph has n vertices, approximately $n^2/4$ edges, a C_3, and no odd cycles strictly between 3 and $k + 2$.

Throughout the paper notation will, unless otherwise specified, follow that found in standard texts. Before giving the proofs of the above results two well known extremal theorems are stated. These two theorems are used frequently in the proofs that follow.

Theorem A [1] (Erdös, Gallai) *A graph G on n vertices with at least $[n(k - 2)+1]/2$ edges contains a path P_k on k vertices. Furthermore, when $n = (k - 1)t$ the graph tK_{k-1} (the union of t vertex disjoint copies of K_{k-1}) contains the maximum number of edges in an n vertex graph with no P_k and is the unique such graph.*

The second extremal result deals with the well known problem of Zarankiewicz. Let $Z(n;t)$ denote the maximal size of a bipartite graph $G(n,n)$ having both parts with n vertices such that $G(n,n)$ contains no $K_{t,t}$. The bound given in the theorem below is an improvement by Znám [3] of the bound proved by Kővári, Sós, and Turán [2].

Theorem B [3] (Problem of Zarankiewicz)
(1) If $2 \leq t < n$, then $Z(n;t) < (t-1)^{1/t}n^{2-1/t}+ (t - 1)n/2$.
(2) If a graph of order n does not contain a $K_{t,t}$ then its size is at most
$$((t - 1)^{1/t}n^{2-1/t} + (t - 1)/2^n)/2.$$

2. Proofs
Before proving the main theorem several lemmas and propositions are needed.

Lemma 1 *Let C_t be a cycle of odd length in a graph G. If some vertex x not on the cycle C_t is adjacent to at least five vertices of C_t, then G contains a cycle C_p of odd length for some $p, t/5 \leq p < t$.*

Proof Let $C_t = (x_1,x_2,...,x_t,x_1)$ and assume x is adjacent to $x_{j_i}, 1 \leq i \leq 5$, where $j_{i-1} < j_i$ for $2 \leq i \leq 5$. Set $j_i - j_{i-1} - 1 = a_{i-1}$ for $2 \leq i \leq 5$ and $t + j_1 - j_5 - 1 = a_5$.

Therefore a_{i-1} counts the number of vertices which are strictly between $x_{j_{i-1}}$ and x_{j_i} along the cycle from $x_{j_{i-1}}$ to x_{j_i} $(2 \leq i \leq 5)$ and a_5 counts the number strictly between x_{j_5} and x_{j_1}. Without loss of generality assume $a_1 = \max_{1 \leq i \leq 5}\{a_i\}$ so that $a_1 \geq (t-5)/5$.

Consider the following possibilities: (1) a_1 is even, (2) a_1 is odd and a_2 or a_5 is even, (3) a_1, a_2, a_5 are all odd and exactly one of a_3 or a_4 is even. Note that since

$$t = \sum_{i=1}^{5} a_i + 5$$

and t is odd, one of the three possibilities occur. If (1) occurs let $C_p = (x, x_{j_1}, x_{j_1+1}, \ldots, x_{j_2}, x)$. If (2) occurs assume without loss of generality that a_2 is even and let $C_p = (x, x_{j_1}, x_{j_1+1}, \ldots, x_{j_3}, x)$. If (3) occurs assume without loss of generality that a_4 is even and let $C_p = (x, x_{j_2}, x_{j_2-1}, \ldots, x_{j_1}, x_{j_1-1}, \ldots, x_{j_5}, x_{j_5-1}, \ldots, x_{j_4}, x)$. It is easy to see that C_p as defined is such that p is odd with $t/5 \leq p < t$. $\square$

In each of the remaining lemmas and propositions that preceeds the proof of the main theorem, similar assumptions are needed. Thus the following conditions are assumed through the proof of Theorem 1. The graph G is of order n and minimal degree $\geq 2n/(k+2)$ where k is a fixed odd integer ≥ 3. Also G contains an odd length cycle $C_l, l > k$, but contains no C_{l-2}. Let $C_l = (x_1, x_2, \ldots, x_l, x_1)$, and for $1 \leq i < j \leq l$, let $A_{ij} = \{v \in V(G) - V(C_l) \mid v$ is adjacent to both x_i and $x_j\}$, $X_i = \{v \in A_{i-1,i+1} \mid v$ has precisely two adjacencies to $C_l\}$, and $Y_i = \{v \in V(G) - V(C_l) \mid v$ is adjacent to precisely x_i on $C_l\}$. For A and B disjoint subsets of $V(G)$, $[A, B]$ will denote the bipartite subgraph of G with parts A and B that contains all edges of G between A and B. Finally assume $l \leq 5k$ for each of the lemmas and propositions in this section (but not in the proof of the theorem).

Lemma 2   *Let $h(n)$ be any unbounded nonnegative function such that $\lim h(n)/n \to 0$. For all $1 \leq i, j \leq l, i \neq j, |i - j| \neq 2$, and n sufficiently large $|A_{ij}| \leq h(n)$ so that $|A_{ij}| = o(n)$.*

Proof   By assumption G contains a C_l but no C_{l-2}, l is odd, and $d_{G-C_l}(x) \geq 2n/(k+2) - l$ for all $x \in V(G) - V(C_l)$. Partition the vertices of A_{ij} into sets B_{ij} and C_{ij} such that $d_{A_{ij}}(x) \geq n/(k+2) - l/2$ for $x \in B_{ij}$ and $C_{ij} = A_{ij} - B_{ij}$. Note that $d_{[A_{ij}, V(G)-(A_{ij} \cup V(C_l))]}(x) \geq n/(k+2) - l/2$ for $x \in C_{ij}$. Suppose $|A_{ij}| > h(n)$. It will be shown that this supposition leads to a contradiction. Two cases are considered.

Case 1: $|B_{ij}| > h(n)/2$. Let $m - 1$ be the distance from x_i to x_j along the cycle C_l. It will be shown for n large that A_{ij} contains a path on $l - m - 2$ vertices. Connecting the end vertices of this path in A_{ij} by disjoint edges to x_i and x_j gives the m vertex path on C_l a cycle C_{l-2}, a contradiction. Thus the proof for this case is completed by showing that A_{ij} contains a path on $5k - m - 2 \geq l - m - 2$ vertices. Note that $l - m - 2 > 0$, since $l \geq k + 2$ implies $l - m - 2 \geq l - ((l+1)/2) - 2 \geq (l-5)/2 \geq (k-3)/2 > 0$. The last inequality $(k-3)/2 > 0$ follows since $k = 3$, $l = 5$, and $m = (l+1)/2 = 3$ means $|i - j| = 2$, contrary to the hypothesis of the lemma. But by definition of B_{ij} each of its vertices are adjacent to at least $n/(k+2) - l/2$ vertices of A_{ij} so that $|E(< A_{ij} >)| > h(n)(n/(k+2) - l/2)/4 \geq n(5k - m - 4)/2 \geq |A_{ij}|(5k - m - 4)/2$ for n sufficiently large. Hence by Erdös–Gallai (Theorem A), A_{ij} contains a path P_{5k-m-2} on $5k - m - 2$ vertices.

Case 2: $|C_{ij}| > h(n)/2$. Observe that the number of vertices on the two paths from x_i to x_j on C_l have opposite parities. Thus choose the one with an even number, say m, vertices. Since $|i - j| \neq 2$, $m \leq l - 3$ which implies $l - m - 2 > 0$. This time an even length path on $l - m - 2$ vertices is found in $[C_{ij}, V(G) - (A_{ij} \cup V(C_l))]$ with its end vertices in C_{ij}. This path on $l - m - 2$ vertices has its end vertices joined by disjoint edges to x_i and x_j so that a C_{l-2} results (using the m vertex path from x_i to x_j on C_l), a contradiction.

The proof is thus completed by showing $[C_{ij}, V(G) - (A_{ij} \cup V(C_l))]$ contains all odd length paths of length at most $5k - m - 2$ with end vertices in $[C_{ij}, V(G) - (A_{ij} \cup V(C_l))]$, i.e. contains a path of length $5k - m - 1$. But $|E([C_{ij}, V(G) - (A_{ij} \cup V(C_l))])| > h(n)(n/(k+2) - l/2)/2 \geq n(5k - m - 3)/2$ for n sufficiently large. Hence by Erdös-Gallai $[C_{ij}, V(G) - (A_{ij} \cup V(C_l))]$ contains the desired length path, so G contains a C_{l-2}. $\quad\square$

Lemma 3 *For all $1 \leq i \leq l$ both $n/(k+2) - o(n) \leq |X_i| \leq n/(k+2) + o(n)$ and $|Y_i| \leq o(n)$.*

Proof By Lemma 2 $|A_{ij}| = o(n)$ for all $i \neq j$, $|i - j| \neq 2$ so that each vertex x_i on C_l is adjacent to at least $2n/(k+2) - o(n)$ vertices of $X_{i-1} \cup X_{i+1} \cup Y_i$, i.e. for each $1 \leq i \leq l$

(1) $|X_{i-1}| + |X_{i+1}| + |Y_i| \geq 2n/(k+2) - o(n)$. This gives

$$2\left|\bigcup_{i=1}^{l} X_i\right| + \left|\bigcup_{i=1}^{l} Y_i\right| \geq l(2n/(k+2)) - o(n) \geq 2n - o(n), \text{ since } l \geq k + 2.$$

Then $\left|\bigcup_{i=1}^{l} X_i\right| + \frac{1}{2}\left|\bigcup_{i=1}^{l} Y_i\right| \geq n - o(n)$, while $\left|\bigcup_{i=1}^{l} X_i\right| + \left|\bigcup_{i=1}^{l} Y_i\right| \leq n$.

Hence

(2) $|Y_i| \leq \left|\bigcup_{i=1}^{l} Y_i\right| = o(n)$ and $\left|\bigcup_{i=1}^{l} X_i\right| = n - o(n)$.

Suppose, for some fixed $\varepsilon > 0$ and n large, that there exists an i such that $|X_i| \geq n/(k+2) + \varepsilon n$. Then $|X_{i+2}| \leq n/(k+2) - \varepsilon'n(0 < \varepsilon' < \varepsilon)$, otherwise $|X_i| + |X_{i+2}| \geq 2n/(k+2) + (\varepsilon - \varepsilon')n$, contrary to (1) and (2). Likewise $|X_{i+2}| \leq n/(k+2) - \varepsilon'n$ implies $|X_{i+4}| \geq n/(k+2) + \varepsilon''n$ for some $0 < \varepsilon'' < \varepsilon'$. Hence if for n large $|X_i| \geq n/(k+2) + \varepsilon n$, then there exists a δ, $0 < \delta < \varepsilon$, such that $|X_{i+2j}| \geq n/(k+2) + \delta n$ for $j = 0,2,4,...,2l - 2$ and $|X_{i+2j}| \leq n/(k+2) - \delta n$ for $j = 1,3,5,...,2l - 1$, where all indices $i + 2j$ are taken modulo l. Since these inequalities are incompatible, it follows that $|X_i| \leq n/(k+2) + o(n)$ for all $1 \leq i \leq l$. Applying (1) gives $|X_{i+2}| \geq n/(k+2) - o(n)$ for all $1 \leq i \leq n$. $\square$

Lemma 4 *For n sufficiently large $(n \geq f_2(k))$ each of the bipartite graphs $[X_i, X_{i+1}], 1 \leq i \leq l$, contain a path P_l on l vertices.*

Proof First observe, using Lemma 3, that $d_{X_i}(x) \leq n/(k+2) + o(n)$ for each $x \in X_i$. Therefore since $l \leq 5k$ and

$$\left| \bigcup_{j=1}^{l} X_i \right| = n - o(n), \text{ the degree } d_{\bigcup_{\substack{j \neq i-1}} X_j}(x) \geq n/(k+2) - o(n) \text{ for all } x \in X_i.$$

Partition X_i into two parts, Z_i and $X_i - Z_i$, where Z_i are those vertices of X_i adjacent to at least $n/(2(k+2))$ vertices of X_{i+1}.

Let G' denote the graph $[X_i, X_j]$ when $j \neq i, i-1, i+1$, and let it denote the graph induced by X_i when $j = i$. Suppose $|X_i - Z_i| \geq |X_i|/2 \geq n/(2(k+2)) - o(n)$. By the pigeonhole principle at least $|X_i|/(2(l-2)) \geq n/(2(l-2)(k+2)) - o(n)$ vertices in $X_i - Z_i$ are adjacent (for some $j \neq i-1, i+1$) to at least $n/(2(l-2)(k+2)) - o(n)$ vertices of X_j. Therefore G' contains at least $n^2/(8(l-2)^2(k+2)^2) - o(n)$ edges. By Erdös-Gallai G' contains, for n large, a path of any fixed length. This means when $i = j$ that vertex x_{i+1} (or x_{i-1}) and a path on $l - 3 \leq 5k - 3$ vertices in G' gives a C_{l-2}, a contradiction. Also if $i \neq j$, then let the even length path from x_i to x_j on C_l contain m vertices. Since $j \neq i-1, i+1, m < l$, the m vertex path from x_i to x_j on C_l can be joined to a path in G' on $l - 2 - m$ vertices with a pair of disjoint edges, one from x_i and another from x_j. But this again gives a C_{l-2}, a contradiction. Hence the supposition that $|X_i - Z_i| \geq |X_i|/2$ is false and $|Z_i| \geq |X_i|/2$.

Since $|Z_i| \geq |X_i|/2$ and each vertex of Z_i is adjacent to at least $n/(2(k+2))$ vertices of X_{i+1}, the graph $[X_i, X_{i+1}]$ contains at least $(n/(2(k+2)) - o(n))(n/(2(k+2)))$ edges. Thus Erdös-Gallai again applies and $[X_i, X_{i+1}]$ contains a P_l for n sufficiently large.

It is easy to check that in all of the usages above (and also those of Lemmas 2 and 3) $o(n)$ depends only on n and k. Thus n sufficiently large, used throughout this proof, means there is an $f_2(k)$ such that $n \geq f_2(k)$. $\square$

Proposition 1 *For n sufficiently large $(n \geq f_2(k))$ the cycle C_l has no diagonals.*

Proof If C_l has a diagonal, then G contain an odd length cycle $C_t, t < l$, such that all but one of the edges of C_t are also edges of C_l. Choose any edge $x_i x_{i+1}$ common to C_t and C_l. By Lemma 4 $[X_i, X_{i+1}]$ contains an even length path on $l - t - 2$ vertices. Join the vertices x_i and x_{i+1} to the appropriate end vertices of this $l - t - 2$ vertex path. But then the C_t cycle can be expanded to a C_{l-2} by replacing edge $x_i x_{i+1}$ by the $l - t - 2$ vertex path, a contradiction. $\square$

Proposition 2 *For n sufficiently large $(n \geq f_2(k))$ each vertex x of G not on the cycle C_l has at most two adjacencies to vertices of the cycle.*

Proof Suppose there exists an $x \in V(G) - V(C_l)$ which is adjacent to at least three vertices of C_l. Then it is clear that G contains an odd length cycle $C_t, t < l$, such that C_t has at least $t - 2 \geq 1$ edges in common with C_l. Let $x_i x_{i+1}$ be any common edge. In the same way as was done in the last proof, Lemma 4 implies the existence of an even length path on $l - t - 2$ vertices in $[X_i, X_{i+1}]$ which can replace edge $x_i x_{i+1}$. This gives a C_{l-2}, a contradiction. $\square$

Proof of Theorem 1

Since G is nonbipartite let C_l be an odd length cycle in the graph. It will be shown that $l \geq k$. Suppose $l < k$ and consider the graph G_A induced by A, where $A = V(G) - V(C_l)$. This graph G_A has at least $(n - l)(\delta(G) - l)/2 \geq (n - l)(n/(k + 2) - l/2)$ edges. By Theorem B there exists a $f_1(k)$ such that for $n \geq f_1(k)$ G_A contains the complete bipartite graph $K_{\lceil \frac{k}{2} \rceil, \lceil \frac{k}{2} \rceil}$ Since G is 2-connected, by Menger's Theorem there exist two vertex disjoint paths P_x and P_y connecting C_l to the graph $K_{\lceil \frac{k}{2} \rceil, \lceil \frac{k}{2} \rceil}$. Let $x(y)$ be the vertex common to the path P_x and $C_l (P_y$ and $C_l)$. Note that x and y are joined by two paths on C_l, one with an even number of vertices and the other with an odd number of vertices. Thus using one of these two paths it is easy to see that G contains an odd length cycle C_t using all vertices of $P_x \cup P_y$ and all but at most one of the vertices of $K_{\lceil \frac{k}{2} \rceil, \lceil \frac{k}{2} \rceil}$. Then $t \geq k + 1$ so G contains a C_l for some odd integer $l, l \geq k$.

For the remainder of this proof let l be the smallest odd positive integer $l \geq k$, such that G contains a C_l. Since it is to be shown that $l = k$ or $G \cong H$, assume throughout that $l \geq k + 2$.

Suppose $l > 5k$. Since l is the length of the smallest odd cycle $\geq k$, each vertex x on C_l is adjacent to at most $2(k - 2)$ vertices of the cycle, the $2(k - 2)$ vertices closest to x. Thus each vertex of the cycle is adjacent to at least $2n/(k + 2) - (2k - 4)$ vertices of $G - C_l$. By Lemma 1 no vertex of $G - C_l$ has as many as five adjacencies to C_l. Hence there are at least $l/(2n/(k + 2) - 2k - 4)$ edges from C_l to $G - C_l$ and there are at most $4(n - l)$ from $G - C_l$. It follows that $4n \geq l(2n/(k + 2) - 2k + 8) \geq 5k(2n/(k + 2) - 2k + 8)$, which leads to a contradiction for n large ($n \geq f_3(k)$). Thus we may assume $l \leq 5k$, giving $k + 2 \leq l \leq 5k$.

The reader can check that G now satisfies all of the conditions assumed uniformly throughout the proofs of Lemmas 2, 3, 4 and Propositions 1 and 2. First apply Proposition 1 to C_l. For n sufficiently large each vertex of C_l has at least $2n/(k + 2) - 2$ adjacencies to vertices of $G - C_l$. Also by Proposition 2 for n sufficiently large each vertex of $G - C_l$ has at most two adjacencies to C_l. Therefore $l(2n/(k + 2) - 2) \leq 2(n - l)$. Since $l \geq k + 2$, a contradiction occurs unless $l = k + 2$ and $k + 2$ divides n. Assume $l = k + 2$ so that each vertex of C_l has precisely $2n/(k + 2) - 2$ adjacencies to vertices of $G - C_l$, and each vertex of $G - C_l$ has precisely two adjacencies to C_l. It is shown under these conditions that $G \cong H$.

Consider $G - C_l$. Note that $x \in V(G) - V(C_l)$ implies
$$x \in \bigcup_{j=1}^{l} X_j.$$

If this were not the case, then x has adjacencies x_m and x_j on C_l, where $|m - j| \neq 2$. This gives a C_t, t odd ($t < l$), using edges xx_m, xx_j and the appropriate $t - 2$ edges of the C_l. But as was done in the proofs of Propositions 1 and 2, some edge $x_i x_{i+1}$ common to C_t and C_l can be replaced by a path with an appropriate number of vertices from $[X_i, X_{i+1}]$ together with two edges joining its ends to x_i and x_{i+1}, respectively, to give a C_{l-2}, a contradiction. Thus
$$V(G) - V(C_l) = \bigcup_{j=1}^{l} X_j.$$

Next for each i consider $d_{G-C_l}(x_i) = 2n/(k + 2) - 2 = |X_{i-1}| + |X_{i+3}|$. Thus $|X_{i-1}| = |X_{i+3}|$ for each $i, 1 \leq i \leq l$ (addition modulo l). But l is odd, so that this implies $|X_i| = |X_{i+1}|$ for each i and thus $|X_i| = n/(k + 2) - 1$.

Finally observe that no vertex of X_i is adjacent to a vertex of X_j for $j \neq i - 1, i + 1$. If this were not the case, an odd cycle $C_t, t < l$, is again obtained which can be lengthened to a C_{l-2} by Lemma 4, a contradiction. Thus all adjacencies of X_i are to vertices of $X_{i-1} \cup X_{i+1} \cup \{x_{i-1}, x_{i+1}\}$. Since $|X_{i-1} \cup X_{i+1} \cup \{x_{i-1}, x_{i+1}\}| =$

$2n/(k + 2)$ and $d(x) \geq 2n/(k + 2)$ for $x \in X_i$, it follows that each $x \in X_i$ is adjacent to precisely $X_{i-1} \cup X_{i+1} \cup \{x_{i-1}, x_{i+1}\}$ for all $1 \leq i \leq l$. This gives $G \cong H$.

Throughout the proof there are places where n must be larger than each of $f_1(k), f_2(k)$, and $f_3(k)$. Thus setting $f(k) = \max_{i=1,3}\{f_i(k)\}$ the theorem holds for all $n \geq f(k)$.

Proof of Theorem 2

Since n is odd and G is regular, G is nonbipartite. The reader will recall that in the proof of Theorem 1 the 2-connectedness is only used to show that G contains an odd cycle C_l for some $l \geq k$. Thus this proof becomes a corollary to the proof of Theorem 1 once it is established that G contains a C_l for some odd integer $l, l \geq k$.

Without loss of generality assume G is connected, otherwise one can simply restrict attention to a component of G. Further if G is not 2-connected, consider an end block B of G.

Since B is an end block, it has at most one cut vertex x. Consider an internal vertex y of B (a non-cut-vertex). Since $d(y) \geq 2n/(k + 2)$, $d_B(y) \geq 2n/(k + 2)$ so that the block B has at least $2n/(k + 2) + 1$ vertices.

Suppose B fails to contain an odd cycle. Let B be bipartite with partite sets R and S and with cut vertex x in S. Let $m = d_{G-B}(x)$ and note $1 \leq m \leq 2n/(k + 2) - 2$. Set $r = |R|$ and $s = |S|$. Counting edges from R to S and then from S to R, it follows that $|E(B)| = r(2n/(k + 2)) = s(2n/(k + 2)) - m$. Hence $m = (s - r)(2n/(k + 2))$ so that $2n/(k + 2)$ divides m, a contradiction. Therefore B contains some odd length cycle C.

Assume $|C| < k$. Now $d_B(y) \geq 2n/(k + 2)$ for all y in $B - \{x\}$ so that there exists $\delta > 0$ such that for n large enough B has at least $\delta |B|^2$ edges. Hence for n large it follows from Theorem B that $B - C$ contains a $K_{\lceil \frac{k}{2} \rceil, \lceil \frac{k}{2} \rceil}$. But then since B is 2-connected, by Menger's theorem there exist two vertex disjoint paths P_x and P_y connecting C to the $K_{\lceil \frac{k}{2} \rceil, \lceil \frac{k}{2} \rceil}$. In the same fashion as argued in the proof of Theorem 1, the graph $C \cup P_x \cup P_y \cup K_{\lceil \frac{k}{2} \rceil, \lceil \frac{k}{2} \rceil}$ contains an odd cycle C_l with $l \geq k + 1$. Hence G contains the desired odd cycle, completing the proof. $\square$

Proof of Theorem 3

The graph L given in the first section (prior to the statement of the theorem) shows that G can be different from H and does not need to contain a C_k, k odd, when $k \geq 11$. Thus in what follows k always has one of the values 3, 5, 7, or 9.

As in the proof of Theorem 2 it is assumed that G is connected. If any block B of G contains a non-cut-vertex, then $|B| \geq 2n/(k+2)+1$ and the same type of argument as given in the proof of Theorem 2 applies. This argument implies, when B contains an odd length cycle and n is large, that it contains an odd length cycle C_l for some $l \geq k$. If B (and hence G) contains any such odd length cycle, then the proof of Theorem 1 shows for n large that G contains the desired cycle C_k or is isomorphic to H. Hence assume for the remainder of this proof that each block B of G which contains an odd cycle has no non-cut-vertices. But G is nonbipartite so some block B contains an odd cycle with each of its vertices cut-vertices. This means B is in fact an odd cycle. It will be shown that this assumption leads to a contradiction.

Thus assume G has block $B = C_l = (x_1, x_2, \dots, x_l, x_1)$, l odd, $l < k$, in which each $x_i (1 \leq i \leq l)$ is a cut-vertex common to both B and some connected subgraph G_i of G. Note that G_i is only assumed connected and is not in general itself a block. Since $k \in \{3, 5, 7, 9\}$ there are several possible values for l all of which are handled similarly. Hence in what follows only the case when $k = 9$ and $l = 3$ will be considered. The reader can check that the remaining possible values for k and l are handled in the same way as this special case.

As described above $B = C_3 = (x_1, x_2, x_3, x_1)$ and each $G_i (1 \leq i \leq 3)$ is a connected graph with x_i a cut-vertex and the only vertex in common to both C_3 and G_i. There are two ways in which a lower bound on $|G_i|$ is determined. First assume G_i is bipartite with partite sets R_i and S_i and with $x_i \in S_i$. Since $d(y) \geq 2n/11$ for all y in G, $d_{G_i}(y) \geq 2n/11$ for y in $G_i - \{x_i\}$ and $d_{G_i}(x_i) \geq 2n/11 - 2$. Thus $|S_i| \geq 2n/11$, $|R_i| \geq 2n/11 - 2$ so that $|G_i| \geq 4n/11 - 2$ for all i where G_i is bipartite. Secondly if G_i contains an odd cycle, let $C = (y_1, y_2, \dots, y_l, y_1)$ be one of longest odd length. Note that $d_{G_i}(y_i) \geq 2n/11$ for $y_i \neq x_i$ and $d_{G_i}(x_i) \geq 2n/11 - 2$. It is easy to see (since G_i contains no $l+2$ cycle) that $|G_i| \geq 2n/11 + (2n/11 - 2 - l)$ for each G_i that contains an odd cycle. Hence in all cases $|G| \geq |G_1| + |G_2| + |G_3| > n$ for n large, a contradiction.

It has been shown that no block can be an odd cycle. Since G is nonbipartite, the block that contains an odd cycle has a non-cut-vertex and thus an odd cycle C_l for some $l \geq k$ or G is isomorphic to H. This completes the proof of the theorem. ❑

The proof of the final result (Theorem 4) will not be given in great detail. The reasons are that a detailed proof would be very lengthy and that the result itself can most likely be improved. Thus simply a sketch of the ideas of the proof will be given.

Proof of Theorem 4 (A Sketch)

Since the graph G is nonbipartite let C_l be a shortest odd cycle in G and consider $G - C_l$. It is possible to show that $G - C_l$ contains a subgraph H_1 of order at least $2(1 - \varepsilon)cn$ or a pair of vertex disjoint subgraphs H_1 and H_2, each of order less than $2(1 - \varepsilon)cn$, such that the following conditions hold. Each H_i is δn-connected ($\delta = \delta(c,\varepsilon)$) and each H_i has minimum degree $\geq (1 - \varepsilon)cn$.

If $G - C_l$ contains the pair of vertex disjoint subgraphs H_1 and H_2, then connect H_1 and H_2 by a pair of vertex disjoint paths with end vertices x_1, x_2 and y_1, y_2, with $x_1, x_2 \in V(H_1), y_1, y_2 \in V(H_2)$. These vertex disjoint paths exist, since G is 2-connected, and each can be assumed to each have length independent of n, since the minimum degree in G is $\geq cn$. But $d_{H_i}(x) \geq (1 - \varepsilon)cn$ and $|H_i| < 2(1 - \varepsilon)cn$ so that both H_i's are panconnected. Hence both x_1, x_2 in H_1 and y_1, y_2 in H_2 are connected by paths of all possible lengths greater than are equal to the distance between them in H_i. This shows that G contains all cycles of length at least $h_1(c,\varepsilon)$ and at most $2(1 - \varepsilon)cn$.

Thus the only case that remains is when $G - C_l$ contains the subgraph H_1 of order at least $2(1 - \varepsilon)cn$ described above. It can be shown by proper application of Theorem B that H_1 contains a subgraph H or order at least $4(1 - \varepsilon)cn/3$ which is the disjoint union of copies of complete bipartite graphs. Each of these bipartite graphs is a $K_{t,t}$ for some $t \geq c'lnn$ where $c' = c'(\varepsilon,c)$. The idea is to order the $K_{t,t}$s of H such that consecutive pairs can be joined by two vertex disjoint paths. Also C_l is joined by two vertex disjoint paths to the first $K_{t,t}$ in H under the given order. Further all the paths linking the C_l to a $K_{t,t}$ and consecutive pairs of $K_{t,t}$s are to be such that they are vertex disjoint. This linking is possible by using the δn-connectivity of H and the fact that the linking paths are each of bounded length independent of n.

Finally the linking is done so that at least some $K_{\alpha\sqrt{lnn},\alpha\sqrt{lnn}}$ in each $K_{t,t}$, α fixed and small, has none of its vertices in a linking path (they are protected from the linking paths). Observe that there are at most $(4(1 - \varepsilon)cn)/(6c'lnn)$ complete bipartite graphs $K_{t,t}$ in H, so that there are at most $2(4(1 - \varepsilon)cn)/(6c'lnn)$ vertices in the linking paths and there are at most $(4(1 - \varepsilon)cn/(6c'lnn))(2\sqrt{lnn})$ protected vertices in the $K_{t,t}$s. But for n large $2(4(1 - \varepsilon)cn)/(6c'lnn) + (4(1 - \varepsilon)cn)(2\sqrt{lnn})/(6c'lnn) < \delta n$, the connectivity of H, so that the linking described is possible. It is now a matter of checking that the linkage described from the C_l through all the $K_{t,t}$s in H give odd cycles of all lengths from $h_1(c,\varepsilon)$ to $4(1 - \varepsilon)n/3$. $\quad\square$

REFERENCES

[1] P. Erdös and T. Gallai, On maximal paths and circuits of graphs, *Acta Math. Acad. Sci. Hungr.* 10 (1959), 337-356.

[2] P. Kövári, V. Sós, and P. Turán, On a problem of K. Zarankiewicz, *Colloq. Math* 3(1954), 50-57.

[3] Š. Znám, On a combinatorial problem of K. Zarankiewicz, *Colloq. Math* 11 (1963), 81-84.

Edge Conditions for the Existence of Minimal Degree Subgraphs

Paul Erdös

Hungarian Academy Science

R.J. Faudree[1]

Memphis State University

C.C. Rousseau

Memphis State University

R.H. Schelp[2]

Memphis State University

ABSTRACT

For $m > n$ and k positive integers, $f(m, n, k)$ denotes the smallest positive integer such that any graph of order m and size $f(m, n, k)$ has a subgraph of order n and minimum degree at least k. The function $f(m, n, k)$ is investigated: upper and lower bounds are give for f, and exact values are determined for f when $m - n$ is small and $k = 2$. Also, for the special case $k = 2$, sharper bounds for $f(m, n, k)$ are proved. Extermal graphs are described when exact results are obtained. Similar results are given for the related function $g(m, n, k)$, which is the smallest positive integer such that any graph of order m, size $g(m, n, k)$, and minimum degree at least k has a subgraph of order n and minimum degree at least k.

[1] This research is partially supported by ONR research grant N000014-88-K-0070.
[2] This research is partially supported by NSF research grant DMS–8603717.

1. Introduction

Turán extremal theory deals with the problem of determining the number of edges in a graph that insures the existence of a fixed subgraph. An excellent survey of such results can be found in [4]. A related Turán type extremal problem of determining the number of edges needed to insure the existence of a (proper) subgraph of minimal degree greater than a fixed bound was considered in [2]. We will give results on the edge extremal problem of determining the number of edges required for the existence of a subgraph of predetermined order and fixed minimal degree.

Notation will generally follow [1], and special notation and terminology will be introduced as needed. For fixed positive integers $k < n$, a finite graph G is in $\mathcal{D}_k(n)$ if G has a subgraph of order n that has minimal degree at least k. The class of graphs in $\mathcal{D}_k(n)$ that have order precisely n will be denoted by $\mathcal{D}_k^*(n)$. If G has order $m \geq n$ and a sufficient number of edges, then $G \in \mathcal{D}_k(n)$. Let $f(m, n, k)$ denote the smallest positive integer such that any graph of order m and size at least $f(m, n, k)$ is in $\mathcal{D}_k(n)$. A graph of order p and size q will be called a (p, q) – graph. Thus, using the given notation, an $(m, f(m, n, k))$ – graph is in $\mathcal{D}_k(n)$. The extremal problem of determining $f(m, n, k)$ will be considered, and the following theorems will be proved.

Theorem 1 *For n sufficiently large,*

$$
f(n + p, n, 2) = \begin{cases}
\dbinom{n-1}{2} + p + 2, & 0 \leq p \leq n - 4, \\[2ex]
\dbinom{n-1}{2} + p + 3, & n - 3 \leq p < \left\lceil \dfrac{3n-6}{2} \right\rceil, \\[2ex]
\dbinom{n-1}{2} + p + 4, & \left\lceil \dfrac{3n-6}{2} \right\rceil \leq p < 2n - 4.
\end{cases}
$$

Also, for $r > 0$, $p = \lfloor rn \rfloor$, and n sufficiently large,

$$
\binom{n-1}{2} + p \leq f(n + p, n, 2) \leq \binom{n-1}{2} + p + 32r^2.
$$

Theorem 2 *For fixed integers $k \geq 2$, $r > 0$ and n sufficiently large,*

$$
\binom{n-1}{2} + (p + 1)(k - 1) < f(n + p, n, k) \leq \binom{n-1}{2} + 2(p + 1)(k - 1),
$$

if $p < n - 2k$, and for $p = rn$,

$$
\binom{n-1}{2} + r(k - 1)n < f(n + p, n, k) \leq \binom{n-1}{2} + \left(\binom{2r+1}{2} + r + 1 \right)(k - 1)n.
$$

If a graph G of order m already has a minimal degree at least k, then the number of edges needed to insure that $G \in \mathcal{D}_k(n)$ may be less than $f(m, n, k)$. Denote by $g(m, n, k)$ the smallest positive integer such that any graph of minimal degree at least k with order m and size at least $g(m, n, k)$ is in $\mathcal{D}_k(n)$. Clearly $g(m, n, k) \le f(m, n, k)$. When m is much larger than n, then the functions f and g are approximately the same. However, the following two theorems verify that this is not true for $k = 2$ when $m - n < n$.

Theorem 3 *For n sufficiently large and $83n/84 < p \le n - 4$,*

$$g(n + p, n, 2) = \binom{p + 1}{2} + 2n - 3.$$

Theorem 4 *For $p > 0$ fixed and n sufficiently large,*

$$g(n + p, n, 2) < n^2 / 8 + p(n + p).$$

2. Graphs with subgraphs of minimal degree two.

In this section the examples and arguments to verify Theorem 1 will be given. The examples that will be used to give lower bounds for the function values $f(n + p, n, 2)$. for $0 \le p < 2n - 4$ all have the same nature. The are obtained from a K_{n-1} by attaching a sparse graph of order $p + 1$. For the different intervals of p we will describe a graph of G of order $n + p$ such that $G \notin \mathcal{D}_2(n)$, but $|E(G)|$ gives the correct lower bound for $f(n + p, n, 2)$. In each case, the verification of the properties of G is straightforward, and will be left to the reader.

Case 1 *For $0 \le p \le n - 4$, $f(n + p, n, 2) \ge \binom{n - 1}{2} + p + 2$.*

The graph G is obtained from a K_{n-1} by attaching a star with $p + 1$ edges that has a vertex of K_{n-1} as the center.

Case 2 *For $n - 3 \le p < \lceil (3n - 6)/2 \rceil$, $f(n + p, n, 2) \ge \binom{n - 1}{2} + p + 3$.*

The graph G is obtained from K_{n-1} by attaching a path with $p + 3$ vertices that has precisely the first and last vertices in common with $K_{n - 1}$. For $p > n - 3$, the end vertices of the path are distinct, but when $p = n - 3$, identify these end vertices to form a $C_{p + 2}$.

Case 3   *For* $\lceil (3n-6)/2 \rceil \leq p < 2n-4$, $f(n+p, n, 2) \geq \binom{n-1}{2} + p + 4$.

The graph G is obtained from K_{n-1} by adding a new vertex v and three vertex disjoint paths from v to three distinct vertices of K_{n-1}, such that the total number of vertices in the paths is $p + 4$ and the length of each pair of these paths differ by at most one.

For small values of p the pattern established by the previous examples and the statement of Theorem 1 indicates that $f(n + p, n, 2)$, considered as a function of p, increases by 1 with each increase of approximately $n/2$ by p. This is not true. For example, consider the graph H obtained from a complete bipartite graph $K_{3,3}$ by subdividing each edge $n/4$ times. Then, H is a $(9n/4 + 6, 9n/4 + 9)$–graph that contains no cycle length less than $n + 4$. The graph G obtained from H by adding a vertex disjoint K_{n-1} and one edge between the K_{n-1} and H is a $(n + p, \binom{n-1}{2} + p + 5)$ for $p = 9n/4 + 5$, and $G \notin \mathcal{D}_2(n)$. Therefore, for $p = 9n/4 + 5$, $f(n + p, n, 2) \geq \binom{n-1}{2} + p + 6$, rather than the value of $\binom{n-1}{2} + p + 5$ indicated by the previous pattern. Similar examples exist for larger values of p as well.

If r is a positive integer and $p = r(n - 1) - 1$, a lower bound for $f(n + p, n, 2)$ can be derived from the graph H that is the vertex disjoint union of a K_{n-1} and r copies of a C_{n-1} with one edge between each copy of a C_{n-1} and the K_{n-1}. The graph $H \notin \mathcal{D}_2(n)$ so $f(n + r(n - 1) - 1, n, 2) \geq \binom{n-1}{2} + rn$. For any number $r > 1$, the previous example can be modified slightly by changing the order of the cycles to get the general inequality $f(n + \lfloor rn \rfloor, n, 2) \geq \binom{n-1}{2} + \lfloor rn \rfloor$.

Before proving Theorem 1, two lemmas will be proved and some additional notation used in the lemmas will be introduced. If H is a subgraph of G, then $G - H$ is the subgraph of G obtained by deleting all the vertices of H. As usual, by $o(n)$ we will mean a function of n such that the $\lim_{n \to \infty} \frac{o(n)}{n} = 0$. The following lemma deals with the general case of $\mathcal{D}_k(n)$ and not just when $k = 2$.

Lemma 5   *For a fixed $r > 1$, and integers $t > k \geq 2$, let G be a $(\lfloor rn \rfloor, \binom{n-1}{2}))$–graph. If n is sufficiently large then either*

(1) $G \in \mathcal{D}_k(n)$, or

(2) there is an integer s such that $sK_{t,t} \subset G$ with $2st = n - o(n)$.

Proof   Suppose that $G \notin \mathcal{D}_k(n)$. Select s maximum such that $sK_{t,t} \subset G$, and let M of order $m = 2st$ be such a maximum subgraph.

We can assume that $m < n$, and since (1) is not true, at most $n - m - 1$ vertices of $G - M$ have degree at least k (and at most m) relative to M. Therefore, there are at most $(k - 1)(\lfloor rn \rfloor - n + 1) + m(n - m - 1)$ edges between M and $G - M$. The maximality of M implies that $G - M$ contains no $K_{t,t}$, so $G - M$ has at most $o((\lfloor rn \rfloor - m)^2) = o(n^2)$ edges [3]. This gives the following inequality:

$$\binom{m}{2} + (k - 1)(\lfloor rn \rfloor - n + 1) + m(n - m - 1) + o(n^2) \geq \binom{n - 1}{2}.$$

It follows immediately from this inequality that

$$m(2n-m) > n^2 - o(n^2) \text{ and so } m \geq n - o(n).$$

This completes the proof of Lemma 5. $\square$

Lemma 6 *Let $r > 0$, $4 \leq t = o(n)$, and $s = o(n)$. If G is a $(\lfloor rn \rfloor + s, \lfloor rn \rfloor + s + 32r^2)$ – graph, and n is sufficiently large, then G contains a C_ℓ for some $\ell \leq n - t$.*

Proof We will suppose that no cycle of length at most $n - t$ exists, and show that this leads to a contradiction. If $r < 1$, the conclusion is obvious, so we will assume $r \geq 1$ for the remainder of the proof. Let C_{k_1} be a cycle of smallest length in G, and sequentially select cycles C_{k_j} of smallest length in $G - C_{k_1} - C_{k_2} - \ldots - C_{k_{j-1}}$, until no more cycles exist. Thus, for some a, there are vertex disjoint cycles $C_{k_1}, \ldots, C_{k_a}$ with $n - t < k_1 \leq k_2 \leq \ldots \leq k_a$. Let $T_{\ell_1}, T_{\ell_2}, \ldots, T_{\ell_c}$ be the trees that form the components of the graph $G - C_{k_1} - C_{k_2} - \ldots - C_{k_a}$. We will assume for $i \geq b$ (b could be 1), that T_{ℓ_i} has at most $n/4 - t$ vertices (these will be called *small trees*), and that the remaining trees (*large trees*) have more than $n/4 - t$ vertices.

By the selection of the cycles and trees, there are no chords in the cycles, and there are no edges between the trees. The edges between the cycles or between cycles and trees will be called *cross* edges, and we will show that there are "few" edges of this type. Three separate cases will be considered: edges between the cycles, edges between a cycle and a "large" tree, and edges between the cycles and all of the "small" trees.

Consider the cycles C_{k_i} and C_{k_j} with $i < j$. If $k_j \leq qk_i$ for some integer q, then there are at most $2q$ edges between the two cycles. If this were not true, then there would be two cross edges between the cycles such that the distance along C_{k_j} between the two end vertices in C_{k_j} would be at most $k_j/(2q + 1)$. Since the distance along C_{k_i} between the other two endvertices in C_{k_i} is at most $k_i/2$, a cycle of length at most $2 + (4q+1)k_i/(4q+2)$ can be formed using the cross edges. This contradicts the minimality of the length of C_{k_i}, and substantiates the claim. Hence, there are at most $2r$ cross edges between C_{k_i}, and the union of the cycles C_{k_j}, for $j > i$. Therefore, there are at most $2ar \leq 2r^2$ cross edges between the cycles.

Consider a cycle C_{k_i}, and a large tree T_{ℓ_j} for $j \geq b$. For any pair of cross edges between T_{ℓ_j} and C_{k_i}, the distance in T_{ℓ_j} between the endvertices of the cross edges in T_{ℓ_j} must be at least $k_i/2 - 2$. Otherwise, there is a cycle of length less the k_i, which contradicts the minimality of the length of C_{k_i}. The number of vertices in T_{ℓ_j} that are pairwise a distance of at least $k_i/2 - 2$ apart is at most $2\ell_j/(k_i/2 - 2)$. Since $\ell_1 + \ell_2 + \ldots + \ell_b \leq \lfloor rn \rfloor$, and $k_i \geq n - t$, there are at most $4r$ edges between C_{k_i} and the union of all of the large trees. This implies that there are at most $4ar \leq 4r^2$ edges between the cycles and the large trees.

We now consider the cross edges between the cycles and the small trees. There can be one edge from each of the $c - b$ small trees to the cycles without generating a cycle, but two edges could generate a cycle. Partition the edges of the cycles into a minimum number of paths, each of length at most $n/4 - t$. Clearly, the number of paths needed in the partition is at most $\lceil 4r \rceil + a$. Two edges between a small tree and one of the paths in the partition gives a cycle of length at most $2(n/4 - t) + 2 < n - t$. Also, if two different small trees both have an edge to each of the same pair of paths in the partition of the cycles, then there is a cycle of length at most $4(n/4 - t) + 4 < n - t$. Thus, there are at most $\binom{\lceil 4r \rceil + a}{2}$ pairs of edges emanating from the small trees into the paths, which implies that there are at most $c - b + 2\binom{\lceil 4r \rceil + a}{2} < c - b + 26r^2$ edges between the cycles and the small trees.

The number of edges in the cycles and the trees is $\lfloor rn \rfloor + s - c$, and the number of cross edges is less than $c - b + 32r^2$. This gives that $|E(G)| < \lfloor rn \rfloor + s + 32r^2$, which is a contradiction that completes the proof of Lemma 6. $\square$

We are now prepared to give the proof of Theorem 1.

Theorem 1. *For n sufficiently large,*

$$f(n+p, n, 2) = \begin{cases} \binom{n-1}{2} + p + 2, & 0 \leq p \leq n-4, \\ \binom{n-1}{2} + p + 3, & n-3 \leq p < \lceil \frac{3n-6}{2} \rceil, \\ \binom{n-1}{2} + p + 4, & \lceil \frac{3n-6}{2} \rceil \leq p < 2n-4. \end{cases}$$

Also for $r > 0$, $p = \lfloor rn \rfloor$, and n sufficiently large,

$$\binom{n-1}{2} + p \leq f(n+p, n, 2) \leq \binom{n-1}{2} + p + 32r^2.$$

Proof The lower bounds for each of the equalities and the inequality for $f(n + p, n, 2)$ follow from the examples that preceded Lemma 5. Suppose that G is a $(n + p, \binom{n-1}{2} + p + d)$–graph for the appropriate number d, and that $G \notin \mathcal{D}_2(n)$. We will show that this leads to a contradiction.

By Lemma 5, $G \supset sK_{4,4}$, for some integer s such that $8s = n - o(n)$. Denote this graph by N. Find a subgraph of M with a maximum number of vertices m such that $M \supseteq N$, $m < n$, and $\delta(M) \geq 2$. For each vertex v of $M - N$, let N_v be the neighborhood of v in N if $|N_v| \leq 2$, and let N_v be an arbitrary but fixed pair of vertices of N adjacent to v if $|N_v| \geq 3$. Since $M - N$ has $o(n)$ vertices, $\bigcup_{v \in M-N} N_v$ has $o(n)$ vertices. In fact, there is a subgraph N' of N that is vertex disjoint from $\bigcup_{v \in M-N} N_v$, is the union of $K_{4,4}$'s, and has $n - \ell$ vertices with $\ell = o(n)$. Note that the deletion from M of any subset S of vertices of N' with the property that each $K_{4,4}$ of N' is entirely in S or has a $K_{2,2}$ outside of S leaves a graph of minimum degree at least two (i. e. $\delta(M - S) \geq 2$). Therefore, for each integer t, $4 \leq t \leq m$, M contains a subgraph of order t and minimal degree at least 2.

Several cases that are dependent upon the nature of the edges in $G - M$ will be considered to complete the proof. Before stating these cases, some special terminology must be introduced. By a *pseudo cycle* of order a in $G - M$, we will mean either a cycle of order a with all of its vertices in $G - M$, or a path with the endvertices (possibly the same vertex) in M, but with the a interior vertices of the path in $G - M$. Hence, pseudo cycles in $G - M$ correspond to cycles in the multigraph obtained from G by shrinking the vertices in M to a single vertex. If $G - M$ has a pseudo cycle of order a, then G has a subgraph of order $m + a$ that has minimum degree at least two. Consider the following three cases.

(1) $G - M$ has no pseudo cycles.

The number of edges in M is at most $\binom{m}{2}$ and, by assumption, there are at most $n + p - m$ edges in G that are not in M. Hence,

$$|E(G)| \le \binom{m}{2} + n + p - m$$

$$\le \binom{n-1}{2} + p + 1.$$

This is a contradiction that completes the proof of this first case.

(2) $G - M$ has a pseudo cycle of order $a \le n - \ell - 3$.

The graph M' obtained from M by adjoining the vertices of the pseudo cycle of order a has order $m + a$ and minimum degree at least 2. The maximality of M implies that $m + a > n$. Since $m + a \le n - \ell + m - 3 \le 2n - \ell - 4$, deletion of the appropriate vertices of N' will yield a subgraph or order n and minimum degree at least 2. This contradicts the fact that $G \notin \mathcal{D}_2(n)$, and completes the proof of this case.

(3)   Each pseudo cycle of $G - M$ has order at least $n - \ell - 2 = n - o(n)$.

Three subcases that depend upon the magnitude of p will be considered.

(i)   Consider $0 \le p \le n - 4$.

Since $G - M$ has order at most $n + o(n)$, it has just one pseudo cycle, since, the existence of two pseudo cycles would imply that there is a pseudo cycle of order smaller than $n - o(n)$. Hence, there are at most $n + p - m + 1$ edges in G not in M, and

$$\binom{n-1}{2} + p + 2 = |E(G)| \le \binom{m}{2} + n + p - m + 1.$$

The previous inequality implies that $m \ge n - 1$. Therefore, we must have $m = n - 1$, M must be a complete graph on $n - 1$ vertices, and $G - M$ has a pseudo length of at most $n - 3$. Clearly, the deletion from G of appropriate vertices from M will yield a subgraph $H \in \mathcal{D}_2^{*}(n)$. This contradicts the fact that $G \notin \mathcal{D}_2(n)$, and completes the proof of this subcase.

(ii)   Consider $n - 3 \le p < 2n - 4$.

For the interval $n - 3 \le p < \lceil (3n - 6)/2 \rceil$ there can be at most 3 pseudo cycles, and the deletion of an appropriate 2 edges will destroy all pseudo cycles. For the interval $\lceil (3n - 6)/2 \rceil \le p < 2n - 4$ there can be at most 6 pseudo cycles, and the deletion of an appropriate 3 edges will destroy all pseudo cycles. However, except for some additional case analysis, the nature of the argument for both parts of this subcase is identical to that used in subcase (i). The details are left to the reader.

(iii) Consider $p = \lfloor rn \rfloor$.

Consider the graph L obtained from $G - M$ by adding one additional vertex, and making this vertex adjacent to precisely those vertices of $G - M$ that are adjacent in G to some vertex M. Thus, the graph L can be considered as a graph obtained from G by shrinking the vertices of M to a single vertex and removing multiple edges. Note that, as mentioned earlier, any cycle in L corresponds to a pseudo cycle in $G - M$. By Lemma 6, L has a cycle length of at most $n - \ell - 3$, so $G - M$ has a pseudo cycle of order at most $n - \ell - 3$. That gives a contradiction to the assumption for this case, and completes the proof of this subcase and of Theorem 1. $\square$

3. Graphs with subgraphs of minimal degree k.

The results of Theorem 1 can be generalized from subgraphs of minimum degree 2 to subgraphs of minimum degree k for any fixed $k \ge 2$. Reasonable bound for $f(n + p, n, k)$ can be found, although they will not be as sharp as those obtained for $k = 2$.

Examples that give lower bounds for $f(n + p, n, k)$ are easy to define. For any integers $k \ge 2$ and $p \ge k - 1$, P_{p+1}^{k-1} will denote the graph obtained from a path P_{p+1} of order $p + 1$ by making adjacent any pair of vertices that are a distance at most $k - 1$ on the path. This graph has $(p + 1)(k - 1) - \binom{k}{2}$ edges and no subgraph of minimal degree at least k. Let H be the graph obtained from $K_{n-1} \cup P_{p+1}^{k-1}$ by adding $k - j$ edges between K_{n-1} and the vertex that is a distance j from one end of the path P_{p+1} for $1 \le j \le k$. The graph H is a $(n + p, \binom{n-1}{2} + (p + 1)(k - 1)) -$ graph and $H \notin \mathcal{D}_k(n)$. This implies that $f(n + p, n, k) > \binom{n-1}{2} + (p + 1)(k - 1)$. Also, for any $p \ge 0$, a K_{n-1} can be expanded by $p + 1$ independent vertices, each adjacent to $k - 1$ vertices of the K_{n-1}, to obtain an $H' \notin \mathcal{D}_k(n)$. This also verifies that

$$f(n + p, n, k) > \binom{n-1}{2} + (p + 1)(k - 1).$$

The following Theorem 2 generalizes Theorem 1 and gives upper bounds on $f(n + p, n, k)$ that correspond to the examples just described.

Theorem 2 *For fixed integers* $k \geq 2$, $r > 0$ *and* n *sufficiently large,*

$$\binom{n-1}{2} + (p+1)(k-1) < f(n+p, n, k) \leq \binom{n-1}{2} + 2(p+1)(k-1),$$

if $p < n - 2k$, *and for* $p = rn$,

$$\binom{n-1}{2} + r(k-1)n < f(n+p, n, k) \leq \binom{n-1}{2} + \left(\binom{2r+1}{2} + r + 1\right)(k-1)n.$$

Proof The lower bounds for f are verified by the examples that were described at the beginning of this section.

We will first suppose that G is a $(n+p, \binom{n-1}{2} + 2(p+1)(k-1))$ – graph, that $G \notin \mathcal{D}_k(n)$, and show that his leads to a contradiction. Just as in the proof of Theorem 1, application of Lemma 5 gives that $G \supset sK_{t,t}$ for some $t \geq 2k$ and with $2st = n - o(n)$. Denote this graph by N, and let M be a subgraph with a maximum number of vertices m such that $M \supset N$, $m \leq n - 1$, and $\delta(M) \geq k$. Also, as in the $k = 2$ case of Theorem 1, there is a subgraph $N' \subset N$ such that N' is the disjoint union of the graphs $K_{t,t}$, has $n - \ell$ vertices with $\ell = o(n)$, and $\delta(M - N) \geq k$.

The graph $G - M$, which has order $n + p - m$, does not contain a subgraph of order less than $n - 2k$ of minimum degree k; for otherwise, the deletion of the appropriate vertices from M would yield a subgraph of G in $\mathcal{D}_k^*(n)$.

We will first deal with the case when $p < n - 2k$. The remainder of the proof for this case will be broken into two subcases that depend upon the magnitude of $n + p - m$.

Consider the first subcase when $n + p - m \leq n - 2k$. Delete a vertex of the smallest degree in $G - M$, and then successively delete a vertex of smallest degree in the remaining subgraph of $G - M$. This will deplete the vertices of $G - M$, and at each step of the deletion process at most $k - 1$ edges will be deleted. Thus, $G - M$ has at most $(n + p - m)(k - 1)$ edges. Also, since the maximality of M implies that each vertex of $G - M$ has degree at most $k - 1$ relative to M, we have

$$|E(G)| < \binom{m}{2} + 2(n + p - m)(k - 1)$$

$$< \binom{m}{2} + 2(n - m - 1)(k - 1) + 2(p + 1)(k - 1)$$

$$< \binom{n-1}{2} + 2(p + 1)(k - 1)$$

This is a contradiction that completes the proof of this subcase.

We now consider the subcase when $n + p - m > n - 2k$. Again, delete a vertex of smallest degree in $G - M$, and then successively delete a vertex of smallest degree in the remaining subgraph of $G - M$. Since $G - M$ has no subgraph of order less than $n - 2k$ of minimum degree at least k, there can only be a few vertices, in fact no more than the first $p - m + 2k$ vertices in this deletion process from $G - M$, that will have degree as large as k when deleted. Also, for $1 \le i \le p - m + 2k$, the ith vertex has degree at most $p - m + 3k - i$ relative to $G - M$. These observations will yield the same contradiction to the number of edges in G as was reached in the last subcase.

We finally consider the second set of inequalities, which is the case when $p = rn$. Partition the vertices of $G - M$ into $2r + 1$ subsets $\{A_i \mid 1 \le i \le 2r + 1\}$, each with at most $(n - 2k)/2$ vertices. Since $G - M$ contains no subgraph of order at most $n - 2k$ and of minimum degree at least k, the number of edges in the subgraph spanned by $A_i \cup A_j$ is at most $(k - 1)|A_i \cup A_j| < (k - 1)n$. There are $\binom{2r+1}{2}$ pairs from the set $\{A_i \mid 1 \le i \le 2r + 1\}$, so $G - M$ has at most $\binom{2r+1}{2}(k - 1)n$ edges. Therefore, since each vertex of $G - M$ has degree at most $k - 1$ relative to M,

$$|E(G)| < \binom{m}{2} + \binom{2r+1}{2}(k - 1)n + (r + 1)(k - 1)n$$

$$< \binom{n-1}{2} + (\binom{2r+1}{2} + r + 1)(k - 1)n.$$

This is a contradiction that completes the proof of this case and of Theorem 2. $\square$

4. Minimal degree graphs

If the graph G has order m and minimal degree at least k, then it is possible that the number of edges needed to insure that $G \in \mathcal{D}_k(n)$ is less that that required if G had no minimum degree restriction. Thus, it is clear that $g(m, n, k) \le f(m, n, k)$. An examination of the examples used to verify the lower bound for $f(m, n, 2)$ in Theorem 1 reveals that they all have minimum degree 2 when $m \ge 2n - 3$. Therefore, for $k = 2$ and $m \ge 2n - 3$, the functions $f(m, n, k)$ and $g(m, n, k)$ are the same for some values of m and n, in general, their difference is bounded by a number independent of n.

Similar results about the relationship between the functions f and g are true for $k > 2$ and m large. For k even and $p \ge n$, let H be the graph $K_{n-1} \cup C_{p+1}^{k/2}$. Then $\delta(H) \ge k$, $G \notin \mathcal{D}_k(n)$, and H is a $(n + p, \binom{n-1}{2} + k(p + 1)/2)$ – graph. Thus, $g(n + p, n, k) > \binom{n-1}{2} + k(p + 1)/2$, and recall from Theorem 2 that $f(n + p, n, k) \le \binom{n-1}{2} + ck(p + 1)$ for some constant c. Therefore, the functions f and g have the same order of magnitude. Note also, the graph H just described can be modified to give the same result when k is odd.

The next two theorems verify that the functions $f(n + p, n, 2)$ and $g(n + p, n, 2)$ can differ significantly for some values of $p < n$. In Theorem 3, the lower bound for p can be improved, but this makes the calculations much more tedious and not worth the effort.

Theorem 3 *For n sufficiently large and $83n/84 < p \leq n - 4$,*

$$g(n+p, n, 2) = \binom{p + 1}{2} + 2n-3.$$

Proof Consider the graph H that is obtained from the vertex disjoint union of a K_{p+2} and C_{n-2} by making a vertex of the complete graph adjacent to $n - p - 3$ consecutive vertices on the cycle. The graph $H \notin \mathcal{D}_2(n)$, and has $\binom{p+2}{2} + 2n - p - 5 = \binom{p+1}{2} + 2n - 4$ edges. Hence, $g(n + p, n, 2) \geq \binom{p+1}{2} + 2n - 3$.

We will suppose that G is a $(n + p, \binom{p+1}{2} + 2n - 3)$ – graph, that $G \notin \mathcal{D}_2(n)$, and show that this leads to a contradiction. Just as in the proof of both Theorem 1 and Theorem 2, we generate subgraphs $N, M,$ and N' of G. The graph N is the union of $K_{4,4}$'s and has order at least $p - o(n)$. The graph $M \supseteq N$, has order at most $n - 1$ and $\delta(M) \geq 2$, and it is maximal with respect to these properties. The graph N' is a subgraph of N that is the disjoint union of $K_{4,4}$'s such that $\delta(M - N') \geq 2$ and N' has order $n - t$ with $t \leq 17(n - p) + o(n)$.

If the graph $G - M$ contains a pseudo cycle of order at most $n - t - 4$, then a subgraph of G in $\mathcal{D}_2(n)$ can be formed using the pseudo cycle, $M - N'$, and an appropriate subgraph of N'. Thus, we can assume that each pseudo cycle of $G - M$ has order at least $n - t - 3$. Clearly $G - M$ cannot contain more that one pseudo cycle, for this would imply that G had $5(n - t - 3)/2$ vertices, which exceeds $n + p$. However, $\delta(G) \geq 2$ implies that $G - M$ has at least one pseudo cycle. Hence, $G - M$ contains either a large cycle or is a long path with both endvertices in M. Let L be the large cycle of $G - M$, if it exists, or in the other case the smallest order subgraph of G of minimum degree at least 2 that contains the long path in $G - M$. Thus L is either a large cycle or two cycles joined by a long path, and if ℓ is the order of L, then $\ell \geq n - t - 3$.

The proof of Theorem 3 will be completed by considering five cases that depend upon the value of ℓ. We will assume that $p < n - 6$ in the following argument. The cases $n - 6 \leq p < n - 4$ will be left to the reader since they can be proved with precisely the same techniques, but require some additional but straightforward case analysis.

Case 1 *Let $\ell \leq n - 4$.*

The graph L can be extended to a graph in $\mathcal{D}_2{}^*(n)$ by adding appropriate vertices of N', a contradiction.

Case 2 *Let $\ell > n$.*

The nature and minimality of L implies that each vertex not in L can be adjacent to at most three vertices of L. If $\ell = n + q$, then

$$|E(G)| \leq n + q + 1 + \binom{p-q}{2} + 3(p - q)$$

$$\leq n + 3 + \binom{p+1}{2} - (q - 2)(p - q - 1)$$

If $q \geq 2$, then G has at most $\binom{p+1}{2} + n + 3$ edges, and if $q = 1$, then G has at most $\binom{p+1}{2} + n + p + 1 \leq \binom{p+1}{2} + 2n - 4$ edges. In either subcase, a contradiction to the size of G is reached

Case 3 *Let $\ell = n - 1$.*

No vertex of $G - L$ can be adjacent to two vertices of L, so

$$|E(G)| \leq n + \binom{p+1}{2} + p + 1$$

$$\leq \binom{p+1}{2} + 2n - 6,$$

a contradiction to the size of G.

Case 4 *Let $\ell = n - 2$.*

At most one vertex of $G - L$ can be adjacent to two or more vertices of L, and a vertex can be adjacent to as many as three vertices of L only if it is adjacent to precisely three consecutive vertices on a path or cycle. Also, if two vertices of $G - L$ are adjacent, they both cannot be adjacent to a vertex of L. It follows that

$$|E(G)| \le n - 1 + \binom{p+2}{2} + 3$$

$$\le \binom{p+1}{2} + n + p + 3,$$

again a contradiction to the size of G.

Case 5 *Let $\ell = n - 3$.*

Clearly, the graph $G - L$ contains no K_3, and thus has at most $(p + 3)^2/4$ edges [4]. Each vertex of $G - L$ is adjacent to at most three vertices of L. Hence,

$$|E(G)| \le n - 2 + (p + 3)^2/4 + 3(p + 3),$$

which is much smaller than $\binom{p+1}{2} + 2n - 3$. This is a contradiction that completes the proof of this case and of Theorem 3. $\square$

When p is small compared to n, the difference between $f(n + p, n, 2)$ and $g(n + p, n, 2)$ is even more dramatic. The next theorem illustrates this.

Theorem 4 *For $p > 0$ fixed and n sufficiently large,*

$$g(n + p, n, 2) < n^2/8 + p(p + n)$$

Proof Suppose that G is a $(n + p, n^2/8 + p(n + p))$ – graph with $\delta(G) \ge 2$ and $G \notin \mathcal{D}_2(n)$. We will show that this leads to a contradiction.

First consider the case when each vertex of G, including any of degree 2, is adjacent to a vertex of degree 2. If $p = 1$, then this is certainly true of G, for otherwise it would be possible to delete a vertex and get the desired subgraph in $\mathcal{D}_2^*(n)$. Each vertex in G of degree 2 is adjacent to at most one vertex of degree greater than 2. Thus, there are at least $\lceil (n + p)/2 \rceil$ vertices of degree 2. Even if the vertices of degree greater than 2 formed a complete graph, we would have

$$|E(G)| \le \binom{\lfloor (n + p)/2 \rfloor}{2} + \frac{3}{2}\lceil (n + p)/2 \rceil < n^2/8 + (n + p),$$

which is a contradiction to the size of G.

For $p = 1$ the theorem is true, so proceed by induction on p. For the earlier observations, we can now assume that G has a vertex v that is not adjacent to any vertex of degree 2. By assumption, the graph $G - v$ is not in $\mathcal{D}_2(n)$ and $\delta(G - v) \ge 2$. Therefore,

$$|E(G)| < |E(G - v)| + n + p < n^2/8 + (p - 1)(n + p - 1) + n + p < n^2/8 + p(n + p)$$

This gives a contradiction that completes the proof of Theorem 4. ❏

The same argument used in the proof of Theorem 4 will verify that $g(n + 1, n, 2) \leq (n + 1)^2/8 + (n + 1)/2$, and in fact this is the correct value of g. We will next describe a class of examples that will give lower bounds for $g(n + p, n, 2)$ and, in particular, confirm that $g(n + 1, n, 2) = (n + 1)^2/8 + (n + 1)/2$. This class of examples will also form a basis for conjecturing values for $g(n + p, n, 2)$.

For $m \geq 2$, fix a Hamiltonian cycle of the complete graph K_{2m}. Subdivide every second edge on the Hamiltonian cycle r times to obtain a graph $G_{m,r}$, which is a $(m(r + 2), \binom{2m}{2} + mr)$ – graph that has minimum degree 2. However, any subgraph of m $G_{m,r}$ that has minimum degree at least 2 must contain either all or none of the vertices on each of the m subdivided edges (suspended paths). Thus, in particular, any proper subgraph of $G_{m,r}$ has at most $m(r + 2) - r$ edges. In addition, if $H \subset G_{m,r}$ and $\delta(H) \geq 2$, then H has order precisely $m(r + 2) - tr - s$ for some integer $t \leq m$ and $0 \leq s \leq 2t$. This last observation comes from the fact that after the interior vertices of a suspended path are deleted, one or both of the endvertices of the suspended may be deleted.

If $r = p + 1$ and $m = (n + p)/(p + 3)$ are integers, then $G' = G_{m,r}$ is a graph of order $n + p$ with $\delta(G') \geq 2$, but $G' \notin \mathcal{D}_2(n)$. For p fixed and n sufficiently large the graph G' has approximately (and also at least) $2n/(p + 3)^2$ edges, so

$$g(n + p, n, 2) > \frac{2n}{(p+3)^2} .$$

Also, if $p = 1$, the graph G' implies $g(n + 1, n, 2) \geq (n + 1)^2/8 + (n + 1)/2$, as indicated earlier.

A more involved version of the previous class of examples gives improved lower bounds for some values of p. We describe this new class now. For a fixed p, select non-negative integers s and t such that $st < p$, $s(t + 1) > p$, and $(s + 2) < p$. If $r = s$ and $m = (n + p)/(s + 2)$ are integers, then $G'' = G_{m,r}$ has order $n + p$, but $G'' \notin \mathcal{D}_2(n)$. If $t = 0$ and $s = p + 1$, then G' and G'' are the same. At the other extreme, s can be selected to be as small as possible such that there is a t with the three inequalities stated earlier being satisfied. For example, if $s = 2t + 2$ and $p = (s + 2)t + 1$, then s and t satisfy the inequalities that insure that $G'' \notin \mathcal{D}_2(n)$. Note that if p is large, then in this example s^2 is approximately $2p$. Therefore for p fixed and n sufficiently large, G'' has approximately $2n^2/s^2$ edges (which is approximately n^2/p edges). This class of examples and the proof techniques of

Theorem 4 indicate that it is possible that the order of magnitude of $g(n + p, n, 2)$ is n^2/p if p is fixed and n is sufficiently large.

In this section we have considered the case $k = 2$. However, there are corresponding results for $g(n + p, n, k)$ with $k \geq 2$ when p is fixed and n is sufficiently large.

REFERENCES

[1]   G. Chartrand and L. Lesniak, *Graphs and Diagraphs*, Wadsworth and Brooks/Cole, Monterey, CA (1986).

[2]   P. Erdös, R. J. Faudree, C. C. Rousseau and R. H. Schelp, Subgraphs of Minimal Degree k, to appear.

[3]   T. Kövári, V. T. Sós and P. Turán, On a Problem of Zarankiewicz. *Mat. Lapok* 3 (1957) 50–57.

[4]   M. Simonovits, *Extremal Graph Theory*. "Selected Topics in Graph Theory 2," Academic Press, New York (1983) 161–200.

A Note on the Largest H–Free Subgraph in a Random Graph

Paul Erdös

Hungarian Academy of Science

John Gimbel

University of Alaska

ABSTRACT

Given a fixed graph H and a random graph G, in Model A with an edge probability of $\frac{1}{2}$, there is a $c > 0$ dependent only on H, such that G almost surely contains no H–free subgraph on $c \log n$ vertices, where n is the order of G.

In one of the first theorems ever proved in random graph theory, it was shown that a random graph on n vertices almost surely (a.s.) has no set of $\frac{2}{\log 2} \log n$ vertices which induce a clique or an empty graph. This result, originally established in [3], can be restated as follows: If G is a random graph on n vertices the order of the largest K_2–free subgraph is a.s. less than $c \log n$ where $c = \frac{2}{\log 2}$ as is the order of the largest $\overline{K_2}$–free subgraph.

We will generalize this notion and show that if H is a fixed graph and G is a random graph, the largest H–free subgraph in G has a.s. fewer than $c \log n$ vertices where c is a constant depending only on H and n is the cardinality of the vertex set of G.

The reader is urged to consult [1] and [5] for undefined terms and concepts. By a subgraph we mean an induced subgraph. Given graphs H and G we will say G *contains* H if G contains an induced isomorphic copy of H. We shall say G is H–*free* if G doesn't contain H. By $\langle U \rangle$ we mean the subgraph induced by U. Our sample space will be Model A of [5] where each edge has a probability of $\frac{1}{2}$ of being present.

The following is a direct result of R. Wilson's work in [6].

Theorem 1 *For each fixed integer m there is a $c_m > 0$ such that any set S of cardinality $n \gg m$ has $c_m n^2$ subsets of size m so that any two of these subsets have at most one element in common.*

Suppose H is a nonempty graph and G is a random graph of order n. From [2] and [4] we know that G a.s. has a set of $\log_2 n$ vertices which induce an empty graph. Hence, G a.s. has an H–free graph of order $\log_2 n$. We note from the same sources that G a.s. has a clique of the same order. Hence the following.

Remark 2 *If H is a fixed graph and G is a random graph on n vertices then G contains an H–free subgraph of order $\log_2 n$.*

We know from [1] (p.85) that for any fixed H, almost every (a.e.) graph G contains H. Thus, on a hueristic level, we see that if K is a "large" subgraph of G then K will a.s. contain H. We will see that "large" simply means $O(\log n)$ where n is the order of G.

Theorem 3 *Given a fixed graph H and a random graph G of order n, the largest H–free subgraph of G a.s. has cardinality less than $c \log n$ where c is dependent only on H.*

Proof Let H be a fixed graph on m vertices and G a random graph on n vertices. Also, let c_m be chosen as in Theorem 1. Choose c large enough so that $(1 - \frac{1}{2^{\binom{m}{2}}})^{cc_m} < e^{-2}$. Now, set $t = c \log n$. Suppose S is a subset of $V(G)$ having cardinality t. Chose subsets of S, say $S_1, S_2, \ldots, S_{c_m t^2}$ so that

$$|S_i| = m \quad (i = 1, 2, \ldots, c_m t^2)$$

and

$$|S_i \cap S_j| \le 1 \quad (1 \le i < j \le c_m t^2)$$

We note that for distinct i and j, the graphs induced by S_i and S_j have no common edge.

For a fixed i, the probability that S_i induces an H is greater than $\frac{1}{2^{\binom{m}{2}}}$. Hence the probability that no S_i induces an H is less than $(1 - \frac{1}{2^{\binom{m}{2}}})^{c_m t^2}$. Also, the probability that the graph induced by S contains no H is less than this expression as well. There are $\binom{n}{t}$ subgraphs of G of size t. Hence, an upper bound on the expected number of H–free subgraphs of size $c \log n$ in G is

$$\binom{n}{t} \left(1 - \frac{1}{2\binom{m}{2}}\right)^{c_m t^2} \leq n^t \left(1 - \frac{1}{2\binom{m}{2}}\right)^{c_m ct \log n} \leq [n(e^{-2})^{\log n}]^t \rightarrow 0 \qquad \Box$$

Clearly, this proof can be altered to show for any fixed H and fixed probability $p > 0$, the existence of a $c > 0$ for which a random graph in Model A with probability p on n vertices a.s. has no H–free subgraph of order $c \log n$.

REFERENCES

[1]   B. Bollabás, *Random Graphs*, Academic Press, London, 1985.

[2]   B. Bollobás and P. Erdös, Cliques in Random Graphs, Math. Proc. Cambridge Philos. Soc. **80** (1976) 419 – 427.

[3]   P. Erdös, Some Remarks on Theory of Graphs, *Bull. Amer. Math. Soc.* **53** (1947), 292 – 294.

[4]   D. Matula, The Largest Clique in a Random Graph, Technical Report, Dept. of Computer Science, Southern Methodist Univ., Dallas (1976).

[5]   E. Palmer, *Graphical Evolution*, John Wiley, New York, 1985.

[6]   R. Wilson, An Existence Theory for Pairwise Balanced Designs, III, *J. Comb. Theory* **18**(1975) 71 – 79.

Graphs Realizing the Same Degree Sequences and their Respective Clique Numbers

P. Erdös
Hungarian Academy of Science

M.S. Jacobson[1]
University of Louisville

J. Lehel
Computer and Automation Institute
Hungarian Academy of Science

ABSTRACT

There are several characterizations of graphical sequences, yet only little work on the possible structure of these graphs. We begin considering the question of the possible clique number attained by graphs with the same degree sequence. In particular, it is shown that the maximum difference between the largest and the smallest possible clique number for graphs realizing the same degree sequence of n elements is $n - cn^{2/3}$ for sufficiently large n. Additional results on sequences having a realization H with $\omega(H) \geq k$ are presented.

1. Introduction

Let $\pi = (d_1, d_2, \ldots, d_n)$ be a sequence of integers with $d_1 \geq d_2 \geq \ldots \geq d_n \geq 0$. The sequence π is a *graphical sequence* if there exists a graph G with degree sequence π. Additionally, G is referred to as a *realization* of π. The sequences that are realizable were characterized by Erdös and Gallai [3] and Havel [7] and Hakimi [4]. These results do not give an indication of the possible variance in the structure, of the graphs realizing the degree sequence.

1 Research supported by O.N.R. contract N00014–85–K–0694

Let $\Gamma(\pi)$ denote the family of graphs realizing π and $\Gamma(P)$ denote the family of graphs with some property P, such as "being connected", "being hamiltonian", or "containing a clique of order k", etc. A graphical sequence π is said to be *potentially (forcibly) P-graphical* if $\Gamma(\pi) \cap \Gamma(P) \neq \varnothing$ ($\Gamma(\pi) \subseteq \Gamma(P)$). In [6], Rao considers the problem of characterizing potentially P–graphical sequences when P is the property "containing a clique of a given order". In the paper a "Havel–Hakimi" type procedure was developed to help determine the maximum clique number of a graph with a given degree sequence.

In this paper we are interested in determining the difference between the largest clique number and the smallest clique number over all graphs in $G(\pi)$. Also, we will study certain conditions on the degree sequence that imply that there is a realization with a clique of a given size.

For convenience we denote the clique number of G, which is the order of the largest complete subgraph in G, by $\omega(G)$. Also we write $G \approx H$ to denote that G and H are two realizations with the same degree sequence.

Two graph operations will be used throughout this paper. If G and H are graphs then $G \cup H$ is the disjoint union of G and H. In the case when $G = H$ then we abbreviate $G \cup H$ as 2G. We denote $G + H$ as the graph with $V(G + H) = V(G) \cup V(H)$ and $E(G + H) = E(G) \cup E(H) \cup \{xy: x \in V(G) \text{ and } y \in V(H)\}$. We denote the degree of x in G by $d_G(x)$, or simply $d(x)$ when no confusion will result. The main results of the paper will be presented in Sections 2 and 3.

In Section 2, the first result we give pertains to regular graphs. If G is r–regular of order n with $r < n - 1$, then $\omega(G) \leq \lfloor n/2 \rfloor$, where $\lfloor x \rfloor$ is the largest integer less than or equal to x. Then we define $\Omega(n;k)$ to be the largest clique number of a graph G of order n so that there exists a graph H with $G \approx H$ and $\omega(H) \leq k$. Proposition 2.2 gives $\Omega(n; 2)$ explicitly while Proposition 2.4 gives an upperbound for $\Omega(n; k)$. Finally, this enables us to determine the order of magnitude of

$$\Omega(n) = \max_{G,H} \{\omega(G) - \omega(H) : G \approx H \text{ and } G \text{ has order } n\}.$$

We show in Theorems 2.7 and 2.8 that $\Omega(n)$ is approximately $n - 2n^{2/3}$, for sufficiently large n.

In Section 3, we investigate the question of potentially P–graphical for various degree sequences when P is the property "containing a clique of a given size". For a degree sequence to have a realization with k–clique, we consider the number of vertices of degree $k - 1$ it must contain. Also the dependency on the number of vertices and the number of edges.

The final section includes a number of questions and unsolved problems pertaining to these ideas.

2. Clique Difference

In this section we will consider the following:

$$\Omega(n) = \max_{G,H} \{\omega(G) - \omega(H) : G \approx H \text{ and } G \text{ has order } n\}.$$

Initially, we will restrict our attention to regular graphs. For convenience, we introduce additional notation. Let G be a graph with A and B two subsets of $V(G)$. Denote by $m_G(A,B)$ the number of edges in G between vertices in A and vertices in B.

Proposition 2.1 *If G is an r–regular graph of order n with $r < n - 1$, then $\omega(G) \leq \lfloor n/2 \rfloor$. Moreover, this bound is sharp.*

Proof Let G be an r–regular graph with $r < n - 1$. Let $A \subseteq V(G)$ which forms a complete subgraph of G and $|A| = \omega(G)$. Let $B = V(G) - A$. It follows that $m_G(A,B) = |A| \cdot (r - (|A| - 1)) \geq |B| \cdot (r - (|B| - 1))$. Since $|A| < n$ and the function $f(x) = x(r + 1 - x)$ is decreasing on the interval $1 \leq x \leq r + 1$, it follows that $|A| < |B|$. Consequently, $\omega(G) \leq \lfloor n/2 \rfloor$ follows from the fact that $|A| + |B| = n$.

To see that this bound is sharp, let n, the order of G, be even. For $n - 2 > r > n/2$, consider any $(r - n/2 + 1)$–regular bepartite graph on $V(G)$ with partite sets A and B. Now put $xy \in E(G)$ if x and y are both in A or both in B. This graph is r–regular with $\omega(G) = n/2$. Similar graphs exist for n odd. $\square$

This result is also given in [2].

Now we give the first result pertaining to the difference between the largest and smallest clique number for a given degree sequence. Recall that $\Omega(n;k)$ is the largest clique number of a graph G of order n so that there exists a graph H with $G \approx H$ and $\omega(H) \leq k$.

Proposition 2.2 $\Omega(n;2) = \lfloor n/2 \rfloor + 1$ *for* $n > 8$.

Proof First we show that $\Omega(n;2) < \lfloor n/2 \rfloor + 2$. Suppose to the contrary that G is a graph of order n with $\omega(G) \geq \lfloor n/2 \rfloor + 2$. This implies that there are at least $\lfloor n/2 \rfloor + 1$. Thus, for every $H \approx G$, H must contain two vertices x and y of degree at least $\lfloor n/2 \rfloor + 1$ which are adjacent. But since the degree of x and y is at least $\lfloor n/2 \rfloor + 1$ they must have a common neighbor, consequently $\omega(H) \geq 3$.

To complete the equality we construct graphs G and H with n vertices such that $\omega(G) = \lfloor n/2 \rfloor + 1$, H is bipartite and $G \approx H$.

Case 1 $n = 4t$, $t \geq 3$. Let $G = K_{2t+1} \cup (C_t + K_{t-1})$ and $H = K_{2t, 2t} - K_{t, t-1}$.

Case 2 $n = 4t + 1$, $t \geq 1$. Let $G = K_{2t+1} \cup K_{t, t}$. Let $H' = K_{t+1, t+1} - (t + 1)K_2$ with partite sets A' and B' and $H'' = K_{t, t-1}$ with partite sets A'' and B'' and $|A''| = t$. Let $H = H' \cup H'' \cup \{$all edges between A' and $A''\}$.

Case 3 $n = 4t + 2$, $t \geq 2$. Let G' be any $(t + 1)$–regular graph of order $2t$, and $G = K_{2t+2} \cup G'$. Let $H = K_{2t+1, 2t+1} - K_{t, t}$.

Case 4 $n = 4t + 3$, $t \geq 1$. Let $G = K_{2t+2} \cup K_{t, t+1}$. Let $H' = K_{t+1, t+1} - (t + 1)K_2$ with partite sets A' and B' and $H'' = K_{t+1, t}$ with partite sets A'' and B'' and $|A''| = t + 1$. Let $H = H' \cup H'' \cup \{$all edges between A' and $A''\}$.

In each of the four cases it is easy to check that the desired graphs have been constructed. $\square$

For the small values of n, the proof above gives $\Omega(n;2) = \lfloor n/2 \rfloor + 1$ for $n = 5,7$. For $n = 3, 4, 6$, and 8, it can easily be checked that $\Omega(n;2) = \lceil n/2 \rceil$.

Before proceeding we give a lemma that will be useful in the results to follow.

Lemma 2.3   *Let G and H be two graphs on the same vertex set V, with $G \approx H$ and $d_G(x) = d_H(x)$ for all $x \in V$. If $A \subseteq V$ and $B = V - A$, then*

$$m_G(A,A) - m_H(A,A) = m_G(B,B) - m_H(B,B).$$

Proof   Let G, H, A, and B be as given in the statement of the lemma. Clearly,

$$0 = 1/2 \cdot \sum_{x \in B} (d_H(x) - d_G(x)) + 1/2 \cdot \sum_{x \in B} (d_G(x) - d_H(x))$$

$$= m_H(A,A) - m_G(A,A) + 1/2 (m_H(A,B) - m_G(A,B))$$
$$+ m_G(B,B) - m_H(B,B) + 1/2 (m_G(B,A) - m_H(B,A)).$$

Hence, we can conlude that

$$m_G(A,A) - m_H(A,A) = m_G(B,B) - m_H(B,B) \text{ since}$$
$$m_G(A,B) = m_G(B,A) \text{ and } m_H(A,B) = m_H(B,A). \square$$

The next result gives a bound for $\Omega(n;p)$ in general. As usual, let $T(n, k + 1)$, the Turan number [10], denote the maximum number of edges in a graph of order n containing no K_{k+1} $(1 < k < n)$.

Theorem 2.4 *Let* $k \geq 3$, *and* $x = t_0$ *be the smallest integer satisfying the inequality*

$$\binom{n - x - T(n - x, k + 1)}{2} \leq \binom{x}{2}$$

Then $\Omega(n;k) \leq n - t_0$ *for sufficiently large* n.

Proof Let G and H be two graphs of order n with the same vertex set such that $G \approx H$. Assume $d_G(x) = d_H(x)$ for every $x \in V$. Let $\omega(G) = \Omega(n;k)$ and $\omega(H) \leq k$. Let $A \subseteq V$ be a subset of vertices defining a maximum clique in G and set $B = V - A$. We shall prove that $t_0 \leq |B|$. By Lemma 2.3

$$m_G(A,A) - m_H(A,A) = m_G(B,B) - m_H(B,B).$$

Since A induces a complete subgraph of G, it follows that $m_G(A,A) = \binom{n - |B|}{2}$. Clearly, $m_G(B,B) - m_H(B,B) \leq \binom{|B|}{2}$, this yields from (2) that

$$n - |B| - m_H(A,A) \leq \binom{|B|}{2}.$$

Since $\omega(H) \leq k$ it must be the case that

$$m_H(A,A) \leq T(n - |B|, k + 1). \qquad \square$$

Theorem 2.5 *If* $k \geq 5$, n *sufficiently large and* t_0 *the smallest integer satisfying* *(1), then* $\Omega(n;k) = n - t_0$.

Proof It only remains to show that there exists a graph G, with $\omega(G) = n - t_0$ and a graph H with $G \approx H$ having $\omega(H) \leq k$. Let $\Omega = \Omega(n;k)$. By the definition there exist graphs G and H with $\omega(G) = \Omega$, $\omega(H) \leq k$ and $G \approx H$. Let $G = K_\Omega \cup G_0$ where G_0 has $n - \Omega$ vertices and $f = \binom{\Omega}{2} - T(\Omega; k + 1)$ edges. By Turan's theorem, there is an F with f edges so that $\omega(K_\Omega - F) \leq k$ and $K_\Omega - F$ is partitioned into k independent sets $I_1, I_2, \ldots, I_p$ of almost equal size $\lfloor \frac{\Omega}{p} \rfloor$ or $\lceil \frac{\Omega}{p} \rceil$. To build the desired H, first we need a bipartite graph H_1 with partite sets of H_1 being $V(K_\Omega)$ and $V(G_0)$ and $d_{H_1}(y) = d_{G_0}(y)$ for each $y \in V(G_0)$ and $d_{H_1}(x) = (\Omega - 1) - d_{K_\Omega - F}(x)$ for each $x \in V(K_\Omega)$. To obtain H_1, we label the vertices of G_0 by $y_1, y_2, \ldots, y_{n-\Omega}$, and the vertices of K_Ω by $x_1, x_2, \ldots, x_\Omega$ with $x_1, x_2, \ldots, x_{m_1}$ the elements of I_1, x_{m_1+1}, $x_{m_1+2}, \ldots x_{m_2}$ the elements of I_2 and so forth.

In the first step we put edges between y_1 and the first $d_{G_0}(y_1)$ vertices of $V(K_\Omega)$ and in the ith step we put edges between y_i and the next subsequence of $d_{G_0}(y_i)$ vertices of $V(K_\Omega)$. Finally, we put $K_\Omega - F$ on the vertices $x_1, x_2, \ldots, x_\Omega$ with $x_1, \ldots, x_{m_1}$ being I_1, $x_{m_1+1}, \ldots, x_{m_2}$ being I_2 etc.

It is clear that $\Omega(G) = \Omega$. Now, simple algebraic manipulation shows that $\frac{nk}{2k-2}$ satisfies (1), thus we can conclude that $\Omega \geq \frac{np}{2p-2}$ and hence it follows that $n - \Omega \leq \Omega - \frac{2\Omega}{k}$. Clearly, since $d_H(y) \leq n - \Omega - 1 < \Omega - \frac{2\Omega}{k}$ for all $y \in V(G_0)$, y cannot be adjacent to more than $k - 1$ of the subsets I_j. Furthermore, since $\omega(K_\Omega - F) = k$ it must be the case that $\omega(H) = k$. By choosing Ω as large as possible, which is the same as choosing t_0 as small as possible, gives the desired result. $\square$

Proposition 2.6 *For fixed k, $k \geq 3$,*

$$\Omega(n;k) \leq \left(\frac{\sqrt{k}}{\sqrt{k+1}}\right)n + O(1)$$

if n is sufficiently large.

Proof Rewrite (1) by substituting

$$y = n - x: \quad 2y(n-1) - n^2 + n \leq 2 \cdot T(y, k+1) \tag{4}$$

By Theorem 2.4, find the largest y satisfying this inequality. Set

$$f(y) = \frac{k-1}{k}y^2 - 2y(n-1) + n^2 - n, \quad \text{since} \quad T(y, k+1) \leq \frac{k-1}{2k}y^2$$

(see [1]) it is enough to show that $f(y) < 0$ for every y. But $\frac{\sqrt{k}}{\sqrt{k+1}}n + \frac{\sqrt{k}}{2} \leq y \leq n$ for sufficiently large n. Since $f(y) < 0$ when $y = n$ and $y = \frac{\sqrt{k}}{\sqrt{k+1}}n + \frac{\sqrt{k}}{2}$ the proposition follows. $\square$

Theorem 2.7 *For sufficiently large n, $\Omega(n) \leq n - cn^{2/3}$ for some $c \geq 1$.*

Proof Let $\Omega = \Omega(n) = M - m$ with $G \approx H$ and $\omega(G) = M$ and $\omega(H) = m$. Suppose $\Omega > n - cn^{2/3}$ that is let $\Omega = n - cn^{2/3-\varepsilon}$. We will show for any $\varepsilon > 0$ this leads to a contradiction.

Since $M - m = n - cn^{2/3-\varepsilon}$ it follows that $n \geq M \geq n - cn^{2/3-\varepsilon}$. From Lemma 2.3 we get

$$m\binom{\frac{M}{m}}{2} \leq \binom{n-M}{2}$$

$$m\left(\frac{M}{m}\right)^2 \leq (n-M)^2$$

$$\frac{(n - cn^{2/3-\varepsilon})^2}{cn^{2/3-\varepsilon}} \leq \frac{M^2}{m} \leq (n - M) \leq (n - (n - cn^{2/3-\varepsilon}))$$

$$(n - cn^{2/3-\varepsilon})^2 \leq (cn^{2/3-\varepsilon})^3.$$

But for n sufficiently large, this yields

$$n^2 \le c^3 n^{2-3\varepsilon}$$

which can be true for no $\varepsilon > 0$, hence a contradiction, thus $\Omega(n) \le n - cn^{2/3}$. Repeating the argument gives $c^3 \ge 1$ or $c \ge 1$. $\square$

Theorem 2.8 *For infinitely many n, $\Omega(n) \ge n - 2n^{2/3}$.*

Proof We exhibit a degree sequence and two realizations attaining this difference. Let $t = (n - 1)^{1/3} > 1$ be an integer with $a = t^3 - t^2 + t$, $b = t^2 - t + 1$, and thus $a + b = t^3 + 1 = n$. Let $G = K_{a+1} \cup K_{b+1}$ with $V(K_{a+1}) = A$ and $V(K_{b+1}) = b$. Construct $H = H_1 \cup H_2$ where $V(H_1) = A$ and H_1 is the complete $(b + 1)$ partite graph with t vertices in each part and H_2 be a bepartite graph with parts A and B so that $d(x) = t - 1$ and $d(y) = b$ for each $x \in A$ and $y \in B$.

It is easy to check that $H \approx G$ and $\omega(G) = a + 1$ and $\omega(H) \le b + 1$. Thus $\Omega(n) \ge a - b \ge n - 2b \ge n - 2n^{2/3}$. $\square$

3. Degree sequences with Realizations having a Given Clique Number

In this section we denote by P_k $(\overline{P_k})$ the property "(not) containing a clique of order $k + 1$" and begin the study of various properties of a degree sequence D so that D is potentially P_k–graphical.

We state the Erdös – Gallai [3] result for realizable degree sequences.

Theorem *A sequence $d_1 \ge d_2 \ge \ldots d_n$ is graphical if and only if*

$$\sum_{i=1}^{t} d_i \le t(t-1) + \sum_{j=t+1}^{n} max\{t, d_j\}$$

for $1 \le t \le n - 1$, and $\sum_{i=1}^{n} d_i$ is even.

We define $f(k)$ to be the smallest number of k's, a graphical degree sequence D, with maximum element k, must have in order for D to be potentially P_k–graphical.

Proposition 3.1 *For $k \ge 2, f(k) = 2k + 2$.*

Proof To see that $f(k) > 2k + 1$, we give a graphical degree sequence that is forcibly $\overline{P_k}$–graphical. Let D be the sequence of $(2k + 1)$ k's and $(k - 2)$ 1's. Let G be the graph $K_{k,k}$ with one edge subdivided and an additional $k - 2$ edges joined to the new vertex. This graph has degree sequence D. Suppose that there was a realization containing a K_{k+1}. It follows that D', the sequence of k k's and $(k - 2)$ 1's would necessarily be graphical. If that were the case then there would need to be at least as many vertices of degree 1 as there are vertices of degree k, since a vertex of degree k can be adjacent to only $k - 1$ vertices of degree k. Hence D is forcibly $\overline{P_k}$–graphical.

To see that $f(k) \leq 2k + 2$, let $D = d_1, d_2, \ldots, d_n$ with $k = d_1 = d_2 = \ldots = d_{2k+2} \geq d_{2k+3} \geq \ldots \geq d_n$ be a graphical sequence. We need only show that $D' = d_1', d_2', \ldots, d_{n-k-1}'$ with $k = d_1' = d_2' = \ldots = d_{k+1}'$ and for $i > k + 1$, $d_i' = d_{i+k+1}$ is a graphical sequence. Clearly, $\sum_{i=1}^{n} d_i'$ is even since both $\sum_{i=1}^{n} d_i$ and $k(k+1)$ are even. Apply the Erdös–Gallai result: If $t < k + 1$;

$$\sum_{i=1}^{t} d_i' = tk \leq tk + \sum_{k+2}^{n-k-1} d_j'$$

$$= t(t - 1) + t(k - t + 1) + \sum_{k+2}^{n-k-1} d_j'$$

$$= t(t - 1) + \sum_{t+1}^{n-k-1} \min \{t, d_j'\}.$$

If $t \geq k + 1$;

$$\sum_{i=1}^{t} d_i' \leq t(t - 1)$$

$$\leq t(t - 1) + \sum_{t+1}^{n-k-1} \min \{t, d_j'\}.$$

Hence D' is realizable and consequently there is a realization of D containing a K_{k+1} as a component. Thus $f(k) = 2k + 2$. $\square$

As a result of the previous proof we get the following:

Corollary 3.2 *Let $D = d_1 \geq d_2 \geq \ldots \geq d_n$ with $n \geq k + 1$ and $\sum_{i=1}^{n} d_i$ even. If $k = d_1 = d_2 = \ldots = d_{k+1}$ then D is a graphical sequence.*

The problem appears to become considerably more difficult if we require a graphical sequence to conain vertices of degree at least p.

Next, we consider sequences of an appropriate length with a sufficient number of k's. First we give the following lemma.

Lemma 3.3 *If $k \geq d_1 \geq d_2 \geq \ldots \geq d_n \geq 1$, $n \geq \left(\frac{k+2}{2}\right)^2$ and $\sum_{i=1}^{n} d_i$ is even then $D = d_1, d_2, \ldots, d_n$ is graphical.*

Proof We need only show that

$$\sum_{i=1}^{t} d_i \leq t(t - 1) + \sum_{t+1}^{n} \min \{d_i, t\}.$$

Case 1 Let $1 \le t \le k$. Suppose

$$\sum_{i=1}^{t} d_i > t(t-1) + \sum_{t+1}^{n} \min \{d_i, t\}$$

This implies

$$tk > t(t-1) + n - t$$

which implies $t(k - t + 2) > n$. But $t(k - t + 2)$ is maximized when $t = \dfrac{k+2}{2}$, which

implies $\left(\dfrac{k+2}{2}\right)^2 > n$, but this contradicts the hypothesis.

Case 2 $k < t \le n - 1$.

$$\sum_{i=1}^{t} d_i \le kt \le t(t-1) \le t(t-1) + \sum_{t+1}^{n} \min \{d_i, t\}. \quad \square$$

We note that this result is sharp. Let $n = \left(\dfrac{k+2}{2}\right)^2 - 1$, and D be the sequence of $\dfrac{k+2}{2}$

p's and $\left(\left(\dfrac{k}{2}\right)\left(\dfrac{k+2}{2}\right) - 1\right)$ 1's. When k is even, D can be seen to be forcibly $\overline{P_k}$–

graphical.

Proposition 3.4 *If $k = d_1 = d_2 = \ldots = d_{k+1} \ge d_{k+2} \ge \ldots \ge d_n \ge 1$ is a graphical sequence with $n \ge \left(\dfrac{k+2}{2}\right)^2 + k + 1$ then the sequence is potentially P_k–graphical.*

Proof This follows since $d_{k+1}, d_{k+2}, \ldots, d_n$ is graphical from Lemma 3.3. $\quad \square$

As previously observed, this result is sharp. Finally, we consider degree sequences that imply a minimum number of edges, and whether they are potentially P_k–graphical. Let $G_k = K_{k-1} + \overline{K_{n-k+1}}$ and D_k be the corresponding degree sequence. Clearly, D_k is forcibly $\overline{P_k}$–graphical since there are only $k - 1$ vertices of degree $\ge k$. We feel that these are the extremal examples, that is, any degree sequence giving more edges than D_k is potentially P_k–graphical. The next result is the case for $k = 3$.

Theorem 3.5 *If $D = d_1 \ge d_2 \ge \ldots \ge d_n \ge 1$, $n \ge 6$ is a graphical degree sequence with $\sum_{i=1}^{n} d_i \ge 2n + 1$ then D is potentially P_3–graphical.*

Proof Let G be a realization of D with smallest possible induced cycle. If G contains an induced cycle of length 6 or more then the edges can be rearranged among the vertices of the cycle to form a graph G' realizing D containing a K_3. Thus we need only to deal with G containing a smallest induced cycle of length 4 or 5 and show that either G contains a K_3 or the edges can be rearranged to define a realization containing a K_3.

Suppose the smallest cycle is a C_4. If $x_1x_2x_3x_4$ form the C_4 and $yx_1 \in E(G)$, then yx_1 and x_2x_3 can be replaced by x_1x_3 and x_2y forcing a K_3. So G contains a C_4: $x_1x_2x_3x_4$ and an edge yz. But x_1x_2, x_2x_3 and yz can be replaced by x_1x_3, x_2y and x_2z. Thus we assume the smallest cycle is a C_5. No vertex off the cycle can be adjacent to more than one vertex on the cycle, otherwise a K_3 or C_4 would result. If $x_1x_2x_3x_4x_5$ form the C_5 and $yx_1 \in E(G)$ then x_1y and x_3x_4 can be replaced by x_1x_4 and x_3y forcing a K_3. So G contains a C_5: $x_1x_2x_3x_4x_5$ and an edge zy. But x_1x_2, x_3x_4 and yz can be replaced by x_1x_4, x_2y and x_3z. In each case, a realization containing a K_3 results. $\square$

Our final results give some related work to Theorem 3.5.

Proposition 3.6   *If $d_1 \geq d_2 \geq \ldots \geq d_n$ is a realizable degree sequence so that for some $3 \leq k \leq n/3$, $d_n > 3k - 5$ then there is a realization H with $\omega(H) \geq k$.*

Proof   Set $v_i \in v(H)$ so that $d(v_i) = d_i$, let $A = \{v_1, v_2, \ldots, v_k\}$ and choose H so that the graph induced on A is as dense as possible. If it is complete then $\omega(H) \geq k$, and we would be done. Suppose $v_iv_j \notin E(H)$ for $i, j \leq k$. Consider the sets

$X = \{v \in v(H) - A$ with $v_iv \in E(H)$ and $v_jv \notin E(H)\}$

$Y = \{v \in v(H) - A$ with $v_jv \in E(H)$ and $v_iv \notin E(H)\}$

$Z = \{v \in v(H) - A$ with v_iv and $v_jv \in E(H)\}$.

If there is $x \in X$ and $y \in Y$ so that $xy \notin E(H)$ then be deleting v_ix and v_jy and adding v_iv_j and xy a realization with an additional edge in A would be attained, thus H must contain all possible edges between X and Y. A similar exchange of edges would imply that at least one of $\langle X \rangle$ or $\langle Y \rangle$ is complete, as well as implying that $\langle Z \rangle$ is complete. Without loss of generality suppose that $\langle X \rangle$ is complete. If $|X| \geq k - 1$ or $|Z| \geq k - 1$ then $\omega(H) \geq k$. If this were the case, since there are only $k - 2$ additional vertices of A possibly adjacent to v_i, $d(v_i) \leq 3k - 6$ but that contradicts the assumption that $d(v_i) \geq 3k - 5$. Thus there is a realization H with $\omega(H) \geq k$. $\square$

Corollary 3.7   *If $d_1 \geq d_2 \geq \ldots \geq d_n$ is a realizable degree sequence with $\sum d_i \geq (4n - 3k + 2)(k - 2) + n$ then the sequence is potentially P_{k-1}-graphical.*

Proof   This condition ensures that $d_k \geq 3k - 5$ and applying Proposition 3.6 gives the result. $\square$

4.   Conclusion and Questions

The results given in this paper suggest several natural questions. First, pertaining to the previous section, determine the smallest degree sum, or number of edges, that yield a potentially P_{k-1}–graphical sequence. Next, to determine exactly the constant c for $\Omega(n)$. The third question that arises is if $G \approx H$ and $\omega(G) > \omega(H)$, for $\omega(G) > m >$ $\omega(H)$, does there exist a graph $F \approx G$ so that $\omega(F) = m$? Finally, what other structural differences between realizations can be studied?

REFERENCES

[1]   B. Bollabás, *Extremal Graph Theory,* Academic Press, London, 1978.

[2]   L. Caccetta and N.J. Pullman, Regular Graphs with Prescribed Chromatic Number, preprint.

[3]   P. Erdös, T. Gallai, Graphen mit Punkten vorge–schriebenen Grades, Mat. Lapok, 11 (1960 264–274.

[4]   S.L. Hakimi, On realizability of a set of integers as degrees of the vertices of a linear graph, I, J. SIAM 10 (1962) 496-506.

[5]   S.L. Hakimi, E.F. Schmeichel, Graphs and their degree sequences: A survey, Lecture Notes in Math. No 642, Springer–Verlag, 1978, 225–235.

[6]   P.L. Hammer, B. Simeone, The splittance of a graph, Univ. of Waterloo Dept. of Combinatorics and Optimization Res. Report, CORR 1977 –39.

[7]   V. Havel, Eine Bemerkung uber die Existenz der Endlichen Graphen, Casopis Pest Mat. 80 (1955), 477–480.

[8]   A.R. Rao, The clique number of a graph with a given degree sequence, Proc. Symposium on Graph Theory (A.R. Rao ed.), MacMillan and Co. India Ltd., I.S.I. Lecture Notes Series 4 (1979), 251–267.

[9]   S.B. Rao, A survey of the theory of potentially P–graphic and forcibly P–graphic degree sequences, Lecture Notes in Math. No 885, Springer–Verlag, 1981, 417–440.

[10] P. Turán, Eine Extremalaufgabe aus der Graphen theorie, Mat. Fiz. Lapok 48 (1941), 436–452.

On Surviving Route Graphs of
Iterated Line Digraphs

M. Escudero

J. Fàbrega

M. A. Fiol

Universitat Politècnica de Catalunya
Barcelona, Spain

N. Homobono

Université Paris–Sud
Orsay Cedex, France

ABSTRACT

A routing ρ on a digraph G assigns to any pair (x, y) of vertices of G a fixed path $\rho(x, y)$ from x to y. Given a set F of "faulty elements" (vertices or arcs), the surviving route graph $R(G, \rho)/F$ of G is the digraph whose vertices are the non–faulty vertices of G, and where x is adjacent to y if the path $\rho(x, y)$ does not contain a faulty element. It is interesting that the diameter of $R(G, \rho)/F$ be as small as possible.

We call a vertex adjacent to and from all other vertices of $R(G, \rho)/F$ a (ρ, F)–central vertex. The existence of this kind of vertices in de Bruijn and Kautz digraphs is proved for a general class of routings. This implies that the diameter of $R(G, \rho)/F$ is at most 2. A simple method to find (ρ, F)–central vertices in these digraphs is also presented. Besides, an upper bound for the diameter of the surviving route graphs of iterated line digraphs is given.

1. Introduction

In the design of interconnection networks several features must be taken into account. One of them is the existence of efficient algorithms for routing messages. If the nodes have no information about the topology of the network, the problem of finding

This work has been supported by the Spanish Research Council (Comision Interministerial de Ciencia y Tecnologia, CICYT) under projects 1180–84 and 0173/86

the paths, along which the messages are to be sent, can be solved using a non–local routing fixed in advance. Thus, given any two nodes x, y, a fixed path is used for all communications from x to y. Usually, one is interested in a routing that assigns shortest paths between nodes.

When some nodes or links in the network fail, some routes become unavailable. However, assuming that the network remains connected, communication is still possible by sending every affected message along a sequence of surviving routes. If the time required to send a message along a route is dominated by the processing time at the endpoints, the total transmission time is roughly proportional to the number of re–routing steps. This number is a measure of the degradation caused by the faults. Hence, one is interested in *fault–tolerant* routings requiring a small number of re–routing steps.

To formalize the problem we consider the network modeled by a *strongly connected* digraph $G = (V, A)$. See [3] for the definitions not given in the paper. Given $x, y \in V$, an $x \to y$ *path* of G is a finite sequence $x = u_0, u_1, ..., u_{n-1}, u_n = y$ where $[u_{i-1}, u_i] \in A$ for $1 \le i \le n$. The number n is called the *length* of the path. If $n = 0$ the path is *trivial*. A *routing* ρ in G is a function $\rho: V \times V \to A^*$ where A^* is the set of all paths of G. See [4], [11]. A *minimal length* routing is a routing such that $\rho(x, y)$ is the shortest $x \to y$ path. Throughout the paper it is supposed that given any two vertices x,y the $x \to y$ path $\rho(x, y)$ has length $|\rho(x, y)|$ satisfying

$$0 \le |\rho(x,y)| \le D(G), \tag{1}$$

where $D(G)$ stands for the diameter of G. If $F = V_F \cup A_F$, $V_F \subset V$, $A_F \subset A$, we say that $\rho(x, y)$ *avoids* F if it contains no vertex of V_F nor arc of A_F.

Following Dolev, Halpern, Simons and Strong, [4], given a strongly connected digraph $G = (V, A)$, a routing ρ in G and a set of vertices and arcs $F = V_F \cup A_F$, we define the *surviving route graph* $R(G,\rho)/F$ as the digraph with set of vertices $V - V_F$ and where vertex x is adjacent to vertex y if, and only if, $\rho(x, y)$ avoids F. Of course, the cardinality of F must be less than the strong vertex–connectivity $\kappa(G)$ of G; otherwise F could disconnect G. By the considerations above, the "faulty" set F contains the nodes and links that fail in the network and the diameter of $R(G, \rho)/F$ is a measure of the worst–case behaviour of the faulty network. Thus, an interesting problem is to obtain graphs with routings that can tolerate many faults without too much increasing the diameter of the corresponding surviving route graph. See also [1], [2], [4], [5], [8], [10], [11], [12], [13], [14] for this and other related problems.

The line digraph technique has proved to be very useful to obtain large digraphs, that is, digraphs that have a large order for fixed maximum degree and diameter, [9]. Line digraphs admit very simple routing algorithms to find short paths between vertices.

Moreover, it has been proved in [7] that iterated line digraphs are also maximally connected if the iteration order is large enough.

This paper is devoted to showing that the line digraph technique is also useful in digraph problems involving routing vulnerability. In the next section we show how to define routings in line digraphs that satisfy condition (1). In particular, minimal length routings in line digraphs can be easily derived. An interesting property of the surviving route graph of the two best–known families of line digraphs, namely *de Bruijn* and *Kautz* digraphs, is studied in Section 3. We call a vertex c adjacent to and from all other vertices of $R(G, \rho)/F$ a *(ρ, F)–central* vertex. The existence of a (ρ, F)–central vertex c is interesting since, in the network modeled by G, any communication from node x to node y could always be carried on through such a vertex with only one re–routing step. This implies that the diameter of $R(G, \rho)/F$ is at most 2. If c is (ρ, F)–central for all ρ satisfying (1) we will refer to it as *F–central*. When $F = V_F \subset V$ (i.e. $A_F = \varnothing$) the existence of this kind of vertices is proved for de Bruijn and Kautz digraphs. We also give a simple algorithm to find F–central vertices in these two families of digraphs. The general case when $F = V_F \cup A_F$, is also considered. Finally, in Section 4, an upper bound for the diameter of the surviving route graphs of iterated line digraphs is given.

2. Routings in Iterated Line Digraphs

In the *line digraph* LG of a digraph G, each vertex represents an arc of G. Thus, $V(LG) = \{uv: [u, v] \in A(G)\}$; and a vertex uv is adjacent to a vertex wz if and only if $v = w$, that is, when the arc $[u, v]$ is adjacent to the arc $[w, z]$ in G. The *k–iterated line digraph*, L^kG, is defined recursively by $L^kG = LL^{k-1}G$ starting with $L^0G = G$. Hence, a vertex x of L^kG may be represented by a sequence $x_0x_1...x_k$ of vertices of G such that $[x_{i-1}, x_i] \in A$, $1 \le i \le k$, i.e., by an $x_0 \to x_k$ path, and x is adjacent to another vertex y if and only if $y = x_1...x_kx_{k+1}$, see [9]. It is well–known that, if G is different from a cycle, then

$$D(L^kG) = D(G) + k \qquad\qquad (2)$$

for any integer $k \ge 1$. See, for instance, [9], [15].

The two best–known families of large d–regular digraphs with diameter D, for any $d > 1$ and $D \ge 1$, are the de Bruijn digraphs $B(d, D)$ and Kautz digraphs $K(d, D)$. They are, respectively, the iterated line digraphs $L^{D-1}K_d^+$ and $L^{D-1}K_{d+1}$, where K_{d+1} denotes the *complete symmetric digraph* on $d+1$ vertices in which each vertex is adjacent to all others, and K_d^+ is obtained from K_d by adding a loop on each vertex. The de Bruijn digraph $B(d, D)$ has order d^D and the Kautz digraph $K(d, D)$ has order

$d^D + d^{D-1}$. Besides, the strong vertex–connectivity of $B(d, D)$ is $d-1$ ($B(d, D)$ has loops) and the strong vertex–connectivity of $K(d, D)$ is d. See [7], [11], [15].

If a routing ρ' satisfying condition (1) is defined in G, then it is very easy to derive a routing ρ in $L^k G$ that also satisfies (1). Given any two vertices $x = x_0 x_1 ... x_k$ and $y = y_0 y_1 ... y_k$ of $L^k G$ let $\rho'(x_k, y_0) = x_k, x_{k+1}, ..., x_{k+n-1}, y_0$, $n = |\rho'(x_k, y_0)|$, $0 \leq n \leq D(G)$, be the $x_k \rightarrow y_0$ path given by the routing in G. Writing $y = y_0 y_1 ... y_k$ as $x_{k+n} x_{k+n+1} ... x_{2k+n}$, $\rho(x, y)$ can be defined as:

$$\rho(x, y) = z_0 \; (=x), \; z_1, ..., z_{n+k-1}, \; z_{n+k} \; (=y),$$

$$z_i = x_i x_{i+1} ... x_{i+k} \; , \; 0 \leq i \leq n+k. \tag{3}$$

Note that $|\rho(x, y)| = n + k \leq D(G) + k = D(L^k G)$. Of course, $\rho(x, y)$ is not in general a shortest $x \rightarrow y$ path.

If $G = K_d^+$ and $\rho'(u, v) = u,v$ for all $u, v \in V(G)$ (when $u = v$ the path $\rho(u, u)$ is not trivial and consists of the loop $[u, u]$), the routing ρ defined in $B(d,D) = L^{D-1} K_d^+$ by (3) assigns to any pair of vertices $x, y \in V(B(d, D))$ the following path $\rho(x, y)$ of length D:

$$\rho(x, y) = z_0 \; (=x), \; z_1, ..., z_{D-1}, \; z_D \; (=y),$$

$$z_i = x_i ... x_{D-1} y_0 ... y_{i-1} \; , \; 0 \leq i \leq D. \tag{4}$$

With such a routing ρ, in the corresponding network any message can be sent from its origin to its destination with precisely D delay time units. See also, [10].

Similarly, if $G = K_{d+1}$ and $\rho'(u, v) = u, v$ when $u \neq v$ and $\rho'(u, u) = u$, the routing (3) in $K(d, D) = L^{D-1} K_{d+1}$ assigns to x, y a path $\rho(x, y)$ of length D if $x_k \neq y_0$ or $D-1$ if $x_k = y_0$.

In order to define a minimal length routing ρ_m in $L^k G$, it seems natural to take advantage of the largest common subsequence of the two sequences $x_0 x_1 ... x_k$ and $y_0 y_1 ... y_k$ representing x and y. More precisely, let t, $1 \leq t \leq k+1$, be the largest positive integer — if it exists — such that $x_{k-t+1} = y_0, x_{k-t+2} = y_1, ..., x_k = y_{t-1}$. In this case, $t = t(x, y)$ is said to be the *intersection length* of x and y. If there is not such a t, let $x_k, x_{k+1}, ..., x_{k+n-1}, x_{k+n} \; (=y_0)$ be a shortest $x_k \rightarrow y_0$ path in G and define the intersection length now as $t(x, y) = -n + 1$. So, $-D(G) + 1 \leq t(x, y) \leq k + 1$ in all cases. In particular, if $L^k G = B(d, D)$ or $L^k G = K(d, D)$, then $0 \leq t(x, y) \leq D$ because $D(K_{d+1}) = D(K_d^+) = 1$.

The intersection length of x and y is used in [6], [7] to find a shortest $x \to y$ path in $L^k G$ using only the information of the structure of G. Writing $y = y_0 y_1 ... y_k$ as $x_{k-t+1} x_{k-t+2} ... x_{2k-t+1}$, $p_m(x, y)$ can be defined as

$$p_m(x, y) = z_0 \, (=x), \, z_1, \, ..., \, z_{k-t}, \, z_{k-t+1} \, (=y),$$

$$z_i = x_i x_{i+1} ... x_{i+k} \, , \, 0 \le i \le k - t + 1. \tag{5}$$

The shortest path thus obtained has length $k - t + 1$. Indeed, the distance $d(x, y)$ from x to y is

$$d(x, y) = |p(x, y)| = k - t + 1. \tag{6}$$

3. F–central Vertices in De Bruijn and Kautz Digraphs

In this section $G^D = (V^D, A^D)$ stands for $B(d,D) = L^{D-1} K_d^+$ or $K(d, D) = L^{D-1} K_{d+1}$. Hence, if V denotes the vertex set of K_d^+ or K_{d+1},

$$V^D = \{x = x_0 x_1 ... x_{D-1}, \, x_i \in V, \, |V| = d\}$$

when $G^D = B(d,D)$, or

$$V^D = \{x = x_0 x_1 ... x_{D-1}, \, x_i \in V, \, x_i \ne x_{i+1}, \, |V| = d + 1\}$$

when $G^D = K(d, D)$. In both cases $|V| = \kappa + 1$ where κ is the strong vertex–connectivity of G^D.

Let ρ be a routing in G^D such that $0 \le |\rho(x, y)| \le D$ for all $x, y \in V^D$, F^D be a subset of V^D with cardinality $|F^D| < \kappa$, and consider the surviving route graph $R(G^D, \rho)/F^D$. As it was stated in the Introduction, a vertex $c \in V^D - F^D$ is F^D–central if the two paths $\rho(x, c)$ and $\rho(c, y)$ avoid F^D for all $x, y \in V^D - F^D$. A sufficient condition for c to be F^D–central is that c be at distance D from and to any vertex f of F^D:

$$d(c,f) = d(f,c) = D. \tag{7}$$

Indeed, since the paths $\rho(x,c)$ and $\rho(c,y)$ have length at most D, they cannot contain vertices of F^D. By (6) (with $k = D-1$) a condition equivalent to (7) is:

$$t(c,f) = t(f,c) = 0 \, \text{ for any } \, f \in F^D. \tag{8}$$

It can be proved that condition (7) is also necessary for c to be F^D–central. However, for a particular routing ρ there could exist a (ρ, F^D)–central vertex not satisfying (7).

Our first result is:

Proposition 1   *Suppose* $F^D \subset V^D$, $|F^D| < \kappa$. *Then there exist in* G^D *two* F^D–*central vertices* $c^1 = c_0^1 c_1^1 \ldots c_{D-1}^1$ *and* $c^2 = c_0^2 c_1^2 \ldots c_{D-1}^2$, *such that* $c_0^1 \neq c_0^2$ *and*

$$t(c^i,f) = t(f,c^i) = 0, \quad i = 1,2, \text{ for any } f \in F^D. \tag{9}$$

Proof   By induction on the diameter D.

(a) Since $B(d,1) = K_d^+$ and $K(d,1) = K_{d+1}$, if $c \in V{-}F^1$, then for any $f \in F^1$:

$$d(c,f) = d(f,c) = 1$$

or equivalently

$$t(c,f) = t(f,c) = 0. \tag{10}$$

Moreover, since $F^1 \subset V$, $|V{-}F^1| = |V| - |F^1| \geq (\kappa+1) - (\kappa-1) = 2$ and the proposition holds when $D = 1$.

(b) Suppose now that the proposition is true for G^{D-1}. Let $f_0 f_1 \ldots f_{D-1}$ be the sequence representing a generic vertex f of F^D, and consider the set

$$F^{D-1} = \{f_1 f_2 \ldots f_{D-1}: \ f_0 f_1 \ldots f_{D-1} \in F^D \} \subset V^{D-1} \tag{11}$$

as the faulty set in G^{D-1}. We claim that for any F^D–central vertex $c_0 c_1 \ldots c_{D-2}$ of G^{D-1} satisfying (9), there is a $c_{D-1} \in V$ such that $c = c_0 c_1 \ldots c_{D-2} c_{D-1}$ is a F^D–central vertex of G^D that also satisfies (9). This will complete the proof because, by the induction hypothesis, there exist in G^{D-1} two F^D–central vertices satisfying (9) such that the first element of their representing sequences is different.

First of all note that c_{D-1} cannot belong to the set

$$A_0 = \{f_0: \ f_0 f_1 \ldots f_{D-1} \in F^D\}; \tag{12}$$

otherwise $t(c,f) \neq 0$ for some $f \in F^D$.

To choose an appropriate c_{D-1} let us consider a set $B \subset V$ defined in the following way: if $c_{D-2} \in A_0$, that is, if $c_{D-2} = f_0^1 = \ldots = f_0^s$ for some $f^i \in F$, $i = 1,2,\ldots,s$, $s < \kappa$, let

$$A_1 = \{f_1^i \ ; \ 1 \leq i \leq s\} \tag{13}$$

and let $B = A_0 \cup A_1$. The cardinality of B is at most

$$|B| \le |A_0| + |A_1| \le (\kappa - s) + s = \kappa. \tag{14}$$

On the other hand, if $c_{D-2} \notin A_0$, let either $B = A_0$ when $G^D = B(d,D)$ so that $|B| \le \kappa - 1$ or $B = A_0 \cup \{c_{D-2}\}$ when $G^D = K(d,D)$ so that $|B| \le \kappa$. As $|V| = \kappa + 1$, in both cases there is at least one element in the set $V - B$.

If $c_{D-1} \in V - B$, then $c = c_0 c_1 ... c_{D-2} c_{D-1}$ is the F^D-central vertex claimed above. Indeed, let $t = t(c,f)$, $f \in F$:

(b1) The intersection length t cannot be greater than 2 since if $t > 2$:

$$c_{D-t} c_{D-t+1} ... c_{D-1} \equiv f_0 f_1 ... f_{t-1} \Rightarrow$$

$$\Rightarrow c_{D-t+1} ... c_{D-2} \equiv f_1 ... f_{t-2} \Rightarrow t(c_0 c_1 ... c_{D-2}, f_1 f_2 ... f_{D-1}) \ne 0$$

which is impossible if $c_0 c_1 ... c_{D-2}$ satisfy (9).

(b2) It must be $t \ne 2$ since if $t = 2$, then

$$c_{D-2} c_{D-1} \equiv f_0 f_1$$

which is impossible because either $c_{D-2} \ne f_0$ or c_{D-1} is different from f_1 when c_{D-2} is equal to f_0 by the choice of c_{D-1}.

(b3) Moreover, $t \ne 1$ since if $t = 1$, then c_{D-1} is equal to f_0 against the choice of c_{D-1}.

Finally, let now $t' = t(f,c)$

(b1') $t' > D-1$; otherwise if $1 \le t' \le D-1$, then

$$f_{D-t'} f_{D-t'+1} ... f_{D-1} \equiv c_0 c_1 ... c_{t'-1} \Rightarrow$$

$$\Rightarrow t(f_1 f_2 ... f_{D-1}, c_0 c_1 ... c_{D-2}) \ne 0$$

which is impossible again.

(b2') Besides, $t' \ne D$ since $t' = D \Rightarrow c = f \Rightarrow t(c,f) = D$ against (b1)☐

A constructive method to obtain F^D-central vertices in G^D follows from this proof:

Algorithm

$A_0 := \{f_{D-1}: f_0f_1...f_{D-1} \in F^D\}$

let $c_0 \in V - A_0$;

$k := 1$

    While $k < D$

        $A_0 = \{f_{D-k-1}: f_0f_1...f_{D-1} \in F^D\}$

        if $c_{k-1} \notin A_0$

            if $G^D = B(d,D)$ *let* $c_k \in V - A_0$

            else   *let* $c_k \in V - (A_0 \cup \{c_{k-1}\})$

        else

            $A_1 := \{f_{D-k}^i : c_{k-1} = f_{D-k-1}^i, 1 \leq i \leq s\}$

            let $c_k \in V - (A_0 \cup A_1)$;

    $k := k + 1$

In the first step of the algorithm there are always two possibilities for the choice of c_0. So, the number of F^D–central vertices generated by the algorithm is at least two. This is exactly the number of vertices obtained when G^D is a Kautz digraph with degree 2, $K(2,D)$, since in this case there is only one feasible element c_k in each step. On the other hand, when $G^D = B(d,D)$ and $c_{k-1} \notin A_0$, there are at least 2 possibilities for the choice of c_k.

Example: Let $G^5 = K(3,5)$, $V = \{0,1,2,3\}$, $F^5 = \{f^1=02310, f^2=12301\}$. The following table shows how the F^5–central vertices are generated.

$k=0$ c_0	$k=1$ c_1	$k=2$ c_2	$k=3$ c_3	$k=4$ c_4	c
2	3	2	0	3	23203
			1	3	23213
3	2	0	1	3	32013
			3	2	32032
		1	0	3	32103
			3	2	32132

Remark 1:   The F^D–central vertices $c_0c_1...c_{D-1}$ generated by the algorithm are such that for $1 \leq k \leq D$, $c_0c_1...c_{k-1}$ is F^k–central in G^k when $F^k = \{f_{D-k}f_{D-k+1}...f_{D-1}: f_0f_1...f_{D-1} \in F^D\} \subset V^k$.

Remark 2:   In the proof of Proposition 1 we have considered the set F^{D-1} whose members are of the form $f_1 f_2 \ldots f_{D-1}$. If we consider the set F^{D-1} formed by the sequences $f_0 f_1 \ldots f_{D-2}$ (that are obtained deleting the last element of the sequences $f \in F^D$), an analogous proposition can be established. In this case, the two F^D–central vertices in G^D, instead of differing in the first element of their representing sequences, differ in the last element. Moreover, an analogous algorithm to generate F^D–central vertices can be given. However, using the fact that de Bruijn and Kautz digraphs are isomorphic to their converse digraphs, it can be proved that the F^D–central vertices thus obtained are those generated by the algorithm given above.

Therefore the F^D–central vertices generated by the algorithm are such that $c_{D-k} c_{D-k+1} \ldots c_{D-1}$ is also F^k–central in G^k when $F^k = \{f_0 f_1 \ldots f_{k-1} : f_0 f_1 \ldots f_{D-1} \in F^D\} \subset V^k$, $1 \le k \le D$ (compare with Remark 1).

Let us now consider the general case when the faulty set F^D contains arcs as well as vertices, that is $F^D = V_F \cup A_F$, $V_F \subset V^D$, $A_F \subset A^D$, $|V_F| + |A_F| < \kappa$. If $A_F^D \neq \varnothing$ the existence of a F^D–central vertex cannot be assured. For instance, let $G^1 = K(3,1)$ and let $F^1 = A_F = \{[0,1], [2,3]\}$. See Figure 1. In $R(G^1,\rho)/F^1$ there is no F^1–central vertex.

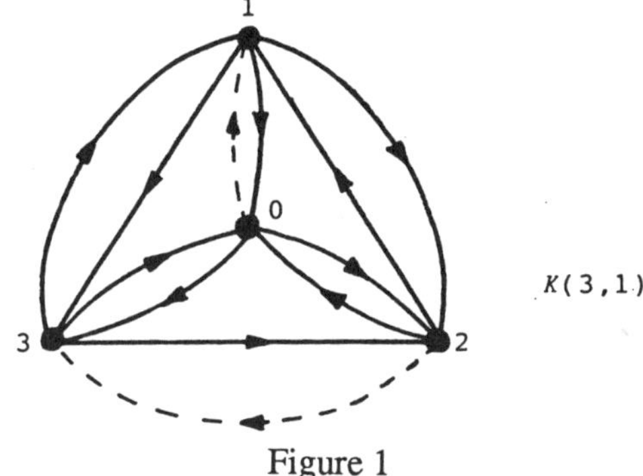

Figure 1

Even so, the diameter of this surviving route graph is 2. This is a general result that can be proved by Proposition 1.

Proposition 2   *Given $F^D = V_F \cup A_F$, $|F^D| < \kappa$, the diameter of $R(G^D,\rho)/F^D$ is 2.*

Proof   We have to prove that given any two vertices $x,y \in V^D - F^D$, the distance from x to y in $R(G^D,\rho)/F^D$ is at most 2.

(a) Consider first the case in which there is not a faulty arc from x to y or from y to x (i.e., if the arcs $[x,y]$ or $[y,x]$ exist, then they do not belong to A_F). Let $F' \subset V^D$ be the set $V_F \cup W_F$ where $W_F = \{u : [u,v] \in A_F, u \neq x, y\} \cup \{v : [u,v] \in A_F, u = x \text{ or } u = y\}$. Proposition 1 implies the existence of a F^D–central vertex c of

$R(G^D,\rho)/F'$. Then, c is such that both $\rho(x,c)$ and $\rho(c,y)$ avoid F' and therefore they also avoid F^D.

(b) Suppose now that there is not a faulty arc from x to y but $[y,x] \in A_F$. In this case let $F' = V_F \cup W_F$ where $W_F = \{u: [u,v] \in A_F\}$, and note that $y \in W_F$. Let c be a F^D–central vertex of $R(G^D,\rho)/F'$ at distance D to and from any vertex of F' (Proposition 1). The path $\rho(x,c)$ avoids F' and so, also F^D. Moreover, as $d(c,f) = D$ for all $f \in F'$ and the length of the path $\rho(c,y)$ is at most D, this path cannot contain any vertex of F' except y. Thus, $\rho(c,y)$ also avoid F^D.

(c) Finally, it remains to consider the case $[x,y] \in A_F$. Let again F' be the set $V_F \cup W_F$ with $W_F = \{u: [u,v] \in A_F\}$, and note that $x \in W_F$. If $x = y$, let $c \in V^D - F'$ be at distance D to and from all vertices of F'. Then $\rho(x,c)$ and $\rho(c,x)$ do not contain any other vertex of F' except x. This implies that both $\rho(x,c)$ and $\rho(c,x)$ avoid F^D. On the other hand, if $x \neq y$, there exist (again by Proposition 1) two F'–central vertices $c^1 = c_0^1 c_1^1 ... c_{D-1}^1$ and $c^2 = c_0^2 c_1^2 ... c_{D-1}^2$, $c_0^1 \neq c_0^2$, which are at distance D to and from all vertices of F'. In particular, $d(x,c^i) = d(c^i,x) = D$, $i = 1,2$. But $[x,y] \in A_F$ so $d(y,c^i)$ must be D or $D{-}1$. Assume that there exists an i (say $i = 1$) such that $d(y,c^1) = D{-}1$. Then,

$$y_{D-1} = c_0^1 \;\Rightarrow\; y_{D-1} \neq c_0^2 \;\Rightarrow\; d(y,c^2) = D.$$

Hence, neither $\rho(x,c^2)$ nor $\rho(c^2,y)$ contain the arc $[x,y]$.$\quad\square$

An inductive proof can also be given for Proposition 2. Thus, given $x,y \in V^D - F^D$, an analogous algorithm to the one presented above to find a vertex c_{xy} such that $\rho(x,c_{xy})$ and $\rho(c_{xy},y)$ avoid F^D can be formulated.

Proposition 2 is also proved in [10] when ρ is the minimal length routing or the routing in which all the paths have length D or $D{-}1$ (Kautz digraph) or D (de Bruijn digraph).

4. Bounds on the Diameter of Surviving Route Graphs of Iterated Line Digraphs.

Throughout this section it is supposed that $G = (V,A)$ is a strongly connected digraph without loops. It follows from the definition of LG that $\delta(LG) = \delta(G) = \delta$ (δ stands for the minimum degree) and thus, $\delta(L^kG) = \delta$ for $k \geq 1$. Besides, it has been proven in [7] that if $k \geq D(G) - 1$, then the strong vertex–connectivity and the strong arc–connectivity of L^kG are maximum, that is, $\kappa(L^kG) = \lambda(L^kG) = \delta$.

Let ρ be a routing in L^kG (satisfying (1)) and F be a faulty set with cardinality less than δ. In this last section we will derive an upper bound for the diameter of $R(L^kG,\rho)/F$ when $k \geq D(G) - 1$. Note that, by the considerations above, $G - F$ is strongly connected.

We need some definitions:

Definition 1   *([7]): Let $\ell = \ell(G)$, $1 \leq \ell \leq D(G)$, be the greatest integer such that, for any two (not necessarily different) vertices $x,y \in V$,*

(a) if $d(x,y) < \ell$, then the shortest $x \to y$ path is unique and there is no $x \to y$ path of length $d(x,y) + 1$;

(b) if $d(x,y) = \ell$, then there is only one shortest $x \to y$ path.

It is proved in [7] that if G is different from a cycle, then

$$\ell(L^kG) = \ell(G) + k. \tag{15}$$

Example: If $G = K_{d+1}$, trivially $\ell(G) = 1$. Then,

$$\ell(K(d,D)) = \ell(L^{D-1}K_{d+1}) = \ell(K_{d+1}) + (D-1) = D.$$

So, for the Kautz digraphs, the parameter ℓ is equal to the diameter D.

Before stating the second definition, let us introduce a little more notation. Given $F \subset V$ and $x \in V$ let $d(x, F)$ denote the minimum of the distances from x to every vertex of F. Define $d(F, x)$ analogously. Let $\overline{p}(x \to y)$ denote a shortest path from x to y. For each $f \in F$ such that $d(x, f) \leq \ell$, the vertex adjacent from $x \in V - F$ in the unique shortest $x \to f$ path is denoted by $v(x \to f)$. Analogously, $v(x \to F) = \{v(x \to f): f \in F, d(x, f) \leq \ell\}$.

From now on it is supposed $\delta(G) > 1$.

Definition 2   *Let $\imath = \imath(G)$, $1 \leq \imath \leq \ell(G)$, be the maximum integer such that given any set $F \subset V$, $|F| < \delta$, there exists at least one vertex $z \in V - F$ such that $d(F, z) \geq \imath$ and $d(z, F) \geq \imath$.*

Note that taking z to be any vertex of $V - F$, we always have $\imath \geq 1$.

Proposition 3   *For all $k \geq 0$,*

$$\imath(L^kG) = \imath(G) + k. \tag{16}$$

Proof   As $L^k G = LL^{k-1} G$ it suffices to consider the case $k = 1$.

Let $r(G) = r$, $\ell(G) = \ell$, $F \subset V(LG)$, $|F| < \delta$. The elements of F have the form $f = f_0 f_1$, where $[f_0, f_1] \in A(G)$. Consider the two subsets of $V(G)$, $F_0 = \{f_0: f_0 f_1 \in F\}$ and $F_1 = \{f_1: f_0 f_1 \in F\}$. By Definition 2, there exists in $V(G) - F_1$ a vertex z such that $d(F_1, z) \geq r$ and $d(z, F_1) \geq r$.

When $r > 1$ and t is any vertex adjacent from z, zt is a vertex of LG such that $d(f, zt) = d(f_1, z) + 1 \geq r + 1$ for any $f \in F$; that is, $d(F, zt) \geq r + 1$. If $r = 1$ the vertex z could belong to F_0. Then, to assure that $zt \notin F$ and so $d(F, zt) \geq r + 1$, let $t \in \Gamma^+(z) - Z$, where $\Gamma^+(z)$ denote the set of vertices adjacent from z and $Z = \{f_1: f_0 f_1 \in F, f_0 = z\}$.

To complete the proof we show that there is at least one vertex $t \in \Gamma^+(z) - Z$ that satisfies

$$d(zt, F) \geq r + 1. \tag{17}$$

To this end, if $r > 1$ let $t \in \Gamma^+(z) - v(z \to F_0)$. As before, when $r = 1$ it could be $z = f_0$ for some $f_0 \in F_0$. In this case, to assure that $zt \notin F$, consider $N = \{v(z \to f_0): f_0 \in F_0, f_0 \neq z\} \cup Z$ (clearly $|N| < \delta$) and let $t \in \Gamma^+(z) - N$. Let us see that for such a t we have $d(t, F_0) > r$ so that (17) holds.

Since f_0 is adjacent in G to f_1, $d(z, F_1) \leq d(z, F_0) + 1$. Hence,

$$d(z, F_0) \geq d(z, F_1) - 1 \geq r - 1. \tag{18}$$

For each $f_0 \in F_0$ the two $z \to f_0$ paths $\overline{p}(z \to f_0) = z, \overline{p}(w \to f_0)$, where $w = v(z \to f_0)$, and $z, \overline{p}(t \to f_0)$ are different because so are the two arcs $[z, w]$ and $[z, t]$ (Of course, if $z = f_0$, then $\overline{p}(z \to f_0)$ is trivial). The path $z, \overline{p}(t \to f_0)$ has length $1 + d(t, f_0)$. Thus, we can conclude:

(a) if $d(z, f_0) > r$, then

$$1 + d(t, f_0) \geq d(z, f_0) \;\Rightarrow\; d(t, f_0) \geq r,$$

(b) if $d(z, f_0) = r$, then

$$1 + d(t, f_0) \geq d(z, f_0) + 1 \;\Rightarrow\; d(t, f_0) \geq r,$$

since, as $r \leq \ell$, there is only one shortest $z \to f_0$ path;

(c) if $d(z, f_0) = r - 1$ then, since $r - 1 < \ell$, there is only one shortest $z \to f_0$ path and no $z \to f_0$ path of length $d(z, f_0) + 1$. Hence,

$$1 + d(t, f_0) \geq d(z, f_0) + 2 \implies d(t, f_0) \geq d(z, f_0) + 1 \geq \tau.$$

Finally, note that, according to Definition 2, $\tau + 1 \leq \ell + 1 = \ell (LG)\square$

Example: As $\tau(K_{d+1}) = 1$, $\tau(K(d, D)) = \tau(L^{D-1}K_{d+1}) = D$, $D \geq 1$. This proves the existence of a F–central vertex in the Kautz digraph $K(d,D)$ such that $d(z,F) = d(F, z) = D$. Moreover, since in K_{d+1} there are two vertices, c_0^1 and c_0^2, at distance 1 to and from F, $|F| < d$, there are also in $K(d,D)$ two F–central vertices, $c^1 = c_0^1 c_1^1 \ldots c_{D-1}^1$ and $c^2 = c_0^2 c_1^2 \ldots c_{D-1}^2$ satisfying (9), as stated in Proposition 1.

Suppose that ρ is a routing in $L^k G$ such that the path $\rho(x,y) = x,y$ when x is adjacent to y. In particular, ρ could be any minimal length routing.

Theorem 1　　*Let $\ell(G) = \ell$, $\tau(G) = \tau$, $D(G) = D$. Let $F \subset V(L^k G)$, $|F| < \delta$. be a faulty set in $L^k G$. If $k > D - \tau - \ell$, then the diameter D_R of $R(L^k G,\rho)/F$ satisfies $D_R \leq 2(D - \tau + 1)$.*

Proof　Let x, y be vertices of $L^k G - F$. It is proved in [7] that if $0 \leq s < \ell (L^k G) = \ell + k$, then there are, in $L^k G - F$, $x \to x'$ and $y' \to y$ paths of length s such that $d(x', F) > s$ and $d(F, y') > s$. By Proposition 3 there exists in $L^k G - F$ a vertex z such that $d(z, F) \geq \tau (L^k G) = \tau + k$ and $d(F, z) \geq \tau (L^k G) = \tau + k$. Besides, $d(x', z) \leq D(L^k G) = D + k$ and $d(z, y') \leq D(L^k G) = D + k$. See Figure 2.

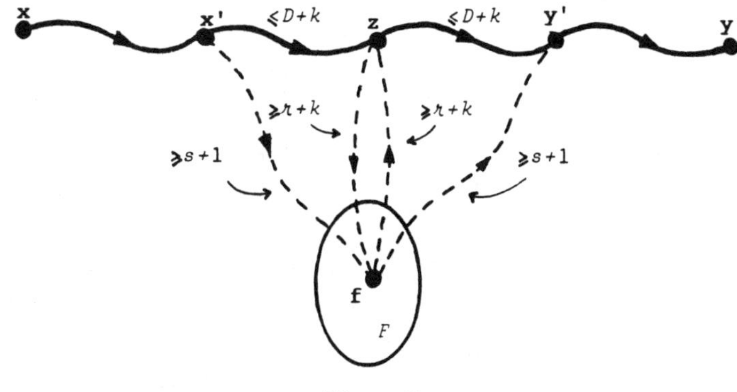

Figure 2

All $x' \to z$ paths [$z \to y'$ paths] containing some vertex $f \in F$ have length at least $s + k + \tau + 1$. Thus, if

$$s + k + \tau + 1 > D + k, \tag{19}$$

then $\rho(x', z)$ avoids F [$\rho(z, y')$ avoids F] because $|\rho(x', z)| \leq D + k$ [$|\rho(z, y')| \leq D + k$]. Let s take the minimum possible value so that (19) holds; that is, $s = D - \ell$. Thus, since s has to be less than $\ell + k$:

$$D - \ell < \ell + k \;\Rightarrow\; k > D - \ell - \ell \qquad (20)$$

So, if the iteration order k satisfies (20), both $\rho(x', z)$ and $\rho(z, y')$ avoid F.

Because of the condition imposed to ρ, the $x \to x'$ path [$y' \to y$ path] in $L^k G - F$ is also in $R(L^k G, \rho)/F$. Besides, x' is adjacent to z [z is adjacent to y'] in $R(L^k G, \rho)/F$. Thus, denoting the distance from x to y in $R(L^k G, \rho)/F$ by $d_R(x, y)$, we have

$$d_R(x, y) \leq 2(s + 1) = 2(D - \ell + 1) \square \qquad (21)$$

Since $\ell \geq 1$, $\ell \geq 1$, if $k \geq D - 1$, then (21) holds.

Remark: If $\ell = \ell = D$, the minimum value of s in (21) is $s = D - \ell = 0$. So, $d_R(x, y) \leq 2$. Moreover, in this case it must be x = x', y = y', and the condition we have imposed on ρ ($\rho(x, y) = x, y$ when x is adjacent to y) is not needed. That is the case when the iterated line digraph is a Kautz digraph.

To prove that Theorem 1 is also valid when the faulty set F contains some arcs, we need to generalize Definition 2.

Definition 3   *Let $\ell' = \ell'(G)$, $1 \leq \ell' \leq \ell(G)$, be the maximum integer such that given any set $F = V_F \cup \{[u, v]\}$, $V_F \subset V$, $[u, v] \in A$, $|F| < \delta$, there exists at least one vertex $z \in V - (V_F \cup \{u, v\})$ such that $d(z, V_F) \geq \ell'$, $d(V_F, z) \geq \ell'$ and $d(z, u) \geq \ell'$, $d(v, z) \geq \ell'$.*

This parameter $\ell'(G)$ is well–defined because taking z to be any vertex of $V - (V_F \cup \{u, v\})$ we always have $\ell' \geq 1$.

Reasoning as in the proof of Proposition 3, a result analogous to (16) can be obtained.

Proposition 4   *For all $k \geq 0$,*

$$\ell'(L^k G) = \ell'(G) + k. \qquad (22)$$

Example: Since $\ell'(K_{d+1}) = 1$, it follows that $\ell'(K(d, D)) = D$.

Let ρ be a routing defined as in Theorem 1 and suppose now that $F = V_F \cup A_F$, $V_F \subset V(L^k G)$, $A_F \subset A(L^k G)$, $|F| < \delta$. When there is not a faulty arc from x to y or from y to x, $x, y \in V(L^k G) - V_F$, it is proved by reasoning as in the proof of Theorem 1, but using now the definition of $\imath'(G)$ and Proposition 4, that

$$d_R(x, y) \le 2(D - \imath' + 1) \quad \text{if} \quad k > D - \imath' - \ell. \tag{23}$$

Let us just examine in more detail the case when there is an arc of A_F from x to y. Let F' be the set $V_F' \cup \{[x, y]\}$ where $V_F' = V_F \cup W_1 \cup W_2$, with $W_1 = \{u: [u, v] \in A_F, u \ne x, y\}$, and $W_2 = \{v: [u, v] \in A_F, u = x \text{ or } u = y, v \ne x, y\}$. From the definition of $\ell(G)$ it follows that, if $0 \le s < \ell(L^k G) = \ell + k$ then there is in $L^k G - F'$ an $x \rightarrow x'$ path [an $y' \rightarrow y$ path] of length s such that $d(x', V_F') \ge s + 1$ and $d(x', x) \ge s$ $[d(V_F', y') \ge s + 1$ and $d(y, y') \ge s]$. Moreover, if there is also a faulty arc from y to x, this $x \rightarrow x'$ path $[y' \rightarrow y$ path] is such that $d(x', y) \ge s + 1$ $[d(x, y') \ge s + 1]$; see [7]. Besides, as $r'(L^k G) = \imath' + k$, there exists in $L^k G - F'$ a vertex z such that $d(z, V_F') \ge \imath' + k$, $d(V_F', z) \ge \imath' + k$ and $d(z, x) \ge \imath' + k$, $d(y, z) \ge \imath' + k$. See Figure 3. Now, reasoning again as in the proof of Theorem 1, (23) is easily derived.

Finally, the case $[y, x] \in A_F$ can be analyzed in a similar way.

Thus, the following result can be stated:

Theorem 1':   *Let* $F = V_F \cup A_F$, $|F| < \delta$. *If* $k \ge D - \imath' - \ell$, *then the diameter* D_R *of* $R(L^k G, \rho)/F$ *satisfies* $D_R \le 2(D - \imath' + 1)$.

The remark after Theorem 1 also applies to Theorem 1'. Indeed, when G^D is a Kautz digraph, Proposition 2 is a corollary of Theorem 1'.

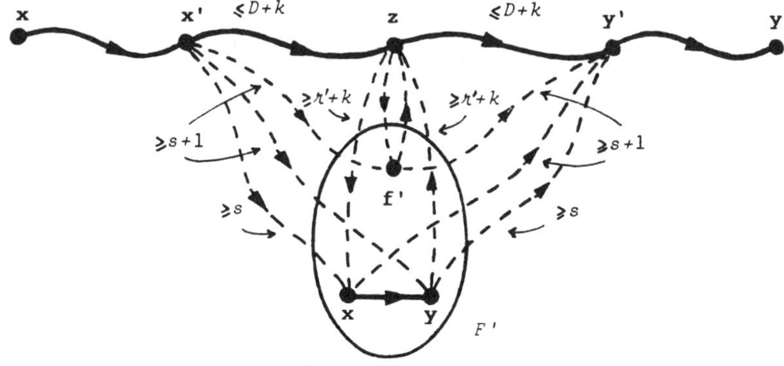

Figure 3.

ACKNOWLEDGMENT: The authors would like to thank the referee for his helpful comments.

REFERENCES

[1]   D. I. Ameter and M. D. Gree, Graphs and Interconnection Networks. To appear.

[2]   A. Broder, D. Dolev, M. Fischer and B. Simons, Efficient fault–tolerant routing in networks, Proc. of ACM 16th. STOC (1984) 536–541.

[3]   G. Chartrand and L. Lesniak, Graphs and Digraphs. Wadsworth, Monterrey (1986).

[4]   D. Dolev, J. Halpern, B. Simons and R. Strong, A new look at fault–tolerant network routing. Proc. of ACM 16th. STOC (1984) 526–535.

[5]   M. Escudero, J. Fàbrega and P. Morillo, Fault–tolerant routings in double–loop networks. *Ars Combinatoria* **25A** (1988) 187–198.

[6]   J. Fàbrega, M. A. Fiol and J. L. A. Yebra, Connectivity and reliable routing algorithms in line digraphs. Proc. III IAESTED International Symp. Applied Informatics, Grindelwald, Switzerland Acta Press (1985) 45–50.

[7]   M. A. Fiol, J. Fàbrega, and M. Escudero, Short paths and connectivity in graphs and digraphs, XII British Combinatorial Conference, Norwich, 3 –7 Jul 1989. Submitted to *Ars Combinatoria*.

[8]   P. Feldman, Fault–tolerance of minimal path routings in a network. Proc. of ACM 17th STOC (1985) 327–334.

[9]   M. A. Fiol, J. L. A. Yebra and I. Alegre, Line digraph iterations and the (d,k) digraph problem. *IEEE Trans Comput.* C–33 (1984) 400–403.

[10] N. Homobono, and C. Peyrat, Fault–tolerant routings in Kautz and de Bruijn networks. Submitted, *LRI Report* **281** (1986).

[11] N., Homobono, *Resistance aux pannes de grans reseaux d'interconnexion. These de docteur 3ieme cycle*, Université de Paris–Sud, Orsay (1987).

[12] M. Imase and Y. Manabe, Fault–tolerant routings in a κ–connected network. Submitted to *Information Processing Letters* (1987).

[13] K. Kawaguchi and K. Wada, Highly fault–tolerant network routing for (k+1)– node (edge) connected graphs, Paper of Technical Group COMP86–16, IECE Japan (1986).

[14] D. Peleg and B. Simons, On fault–tolerant routings in general graphs. *IBM Research Report RJ5194* (1986).

[15] S. M. Reddy, J. G. Kuhl, S. H. Hosseini and H. Lee, On digraphs with minimum diameter and maximum connectivity. Proc. 20th Annual Allerton Conference (1982) 1018–1026.

Hamiltonian Properties and Adjacency Conditions in K(1, 3)–Free Graphs

Ralph J. Faudree*
Memphis State University

Ronald J. Gould*
Emory University

Terri E. Lindquester
Rhodes College

ABSTRACT

We investigate several hamiltonian related properties in K(1, 3)–free graphs by imposing a bound on the neighborhood union of pairs of nonadjacent vertices. We show that the basic results concerning neighborhood unions and hamiltonian properties in graphs can be improved for graphs containing no induced K(1, 3) as a subgraph.

A graph is said to be *K(1, 3)–free* if it does not contain a copy of the complete bipartite graph, K(1, 3), as an induced subgraph. There have been many results in recent years dealing with K(1, 3)–free graphs. For example, see [6] and [11]. The assumption that a graph is K(1, 3)–free is quite strong, and it provides the structure needed to obtain some interesting results involving longest paths and cycles in graphs. Terms not defined here can be found in [5].

In [10], M. Matthews and D. Sumner proved that if G is a connected K(1, 3)–free graph with $\delta \geq (p - 2)/3$, then G is traceable. They also showed that any 2–connected K(1, 3)–free graph with the same minimum degree condition is hamiltonian. This imposition of a lower bound on δ accounts for a more even edge distribution in the graph.

* Supported by O.N.R. contract No. N000014–88–K–0070.

Recently, a different approach was taken by Zhang [14]. With a lower bound on the degree sums of sets of vertices, he was able to avoid restricting the graph with a minimum degree condition. More specifically, he proved:

Theorem   (Zhang [14])   *If G is a k–connected $(k \geq 2)$ $K(1, 3)$–free graph of order p such that*

$$\sum_{v \in I} deg(v) \geq p - k$$

for any independent $(k + 1)$–set I, then G is hamiltonian.

To further reduce the edge density of the graph, we consider the neighborhood unions of pairs of nonadjacent vertices in $K(1, 3)$–free graphs. We find the basic results concerning neighborhood unions and hamiltonian properties in graphs [2] can be improved for graphs containing no induced $K(1, 3)$. As in [1], we define

$$NC = \min |N(u) \cup N(v)|,$$

where the minimum is taken over all pairs of nonadjacent vertices u, v. The first result is one involving traceability.

Theorem 1   *If G is a 2–connected $K(1, 3)$–free graph of order p such that*
$$NC \geq (p - 2)/2,$$

then G is traceable.

Proof   Assume that G is not traceable. Let $P : x_1, x_2, \ldots, x_t$ be a longest path in G with $t \leq p - 1$, and let $x \in V(G - P)$. Since G is 2–connected, there exist at least two paths, disjoint except at x, from x to P. Let $x_i \in V(P)$ be the end vertex of a path, P_1, from x to P such that i is minimized. Then choose the path, P_2, from x to P so that if x_j is the end vertex of P_2, $i < j$, there are no paths from x to P with end vertices in $\{x_{i+1}, \ldots, x_{j-1}\}$. Observe that this is posssbile, since x cannot be the end vertex of two paths in $V(G - P)$ that have consecutive end vertices on P.

Since P is a longest path in G, certainly x_1 and x_t have no adjacencies off P. Also since G is $K(1, 3)$–free, the edges $x_{i-1}x_{i+1}$ and $x_{j-1}x_{j+1}$ are in $E(G)$. (Note that there exists no path—disjoint from $V(P - \{x_2, x_{i+2}, x_{t-1}\})$—from x to x_2, x_{i+2}, or x_{t-1}, since these paths and the forbidden subgraph property would force the existence of the edges x_1x_3, $x_{i+1}x_{i+3}$, or $x_{t-2}x_t$, respectively. Any of these situations would create a path longer than P.)

We consider the following adjacencies of x_1:
$$S_1 = \{x_{k-1} : x_k \in N(x_1) \cap V(P)\}.$$
We will show that $S_1 \cap N(x_{j-1}) = \emptyset$ and $S_1 \cap N(x_t) = \emptyset$.

Now for $x_k \in N(x_1)$, $j + 3 \leq k \leq t - 1$, (observe $k \neq j + 1$, $j + 2$, and t) we see that $x_{k-1} \notin N(x_{j-1})$, or the path

$$x_t, ..., x_k, x_1, ..., x_{j-1}, x_{k-1}, x_{k-2}, ..., x_j, ..., x,$$

is longer than P. Similarly, $x_{k-1} \notin N(x_t)$ or the path

$$x, ..., x_j, x_{j-1}, ..., x_1, x_k, x_{k+1}, ..., x_t, x_{k-1}, ..., x_{j+1},$$

is longer than P. Also for $x_k \in N(x_1)$, $i + 3 \leq k \leq j - 3$ and $2 \leq k \leq i - 1$, $x_{k-1} \notin N(x_t)$ or the path

$$x, ..., x_j, ..., x_t, x_{k-1}, ..., x_1, x_k, ..., x_{j-1},$$

is a path longer than P. Lastly, $N(x_{j-1}) \cap S_1 = \varnothing$ for $i + 3 \leq k \leq j - 3$ and $2 \leq k \leq i - 1$ since the paths

$$x_t, ..., x_j, ..., x, ..., x_i, ..., x_{k-1}, x_{j-1}, ..., x_k, x_1, ..., x_{i-1}, \text{ and}$$

$$x_t, ..., x_j, ..., x, ..., x_i, ..., x_{j-1}, x_{k-1}, ..., x_1, x_k, ..., x_{i-1},$$

are longer than P.

Similarly, we now consider the set S_2:
$$S_2 = \{y : y \in N(x) - V(P)\} \cup \{x_{k-1} : x_k \in N(x) \cap V(P), j + 3 \leq k \leq t - 2\}.$$
Then clearly $S_2 \cap N(x_{j-1}) = \varnothing$ and $S_2 \cap N(x_t) = \varnothing$. (Note that x is not adjacent to x_{j+2}.) By the choice of P_1 and P_2 there are no other adjacencies of x.

We define the function f: $N(x) \cup N(x_1) \rightarrow N(x_{j-1}) \cup N(x_t)$ by:
$$f(y) = y, \text{ for } y \notin V(P);$$
$$f(x_{k-1}) = x_k, \text{ for } 2 \leq k \leq t - 1.$$
From the previous arguments, clearly f is injective; therefore, we have
$$f(N(x) \cup N(x_1)) \cap (N(x_{j-1}) \cup N(x_t)) = \varnothing.$$
Also
$$f(N(x) \cup N(x_1)) \cup (N(x_{j-1}) \cup N(x_t)) \subseteq V(G) - \{x_i, x_t, x\}.$$
Since $NC \geq (p - 2)/2$, this implies that
$$(p - 2)/2 + (p - 2)/2 \leq p - 3, \text{ a contradiction.}$$
Therefore, G is traceable. $\square$

Figure 1 illustrates that the connectivity condition in Theorem 1 cannot be lowered.

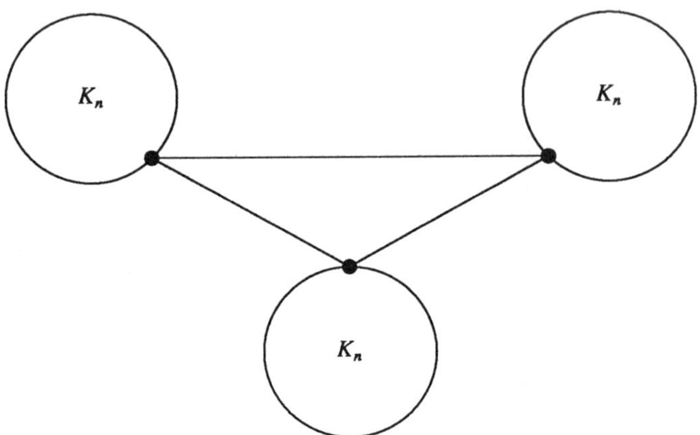

Figure 1   A 1–connected K(1, 3)–free graph that is not traceable.

The bipartite graph $K(n, n-2)$ has order $p = 2n - 2$, while $NC \geq (p - 2)/2$. However, $K(n, n-2)$ is not traceable since induced $K(1, 3)$'s abound.

We next present a series of lemmas which will allow us to prove analogous results for the hamiltonian property and a "highly hamiltonian" property–pancyclicity. A graph of order p is *pancyclic* if it contains a k–cycle for every k such that $3 \leq k \leq p$. It was shown in [3] that if G is a 2–connected graph of order p such that $NC \geq \frac{2p-1}{3} + 2$, then G is pancyclic. We were able to improve this bound by assuming that G contains no induced $K(1, 3)$.

First, it will be useful to define a related concept. A connected graph G of order p is said to be *panconnected* if for each pair u, v of distinct vertices of G there exists a uv–path of length t, for each t satisfying $dist(u, v) \leq t \leq p - 1$. In [13], Williamson proved that by restricting δ in G, panconnectedness can be achieved.

Theorem   (Williamson [13])   *If G is a graph of order $p \geq 4$ such that $deg(v) \geq p/2 + 1$ for every vertex v of G, then G is panconnected.*

First we grasp some basic structural ideas:

Lemma 1   *Let H be a graph of order $p \geq 2$ with independence number $\alpha(H) \leq 2$, and let $G = K_1 + H$. Then either G is panconnected or $H = K_r \cup K_{p-r}$ for some $1 \leq r < p$.*

Proof   If $p \leq 5$, the result holds trivially, so assume $p \geq 6$ and proceed by induction on $p + 1$, the order of G. Select a vertex v of maximum degree in H and consider the graph $G - v$ which is either panconnected or $H - v = K_r \cup K_{p-r}$ for some $1 \leq r < p - 1$.

If $G - v$ is panconnected, then for any pair of vertices x and y, there are xy–paths of each length from 2 to $p - 1$. Let $P = (x = x_1, x_2, \ldots, x_p = y)$ be an xy–path of length $p - 1$. Since v has maximum degree in H, $\deg(v) \geq 3$. Thus, $vx_i, vx_j \in E(G)$ with $i < j < p$. Since $\alpha(H) \leq 2$, one of the edges vx_{i+1}, vx_{j+1}, or $x_{i+1}x_{j+1} \in E(G)$, which implies that there exists an xy–path in G of length p.

If $H - v = K_r \cup K_{p-r-1}$, then v must be adjacent, without loss of generality, to each vertex of K_{p-r-1}, for otherwise there would be three independent vertices. If v has an adjacency in K_r, it is easily verified that G is panconnected. If not, then $H = K_r \cup K_{p-r}$, and the proof is complete.   □

Lemma 2   *If G is a 2–connected $K(1,3)$–free graph of order $p \geq 11$ that satisfies*

$$NC \geq \frac{(2p-1)}{3} \quad \text{and} \quad \delta(G) \leq \frac{(p-5)}{3},$$

then G is pancyclic.

Proof   Let v be a vertex of degree $\delta = \delta(G)$, and let $H = G - v - N(v)$. Then $\delta(H) \geq (2p - 1)/3 - \delta$. Since $\delta \leq (p - 5)/3$, we have

$$\delta(H) \geq (p - \delta - 1)/2 + 1 = |H|/2 + 1,$$

so that H is panconnected by [13]. Therefore, H has cycles of each length from 3 to $p - \delta - 1$. Let L be the graph spanned by $v + N(v)$. Since G is $K(1,3)$–free, $\alpha(L) \leq 2$. Because G is 2–connected, there are distinct vertices $u_1, u_2 \in V(L)$ and $v_1, v_2 \in V(H)$ such that $u_i v_i \in V(G)$ for $i = 1, 2$. From Lemma 1, we consider two possibilities.

If L is panconnected, then there are paths in L between u_1 and u_2 of each length from 2 to δ. Also, there are paths in H of each length from 2 to $p - \delta - 2$ between v_1 and v_2. Using those paths, and the edges $u_1 v_1$ and $u_2 v_2$, cycles of each length from 6 to p can be obtained which implies that G is pancyclic.

Suppose that L is not panconnected. Then $L = v + (K_r \cup K_s)$ with $r + s = \delta$ and $r, s \geq 1$. Without loss of generality, assume $u_1 \in K_r$ and $u_2 \in K_s$. Again paths in L between u_1 and u_2 exist of each length from 2 to δ. Thus, as in the previous case, G is pancyclic.   □

As a corollary to the proof of Lemma 2, using the result of Ore [12] on hamiltonian–connected graphs, we have the following

Corollary 1 *If G is a 2–connected $K(1,3)$–free graph of order $p \geq 11$ that satisfies*

$$NC > (2p - 3)/3 \text{ and}$$
$$\delta(G) \leq (p - 5)/3,$$

then G is hamiltonian.

Lemma 3 *Let G be a 2–connected $K(1,3)$–free graph of order $p \geq 14$ that satisfies*

$$NC \geq (2p - 1)/3 \text{ and}$$
$$\delta(G) \geq (p - 4)/3.$$

If G contains a C_m for some $m \geq (2p + 5)/3$, then G contains cycles of each length from m to p.

Proof   Select t such that G contains a C_t but no C_{t+1} for some $t \geq m$. Let $C_t = (x_1, x_2, \ldots, x_t, x_1)$ and let $H = G - C_t$. Note that if $y \in V(H)$ and $yx_i \in E(G)$, then $yx_{i-1}, yx_{i+1} \notin E(G)$, otherwise a C_{t+1} would exist. The $K(1,3)$–free property implies $x_{i-1}x_{i+1} \in E(G)$. Using this, it is easy to verify that $yx_i \in E(G)$ implies that yx_{i+2} and $yx_{i+3} \notin E(G)$, for otherwise a C_{t+1} would exist in G.

Consider the case when H contains nonadjacent vertices y_1 and y_2. By the previous remarks, each of y_1 and y_2 can be adjacent to at most $t/4$ vertices of C_t, so that

$$\frac{(2p - 1)}{3} \leq |N(y_1) \cup N(y_2)| \leq t/2 + (p - t - 2).$$

This implies $t \leq (2p - 10)/3$, a contradiction. Therefore H is the complete graph K_{p-t}.

If $t = p - 1$, then the vertex in H has degree at most $(p - 1)/4$. Thus we get $(p - 4)/3 \leq (p - 1)/4$, which is a contradiction for $p \geq 14$. Hence, H has at least two vetices. Each vertex of H has at least $s \geq (p - 4)/3 - (p - t - 1) \geq t - (2p + 1)/3$ adjacencies on C_t, so clearly $s \geq 2$.

If $yx_i \in E(G)$ and $y'x_j \in E(G)$ for $y, y' \in V(H)$, then $j > i + p - t + 2$, for otherwise G would contain a C_{t+1}. Using the fact that y has at least s adjacencies on C_t and that each pair of these adjacencies are at a distance at least four on C_t, there are at least $4(s - 1) + 2(p - t + 2)$ vertices of C_t that are not adjacent to y'. Since y' has s adjacencies on C_t, we have

$$s + 4(s - 1) + 2(p - t + 2) \leq t.$$

This implies $t \leq (4p + 5)/6$, a contradiction. ◻

As a corollary to the proof of Lemma 3, using Corollary 1, we get the following

Lemma 4 *Let G be a 2–connected $K(1,3)$–free graph of order $p \geq 14$ that satisfies*

$$NC > (2p - 3)/3 \text{ and}$$
$$\delta(G) > (p - 5)/3.$$

If G contains a C_m for some $m \geq (2p + 4)/3$, then G is hamiltonian.

Lemma 5 *If G is a 2–connected graph of order $p \geq 9$ with $\alpha(G) < 3$, then G is pancyclic.*

Proof Since the Ramsey number $r(K_3, C_t) = 2t - 1$ (c.f. [5] pp. 272), G contains a cycle of each length $t \leq (p + 1)/2$. Select m such that G contains a cycle of each length from 3 to m, but no cycle of length $m + 1$. Let $C = (x_1, x_2, ..., x_m, x_1)$ be a cycle of length m and let $H = G - C$.

If $y \in V(H)$ and $yx_i, yx_j \in E(G)$, then yx_{i+1}, yx_{j+1}, and $x_{i+1}x_{j+1} \notin E(G)$ or we create a C_{m+1}. Since $\alpha(G) < 3$, each vertex of H must be adjacent to at most one vertex of C, and H is complete.

Similarly, $x_ix_j \in E(G)$ unless each vertex of H is adjacent to either x_i or x_j. Therefore, either C forms a complete graph K_m or a $K_m - e$, or there exists a vertex x of C which is adjacent to each vertex of H, while $C - x$ forms a complete graph. However, in all cases, there are two disjoint subgraphs that are complete (or one has a missing edge) that span G. Since G is 2–connected there are two independent edges between these subgraphs. Under those conditions it is easily verified that G is pancyclic.

With these lemmas at hand, we can now prove the main theorem.

Theorem 2 *If G is a 2–connected $K(1, 3)$–free graph of order $p \geq 14$ with*
$$NC > (2p - 2)/3,$$
then G is pancyclic.

Proof Assume G is not pancyclic. By Lemma 2 we can assume that $\delta(G) \geq (p - 4)/3$ and by Lemma 5 that $\alpha(G) \geq 3$.

Consider pairwise nonadjacent vertices x, y, and z and assume that $\deg(z) \geq \deg(x) \geq \deg(y)$. Then from the hypothesis, $\deg(z) \geq \deg(x) \geq (2p - 1)/6$. We claim that z is adjacent to at least two vertices in $N(x)$ or $N(y)$. If not then,
$$(2p - 1)/3 + (2p - 1)/6 - 2 \leq p - 3,$$
a contradiction. This proves the claim. Therefore we can assume there exist subgraphs A and B with $A \cap B = \{u, v\}$ and with $|A \cup B| \geq (2p - 1)/3$. Also x is adjacent to each vertex of A and z is adjacent to each vertex of B.

We will subsequently show that G contains cycles of each length from 3 to $\lceil (2p+5)/3 \rceil$. We can then conclude that G is pancyclic from Lemma 4. To prove the existence of these cycles we must consider the graph H spanned by $A \cup B \cup \{x, y\}$.

Note that from Lemma 1, $A + x$ (respectively $B + z$) is panconnected or $A = A_1 \cup A_2$, where A_1 and A_2 are disjoint complete graphs (respectively $B = B_1 \cup B_2$, where B_1 and B_2 are disjoint complete graphs). We now consider various cases depending on the structure of A and B and the location of u and v, and show the existence of the required cycles.

Case 1   Suppose $A + x$ is panconnected or $u \in V(A_1)$ and $v \in V(A_2)$, and $B + z$ is panconnected or $u \in V(B_1)$ and $v \in V(B_2)$.

In $A + x$ (and $B + z$) there is a path from u to v of each length from 2 to $|A|$ (and $|B|$), thus H has cycles of each length from 4 to $|H| \geq (2p + 5)/3$. Clearly, $A + x$ (and $B + z$) contains a triangle.

Case 2   Suppose $A + x$ is panconnected and $u, v \in B_2$.

Since G is 2–connected, there is a path in $G - y$ between B_1 and $B_2 \cup A$. Let $P = (w = w_0, w_1, \ldots, w_t = w')$ be such a path of minimal length. We claim that $t \leq 2$. If this is not true, then consider the independent vertices w_0, w_2, and x, and observe that $N(x) \cap (N(w_0) \cup N(w_2)) \neq \varnothing$. This verifies the claim. It is easily seen that $H \cup P$ contains cycles of each length from 3 to $|H| + t - 1$, independent of whether w' is in A or B_2. Note that all cycles of length greater than $|A \cup B_2| + 2$ will use the path P.

If $u \in A_1 \cap B_1$ and $v \in A_2 \cap B_2$, then we get the same result. That is, $H \cup P$ contains cycles of each length from 3 to $|H| + t - 1$, independent of whether w' is in $A_1 \cup A_2$ or B_2.

Case 3   The vertices $u, v \in A_2 \cap B_2$.

Since G is 2–connected, there is a path in $G - y$ between B_1 and $A \cup B_2$. Let $P = (w = w_0, w_1, \ldots, w_t = w')$ be such a path of minimal length. We claim that $t \leq 2$. If this is not the case, then consider the independent vertices w_0, w_2, and x, and observe that $N(x) \cap (N(w_0) \cup N(w_2)) \neq \varnothing$. This verifies the claim.

Subcase i   Let $w' \in A_1$.

Just as in the previous cases, $H \cup P$ contains cycles of each length from 3 to $|H| + t - 1$, since $p \geq 14$. The remaining cycles of length ≥ 6 can be constructed using the path P.

Subcase ii   Let $w' \in A_2 \cup B_2$.

A repeat of the argument exhibiting P implies the existence of a corresponding path P' from A_1 to $A_2 \cup B_2$ of length $t \leq 2$. There are several possibilities that must be considered, but it is straightforward to verify that $H \cup P \cup P'$ contains cycles of each length from 3 to $|H| + t + t' - 2$.

This completes the proof of Theorem 2. $\square$

Theorem 3   *If G is a 2–connected $K(1,3)$–free graph of order $p \geq 14$ with $NC > (2p - 3)/3$, then G is hamiltonian.*

Proof   Consider nonadjacent vertices x and y. Then $|N(x) \cup N(y)| > (2p - 3)/3$. Thus there exist disjoint subgraphs A and B with $|A \cup B| > (2p - 3)/3$ and with x adjacent to each vertex of A and y adjacent to each vertex of B.

Since G is 2–connected, there exist vertex disjoint paths P_1 and P_2 from A to B (avoiding x and y). By Lemma 1, either $A + x$ is panconnected or $A = A_1 \cup A_2$ where A_1 and A_2 are disjoint complete graphs. The same is true of B with disjoint subgraphs B_1 and B_2. Also if $A + x$ $(B + y)$ is not panconnected, then the paths P_1 and P_2 come from A_1 and A_2 $(B_1$ and $B_2)$, respectively.

Clearly, G contains a cycle of length at least $|(A \cup B) \cup (P \cup P')| + 2 \geq (2p + 4)/3$. From Lemma 5, we see that G is hamiltonian. $\square$

We note here that we have obtained this same result for $3 \leq p \leq 13$ with other techniques.

A variation of a graph used in [10] illustrates the sharpness of our theorem. Let G be the graph of order $p = 3n + 6$ in Figure 2. Then G is 2–connected and $K(1,3)$–free. Notice also that G is not hamiltonian. For nonadjacent vertices u and v in G,

$$|N(u) \cup N(v)| = 2n + 2 = 2((p - 6)/3) + 2 = 2p/3 - 2.$$

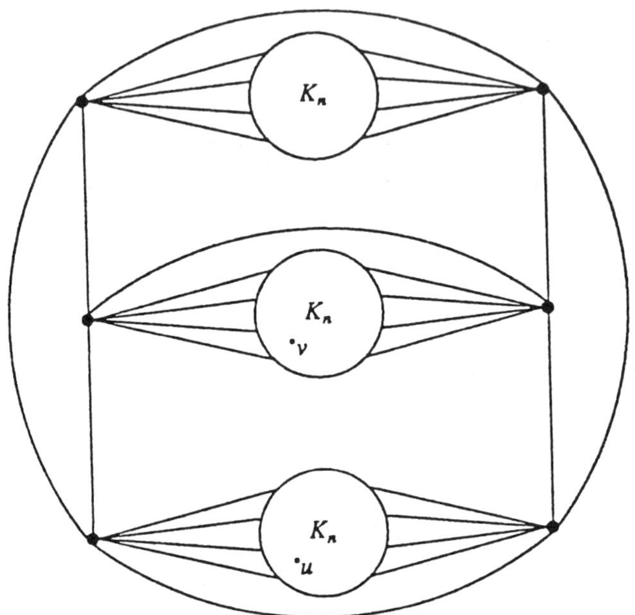

Figure 2   A nonhamiltonian K(1,3)–free graph illustrating
the sharpness of Theorem 2.

We feel that from the nature of the proof techniques and related results in this area that a generalization of Theorem 3 to sets of more than two vertices is imminent. Theorem 3 and a result of Fraisse [4] lead us to the following conjecture:

Conjecture   *Let G be a K(1,3)–free graph of order p and connectivity k. Suppose there exists some t, $t \leq k$, such that for every independent set $S \subseteq V(G)$ of cardinality t we have:*

$$|N(S)| \geq t(p-1)/(t+1)$$

then G is hamiltonian.

From Theorem 3 we get the following result for connected K(1,3)–free graphs:

Theorem 4   *If G is a connected K(1,3)–free graph of order $p \geq 14$ with $NC > (2p-5)/3$, then G is traceable.*

Proof   Consider the graph $H = G + v$ for some vertex $v \notin V(G)$. Then

$$NC(H) > (2p - 4)/3 + 1 = \frac{2(p+1) - 3}{3}.$$

Since $|H| \geq 14$, and H is 2–connected, then H is hamiltonian which implies that G is traceable. $\square$

A graph is *homogeneously traceable* if for every $x \in V(G)$ there exists a path beginning at x and containing all of the vertices of G. In general, there are not many results dealing with homogeneously traceable graphs. It appears that sufficient conditions for a nonhamiltonian graph to have this property are not easy to find due to the theoretical closeness of homogeneous traceability and hamiltonicity. As with the hamiltonian–connected property, we need the 3–connectivity stipulation to prove this result. We also point out that Matthews [8] has proven that if G is a 3–connected $K(1, 3)$–free graph of order $p < 20$, then G is hamiltonian. If G is hamiltonian, then G is homogeneously traceable, hence our result for $K(1, 3)$–free graphs is only interesting when $p \geq 20$.

Theorem 5 *If G is a 3–connected $K(1, 3)$–free graph of order p such that*
$$NC > (2p - 5) / 3,$$
then G is homogeneously traceable.

Proof We proceed by contradiction. First, we define an x_m–*hamiltonian path* to be a path in G that has x_m as an end vertex, and that contains all of the vertices of $V(G)$. Suppose there is some vertex x_m such that there exists no x_m–hamiltonian path in G. Let P: $x_1, x_2, ..., x_m$ be a longest x_m–path.

Since G is 3–connected, for some vertex x not on P, there exist two disjoint (except at x) paths from x to P such that the end vertices x_i, x_j, of these paths satisfy $i \neq j \neq m$. Moreover, x_i can be chosen so that there are no endpoints of x–paths between x_1 and x_i. Once x_i is chosen, then we select x_j so that there are no endpoints of x–paths between x_i and x_j. (Note that $x_1 x, x_{i+1} x, x_{i+2} x \notin E(G)$ by the maximality of P and since G is $K(1, 3)$–free.)

Consider the nonadjacent pair of vertices x_1 and x_{j+1}. Let
$$S = \{x_k : x_k \in N(x_1) \cup N(x_{j+1})\}.$$
Then since G is $K(1, 3)$–free and $N(x) \cap V(P) \subseteq \{x_i, x_j, x_{j+1}, ..., x_m\}$, we see that $|S \cap N(x)| \leq 2$. (It could be that $S \cap N(x) = \{x_j, x_m\}$.) Hence,
$$|N(x)| < (p - 1) - ((2p - 5) / 3 - 2) - 3 = (p - 1) / 3.$$
We subtract three at the end since $x_{j+1} \notin N(x)$, $x_{i+1} \notin N(x)$, and $x_1 \notin N(x)$. (Note that $x_{j+1} \notin N(x_{i+1})$ and $x_1 \notin N(x_{i+1})$ so x_{i+1} is not in $N(x_{j+1}) \cup N(x_1)$.)

In addition, x and x_1 are nonadjacent which implies that
$$|N(x_1)| > (2p - 5) / 3 - (p - 1) / 3 = (p - 4) / 3.$$

We use this fact to contradict the cardinality of $N(x_{i+1}) \cup N(x)$.

For every $x_k \in N(x_1)$, $j + 3 \le k \le m - 1$, $x_{k+1} \notin N(x_{i+1})$ or the path

$$x_m, \ldots, x_{k+1}, x_{i+1}, \ldots, x_k, x_1, \ldots, x_i, \ldots, x$$

is a longer x_m–path. Analogously, $x_{k+1} \in N(x)$ or we obtain

$$x_m, \ldots, x_{k+1}, x, \ldots, x_j, \ldots, x_k, x_1, \ldots, x_{j-1}$$

which is also a longer x_m–path. Observe that x_1 is not adjacent to any vertex in

$$\{x_i, x_{i+1}, x_{i+2}, x_{j-2}, x_{j-1}, x_j, x_{j+1}, x_{j+2}\}$$

because G is $K(1, 3)$–free.

For $x_k \in N(x_1)$, $i + 3 \le k \le j - 3$, then $x_{k+1} \notin N(x_{i+1})$ or a longer x_m–path is the following:

$$x_m, \ldots, x_{k+1}, x_{i+1}, \ldots, x_k, x_1, \ldots, x_i, \ldots, x.$$

Lastly, for $x_k \in N(x_1)$, $2 \le k \le i - 1$, $x_{k-1} \notin N(x_{i+1})$. Since we have no corresponding nonadjacency to subtract from $|N(x) \cup N(x_{i+1})|$ if perhaps $x_1 x_m \in E(G)$, we have that

$$|N(x) \cup N(x_{i+1})| < p - ((p - 4) / 3 - 1) - 4 = (2p - 5) / 3.$$

We subtract four at the end since $x_{j+2}, x_{j+1}, x_{i+1}$, and x have not been counted and are not in $N(x) \cup N(x_{i+1})$. Since $NC > (2p - 5) / 3$ we have our contradiction. Therefore, G is homogeneously traceable. ❏

From the nature of homogeneously traceable graphs and the strength of the neighborhood condition, we believe the following is true:

Conjecture *If G is a 3–connected $K(1, 3)$–free graph of order p such that*

$$NC > (2p - 5) / 3,$$

then G is hamiltonian.

REFERENCES

[1] R.J. Faudree, R.J. Gould, M.S. Jacobson, R.H. Schelp, Extremal Problems Involving Neighborhood Unions, *J. Graph Theory*, 111 (1987), 555 – 564.

[2] R.J. Faudree, R.J. Gould, M.S. Jacobson, R.H. Schelp, Neighborhood Unions and Hamiltonian Properties in Graphs, *J. Combinatorial Theory, Ser. B,* to appear.

[3] R.J. Faudree, R.J. Gould, M.S. Jacobson, L.M. Lesniak, Neighborhood Unions and Highly Hamiltonian Graphs, preprint.

[4] P. Fraisse, A New Sufficient Condition for Hamiltonian Graphs, *J. Graph Theory*, 10 (1986), 405–409.

[5] R.J. Gould, *Graph Theory*, Benjamin/Cummings Publishing Co., Menlo Park, Calif., 1988.

[6] R.J. Gould, M.S. Jacobson, Forbidden Subgraphs and Hamiltonian Properties in the Square of a Connected Graph, *J. Graph Theory*, 8 (1984), 147 – 154.

[7] T.E. Lindquester, The Effects of Distance and Neighborhood Union Conditions on Hamiltonian Properties in Graphs, to appear in J. Graph Theory.

[8] M.M. Matthews, Every 3–connected K(1, 3)–free Graph with Fewer than Twenty Vertices is Hamiltonian, *Technical Report 82–004*, Department of Computer Science, University of South Carolina.

[9] M.M. Matthews, D.P. Sumner, Hamiltonian Results in K(1, 3)–Free Graphs, *J. Graph Theory*, 8 (1984), 139 – 146.

[10] M.M. Matthews, D.P. Sumner, Longest Paths and Cycles in K(1, 3)–Free Graphs, *J. Graph Theory*, 9 (1985), 269 – 277.

[11] D.J. Oberly, D.P. Sumner, Every Connected, Locally Connected Nontrivial Graph with No Induced Claw is Hamiltonian, *J. Graph Theory*, 3 (1979), 351 – 356.

[12] O.Ore, Note on Hamiltonian Circuits, *Amer. Math. Monthly*, 67 (1960), 55.

[13] J.E. Williamson, Panconnected Graphs II, *Perod. Math. Hungar.*, 8 (1977), 105 – 116.

[14] C.Zhang, Hamiltonian Cycles in Claw–Free Graphs, *J. Graph Theory*, 12 (1988), 209–216.

Lexicographically Factorable Extensions
of Irreducible Graphs

Joan Feigenbaum

AT&T Bell Laboratories

ABSTRACT

The lexicographic product *G[H] of graphs G and H is defined by*
$$V(G[H]) = V(G) \times V(H) \text{ and}$$
$$E(G[H]) = \{(x, y) - (u, v): x - u \in E(G) \text{ or } (x = u \text{ and } y - v \in E(H))\}.$$
*We show that minimal lexicographically factorable extensions of irreducible
graphs are not unique and that finding them is NP–Hard.*

1. Introduction

Let G and H be finite, simple, undirected graphs. The *lexicographic product*
G[H] of G and H is defined by
$$V(G[H]) = V(G) \times V(H) \text{ and}$$
$$E(G[H]) = \{(x, y) - (u, v): x - u \in E(G) \text{ or } (x = u \text{ and } y - v \in E(H))\}.$$
Informally, form the lexicographic product G[H] by substituting a *copy* of H for each
node in G and drawing in all possible edges between pairs of copies that correspond to
adjacent nodes. Figure 1 shows the lexicographic product of a path of length two and a
triangle.

The lexicographic product G[H] is sometimes called the *composition* of G and H.
The trivial graph K_1 acts as a unit for this operator; that is,
$$K_1[G] \cong G[K_1] \cong G, \text{ for all graphs G.}$$
A graph G is called *irreducible* if $G \cong G_1[G_2]$ implies that at least one of G_1 or G_2
is K_1. If G is not irreducible, then we say that it is *factorable*.

Lexicographic product graphs have been used to bound the computational complexity
of one graph problem in terms of the complexity of another. For example, Papadimitriou
and Yannakakis used them to show that the EXACT CLIQUE problem is complete for
the complexity class D^P ([10]), and Garey and Johnson used them to show that no

481

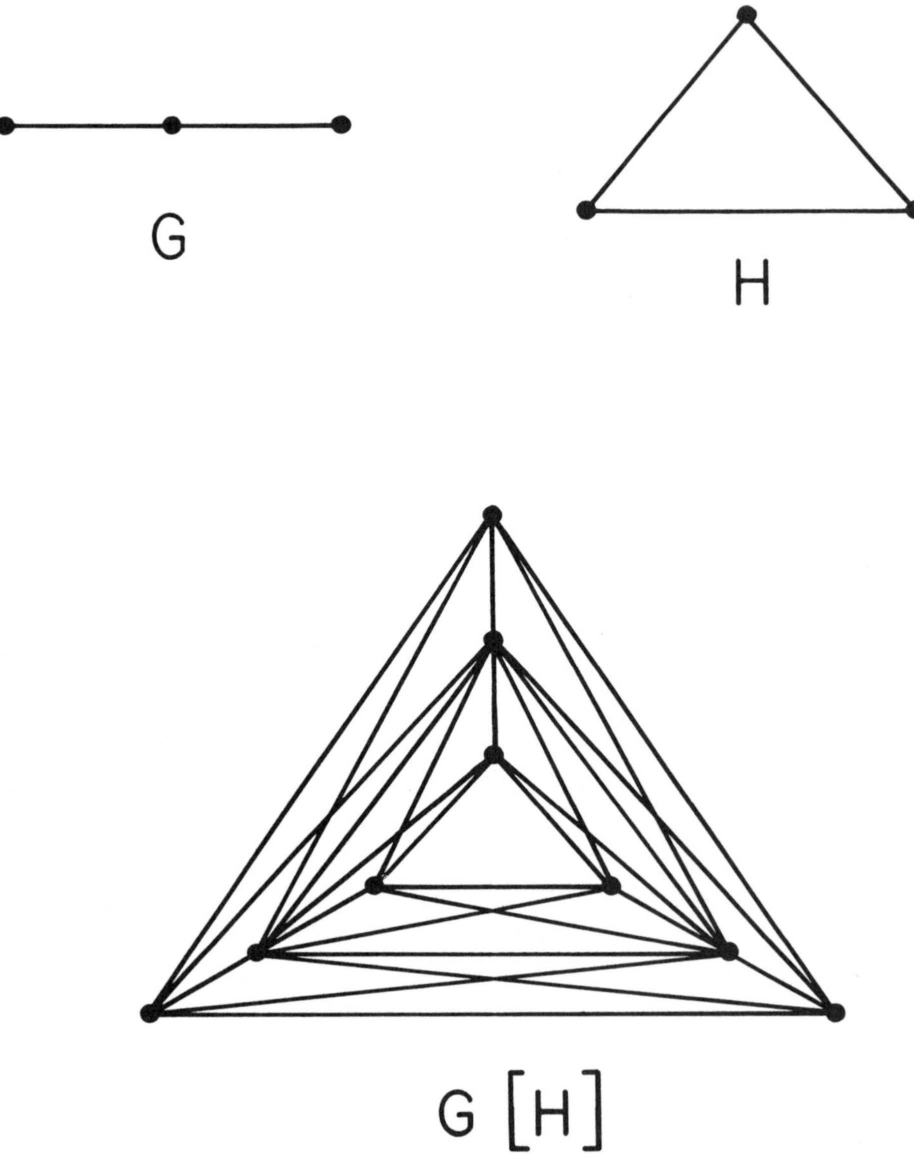

Figure 1. Lexicographic product of a path of length two and a triangle

polynomial–time algorithm can always approximate the chromatic number to within a factor of two, assuming that $P \neq NP$ ([7]).

A simple proof that most graphs are irreducible is given in [3]. The lexicographic product operator is *cancellable*, i.e.,

$$G[H] \cong G[J] \text{ implies } H \cong J, \text{ for all } G, H, J, \text{ and}$$
$$H[G] \cong J[G] \text{ implies } H \cong J, \text{ for all } G, H, J,$$

but it does not have the *unique factorization property*; see [2] and for a complete characterization of the graphs with two or more inequivalent factorizations.[1] The computational problem of finding all the factorizations of a graph G (or proving that G is irreducible) is polynomially equivalent to testing graph isomorphism ([5]). See [3] or [5] for a more thorough discussion of the properties of the lexicographic product operator and a more complete bibliography.

In this paper, we consider *minimal factorable extensions* of irreducible graphs. A factorable extension H of G is a factorable graph H of which G is a subgraph (denoted $G \subset H$). A minimal factorable extension H is one with the fewest number of nodes and, among all factorable extensions with that number of nodes, the fewest number of edges. Becuase the complete graph K_n is factorable if n is composite, the number of nodes of H is $n = |V(G)|$ if n is composite, and it is $n + 1$ if n is prime. Thus, the interesting comparison is really the density of H with the density of G.

In Section 2 below, we show that, in general, an irreducible graph G can have nonisomorphic minimal facotrable extensions H and H'. In Section 3, we prove that it is NP–Hard to find a minimal factorable extensions of a graph G. In Section 4, we pose the general question "What is the maximum possible density of G's minimal factorable extension H, as a function of the density of G," and derive a partial result.

The analogous problems for *cartesian–factorable* extensions are solved in [4].

2. Nonuniqueness

We exihibit an infinite class C of irreducible graphs with nonunique minimal factorable extensions. For each prime integer $p \geq 5$, we construct a member G_p of C on p^2 nodes. The node set $V(G_p)$ consists of p disjoint sets $S_1, ..., S_p$ of p nodes each. Construct $E(G_p)$ as follows. Between node sets S_i and S_j, where $i \neq j$, all possible edges are present. The subgraph induced by S_1 contributes two edges, in the form of a path of length two. Each of the subgraphs induced by S_2 through S_p also contributes two edges, but in the form of two edges with no endpoints in common, rather than a path of length two. Figure 2a show G_5.

---

1. This characterization was later rediscovered and appears in English in [1].

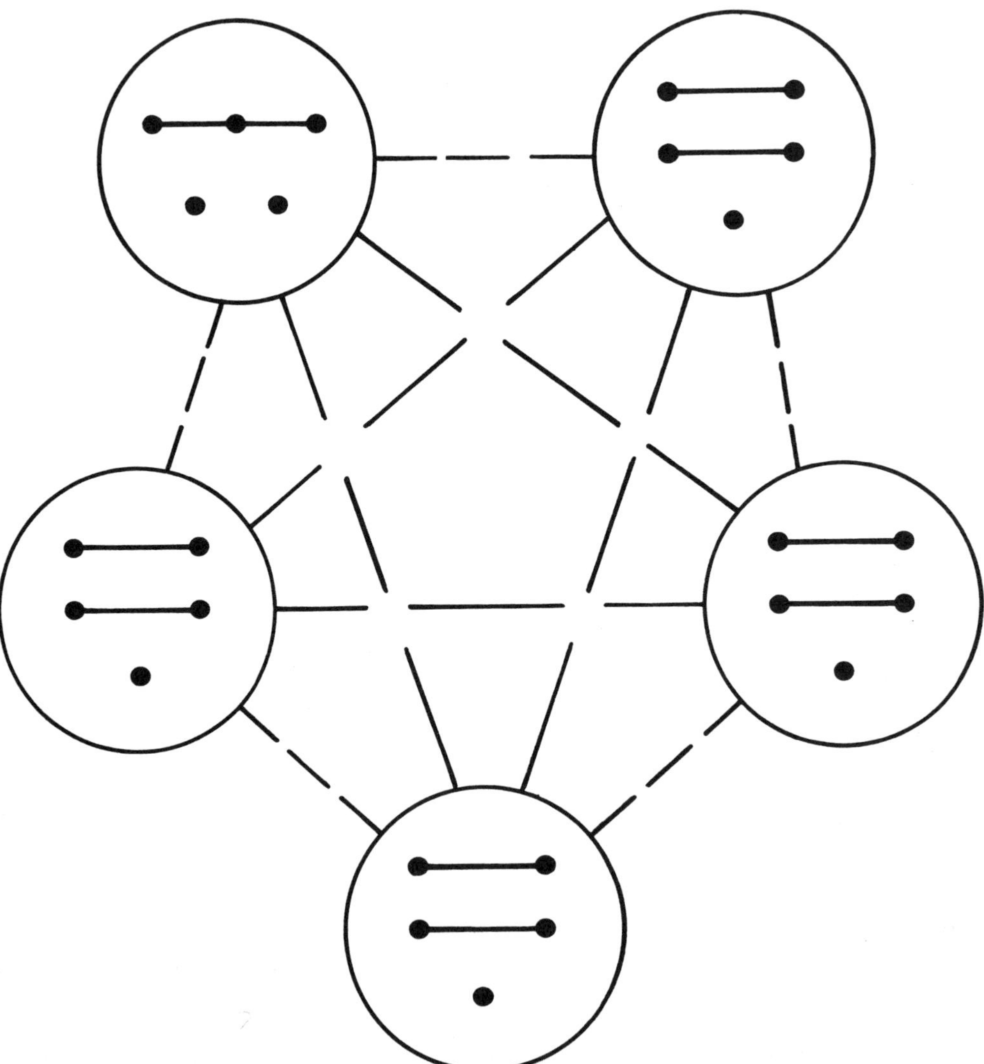

Figure 2a. Irreducible graph G_5 of Section 2. The dotted lines between disjoint subgraphs indicate that all possible edges between those subgraphs are present.

For any p, the graph G_p is irreducible. To see this, note that, if G_p were factorable, both of its factors would have to have p nodes; assume a right factor with degree sequence $d_1 \leq d_2 \leq ... \leq d_p$. Then, by definition of the lexicographic product, any integer that occurred as the degree of a node of G_p would be congruent to $d_i \bmod p$, for some i, $1 \leq i \leq p$, and the number of nodes with degrees congruent to d_i would be a multiple of p. These conditions do not hold for G_p, which has a unique node whose degree is congruent to 2 mod p. Let $H_p = K_p[A_p]$ and $H'_p = K_p[B_p]$, where A_p is a graph on p nodes whose edge set is a path of length three, B_p is a graph on p nodes whose edge set is a path of length two and an edge neither of whose endpoints is on the path, and K_p is the complete graph on p nodes. Then H_p and H'_p are both minimal factorable extensions of G_p and $H_p \not\equiv H'_p$. Figure 2b shows H_5 and H'_5.

This construction shows

Remark: *Minimal factorable extensions are not unique.*

3 . NP–Hardness

The terminology and conventions of this section follow those of [6].

We are interested in the computational complexity of constructing a minimal factorable extension H of a graph G. An algorithm that solved the construction problem could clearly be used to solve the following decision problem:

Minimal Factorable Extension (MFE):

Input: A graph G and positive integers N and E.

Question: Is there a lexicographically factorable graph H such that G is a subgraph of H, $|V(H)| \leq N$, and $|E(H)| \leq E$?

Therefore, we prove that the construction problem is intractable (unless P is equal to NP) by proving the NP–Completeness of the decision problem.

The MFE problem is clearly in NP: given a graph H, two nontrivial factors H_1 and H_2, an injection ϕ_1 from V(G) into V(H), and a bijection ϕ_2 from V(H) onto $V(H_1[H_2])$, one can verify in polynomial time that ϕ_1 and ϕ_2 are isomorphisms, that $|V(H)| \leq N$, and that $|E(H)| \leq E$.

The proof of completeness is by many–to–one polynomial–time reduction from CLIQUE. Let (J, k) be an arbitrary instance of CLIQUE; so J is, without loss of generality, a connected graph, and k is an integer between 1 and $|V(J)|$. We construct an equivalent instance (G, N, E) of MFE as follows. Let $n = |V(J)|$, and assume that $V(J) = \{v_1, ..., v_n\}$. Choose a prime integer p between n^3 and $2n^3$; such a prime must exist, by a theorem of Chebyshev ([8]), and one can be found deterministically in

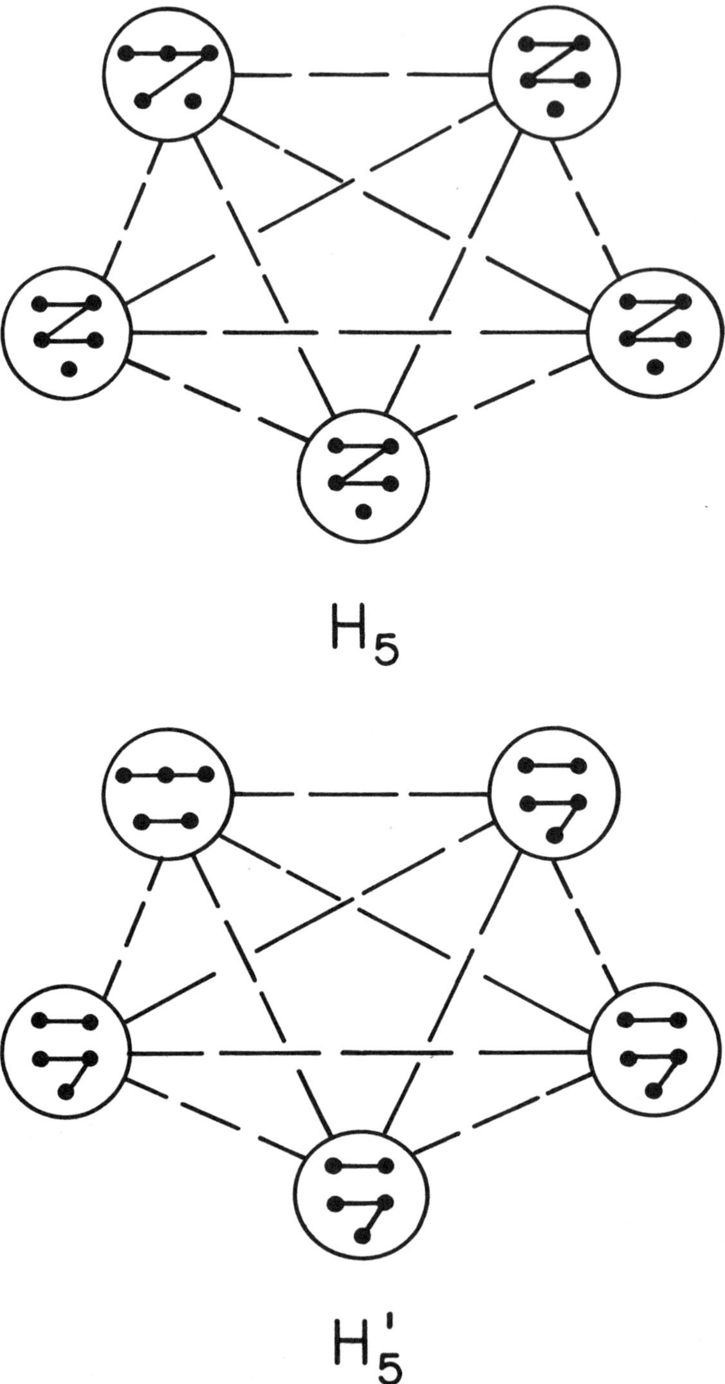

Figure 2b. Nonisomorphic minimal factorable extensions H_5 and H'_5 of the irreducible graph G_5 shown in Figure 2a.

time polynomial in n by trial division of successive odd integers.[1] Let $V_1 = \{v_{n+1}, ...,$ $v_p\}$ and $V_2 = \{w_1, ..., w_p\}$ be two sets of nodes that are disjoint from each other and both disjoint from $V(J)$, and let $V(G) = V(J) \cup V_1 \cup V_2$. The edge set $E(G)$ consists of $E(J)$, all possible edges $v_i - v_j$ such that $1 \le i \le n$ and $n + 1 \le j \le p$, all possible edges $v_i - v_j$ such that $n + 1 \le i, j \le p$ and $i \ne j$, all possible edges $w_i - w_j$ such that $n - k + 1 \le i, j \le p$ and $i \ne j$, and all possible edges $w_i - w_j$ such that $1 \le i \le n - k$ and $n + 1 \le j \le p$. (See Figure 3.) Finally, let $N = 2p$ and

$$E = 2|E(J)| + 2\binom{p - n}{2} + 2n(p - n).$$

Note that

$$|E(G)| = |E(J)| + \binom{k}{2} + 2\binom{p - n}{2} + 2n(p - n).$$

The size of (G, N, E) is polynomial is n and the reduction takes polynomial time.

If (J, k) is a yes–instance of CLIQUE, then there are k nodes in $V(J)$ that induce a complete subgraph; assume without loss of generality that they are v_{n-k+1} through v_n. Let H be a graph with node set $V(H) = V(G)$ and edge set

$$E(H) = E(G) \cup \{w_i - w_j : 1 \le i, j \le n \text{ and } v_i - v_j \in E(J)\}.$$

Then $|V(H)| = 2p$, $|E(H)| = E$, $G \subset H$, and H has the nontivial factorization $H_1[H_2]$, where H_1 is the graph on two nodes and no edges, and H_2 is the subgraph of G induced by $V(J) \cup V_1$. Thus, (G, N, E) is a yes–instance of MFE.

Now suppose that (G, N, E) is a yes–instance of MFE and that H is factorable extension of G with the desired size. Let $H_1[H_2]$ be a nontrivial factorization of H. Because $|V(H)| = 2p$, and p is prime, the only possible sizes of H's factors are $|V(H_1)| = 2, |V(H_2)| = p$ and $|V(H_1)| = p, |V(H_2)| = 2$. Assume first that $|V(H_1)| = 2$ and $|V(H_2)| = p$. By definition of the lexicographic product, $V(H)$ can be partitioned into two sets S and T, both of which induce subgraphs isomorphic to H_2. We show that the only possibilities for S and T are $V(J) \cup V_1$ and V_2, respectively. If $S = V(J) \cup V_1$ and $T = V_2$, then $V(J)$ must contain k nodes that induce a complete subgraph of J, by the following argument. The subgraph induced by T contains a clique of size $p - n + k$. It is isomorphic to the subgraph induced by S, and hence the subgraph induced by S also contains a clique of size $p - n + k$; at most $p - n$ of the nodes of this clique are in V_1, and thus $V(J)$ contains k nodes that induce a clique of J.

If $S \ne V(J) \cup V_1$ and $T \ne V_2$, then there are positive integers a and b, where $p = a + b$, such that $V(J) \cup V_1$ contributes a nodes to S and b nodes to T, and V_2

---

1. Note that this does not assume that we can test the primality of a integer i deterministically in time polynomial in the number of bits in the binary representation of i (the important unsolved problem in computational number theory that may spring to mind), but merely that we can test it deterministically in time polynomial in i, which is trivially true.

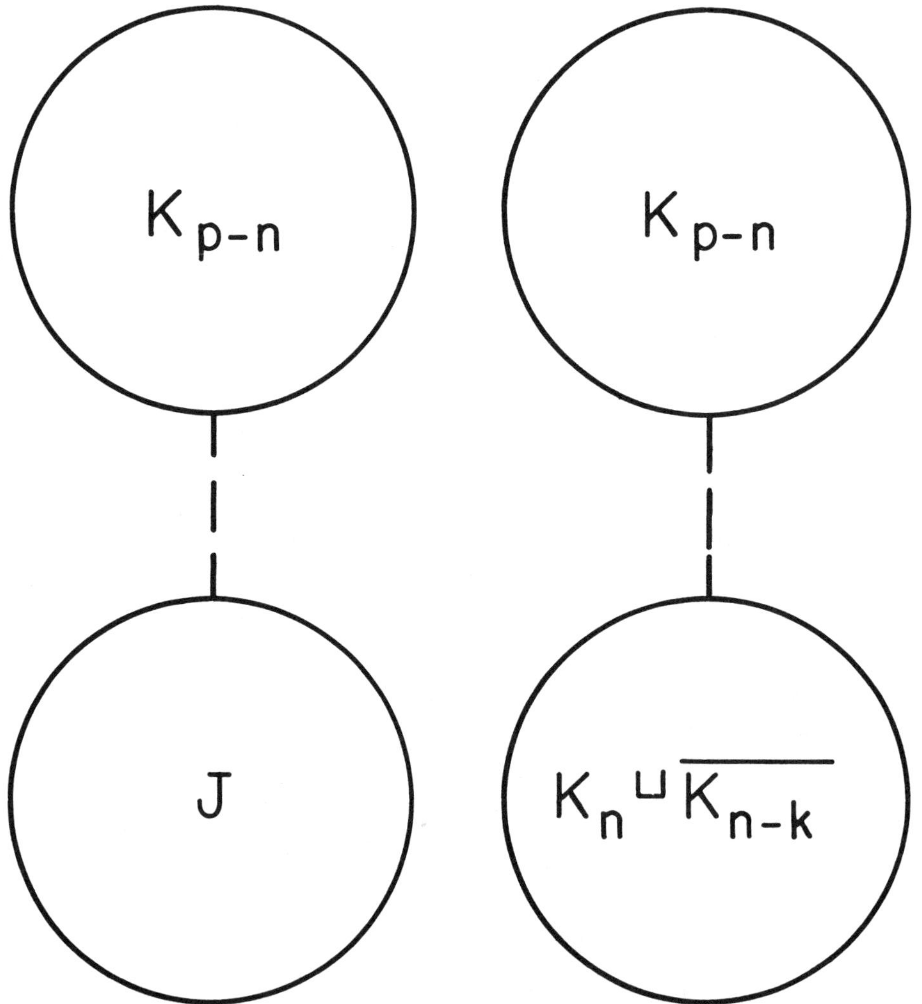

Figure 3. Graph G of the MFE instance (G, N, E) that corresponds to the CLIQUE instance (J, k) is Section 3. The symbol U indicates disjoint union.

contributes b nodes to S and a nodes to T. Also by definition of the lexicographic product, if there is at least one edge of H between S and T, then all possible edges between S and T are present. By construction, $V(J) \cup V_1$ and V_2 induce connected subgraphs of G; thus, if V_2, say, intersects both S and T, all possible edges between S and T must be present in H. In G, there are no edges between $V(J) \cup V_1$ and V_2; hence, filling out G to H entails adding at least $a^2 + b^2$ edges (i.e., all edges between $S \cap (V(J) \cup V_1)$ and $T \cap V_2$ and all edges between $T \cap (V(J) \cup V_1)$ and $S \cap V_2$). The quantity $a^2 + b^2$ is at least $(p^2 + 1)/2$, becuase $a + b = p$. However, the maximum number of edges we are allowed to add in order to go from G to H is

$$E - |E(G)| \leq |E(J)| \leq \binom{n}{2} < p < \frac{p^2 + 1}{2},$$

because p was chosen between n^3 and $2n^3$. Thus, if the left factor H_1 of H has two nodes, S must be $V(J) \cup V1$, T must be $V2$, and (J, k) must be a yes–instance of CLIQUE.

Finally, we show that the left factor of H must have two nodes. The only other possibility is that it has p nodes. If H_1 has p nodes, then $V(H)$ can be partitioned into p pairs of nodes, each of which induces a copy of H_2. At least one of these pairs, say $\{s, t\}$, must intersect both $V(J) \cup V_1$ and V_2, because $V(J) \cup V_1$ and V_2 have odd cardinality and hence can't be partitioned into pairs. If $s \in V(J) \cup V_1$, then it is adjacent (in G) to at least $p - n$ other nodes, say $s_1, ..., s_{p-n}$ in $V(J) \cup V_1$, none of which is in the copy of H_2 induced by $\{s, t\}$. By definition of the lexicographic product, all edges $t - s_i$, $1 \leq i \leq p_n$, must be present in H if the edges $s - s_i$ are present and $\{s, t\}$ induces a copy of H_2. None of the edges $t - s_i$ are in $E(G)$. This means that filling out G to H entails adding at least

$$p - n > n^3 - n > n^2 > |E(J)|$$

edges, which is too many, as in the previous case.

This completes the proof of

Theorem: *The Minimal Factorable Extension problem is NP–Complete.*

4. Density

The NP–Completeness result of the previous section indicates that it is probably computationally intractable to find the *exact* value of $|E(H)|$, where H is a minimal factorable extension of a given graph G. In this section, we seek bounds on $|E(H)|$ as a function of $|E(G)|$.

Part of the motivation for this question is the following. The *cartesian product* of graphs G_1 and G_2 is formed by substituting a copy of G_2 for each node in G_1 and drawing in the trivial isomorphism between each pair of adjacent copies. In [4], it is shown that, if G is a cartesian–prime graph on n nodes and at most $c_1n^{2-1/k}$ edges, then the minimal cartesian–factorable extension of G has at most $c_2n^{2-1/2k}$ edges. (The factors c_1 and c_2 are unimportant constants.) These bounds are tight: for each k, there is an infinite family of graphs for which this many additional edges are required to achieve factorability. Thus, "maximum fill–in" is never required in order to make a graph cartesian–factorable: if G does not have a constant fraction of all possible edges, then neither does its minimal cartesian–factorable extension.

Not too surprisingly, maximum fill–in may be required in order to make a graph lexicographically factorable, as the following fact shows.

Fact: *There is an infinite family of graphs $G = \{G_n\}$ such that $|V(G_n)| = n$, $|E(G_n)| = O(n^{3/2}\log n)$, and the minimal lexicographically factorable extension H_n of G_n satisfies $|V(H_n)| = n$, $|E(H_n)| = \Omega(n^2)$.*

Proof For each $G_n \in G$, the number n is of the form p^2, where p is a prime. thus any minimal factorable extension H_n of G_n has n nodes and is of the form $H_n = H_n^1[H_n^2]$, where $|V(H_n^1)| = |V(H_n^2)| = p$. We would be done if we could show that H_n^1 would have to be K_p, because in that case the edges between copies of H_n^2 would contribute $\binom{p}{2}p^2 = \Omega(n^2)$ edges to $E(H_n)$. Thus it suffices to demonstrate the existence of a family of graphs $\{G_{p^2}\}$ with this property: For every partition of $V(G_{p^2})$ into p sets of size p, for each pair S, T of sets in the partition, there is at least one edge in $E(G_{p^2})$ with one endpoint in S and the other in T. Existence follows from Pippenger's result on expanding graphs ([10]): a graph is called *a–expanding* if any two disjoint sets of nodes, each of size at least $a + 1$, are joined by an edge. A straightforward probabilistic argument shows

Lemma (Pippenger): *For all $1 \leq a \leq n$, for all sufficiently large n, there is an a–expanding graph with n nodes and $O((n^2\log n)/a)$ edges.*

Putting $n = p^2$ and $a = p - 1$ gives the desired family $\{Gn\}$. $\square$

This fact answers a special case of the following general question.

Question: If $|V(G)| = n$ and $|E(G)| = m$, how large can $|E(H)|$ be in relation to n and m, where H is a minimal lexicographically factorable extension of G? Find an infinite family of irreducible graphs that require this many edges.

REFERENCES

[1] D. Coppersmith and J. Feigenbaum. Finite Graphs with Two Inequivalent Factorizations under the Composition Operator, IBM Research Report RC11149, Yorktown Heights, 1985.

[2] W. Dörfler and W. Imrich. Das Lexikographische Produkt Gerichteter Graphen, Monats. Math. 76 (1972), 21 – 30.

[3] J. Feigenbaum. Product Graphs: Some Algorithmic and Combinatorial Results, Stanford University Technical Report STAN–CS–86–1121, PhD Thesis, 1986.

[4] J. Feigenbaum and R.W. Haddad. On Factorable Extensions and Subgraphs of Prime Graphs, SIAM J. on Disc. Maths. (2), 1989, 197 – 218.

[5] J. Feigenbaum and A.A. Schäffer. Recognizing Composite Graphs is Equivalent to Testing Graph Isomorphism, SIAM J. on Comp. 15 (1986), 619 – 627.

[6] M. Garey and D. Johnson. Computers and Intractability: A Guide to the Theory of NP–Completeness, Freeman, San Francisco, 1979.

[7] M. Garey and D. Johnson. The Complexity of Near–Optimal Graph Coloring, JACM (23), 1976, 43 – 49.

[8] I. Niven and H.S. Zuckerman. An Introduction to the Theory of Numbers, 3rd Edition, John Wiley & Sons, New York, 1972.

[9] C. Papadimitriou and M. Yannakakis. The Complexity of Factes (and some Facets of Complexity), JCSS (28), 1984, 244 – 259.

[10] N. Pippenger, Sorting and Selecting in Rounds, SIAM J. on Comp. (16), 1987, 1032–1038.

A Boolean Algebra Approach to the Construction of Snarks

M.A. Fiol[1]

Universitat Politècnica de Catalunya

ABSTRACT

This work deals with the construction of snarks, that is, cubic graphs that cannot by 3–edge–colored. A natural generalization of the concept of "color", that describes in a simple way the coloring ("0" or "1") of any set of (semi)edges, is introduced. This approach allows us to apply the Boolean logic theory to find an ample family of snarks, which includes many of the previously known constructions and also some interesting new ones.

1. Introduction

Let G be a graph with maximum degree Δ. The *chromatic index* of G, denoted by $\chi'(G)$, is the minimum integer k such that G is k–edge–colorable. By the well-known theorem of Vizing [25],

$$\Delta \leq \chi'(G) \leq \Delta + 1.$$

When $\chi'(G) = \Delta$, the graph G is said to be of *class 1*. Otherwise, i.e. when $\chi'(G) = \Delta + 1$, G is said to be of *class 2*.

The term *Tait coloring* of G is used to mean a 3–edge–coloring of G when such a graph is cubic [24]. For a general textbook on edge–coloring we refer the reader to [11].

The main concern of this paper is the construction of snarks. Following [3], we define a *snark* as a cubic graph that cannot be Tait colored (i.e., with chromatic index 4). This name was proposed by M. Gardner [13] who borrowed it from the Lewis Carroll ballad *"The Hunting of the Snark"*. Usually, and in order to avoid "trivial cases", a class 2 cubic graph is called snark only if it is cyclically 4–edge–connected and with girth at least 5. See, for instance, [6], [18] or [26]. However, the decomposition results

1 This work has been supported by the Spanish Research Council (Comision Interministerial de Ciencia y Tecnologia, CICYT) under projects 1180/84 and 0173/86.

given in [3] and [15] showed that this notion of nontriviality may not be appropriate. Hence, we adopt the most simple definition given above. In the next section we will study this question in detail.

To the author's knowledge, the history of the hunt of (nontrivial) snarks may be summarized as follows. In 1973 only four snarks were known, the earliest one being the ubiquitous Petersen graph P [22]. The other three, on 18, 210, and 50 vertices, were found by D. Blanuša [2], B. Descartes [7] and G. Szekeres [23] respectively. Quoting A. Chetwynd and R. Wilson [6], "In 1975 the art of snark hunting underwent a dramatic change when R. Isaacs [18] described two infinite families of snarks." One of these families, called the *BDS class*, included all (three) snarks previously known. In fact, this family is based on a construction also discovered independently by G.M. Adelson–Velski and A. Titov in [1]. The members of the other family are the so–called *flower snarks*. They were also found independently by Grinberg in 1972, although he never published his work.

In [20], Jakobsen proposed a method, based on the well–known Hajós–union [16], to construct class 2 graphs. As it was pointed out by M.K. Goldberg in [15], some snarks of the BDS class can also be obtained by using this approach.

Later, R. Isaacs [19] described two new infinite sets of snarks found by F. Loupekine.

In [8], the author proposed a new method of generating snarks, based on Boolean algebra. This method led to a new characterization of the BDS class and also to a significant enlargement of it. For instance, Loupekhine's graphs [19] and most of the Goldberg's snarks [14], [15] can be viewed as members of this class. This paper is mainly devoted to the study of such a method.

In [8], infinitely many snarks of another family, called by R. Isaacs the *Q class*, were also given. Apart from the Petersen graph P and the flower snark J_5, in [18] Isaacs had given a further snark of this class: the *double star graph*. The graphs of this class are all cyclically 5–edge–connected. Recently, P. Cameron, A.G. Chetwynd and J.J. Watkins [3] gave a method to construct new snarks belonging to such a family.

Other constructions of snarks, most of them belonging to the BDS class, have been proposed by several authors. See, for instance, the papers of U.A. Celmins and E.R. Swart [5], J.L. Fouquet, J.L. Jolivet and M. Rivière [12], and J.J. Watkins [26].

2. Multipoles

In the study of snarks it is useful to think of them as made up by joining two or more graphs with "dangling edges". We call these graphs multipoles. More precisely, a *multipole* or *m–pole* $Gp = (V, E, X)$ consists of a (finite) set of vertices $V = V(Gp)$, a set of edges $E = E(Gp)$ or unordered pair of vertices, and a set $X = X(Gp)$, $|X| = m$, whose elements x_i are called *semiedges*. Each semiedge is associated either with one vertex or with another semiedge making up what we call an *isolated edge*. An example of multipole is depicted in Fig. 1a. Notice that, according to our definition, a multipole can be disconnected or even be "empty" in the sense that it can have no vertices. The diagram of a generic m–pole will be as shown in Fig. 1b.

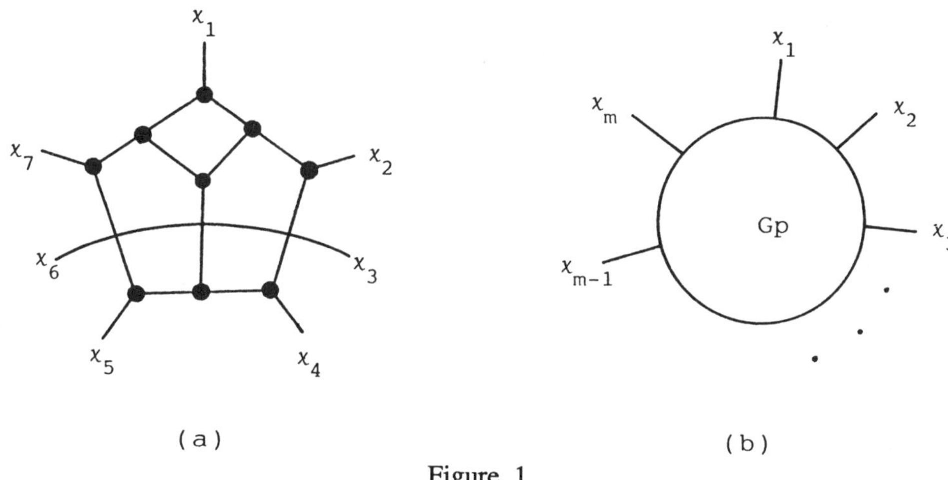

(a)   (b)

Figure 1

The behavior of the semiedges is as expected. For instance, if the semiedge x is associated with vertex u, we say that x is *incident* to u. Then we write $x = (u)$ following Goldberg's notation [15]. By joining the semiedges (u) and (v) we obtain the edge (u, v). The *degree* of u, d(u) is defined as the number of edges plus the number of semiedges incident to it. Throughout this paper, a multipole will be supposed to by cubic, i.e. d(u) = 3 for all $u \in V$.

Given a multipole Gp, we denote by Gp* the graph (with maximum degree 3) obtained from Gp by leaving out all its semiedges. Then, Gp is said to be *contained* in a cubic graph G if Gp* is a (proper) subdigraph of G. Notice that, in this case, Gp can be thought of as being obtained from G by cutting (in one or more points) some of its edges.

Let $C = \{1, 2, 3\}$ be a set of "colors". A *Tait coloring* of a m–pole (V, E, X) is an assignment of colors to its edges and semiedges, i.e. a mapping $\phi : E \cup X \to C$, such that in each vertex incide edges and/or semiedges with different color and each isolated edge has both semiedges of the same color. For example, Fig. 2 shows a Tait coloring of the 7–pole of Fig. 1a. Note that the numbers of semiedges with equal color have the same parity. The following basic lemma states that this is always the case.

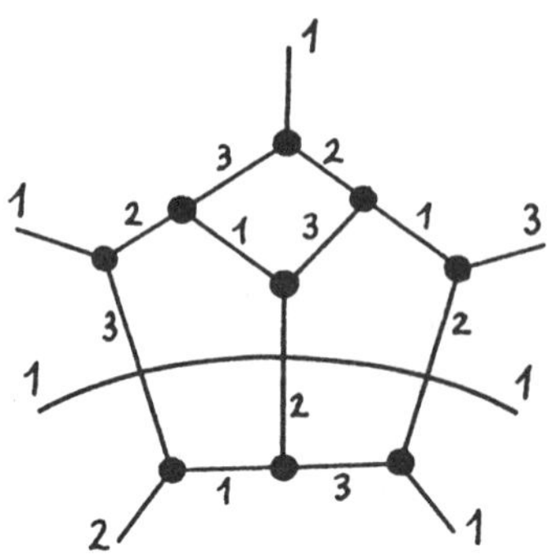

Figure 2

The Parity Lemma *Let m_i be the number of semiedges with color i, $i = 1, 2, 3$, in a Tait colored m–pole. Then*

$$m_1 \equiv m_2 \equiv m_3 \equiv m \ (mod\ 2). \tag{1}$$

This result has been used extensively in the literature on the subject. See for instance [2], [7], [15] or [18]. Although in these references isolated edges are not allowed, the proof is basically the same and hence we refer the reader to them.

Given a m–pole Gp with semiedges $x_1, ..., x_m$, we define its *set $C(Gp)$ of semiedge colorings* as

$$C(Gp) = \{(\phi(x_1), \phi(x_2), ... \phi(x_m)) : \phi \text{ is a Tait coloring of } Gp\}.$$

Note that $C(Gp)$ depends on the order in which the semiedges are considered. Thus, when referring to such a set we will implicitly assume that this ordering is given.

Of course, $C(Gp) = \varnothing$ iff Gp is not Tait colorable. In this case it is trivial to obtain a class 2 graph from Gp. Indeed, we can either remove all its semiedges or join them properly in order to achieve regularity (using additional vertices if necessary). By the parity lemma, the simplest example of non–Tait–colorable m–pole is when $m = 1$, so that any cubic graph with a bridge is trivially of class 2.

In the other extreme, we will say that Gp is *c–complete* if $C(Gp)$ has maximum cardinality. In other words, Gp is c–complete if it can be Tait colored so that its semiedges have any combination of colors satisfying the parity lemma. For instance, all Tait colorable 2–poles and 3–poles are c–complete because, according to (1), the only possibilities, up to permutation of the colors, are $(\phi(x_1), \phi(x_2)) = (a, a)$ and $(\phi(x_1), \phi(x_2), \phi(x_3)) = (a, b, c)$ respectively —here, and henceforth, the letters a, b, c stand for the colors 1, 2, 3 in any order. Clearly, the simplest c–complete 2–pole and 3–pole are respectively an isolated edge and a single vertex with 3 semiedges incident to it. They will be denoted by **e** and **v** respectively. On the other hand, a c–complete

4–pole has four different values of $(\phi(x_1), \phi(x_2), \phi(x_3), \phi(x_4))$— namely, (a, a, a, a), (a, a, b, b), (a, b, a, b) and (a, b, b, a). In general, a c–complete multipole will be denoted by **Z**.

Other useful definitions related with the semiedge coloring set follow. Let Gp_1 and Gp_2 be two m–poles. Then:

Gp_1 and Gp_2 are said to be *c–equivalent* if $C(Gp_1) = C(Gp_2)$;

Gp_1 is said to be *c–contained* in Gp_2 if $C(Gp_1) \subseteq C(Gp_2)$;

A multipole Gp is said to be *c–reducible* if there exists another multipole Gp', c-contained in Gp, such that $V(Gp') < V(Gp)$. In such a case we also say that Gp is *c–reducible to Gp'*.

Let us assume that a snark U contains the multipole Gp which c–contains the multipole Gp'. Then, it is clear that Gp can be replaced by Gp' —in the obvious way— without affecting the non–Tait–colorability of the resulting graph. Moreover, if Gp is c–reducible to Gp', such a graph will have fewer vertices than U.

By the above, the conditions of nontriviality for snarks, given in the Introduction, are a consequence of the following statements:

A1 Any 1–pole is not Tait colorable.

A2 Any 2–pole different from **e** is either not Tait colorable or c–reducible to **e**.

A3 Any 3–pole different from **v** is either not Tait colorable or c–reducible to **v**.

As A square, i.e. a 4–cycle with one semiedge incident to each vertex, is c–reducible to two parallel isolated edges (the cyclic orderings of the semiedges being induced by the drawings).

Let U denote a snark. M.K. Goldberg [15] and P.J. Cameron, A.G. Chetwynd and J.J. Watkins [3] implicitly proved the following result.

A4 Any 4–pole Gp contained in U with $V(Gp) > 2$ is either not Tait colorable or c–reducible.

In the latter paper the following result was also proved.

A5 Any 5–pole contained in U with $V(Gp) > 5$ is either not Tait colorable or c–reducible.

In general, since the number of possible semiedge colorings $(\phi(x_1), \phi(x_2), ..., \phi(x_m))$ is finite, there exists a positive integer–valued function $v(m)$ such that the following result holds.

Am   Any m–pole Gp contained in U with V(Gp) > v(m) is either not Tait colorable or c–reducible.

The exact value of v(m) is unknown for m ≥ 6. Even so, the above statement shows that any snark U with a cutset of m edges and V(U) > 2v(m) can be "reduced" to another snark with fewer vertices. See [3] for the cases m = 4, 5.

Let Gp_1 and Gp_2 be two m–poles with semiedges x_i and y_i, i = 1, 2, ..., m, respectively, and assume that by joining x_i with y_i for all i we obtain the cubic graph G. Then we will say that Gp_1 and Gp_2 are *complementary (with respect to G)*, or that Gp_2 is the *complement of* Gp_1, written $Gp_2 = Gp'_1$.

On the other hand, the m–poles Gp_1 and Gp_2 are said to be *c–disjoint* if $C(Gp_1) \cap C(Gp_2) = \varnothing$. In particular that is the case when one of the m–poles is not Tait colorable.

The analysis and synthesis of snarks is based on the following straightforward result.

Proposition   *Let Gp and Gp' be two complementary multipoles of graph G. Then G is a snark iff Gp and Gp' are c–disjoint.*

Thus, the problem of constructing snarks can be reduced to the problem of finding pairs of c–disjoint multipoles. The main problem to proceed in this way is that, when the number of semiedges increases, the characterization of the set C(Gp) becomes more and more difficult. To overcome this drawback the idea is to group the semiedges in different sets and give a proper characterization of the "global" coloring of their elements, as we do in the next section.

3. Multisets and Boole Colorings

A *mulitset* or *n–set* is simply an n'–pole Gp whose n' semiedges are grouped in n ≤ n' sets, say X_i, i = 1, 2, ..., n, with $n_i = |X_i|$. See Fig. 3.

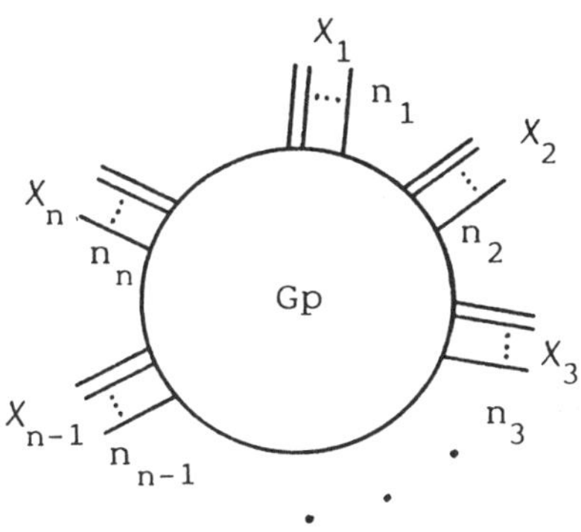

Figure 3

The main concern of this section is to characterize in a useful and simple way the coloring of the sets X_i for a given Tait coloring ϕ of Gp. To this end, let X be a generic set of m semiedges, m_i of which have color i, $i = 1, 2, 3$. Then, depending upon the parity of these numbers, we basically distinguish two cases:

Case 0 *They have the same parity:*

We say that the set X has *Boole coloring* 0, denoted by $\phi(X) = X = 0$, iff

$$m_1 \equiv m_2 \equiv m_3 \equiv m \pmod 2. \tag{2}$$

Case 1 *They have different parity:*

We say that the set X has *Boole coloring* 1 (or, more specifically, 1_a), denoted by $\phi(X) = X = 1(1_a)$, iff

$$m_a + 1 \equiv m_b \equiv m_c \equiv m + 1 \pmod 2. \tag{3}$$

This characterization (as well as a generalization of it involving more than 3 colors) was introduced in [8] to construct snarks. Other related applications are discussed in [9] and [10].

Note that if X coincides with the semiedge set of a Tait colored multipole (1–set), the parity lemma holds and hence $X = 0$.

Clearly, the above definitions can also be used to characterize the Boole coloring of any 3–colored set of edges.

By way of example, Table 1 shows the Boole coloring of X for different values of the coloring–vector (m_1, m_2, m_3) when $1 \leq m \leq 3$.

| Boole coloring X | coloring–vector (m_1, m_2, m_3) | | |
	$m = 1$	$m = 2$	$m = 3$
0		(2, 0, 0) (0, 2, 0) (0, 0, 2)	(1, 1, 1)
1_1	(1,0,0)	(0,1,1)	(1,0,2) (1,2,0)
1_2	(0,1,0)	(1,0,1)	(0,1,2) (2,1,0)
1_3	(0,0,1)	(1,1,0)	(0,2,1) (2,0,1)

Table 1

It is interesting to note the following remarks:

B1 When $m = 1$, the only possible Boole coloring of (the semiedge of) X is 1.
Moreover, $X = 1_i$ iff such a semiedge has color i, $i = 1, 2, 3$.

B2 When $m = 2$, we have a Boole coloring 0 (resp. 1) iff the two semiedges of X
have the same (resp. different) color. This characterization was independently
used by M. Goldberg [14], [15]; and by I. Holyer [17] whose values T ("true")
and F ("false") correspond to our 0 and 1 respectively. More specifically, note
that the Boole coloring is 1_a iff the missing color is a.

B3 When $m = 3$, the Boole coloring of X is 0 iff the three colors of the semiedges
are all different. Thus, we can say that a cubic graph is Tait colored iff the set of
edges incident to each vertex has Boole coloring 0. Otherwise, if two semiedges
have the same color, the Boole coloring of X is 1_a where a is the color of the
third semiedge. An equivalent characterization, but without using the 1's, was
used by B. Descartes in [7] to construct his graph.

A natural definition of the sum of Boole colorings is now the following. Let X and
y be two sets of semiedges with Boole colorings X and Y respectively. Then we
define the sum X + Y as the Boole coloring that, according to (2) and (3),

corresponds to the set $X \cup Y$. It is very easy to check that this definition leads to Table 2, so that we obtain the "Klein group of Boole colorings".

+	0	1_1	1_2	1_3
0	0	1_1	1_2	1_3
1_1	1_1	0	1_3	1_2
1_2	1_2	1_3	0	1_1
1_3	1_3	1_2	1_1	0

Table 2

Note that, as each element coincides with its inverse, $m1_i = 1_i + \ldots + 1_i$ is 0 when m is even and 1_i when m is odd. The following result, based on this simple fact, is of fundamental importance in our study. Because of **B1**, it can be seen as a generalization of the parity lemma.

Lemma 1 *Let Gp be a Tait colored n–set with $m_i \geq 0$ sets of semiedges having Boole coloring 1_i, $i = 1, 2, 3$, $m_1 + m_2 + m_3 = m \leq n$. Then,*

$$m_1 \equiv m_2 \equiv m_3 \equiv m \pmod 2.$$

Proof As stated before, the Boole coloring of the whole set of semiedges of Gp must be 0. So we have

$$\sum_{i=1}^{3} m_i 1_i + (n - m)0 = \sum_{i=1}^{3} m_i 1_i = 0.$$

But this equality only holds if either $m_i 1_i = 0$ or $m_i 1_i = 1_i$ for all i. Since $m_1 + m_2 + m_3 = m$, the lemma follows. $\square$

Table 3 shows the feasible Boole colorings of (the semiedge sets of) a Tait colored n–set, $1 \leq n \leq 3$, according to the above result.

n	X_1	X_2	X_3
1	0		
2	0	0	
	1_a	1_a	
3	0	0	0
	0	1_a	1_a
	1_a	0	1_a
	1_a	1_a	0
	1_a	1_b	1_c

Table 3

Notice that, as pointed out before, if $n = 1$ it must be $X_1 = 0$, which is just a reformulation of the parity lemma.

Leaving out the subindexes of the Boole colorings 1, the entries in Table 3 can be thought of as being the possible values of the "logic variables" X_i. This suggests the possibility of using Boolean algebra in order to characterize the possible Boole colorings of a given multiset.

For any value of n we have the following corollary of Lemma 1.

Corollary 1 *A Tait colored n–set, $n \geq 1$, cannot have only one set with Boole coloring 1, i.e., $X_j = 1$ and $X_i = 0$ for all $i \neq j$.*

Given an n–set Gp with semiedge sets X_j, $j = 1, 2, ..., n$, we define its *set B(Gp) of Boole coloring vectors*, or simply *Boole colorings*, as

$$B(Gp) = \{(\phi(X_1), \phi(X_2), ..., \phi(X_n)) : \phi \text{ is a Tait coloring of Gp}\}.$$

Of course, this set depends upon the subindexes of the Boole colorings 1 begin considered or not, but that will be either immaterial or clear from the context. Hence the unified notation.

Most of the definitions and remarks involving the semiedge coloring set C(Gp) apply also to the set B(Gp) with minor and trivial changes. For instance, if B(Gp) has maximum cardinality, the multiset Gp is said to be c–complete. It is readily seen that if a multipole is c–complete any multiset obtained from it —with sets of at least two semiedges— is c–complete (considering subindexes or not). However, the converse does not hold in general.

Let us now consider a family of multisets which are joined in such a way that each set X_i of n_i semiedges is joined to exactly one set X_j of n_j semiedges, so making up a set of edges that we denote by (X_i, X_j) —for simplicity we can now assume that $n_i = n_j$, but later we shall see that a junction can be easily done in general. If the pairs of semiedges to be joined are not specified, this structure, which will be called a *logic network*, represents a family of cubic graphs.

Let $B = \{0, 1_1, 1_2, 1_3\}$ or $\{0, 1\}$, depending upon the case, be a set of "Boole colors". Then, a *Boole coloring* of a logic network is defined to be an assignment of Boole colors to its edge sets (X_i, X_j) such that the induced Boole coloring of each multiset Gp belongs to B(Gp).

Obviously, the same ideas above apply to the construction, from some multisets, of a *"logic multiset"*, as well as to its Boole colorability. Besides, note that from a non–Boole–colorable (logic) multiset we can readily obtain a non–Boole–colorable logic network.

In this context, it is clear that if a logic network is not Boole colorable, its "underlying" graphs are not Tait colorable. So, we will manage to construct snarks if we know how to construct such logic networks.

The most obvious way to construct a non–Boole–colorable logic network is by joining two c–disjoint (in terms of Boole colorings) multisets. Some methods to obtain them are explained in the last section, but first we need to consider some useful configurations which are the concern of the next section.

4. Multisets and Boolean Algebra

As said before, in our study we make ample use of a (2–valued) Boolean algebra. So, we have a set $B = \{0, 1\}$ jointly with two binary opreations, $+$ and $\cdot$, called *logic sum* and *logic product* respectively, satisfying some well–known axioms. It will be clear from the context when "$+$" denotes logic sum or sum of Boole colorings. Given $x \in B$, we use $\bar{x}$ to mean the complement of x, i.e., $\bar{0} = 1$ and $\bar{1} = 0$. The remaining notation used below is, I hope, self–explanatory.

The Identity Operator

According to Table 3, the Boole coloring (X_1, X_2) of a Tait colorable 2–set must be $(0, 0)$ or $(1, 1)$. Therefore we can write

$$X_2 = X_1,$$

which corresponds to the *identity* logic function.

Then a 2–set will allow us to join two sets with any number, say n_1 and n_2, of semiedges without changing their Boole colorings, see Fig. 4a. The symbol used for this "operator" is shown in Fig. 4b, where bold lines represent semiedge sets.

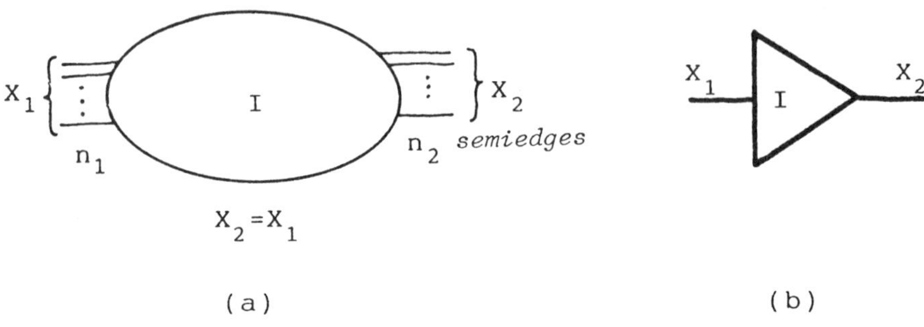

Figure 4

In particular, when one set of a Tait colored 2–set, say X_1, has only one semiedge (with color a) we will have, by **B1**, $X_2 = X_1 = 1$ (1_a).

The Truth Operator

A method to obtain a 2–set with the only possible Boole coloring $(X_1, X_2) = (1, 1)$ —and sets with more than one semiedge— is the following. Assume that a snark U contains the multipole (2–set) shown in Fig. 5a jointly with its only possible Boole coloring, and denoted by $\{Z_1, Z_2\}$ or, simply, $\{Z, Z\}$. Then it is clear that its complement, $\{Z, Z\}'$, see Fig. 5b, cannot have the Boole coloring $(0, 0)$. Otherwise, since Z_1 and Z_2 are c–complete, the colorings of the semiedges of X_1 and X_2 would be elements of $C(Z_1)$ and $C(Z_2)$ repsectively, giving in this way a Tait coloring of U. Hence, the 2–set $\{Z, Z\}'$ can only have the Boole coloring

$$X_1 = X_2 = 1,$$

which corresponds to the *"truth"* logic function (with 2 variables). Fig. 5c shows the symbol used for such a 2–set.

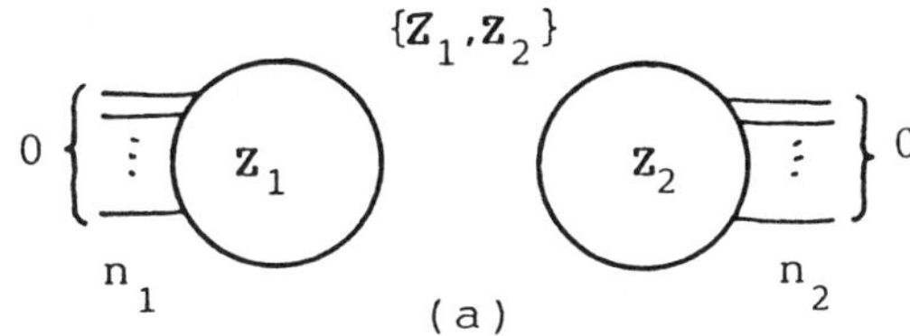

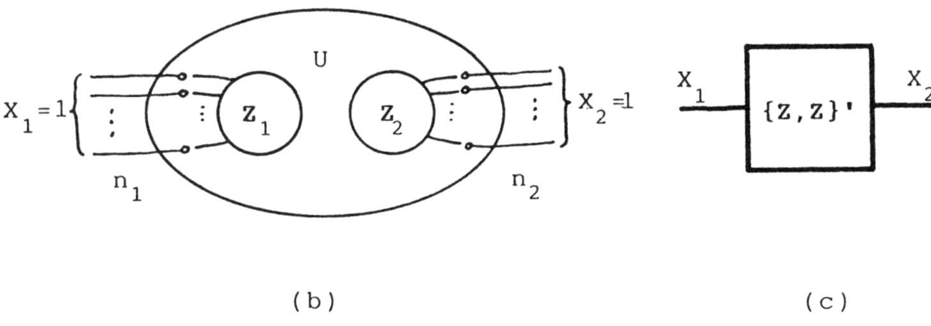

Figure 5

In particular, the simplest cases are obtained when the Z_i's are **e** ($n_i = 2$) or **v** ($n_i = 3$). For instance, the case $Z_1 = Z_2 = $ **e** was considered by R. Isaacs in [18] and by G.M. Adelson–Velski and A. Titov in [1]; and the case $Z_1 = Z_2 = $ **v**, $U = P$, was used by B. Descartes in [7] to obtain his graph. In the next section we will discuss the case $n_i \geq 4$.

The Untruth Operator

Let us now consider a snark U containing the 2–set shown in Fig. 6a, that we denote by $\{ZeZ\}$. Then, since the Boole coloring of the edge **e** must be 1, its only possible Boole coloring is $(X_1, X_2) = (1, 1)$. Hence, reasoning as in the preceding subsection, we conclude that the 2–set $\{ZeZ\}'$, shown in Fig. 6b, can only have the Boole coloring

$$X_1 = X_2 = 0,$$

which corresponds to the *"untruth"* logic function (with 2 variables). The symbol used for this 2–set is shown in Fig. 6c.

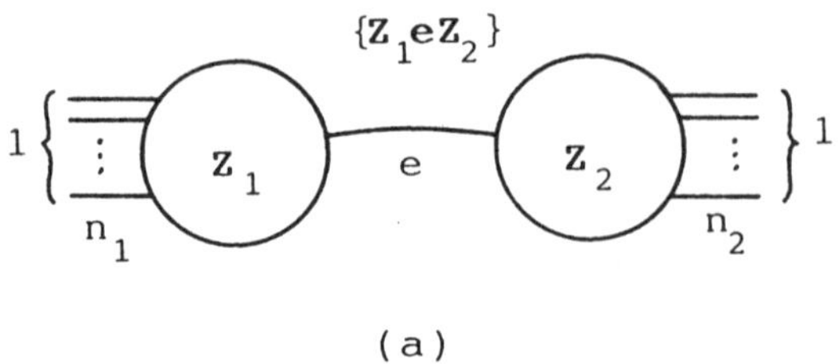

(a)

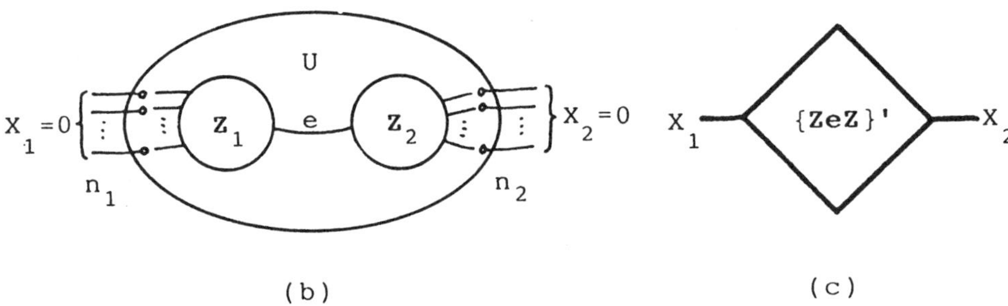

(b)   (c)

Figure 6

Apart from the trivial cases $Z_1 = Z_2 = e$ and $Z_1 = e, Z_2 = v$, the most simple 2-set of this type is obtained when $Z_1 = Z_2 = v$ ($n_1 = n_2 = 2$), which was also dealt with in the above references [1] and [18].

The Or and Exclusive–Or Operators

Reasoning as above we can obtain 3–sets with only some of the possible Boole colorings shown in Table 3. As before, all these configurations are derived by taking the complement, with respect to a snark U, of a suitable multiset made up by some c-complete multipoles. The 3–sets thus obtained are shown in Table 4 together with their truth table, the corresponding logic function if any, and the symbol —usually borrowed from logic circuit theory— we use to represent them.

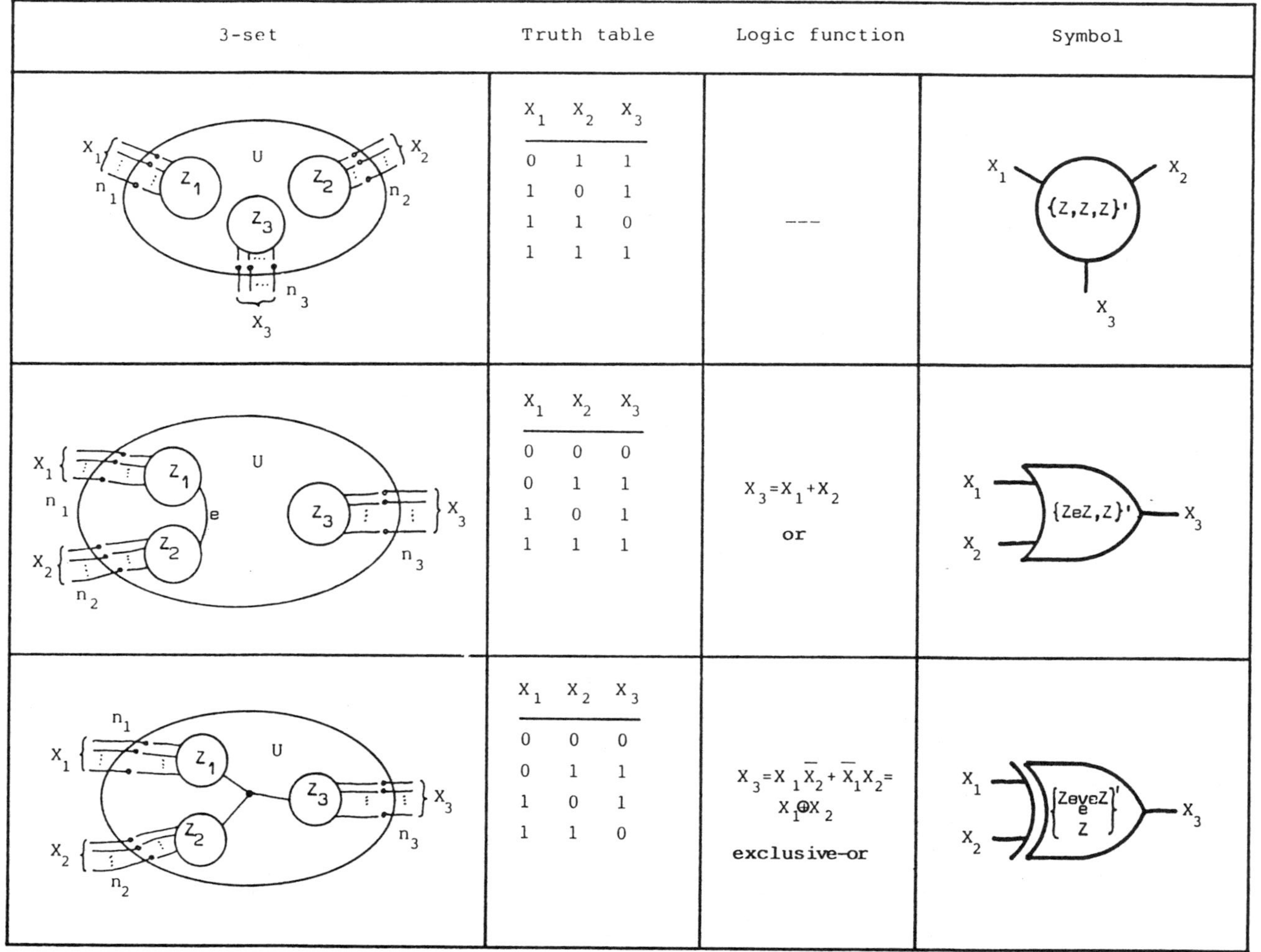

3-set	Truth table	Logic function	Symbol
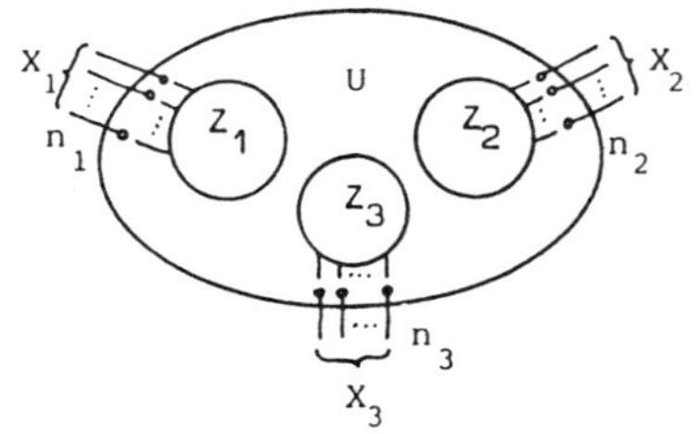X_1 n_1 U Z_1 Z_2 Z_3 X_2 n_2 n_3 X_3	X_1 X_2 X_3 0 1 1 1 0 1 1 1 0 1 1 1	---	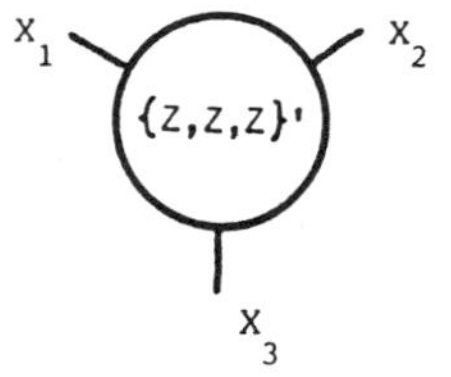X_1 X_2 {Z,Z,Z}' X_3
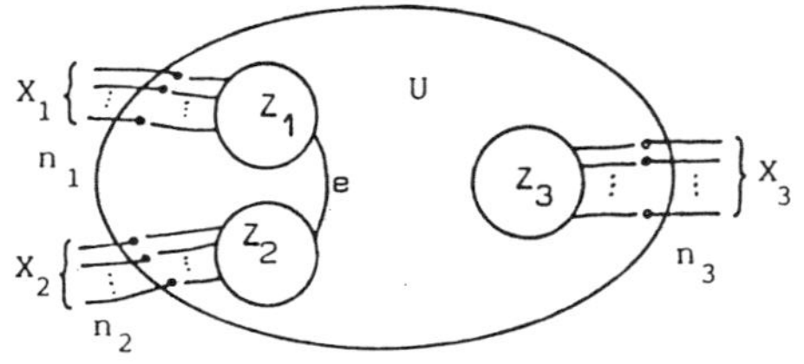X_1 n_1 X_2 n_2 U Z_1 Z_2 e Z_3 X_3 n_3	X_1 X_2 X_3 0 0 0 0 1 1 1 0 1 1 1 1	$X_3=X_1+X_2$ or	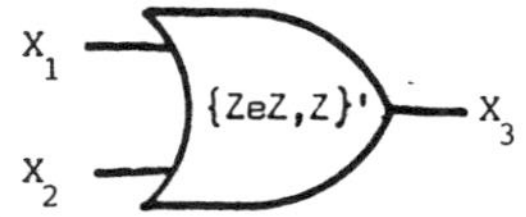X_1 X_2 {ZeZ,Z}' X_3
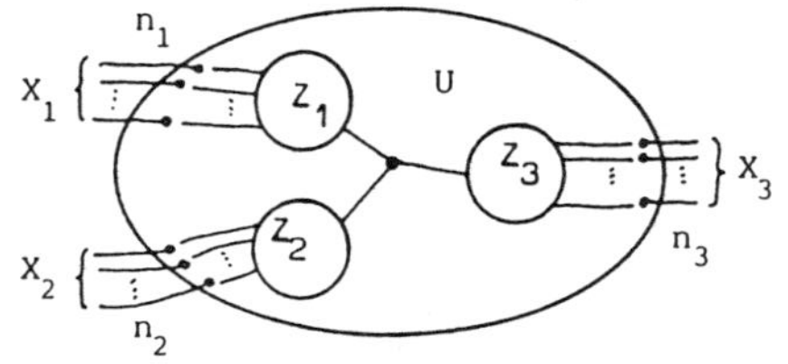n_1 X_1 U Z_1 Z_2 Z_3 X_3 n_3 X_2 n_2	X_1 X_2 X_3 0 0 0 0 1 1 1 0 1 1 1 0	$X_3=X_1\bar{X}_2+\bar{X}_1X_2=$ $X_1 \oplus X_2$ exclusive-or	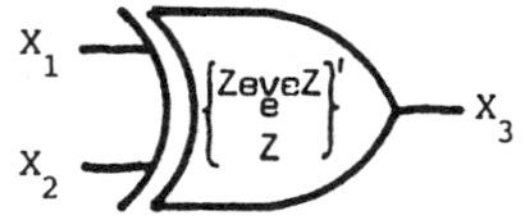X_1 X_2 {ZeveZ, Z}' X_3

Table 4

From these configurations, many particular cases may be derived. For instance, the *exclusive–or* 3–set obtained by taking $Z_1 = Z_2 = Z_3 = v$ ($n_1 = n_2 = n_3 = 2$) was independently considered in [15] and [5]. In [15], M.K. Goldberg proposed it as an example of *"even cell"*, i.e. a multiset with exactly two semiedges in each set and all Boole coloring vectors having an even number of 1's, see Section 5.

It should be pointed out that, for each general configuration, the truth table gives only its <u>possible</u> Boole colorings, Thus, in some particular cases the multiset we obtain may have not all of such colorings, or even it may have none of them (if it is not Tait colorable). For instance, when some X_i has only one element the value of X_i cannot be 0, making some Boole colorings in the truth table to be not possible. The two following particular cases of the *or* and *exclusive–or* operators are based on this fact.

The Not Operator

If in the 3–set $\left\{ \begin{smallmatrix} \mathbf{Z}e\mathbf{v}e\mathbf{Z} \\ e \\ \mathbf{z} \end{smallmatrix} \right\}'$ we consider a c–completé multipole, say Z_3, equal to e, we obtain the structure shown in Table 5 jointly with its truth table and symbol (*"not gate"* with an additional semiedge x). Notice that, disregarding X_3 (= 1), the other variables satisfy

$$X_2 = \overline{X}_1,$$

which corresponds to the *not* logic function.

By Corollary 1, we know that there is no 2–set with the Boole coloring $(0, 1)$ or $(1, 0)$. Hence, our *not* operator must have an additional semiedge —or semiedge set— whose Boole coloring is, in general, immaterial. For this reason it will be referred as a *neutral* semiedge.

The simplest particular case $Z_1 = Z_2 = v$ ($n_1 = n_2 = 2$), denoted by $\{vevev\}'$, is the basic structure (5–pole) of Loupekine's snarks [19].

3-set	Truth table	Logic function	Symbol
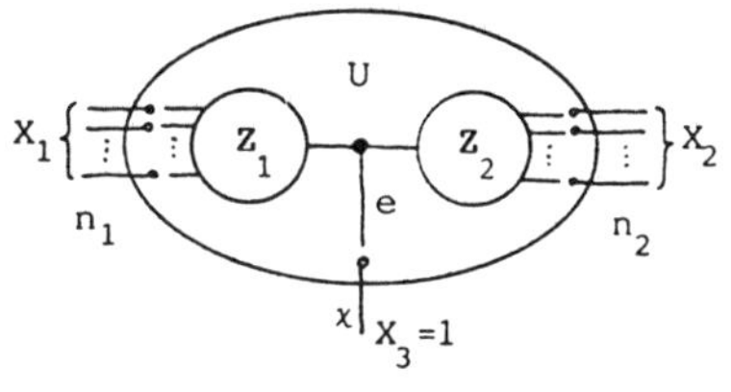 $X_1\{$ ... Z_1 ... Z_2 ... $\}X_2$, U, n_1, e, n_2, x $X_3=1$	$X_1 \quad X_2$ — $0 \quad 1$ $1 \quad 0$	$X_2 = \overline{X}_1$ not	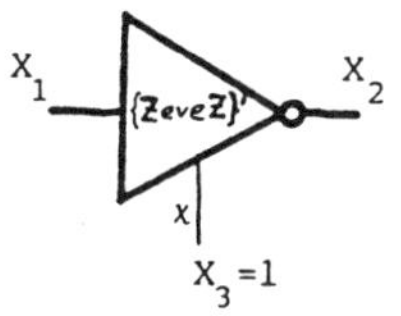 X_1 —$\{Z\,e\,veZ\}'$— X_2, x, $X_3=1$
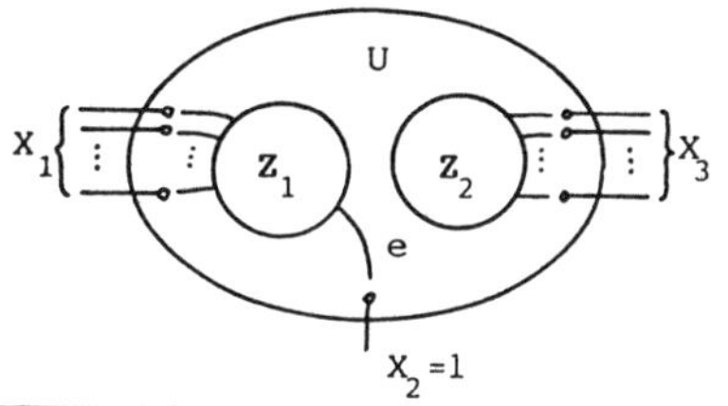 $X_1\{$ ... Z_1 ... Z_2 ... $\}X_3$, U, e, $X_2=1$	$X_1 \quad X_2 \quad X_3$ — $0 \quad 1_a \quad 1_a$ $1_a \quad 1_b \quad 1_c$	$---$ ("partial-or")	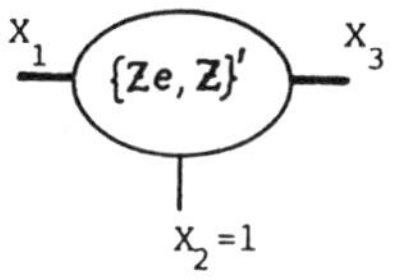 X_1 —$\{Ze, Z\}'$— X_3, $X_2=1$

Table 5

The Partial–Or Operator

This 3–set is obtained by considering the *or* operator $\{Z_1 e Z_2, Z_3\}'$ with $Z_2 = e$.
It is also shown in Table 5 together with its symbol and truth table where the subindexes
corresponding to the 1's have been explicitly written.

As a summary, Table 6 shows some particular cases of the above multisets when
$U = P$.

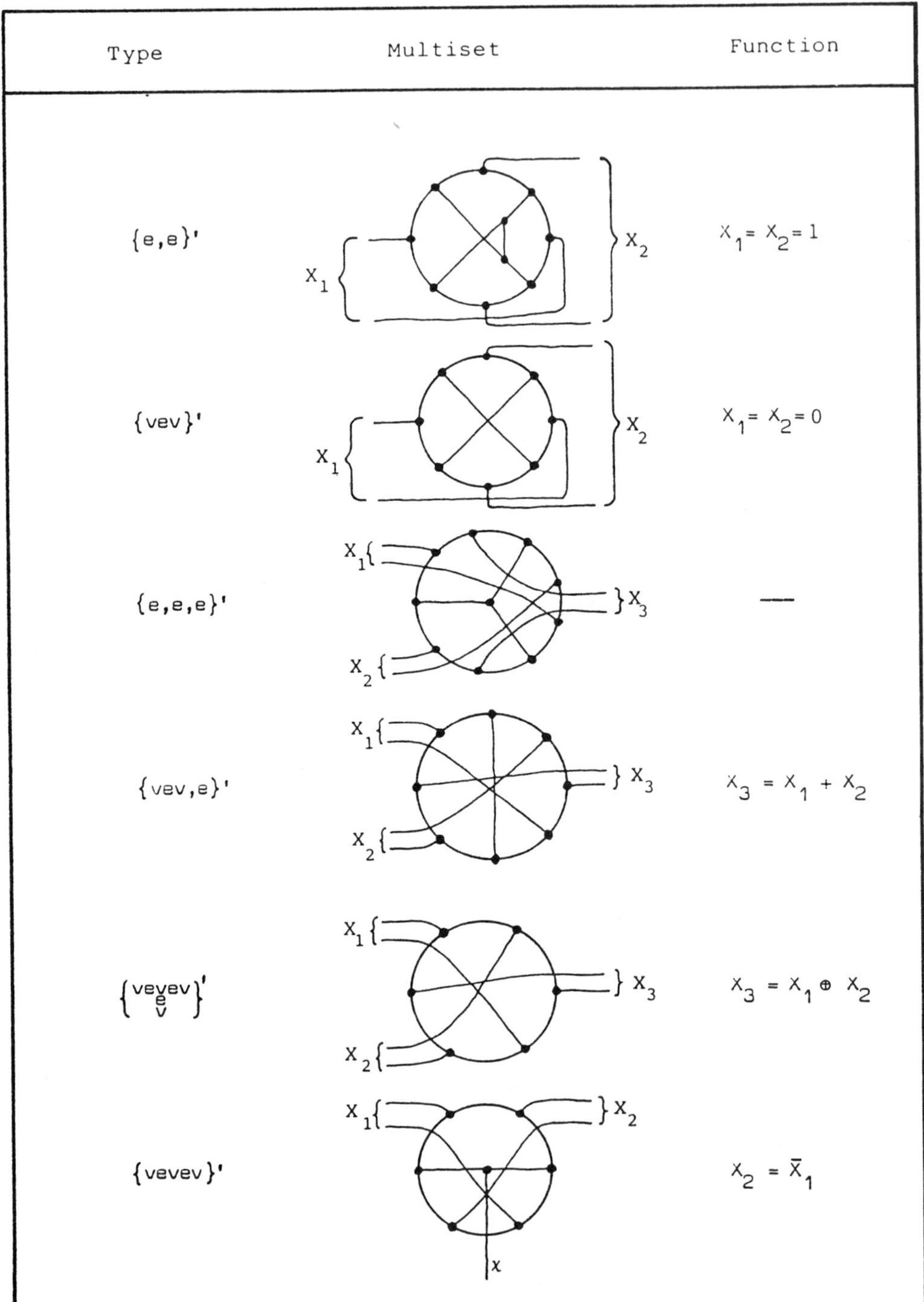

Type	Multiset	Function
$\{e,e\}'$		$X_1 = X_2 = 1$
$\{vev\}'$		$X_1 = X_2 = 0$
$\{e,e,e\}'$		—
$\{vev,e\}'$		$X_3 = X_1 + X_2$
$\left\{\begin{array}{c}vevev\\e\\v\end{array}\right\}'$		$X_3 = X_1 \oplus X_2$
$\{vevev\}'$		$X_2 = \bar{X}_1$

Table 6

Other Operators

From the logic operators obtained before it is now easy to construct multisets associated with other basic logic functions. For instance, by De Morgan's laws the *and* function can be written as

$$X_3 = X_1 X_2 = \overline{\overline{X_1 X_2}} = \overline{\overline{X}_1 + \overline{X}_2},$$

which lead us to the structure shown in Fig. 7 jointly with its truth table.

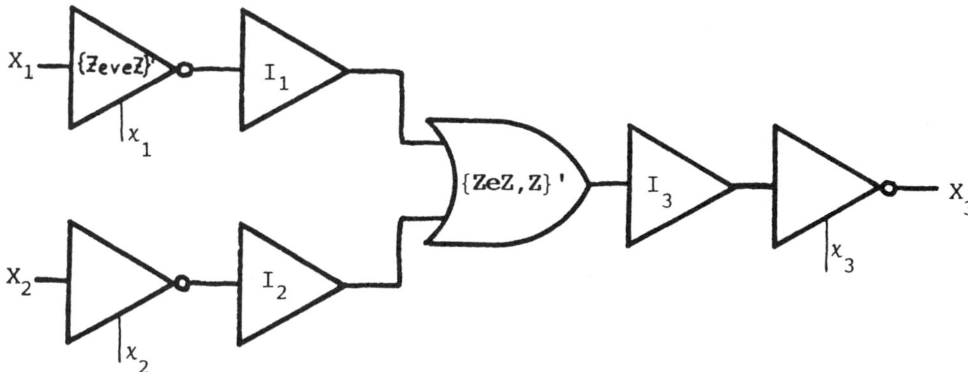

X_1	X_2	X_3
0	0	0
0	1	0
1	0	0
1	1	1

Figure 7

Notice the presence of the 2–sets I_i which, as said before, are used to join two sets with different numbers of semiedges, "transferring" the Boole coloring from one to the other. As the presence of such 2–sets is obvious, we will not draw them henceforth.

When the neutral semiedges x_1, x_2 and x_3 are joined to a common vertex, we obtain a 3–set which, by Corollary 1, cannot have the Boole colorings (0, 1, 0) and (1, 0, 0). Then, since the remaining Boole colorings satisfy $X_1 = X_2 = X_3$ (*identity operator with 3 variables*), we call this configuration a *meeting point*. Notice that it might also be obtained by using other 3–sets (e.g. a c–complete 3–set) instead of the *or* operator.

Besides, if in such a structure we replace the *or* operator by the *exclusive–or* operator, we obtain a "*truth* operator with 3 variables", see Fig. 8. Note the interesting "color isomorphism" $i \equiv 1_i$, $i = 1, 2, 3$, between this 3–set and the most simple 3-pole **v**.

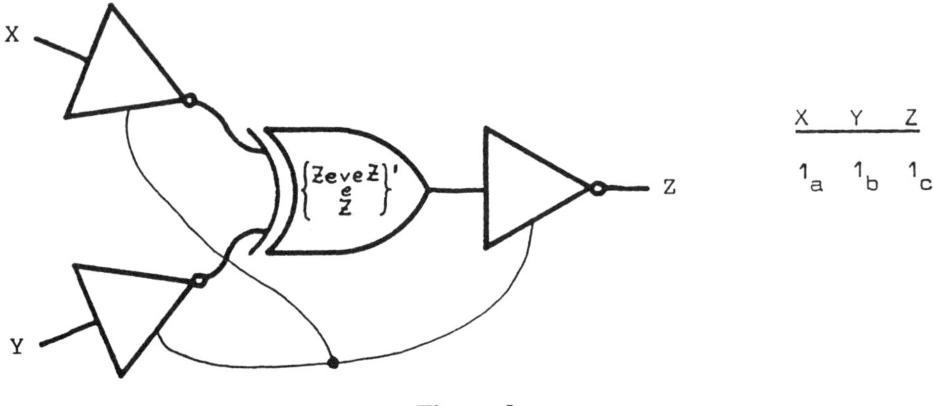

Figure 8

Actually, because of Corollary 1, it is readily seen that our *not* operator is "functionally complete", that is, any other logic operator can be derived from it. For instance, Table 7 shows how we can obtain the *exclusive–or* and *equivalence* operators, as well as other "3–sets" (disregarding neutral semiedges) whose truth table does not correspond to any logic function. Note that in obtaining such an *exclusive–or* operator, A may be any c–complete 3–set. Then, when the most simple of them —made up by 3 isolated edges— is used, and the *not* operators are of type {**vevev**}', the 3–set of Fig. 9 appears.

Diagram \ A	c-complete 3-set	$\{Z_1 e Z_2, Z_3\}'$ or operator	$\{Z_1, Z_2, Z_3\}'$
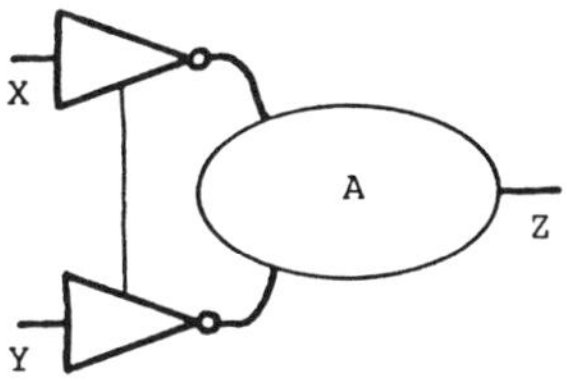	X Y Z 0 0 0 0 1 1 1 0 1 1 1 0 exclusive-or	X Y Z 0 1 1 1 0 1 1 1 0 —	X Y Z 0 0 0 0 1 1 1 0 1 —
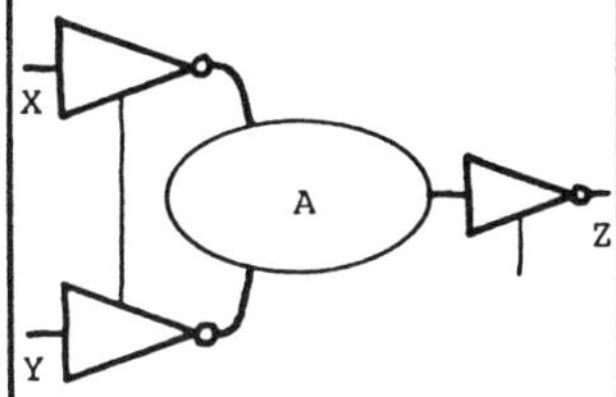	X Y Z 0 0 1 0 1 0 1 0 1 1 1 1 equivalence	X Y Z 0 1 0 1 0 0 1 1 1	X Y Z 0 0 1 0 1 0 1 0 0

Table 7

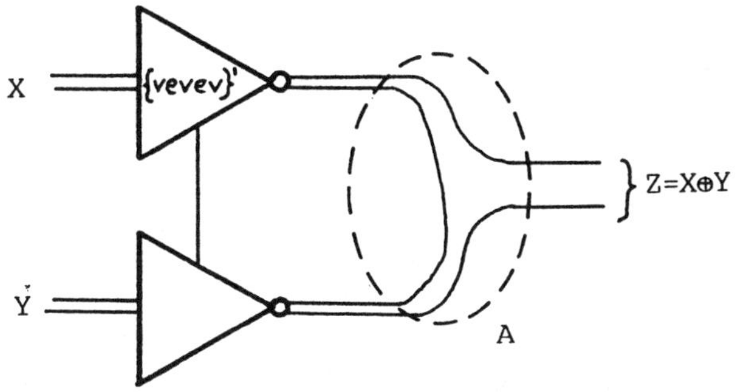

Figure 9

In order to construct a meeting point with more than 3 variables, we use a 3–set made up by joining "in series" an even number p of *not* operators and grouping all their neutral semiedges x_i, $1 \leq i \leq p$, into a set W, see Fig. 10a, its symbol being shown in Fig. 10b. From Corollary 1 and the parity of p it is trivial to check that $X = 0 \Rightarrow Y = W = 0$; $Y = 0 \Rightarrow X = W = 0$; and $W = 1 \Rightarrow X = Y = 1$. Hence it follows that the n–set shown in Fig. 10c has only the Boole colorings $(0, 0, ..., 0)$ or $(1, 1, ..., 1)$ as claimed.

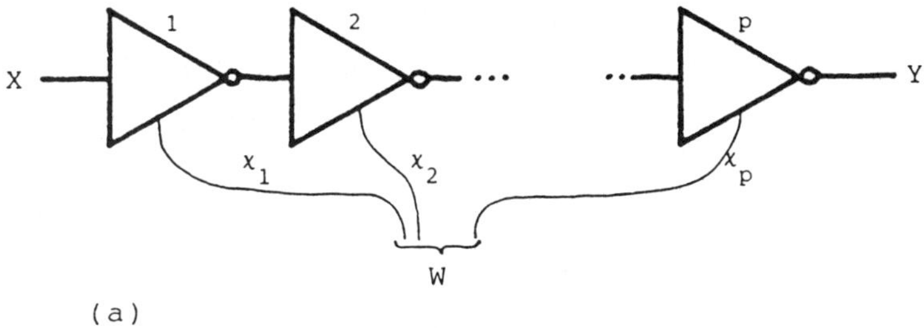

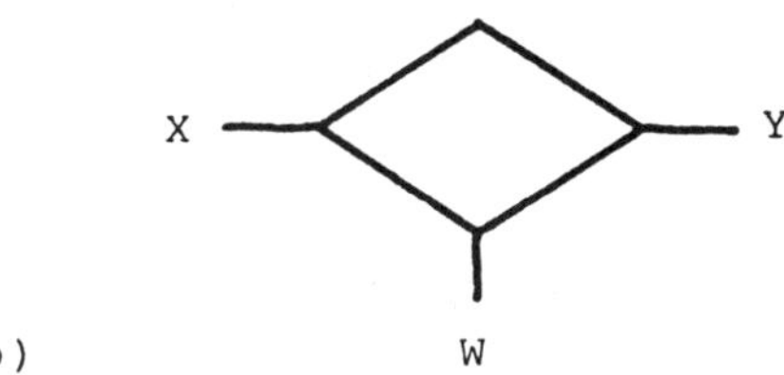

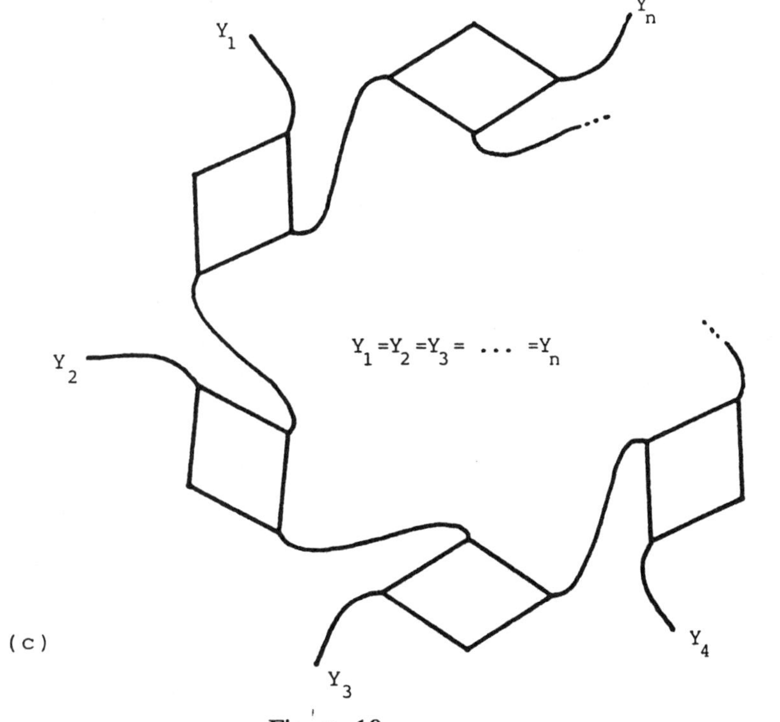

Figure 10

A particular case of this configuration, obtained from pairs of *not* operators $(p = 2)$ like the one shown in Table 6, was independently used by I. Holyer [17] to prove that the problem of finding the chromatic index of an arbitrary cubic graph is NP–complete.

By joining conveniently the logic operators and the meeting points studied throughout this section, we can now obtain multisets implementing any logic function $f : B^n \to B^m$, $B = \{0, 1\}$. Also, by Corollary 1, many interesting configurations can be derived using only *not* operators. As an example, we have chosen the 4–set A_2 depicted in Fig. 11a, where A_0 represents a c–complete 4–set. Since A_1 and A_2, as well as A_0 itself, must satisfy Corollary 1, it is readily seen that the resulting 4–pole A_2 has the truth table shown in Fig. 11b, where subindexes have been included (just obtain the possible Boole coloring vectors of A_i from those of A_{i-1}, $i = 1, 2$, and, after each step, leave out the vectors with only one 1). Analogously to the case of the *truth* operator with 3 variables, it is worth noting the color isomorphism between this 4–set and a vertex with 4 semiedges (colored with, say, 0, 1, 2, 3).

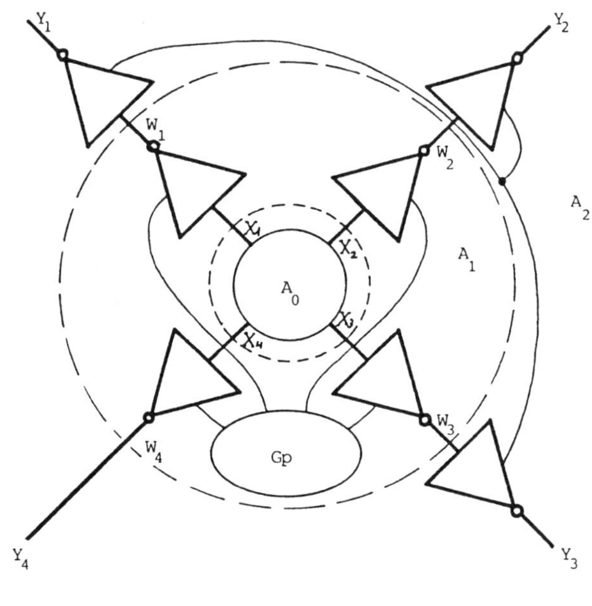

(a)

Y_1	Y_2	Y_3	Y_4
0	1_a	1_b	1_c
1_a	0	1_b	1_c
1_a	1_b	0	1_c
1_a	1_b	1_c	0

(b)

Figure 11

5. The BBDS Class of Snarks

We propose to slightly change the name of the BDS class of snarks and call it the *BBDS class*. The capital letters would, of course, stand for Boole and Blanuša & Descartes & Szekeres whose logic and graphs respectively inspired its construction.

Roughly speaking, we say that a snark U belongs to the BBDS class, denoted by U ∈ {BBDS}, if it derives from a non–Boole–colorable logic network. Hence, a basic characteristic of such a graph is that, when constructing it, we can insert an arbitrary 2-set (e.g. the *identity* operator) between every pair of semiedge sets to be joined. This fact, and some further minor considerations, led the author [8] to the following more precise definition.

The graph U belongs to {BBDS} iff it contains at least one m–pole Gp, m > 3, such that it can be replaced by a c–complete m–pole **Z** without affecting non–Tait–colorability; and there is at most one semiedge incident to each vertex of the complementary m–pole Gp'.

Note that, by the proposition in Section 2, Gp' must be non–Tait–colorable or, what is the same, the (non–regular) graph Gp'* is of class 2. Thus an alternative definition is to say that U ∈ {BBDS} iff it contains a subgraph of class 2 with m > 3 vertices of degree 2 (and none of degree 1). For instance, in the Blanuša graph and also in Loupekine's snark L_3 one can find such subgraphs with m = 5 and 4 respectively, see Figs. 12a and 12b (the corresponding m–poles Gp are drawn in dashed lines). The subdigraph of L_3 is a counter–example to the critical graph conjecture, see [4], [9], [15] and [21].

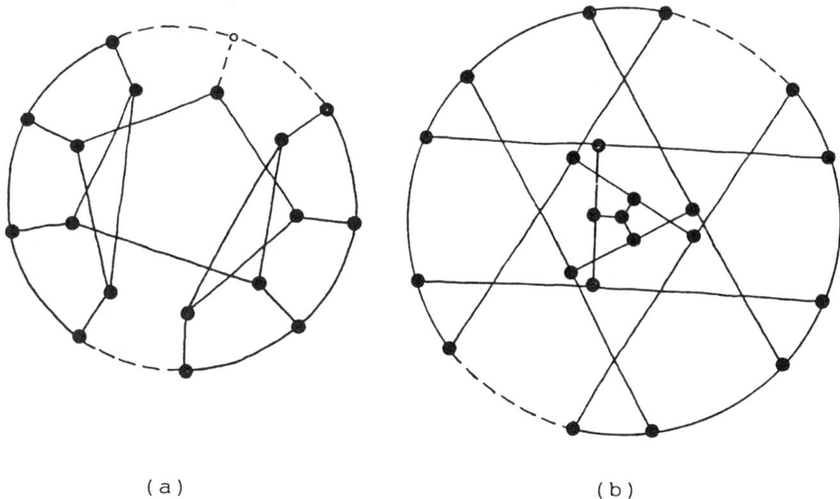

(a)   (b)

Figure 12

The two snarks above are particular instances of two simple non–Boole–colorable logic networks. Namely, those shown in Figs. 13a and 13b (k odd), which correspond to the constructions of R. Isaacs [18], called the *dot product*, and F. Loupekine [19]. The non–Boole–colorability of both structures follows trivially from the truth tables of the operators involved.

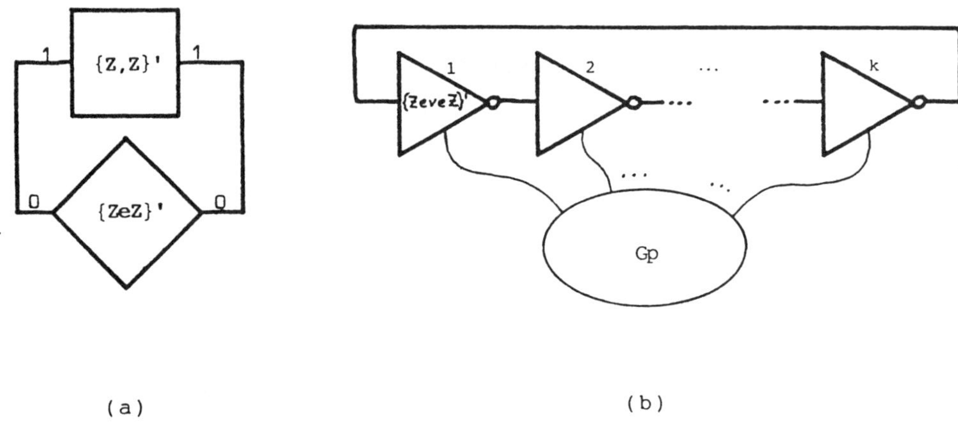

(a)   (b)

Figure 13

Let $U \in \{BBDS\}$. By the considerations above, it can be assumed that this graph contains some c–complete m–poles with $m \geq 4$. Then, applying the method of Section 4, we can obtain logic operators with larger semiedge sets than those obtained there. For example, if U is the "Star of David" of Fig. 12b (as we called L_3 before knowing of Loupekine's work), the *not* operator $\{Z_1eveZ_2\}'$ with Z_1 a c–complete 4–pole $(n_1 = 3)$ and $Z_2 = v$ $(n_2 = 2)$ is shown in Fig. 14.

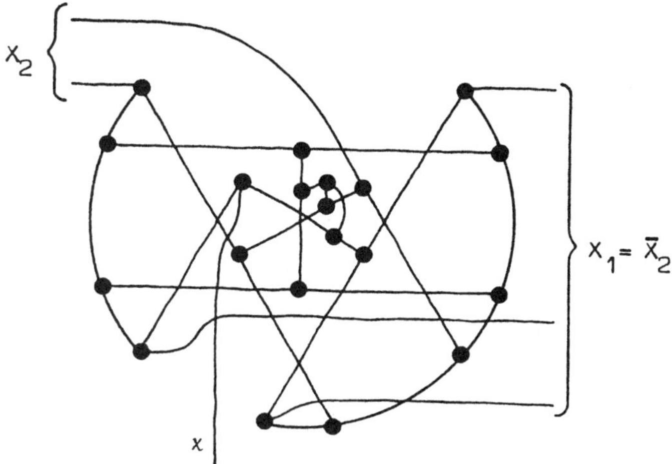

Figure 14

In what follows several methods of constructing snarks, based on the "Boole coloring theory", are given. As expected, most of the graphs obtained belong to the BBDS class.

Snarks from Logic Functions

Let us begin with an ample family of non–Boole–colorable logic networks, which include those given above. Let $f = (f_1, f_2, ..., f_m)$ and $\overline{f} = \overline{f}_1, \overline{f}_2, ..., \overline{f}_m)$ be two logic functions from B^n to B^m, $B = \{0, 1\}$, such that $y_i = f_i(x_1, x_2, ..., x_n), 1 \leq i \leq m$, $z_i = \overline{f}_i(x_1, x_2, ..., x_n) = \overline{y}_i$ if $i \in I \subset \{1, 2, ..., m\}, I \neq \varnothing$, and $z_i = y_i$ otherwise. Then, as the values of y_i and z_i, $i \in I$, mismatch for any given $(x_1, x_2, ..., x_n)$, it is clear that any pair of multisets implementing such functions are c–disjoint and, as said before, they can be used to derive a non–Boole–colorable logic network.

Snarks from Even and Odd Multisets

When the truth table of the considered multisets do not correspond to any logic function we can use other methods to find pairs of c–disjoint multisets. The method described here, based on the parity of the number of Boole colorings 1 (or 0), was proposed in [8]. The same method was independently developed in [15] for the above–mentioned cells.

A Tait colorable n–set Gp is said to be *even* (resp. *odd*) if all its Boole coloring vectors have an even (resp. odd) number of 1's. Note that some of the multisets studied in the preceding section belong to one of these categories. For instance the meeting point with an even number of variables and the $(n + 1)$–set corresponding to the *exclusive–or* function $y = x_1 \oplus x_2 \oplus ... \oplus x_n$ are even, whereas the 4–set of Fig. 11a is odd. Other interesting examples can easily be found applying the proposed methods.

Of course, any logic network obtained by joining two n–sets with different parity is not Boole colorable. More general results are obtained from the following statement, which is easily proved using a simple parity argument.

C *Let Gp a (logic) n–set obtained by joining n_E even multisets and n_O odd mulitsets. Then the "parity" of Gp coincides with the parity of n_O.*

Considering a logic network as a (even) 0–set and applying Corollary 1, we have the following corollaries.

C1 *Any logic network made up by joining an odd number of odd multisets and any number of even mulisets is not Boole colorable.*

C2 *Any 1–set or 2–set made up as in* **C1** *is not Boole colorable.*

C3 *Any 3–set constructed as in* **C1** *has the only possible Boole coloring $(1, 1, 1)$ —truth operator with 3 variables.*

C4 *Any 4–set constructed as in* **C1** *has only the possible Boole colorings shown in Fig. 11b.*

Snarks from Corollary 1

The method described at the end of Section 4, which is based on the use of *not* operators and Corollary 1, applies also the the construction of non–Boole–colorable logic networks or multisets. For instance, if each semiedge set $\mathcal{Y}_i$ of the 4–set of Fig. 11a is joined to a *not* operator (the neutral terminals being joined to any multipole), we obtain a 4–set with semiedge sets Z_i such that $Z_i = \overline{Y_i}$, $1 \le i \le 4$. Therefore, from the truth table of Fig. 11b, its possible Boole colorings would have only one 1, which is impossible.

Snarks from Color Isomorphisms

Let us now consider the subindexes of the Boole colorings 1. Then, as said before, some interesting color isomorphisms between multisets and multipoles appear. The first of them is based on the equivalence $i \equiv 1_i$, $i = 1, 2, 3$, between colors and Boole colorings (see **B1**). Thus, considering an m–pole as a special case of m–set, we can say that the *truth* operator of 2 (resp. 3) variables is c–equivalent to the 2–set **e** (resp. 3–set **v**), i.e., they have the same set of Boole coloring vectors. This fact allows us to construct easily a non–Boole–colorable logic network from a snark U. For instance, we can replace all the edges and vertices of U by *truth* operators of 2 variables and arbitrary 3–sets respectively —using *identity* operators if necessary. An example of this construction is the graph of B. Descartes [7]. Another possibility is to replace all the vertices of U by *truth* operators of 3 variables. Many other variations can also be considered.

Let us now consider the "c–equivalence" which exists between the 4–set of Fig. 11a or **C4** and a vertex with 4 semiedges, being "colored" with $\{0, 1_1, 1_2, 1_3\}$ and $\{0, 1, 2, 3\}$ respectively. In this case, we can derive, in the obvious way, a non–Boole–colorable logic network from a non–edge–colorable 4–regular graph. Examples of such 4–regular graphs are the line graphs of snarks and those graphs having an odd number of vertices (note that, in this latter case, the proposed construction consists in joining an odd number of even 4–sets). Analogously, using also *not* operators and *truth* operators of 3 variables to replace vertices of degree 2 and 3 respectively, non–Boole–colorable logic networks can be obtained from class 2 (non–regular) graphs with maximum degree 4.

Snarks from c–Equivalent Multipoles

As said in Section 2, a multipole Gp contained in a snark U can be replaced by a multipole Gp', c–contained in Gp, giving rise to another snark U'. In this subsection we focus on the case when Gp and Gp' are c–equivalent.

The two following statements show that some of the configurations studied before are c–equivalent to some simple multipoles contained in any snark.

D1   The "untruth cell" of n variables, seen as a $2n$–set, is c–equivalent to n isolated edges, see Fig. 15a.

D2   The truth cells of $n = 2$ and 3 variables are c–equivalent to the multipoles $\{vev\}$ and $\left\{\begin{smallmatrix} vevev \\ e \\ v \end{smallmatrix}\right\}$ respectively, see Figs. 15b and 15c.

The proofs are simple consequences of remark **B2**, Lemma 1 and standard Kempe–chain arguments. The cases $n = 2$ are well–known since the above–mentioned dot product [1], [18] is based on them. The case $n = 3$ in **D2** was implicitly considered by M.K. Goldberg in [15] to construct his (even) "hooking–cells".

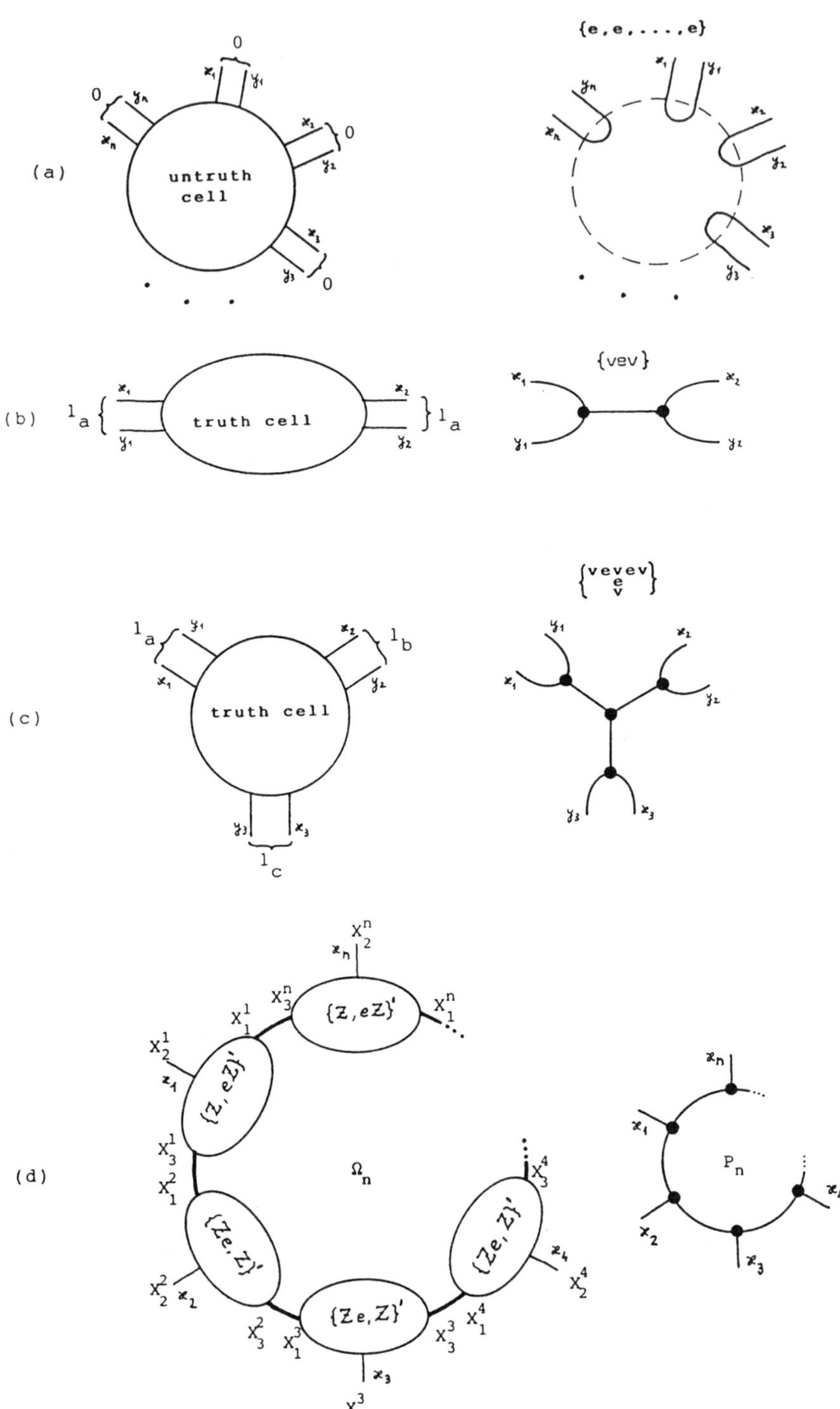

Figure 15

Another interesting construction of c–equivalent multipoles is as follows. Let Ω_n be the n–pole obtained by joining n *partial–or* operators with semiedge sets x_1^i, x_2^i, x_3^i, $i \in Z_n$ —see Table 5— in such a way that $X_3^i = X_1^{i+1}$, as shown in Fig. 15d.

Then,

D3   The multipole Ω_n is c–equivalent to an "n–gon", i.e., an n–cycle with one semiedge incident to each vertex. See the above figure.

Indeed, as $X_1^{i+1} = X_3^i = 1$, the only possible Boole coloring of each 3–set $\{Ze, Z\}'$ is $(X_1^i, X_2^i, X_3^i) = (1_a, 1_b, 1_c)$.

ACKNOWLEDGEMENTS

I am indebted to Prof. J.L.A. Yebra for helpful comments, and also to the late Prof. Rufus Isaacs from whom I received the first encouragements to publish this work.

I thank Dr. A.G. Chetwynd and Prof. M.K. Goldberg for supplying much information on the subject.

REFERENCES

[1]   G.M. Adelson–Velski and A. Titov, On 4–chromatic cubic graphs. *Voprosu Kibernetiki* **1** (1974).

[2]   D. Blanuša, Problem ceteriju boja (The problem of four colors). *Hrvatsko Prirodoslovno Društvo Glasnik Mat.–Fiz. Astr. Ser. II,* **1** (1946) 31 – 42.

[3]   P.J. Cameron, A.G. Chetwynd and J.J. Watkins, Decomposition of snarks. *J. Graph Theory* **11** (1987) 13 – 19.

[4]   A.G. Chetwynd and R.J. Wilson, The rise and fall of the critical graph conjecture. *J. Graph Theory* **7** (1983) 153 – 157.

[5]   U.A. Celmins and E.R. Swart, The construction of snarks. *Research Report CORR 79–18*, Dpt. Of Combin. and Optim., Univeristy of Waterloo, Waterloo, Ontario, Canada.

[6]   A.G. Chetwynd and R.J. Wilson, Snarks and supersnarks. In *Theory and Application of Graphs.* Wiley, New York (1981) 215 – 241.

[7]   B. Descartes, Network colourings. *Math. Gazette* **32** (1948) 67 – 69.

[8]   M.A. Fiol, *Contribución a la Teoria de Grafos Regulares.* P.F.C. (Master Dissertation), Universitat Politècnica de Barcelona, 1979.

[9]   M.A. Fiol, 3'–grafos críticos. *Research Report*, Dpt. de Matemàtica, Universitat Politècnica de Barcelona, 1979.

[10] M.A. Fiol and M.L. Fiol, Coloracions: un nou concepte dintre la teoria de coloració de graphs. *L'Escaire (Barcelona)* **11** (1984) 33 – 44.

[11] S. Fiorini and R.J. Wilson, *Edge-Colourings of Graphs*. Research Notes in Mathematics, **16**, Pitman Publ., London, 1977.

[12] J.L. Fouquet, J.L. Jolivet and M. Rivière, Graphes cubiques d'indice trois, graphes cubiques isochromatiques, graphes cubiques d'indice quatre. *J. Combin. Theory Ser.* *B31* (1981) 262 – 281.

[13] M. Gardner, Mathematical Games: Snarks, Boojums and other conjectures related to the four–color–map theorem. *Sci. Amer.* **234** (1976) 126 – 130.

[14] M.K. Goldberg, On graphs of degree 3 with chromatic index 4. *Bull. Acad. Sci. GSSR* **93** (1979) 29 – 31.

[15] M.K. Goldberg, Construction of class 2 graphs with maximum vertex degree 3. *J. Combin. Theory Ser. B* **31** (1981) 282 – 291.

[16] G. Hajós, Über eine konstruction nicht n–färbarer graphen. *Wiss. Z. Martin Luther Univ. Halle Wittenberg, Math. Naturwiss. Reihe* **10** (1961) 116 – 117.

[17] I. Holyer, The NP–completeness of edge–coloring. *SIAM J. Comput.* **10** (1981) 718 – 720.

[18] R. Isaacs, Infinite families of non–trivial trivalent graphs which are not Tait colorable. *Amer. Math. Monthly* **82** (1975) 221 – 239.

[19] R. Isaacs, Loupekhine's snarks: a bifamily of non–Tait–colorable graphs. *Technical Report 263,* Dpt. of Math. Sci., The Johns Hopkins University, Maryland, U.S.A., 1976.

[20] I.T. Jakobsen, Some remarks on the chromatic index of a graph. *Arch. Math. (Basel)* **24** (1973) 440 – 448.

[21] I.T. Jakobsen, On critical graphs with chromatic index 4. *Discrete Math.* **9** (1974) 265 – 276.

[22] J. Petersen, Die Theorie der regulären Graphen. *Acta Math., Stockholm,* **15** (1891) 193 – 220.

[23] G. Szekeres, Polyhedral decomposition of cubic graphs. *Bull. Austral. Math. Soc.* **8** (1973) 367 – 387.

[24] P.G. Tait, Remarks on the colouring of maps. *Proc. Roy. Soc. Edimburgh* **10** (1880) 501 – 503, 729.

[25] V.G. Vizing, The chromatic class of a multigraph. *Kibernetika (Kiev)* **3** (1965) 29 – 39.

[26] J.J. Watkins, On the construction of snarks. *Ars Combin.* **16–B** (1983) 111 – 123.

On the Cycle Structure of Multipartite Tournaments

Wayne D. Goddard
Massachusetts Institute of Technology

Ortrud R. Oellermann*
University of Natal and Western Michigan University

ABSTRACT

It is shown that if T is a strongly connected n–partite tournament, $n \geq 3$, then every vertex of T lies on a cycle that contains vertices from exactly m partite sets for every m with $3 \leq m \leq n$. Furthermore, T has a cycle of every length m for $3 \leq m \leq n$. Finally, every strongly connected n–partite tournament, $n \geq 3$, contains at least $n - 2$ 3–cycles, and the sharpness of this result is discussed.

1. Introduction

Tournaments are directed graphs obtained by assigning directions to the edges of a complete graph, while *bipartite tournaments* are produced by directing the edges of a complete bipartite graph. More generally, an n–*partite tournament* (n $\geq$ 2) or *multipartite tournament* results when each edge of a complete n–partite graph is assigned a direction. Thus, a tournament of order p is a p–partite tournament each of whose partite sets consists of a single vertex. A digraph D that is obtained by assigning directions to the edges of a graph G is called an *orientation* of G. If, in addition, D is strongly connected, we say that D is a *strongly connected orientation* of G.

Tournaments have been studied extensively in the literature and very nice accounts of tournaments can be found in [11] and [12]. Furthermore, a theory of bipartite tournaments has been developed (see, for example, [1] – [8]). In [10] several

* Research supported in part by Office of Naval Research Contract N00014 – 88 – K – 0018. Western Michigan University is thanked for supporting the second author's attendance at this conference.

525

properties of multipartite tournaments are established. The following useful result was established in [10].

Theorem A   *If T is a strongly connected n–partite tournament, $n \geq 3$, then T contains a 3–cycle.*

The goal here is to further investigate the cycle structure of multipartite tournaments and to settle a conjecture made in [10]. For basic graph theory terminology, we follow [9].

2. Cycles in Multipartite Tournaments

If D is a digraph and v is a vertex of D, then $N^+(v)$ and $N^-(v)$ denote the set of vertices adjacent from and adjacent to v, respectively. Furthermore, the *score* of v is defined by $s(v) = |N^+(v)|$. A digraph D of order $n \geq 3$ is *vertex–pancyclic* if each vertex of D lies on a cycle of length m for each $m = 3, 4, ..., n$. Moon [12] showed that every strongly connected tournament is vertex–pancyclic. Equivalently, this result states that if T is a strongly connected n–partite tournament $(n \geq 3)$, each of whose partite sets consists of a single vertex, then every vertex of T lies on a cycle that contains vertices from exactly m partite sets for every $m = 3, 4, ..., n$. The following theorem extends this result to all multipartite tournaments with at least three partite sets.

Theorem 1   *Let T be a strongly connected n–partite tournament, $n \geq 3$. Then every vertex of T lies on a cycle that contains vertices from exactly m partite sets for every $m = 3, 4, ..., n$.*

Proof   Let v be a vertex of T. We proceed by induction. Since T is strongly connected, there is a shortest cycle that contains v, say $C : v = v_1, v_2, v_3, ..., v_k, v_1$. If v_1 and v_3 belong to distinct partite sets of T, then $(v_3, v_1) \in E(T)$; otherwise, $v_1, v_3, ..., v_k, v_1$ is a shorter cycle than C that contains v. However, then v_1, v_2, v_3, v_1 is a cycle that includes v and contains vertices from exactly three partite sets. Assume now that v_1 and v_3 belong to the same partite set of T. Then $k \geq 4$ and v_1 and v_4 are in distinct partite sets of T. If $(v_1, v_4) \in E(T)$, then $v_1, v_4, ..., v_k, v_1$ is a cycle that includes v and is shorter than C, which is not possible. Hence, $(v_4, v_1) \in E(T)$ so that v_1, v_2, v_3, v_4, v_1 is a cycle that includes v and contains vertices from at most three partite sets of T.

Assume now that C is a cycle that includes v and that contains vertices from exactly m partite sets $2 \leq m < n$. We shall show that there exists a cycle C' which includes v and contains vertices from exactly $m + 1$ partite sets. Let S be the set of all vertices that belong to partite sets that are not represented on C. If $u \in S$ and if there exists a pair of vertices on C such that one of these is adjacent to u and the other one is adjacent from u, then there exist adjacent vertices x and y on C such that (x, y), (x, u) and (u, y) are arcs of T. Let C' be the cycle obtained by replacing the arc (x, y) of C by the path x, u, y. Then C' includes v and contains vertices from exactly $m + 1$ partite sets.

Suppose now that $S = S_1 \cup S_2$ where S_1 is the set of vertices of S each of which is adjacent from every vertex of C, while S_2 is the collection of vertices in S each of which is adjacent to every vertex of C. At least one of the two sets S_1 and S_2 is nonempty. Assume that S_1 is nonempty.

Since T is strongly connected there is a path from every vertex of S_1 to a vertex of C. Let P be a shortest such path, and assume that P is a $z - y$ path. Suppose that x precedes y on C. If P does not contain vertices of S_2, then the cycle obtained from C by replacing the arc (x, y) by the $x - y$ path $x, (x, z), P$, is a cycle that includes v and vertices from exactly $m + 1$ partite sets.

Assume thus that P contains a vertex of S_2. Since P is a shortest path from S_1 to C and since every vertex of S_2 is adjacent to every vertex of C it follows that the penultimate vertex of P, say w, belongs to S_2 and that all other vertices of P different from z do not belong to S. Suppose that the $z - w$ subpath of P has length at least 2. Let u be an internal vertex of this subpath. Since P is a shortest path from S_1 to C and since C contains vertices from at least two partite sets, some vertex a of C is adjacent to u. Let b be the vertex that follows a on C. Let P' be the $a - b$ path obtained from the path a, u followed by the $u - w$ subpath of P and then the path w, b. If we now replace the arc (a, b) of C by the path P', then the resulting cycle C' includes v and contains vertices from exactly $m + 1$ partite sets.

Suppose now that the $z - w$ subpath of P has length 1. Let H be a $v - v'$ subpath of C that contains vertices from exactly $m - 1$ partite sets of T. Then H followed by the path v', z, w, v produces the desired cycle C'.

Similarly it can be shown if $S_2 \neq \varnothing$, that there exists a cycle C' that includes v and vertices from exactly $m + 1$ partite sets. $\square$

Since a strongly connected tournament of order $n \geq 3$ is vertex–pancyclic, such a tournament contains a cycle of every length l, $3 \leq l \leq n$. Thus, if T is a strongly connected n–partite tournament, $n \geq 3$, each of whose partite sets consists of a single vertex, then T contains a cycle of every length l, $3 \leq l \leq n$. We now present a lemma that will aid us in extending this result to more general n–partite tournaments.

Lemma *Let T be a strongly connected n–partite tournament, $n \geq 3$. Let C be a cycle of minimum length in T that contains vertices from exactly j partite sets,*

$4 \leq j \leq n$. Now let C' be a cycle, in the strongly connected j–partite tournament $T' = \langle V(C) \rangle$, that contains vertices from exactly $j-1$ partite sets. If C has length s and C' has length t, then T' contains cycles of every length l, $t \leq l \leq s$.

Proof If $t = s - 1$, then the lemma follows. Assume thus that $t < s - 1$. Let v be a vertex from a partite set of T' that is not represented in the cycle C' : $v_1, v_2, \ldots, v_t$, v_1. If v is adjacent to some vertex of C' and adjacent from some vertex of C', then there exist adjacent vertices on C', say v_i and v_{i+1} (indices expressed modulo t) such that (v_i, v) and (v, v_{i+1}) are arcs of T'. However, then $v_1, v_2, \ldots, v_i, v, v_{i+1}, \ldots, v_t, v_1$ is cycle of length $t + 1 < s$ that contains vertices from exactly j partite sets. This contradicts the fact that C is a cycle of minimum length that contains vertices from exactly j partite sets. Thus the vertices of C' are either all adjacent to v or all adjacent from v; we may assume, without loss of generality, that all vertices of C' are adjacent to v.

Let $P : v = w_1, w_2, \ldots, w_r = v_i$ be a shortest path (in T') from v to a vertex of C'. Then $C'' : v = w_1, w_2, \ldots, w_r (= v_i), v_{i+1}, \ldots, v_{i-1}, v$ (indices expressed modulo t) is a cycle in T that contains vertices from exactly j partite sets. Thus C'' contains all the vertices of C and $r = s - t + 1 \geq 3$. From our choice of P it follows that v is not adjacent to w_m for all $m = 3, 4, \ldots, r$. Also, since C is a shortest cycle that contains vertices from exactly j partite sets, the cycle $v = w_1, w_2, \ldots, w_r (= v_i)$, $v_{i+1}, \ldots, v_{i-2}, v$ (indices expressed modulo t) contains vertices from exactly $j - 1$ partite sets. Thus v_{i-1} is the unique vertex from its partite set in C. Therefore, since P is a shortest path from v to a vertex of C', it follows that (v_{i-1}, w_m) is an arc of T for $m = 1, 2, \ldots, s - t - 1$.

Observe now that $v_{i-1}, w_m, w_{m+1}, \ldots, w_r, v_{i+1}, \ldots, v_{i-1}$ is a cycle of length $s - m + 1$ for $m = 2, 3, \ldots, s - t - 1$. It remains to show that there is a cycle of length $t + 1$.

First we show that $w_{r-1} = w_{s-t}$ does not belong to the same partite set as v. Since $r \geq 3$ it follows that $w_{r-1} \neq v$. If w_{r-1} is in the same partite set as v, then $r \geq 4$ and $w_{r-2}, w_{r-1}, w_r (= v_i), v_{i+1}, \ldots, v_{i-1}, w_{r-2}$ is a cycle that contains vertices from exactly j partite sets, which contradicts our choice of C.

Since $C'' : w_2, w_3, \ldots, w_r (= v_i), v_{i+1}, \ldots, v_{i-1}, w_2$ is a cycle that does not contain v, but contains all vertices of C', it follows from the choice of C, that C'' contains vertices from exactly $j - 1$ partite sets. Therefore C' contains a vertex that is in the same partite set as w_{r-1}. Therefore, there exists a vertex v_k on C such that w_{r-1} is adjacent to v_k and w_{r-1} is not adjacent to v_{k-1} (possibly v_{k-1} is in the same partite set as w_{r-1}). Since $v, w_2, \ldots, w_{r-1}, v_k$ is also a shortest path from v to a vertex of C, it follows as in case of v_{i-1}, that v_{k-1} is the unique vertex from its partite set in C.

Therefore v_{k-1} and w_{r-1} belong to distinct partite sets and $(v_{k-1}, w_{r-1}) \in E(T)$. Hence $v_{k-1}, w_{r-1}, v_k, v_{k+1}, \ldots, v_{k-1}$ is a cycle of length $t + 1$. ❏

Theorem 2 *Let T be a strongly connected n–partite tournament, $n \geq 3$. Then T contains a cycle of every length l, $3 \leq l \leq n$.*

Proof If $n = 3$, then the result follows from Theorem A. Assume thus that $n \geq 4$. Let $C^{(n)}$ be a cycle of minimum length that contains vertices from n partite sets. Let T_n be the tournament induced by the vertices of $C^{(n)}$. Suppose now that T_{n-k} and $C^{(n-k)}$ have been defined where $0 \leq k < n - 3$. Let $C^{(n-k-1)}$ be a cycle of minimum length in T_{n-k} that contains vertices from exactly $n - k - 1$ partite sets and set $T_{n-k-1} = \langle V(C_{n-k-1}) \rangle$.

Since $C^{(3)}$ is a 3–cycle (by Theorem A) and since $C^{(n)}$ contains at least n vertices, it follows by repeated application of Lemma 1 that T_n and therefore T contains a cycle of every length l, $3 \leq l \leq n$. ❏

3. Multipartite Tournaments and 3–cycles

It is well-known that every strongly connected tournament of order $n \geq 3$ (i.e., a strongly connected n–partite tournament each of whose partite sets consists of a single vertex) contains at least $n - 2$ 3–cycles (see [11]).

From Theorem A it follows that every strongly connected 3–partite tournament contains at least one 3–cycle. In [10] it is further shown that there exist strongly connected 3–partite tournaments with exactly one 3–cycle. It was conjectured in [10] that every strongly connected n–partite tournament contains at least $n - 2$ 3–cycles. We settle this conjecture next.

Theorem 3 *Let T be a strongly connected n–partite tournament $n \geq 3$. Then T contains at least $n - 2$ 3–cycles.*

Proof By induction on n. If $n = 3$, then the result follows from Theorem A.

Suppose now that $n \geq 4$ and that every strongly connected $(n - 1)$–partite tournament contains at least $(n - 1) - 2$ 3-cycles. Let C be a cycle that contains vertices from exactly $n - 1$ partite sets. Let $H = \langle V(C) \rangle$. Then H is a strongly connected $(n - 1)$–partite tournament. By the inductive hypothesis H contains at least $n - 3$ 3–cycles. Let w be a vertex from a partite set that is not represented on C. By Theorem 1, w belongs to a cycle C' that contains vertices from exactly three partite sets. Let $T' = \langle V(C') \rangle$. Then T' is strongly connected and therefore contains a 3–cycle that is different from the 3–cycles in H. Hence T contains at least $n - 2$ 3–cycles. ❏

It was shown in [10], that if G is a complete 3–partite graph that is not isomorphic to $K_{2,2,2}$, then there exists a strongly connected orientation of G that contains exactly one 3–cycle. We next extend this result to more general n–partite tournaments, $n \geq 3$. To aid us in our dicussion we first introduce some additional terminology.

Let S be a sequence of $n \geq 3$ positive integers. If the elements of S can be (re)ordered as $a_1, a_2, \ldots, a_n$ such that $\sum_{i=1}^{r} a_i - \sum_{i=r+1}^{n} a_i \in \{0,1\}$, then we say that S is *balanced*. Further, we say that $(a_1, a_2, \ldots, a_r; a_{r+1}, a_{r+2}, \ldots, a_n)$ is a *balanced split* of S. Let D be a digraph and v a vertex of D. A digraph H is said to be obtained from D by *expanding* v *to a set* S if $V(H) = V(D - v) \cup S$ where S is an independent set of vertices such that each vertex in S has the same adjacencies in H as v in D.

The following lemma will prove to be useful.

Lemma 2 *Let S be a sequence of $n \geq 3$ positive integers. Then one of the following occurs.*

    (a) S is balanced.

    (b) The elements of S can be ordered as $a_1, a_2, \ldots, a_n$ such that there exist positive integers k and r, $2 \leq k < a_1$ such that $(k, a_2, \ldots, a_r; a_{r+1}, a_{r+2}, \ldots, a_n)$ is a balanced split of the sequence $k, a_2, \ldots, a_r, a_{r+1}, \ldots, a_n$.

    (c) The elements of S can be ordered as $a_1, a_2, \ldots, a_n$ such that the following hold (i) There exists an integer $r \geq 2$ such that $(2, a_2, \ldots, a_r - 1; a_{r+1}, a_{r+2}, \ldots, a_n)$ is a balanced split of the sequence $2, a_2, \ldots, a_r - 1, a_{r+1}, \ldots, a_n$; and (ii) $a_1 = a_2 = \ldots = a_r \geq 3$.

    (d) n is odd and every element of S equals 2.

Proof   Among all partitions of S into two nonempty subsequences let $\{A, B\}$ be a partition such that $d = \left| \sum_{b \in B} b - \sum_{a \in A} a \right|$ is as small as possible.

If $d \in \{0,1\}$, then S is balanced. Assume thus that $d \geq 2$, and assume without loss of generality that $\sum_{b \in B} b > \sum_{a \in A} a$.

We show next that $b \geq d$ for all $b \in B$. If $|B| = 1$, then the result follows. Assume thus that $|B| \geq 2$. Assume, to the contrary, that there exists $b' \in B$ such that $b' < d$. If $b' \leq d/2$, then $\{B - \{b'\}, A \cup \{b'\}\}$ is a partition of S into two nonempty subsequences such that $0 \leq \sum_{b \in B - \{b'\}} b - \sum_{a \in A \cup \{b'\}} a = d - 2b'$, which contradicts our choice of the partition $\{A, B\}$. Assume now that $d/2 < b' \le d$. Then $0 \le \sum\limits_{a \in A \cup \{b'\}} a - \sum\limits_{b \in B-\{b'\}} b = 2b' - d < d$, which again contradicts our choice of the partition $\{A, B\}$.

Suppose now that the elements of B have been ordered in nonincreasing order as $a_1, a_2, \ldots, a_r$. If $a_1 > d$, then let $k = a_1 + 1 - d$, and observe that for such k and r, (b) holds.

Assume thus that $b = d$ for all $b \in B$. Since A is nonempty, it follows that $r = |B| \ge 2$. Once again suppose that the elements of B are $a_1, a_2, \ldots, a_r$. If $d \ge 3$, then (c) holds. Assume thus that $d = 2$. Then it follows from our choice of $\{A, B\}$, that every element of A is also equal to 2 and that $|B| = |A| + 1$. Hence n is odd and (d) holds. $\square$

Theorem 4 *Let G be a complete n–partite graph, $n \ge 3$, such that $G \not\equiv K_{2, 2, \ldots, 2}$ for odd n. Then there exists a strong orientation of G with exactly $n - 2$ 3–cycles.*

Proof Let S be the sequence of the cardinalities of the partite sets of G. Then S satisfies one of the conditions (a), (b) or (c) of Lemma 2.

Case 1. Suppose that S satisfies condition (a) of Lemma 2. Suppose that S has been ordered as $a_1, a_2, \ldots, a_n$ such that $(a_1, a_2, \ldots, a_r; a_{r+1}, a_{r+2}, \ldots a_n)$ is a balanced split of S. Let $p = \sum\limits_{i=1}^{n} a_i$. Label the vertices from the n partite sets $V_1, V_2, \ldots, V_n$ of G as follows:

$$
\begin{aligned}
V_1 &= \{v_1, v_3, \ldots, v_{2a_1-1}\} \\
V_2 &= \{v_{2a_1+1}, \ldots, v_{2(a_1+a_2)-1}\} \\
&\quad\vdots \\
V_r &= \{v_{2(a_1+\ldots+a_{r-1})+1}, v_{2(a_1+\ldots+a_{r-1})+3}, \ldots, v_{2(a_1+a_2\ldots+a_r)-1}\} \\
V_{r+1} &= \{v_2, v_4, \ldots, v_{2a_r+1}\} \\
&\quad\vdots \\
V_n &= \{v_{2(a_{r+1}+a_{r+2}+\ldots+a_{n-1})+2}, \ldots, v_{2(a_{r+1}+a_{r+2}+\ldots+a_n)}\}
\end{aligned}
$$

We now assign directions to the edges of G to produce an n–partite tournament T as follows:

$$E(T) = \{(v_i, v_j) \mid i = j + 1 \text{ for } i = 2, 3, \ldots, p \text{ or}$$

$$j \ge i + 2 \text{ and } v_i \text{ and } v_j \text{ belong to different partite sets of } G \text{ for } 1 \le i \le p - 2\}.$$

(See Figure 1. All arcs of T that are not shown in the figure are directed down.)

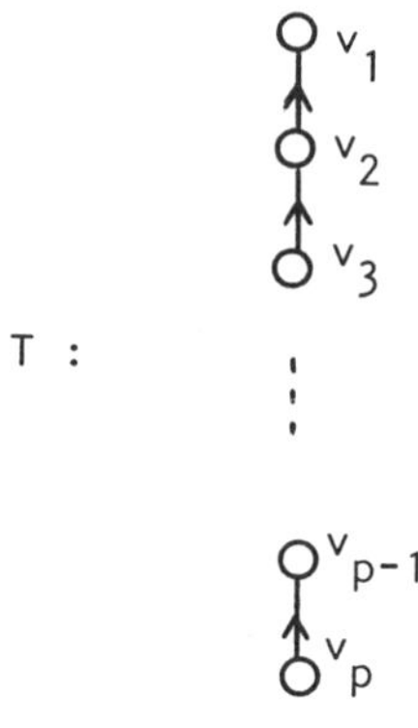

Figure 1

Observe that T is strongly connected. Further, all 3–cycles of T are of the type v_i, v_{i+2}, v_{i+1}, v_i for some i. However, v_i, v_{i+2}, v_{i+1}, v_i is a 3–cycle of T if and only if v_i and v_{i+2} belong to distinct partite sets of G. But $|\{i \mid v_i, v_{i+2}$ belong to different partite sets of $G\}| = n - 2$.

Case 2. Suppose S satisfies condition (b) of Lemma 2. Since the sequence S': k, a_2, ..., a_r, a_{r+1}, ..., a_n is balanced and $(k, a_2, ..., a_r; a_{r+1}, ..., a_n)$ is a balanced split of S; the complete n–partite graph $H = K_{k, a_2, ..., a_r, a_{r+1}, ..., a_n}$ has, by Case 1, a strong orientation with exactly $n - 2$ 3–cycles. Label the vertices of H as in Case 1 and let T' be the tournament obtained from H as in Case 1. Then T' has exactly $n - 2$ 3–cycles. Let T be the strongly connected n–partite tournament obtained by expanding v_1 to a set S of $a_1 - k + 1$ vertices. Since $k \geq 2$, the tournament T has the same number of 3–cycles as T'.

Case 3. Suppose S satisfies conditions (c) of Lemma 2. Since the sequence S': 2, a_2, a_3, ..., a_{r-1}, $a_r - 1$, a_{r+1}, a_{r+2}, ..., a_n is balanced and $(2, a_2, a_3, ..., a_{r-1}, a_r-1; a_{r+1}, a_{r+2}, ..., a_n)$ is a balanced split of S', the complete n–partite graph $H = K_{2, a_2, a_3, ..., a_{r-1}, a_r-1, a_{r+1}, a_{r+2}, ..., a_n}$ has, by Case 1, a strong orientation with exactly $n - 2$ 3–cycles. Since $a_1 = a_2 = ... = a_r = d$ (where d is as in Lemma 2) and $\sum_{i=1}^{r} a_i - \sum_{i=r+1}^{n} a_i = d$, it follows that $2 + a_2 + a_3 + ... + a_{r-1} + (a_r - 1) = (r - 1)d + 1$ and $a_{r+1} + a_{r+2} + ... + a_n = (r - 1)d$. Therefore, if we label the vertices of H as in Case 1 and let T' be the tournament obtained from H as in Case 1, then $v_1 \in V_1$ and $v_p \in V_r$. Let T be obtained from T' by first expanding v_1 to a set of $a_1 - 1$ vertices and then expanding v_p to a set of two vertices. Then T is strongly connected and contains the same number of 3–cycles as T', i.e., T contains exactly $n - 2$ 3–cycles. $\square$

REFERENCES

[1] K. S. Bagga, On upsets in bipartite tournaments. *Graph Theory with Applications to Algorithms and Computer Science* (eds. Y. Alavi, G. Chartrand, L. Lesniak, D. R. Lick and C. E. Wall). Wiley Interscience, New York (1985) 37 – 45.

[2] K. S. Bagga and L. W. Beineke, Some results on binary matrices obtained via bipartite tournaments. *Graph Theory with Applications to Algorithms and Computer Science* (eds. Y. Alavi, G. Chartrand, L. Lesniak, D. R. Lick and C. E. Wall). Wiley Interscience, New York (1985) 47 – 55.

[3] K. S. Bagga and L. W. Beineke, On superstrong bipartite tournaments. *Congressus Numerantium* **53** (1986) 113 – 119.

[4] K. S. Bagga and L. W. Beineke, On the numbers of some subtournaments of a bipartite tournaments. *Proceedings of the Third New York Conference on Combinatorial Mathematics,* to appear.

[5] K. S. Bagga and L. W. Beineke, Uniquely realizable score lists in bipartite tournaments. *Czech. Math. J.* **37** (1987) 323–333.

[6] L. W. Beineke, A tour through tournaments or bipartite and ordinary tournaments: A comparative survey. *Combinatorics,* London Mathematical Society Lecture Note Series 52, Cambridge University Press (1981) 41 – 55.

[7] L. W. Beineke and C. H. C. Little, Cycles in bipartite tournaments. *J. Combinatorial Theory* **32B** (1982) 140 – 145.

[8] L. W. Beineke and J. W. Moon, On bipartite tournaments and scores. *The Theory and Applications of Graphs* (eds. Y. Alavi, G. Chartrand, D. Goldsmith, L. Lesniak–Foster and D. R. Lick) Wiley Interscience, New York (1981) 55 – 71.

[9] G. Chartrand and L. Lesniak, *Graphs & Digraphs 2nd Edition*, Wadsworth & Brooks/Cole, Monterey (1986).

[10] W. D. Goddard, G. Kubicki, O. R. Oellermann and S. Tian, On multipartite tournaments. Submitted for publication.

[11] F. Harary and L. Moser, The theory of round robin tournaments. *Amer. Math. Monthly* **73** (1966) 231 – 246.

[12] J. W. Moon, *Topics on Tournaments*, Holt, Rinehart and Winston, New York (1968).

ON SOME EXTREMAL PROBLEMS IN CONNECTIVITY

W. D. Goddard

Henda C. Swart

University of Natal at Durban

ABSTRACT

In this paper are considered the three parameters integrity, toughness and binding number and three extremal problems: the characterisation of maximal graphs; the problem of maximising the value of the parameter given the size (and order); and Nordhaus-Gaddum type results. Main results include the characterisation of maximal graphs, and for binding number, complete solutions of the latter two problems.

1.   Introduction. The Three Parameters.

We shall use the notation of [3]. Further we use $\subset$ to denote strict containment (in the comparison of sets) and *cut-set* to mean a set $S \subset V(G)$ such that $G - S$ is disconnected. Integrity was introduced by Barefoot, Entringer and Swart [1] as an alternative measure of the vulnerability of a graph. This was further explored in [2] and [7]. The concept of the binding number of a graph was introduced in 1973 by Woodall [15]. Toughness was introduced in 1973, by Chvátal [5] in his dealings with hamiltonicity. The general theory was partly developed by Chvátal in his original paper and by Pippert [13]. We offer a slightly altered definition, though this change affects only the values of the toughness of the complete graph.

The relevant definitions are:

[1] For any graph G, the *integrity* of G, denoted $I(G)$ is defined by

535

$$I(G) := \min_{S \subseteq V(G)} \{|S| + m(G - S)\}$$

where $m(G - S)$ denotes the maximum order of a component of $G - S$.

[15] For any graph G, the *binding number* of G, denoted by $bind(G)$ is defined by

$$bind(G) := \min_{S \in \mathcal{F}(G)} \frac{|N(S)|}{|S|}$$

where $\mathcal{F}(G) = \{S \subseteq V(G) : S \neq \phi \,\&\, N(S) \neq V(G)\}$.

[5] For any graph G, the *toughness* of G, denoted by $t(G)$, is defined by

$$t(G) := \min_{S \in J(G)} \{|S|/k(G - S)\}$$

where $J(G) = \{S \subset V(G) : \kappa(G - S) = 0\}$.

This yields immediately the following alternative definition:

$$t(G) = \begin{cases} \min\{|S|/k(G - S) : S \text{ is a cut-set of } G\} & G \text{ noncomplete} \\ p(G) - 1 & G \text{ complete.} \end{cases}$$

Chvátal's original definition was that $t(G) = \min\{|S|/k(G - S) : S$ is a cut-set of $G\}$ for all G so that $t(K_n) = \min \phi = \infty$ for all positive integers n.

Further, an appropriate S for which the minimum is attained is known as an *I-set, binding set* or *tough set* of G as the case may be. We shall need the following:

Proposition 1.1   For all graphs G:

a)   [7] $I(G) \leq \alpha(G) + 1$;

b)   [15] $bind(G) \leq (p(G) - 1)/(p(G) - \delta(G))$;

c)   [5], [11] If G is noncomplete then $t(G) \leq \kappa(G)/2$ with equality if G is $K(1, 3)$-free;

d)   [7] $I(G) = p(G) - 1$ iff $\overline{G}$ is non-empty and has girth at least 5.

2. Maximal Graphs

If μ is a graphical invariant, then a graph G is μ-*maximal* iff $\mu(G+e) > \mu(G)$ for every edge e of $\overline{G}$.

It is easily verified for each parameter that such graphs either are such that every component is complete, or result from the join of such a graph with a complete graph. For toughness this is sufficient:

Theorem 2.1 A graph G is tough-maximal iff G is i) complete, ii) the union of two complete graphs, or iii) the join of a complete graph with one whose every component is complete.

By considering the effect of adding an edge between the two largest or the two smallest components of $G - S$ where S is an I-set of G, we get the following result for integrity:

Theorem 2.2   A graph G is I-maximal iff either

i) G is complete, or

ii) $G \cong K_s + (K_{g1} \cup K_{g2} \cup ... \cup K_{gr})$ where $r \geq 2, g_1 = g_2 \geq g_3 \geq ... \geq g_r \geq 1$ and $g_{r-1} + g_r \geq g_1 + 1$.

The situation for the binding number is more complicated. This springs in part from the fact that the formula for the binding number of the join of a complete graph and one whose every component is complete is not as simple as in the case of the other two parameters. This is given by:

Theorem 2.3   [6] Let G be a disconnected graph, of order m, in which every component is complete, having b trivial components and a smallest non-trivial component (if one exists) or order a. Then

$$bind(G + K_n) = \begin{cases} n/b & \text{if } n \leq b \neq 0, \\ (m+n-b)/m & \text{if } n \geq b > 0, \\ \min\left\{\frac{m-a+n}{m-a}, \frac{m-1+n}{m-a+1}\right\} & \text{if } b = 0. \end{cases}$$

We may proceed via the concept of bind-quasi-maximal graphs:

A graph G is bind-quasi-maximal with respect to a binding set S (of G) iff for every $e \in E(\overline{G})$ it holds that $bind(G + e) > bind(G)$ or S is not a binding set of $G + e$.

We observe immediately that every bind-maximal graph G is bind-quasi-maximal with respect to every binding set of G. We shall now categorise the set Q^* of graphs that are bind-quasi- maximal with respect to at least one of their binding sets.

Lemma 2.4   A graph G belongs to Q^* iff $G \cong G(p,n,s,c) := K_{n-c} + (K_{p+c-s-n}) \cup K_c \cup \overline{K}_{s-c}$ such that

a) $1 \le s \le p, 0 \le n < .p, 0 \le c \le \min\{n,s\}, c \ne 1$ and $s + n - c \le p$;

b) $bind(G) = n/s$.

Proof   Let G be bind-quasi-maximal with respect to S. Then it is easily seen that if $S \ne V(G)$ then $G - S$ is complete; that each vertex in $S - N(S)$ is adjacent to every vertex in $N(S) - S$ and to no other vertex; and that each vertex in $S \cap N(S)$ is adjacent to every vertex in $N(S)$ and to no other vertex. Letting $|S| = s, |N(S)| = n$ and $|S \cap N(S)| = c$, we get the necessity of the conditions.

Conversely, let $G = G(p,n,s,c)$ such that the conditions of a) and b) hold. Clearly S given by $S = V(K_c) \cup V(\overline{K}_{s-c})$ is a binding set of G, and further G is bind- quasi-maximal with respect to S.

Hence we obtain the following theorem:

Theorem 2.5   The graphs of Q^* are given by

(i) $K_n + (K_{p-s-n} \cup \overline{K}_s)$ provided $n < s, p - n - s \ge 0$ and $p - n - s \ne 1$;

(ii) $K_{s-c} + (K_{p+c-2s} \cup K_c \cup \overline{K}_{s-c})$ provided $c \ne 1$ and $p + c - 2s \ne 1$;

(iii) $K_{n-c} + (K_c \cup \overline{K}_{s-c}$ provided $n > s$ and $c \ne 1$ and $c < s$;

(iv) $K_{n-s} + (K_{p-n} \cup K_s)$ provided $n > s$ and $p - n \ge 2$ and $n(s+1) \le s(p-1)$.

Now we note that Kane et al. [10] determined for each p the set of all pairs (n,s) such that there exists a graph G on p vertices possessing a binding set S such that $|S| = s, |N(S)| = n$ (and $bind(G) = n/s$). The above theorem recovers this result since any graph G with binding set S is a spanning subgraph of graph G' with binding set S and $N_G(S) = N_{G'}(S)$ but with G' bind-quasi-maximal with respect to S. The characterisation of bind-maximal graphs also follows easily:

Theorem 2.6 A graph G is bind-maximal iff one of the following conditions holds:

(a) $G \cong K_1 \cup K_{p-1}$;

(b) $G \cong K_m + (K_n \cup \overline{K}_m)$ where $m \geq 1$ and $n \geq 0, n \neq 2$;

(c) $G \cong K_{p-s} \cup K_s$ where $s, p - s \geq 3$;

(d) $G \in Q^*, bind(G) \notin \{0,1\}$, and if G is given by (iv) of theorem 2.5 then $n(s + 1) < s(p - 1)$ in the notation of that theorem.

Proof Note that $G \in Q^*$ is certainly bind-maximal if it has a unique binding set. Consider then the classes of Q^* given by theorem 2.5. Then in (i), $n > 0$ implies that the graph has a unique binding set (yielding case (d)) while if $n = 0$ then the graph is bind- maximal iff $s = 1$ (case (a)). In (ii), if $c < s$ then as $\overline{K}_{s-c}$ is a binding set, the graph is bind-maximal iff at least one of $c, p + c - 2s$ is zero and neither equal to 2 (case (b)). If $c = s$ then the graph is bind-maximal iff $s, p - s \neq 2$ (case (c)). In (iii) every graph has a unique binding set and in (iv), the graph is bind-maximal iff $n(s + 1) < s(p - 1)$. Both lead to case (d) of the theorem.

For μ a graphical invariant, one may also define a graph G to be μ-*minimal* iff $\mu(G-e) < \mu(G)$ for every edge e of G. However, the investigation and characterisation of such minimal graphs appears very difficult especially with integrity. While it is easy to see that certain general classes of graphs are minimal and that some constructions yield minimal graphs, we have not been able to characterise the minimal graphs for any of the parameters. Some results (including characterisations for complete multipartite graphs) are to be found in [6].

3. Maximum Parameter for Graphs of Given Order and Size

The most general question one may ask is, given order p and size q of a graph, what is the maximum possible value of a parameter? For integrity we cannot say much. In [2] it is shown that paths and cycles have the highest value of integrity given their order and size. However, the question appears difficult. For example, by proposition 1.1d it follows that the question of determining the minimum size needed for $I(G) = p(G) - 1$ is equivalent to determining the maximum number of edges in a graph of girth at least five.

We shall need the Harary graphs $H_{m,n}$ which are used in the solution to the above problem with connectivity as the parameter. For $C_n = v_0 v_1 \cdots v_{n-1}, H_{m,n} = C_n^{m/2}$ if m is even, $H_{m,n} = H_{m-1,n} + \{v_i v_{i+n/2} : i = 0, \cdots, n/2 - 1\}$ if m is odd and n even, and $H_{m,n} = H_{m-1,n} + \{v_i v_{i+(n+1)/2} : i = 0, \cdots, (n-3)/2\} + \{v_0 v_{(n-1)/2}, v_0 v_{n+1}/2\}$ for m and n odd. The important observation is that $H_{m,n}$ is m- connected.

For toughness of noncomplete graphs G, we have an immediate upper bound; for, in this case, $t(G) \le \kappa(G)/2$ while further $\kappa(G) \le \lfloor 2q/p \rfloor$ so that:

$$t(G) \le \lfloor 2q/p \rfloor /2. \tag{3.1}$$

For equality to hold in 3.1 above, we need graphs with equality in both the inequalities preceding it. Equality in the latter of these two is attained by the Harary graphs. Unfortunately, these do not always have the desired toughness. We consider the (simpler) question of whether, given $2 \le r < p$, there exists an r-regular graph on p vertices with toughness (at least) $r/2$. The case with $r = p - 1$ corresponds to the complete graphs and thus the answer there is in the affirmative. Further, we note that affirmative answers to this question provide solutions to the more general question posed above.

Now if r is even, then the question *is* answered in the affirmative. Here the Harary graph $C_p^{r/2}$ provide the required extremal graphs; they have $\kappa = r$ while because they are $K(1,3)$-free, $t = \kappa/2$.

The case when r is odd is much more complicated and was briefly considered by Chvátal in his original paper [5]. We note immediately that, of course, in this case p must be even. Further, the Harary graphs in this case are not $K(1,3)$-free for r small with respect to p so that there is no immediate guarantee that such graphs have the desired toughness. Indeed, $C_p < 1, l >$ is bipartite and has toughness only 1!

Nevertheless, Chvátal showed that in certain circumstances one does have the existence of r- regular graph on p vertices with toughness $r/2$. If $p = mr$, where m is even and $m \geq r + 1$ and G is an r-regular graph on m vertices with $\kappa_1(G) = r$, then Chvátal showed that $L(S(G))$ is suitable. Thus for r odd, if p is an *even multiple* of r with $p \geq r(r + 1)$ then there exists an r-regular graph with p vertices which is $r/2$-tough. It is not necessary that $p \geq r(r + 1)$ (for $r = 3$ and $p = 6$ take $K_2 \times K_3$). However Chvátal went on to prove that in the case of $r = 3$ at least, r must be a divisor of p and conjectured that this situation is typical of odd r. He used a result which we may (improve slightly and) state as follows:

Theorem 3.2   [5] If G is a noncomplete cubic graph then $t(G) \leq p/(p - \lfloor p/3 \rfloor)$.

In the light of his construction we conjecture the following:

Conjecture 3.3   If G is r-regular and $t(G) = r/2$ then G is $K(1,3)$-free.

Certainly the result is not true if we require only that the graph be tough-minimal and have toughness $\kappa(G)/2$.

We now consider the results for the binding number. In [8] is introduced the concept of the *excess* of a subset S of $V(G)$; denoted by $Exc(S)$, this is given by

$Exc(S) := |N(S)| - |S|$. Then the following hierarchy of conditions for graphs without isolated vertices is defined:

E1: $bind(G) = (p(G) - 1)/(p(G) - \delta(G))$.

E2: For all $S \in \mathcal{F}(G)$ it holds that $Exc(S) \geq \delta(G) - 1$.

E3:   E2 holds with equality if and only if $S = \{v\}$ or $S = V(G) - N(v)$ where v is a vertex of degree $\delta(G)$.

The following result can be established in a straightforward manner:

Theorem 3.4   [8] If $\delta(G) \geq 1$, then E3 implies E2 and E2 implies E1.

Further we shall need the following two results:

Theorem 3.5   [8] Let G be a graph which satisfies E3. If G' is a spanning supergraph of G such that $\delta(G') = \delta(G) + 1$, then G' satisfies E2.

Theorem 3.6   [8] Let G be a $K(1,3)$-free graph of order p and minimum degree $\delta \geq 3$.

    a) If $\kappa(G) \geq \delta - 1$ and $p \neq \delta + 2$, then G satisfies E2.

    b) If $\kappa(G) = \delta$ and $p \neq \delta+2, p \neq \delta+3$, and $(\delta, p) \notin \{(3,8); (3,10); (4,9)\}$, then G satisfies E3.

Following straight from this theorem together with observations about cycles inter alia, we have the following result on the noncomplete powers of a cycle:

Theorem 3.7

    a) If $2 \leq a \leq (p-4)/2$ then C_p^a satisfies E3 unless $a = 2$ and $p = 9$.

    b) If $2 \leq a \leq (p-3)/2$ or if $a = 1$ and p is odd, then C_p^a satisfies E2.

    c) [1] If $2 \leq a \leq (p-3)/2$ or if $a = 1$ and p is odd, then C_p^a satisfies E1.

The following theorem considers the Harary graphs when m is odd.

Theorem 3.8   Let m be odd. If $m = n - 1$ or, if $5 \leq m \leq n - 3$ or if $m = 3$ and $4|n$, then $bind(H_{m,n}) = (n-1)/(n-m)$; moreover, unless $m = 5$ and $n = 9$ then $H_{m,n}$ satisfies E2.

Proof

The case $m = n - 1$ corresponds to complete graphs. Consider therefore $H_{m,n}$ for $5 \leq m \leq n - 3$. Then as $H_{m,n}$ is formed from $C_n^{(m-1)/2}$ by adding edges such that the minimum degree is raised by 1, the result follows from theorem 3.7a and theorem 3.5 except when $m = 5$ and $n = 9$. Further, it is easily checked that $bind(H_{5,9}) = 2$, as required and that when $m = 3$ and $4|n$ then $H_{3,n}$ satisfies E2.

It is to be noted that the Harary graphs not mentioned above do not have the desired binding number. For completeness and later use we include the following:

Theorem 3.9   If $1 \leq a < (p-1)/2$ then $\overline{C}_p^a$ satisfies E2.

Proof   Let $G = \overline{C}_p^a, S \in \mathcal{J}(G)$ and let $N_1, N_2, \cdots, N_r$ be the maximal sets of consecutive (in C_p) vertices in $N(S)$. Then for $i = 1, \cdots, r, N_i$ corresponds to a maximal set X_i of consecutive vertices such that $N_i = N(X_i)$ and such that $|N_i| = |X_i| + \delta(G) - 1$. Thus $S \subseteq \cup_{i=1}^r X_i$ and so

$$|N(S)| = \sum_i |N_i| = r(\delta(G) - 1) + \sum_i |X_i| \geq r(\delta(G) - 1) + |S|$$

proving the result.

Note that, in general, such graphs contain many copies of $K(1,3)$ and do not satisfy property E3.

For binding number we have an immediate upper bound. To see this, recall that $bind(G) \leq (p-1)/(p - \delta(G))$ and that $\delta(G) \leq \lfloor 2q/p \rfloor$ so that:

$$bind(G) \leq (p-1)/(p - \lfloor 2q/p \rfloor). \tag{3.10}$$

It is therefore sensible to consider graphs that are regular or 'almost' regular:

Theorem 3.11   If $1 \leq r < p$ then the following conditions are equivalent:
1. there exists an r-regular graph G of order p satisfying E1;
2. there exists an r-regular graph G of order p satisfying E2;
3. pr is even, $r \neq p - 2$ and if $r = 2$ then p is odd.

Proof   That (2) implies (1) is obvious (as E2 implies E1). To see that (1) implies (3) note that pr must be even for an r-regular graph on p vertices to exist. Further it is easily seen that if $\delta(G) = p - 2$ then G does not satisfy E1. Also, the only 2-regular graphs with binding number $(p-1)/(p-2)$ are the odd cycles. What remains is to show that (3) implies (2). For $4 \le r \le p$ we know that the relevant Harary graphs $H_{r,p}$ satisfy the requirements (theorems 3.7b and 3.8). Now, if $r = 1$ then take $G \cong \frac{p}{2}K_2$ and if $r = 2$, then C_p. For $r = 3$, take $H_{3,p}$ if $4|p$ (theorem 3.8) and $K_2 \times C_{p/2}$ otherwise ([14]).

Theorem 3.12   Let $1 \le r < p$ where r and p are odd integers. Then the following are equivalent:

1. there exists a graph G satisfying E1 with order pl, minimum degree r and size $\lceil rp/2 \rceil$;

2. there exists a graph G satisfying E2 with order p, minimum degree r and size $\lceil rp/2 \rceil$;

3. $r \ne p - 2, r \ne 1$ and if $r = 3$ then $p \ne 7$.

Proof   As before, (2) obviously implies (1). Further, it is easily checked that (1) implies (3). (The check for $r = 3$ and $p = 7$ is simplified by noting that, as the binding number must be $3/2$, the graph must be hamiltonian.) We now show that (3) implies (2),.

If $r \ge 5$, then the Harary graphs $H_{r,p}$ detailed earlier suffice (theorem 3.8) unless $r = 5$ and $p = 9$. If $r = 3$ then we know that the Harary graphs are *not* suitable. However the following construction yields graphs which are. For $p \ge 9$, form G_p from C_p as follows. Label the vertices of C_p with $-s, -s + 1, \cdots, -1, 0, 1, \cdots, s, -s$. Then join i to $-i$ for $i = 1, \cdots, s - 3$; join $s - 1$ to $-s + 2$. Then it is easily verified that such graphs have the requisite binding number and indeed satisfy E2. For $p = 9$ and $r = 5$ take $G = G_9 \oplus C_9^2$.

We now consider the situations excluded by condition (3) of the previous two theorems. We ask (i) what is the largest value of the binding number that can occur, and (ii) how many edges are needed to attain the binding number $(p - 1)/(p - \delta)$? For the excluded graphs from the first theorem with $r = 2$

and p even, the maximum binding number is 1; to get the bound of (ii) take $C_p + e$ where e creates an odd cycle. In either theorem, with $r = p - 2$, the maximum binding number is $(p-2)/2$ and only complete graphs have higher binding number. If $r = 1$ and p is odd then the maximum value is $1/2$ and to get the bound of 1 take one extra edge to form $K_3 \cup \frac{p-3}{2} K_2$. On the other hand, when $r = 3$ and $p = 7$ then the maximum binding number is $4/3$, while one needs at least two extra edges (yielding a graph of size 13) to obtain the bound.

4. Nordhaus-Gaddum-type Results

Nordhaus-Gaddum type results, named after the classic paper [12] on inequalities involving the chromatic number $\chi(X)$ of a graph G, entail bounds, as functions of p (the order of G), on the sum or product of $\mu(G)$ and $\mu(\overline{G})$ for some graphical parameter μ. We present first the lower bounds for $\mu = I$.

Theorem 4.1   For all graphs G of order p,

a) $I(G) + I(\overline{G}) \geq p + 1$

b) $I(G) \cdot I(\overline{G}) \geq p$.

Proof   a) We may assume that G (say) is connected. If G is complete then $I(G) + I(\overline{G}) = p + 1$ and the hypothesis holds; therefore we may assume that G is noncomplete. Let H be a graph which is I-maximal with $I(G) = I(H)$ and which contains G as a spanning subgraph; then $I(\overline{G}) \geq I(\overline{H})$. How by theorem 2.2, $H \cong K_s + (K_{g1} \cup K_{g2} \cup \cdots \cup K_{gr})$ where $I(G) = I(H) = s + gl$ and $g_1 \geq g_2 \geq \cdots \geq g_r$; hence $\overline{H} \cong sK_1 \cup K(g_1, g_2, \cdots, g_r)$ yielding $I(\overline{H}) = (p - s) - g_1 + 1$. Therefore, $I(G) + I(\overline{G}) \geq I(H) + I(\overline{H}) = p + 1$ proving the result. b) This is a simple consequence of a).

It is immediately evident that equality is attained for the product only for $\{G, \overline{G}\} = \{K_p, \overline{K}_p\}$. For the sum, equality is reached for the H of the proof of the theorem (any graph which can be expressed as the join of a complete graph and one in which every component is complete) and others; for example consider $G = K(2, 4) - e$ where $I(G) = 3, I(\overline{G}) = 4$ and $p(G) = 6$.

The best upper bounds we have been able to find are derived from the Nordhaus-Gaddum type results for independence number. These were obtained by Chartrand and Schuster in 1974 [4]. We require the lower bounds.

If $r(m, n)$ denotes the appropriate Ramsey number, then define for each positive integer p,

$$\sigma_p := \min\{(m - 1) + (n - 1) : r(m, n) > p\},$$

$$\mu_p := \min\{(m - 1) \cdot (n - 1) : r(m, n) > p\}.$$

Chartrand and Schuster then proved the following result:

Theorem 4.2 [4] For each graph G of order p, $\beta(G) + \beta(\overline{G}) \geq \sigma_p$ and $\beta(G) \cdot \beta(\overline{G}) \geq \mu_p$. Moreover these bounds are best possible.

We are now able to determine upper bounds for $I(G) + I(\overline{G})$ and $I(G) \cdot I(\overline{G})$:

Theorem 4.3 For every graph G of order p,

$$\text{a)} \quad I(G) + I(\overline{G}) \leq S_p := 2(p + 1) - \sigma_p,$$

$$\text{b)} \quad I(G) \cdot I(\overline{G}) \leq P_p := \lfloor S_p/2 \rfloor \lceil S_p/2 \rceil.$$

Proof

a) Since $I(G) \leq \alpha(G) + 1 = p + 1 - \beta(G)$, it follows from theorem 4.2 that

$$I(G) + I(\overline{G}) \leq 2(p + 1) - (\beta(G) + \beta(\overline{G})) \leq 2(p + 1) - \sigma_p.$$

b) This follows from a).

We note that the bound given in b) appears to be better than that based on $I(G) \cdot I(\overline{G}) \leq (p + 1 - \beta(G)) \cdot (p + 1 - \beta(\overline{G}))$. Now, while we have not in general determined the relevant Nordhaus-Gaddum type bounds, some data is expressed in table 1 where I and $\overline{I}$ denote $I(G)$ and $I(\overline{G})$ respectively.

A graph on 10 vertices with $I(G) = I(\overline{G}) = 8$ may be constructed by taking C_9^2 and introducing a new vertex v such that if the vertices of C_9^2 are numbered consecutively with 1 through 9 then $N(v) = \{2, 4, 6, 8\}$ say. Verifying the integrity of G and $\overline{G}$ is simplified by noting that $I(G - v) = I(\overline{G} - v) = 7$ so that if the integrity were below eight in either case, the I-set could not include v. It is not known whether this example is unique.

p	S_p	$I + \bar{I}$	P_p	$I \cdot \bar{I}$	Extremal graphs G or $\bar{G}$
1	2	2	1	1	K_1
2	3	3	2	2	K_2
3	4	4	4	4	K_3 (sum only) and P_3
4	6	6	9	9	P_4
5	8	8	16	16	C_5
6	9	9	20	20	C_6 inter alia
7	11	11	30	30	C_7
8	13	12	42	36	C_8^2 inter alia
9	14	14	49	49	C_9^2 inter alia
10	16	16	64	64	See Text

Table 1

Some further discussion is in order. The case $p = 8$ is important, as for this value the upper bound S_8 (and hence P_8) is not achieved. For, to achieve this bound, one would (as in the proof of the bound) require a graph G on eight vertices with $\beta(G) = 2$ and $\beta(\bar{G}) = 3$ (or vice versa). By Ramsey theory, the only possibility is $G = C_8^2$ but we know that for this graph, $I(G) = I(\bar{G}) = 6$.

To prove that C_7 is the unique extremal graph (up to complementation) for $p = 7$, we argue that we need a graph with integrity five and whose complement has integrity six (or vice versa). Thus we need a graph with integrity five and girth at least five (by proposition 1.1d) whence, by simple enumeration, C_7 is the only one which meets the requirements.

The best guaranteed sum that we have found is the following:

Theorem 4.4   Let $p \geq 4$ be given. Then there exists a graph G of order p such that $I(G) + I(\bar{G}) = \lceil 3p/2 \rceil$.

Proof   If $p = 4$ then such a G is P_4. For $p \geq 5$ consider the graph $G = C_p^a$ where $a = \lfloor p/4 \rfloor$.

Then, by results of [2] and[7] (respectively), it holds that $(G) = a + \lceil p/2 \rceil$ and that $I(\bar{G}) = p - a$ and thus $I(G) + I(\bar{G}) = p + \lceil p/2 \rceil = \lceil 3p/2 \rceil$, proving our result. Pippert [13] was the first to consider results for toughness.

Theorem 4.5   [13] For all non-trivial graphs G of order p,

$$\frac{1}{p-1} \le t(G) + t(\overline{G}) \le p - 1.$$

Moreover, if both G and $\overline{G}$ are noncomplete then $t(G) + t(\overline{G}) \le (p-1)/2$.

The only 'interesting' bound is that of the upper bound of the product. The following is an improvement of a result of Pippert and is based directly on the upper bounds for δ:

Theorem 4.6   For all graphs G,

$$4 \cdot t(G) \cdot t(\overline{G}) \le \begin{cases} p(p-2)/2 & p \text{ even,} \\ (p-1)^2/4 & p \equiv 1 \pmod 4, \\ (p+1)(p-3)/4 & p \equiv 3 \pmod 4. \end{cases}$$

The actual values (as far as they have been determined) and the bounds for small values of p are presented in table 2. The results for $p = 8$ are important. To prove this value, let us assume that $\delta(G) \le \delta(\overline{G})$ where, of course $\delta(G) + \delta(\overline{G}) \le 7$. If $\delta(G) < 3$ or $\delta(\overline{G}) < 4$ then $\delta(G) \cdot \delta(\overline{G}) \le 5/2$. On the other hand, if $\delta(G) = 3$ and $\delta(\overline{G}) = 4$ then G is a cubic graph on eight vertices and thus has toughness at most $4/3$ by theorem 3.2. Clearly, $t(C_8^2) = 2$ and $t(\overline{C}_8^2) = 4/3$ so that the result follows.

For large values of p this approach is not optimal as the following upper bound shows. Define

$$\theta_p = \max_{(m,n)\in R_p} \left\{ \left(\frac{p}{m-1} - 1\right) \cdot \left(\frac{p}{n-1} - 1\right) \right\}.$$

Then it is easily shows that θ_p is an upper bound on $(p/\beta(G)-1)\cdot(p/\beta(\overline{G})-1)$ and hence an upper bound on $t(G).t(\overline{G})$ where G is any graph of order p. Now for small $p \ge 4$, this bound is not attainable and indeed the previous upper bound is better. However, for large p, this bound will be better (certainly for $p \ge r\ (4,6)$). It remains to be seen whether this bound will be attainable or even good. The actual values for large p thus remain open.

For binding number the Nordhaus-Gaddum problem has been solved:

p	Bound of theorem 4.6	Actual Max of $t(G) \cdot t(\bar{G})$	Extremal graphs G or $\bar{G}$
1	0	0	K_1
2	0	0	K_2
3	0	0	K_3 and P_3
4	1/2	1/2	P_4
5	1	1	C_5
6	3/2	3/2	C_6
7	2	2	C_7
8	3	8/3	C_8^2 (inter alia?)

Table 2

Theorem 4.7   Let G have order p where $p \geq 2$; then

a) $1/(p-1) \leq bind(G) + bind(\bar{G}) \leq p - 1$;

$$0 \leq bind(G) \times bind(\bar{G}) \leq \begin{cases} \frac{1}{2}(p-2) & \text{if } p \text{ is even,} \\ \frac{1}{3}(p-1)^2/(p-2) & \text{if } p \text{ is odd but } p \neq 3, \\ 0 & \text{if } p = 3; \end{cases}$$

with equality possible in all four bounds for all p.

Proof

a) To prove the lower bound, we note that at least one of G and $\bar{G}$ (say G) is connected so that $bind(G) \geq 1/(p-1)$ and the result follows. Equality is possible for G (say) a star, uniquely.

On the other hand, if both G and $\bar{G}$ are noncomplete then $bind(G) + bind(\bar{G}) \leq 2 \cdot bind(K_p - e) = p - 2$; hence for all G, $bind(G) + bind(\bar{G}) \leq bind(K_p) + bind(\bar{K}_p) = p - 1$ with equality for G (say) complete, uniquely.

b) The lower bound is obvious. Equality is attained iff G or $\bar{G}$ contains an isolated vertex. To prove the upper bound we let $\delta(\Delta)$ and $\bar{\delta}(\bar{\Delta})$ denote the smallest (largest) degree in G and $\bar{G}$ respectively. Recalling the bound $bind(G) \leq (p-1)/(p-\delta)$ and noting that $\delta \leq p - 1 - \bar{\Delta}$ (and mutatis mutandis) we obtain that:

$$bind(G) \times bind(\overline{G}) \leq \frac{(p-1)^2}{(\Delta+1)\cdot(\overline{\Delta}+1)}.$$

We note that $\Delta + \overline{\Delta} \geq p - 1$, and we assume that $\Delta \leq \overline{\Delta}$. Consider three cases:

1. If $\Delta = 0$, then $bind(G) \times bind(\overline{G}) = 0$.

2. If $\Delta = 1$, then $(p \geq 3$ and$)$ to attain $bind(G) > 0$ we must have $\delta = 1$ so that p is even and $G \cong \frac{p}{2}K_2$. Thus for p odd and $\Delta = 1$, it holds that $bind(G) \times bind(\overline{G}) = 0$. However, for p even, it holds that $bind(G) \times bind(\overline{G}) \leq 1 \cdot \frac{1}{2}(p-2)$ with equality uniquely for $G \cong \frac{p}{2}K_2$ (say).

3. If $\Delta \geq 2$, then $(p \geq 5$ and$)$ $(\Delta+1)\cdot(\overline{\Delta}+1) \geq 3(p-2)$ so that

$$bind(G) \times bind(\overline{G}) \leq (p-1)^2/(3(p-2)).$$

Equality requires (inter alia) $\delta = \Delta = 2$ and $bind(G) = (p-1)/(p-2)$. If p is odd then this is satisfied by C_p (only); further, from theorem 3.9 it follows that $bind(\overline{C}_p) = (p-1)/3$ so that we do indeed obtain equality (for p odd). If p is even and $p \geq 8$, then the (attainable) bound $(p-2)/2$ obtained in case 2 is larger than the above upper bound. It is easily checked for $p = 6$ and $\Delta = 2$ that $bind(G) \times bind(\overline{G}) \leq 5/3 < (6-2)/2$.

REFERENCES

[1] C. A. Barefoot, R. Entringer and H. C. Swart, "Vulnerability in Graphs - A comparative survey," *J. Combin. Math. Combin. Comp.* 1 (1987), 13-22.

[2] C. A. Barefoot, R. Entringer and H. C. Swart, "Integrity of trees and powers of cycles, *Congress. Numer.* 58 (1987), 103-114.

[3] G. Chartrand and L. Lesniak, Graphs and Digraphs (Second Edition), Wadsworth (1986).

[4] G. Chartrand and S. Schuster, "On the Independence numbers of Complementary Graphs," Transactions of New York Academy of Sciences, Series II, 30 (1974), 247-251.

[5] V. Chvátal, "Tough graphs and Hamilton Circuits," *Discrete Mathematics* 5 (1973), 215-228.

[6] W. D. Goddard, "On the Vulnerability of Graphs," M. Sc thesis, University of Natal at Durban (Submitted).

[7] W. D. Goddard and Henda C. Swart, "Integrity in Graphs: Bounds and Basics," Submitted.

[8] W. D. Goddard and Henda C. Swart, "The Binding Number of a Graph Revisited," Submitted.

[9] F. Harary, "The Maximum Connectivity in a Graph." *Proc. Nat. Acad. Sci. USA* 48 (1962), 1142-1146.

[10] V. G. Kane, S. P. Mohanty and E. G. Strauss, "Which Rational Numbers are Binding Numbers?", *J. Graph Theory* 5 (1981), 379-384.

[11] M. M. Matthews and D. P. Sumner, "Hamiltonian results in $K_{1,3}$-free graphs," *J. Graph Theory* 8 (1984), 139-146.

[12] E. A. Nordhaus and J. W. Gaddum, "On complementary graphs," *American Math. Monthly* 63 (1956), 175-177.

[13] R. E. Pippert, "On the Toughness of a Graph," Lecture Notes in Mathematics 303 (1972), 225-233.

[14] J. Wang, S. Tian and J. Liu, "The Binding Number of Product Graphs," Lecture Notes in Mathematics 1073 (1984), 119-128.

[15] D. R. Woodall, "The Binding Number of a Graph and its Anderson Number," *J. Combin. Th. (B)* 15 (1973), 225-255.

Bounds on the Number of Isolated Vertices
in Sum Graphs

Ronald J. Gould[1]

Vojtech Rödl

Emory University
Atlanta, GA 30322

ABSTRACT

A graph G is called a <u>sum graph</u> if the vertices of G can be labeled with distinct positive integers so that e = uv is an edge of G if, and only if, the sum of the lables on vertex u and vertex v is also a label in G. It is clear that if G is a properly labeled sum graph, then the vertex receiving the highest label cannot be adjacent to any other vertex. Thus, every sum graph must contain isolated vertices. We consider the problem of finding a general upper bound on the number of isolated vertices present in a sum graph, as well as the problem of finding a lower bound, at least for certain graphs.

1. Introduction

All terms not defined here can be found in [3]. Throughout this paper we will consider (p, q)–graphs, that is, graphs with p vertices and q edges. We denote the vertex set of G as V(G), the edge set as E(G) and a label attached to a vertex v as L(v).

Many different graph labeling problems have been studied (see for example [1]). Recently, Harary introduced a new variation, called sum graphs. A *sum graph* is a graph G in which each vertex x is labeled with a distinct positive integer L(x) such that $e = uv$ is an edge of G if, and only if, $L(u) + L(v) = L(w)$, for some $w \in V(G)$. It is obvious that if G is properly labeled as a sum graph, then the vertex assigned the highest label cannot be adjacent to any other vertex of G; thus, every sum graph must

1 Research supported in part by O.N.R. contract number N000014–88–K–0070.

553

contain at least one isolated vertex. We denote by $s(G)$ the number of isolated vertices we must add to a given graph G in order to be able to label it as a sum graph. Harary [4] posed the problem of finding an upper bound for $s(G)$. The object of this paper is to present a general upper bound and, at least for certain graphs, a lower bound on $s(G)$.

In studying sum graphs, it becomes obvious that if we wish to minimize the number of isolated vertices necessary for a proper labeling, then we must maximize the number of duplicate sums; that is, the number of edges $e = uv$, where $L(u) + L(v) = k$, for some fixed value of k.

2. Main Results

In this section we concentrate on obtaining general upper and lower bounds on $f(G)$, where $f(G) = |V(G)| + s(G)$. Bounds for $s(G)$ are then immediate. We note that it is shown in [2] that $s(K_n) \leq 2n - 3$, with equality holding when $n \geq 6$. This can be achieved by labeling the vertices of K_n using consecutive elements of an arithmetic sequence, a fact that will prove useful to us later. For example, the labeling of 11, 21, 31 ,41 on K_4 produces the need for five additional labels, namely 32, 42, 52, 62, 72. A *clique cover* of G is a partition of $V(G)$ into cliques. The set of all clique covers for G will be denoted $Cl(G)$, while for a particular cover $\hat{P} \in Cl(G)$, the number of cliques in $\hat{P}$ will be denoted $|\hat{P}|$.

Theorem 1 *Let G be a (p, q)–graph. Then G can be labeled as a sum graph with*

$$f(G) \leq q + 3p - 3\,|\hat{P}| - \left| \bigcup_{i=1}^{|\hat{P}|} E(C_i) \right|,$$

where $\hat{P}$ is a clique cover in $Cl(G)$ containing the maximum number of edges of G within its cliques $C_1, C_2, ..., C_{|\hat{P}|}$.

Proof Let G be a (p, q)–graph. We begin by decomposing $V(G)$ into cliques. We denote the classes of vertices in this cover by $C_1, C_2, ..., C_k$. Further, we assume that $|V(C_i)| = t_i$, $1 \leq i \leq k$. Thus, each class of vertices in this partition has order t_i and size $\binom{t_i}{2}$.

Select integers a_i and d_i and label vertex v_j^i of C_i with

$$L(v_j^i) = a_i + jd_i.$$

By labeling in this manner, we introduce at most $2t_i - 3$ new labels (hence, new isolated vertices) relative to each clique C_i (c.f. [2]). Further, the new labels necessary are

$$2a_i + 3d_i, 2a_i + 4d_i, ..., 2a_i + (2t_i - 1)d_i.$$

In general, such a labeling is done for each class C_i, $1 \leq i \leq k$.

In performing this labeling, we select the integers a_i and a_j so that the labels assigned to the vertices along with their subsequent implied sums for the edges between cliques are all distinct. More precisely, for $i \neq j$, we want

$$a_i + xd_i \neq a_j + yd_j \tag{1}$$

whenever $x \leq 2t_i - 3$ and $y \leq 2t_j - 3$. Moreover, for $\{i, j\} \neq \{r, s\}$, we also want

$$(a_i + xd_i) + (a_j + yd_j) \neq (a_r + zd_r) + (a_s + wd_s). \tag{2}$$

To see that such a labeling is possible, we may for example, select the integers using the following rules:

1. $d_{i+1} > 2td_i$ for $i = 1, 2, ..., k - 1$ and $t = \max_{1 \leq i \leq k} t_i$.

2. $a_i - a_j \gg d_k$, for every i, j, $1 \leq i$, $j \leq k$.

3. $a_1 \ll a_2 \ll a_3 \ll ... \ll a_k$.

Rule 2 insures us that equation (1) holds and that we never have the situation that for $\{i, j\} \neq \{r, s\}$

$$a_i + a_j = a_r + a_s \text{ holds.}$$

Further, it follows from Rule 3 that

$$(a_i + xd_i) + (a_j + yd_j) = (a_r + zd_r) + (a_s + wd_s) \tag{3}$$

implies that

$$a_i + a_j = a_r + a_s.$$

Thus, by our previous statements, $i = r$ and $j = s$. But then,

$$xd_i + yd_j = zd_r + wd_s,$$

and in fact, we have that

$$(x - z)d_i = (w - y)d_j. \tag{4}$$

But, using Rule 1 and the fact that $1 \leq w - y \leq 2t$, and assuming without loss of generality that $i < j$, we see that

$$(w - y)d_j \geq d_j > 2td_i \geq (x - z)d_i,$$

contradicting equation (4). But, then (2) must hold. Thus, the labeling is as claimed.

Now, with this labeling in mind, how large is $f(G)$? The total number of vertices (and hence labels) necessary for this labeling is

$$p + q + \sum_{i=1}^{k} [(2t_i - 3) - \binom{t_i}{2}].$$

Thus, if $\hat{P} \in Cl(G)$ contains the maximum number of edges of G within its cliques, then

$$f(G) \leq q + 3p - 3|\hat{P}| - \left| \bigcup_{i=1}^{|\hat{P}|} E(C_i) \right|,$$

completing the proof. $\square$

We now trun our attention to obtaining a lower bound on f(G). We begin with a Lemma.

Lemma 1   *Let p and d be positive integers satisfying*

$$\frac{\log (ep^3)}{\log (5/4)} < d < \frac{p}{10},$$

then

$$\frac{p}{2d} \left(\frac{1}{ep^3}\right)^{1/d} > 4. \tag{*}$$

Proof   The fact that

$$\frac{\log (ep^3)}{\log (5/4)} < d$$

implies that

$$(5/4)^d > ep^3$$

and hence that

$$\frac{1}{ep^3} (10d)^d > (8d)^d.$$

Now, using the fact that $10d < p$, we infer that

$$\frac{1}{ep^3} p^d > (8d)^d$$

which is equivalent to (*). ❑

Theorem 2   *Let*

$$\frac{\log (ep^3)}{\log (5/4)} < d < \frac{p}{10},$$

then there exists a (p, q)–graph G with $q = \binom{p}{2} - j$ where $j \leq pd$ and

$$s(G) \geq q - \frac{(p+1)(p-1)}{2d} \ln(ep^3) - p.$$

Proof   Here we consider $\hat{K}$, the class of graph with p vertices and
$q = \binom{p}{2} - j$ edges ($j \leq pd$). To each member of $\hat{K}$ we assign the most economical labeling possible, that is, a labeling that minimizes f(G). This labeling induces a system of equations of type

1. $L(v_i) + L(v_j) = L(v_k)$ and
2. $L(v_i) + L(v_j) = L(v_{i'}) + L(v_{j'})$.

For otherwise, we would have $p + q$ labels, which is clearly not the best possible labeling.

Now, for each graph $G \in \hat{K}$, only certain of the equations of type (1) or (2) actually hold. Suppose we view this system of equations as a homogeneous system of linear equations for the variables $L(v_i)$. Then, it follows from the fact that this system has a nonzero solution, that there are at most $p - 1$ equations with the property that all others are consequences of these equations (i.e. a linear combination of these equations). Thus, we may assign to each $G \in \hat{K}$, a system of at most $p - 1$ equations of type (1) and (2).

On the other hand, there are at most

$$\binom{p}{2} (p - 3)$$

equations of type (1) and at most

$$\frac{1}{2} \binom{p}{2} \binom{p-2}{2}$$

equations of type (2). Thus, the total number of possible equations of type (1) and (2) is clearly less than p^4. Hence, the number of ways in which we can choose this collection of at most $p - 1$ equations is

$$b = \sum_{j \le p-1} \binom{p^4}{j} < \binom{p^4}{p} < (ep^3)^p. \tag{2.1}$$

We partition the graphs in $\hat{K}$ into b classes, say

$$K_1, K_2, ..., K_b$$

according to which tuple of equations was assigned to K_i $(1 \le i \le b)$. As a notational convenience, in what follows let

$$m = \binom{p}{2},$$

and let

$$N = \sum_{i=1}^{b} |K_i|.$$

Since

$$\sum_{i=1}^{b} |K_i| = N = \sum_{j \le pd} \binom{m}{j}, \tag{2.2}$$

then the average number of graphs in any of the classes is $\frac{N}{b}$. Without loss of generality, suppose K_1 is a class containig at least the average number of graphs. Then we have

$$|K_1| \ge \frac{N}{b}. \tag{2.3}$$

Equations of type (1) and (2) satisfied in K_1 induce an equivalence relation $P_1, P_2, ..., P_k$ on the set of all pairs of vertices. The edge set of each graph $G \in K_1$ is equal to

$$\bigcup_{i=1}^{v} P_{j_i}$$

where $P_{j_1}, P_{j_2}, ..., P_{j_v}$ is a subset of $P_1, ..., P_k$ that satisfies

$$\sum_{i=1}^{v} |P_{j_i}| = q \ge \binom{p}{2} - pd. \tag{2.4}$$

As all labels of vertices of G need not correspond to sums of edge labels (for example, the minimum label), the number of additional isolated vertices needed to label G is at least $v - p$. If we set $|P_{j_i}| = t_i$ $(1 \le i \le k)$, then equation (2.4) implies that

$$v = q - \sum_{i=1}^{v} (|P_{j_i}| - 1) \ge q - \sum_{i=1}^{k} t_i + k,$$

and hence,

$$f(G) = s(G) + p \ge v \ge q - \sum_{i=1}^{k} t_i + k = q - x \tag{2.5}$$

where, $x = \sum_{j=1}^{k} t_j - k$.

It is clear that

$$\sum_{j \le pd} \binom{m-x}{j} \ge |K_1|,$$

and hence, from equation (2.3), it is clear that

$$\sum_{j \le pd} \binom{m-x}{j} \ge |K_1| \ge \frac{N}{b}. \tag{2.6}$$

Using equations (2.1) and (2.2) we also see that

$$\frac{N}{b} \ge \sum_{j \le pd} \binom{m}{j} (ep^3)^{-p} \ge \binom{m}{pd} (ep^3)^{-p}. \tag{2.7}$$

Therefore, (2.6) and (2.7) imply that

$$\sum_{j \le pd} \binom{m-x}{j} \ge \binom{m}{pd} (ep^3)^{-p}. \tag{2.8}$$

Now, using the fact that $\binom{r}{s} > \left(\frac{r}{s}\right)^s$ and the fact that $\binom{p}{2} \approx \frac{p^2}{2}$, we see that

$$\binom{m}{pd} (ep^3)^{-p} \ge \left[\left(\frac{p}{2d}\right)^d \frac{1}{ep^3}\right]^p.$$

Comparing this and an obvious upper bound for the left hand side of equation (2.8), we infer that

$$2^{m-x} \ge \sum_{j \le pd} \binom{m-x}{j} \ge \left[\left(\frac{p}{2d}\right)^d \frac{1}{ep^3}\right]^p,$$

and applying $\log_2$, we see that

$$m - x \ge pd \log_2 \left(\frac{p}{2d} \left(\frac{1}{ep^3}\right)^{1/d}\right).$$

As the assumptions of Lemma 1 are satisfied, we infer that

$$\log_2 \left(\frac{p}{2d} \left(\frac{1}{ep^3}\right)^{1/d}\right) > 2$$

and hence, $m - x \ge 2pd$ and thus,

$$pd \leq 1/2(m - x).$$

Using this new upper bound for pd, we obtain a new upper bound for $\binom{m}{pd}(ep^3)^{-p}$, namely

$$(1 + pd)\binom{m - x}{pd} > \sum_{j \leq pd}\binom{m - x}{j} \geq \binom{m}{pd}(ep^3)^{-p}. \qquad (2.9)$$

Dividing both sides of (2.9) by $(1 + pd)\binom{m}{pd}$ we see that

$$(1 - \frac{x}{m})(1 - \frac{x}{m - 1}) \ldots (1 - \frac{x}{m - pd + 1}) \geq \frac{1}{1 + pd}(ep^3)^{-p}.$$

Then, using the largest of the factors, we obtain the fact that

$$(1 - \frac{x}{m})^{pd} \geq \frac{1}{1 + pd}(ep^3)^{-p}.$$

However, since $(1 - y)^z = e^{z\ln(1-y)} < e^{-zy}$, if we let

$$y = \frac{x}{m} \text{ and } z = pd,$$

we see that

$$e^{-pdx/m} > (1 - \frac{x}{m})^{pd} \geq \frac{1}{1 + pd}(ep^3)^{-p} > (ep^3)^{-p-1} = e^{-(p+1)\ln(ep^3)}.$$

Now since e^w is an everywhere increasing function, we see that

$$\frac{-pdx}{m} > -(p + 1) \ln (ep^3)$$

or

$$\frac{pdx}{m} < (p + 1) \ln (ep^3).$$

Thus,

$$x \leq \frac{p + 1}{p}\frac{m}{d} \ln (ep^3). \qquad (2.10)$$

Now, from (2.5) and (2.10) we obtain the desired bound. $\square$

Our next goal is to restate Theorem 1 in order to show that in many cases the bound of Theorem 2 is close to best possible. Recall that we denoted by $\hat{K} = \hat{K}(p, d)$, the class of all graphs with p vertices and $\binom{p}{2} - j$ edges, where $j \leq pd$.

In order to esitmate the upper bound in Thoerem 1, we will actually find a lower bound for

$$3|\hat{P}| + \left| \bigcup_{i=1}^{|\hat{P}|} E(C_i) \right|$$

where $\hat{P}$ is a clique cover of G containing the maximum number of edges of G within its cliques. In order to do this, we find it convenient to consider $\overline{G}$, the complement of G, and we shall consider colorings of the vertices of G with color classes $C_1, C_2, \ldots, C_{|\hat{P}|}$.

Let $g(p, d)$ be the minimum of

$$\max \left(3|\hat{P}| + \sum_{i=1}^{|\hat{P}|} |E(C_i)| \right),$$

where the minimum is taken over all graphs G with at least p vertices and at most pd edges; while the maximum is taken over all colorings of G with color classes C_1, $...,C_{|\hat{P}|}$.

We will show that

$$g(p, d) \geq \frac{p^2}{6d + 6} + d - \frac{p}{2} = f(p, d) \tag{0}$$

Our proof is base on induction on p. For $p = 2$ the statement trivially hold. Now, let G be a graph with p vertices and at most pd edges. Let the largest independent set in G be C_0 (where $|C_0| = x$) and consider the graph $G - C_0$. Since each vertex of $V(G) - C_0$ is joined to some vertex of C_0 (due to the maximality of C_0), the subgraph G' induced on $V(G) - C_0$ has at most $pd - (p - x)$ edges; hence its average degree is at most

$$d' = \frac{pd}{p - x} - 1.$$

Using C_0 as one partition class and partitioning G' according to the induction assumption we see that

$$g(G) \geq 3 + \binom{x}{2} + g(G') \geq 3 + \binom{x}{2} + f(p - x, \frac{pd}{p - x} - 1).$$

Thus, it remains to show that

$$3 + \binom{x}{2} + f(p - x, \frac{pd}{p - x} - 1) \geq f(p, d). \tag{1}$$

We will prove (1) by induction on x.

Note that $x \geq \frac{p}{2d + 1}$ by Turan's Theorem [5] and thus, to verify the anchor of the induction, we will show that the inequality holds for $x = \frac{p}{2d + 1}$. (Note that we do not assume that $\frac{p}{2d + 1}$ is necessarily an integer.) Thus,

$$3 + \binom{p / (2d + 1)}{2} \geq f(p, d) - f\left(p - \frac{p}{2d + 1}, \frac{pd}{p - \frac{p}{2d + 1}} - 1\right)$$

$$= f(p, d) - f\left(\frac{2d}{2d + 1} p, \frac{2d - 1}{2}\right). \tag{2}$$

Substituting we see that

$$\frac{p^2}{6d + 6} - \frac{\left(\frac{2d}{2d + 1} p\right)^2}{6d + 3} + \frac{1}{2} - \frac{p}{2(2d + 1)} < \frac{p^2}{3} \left(\frac{1}{2d + 1} - \frac{2d}{(2d + 1)^2}\right) + \frac{1}{2} - \frac{p}{2(2d+1)}$$

$$< \frac{p^2}{2} \frac{1}{(2d + 1)^2} - \frac{1}{2} \frac{p}{2d + 1} + \frac{1}{2},$$

which is obviously smaller than the left hand side of (2), verifying the anchor of the induction.

Now we show the inductive step. This amounts to showing that

$$\left(\begin{array}{c} x+1 \\ 2 \end{array}\right) + f(p-x-1, \frac{pd}{p-x-1}-1) \geq \left(\begin{array}{c} x \\ 2 \end{array}\right) + f(p-x, \frac{pd}{p-x}-1). \quad (3)$$

Equation (3) is clearly equivalent to

$$x \geq f(p-x, \frac{pd}{p-x}-1) - f(p-x-1, \frac{pd}{p-x-1}-1), \quad (4)$$

which we now verify.

First, we esitmate the right hand side of equation (4) as follows:

$$\frac{(p-x)^2}{6\frac{pd}{p-x}} - \frac{(p-x-1)^2}{6\frac{pd}{p-x-1}} + \frac{pd}{p-x} - \frac{pd}{p-x-1} - \frac{1}{2} \leq \frac{(p-x)^3}{6pd} - \frac{(p-x-1)^3}{6pd}$$

$$\leq \frac{3(p-x)^2}{6pd}. \quad (5)$$

We will show that (5) is smaller than x. To this end it is enough to verify that

$$3(p-x)^2 \leq 6pdx. \quad (6)$$

Since $p \geq x \geq 0$ and $d \geq 0$, then the right hand side of equation (6) is increasing in x while the left hand side of equation (6) is decreasing in x. Thus, as $x \geq \frac{p}{2d+1}$, it is enough to verify equation (6) for $x = \frac{p}{2d+1}$. We leave this easy computation to the reader. This completes the induction on x and hence verifies equation (1). Therefore, our proof of the desired bound is completed.

Considering $G \in \hat{K}$, and substituting the just derived bound into the inequality of Theorem 1 yields

$$f(G) \leq q + 3p - \left(\frac{p^2}{6d+6} + d - \frac{p}{2}\right). \quad (7)$$

At the same time, Thoerem 2 tells us that there is $G \in \hat{K}$ with

$$f(G) \geq q - \frac{3(p+1)(p-1)}{6d} \ln(ep^3). \quad (8)$$

Thus, we see these two bounds are reasonably close.

Finally, we set

$$f(p) = \max \{f(G) \mid G \text{ has } p \text{ vertices}\}.$$

Using (7) and Theorem 2 we infer the following.

Theorem 3

$$\left(\begin{array}{c} p \\ 2 \end{array}\right) - \sqrt{2 \ln ep^3}\, p^{3/2} \leq f(p) \leq \left(\begin{array}{c} p \\ 2 \end{array}\right) - \sqrt{2/3}\, p^{3/2} + \frac{9p}{2}.$$

Proof   First we outline the upper bound. Noting that $q = \left(\begin{array}{c} p \\ 2 \end{array}\right) - pd$ and maximizing formula (7) over all d, we infer that the minimum is attained for

$$d + 1 = \sqrt{\frac{p}{6}}.$$

On the other hand, using the fact that formula (8) attains its maximum for

$$d = \sqrt{\frac{(p+1)(p-1)}{2p} \ln(ep^3)} \ ,$$

we infer the lower bound. □

REFERENCES

[1] Chung, F.R.K., *Some Problems and Results in Labelings of Graphs*, The Theory and Applications of Graphs, ed. by Chartrand, Alavi, Goldsmith, Lesniak–Foster, and Lick, Wiley Interscience, New York, NY, 1981.

[2] Faudree, R., Gould, R., Gyráfás, A., Lesniak, L., Schelp, R., *On Sum Graphs*, preprint.

[3] Gould, R., *Graph Theory*, Benjamin/Cummings Publishing Co., Menlo Park, CA, 1988.

[4] Harary, F., personal communication.

[5] Turán, P., *On an Extremal Problem in Graph Theory*. Mat. Fiz. Lapok, 48 (1941), 436 – 452.

ON THE MINIMUM INTERSECTION OF MINIMUM DOMINATING SETS IN SERIES-PARALLEL GRAPHS*

D. L. Grinstead

P. J. Slater

University of Alabama in Huntsville

ABSTRACT

Recently the authors introduced a type of graph-theoretic problem "intermediate" between finding a single vertex set with a specified property (such as being a minimum dominating set) and finding a partition of the entire vertex set of the graph into such sets (for example, a partition into a maximum number of dominating sets). Specifically, we defined $M_\gamma(G)$ to be the minimum cardinality of the intersection of two minimum dominating sets of G. We have shown that simply determining whether or not $M_\gamma(G) = 0$ is NP-hard for arbitrary bipartite graphs.

Here we present algorithmic results for the class of series-parallel graphs.

1. Introduction

Much of the extensive research in graph algorithms has been concerned with developing polynomial time algorithms for NP-complete problems restricted to appropriate classes of graph. Typically, the problems investigated have been of two types: (1) finding a single vertex subset with

---

*Research supported in part by U.S. Office of Naval Research under Grant N00014-86-K-0745.

a specified property (such as being a minimum dominating set) or (2) finding a partition of the vertex set into such sets (for example, a partition into a maximum number of dominating sets). Recently, in [7], we introduced a "two-set/single property" problem, a type intermediate between the other two. (A "single-set/two property" problem is introduced in these proceedings [15].) Specifically, the example we used as a model for problems of this type was that of finding two minimum dominating sets with an optimization criterion of minimizing the cardinality of their intersection.

Given a graph $G = (V, E)$, a vertex subset $D \subset V$ is a dominating set if each $v \in V - D$ is adjacent to at least one vertex in D; $\gamma(G)$ here denotes the minimum number of vertices in a dominating set. Problems involving dominating sets have been widely studied, and an extensive bibliography on the subject has been compiled by S. T. Hedetniemi and R. Laskar (Clemson University). Examples of papers on theoretical and algorithmic aspects of domination include [1, 3, 6, 11, 13].

We let $M_\gamma(G)$ denote the minimum cardinality of the intersection of two minimum dominating sets in G. In [7], we showed that simply determining if there exist two disjoint minimum dominating sets (that is, determining if $M_\gamma(G) = 0$) is NP-hard for arbitrary bipartite graphs, and we presented a linear algorithm for determining $M_\gamma(T)$ for a tree T.

In this paper we describe a linear algorithm for determining $M_\gamma(G)$ for a series-parallel graph G. To do this we will first discuss series-parallel graphs and their structure (Section 2). Then in Section 3 we introduce nine single-set γ-related parameters giving initial values, recursive formulations, and final conditions for computing $\gamma(G)$ for a series-parallel graph G. These nine single-set parameters provide necessary information for the two-set γ-related parameters described in Section 4. In Section 4 we introduce 45 of these two-set parameters and describe their initial values, recursive formulations, and final conditions for series-parallel graphs. Although the algorithm is linear, its length prohibits us from presenting it here in its entirety; to do so would require more than one-hundred pages of detailed recurrence equations. Instead we will describe it, in Section 5, and show the order in which parameters would be computed for a series-parallel graph G. The complete algorithm may be found in Grinstead [9].

Another two-set/single property problem involving maximum independent sets is described in [8], in which a general discussion of various optimization criteria for this type of problem is included.

2. Series-Parallel Graphs

A series-parallel graph G is a graph which contains no homeomorph of K_4. This definition is not easily implemented algorithmically, but it is equivalent to the following recursive definition, which is presented essentially in the form introduced by Wimer [18].

Definition (Series-parallel graph):

1) K_2 is series-parallel with its two vertices, say u and v, as terminals.

2) If $G1$ is series-parallel with terminals u_1 and v_i and $G2$ is series-parallel with terminals u_2 and v_2, then:

   a) the *Series 1* composition of $G1$ and $G2$ obtained by identifying the right terminal of $G1$ (i.e. v_1) with the left terminal of $G2$ (ie.e u_2) is series-parallel with terminals u_1 and v_2 (see Figure 2.1.),

   b) The *Series 2* composition of $G1$ and $G2$ obtained by identifying the right terminal of $G1$ with the left terminal of $G2$ is series-parallel with terminals u_1 and $v_1 = u_2$ (see Figure 2.2.), and

   c) The *Parallel* composition of $G1$ and $G2$ obtained by identifying the left terminal of $G1$ with the left terminal of $G2$ and identifying the right terminal of $G1$ with the right terminal of $G2$ is series-parallel with terminals $u_1 = u_2$ and $v_1 = v_2$ (see Figure 2.3).

3) All and only series-parallel graphs may be obtained by a finite number of applications of 1) and 2).

Thus, for each series-parallel graph H there exists a labelled binary tree (called a parse tree of H) which represents it. Each leaf of this tree represents

Figure 2.1. The Series 1 composition of G1 and G2.
(Darkened vertices denote terminals.)

Figure 2.2. The Series 2 composition of G1 and G2.
(Darkened vertices denote terminals.)

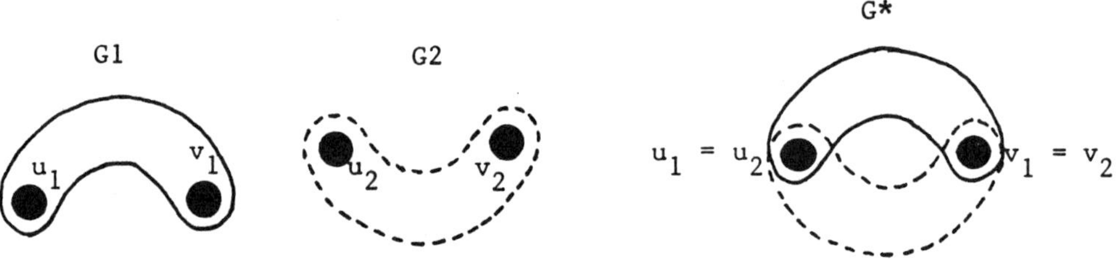

Figure 2.3. The Parallel composition of G1 and G2.
(Darkened vertices denote terminals.)

K_2, the basis graph for the class of series- parallel graphs. Each internal node of the parse tee is labelled either $S1, S2$ or P (for Series 1, Series 2, or Parallel) and represents the graph G^* obtained by combining the graph $G1$ (represented by the left child of the node) and the graph $G2$ (represented by the right child of the node) through the appropriate type of composition. Hence each node of the parse tree represents a series- parallel graph, and the graph represented by the root node is the graph H. A parse tree for a series-parallel graph may be found in linear time using a procedure similar to the one in [11], so requiring a parse tree as input does not increase the order of the execution time of our algorithm. In fact, apparently the first detailed parsing algorithm for series-parallel graphs is due to J. Valdes [16, 17].

3.   Computation of Some Single-Set γ-Related Parameters for Series-Parallel Graphs

Let G^* be a series-parallel graph obtained by joining $G1$ and $G2$ (each series-parallel) by one of the three possible connections, for example, Series 1. Next let D be a dominating set of G^*. Then, if we consider the sets $D1 = D \cap V(G1)$ and $D2 = D \cap V(G2)$, there are three possibilities. (1) If $v_1 = u_2 \in D$, then $D1$ is a dominating set of $G1$ containing v_1 and D_2 is a dominating set of $G2$ containing u_2. (2) If $v_1 = u_2 \notin D$, then there must be some vertex $w \in V(G^*)$ such that $w \in D$ and w is adjacent to $v_1 = u_2$. (2a) If $w \in V(G1)$, then $D1$ is a dominating set of $G1$ and $D2$ is a dominating set of $G2 - u_2$ (not containing u_2). Note that the vertex u_2 may or may not be dominated by the set $D2$. (2b) If $w \in V(G2)$, then $D1$ is a dominating set of $G1 - v_1$ (not containing v_1) and $D2$ is a dominating set of $G2$ (not containing u_2).

Because of these three cases we introduce nine domination parameters:

$$\gamma_{yy}(G) \quad \gamma_{ny}(G) \quad \gamma_{\bar{n}y}(G)$$

$$\gamma_{yn}(G) \quad \gamma_{nn}(G) \quad \gamma_{\bar{n}n}(G)$$

$$\gamma_{y\bar{n}}(G) \quad \gamma_{n\bar{n}}(G) \quad \gamma_{\bar{n}\bar{n}}(G)$$

In these parameters the two subscripts describe the situation of the two terminals of series parallel graph G in the dominating set. A subscript of y indicates that the vertex is in the set, a subscript of n indicates that the vertex is not in the set but is dominated by the set, and a subscript of $\bar{n}$ indicates that the vertex is not in the set and may or may not be dominated by the set. That is, $\gamma_{yy}(G)$ denotes the minimum cardinality of a dominating set of G which contains both terminals, $\gamma_{yn}(G)$ denotes the minimum cardinality of a dominating set of G which contains the left terminal and not the right terminal, $\gamma_{y\bar{n}}(G)$ denotes the minimum cardinality of a dominating set of G minus the right terminal which contains the left terminal and not the right terminal, etc.

We first initialize these nine parameters for $G = K_2$, the basis graph for the class of series-parallel graphs. Then we will present recurrences for the parameters for each of the three types of series-parallel compositions (Series 1, Series 2, Parallel). Finally, when we reach the root node of the parse tree of G we will be able to determine $\gamma(G)$.

To see how these parameters should be initialized, consider the series-parallel basis graph K_2. The dominating set of K_2 which contains both terminals has cardinality two, so $\gamma_{yy}(K_2) = 2$. A set D containing exactly one of the terminals dominates the entire graph, so $\gamma_{yn}(K_2) = \gamma_{ny}(K_2) = \gamma_{y\bar{n}}(K_2) = \gamma_{\bar{n}y}(K_2) = 1$. Any time neither vertex is contained in the dominating set but domination is required we have an impossible situation so those parameters will be initialized to infinity. Thus $\gamma_{nn}(K_2) = \gamma_{n\bar{n}}(K_2) = \gamma_{\bar{n}n}(K_2) = \infty$.

For the final parameter $\gamma_{\bar{n}\bar{n}}$, neither terminal is included in the dominating set but domination is not required, so $\gamma_{\bar{n}\bar{n}}(K_2) = 0$.

Each of the nine γ-parameters must have a recurrence for each of the three types of series- parallel compositions, for a total of twenty-seven recurrences. Here we will take $\gamma^*_{y\bar{n}}$ as an example, writing $\gamma^*_{y\bar{n}}$ for $\gamma_{y\bar{n}}(G^*)$.

Suppose G^* is obtained by a Series 1 composition of series-parallel graphs $G1$ and $G2$. If we are looking for $\gamma^*_{y\bar{n}}$ we want to consider sets which contain the left terminal of G^*, do not contain the right terminal of G^*, and may or may not dominate the right terminal of G^*. Such sets with minimum cardinality are called $\gamma_{y\bar{n}}$-sets. If we consider such a set D in relation to the graphs $G1$ and $G2$, we see that D contains the left terminal of $G1$ (which becomes the left terminal of G^*) and that D does not contain, and possibly does not dominate, the right terminal of $G2$ (which becomes the right terminal of G^*). So now we are only concerned with the right terminal v_1 of $G1$ and the left terminal u_2 of $G2$ which are identified in the Series 1 composition.

One possibility is that D contains $v_1 = u_2$. Then in $G1$ we are looking at a set which contains both terminals (see γ^1_{yy} below), and in $G2$ we are looking at a set which contains the left terminal and does not contain, or necessarily dominate, the right terminal (see $\gamma^2_{y\bar{n}}$ below). Thus $\gamma^*_{y\bar{n}} \leq \gamma^1_{yy} + \gamma^2_{y\bar{n}} - 1$. The minus one must be included because the vertex $v_1 = u_2$ is counted in the dominating sets of both $G1$ and $G2$. A second possibility is that D does not contain $v_1 = u_2$. Since this vertex is not a terminal of the graph G^* we must require domination. Here there are two choices: (1) We may require that v_1 be dominated in $G1$ and not care if domination of u_2 occurs in $G2$, or (2) we may not care if domination of v_1 occurs in $G1$ and require that u_2 be dominated in $G2$. Note that either of these two cases will accommodate the situation in which $v_1 = u_2$ is dominated in both $G1$ and $G2$. From case (1) we see that $\gamma^*_{y\bar{n}} \leq \gamma^1_{yn} + \gamma^2_{n\bar{n}}$ and from case (2) we see that $\gamma^*_{y\bar{n}} \leq \gamma^1_{y\bar{n}} + \gamma^2_{nn}$.

Observing that these are the only possibilities and combining the three inequalities we get the Series 1 recurrence $\gamma^*_{y\bar{n}} = \min\{\gamma^1_{yy} + \gamma^2_{y\bar{n}} - 1, \gamma^1_{yn} + \gamma^2_{n\bar{n}}, \gamma^1_{y\bar{n}} + \gamma^2_{nn}\}$ (see line (3) of Table 3.1). The remaining Series 1 recurrences may be found in a similar manner and are presented in Table 3.1.

$$\gamma^{*}_{yy} = \min\left(\gamma^{1}_{yy} + \gamma^{2}_{yy} - 1,\ \gamma^{1}_{yn} + \gamma^{2}_{\bar{n}y},\ \gamma^{1}_{y\bar{n}} + \gamma^{2}_{ny}\right) \tag{1}$$

$$\gamma^{*}_{yn} = \min\left(\gamma^{1}_{yy} + \gamma^{2}_{yn} - 1,\ \gamma^{1}_{yn} + \gamma^{2}_{\bar{n}n},\ \gamma^{1}_{y\bar{n}} + \gamma^{2}_{nn}\right) \tag{2}$$

$$\gamma^{*}_{y\bar{n}} = \min\left(\gamma^{1}_{yy} + \gamma^{2}_{y\bar{n}} - 1,\ \gamma^{1}_{yn} + \gamma^{2}_{\bar{n}\bar{n}},\ \gamma^{1}_{y\bar{n}} + \gamma^{2}_{n\bar{n}}\right) \tag{3}$$

$$\gamma^{*}_{ny} = \min\left(\gamma^{1}_{ny} + \gamma^{2}_{yy} - 1,\ \gamma^{1}_{nn} + \gamma^{2}_{\bar{n}y},\ \gamma^{1}_{n\bar{n}} + \gamma^{2}_{ny}\right) \tag{4}$$

$$\gamma^{*}_{nn} = \min\left(\gamma^{1}_{ny} + \gamma^{2}_{yn} - 1,\ \gamma^{1}_{nn} + \gamma^{2}_{\bar{n}n},\ \gamma^{1}_{n\bar{n}} + \gamma^{2}_{nn}\right) \tag{5}$$

$$\gamma^{*}_{n\bar{n}} = \min\left(\gamma^{1}_{ny} + \gamma^{2}_{y\bar{n}} - 1,\ \gamma^{1}_{nn} + \gamma^{2}_{\bar{n}\bar{n}},\ \gamma^{1}_{n\bar{n}} + \gamma^{2}_{n\bar{n}}\right) \tag{6}$$

$$\gamma^{*}_{\bar{n}y} = \min\left(\gamma^{1}_{\bar{n}y} + \gamma^{2}_{yy} - 1,\ \gamma^{1}_{\bar{n}n} + \gamma^{2}_{\bar{n}y},\ \gamma^{1}_{\bar{n}\bar{n}} + \gamma^{2}_{ny}\right) \tag{7}$$

$$\gamma^{*}_{\bar{n}n} = \min\left(\gamma^{1}_{\bar{n}y} + \gamma^{2}_{yn} - 1,\ \gamma^{1}_{\bar{n}n} + \gamma^{2}_{\bar{n}n},\ \gamma^{1}_{\bar{n}\bar{n}} + \gamma^{2}_{nn}\right) \tag{8}$$

$$\gamma^{*}_{\bar{n}\bar{n}} = \min\left(\gamma^{1}_{\bar{n}y} + \gamma^{2}_{y\bar{n}} - 1,\ \gamma^{1}_{\bar{n}n} + \gamma^{2}_{\bar{n}\bar{n}},\ \gamma^{1}_{\bar{n}\bar{n}} + \gamma^{2}_{n\bar{n}}\right) \tag{9}$$

Table 3.1. γ recurrences for Series 1 composition.

$$\gamma^{*}_{yy} = \min \{ \gamma^{1}_{yy} + \gamma^{2}_{yy} - 1, \ \gamma^{1}_{yy} + \gamma^{2}_{yn} - 1 \} \tag{10}$$

$$\gamma^{*}_{yn} = \min \{ \gamma^{1}_{yn} + \gamma^{2}_{\bar{n}y}, \ \gamma^{1}_{y\bar{n}} + \gamma^{2}_{ny}, \ \gamma^{1}_{yn} + \gamma^{1}_{\bar{n}n}, \ \gamma^{1}_{y\bar{n}} + \gamma^{2}_{nn} \} \tag{11}$$

$$\gamma^{*}_{y\bar{n}} = \min \{ \gamma^{1}_{y\bar{n}} + \gamma^{2}_{\bar{n}y}, \ \gamma^{1}_{y\bar{n}} + \gamma^{2}_{\bar{n}n} \} \tag{12}$$

$$\gamma^{*}_{ny} = \min \{ \gamma^{1}_{ny} + \gamma^{2}_{yy} - 1, \ \gamma^{1}_{ny} + \gamma^{2}_{yn} - 1 \} \tag{13}$$

$$\gamma^{*}_{nn} = \min \{ \gamma^{1}_{nn} + \gamma^{2}_{\bar{n}y}, \ \gamma^{1}_{n\bar{n}} + \gamma^{2}_{ny}, \ \gamma^{1}_{nn} + \gamma^{2}_{\bar{n}n}, \ \gamma^{1}_{n\bar{n}} + \gamma^{2}_{nn} \} \tag{14}$$

$$\gamma^{*}_{n\bar{n}} = \min \{ \gamma^{1}_{n\bar{n}} + \gamma^{2}_{\bar{n}y}, \ \gamma^{1}_{n\bar{n}} + \gamma^{2}_{\bar{n}n} \} \tag{15}$$

$$\gamma^{*}_{\bar{n}y} = \min \{ \gamma^{1}_{\bar{n}y} + \gamma^{2}_{yy} - 1, \ \gamma^{1}_{\bar{n}y} + \gamma^{2}_{yn} - 1 \} \tag{16}$$

$$\gamma^{*}_{\bar{n}n} = \min \{ \gamma^{1}_{\bar{n}n} + \gamma^{2}_{\bar{n}y}, \ \gamma^{1}_{\bar{n}\bar{n}} + \gamma^{2}_{ny}, \ \gamma^{1}_{\bar{n}n} + \gamma^{2}_{\bar{n}n}, \ \gamma^{1}_{\bar{n}\bar{n}} + \gamma^{2}_{nn} \} \tag{17}$$

$$\gamma^{*}_{\bar{n}\bar{n}} = \min \{ \gamma^{1}_{\bar{n}\bar{n}} + \gamma^{2}_{\bar{n}y}, \ \gamma^{1}_{\bar{n}\bar{n}} + \gamma^{2}_{\bar{n}n} \} \tag{18}$$

Table 3.2. γ recurrences for Series 2 composition.

$$\gamma^{*}_{yy} = \gamma^{1}_{yy} + \gamma^{2}_{yy} - 2 \tag{19}$$

$$\gamma^{*}_{yn} = \min\left(\gamma^{1}_{yn} + \gamma^{2}_{y\bar{n}} - 1,\ \gamma^{1}_{y\bar{n}} + \gamma^{2}_{yn} - 1\right) \tag{20}$$

$$\gamma^{*}_{y\bar{n}} = \gamma^{1}_{y\bar{n}} + \gamma^{2}_{y\bar{n}} - 1 \tag{21}$$

$$\gamma^{*}_{ny} = \min\left(\gamma^{1}_{ny} + \gamma^{2}_{\bar{n}y} - 1,\ \gamma^{1}_{\bar{n}y} + \gamma^{2}_{ny} - 1\right) \tag{22}$$

$$\gamma^{*}_{nn} = \min\left(\gamma^{1}_{nn} + \gamma^{2}_{\bar{n}\bar{n}},\ \gamma^{1}_{n\bar{n}} + \gamma^{2}_{\bar{n}n},\ \gamma^{1}_{\bar{n}n} + \gamma^{2}_{n\bar{n}},\ \gamma^{1}_{\bar{n}\bar{n}} + \gamma^{2}_{nn}\right) \tag{23}$$

$$\gamma^{*}_{n\bar{n}} = \min\left(\gamma^{1}_{n\bar{n}} + \gamma^{2}_{\bar{n}\bar{n}},\ \gamma^{1}_{\bar{n}\bar{n}} + \gamma^{2}_{n\bar{n}}\right) \tag{24}$$

$$\gamma^{*}_{\bar{n}y} = \gamma^{1}_{\bar{n}y} + \gamma^{2}_{\bar{n}y} - 1 \tag{25}$$

$$\gamma^{*}_{\bar{n}n} = \min\left(\gamma^{1}_{\bar{n}n} + \gamma^{2}_{\bar{n}\bar{n}},\ \gamma^{1}_{\bar{n}\bar{n}} + \gamma^{2}_{\bar{n}n}\right) \tag{26}$$

$$\gamma^{*}_{\bar{n}\bar{n}} = \gamma^{1}_{\bar{n}\bar{n}} + \gamma^{2}_{\bar{n}\bar{n}} \tag{27}$$

Table 3.3. γ recurrences for Parallel composition.

If G^* is obtained by a Series 2 composition of $G1$ and $G2$ then the terminals of G^* were the terminals of $G1$. Thus a $\gamma_{y\bar{n}}$-set of G^* must contain a $\gamma_{y\bar{n}}$-set of $G1$. Also the left terminal of $G2$ is the same as the right terminal of $G1$, and so it is in an $\bar{n}$ situation. This only leaves the right terminal of $G2$ at which we require domination, so our only two choices are that v_2 is contained in the set or that v_2 is not contained in, but is dominated by, the set. Thus $\gamma_{y\bar{n}}^* = \min\{\gamma_{y\bar{n}}^1 + \gamma_{\bar{n}y}^2, \gamma_{y\bar{n}}^1 + \gamma_{\bar{n}n}^2\}$ is the Series 2 recurrence for $\gamma_{y\bar{n}}$. This recurrence, along with the remaining Series 2 recurrences, is shown in Table 3.2.

Next suppose that G^* is obtained by a Parallel composition of $G1$ and $G2$. Then the terminals of $G1$ and $G2$ are the terminals of G^* so a $\gamma_{y\bar{n}}$-set of G^* must contain both a $\gamma_{y\bar{n}}$-set of $G1$ and a $\gamma_{y\bar{n}}$-set of $G2$. Hence the Parallel recurrence for $\gamma_{y\bar{n}}^*$ is simply $\gamma_{y\bar{n}}^* = \gamma_{y\bar{n}}^1 + \gamma_{y\bar{n}}^2 - 1$. The Parallel recurrences are all presented in Table 3.3. Note that the Parallel recurrences with more than one choice arise when domination is required at a terminal which is not to be used in the set, thus domination of the terminal is required in either $G1$ or $G2$ but not necessarily both.

When the root node of the parse tree is reached, we have the value of all nine parameters for G and $\gamma(G) = \min\{\gamma_{yy}(G), \gamma_{yn}(G), \gamma_{ny}(G), \gamma_{nn}(G)\}$. Note that any parameter containing an $\bar{n}$ as a subscript does not require domination of the corresponding terminal, and thus it cannot be included in this minimum for determining $\gamma(G)$.

4. Computation of Some Two-Set Parameters

In this section we define 45 two-set parameters which, in conjunction with nine γ-related single-set parameters, can be used in determining $M_\gamma(G)$, the minimum cardinality of the intersection of two minimum dominating sets in a series-parallel graph G. Although the problem of finding two MDS's, D_1 and D_2, for which $|D_1 \cap D_2| = M_\gamma(G)$ can be solved in linear time, we only consider here the problem of determining the value of $M_\gamma(G)$. Since each of the 45 two-set parameters must have a recurrence for each of the three types of series- parallel composition, we must specify, in principle, 135 recurrences. In the interest of space, however, we will only present three of the recurrences here; all of the others may be found in a similar manner. A complete set of recurrences, along with the details for finding two sets D_1 and D_2, is available in [9].

Again we will use the notation that a subscript of y indicates that the corresponding terminal is in the dominating set, n indicates that the terminal is not in the dominating set but it has been dominated, and $\bar{n}$ indicates that the terminal is not in the dominating set and it may or may not be dominated. The 45 parameters for measuring the minimum cardinality of the intersection of various combinations of types of sets are as follows:

$\lambda_{yy,yy}$

$\lambda_{yy,yn}\lambda_{yn,yn}$

$\lambda_{yy,y\bar{n}}\lambda_{yn,y\bar{n}}\lambda_{y\bar{n},y\bar{n}}$

$\lambda_{yy,ny}\lambda_{yn,ny}\lambda_{y\bar{n},ny}\lambda_{ny,ny}$

$\lambda_{yy,nn}\lambda_{yn,nn}\lambda_{y\bar{n},nn}\lambda_{ny,ny}\lambda_{nn,nn}$

$\lambda_{yy,n\bar{n}}\lambda_{yn,n\bar{n}}\lambda_{y\bar{n},n\bar{n}}\lambda_{ny,n\bar{n}}\lambda_{nn,n\bar{n}}\lambda_{n\bar{n},n\bar{n}}$

$\lambda_{yy,\bar{n}y}\lambda_{yn,\bar{n}y}\lambda_{y\bar{n},\bar{n}y}\lambda_{ny,\bar{n}y}\lambda_{nn,\bar{n}y}\lambda_{n\bar{n},\bar{n}y}\lambda_{\bar{n}y,\bar{n}y}$

$\lambda_{yy,\bar{n}n}\lambda_{yn,\bar{n}y}\lambda_{y\bar{n},\bar{n}n}\lambda_{ny,\bar{n}n}\lambda_{nn,\bar{n}n}\lambda_{n\bar{n},\bar{n}n}\lambda_{\bar{n}y,\bar{n}n}\lambda_{\bar{n}n,\bar{n}n}$

$\lambda_{yy,\bar{n}\bar{n}}\lambda_{yn,\bar{n}\bar{n}}\lambda_{y\bar{n},\bar{n}\bar{n}}\lambda_{ny,\bar{n}\bar{n}}\lambda_{nn,\bar{n}\bar{n}}\lambda_{n\bar{n},\bar{n}\bar{n}}\lambda_{\bar{n}y,\bar{n}\bar{n}}\lambda_{\bar{n}n,\bar{n}\bar{n}}\lambda_{\bar{n}\bar{n},\bar{n}\bar{n}}$

Here the first two subscripts describe the situation of the two terminals in one dominating set, while the second two subscripts describe the situation of the two terminals in a second dominating set. Thus $\lambda_{yn,n\bar{n}}(G)$ denotes the minimum cardinality of the intersection of sets D_1 and D_2 where D_1 is a dominating set of G of minimum cardinality among those than contain the left terminal of G and not the right terminal of G, and D_2 is a dominating set of G minus the right terminal of minimum cardinality among those which contain neither terminal. Note that there would be 81 combinations of set types but that $\lambda_{ab,cd}(G) = \lambda_{cd,ab}(G)$ because the order of the two sets does not matter, so we are left with only forty- five distinct combinations. As in Section

3, we will present here initialization for these 45 parameters on K_2, a sample recurrence for the parameters under each of the three types of series-parallel connections, and final conditions for determining $M_\gamma(G)$. Initialization of these parameters for $G = K_2$ is straightforward as follows:

$$\lambda_{yy,yy}(K_2) = 2$$

$$\lambda_{yy,yn}(K_2) = \lambda_{yy,ny}(K_2) = \lambda_{yy,y\bar{n}}(K_2) = \lambda_{yy,n\bar{y}}(K_2)$$

$$= \lambda_{yn,yn}(K_2) = \lambda_{yn,y\bar{n}}(K_2) = \lambda_{ny,ny}(K_2) = \lambda_{ny,\bar{n}y}(K_2)$$

$$= \lambda_{y\bar{n},y\bar{n}}(K_2) = \lambda_{\bar{n}y,\bar{n}y}(K_2) = 1$$

$$\lambda_{yy,\overline{nn}}(K_2) = \lambda_{yn,ny}(K_2) = \lambda_{yn,\bar{n}y}(K_2) = \lambda_{yn,\overline{nn}}(K_2)$$

$$= \lambda_{y\bar{n},ny}(K_2) = \lambda_{ny,\overline{nn}}(K_2) = \lambda_{y\bar{n},\bar{n}y}(K_2) = \lambda_{y\bar{n},\overline{nn}}(K_2)$$

$$= \lambda_{\bar{n}y,\overline{nn}}(K_2) = \lambda_{\overline{nn},\overline{nn}}(K_2) = 0$$

The remaining 24 λ parameters must be initialized to $+\infty$ (as in the case of $\gamma_{nn}, \gamma_{\bar{n}n}, \gamma_{n\bar{n}}$) since they pair an n with another n or an $\bar{n}$ in one of the two sets, thus requiring domination but not dominating.

For each type of series-parallel composition we must have recurrences for each of the 45 λ- parameters, for a total of 135 recurrences. Here we will consider $\lambda_{yn,n\bar{n}}$ and explain the three recurrences for this parameter.

Recall that $\lambda_{yn,n\bar{n}}(G^*)$ is the minimum cardinality of the intersection of two sets D_1 and D_2 where D_1 is a γ_{yn}-set (that is, D_1 contains the left terminal of G^* but not the right terminal, D_1 dominates all of the vertices of G^*, and $|D_1| - \gamma_{yn}(G^*)$), and D_2 is a $\gamma_{n\bar{n}}$-set (i.e., D_2 contains neither terminal of G^*, it dominates G^* minus the right terminal, and $|D_2| = \gamma_{n\bar{n}}(G^*)$). In the discussion that follows we will write $\lambda^*_{yn,n\bar{n}}$ for $\lambda_{yn,n\bar{n}}(G^*)$, $\lambda^1_{yn,n\bar{n}}$ for $\lambda_{yn,n\bar{n}}(G1)$, and $\lambda^2_{yn,n\bar{n}}$ for $\lambda_{yn,n\bar{n}}(G2)$.

To ensure that D_1 is a γ_{yn}-set and D_2 is a $\gamma_{n\bar{n}}$-set we must refer to the recurrences presented in Section 3. We will utilize equations (2) and (6) of Table 3.1 for the Series 1 composition, equations (11) and (15) of Table 3.2 of the Series 2 composition, and equations (20) and (24) of Table 3.3 for the

Parallel composition.

Suppose, for example, that G^* is obtained by a Series 1 composition of series-parallel graphs $G1$ and $G2$. Note that, from (2), there are three possibilities to be considered for obtaining a γ_{yn}-set D_1. And, from (6), there are three possibilities to be considered for a $\gamma_{n\bar{n}}$-set D_2. The nine resulting combinations of possibilities for $\lambda_{yn,n\bar{n}}$ can be examined as follows. First suppose that $\gamma_{yn}^* = \gamma_{yy}^1 + \gamma_{yn}^2 - 1$ (the first choice from equation (2)) and that $\gamma_{n,\bar{n}}^* - \gamma_{ny}^1 + \gamma_{y\bar{n}}^2 - 1$ (the first choice from equation (6)). Then one viable alternative involves a γ_{yy}-set and a γ_{ny}-set from $G1$ (see the $\lambda_{yy,ny}^1$ below), and a γ_{yn}-set and a $\gamma_{y\bar{n}}$-set from $G2$ (see $\lambda_{yn,y\bar{n}}^2$ below). The fact that the sets may be put together to form a γ_{yn}-set D_1 and a $\gamma_{n\bar{n}}$-set D_2 of G^* shows that $\lambda_{yn,n\bar{n}}^* \leq \lambda_{yy,ny}^1 + \lambda_{yn,y\bar{n}}^2 - 1$. The minus one is included because the right terminal of $G1$ is identified with the left terminal of $G2$ and gets counted twice. The right hand side of this inequality is set equal to a variable A (see line (28) below) and then in equation (29) $\lambda_{yn,n\bar{n}}^*$ is set equal to the minimum of the nine possibilities (one of which is A). If, however, $\gamma_{yn}^* \neq \gamma_{yy}^1 + \gamma_{yn}^2 - 1$ or $\gamma_{n\bar{n}}^* \neq \gamma_{ny}^1 + \gamma_{y\bar{n}}^2 - 1$ then the value of A will be infinity, and thus it will not be chosen as the minimum in equation (29). All nine possibilities are presented in the Series 1 recurrence for $\lambda_{yn,n\bar{n}}^*$ which follows.

If $\gamma_{yn}^* = \gamma_{yy}^1 + \gamma_{yn}^2 - 1$

    then

  if $\gamma_{n,\bar{n}}^* = \gamma_{ny}^1 + \gamma_{y\bar{n}}^2 - 1$

$$\text{then } A := \lambda_{yy,ny}^1 + \lambda_{yn,y\bar{n}}^2 - 1$$

$$\text{else } A := \infty$$

  if $\gamma_{n\bar{n}}^* = \gamma_{nn}^1 + \gamma_{\bar{n}\bar{n}}^2$

$$\text{then } B := \lambda_{yy,nn}^1 + \lambda_{yn,\overline{nn}}^2$$

$$\text{else } B := \infty$$

(28)

if $\quad \gamma^*_{n\bar{n}} = \gamma^1_{n\bar{n}} + \gamma^2_{n\bar{n}}$

$$\text{then } C := \lambda^1_{yy,n\bar{n}} + \lambda^2_{yn,n\bar{n}}$$

$$\text{else } C := \infty$$

If $\gamma^*_{yn} = \gamma^1_{yn} + \gamma^2_{\bar{n}n}$

    then

if $\quad \gamma^*_{n\bar{n}} = \gamma^1_{ny} + \gamma^2_{y\bar{n}} - 1$

$$\text{then } D := \lambda^1_{yn,ny} + \lambda^2_{\bar{n}n,y\bar{n}}$$

$$\text{else } D := \infty$$

if $\quad \gamma^*_{n\bar{n}} = \gamma^1_{nn} + \gamma^2_{\bar{n}n}$

$$\text{then } E := \lambda^1_{yn,ny} + \lambda^2_{\bar{n}n,\overline{nn}}$$

$$\text{else } E := \infty$$

if $\quad \gamma^*_{n\bar{n}} = \gamma^1_{n\bar{n}} + \gamma^2_{n\bar{n}}$

$$\text{then } F := \lambda^1_{yn,n\bar{n}} + \lambda^2_{\bar{n}n,n\bar{n}}$$

$$\text{else } F := \infty$$

If $\gamma^*_{yn} = \gamma^1_{y\bar{n}} + \gamma^2_{nn}$

    then

if $\quad \gamma^*_{n\bar{n}} = \gamma^1_{ny} + \gamma^2_{y\bar{n}} - 1$

$$\text{then } G := \gamma^1_{y\bar{n},ny} + \lambda^2_{nn,y\bar{n}}$$

$$\text{else } G := \infty$$

if $\quad \gamma^*_{n\bar{n}} = \gamma^1_{nn} + \gamma^2_{\bar{n}n}$

$$\text{then } H := \lambda^1_{y\bar{n},nn} + \lambda^2_{nn,\overline{nn}}$$

$$\text{else } H := \infty$$

if $\quad \gamma^*_{n\bar{n}} = \gamma^1_{n\bar{n}} + \gamma^2_{n\bar{n}}$

$$\text{then } I := \lambda^1_{y\bar{n},n\bar{n}} + \lambda^2_{nn,n\bar{n}}$$

$$\text{else } I := \infty$$

$$\lambda^*_{yn,n\bar{n}} = \min\{A, B, C, D, E, F, G, H, I\} \tag{29}$$

The Series 2 recurrence for $\lambda^*_{yn,n\bar{n}}$ which follows is very similar to the recurrence for Series 1. The main difference is that here there are four possibilities to be considered for the γ_{yn}-set D_1 and two possibilities for the $\gamma_{n\bar{n}}$-set D_2 (see equations (11) and (15) of Table 3.2), for a total of eight resulting combinations.

If $\gamma^*_{yn} + \gamma^2_{\bar{n}y}$

    then

  if $\gamma^*_{n\bar{n}} = \gamma^1_{n\bar{n}} + \gamma^2_{\bar{n}y}$

        then $A := \lambda^1_{yn,n\bar{n}} + \lambda^2_{\bar{n}y,\bar{n}y}$

        else $A := \infty$

  if $\gamma^*_{n\bar{n}} = \gamma^1_{n\bar{n}} + \gamma^2_{\bar{n}n}$

        then $B := \lambda^1_{yn,n\bar{n}} + \lambda^2_{\bar{n}y,\bar{n}n}$

        else $B := \infty$

If $\gamma^*_{yn} = \gamma^1_{y\bar{n}} + \gamma^2_{ny}$

    then

  if $\gamma^*_{n\bar{n}} = \gamma^1_{n\bar{n}} + \gamma^2_{\bar{n}y}$

        then $C := \lambda^1_{y\bar{n},n\bar{n}} + \lambda^2_{ny,\bar{n}y}$

        else $C := \infty$

  if $\gamma^*_{nn} = \gamma^1_{n\bar{n}} + \gamma^2_{\bar{n}n}$

        then $D := \lambda^1_{y\bar{n},n\bar{n}} + \lambda^2_{ny,\bar{n}n}$

        else $D := \infty$

If $\gamma^*_{yn} = \gamma^1_{yn} + \gamma^2_{\bar{n}n}$

    then

  if $\gamma^*_{n\bar{n}} = \gamma^1_{n\bar{n}} + \gamma^2_{\bar{n}y}$

$$\text{then } E := \lambda^1_{yn,n\bar{n}} + \lambda^2_{\bar{n}n,\bar{n}y}$$

$$\text{else } E := \infty$$

if $\quad \gamma^*_{n\bar{n}} = \gamma^1_{n\bar{n}} + \gamma^2_{\bar{n}n}$

$$\text{then } F := \lambda^1_{yn,n\bar{n}} + \lambda^2_{\bar{n}n,\bar{n}n}$$

$$\text{else } F := \infty$$

If $\gamma^*_{yn} = \gamma^1_{y\bar{n}} + \gamma^2_{nn}$

    then

if $\quad \gamma^*_{n\bar{n}} = \gamma^1_{n\bar{n}} + \gamma^2_{\bar{n}y}$

$$\text{then } G := \lambda^1_{y\bar{n},n\bar{n}} + \lambda^2_{nn,\bar{y}}$$

$$\text{else } G := \infty$$

if $\quad \gamma^*_{n\bar{n}} = \gamma^1_{n\bar{n}} + \gamma^2_{\bar{n}n}$

$$\text{then } H := \lambda^1_{y\bar{n},n\bar{n}} + \lambda^2_{nn,\bar{n}n}$$

$$\text{else } H := \infty$$

$$\lambda_{yn,n\bar{n}} = \min\{A,B,C,D,E,F,G,H\}.$$

Finally, we present the Parallel recurrence for $\lambda^*_{yn,n\bar{n}}$. Here there are two possibilities to be considered for each of the γ_{yn}-set D_1 and the $\gamma_{n\bar{n}}$-set D_2 of G^*, resulting in four combinations

If $\gamma^*_{yn} = \gamma^1_{yn} + \gamma^2_{\overline{yn}}$

    then

if $\quad \gamma^*_{n\bar{n}} = \gamma^1_{n\bar{n}} + \gamma^2_{\overline{nn}}$

$$\text{then } A := \lambda^1_{yn,n\bar{n}} + \lambda^2_{\overline{yn},\overline{nn}}$$

$$\text{else } A := \infty$$

if $\quad \gamma^*_{n\bar{n}} = \gamma^1_{\overline{nn}} + \gamma^2_{n\bar{n}}$

$$\text{then } B := \lambda^1_{yn,\overline{nn}} + \lambda^2_{\overline{yn},n\bar{n}}$$

$$\text{else } B := \infty$$

If $\gamma^{*}_{yn} = \gamma^{1}_{y\overline{n}} + \gamma^{2}_{yn} - 1$

   then

  if $\gamma^{*}_{n\overline{n}} = \gamma^{1}_{n\overline{n}} + \gamma^{2}_{\overline{nn}}$

      then $C := \lambda^{1}_{y\overline{n},n\overline{n}} + \lambda^{2}_{yn,\overline{nn}}$

      else $C := \infty$

  if $\gamma^{*}_{n\overline{n}} = \gamma^{1}_{\overline{nn}} + \gamma^{2}_{n\overline{n}}$

      then $D := \lambda^{1}_{y\overline{n},\overline{nn}} + \lambda^{2}_{yn,n\overline{n}}$

      else $D := \infty$

$\lambda_{yn,n\overline{n}} = \min\{A, B, C, D\}$

We note here that in determining $M_{\gamma}(G)$ for the final series-parallel graph G we must disregard any of the λ-parameters that contain an $\overline{n}$ as one of the subscripts because these parameters consider sets which do not necessarily dominate the entire graph G. As will be described in Section 5, the remaining λ-parameters will be considered in determining $M_{\gamma}(G)$ only if the corresponding sets D_1 and D_2 are *minimum* dominating sets of G.

5. A Description of the Algorithm

In Sections 3 and 4 we have indicated 27 γ-recurrences and 135 λ-recurrences. However, not all of these 162 recurrences will be calculated at each node of the parse tree. Because parse tree nodes represent a specific type of series-parallel composition, we merely choose the appropriate set of nine γ-parameters and 45 λ-parameters. This is still a total of 54 recurrences that are associated with each node of the parse tree; so, even though our algorithm is linear, the coefficient is quite large.

A description of our algorithm for $M_{\gamma}(G)$ appears below. Note that the γ- and the λ-parameters are *both* determined in a single pass through the parse tree. Also note that the only λ-parameters that can be used in the final determination of $M_{\gamma}(G)$ are those for which there is an associated *minimum* dominating set of G.

6.   Algorithm Minintersection

Step 1   Determine a parse tree T for the series-parallel graph G given as input. As previously noted, we "identify" each node of T with the series-parallel graph it represents. Thus, the root node corresponds to G.

Step 2   Initialize the nine γ-parameters and 45 λ-parameters for each leaf of the parse tree (which represents the series-parallel basis graph K_2) as described in Sections 3 and 4.

Step 3   Working from the leaves up to the root, at each node in the parse tree note the type of series-parallel composition associated with it, and calculate the appropriate set of nine γ-parameters and the appropriate set of 45 λ-parameters.

Step 4   Determine $M_\gamma(G)$ as follows:

$$\gamma(G) = \min\ \{\ \gamma_{yy}(G), \gamma_{yn}(G), \gamma_{ny}(G), \gamma_{nn}(G)\ \}$$

(* Initialize A=B=C=D=E=F=G=H=I=J= ∞ *)

If $\gamma(G) = \gamma_{yy}(G)$

    then

  if $\gamma(G) = \gamma_{yn}(G)$

           then $A := \lambda_{yy,yn}(G)$

           else $A := \infty$

  if $\gamma(G) = \gamma_{ny}(G)$

           then $B := \lambda_{yy,ny}(G)$

           else $B := \infty$

  if $\gamma(G) = \gamma_{nn}(G)$

           then $C := \lambda_{yy,nn}(G)$

           else $C := \infty$

D: $= \lambda_{yy,yy}(G)$

else $D := \infty$

If $\gamma(G) = \gamma_{yn}(G)$

    then

   if $\gamma(G) = \gamma_{ny}(G)$

                then $E := \lambda_{yn,ny}(G)$

                else $E := \infty$

   if $\gamma(G) = \gamma_{nn}(G)$

                then $F := \lambda_{yn,nn}(G)$

                else $F := \infty$

  $G := \lambda_{yn,yn}(G)$

else $G := \infty$

If $\gamma(G) = \gamma_{ny}(G)$

    then

   if $\gamma(G) = \gamma_{nn}(G)$ then $H := \lambda_{ny,nn}(G)$

  $I := \lambda_{ny,ny}(G)$

    else $I := \infty$

If $\gamma(G) = \gamma_{nn}(G)$ then $J := \lambda_{nn,nn}(G)$

              else $J := \infty$

$M_\gamma(G) = \min\{\ A, B, C, D, E, F, G, H, I, J\ \}$

Remarks

Much research is being done in the developing theory of polynomial/linear algorithms for graph theoretic problems. Some quite recent work is concerned with predicting the nature of problems for which there will exist linear time algorithms on recursive families of graphs. Such work includes that of Bern, Lawler and Wong [2], Bodlaender [4], Borie, Parker and Tovey [5], Mahajan and Peters [12], and Seese [14].

We emphasize that the multiset problem discussed here is a new type of problem. In fact, Gary Parker in a personal correspondence, informed us that our minimum intersection of dominating sets problem can apparently not be formulated in such a way that the results in [5] can apply. Also, in [7] we showed that the Disjoint Minimum Dominating Sets (DMDS) problem of deciding if a graph has two disjoint minimum dominating sets is NP-hard, and the problem is in NP^{NP}. However, DMDS appears not to be in NP because verification of a proposed solution {D1, D2} of minimum dominating sets includes verifying that each of D1 and D2 is, in fact, *minimum*.

REFERENCES

[1] S. Arnborg and A. Proskurowski, "Linear-time algorithms for NP-hard problems restricted to partial k-trees", TRITA-NA-8404, The Royal Institute of Technology, Sweden, 1984.

[2] M. W. Bern, E. L. Lawler, and A. L. Wong, "Why certain subgraph computations require only linear time", 26th Annual Symposium on the Foundations of Computer Science, Portland, OR, 1985, 117-125.

[3] T. Beyer, A. Proskurowski, S. Hedetniemi, and S. Mitchell, "Independent domination in trees", Proc. 8th S. E. Conference on Combinatorics, Graph Theory and Computing, *Utilitas Mathematica*, Winnipeg, 1977, 321-328.

[4] H. L. Bodlaender, "Polynomial algorithms for chromatic index and graph isomorphism on partial k-trees,", Tech. Rept. RUU-cs-87-17, Dept. of Computer Science, Univ. of Utrecht, Netherlands, October 1987.

[5] R. B. Borie, R. Gary Parker and Craig A. Tovey, "Automatic generation of linear algorithms from predicate calculus descriptions of problems on recursively constructed graph families", preprint, Georgia Institute of Technology, July 1988.

[6] E. J. Cockayne, S. Goodman, and S. T. Hedetniemi, "A linear algorithm for the domination number of a tree", *Information Processing Letters 4*, 1975, 41-44.

[7] D. L. Grinstead and P. J. Slater, "On minimum dominating sets with minimum intersection", *Annals of Discrete Math.*, to appear.

[8] D. L. Grinstead, "A multiset single property problem: maximum independent sets with minimum intersection", in preparation.

[9] D. L. Grinstead, "Algorithmic Templates and Multiset Problems in Graphs", Ph.D. thesis, in preparation.

[10] S. T. Hedetniemi and R. Laskar, "A bibliography of papers on domination in graphs", personal communication.

[11] T. Kikuno, N. Yoshida, and Y. Kakuda, "A linear algorithm for the domination number of a series-parallel graph", *Discrete Appl. Math.* *5*, 1983, 299-311.

[12] S. Mahajan and J. G. Peters, "Algorithms for regular properties in recursive graphs", Twenty-fifth Annual Allerton Conf. on Communications, *Control, and Computing*, 1987, 14-23.

[13] J. Pfaff, R. Laskar, and S. T. Hedetniemi, "Linear algorithms for independent domination and total domination in series-parallel graphs", *Congressus Numerantium 45*, 1985, 71-82.

[14] D. Seese, "Tree-partite graphs and the complexity of algorithms", Tech. Rept. *P- MATH-08-86*, Akademie der Wissenschaften der DDR, Karl-Weierstrass-Institut fur Mathematik, Berlin, 1986.

[15] W. J. Selig and P. J. Slater, "Minimum dominating, optimally independent vertex sets in graphs", these proceedings.

[16] J. Valdes, "Parsing flowcharts and series-parallel graphs", Tech. Rept. STAN-CS-78- 682, Computer Science Dept., Stanford Univ., 1978.

[17] J. Valdes, R. E. Tarjan and E. L. Lawler, "The recognition of series-parallel digraphs", *SIAM J. Comput.*, 11 (2), 1982, 298-313.

[18] T. V. Wimer, "Linear Algorithms on k-terminal graphs", Ph.D. dissertation, Computer Science Department, Clemson University, 1987.

To Chakravarity
my friend.

J. Olani
4/15/93

Graph Theory, Combinatorics, and Applications

Volume 2

PROCEEDINGS OF THE SIXTH QUADRENNIAL
INTERNATIONAL CONFERENCE ON THE THEORY
AND APPLICATIONS OF GRAPHS

Western Michigan University

Edited by

Y. Alavi
G. Chartrand
O.R. Oellermann
A.J. Schwenk

A Wiley-Interscience Publication
JOHN WILEY & SONS, INC.
New York / Chichester / Brisbane / Toronto / Singapore

Copyright © 1991 by John Wiley & Sons, Inc.

Library of Congress Cataloging in Publication Data:

International Conference on the Theory and Applications of Graphs (6th
 : 1988 : Western Michigan University)

Graph theory, combinatorics, and applications : proceedings of the
Sixth Quadrennial International Conference on the Theory and
Applications of Graphs / edited by Y. Alavi . . . [et al.].
 p.   cm.
"A Wiley-Interscience publication."
Includes bibliographical references.

ISBN 0-471-53245-2 (set). —ISBN 0-471-60917-X (v. 1)
               —ISBN 0-471-53219 (v. 2)
 1. Graph theory—Congresses. 2. Combinatorial analysis—
Congresses. I. Alavi, Y. II. Title.
QA166.I55   1988
511′.5—dc20   90-28082
                                               CIP

Printed in the United States of America

10 9 8 7 6 5 4 3 2 1

Contents

Volume 2

*G. Gunther and B. Hartnell**
On m-Connected and k-Neighbour-Connected Graphs 585

R.K. Guy
Graphs and the Strong Law of Small Numbers 597

Y. Hamidoune
Combinatorial Additive Number Theory and Cayley Graphs 615

F. Harary
Recent Results and Unsolved Problems on Hypercube Theory 621

*F. Harary and M. Lewinter**
Spanning Subgraphs of a Hypercube VI: Survey and Unsolved Problems 633

H. Harborth and I. Mengersen*
The Ramsey Number of $K_{3,3}$ 639

M. Henning, H.C. Swart, and P.A. Winter*
On Graphs with (I,n)-Regular Induced Subgraphs 645

*D. Holton and M.D. Plummer**
Matching Extension and Connectivity in Graphs II 651

D.F. Hsu
Orthomorphisms and Near Orthomorphisms 667

S. Huang
Cartan Matrices and Strong Perfect Graph Conjecture 681

B.W. Jackson
Shortness Parameters of r-Regular r-Connected Graphs 687

M.S. Jacobson, F.R. McMorris, and H.M. Mulder
An Introduction to Tolerance Intersection Graphs 705

M. Johnson
Graph Transforms: A Formalism for Modeling Chemical Reaction Pathways 725

---

* Indicates the speaker at the Conference.

*H.A. Kierstead and W.T. Trotter**
A Note on Removable Pairs 739

F. Lazebnik
On the Number of Maximal Independent Sets in Some (v,e)-Graphs 743

S.-M. Lee and A.-Y. Wang*
On Critical and Cocritical Diameter Edge-invariant Graphs 753

J. Lehel
Facts and Quests on Degree Irregular Assignments 765

L. Lesniak
Neighborhood Unions and Graphical Properties 783

H. Levinson
An Algorithm to Decide if a Cayley Diagram Is Planar 801

W. Mader
Critically n-Connected Digraphs 811

B. Manvel and R. Osborne
Walks in Covering Spaces 831

T.A. McKee
Chordal and Interval Multigraphs 841

K.A. McKeon
The Expected Number of Symmetries in Locally-Restricted Trees I 849

Z. Miller
Multidimensional Bandwidth in Random Graphs 861

B. Mohar
The Laplacian Spectrum of Graphs 871

C.L. Monma and S. Suri*
**Partitioning Points and Graphs to Minimize the Maximum
or the Sum of Diameters** 880

J.W. Moon
On the Number of Well-Covered Trees 913

L.S. Moss
The Universal Graphs of Fixed Finite Diameter 923

C.M. Mynhardt
**On the Difference Between the Domination and Independent
Domination Numbers of Cubic Graphs** 939

W. Myrvold
A Report on the Ally Reconstruction Problem 949

E.M. Palmer
Random Superposition: Multigraphs 957

*T.D. Parsons and T. Pisanski**
Graphical Designs   971

B.L. Piazza, R. Ringeisen, and S.K. Stueckle*
Properties of Non-Minimum Crossings for Some Classes of Graphs   975

D.L. Powers
Partially Distance-Regular Graphs   991

M. Randić
On Enumeration of Complete Matchings in Hexagonal Lattices   1001

R.C. Read and G.F. Royle*
Chromatic Roots of Families of Graphs   1009

F.S. Roberts
**From Garbage to Rainbows: Generalizations of Graph Coloring
and Their Applications**   1031

S. Schuster
Edge Dominating Numbers of Complementary Graphs   1053

*W.J. Selig and P.J. Slater**
Minimum Dominating, Optimally Independent Vertex Sets in Graphs   1061

F. Shahrokhi
Duality Theorems for the Maximum Concurrent Flow Problem   1075

P.K. Stockmeyer
**Who is the Best Doubles Tennis Player? An Introduction
to k-Tournaments**   1083

C. Thomassen
Recent Results on Graph Embeddings   1093

T.W. Tucker
Symmetric Embeddings of Cayley Graphs in Nonorientable Surfaces   1105

P.D. Vestergaard
**The Number of Isomorphism Classes of Spanning Unicyclic
Subgraphs of a Graph**   1121

J.J. Watkins and R.J. Wilson*
A Survey of Snarks   1129

H.S. Wilf
**Graphical Combinatorial Families and Unique Representations
of Integers**   1145

R.J. Wilson
Number Theory for Graphs   1151

P.A. Winter
The Destructibility Number of a Vertex   1161

ON M-CONNECTED AND K-NEIGHBOUR-CONNECTED GRAPHS

G. Gunther

Sir Wilfred Grenfell College

B. L. Hartnell

Saint Mary's University

ABSTRACT

A graph is k-neighbour-connected (k-NC) if the removal of fewer than k vertices and all their neighbours does not disconnect the graph nor result in a complete graph. We shall examine briefly a number of problems dealing with k-neighbour-connected graphs and their relationship with the usual notion of connectivity. For instance, in an attempt to examine those graphs which, in some sense, are "between" k-NC and m-connected, we initiate an investigation into graphs which are still k-NC after the removal of any set of m vertices.

1.   Introduction

The problems we look at in this paper originated out of a study of dominating sets. There, one wishes to look at properties of sets of vertices that dominate the entire graph. Our concern, however, has been to look for graphs that are in some sense difficult to dominate. With this end in mind, we have concentrated our efforts on studying properties of the set of vertices that remain undominated; that is, we have looked at the survivor set of vertices left remaining in a graph G after a set of closed neighbourhoods has been deleted from G.

The details of some of this work can be found in [5] - [14], as well as in [16]. Clearly, these types of questions are also relevant in the context of network reliability. See for example [1], [2], [3], [4] and [15].

One question we have pursued with some vigour in the last few years is concerned with determining those graphs which remain connected after the deletion of a number of closed neighbourhoods. We define a graph G to be *k-neighbour-connected*, or *k-NC*, if for any $X \subseteq V(G)$ with $|X| \leq k - 1$, the graph $G - N[X]$ is a connected graph which is not complete. (**Note:**

$N[X]$ denotes the closed neighbourhood of X, and is defined by $N[X] = \cup N[x]$.) The possibility of a complete graph is excluded as

$$x \in X$$

the deletion of any further closed neighbourhood would completely destroy the graph.

It is not difficult to see that if G is k-NC, then G must be k-connected (in the standard sense of connectivity). See Corollary 1 of [8] for this result. It is also at once apparent that the converse of this is false - one needs only to consider any complete graph K_n, which is very highly connected, but which is not even 1-NC. Indeed, the property of being k-NC is very sensitive with respect to the number of edges present in the graph. As the example of the complete graph above shows, if there are too many edges present in the graph, the neighbour-connectivity actually decreases. It is unfortunately not at all clear what constitutes "too many" edges. The complicated relationship that holds between neighbour-connectivity and the number of edges present is manifested by the sequence of diagrams in Figure 1.

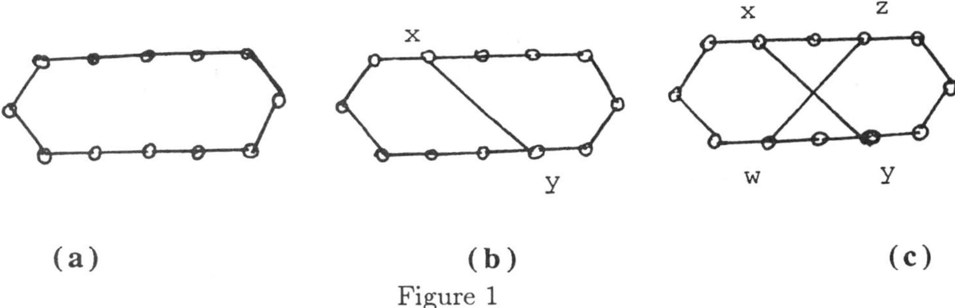

(a)   **(b)**   **(c)**

Figure 1

Diagram (a) shows C_{12}, which is 2-NC. In diagram (b), we have added edge xy to C_{12} - this graph is no longer 2-NC, for deleting $N[x]$ (i.e. x and its neighbours) disconnects the survivor graph. If, however, we add the additional edge zw (illustrated in (c)), then we have again re-established 2-neighbour-connectivity in the resultant graph.

2.   Some Results and Open Problems

A number of families of k-NC graphs are described in [5], [6], [8] and [9]. All of these are k-regular in addition to being k-NC. It turns out that one parameter which governs the types of minimal k-regular, k-NC graphs G that are known is the size m of maximum cliques that one can find in G. In particular, in [9] one finds a classification of all minimal k-regular, k-NC graphs which contain k- cliques. These turn out to be of order $k(k+1)$. With these as seed graphs, the following construction yields new examples of k-NC graphs.

Lemma 1   Let G be a k-regular, k-NC graph which contains a k-clique $C = c_1, \cdots, c_k\}$. Form a new graph G' from G by adding an m-clique $D = \{d_1, d_2, \cdots, d_m\}$ where each d_i is joined to every $c_j \in C$ but to no other vertex in G. That is, C has been replaced by a $(k+m)$-clique. Then the new graph G' obtained from G in this way is k-NC.

Proof   Suppose that we delete the $k-1$ closed neighbourhoods $N[x_1]$, $N[x_2], \cdots, N[x_{k-1}]$ with all $x_i \in G'$. If x_i is a vertex of C, then deletion of

$N[x_i]$ deletes all the new vertices in D, thus leaving the same survivor set in G' as would have been left in G. If all the vertices x_i belong to $G - C$, one must worry about the possibility that all of C might be conceivably be contained within $\cup N[x_i]$. If this were the case, then deleting all the $N[x_i]$ would disconnect D from the remainder of the survivor graph. However, Lemma 3.3 of [9] assures us that $|N[x] \cap V(C)| \leq 1$ and hence this particular situation cannot occur. A second potential problem might arise in the event where one of the vertices x_i belongs to D. In this case, deleting $N[x_i]$ will delete all of C, but will not affect any other vertex in the original vertex set of G. The only way we were able, in G, to delete all of C was by choosing some $c_j \in V(C)$ and deleting $N[c_j]$. Such a deletion, however, always removed exactly one vertex $z \in G - C$. The difference now is that we can delete C (from $d_i \in V(D)$) without touching vertex z. A possible concern is that there is some choice of the $k - 1$ vertices $x_1, \cdots, x_{k-1}$ which results in the deletion of all k vertices in $N(z)$ without touching z itself. But for any $y \notin N[z]$, we have $|N(y) \cap N(z)| \leq 1$, for otherwise we could find some $s, t \in N(y) \cap N(z)$ so that $G[\{z, s, y, t\}]$ would be either a 4-cycle or a 4-cycle with one diagonal, both of which are impossible configurations in a k-regular, k-NC graph (Lemma 3.2 in [9]). This completes the proof.

We describe without proof, a second construction which yields new examples of k-NC graphs.

Lemma 2   Let G be a k-NC graph. Choose a vertex $v \in G$, and adjoin to G a new vertex v' which is joined to exactly all vertices in $N[v]$. Then the new graph G' for which $V(G') = V(G) \cup \{v'\}$ is also k-NC.

These constructions allow us to conclude the following result.

Proposition 1   For all positive integers k and all integers $p \geq k(k + 1)$, there exist k-NC graphs on p vertices.

Proof   By [9], there exists a k-regular, k-NC graph on $k(k + 1)$ vertices which has a k-clique. By Lemma 1, the result follows.

In an attempt to encourage investigation that would lead to a better understanding of the relation that holds between the concepts of k-connectivity and k-neighbour-connectivity, we pose three problems.

First, observe that with traditional k-connectivity, the allowable survivor set of vertices after the deletion of $k-1$ or fewer vertices is a connected graph which is not a single point (considered trivial since removing one more point destroys the graph), whereas with k-neighbour-connectivity, the allowable survivor set is a connected graph which is not a clique (again considered trivial since removing any further neighbourhood completely annihilates the graph).

Problem 1. Call a graph G, a *weak-k-neighbour-connected* graph if the deletion of fewer than k closed neighbourhoods neither disconnects G nor leaves only a single vertex. For each k, determine the minimum number, p, of vertices required to have a weak-k-neighbour- connected graph on p vertices.

We observe that the minimum weak-2-NC graph is the 5-cycle (not the 6-cycle as it was for 2-NC). In general the numbers achieved by k-NC graphs give an upper bound for weak-k-NC graphs. By Proposition 1 this shows $k^2 + k$ to be achievable. The construction given in the following proposition improves this bound.

Proposition 2   For all positive integers k there exists a weak-k-NC graph on $k^2 + 1$ vertices.

Proof   For any positive integer k consider a k-regular graph G on $k^2 + 1$ vertices where G consists of k cliques, say $C_1, C_2, \cdots, C_k$, each of size k and one additional point, say x. There is exactly one edge between each pair of cliques C_i and C_j and between each clique and x. As G is k-regular each vertex in a clique C_i has precisely one neighbour not in that clique.

To see that G is a weak-k-NC graph we first observe that since G is k-regular the deletion of fewer than k closed neighbourhoods leaves more than a single point.

Furthermore we note that if u and v are any two vertices which remain after such a deletion they must still be connected. This follows since it is impossible to destroy the link between any two cliques without entirely destroying one of the cliques. This is also true for the edge between x and any of the cliques (x or the clique itself must be removed else the edge remains).

Figure 2 illustrates the preceding construction for $k = 2$ (the 5-cycle) and $k = 3$. In these cases it is straightforward to verify that these are in fact the minimum order graphs for weak-k-NC. We conjecture that the bound $k^2 + 1$ is sharp for all k.

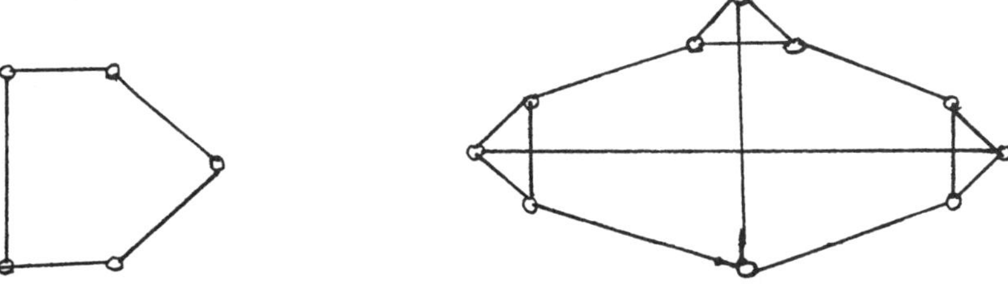

Figure 2

We would like to comment briefly on another subtlety that distinguishes neighbour-connectivity from regular connectivity. In dealing with k-connectivity, one worries about the effects that the deletion of a number of vertices might have. It is clearly immaterial in which order these vertices are deleted. When one deletes entire neighbourhoods, however, this question of order becomes relevant. Indeed, in the definition of k-NC, it implicitly follows that all the closed neighbourhoods are to be deleted *simultaneously*. This means, in particular, that two neighbourhoods $N[x]$ and $N[y]$ may be deleted even when x and y are themselves adjacent to each other.

If we weaken this definition, but in a manner different from that of Problem 1, we might wonder about graphs that remain connected after the successive deletions of a number of neighbourhoods. This forces the deleted vertices to be independent. In particular, let us call a graph G *k-I-neighbour- connected* (k-INC) if for any independent set $X \subseteq V(G)$ with $|X| \leq k - 1$, the graph

$G - N[X]$ is connected but not complete.

Problem 2. Determine the minimum number of vertices required in a k-I-neighbour- connected graph. Note that now, of course, we will no longer have the possibility of deleting N[x] and N[y] for adjacent vertices.

Comment Clearly, this distinction between deleting simultaneously or successively is meaningless in the case of 2-INC graphs, since there we only allow the deletion of one closed neighbourhood. The minimum graph here is still C_6, the cycle on six vertices. If we look for 3- INC graphs, i.e., where the neighbourhoods are to be deleted in succession, we quickly realize that our results change. A simple argument shows that the smallest 3-INC graph is the familiar Peterson graph of order 10, which we show in Figure 3.

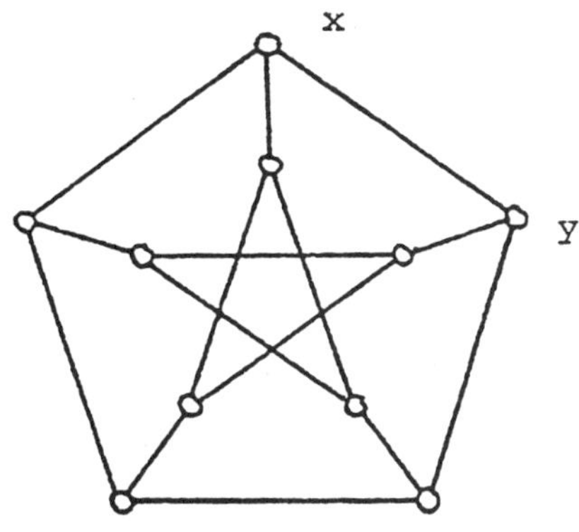

Figure 3

This graph is 1-transitive on its vertices, so we may pick any vertex to begin our deletion. Note that deleting N[x] leaves as a survivor set the 6-cycle, which we know is 2-NC. If, however, we delete N[x] and N[y] simultaneously, we end up with a disconnected survivor set. Thus, this graph provides us with an example of a graph which is not 3-NC (i.e., under simultaneous deletions), but which is 3-INC (i.e., under successive deletions).

Let us try to extend this pattern by searching for a smallest 4-INC graph H which has the property that for any vertex $x \in V(H)$, the survivor graph

H-N[x] is isomorphic to the Petersen graph. That is, we wish to adjoin a new vertex z with $t+3$ neighbours each of which is joined to $t+2$ vertices in the Petersen graph. Each vertex in the Petersen graph has t new edges to the neighbours of z. Hence

$$10t = (t+3)(t+2)$$

$$10t = t^2 + 5t + 6$$

$$t^2 - 5t + 6 = 0$$

$$\therefore t = 2 \text{ or } t = 3$$

For the smallest case we consider $t = 2$.

How should the new edges connect $N(z)$ to the Petersen graph? We illustrate the situation in Figure 4, numbering the vertices from 1 to 16 for convenience.

We ask first whether vertex 2 may be joined to two neighbouring vertices x, y in the Petersen graph?

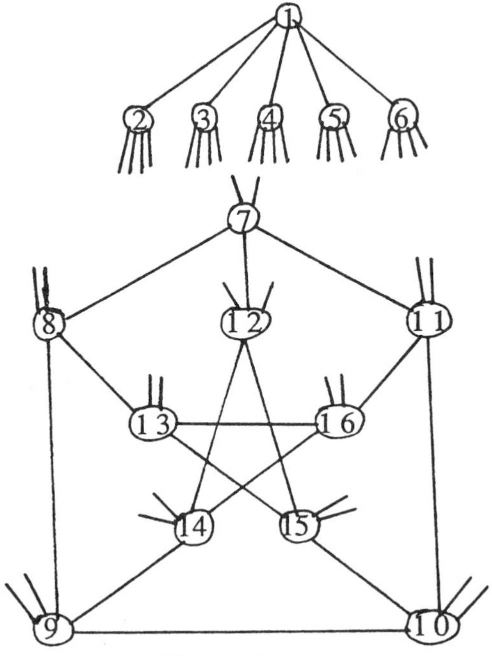

Figure 4

Assume that this happens. Then a simple count shows that there must exist at least two vertices u, v in the Petersen graph, neither of which is connected to 2 or x or y. But then if we delete N[u] and look at the survivor graph H-N[u], we note that this graph would contain the 3-cycle $2 - x - y$, which cannot happen as the Petersen graph does not contain any 3-cycles.

Consequently, the four edges out of vertex 2 must go to four vertices in the Petersen graph which are pairwise not connected. But there exist precisely five such sets of four vertices in the Petersen graph which are pairwise not connected. These are, using the numbering of Figure 4, the five quadruples 7-10-13-14, 7-9-15-16, 9-11-12-13, 8-11-14-15 and 8-10-12-16. Since the numbering of vertices 2,3,4,5,6 is arbitrary, we may without any loss of generality connect these five vertices to the five quadruples listed, so that 2 is connected to each of 7,10,13,14 and so on. It is easy to check that the resultant graph H on 16 vertices is 5-regular and has the property that the deletion of any closed neighbourhood leaves a Petersen graph as the survivor graph. Consequently, H is the smallest 4-INC graph.

We show a symmetric representation of H in Figure 5.

Finally, we observe that, in general, k-NC graphs, which we know exist for $k(k + 1)$ vertices, give us a bound for k-INC graphs.

We conclude by describing a family of graphs in which the deletion of m vertices and k closed neighbourhoods still leaves an allowable survivor set. Again, there are several possibilities for "allowable". Here we will demand that the survivor graph be connected and not complete.

Lemma 3   Let G be k-NC. Let s, t be two natural numbers with $s + t < k$. If we delete s closed neighbourhoods and t additional vertices in G, then the survivor graph remains connected and is not a complete graph.

Proof   Deleting s neighbourhoods results in a graph which is still (k - s)-NC, and hence (k - s)-connected. But, $k - s > t$, and thus we can still delete t vertices without jeopardizing connectivity. Also, since the survivor set we would leave after the deletion of $s+1$ *neighbourhoods* is not a clique, it follows that the survivor set now cannot be a clique either.

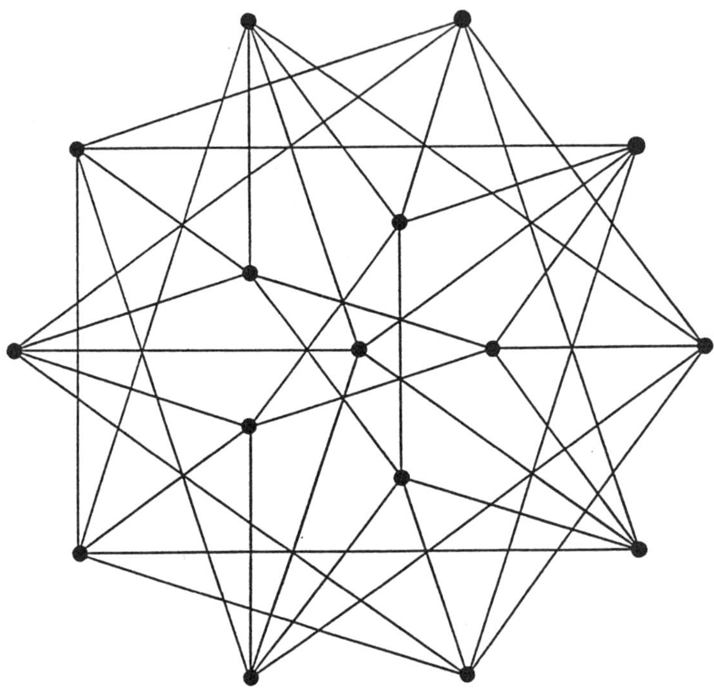

Figure 5

It turns out that we can find smaller graphs which remain connected in the allowable way after the deletion of a mixture of neighbourhoods and vertices. For example, the smallest graph which remains connected (but is not complete) after the deletion of one neighbourhood and one further vertex, is the graph on nine vertices shown in Figure 6.

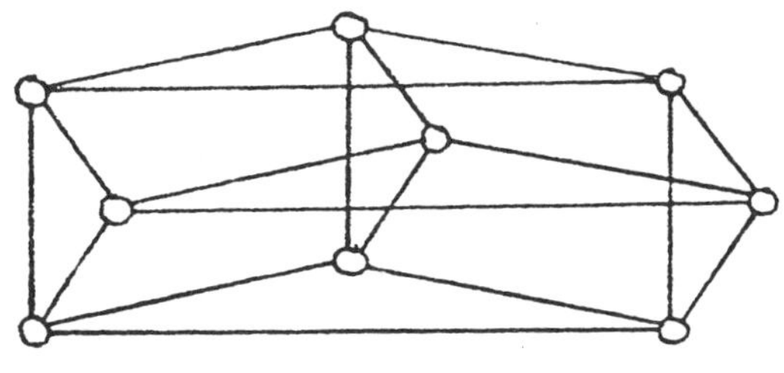

Figure 6

We recognize in this graph the product $K_3 \times K_3$.

This example generalizes nicely, as we note in the following lemma.

Lemma 4   Let $G = K_k \times K_m$. Then the following hold for G.

a) G is a $(k + m - 2)$-regular graph of order km

b) G is $(k + m + 2)$-connected

c) G is p-NC, where $p = \min\{m - 1, k - 1\}$

d)   If in G we delete s neighbourhoods, where $s < p$, then the survivor

graph is still at least $(k + m - 2 - 2s)$-connected.

Proof   Part (a) is immediate and part (b) follows from Menger's Theorem.
Part (c) is a consequence of the fact that if $x \in G$, then $G - N[x] = K_{k-1} \times K_{m-1}$. This result, in combination with (b), is sufficient to force (d) as well.
We conclude with,

Problem 3   Determine the minimum number, p, of vertices such that there
exists a graph on p vertices such that the deletion of no more than k closed
neighbourhoods and m additional vertices leaves a connected graph which is
not a clique.

REFERENCES

[1] K.S. Bagga, L. W. Beineke, M. J. Lipman and R. E. Pippert, "On
the edge-integrity of graphs," *Congressus Numerantium* 60 (1987),
141-144.

[2] F. T. Boesch and A. P. Felzer, "A general class of invulnerable graphs,"
Networks 2 (1972) 261-283.

[3] F. T. Boesch and C. L. Suffel, "An overview of discrete methods in the synthesis of reliable networks," *Large Scale Systems* 7 (1984), no. 2-3, 191-196.

[4] S. Goodman and D. Shier, "On designing a reliable hierarchical structure," *SIAM J. Appl Math.* 32(2) (1977) 418-430.

[5] G. Gunther, "Minimum k-neighbour-connected graphs and affine flags," *Ars Combinatoria* 26(1988) 83-90.

[6] G. Gunther and B. L. Hartnell, "Flags and neighbour-connectivity," *Ars Combinatoria* 24 (1987) 31-38.

[7] G. Gunther, "On the existence of neighbour-connected graphs", *Congressus Numerantium* 54 (1986) 105-110.

[8] G. Gunther, B. L. Hartnell and R. Nowakowski, "Neighbour-connected graphs and projective planes," *Networks* 17 (1987) 241-247.

[9] G. Gunther and B. L. Hartnell, "Neighbour-connectivity in regular graphs," *Disc. Appl. Math.* 77 (1985) 233-243.

[10] G. Gunther and B. L. Hartnell, "Optimal k-secure graphs," *Disc. Appl. Math* 2 (1980) 225-231.

[11] G. Gunther and B. L. Hartnell, "On minimizing the effects of betrayals in a resistance movement," *Proc. Eighth Manitoba Conference on Numerical Mathematics and Computing* (1978) 285-306.

[12] B. L. Hartnell, "The Optimum defence against random subversions in a network," *Proc. Tenth Southeastern Conference on Combinatorics, Graph Theory and Computing*, Boca Raton (1979) 293-299.

[13] B. L. Hartnell, "Some problems on minimum dominating sets," *Proc. Eighth Southeastern Conference on Combinatorics, Graph Theory and Computing*, Baton Rouge (1977) 317- 320.

[14] B. L. Hartnell and R. Nowakowski, "Indivisible graphs," *Congressus Numerantium* 33 (1981) 55-62.

[15] H. Heffes and A. Kumar, "Incorporating dependent node damage in deterministic connectivity analysis and synthesis of networks," *Networks* 16 (1986) 51-65.

[16] A. Meir and J. Moon, "Survival under random coverings of trees," *Graphs and Combinatorics* 4 (1988) 49-65.

GRAPHS AND THE STRONG LAW OF SMALL NUMBERS

Richard K. Guy

ABSTRACT

Graphs furnish many examples of the fickle nature of The Strong Law of Small Numbers [12,14]: "There aren't enough small numbers to carry out the many tasks demanded of them."

It is the enemy of mathematical discovery. On the one hand, a few early exceptions can hide an underlying pattern. On the other hand, spurious similarities can cause careless conjectures. The Second Strong Law of Small Numbers [15] states that: "If two numbers look equal, they aren't necessarily so."

A common form of puzzle, or intelligence test, is to give a few members of a sequence and to ask for the next few members. A mathematician knows that it is possible to make a rule whereby any given finite sequence can be continued with any other given finite sequence of additional terms, so such exercises are not always fair. However, in particular cases, some continuation rules are very much more plausible than others. As an example, give the next three terms of the sequence

(1), 2, 3, 5, 7, 11, 13, 17, 19, 23, 29, 31, 37, 41, 43, 47, 53, ...

But in this paper we give you the rules. An earlier draft had the format of [14]: a long list of examples, followed by a long list of answers. Here, at the referee's suggestion, we give six sets of examples of the appearance of six different sequences in six different sections. The reader is invited to decide, within each set, whether the coincidences are genuine or not. Are there, in fact, many more sequences? At the end of each section we give the answers,

597

insofar as we know them.

1.   Introduction

The first sequence

$$1, \quad 1, \quad 2, \quad 5, \quad 14, \quad \ldots$$

is exemplified by

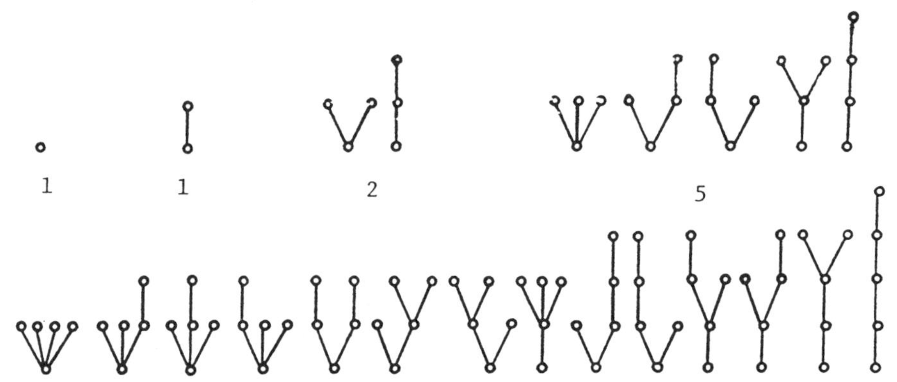

Example 1A   The numbers of rooted plane trees with n edges and $n+1$ vertices $(n = 0, 1, 2, \ldots)$, where by *plane* we mean to distinguish between an embedding of a tree in the plane and its reflextion in a vertical axis by:

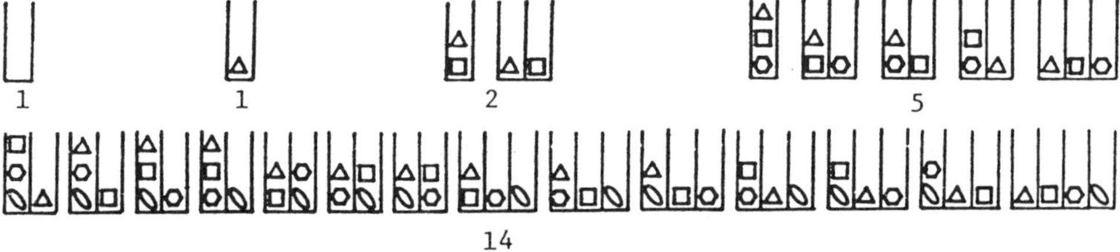

Example 1B   The numbers of distributions of n distinguishable objects in indistinguishable boxes, with at most 3 objects in a box by:

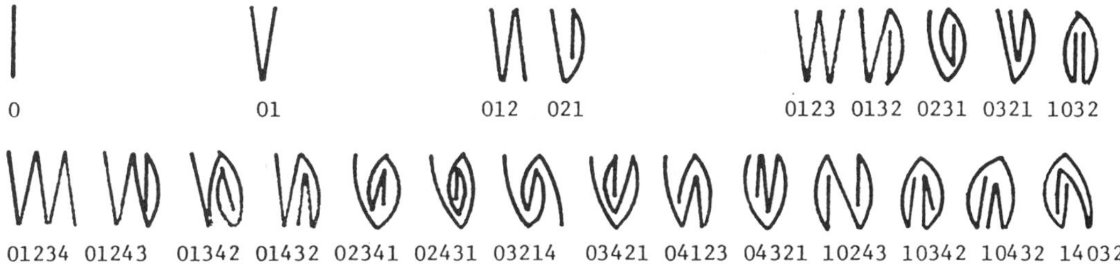

Example 1C   The numbers of ways of making n folds in a strip of $n+1$ postage stamps, where we don't distinguish between the back and front, or the top and bottom, or the left and right of a stamp by:

Example 1D   The numbers of plane binary trees with n decisions and $n+1$ outcomes; again the word *plane* means that we distinguish between left and right, and the dots at the end of the figure indicate that there are seven more trees, the reflextions of the first seven; notice that when n is even and positive, the number of trees will be even: and, in addition by:

Example 1E   The numbers of different groups, up to isomorphism, of order 2^n.

Although the one-one correspondence between them is not quite immediate, the objects in Examples 1A and 1D, the rooted plane trees and the plane binary trees, are both manifestations of the ubiquitous Catalan numbers [2,

13, 37, #577], $\binom{2n}{n}/(n+1), n = 0, 1, 2, \ldots$

$$1, 1, 2, 5, 14, 42, 132, 429, 1430, 4862, 16796, 58786, 208012, 742900, \ldots$$

Kutchiniski [23] relates them by putting each in correspondence with certain sequences of zeros and ones, while Conway and Guy [7] do so via the various ways of parenthesizing a repeated exponentiation.

On the other hand, the sequence defined by Example 1B [27, 37 #579]

$$1, 1, 2, 5, 14, 46, 166, 652, 2780, 12644, 61136, 312676, 1680592, 9467680, \ldots$$

and that in example 1C [22, 37 #576], see also [38],

$$1, 1, 2, 5, 14, 39, 120, 358, 1176, 3527, 11622, 36627, 121622, 389560, \ldots$$

(the sixth member is given erroneously as 38 in [11]) are not connected with the Catalan numbers, nor with one another. Also, the number of groups of order $2^n, n = 0, 1, \ldots$ [37 #581]

$$1, 1, 2, 5, 14, 51, 267, \ldots$$

is again different, and perhaps not known beyond $n = 6$.

The nth term of the sequence in Example 1B is given by $n!\Sigma 1/a.!b!c!2^b6^c$, where the sum is taken over all triples of positive integers satisfying $a+2b+3c = n$.

2. The Second Sequence,

$$1, \quad 1, \quad 1, \quad 2, \quad 3, \quad 6, \quad 11, \quad \cdots$$

manifests itself as

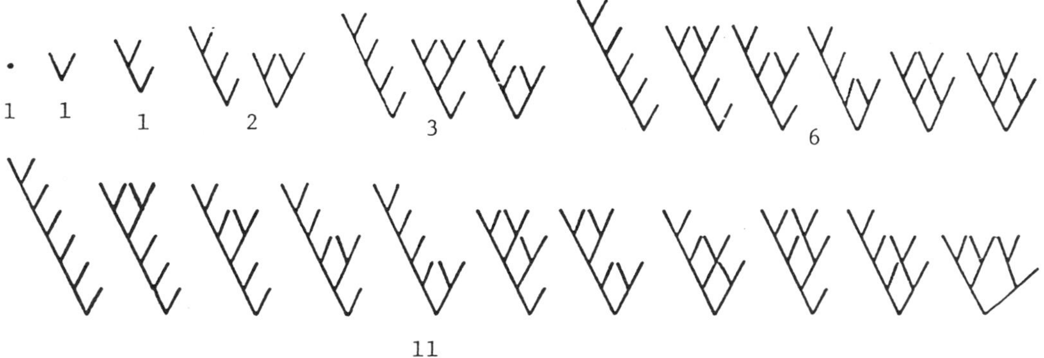

Example 2A   The numbers of binary trees with $0, 1, 2, 3, 4, 5, 6$ branchings, where we no longer distinguish between left and right;

Example (and Figure) 2B   The number of ways of arranging a knock-out tournament, among $1, 2, 3, 4, 5, 6, 7$ players, by numbers of matches in each round: at first sight, this appears to be isomorphic to 2A (read the numbers of branchings at each level, from the top downwards, the winner being the root of the tree):

$$\phi, \quad 1, \quad 11, \quad 111 \text{ or } 21, \quad 1111 \text{ or } 211 \text{ or } 121, \cdots$$

but with 6 players, i.e. 5 matches, we have 11111 or 2111 or 1211 or 1121 or the first three arrangements of Figure 2B, where the first two are the same binary tree, but different arrangements of matches (1 winner in the first round of the first arrangement gets a bye in the second round, while 2 players get byes in the first round of the second arrangement) and the last two are counted as the same arrangement, 221, although they are different binary trees: similarly for the second and third sets of three arrangements in Figure 2B, involving 7 players; the eleven different arrangements for this number of players being 111111, 21111, 12111, 11211, 11121, 2211, 2121, 1221, 321, 1311 and 3111;

Figure 2B. Some arrangements of knock-out tournaments with 6 or 7 players

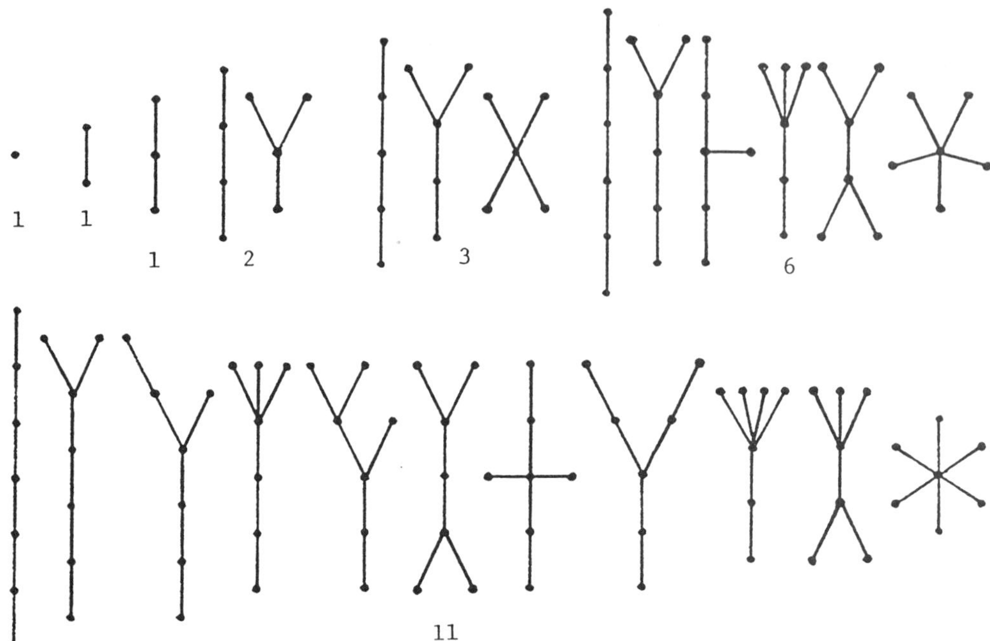

Example 2C   The numbers of trees with 1,2,3,4,5,6,7 vertices, where the trees are now neither rooted nor labelled in any way; and as

Example 2D   The coefficients of $x^n (0 \leq n \leq 6)$ in the expansion of the generating function

$$\frac{1}{2}\Big\{\, 1 + \frac{1 - 3x^2}{(1 - 2x)(1 - x^2)} \,\Big\},$$

which can also be described recursively, after the initial values, by the rule: alternately double, and double and subtract one.

No two of the sequences in section two are connected. The binary trees in Example 2A are a manifestation of the Wedderburn-Etherinton numbers [9,40, 37 #298, 6, p. 54]. They are

$$1, 1, 1, 2, 3, 6, 11, 23, 46, 98, 207, 451, 983, 2179, 4850, 10905, 24631, 56011, \ldots$$

The arrangements, by numbers of matches in the various rounds of a knock-out tournament, could be called Capell-Narayana numbers [3, 37 #297]:

$$1, 1, 1, 2, 3, 6, 11, 22, 42, 84, 165, \ldots$$

and don't appear to have been calculated any further; perhaps because the arrangements are mostly of interest to the organizers of the tournament. What the general public wants to know, in addition, is the order in which the players appear in the arrangement, who is "seeded", who gets a bye, and so on. The numbers of *labelled* knock-out tournaments for 1,2,3,... players is

$$1, 1, 3, 15, 105, 945, 10395, 135135, 2027025, 34459425, 654729075, \ldots$$

These are known as double factorials [37 #1217]: if $m(= n-1)$ is the number of matches, they may be written as $(2m)!/2^m m!$ Even more interesting, of course, is: who is the winner? The number of ways in which he emerges is

$$1, 2, 12, 120, 1680, 30240, 665280, 17297280, 518918400, 17643225600, \ldots,$$

i.e. $(2m!)/m!$ [37, #808]. These numbers appear as coefficients when representing powers in terms of Hermite polynomials [36]. Is there some inversion formula whereby we can recapture the Capell-Narayana numbers from these more familiar sequences?

The number of trees with n vertices, $n = 1, 2, \cdots$ (Example 2C)

$$1, 1, 2, 3, 6, 11, 23, 47, 106, 235, 551, 1301, 3159, 7741, 19320, 48629, 123867, \ldots$$

was first obtained by Otter [30], see also [37 #299; 33 pp. 135-138; 16, p. 232; 26].

The generating function given as Example 2D was suggested by Jonathan Schaer. It is a red herring, swimming in the narrow straits between the Wedderburn-Etherington and Capell-Narayana numbers. The coefficients of the power series are

$$1, 1, 1, 2, 3, 6, 11, 22, 43, 86, 171, 342, 683, 1366, 2731, 5462, 10923, 21846, \ldots$$

and can be simply described after examining them briefly.

3. The Third Sequence

$$1, \quad 1, \quad 2, \quad 4, \quad 9, \cdots$$

appears in

Example 3A   The number of possible score sequences in a round-robin tournament between 1,2,3,4,5 players, there being no ties:

1	0	1
2	10	1
3	210,111	2
4	3210,3111,2220,2211	4
5	43210,43111,42220,42211,33310,33220,33211,32221,22222	9

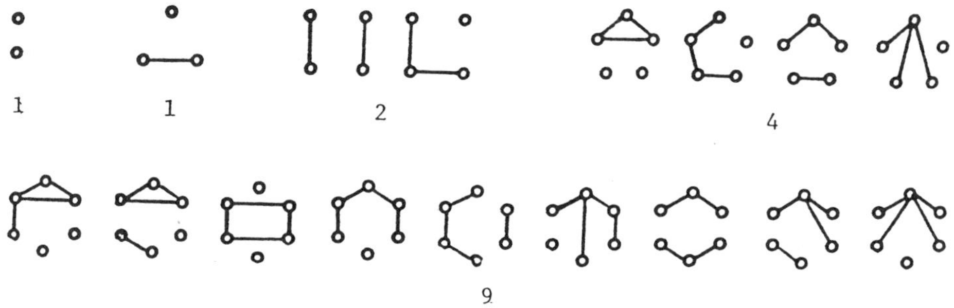

in Example 3B, graphs with $n+1$ vertices and $n-1$ edges, $n = 1,2,3,4,5$:

in Example 3C, the values of k for which $(15 \times 2^k) + 1$ is prime in:

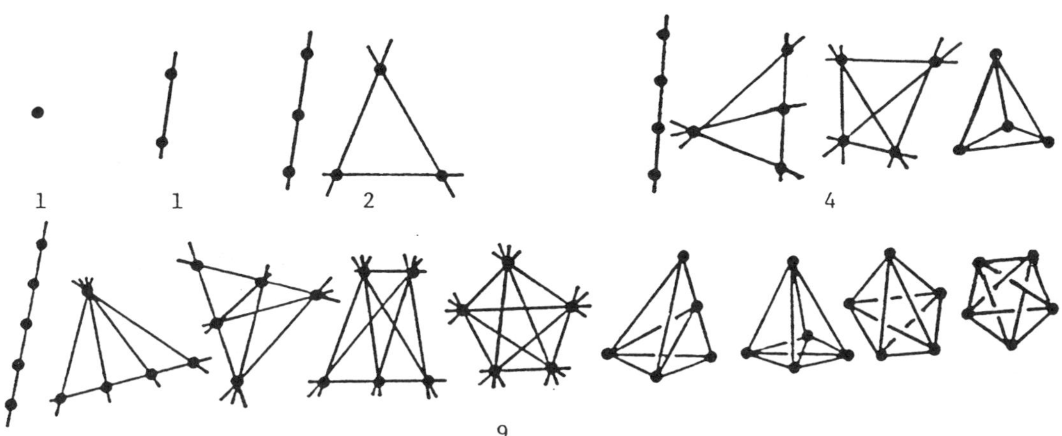

Example 3D   Combinatorial geometries with 1,2,3,4,5 points; notice that the last two diagrams with four points differ in that they have dimensions 2 and 3; the nine diagrams with five points are one of dimension 1, four of dimension 2, three of dimension 3, and one of dimension 4: and finally in

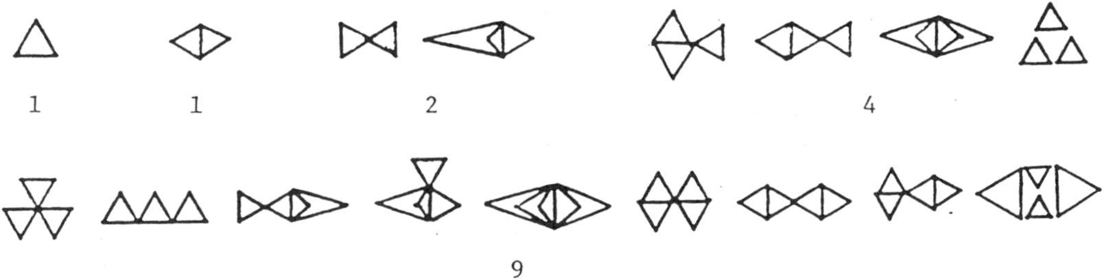

Example 3E   Minimal triangle graphs with $n+2$ vertices, $n = 1, 2, 3, 4, 5$; notice that the last examples with 6 and 7 vertices have only 3 and 4 triangles respectively, not 4 and 6. It may be easier to visualize these as minimal (but not necessarily minimum) sets of triples which cover a set of $n+2$ elements (vertices).

All five of the examples beginning 1,1,2,4,9 are different sequences! The numbers of score sequences in Example 3A are

$$1, 1, 2, 4, 9, 22, 59, 167, 490, 1486, 4639, 14805, 48107, 158808, 531469, 1799659, \ldots$$

and might be called Narayana-Bent numbers [27]; see also [28, p. 66; 37 #459]. The numbers of graphs with $n + 1$ vertices and $n - 1$ edges (necessarily disconnected; Example 3B) are

$$1, 1, 2, 4, 9, 21, 56, 148, 428, 1305, 4191, 14140, 50159, 185987, 720298, 2905512, \ldots$$

and are listed in [33, p. 146; 37 #458]. The expression of $15 \times 2^k + 1$ is prime for $k =$

$$1, 2, 4, 9, 10, 12, 27, 37, 38, 44, 48, 78, 112, 168, 229, 297, 339, \ldots$$

[37 #445]. The numbers of combinatorial geometries with $1, 2, \ldots$ points are

$$1, 2, 4, 9, 26, 101, 950, \ldots$$

[8, 37 #462] and may not have been calculated for more than 7 points. The numbers of minimal triangle graphs with $3, 4, 5, \ldots$ vertices,

$$1, 1, 2, 4, 9, 48, 117, 307, 821, 2277, \ldots$$

[1, 37 #450] also do not appear to be known for more than 12 vertices.

4. The Fourth Sequence

$$1, \quad 1, \quad 2, \quad 3, \quad 5, \quad 7, \quad 11, \quad \ldots$$

features in

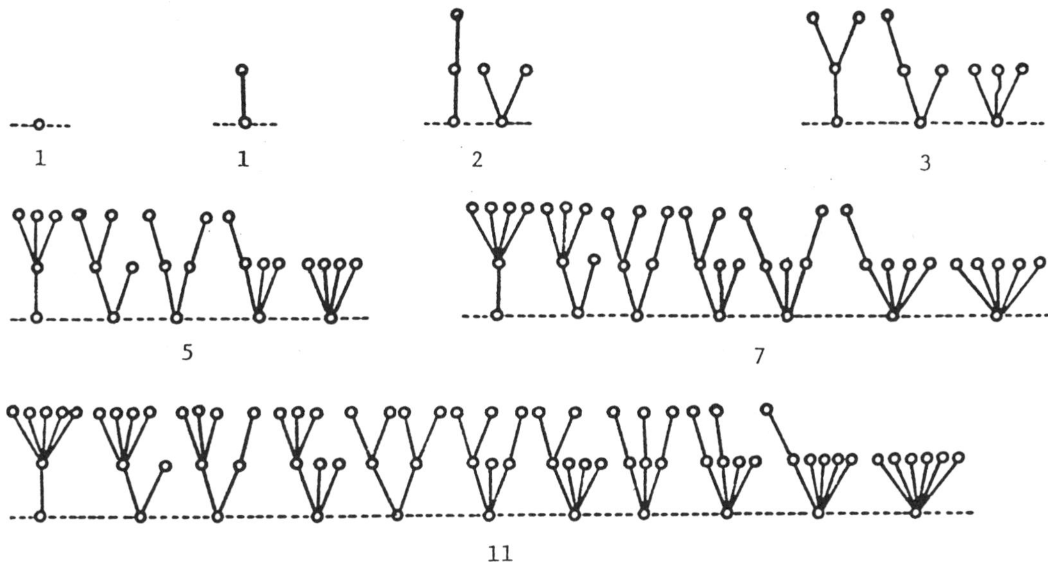

Example 4A   Trees of height at most two, with 0, 1,2,3,4,5,6 edges:

Example 4B   The number of partitions of n (it is conventional to define $p(0) = 1$):

1	1
2,11	2
3,21,111	3
4,31,22,211,1111	5
5,41,32,311,221,2111,11111	7
6,51,42,411,33,321,3111,222,2211,21111,111111	11

Example 4C   (Apart from the initial ones), the primes: and

Example 4D   The integer part of $(3/2)^n$

$n =$	0	1	2	3	4	5	6
$(3/2)^n =$	1	1.5	2.25	3.375	5.0625	7.59375	11.390625

There is a neat correspondence between the trees of height at most two with n edges and the partitions of n (Examples 4A and 4B): after defining

$p(0) = 1$, we simply ask what are the valences of the vertices at height one in each tree:

$$1 \ ; \ 2, 11 \ ; \ 3, 21, 111 \ ; \ 4, 31, 22, 211, 1111 \ ; \ 5, 41, ...$$

Conversely, to construct a tree corresponding to a given partition of n, take as many vertices at height one as there are parts, connect them to a root vertex, and add enough vertices at height two to make the total $n + 1$, and connect them to the height one vertices to make the valences correspond to the sizes of the parts. Of course, the values of the partition function [37 #244]

$$1, 1, 2, 3, 5, 7, 11, 15, 22, 30, 42, 56, 77, 101, 135, 176, 231, 297, 385, 490, ...$$

are rarely prime, after the first few; nor does the sequence of integer parts of powers 1.5 coincide any longer [37 #245]:

$$1, 1, 2, 3, 5, 7, 11, 17, 25, 38, 57, 86, 129, 194, 291, 437, 656, 989, 1477, 2216, 3325, ...$$

5. The Members of the Fifth Sequence

$$1, \quad 2, \quad 4, \quad 6, \quad 10, \quad 14, \quad ...$$

are just one less than those in the fourth. They appear in Example 4A $(n = 0, 1, 2, 3, 4)$ as numbers of trees of height two with $n + 2$ edges; for each value of n, omit the trees of height one; and in Figure 5A $(n = 5)$:

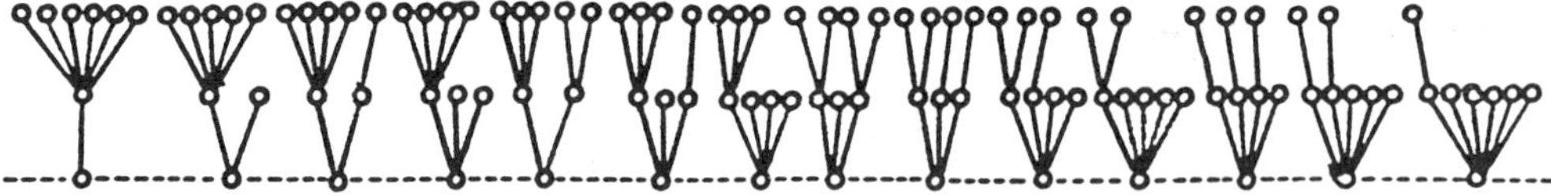

Figure 5A. Fourteen trees of height two, with 7 edges

in Example 5B   As numbers of partitions of $2n$ into parts which are powers of two (define this as 1 for $n = 0$): exponents indicate repetitions of part size:

n = 1	$2,1^2$	2
2	$4,2^2,21^2,1^4$	4
3	$42,41^2, 2^3, 2^21^2, 21^4, 1^6$	6
4	$8,4^2,42^2,421^2,41^4,2^4,2^31^2,2^21^4,21^6,1^8$	10
5	$82,81^2,4^22,4^21^2,42^3,42^21^2,421^4,41^6,2^5,2^41^2,2^31^4,2\ ^21^6,21^8,1^{10}$	14

Example 5C   Here, the values of k which make $k^2 + 1$ prime,

$(0^2 + 1), 1^2 + 1 = 2,\ 2^2 + 1 = 5,\ 4^2 + 1 = 17,\ 6^2 + 1 = 37, 10^2 + 1 = 101,$

$14^2 + 1 = 197, ...;$ and

1

2

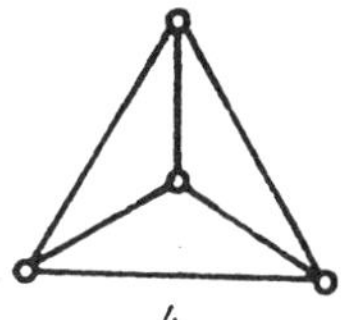

4

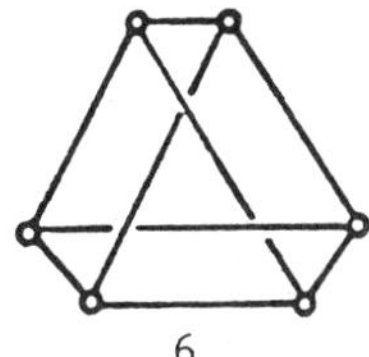

6

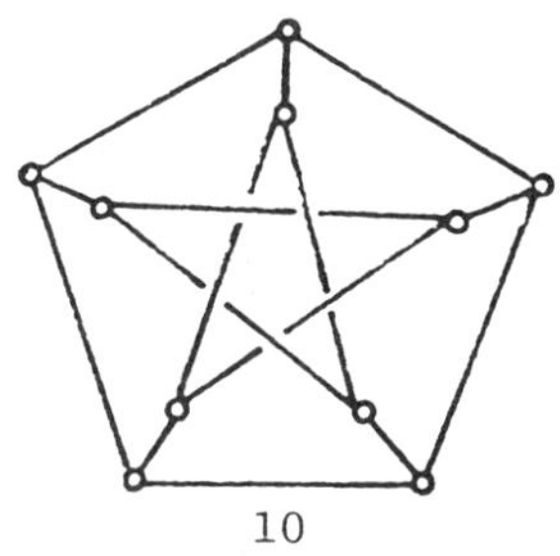

10

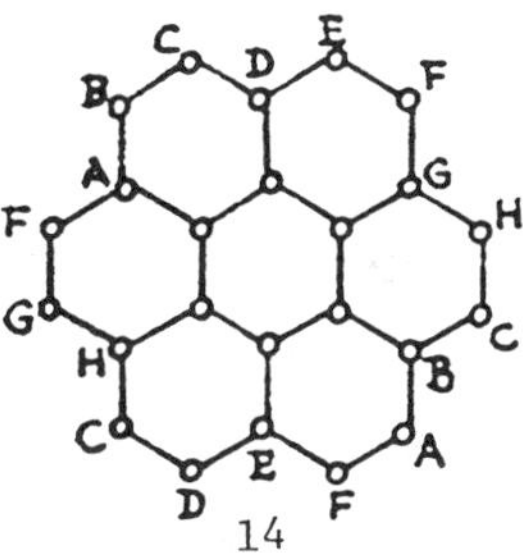

14

Example 5D   The least number of vertices in a regular trivalent graph of
girth $n + 1$. For $n = 0$ and 1 we have to use non-Michigan graphs, and count
a loop as contributing only 1 to the valence. But for $n = 2, 3, 4, 5$ we have
those standbys without which no paper on graph theory is complete: $K_4, k_{3,3},$
the Petersen graph and the Heawood map on the torus. In this last, identify
vertices bearing the same label.

The trees of height two (Figure 5A) are just one less numerous than the
trees of height at most two, so the number with n edges is $p(n) - 1$. However,
the numbers of partitions of $2n$ into parts which are powers of 2,

$(1), 2, 4, 6, 10, 14, 20, 26, 36, 46, 60, 74, 94, 114, 140, 166, 202, 238, 284, 330, 390, 450, ...$

are not the same thing [4, 5, 37 #378]. Nor, of course, are the values of k making $k^2 + 1$ prime:

$$1, 2, 4, 6, 10, 14, 16, 20, 24, 26, 36, 40, 54, 66, 74, 84, 90, 94, 110, 116, 120, 124, 126, \ldots$$

nor the numbers of vertices in the minimal regular trivalent graphs of girth $n + 1$.

$$1, 2, 4, 6, 10, 14, 24, 30, \ldots$$

which have only been determined up to girth 8[39,34,35,25,37 #380].

6.   The Sixth and Final Sequence

$$1, \quad 2, \quad 5, \quad 15, \quad \ldots$$

occurs as

Example 6A   The number of arrangements of 1,2,3,4 different objects in indistinguishable boxes, at most four in a box; as in Figure 1B, but add a box containing four objects to the set of fourteen; as

Example 6B   The first members of a sequence of "convergent sets" for a ternary continued fraction, the ratio of alternate members being good approximations to the real root of $x^3 - 7x^2 + 3x - 1 = 0$; and as

Example 6C   The number of ways of answering the questions in this paper; i.e. the number of equivalence relations on 1,2,3,4 objects, the number of rhyme schemes for verses with 1,2,3,4 lines:

1		1
12;	1,2	2
123;	1,23; 2,13; 3,12; 1,2,3	5
1234;	1,234; 2,134; 3,124; 4,123; 12,34; 13,24	
	14,23; 1,2,34; 1,4,23; 2,3,14; 2,4,13; 3,4,12; 1,2,3,4	15

The final sequence, in the forms of Example 6C, is the sequence of Bell numbers [13,37 #585]

$$1, 2, 5, 15, 52, 203, 877, 4140, 21147, 115975, 678570, 4213597, 276544437, \ldots$$

One way of describing them is the number of ways of arranging n distinguishable objects in indistinguishable boxes, with no restriction on the number of objects in a box. So the members of Example 6A fall short by increasing numbers:

$$1, 2, 5, 15, 51 = 52 - 1, 196 = 203 - 7, 827, 3795, 18755, 99146, 556711, 3305017,$$

[27, 37 #584]. Example 6B is

$$1, 2, 5, 15, 32, 99, 210, 650, 1379, 4268, 9055, 28025, 59458, 184021, 390420, \ldots$$

arising in an investigation [24,37 #582] into a problem of Jacobi.

7. Conclusion

There are still many unsolved problems in enumerative combinatorics, in spite of the recent advances that have been made [17,32,41]. Take a look at three problems which have a superficial look of similarity. First the now classical one of enumeration of polyominoes: you can adjoin four squares to a single quare in the obvious way; then three more to one of these, but not to all four of them, because of overlap. Second, the problem of the number of different deltahedra we can make from regular tetrahedra: we can again adjoin four tetrahedra to the faces of a single one; then twelve more, one to each of the three exposed faces on these four; but at the third stage we run into restrictions, since at most five tetrahedra will fit round an edge. Third, the problem of the number of isomers of the saturated hydrocarbons, $C_n H_{2n+2}$ [18-21,31] has been tackled as a mathematical, but not as a chemical problem. A tetravalent carbon atom can be associated with four other such: can each of

these four be associated with another three, and each of these twelve with...?
Clearly there is a physical bound.

But even as mathematical problems, many enumerations have not always
been clearly defined: the literature contains inconsistencies, eg. [10]: exact formulas remain beyond our grasp: and even the best known asymptotic formulas
are often weak.

[Note: The next three terms of the sequence in the Introduction are, of
course, 59, 60, 61: the orders of the simple groups!]

REFERENCES

[1] Robert Bowen "The generation of minimal triangle graphs," *Math. Comput.* 21 (1967) 248-250.

[2] W. G. Brown, "Historical note on a recurrent combinatorial problem," *Amer. Math. Monthly* 72 (1965) 973-977.

[3] P. Capell and T. V. Narayana, "On knock-out tournaments," *Canad. Math. Bull.* 13 (1970) 105-109.

[4] R. F. Churchhouse, "Binary partitions," in Atkin, Birch (eds) Computers in Number Theory, Atlas Symposium, Oxford, 1969, Academic Press, London, 1971, 397-400.

[5] R. F. Churchhouse, "Congruence properties of the binary partition function," *Proc. Cambridge Philos. Soc.*, 66 (1969)371-376.

[6] L. Comtet, "Advanced Combinatorics," D. Reidel, Drodrecht, Holland, 1974.

[7] John Conway and Richard Gury, "The Book of Numbers," Scientific American Library (in preparation).

[8] Henry H. Crapo and Gian-Carlo Rota, "On the foundations of combinatorial theory II."Combinatorial geometries, *Stud. Appl. Math.* 49 (1970) 109-133.

[9] I. M. H. Etherington, "Non-associate powers and a functional equation," *Math. Gaz.* 21 (1937) 36-39.

[10] R. A. Fisher, "Some combinatorial theorems and enumerations connected with the numbers of diagonal types of a latin square," *Ann. Eugenics*, 11 (1942) 395-401; see review by Coxeter, MR 4, 183e.

[11] Martin Gardner, "Mathematical games: permutations and paradoxes in combinatorial mathematics," *Sci. Amer.* 209 #2 (Aug. 1963) 112-119; esp. p. 114; see also #3 (Sept.) p. 262.

[12] Martin Gardner, "Mathematical games: patterns in primes are a clue to the strong law of small numbers," *Sci. Amer.* 243 #6 (Dec. 1980) 18-28.

[13] Henry W. Gould, "Bell and Catalan Numbers: Research Bibliography of Two Special Number Sequences," Morgantown, WV, 6th edition, 1985.

[14] Richard K. Guy, "The strong law of small numbers," *Amer. Math. Monthly*, 95 (1988).

[15] Richard K. Guy, "The second strong law of small numbers" (in preparation)

[16] F. Harary, "Graph Theory," Addison-Wesley, 1969.

[17] Frank Harary and Edgar M. Palmer, "Graphical Enumeration," Academic Press, New York, 1973.

[18] Henry R. Henze and Charles M. Blair, "The number of structurally isomeric alcohols of the methanol series," *J. Amer. Chem. Soc.* 53 (1931) 3042-3046.

[19] Henry R. Henze and Charles M. Blair, "The number of isomeric hydrocarbons of the methane series," *J. Amer. Chem. Soc.* 53 (1931) 3077-3085.

[20] Henry R. Henze and Charles M. Blair, "The number of structurally isomeric hydrocarbons of the ethylene series," *J. Amer. Chem. Soc.* 55 (1933) 680-686.

[21] Henry R. Henze and Charles M. Blair, "The number of structural isomers of the more important types of aliphatic compounds, *J. Amer. Chem. Soc.* 56 (1934) 157.

[22] John E. Koehler, "Folding a strip of stamps," *J. Combin. Theory*, 5 (1968) 135-152.

[23] Mike Kuchinski, "Catalan Structures and Correspondences," M.Sc. thesis, W. Virginia University, 1977.

[24] D. N. Lehmer, "On ternary continued fraction," *Tôhoku Math. J.* 37 (1933) 436-445.

[25] W. F. McGee, "A minimal cubic graph of girth seven," *Canad. Math. Bull.* 3 (1960) 149-152; MR 23 #A1550.

[26] Z. Melzak, "A note on homogeneous dendrites," *Canad. Math. Bull.* 11 (1968) 85-93; MR 37 #3954.

[27] F. L. Miksa, L. Moser and M. Wyman, "Restricted partitions of finite sets," *Canad. Math. Bull.* 1 (1958) 87-96; MR 20 #1636.

[28] J. W. Moon, "Topics on Tournaments," Holt, New York, 1968.

[29] T. V. Narayana and D. H. Bent, "Computation of the number of score sequences in round-robin tournaments," *Canad. Math. Bull.* 7 (1964) 133-136.

[30] Richard Otter, "The number of trees," *Annals of Math.* (2) 49 (1948) 583-599; MR 10, 53.

[31] Douglas Perry, "The number of structural isomers of certain homologs of methane and methanol," *J. Amer. Chem. Soc.* 54 (1932) 2918-2920.

[32] George Pólya and R. C. Read, "Combinatorial Enumeration of Groups, Graphs and Chemical Compounds," Springer-Verlag, New York, 1987.

[33] John Riordan, "An Introduction to Combinatorial Analysis," Wiley, New York, 1958.

[34] H. Sachs, "On regular graphs with given girth," in M. Fiedler (ed) Theory of Graphs and its Applications, Smolenice, 1963, 91-97, MR 30 #3467.

[35] H. Sachs, "Regular graphs with given girth and restricted circuits," *J. London Math. Soc.* 38 (1963) 423-429; MR 28 #1613.

[36] Herbert E. Salzer, "Coefficients for expressing the first thirty powers in terms of the Hermite polynomials, *Math. Tables Aids Comput.* 3 (1948) 167-169.

[37] Neil J. A. Sloane, "A Handbook of Integer Sequences," Academic Press, New York, 1973.

[38] Jacque Touchard "Contributions á l'étude du problème des timbres-poste," *Canad. J. Math.* 2 (1950) 385-398.

[39] W. T. Tutte, "A family of cubical graphs," *Proc. Cambridge Philos. Soc.* 43 (1947) 459-474; MR 9, 97.

[40] J. H. M. Wedderburn, "The functional equation $g(z^2) = 2ax + [g(x)]^2$," *Ann. of Math.* 24 (1923) 121-140.

[41] I. P. Goulden and David M. Jackson, "Combinatorial Enumeration," John Wiley, Chichester, 1983.

Combinatorial Additive Number Theory and Cayley Graphs

Yahya Ould Hamidoune
Université Pierre et Marie Curie

ABSTRACT

We obtain the following generalization of the Cauchy–Davenport Theorem.

Let G be an Abelian group and let S be a finite subset of G. There exists $s \in S$ such that the subgroup H generated by s has the following property. For every $A \subset G$ such that $A \cap H \neq \varnothing$, either $|A \cup (A + S)| \geq |A| + |S|$ or $H \subset A \cup (A + S)$.

0. Introduction

Addition theorems have many applications in Analysis, Number Theory and Combinatorics.

These theorems give some relations between the cardinality of the sum of two subsets of an Abelian group and the cardinalities of these sets. The book of H.B. Mann [12] is an appropriate reference for these questions

In this paper we mention a relation between addition theorems and the strong–connectivity of Cayley directed graphs.

By a *directed graph* we mean an ordered pair (V, E), where V is a set and E is a subset of V x V. We consider only finite directed graphs.

Let X be a directed graph. The vertex (resp. edge)–set of X will be denoted by V(X) (resp. E(X)).

Let A be a subset of V(X). The set of vertices incident from A will be denoted by $N_X^+(A)$, or simply $N^+(A)$. More precisely

$$N_X^+(A) = \{y \in V(X)\backslash A : \exists\, x \in A \text{ such that } (x, y) \in E(X)\}.$$

Let X be a directed graph. The *strong–connectivity* of X, denoted by $\kappa(X)$, is defined as follows.

$$\kappa(X) = \min\{|N^+(A)| : |A| = 1 \text{ or } A \cup N^+(A) \neq V(X)\}.$$

615

It is easy to see that $\kappa(X)$ is the minimum number of vertices whose deletion reduces X to a single vertex or to a nonstrongly connected directed graph.

In [6, 7, 8, 9], we have only considered directed graphs without loops. We allow loops in this paper in connection with additive number theory. But the results of [6, 7, 8, 9] apply in this situation since the strong–connectivity is not affected if we delete the loops.

More details about the strong–connectivity of directed graphs and its relation with groups can be found in [6, 7, 8, 9].

1. Theorems of Lagrange and Cauchy

The following theorem is due to Lagrange, cf. [3]. It was a piece in the proof of his famous theorem that any integer is the sum of 4 squares.

Theorem 1.1 (Lagrange [11])   *Given a prime number p, $x, y \not\equiv 0 \bmod p$. There are $t, u \in \mathbf{Z}$ such that $y \equiv t^2 + xu^2 \bmod p$.*

Cauchy found a combinatorial proof to the following more general result.

Let $X_1, X_2, \ldots, X_r$ be distinct variables and $f \in \mathbf{Z}[X_1, \ldots, X_r]$. Let $A(f)$ be the number of distinct residues $\bmod p$ among all the values of f.

Theorem 1.2 (Cauchy [3])   *Let $X_1, X_2, \ldots, X_r, \ldots, X_{r+s}$ be distinct variables. Let $f \in \mathbf{Z}[X_1, \ldots, X_r]$, $g \in \mathbf{Z}[X_{1+r}, \ldots, X_{r+s}]$. Then either $|A(f + g)| = p$ or $|A(f + g)| \geq |A(f)| + |A(g)| - 1$.*

Let $L : \mathbf{Z}_p \times \mathbf{Z}_p \to \mathbf{Z}_p$, be the mapping $L(t, u) = t^2 + xu^2$. Lagrange's Theorem states that L is surjective. Cauchy's Theorem easily implies this fact.

2. The Cauchy–Davenport Theorem

Let G be an Abelian group and let $A, B \subset G$. We write $A + B = \{x + y \mid x \in A$ and $y \in B\}$.

Theorem 2.1 (The Cauchy–Davenport Theorem)   *Let p be a prime number, and let A and B be two proper subsets of $\mathbf{Z}_p$ such that $B + A \neq \mathbf{Z}_p$. Then $|A + B| \geq |A| + |B| - 1$.*

This result is proved by Davenport [4]. Davenport mentioned in his historical note [5] that this theorem is contained in [3]. Theorem 2.1 is known as the Cauchy–Davenport Theorem.

The equivalence between this theorem and Cauchy's Theorem is due to the well known lemma saying that every subset of $\mathbf{Z}_p$ can be represented as the set of values of a polynomial.

We shall show now how to reduce this problem to a connectivity problem on Cayley directed graphs.

Let G be a finite group and S be a subset of G. We shall use the additive notation. The *Cayley directed graph* of G with respect to S is the graph $\text{Cay}(G, S) = (G, E)$, where $E = \{(x, y) \mid -x + y \in S\}$.

Remark 1 Let $X = \text{Cay}(G, S)$. Then $N^+(A) = (A + S)\backslash A$.

Theorem 2.2 (Hamidoune [6]) *Let $S \subset \mathbf{Z}_p$. Then $\kappa(Cay(\mathbf{Z}_p, S)) = \mid S\backslash 0 \mid$.*

Proof of the Cauchy–Davenport Theorem using Theorem 2.2. We may assume without loss of generality that $0 \in B$. Let $X = \text{Cay}(\mathbf{Z}_p, B)$. Since $0 \in B$, we have $A \subset A + B$. By Remark 1, we have

$$A + B = A \cup N^+(A). \tag{1}$$

It follows that $A \cup N^+(A) \neq V(X)$. By the definition of $\kappa(X)$ and Theorem 2.2, we have

$$|N^+(A)| \geq \kappa(X) = |B\backslash 0| = |B| - 1. \tag{2}$$

By (1) and (2), we have $|A + B| = |A \cup N^+(A)| \geq |A| + |B| - 1$. $\square$

3. Theorems of Shepherdson and Alon

Using the Davenport transfer argument, Shepherdson proved the following.

Theorem 3.1 (Shepherdson [14]) *Let G be a finite Abelian group and let A, S be two subsets of G such that $S \subset A$. Then either $\mid A \cup (A + S) \mid \geq \mid A \mid + \mid S \mid$ or there is $s \in S$ such that $A \cup (A + S)$ contains the subgroup generated by s.*

As an application of Theorem 3.1, Shepherdson obtained the following.

Theorem 3.2 (Shepherdson [13]) *Let G be a finite Abelian group and let A and let S be a subset of G. Let $k \in N$ be such that $k \mid S \mid \geq \mid G \mid$. Then there is a sequence of natural numbers $(n_s; s \in S)$ such that $\sum n_s s = 0$ and $1 \leq \sum n_s \leq k$.*

This result is only stated by Shepherdson for $\mathbf{Z}_n$ and $\mathbf{Z}_n^*$ (the group of units of $\mathbf{Z}_n$). But Shepherdson's method works for all finite Abelian groups.

We proved in [8] the following.

Theorem 1.D (Hamidoune [8])   *Let G be a finite group and let $S \subset G$. Then there is a nonvoid sequence of elements of S with product $= 1$ and length $\leq \lceil |G|/|S| \rceil$.*

We did not realize in [8] that Theorem 1.D extends Shepherdson's Theorem (3.2) to the case where the group is not necessarily Abelian. Indeed we were unaware of Shepherdson's Theorem.

We obtained in [8] a more general lower bound for the length of a directed cycle in a directed graph with a transitive group of automorphisms.

Recently, Theorem 3.2 has been rediscovered by Alon [1]. The proof of Alon uses a result of Scherk [13].

Let $n, m \in \mathbf{N}$. We put $[n] = \{1, \ldots, n\}$. Define $f(n, m)$ as the smallest integer k such that $\forall A \subset [n]$ $(|A| > k \Rightarrow (\exists B \subset A ; B \neq \varnothing$ and $\sum_{x \in B} x = m))$.

It was conjectured by Erdös and Graham that $f(n, 2n) = (\frac{1}{3} + o(1))n$. This conjecture was proved by Alon [1], as a corollary of the following.

Theorem 3.3 (Alon [1])   *For every $\varepsilon > 0$ and $k > 1$, there is an n_0 such that for every $n > n_0$, and $A \subset \mathbf{Z}_n$ with $|A| > (\frac{1}{k} + \varepsilon)n$, there is a nonvoid subset of A with sum $= 0$ and cardinality $\leq k$.*

Alon proved a similar statement for Abelian groups of odd order.

Theorem 3.4 (Alon [1])   *For every $\varepsilon > 0$ and $k > 1$, there is an n_0 such that for every Abelian group G of odd order $n > n_0$, and $A \subset G$ with $|A| > (\frac{1}{k} + \varepsilon)n$, there is a nonvoid subset of A with sum $= 0$ and cardinality $\leq k$.*

4. A New Addition Theorem

The proofs of the results announced in this section can be found in [10]. These proofs use the theory of atoms of Cayley directed graphs [6, 7, 8, 9]. Our main result is the following.

Theorem 4.1 [10]   *Let G be an Abelian group and let S be a finite subset of G. There exists an $s \in S$ such that the subgroup H generated by s has the following property. For every $A \subset G$ such that $A \cap H \neq \varnothing$, either $|A \cup (A + S)| \geq |A| + |S|$ or $H \subset A \cup (A + S)$.*

Theorem 4.1 generalizes the Cauchy–Davenport Theorem and Shepherdson's Theorem (3.1). As applications we obtain the following generalizations to Alon's Theorem (3.3).

Theorem 4.2 *For every $\varepsilon > 0$ and $k > 1$, there is an n_0 such that for every $n > n_0$, and $A \subset Z_n$ with $|A| > (\frac{1}{k} + \varepsilon)n$, there exists an $a \in A\backslash\{0\}$ such that any element of the subgroup generated by a is a sum of a nonvoid subset of A cardinality $\leq k$.*

Theorem 4.3 *For every $\varepsilon > 0$ and $k > 1$, there is an n_0 such that for every group G of odd order $n > n_0$, and for every $A \subset G$ with $|A| > (\frac{1}{k} + \varepsilon)n$, there exists an $a \in A\backslash\{0\}$ such that any element of the subgroup generated by a is a sum of a nonvoid subset of A with cardinality $\leq k$.*

Theorem 4.2 implies the following result proved by Alon [2].

$$\forall \varepsilon > 0, \ \exists \ \text{an } n_0, \text{ such that } \ \forall \ \text{prime } p > n_0, \ x \in Z_p \text{ and } A \subset Z_p \text{ with } |A| > (\tfrac{1}{k} + \varepsilon)p, \text{ there is } A \subset Z_p \text{ such that } 1 \leq |A| \leq k \text{ and } \sum_{y \in A} y = x.$$

REFERENCES

[1]. N. Alon, Subset sums, *J. Number Theory* **27** (1987), 196 – 205.

[2]. N. Alon, private communication.

[3]. A. Cauchy, Recherches sur les nombres, *J. Ecole Polytechnique* **9** (1813), 99 – 116.

[4] H. Davenport, On the addition of residue classes, *J. London Math. Soc.* **10** (1935), 30 – 32.

[5] H. Davenport, A historical note, *J. London Math. Soc.* **22** (1947), 100 – 101.

[6] Y.O. Hamidoune, Sur les atomes d'un graphe orienté, *C.R. Acad. Sc. Paris A* **284** (1977), 1253 – 1256.

[7] Y.O. Hamidoune, Sur la séparation dans les graphes de Cayley abeliens, *Discrete Math.* **55** (1985), 323 – 326.

[8] Y.O. Hamidoune, An application of connectivity theory in graphs to factorization of elements in groups, *Europ. J. of Combinatorics* **2** (1981), 108 – 112.

[9] Y.O. Hamidoune, On the connectivity of Cayley digraphs, *Europ. J. of Combinatorics* **5** (1984), 309 – 312.

[10] Y.O. Hamidoune, On subset products in finite groups, preprint.

[11] J.L. Lagrange, *Nouv. Mémoires* Acad. Berlin (1770), 123 – 133, Ouvres 3, 189 – 201.

[12] H.B. Mann, Addition theorems, *Interscience tracts* **18** John Wiley and Sons, (1965)

[13] P. Scherk, *Amer. Math. Monthly* **62** (1955), 46 – 47.

[14] J.C. Shepherdson, On the addition of elements of a sequence, *J. London Math. Soc.* **22** (1947), 85 – 88.

Recent Results and Unsolved Problems
on Hypercube Theory

Frank Harary

New Mexico State University

ABSTRACT

The study of graph theoretic properties of hypercubes is relatively uninvestigated and offers a rich field for original research. After reviewing seven presentations of a hypercube Q_n, we look into various properties. First we characterize the edge sums of a hypercube, i. e., the numbers which can arise as the sum of the numerical labels on two adjacent nodes. Then we study cubical graphs which are the subgraphs of some hypercube. The cubical dimension of a cubical graph G is the smallest n such that G is contained in Q_n. The topological cubical dimension of an arbitrary graph G is the smallest n such that some subdivision of G is a subgraph of Q_n. There are several unsolved problems involving these two dimensions. For practical computer-reasons, it would be helpful to have a characterization of the structure of those trees and other graphs which span a hypercube. Partial progress toward this difficult problem is reported. Finally several new elementary problems on changing and unchanging the diameter of a hypercube are given.

000. Introduction

The utilization of hypercube architectures in massively parallel computers has motivated considerable research activity on hypercube theory. While much is known about the structure of hypercubes, there is very much more which has not yet been investigated. This is the inspiration for the survey of hypercube theory in [11], and for my present research.

My current doctoral student, Niall Graham, is also fascinated by hypercube theory. I am grateful for his kind assistance in the preparation of this manuscript.

621

Table of Contents

000. Introduction

001. Presentations

010. Edge Sums

011. Cubical Graphs

100. Cubical Dimensions

101. Embedding

110. Spanning Subgraphs

111. Diameter

001.   Presentations

There are many logically equivalent ways to present the information that a given graph G is a hypercube, or more specifically that G and Q_n are isomorphic. These are summarized in [9] where the following seven presentations, P1 to P7, of a hypercube are given in detail, and others.

P1. Binary Sequences

The n–dimensional *hypercube* $Q_n = (V_n, E_n)$ is the graph whose node set V_n consists of all the 2^n binary n–strings. A pair of nodes $x = (x_1, ..., x_n)$ and $y = (y_1, ..., y_n)$ are adjacent in Q_n if and only if $\Sigma |x_i - y_i| = 1$, i. e., the binary words x and y differ in exactly one place. Figure 1.1 shows the first four hypercubes Q_1 to Q_4 with this binary labeling of their nodes.

Thus Q_n is regular of degree n, is bipartite, has $p = 2^n$ nodes, and hence has $q = n2^{n-1}$ edges, a quantity which has been jokingly described as "the derivative of 2^n with respect to 2."

P2. Cartesian Products

Referring to [7, p. 21] for the definition of the cartesian product $G \times H$ of two graphs, we have the following recursive definition of Q_n:

$$Q_1 = K_2 ; \quad Q_n = Q_{n-1} \times Q_1 \qquad (1.1)$$

Thus in particular $Q_2 = C_4$, the 4–cycle seen in Figure 1.1, and $Q_4 = C_4 \times C_4$ which naturally embeds on the surface of a torus. Figure 1.2 brings out the toroidal nature of Q_4 as the cartesian product of two cycles.

A company which manufactures hypercube computers has distributed a poster which vividly displays an extension of the drawing in Figure 1.2 to the next hypercube Q_5.

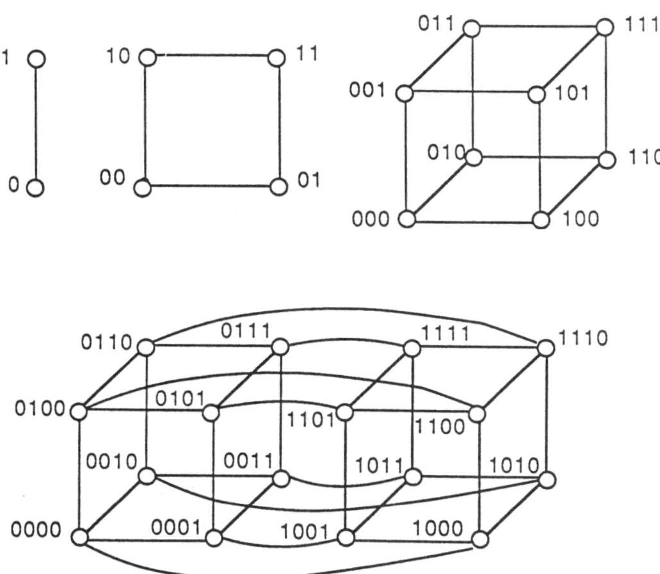

Figure 1.1 The smallest hypercubes

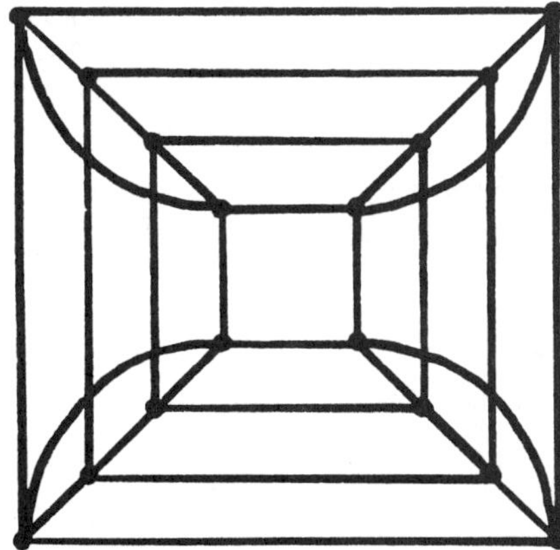

Figure 1.2 A drawing of Q_4 with crossing number 8

P3. *Unit Cube in R^n*

In n–dimensional euclidean space R^n, the hypercube Q_n may be regarded as the expected extension of the unit interval $[0,1] = Q_1$, the unit square Q_2, the Platonic solid Q_3, and so forth. Thus Q_n is the unit hypercube in R^n with coordinates which are the binary sequences stated in presentation P1 and shown in Figure 1.1. Hence it is meaningful to consider hyperplanes in R^n which "cut through" Q_n without passing through any vertex of Q_n. The set of all nodes on either side of such a cutting hyperplane has been called a *linearly separable boolean function*. Such functions have been studied extensively. The natural extension of the boolean functions determined by two or more cutting planes has not yet been investigated.

P4. *The Power Set of an n–Set*

Writing $S = N_n = \{1, 2, ..., n\}$, the *power set* 2^S is the collection of all the 2^n subsets of S. This gives the node set $V_n = 2^S$ of Q_n. The obvious 1–1 correspondence between a subset $X \subset S$ and a binary sequence is given by the characteristic function of X. The edge set E_n is then defined as in P1.

P5. *Lattices*

The collection of all subsets of a finite set N_n gives the complete lattice L_n whose covering relation, the Hasse diagram $H(L_n)$, is precisely Q_n. It is this presentation of a hypercube which proved useful in determining (in our last section) the maximum number of new edges which can be added to Q_n without decreasing the diameter.

P6. *Finite Boolean Algebras*

Whether defined by any one of literally dozens of axiom systems or in any other correct manner, a finite boolean algebra with n atoms is precisely realized by Q_n. Its n atoms correspond to the nodes $(\delta_{i1}, \delta_{i2}, ..., \delta_{in})$, for i = 1 to n, where δ_{ij} is the familiar Kronecker delta. Each binary sequence label on a node stands for the subset of N_n as in presentation P4.

In the light of this presentation, every theorem about finite boolean algebras is a theorem about hypercubes, *and conversely!* Thus each new result about hypercubes is a disguised form of a new theorem about finite boolean algebras, a subject which logicians and algebraists have long regarded as closed.

Let $u = u_1, u_2, ..., u_n$ be a binary sequence, i. e., a node of Q_n. Then u can be regarded as one term in the disjunctive normal form of a boolean function on n variables x_i as in the following illustration for n = 4:

$$u = 1101 \rightarrow \text{term } x_1 \, x_2 \, x'_3 \, x_4$$

Thus a subset S of $V(Q_n)$ determines a boolean function in a $1-1$ manner. The automorphism group of Q_n provides the equivalence relation which defines the type of a boolean function. Further, each boolean function written $S \subset V(Q_n)$ yields the induced subgraph $\langle S \rangle$, called "the graph of a boolean function."

P7. Adjacency matrices

It is well known that an adjacency matrix A of a graph G specifies G up to isomorphism. When G has p nodes, A is a $p \times p$ matrix with zero diagonal. Referring to the labeling in Figure 1, we have the illustration:

$$A(Q_2) = \begin{array}{c} \\ 00 \\ 01 \\ 10 \\ 11 \end{array} \overset{\begin{array}{cccc} 00 & 01 & 10 & 11 \end{array}}{\begin{bmatrix} 0 & 1 & 1 & 0 \\ 1 & 0 & 0 & 1 \\ 1 & 0 & 0 & 1 \\ 0 & 1 & 1 & 0 \end{bmatrix}}$$

When G is bipartite, a more concise matrix $B = B(G)$ contains all the information in $A(G)$. Following [1, 6], the *bipartite adjacency matrix* $B = [b_{ij}]$ of graph G with node color sets U and W has $b_{ij} = 1$ when nodes u_i and w_j are adjacent; and $b_{ij} = 0$ otherwise. Thus for Q_2, we have

$$B(Q_2) = \begin{array}{c} \\ 00 \\ 11 \end{array} \overset{\begin{array}{cc} 01 & 10 \end{array}}{\begin{bmatrix} 1 & 1 \\ 1 & 1 \end{bmatrix}}$$

We observed in [3] that writing $B_n = B(Q_n)$, we have the recursive formulation:

$$B_1 = [1] \qquad B_{n+1} = \begin{bmatrix} B_n & I \\ I & B_n \end{bmatrix} \tag{1.2}$$

We then used this recursion to evaluate the number of perfect matchings in Q_n, written $f_1(Q_n)$, with the symbol f_1 serving as a mnemonic for 1–factor, by applying the straightforward permanent formula,

$$f_1(Q_n) = \operatorname{per}(B_n).$$

Other Presentations

This is a dynamic field of research and various other presentations and characterizations of a hypercube are currently being developed.

010.　Edge Sums

The usual binary sequences associated as in Figure 1.1 with the nodes of Q_n can be regarded either as binary numbers or as the numbers from 0 through $2^n - 1$. It is natural to ask which numbers are obtained when one writes on each edge the sum of the labels of its two incident nodes. A positive integer k is an *edge sum of a hypercube* if it appears in this way on an edge of some Q_n. We show in Figure 2.1 the edge sums of Q_2 and Q_3.

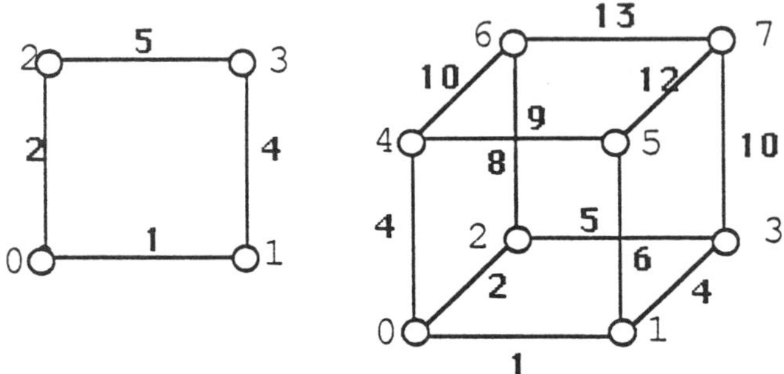

Figure 2.1 The edge sums of two hypercubes

We determined in [4] exactly which numbers are edge sums:

Theorem 2A   *A natural number x is an edge sum if and only if $x \not\equiv 3(mod\ 4)$,   i.e., the binary expression for x does not end in 11.*

To prove this sufficiency, note that for arbitrary binary substrings, α and β, the sum $\alpha 00\beta + \alpha 10\beta = \alpha 010\beta 0$. Thus any positive even number may be an edge sum. The sum $\alpha 0 + \alpha 1 = \alpha 01$ generates any binary number ending in 01. Hence every binary number not ending in 11 is indeed an edge sum.

For the necessity we observe that to achieve a sum that ends in 11 which is odd, we must vary the last digit. But we have just seen that such a sum is of the form $\alpha 01$, affirming that only strings of the form $\alpha 11$ cannot arise as edge sums.

This modest exercise illustrates the fact that there exist new, elementary, interesting results about hypercubes to be found.

011.   Cubical graphs

An *n–cubical graph* G is isomorphic to a subgraph of Q_n. A *cubical graph* is n–cubical for some n. Garey and Graham [2] proved that the problem of deciding whether a given graph G is n–cubical is NP–complete. They accomplished this by the study of minimal noncubical graphs G, that is, graphs G which are not cubical but the removal of any one edge results in a cubical graph.

As Q_n is bipartite, we know that every cubical graph is bipartite. However the converse does not hold: the smallest counterexample is provided by the complete bipartite graph $K_{2,3}$ as mentioned in [8].

It is well known that every tree is cubical and so is every unicyclic graph whose unique cycle is even. Every square–cell animal is cubical and so is every polyhex. The only cubical graphs among the complete bipartite graphs $K_{r,s}$ with $r \le s$ are the stars $K_{1,s}$ and the 4–cycle $K_{2,2} = C_4$.

It is easy to see that every cubical graph G can be embedded in some hypercube Q_n as an induced subgraph.

100. Cubical dimensions

The *cubical dimension* $cd(G)$ of a cubical graph G is the minimum n such that $G \subset Q_n$ (subgraph). This concept and the above notation were the subject of [8]. Various easily verified observations which give exact values of $cd(G)$ for several graphs G include the following formulas for stars, paths, even cycles, meshes, full binary trees T_h of height h, and matchings rK_2.

(4.1) $cd(K_{1,s}) = s$

(4.2) $cd(P_r) = \lceil \log_2 r \rceil$

(4.3) $cd(C_k) = \lceil \log_2 k \rceil$ for k even

(4.4) $cd(M_{r,s}) = \lceil \log_2 r \rceil + \lceil \log_2 s \rceil$

(4.5) $cd(T_h) = h + 2, \ h \ge 2$

(4.6) $cd(rK_2) = 1 + \lceil \log_2 r \rceil$

Unsolved problems

1. Among all trees T with p nodes, the star has the largest cubical dimension and the path has the smallest. What is the distribution of cubical dimensions among all other trees? What are the mean, median and mode among all these values of $cd(T)$?

2. What is the exact value of $cd(T)$ for the following two special classes of trees T:
   (a) starlike trees $S(a_1, a_2, \ldots, a_n)$ in which branch i is a path of length a_i
   (b) caterpillars $C(a_1, a_2, \ldots, a_k)$ where the spine is the path $P_k = v_1, v_2, \ldots, v_k$
       and a_i is the number of endnodes adjacent to spinal node v_i

When a graph G is cubical, say cd(G) = n, then G is necessarily isomorphic to an induced subgraph of some hypercube Q_m, with m > n unless G is already an induced subgraph of Q_n. This fact immediately suggests defining the *induced cubical dimension* of G, written icd(G), as the smallest such n. The exact determination of icd(G) for various cubical graphs G, including paths and cycles, appears to offer an intrinsically intractable problem area, as we shall see in the last section.

101. Embedding

For practical computing purposes, it is necessary to embed graphs G into hypercubes even when G is not cubical, as reported in Livingston and Stout [13]. This is done by embedding not G itself but a subdivision of G.

Folklore Theorem   *Every graph has a cubical subdivison.*

Folklore Proof   It is sufficient to verify that this holds for every complete graph, K_p, as every graph G is a spanning subgraph of the complete graph $G \cup \overline{G}$. Recall that the *weight* of a node $u = (u_1, \ldots, u_n)$ of Q_n is the sum Σu_i. Thus the origin $(0, \ldots, 0)$ is the only node of weight 0 and there are just n nodes of weight 1.

To construct a subdivision S of the complete graph K_{n+1} which is a subgraph of Q_n, take the n + 1 nodes of weight at most 1 as the node set of K_{n+1} and define S to be the subgraph of Q_n induced by all nodes of weight 0, 1 or 2.

The graph S has the n edges incident with the origin as unsubdivided edges of K_{n+1}, and each of the remaining $\binom{n}{2}$ edges of K_n is subdivided by one node of degree 2. These degree-2 nodes of S are just the nodes of weight 2 in Q_n, as illustrated in Figure 5.1 for n = 3.

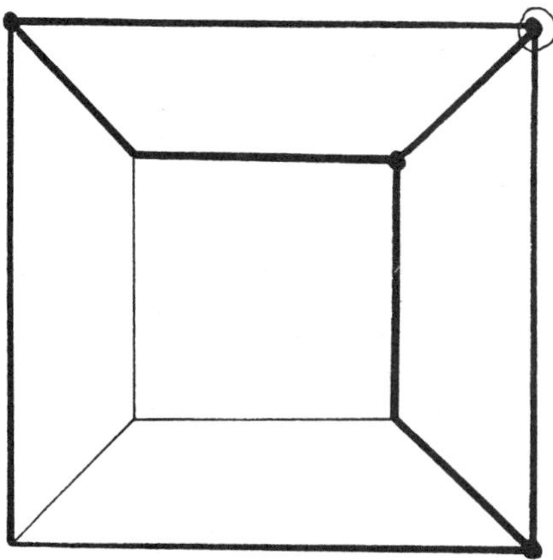

Figure 5.1 Topological embedding of K_4 in Q_3

It follows at once that for every graph G with p nodes, some subdivision SG is $(p-1)$-cubical.

The *topological cubical dimension* $tcd(G)$ of an arbitrary graph G is the smallest n such that some subdivision SG satisfies $SG \subset Q_n$. When G is cubical, it is obvious that $tcd(G) = cd(G)$. The invariant $tcd(G)$ was studied in [10].

110. Spanning subgraphs

The investigation of various families of spanning subgraphs of a hypercube is the subject of a series of papers with M. Lewinter. Our progress to date is summarized in the survey article [12] in this volume. Of course, a tree T which spans Q_n has 2^n nodes. As Q_n has 2^{n-1} nodes of even weight and 2^{n-1} of odd weight, it is a bipartite graph with the same number of nodes in each of its two color classes. A bipartite graph with this property is called *equitable*. Obviously, every connected spanning subgraph of Q_n is an equitable bipartite graph. A *starlike tree* is a subdivision of a star $K_{1,k}$ with $k \geq 3$. Just these two necessary conditions prove to be sufficient for starlike trees:

A starlike tree T spans Q_n if and only if T has 2^n nodes and is equitable.

The general problem of characterizing the structure of trees which span Q_n appears quite intractable. There is an adage which recommends that if one cannot solve a problem for graphs, he should attempt to settle it for trees. This maxim may be modified to suggest that a difficult problem for trees should first be tried for caterpillars. Even this restricted problem has only yielded partial progress.

However, we did succeed in characterizing those meshes which span Q_n. Recall that a *mesh* is the cartesian product of a finite number d of nontrivial paths. As usual, P_h denotes the path with h nodes. We call P_h a *binary path* if $h = 2^k$ for some k. Then a *binary mesh* M is the cartesian product of binary paths, as illustrated in Figure 6.1. Our result is:

A mesh M spans Q_n if and only if M is a binary mesh with 2^n nodes.

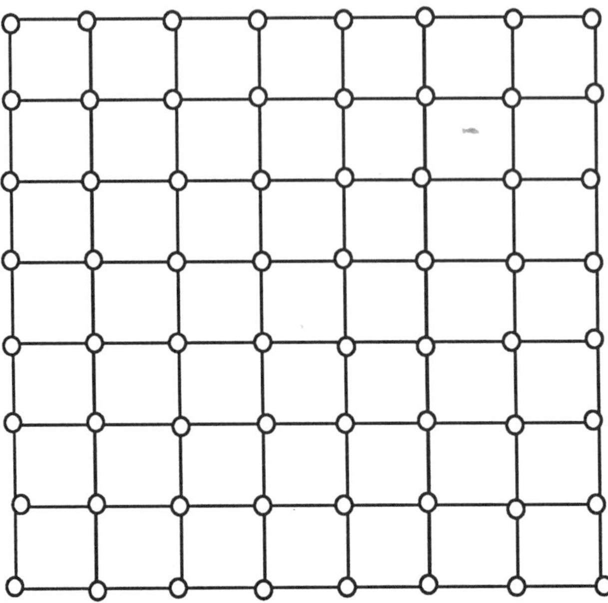

Figure 6.1 A binary mesh which spans Q_6

111. Diameter

In a parallel processor the diameter of the underlying graph architecture imposes a bound on the communication time between an arbitrary pair of processors. Motivated by an investigation of the diameter in the presence of edge faults, the following extremal invariants of a hypercube suggest themselves. These changing and unchanging invariants were named and studied in [5].

Let $ch^+(Q_n)$ be the least number of edges whose addition to Q_n decreases the diameter, and let $ch^-(Q_n)$ be the least number of edges whose deletion increases the diameter, ch standing for changing. Similarly, let $un^+(Q_n)$ be the maximum number of edges which can be added to Q_n without changing (decreasing) the diameter, and let $un^-(Q_n)$ be the greatest number of edges whose removal from Q_n does not affect (increase) the diameter, un standing for unchanging. The following results were found in [5] concerning these invariants.

$$ch^-(Q_n) = n - 1 \qquad\qquad (7.1)$$

In words, the removal of $n - 1$ edges and no fewer is sufficient to increase the diameter. This is only achieved by removing all but one edge incident to the same node.

$$ch^+(Q_n) = 2 \qquad\qquad (7.2)$$

Two additional edges are required to decrease the diameter. There are several solutions. The simplest is to augment any subcube Q_2 in Q_n into a copy of K_4 by adding two diagonal edges.

$$(n-3)2^{n-1} + 2 \leq un^-(Q_n) < (n-2)2^{n-1} \tag{7.3}$$

The lower bound was achieved by constructing a family of graphs defined in terms of spanning trees of Q_n. More recently, a lower bound asymptotic to $(n - 20/7)2^{n-1}$ was found, offering a slight improvement. The upper bound is a consequence of the above mentioned fact that the minimum degree $\delta \geq 2$. A consequence of the lower bound is that

$$\lim_{n \to \infty} un^-(Q_n)/n2^{n-1} = 1 \tag{7.4}$$

In words "almost all" the edges of a hypercube may be removed without altering the diameter!

$$un^+(Q_n) = \binom{2n}{n-1} + \frac{1}{2}\binom{2n}{n} - (n+1)2^{n-1} \tag{7.5}$$

The derivation of this quantity is routine as exactly one pair of nodes, namely, $00 \ldots 0$ and $11 \ldots 1$, remain at distance n apart after the addition of all the new edges which join nodes whose weights differ by 1.

The minimum diameter d of a spanning tree T of Q_n is given by:

$$\min(d : T \text{ spans } Q_n) = 2n - 1$$

As every node in Q_n has eccentricity n, the center of a spanning tree must lie on a path joining two nodes whose distances from the center are respectively at least n and $n - 1$. A family of spanning trees with this property are easily constructed.

REFERENCES

[1]   R. Brualdi, F. Harary and Z. Miller, Bigraphs versus digraphs via matrices. *J. Graph Theory* **4** (1980) 51–57.

[2]   M. R. Garey and R. L. Graham, On cubical graphs. *J. Combin. Theory* **B18** (1975) 84–95.

[3]   N. Graham and F. Harary, The number of perfect matchings in a hypercube. *Applied Math. Letters* **1** (1988) 41–44.

[4]   N. Graham and F. Harary, Edge sums in hypercubes. *Irish Math. Soc. Bull.* **2** (1989) 8–12.

[5]   N. Graham and F. Harary, Changing and unchanging the diameter of a hypercube, to appear.

[6]   F. Harary, Permanents, determinants and bipartite graphs. *Math. Mag.* **42** (1969) 146–148.

[7]   F. Harary, *Graph Theory*. Addison–Wesley, Reading (1969).

[8]   F. Harary, Cubical graphs and cubical dimensions. *Comput. Math. Appl.* **15** (1988) 271–275.

[9]   F. Harary, Presentations of a hypercube and their roles. *Proc. Intl. Conf. on Computer Science, Hong Kong* (1988).

[10] F. Harary,The topological cubical dimension of a graph, to appear.

[11] F. Harary, J. P. Hayes and H. J. Wu, A survey of the theory of hypercube graphs. *Comput. Math. Appl.* **15** (1988) 277–289.

[12] F. Harary and M. Lewinter, Spanning subgraphs of a hypercube VI: Survey and unsolved problems. This volume.

[13] M. Livingston and Q. F. Stout, Embeddings in hypercubes. *Math. Comput. Modeling* **11** (1988) 222–227.

Spanning Subgraphs of a Hypercube VI:
Survey and Unsolved Problems

Frank Harary
New Mexico State University

Martin Lewinter
SUNY at Purchase

ABSTRACT

The hypercube Q_n is defined recursively by $Q_1 = K_2$, while $Q_n = Q_{n-1} \times K_2$. A characterization of the spanning trees of Q_n is at present unknown. We present a survey of results in this area and propose various unsolved or partially solved problems. The n–dimensional mesh $M(a_1, ..., a_n)$ is the cartesian product of the paths P_{a_i}, $i = 1, ..., n$. While we have shown that this mesh spans a hypercube if and only if each a_i is a power of 2, there are several open questions concerning embedding meshes in hypercubes.

1. Introduction

Hypercubes are of considerable current interest in computer science, particularly in computer architecture, as they are being successfully utilized in parallel processing.

The *hypercube* Q_n is defined recursively by $Q_1 = K_2$ while $Q_n = Q_{n-1} \times K_2$. We employ the graph–theoretic terminology of [1] except that we use *nodes* and *edges* rather than points and lines. The node set $V_n = V(Q_n)$ is the set of all binary n–tuples, so that $|V_n| = 2^n$. Let nodes x, y $\in V(Q_n)$ have binary labels $(x_1, ..., x_n)$ and $(y_1, ..., y_n)$. Then xy $\in E_n = E(Q_n)$ if and only if $\Sigma |x_i - y_i| = 1$. More generally, the distance $d(x, y) = \Sigma |x_i - y_i|$. Note that $|E_n| = n2^{n-1}$.

We require several definitions. A tree is *equitable* if in the 2–coloring of V(T), both colors sets have the same cardinality. If the cardinalities differ by one, it is *nearly equitable;* otherwise it is called *skewed*. A *caterpillar* is a tree which becomes a path when its endnodes are removed. This path is called the *spine* of the caterpillar.

633

The path P_k has k nodes. Denoting, as usual, the maximum degree of a graph G by $\Delta(G)$, a *k–legged* caterpillar T satisfies $\Delta(T) = k + 2$. When the degree set of a caterpillar T is $\{1, 2, k + 2\}$, $k \geq 2$, it is called *strictly* k–legged. A *starlike* tree T is homeomorphic to a star. If $\Delta(T) = k$, the starlike tree is called *k–branched*. A *double star* is a tree with exactly two nodes which are not endnodes, i.e., it is a caterpillar with spine K_2. A *double starlike* tree is a subdivision of a double star S in which the edge joining the two central nodes of S is not subdivided.

The *n–dimensional mesh* denoted by $M = M(a_1, ..., a_n)$ has node set $V(M)$ consisting of all the ordered n–tuples $x = (x_1, ..., x_n)$ such that $1 \leq x_i \leq a_i$ for each i between 1 and n. As for hypercubes, two nodes $x, y \in V(M)$ are adjacent if and only if $\Sigma |x_i - y_i| = 1$. If each $a_i = 2$, then $M = Q_n$. Obviously $M(a_1, ..., a_n)$ is the cartesian product of the paths $P_{a_1}, ..., P_{a_n}$. If each a_i is a power of 2, then M is a *binary mesh*.

An integer–valued function f on a set S of trees *interpolates* if given trees T, $T' \in S$ and an integer m such that $f(T) < m < f(T')$, then there exists a tree $T'' \in S$ with $f(T'') = m$. If m must be an even (odd) number, then f has the *even (odd) interpolation property*. In Section 2, we summarize various results on spanning trees of hypercubes and present several open questions. Section 3 deals with embedding meshes in hypercubes.

2. Spanning Trees of a Hypercube

An open question is to characterize the spanning trees of Q_n. In [3] we showed that equitable k–branched starlike trees on 2^n nodes with $3 \leq k \leq n$ spans Q_n, thereby extending a result of Nebesky [10], who proved this for $k = 3$. We proved in [4] that if an equitable double starlike tree T on 2^n nodes satisfies $\Delta(T) = 3$, then T spans Q_n. This result has not yet been settled for $\Delta(T) > 3$, although it is conjectured to hold.

Havel and Liebel in [9] showed that equitable 1–legged caterpillars on 2^n nodes span Q_n. It is not known exactly which k–legged caterpillars span Q_n. In [8], we presented several classes of strictly 2–legged caterpillars which span Q_n and gave a tight upper bound for the number of nodes of degree 4 in the next result.

Theorem A *There exist strictly 2–legged caterpillars T which span Q_n having the following properties:*

    (1) $n \geq 5$ is odd and T has $(2^n - 2)/3$ nodes of degree 4.

    (2) *$n \geq 6$ is even and T has $(2^n - 4)/3$ nodes of degree 4.*

We also proved in [7] that the number of nodes of degree 4 in hypercube–spanning strictly 2–legged caterpillars is even and has the even interpolation property.

We ask the following question: Given $n \geq 5$, what is the shortest possible spine among all the caterpillars which span Q_n? The 3–legged caterpillar of Figure 1 spans Q_5 and its spine is P_{10}. Note that the sequence which lists the number of endnodes (consecutively) of the spinal nodes, i.e., the *caterpillar code*, is the palindrome $(3, 3, 3, 2, 0, 0, 2, 3, 3, 3)$. The construction involves considering Q_5 as four copies of Q_3 and using the spine to span the two "opposite" copies of Q_2 in the upper left and lower right copies of Q_3. This requires two nodes (with code 0) to connect these spinal segments.

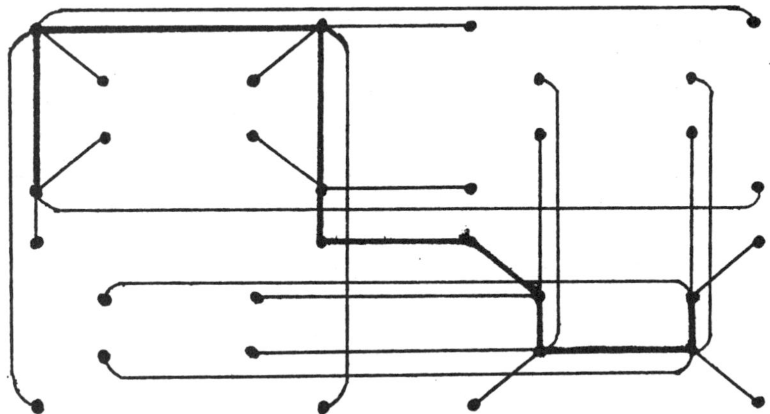

Figure 1. A 3–legged caterpillar which spans Q_5 in which Q_5 is represented as $Q_3 \times Q_2$.

For $n \geq 5$, a similar construction yields a 3–legged caterpillar with a spine of order $2^{n-2} + 2$, whose code is $(3, ..., 3, 2, 0, 0, 2, 3, ..., 3)$. We conjecture that this is the shortest such spine.

Observe that the number of endnodes of this shortest caterpillar is $2^n - (2^{n-1} + 2) = 3 \cdot 2^{n-2} - 2$ which asymptotically equals 75% of $|V(Q_n)|$. This raises the open question: What is the maximum possible number of endnodes of a spanning tree of Q_n? If the above observation always gives the maximum, then the answer for caterpillars is $3 \cdot 2^{n-2} - 2$, since the number of endnodes plus the number of spinal nodes equals 2^n.

In [7], we call a subgraph H of Q_n a *subcube of codimension* k if $H = Q_{n-k}$. Given an embedded spanning tree T of Q_n and a subcube H where $T \cap H$ is connected and is, in fact, a spanning tree of H, then H is called a *spanned subcube* of Q_n relative to T. We exhibited in [7], using techniques developed in [6], a method of determining whether a copy of Q_n with an embedded spanning tree T contains a spanned subcube of codimension 1. Each spanning tree of Q_3 determines at least one spanned subcube of codimension 1, while for $n \geq 4$, there exists spanning trees which do not. The analogous situation for spanned subcubes of codimension greater than 1 remains to be examined.

3. Embedding Meshes in Q_n

We characterized in [5] the meshes which span a hypercube:

Theorem B   *The m-mesh $M(a_1, ..., a_m)$ spans Q_n if and only if $a_1 a_2 ... a_m = 2^n$. In words, a mesh M spans some hypercube if and only if M is a binary mesh.*

As a consequence, if a tree spans a binary mesh, then it spans a hypercube. In particular, if a tree T spans a binary ladder $M(2, 2^{n-1})$, then T spans Q_n.

For the mesh $M(k, k)$ to be contained in Q_n, we must have $k^2 \leq 2^n$, from which we obtain $\lceil 2 \log_2 k \rceil \leq n$. Now since Q_m contains a hamiltonian path P_{2^m}, it follows that the mesh $P_{2^m} \times P_{2^m}$ spans $Q_m \times Q_m = Q_{2m}$. As $M(k, k) = P_k \times P_k$, we observe that $M(k, k)$ embeds in $Q_s \times Q_s = Q_{2s}$, where $s = \lceil \log_2 k \rceil$. Thus the minimum n for which $M(k, k) \subset Q_n$ satisfies

$$(1) \qquad\qquad \lceil 2 \log_2 k \rceil \leq n \leq 2 \lceil \log_2 k \rceil.$$

Now for any positive number x, we have $2\lceil x \rceil - \lceil 2x \rceil \leq 1$, hence the left and right expressions in (1) differ by at most 1. It is noted in the survey article [2] that the upper bound in (1) is exactly the smallest n.

By a similar argument, the minimum n such that the d-dimensional mesh $M(k,k,...,k)$ embeds in Q_n satisfies

$$(2) \qquad\qquad \lceil d \log_2 k \rceil \leq n \leq d \lceil \log_2 k \rceil.$$

Again, the upper bound gives the exact value.

REFERENCES

[1]   F. Harary, *Graph Theory.* Addison–Wesley, Reading (1969).

[2]   F. Harary, J.P. Hayes and H.J. Wu, A survey of the theory of hypercube graphs. *Comput. Math. Appl.* 15 (1988) 277 – 289.

[3]   F. Harary and M. Lewinter, The starlike trees which span a hypercube. *Comput. Math. Appl.* 15 (1988) 299 – 302.

[4] F. Harary and M. Lewinter, Spanning subgraphs of a hypercube II: Double starlike trees, *Math Comput. Modelling* 11 (1988) 216 – 217.

[5] F. Harary and M. Lewinter, Spanning subgraphs of a hypercube III: Meshes, *International Journal of Computer Mathematics,* 25 (1988) 1 – 4.

[6] F. Harary and M. Lewinter, Spanning subgraphs of a hypercube IV: Rooted trees, *Comput. Math. Appl.,* to appear.

[7] F. Harary and M. Lewinter, Spanning subgraphs of a hypercube V: Spanned subcubes, *Proc. First China–USA Conf. on Graph Theory*, to appear.

[8] F. Harary, M. Lewinter and W. Widulski, On two–legged caterpillars which span hypercubes. *Congress Numerantium*, to appear.

[9] I. Havel and P. Liebl, One–legged caterpillars span hypercubes. *J. Graph Theory* 10 (1986) 69 – 77.

[10] L. Nebesky, On quasistars in n–cubes, *Casopis Pest. Mat.* 109 (1984) 153 – 156.

The Ramsey Number of $K_{3,3}$

Heiko Harborth

Ingrid Mengersen

Technische Universität Braunschweig, West Germany

ABSTRACT

The Ramsey numbers $r(K_{2,3}, K_{3,3}) = 13$ and $r(K_{3,3}) = 18$ are proved.

The Ramsey number $r = r(G, H)$ is the smallest number r such that in every two–coloring of the edges of the complete graph K_r there is a copy of G with all edges of the first color (green) or a copy of H with all edges of the second color (red). If $G = H$ then $r(G, G) = r(G)$ denotes the so–called diagonal Ramsey number.

Burr [1] has listed the diagonal Ramsey numbers $r(G)$ for all graphs G with up to six edges and Hendry [5] has extended this list to all graphs with up to seven edges. Except for the complete graph of order five, $r(G)$ is known for every graph G with at most five vertices (see [1], [2], [3], [6], [7]). The proof of $r(K_5 - e) = 22$ was found recently [3]. At present it is known that $43 \le r(K_5) \le 55$.

So the Ramsey number is unknown for one of the two Kuratowski graphs, K_5. What about the other one, $K_{3,3}$?

The case where G is a complete bipartite graph is of fundamental interest. In only a few of the cases is $r(G)$ known exactly. It is known that

$$r(K_{1,n}) = \begin{cases} 2n & \text{if } n \text{ is odd,} \\ 2n - 1 & \text{if } n \text{ is even.} \end{cases}$$

Also, $r(K_{2,2}) = 6$ and $r(K_{2,3}) = 10$ are found in Burr's table [1]. Some general estimations are given in [2]. It is the purpose of this note to prove $r(K_{3,3}) = 18$.

Let $ge(v)$ or $re(v)$ denote the numbers of green or red edges, respectively, incident to a vertex v in a two–coloring of G. The sets of neighbors of v incident to v in green or red will be denoted by G_v or R_v, respectively. The numbers of green or red edges joining two vertex sets S and T are denoted ge(S,T) or re(S,T), respectively. If S consists of a single vertex x, then ge(x,T) or re(x,T) is used.

Theorem 1 $r(K_{1,3},K_{m,n}) = m + n + 2.$

Proof A green cycle graph C_{m+n+1} proves $r(K_{1,3},K_{m,n}) > m + n + 1$. In K_{m+n+2} without a green $K_{1,3}$ at most two green edges are incident to any vertex. Then m vertices of K_{m+n+2} can be chosen such that at most two of the remaining vertices are adjacent each by one green edge to one of the m vertices. The remaining n vertices together with these m vertices determine a red $K_{m,n}$, and $r(K_{1,3},K_{m,n}) \leq m + n + 2$ is proved. ❏

Theorem 2 $r(K_{2,3},K_{3,3}) = 13.$

Proof Figure 1 shows the green edges of a two–colored K_{12} without a green $K_{2,3}$ and without a red $K_{3,3}$. This proves $r(K_{2,3},K_{3,3}) > 12$.

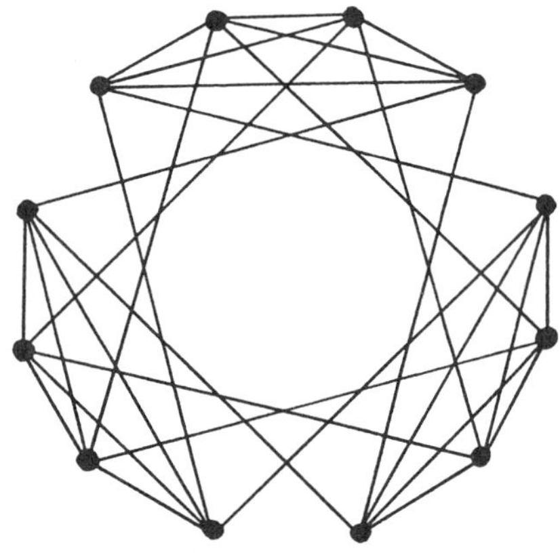

Figure 1

The following sequence of conclusions guarantees a green $K_{2,3}$ or a red $K_{3,3}$ in every two–coloring of the edges of K_{13}.

If $re(v) \leq 4$ in a two–coloring of K_{13}, then $ge(v) \geq 8$, and together with $r(K_{1,3},K_{3,3}) = 8$ a green $K_{2,3}$ or a red $K_{3,3}$ occurs. If $re(v) \geq 10$, then $r(K_{2,3},K_{2,3}) = 10$ guarantees a green $K_{2,3}$ or a red $K_{3,3}$.

Assume $5 \leq re(v) \leq 6$ for all vertices v in a two–coloring of K_{13}. Then a green $K_{2,3}$ exists, since otherwise $re(R_v,G_v) \geq re(v)(ge(v) - 2) \geq 24$ for a fixed vertex v, $re(w,G_v) \geq ge(v) - 3 \geq 3$ for every vertex $w \in G_v$, and thus $re(w_0) = re(w_0,R_v) + re(w_0,G_v) \geq 7$ for at least one vertex $w_0 \in G_v$, which contradicts the assumption.

In the remaining case a vertex v exists with $7 \leq re(v) \leq 9$.

Assume first $ge(R_v,G_v) > \min\{2re(v), re(v) + \binom{ge(v)}{2}\}$, that is, $ge(R_v,G_v) \geq 15$, 15, or 13 for $re(v) = 12 - ge(v) = 7, 8$, or 9, respectively. Then either one vertex of R_v is joined by three green edges to G_v, or two vertices of R_v together with two vertices of G_v determine a green $K_{2,2}$. In both cases together with vertex v a green $K_{2,3}$ occurs.

Assume second $re(R_v,G_v) > 4ge(v)$, that is $re(R_v,G_v) \geq 21, 17$, or 13 for $re(v) = 12 - ge(v) = 7, 8$, or 9, respectively. Then $re(w,R_v) \geq 5$ for at least one vertex $w \in G_v$. If w is adjacent by six red edges to a set S of six vertices of R_v, then either a red $K_{3,3}$ occurs, or two vertices of G_v are joined by at least eight green edges to S. These two vertices together with v and two vertices of S determine a green $K_{2,3}$. If w is adjacent by exactly five red edges to a set T of five vertices of R_v, then either v or w are joined by three green edges to a set, say C, of three vertices. If no red $K_{3,3}$ occurs, then $ge(T,C) \geq 9$, and v together with one vertex of T and all vertices of C, or with two vertices of T and two vertices of C determine a green $K_{2,3}$.

It remains
$$re(R_v,G_v) + ge(R_v,G_v) = re(v)ge(v) \leq 4ge(v) + \min\{2re(v), re(v) + \binom{ge(v)}{2}\},$$

which, however, is impossible for $re(v) = 7, 8$, and 9. Thus the proof of $r(K_{2,3},K_{3,3}) \leq 13$ is complete. ❑

Theorem 3 *$r(K_{3,3}) = 18$.*

Proof For the proof of $r(K_{3,3}) > 17$ it can be checked that the well–known unique two–coloring of K_{17} without a monochromatic K_4 does not contain a monochromatic $K_{3,3}$.

The proof of $r(K_{3,3}) \leq 18$ is partitioned into the following Lemmas.

Lemma 1 *If a two–colored complete graph contains a $K_{2,4}$ of one and a $K_{2,5}$ of the other color such that at most the sets of two vertices intersect one another, then there is a monochromatic $K_{3,3}$.*

Proof To avoid a monochromatic $K_{3,3}$ the four vertices of $K_{2,4}$ are joined by at least 12 edges of one color to the five vertices of $K_{2,5}$, and these are joined by at least 10 edges of the other color to the four vertices of $K_{2,4}$, however, there exist only 20 edges. ❑

Lemma 2 *If for some vertex v in a two–colored complete graph and for some $w \in G_v$, one has $ge(v) \geq 8$ and $re(w,R_v) \geq 5$ or else $ge(v) \geq 6$ and $re(w,R_v) \geq 6$, then there is a monochromatic $K_{3,3}$. — The same conclusion holds if colors green and red are interchanged.*

Proof Assume $ge(v) \geq 8$, and let v be joined by green edges to w and to a_1, a_2, ..., a_7, and v and w both be joined by red edges to b_1, b_2, ..., b_5. Either a red $K_{3,3}$ occurs or $ge(\{a_1, ..., a_7\},\{b_1, ..., b_5\}) \geq 21$. Then one vertex of $\{b_1, ..., b_5\}$, say b_5, is adjacent by green edges to five vertices of $\{a_1, ..., a_7\}$. Together with v there exists a green $K_{2,5}$. Moreover, v, w and $\{b_1, ..., b_4\}$ determine a red $K_{2,4}$, and Lemma 1 can be used. — The corresponding proofs of the second part of Lemma 2 and with colors interchanged are left to the reader. $\square$

In the following, $s_i(v)$ denotes the number of vertices $w \in G_v$ with $re(w,R_v) = i$, and $t_i(v)$ denotes the number of vertices $w \in R_v$ with $ge(w,G_v) = i$.

Lemma 3 *If*

$$\sum_{i \geq 3} \binom{i}{3} s_i(v) > \binom{re(v)}{3},$$

or

$$\sum_{i \geq 3} \binom{i}{3} s_i(v) = \binom{re(v)}{3}, \text{ and } re(w,R_v) \geq 3 \text{ for one } v \in R_v,$$

or the corresponding expressions with colors green and red interchanged are valid, then there is a monochromatic $K_{3,3}$.

Proof This is an easy consequence of the definitions of $s_i(v)$ and $t_i(v)$. $\square$

Lemma 4 *If for some vertex v in a two–colored K_{18} one has $ge(v) = 9$ or $re(v) = 9$, then there is a monochromatic $K_{3,3}$.*

Proof Assume $ge(v) = 9$, and thus $re(v) = 8$. Either $a \in G_v$ exists with $re(a,R_v) \geq 5$, or $b \in R_v$ with $ge(b,G_v) \geq 5$, and then Lemma 2 can be used. — For $re(v) = 9$ interchange colors green and red. $\square$

Lemma 5 *If for some vertex v in a two–colored K_{18} one has $ge(v) = 11$ or $re(v) = 11$, then there is a monochromatic $K_{3,3}$.*

Proof Assume $ge(v) = 11$, and therefore $re(v) = 6$. Furthermore assume that no monochromatic $K_{3,3}$ occurs.

Lemma 2 implies $t_i(v) = 0$ for $i \geq 6$, and thus $re(R_v,G_v) \geq 36$. Also Lemma 2 implies $s_i(v) = 0$ for $i \geq 5$. By Lemma 3 then $s_3(v) + 4s_4(v) \leq 20$ can be assumed. This together with $re(R_v,G_v) \geq 36$ leaves the unique possibility $s_4(v) = 3$, $s_3(v) = 8$, $s_3(v) + 4s_4(v) = 20$ and $re(R_v,G_v) = 36$. It follows $ge(R_v,G_v) = 30$, $ge(w,G_v) = 5$ and $re(w,G_v) = 6$ for all $w \in R_v$, and Lemma 3 yields $re(w,G_v) \leq 2$ for all $w \in R_v$.

For all $w \in R_v$ one has $re(w) = 1 + re(w,G_v) + re(w,R_v) = 7, 8,$ or 9. Lemma 4, however, leaves only $re(w) = 7$, and thus $re(w,R_v) = 0$ for all $w \in R_v$. This means, R_v is a green K_6 which contains a green $K_{3,3}$, in contradiction to the assumption. — For $re(v) = 11$ interchange colors green and red.

Lemma 6 *If for some vertex v in a two–colored K_{18} one has $ge(v) = 10$ or $re(v) = 10$, then there exists a monochromatic $K_{3,3}$.*

Proof Assume $ge(v) = 10$, and therefore $re(v) = 7$. Furthermore assume that no monochromatic $K_{3,3}$ occurs.

Lemma 2 implies $s_i(v) = 0$ for $i \geq 5$. By Lemma 3 it follows $s_3(v) + 4s_4(v) \leq 35$, and thus $s_4(v) \leq 8$, $re(G_v,R_v) \leq 8 \cdot 4 + 2 \cdot 3 = 38$, $ge(R_v,G_v) = 7 \cdot 10 - re(G_v,R_v) \geq 32$, and together with Lemma 2 then $t_i(v) = 0$ for $i \geq 6$ and $t_5(v) \geq 4$.

Let w_i for $1 \leq i \leq 7$ denote the vertices of R_v, and let $re(w_i,G_v) = 5$, that is $ge(w_i,G_v) = 5$, for $1 \leq i \leq t_5(v)$. If $re(w_i) \geq 10$ for some $i \leq t_5(v)$, then v and w_i are the sets of two vertices of a green $K_{2,5}$ and a red $K_{2,4}$, and Lemma 1 can be used. If $re(w_i) = 9$ or $re(w_i) = 8$ for some $i \leq t_5(v)$, then Lemma 4 works. If $re(w_i) = 6$ for some $i \leq t_5(v)$, then Lemma 5 can be applied. Since $re(w_i) \geq re(w_i,G_v) + 1 = 6$ for $i \leq t_5(v)$, it remains $re(w_i) = 7$, and thus $re(w_i,R_v) = 1$ for all $1 \leq i \leq t_5(v)$.

Now in R_v two vertices w_i and w_j with $i, j \leq t_5(v)$, which are joined in red, together with w_k, $k \leq t_5(v)$, determine a green $K_{3,3}$. Thus all edges (w_i,w_j), $i, j \leq t_5(v)$, can be assumed green.

If $t_5(v) \geq 5$, then w_6 or w_7, say w_7, is connected by at least three red edges to vertices w_i, $i \leq t_5(v)$, say to w_1, w_2, w_3. Then w_1, w_2, w_3 together with w_4, w_5, w_6 determine a green $K_{3,3}$.

It remains $t_5(v) = 4$. Then $ge(G_v,R_v) \geq 32$ implies $4t_4(v) + 3t_3(v) + \dots \geq 12$, and this is possible only for $t_4(v) = 3$ and $ge(G_v,R_v) = 32$. It follows $re(w_i,G_v) = 6$ for $i = 5, 6, 7$. Three red edges from w_5, w_6 or w_7 to $\{w_1,w_2,w_3,w_4\}$ force a green $K_{3,3}$ as before. Thus for example w_5 is joined by exactly two red edges, say to w_1 and w_2. Then w_1, w_2 both are joined by green edges to w_3, w_4, w_6, and w_7. Any green edge from w_5 to w_6 or to w_7 forces a green $K_{3,3}$. Two red edges (w_5,w_6) and (w_5,w_7), however, imply $re(w_5) = 11$, which is covered by Lemma 5. — For $re(v) = 10$ interchange colors green and red. ❏

Lemma 7 *If in a two–colored K_{18} for all vertices v either $ge(v) = 12$ or $re(v) = 12$, then there exists a monochromatic $K_{3,3}$.*

Proof Assume that $re(v) = 12$ for at least 9 vertices v of K_{18}. If every pair of them is connected by a red edge, then a red K_9 occurs which contains a red $K_{3,3}$. If at least one pair is connected by a green edge, then both vertices are joined by red edges to at least eight other vertices. Then, however, $r(K_{3,3}, K_{1,3}) = 8$ (Theorem 1) guarantees a monochromatic $K_{3,3}$. ❏

Now the proof of $r(K_{3,3}) \leq 18$ can be finished. By Lemmas 4 to 7 one vertex exists in a two–coloring of K_{18} with $ge(v) \leq 4$ or $ge(v) \geq 13$. Then $r(K_{3,3}, K_{2,3}) = r(K_{2,3}, K_{3,3}) = 13$ (Theorem 2) forces a monochromatic $K_{3,3}$, and Theorem 3 is proved. ❏

Thus $K_{3,3}$ is another graph which has 18 as its diagonal Ramsey number just as the path P_{13}, the star $K_{1,9}$, the book B_4, the complete graph K_4, and some other graphs. How large is the set of all graphs G with $r(G) = 18$?

A final remark: If the edges of a complete bipartite graph $K_{s,t}$ instead of K_r are two–colored, then all pairs (s, t) are known [4] such that any two–coloring of $K_{s,t}$ contains a monochromatic $K_{3,3}$, but $K_{s-1,t}$ and $K_{s,t-1}$ do not $(s \leq t)$, and these pairs are (5,41), (7,29), (9,23) and (13,17).

REFERENCES

[1]   S.A. Burr, Diagonal Ramsey numbers for graphs, *J.Graph Theory* **7** (1983), 57 – 69.

[2]   F.R.K. Chung and R.L. Graham, On multicolor Ramsey numbers for complete bipartite graphs, *J. Combinatorial Theory* **18** (B) (1975), 164 – 169.

[3]   C. Clapham, G. Exoo, H. Harborth, I. Mengersen and J. Sheehan, The Ramsey number of $K_5 - e$, *J. Graph Theory* **13** (1989), 7 – 15.

[4]   F. Harary, H. Harborth and I. Mengersen. Generalized Ramsey theory for graphs XII: Bipartite Ramsey sets, *Glasgow Math. J.* **22** (1981), 31 – 41.

[5]   G.R.T. Hendry, Diagonal Ramsey numbers for graphs with seven edges, *Utilitas Mathematica* **32** (1987), 11 – 34.

[6]   G.R.T. Hendry, Ramsey numbers for graphs with five vertices, *J. Graph Theory* **13** (1989), 245 – 248.

[7]   H. Harborth and I. Mengersen, All Ramsey numbers for five vertices and seven or eight edges, *Discrete Math.* **73** (1988/89), 91–98.

ON GRAPHS WITH (I, n)-REGULAR INDUCED SUBGRAPHS

M. A. Henning

Henda C. Swart

P. A. Winter

University of Natal, Durban

ABSTRACT

For $n \geq 2$, a graph G is said to be (I, n)-regular of degree k if each vertex of G is an end vertex of exactly k non-trivial paths of length vertex of G is an end vertex of exactly k non-trivial paths of length at most $n-1$. For $n \geq 3$, a complete characterization of connected graphs of girth at least n for which every vertex-induced, connected, non-trivial, proper subgraph is (I, n)-regular is obtained.

1.   Introduction

Th terminology and notation of [1] will be used throughout. In particular G will denote a connected graph with vertex set V, edge set E, order p, size q and girth g. If $v \in V$, the degree of v in G is written as deg v and $\delta = $ min $\{deg\, v : v \in V\}, \Delta = \max\{deg\, v : v \in V\}$. The degree set of G is denoted by D_G.

If n is an integer, $n \geq 2$ and $v \in V$, we denote by $p_n(v)$ the number of paths of length $n-1$ in G which have v as an end vertex and by $deg(n, v)$ the number of non- trivial paths of length $\leq n-1$ with v as an end vertex; hence $deg(n, v) = \sum_{m=2}^{n} p_m(v)$. The graph G is said to be (I, n)-*regular of degree* k if $deg(n, v) = k$ for all v in V. The graph G is said to be *highly* (I, n)-*regular* if G and all its non-trivial, vertex-induced, connected subgraphs are (I, n)-regular.

It is our purpose to characterize connected graphs of girth $\geq n$ for which every vertex-induced, connected, non-trivial, proper subgraph is (I, n)-regular for all integers $n \geq 2$, and also to characterize highly (I, n)-regular graphs of girth $\geq n$.

These concepts find application in the fields of Sociology, Epidemiology, etc., as follows: Given that an idea (infection, etc.) may be transferred from the originator and transmitted sequentially to $n - 1$ others, without losing its integrity, design experiments to test the potency of ideas such that all participants in the experiment have equal potential for disseminating their ideas. We associate the participants in the experiment with the vertices of a graph G, two vertices of G being adjacent if and only if ideas may be transferred directly between the corresponding participants. The demands of the experiment are met if G if (I, n)-regular, where G is to be of girth at least n to avoid feed-back. However, since such experiments may be conducted over a long period, in which some participants are expected to withdraw and the results processed only for a proper subset of participants (those remaining to the end), it is preferable that G should have the property that each of its proper, connected, non- trivial, vertex-induced subgraphs be (I, n)-regular for all integers $n \geq 3$. Highly (I, n)-regular graphs would be preferable, but their use is not always feasible.

Obviously a graph is (I, n)-regular if and only if it is regular. The authors [2] obtained a complete characterization of (I, n)-regular graphs of girth at least n.

Theorem 1 For $n \geq 3$, a connected graph G of girth $\geq n$ is (I, n)-regular if and only if it satisfies one of the following conditions:

(a) G is r-regular.

(b) G is a tree of diameter $\leq n - 1$.

(c) G is a bipartite graph with bipartition (X, Y), such that $deg\, x = \Delta$ and $deg\, y = \delta$ for all vertices $x \in X$ and $y \in Y$ and n is odd.

(d) If j is a proper factor of $n - 1, G$ is obtained from a graph H satisfying condition (a) or (c) above by the replacement of each edge of H by a path of length j.

(e) If $j = n - 1, G$ is obtained from a Δ-regular multigraph H by the replacement of each edge of H by a path of length j.

In particular, a connected graph G is $(I, 3)$-regular if and only if it satisfies (a) (and $r^2 = k$) or (c) (and $k = \delta\Delta$).

We note that, if G satisfies condition (d) or (e), then j is the shortest distance between a vertex of degree Δ and another vertex of degree > 2 in G.

Theorem 2   A connected graph G of order $p \geq 3$ has the property that every vertex-induced, connected, non-trivial, proper subgraph is $(I, 3)$-regular if and only if it satisfies one of the following conditions:

(i) G is a connected graph of order 3 of 4;

(ii) $p \geq 5$ and G is Kp of $K(p - t, t)$ for $1 \leq t \leq p - 1$.

Proof   If G is one of the graphs listed in (i) or (ii) above, then, clearly, every vertex-induced, connected, non-trivial, proper subgraph of G is $(I, 3)$-regular.

Let G be a graph which satisfies the hypotheses of the theorem. Since there is nothing to show if $p \leq 4$, we may assume that $p \geq 5$.

If G is a tree, let v be an end-vertex of G. From the comment following Theorem 1 and the observation that G contains no path of length 3 as a vertex-induced subgraph, we obtain $G - v = K(p - 2, 1)$ and $G = K(p - 1, 1)$.

For the remainder of the proof we shall assume that G is not a tree. Let C be a cycle in G of minimum order. Necessarily C is a vertex-induced subgraph of G. If C is a cycle of order ≥ 5, then G would contain a vertex-induced proper subgraph which is a path of length > 2 and is not $(I, 3)$-regular. Thus $C = C_3$ or $C = C_4$.

We next show that G is 2-connected. If this is not the case, then there is a cut-vertex v of G. Since G contains the cycle C, at least one component of $G - v$ is of order ≥ 2; let G_1 denote such a component and G_2 any other component of $G - v$. Then there exist vertices $u_1 \in V(G_1)$ and $u_2 \in V(G_2)$, each of which is adjacent to v in G. Let w be any neighbour of u_1 in G_1. Necessarily the vertex-induced subgraph on $\{u_1, u_2, v, w\}$ is either P_4 or $K_1 + (K_1 \cup K_2)$, neither of which is $(I, 3)$-regular, which contradicts our choice of G. Hence G is 2-connected.

We proceed by induction on p.

Case $C = C_3$: Suppose that $p = 5$ and $V(G) = V(C) \cup \{u_1, u_2\}$. Since G is 2-connected, each of u_1 and u_2 is adjacent to some vertex of C. Further, since the vertex-induced subgraph on $V(C) \cup \{u_1\}$ contains a triangle, the comment following Theorem 1 implies that this subgraph is K_4. Similarly, the vertex-induced subgraphs on $V(C) \cup \{u_2\}$ and $V(G) - \{v\}$, where $v \in V(C)$, are K_4. Consequently, G is K_5.

Suppose $p > 5$ and assume validity of the theorem for all connected graphs of order $p - 1$. This assumption, together with the fact that G contains a triangle and the observation that $G - v$ is connected for all vertices $v \in V$, implies that the vertex-induced subgraph on $V - \{v\}$ is K_{p-1} for every vertex $v \in V$. Therefore, obviously, $G = K_p$.

Case $C = C_4$: Suppose that $p = 5$ and that $V(G) = V(C) \cup \{u\}$. In view of the fact that G is 2-connected and triangle free, u must be adjacent to exactly two non-adjacent vertices of C. Therefore $G = K(2, 3)$.

Suppose $p > 5$ and assume validity of the theorem for all connected graphs of order $p - 1$. This assumption, together with the fact that G is triangle free and the observation that $G - v$ is connected for all vertices $v \in V$, implies that the vertex-induced subgraph on $V - \{v\}$ is $K(p - 1 - t_1, t_1)$ or $K(p - 1 - t_2, t_2)$ where $1 < t_1 < t_2 < p - 2$ and $t_2 - t_1 = 1$. Thus the complement of G is disconnected and G is reconstructible. Therefore G is $K(p - t_2, t_2)$, as required.

Theorem 3   For $n \geq 4$, a connected graph G of girth $\geq n$ has the property that every vertex-induced, connected, non-trivial, proper subgraph of G is (I, n)- regular if and only if it satisfies one of the following conditions:

(i) $G = K(2, 3)$ and $n = 4$;

(ii) G is a tree of diameter $\leq n - 1$;

(iii) $G = P_{n+1}$;

(iv) $G = C_n$;

(v) $G = C_{n+1}$;

(vi) $G = C_n \cup <u> +uw$, where $w \in V(C_n)$.

Proof   If G is one of the graphs listed in (i) to (vi) above, then, clearly, every vertex-induced, connected, non-trivial, proper subgraph of G is (I, n)-regular.

Let G be a graph which satisfies the hypotheses of the theorem and which does not satisfy conditions (ii), (iii), (iv) or (v). Since G does not satisfy (ii) or (iii), Theorem 1 implies that G is not a tree. Let C be a cycle of minimal order in G. Note that since C is minimal it is a vertex-induced subgraph of G. Any vertex-induced subgraph of C is a vertex-induced subgraph of G and cannot be P_{n+1}, whence $|V(C)| \leq n + 1$. Since the girth of $G \geq n$, it follows that $|V(C)| \in \{n, n + 1\}$. Because G does not satisfy (iv) or (v), $C \neq G$.

Let u be a vertex in $V(G) - V(C)$ which is adjacent to a vertex of C. If u is adjacent to more than one vertex of C, then $C = C_4$ and the vertex-induced subgraph G_1 on $V(C) \cup \{u\}$ is $K(2, 3)$ with $n = 4$. If $G_1 \neq G$, then G_1 must be (I, n)-regular, which is false by Theorem 1. Therefore $G = G_1 = K(2, 3)$.

Assume u is adjacent to only one vertex of C. Let G_1 be the vertex-induced subgraph of G on $V(C) \cup \{u\}$. Then $C \neq C_{n+1}$, since otherwise G_1 would contain a vertex-induced path of length n which is not (I, n)-regular. Thus $G_1 = C_n \cup <u> +uw$ for some $w \in V(C)$. If $G_1 \neq G$, then G_1 must be (I, n)-regular which is false by Theorem 1. Therefore $G = G_1$.

The above theorems now yield a complete characterization of highly (I, n)-regular graphs of girth $\geq n$:

Corollary 1 For $n \geq 3$, a connected graph G of girth $\geq n$ is highly (I, n)-regular if and only if it satisfies one of the following conditions:

(a) $G = K_p$ and $n = 3$;

(b) $G = K(p - t, t)$ where $1 \leq t \leq p - 1$ and $n = 3$;

(c) G is a tree of diameter $\leq n - 1$;

(d) $G = C_n$;

(e) $G = C_{n+1}$.

The authors thank the referee for improvements to the paper.

REFERENCES

[1] G. Chartrand and L. Lesniak, *Graphs and Digraphs* (Second Edition), Wadsworth, Monterey, 1986.

[2] M. A. Henning, Henda C. Swart and P. A. Winter, "On (I,n)-regular graphs," *Ars Combinatoria* 23A (1987), 141-152.

MATCHING EXTENSION AND CONNECTIVITY IN GRAPHS ii

Derek Holton*

University of Otago - New Zealand

M. D. Plummer**

Vanderbilt University

1. Introduction

All graphs considered are finite, undirected, connected and simple (i.e., they have no loops or parallel lines). Let n and p be positive integers with $n \leq (p-2)/2$ and let G be a graph with p points having a perfect matching. Graph G is said to be n-*extendable* if every matching of size n in G extends to a perfect matching. (We shall call graph G 0-*extendable* if it contains a perfect matching.) For a discussion of the role of the concept of n-extendability within the general framework of matching theory in graphs and for an historical résumé of the development of n-extendability, the reader is referred to the book [3], to [6] or to [7].

---

*Work supported by Grant UGC 32-635
** Work supported by ONR Contract #N00014-85-K-0488, NSF New Zealand-U.S.A. Cooperative Research Grant INT-8521818, and the University of Otago

In this paper we continue work begun in [6] and [7]. In Section 2 of this paper, we improve a general connectivity result as follows. In [7] it was shown that if one has a point cutset S of size $n+1$ in an n-extendable graph G where $n \geq 2$, then the graph $G - S$ has at most $n+1$ components with equality holding if and only if $G = K_{n+1,n_1}$. We sharpen this result to show that for $n \geq 2$, if G is n-extendable and $\neq K_{n+1,n+1}$ and if S is a point cutset of size $n+1$, then $G - S$ must have *exactly two* components.

In Section 3, we address a problem on matching extension in bipartite graphs. In [5], it was shown that no planar graph can be 3-extendable. In light of that result, it seemed natural to investigate 2-extendable planar graphs and this study was begun in [1]. In that paper it was shown, in particular, than any cubic 3-connected planar graph which is cyclically 5-connected must be 2-extendable. The planarity property played a strong role in the proof and it occurred to the authors to ask whether in a *non*-planar $n+1$-regular $n+1$-connected graph there was *any* lower bound on cyclic connectivity which would guarantee that the graph be n-extendable. We now show that the answer is yes, at least in the case when the graph is bipartite. Moreover, the bound we obtain is sharp by a recent result of Lou and Holton [2].

2. A Point-Connectivity Result

There are two basic theorems about n-connected graphs which we will need repeatedly in this paper, and we list them for the reader's convenience. Proofs of each may be found in [4].

1980A. Theorem   If $n \geq 2$ and G is n-extendable, then G is $(n-1)$-extendable.

1980 B. Theorem   If $n \geq 1$ and G is n-extendable, then G is $(n+1)$-connected.

We now present the main result of this section.

Theorem 2.1   Suppose n is a positive integer. Suppose further that G is an n-extendable graph and S is a point cutset of G with $|S| = n+1$. then:

(a) $G = K_{n+1,n+1}$, or

(b)   if $n = 1$, then $G - S$ has at least 2 even components, but no odd components, or exactly 2 odd components, but no even components;

(c)   if $n \geq 3$ and n is odd, then $G - S$ has exactly 2 odd components and no even components, or exactly 2 even components and no odd components;

or

(d)   if $n \geq 2$ and n is even, then $G - S$ has exactly 2 components, one of which is odd and the other, even.

Proof   In this proof, in addition to Theorems 1980A and 1980B stated above, we will also make use of the simple observation that if $n \geq 1$ if G is *any* n-extendable graph and if $e = ab$ is any line in G, then the graph $G - a - b$ is $(n - 1)$-extendable. The proof will proceed by induction on n. First we deal with the "anomalous" case when $n = 1$.

Suppose $G - S$ has an even component C. Let e_1 be a line joining C to S. Let M be a perfect matching containing e_1. Then by parity, M must match S into C. Thus $G - S$ has no odd components and at least 2 even components, as claimed.

Now suppose $G - S$ has no even components and hence at lest 2 odd components. But then since G is 1-extendable, $G - S$ must have exactly 2 odd components.

Now suppose $n = 2$. First suppose $G - S$ has at least 2 odd components (and hence by parity at least 3 odd components). Furthermore, let us suppose that $G - S$ has an even component C_e. Let e_1 be a line joining S to C_e. Then e_1 does not extend to a perfect matching, a contradiction. Hence $G - S$ has only odd components and at least of 3 of them.

Suppose one of these odd components, C_0, has at least 3 points. Let e_1 and e_2 be a matching from C_0 to S. Then $\{e_1, e_2\}$ does not extend to a perfect matching and again we have a contradiction. Thus all components of $G - S$ are singletons. Since G has a perfect matching, $G - S$ must consist of exactly 3 singletons. But since G is 3- connected, all of these singletons must have degree 3. Also recall from Theorem 2.2 of [7] that set S is independent. Thus $G = K_{3,3}$.

Suppose now that $G - S$ has exactly 1 odd component and hence at least 1 even component.

First we shall suppose that $G - S$ has at least 2 even components, say C_1 and C_2. Let $\{e_1, e_2\}$ be a matching of one point of C_1 and one of C_2 into S. Then the matching $\{e_1, e_2\}$ does not extend, a contradiction. Thus $G - S$ has exactly one even component as claimed.

Now let us suppose that $n = 3$. Suppose $G - S$ has at least 2 even components. Let C_3, C_4 be two such. Let $\{e_3, e_4\}$ be a matching of size 2 where e_i matches a point of C_i into S. Then $\{e_3, e_4\}$ extends to a perfect matching which must match a point in each of C_3 and C_4 to the remaining two points of S. Hence $G - S$ has no odd components.

Suppose now that $G - S$ has at least 3 even components. Let C_5, C_6 and C_7 be 3 such even components and let $\{e_5, e_6, e_7\}$ be a matching of size 3 of C_5, C_6 and C_7 respectively into S. Then $\{e_5, e_6, e_7\}$ does not extend to a perfect matching, a contradiction.

Thus G must have exactly 2 even components and hence no odd components. Note further here that it is impossible in this case that both even components are K_2's for if both were K_2's, graph G would not be 2-extendable. Hence by Theorem 1980A, G would not be 3-extendable either and this is a contradiction.

Suppose next that $G - S$ has no even components. Thus $G - S$ has at least 2 odd components.

Suppose one of these odd components has at least 3 points. Call it C_8. Let e_8, e_9 and e_{10} be a matching from components C_8 into S. Since this matching extends to a perfect matching, it follows that $G - S$ has precisely 2 odd components.

Now suppose that all the odd components of $G - S$ are singletons. Then again, since S is independent, it follows that $G = K_{4,4}$.

Finally, suppose that $G-S$ has exactly 1 even component C_e. (Hence G has at least 2 odd components.) Suppose that $|V(C_e)| \geq 4$. Then let e_{11}, e_{12} and $\{e_{13}\}$ be a matching of size 3 from C_e into S. Extending $\{e_{11}, e_{12}, e_{13}\}$ to a perfect matching is then impossible, contrary to the assumption that G is 3-extendable.

So $|V(C_e)| = 2$ and hence $C_e = K_2$.

Now suppose $G - S$ has at least 3 odd components. Let C_9, C_{10} and C_{11} be 3 such. Let $\{e_{14}, e_{15}, e_{16}\}$ be a matching where e_{14} joins C_9 to S, e_{15} joins C_{10} to S and e_{16} joins C_e to S. Then $\{e_{14}, e_{15}, e_{16}\}$ does not extend and again we have a contradiction.

Thus $G-S$ has exactly 2 odd components. Call these 2 components C_{12} and C_{13}. (Let us denote the even components C_e which is a single line by f.) Let $\{e_{17}, e_{18}\}$ be a matching of C_{12} (respectively C_{13}) into S. Then $\{e_{17}, e_{18}, f\}$ does not extend to a perfect matching, a contradiction.

This completes the proof for $n = 3$.

1. First suppose that n is even, $n \geq 4$ and that G is n-extendable.

1.1 Suppose $G - S$ has an even component C_e. Let $e = ab$ be a line joining S and C_e, where $a \in S$ and $b \in V(C_e)$. Let $H = G - a - b$. Then H is $(n-1)$-extendable (by Theorem 1980A) and $n-1$ is odd. Also $S' = S - a$ is a (minimum) cutset of H (by Theorem 1980B). Thus by the induction hypothesis, there are three possibilities: either $H = K_{n,n}, H - S'$ consists of 2 odd components and no even components, or $H - S'$ consists of 2 even components and no odd components.

1.1.1 Suppose $H = K_{n,n}$. Let $T = V(G) - S - b$. Now we already know that $S' = S - a$ is independent has cardinality n. Hence S' must be one of the two color classes of $H = K_{n,n}$. But then T must be the other color class of $H = K_{n,n}$ and hence T is independent in H. It now follows that all odd components of $G - S$ are singletons and that C_e is the only even component of $G - S$. Furthermore, since C_e is connected and $|V(C_e)| \geq 2$, component C_e must contain a line which must be of the form bc, where

$c \in V(C_e)$. Both line bc does not extend to a perfect matching and so G is not 1-extendable. Hence, yet again by Theorem 1980A, graph G is not n-extendable, a contradiction.

1.1.2. Next suppose that $H - S'$ consists of precisely two odd and no even components. Call the odd components A and B. Without loss of generality, suppose that $B \subseteq C_e$. Now $b \in V(C_e)$ and since S is a cutset of G, b is not adjacent to any point of A. Thus $G - S$ consists of precisely 2 components, one odd, the other even, as claimed.

1.1.3. Now suppose that $H - S'$ has precisely 2 even and no odd components. Call the even components D and E. Again, since S is a cutset in G, point b is adjacent to at most one of D and E and hence to exactly one of D and E, say E. But then $C_e = G[V(E) \cup \{b\}]$ is an odd component of $G - S$, a contradiction.

1.2. Let us now suppose that $G - S$ has no even components.

1.2.1. First suppose that all odd components of $G - S$ are singletons. Then since G is $(n+1)$-connected, we must have $G = K_{n+1,n+1}$.

1.2.2. Next suppose that at least one of the odd components of $G - S$ has at least 3 points, say component C_0. Let $ab = e$ be a line joining C_0 to S. Let $H = G - a - b$. Then H is $(n-1)$-extendable), and $S' = S - a$ is a minimum cutset of H, as before. So by the induction hypothesis, either $H = K_{n,n}$, $H - S'$ has precisely 2 odd components and no even components, or $H - S'$ has precisely 2 even components and no odd components.

1.2.2.1. Suppose $H = K_{n,n}$. It then follows that $G = K_{n+1,n+1}$.

1.2.2.2. Suppose $H - S'$ has 2 odd components and no even ones. Let the odd components be called D and E. Again, as before, point b is adjacent to exactly one of D and E, say E. This $G - E$ has an even component, namely $G[V(E) \cup \{b\}]$, contradiction.

1.2.2.3. Now suppose that $H - S'$ has 2 even components and no odd components. Again, point b is adjacent to exactly one of D and E, say E. Thus $G - S$ has an even component, namely D, and once more we have a contradiction.

This completes the proof in Case 1.

2.   Now let us suppose that n is odd, $n \geq 5$ and G is n-extendable.

2.1   Suppose $G - S$ has an even component, C_e. Let H, a, b and $S\prime$ be as in Case 1.1. Then by the induction hypothesis, either $H = K_{n,n}$ or $H - S'$ has exactly two components, one odd and the other, even.

2.1.1.   Suppose $H = K_{n,n}$. As before, S' is an independent set of size n in H and as such, must be one of the two color classes of $H = K_{n,n}$. But then T must be the other color class of $H = K_{n,n}$ of size n and thus again $G - S$ has exactly one even component C_e (and $C_e = K_2$) and $n - 1$ singleton (odd) components. But since S is independent, the line constituting C_e cannot extend to a perfect matching. Hence G is not 1-extendable, and hence by Theorem 1980A, G is not n-extendable, a contradiction.

2.1.2.   Now suppose $H - S'$ has one odd component-call it A-and the other even - call it B. As before, point b is adjacent only to points of one of A and B, and since $b \in C_e$, b is adjacent to points of the odd component A. So $G[V(A) \cup \{b\}] = C_e$ and $G - S$ has exactly 2 even components.

2.2   So, finally, suppose $G - S$ has no even components, and hence at least 2 odd components.

2.2.1.   Suppose one of these odd components, C_0, has at least 3 points. Choose a line $e = ab$ joining a point a in S to a point b in C_0. Let $H = G - a - b$ as before. By the induction hypothesis, graph H has precisely one odd component and one even. Let the odd component be A and the even, B. But then $G - S$ has exactly 2 odd components; namely A and $G[V(B) \cup \{b\}] = C_0$.

2.2.2.   So we may suppose that all the odd components of $G - S$ are singletons. But then $G = K_{n+1,n+1}$ and the proof of the theorem is complete.

In closing this section, we remark that each of the classes (b), (c) and (d) of n-extendable graphs mentioned in the conclusion of the above theorem is non-empty. To see this, let us introduce some notation. Let G and H

denote any 2 graphs. Then by $G + H$ we mean the *join* of G and H. The join is defined to be the graph obtained from graphs G an H by joining each point of G to each point of H. Now let F, G and H be any three graphs. Let us denote by $F + G + H$ the join $G + (F \cup H)$. Finally, let K_2 denote the graph consisting of r disjoint liens. With this notation, we note that class (b) contains the infinite family $\overline{K_2} + rK_2$, for $r \geq 2, K_{2_r} + \overline{K_2} + K_{2s}$, for $r \geq 2$ and $s \geq 2$ and $K_{2r+1} + \overline{K_2} + K_{2s+1}$, for $r, s \geq 2$.

In class (c) for $n \geq 3, n$ odd, we have $K_{r+2} + \overline{K_{n+1}} + K_s, r, s \geq n$ and both odd, as well as $K_{r+3} + \overline{K_{n+1}} + K_{s+1}, r, s \geq n$, and both odd.

Finally, in class (d) for $n \geq 2, n$ even, we have $K_{r+1} + \overline{K_{n+1}} + K_{s+2}, r, s, \geq n$ and both even.

In closing our discussion of Theorem 2.1, it is worth observing that none of the conditions (b), (c) or (d) implies that G is n-extendable, so that a converse of Theorem 2.1 of this type fails to hold.

3. A Cyclic Connectivity Result for Regular Bipartite Graphs

In this section we will be concerned with extending matchings in regular bipartite graphs. Let us begin by recalling the concept of cyclic connectivity. The *cyclic connectivity* of a connected graph G is the cardinality of a smallest cutset of lines in G the deletion of which leaves 2 components each containing a cycle. We denote this number by $c\lambda(G)$.

Before presenting our main result, we state the next Lemma, the proof of which is trivial and is left to the reader.

Lemma 3.1   Let F be a forest with no isolated points. Suppose the bipartition of $V(F)$ is $A \cup B$, where $|A| = a, |B| = b$ and $a > b$. Then F contains a tree with at least 2 endpoints in A.

Now we are prepared for the main result of this section.

Theorem 3.2   Let n be a positive integer and let G be an $(n + 1)$-regular, $(n + 1)$-connected bipartite graph. Then if $c\lambda(G) \geq n^2$, G is n-extendable.

Proof   If $n = 1$, the conclusion is immediate since G is just an even cycle.

So let us assume that $n \geq 2$. Suppose graph G satisfies the hypotheses of this theorem and suppose further that the bipartition of its points is $V(G) = A \cup B$. Now suppose that G is not n-extendable. Then there are n independent lines $e_1 = a_1 b_1, \cdots, e_n = a_n b_n$ where each point $ai \in A$ and $b_i \in B$ and such that $G' = G - a_1 - \cdots - a_n - b_1 - \cdots - b_n$ has no perfect matching. Thus by the well-known theorem of Philip Hall on bipartite matching, we may assume, without loss of generality, that there exists a point set $A_1 \subseteq A$ with $|\Gamma_{G'}(A_1)| < |A_1|$. (Here $\Gamma_{G'}(A_1)$ denotes the set of neighbors of set A_1 in graph G'.) Moreover, since graph G is $(n+1)$-connected, $\{b_1, ..., b_n\}$ is not a cutset of G and hence A_1 contains no isolates in G'. Let us denote $\Gamma_{G'}(A_1)$ by B_1. Let $G_1 = G[A_1 \cup B_1]$ and let $G_0 = G[A_0 \cup B_0]$ where $A_0 = A - (A_1 \cup \{a_1, ..., a_n\})$ and $B_0 = B - (B_1 \cup \{b_1, ..., b_n\})$.

Note that here at the outset that the pair (A_1, B_1) where $\Gamma_{G'}(A_1) = B_1$, point set A_1 contains no isolates and $|\Gamma_{G'}(A_1)| = |B_1| < |A_1|$ is certainly not necessarily unique. Let us call such a pair (A_1, B_1) a *Hall barrier* in G'. (For more on the barrier concept in general, see [3].)

Note that $B_1 \neq \phi$ for if $B_1 = \phi$, then $\{b_1, ..., b_n\}$ separates A_1 from $G - A_1$, thus contradicting the assumption that G is $(n+1)$-connected. Thus $|B_1| \geq 1$ and hence $|A_1| \geq 2$. (Similarly, $B_0 \neq \phi$ and since $\{a_1, ..., a_n\}$ is not a cutset, we have $|A_0| \geq 1$ and hence $|B_0| \geq 2$.)

For the rest of this proof, let us adopt the following terminology:

$$m = \text{ the number of lines from } A_1 \text{ to } \{b_1, ..., b_n\},$$

$$n_0 = \text{ the number of lines from } B_0 \text{ to } \{a_1, ..., a_n\},$$

$$n_1 = \text{ the number of lines from } B_1 \text{ to } \{a_1, ..., a_n\},$$

$$n_2 = \text{ the number of lines from } B_1 \text{ to } \{a_1, ..., a_n\},$$

$$n_3 = \text{ the number of lines from } \{b_1, ..., b_n\} \text{ to } A_0, \text{ and}$$

$$n_4 = \text{ the number of lines in } G[S],$$

where $S = \{a_1, ..., a_n\} \cup \{b_1, ..., b_n\}$. (See Figure 3.1 below. Line cut L shown in that figure will be discussed later in this proof.)

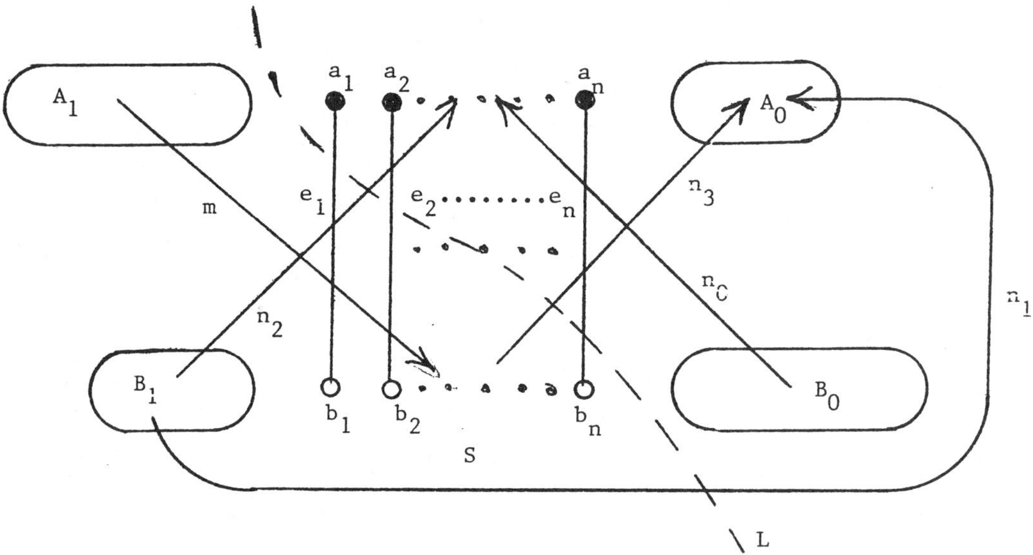

Figure 3.1

Now counting lines in G_1, we have $|A_1|(n+1) - m = |B_1|(n+1) - n_1 - n_2$, or

$$|A_1|(n+1) = |B_1|(n+1) - n_1 - n_2 + m. \tag{1}$$

Also $|A_1| \geq |B_1| + 1$, so

$$|A_1|(n+1) \geq (|B_1|+1)(n+1) = |B_1|(n+1)+n+1. \tag{2}$$

Combining (1) and (2), we get $|B_1|(n+1)+n+1 \leq |B_1|(n+1)-n_1-n_2+m$, or

$$n_1 + n_2 \leq m - n - 1. \tag{3}$$

Claim 1 If G_0 is a forest, then $|A_0| + |B_0| \leq 2n$. Since G_0 is a forest, $|V(G_0)| \geq |E(G_0)| + 1$, and so

$$
\begin{aligned}
2(|V(G_0)|) &= 2(|A_0| + |B_0|) \\
&\geq 2(|E(G_0)| + 1) \\
&= \sum_{v \in V(G_0)} deg\,v + 2 \\
&= |A_0|(n+1) - n_1 - n_3 + |B_0|(n+1) - n_0 + 2
\end{aligned}
$$

or

$$(|A_0| + |B_0|)(n-1) \leq n_0 + n_1 + n_3 - 2. \tag{4}$$

Now we also have

$$n_0 \leq n^2 - n_2, \tag{5}$$

and

$$n_3 \leq n^2 - m. \tag{6}$$

Thus substituting (3), (5) and (6) into (4), we obtain

$$(|A_0| + |B_0|)(n-1) \leq n^2 - n_2 + n_1 + n_3 - 2$$

$$\leq n^2 - n_2 + m - n - 1 - n_2 + n_3 - 2$$

$$\leq n^2 - 2n_2 + m - n - 3 + n^2 - m$$

$$= 2n^2 - n - 3 - 2n_2$$

$$\leq 2n^2 - n - 3$$

$$= n^2 - n + n^2 - 3$$

So

$$|A_0| + |B_0| \leq n + \frac{n^2 - 3}{n - 1}$$

$$< n + \frac{n^2 - 1}{n - 1}$$

$$= 2n + 1.$$

So $|A_0| + |B_0| \leq 2n$, as claimed.

Claim 3 $n_1 + n_2 + n_3 + n_4 \leq n^2 - 1.$

For we have

$$n_1 + n_2 + n_3 + n_4 = n_1 + n_2 + n_3 + n(n+1) - m - n_3$$

$$\leq m - n - 1 + n(n+1) - m$$

$$= n^2 - 1.$$

Now among all Hall barriers in G', let (A_1, B_1) be one with the smallest number of points. Without loss of generality, we may assume that $A_1 \subseteq A, B_1 \subseteq B$ and $|A_1| \geq |B_1| + 1 \geq 2$. Let $G_1 = G[A_1 \cup B_1]$. We now define $H_0 = G[A_0 \cup B_0 \cup \{a_1, ..., a_n\}]$ and $H_1 = G[A_1 \cup B_1 \cup \{b_1, ..., b_n\}]$.

Claim 3   Subgraph H_1 contains a cycle.

Suppose not. Then H_1 is a forest and hence so is G_1. Since G_1 is a Hall barrier, it contains no isolates. Since $|A_1| > |B_1|$, we may apply Lemma 3.1 to conclude that G_1 contains a tree T_1 with at least 2 endpoints in A_1. But since G is $(n+1)$- regular, each of these endpoints is adjacent to all of

$\{b_1, ..., b_n\}$ and hence, since $n \geq 2$, subgraph H_1 must contain a 4-cycle, a contradiction.

Claim 4 $|A_0| = |B_0| - 1$.

We will show that $|A_1| = |B_1| + 1$, and the claim will follow. Suppose $|A_1| \geq |B_1| + 2$.

Suppose $\alpha \in A_1$. Furthermore, suppose that every neighbor in B_1 of α is adjacent to some point in $A_1 - \alpha$. Then $(A_1 - \alpha, B_1)$ is a smaller Hall barrier than A_1, B_1), a contradiction.

So for every $\alpha \in A_1$ there must exist a $\beta \in B_1$ such that α is adjacent to β, but β is adjacent to no other point in A_1. But then we must have a matching of A_1 into B_1 and hence $|A_1| \leq |B_1|$ which is again a contradiction.

Claim 5 H_0 contains a cycle.

Let $\ell = |B_0|$. Then $|A_0| = \ell - 1$ by Claim 4. Now suppose H_0 does not contain a cycle. Then H_0 is a forest (and hence so is G_0). Hence

$$
\begin{aligned}
|V(H_0)| &= 2\ell - 1 + n \\
&\leq |E(H_0)| + 1 \\
&= (n + \ell - 1)(n + 1) - (n_1 + n_2 + n_3 + n_4) + 1 \\
&\leq (n + \ell - 1)(n + 1) - (n_2 - 1) + 1 \\
&= \ell n + \ell + 1.
\end{aligned}
$$

So, in particular,

$$
\begin{aligned}
\ell n &\leq 2\ell + n - \ell - 2 \\
&= \ell + n - 2 \\
&< 2n - 2
\end{aligned}
$$

where we have used Claims 2 and 1 respectively.

But $\ell \geq 2$ and we have $2n < 2n - 2$, a contradiction.

Thus by Claims 3 and 5, if L is the set of lines counted by $n_1+n_2+n_3+n_4$ (see Figure 3.1), then L is a *cyclic* line cut and $c\lambda(G) \leq |L| \leq n^2 - 1$, contradicting the hypothesis and completing the proof of the theorem.

Remark 1   Figure 3.2 below shows a 3-regular 3-connected bipartite graph G with $c\lambda(G) = 3$ which is not 2-extendable, thus showing that our lower bound in Theorem 3.2 is sharp, at least for the value $n = 2$. Very recently, Lou and Holton [2] have shown that the lower bound in Theorem 3.2 is sharp for *all* $n \geq 2$ by constructing, for all such n, $(n+1)$-regular $(n+1)$-connected bipartite graphs which have $c\lambda(G) = n^2 - 1$, but which are not n-extendable.

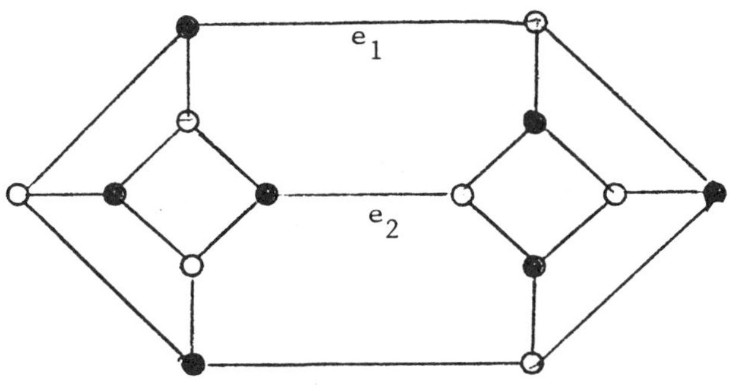

Figure 3.2

Remark 2   It is also interesting to compare the case $n = 2$ in Theorem 3.2 above with Theorem 3.1 of [1] which says that if G is a 3-regular 3-connected planar (but not necessarily bipartite) graph with $c\lambda(G) \geq 4$ and if G has no faces of size 4, then G is 2-extendable.

REFERENCES

[1] D. A. Holton and M. D. Plummer, "2-extendability in 3-polytopes," *Proceedings of the Seventh Hungarian Symposium on Combinatorics (Eger, 1987)* 52, Colloq. Math. Soc. János Bolyai, Akadémiai Kiadó, Budapest, 281-300.

[2] Dingjun Lou and D. A. Holton, "Lower bound of cyclic edge connectivity for n- extendability of regular graphs, preprint, 1988.

[3] L. Lovász and M. D. Plummer, *Matching Theory*, Ann. Discrete Math. 29, North- Holland, Amsterdam, 1986.

[4] M. D. Plummer, "On n-extendable graphs," Discrete Math. 31, 1980, 201-210.

[5] —————————, "A theorem on matchings in the plane," *Graph Theory in Memory of G. A. Dirac*, Eds.: Andersen, et al., Ann. Discrete Math. 41, North-Holland, Amsterdam, 1989, 347-354.

[6] —————————, "Matching extension in bipartite graphs," *Proceedings of the Seventeenth Southeastern Conference on Combinatorics, Graph Theory and Computing*, Congress. Numer. 54, Utilitas Math., Winnipeg, 1986, 245-258.

[7] —————————, "Matching extension and connectivity in graphs," *Proceedings of the 250th Anniversary Conference on Graph Theory*," Eds.: K. S. Bagga, et al., Congress. Numer. 63, Utilitas Math., Winnipeg, 1987, 147-160.

ORTHOMORPHISMS AND NEAR ORTHOMORPHISMS

D. Frank Hsu

Fordham University

ABSTRACT

*We briefly survey the development of orthomorphism, near orthomorphism,
orthomorphism graph and their applications. Using a classification scheme
similar to that used in the study of neofields, we exhibit orthomorphisms
and near orthomorphisms of cyclic group of small order.*

1.   Introduction

Let $|G|$ be the order of a finite group $(G, .)$. We define a *K-Complete Mapping* where $K = \{k_1, ..., k_n\}$ to be an arrangement of the non-identity elements of G into s cyclic sequences of lengths $k_1 ..., k_s : (g_{11} \cdots g_{1k_1}), (g_{21} \cdots g_{2k_2}), \cdots, (g_{s1} \cdots g_{sk_s})$, so that the elements $g_{i,j}^{-1} g_{i,j+1})$, together with the elements range over all the non-identity elements of G when i ranges from 1 to the positive integer s. It is obvious that $\sum_{i=1}^{s} k_i = |G| - 1$. Similarly, a *K-Near Complete Mapping* where $K = \{k; k_1, ..., k_s\}$ is an arrangement of the elements of $|G|$ into a sequence with length h and s cyclic sequences of length $k_1, \cdots, k_s : [g_1' \cdots g_h'](g_{11} \cdots g_{1k_1}) \cdots (g_{s_1} \cdots g_{sk_s})$ such that the elements $(g_i')^{-1} g_{1+1}'$ and $g_{i,j}^{-1} g_{i,j+1}$ together with $g_{i,k_i}^{-1} g_{i,1}$ consist of each of the non-identity elements of G exactly once. Here we have $h + \sum_{i=1}^{s} k_1 = |G|$. A *Generalized Complete Mapping* is either a K-complete Mapping or a K-Near Complete Mapping.

Generalized Complete Mapping was defined and studied in Hsu and Keedwell [14] and [15] in a more general context. The concept of a generalized

complete mapping subsumes those of a complete mapping, a starter, an even
starter, a sequencing, a near-sequencing (or R-sequencing) and a left (or right)
neofield. Generalized complete mappings have been used to characterize left
neofields (see Hsu and Keedwell [14], to provide new constructions of block
designs of Mendelsohn type (see Hsu and Keedwell [15] and to give graceful
labelings of directed graphs (see Bloom and Hsu [1]). Reprints have been col-
lected in Hsu [12] and a survey on the aspect of neofields and combinatorial
designs is given in the collection. For other works of generalized complete mapping and its applications, the reader is referred to Hsu [13]. A K- Complete
Mapping has been also studied under the notion of an orthomorphism in the
construction of mutually orthogonal latin squares and finite projective planes.
Since most of the constructions are in abelian group (especially in cyclic group),
orthomorphism is defined addictively as follows: Let G be a group of order n
written addictively whether abelian or not, an orthomorphism of G is a bijection $f : G \to G$ such that the mapping defined by $g(x) = f(x) - x$ is also a
bijection. Without loss of generality, we can assume $g(0) = f(0) = 0$, when 0
is the identity of G.

In the following section, we briefly survey the development of orthomorphism, its structure and applications. We also define the notion of a near
orthomorphism and show its relation to K-near complete mappings. In section
3, we introduce a classification scheme for the class of orthomorphisms and
near orthomorphisms on cyclic groups. This scheme is analogous to that used
in Hsu [11] for neofield with cyclic multiplicative group. Using this classification, we exhibit orthomorphisms and near orthomorphisms of cyclic groups of
order up to $n = 9$.

2. Orthomorphism and Near Orthomorphism

The notion of an orthomorphism was first defined and studied in Johnson,
Dulmage and Mendelsohn [16]. It is equivalent to complete mapping defined
and studied earlier by Hall and Paige [10]. In fact, Mann [18] was the first to
define and study complete mappings. More specifically, if $x \to f(x)$ is an

orthomorphism, then $x \rightarrow -f(x)$ is a complete mapping. Hall and Paige [10] showed that a group G with a nontrivial cyclic Sylow 2-subgroup does not admit orthomorphisms. Although they did not prove the converse, they have shown that that converse is true for certain classes of groups, such as non-cycle 2-groups and the symmetric group $S_n, n > 3$.

In [11], Hsu studied neofields which have cyclic multiplication group and combinatorial designs using orthomorphism. Algorithm was given and a FOR-TRAN program was run on the Michigan Terminal System which lists all orthomorphisms of cyclic group of odd order. Orthomorphism does not exist in cyclic group of even order. However, certain mapping very similar to an orthomorphism does exist in cyclic group of even order. In general, we have the following definition.

Definition 1   Let G be a group written additively with 0 as identity. Let f be a bijection from G to G. Then f is called a *near orthomorphism* if $\{f(x) - x \mid x \in G\} = G - \{h\}$ where h is some element in G.

Just as an orthomorphism may be thought of as a K-complete mapping, so a near orthomorphism may be regarded as a K-near complete mapping. Moreover, given a K-near complete mapping, one can construct a near orthomorphism. Let f be a K-near complete mapping written addictively as follows:

$$[g_1' g_2' \cdots g_h'](g_{11} g_{12} \cdots g_{1k_1})(g_{21} g_{22} \cdots g_{2k_2}) \cdots (g_{s1} g_{s2} \cdots g_{sk_s})$$

where the elements $g_{i+1}' - g_i'$ and $g_{i,j+1} - g_{ij}$ together with the elements $g_{i,1} - g_{i,k_l}$ comprise the non-identity elements of G. In other words, the mappings $f(g_i') = g_{i+1}'$ and $f(g_{i,j}) = g_i, j+1$ together with $f(g_i, k_i) = g_{i,1}$ map $G - \{g_h\}$ one-to-one onto $G - \{0\}$ and $\{f(x) - x / x \in G - \{g_h'\}\} = G - \{0\}$. Without loss of generality, we assume $g_1' = 0$. Let the mapping g be defined as follows:

$$\begin{cases} g(x) = f(x) \ +g_h' \cdot \forall x \neq g_h' \\ g(g_h') = gg_h' \end{cases}$$

Hence the mapping g from G to G is one-to-one. For if $g(x) = g(y)$, we have either $f(x) = f(y)$ or $f(x) = 0$. Since $f(x) \neq 0$, it follows that $f(x) = f(y)$. This is impossible since f is one-to-one. Since $\{f(x) - x \mid x \in G - \{g'_h\}\} = G - \{0\}$, we have $\{g(x) - x \mid x \in G - \{g'_h\}\} = G - \{g'_h\}$. Therefore, $g(x) - x \mid x \in G\} = G - \{g'_h\}$ since $g(g'_h) - g'_h = g'_h - g'_h = 0$. Hence g is a near orthomorphism.

Using Theorem 3.5 of Hsu and Keedwell [14], we have

Theorem 1   For abelian groups, the existences of an orthomorphism and a near orthomorphism are mutually exclusive. An abelian group $(G, +)$ has an orthomorphism if and only f the sum of all its elements is 0. It has a near orthomorphism if and only if it has a unique element of order 2 and then the sum of all its elements is this unique element of order 2.

The proof of Theorem 3.5 in Hsu and Keedwell [14] is based on the results of L.J. Paige [21] and B. Gordon [9].

If an abelian group $(G, +)$ has a near orthomorphism f and equivalently K-near complete mapping with $g'_1 = 0$, then the element g'_h is the unique element of order 2 in G. Here we give two examples of near orthomorphism.

Example (1)   The following mapping g is a near orthomorphism of Z_6 with its corresponding K-near complete mapping f where $K = \{3 : 3\}$:

x	0	1	2	3	4	5
g(x)	1	5	2	3	0	4
g(x)-x	1	4	0	0	2	5

x	0	1	2	3	4	5
f(x)	4	2	5	*	3	1
$f(x) - x$	4	1	3	*	5	2

Example (2)   A K-near complete mapping f where $K = \{2, 3, 3, 3, 3\}$ is given in the dihedral group $D_7 = gp\{a, b : a^7 = b^2 = \in, ab = ba^{-1}\}$ together with its corresponding near orthomorphism g. For convenience, the group is wither multiplicatively.

x	e	a	a^2	a^3	a^4	a^5	a^6	b	ba	ba^2	ba^3	ba^4	ba^5	ba^6
$f(x)$	ba^3	a^4	a	ba	a^2	ba^2	ba^6	a^6	ba^5	ba^4	$*$	a^5	a^3	b
$x^{-1} \times f(x)$	ba^3	a^3	a^6	ba^4	a^5	b	ba^5	ba^6	a^4	a^2	$*$	ba^2	ba	a

x	a	a	a^2	a^3	a^4	a^5	a^6	b	ba	ba^2	ba^3	ba^4	ba^5	ba^6
$g(x)$	e	ba^6	ba^2	a^2	ba	a	a^4	ba^4	a^5	a^6	ba^3	ba^5	b	a^3
$x^{-1} \times g(x)$	e	b	ba^4	a^6	ba^5	a^3	a^5	a^4	ba^6	ba	e	a	a^2	ba^2

The *orthomorphism graph* $Orth(G)$ *of a group* G is the graph with vertex set consisting of all orthomorphisms of G, two orthomorphisms α, β of G being adjacent if and only if the mapping $\delta : G \to G$ defined by $\delta(x) = \alpha(x) - \beta(x)$ is a bijection. The notion of an orthomorphism graph was first defined by Evans ([4], [5] and [6]) and used to study mutually orthogonal latin square and to construct affine planes (see also Johnson, Dulmage and Mendelsohn [16]. More specifically, an n-clique of the orthomorphism graph can be used to construct a set of $n+1$ mutually orthogonal latin squares from the Cayley table of G by permutating its columns while an affine plane of order n can be constructed from an $(n-2)$-clique of the orthomorphism graph of the group G. Recently, orthomorphism graphs have been studied extensively. However, the proofs have tended to use computer searches rather than cohesive theoretical approaches. A theory of orthomorphism graphs of groups needs to be developed.

Among the best examples of orthomorphism graphs studied are two particular finite groups Z_{11} and $Z_2 \times Z_6$. The clique number of Orth(Z_{11}) is 9 since the mappings $x \to ax.2 \leq a \leq 10$ form a 9-clique. In 1961 Johnson, Dulmage and Mendelsohn [16] showed that Orth(Z_{11}) has only one 9-clique and found that the number of vertices of Orth(Z_{11}) is 3441 by using a computer. This was confirmed by Hsu in 1980 and Evans and McFarland in 1984. In his study of neofields, Hsu [11] showed that the number of cyclic neofields (i.e. neofields with a cyclic multiplicative group $G = Z_{11}$) is 3441. While in 1984, Evans and McFarland [8] found Orth (Z_{11}) to have 660 vertices of degree 3, 135 vertices of degree 8 and 6 vertices of degree 162, the remaining 2640 vertices are isolated.

The clique number of $\mathrm{Orth}(Z_2 \times Z_6)$ is 4. The constructions of the 4-cliques were done independently by Bose. Chakravarti and Knuth [3] in 1960 using a Hadamard Matrix of order 12 and a computer and by Johnson, Dulmage and Mendelsohn [16] in 1961 using a search algorithm for computation. These three authors in 1961 also mentioned that Parker and Van Duren proved that Orth $(Z_2 \times Z_6)$ contained no 5-cliques. This was reconfirmed in 1973 by Baumert and Hall [2] using computers. Construction of a 4-clique in Orth $(Z_2 \times Z_6)$ was given by Mills [20] in 1977. However, no proof of the non-existence of a 5-clique has ever been given. A survey of the work done on $\mathrm{Orth}(Z_p)$, p an odd prime was given by Evans [6]. Other works on orthomorphisms, orthomorphism graphs and relation to difference matrices and generalized Hadamard Matrices can be found in Evans ([4] and [5]) and Jungnickel ([17] and [18]). Recently Evans and Hsu [7] used orthomorphisms to construct various types of triple systems such as directed triple systems and steiner triple systems.

3. Existence and Enumeration

All orthomorphisms and near orthomorphisms of $Z_n, 2 \leq n \leq 9$ are exhibited in the Appendix. As stated in Theorem 1, orthomorphisms are mutually exclusive for abelian groups. For cyclic groups, it is known that orthomorphism exists in every Z_n, n odd and near orthomorphism exists in every Z_n, n even. Johnson, Dulmage and Mendelsohn [15] gave an analysis of some special cases for orthomorphisms when the order n of the group is small. We briefly discuss orthomorphisms and near orthomorphisms for non-cyclic groups.

For $n = 4$, the group $Z_2 \times Z_2$ has 3 orthomorphisms. For $n = 8$, the group $Z_2 \times Z_4$ has no orthomorphisms which are automorphisms but has 49 orthomorphisms. The group $Z_2 \times Z_2 \times Z_2$ does not have orthomorphisms which are not automorphisms. However, the automorphisms group of $Z_2 \times Z_2 \times Z_2$ is the simple group of order 168. Following Sylow's theorem there exist 8 subgraphs of order 7 and each of these subgroups consists of elements which are orthomorphisms. For $n = 9$, other than the 225 orthomorphisms of Z_9 listed in the Appendix, the group $Z_3 \times Z_3$ has 28 orthomorphisms among the 48 automorphisms it has. There are plenty of other orthomorphisms of $Z_3 \times Z_3$ which are not part of a clique in $\mathrm{Orth}(Z_3 \times Z_3)$. For the case $n = 12$, the group $Z_2 \times Z_6$ has only two orthomorphic automorphisms other than the identity. Other calculations using transversals shows that there exist 4-cliques in $\mathrm{Orth}(Z_2 \times Z_6)$ as stated before. Apart from the identity, it was shown that there are exactly 16.512 orthomorphisms of $Z_2 \times Z_6$.

In general, the quest for the construction of orthomorphisms and near orthomorphisms is a challenging job. A theory together with applications has to be developed. It is known that the only finite abelian groups which have no orthomorphism are those with a unique element of order 2. It is also known that if the finite group G has an orthomorphism then G does not contain a cyclic Sylow 2-subgroup. It is commonly conjectured that the converse is true. In the case of near orthomorphism, by Theorem 1 and the discussion above, an abelian group G can not have a near orthomorphism unless it has a unique element of order 2. The converse of this is true since this kind of abelian group has a sequencing which is a near orthomorphism. One natural question to ask is: which finite non-abelian groups have orthomorphisms, which have near orthomorphism? Most research has been centered on dihedral groups. The reader is referred to Hsu [12] and Hsu [13] and references listed for further reading. The following orthomorphism is given in Hsu and Keedwell [14], where G is the non-abelian group of order 21 $\mathrm{gp}\{a, b : a^7 = b^3 = \epsilon, ab = ba^2\}$.

$$(a \; a^2 \; a^4) \, (a^3 \; a^6 \; a^5) \, (b \; b^2) \, (ba \; b^2 \; a^3) \, (ba^2 \; b^2 \; a^6) \, (ba^3 \; b^2 \; a^2)$$
$$(ba^4 \; b^2 \; a^5) \, (ba^5 \; b^2 \; a) \, (ba^6 \; b^2 \; a^4).$$

This orthomorphism was used to construct a neofield of order 22 when the night distributive law holds.

The updated information for the number $\sharp(n)$ of orthomorphisms or near orthomorphisms other than the identity in cyclic group Z_n is:

n	3	4	5	6	7	8	9	10	11	12	13	14	15
$\sharp(n)$	1	2	3	8	19	64	225	928	3441	*	79.259	*	2.424.195

Here we introduce a classification scheme which is different from the study of orthomorphism graphs. Rather, this classification is analogous to the study of cyclic neofields. This method is more algebraic than combinatorial. Consequently, it may help in the quest for a solution to the enumeration problem, although we are not clear about its significance in the construction of mutually orthogonal latin squares.

Let G_n be the set of all orthomorphisms or near orthomorphisms of cyclic group Z_n of order n (we restrict to cyclic group for the purpose of explaining the Appendix). We define three operations dr, dc and tp on the elements of G_n as follows:

For every $f \in G_n : f(k) = t$ implies

$$f(k - t + h) = -t \quad \text{in} \quad dr(f)$$

$$f(-k) = t - k \quad \text{in} \quad dc(f)$$

$$f(t + h) = k + h \quad \text{in} \quad tp(f),$$

where

$$h = \begin{cases} 0 & \text{if } n \text{ is odd} \\ n/2 & \text{if } n \text{ is even} \end{cases}$$

Define $\Delta = tp * dc$. Then it is easily shown that $\Delta^3 = (dr)^2 = (dc)^2 = (tp)^2 = i$, the identity mapping. Since dc, dr, tp and Δ are well-defined mapping from G_n to G_n we consider G_n as a graph with elements of G_n as vertices. Two elements f and g in G_n are adjacent to each other if f is related to g by one of the four mappings dc, dr, tp and Δ. Moreover, $\{i.dc, dr, tp, \Delta, \Delta^2\}$ has subgroups $\{i\}, \{i, di\}, \{i, dr\}, \{i, tp\}$ and $\{i, \Delta, \Delta^2\}$. In 1980, Hsu [10] showed that the graph G_n consists of isolated cliques of $K_1, K_2, K_3,$ and K_6 in the study of cyclic neofields.

In the case of K_3, three vertices of G_n are connected by dc.tp or dr. For K_2, two vertices of G_n are adjacent through dc, dr and tp where $dc(f) = dr(f) = tp(f)$. For the case of K_1, the vertex f is fixed by all mappings. In other words, $dr(f) = dc(f) = tp(f) = \Delta(f) = i(f)$. Therefore, the total number of orthomorphisms and near orthomorphisms, i.e. the number of vertices in G_n, is equal to $g(n) = 1 \times a_1 + 2 \times a_2 + 3 \times a_3 + 6 \times a_6$ where a_i is the total number of complete subgraphs K_i in G_n. In the appendix, AB, BC, CD, DE, EF, FA, AC, and AD represent the adjacency dr, dc, dr, dc, dr, dc, Δ and dt respectively.

Appendix

All orthomorphisms and near orthomorphisms of $Z_n, 3 \le n \le 9$ are listed here in the form of a generalized complete mapping. Following Theorem 1, all listed mappings are orthomorphisms if n is odd, and are near orthomorphisms if n is even. In the appendix, AB, BC, CD, DE, EF, FA, AC and AD represent the adjacency dr, dc, dr, dc, dr, dc, dr, dc, Δ and dt respectively.

	A	B	C	D	E	F
z_3	(12)					
z_4	[013] [031]					
z_5	(14)(23)	(13 42)		(12 43)		
z_6	[021453] [045213] [015243] [051423]	[043](125) [023](154)		[013](254) [053](124)		
z_7	(13)(26)(45) (15)(46)(23) (13 26 45) (16)(25)(43) (145 326) (154236)	(154623) (142)(356) (156243) (132564)	(134265) (146523)	(124)(365) (125463) (126435)	(136452) (153462)	(162354) (163245)
z_8	[04](127)(365) [04](176)(235) [04](1276)(35) [04](167)(253) [04](172)(356) [05762314] [03126574] [05326714] [03562174] [01362754] [07526134] [01726354] [07162534] [04](17)(2563) [0254](1673) [0634](1572) [06127354] [02761534]	[04](2365)(17) [04](13)(27)(56) [04](16)(23)(57) [06735214] [02153674] [027314](56) [061574](32) [06371254] [02517634] [023754](16) [065134](72) [04](132756) [054](13)(267) [034](162)(57) [054](17)(236) [034](17)(265)	[04](136752) [014](263)(57) [074](13)(256) [014](276)(35) [074](126)(35)	[04](12)(36)(57) [04](13)(25)(67) [05617324] [03271564] [057364](12) [031524](76) [01653724] [07235164] [013764](25) [075124](63) [04](125763) [0214](3756) [0674](1325) [06523714] [02365174]	[04](165723) [0564](1372) [0324](7516) [05376124] [03512764]	[04](1672)(35) [0164](2573) [0724](6315) [01736524] [07152364]

	A	B	C	D	E	F
Z_9	(12)(36)(48)(57)	(187245)(36)		(154278)(36)		
	(12)(37)(46)(58)	(187346)(25)		(164378)(25)		
	(18)(27)(36)(45)	(157842)(36)		(124875)(36)		
	(14)(26)(35)(78)	(126538)(47)		(183562)(47)		
	(15)(24)(36)(78)	(127548)(36)		(184572)(36)		
	(13)(25)(48)(67)	(137645)(28)		(154673)(28)		
	(16)(28)(34)(57)	(167243)(58)		(134276)(58)		
	(17)(24)(38)(56)	(14)(275683)		(14)(238657)		
	(15)(23)(47)(68)	(17)(235486)		(17)(268453)		
	(15278)(364)	(16378)(254)	(18736)(245)	(12)(375846)	(12)(364857)	(18725)(346)
	(164)(25783)	(17256438)	(18346527)	(142)(37)(568)	(124)(37)(586)	(146)(23875)
	(143)(26)(578)	(16534728)	(18274356)	(16742)(358)	(12476)(385)	(134)(26)(587)
	(15463)(278)	(15428)(367)	(18245)(376)	(136752)(48)	(125763)(48)	(13645)(287)
	(14652783)	(1428)(37)(56)	(1824)(37)(56)	(146752)(38)	(125764)(38)	(13872564)
	(146)(27853)	(17342658)	(18562437)	(16834752)	(12574386)	(164)(23587)
	(178325)(46)	(17325648)	(18465237)	(1452)(37)(68)	(1254)(37)(68)	(152387)(46)
	(18)(265)(347)	(1462)(3785)	(1264)(3587)	(12538746)	(16478352)	(18)(256)(374)
	(15647328)	(1783)(2546)	(1387)(2645)	(12537684)	(14867352)	(18237465)
	(17326458)	(17846)(253)	(16487)(235)	(125)(374)(68)	(152)(347)(68)	(18546237)
	(12748)(365)	(142635)(78)	(153624)(78)	(12638)(475)	(18362)(457)	(18472)(356)
	(1658)(2374)	(14352786)	(16872534)	(12643857)	(17583462)	(1856)(2473)
	(126538)(47)	(17826435)	(15346287)	(12685473)	(13745862)	(18476)(235)
	(1438)(2675)	(16537824)	(14287356)	(12746583)	(13856472)	(1834)(2576)
	(17548)(236)	(15)(247863)	(15)(236874)	(127)(35486)	(172)(36845)	(18457)(263)
	(1758)(26)(34)	(16)(247853)	(16)(235874)	(12743586)	(16853472)	(1857)(26)(34)
	(14375628)	(16524783)	(13874256)	(12764)(358)	(14672)(385)	(18265734)
	(14537628)	(13)(265)(478)	(13)(256)(487)	(12835)(476)	(15382)(467)	(18267354)
	(158)(27634)	(136)(27854)	(163)(24587)	(128436)(57)	(163482)(57)	(185)(24367)
	(15238)(476)	(13786254)	(14526873)	(12846537)	(17356482)	(18325)(467)
	(15347628)	(13)(26)(4785)	(13)(26)(4587)	(128476)(35)	(167482)(35)	(18267435)
	(13826754)	(153824)(67)	(142835)(67)	(13)(2748)(56)	(13)(2847)(56)	(14576283)
	(173)(268)(45)	(15328467)	(17648235)	(1325)(4867)	(1523)(4768)	(137)(286)(45)
	(1682)(3745)	(18435267)	(17625348)	(13264785)	(15874623)	(1286)(3547)
	(16823754)	(15243867)	(17683425)	(1327)(4856)	(1723)(4658)	(14573286)
	(175)(268)(34)	(167)(24853)	(176)(23584)	(13274586)	(16854723)	(157)(286)(34)
	(1534)(2768)	(13854267)	(17624583)	(13284756)	(16574823)	(1435)(2867)
	(1756)(2438)	(16758324)	(14238576)	(134)(27)(586)	(143)(27)(568)	(1657)　(2834)
	(176)(24538)	(13267584)	(14857623)	(13472865)	(15682743)	(167)(28354)
	(15764)(238)	(13867254)	(14527683)	(137)(284)(56)	(173)(248)(56)	(14675)(283)
	(16457)(283)	(17256843)	(13486527)	(14237685)	(15867324)	(17546)(238)
	(178365)(24)	(148)(27563)	(184)(23657)	(14572)(386)	(12754)(368)	(156387)(24)
	(175846)(23)	(17324586)	(16854237)	(1527)(34)(68)	(1725)(34)(68)	(164857)(23)

REFERENCES

[1] G. S. Bloom and D. F. Hsu: "On graceful directed graphs. *SIAM J. of Algebraic and Discrete Methods.* v.6 (1985) p. 519-536.

[2] L. Baumert and M. Hall, Jr. "Nonexistence of Certain Planes of Order 10 and 12," *J. of Combinatorial Theory* A14 (1973) p. 273-280.

[3] R. C. Bose, I. M. Chakravarti and D. E. Knuth, "On Methods of Constructing Sets of Mutually Orthogonal Latin Squares using a computer I," *Technometries* 2 (1960) p. 507-516.

[4] A. B. Evans, "Generating Orthomorphisms of GF(q)$_+$, *Discrete Math.* 63 (1987) p. 21-26.

[5] A. B. Evans, "Difference Matrices, Generalized Hadamard Matrices and Orthomorphism Graphs of Groups," *JCMCC,V I* (1987), p. 97-105.

[6] A. B. Evans, "Orthomorphism Graphs of Z_p," *ARS Combinatoria* 25B (1988) p. 141-152.

[7] A. B. Evans and D. F. Hsu, "Triple Systems and Orthomorphisms," to appear.

[8] A. B. Evans and R. L. McFarland, "Planes of Prime Order with Translations," *Congressus Numerantium* 44 (1984) p. 41-46.

[9] B. Gordon, "Sequences in Groups with Distinct Partial Products," *Pacific J. Math.* 11 (1961) p. 1309-1313.

[10] M. Hall, Jr. and L. J. Paige, "Complete Mappings of Finite Groups," *Pacific J. of Math.* 5(1955), p. 541-549.

[11] D. F. Hsu, "Cyclic Neofields and Combinatorial Designs," LNM #824, Springer-Verlag, Berlin, 1980.

[12] D. F. Hsu, "Advances in Discrete Mathematics," Volume 1, *Neofields and Combinatorial Designs (edited)*, Hadronie Press, Nonantum, MA, 1985.

[13] D. F. Hsu, "Advances in Discrete Mathematics, Volume 2," *Generalized Complete Mappings (edited)*, Hadronie Press, Nonantum, MA, 1987.

[14] D. F. Hsu and A. D. Keedwell, "Generalized Complete Mappings, Neofields, Sequenceable Groups and Block Designs I," *Pacific J. Math.* 111 (1984), p. 317-332.

[15] D. F. Hsu and A. D. Keedwell, "Generalized Complete Mappings, Neofields, Sequenceable Groups and Block Designs II," *Pacific J. Math.* 117 (1985), p. 291-312.

[16] D. M. Johnson, A. L. Dulmage and N. S. Mendelsohn, "Orthomorphisms of Groups and Orthogonal Latin Squares. I.," *Canadian J. Math.* 13 (1961), p. 356-372.

[17] D. Jungnickel, "On Difference Matrices, Resolvable Transversal Designs and Generalized Hadamard Matrices," *Math. Z.*, 167 (1979), p. 49-60.

[18] D. Jungnickel, "On Difference Matrices and Regular Latin Squares," *Abh. Math. Sem. Univ. Hamburg* 50 (1980), p. 219-231.

[19] H. B. Mann, "The Construction of Orthogonal Latin Squares," *Annals of Math. Statistics* 12 (1942), p. 418-423.

[20] W. H. Mills, "Some Mutually Orthogonal Latin Squares," Proc. 8th SE Conf. on Combinatorics, Graph Theory and Computing," (1977), p. 473-487.

[21] L. J. Paige, "A Note on Finite Abelian Groups," *Bull. Amer. Math. Soc.* (A) 53 (1947), p. 590-593.

Cartan Matrices and Strong
Perfect Graph Conjecture

Siming Huang

The University of Iowa

ABSTRACT

In this paper, we introduce the Cartan matrices of graphs. We prove that if G is a critical imperfect graph and its Cartan matrix is affine type, then $G = C_{2m+1}$, that is, an odd circuit. Furthermore, there exists a polynomial algorithm to check whether a Cartan matrix is affine type.

1. Introduction

In this paper, we only deal with simple graphs, that is, no multiple edges and no loops. A graph G is said to be *perfect* if G and each of its induced subgraphs have the property that the chromatic number χ equals the size of a maximum clique ω. A graph is *critically imperfect* if it is not perfect but each of its proper induced subgraphs is perfect. The Strong Perfect Graph Conjecture, posed by Berge in 1960, is that a graph

G is perfect if and only if neither G nor its complement $\overline{G}$ contains as an induced subgraph an odd circuit of length greater than 3. An equivalent version of the Strong Perfect Graph Conjecture can be formulated as follows: The only critically imperfect graphs are the odd circuits and their complements.

In this paper, we introduce the Cartan matrices of graphs. We will state some of its properties and classifications in section 2, and prove that there exists a polynomial algorithm to determine whether the Cartan matrix of a graph is affine type. In section 3, we will prove the main theorem of this paper: If G is a critical imperfect graph and its Cartan matrix is affine type, then $G = C_{2m+1}$. The converse is also true.

2. Cartan Matrices of Graphs

Now we introduce the Cartan matrix of a graph G. More generally, we have the following definition.

Definition   The $n \times n$ matrix A is called a generalized *Cartan matrix* if it satisfies the following conditions:

(C1): $a_{ii} = 2$ for $i = 1, 2, \ldots , n$;

(C2): $a_{ij} \leq 0$ i, j integers, $i \neq j$;

(C3): $a_{ij} = 0$ implies $a_{ji} = 0$.

A matrix A is called *decomposable* if, after reordering the indices (i.e., a permutation of its rows and the same permutation of the columns), A decomposes into the form

$$\begin{bmatrix} A_1 & 0 \\ 0 & A_2 \end{bmatrix}$$

Let G be any simple graph, and A be its adjacent matrix, we define $C(G) = 2I - A$ the Cartan matrix associated to G. For convenience, we denote $C(G) = C$.

The following theorem is obvious.

Theorem 2.1   *C is indecomposable if and only if G is a connected graph.*

We will give a complete classification of Cartan matrices. More generally, we consider the following matrices:

Let $A = (a_{ij})$ be a real $n \times n$ matrix which satisfies the following three properties:

(P1): A is indecomposable;

(P2): $a_{ij} \leq 0$ for $i \neq j$;

(P3): $a_{ij} = 0$ implies $a_{ji} = 0$.

The following theorem [1] gives a complete classification of all the matrices satisfying (P1) – (P3), hence of all Cartan matrices of connected graphs.

Theorem 2.2   *Let A be a real $n \times n$ matrix satisfying (P1) – (P3). Then one and only one of the following three possibilities holds for A:*

(Fin.) $\det A \neq 0$; there exists $u > 0$ such that $Au > 0$; $Av \geq 0$ implies $v > 0$ or $v = 0$;

(Aff.) $rank (A) = n - 1$; there exists $u > 0$ such that $Au = 0$; $Av \geq 0$ implies $Av = 0$;

(Ind.)   there exists $u > 0$ such that $Au < 0$; $Av \geq 0$, $v \geq 0$ implies $v = 0$.

Recall that a matrix of the form $(a_{ij})_{i,j\in S}$, where $S \subset \{1, 2, \ldots , n\}$, is called a *principal submatrix* of $A = (a_{ij})$; we will denote it by A_S. The determinant of a principal submatrix is called a *principal minor*.

The following two lemmas are essentially the Lemma 4.4 and Lemma 4.5 of [1].

Lemma 2.3 *If A is of finite or affine type, then any proper principal submatrix of A decomposes into a direct sum of matrices of finite type.*

Lemma 2.4 *A symmetric matrix $A = (a_{ij})$ satisfying (P1) – (P3) is of finite (resp. affine) type if and only if A is positive definite (resp. positive semidefinite of rank $n - 1$).*

Theorem 2.5 *There exists a polynomial algorithm to determine whether a symmetric matrix $A = (a_{ij})$ satisfying (P1) – (P3) is of finite, affine, or indefinite type.*

Proof Let A_k denote the principal submatrix of A, where

$$A_k = \begin{bmatrix} a_{11} & \cdots & a_{1k} \\ & & \\ a_{k1} & \cdots & a_{kk} \end{bmatrix}, \quad k = 1, 2, \ldots, n.$$

Then by Lemma 2.4, A is of finite (resp. affine) type if and only if A is positive definite (resp. positive semidefinite of rank $n - 1$). By the well–known results of linear algebra, A is positive definite (resp. positive semidefinite of rank $n - 1$) if and only if $\det (A_k) > 0$ for $k = 1, 2, \ldots, n$ (resp. $\det (A_k) > 0$, $k = 1, 2, \ldots, n - 1$, $\det (A) = 0$). The complexity for computing $\det (A_k)$ is k^3, so the total complexity for determining whether A is positive definite (resp. positive semidefinite of rank $n - 1$) is $\Sigma k^3 = O(n^4)$. By Theorem 2.2, these three types are exclusive, hence we have proved the theorem. ❑

Corollary 2.6 *There exists a polynomial algorithm to determine whether the Cartan matrix of a connected graph G is of finite, affine, or indefinite type.*

Proof Since the Cartan matrix of a connected graph G is symmetric and satisfies (P1) – (P3), Theorem 2.5 applies. ❑

The following theorem will play a crucial rule in proving our main theorem in section 3.

Theorem 2.7 *Let G be a simple connected graph which contains a circuit of length greater than or equal to 3 and let C be its Cartan matrix. If C is of finite or affine type, then G is a circuit.*

Proof Suppose G contains a circuit with length ≥ 3, then there exists a principal submatrix B of C of the form

$$\begin{bmatrix} 2 & -1 & 0 & \dots & 0 & -1 \\ -1 & 2 & -1 & \dots & & 0 \\ & & \dots\dots\dots\dots\dots\dots\dots & & & \\ 0 & 0 & & \dots & 2 & -1 \\ -1 & 0 & & \dots & -1 & 2 \end{bmatrix}$$

By Lemma 2.3, B is of finite or affine type. As in this case $\det B = 0$, Lemmas 2.3 and 2.4 imply that $B = C$, hence G is a circuit. $\square$

3. The Main Theorem

A graph G of order $\alpha\omega + 1$ ($\alpha,\omega \geq 2$) is called an (α,ω)–graph if, for every vertex v, the vertices of $G - v$ can be partitioned into α cliques of order ω and into ω independent sets of order α.

Theorem 3.1 [3]   *Every critically imperfect graph G is an (α,ω)–graph with $\alpha = \alpha(G)$ and $\omega = \omega(G)$; where $\alpha(G)$ denotes the independence number of G and $\omega(G)$ denotes the clique number of G.*

The properties given in the next theorem were first established for critically imperfect graphs by Padberg [4], and then shown to apply to all (α,ω)–graphs by Bland, Huang, and Trotter [5].

Theorem 3.2 [4], [5]   *Let G be an (α,ω)–graph of order n. Then*
   (1) G has exactly n ω–cliques;
   (2) G has exactly n independent α–sets;
   (3)   every vertex is in exactly ω ω–cliques;
   (4)   every vertex is in exactly α independent α–sets;
   (5)   every ω–clique is disjoint from exactly one independent α–set;
   (6)   every independent α–set is disjoint from exactly one ω–clique.

Now we are ready to prove our main theorem.

Theorem 3.3   *G is critical imperfect with $C(G)$ affine if and only if $G = C_{2m+1}$, $m \geq 1$.*

Proof   If $G = C_{2m+1}$, then G is critically imperfect and it is easy to check that $C(G)$ is affine type.
   If G is critically imperfect and $C(G)$ is affine, then G is an (α,ω)–graph by

Theorem 3.1. We consider two cases:

1) If $\omega \geq 3$, then G has a cycle of length ≥ 3 by Theorem 3.2 (1). So G is a cycle by Theorem 2.7 since $C(G)$ is affine. Furthermore, G is an odd cycle since G is imperfect.

2) If $\omega = 2$, then by properties of an (α, ω)–graph in Theorem 3.2, it is easy to prove that G is an odd cycle, hence $G = C_{2m+1}$. ❑

From Theorem 3.3, we can state an equivalent version of the Strong Perfect Graph Conjecture.

Strong Perfect Graph Conjecture *If G is critically imperfect, then either the Cartan matrix of G or its complement $\overline{G}$ is affine type.*

The following theorem states that in order to prove the above conjecture we may try to consider both the Cartan matrices $C(G)$ and $C(\overline{G})$.

Theorem 3.4 *Let G be a graph, and $C = C(G)$, $\overline{C} = C(\overline{G})$ be the Cartan matrices of G and $\overline{G}$ respectively, then we have*

a) C and $\overline{C}$ cannot be both finite type if $n > 4$;

b) C and $\overline{C}$ cannot be both affine type if $n > 5$; C and $\overline{C}$ both are affine type for critically imperfect graphs if and only if $G = C_5$;

c) It cannot be the case that C is finite and $\overline{C}$ is affine if $n > 4$ or vice versa.

Proof a) Suppose C and $\overline{C}$ are finite type, and let $e = (1, \dots , 1)^T$ be a vector with all components one. Then by Lemma 2.4, we have

$$e^T C e = 2n - \Sigma d_G(v_i) > 0$$

$$e^T \overline{C} e = 2n - \Sigma d_{\overline{G}}(v_i) > 0$$

which implies that

$$0 < e^T C e + e^T \overline{C} e = 4n - \Sigma \{ d_G(v_i) + d_{\overline{G}}(v_i) \}$$

$$= 4n - \Sigma(n - 1)$$

$$= n(5 - n).$$

Hence we get n < 5, a contradiction.

The proofs of b) and c) are similar to the proof of a), so we will not state them here. ❏

From Theorem 3.4, we can conclude that if we are able to prove that C and $\overline{C}$ cannot be both indefinite or one is indefinite and the other is finite for critically imperfect graphs, then we have proved the above conjecture, i.e., the Strong Perfect Graph Conjecture. It needs further research and work.

REFERENCES

[1]   Victor G. Kac, *Infinite Dimensional Lie–Algebras*, Second Edition, Cambridge University Press 1985.

[2]   C. Berge and V. Chvatal, Topics on perfect graphs. *Annals of Discrete Mathematics* **21**.

[3]   L. Lovasz, Perfect graphs, in *Selected Topics in Graph Theory 2* (eds. Lowell W. Beineke and Robin J. Wilson).

[4]   M.W. Padberg, Perfect zero–one matrices. *Math. Programming* **6** (1974), 180 – 196.

[5]   R.G. Bland, H.C. Huang and L.E. Trotter, Jr., Graphical properties related to minimal imperfection. *Disc. Math* **27** (1979), 11 – 22.

SHORTNESS PARAMETERS OF R-REGULAR R-CONNECTED GRAPHS

Bradley W. Jackson

San Jose State University

ABSTRACT

A graph G is said to be r-regular if every vertex of G has degree r. We also say that G is r-connected if it cannot be reduced to a disconnected graph (or a single vertex) by the removal of fewer than r vertices. In addition, we say that G is s-cyclically-edge- connected if it cannot be disconnected into two components, both containing a cycle, by the removal of fewer than s edges.

1. Introduction

In this paper we discuss techniques for measuring longest cycles in r-regular r-connected graphs. Known results for various families of r-regular r-connected s-cyclically-edge-connected are presented and several open problems are discussed.

2. Hamiltonian Cycles in Classes of 3-regular 3-connected Graphs

A Hamiltonian cycle in a graph G is a cycle that contains every vertex of G. If G has a hamiltonian cycle, we say that G is a hamiltonian graph has and otherwise we say that G is nonhamiltonian.

We denote by F_1 the family of 3-regular 3-connected graphs. It is well known that the family F_1 has nonhamiltonian members. The smallest nonhamiltonian 3-regular 3-connected graph has 10 vertices. It is known as the Petersen graph and is pictured in figure 1.

"

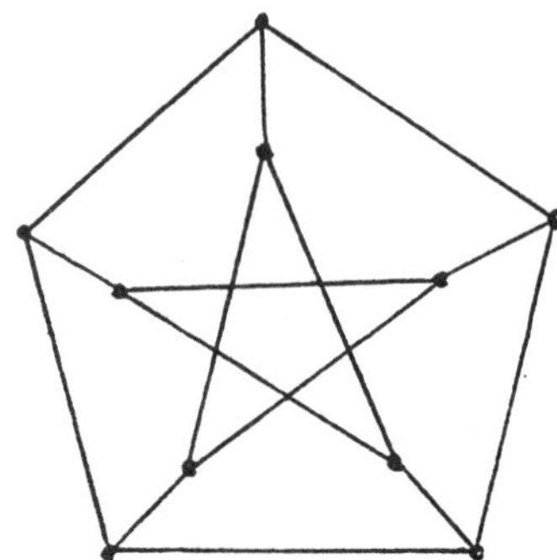

Figure 1. The Petersen graph

We denote by F_2 the family of 3-regular 3-connected planar graphs. The problem of finding hamiltonian cycles in 3-regular 3-connected planar graphs was first studied because of its relationship to the four color problem [18]. In 1946, Tutte presented a nonhamiltonian member of the family F_2 containing 46 vertices [19]. The smallest known nonhamiltonian 3-regular 3-connected planar graph has 38 vertices. It was discovered by Lederberg [14] in 1967. It is also known that any 3-regular 3- connected planar graph with 34 vertices or fewer has a hamiltonian cycle [1].

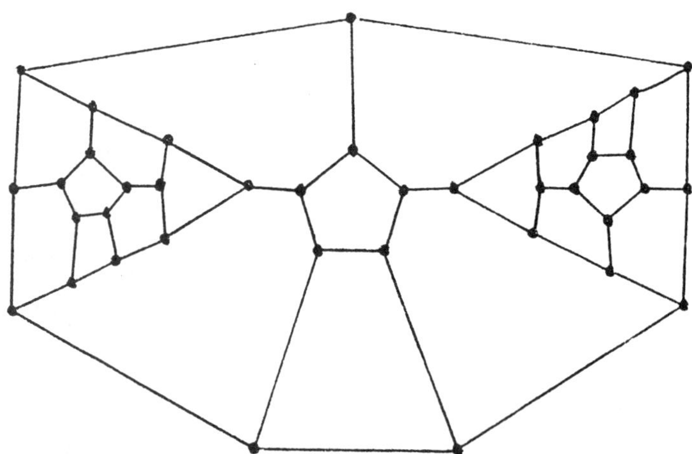

Figure 2. The Lederberg graph

We denote by F_3 the family of 3-regular 3-connected bipartite graphs. It was conjectured by Tutte [21] that every member of F_3 had a hamiltonian cycle, but Horton discovered a member of F_3 had a hamiltonian cycle, but Horton discovered a nonhamiltonian member of this family containing 96 vertices [2]. The smallest known nonhamiltonian 3-regular 3-connected bipartite graph has 54 vertices. It was discovered by Ellingham and Horton [5] and is pictured in figure 3.

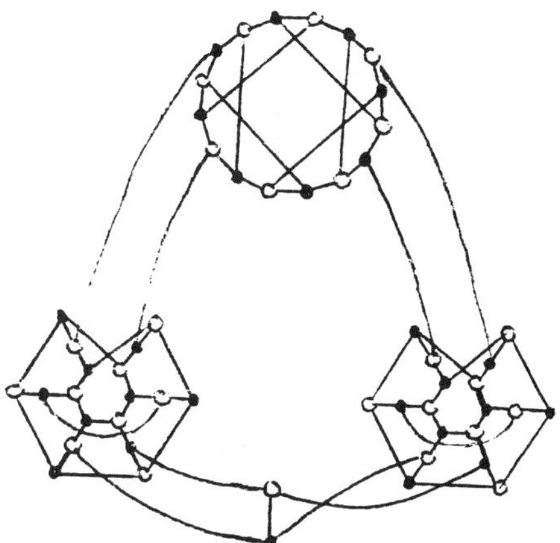

The Ellingham-Horton Graph

Finally we consider F_4, the family of 3-regular 3-connected bipartite planar graphs. Barnette has made the following conjecture.

Problem 1   Every 3-regular 3-connected bipartite planar graph has a hamiltonian cycle.

Figure 4. A counterexample to Barnette's conjecture

(to be supplied by the reader)

Although Barnette's conjecture remains open, it has been shown [9] that any counterexample must have at least 66 vertices.

2. Shortness Parameters of Families of Graphs

Let G be any graph. We denote by $c(G)$, the number of vertices in a longest cycle of G, and we let $n(G)$ represent the total number of vertices in G. We say that $a(G) = \frac{c(G)}{n(G)}$, is the shortness coefficient of G. For a given family of graphs F, we define the shortness coefficient $a(F)$ by $a(F) = \lim\inf \{a(G)|G \in F\}$. If $S \subseteq F$ is a subfamily of F and $\lim\inf \{a(G) \mid G \subseteq S\} = c_1$, then we way that $a(F) \leq c_1$. On the other hand, if $a(G) \geq c_2$, for all G in F, we say that $a(F) \geq c_2$. We say that F has a nontrivial shortness coefficient if $a > c_1 \geq a(F) \geq c_2 > 0$.

In a similar way we define the shortness exponent of a graph G to be $b(G) = \frac{\log c(G)}{\log n(G)}$. As above, we can also discuss the shortness exponent of a family of graphs. Note that we have the following relationships between shortness exponents and shortness coefficients, $a(F) > 0$ implies that $b(F) = 1$ and $b(F) < 1$ implies that $a(F) = 0$.

3. Constructing Families of Graphs with Small Shortness Parameters

One operation that has been successfully exploited in several constructions of families of r-regular r- connected graphs with small shortness parameters is the removal and replacement of vertices. Suppose that H is an r-regular r-connected graph and let v be any vertex of H. We denote by $H \backslash v$ the topological set obtained by removing v from H but leaving the r free edges that were incident to v.

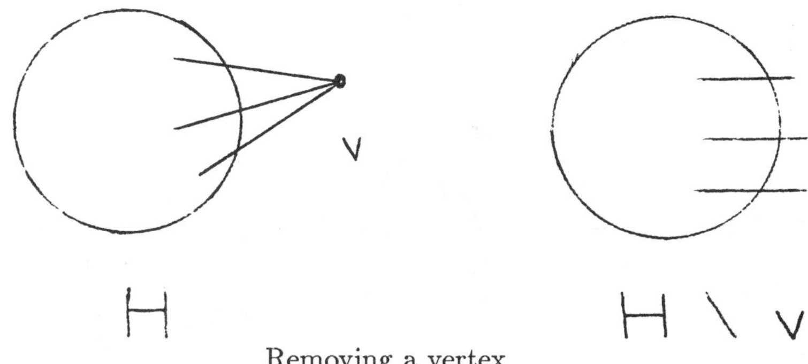

Removing a vertex

If G is an r-regular r-connected graph then $H \backslash v$ can be used to replace any vertex u in G by removing that vertex and identifying the r free edges of $H \backslash v$ with the r free edges of G/u in some order. It is well known [12] that the resulting graph is also r-regular and r-connected. We denote by $G[H]$ any graph obtained from G by replacing each vertex of G with $H \backslash v$ in this manner. Note that the graphs obtained in this manner have cyclic connectivity r but no higher.

In 1980, Bondy and Simonvits [3] described the following recursive construction. Let $G_0 = P$ be the Petersen graph and for $i \geq 1$, let $G_i = G_{i-1}$ [P].

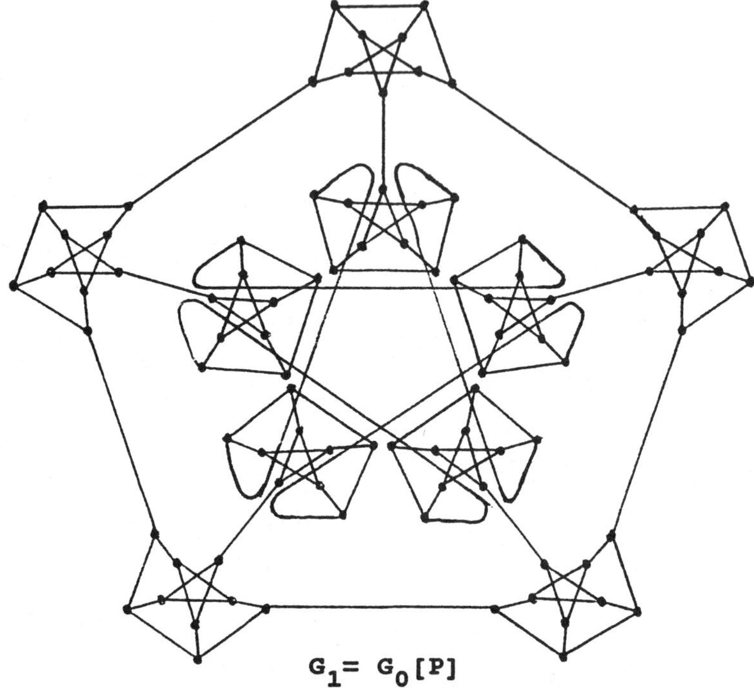

$$G_1 = G_0[P]$$

It is easy to see that $g(G_i) = 10(9^i)$ and $c(G_i) = 9(8^i)$. Thus $b(F_1) \leq \frac{\log 8}{\log 9}$ for the family of 3-regular 3-connected graph.

For the family F_2 a similar construction starting with the Lederberg graph (38 vertices) shows that $b(F_2) \leq \frac{\log 36}{\log 37}$. For the family of 3-regular 3-connected bipartite graphs a construction starting with the Ellingham-Horton graph (54 vertices) shows that $b(F_3) \leq \frac{\log 52}{\log 53}$. Likewise any counterexample to Barnette's conjecture (see figure 4) could be used to construct a family of 3-regular 3-connected bipartite planar graphs with shortness exponent $b < 1$.

In a recent paper, Bill Jackson [10] has shown that $b(F_1) \geq \log_2(1 + \sqrt{5}) - 1$. Thus showing that F_1 has a nontrivial shortness exponent. His result also gives a lower bound for the shortness exponents of the other families, but no further improvements in his bound for $F_2, F_3,$ or F_4 have been computed. To construct families of 3-regular 3-connected graphs with cyclic-edge-connectivity greater than 3, different constructions are needed. One such construction involves the removal of an edge. Suppose that H is a 3-regular 3-connected graph and let e be any edge of H. We denote by $H \backslash e$ the topological set obtained from H by removing e and its endpoints. If x and y are the endpoints of e in H, then we label by x the two free edges of $H \backslash e$ that were incident to x and by y the two free edges of $H \backslash e$ that were incident to y.

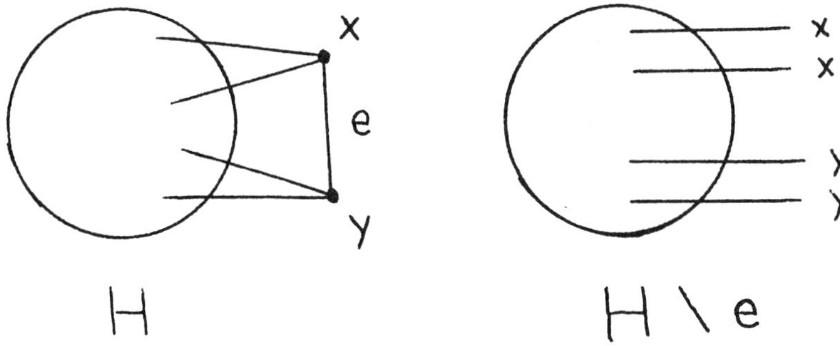

Figure 7. Removing an edge

We denote by $m[H]$ the graph obtained from m copies of $H \backslash e$ by identifying both x-edges of the ith copy H_i with both y-edges of the i+1st copy H_{i+1} mod m. Note that if H is 4-cyclically-edge-connected, then so is $m[H]$.

Consider the graph $m[P]$ in figure 8 which is constructed from m copies of the Petersen graph. Since the Petersen graph is 5-cyclically-edge-connected than $m[P]$ is 4-cyclically-edge-connected and since each copy of $P \backslash e$ has 8 vertices then $n(m[P]) = 8m$. If C is a cycle in $m[P]$ that contains all of the vertices of the ith copy P_i then either C enters P_i once using both x-edges and a second time using both y-edges, or it enters P_i once using both x-edges and a second time using both y-edges. From this it is easy to check that C contains all of the vertices of the ith copy P_i then either C enters P_i once using both x-edges or both y-edges, or it enters P_i once using both x-edges and a second time using both y-edges. From this it is easy to check that C contains all of the vertices in at most two copies of P and that $c(m[P]) = 7m + 2$. Thus the shortness coefficient of the family of 3-regular 3-connected 4-cyclically-edge-connected graphs is at most 7/8.

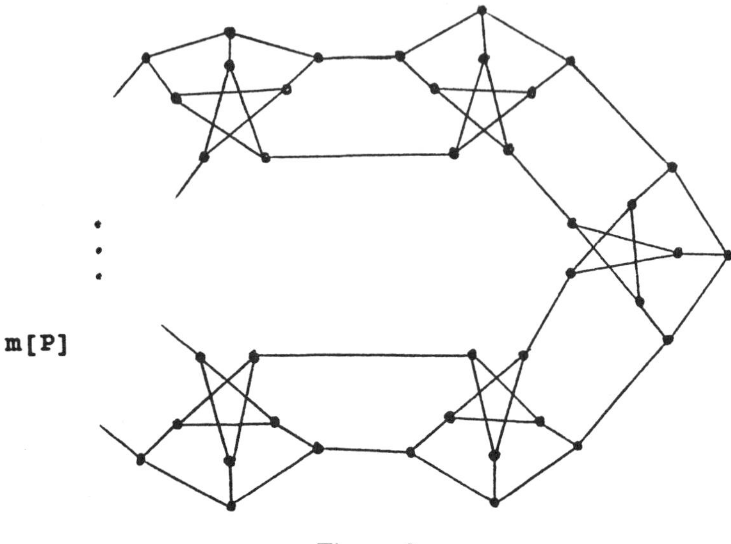

Figure 8

Bondy has made the following conjecture.

Problem 2   The shortness coefficient for the class of 3-regular 3-connected 4- cyclically-edge-connected graphs has a nontrivial lower bound $k > 0$.

For planar graphs the shortness coefficient has been well studied. In 1974, Grunbaum and Malkevitch [7] showed that the shortness coefficient of any 3-regular 3-connected 4-cyclically-edge-connected planar graph is at least 3/4. The smallest known nonhamiltonian 3-regular 3-connected 4-cyclically-edge-connected planar graph has 42 vertices. It was discovered by Honsberger in 1973 and is pictured in figure 9. By removing an edge from this graph and joining m copies as above one can construct a family of 3-regular 3-connected 4-cyclically-edge-connected planar graphs with shortness coefficient $a = 39/40$. Thus this family of graphs has a nontrivial shortness coefficient.

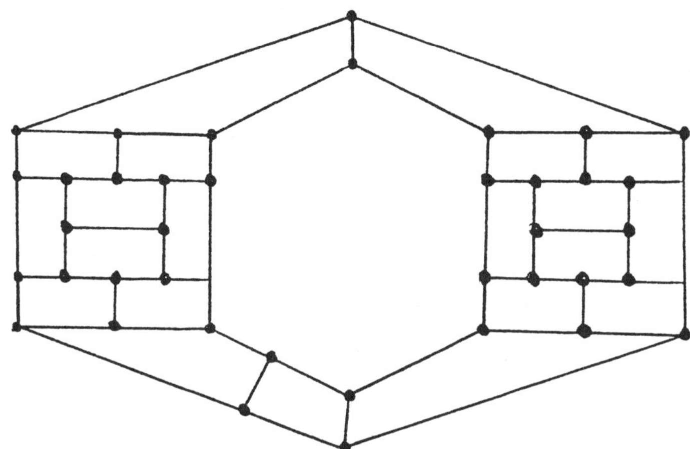

Figure 9. The Honsberger graphs

The Ellingham-Horton graph in figure 3 is a 3-regular 3-connected 4-cyclically-edge-connected bipartite graph with 54 vertices. By removing an edge from this graph and joining m copies one can construct a family of 3-regular 3-connected 4-cyclically-edge connected bipartite graphs with shortness coefficient $a = 51/52$.

There is also a 3-regular 3-connected 5-cyclically-edge connected planar graph with 44 vertices. It was discovered by Tutte in 1972 and is pictured in figure 10.

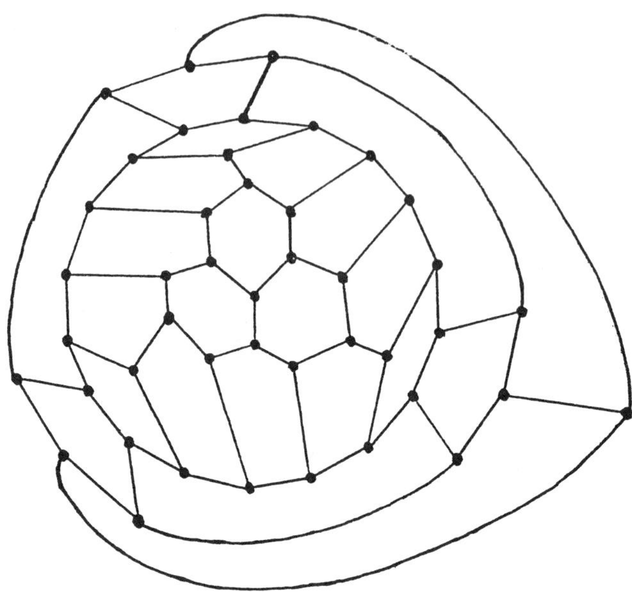

Figure 10. A Tutte graph with cyclic connectivity 5

In 1982, Zaks [22] constructed a family of 3-regular 3-connected 5-cyclically-
edge connected planar graphs, containing only faces with 5,14,20 or 35 sides,
with shortness coefficient equal to $a = 84/85$. His construction uses the subgraph Q_1 in figure 11 with 85 vertices. When Q_1 is a subgraph of a graph
G, connected to the rest of G by the edges e_1, e_2, e_3, e_4, e_5, then no cycle in
G contains all the vertices of Q_1.

A similar construction for arbitrary 3-regular 3-connected 5-cyclically-
edge-connected graphs can b used to construct a family of graphs with shortness
coefficient $a = 34/35$. It uses the subgraph Q_2 in figure 12 which has 35
vertices. As above when Q_2 is a subgraph of a graph G, connected to the
rest of G by the edges e_1, e_2, e_3, e_4, e_5, then no cycle in G contains all the
vertices of Q_2. There are no known 3-regular 3-connected bipartite graphs
with cyclic connectivity equal to five.

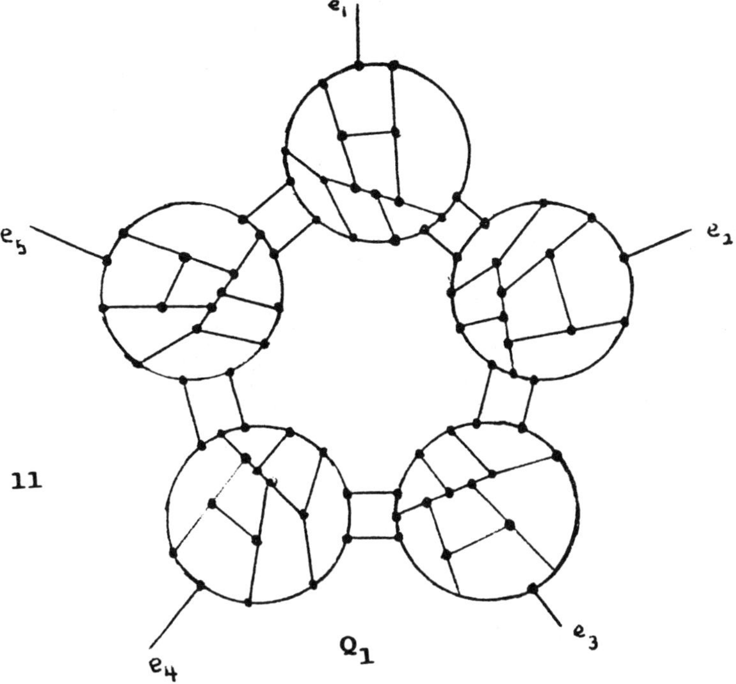

Figure 11

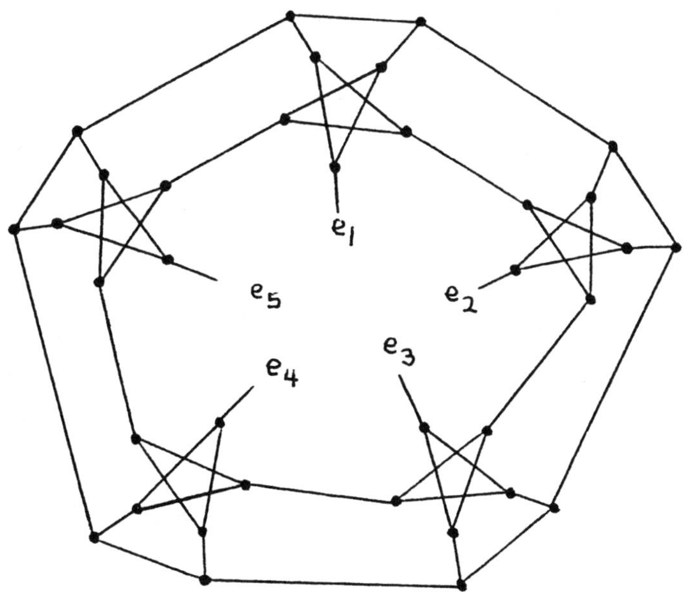

Figure 12

Little is known about nonhamiltonian 3-regular 3-connected graphs with cyclic connectivity greater than five. Note that there are no 3-regular 3-connected planar graphs with cyclic connectivity greater than five. However, for the other two classes of 3-regular 3-connected graphs that we have encountered we have the following problems.

Problems 3-4   For the class of 3-regular 3-connected graphs F_1 (bipartite graphs F_3) is there an s_i sufficiently large, so that all s_i-cyclically-edge-connected graphs in F_i have a hamiltonian cycle.

Note that there is a 3-regular 3-connected 7-cyclically-edge-connected nonhamiltonian graph with 28 vertices. It is known as the Coxeter graph and is pictured in figure 13.

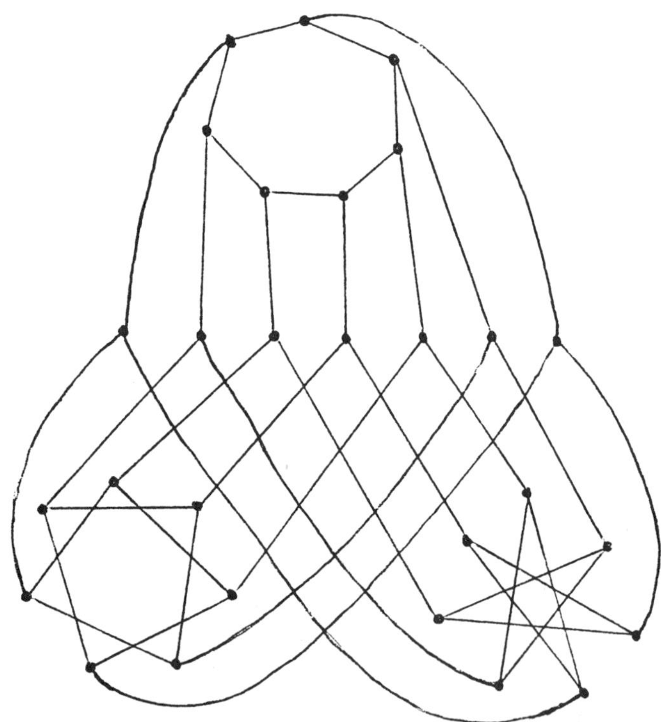

Figure 13. The Coxeter graph

4. Hamiltonian Cycles in Classes of 4-regular 4-connected Graphs In 1971 C. Nash-Williams [17] conjectured that every 4-regular 4-connected graph was hamiltonian. However in 1973, Meredith [16] constructed nonhamiltonian r-regular r-connected graphs, for every $r > 3$. The construction starts with a nonhamiltonian r-regular r-edge-connected multigraph obtained by adding edges to the Petersen graph. Each vertex is then replaced by a copy of $K_{r,r}$ with one vertex removed. The resulting graph is an r-regular r-connected non-hamiltonian graph with $10(2r - 1)$ vertices. A 4- regular 4-connected non-hamiltonian graph with 70 vertices is pictured in figure 14.

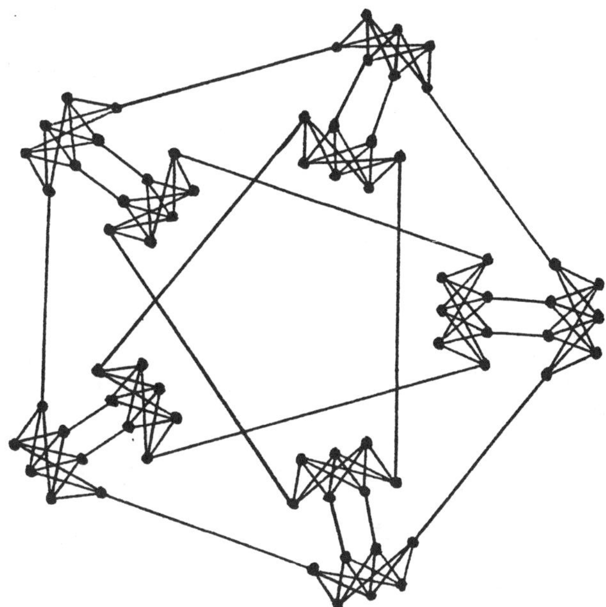

Figure 14. The Meredith graph for $r = 4$

The smallest known 4-regular 4-connected nonhamiltonian graph has 52 vertices. It was discovered in 1987 by Brad Jackson [11] and is pictured in figure 15.

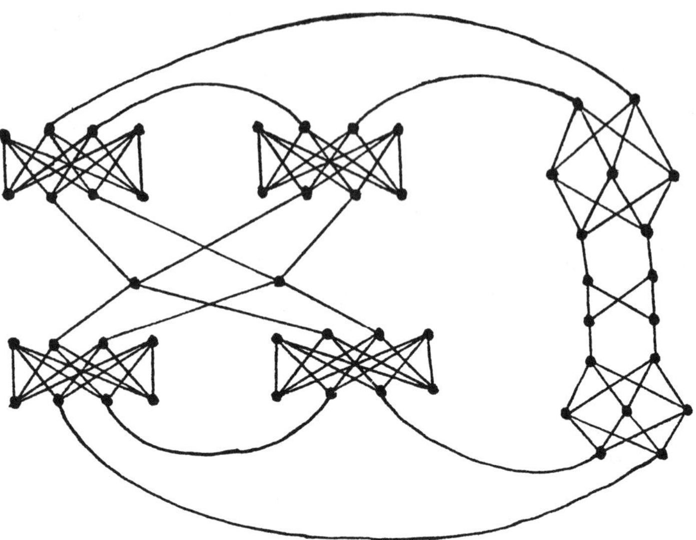

Figure 15. The Jackson graph

As for other classes of 4-regular 4-connected graphs, Tutte [19] proved in 1956 that every 4-connected planar graph has a hamiltonian cycle. By starting with a nonhamiltonian 3-regular 3-connected bipartite graph and adding edges to obtain a nonhamiltonian r-regular r-edge-connected bipartite multigraph, Meredith's construction can also be used to obtain nonhamiltonian r-regular r-connected bipartite graphs, for every $r > 3$. The smallest known 4-regular 4-connected nonhamiltonian bipartite graph was constructed by McCuaig and Rosenfeld [15] in 1984. It has 84 vertices and is pictured in figure 16.

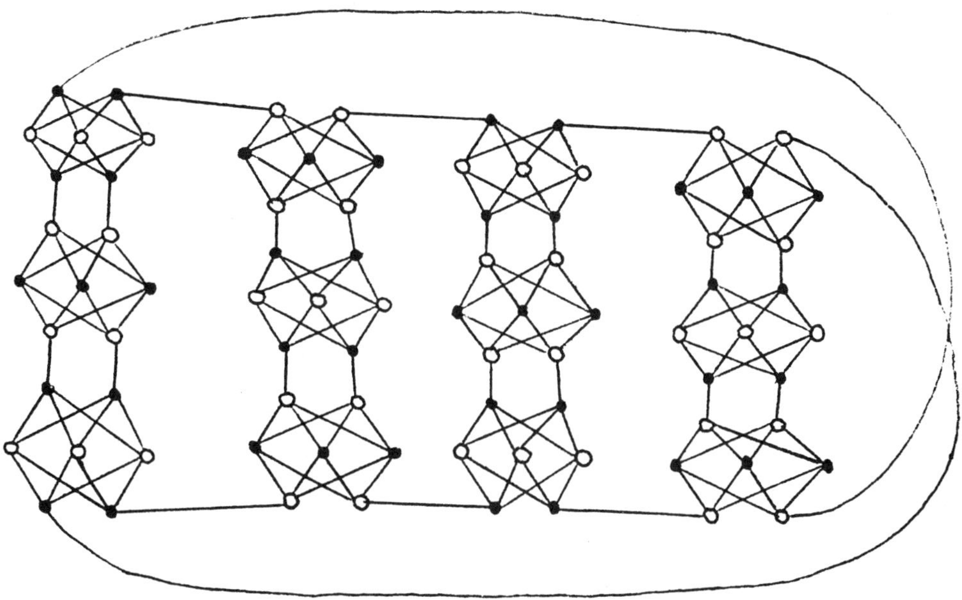

Figure 16. The McCuaig-Rosenfeld graph

5.   Shortness Parameters of 4-regular r-connected Graphs

In 1981, Jackson and Parsons [13] found a family of r-regular r-connected graphs with a nontrivial shortness exponent. Starting with a r-regular r-edge-connected multigraph with 10 vertices their construction would show that the shortness exponent of the family of r-regular r-connected graphs is at most $\frac{\log(10(r-1)-1)}{\log(10(r-1))}$. Note that these graphs have cyclic connectivity r but no greater since the construction involves vertex replacement. In a similar way it is possible to find an upper bound for the shortness exponent r-regular r-connected bipartite graphs. This leads to the following problems.

Problem 5-6   Find a lower bound for the shortness exponent of r-regular r- connected (bipartite) graphs.

Problem 7   Show that the class of r-regular r-connected $r+1$-cyclically-edge- connected graphs has a nontrivial shortness coefficient. For s even and $s < \frac{4}{3}r$ there are r-regular r-connected s-cyclically-edge-connected graphs with no hamiltonian cycle. A 5-regular 5-connected 6-cyclically-edge-connected non-hamiltonian graphs is pictured in figure 17.

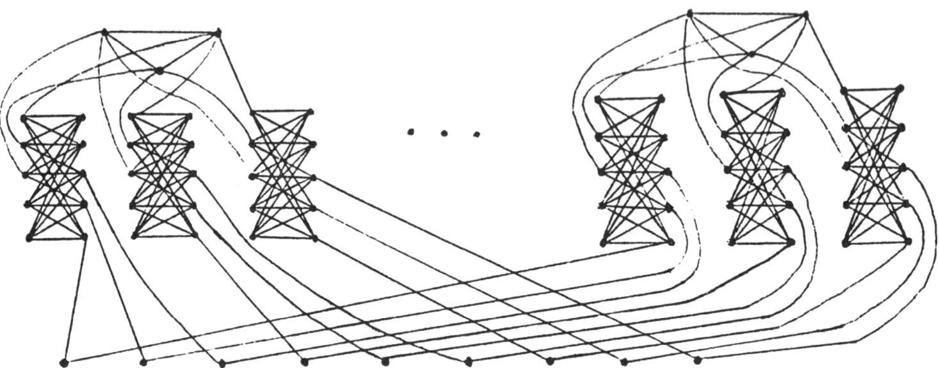

Figure 17. A 5-regular 5-connected nonhamiltonian graph

We conclude with the following set of problems.

Problems 8 - ∞   For a given $r > 3$ and s sufficiently large, show that all r-regular r-connected s-cyclically-edge-connected graphs are hamiltonian.

REFERENCES

[1] D. Barnette, "Every Simple 3-Polytope with 34 vertices is Hamiltonian," *Discrete Math.* 62(1986) 1-20.

[2] J. Bondy and U. Murty, "Graph Theory with Applications," Macmillan and Co., London, 1976.

[3] J. Bondy and M. Simonovits, "Longest cycles in 3-connected 3-regular graphs," *Canad. J. Math.*, 32(1980) 987-992.

[4] M. Capobianco and J. Molluzo, "Examples and Counterexamples in Graph Theory," Chapter 9 *(Traversability)*, North-Holland, New York, 1978.

[5] M. Ellingham and J. Horton, "Honhamiltonian 3-connected cubic bipartite graphs," *J. Combinatorial Th. Ser. B4* (1983) 350-353.

[6] G. Faulkner and D. Younger, "Nonhamiltonian cubic planar graphs," *Discrete Math.*, (1974)67-74.

[7] B. Grunbaum and J. Malkevitch, "Pairs of edge-disjoint hamiltonian circuits," *Aequationes Math.* 14(1976) 191-196.

[8] B. Grunbaum and H. Walther, "Shortness exponents of families of graphs," *J. Combinatorial Theory Ser. A*, 14 (1973), 364-385.

[9] D. Holton, B. Manvel, B. McKay, "Hamiltonian cycles in Cubic 3-connected bipartite planar graphs," *J. Combinatorial Th. Ser. B*, 38 (1985) 279-297.

[10] Bill Jackson, "Longest Cycles in 3-connected cubic graphs," *J. Combinatorial Theory Ser. B*, 41 (1986), 17-26.

[11] Brad Jackson, "Small r-regular r-connected nonhamiltonian graphs," *Ars Combinatoria* 24(1987) 77-83.

[12] Brad Jackson and T. Parsons, "On r-regular r-connected nonhamiltonian graphs," *Bull. Australian Math. Soc.* 24 (1981) 205-220.

[13] Brad Jackson and T. Parsons, "A shortness exponent for r-regular r-connected graphs, *J. Graph Theory*, 6(1982) 169-176.

[14] J. Lederberg, "Hamiltonian circuits of convex trivalent polyhedra (up to 18 vertices)," *American Math. Monthly* 74 (1967) 522-527.

[15] W. McCuaig and M. Rosenfeld, "Cyclability of r-regular r-connected graphs," *Bull. Australian Math. Soc.*, 29 (1984) 1-11.

[16] G. Meredith, "Regular n-valent n-connected nonhamiltonian non-n-edge-colorable graphs," *J. Combinatorial Theory Ser. B*, 14 (1973) 55-60.

[17] C. Nash-Williams, "Possible directions in graph theory," *Combinatorial Mathematics and its Applications*, D. Welsh, editor, Academic Press, New York, 1971.

[18] P. Tait, "Remarks on the coloring of maps," *Proc. Royal Society Edinburgh Ser. A.*, 10(1980) 729.

[19] W. Tutte, "On hamiltonian circuits," *J. London Math. Soc.* 21 (1946) 98-101.

[20] W. Tutte, "A theorem on planar graphs," *Trans. American Math. Soc.* 82 (1956) 99-116.

[21] W. Tutte, "On two-factors of bipartite regular graphs," *Discrete Math.* 41(1982) 35-41.

[22] J. Zaks, "Shortness Coefficient of Cyclically 5-connected cubic planar graphs," *Aequationes Math.* 24 (1982) 97-102.

AN INTRODUCTION TO TOLERANCE INTERSECTION GRAPHS

M. S. Jacobson

F. R. McMorris

University of Louisville

H. M. Mulder

Erasmus Universiteit

ABSTRACT

A masterplan of tolerance intersection graphs is proposed. Any class of intersection graphs can be generalized by introducing tolerances. As an example, nested families of sets with tolerances are studied, by which new characterizations of threshold graphs, interval graphs and coTT graphs are obtained. Some questions and problems are posed.

1. Introduction

The Masterplan of Tolerance Intersection Graphs   An intriguing theme in graph theory is that of the intersection graph of a family of subsets of a set: the vertices of the graph are represented by the subsets of the family and adjacency is defined by a non-empty intersection of the corresponding subsets. By specifying conditions on the family of subsets one gets interesting classes of intersection graphs. Two notable examples are interval graphs and chordal graphs. An interval graph is the intersection graph of a family of closed intervals on the real line. These graphs have many nice characterizations, such as by forbidden subgraphs [8]. Chordal graphs are defined as graphs having no induced cycle C_n of length $n \geq 4$. They were proven to be the intersection graphs of subtrees of a tree [2].

Golumbic and Monma in [5] introduced a generalization of interval graphs using tolerances (see also [6]). In this case, we have a family of intervals $\{I_v\}_{v \in V}$ and each interval I_v is assigned a positive real number t_v, its tolerance. Now we construct a graph with vertex set V by joining two vertices u an v whenever $|I_u \cap I_v| \geq \min\{t_u, t_v\}$, where $|I|$ denotes the length of interval I. A full characterization of these graphs is still open. Besides this problem some other natural questions present themselves. What happens when all tolerances are equal or when $t_v = |I_v|$, for each vertex v? In [5] it was shown that with constant tolerance c we get precisely the interval graphs and with tolerances $t_v = |I_v|$ we get precisely the permutation graphs.

This idea of tolerances on intervals suggests a *Masterplan*.

Let Z be a set and let μ be a measure on Z assigning to each non-empty subset S of Z a positive real number $\mu(S)$. Let $S = \{S_v\}_{v \in V}$ be a (finite) family of non-empty subsets of Z, and let $t : S \rightarrow R+$ be a mapping, which assigns to each subset S_v a *tolerance t_v*. Finally, let $\phi : R^+ \times R^+ \rightarrow R^+$ be a function assigning a positive real number to each pair of positive reals. Then the *tolerance intersection graph* of (S, μ) and (t, ϕ) is the graph $G = (V, E)$ with vertex-set V and

$$uv \in E \text{ if and only if } \mu(S_u \cap S_v) \geq \phi(t_u, t_v).$$

By specifying conditions on S, μ and ϕ we get special classes of tolerance graphs, which could be termed ϕ-*tolerance (S, μ)-graphs*. For each such class we can try to answer the canonical natural questions of finding a characterization of the class, and establishing the prototype theorems (I) on constant tolerances and (II) on tolerances $t_v = \mu(S_v)$, for v in V. It is important to note that being a tolerance intersection graph is hereditary; that is, every induced subgraph is also a tolerance intersection graph of the same type. Note also that this implies that a forbidden subgraph list exists for each such class.

A characteristic line of approach in the spirit of the Masterplan is to start with an intersection representation of a nice class P of graphs. Generalize this class to that of (ϕ-)*tolerance P graphs* by introducing tolerances, and then establish prototype theorems I and II and try to find characterizations of this class. From this point of view it is more appropriate to name the above graphs introduced in [5] by Golumbic and Monma, (min-) tolerance interval graphs instead of "tolerance graphs".

The aim of this paper is to show some of the possibilities of the Masterplan, to open up new vistas, and to present some questions and problems. In Section 2 we consider intersection representations for split graphs and chordal graphs. Apparently these representations are not the right ones to yield interesting generalizations, because in all cases, by introducing tolerances we obtain all graphs. In Section 3 we choose a different line of approach. The family of subsets is in all cases a family of nested sets, and we vary the function on the tolerances. Although prototype theorems I and II reduce to trivia, we obtain some nice new characterizations of known classes of graphs. Finally, in Section 4, we state some problems and suggest several promising generalizations.

In the following we will restrict ourselves to families consisting of intervals on the real line, finite sets, or subgraphs. As a measure of the respective families we use length of intervals, the cardinality of the finite sets, or the number of vertices of the subgraphs.

2. Subtrees of a Tree with Tolerances Chordal graphs, split graphs, and threshold graphs all have intersection representations involving subtrees of a tree. In this section we study the effects of introducing tolerances, where the measure is the number of vertices in the subtree and ϕ is the minimum of the tolerances. First we recall some relevant definitions and results that are needed here and in the next section. They can all be found in [3].

A graph $G = (V, E)$ is a *threshold graph* if we can associate a *threshold* Θ with G and assign weights w_v to the vertices v in V such that

$$uv \in E \text{ if and only if } w_u + w_v > \Theta.$$

The typical structure of a threshold graph is depicted in Figure 1. The vertex-set of G is partitioned into $D_0, D_i, ..., D_m$, where D_0 may be empty and $D_{\lceil m/2 \rceil}$ only exists if m is odd. The sets D_i are indicated by cells in Figure 1, and a line between cells D_i and D_j means that each vertex in D_i is adjacent to each vertex in D_j. Moreover, the cells on the left hand side consist of independent sets and the cells on the right hand side are cliques. So the left hand side is an independent set and the neighborhoods of its vertices are ordered by inclusion downwards. And the right hand side is a clique and the closed neighborhoods of its vertices are ordered by inclusion upwards.

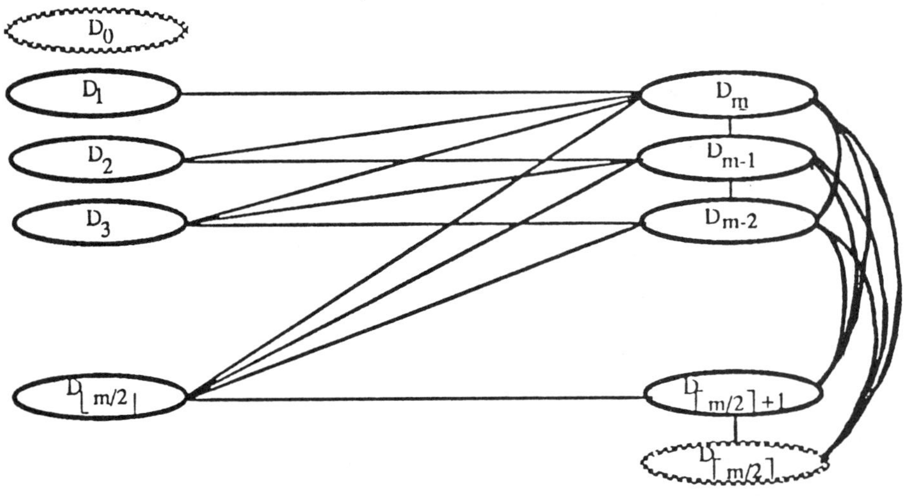

Figure 1

A *split graph* $G = (V, E)$ is a graph whose vertex-set V can be partitioned into an independent set I and a clique K. So, in particular, a threshold graph is split. Clearly, threshold graphs and split graphs are special instances of chordal graphs.

In [9] it was observed that a graph is split if and only if it is the intersection graph of a family of substars of a star $K_{1,n}$ (for some n). The substars in the family consisting of only a pendant vertex of the large star constitute the independent set of the split graph. All other substars in the family contain the center of the large star and hence constitute a clique of the split graph, yielding the desired partition. However, when we introduce tolerances in this case we do not get a very interesting generalization of split graphs. This is an immediate consequence of the following lemma which is known [4], but we include a proof for completeness.

Lemma 1 Let $G = (V, E)$ be a graph. Then G is the edge-intersection graph of substars of a star.

Proof Set $m = |E|$ and label the pendant vertices of the star $K_{1,m}$ with the edges of G. Let c be the center of $K_{1,m}$. For each vertex v of G, let S_v be the substar of $K_{1,m}$ consisting of the center of $K_{1,m}$ and the pendant vertices labeled with the edges incident with v. Then uv is an edge in G if and only if S_u and S_v have an edge in common, viz. the edge between the center c and the pendant vertex labeled with uv in E.

We proceed by considering prototype theorems I and II for the substars of a star case.

Theorem 2.1 Every graph G is a min-tolerance intersection graph of substars of a star for some constant tolerance.

Proof From Lemma 1, using constant tolerance 2, we get any graph from substars of a star.

Recall that a comparability graph is obtained from a partially ordered set $(V, \leq)$, with u and v adjacent if and only if $u < v$ or $v < u$.

Theorem 2.2 A graph G is a min-tolerance intersection graph of substars of a star with tolerances equal to the number of vertices in substars if and only if it is a comparability graph.

Proof If the tolerances are equal to the number of vertices of the substars, then adjacency is equivalent to substar containment. Thus we have a containment graph of substars of a star, which is known to be equivalent to being a comparability graph [7].

Chordal graphs are the intersection graphs of subtrees of a tree. Theorems 2.1 and 2.2 imply that in this case we do not get a nice generalization of chordal graphs. We simply state prototype theorems 1 and 2.

Theorem 3.1   Any graph is the min-tolerance intersection graph of subtrees of a tree for some constant tolerance.

Theorem 3.2   A graph is the min-tolerance intersection graph of subtrees with tolerances equal to the number of vertices in the subtree if and only if it is a comparability graph.

Finally, a threshold graph is an intersection graph of substars of a star, where the substars containing the center of the large star are nested when restricted to pendant vertices that are substars. In this case we get the same results. All graphs in prototype theorem 1 and the comparability graphs in prototype theorem 2.

We also have considered the continuous versions of the above. That is, each edge is replaced by a line segment. Then the size of an intersection is the sum of the lengths of the line segments in the intersection.

Even in this case, however, we get precisely the same classes of graphs as previously obtained in prototype theorems 1 and 2.

3.   Nested Families with Tolerances

In this section we restrict ourselves to nested families of sets (i.e. linearly ordered by inclusion), and we vary the function ϕ on the tolerances. For ϕ we consider the minimum, the maximum, and the sum of the two tolerances. Other such functions will be considered in future work. Although prototype theorems 1 and 2 reduce to trivia, we attain new characterizations of some important classes of chordal graphs when arbitrary tolerances are considered. For the nested families we consider intervals on the real line, with their lengths as measure, and finite sets, with their cardinalities as measure. We will refer to the measure of a set as the *set-size*. First we observe that without loss of generality we may choose as nested intervals $N = \{[0, r_i] : i = 1, 2, ..., \}$

with $0 < r_1 \leq r_2 \leq \ldots \leq r_n$, and in the case of finite sets we may choose them to be of the form $\{1, \ldots, k_i\}$ with positive integers k_i satisfying $1 \leq k_1 \leq k_2 \leq \ldots \leq k_n$. Furthermore, in the case of finite sets $\{1, .. k_i\}$ we may convert them into intervals $[0, k_i]$, and vice versa, if we have intervals $[0, r_i]$ with integer r_i, then we may convert them into finite sets $\{11, \ldots, r_i\}$ without changing the tolerance-intersection structure. Let us call the associated tolerance intersection graphs ϕ *tolerance chain graphs*, depending on the choice of ϕ.

A graph is said to admit a *P-elimination scheme* if its vertices can be ordered $v_1, v_2, \cdots, v_n$ such that each v_i is of type P in the subgraph induced by $v_i, v_{i+1}, \cdots, v_n$. The ordering $v_1, v_2, \cdots, v_n$ is a *P-elimination ordering* of the vertices. A *simplicial vertex* is a vertex having a clique (a set inducing a complete graph) as its neighborhood. A nice and well-known characterization of chordal graphs (see [3]) is that they are the graphs having a simplicial elimination scheme, where any simplicial vertex may start a simplicial elimination ordering. An equivalent formulation of this property is that every induced subgraph of a chordal graph contains a simplicial vertex. A *central vertex* is a vertex adjacent to all other vertices of the graph. The following fact follows immediately from the structure of threshold graphs as given in Figure 1.

Lemma 4   A graph G is a threshold graph if and only if G admits an (isolated or central)-elimination ordering. Here any isolated or central vertex may start an elimination ordering.

First we consider the min-tolerance chain graphs. As we noted earlier, the prototype theorems reduce to trivia.

Theorem 5.1   A graph G is a min-tolerance chain graph with constant tolerances if and only if G is the disjoint union of a complete graph and an edge-less graph.

Theorem 5.2   A graph G is a min-tolerance chain graph with tolerances equal to the set-sizes if and only if G is a complete graph.

With arbitrary tolerances we get a new characterization of threshold graphs.

Theorem 6   A graph is a threshold graph if and only if it is a min-tolerance chain graph.

Proof   First let $N = \{[0, r_i] : i = 1, 2, ..., n\}$ be a nested family of intervals with $0 < r_1 \leq r_2 \leq ... \leq r_n$, where interval $[0, r_i]$ has tolerance t_i, for $i = 1, ..., n$. Let G be the associated min-tolerance chain graph with v_i represented by $[0, r_i]$, for $i = 1, ... n$. In view of the fact observed above, it suffices to show that G has a central vertex or an isolated vertex. If $t_1 \leq r_1$, then v_1 is adjacent to all other vertices. If $t_1 > r_1$ and v_1 is not isolated, then let v_j be a neighbor of v_1. This implies that $t_j \leq r_1$, whence v_j is a central vertex. For the converse we have indicated in Figure 2 a min-tolerance chain representation for the typical threshold graph. For every cell the representing set and tolerance for all vertices in the cell are given.

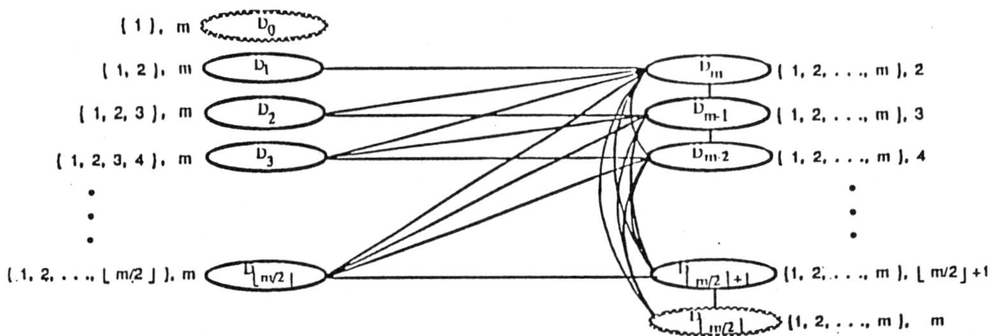

Figure 2

Next we consider the max-tolerance chain graphs.

Theorem 7.1   A graph G is a max-tolerance chain graph with constant tolerances if and only if G is the disjoint union of a complete graph and an edge-less graph.

Theorem 7.2   A graph G is a max-tolerance chain graphs with tolerances equal to set-sizes if and only if G is a disjoint union of complete graphs.

With arbitrary tolerances we get a new characterization of interval graphs. This can be proved by a transformation of the nested intervals with tolerances into arbitrary intervals with intersections, and vice versa. We also include a second proof using the tolerances. It is somewhat longer, but it is more illuminating with respect to the effects of the tolerances.

Theorem 8 A graph is a max-tolerance chain graph if and only if it is an interval graph.

First Proof Let $N = \{[0, r_i] : i = 1, 2, ..., n\}$ be a nested family of intervals with $0 < r_1 \leq r_2 \leq ... \leq r_n$ and with respective tolerances t_i. Let $G = (V, E)$ be the associated max-tolerance chain graph, where v_i is represented by $[0, r_i]$, for $i = 1, ..., n$.

We may assume that $t_i \leq r_i$, for $i = 1, ..., n$, for if $t_i > r_i$, then v_i is isolated and we can disregard v_i. consequently, v_i and v_j are adjacent if and only if

$$\min\{r_i, r_j\} \geq \max\{t_i, t_j\}.$$

This is true if and only if the two intervals $[t_i, r_i]$ and $[t_j, r_j]$ have a non empty intersection.

Thus, the max-tolerance chain graph of N is the interval graph of the intervals $\{[t_i, r_i] : i = 1, 2, ..., n\}$, and vice versa.

Second Proof Let N, t_i and G be as in the first proof. We may assume that G is connected, so that in particular $t_i \leq r_i$, for $i = 1, ..., n$. We show that v_1 is simplicial. Let v_1 and v_j be distinct neighbors of v_1. Then we have $t_i, t_j \leq r_1 \leq r_i, r_j$, so that v_i and v_j are also adjacent. Clearly, any neighbor of v_1, except v_2, is adjacent to v_2, since $r_1 \leq r_2$. So we have a simplicial elimination ordering $v_1, v_2, ..., v_n$ for G such that, for each i, all neighbors of v_i among $v_{i+2}, ..., v_n$ are adjacent to v_{i+1}. It is easily checked that all forbidden subgraphs for interval graphs (see [8]) fail to have such an elimination ordering. Hence G is an interval graph.

Conversely, an interval graph H has the property that its maximal cliques can be linearly ordered as $C_1, ..., C_m$ such that any vertex of H occurs in consecutive maximal cliques. Now, if vertex v is in cliques $C_i, C_{i+1}, ..., C_j$, then we associate with v the set $\{1, 2, ..., j\}$ and tolerance i. It is easily checked that the chain of sets with tolerances thus obtained is a max-tolerance chain representation of H.

In the second proof we have shown that a graph is an interval graph if and only if it admits a simplicial elimination ordering $v_1, v_2, ..., v_n$ with the additional property that for each i, the neighbors of v_i among $v_{1+2}, ..., v_n$ are all adjacent to v_{i+1}. As far as we know this seems to be another new characterization of interval graphs.

Now we turn to the sum-tolerance case. Again we state prototype theorems I and II.

Theorem 9.1   A graph G is a sum-tolerance chain graph with constance tolerances if and only if G is the disjoint union of a complete graph and an edge-less graph.

Theorem 9.2   A graph G is a sum-tolerance chain graph with tolerances equal to set-sizes if and only if G is an edge-less graph.

In a recent paper Monma, Reed and Trotter [11] have introduced what they call threshold tolerance graphs, which generalize threshold graphs. With each vertex v of a graph we associate a weight t_v and a tolerance r_v. Then in u and v are adjacent if and only if

$$t_u + t_v > \min \{r_u, r_v\}.$$

The main results in [11] concern complements of threshold tolerance graphs, called *co TT- graphs*. Here adjacency is defined by

$$t_u + t_v \leq \min\{r_u, r_v\}.$$

Hence, coTT graphs are equivalent to the sum-tolerance chain graphs with nested intervals $N = \{[0,_i] : i = 1, 2, ..., n\}$ and tolerances t_i. We now analyze coTT from our perspective. Because our proofs are considerably different, we will reprove some results in [11] as Lemmas 12 and 13.

Lemma 10 Let G b the sum-tolerance chain graph of the nested intervals N with tolerances t_i. Then any vertex with maximum tolerance is simplicial in G.

Proof Let v_j be a vertex with maximum tolerance, and let v_i and v_k be distinct neighbors of v_j. Then we have

$$t_i + t_k \leq t_i + t_j \leq \min\{r_i, r_j\} \leq r_i$$

and

$$t_i + t_k \leq t_j + t_k \leq \min\{r_j, r_k\} \leq r_k,$$

so that

$$t_i t_k \leq \min\{r_i, r_k\}.$$

Hence v_i and v_k are adjacent.

Thus, sum-tolerance chain graphs are chordal.

Lemma 11 lt $u_1, u_2, ..., u_p$ be an induced path in a sum-tolerance chain graph, where s_i is the tolerance of u_i, for $i = 1, .., p$. Then either

$$s_1 = s_2 < s_3 < ... < s_p,$$

or

$$s_1 > s_2 > ... > s_{p-1} = s_p,$$

or

$$s_1 > s_2 > ... > s_j < s_{j+1} < ... < s_p,$$

for some j with $1 \leq j \leq p$.

Proof   Let $[0, x_i]$ be the interval associated with u_i, for $i = 1, ..., p$. Assume that $s_i \leq x_{i+1}$, for some i. Since u_{i+1} and u_{i+2} are adjacent, we have

$$x_{i+2}, x_{i+1} \geq s_{i+2} + s_{i+1} \geq s_{i+2} + s_i$$

Since u_{i+2} and u_i are not adjacent, it follows that

$$s_{i+2} + s_i > x_i.$$

Now u_{i+1} and u_i are adjacent, so that $x_i \geq s_{i+1} + s_i$. This implies that

$$s_{i+2} > s_i.$$

Now suppose that $s_i = s_{i+1}$, for some i with $1 < i < p - 1$. Then, by the previous argument, we have $s_{i-1} > s_i = s_{i+1} < s_{i+2}$, with, say, $s_{i+2} \leq s_{i-1}$.

Then we have

$$x_{i+2}, x_{i+1} \geq s_{i+2} + s_{i+1} = s_{i+2} + s_i > x_i \geq s_{i-1} + s_i,$$

which is impossible. These observations immediately imply the assertion in the lemma.

An m-*trampoline*, or m-*sun*, with $m \geq 3$ consists of a complete graph with vertex set $\{1, 2, ..., m\}$ together with vertices $m+1, m+2, ..., 2m$, where $m + 1$ is adjacent to i and $i + 1 \pmod{m}$. The *strongly chordal graphs* are the chordal graphs without induced m-suns, for $m \geq 3$ (see [1]).

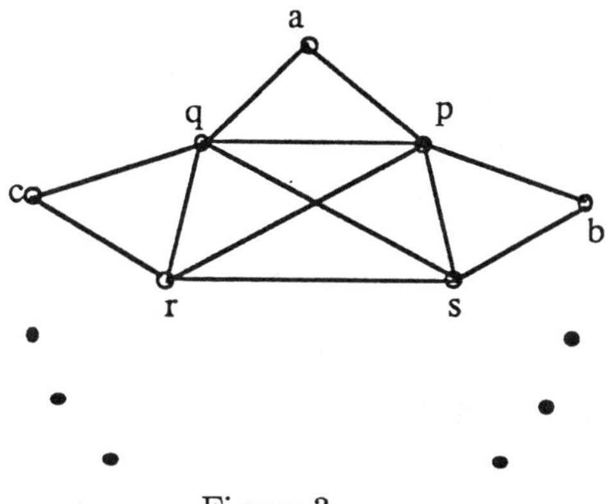

Figure 3

Lemma 12   A sum-tolerance chain graph is strongly chordal.

Proof   It suffices to show that m-suns do not have a sum-tolerance chain representation. Assume the contrary. In Figure 3 an m-sun is depicted, where $r = s$ in the case $m = 3$. The vertices are labeled with their tolerances, and the intervals of vertices $b, p, q,$ and c are, respectively, [0,B], [0P], [0,Q], and [0,C]. Without loss of generality, we may assume that in Figure 3 we have $a \geq b \geq c$. Then we have

$$P \geq p + b \geq p + c > C \geq q + c,$$

so $p > q$. On the other hand, we have

$$Q \geq a + q \geq b + q > B \geq b + p,$$

whence $q > p$. This contradiction completes the proof. In Figure 4 we see two other graphs that are forbidden in sum-tolerance chain graphs.

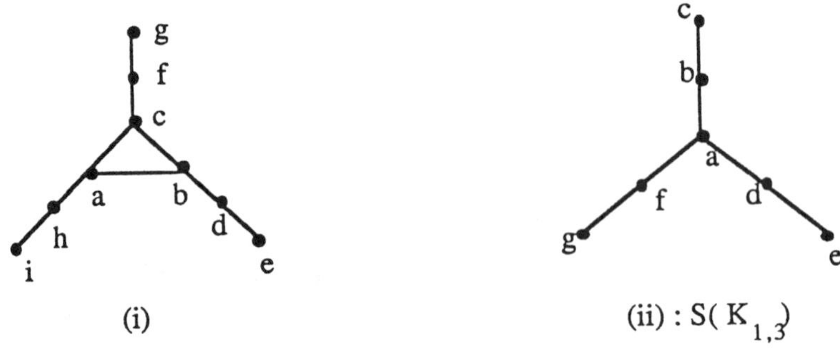

Figure 4

Lemma 13 The graphs in Figure 4 do not occur as induced subgraphs in a sum- tolerance chain graph.

Proof Assume that the graphs of Figure 4 each have a sum-tolerance chain representation. Again the vertices are labeled with their tolerances and the associated intervals are of the form $[0,X]$, where X is associated with vertex and tolerance x. First consider the graph of Figure 4(i). Since the edges of the triangle are each internal edges of some induced path, the tolerances on the triangle are all distinct, say, $a < b < c$ by Lemma 11. Again by Lemma 11, we have $b < d < e$ and $c < f < g$. Now, if $d \leq c$, then we get

$$A, C \geq c + a \geq d + a > D \geq d + b > d + a,$$

which is impossible. Hence $d > c$ and we have

$$D \geq d + e > d + c. \qquad (^*)$$

On the other hand, we have

$$F, C \geq f + c > f + b > B \geq d + b,$$

so $f > d$. Hence we have

$$C \geq f + c > d + c. \qquad (^{**})$$

Now (*) and (**) force d and c to be adjacent. This impossibility shows that Figure 4(i) depicts a forbidden subgraph for sum-tolerance chain graphs.

Second consider the graph of Figure 4(ii). By Lemma 10, one of the pendant vertices must have maximal tolerance, say c is maximal. There are three candidates for the next maximal tolerance, viz. b, e, and g. Assume that $b \leq e$. Then we have

$$D \geq d + e \geq d + b > B \geq c + b \geq d + b,$$

which is impossible. So we have $b > e$, and similarly $b > g$. Now assume that $a \leq d$. Then we get

$$E \geq d + e \geq a + e > A \geq a + b > a + e,$$

which is impossible. So we have $a > d$. Similarly, we infer that $a > f$. But now we are in conflict with Lemma 11, which completes the proof. The graphs of Figure 4 are minimal forbidden subgraphs as is shown by the graphs in Figures 5 and 6. Note that the graph of Figure 4(ii) minus a pendant vertex is induced in the graph of Figure 6. This last graph shows how we can realize arbitrary long paths and caterpillars (with extra backbone) using Fibonacci's sequence. This will be used for Theorem 15.

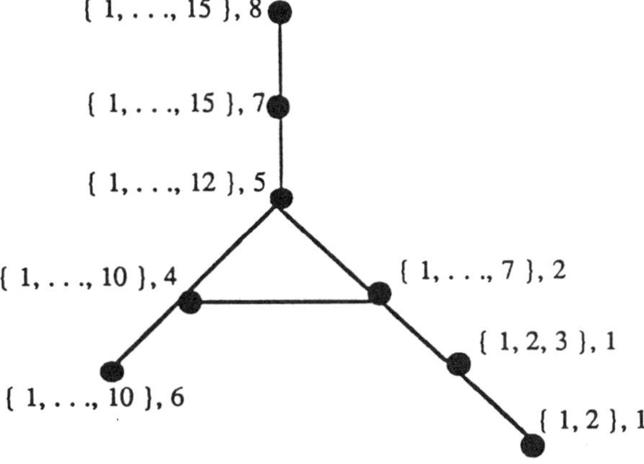

Figure 5

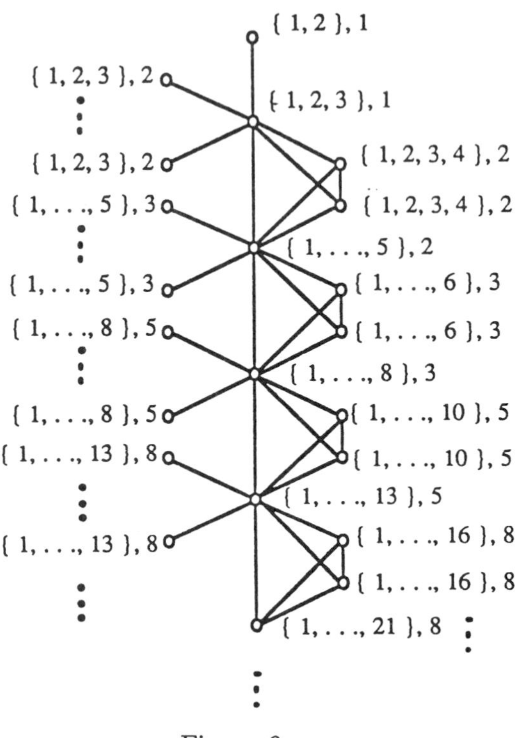

Figure 6

Theorem 14   A sum-tolerance chain graph is a strongly chordal graph without any of the graphs of figure 4 as induced subgraphs.

Proof   Use Lemmas 12 and 13.

We conclude this section showing that trees which are sum-tolerance chain graphs are precisely those with no induced $S(K_{1,3})$.

Theorem 15   A tree T is a sum-tolerance chain tree if and only if T contains no induced subgraph isomorphic to $S(K_{1,3})$.

Proof   We only need show that if T is a tree with no induced $S(K_{1,3})$'s then T is a sum-tolerance chain tree. We proceed by induction. Assume the result is true for all such trees with less than n vertices (it is easy to derive a representation for "small" trees) and let T be a tree with no induced $S(K_{1,3})$ of order n. It T were a path, as previously noted, T would be a sum-tolerance graph. Thus, we may assume that T contains a vertex x of degree at least 3. Furthermore, since T contains no induced $S(K_{1,3})$ we may assume that x is adjacent to at least one pendant vertex v. By the induction hypothesis $T - v$ has a sum-tolerance representation.

Choose z and y neighbors of x in $T - v$, so that t_y and t_z are the smallest two values over all neighbors of x. If both t_y and t_z were at most t_x, then

$$t_y + t_z \leq t_x + t_z \leq \min\{r_x, r_z\} \leq r_z,$$

$$t_y + t_z \leq t_y + t_x \leq \min\{r_y, r_x\} \leq r_y,$$

which implies $t_y + t_z \leq \min\{r_y, r_z\}$, hence $yz \in E(T - v)$ which is a contradiction. Thus at least one of t_z or t_y is greater than t_x. Without loss of generality, say $t_y > t_x$. Since $t_y > t_x$. Since $t_x + t_y \leq r_x$, it follows that $t_y \leq r_x - t_x$. To complete the sum-tolerance representation for T, let $t_v = r_x - t_x$ and $r_v = r_x$. Clearly, if t_z were also greater than t_x, then it would follow that v would only be adjacent to x, and T would result. The only possibility remaining is to suppose $t_y > t_x \geq t_z$. Since $t_v = r_x - t_x \geq t_y > t_x$, it follows that all vertices non-adjacent to x would also be non-adjacent to v. Hence we need only show that z and v are not adjacent.

First, suppose $r_z \geq r_z$. This implies $t_z + t_y \leq t_x + t_y \leq \min\{r_x, r_y\} \leq \min\{r_2, r_y\}$ which would imply that zy is an edge. Since this cannot be the case, we can conclude that $r_z < r_x$.

Now suppose $t_v + t_z \leq \min\{r_v, r_z\}$. If $r_z \leq r_y$, then

$$r_z = \min\{r_z, r_y\} < t_z + t_y \leq t_z + t_v = t_z + r_x - t_x \leq r_z$$

Thus $r_y < r_z$, which implies $t_x + t_y \leq \min\{r_x, r_y\} = r_y = \min\{r_z, r_y\} < t_z + t_y$, which yields the contradiction $t_z > t_x$.

Consequently, $t_v + t_z > r_z$ which implies that $t_v + t_z > \min\{r_v, r_z\}$ and hence v and z are not adjacent so that the resulting representation gives precisely T.

Note that in [6], the trees of Theorem 15 were shown to be bounded min-tolerance interval trees.

4. Open Problems and Questions

We have seen that in the case of interval graphs by introducing tolerances one gets a very interesting class of graphs ([5], [6]). In section 2 we tried to obtain similar classes for threshold split and chordal graphs with somewhat unsatisfactory results. We pose the problem of searching for intersection representations for these classes so that one gets meaningful and interesting classes of tolerance threshold graphs, tolerance slit graphs, and tolerance chordal graphs.

A question that presents itself after section 3 is whether there are functions ϕ on the tolerances such that the ϕ-tolerance chain graphs are not chordal.

Along the lines of the previous section we may consider more general families than chains of intervals or finite sets. A natural candidate is a family of sets called a "tree of subsets" (with respect to set inclusion, the sets form a tree ordered set.) Every chain in a tree of subsets is characterized by the results in the previous section. In particular, it is easy to show that with min-tolerance and max-tolerance the smallest sets correspond to simplicial or isolated vertices. Hence in these cases we have again classes of chordal graphs. In section 3 the prototype theorems I and II reduced to trivia, but in this situation we expect non-trivial results for prototype theorems I and II.

Another promising tolerance intersection problem is the following. In [10] McMorris and Scheinerman observed that the chordal graph are precisely the intersection graphs of families of leaf-generated subtrees of some full binary tree. Here a full binary tree is a rooted tree with root of degree 2, all leaves (pendant vertices) at equal distance from the root and all other vertices have degree 3. A leaf-generated subtree is obtained by choosing some leaves of the binary tree and then taking the subtree having precisely the chosen leaves as pendant vertices. Surprisingly, it turns out that different constant tolerances yield different classes of graphs, and this will be reported elsewhere. However, much work needs to be done on this representation.

The authors hope that the above problems and questions will stimulate much work in the near future.

REFERENCES

[1] M. Farber, "Characterizations of strongly chordal graphs," *Discrete Math.* 43 (1983), 173-189.

[2] F. Gavril, "The intersection graph of subtrees in trees are exactly the chordal graphs," *J. Combin. Theory B* 16 (1974), 47-56.

[3] M. C. Golumbic, Algorithmic Graph Theory and Perfect Graphs, Academic Press, New York, 1980.

[4] M. C. Golumbic and R. E. Jamison, "The edge intersection graphs of paths in a tree, *J. Combin. Theory* B 38 (1985), 8-22.

[5] M. C. Golumbic and C. L. Monma, "A generalization of interval graphs with tolerances," *Congressus Numerantium* 35 (1982), 321-331.

[6] M. C. Golumbic, C. L. Monma and W. T. Trotter, "Tolerance graphs," *Discrete Appl. Math.* 9(1984), 157-170.

[7] M. C. Golumbic and E. R. Scheinerman, "Containment graphs, posets and related classes of graphs," to appear in *Proc. 3rd International Conf. on Combinatorial Math.,* New York Academy of Sciences.

[8] C. G. Lekkerkerker and J. C. Boland, "Representation of a finite graph by a set of intervals on the real line," *Fund. Math.* 51(1962), 45-64.

[9] F. R. McMorris and D. Sheir, "Representing chordal graphs on $K_{1,n}$," *Comm. Math. Univ. Carolinae* 24(1983), 489-494.

[10] F. R. McMorris and E. R. Scheinerman, "Connectivity threshold for random chordal graphs," to appear in *Graphs and Combin.*

[11] C. L. Monma, B. Reed and W. T. Trotter, "Threshold tolerance graphs," to appear in *J. Graph Theory.*

GRAPH TRANSFORMS:
A FORMALISM FOR MODELING CHEMICAL REACTION PATHWAYS

Mark Johnson

The Upjohn Company
Kalamazoo, Michigan

ABSTRACT

A goal of computer-assisted organic synthesis is the construction of a kit of reaction rules that enables the computer to suggest chemical paths of compounds leading to or from a compound of interest. Implicit in computer-aided organic synthesis the modeling of chemical reaction pathways. A metadigraphs is introduced as a graph-theoretic counterpart of a chemical reaction pathway and is defined to be a digraphs whose vertex set Γ is a set of graphs. A graph transform t is defined such that for each ordered pair (G, G') of graphs one can determine if t can transform G into G'. A iransform kit is defined to be an ordered pair $(T^i T^b)$ of sets T^i of "inducing" transforms and T^b of "blocking" transforms. A set Γ of graphs and a kit K defines a unique metadigraph $M(\Gamma, K)$. The ordered pair (Γ, K) is referred to as a specification of the metadigraph M if $M = M(\Gamma, K)$. It is shown that every metadigraph M admits a specification of the form $(V(M), K)$ for some kit K where $V(M)$ denotes the vertex set of M.

1.   Introduction

Chemical graphs are labeled graphs in which letters assigned to the vertices denote the atoms of a molecule and letters assigned to the edges denote the bonds. They are widely used in chemistry as representations of molecular structures. A chemical reaction type can be viewed as a transform which, when applied to a compound, generates new compounds. For example, benzene undergoes hydroxylation to form hydroxybenzene. This idea has been made algorithmically explicit in a variety of ways with respect to chemical graphs in computer-assisted organic synthesis (1,10,11) and computer-assisted drug metabolism (6). The goal in these approaches is the construction of a kit of reaction rules that enables the computer to suggest chemical paths of compounds leading to or from a compound of interest. Recently, Koća et al. (10) have re-examined the approach by Ugi et al. (11) using graph-theoretic concepts. This paper develops a general graph-theoretic formalism analogous to the chemical concepts of reactions and kits of reaction rules, but freed from terminology and assumptions that limit the applications of this formalism to a particular scientific discipline. Such a mathematical formalism should eventually lead to improvements in computer-assisted organic synthesis.

Chemical pathways are diverse. They may consist of a single reaction or a complex biochemical pathway. Figure 1 presents part of the "shikimic acid" metabolic pathway (3). There we see cpd. 2 was hydroxylated to give cpd. 3. Similarly, cpd. 4 was hydroxylated to give cpd. 5 and cpd. 5 was further hydroxylated to give cpd. 6. In our analysis, "hydroxylation", which performs a particular structural change, will correspond to the graph-theoretic concept of a transform. In hydroxylation, a hydrogen atom is deleted and the atoms and bonds of an hydroxyl group are added. The graph elements (vertices or edges corresponding to the atoms and bonds) to be deleted by a transform will subsequently be indicated by assigning each such element $a-1$, and the graph elements to be added will be indicated by assigning each such element $a+1$. Note the asymmetry here. The reverse change, deleing an hydroxyl group and adding a hydrogen atom is called dehydroxylation. This asymmetry must be incorporated into our concept of a transform. In addition, we see that a transform acts on some compounds and not on others. In Figure 1, compounds 2,

4 and 5 were hydroxylated, but compounds 1, 3, and 6 were not. Those graph elements used to define the local environment where a structural change will take place, but which are neither to be added nor deleted, will subsequently be indicated by assigning each such element a 0. Finally, Figure 1 indicates that compound 2 is "deanimated" to give compound 5. The "deanimation" transform perform a different structural change than the "hydroxylation" transform. Thus, a solution to the problem of modeling chemical reaction pathways will require a specification of a broad class of possible transforms and a set of rules for deciding which transform is to operate where on each graph.

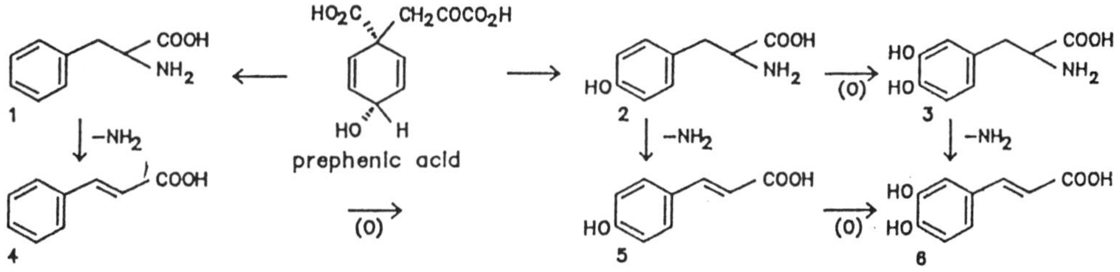

Figure 1. Part of the shikimic acid pathway

Figure 2 presents a graph-theoretic counterpart of a chemical reaction pathway. This figure should be viewed as involving two digraphs M_1 and M_2 whose vertices are themselves graphs. We shall refer to such digraphs as metadigraphs. The arcs of the metadigraphs are the graph-theoretic counterparts of chemical reactions. Those operations which convert one graph into another, the counterparts of "hydroxylation" and "deamination", will be called transforms. A transform kit K will consist of a set T^i of "inducing" transforms and set T^b of "blocking" transforms. Rules are given whereby an arc of a metadigraph can be considered to be an action of an "inducing" transform in T^i not blocked by a "blocking" transform in T^b. It will be seen that metadigraph M_1 in Figure 2 can be simply specified by an ordered pair $(\{P_1\}, K)$ where K consists of a single inducing transform and a single blocking transform. It will also be shown that any metadigraph M admits a specification of the form $(V(M), K)$ for some transform kit K where $V(M)$

denotes the set of graphs that comprise the vertex set of M.

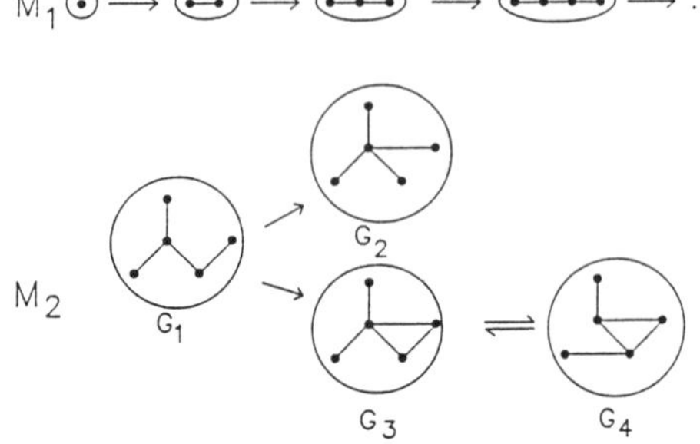

Figure 2. Two metadigraphs

2.   Semitransforms and semiactions

The preceding notions will now be developed rigorously following the notation in Chartrand and Lesniak (4). A graph $C = (V, E)$ together with an assignment of the integers -1,0 and 1 to the elements (vertices and edges) of C, one integer per element, will be called a *semitransform* on C if zero edges (edges assigned 0's) are adjacent to only zero vertices, nonpositive (≤ 0) edges are adjacent only to nonpositive vertices, and nonnegative (≥ 0) edges are adjacent only to nonnegative vertices.

A semitransform will be illustrated by indicating negative, zero and positive edges with dashed, solid and dotted lines, and indicating negative, zero and positive vertices with x's, solid dots and circles, respectively. Figure 3 shows three semitransforms.

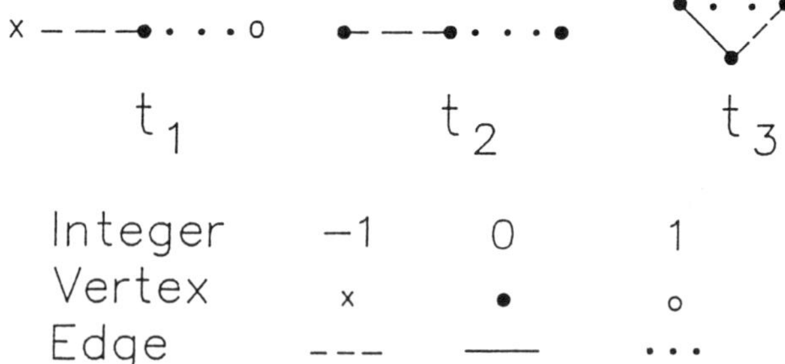

Figure 3. Some semitransforms

Let t be a semitransform on C. Four graphs will be commonly associated with t. Graph C is the *covering graph* of t. The nonpositive elements of t

define the *ingraph* I of t; the nonnegative elements define the *outgraph* O of t; and the zero elements define the *linking graph* L of t. The restrictions on the assignment function assure that I, O and L are well-defined. In Figure 3, the covering graph of t_3 is the cycle C_3, the ingraph and outgraph are both P_3 paths. The linking graph is the union of K_1 and P_2. The graphs C, I, O and L will be called the *defining* graphs of t. These letters, with appropriate superscripts and subscripts, will denote the defining graphs of a semitransform with corresponding superscripts and subscripts. Clearly, if any three of the defining graphs of a semitransform, when viewed as labeled subgraphs of C, are known or if the ingraph and outgraph are known, all of the defining graphs are completely determined. Define the *cardinality* $|G|$ of the graph $G = (V, E)$ by $|G| = |V| + |E|$. We obviously have

$$|I| + |O| = |C| + |L|, \qquad (1)$$

$$p(I) + p(O) = p(C) + p(L), \qquad (2)$$

and

$$q(I) + q(O) = q(C) + q(L), \qquad (3)$$

where $p(G)$ and $q(G)$ denote the order and size of graph G.

There are many ways to form a semitransform having G for its ingraph and G' for its outgraph. To see this, let C be any common supergraph of G and G'. Let I and O denote labeled subgraphs of C that are isomorphic to G and G', respectively. Then the semitransform having I and O as the ingraph and outgraph is such a transform. Note that C will be the covering graph of that transform only if I and O cover C, i.e. every element of C is an element of either I or O. Let t be a semitransform on $C = (V, E)$ with assignment function $g : V \cup E \to \{-1, 0, 1\}$ and let t' be a semitransform on $C' = (V', E')$ with assignment function $h : V' \cup E' \to \{-1, 0, 1\}$. Then t and t' are isomorphic, and we write $t \simeq t'$, if there exists an isomorphism $f : V \to V'$ between C and C' such that $g(v) = h(f(v))$ for all $v \in V$ and $g(uw) = h(f(u)f(w))$ for all $uw \in E$.

Let t be a semitransform on C with assignment function g. Let C' be a labeled subgraph of C and let g' be the restriction of g to the elements of C'. The graph C' together with g' is a semitransform t' called a *subtransform* of t. If, in addition, t' has as many nonzero elements as t, then t' is a *reduction* of t. If t'' is any other semitransform isomorphic to t', then t'' is also called a subtransform (or reduction if such is the case) of t. The semitransform t_2 is a reduction of the semitransform t_3 in Figure 3. If t' is a reduction of t, then t is an *extension* of t'. Clearly, if t' is a reduction of t, then

$$|X| - |Y| = |X'| - |Y'|, \qquad (4)$$

$$p(X) - p(Y) = p(X') - p(Y') \qquad (5)$$

$$q(X) - q(Y) = q(X') - q(Y') \qquad (6)$$

where X and Y denote any pair of defining graphs of t and X' and Y' denote the corresponding pair of defining graphs of t'.

Let (G, G') be an ordered pair of graphs, and let t be a semitransform. We shall say (G, G') is a *semiaction* t if there exists an extension t' of t such that G and G' are isomorphic to the ingraph and outgraph of t'. The extension t' is called a *semioverlay* of (G, G') with respect to t. Clearly, every semitransform t is a semioverlay of (I, O) with respect to t itself where I and O denote the ingraph and outgraph of t. Let (G, G') be any semiaction of t having ingraph I and outgraph O. It follows directly from equations 5 and 6 that

$$p(G') = p(G) + p(O) - p(I), \qquad (7)$$

$$q(G') = q(G) + q(O) - q(I), \qquad (8)$$

$$|G'| = |G| + |O| - |I|. \qquad (9)$$

Example 1   Let (G, G') be any semiaction of t_1 in Figure 3. Let t be any overlay of (G, G') with respect to t_1 and let C, I, O, L denote the defining graphs of t. Then the linking graph L of t contains all the vertices and edges in I and O except for one edge, say uy, of I where $u \in V(L)$ and $y \notin V(L)$ and one edge of O adjacent to u, say uz, where $z \notin V(L)$. Define the function $f : V(I) \to V(O)$ by $f(v) = v$ if $v \neq y$ and $f(v) = z$ if $v = y$. Clearly, f defines an isomorphism between I and O. Thus, if (G, G') is a semiaction of t_1, then $G \simeq G'$. It might be pointed out that equations (7) and (8) imply that G and G' have identical orders and sizes since I and O have identical orders and sizes.

A graph G is transformable into G' by an *edge rotation* if G contains distinct vertices u, v, w such that $uv \in E(G)$, $uw \notin E(G)$ and $G' \simeq G - uv + uw$ (5). An edge rotation is an *edge shift*, if, in addition, $vw \in E(G)$ (9). It is a straightforward matter to show that G is transformable into G' by an edge rotation (edge shift) if and only if (G, G') is a semiaction of $t_2(t_3)$ in Figure 3.

3.   Transforms and Actions

Let G and G' be two graphs. A *maximum common subgraph* of G and G' is a graph H of maximum cardinality which is isomorphic to a subgraph of both G and G'. A *minimum common supergraph* of G and G' is a graph H of minimum cardinality which is isomorphic to a supergraph of G and G'. (See Johnson (8) for other uses of maximum common subgraphs and minimum common supergraphs, referred to there as maximum intersections and minimum unions, for comparing graphs of arbitrary order and size.) A semioverlay t' of (G, G') with respect to a semitransform t is an *overlay* of (G, G') with respect to t if the linking graph of t' is a maximum common subgraph of G and G'. It follows directly from equation 1 that the linking graph of t' is a maximum common subgraph of G and G' if and only if the covering graph of t' is a minimum common supergraph of G and G'.

Figure 4 gives three overlays of graphs G_5 and G_6. Note that overlays o_1 and o_2 have isomorphic linking graphs but nonisomorphic covering graphs while overlays o_2 and o_3 have nonisomorhic linking graphs.

Various overlays of (G_5, G_6).

A semiaction (G, G') of a semitransform t is an *action* of t if t is a reduction of an overlay of (G, G'). A semitransform t is a *transform* if there exists an action of t. The following theorem establishes a consistency relationship between actions and overlays.

Theorem 1   A semioverlay of (G, G') with respect to t is an overlay with respect to t if and only if (G, G') is an action of t.

Proof   By definition, it t' is an overlay of (G, G') with respect to t, then (G, G') is an action of t. To establish the only if relation, assume (G, G') is an action of t and t' is a semioverlay of (G, G') with respect to t By definition, there exists an overlay t'' of (G, G') with respect to t. Using equation 4 with $X = I$ and $Y = L$, we have

$$|G| = |I'|$$
$$= |L'| + |I'| - |L'| \tag{10}$$
$$= |L'| + |I| - |L|$$

and

$$|G| = |I''|$$
$$= |L''| + |I''| - |L''| \tag{11}$$
$$= |L'| + |I| - |L|.$$

Equating the right hand sides of equations (10) and (11), we have $|L'| = |L''|$. Since L' is a subgraph of both G and G' of cardinality $|L|$, L' must be a maximum common subgraph.

Not all semitransforms are transforms. In particular, the semitransform given by t_1 of Figure 3 is not. To see this, suppose (G, G') were an action of t_1. By definition, there exists an extension t of g_1 which is an overlay of G, G'). Let C, I, O, L be the defining graphs of t. Then $I \simeq G$ and $' \simeq O$. However, we showed in example 1 that $G \simeq G'$. Since t is an overlay, L is a maximum common subgraph of I and O, and consequently, $L = I = O$, i.e. t contains only zero elements. Since t_1 is a reduction of t, t_1 contains only zero elements, a contradiction. Thus t_1 cannot be a transform.

Let ST, SO, T and O denote the set of semitransforms, semioverlays, transforms and overlays. When defining semioverlays, it was shown that ST=SO. It has just been shown that T is a proper subset of ST. To see that O is a proper subset of T, note that the linking graph of t_2 in Figure 3 is the complement of K_3 while the maximum common subgraph of the ingraph and outgraph of t_2 is $P_2 \cup K_1$. Thus, t_2 is not an overlay. To show that t_2 and t_3 of Figure 3 are transforms, let $G \simeq K(1,3)$ and $G' \simeq P_4$. Let t be defined by t_4 of Figure 5. Let C, I, O and L be the defining graphs of t. Since $G \simeq I$ and $G' \simeq O, t$ is a semioverlay of (G, G'). Clearly, $L \simeq P_3 \cup K_1$ is a maximum common subgraph of G and G'. Thus, t is an overlay. Now we simply note that both t_2 and t_3 are reductions of t.

It should also be noted that not all semiactions of a transform are actions of that transform. To see this, let t' be defined by t_5 of Figure 5. Clearly, t' is a semioverlay of (P_4, P_4). Since t_2 and t_3 of Figure 3 are reductions of $t', (P_4, P_4)$ is a semiaction of both t_2 and t_3. However, the linking graph of t' is $P_2 \cup P_2$, which is not a maximum common subgraph of P_4 and P_4. Thus, t' is not an overlay of t_3, and consequently not of t_2. It follows from Theorem 1 that (P_4, P_4) is not an action of either t_2 or t_3.

Figure 5. An overlay and a semioverlay with respect to t_3

Let $S(t)$ and $A(t)$ denote the sets of semiactions and actions of t. It would be nice to be able to prove that $A(t) = A(t')$ implies $t \simeq t'$, but I have been unable to do so. However, the proof of the following theorem is straight-forward.

Theorem 2   Let t be a transform and let t' be a reduction of t. Then t' is a transform. Moreover, $A(t) \subset A(t') \subset S(t')$.

3.   Specifications of Metadigraphs

A *metadigraph* M is a, possibly infinite, digraph $D = (V, E)$ together with an injective function $g : V \to \Gamma_0$ that assigns a unique graph in Γ_0 to each vertex of D. Thus, a metadigraph can be considered to be a labeled digraph in which the set of labels is a set of unlabeled graphs. It follows that we can uniquely and unambiguously denote M by (Γ, E^g) where Γ is the image set $g(V)$ of V and $E^g = \{g(u)g(v)|uv \in E\}$, i.e. we can uniquely represent M by (Γ, E^g). Examples of two metadigraphs were given in Figure 2. Let M' be another metadigraph $D' = (V', E')$ with assignment function h, and let (Γ', E^h) denote its corresponding labeled graph representation. M is isomorphic to M' if there exists an isomorphism $f : V \to V'$ between D and D' such that $g(v) = h(f(v))$ for $v \in V$. Clearly, M is isomorphic to M' if and only if their corresponding labeled graph representations (Γ, E^g) and (Γ', E^h) are identical, i.e. $\Gamma = \Gamma'$ and $E^g = E^h$. Thus, we shall say M is a submetadigraph of M' if $\Gamma \subset \Gamma'$ and $E^g \subset E^h$.

An ordered pair (T^i, T^b) of, possibly infinite, sets of transforms is a *transform kit*. An action (G, G') of t is *blocked* by t' if (a) t' is an extension of t, (b) (G, G') is an action of t' and (c) $t' \in T^b$. Let $A(t|T^b)$ denote the actions of t that are not blocked by t' for any $t' \in T^b$. Clearly,

$$A(t|T^b) = A(t)\backslash \cup_t' A(t')$$

where the union is over those t' in T^b which are extensions of t. An ordered pair (G, G') of graphs is an *action* of $K = (T^i, T^b)$ is $(G, G') \in A(t|T^b)$

form some $t \in T^i$. A graph G' will be said to be *K-reachable* from G if either $G' \simeq G$ or there exists a sequence $G_1, \cdots, G_n$ such that $G \simeq G_1, G' \simeq G_n$ and (G_i, G_{i+1}) is an action of K for $i = 1, ..., n-1$. Let Γ be a set of graphs and let K be a transform kit. Define the metadigraph $M(\Gamma, K)$ as the metadigraph (Γ', E) where Γ' is the set of graphs which are K- reachable from a graph in Γ and where E is the subset of $\Gamma' \times \Gamma'$ whose members are actions of K. Given a metadigraphs M, we shall say (Γ, K) is a *specification* of M if $M(\Gamma, K) = M$.

Example 2 Let t_4 and t_5 be defined by Figure 6. Let $K = (\{t_4\}, \{t_5\})$ and let $M = (\Gamma, E)$ denote the metadigraph M_1 in Figure 2, i.e. $\Gamma = \{P_n | n = 1, ...\}$ and $E = \{(P_i, P_{i+1}) | i = 1, ...\}$. Then $(\{P_1\}, K)$ is a specification of M. To see this, we simply note that for every graph G and every vertex $v \in V(G), (G, G + u + uv)$ is an action of t_4 were $u \notin V(G)$. Similarly, $(G, G + u + uv)$ is an act of t_5 if and only if the deg $v > 1$. It follows that if (G, G') is an actio of K, then $G' \simeq G + u + uv$ where deg $v < 2$. Thus, if G is a path P_i, then $G' \simeq P_{i+1}$. Since every path is K-reachable from P_1, it follows that $(\{P_1\}, K)$ is a specification of M.

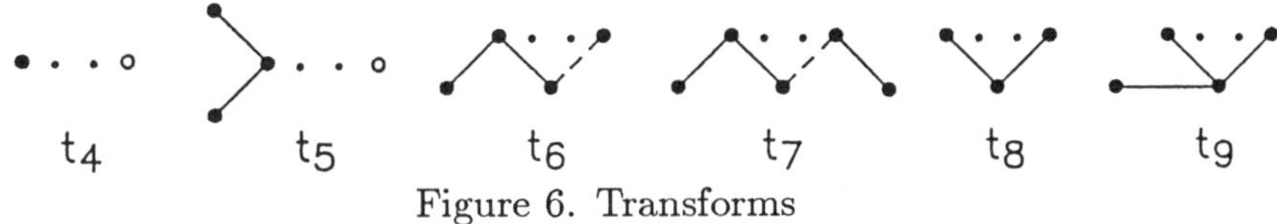

$t_4 \qquad t_5 \qquad t_6 \qquad t_7 \qquad t_8 \qquad t_9$

Figure 6. Transforms

Example 3 Let M be metadigraph M_2 in Figure 2, and let $t_i, i = 6, ..., 9$ be transforms defined by Figure 6. Define $T^i = \{t_6, t_8\}$ and $T^b = \{t_7, t_9\}$, and let K be the transform kit (T^i, T^b). Then $(\{G_1\}, K)$ is a specification of M. To see this, note that t_6 performs an edge shift of the form $G - uv + uw$ where v is adjacent in G to a vertex w with deg $w > 1$, while t_7 blocks those cases in which u is not a terminal vertex. Likewise, t_8 adds an edge between vertices u and v which are not adjacent, but which are both adjacent to another vertex w, while t_9 blocks those cases in which deg $w > 2$.

A theorem will now be proven which implies, as a corollary, that any metadigraph admits a specification for some K. The proof requires the following lemma.

Lemma 1   If t is a reduction of a semitransform t' for which $I \simeq I'$ and $O \simeq O'$, then $t \simeq t'$.

Proof   To see that the reduction condition is essential, let t be defined by transform t_5 in Figure 5, and let t' be the transform for which $C' \simeq P_4 \simeq L'$. Since $L \simeq P_2 \cup P_2$, t and t' are not isomorphic. On the other hand, $I \simeq O \simeq P_4 \simeq I' \simeq O'$. Thus, the lemma fails without the reduction condition.

By definition, t is isomorphic to a reduction t'' of t' for which I'' and O'' are labeled subgraphs of I' and O'. Since $I'' \simeq I \simeq I'$ and $O'' \simeq O \simeq O$, it follows that $I'' = I'$ and $O'' = O'$. Thus, $t'' = t'$. Since $t \simeq t''$, transitivity implies $t \simeq t'$. Let $\mathcal{T}_0$ denote the set of all transforms. Theorem 3 and its corollary show how $\mathcal{T}_0$ can be used to prove that all metadigraphs admit a specification with respect to some transform kit.

Theorem 3   Let $K = (T^i, T^b)$ be the transform kit based on any partitioning $\{T^i, T^b\}$ of $\mathcal{T}_0$ such that every member of T^i is an overlay. Then (G, G') is an action of K if and only if there exists an overlay of (G, G') in T^i.

Proof   Assume there exists an overlay t in T^i of (G, G'), i.e. $I \simeq G$ and $O \simeq G'$. If (G, G') is not an action of K, then the action (G, G') of t must be blocked by some $t' \in T^b$. This implies that I is a subgraph of I' and O is a subgraph of O'. By definition, (G, G') must be an action of t', i.e. t' is a reduction of an overlay of (G, G'). This implies that I' is a subgraph of G and that O' is a subgraph of G'. It follows that $I \simeq I'$ and $O \simeq O'$. By Lemma 1, $t \simeq t'$, a contradiction since T^i and T^b are disjoint sets.

On the other hand, assume (G, G') is an action of K. Then $(G, G') \in A(t|T^b)$ for some $t \in T^i$. Let t' be any extension of t such that (G, G') is an action of t'. If $t' \in T^b$, then $(G, G') \notin A(t|T^b)$. Thus, all extensions of t having (G, G') as an action must lie in T^i. But $(G, G') \in A(t|T^b)$ implies the existence of an overlay t' of (G, G') with respect to t. As an extension of t having (G, G') as an action, t' must lie in T^i, i.e. there exists an overlay of $(G, G') \in T^i$.

Corollary 1 For any metadigraph M, there exists a transform kit K such that $(V(M), K)$ is a specification of M.

Proof Write $M = (\Gamma, E)$. For every edge $GG' \in E$, form an overlay of (G, G') as noted in the section on semitransforms by letting the covering graph be a minimum common supergraph of G and G'. Let T^i denote the st of overlays so formed, and let $K = (T^i, \mathcal{T}_0 \setminus T^i)$. By Theorem 3, GG' is an action of K if and only if GG' is an edge of M. Since M and $M(\Gamma, K)$ have identical vertex set, $M = M(\Gamma, K)$.

As there generally exists many overlays of a pair (G, G') of graphs, there will generally be many kits K that satisfy the conditions of the preceding proof.

Acknowledgements

I would like to thank Gerald Maggiora for his encouragement, ideas an perspectives. I also want to thank the reviewers for their careful reading of the original manuscript and numerous helpful and detailed suggestions and clarifications. This work was supported by The Upjohn Company.

REFERENCES

[1] R. Barone and M. Chanon, "Computer-Aided Organic Synthesis (CAOS), In *Computer Aids to Chemistry*, Ed. by G. Vernin and M. Chanon, John Wiley and Sons, (1986) 19- 102.

[2] A. H. Beckett and D. A. Cowan, "Pitfalls in Drug-Metabolism Methodology," In *Drug Metabolism in Man*, ed. by J. W. Gorrod and A. H. Beckett, Taylor and Francis Ltd., London (1978) 237-257.

[3] J. D. Bu'lock, "The Biosynthesis of Natural Products: An Introduction to Secondary Metabolism," McGraw-Hill Pub. Co., New York (1965).

[4] G. Chartrand and L. Lesniak, "Graphs and Digraphs," Second Edition, Wadsworth and Brooks, Monterey, 1986.

[5] G. Chartrand, F. Saba and H.-B. Zou, "Edge Rotations and Distance between Graphs. *Casopis Pro Pest. Mat.* 110 (1985) 87-91.

[6] F. Darvas, "Predicting Metabolic Pathways by Logic Programming," *J. Mol. Graphics* 6 (1988) 80-86.

[7] S. Fujita, "Description of Organic Reactions Based on Imaginary Transition Structures. 1. Introduction of New concepts," *J. Chem. Inf. Comput. Sci.* 26 (1986) 205-212.

[8] M. Johnson, "Relating Metrics Lines and Variables Defined on Graphs to Problems in Medicinal Chemistry," In *Graph Theory and Its Applications to Algorithms and Computer Science*, ed. by Y. Alavi, G. Chartrand, L. Lesniak, D., R. Lick and C. E. Wall, John Wiley and Sons, Inc. (1985) 457-470.

[9] M. Johnson, "An Ordering of Some Metrics Defined on the Space of Graphs," *Czech. Math. J.* 37 (1987) 75-85.

[10] J. Koca, M. Kratochvíl, V. Kvasnička, L. Matyska and J. Pospíchal, "Synthon Model of Organic Chemistry an Synthesis Design," Springer-Verlag, Berlin, (1989) 207 p.

[11] I. Ugi, J. Bauer, J. Brandt, J. Friedrich, J. Gasteige, C. Jochum and W. Schubert, "New Applications of Computers in Chemistry," *Angew. Chem. Int. Ed. Engl.*, 18 (1979) 111-123.

[12] T. H. Varkony, D. H. Smith and C. Djerassi, "Computer-Assisted Structure Manipulation: Studies in the Biosynthesis of Natural Products," *Tetrahedron*, 34, (1978) 841-852.

A Note on Removable Pairs

H.A. Kierstead[1]
W.T. Trotter[2]

Department of Mathematics
Arizona State University

ABSTRACT

A long standing conjecture in the dimension theory for finite ordered sets asserts that every ordered set (of at least three points) contains a pair whose removal decreases the dimension at most one. Two stronger conjectures have been made:

(1) If (x, y) is a critical pair, then $dim(P) \leq 1 + dim(P - \{x, y\})$.

(2) For every $x \in P$, there exists $y \in P - \{x\}$ so that $dim(P) \leq 1 + dim(P - \{x, y\})$.

K. Reuter has disproved conjecture 1 by constructing a four–dimensional poset P containing a critical pair (x, y) so that $dim(P - \{x, y\}) = 2$. In this note, we construct for every $n \geq 5$ an n–dimensional poset P_n containing a critical pair (x, y) so that $dim(P_n - \{x, y\}) = n - 2$. Point y is a maximal point of P_n.

1. Preliminaries

Recall that the *dimension* of a finite ordered set P in the least positive integer t so that there exist t linear extensions $L_1, L_2, ..., L_t$ so that $P = L_1 \cap L_2 \cap ... \cap L_t$. An incomparable pair (x, y) is called a *critical* pair if any point less than x is less than y and any point greater than y is greater than x. The dimension of P is the least t for which there exist t linear extensions of P so that for every critical pair (x, y), there is at least one i for which $y < x$ in L_i. We refer the reader to the survey article [3] by D. Kelly and W.T. Trotter and the chapters [6], [7] by Trotter for additional background information on dimension theory.

---

1. Research supported in part by the Office of Naval Research.
2. Research supported in part by the National Science Foundation.

2. Removable Pairs

The following conjecture is one of the best known open problems in dimension theory and is a featured problem in **ORDER**. We believe the first reference to the conjecture is [1].

Conjecture 0 *If P is an ordered set having at least three points, then P contains a distinct pair (x, y) so that dim(P) ≤ 1 + dim(P − {x, y}).*

A pair x, y ∈ P for which dim(P) ≤ 1 + dim(P − {x, y}) is called a *1−removable* pair, so that Conjecture 0 asserts that every poset contains a 1−removable pair.

The first reference to the following conjecture is apparently [5].

Conjecture 1 *Every critical pair is 1−removable.*

In [2], D. Kelly made the following conjecture which is also stronger than Conjecture 0.

Conjecture 2 *For every x ∈ P, there is a point y ∈ P − {x} so that x, y is a 1-removable pair.*

K. Reuter [4] has disproved Conjecture 1 by constructing the ordered set shown in Figure 1. This ordered set P has dimension 4, (x, y) is a critical pair, and dim(P - {x, y}) = 2. Note that y is a maximal point.

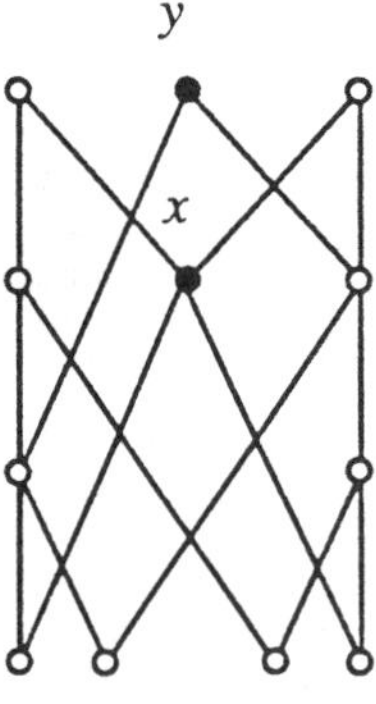

Figure 1

The purpose of this note is to show that Reuter's example is not an isolated phenomenon. To accomplish this, we will establish the following result.

Theorem *For every* $n \geq 4$*, there exists an* n*–dimensional ordered set* P_n *containing a critical pair* (x, y) *so that* y *is a maximal element in* P_n*, but* (x, y) *is not 1-removable, i.e.,* $dim(P - \{x, y\}) = n - 2$.

Proof For $n = 4$, we have Reuter's example shown in Figure 1. For $n \geq 5$, the point set of P_n contains $4n - 4$ points labelled $\{a_i : 1 \leq i \leq n - 2\} \cup \{b_i : 1 \leq i \leq n - 2\} \cup \{c_i : 1 \leq i \leq n - 2\} \cup \{d_i : 1 \leq i \leq n - 2\} \cup \{x, y, z, w\}$. For all i, j with $1 \leq i$, $j \leq n - 2$ and $i \neq j$, we have the cover relations $a_i <: b_j$ and $c_i <: d_j$. For each i with $1 \leq i \leq n - 2$, we have $a_i <: y$, $c_i <: y$, $c_i < x$, $z <: b_i$, $w <: b_i$, $w < d_i$, $x <: b_i$, and $z < d_i$. We also have $w <: y$. We illustrate this definition with a diagram for P_n when $n = 5$.

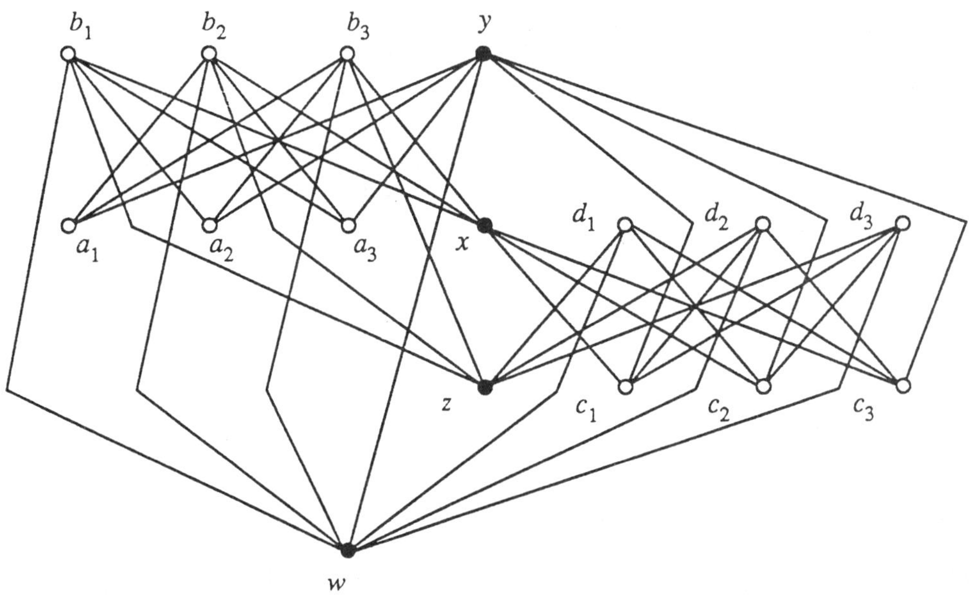

Figure 2

We first show that $dim(P_n) \geq n$. To the contrary, suppose $dim(P_n) \leq n - 1$, and let $L_1, L_2, \ldots, L_{n-1}$ be linear extensions whose intersection is P_n. Without loss of generality, we may assume that $b_i < a_i$ in L_i for $i = 1, 2, \ldots, n - 2$. Thus we must have $x > y$ in L_{n-1} and $z > y$ in L_{n-1}. However, this implies that for each $i = 1, 2, \ldots, n - 2$, there exists a unique $j_i \in \{1, 2, \ldots, n - 2\}$ so that $c_i > d_i$ in L_{j_i}. Hence

$w > x$ in L_{n-1} also. But this implies that $w > x > y$ in L_{n-1} which is impossible since $w < y$ in P_n. The contradiction completes the proof that $\dim(P_n) \geq n$.

We observe that y is maximal element of P_n and that (x, y) is a critical pair. We now show that $\dim(P_n - \{x, y\}) \leq n - 2$. To accomplish this, consider the poset $Q_n = P_n - \{x, y\}$. In Q_n, we observe that z and w have duplicated holdings so that $\dim(Q_n - \{z\}) = \dim(Q_n)$. Let $Q'_n = Q_n - \{z\}$. We show that Q'_n has $n - 2$ linear extensions which intersect to give Q'_n. Let $A = \{a_1, a_2, ..., a_{n-2}\}$, $B = \{b_1, b_2, ..., b_{n-2}\}$, $C = \{c_1, c_2, ..., c_{n-2}\}$ and $D = \{d_1, d_2, ..., d_{n-2}\}$.

For $i = 1, 2$, L_i is any linear extension of Q'_n so that $A - \{a_i\} < C - \{c_i\} < w < d_i < c_i < b_i < a_i < B - \{b_i\} < D - \{d_i\}$. For $i = 3, 4, ..., n - 2$, L_i is any linear extension of Q'_n so that $w < C - \{c_i\} < d_i < c_i < D - \{d_i\} < A - \{a_i\} < b_i < B - \{b_i\}$. It is easy to see that any family constructed by these rules forms a realizer. With this observation, our proof is complete. $\square$

We pause to note that the construction given in the preceding family for $\{P_n : n \geq 5\}$ does not work for $n = 4$. In this case, $\dim P_n = 4$, but $\dim(P_n - \{x, y\}) = 3$. Thus to handle the case $n = 4$, we need a special example, and Reuter's construction suffices.

3. Concluding Remarks

We view the results of this note as providing additional evidence as to the difficulty of Conjecture 0, but we are unable to decide whether our theorem argues for or against the conjecture. It is easy to see that the examples satisfy Conjecture 2, so at least this stronger form of the original conjecutre remains open.

REFERENCES

[1]   K. Bogart and W.T. Trotter, Irreducible Posets with Arbitrarily Large Height Exist, *J. Comb. Theory (A)* **17** (1974), 337 – 344.

[2]   D. Kelly, Removable Pairs in Dimension Theory, *Order 1* (1984) 217 – 218.

[3]   D. Kelly and W.T. Trotter, Dimension Theory for Ordered Sets, *Proceedings of the Symposium on Ordered Sets*, I. Rival et. al., Reidel Publishing (1982), 171 – 212.

[4]   K. Reuter, personal communication.

[5]   W.T. Trotter, A Generalization of Hiraguchi's Inequality for Posets, *J. Comb. Theory (A)* **20** (1976), 114 – 123.

[6]   W.T. Trotter, Graphs and Partially Ordered Sets, in *Selected Topics in Graph Theory II*, R. Wilson and L. Beineke, eds. Academic Press, (1983), 237 – 268.

[7]   W.T. Trotter, Partially Ordered Sets, A chapter in *Handbook of Combinatorics*, R. Graham, M. Groetschel, L. Lovasz, eds., to appear.

On the Number of Maximal Independent Sets in Some (v,e)-Graphs

Felix Lazebnik

University of Delaware

ABSTRACT

Let V(G) be the set of vertices of a simple undirected graph G and S be a subset of V(G). S is an independent *set in G if no two vertices of S are joined by and edge of G. S is a* maximal independent *set (m.i.s.) in G if S is independent and is not a subset of any other independent set. Let $\mu(G)$ denote the number of m.i.s. of a graph G, and $\mu(v, e) = max \{\mu(G) \mid G has v vertices and e edges\}. For $0 \le e \le v$, new bounds for $\mu(v,e)$ are found. For some subranges of the parameters, $\mu(v, e)$ is determined and extremal graphs are described. The results refine some known theorems from extremal graph theory and an upper bound for the running time of an algorithm of E. L. Lawler for determining the chromatic number of a graph.*

1. Introduction

The definitions in this paper are based on [Bo76]. All graphs we consider are undirected labelled graphs without loops and multiple edges. V(G) and E(G) denote sets of vertices and edges of G respectively. The number of elements of a finite set A is denoted by |A|. We write $v = v(G) = |V(G)|$ and $e = e(G) = |E(G)|$ and call G a (v, e)–graph. Let $\{x;y\}$ be an edge of G. Then by $G - \{x;y\}$ we mean the graph obtained from G by deleting $\{x;y\}$. By K_v, K_v^c, T_v and $K_{m,n}$ we denote correspondingly the complete graph on v vertices (any two vertices are joined by an edge), the completely disconnected graph on v vertices (no edges at all), a tree on v vertices, and the complete bipartite graph whose vertex classes contain m and n vertices. By $G + H$ we denote the disjoint union of graphs G and H. For a given x $\in$ V(G), by $N_G(x)$ we denote the set of all neighbors of x in G, i. e. the set of all y $\in$

---

1   This paper is based on a part of a Ph. D. Thesis written by the author under the supervision of Prof. H. S. Wilf at the University of Pennsylvania.

V(G) such that {x;y} ∈ E(G). A set S, $S \subseteq V(G)$, is an *independent set* in G if no two vertices of S are joined by an edge of G. S is a *maximal independent set* (m.i.s.) in G if S is independent and is not a subset of any other independent set of G. Let $\mathcal{M}(G)$ denote the set of all m.i.s. of vertices in G, $\mu(G) = |\mathcal{M}(G)|$, $\mu(v) = \max\{\mu(G): |V(G)| = v\}$, $\mu(v, e) = \max\{\mu(G):$ G is a (v, e)–graph}. Obviously, for any (v, e)–graph G, $\mu(G) \le \mu(v,e) \le \mu(v)$. A *clique* in G is a maximal complete subgraph of G. Let cl(G) denote the number of cliques of graph G. Let cl(v) = max{cl(G): |V(G)| = v}, cl(v,e) = max{cl(G): G is a (v,e)–graph}. Let G^c denote the complement of graph G. It is easy to see that $\mu(G) = cl(G^c)$, $\mu(v) = c(v)$ and $\mu(v, e) = cl(v, v(v-1)/2 - e)$. The following problem was formulated by H. S. Wilf:

> For the given pair of positive integers (v,e), find $\mu(v, e)$ or give a non–trivial upper bound of $\mu(v, e)$.

In this paper we present some partial results. Problems similar to this, but for different families of graphs, were considered by several authors. The value of cl(v) was determined by Miller and Muller [MiMu60] and independently by a different method by Moon and Mosser [MoMo65] in which they characterized the extremal graphs. The found that

$$\mu(v) = cl(v) = \begin{cases} 3^t & \text{if } v = 3t \ge 3; \\ 4 \cdot 3^{t-1} & \text{if } v = 3t+1 \ge 4; \\ 2 \cdot 3^t & \text{if } v = 3t+2 \ge 2 \end{cases} \tag{1.1}$$

and that the extremal graphs (for the number of cliques) are Turan graphs $T_t(v)$.

It turned out that the problem of finding cl(v) can be shown to be equivalent (Yao [Y76] attributes this result to D. E. Muller) to the problem of Katona on minimal separating systems:

Given the set [n] = {1, 2, ..., n}. Find the smallest number f(n) for which there exists a family of subsets of [n] {$A_1, A_2, ..., A_{f(n)}$} with the following property: given any two elements x, y ∈ [n] (x ≠ y), there exist k, ℓ such that $A_k \cap A_\ell = \varnothing$, and x ∈ A_k, y ∈ A_ℓ.

Katona's problem was solved by Yao [Y76] and independently by a similar method by Cai [Ca83].

There were several papers in which the authors restricted their attention to a subset $\mathcal{F}$ of all the graphs on v vertices and determined either cl(v, $\mathcal{F}$) = max {cl(G): G ∈ $\mathcal{F}$} or $\mu(v, \mathcal{F})$ = max {$\mu(G)$: G ∈ $\mathcal{F}$}.

Hedman [He85,He] found the maximal number of cliques for the family $\mathcal{F}$ of all graphs on v vertices with the given clique number ω (the clique number ω of a graph G is the greatest number of vertices in a clique of G among all cliques of G).

Wilf [W86] found the larges number of m.i.s. vertices that any tree of v vertices can have. The same results were obtained by Cohen [Co84] and Sagan [Sa88] by different methods. Sagan's paper completely describes all extremal graphs. If we denoted this number by $\mu(v, \text{Tree})$, then the result is

$$\mu(v, \text{Tree}) = \begin{cases} 2^{t-1} + 1 & \text{if } v = 2t > 0; \\ 2^t & \text{if } v = 2t+1 ; \\ 1 & \text{if } v = 0 \end{cases} \tag{1.2}$$

Furedi [Fu87] gave a new proof of (1.1) and established an exact upper bound for $\mu(G)$ for a non–extremal graph G. In the same paper he found $\mu(v, \text{Conn}) =$ the maximum number of m.i.s. that a connected graph on v vertices can have (for v > 50) and described all extremal graphs. Independently, Griggs, Grinstead and Guichard [GGG88] determined $\mu(v, \text{Conn})$ for all $v \geq 6$ and described all the extremal graphs. Their result is

$$\mu(v, \text{Conn}) = \begin{cases} 2 \cdot 3^{t-1} + 2^{t-1} & \text{if } v = 3t > 6; \\ 3^t + 2^{t-1} & \text{if } v = 3t+1 > 6; \\ 4 \cdot 3^{t-1} + 3 \cdot 2^{t-2} & \text{if } v = 3t+2 > 6 \end{cases} \tag{1.3}$$

Harary and Lempel [HL74] studied the extremal graphs for the family of all graphs on v vertices with e edges. They developed some standard forms for such graphs and suggested a transformation which brings an extremal graph into this form. Similar results were obtained independently by the author. Unfortunately they have not helped much in finding $\mu(v, e)$.

Another motivation for the present work was an article by E. Lawler [La76] in which an algorithm for determining the chromatic number of a graph is discussed, and it is shown that its run time, in the worst case, is $0[ev(1+3^{1/3})^v]$ for graphs of e edges and v vertices. The appearance of $3^{1/3}$ derives from (1.1) because of the fact that a graph on v vertices has at most $3^{v/3}$ maximal independent sets. The fact that the graph has e edges is not used when the greatest number of maximal independent sets is estimated.

In Section 2 we give new bounds for $\mu(v, e)$ and determine $\mu(v,e)$ exactly for some ranges of v and e. The main results are in Theorems 2.1, 2.6 and Corollary 2.3.

2. Results

In this section we determine explicitly or find bounds for $\mu(v, e)$, for $0 \le e \le v$. We start with the following.

Theorem 2.1 *Let v, e be non-negative integers, $0 \le e \le v$, and $m(v, e) = 2^{v-e} \cdot 3^{(2e-v)/3}$. Then*

$$\mu(v, e) \le m(v, e) \qquad (2.1)$$

The equality in (2.1) occurs if and only if $v/2 \le e \le v$ and $2e - v = 3t$ for some non-negative integer t. The only graph G, for which $\mu(G) = \mu(v, e) = m(v, e)$ is $G = (v - e)K_2 + tK_3$.

Proof We notice that for $e > v$, our upper bound $m(e, v)$ is worse than $\mu(v)$ given by (1.1). This explains the restriction $0 \le e \le v$. If $v = e = 0$, then $\mu(\varnothing) = m(0, 0) = 1$. If $v = 1$, $e = 0$, then $\mu(1, 0) = \mu(K_1) = 1 < m(1, 0) = 2/(3^{1/3})$. Let G be an extremal graph and $G_1, G_2, \ldots, G_n$ be connected components of G. Suppose G_i is a (v_i, e_i)–graph, $1 \le i \le n$. Then $\Sigma v_i = v$ and $\Sigma e_i = e$. Since $\mu(v, e) = \mu(G) = \Pi\mu(G_i)$ and $m(v, e) = \Pi\mu(v_i, e_i)$, Then in order to prove the theorem it is sufficient to show that for all i, $1 \le i \le n$, $\mu(G_i) \le m(v_i, e_i)$.

Lemma 2.2 *For any connected (v, e)–graph G, $\mu(G) \le m(v, e)$. $\mu(G) = m(v, e)$ if and only if $G = \varnothing$, $G = K_2$ or $G = K_3$.*

Proof It is enough to show that

$$\mu(v) \le m(v, e). \qquad (2.3)$$

Then the first statement will be proved. Let $e = v + p$. Then

$$m(v, v+p) = 2^{-p} \cdot 3^{(v+2p)/3} = (9/8)^{p/3}3^{v/3},$$

and (2.3) can be easily checked by using the table from Figure 1. Entries in the $\mu(v)$ column come from (1.1) and (1.2) (the only connected (v, e)–graphs with $e \le v - 1$ are trees, and $e = v - 1$.

e	v	$\mu(v)$	$m(v,e)$
$e = v + p,\ p \geq 0$	$v = 3t,\quad t \geq 1$	3^t	$(9/8)^{p/3}3^t$
	$v = 3t+1,\ t \geq 1$	$4 \cdot 3^{t-1}$	$3^{1/3}(9/8)^{p/3}3^t$
	$v = 3t+2,\ t \geq 0$	$2 \cdot 3^t$	$3^{2/3}(9/8)^{p/3}3^t$
$e = v-1$	$v = 2t,\ t \geq 1$	$2^{t-1}+1$	$(8/9)^{1/3}3^{2t/3}$
	$v = 2t+1,\ t \geq 0$	2^t	$(2/3^{1/3})3^{2t/3}$

Figure 1

Comparing the entries in the table on Figure 1, we conclude that equality occurs if and only if $v = 3t$, $p = 0$ or $v = 2$, $p = -1$. For $v = 3t$, $p = 0$, we have $e = 3t$, $\mu(3t) = 3^t$. As it follows from [MiMu60} and [MoMo65], the only extremal graph in this case is tK_3. This graph is connected for $t = 1$. Therefore $G = K_3$. For $v = 2$, $p = -1$, we have $e = 1$ and $G = K_2$. This proves the lemma. $\square$

Thus (2.2) is true for each connected component of G and the bound (2.1) is proved. In order to get an equality in (2.1), each connected component of G has to be either K_3 or K_2. Suppose $G = tK_3 + sK_2$, for some non–negative integers t, s. then

$$3t + 2s = v \quad \text{and} \quad 3t + s = e \tag{2.4}$$

The only solution of (2.4) is $s = v - e$, $t = (2e - v)/3$, and this concludes the proof of the theorem.

Corollary 2.3 *Let v, e be non–negative integers, $v/2 \leq e \leq v$. If $2e - v = 3t + 1$, then $3^{-1/3}m(v, e) \leq \mu(v, e) < m(v, e)$. If $2e - v = 3t + 2$, then $\mu(v, e) = (1/2)(3^{1/3})m(v, e)$ for $v = 4, 6$; $(5/3^{5/3})m(v, e) \leq \mu(v, e) < m(v, e)$ for $v = 5$, $v \geq 7$.*

Proof The upper bounds follow from Theorem 2.1. In the case $2e - v = 3t + 1$, the lower bound comes from the graph $tK_3 + (v - e - 1)K_2 + P_2$, where P_2 is a path with two edges. If $2e - v = 3t + 2$, then for $v = 4$, the only possible value of e is 4. In both cases $\mu(v, e) = (1/2)(3^{1/3})m(v, e)$. For $v = 5$ or ≥ 7, the lower bound comes from the graph $(t - 1)K_3 + (v - e)K_2 + H$, where H is $(5, 5)$–graph shown on Figure 2. The lower bounds seem to be the best possible, but the author has been unable to prove it.

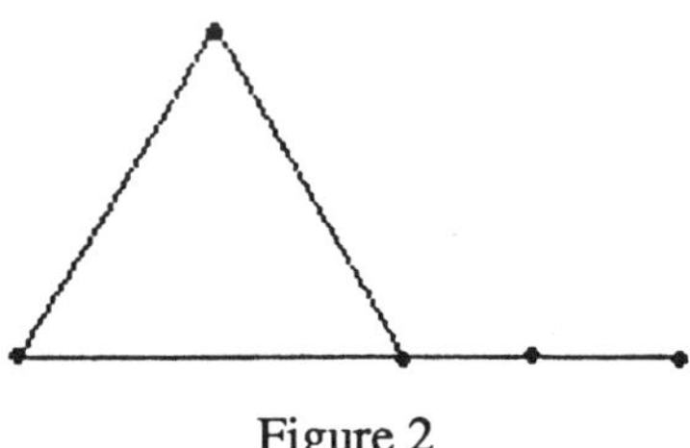

Figure 2

It turns out that for e, $0 \leq e \leq v/2$, the result of Theorem 2.1 can be substantially improved. The following lemma is the main step in this direction. It is also interesting on its own.

Lemma 2.4 *Let* G *be a graph and* $\{x; y\}$ *be an edge of* G. *Then*

$$\mu(G) \leq 2\mu(G - \{x; y\}). \tag{2.5}$$

The equality in (2.5) occurs if and only if $\{x; y\}$ is a connected component of G.

Proof The idea of the proof is to partition both $\mathcal{M}(G)$ and $\mathcal{M}(G-\{x;y\})$ into several classes and to compare numbers of elements in the corresponding classes. The description of the partitions is rather long, but the comparison will be easy. Figure 3 illustrates both stages of the proof. We divide $\mathcal{M}(G)$ into the following 7 classes some of which can be empty ($\overset{\bullet}{\cup}$ stands for the disjoint union of sets):

$$\mathcal{M}_{x,1}(G) = \{M \in \mathcal{M}(G): M = \{x\} \overset{\bullet}{\cup} M', M' \neq \varnothing, M' \cap N_G(y) = \varnothing\};$$

$$\mathcal{M}_{y,1}(G) = \{M \in \mathcal{M}(G): M = \{x\} \overset{\bullet}{\cup} M', M' \neq \varnothing, M' \cap N_G(x) = \varnothing\};$$

$$\mathcal{M}_{x,2}(G) = \{M \in \mathcal{M}(G): M = \{x\} \overset{\bullet}{\cup} M', M' \cap N_G(y) \neq \varnothing\};$$

$$\mathcal{M}_{y,2}(G) = \{M \in \mathcal{M}(G): M = \{x\} \overset{\bullet}{\cup} M', M' \cap N_G(x) \neq \varnothing\};$$

$$\mathcal{M}_3(G) = \{M \in \mathcal{M}(G): M \neq \varnothing, x \notin M, y \notin M\};$$

$$\mathcal{M}_x(G) = \begin{cases} \{x\}, & \text{if } \{x\} \in \mathcal{M}(G) \\ \varnothing, & \text{otherwise} \end{cases}$$

$$\mathcal{M}_y(G) = \begin{cases} \{y\}, \text{ if } \{y\} \in \mathcal{M}(G) \\ \varnothing, \text{ otherwise} \end{cases}$$

It is easy to check that $\mathcal{M}(G)$ is the disjoint union of these classes. Similarily we divide $\mathcal{M}(G - \{x; y\})$ into the following 5 classes:

$$\mathcal{M}_{x,y,1}(G - \{x; y\}) = \{M \in \mathcal{M}(G - \{x; y\}): M = \{x\} \cup \{y\} \cup M', M' \neq \varnothing\};$$

$$\mathcal{M}_{x,2}(G - \{x; y\}) = \{M \in \mathcal{M}(G - \{x; y\}): x \in M, M \cap N_{G-\{x; y\}}(y) \neq \varnothing\};$$

$$\mathcal{M}_{y,2}(G - \{x; y\}) = \{M \in \mathcal{M}(G - \{x; y\}): y \in M, M \cap N_{G-\{x; y\}}(x) \neq \varnothing\};$$

$$\mathcal{M}_3(G - \{x; y\}) = \{M \in \mathcal{M}(G - \{x; y\}): M \neq \varnothing, x \notin M, y \notin M\};$$

$$\mathcal{M}_{x,y}(G - \{x; y\}) = \begin{cases} \{x, y\}, \text{ if } \{x, y\} \in \mathcal{M}(G - \{x; y\}) \\ \varnothing, \text{ otherwise} \end{cases}$$

It is easy to check that $\mathcal{M}(G-\{x;y\})$ is the disjoint union of these classes.

The following bijections between some of these classes are obvious:

$$f_{x,1} : \mathcal{M}_{x,1}(G) \to \mathcal{M}_{x,y,1}(G-\{x;y\}), \; f_{x,1}(\{x\} \cup M') = \{x, y\} \cup M';$$

$$f_{y,1} : \mathcal{M}_{y,1}(G) \to \mathcal{M}_{x,y,1}(G-\{x;y\}), \; f_{y,1}(\{y\} \cup M') = \{x, y\} \cup M';$$

$$f_{x,2} : \mathcal{M}_{x,2}(G) \to \mathcal{M}_{x,2}(G-\{x;y\}), \; f_{x,2}(M) = M;$$

$$f_{y,2} : \mathcal{M}_{y,2}(G) \to \mathcal{M}_{y,2}(G-\{x;y\}), \; f_{y,2}(M) = M;$$

$$f_3 : \mathcal{M}_3(G) \to \mathcal{M}_3(G-\{x;y\}), \; f_3(M) = M;$$

$$f_x : \mathcal{M}_x(G) \to \mathcal{M}_{x,y}(G-\{x;y\}), \; f_x(\{x\}) = \{x, y\}, \text{ if } \{x\} \text{ is m.i.s.};$$

$$f_x(\{\varnothing\}) = \varnothing, \text{ if } \{x\} \text{ is not m.i.s.};$$

$$f_y : \mathcal{M}_y(G) \to \mathcal{M}_{x,y}(G-\{x;y\}), \; f_y(\{y\}) = \{x, y\}, \text{ if } \{y\} \text{ is m.i.s.};$$

$$f_y(\{\varnothing\}) = \varnothing, \text{ if } \{y\} \text{ is not m.i.s.}$$

(Notice that for each of these mappings the domain and the range are non–empty simultaneously.)

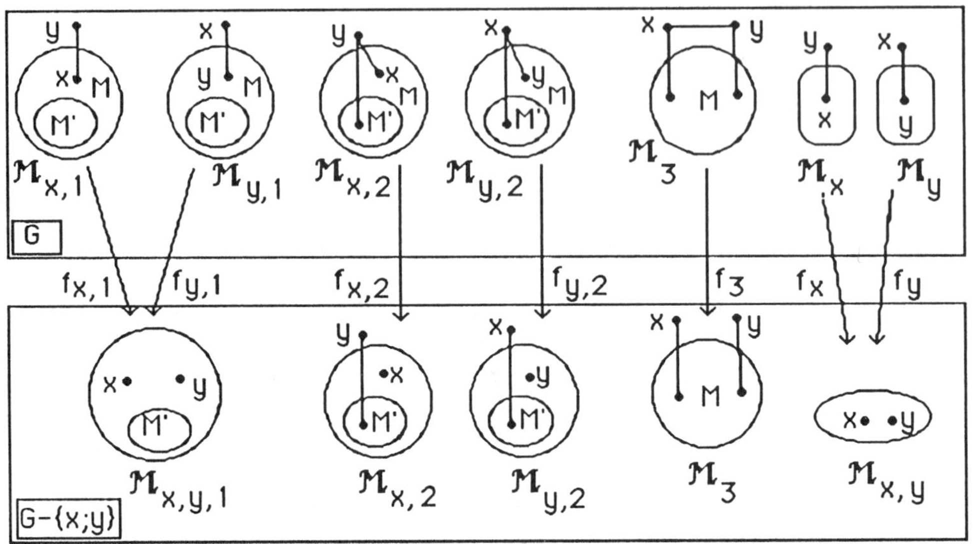

Figure 3

Finally we denote $|\mathcal{M}_{x,1}(G)| = |\mathcal{M}_{y,1}(G)| = |\mathcal{M}_{x,y,1}(G-\{x;y\})| = i_1,$

$|\mathcal{M}_{x,2}(G)| = |\mathcal{M}_{x,2}(G-\{x;y\})| = i_{x,2},\ |\mathcal{M}_{y,2}(G)| = |\mathcal{M}_{y,2}(G-\{x;y\})| = i_{y,2},$

$|\mathcal{M}_3(G)| = |\mathcal{M}_3(G-\{x;y\})| = i_3,\ |\mathcal{M}_x(G)| = i_x\ (=1\text{ or }0),$

$|\mathcal{M}_y(G)| = i_y\ (=1\text{ or }0),\ |\mathcal{M}_{x,y}(G-\{x;y\})| = i_{x,y}\ (=1\text{ or }0).$

Then

$$2\mu(G-\{x;y\}) = 2(i_1 + i_{x,2} + i_{y,2} + i_3 + i_{x,y}),$$

$$\mu(G) = 2i_1 + i_{x,2} + i_{y,2} + i_3 + i_x + i_y.$$

Hence, $\mu(G) = 2\mu(G-\{x;y\}) - (i_{x,2} + i_{y,2} + i_3) - (2i_{x,y} - i_x - i_y)$. Obviously $i_{x,2} + i_{y,2} + i_3 \geq 0$. If at least one of i_x or $i_y = 1$, then $i_{x,y} = 1$, and $2i_{x,y} - i_x - i_y \geq 0$. If $i_x = i_y = 0$, then again $2i_{x,y} - i_x - i_y \geq 0$. Therefore we get

$$\mu(G) \leq 2\mu(G-\{x;y\}) \tag{2.6}$$

The equality sign in (2.6) occurs if and only if $i_{x,2} + i_{y,2} + i_3 = 0$ and $2i_{x,y} - i_x - i_y = 0$. The first of the equalities implies $i_{x,2} = i_{y,2} = i_3 = 0$. If $i_{x,2} = i_{y,2} = 0$, then vertices x and y have the same set of neighbors in G (each independent set of a graph is a subset of at least one m.i.s.). But $i_3 = 0$ implies that this set of neighbors is empty. Therefore the edge $\{x;y\}$ is a connected component in G. If this is the only connected component of G, i. e. $G = K_2$, then $i_{x,y} = i_x = i_y = 1$ and $2i_{x,y} - i_x - i_y = 0$. If G has more than one connected component, then $i_{x,y} = i_x = i_y = 0$ and again $2i_{x,y} - i_x - i_y = 0$. The lemma is proved. ❑

Corollary 2.5 *If an extremal (v, e)–graph G has two isolated vertices, then it is a disjoint union of edges and isolated vertices.*

Proof Let G have a connected component H with 2 or more edges and two isolated vertices a and b. By deleting and edge $\{x;y\}$ in H and joining vertices a and b we obtain a (v, e)–graph G'. Since $\mu(H + \{a\} + \{b\}) = \mu(H)$ and $\mu(\{a;b\} + (H - \{x;y\})) = 2\mu(H - \{x;y\})$, and by Lemma 2.4, $\mu(H) < 2\mu(H - \{x;y\})$, then $\mu(H + \{a\} + \{b\}) < \mu(\{a;b\} + (H - \{x;y\}))$. All other connected components (with the vertices in $V(G) - \{a, b\} - V(H)$) of G and G' are the same. Therefore $\mu(G) < \mu(G')$, which contradicts the extremality of G. ❑

The following theorem gives the exact value of $\mu(v,e)$ and describes the extremal graphs for $0 \le e \le v/2$.

Theorem 2.6 Let $0 \le e \le v/2$. Then $\mu(v,e) = 2^e$, and the only extremal graph is $eK_2 + (v - 2e)K_1$.

Proof The greatest number of vertices in a graph which are incident to e edges is 2e and this happens *only* if the graph is eK_2. Therefore if $v - 2e \ge 1$, then the statement of the theorem follows from Corollary 2.5. so we assume that $v = 2e$. Let G be extremal graph and $G_1, G_2, \ldots, G_n$ be connected components of G. Suppose $v(G_i) = v_i$ and $e(G_i) = e_i$, $1 \le i \le n$. Since for all i, $1 \le e_i \le v_i - 1$, then

$$\Sigma e_i = e \ge v - n \quad \text{and} \quad n \ge v - e = 2e - e = e.$$

If G had two isolated vertices, then, due to Corollary 2.5, it would have at least 2e + 2 vertices and this is not the case.

Suppose G has no isolated vertices. Then each component must have at least one edge and $n \le e$. So $n = e$, $G = eK_2$ and the theorem is proved.

The only case left is when G has only one isolated vertex. Then each of the remaining $n - 1$ components has at least one edge. If each of them has exactly one edge, then $G = eK_2 + K_1$ and $v(G) = 2e + 1$, but the latter is false. So there should be a component with at least two edges. It cannot have three edges, since in this case $e(G) \ge$

$3 + (n - 2) = n + 1 > e$. Thus $G = P_2 + (e - 2)K_2 + K_1$ (P_2 is a path with two edges), and $\mu(G) = 2 \cdot 2^{e-2} = 2^{e-1}$. But this is less than $2^e = \mu(eK_2)$ which contradicts the extremality of G. Therefore the theorem is proved.

Acknowledgement   The author wishes to thank the referee for his numerous comments, suggestions and corrections which resulted in the improvement of the original version of this paper.

REFERENCES

[Bo78]   B. Bollobas, Extremal Graph Theory. *Academic Press* (1978).

[Ca83]   Cai Mao–cheng, Solution to Edmonds' and Katona's problems on families of separating subsets. *Discrete Math* **47** (1983) 13–21.

[Co84]   D. I. A. Cohen, Counting stable sets in trees. *Seminaire Lotharingien de Combinatoire,* 10eme session, R. Konig ed, L'Institute de Recherche Mathematique, Avancee pub. Strasbourg, France (1984) d 48–52.

[Fu87]   Z. Furedi, The number of maximal independent sets in connected graphs. *Journal of Graph Theory*, Vol. 11, No. 4 (1987) 463–470.

[GGG88]   J. R. Griggs, C. M. Grinstead, D. R. Guichard, The number of maximal independent sets in a connected graph. *Discrete Math* **68** (1988) 211–220.

[HL74]   F. Harary and A. Lempel, On clique–extremal (p, q)–graphs. *Networks* **4** (1974) 371–378.

[He]   B. Hedman, Another extremal property of Turan graphs, to appear.

[He85]   B. Hedman The maximal number of cliques in deuse graphs. *Discrete Math* **54** (1985) 161–166.

[La76]   E. L. Lawler, A note on the complexity of the chromatic number problem. *Information Processing Letters*, v. 5, nom. 3 (1976) 66–67.

[MiMu60]   R. E. Miller and D. E. Muller, A problem of maximum consistant set. *IBM Research Report RC–240*, J. T. Watson Res. Center, Yorktown Hights, N. Y. (1960).

[MoMo65]   J. W. Moon and L. Moser, On clique graphs. *Israel J. Math* **3** (1965) 23–28.

[Sa88]   B. Sagan, A note on independent sets in a tree, *SIAM J. Discrete Math.* **1** (1988) no. 1, 105–108.

[W86]   H. S. Wilf, The number of maximal independent sets in a tree. *SIAM J. Alg. Disc. Math.* **7** (1986) 125–130.

[Y76]   A. C. – C. Yao, On problem of Katona on minimal separating systems. *Discrete Math.* **15** (1976) 193–199.

On Critical and Cocritical Diameter Edge–invariant Graphs

Sin–Min Lee

San Jose State University

Ai–Yun Wang

Santa Clara University

ABSTRACT

The concepts of critical and cocritical diameter edge–invariant graphs are introduced. It is shown that every connected graph is an induced subgraph of a cocritical diameter edge–invariant graph.

1. Introduction

We can model a communication network by a graph $G = (V,E)$ consisting of a set V of vertices, or communication sites, and a set E of edges, or communication lines. The distance between two vertices x and y is defined as the length of the shortest (x,y)–path in G. The diameter $D(G)$ of a graph G is defined by $D(G) = \max\{d(x,y): x,y \in V(G)\}$. The diameter of a graph corresponds to the maximum number of links over which a message between two vertices of the network must travel. In networks where the time delay of signal degradation is approximately proportional to the length of the path being used, the diameter of the network is an important parameter which measures the efficiency and reliability of the network. ([1], [2], [3], [4])

Ore [14] considered graph G with $D(G) > D(G+e)$ for all e not in $E(G)$. Glivjak [5] in the sixties initiated the study of graphs G with property $D(G) < D(G \setminus e)$ for all e in $E(G)$. There is a vast literature devoted to the above two concepts. (See [6], [7], [8], [9], [18].)

Recently, the first author [12], [13] investigated the so–called diameter edge–invariant networks. A graph $G = (V,E)$ is said to be *diameter edge–invariant* if by deleting any edge e of G, its diameter does not change, i.e., $D(G) = D(G \setminus e)$ for all e in E. We will use an abbreviation "dei" for the diameter edge–invariant. The dei graph should be c–connected for c ≥ 2, i.e., any two nonadjacent nodes of G are joined by c internally disjoint paths and cannot be separated by the removal of fewer than c vertices. However, c–connectedness is not sufficient for G to be dei; for example, $C_3 \times K_2$ is 3–connected but not diameter edge–invariant.

We shall denote the class of dei graphs by DEI. For any integer k ≥ 2, we will denote the class of all dei graphs with diameter k by DEI(k).

A diameter edge–invariant graph G = (V,E) is said to be *critical* (respectively *cocritical*), if the deletion of any vertex v in V results in a graph G \ v which is not dei (respectively G \ v is dei).

The potential importance of the cocritical graphs to the design of interconnection network is indicated by their high reliability.

Several methods of construction of dei graphs are proposed in [12]. It is the purpose of this paper to investigate the class of critical dei graphs and cocritical dei graphs that arise from such constructions.

Not every dei graph is critical dei or cocritical dei. The graph in Figure 1 (a) is in DEI(2). However, we see that G \ A is in DEI(2) but G \ B is not in DEI. Hence G is neither critical nor cocritical.

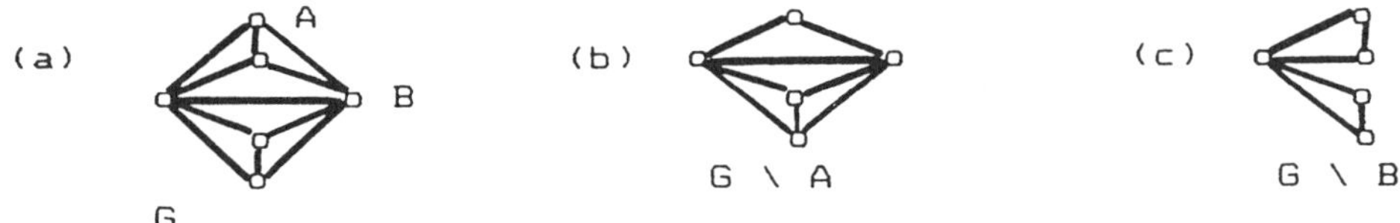

Figure 1

The second section of this paper is devoted to the construction of critical dei graphs by edge expansion of graphs. The critical dei graphs of diameter k with minimum number of nodes are completely described.

Section 4 presents the Sabidussi sum of graphs. This construction will lead to a large class of cocritical dei graphs. We show that every connected graph is an induced subgraph of a cocritical dei graph. The result not only indicates that there is an abundance of cocritical dei graphs but, as a consequence, also shows that it is impossible to have a characterization of cocritical dei graphs by forbidden subgraphs. For a fixed k ≥ 2, we completely describe the structure of the cocritical dei graphs of diameter k with minimum number of nodes.

2. Critical dei graphs constructed by edge expansion of graphs.

Let *Gph* be the class of all undirected graphs. Denote by 2–*Gph* the class of all graphs of the form ⟨G; u,v⟩ where u, v in V(G), i.e. graphs with 2 terminals.

Given a directed graph G = (V,E) without loops and a mapping f: E→ 2–*Gph*, we construct a new undirected graph (G,f) which is called the *edge expansion* of G by f in [12] as follows:

 if f((a,b)) = ⟨H; u,v⟩, then we replace the edge (a,b) in G by the graph H which
  identifies u with a and v with b.

An illustration of the construction of the edge expansion G from a mapping f is given in Example 1.

Example 1 Suppose G is given by Figure 2(a) and mapping f is given by Figure 2(b). Then (G,f) is shown in Figure 2(c).

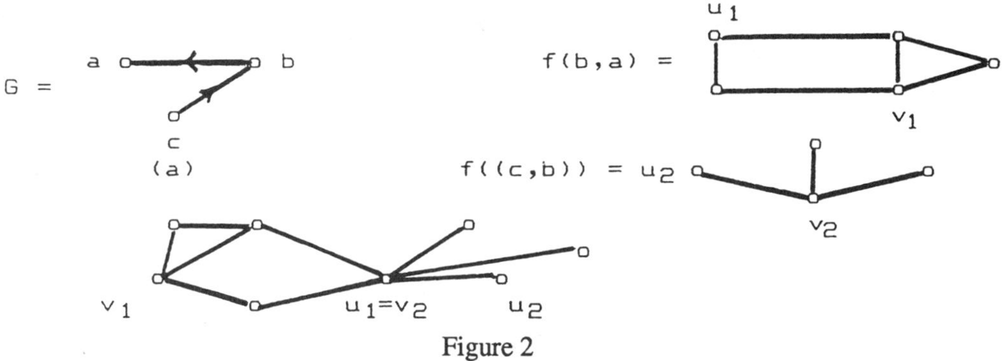

Figure 2

A graph G is a *minimum DEI(k)–graph* if it is a dei graph of diameter k with minimum number of nodes. It is clear that every minimum DEI(k)–graph is a critical dei graph.

The only minimum DEI(2)–graph is the "diamond" D of Klee and Quaife [10, 11], it is $P_3 + K_1$. There are two minimum DEI(3)–graphs J_1 and J_2 (Figure 3).

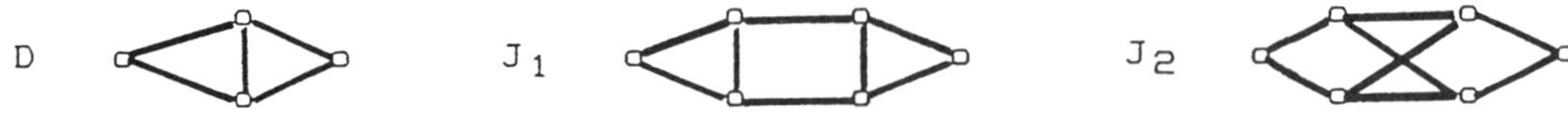

Minimum DEI(k)–graphs for k = 2 and 3.

Figure 3

The minimum DEI(k)–graphs are completely described by the first author in [12].

Theorem 1   *A graph G is a minimum DEI(k)–graph,*

(1)   for $k = 2m$ iff it is the edge expansion of P_{m+1} by f, where $f(e_i) = \langle D; u,v \rangle$ for $i = 1$, m and $f(e_i) \in \{ \langle D; u,v \rangle, \langle C_4; a,b \rangle \}$ for $1 < i < m$ where

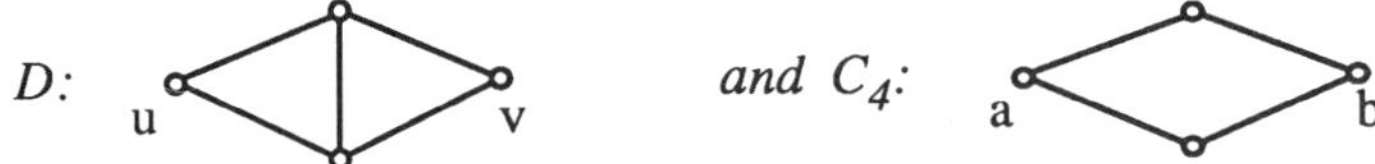

(2)   for $k = 2m + 1$ iff it is the edge expansion of P_{m+1} by f where

(a) $m = 1$.

$$f(e_1) = J_1 \ or \ J_2$$

(b) $m > 1$, $P_{m+1} = (e_1, ..., e_m)$, and for a fixed k,

where $1 \le k \le m$,

$$f_k(e_i) = \begin{cases} J_1 \ or \ J_2 & for \ i = k \\ D & for \ i = m \\ D \ or \ C_4 & otherwise. \end{cases}$$

Example 2   The following configurations are minimum DEI(6) and DEI(7) graphs (Figure 4).

For $n = 3$ and $m \ge 3$, we denote by SF(n,m) the edge expansion of C_n by the complete graph K_m. The graph SF(n,m) is called the *generalized sunflower*. For m = 3, the graphs SF(n,3) for $n = 3, 4, 6$ are depicted in Figure 5. The graph SF (n,3) is called a sunflower graph.

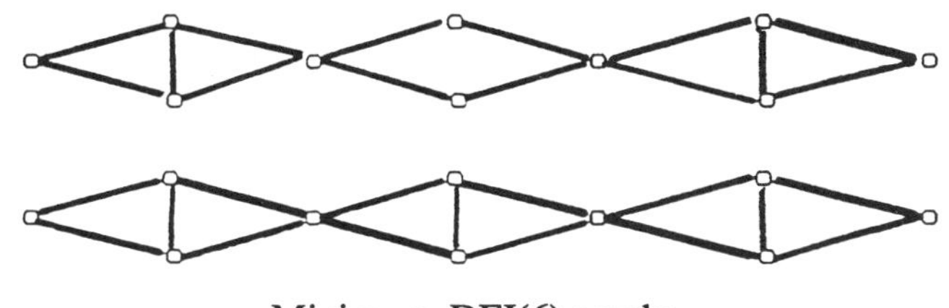

Minimum DEI(6) graphs.

(P_3, f_2)

(P_3, f_1)

Minimum DEI(7) graphs.

Figure 4

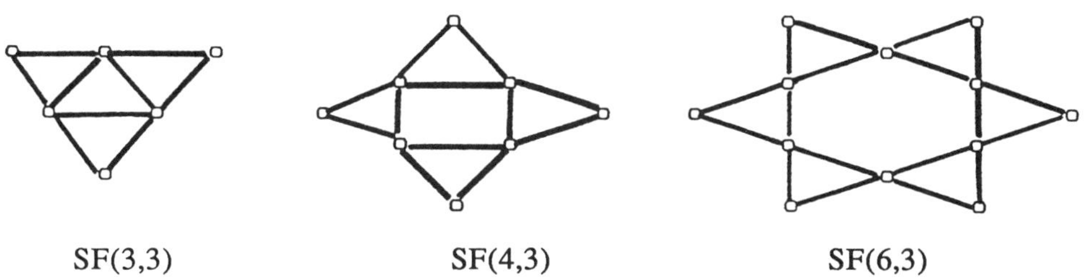

SF(3,3)   SF(4,3)   SF(6,3)

Sunflower Graphs.

Figure 5

Theorem 2   *The generalized sunflower SF(n,m) is critical dei if and only if m = 3,
and n is an even number greater than 4.*

Proof   It is clear that the graph SF(n,m) is dei if and only if n is even.
   If m ≥ 4 and n is even, the graph SF(n,m) is not critical.

Assume m = 3 and n = 4, then the subgraph of SF(4,3) obtained by deleting an outer–vertex is diameter edge–invariant (Figure 6). Therefore, the graph SF(4,3) is not critical diameter edge invariant.

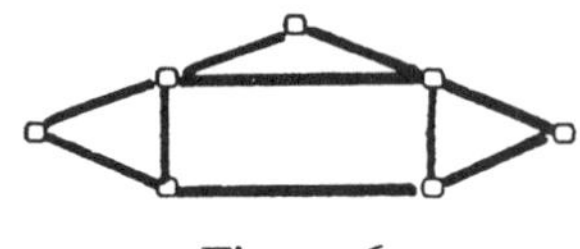

Figure 6

For n = 2k ≥ 6, there are two types of vertex deleted subgraphs (Figure 7 for k = 3).

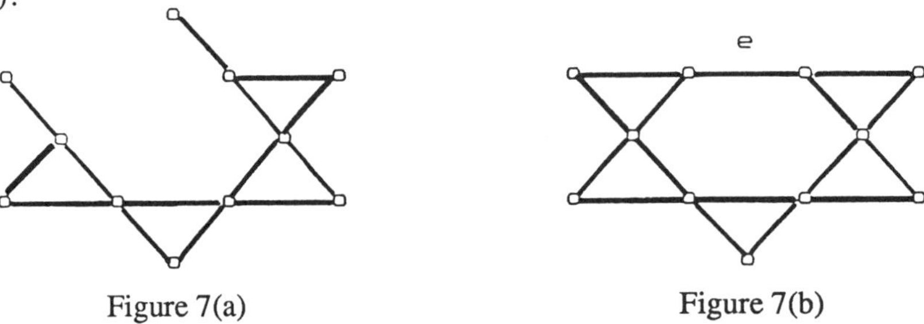

Figure 7(a)   Figure 7(b)

We observe that Figure 7(a) is clearly not dei and the graph of Figure 7(b) has diameter (n/2) + 1. If we delete the edge e, then the resulting graph has diameter n − 1. Clearly n − 1 > (n/2) + 1 for n = 2k ≥ 6. Therefore, SF(2k,3) is critical. ❑

Theorem 3 *If G is critical dei and f: E(G) → {Diamond}, then the edge expansion (G,f) is critical dei.*

Example 3   The graph which is constructed by the edge expansion of SF(6,3) by the diamond D is shown as follows (Figure 8):

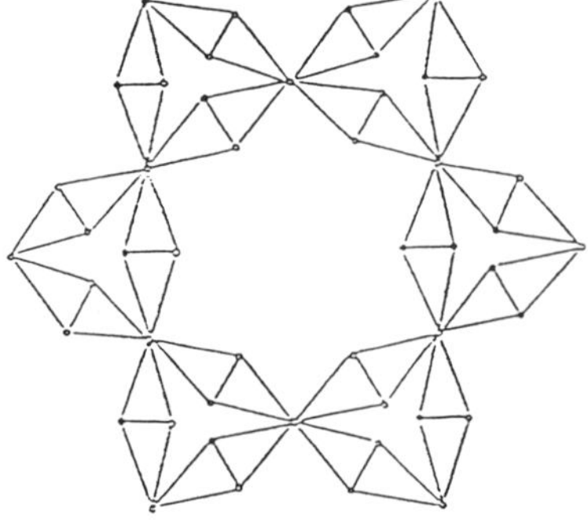

Edge expansion of SF(6,3) by D.

Figure 8

3. Critical dei graphs constructed by Cartesian product.

The operation of forming Cartesian products affords the possibility of constructing many dei graphs [12], [13]. It may be worthwhile to note that not all such graphs are critical.

The Cartesian product $G \times H$ of two graphs is the graph with $V(G) \times V(H)$ and two vertices (g_1, h_1) and (g_2, h_2) are adjacent if $g_1 = g_2$ and $(h_1, h_2) \in E(H)$ or $(g_1, g_2) \in E(G)$ and $h_1 = h_2$.

The 2–dimensional grid is the Cartesian product of two paths (see Figure 9). It is clear that $P_2 \times P_n$ is not dei for any $n \geq 2$. But $P_m \times P_n$ is dei if $\min\{m,n\} \geq 3$.

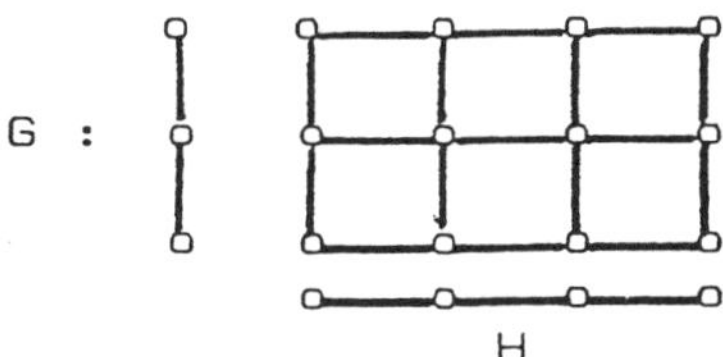

The products of two paths

Figure 9

The following theorem characterizes critical dei graphs which are 2–dimensional grids.

Theorem 4   *The 2–dimensional grid $P_n \times P_m$ is critical dei if and only if $(n,m) =$ (3,3) or (3,4).*

Theorem 5   *The prism $C_n \times K_2$ is critical dei, if and only if $n > 3$.*

Since C_3 is a complete graph, one may conjecture that if G is a non–complete 2–connected graph, then $G \times K_2$ is critical diameter e–invariant.

However, the smallest counter–example is provided by the "diamond" D of Klee and Quaife. We see that $D \times K_2$ is dei. However, its two subgraphs which are depicted in Figure 10 are not both dei graphs.

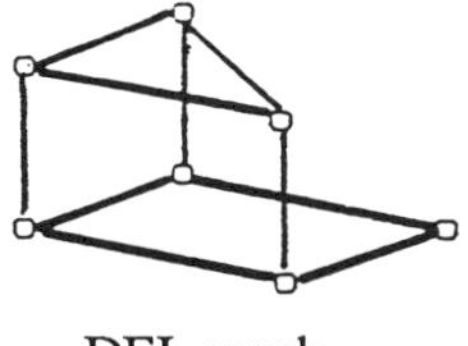 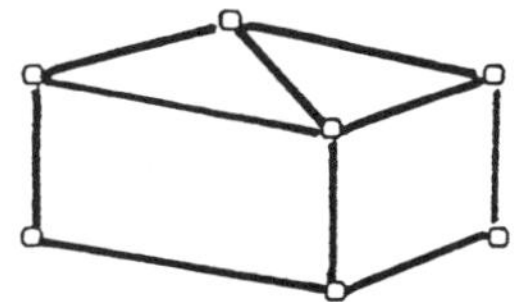

DEI–graph   not DEI–graph

Figure 10

4. Cocritical dei graphs which are Sabidussi sum of graphs.

Let G be a connected graph and f: V(G) $\rightarrow$ *Gph* be a nontrivial mapping; we construct a graph $S^+(G,f)$ as follows:

$$V(S^+(G,f)) = U \{ f(v):\ v \in V(G) \}$$

$$E(S^+(G,f)) = \{ (x,y):\ x \in V(f(v)),\ y \in V(f(u)),\ (v,u) \in E(G) \}$$

The graph is connected and it is called the *Sabidussi sum* of { f(v): v $\in$ V(G) } in [12]. It is clear that $D(S^+(G,f)) = D(G)$, for $D(G) \neq 1$. For $D(G) = 1$, we have $D(S^+(G,f)) = 2$. This construction is a generalization of Sabidussi's composition of lexicographic product [15], [16], [17]. The Zykov sum G + H of two graphs can be considered as $S^+(K_2,f)$ where f(u) = G and f(v) = H. (Figure 11)

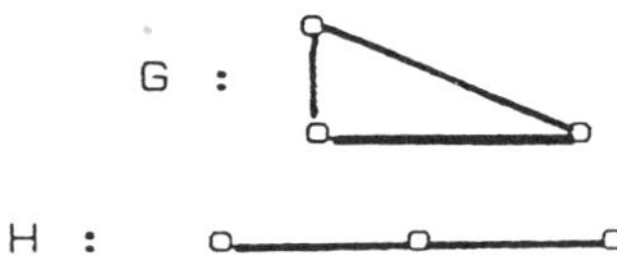

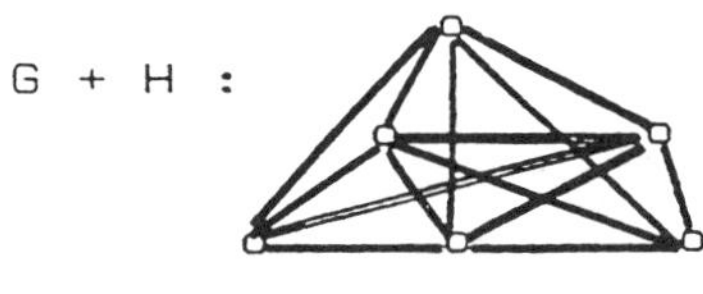

Figure 11

The lexicographic product of the graphs G_1 and G_2 is a graph denoted $G_1[G_2]$ whose vertex set is $V(G_1 \times G_2)$. For vertices $u = (u_1, v_2)$ and $v = (v_1, v_2)$ we have $uv \in E(G_1[G_2])$ if and only if either (a) $u_1 = v_1$ and $u_2v_2 \in E(G_2)$ or (b) $u_1v_1 \in E(G_1)$.

We see that $G_1[G_2] = S^+(G_1,f)$ where f(v) = G_2 for all v in $V(G_1)$.

Now we present the main innovation of this section.

Theorem 6   *For a 2–connected graph G, and f: V(G) $\rightarrow$ Gph with the property that f(v) is connected and $|V(f(v))| \geq 2$, the Sabidussi sum $S^+(G,f)$ is cocritical dei.*

We illustrate the above result by the following example:

Example 4   Let $C = C_4$ and f: $V(C_4) \rightarrow Gph$ be f(i) = K_2, then $S^+(G,f)$ is a graph which is cocritical dei. (Figure 12)

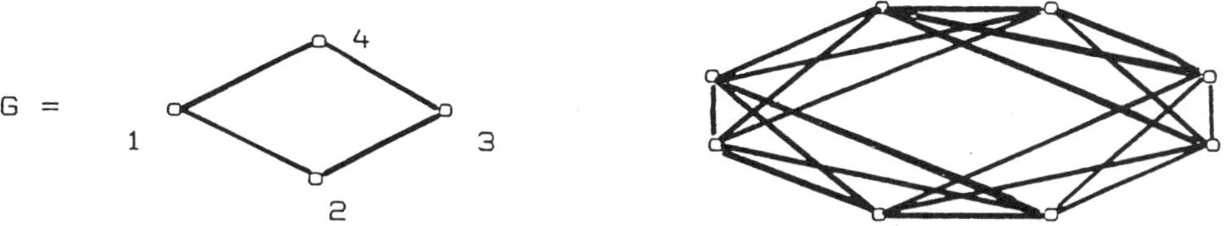

Cocritical dei graph with diameter 2.

Figure 12

Remark 1   The 2–connectedness in Theorem 6 is necessary. There exist connected graphs G such that f: V(G) $\rightarrow$ *Gph* with f(v) connected and $|V(f)| \geq 2$ but $S^+(G,f)$ is not cocritical dei. Let G = P_3, f(v_1) = P_3 and f(v_2) = f(v_3) = P_2 then the Sabidussi sum $S^+(G,f)$ (Figure 13) is not cocritical diameter e–invariant.

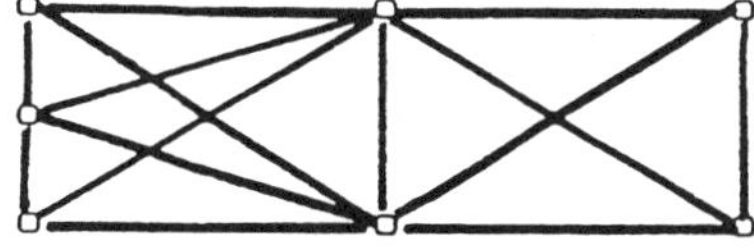
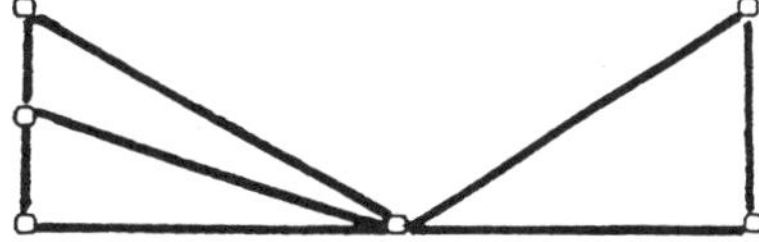

Figure 13

Remark 2   Let G be C_4 and f(v_1) = f(v_2) = f(v_4) = K_2 and f(v_3) = K_1, then $S^+(C_4,f)$ is not even cocritical (Figure 14). This example illustrates that the condition $|V(f(v))| \geq 2$ for all v is necessary.

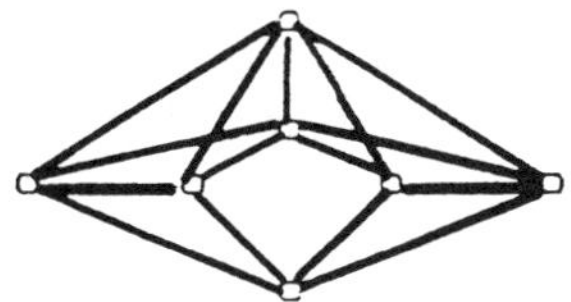

Figure 14

Corollary 7   *Every connected graph H is an induced subgraph of a cocritical dei graph.*

Proof   If $|H| = 2$, the result is trivial true. Assuming $|H| \geq 3$, take any 2–connected graph G. Consider the lexicographic product G[H], then by Theorem 6 it is cocritical dei.◻

Our characterization of minimum cocritical DEI(k) graphs is

Theorem 8   *The minimum cocritical DEI graph with diameter   k   is given by $S^+(C_{2k},f)$ where $f(v) = P_2$ for all   v   in $V(C_{2k})$.*

Example 5   The following graphs (Figure 15) are minimum cocritical DEI graphs of diameter 2 and 3.

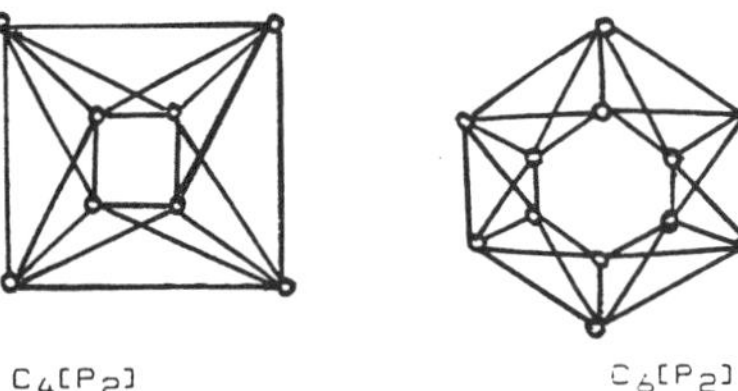

Figure 15

Acknowledgement   We appreciate the referee for his helpful suggestions which improve the style of the paper.

REFERENCES

[1]   J. C. Bermond and B. Bolobas, Diameters in graphs: a survey. *Congressus Numerantium* **32**, (1981), 3–27.

[2]   J. C. Bermond, J. Bond, M. Paoli and C. Peyrat, Graphs and interconnection networks: Diameter and vulnerability, *London Math Society Lecture Notes Ser. 82*, Cambridge University Press, (1983), 1–30.

[3]   J. C. Bermond, C. Delorme and G. Farhi, Large graphs with given degree and diameter III, *Annals of Discrete Mathematics* **13**, (1982), 23–32.

[4]   F. R. K. Chung, Diameters of communication networks, in "Math. of information Processing", *Proc. Symposium in Applied Math.* **34**, ed. by Michael Anshel and William Gewirtz, American Math Society, (1986), 1–18.

[5]   F. Glivjak, On certain classes of graphs of diameter two without superfluous edges, *Acta Fac. Rer. Nat. Univ. Comenian Math.* **21**, (1968), 39–48.

[6]   F. Glivjak, On certain edge–critical graphs of a given diameter, *Mat. Casopis Sloven Akad Vied.* **25**, (1975), 249–263.

[7]   F. Glivjak, On the impossibility to construct diametrically critical graphs by extensions, *Arch. Math. (Brno)* **11**, (1975), 131–137.

[8]   D. Greenwill and P. Johnson, On subgraphs of critical graphs of diameter k, *Proc. 10th S–E Conf. Combinatorics, Graph Theory and Computing*, 465–467.

[9] J. Hartman and I Rubin, "On Diameter Stability of Graphs," *Theory and Applications of Graphs,* Eds. Y. Alavi and D. R. Lick, Springer Lecture Notes No. 642, (1987), 247–254 (1987).

[10] V. Klee and H Quaife, Minimum graphs of specified diameter, connectivity and valence 1. *Mathematics of Operations Research,* **1** (1976), 28– 31.

[11] V. Klee and H. Quaife, Classifications and enumeration of minimum (d,1,3)–graphs and minimum (d,2,3)–graphs, *J. Comb. Theory (B)* **23**, (1977), 83–93.

[12] Sin–Min Lee, Design of diameter e–invariant networks, *Congressus Numerantium* **65** (1988), 89–102.

[13] Sin–Min Lee and Rudy Tanoto, Three classes of diameter edge–invariant graphs. *Commentations Mathematicae Universatis Carolinae* **28**, (1987), 227–232.

[14] O. Ore, Diameters in graphs, *Journal of Comb. Theory* **5**, (1968), 75–81.

[15] G. Sabidussi, The composition of graphs, *Duke Math. Journal* **26**, (1959), 693–696.

[16] G. Sabidussi, The lexicographic product of graphs, *Duke Math. Journal* **28**, (1961), 573– 578.

[17] G. Sabidussi, Graph derivatives, *Math Zeitschrift* **76**, (1961), 385–401.

[18] A. A. Schone, H. L. Bodlaender and J. van Leeuwen, Diameter increase caused by edge deletion, *Journal of Graph Theory* **11** (1987), 409–427.

Facts and Quests on Degree Irregular Assignments

J. Lehel[1]

Computer and Automation Institute
Hungarian Academy of Sciences

ABSTRACT

We summarize the status of research on irregularity recently done by many people: G. Chartrand, R.J. Faudree, A. Gyárfás, K.S. Holbert, M.S. Jacobson, L. Kinch, J. Lehel, O.R. Oellermann, S. Ruiz, F. Saba, R.H. Schelp, H.C. Swart and T. Szônyi. The problem of irregular degree assignments is formulated in terms of set systems. The paper presents 27 conjectures and questions for particular structures together with the related facts and results.

Introduction

In [CJLORS], the following problem was proposed by Chartrand, Jacobson, Lehel, Oellermann, Ruiz and Saba. Assign positive integer weights to the edges of a simple connected graph of order at least 3 in such a way that the graph becomes irregular, i.e., the weight sums at each vertex are distinct. What is the minimum of the largest weight over all such irregular assignments?

The basic problem on irregular assignments is formulated here as a general extremal problem in terms of hypergraphs, i.e., set systems. Standard graph and hypergraph terminology will be assumed or may be found in [B].

A hypergraph H is a pair (V(H),E(H)), where V(H) is a finite non–empty set of elements called vertices, and E(H) is a finite collection of subsets of V(H) called (hyper)edges. A hypergraph is *vertex–distinguishable* if for each pair of vertices there is an edge of H which contains precisely one of them. We consider here simple hypergraphs which contain no multiple edges (that is the collection of E(H) is a set).

1 The work was done in part at the University of Louisville, Louisville, Kentucky, and was partially supported by the Hungarian Academy of Sciences under AKA grant 1–3–86–264.

Assign positive integer weights to the edges of a hypergraph H. This assignment is *irregular* if for all $x \in V(G)$ the weighted degree of x (the sum of the weights of the edges containing x) are distinct. The minimum of the largest weights assigned over all such irregular assignments of H is called the *irregularity strength* of H and is denoted by s(H). We say that s(H) is undefined (or infinite) if H is not vertex–distinguishable.

The basic problem of irregular assignments consists in determining or bounding the irregularity strength of finite set systems.

The material is arranged into sections according to the particular structure investigated or to the methods used as follows:

1. Irregularity Strength of Graphs
1.1 General bounds
1.2 Trees, cycles
1.3 Complete graphs

2. Regular and Dense Graphs
2.1 Regular graphs
2.2 Dense graphs
2.3 Graphs with $s(G) = 2$

3. Hypergraphs
3.1 Irregular hypergraphs
3.2 Irregularity strength
3.3 Complete hypergraphs

4. Irregular Embeddings
4.1 Consecutive degree sequences
4.2 F–degrees
4.3 Embeddings

5. Dual Strength
5.1 Graphs
5.2 Sum–distinct sequences
5.3 Projective planes

The text presents 22 propositions and 27 theorems (assembled from 15 manuscripts), furthermore, there are 27 problems which formulate the related conjectures and quests.

I am grateful to the many people for their explicit or indirect contribution to this paper. In particular I would like to thank Gary Chartrand, Paul Erdös, Ralph Faudree, András Gyárfás, Mike Jacobson, Lael Kinch, Dick Schelp and an anonymous referee.

Special thanks go to Dick, Mike, Ralph and Buck McMorris for their kind support making possible our collaboration at the University of Louisville and at Memphis State University in 1986 and 1987. Many problems presented here were proposed with Mike and Lael of Louisville while I spent three semesters at U of L.

1. Irregularity Strength of Graphs

The problem of studying $s(G)$, the irregularity strength of a graph G was proposed in [CJLORS]. Determining the strength proved to be rather hard, even for very simple graphs.

A graph in which each edge is assigned a positive integer weight will be called a network, and the largest weight s is the strength of the network (c.f. [CJLORS], [JL]). The same structure is Berge's s–graph which is a multigraph with edge multiplicity at most s (c.f. [B]).

An irregular network is one with distinct (weighted) degrees, and the irregularity strength $s(G)$ is the minimum strength of an irregular network with underlying graph G.

1.1. General Bounds Note that the graphs of order 2 (either a single edge or two isolated vertices) are the only minimal graphs which are not vertex–distinguishable. Obviously, $s(G)$ is finite if and only if G contains at most one isolated vertex and no connected component of order 2.

Any known lower bound on $s(G)$ of general use is based on the following observation.

Proposition 1 *([CJLORS])* *If G contains n_i vertices of degree i, for some positive integer i, then $s(G) \geq (n_i + i - 1)/i$.*

A straightforward procedure which assigns non–unit weights to the edges of a spanning forest of the graph results in the following general upper bound.

Theorem 2 *(c.f. [CJLORS])* *If G is a graph of order $n \geq 3$ with finite irregularity strength, then $s(G) \leq 2n - 3$.*

For connected graphs this bound can be lowered.

Theorem 3 *([JL])   If G is a connected graph of order $n \geq 4$ with finite irregularity strength then $s(G) \leq n - 1$.*

Theorem 2 is sharp only for a 3–clique, while there is equality in Theorem 3 for every star.

Problem 4. Does there exist a real α, $0 < \alpha < 1$, such that $s(G) \leq \alpha n$ holds, if n is sufficiently large, for every (connected) graph G of order n containing no vertices of degree one?

Problem 5. Find the smallest real $\alpha = \alpha(d)$ such that $s(G) \leq \alpha n$ holds, if n is sufficiently large, for every (connected) graph G of order n containing no vertices of degree less than d.

One should note that the impact of connectedness on the irregularity strength of the graph is not at all clear.

1.2   Trees, cycles

Proposition 6 *(c.f. {CJLORS})   If T is a tree of order $n \geq 3$ then $s(T) \geq \lceil (n + 2)/3 \rceil$, and the bound is sharp.*

The construction and the proof of Proposition 6 in [CJLORS] shows that the trees with $n' = \lceil (n + 2)/3 \rceil$ vertices of degree 1 and n' vertices of degree 2 are expected to approach the lower bound. As far as we know no attempt of characterizing extremal trees (i.e. trees of order n with irregularity strength n') has been made.

Proposition 7 *([CJLORS])   If G is a unicyclic graph of order n (that is connected and has exactly one cycle) then $s(G) \geq n/3$, and the bound is tight.*

Proposition 8 *(c.f. [CJLORS])   For the path P_n with n vertices*

$$s(P_n) = \begin{cases} n/2 & \text{for } n \equiv 0 \ (mod \ 4) \\ \lfloor n/2 \rfloor + 1 & \text{otherwise} \end{cases} .$$

One can verify that the irregularity strength of the complete binary tree is equal to $(n + 1)/2$, thus its strength is determined by the number of pendant vertices. This observation generates the following problems, the first is due to Chartrand (personal communication), the second one to Schelp ([CSS]).

Problem 9. Find the largest integer $n(p)$ such that $s(T) = p$ holds for every tree T of order $n \le n(p)$ with p vertices of degree 1 (e.g. $n(2) = 4$).

Problem 10. If T is a tree with no vertices of degree 2 then $s(T)$ equals the number of vertices of degree 1.

Note that in the special case of a full binary tree (when it has no vertices of degree 2) Problem 10 is settled in [CSS].

Problem 11. Find the smallest positive real $\alpha = \alpha(D)$ such that $s(T) \le \alpha$ holds, if n is sufficiently large, for every tree of order n with maximum degree $D \ge 2$.

Proposition 12 *([FJLS])   For the cycle C_n with n vertices*

$$s(C_n) = \begin{cases} \lceil n/2 \rceil & \textit{for } n \equiv 1 \ (mod\ 4) \\ \lceil n/2 \rceil + 1 & \textit{otherwise} \end{cases}$$

Proposition 13 *([KLE])   If G is the disjoint union of cycles and paths of order at least 4 then $s(G) \le n/2 + 10$, where n is the order of G.*

The exclusion of the P_3 components in the previous result is justified with the following.

Theorem 14 *([KLE])   Let G be the disjoint union of t paths of order 3. If G has n vertices, then*

$$\lceil (5n - 1)/7 \rceil \le s(G) \le \lceil (5n - 1)/7 \rceil + 1.$$

Kinch also conjectures that $s(G) = \lceil (5n - 1)/7 \rceil$ holds for every n. If $G = tP_3$ as in Theorem 14 then $n = 3t$, $n_i = 2t$ and with $i = 1$ Proposition 1 gives the weaker lower bound $s(G) \ge 2t = 2n/3$. It is worth noting that this is the only example we know where Proposition 1 (and also its slight extension Proposition 44) gives no tight bound on the irregularity strength.

1.3. Complete Graphs   Let K_n denote the complete graph (clique) of order n. In ([CJLORS]), $s(K_n) = s(K_{2k,2k}) = 3$ was proved for every n and for $k \ge 1$. These results are contained as special cases in the results mentioned in this section. Let $K_{p,q}$ denote the complete p x q bipartite graph. A complete multipartite graph is defined as the complement of some pairwise vertex disjoint cliques.

Theorem 15 *([FJLS])   If G is a complete k-partite graph with $k \ge 3$ having the same number of vertices in each vertex class, then $s(G) = 3$.*

In the case when G is a clique Theorem 15 yields the result in [CJLORS] that $s(K_n) = 3$ for every $n \geq 3$.

Theorem 16 *([FJLS])*

$$s(K_{p,q}) = \begin{cases} \lceil (q + p - 1)/p \rceil & \text{for } q \geq 2p \\ 3 & \text{if either } p = q \text{ is even or } 1 < q/2 \leq p < q \end{cases}$$

The missing case when $p = q$ is odd was proposed by Chartrand et al. and settled by Gyárfás recently:

Theorem 17 *([GY])*

$$s(K_{2k+1,2k+1}) = 4.$$

2. Regular and Dense Graphs

The main conjectures are that every r–regular graph of order n has irregularity strength about n/r; and the irregularity strength of graphs of order n with minimum degree proportional to n does not depend on n. These questions are formulated in this section together with some related results.

2.1. Regular graphs   Every r–regular graph G of order n satisfies $s(G) \geq (n + r - 1)/r$ by Proposition 1.

Proposition 18   *If G is regular, then $s(G) \geq 3$.*

The main problem in the background of all results pertaining to the irregularity strength of regular graphs is the following problem due to Jacobson.

Problem 19.   Does there exist an absolute constant c such that $s(G) \leq n/r + c$ holds for every r–regular graph G of order n ?

One can find infinitely many regular graphs with irregularity strength indicated in the lower bound. The construction presented in [FJLS] is based on the vertex multiplication of a graph. Let $G^{(k)}$ denote the graph obtained from G by replacing each vertex of G by an independent set of k new vertices.

Theorem 20 *([FJLS])   Let G be an r–regular graph with n vertices obtained from the cycle C_t by vertex multiplication (that is $G = C_t^{(k)}$, $t \geq 3$, $n = tk$, $r = 2k$ and $k \geq 2$). Then $s(G) = \lceil n/r \rceil + 1$.*

Proposition 21 *([FJLS])   Let G be a 2–regular graph of order n. Then* $\lfloor n/2 \rfloor \leq$ $s(G) \leq \lfloor n/2 \rfloor + 2$. *Moreover, if G is the disjoint union of triangles, then*

$$s(G) = \begin{cases} \lceil n/2 \rceil + 1 & \text{for } n = (4k+1)3 \\ \lceil n/2 \rceil + 2 & \text{otherwise} \end{cases};$$

if G has no triangle components, then

$$s(G) = \begin{cases} \lceil n/2 \rceil & \text{for } n \equiv 1 \ (mod \ 4) \\ \lceil n/2 \rceil + 1 & \text{otherwise} \end{cases}.$$

It has been observed by Jacobson that $s(G)$ is about $n/2$ or less if G has a Hamiltonian cycle. This leads to the following.

Proposition 22 *([FJLS])   If G is an r–regular graph of order n, with $r \geq n/2$ and $n \geq 3$, then $s(G) \leq \lceil n/2 \rceil + 1$.*

A result in [J] yields a similar theorem: *If G is a 2–connected r–regular graph of order n, with $r \geq n/3$, then $s(G) \leq \lceil n/2 \rceil + 1$.*

The disjoint union of t cliques of order p is denoted by tK_p. The irregularity strength of these structures was determined (up to an additive constant) with Kinch.

Theorem 23 *([FJKL])   If $p \geq 3$, $r = p - 1$ and $n = tp$, then $G = tK_p$ is an r–regular graph of order n having irregularity strength $s(G) \leq \lceil (n+1)/r \rceil + 2$ for every t.*

Theorem 24 *([FL])   If G is a regular graph of order n having finite irregularity strength, then $s(G) < n/2 + 9$.*

Problem 25.   Does there exist a constant c such that $s(G) \leq n/3 + c$ holds for every 3–regular graph of order n ?

2.2. Dense graphs   A graph is considered here dense if the degree of each vertex is proportional to the order of the graph. The main problem concerning dense graphs is the following:

Problem 26.   Let α be a real with $0 < \alpha < 1$. Does there exist a constant $c = c(\alpha)$ such that $s(G) \leq c$ for every graph G with a minimum degree $\delta(G) \geq (1 - \alpha)n$?

The upper bound in the next theorem improves on the general bound $s(G) \leq 2n - 3$ of Theorem 2, however, it still depends on n.

Theorem 27 *([FJKL])   Let $0 < \alpha < 1$ and let G be a $(1 - \alpha)n$–regular graph of order n. Then $s(G) \leq (1/(\lfloor 1/\alpha \rfloor - 1)) n + 5$ holds for sufficiently large n.*

More is known about the irregularity strength of graphs of order n having minimum degree $n - t$, with t constant.

Theorem 28 *([FJKL])   If t is a fixed positive integer and n is sufficiently large, then $s(G) \leq 3$ holds for each graph of order n with minimum degree $n - t$.*

In some sense this theorem gives a sharp result on the minimum degree of a graph of irregularity strength 3 or less, since for fixed $t \geq 0$ the strength of infinitely many graphs of minimum degree $n - t$ is equal to 3 (e.g. regular graphs according to Proposition 18).

Problem 29.   Find the largest integer $t = t(n)$ such that $s(G) \leq 3$ holds, if n is sufficiently large, for every graph G of order n with minimum degree $n - t$.

Theorem 30 *([FJKL])   Let G be a graph of order n and minimum degree $\delta(G) = n - t$. If $1 \leq t \leq \sqrt{n/18}$, then $s(G) \leq 3$.*

If $t = t(n) > 2n/3$, it follows that $s(G) > 4$ for every $(n - t)$–regular graph G of order n. By lowering $2n/3$ (or by pushing up the bound $\sqrt{n/18}$ in Theorem 30) one could try to find the threshold between the minimum degree of graphs of irregularity strength 3 and 4.

Problem 31.   Find the smallest integer $t = t(n)$ such that there exist infinitely many graphs of order n with minimum degree $n - t$ and having irregularity strength at least 4.

Problem 32.   Find the smallest constant c such that $s(G) \leq c$ holds for every graph G of order n having minimum degree $n{-}3$.

2.3. Graphs with $s(G) = 2$ Some extremal problems pertaining to the number of edges in a graph of irregularity strength 2 were thoroughly investigated in [FGYS]. Note that in most of the cases of sharp results the non–constructive part of the proof is based on a supply–demand type counting argument developed by Gyárfás (c.f. [GY],[GYA]). Let n and m denote, respectively, the number of vertices and edges of a graph G.

Theorem 33 *([FGYS])   If G has irregularity strength 2, then $m \geq \lceil (n^2 - 1)/8 \rceil$, moreover, $m \geq (n^2 - 1)/8 + 1$ of $n \equiv 3 \pmod 4$. Furthermore, there exist graphs for which equality holds.*

Theorem 34 *([FGYS])   If G has irregularity strength 2, then*

$$m \leq \binom{n}{2} - (n - 1)/4.$$

A related result is the following:

Theorem 35 *([FGYS])   Let $N \subset K_n$ be subgraph of at most n/2 vertices and at least (n − 1)/4 edges. Then $s(K_n - N) = 2$.*

The results above have the following corollary: *$\lceil (n - 1)/4 \rceil$ is the minimum number of edges which need to be removed from K_n in order to decrease its irregularity strength from 3 to 2.*

The question of determining the irregularity strength of $K_n - F_k$, where F_k is a matching of k edges, was proposed in [FJKL].

Proposition 36 *([FJKL])   For every $k \geq 0$ and $n \geq 3$, $s(K_n - F_k) \leq 3$.*

Theorem 37 *([GYA]) $s(K_n - F_k) = 2$ if and only if $n = 4t \pm 1$ or 4t, and $k = t$.*

A graph whose vertex set is the union of a clique and an independent (or stable) set is called a split graph. Then $G = K_n - K_p$ is a complete split graph $(n \geq p)$. It is not hard to verify that $s(K_n - K_p) = 3$ for $n \geq 2p^2 - 2p + 2$.

Proposition 38 *(c.f. [FGYS]) $s(K_n - K_p) = 2$ if and only if $2p - 1 \leq n \leq 2p^2 - 2p + 1$.*

3. Hypergraphs

The starting point is that there do exist non–trivial irregular hypergraphs (H is irregular if the degrees of each vertex are distinct or equivalently, if $s(H) = 1$). The rank of a hypergraph is the maximum cardinality of its hyperedges. A hypergraph is called r–uniform if each edge contains r vertices.

3.1. Irregular hypergraphs

Proposition 39 *([GYJKLS])   If $r \geq 3$ and $n \geq r + 3$, then there exist irregular r–uniform hypergraphs of order n.*

Faudree, Jacobson and Schelp proposed the question of the irregularity behavior of a typical (random) hypergraph.

Theorem 40 *([GYJKLS])   Almost all r–uniform hypergraphs are irregular for $r \geq 6$.*

For the remaining cases Erdös and Gyárfás ask (personal communication):

Problem 41. Is it true that almost all 3– and 4–uniform hypergraphs have 2 vertices with the same degree?

Problem 42. Does it have positive probability that a 5–uniform hypergraph is irregular?

An $\langle r \times m \rangle$–hypergraph is an r–partite, r–uniform hypergraph with m vertices in each vertex class.

Theorem 43 *([GYJKLS])   For either fixed $m \geq 2$ or fixed $r \geq 6$ almost all $\langle r \times m \rangle$–hypergraphs are irregular.*

3.2. Irregularity strength   The lower bound given in [JL] (and essentially in Proposition 1) generalizes to hypergraphs.

Proposition 44 *([GYJKLS])   If H has n_i vertices of degree i, then*

$$s(G) \geq max \left\{ \left(\left(\sum_{k=i}^{j} n_k \right) + i - 1 \right) / j \; : \; i \leq j \right\} .$$

Theorem 45 *([GYJKLS])   If H is a vertex distinguishable hypergraph of rank r, then $s(H) \leq r|V(H)| - r + 1$.*

Proposition 46 *([GYJKLS])   The dual of a complete graph having n edges is a 2– regular hypergraph of order n with irregularity strength $2n - o(n)$.*

Most likely, there is no absolute constant c with $s(H) \leq c|V(H)|$. However, examples with high irregularity strength are unknown.

Problem 47. Find hypergraphs of order n with irregularity strength at least 2n.

3.3 Complete Hypergraphs   Denote by $K_n^{(r)}$ the complete r–uniform hypergraph of order n and by $K\langle r \times m \rangle$ the complete r–partite, r–uniform hypergraph with m vertices in each vertex class.

For the most part the hypergraphs in each of the two mentioned classes have irregularity strength 2. Note that this is equivalent to showing the existence of an irregular hypergraph of the given type.

Proposition 48 *([GYJKLS]) For $r \geq 3$,*

$$s(K_h^{(r)}) = \begin{cases} n & \text{if } n = r + 1 \\ 3 & \text{if } n = r + 2 \\ 2 & \text{if } n \geq r + 3 \end{cases}.$$

Theorem 49 *([GYJKLS]) For $r \geq 3$ and $m \geq 2$,*

$$s(K\langle r \times m \rangle) = \begin{cases} 3 & \text{if } r + m \leq 6 \\ 2 & \text{otherwise} \end{cases}.$$

4. Irregular Embeddings

This section contains problems and results pertaining to the existence of irregular networks under various restrictions or with additional constraints.

4.1. Consecutive degree sequences   In [FJKL], irregular complete networks were studied with degree sequence consisting of consecutive or almost consecutive integers. (A network on a complete graph is called a complete network.)

Theorem 50 *([FJKL]) If $D = (d, d{-}1, ..., d{-}n{+}1)$ is a sequence of n consecutive integers with even sum, then there exists a complete network of strength $\lceil d/(n-1) \rceil$ with degree sequence D for each $n \geq 5$ and $d > 2(n-1)$.*

Note that $\lceil d/(n-1) \rceil$ is obviously the smallest possible strength a network can have with the degree sequence D.

Let $a \geq 0$, $a + 1 \leq b \leq a + 2$ and $b + n - 2 \leq c \leq b + n - 1$. Then $D = (a, b, b{+}1, ..., b{+}n{-}3, c)$ with even sum is called a quasi–consecutive sequence. If D is the degree sequence of some irregular network G of order n, then obviously $s(G) \geq \lceil c/(n-1) \rceil$

Theorem 51 *([FJKL]) If D is a quasi–consecutive sequence of at least 5 integers, then there exists a network with degree sequence D and having irregularity strength*

$$s = \begin{cases} \lceil c/(n-1) \rceil & \text{if } c \neq n - 1 \\ 2 & \text{if } c = n - 1 \end{cases}.$$

4.2. F–degrees   Chartrand, Holbert, Oellermann and Swart in [CHOS] introduced the generalized degree of a graph as follows. Let F and G be graphs; the F-degree of a vertex x in G is the number of subgraphs of G isomorphic to F that contain x. A graph is called *F–irregular* (F–regular) if the F–degrees of the vertices of G are all distinct (equal).

Theorem 52 *([CHOS])   There exist $K_{1,n}$–irregular and K_{n+1}–irregular graphs for every $n \geq 2$.*

Erdös, Székely and Trotter (c.f. [CHOS]) have the remark that there are infinitely many K_3–irregular graphs.

Problem 53.   Do there exist infinitely many $K_{1,n}$–irregular and K_n–irregular graphs?

Chartrand et al. in [CHOS] have the following:

Problem 54.   Does there exist an F–irregular graph for every connected graph F of order at least 3?

4.3. Embeddings   On the analogy of the embedding of a simple graph into a regular one, investigated by König ([K]), Chartrand has proposed the problem of the irregular embedding of networks (personal communication, c.f. [CJLORS]).

Proposition 55 *([CJLORS])   If G is a network of strength at least 2, then there exists an irregular network H containing G as an induced subnetwork and such that $s(H) = s(G)$.*

It was also mentioned in [CJLORS] that if G is bipartite, then H might be bipartite as well. This result is an instance of the following general problem:

Problem 56 [CJLORS]. Given a network G of strength at least 2, and possessing a specified graphical property P, does there exist an irregular network H having property P such that H contains G as an induced subnetwork and $s(H) = s(G)$?

This question has an affirmative answer in the case when P is the property of being k–chromatic.

Theorem 57 *([JLE])   If G is a k–chromatic network of strength at least 2, then there exists an irregular k–chromatic network H with $s(H) = s(G)$ such that H contains G as an induced subnetwork.*

The following general problem on the F–irregular embedding of graphs contains the question on the existence of infinitely many F–irregular graphs (Problems 53 and 54).

Problem 58. Given the graphs F and G (F is connected and has at least 3 vertices). Does there exist an F–irregular graph H which contains G as an induced subgraph?

Note that Problem 58 has a positive answer when $F = P_3$, a path on 3 vertices (Jacobson, personal communication).

If $F = K_r$ then the F–irregular embedding of a graph with maximum clique r can be related in a natural way to the irregular embedding of a conform r–uniform hypergraph. Here we formulate a slightly weaker problem (for hypergraph terminology, see [B]).

Problem 59. Let H be an r–uniform hypergraph. Does there exist an irregular r–uniform hypergraph containing H as a subhypergraph?

5. Dual Strength

If H* stands for the dual of the hypergraph H, and H is simple then obviously H* is vertex–distinguishable. So it is quite natural to introduce s*(H), the dual notion of the irregularity strength; it is the minimum of the largest label used in any *irregular labelling of the vertices* with (not necessarily distinct) positive integers such that for each $e \in E(H)$ the sum of the labels of the vertices of e are distinct. Then s*(H) = s(H*), and it is called the dual (irregularity) strength of H.

5.1. Graphs The study of the dual strength has been initiated in [GYJKLS]. The basic problem pertaining to the dual strength of graphs is: how large s*(G) can be compared to the number of edges. The following conjecture is due to Gyárfás (personal communication).

Problem 60. Does there exist a constant c such that $s^*(G) \leq cm$ for every graph G with m edges?

Note that the question above is open even for bipartite graphs. The case of trees, however, is an easy corollary of the following observation due to Gyárfás (personal communication).

Proposition 61 *If G is a graph of order n such that every induced subgraph contains a vertex of degree at most k, then $s^*(G) \leq kn + 0(1)$.*

For a complete graph the following holds true.

Proposition 62 *([GYJKLS])* *If* G *is an* n–*clique, then*

$$s^*(G)/\binom{n}{2} \to 2 \quad as \quad m \quad approaches \quad infinity.$$

For the dual strength of the complete bipartite graph Gyárfás and Schelp [GYS] have the following conjecture settled in the case of $p = 3$.

Problem 63. Let $p \leq q$ and $G = K_{p,q}$. Is it true that

$$s^*(G) = \begin{cases} p(q+1)/2 & for \ p \ even \\ q(p+1)/2 & for \ q \ even, \ p \ odd \ ? \\ q(p+1)/2 + (p-1)/2 & for \ p \ and \ q \ odd \end{cases}$$

5.2. Sum–distinct sequences   The notation of a B_2–sequence was introduced in additive number theory by Erdös and Turán ([ET]) to denote a sequence $a_1 < a_2 < \ldots < a_m$ of positive integers with distinct pairwise sums. They ask the following question.

Problem 64   ([ET]). Is there a constant c such that $a_m \geq m^2 + cm$?

Partial results to this well studied extremal problem on B_2–sequences have some interesting corollaries for the dual strength of complete graphs (see Propositions 46 and 62).   Note that since cliques contain no loops, the notion of a B_2–sequence is to be modified by considering only sums of distinct elements.

Positive integers $a_1 < \ldots < a_m$ form a sum–distinct sequence if all the 2^m subsums are distinct. An example is the sequence $a_i = 2^{i-1}$, $i = 1, \ldots, m$. A problem due to Erdös and Moser, still open since 1955, is the following. (For answering any of these problems Erdös offers a good price.)

Problem 65   ([E]). Prove or disprove the existence of an absolute constant c such that $a_m > c2^m$ holds for every sum–distinct sequence.

This question is clearly related to the problem of estimating the dual strength of a totally complete hypergraph of order m.

The concept of a sum–distinct sequence is extended to more than one sequence and their generation by 'greedy' procedures in [KL].

Two sequences $A = \{a_1, a_2, ...\}$ and $B = \{b_1, b_2, ...\}$ are said to be compatible sum–distinct sequences if both A and B are sum–distinct, and the subsums of A differ from that of B.

Denote $F(n)$ the nth Fibonacci number, that is $F(0) = F(1) = 1$ and if $n \geq 2$, then $F(n) = F(n-1) + F(n-2)$.

Proposition 66 *([KL])   Create two compatible sum–distinct sequences by the first-fit procedure as follows. Start with $a_1 = 1$ and $b_1 = 2$; if elements a_i and b_i are defined for $i = 1, 2, ..., k$, then choose the smallest positive integer x as a_{k+1}, then the smallest positive integer y as b_{k+1} such that $\{a_i : 1 \leq i \leq k\} \cup \{x\}$ remains sum-distinct and compatible with $\{b_i : 1 \leq i \leq k\}$, and $\{b_i : 1 \leq i \leq k\} \cup \{y\}$ remains sum–distinct and compatible with $\{a_i : 1 \leq i \leq k+1\}$. Then $a_i = F(2i-1)$ and $b_i = F(2i)$ for $i = 1, 2, ...$.*

We say that t sum–distinct sequences are compatible if they are pairwise compatible ($t > 1$). No formula is known so far for the compatible sequences created by the first–fit procedure for $t \geq 3$. Note that in [KL] it is shown that by introducing a minor change in the procedure the sequence of generalized Fibonacci numbers are obtained.

For $n = 2$ one can verify easily the following:

Proposition 67   *The first–fit procedure creates t compatible sum–distinct pairs with maximum element $\lfloor 5t/2 \rfloor$.*

Let $s(t,n)$ denote the minimum of the largest element of t compatible sum–distinct sequences.

Theorem 68 *([KLE]) $s(t,n) \geq (15t-1)/7$ and the bound is sharp (up to an additive constant) for every t.*

As in the case of $t = 1$, one can propose the following general extremal problem.

Problem 69.   Determine the order of magnitude of $s(t,n)$.

Problem 70.   Is there a constant $c = c(n)$ such that $s(t,n) \leq ct$ for every t?

5.3. Projective planes   A projective plane of order q is considered as a $(q+1)$–uniform, regular hypergraph defined by the $q2 + q + 1$ lines as edges. For the terminology pertaining to infinite geometries see [H]. Jacobson asked (personal communication) whether the finite projective plane has large irregularity strength compared to the number of points. It is more convenient to formulate the results and the problems in terms of the dual strength.

Theorem 71 *([LSZ])*   *If P is a projective plane of order q and q $\geq$ 3, then s*(P) $\leq$ q^2.*

Note that the lower bound from Proposition 44 is linear in q. For special q there are sharper results.

Proposition 72 *([LSZ])*   *If P is the Fano plane (q = 2), then s*(P) = 5. If P is the projective plane of order 3, then 6 $\leq$ s*(P) $\leq$ 8.*

Proposition 73 *([LSZ])*   *If P is a Desarguesian projective plane of square order q, then s*(P) $\leq$ q $\sqrt{q}$ + 0(q).*

The following result improves on the lower bound s*(P) $\geq$ q + 2 resulting by proposition 44, if q is sufficiently large.

Theorem 74 *([LSZ])*   *If P is a projective plane of order q, then s*(P) $\geq$ 7q/4 for sufficiently large q.*

Problem 75. Disprove that s*(P) $\leq$ 0(q) for every projective plane P of order q.

There are further structures of geometric nature for which the problem of determining the dual strength seems to have some interest, for instance:

Problem 76. Determine the dual strength of uniform interval hypergraphs.

REFERENCES

[B]   C. Berge, Graphs and hypergraphs (North Holland, Amsterdam) (1973).

[CEO]   G. Chartrand, P. Erdös, O. R. Oellermann, How to define an irregular graph, *College Math. J.* **19** (1988) 36-42.

[CHOS]   G. Chartrand, K. S. Holbert, O. R. Oellermann, H. C. Swart, F–degrees in graphs, *Ars Combinatoria* **24** (1978) 133-148.

[CJLORS]   G. Chartrand, M. S. Jacobson, J. Lehel, O. R. Oellermann, S. Ruiz, F. Saba, Irregular networks, to appear in the Proceedings of the 250th Anniversary Conference on Graph Theory, Fort Wayne, IN, 1986.

[CL]   G. Chartrand, L. Lesniak, Graphs and Digraphs, 2nd Edition, Wadsworth and Brooks/Cole, (1986).

[CSS]   L. A. Cammack, R. H. Schelp, G. C. Schray, Irregularity strength of full d–ary trees, In preparation.

[E]   P. Erdös, Problems and results in additive number theory, in "Colloque sur la Théorie des Nombres, Bruxelles 1955", Liége et Paris, (1956), 127-137.

[ET]   P. Erdös, P. Turán, On a problem of Sidon in additive number theory and some related problems, *J. London Math Soc.* **16** (1941) 212-215.

[FJLS]   R. J. Faudree, M. S. Jacobson, J. Lehel, R. H. Schelp, Irregular networks, regular graphs and integer matrices with distinct row and column sums, to appear in *Discrete Math.*

[FGYS]   R. J. Faudree, A. Gyárfás, R. H. Schelp, On graphs of irregularity strength 2, to appear in the Proceedings of the Hungarian Conference in Eger 1987.

[FL]   R. J. Faudree, J. Lehel, Bound on the irregularity strength of regular graphs. to appear in the Proceedings of the Hungarian Conference in Eger 1987.

[FJKL]   R. J. Faudree, M. S. Jacobson, L. Kinch, J. Lehel, Irregularity of dense graphs. Submitted.

[GY]   A. Gyárfás, The irregularity strength of $K_{m,m}$ is 4 for odd m, *Discrete Math.* **71** (1988) 273-274.

[GYA]   A. Gyárfás, The irregularity strength of $K_n - mK_2$, to appear in *Utilitas Mathematica.*

[GYJKLS]   A. Gyárfás, M. S. Jacobson, L. Kinch, J. Lehel, R. H. Schelp, Irregularity strength of uniform hypergraphs. Submitted.

[GYS]   A. Gyárfás, R. H. Schelp, A matrix labelling problem. Submitted.

[H]   J. W. P. Hirschfeld, Projective Geometries over Finite Fields, Clarendon Press, Oxford, (1979).

[J]   B. Jackson, Hamiltonian cycles in regular 2–connected graphs, *Journal of Combinatorial Theory,* **B 29** (1980) 27-46.

[JL]   M. S. Jacobson, J. Lehel, Upper bound on the irregularity strength of a simple graph. Submitted.

[JLE]   M. S. Jacobson, J. Lehel, Irregular embeddings. In preparation.

[KL]   L. Kinch, J. Lehel, Sum–distinct sequences and Fibonacci numbers. Submitted.

[KLE]   L. Kinch, J. Lehel, The irregularity strength of tP_3. Submitted.

[K]   D. König, Theorie der endlichen und unendlichen Graphen, Leipzig, (1936) (Reprinted Chelsea, New York, 1950)

[LSZ]   J. Lehel, T. Szônyi, Irregular labeling of the finite projective planes. Unpublished.

Neighborhood Unions and Graphical Properties

Linda M. Lesniak

Drew University

ABSTRACT

If S is a set of vertices of a graph G, then the neighborhood N(S) of S is defined to be the set of all vertices of G which are adjacent to at least one vertex of S. A survey is presented of recent results of the following form:

Given a graph G, if |N(S)| is at least k for every subset S of V(G) of a specified type, then G has property P.

We look, for example, at a variety of hamiltonian properties P, where the sets S are all pairs of nonadjacent vertices and also where the sets S are all pairs of vertices.

1. Introduction

Graph theory literature is rich with sufficient conditions of the following type:

If deg $v \geq k$ for every vertex v of a graph G, then G has property P.

Perhaps one of the best-known is a result of Dirac [8] giving a sufficient condition for a graph to be hamiltonian.

Theorem 1 *If G is a graph of order $p \geq 3$ such that deg $v \geq p/2$ for every vertex v, then G is hamiltonian.*

In many instances, minimum degree conditions have been improved by results involving degree sums of nonadjacent vertices. (See [6] for an excellent survey.) Ore [25], for example, extended Dirac's result in this fashion.

Theorem 2 *If G is a graph of order $p \geq 3$ such that deg u + deg $v \geq p$ for each pair u, v of nonadjacent vertices, then G is hamiltonian.*

Theorem 1 was yet generalized further by Bondy [4].

Theorem 3 *If G is a 2–connected graph of order p such that   deg u + deg v + deg w $\geq$ 3p/2 for each triple u, v, w of mutually nonadjacent vertices, then G is hamiltonian.*

A condition of the form " deg v $\geq$ k for every vertex v of G" could alternatively be described as "$|N(S)| \geq k$ for every subset S of V(G) with $|S| = 1$", where N(S) denotes the *neighborhood of the set S* and is defined to be the set of all vertices of G which are adjacent to at least one vertex of S. This rather more complicated way of describing a minimum degree condition suggests obvious variations obtained by changing the sets S under consideration. Such results, unlike Theorems 2 and 3, would reflect overlapping neighborhoods of the vertices in S.

Although this idea of a neighborhood condition is not new, only a few such results appeared in the literature until recently. Hall [21] showed that if G is a bipartite graph with partite sets V_1 and V_2, then V_1 can be matched to a subset of V_2 if and only if $|N(S)| \geq |S|$ for every nonempty subset S of V_1. Anderson [2] showed that if G is a graph of even order p satisfying

$$|N(S)| \geq (p + 2|S| + 1)/4$$

for every nonempty subset S of V(G), then G has a 1–factor. Woodall [28] established that if G is a 2–connected graph of order p satisfying

$$|N(S)| \geq (p + |S| - 1)/3$$

for every nonempty subset S of V(G) and

$$\deg v \geq (p + 2)/3$$

for every vertex v, then G is hamiltonian. Posa [26] showed that for a fixed integer k, if G is a graph such that $|N(S) - S| \geq 2|S| - 1$ for each set S of at most k vertices, then G contains a path of length $3k - 2$.

In this paper we present a survey of the many recent neighborhood union results. In all cases, these results are distinct from known results based on degree conditions. For hamiltonian properties we will compare neighborhood union and degree sum results. For the sake of brevity, we leave other such comparisons as an exercise for the reader.

The following notations will be useful in presenting examples. If G and H are two graphs with disjoint vertex sets, then $G \cup H$ denotes the *union* of G and H and is the graph with vertex set $V(G) \cup V(H)$ and edge set $E(G) \cup E(H)$. The union of n copies of G is denoted by nG. The *join* of G and H, denoted G + H, is the graph with vertex set $V(G) \cup V(H)$ and edge set

$$E(G) \cup E(H) \cup \{ uv \mid u \in V(G) \text{ and } v \in V(H)\}.$$

For all other terms and notation not specifically defined here, see [7].

2. All Pairs of Nonadjacent Vertices

In this section we examine the relationship between the cardinality of the union of the neighborhoods of an arbitrary pair of nonadjacent vertices of a graph and various hamiltonian properties.

A graph G is *hamiltonian* (*traceable*) if it has a cycle (path) containing all of the vertices of G. Such a cycle (path) is called a *hamiltonian cycle* (*hamiltonian path*). A graph G is *hamiltonian–connected* if each pair of vertices of G are the endvertices of a hamiltonian path in G.

Ore's original paper on a degree sum condition for a graph to be hamiltonian contained a similar condition (as a corollary) for a graph to be traceable.

Theorem 4   *If G is a graph of order p such that $deg\ u\ +\ deg\ v\ \geq p - 1$ for each pair $u,\ v$ of nonadjacent vertices, then G is traceable.*

In a later paper [24] he established a similar result for hamiltonian–connected graphs.

Theorem 5   *If G is a graph of order p such that $deg\ u\ +\ deg\ v\ \geq p + 1$ for each pair $u,\ v$ of nonadjacent vertices, then G is hamiltonian–connected.*

If G is a (noncomplete) graph with $p \geq 2$ vertices such that $|N(u) \cup N(v)| \geq k$ for each pair $u,\ v$ of nonadjacent vertices, then necessarily $0 \leq k \leq p - 2$. Since the graph $K_1 \cup K_{p-1}$ is disconnected with $k = p - 2$, it is clear that such a neighborhood condition will not force any connectivity conditions. However, if G is traceable, hamiltonian, or hamiltonian–connected, then G is necessarily connected, 2–connected, or 3–connected, respectively. Thus, in the case of hamiltonian properties, a minimum connectivity condition as well as a neighborhood condition must be assumed.

In [14], the following "hamiltonian type" neighborhood conditions were established.

Theorem 6   *If G is a 2–connected graph of order p such that $|N(u) \cup N(v)| \geq (2p - 1) / 3$ for each pair $u,\ v$ of nonadjacent vertices, then G is hamiltonian.*

Corollary   *If G is a connected graph of order p such that $|N(u) \cup N(v)| \geq (2p - 2) / 3$ for each pair $u,\ v$ of nonadjacent vertices, then G is traceable.*

Theorem 7   *If G is a 3–connected graph of order p such that $|N(u) \cup N(v)| > 2p/3$ for each pair $u,\ v$ of nonadjacent vertices, then G is hamiltonian–connected.*

If G is the graph $3K_n + K_2$, then G is 2–connected of order $p = 3n + 2$. Furthermore, $|N(u) \cup N(v)| = 2n = (2p - 1)/3 - 1$ for each pair $u,\ v$ of nonadjacent vertices. However, G is not hamiltonian. Thus, for graphs of order $p = 3n + 2$, Theorem 6 is best possible.

If H is the graph obtained by taking three copies of K_n ($n \geq 3$), and joining corresponding vertices in each copy by an edge, then H is a 2–connected graph of order $p = 3n$ which satisfies $|N(u) \cup N(v)| = 2n = 2p/3$ for each pair u, v of nonadjacent vertices. Thus Theorem 6 states that H is hamiltonian. However, each vertex of H has degree $n + 1 = p/3 + 1$, and so neither of Theorems 2 nor 3 applies.

Similar constructions exist for hamiltonian–connected graphs. In particular, if G is the graph $3K_n + K_3$, then G is 3–connected of order $p = 3n + 3$ and G satisfies the condition that $|N(u) \cup N(v)| = 2n + 1 = 2p/3 - 1$ for each pair u, v of nonadjacent vertices. However, G is not hamiltonian–connected. On the other hand let H be the graph obtained from three disjoint copies of K_n, $n \geq 10$, where the vertices of each copy are ordered 1, 2, ..., n. Add edges between the copies as follows. The first vertex of each copy is adjacent to the n–th, first, and second vertices of the other two copies. The second vertex of each copy is adjacent to the first, second, and third vertices of each copy. In general, vertex i of each copy is adjacent to vertices $i - 1, i$, and $i + 1$ of the other two copies. Then H is a 3–connected graph of order $p = 3n$ in which each vertex has degree $n + 5 = p/3 + 5$. Since $|N(u) \cup N(v)| \geq 2n + 3 = 2p/3 + 3$ for each pair u, v of nonadjacent vertices, Theorem 7 may be applied to conclude that G is hamiltonian–connected. However, Theorem 5 gives no information since $2p/3 + 10 < p + 1$.

The neighborhood condition for a graph to be traceable can be decreased substantially if we assume that G is 2-connected [14].

Theorem 8 *If G is a 2–connected graph of order p such that $|N(u) \cup N(v)| \geq (p - 1)/2$ for each pair u, v of nonadjacent vertices, then G is traceable.*

If G is the graph $3K_n + K_1$, $n \geq 2$, then G is connected, has order $p = 3n + 1$ and satisfies $|N(u) \cup N(v)| = 2n - 1 \geq (p - 1)/2$ for each pair u, v of nonadjacent vertices. Since G is not traceable we see that the connectivity condition of Theorem 8 cannot be dropped. Secondly, if we consider the complete bipartite graph $K_{n-2,n}$, $n \geq 4$, we have a 2–connected graph of order $p = 2n - 2$ satisfying $|N(u) \cup N(v)| \geq n - 2 = (p - 2)/2$ for each pair u, v of nonadjacent vertices, and this graph is not traceable. Thus the neighborhood condition of Theorem 8 is sharp. Finally, if H is the graph $3K_n + K_2$, $n \geq 2$, then H is 2–connected of order $p = 3n + 2$ and $|N(u) \cup N(v)| = 2n \geq (p - 1)/2$ for each pair u, v of nonadjacent vertices. Thus Theorem 8 states that H is traceable. Since, however, the degree sum of any pair of nonadjacent vertices is $2n + 2 < p - 1$, Theorem 4 cannot be applied.

Theorem 8 suggests that neighborhood conditions may be relaxed somewhat by imposing other conditions (higher connectivity, for example) on the graphs under consideration. Results of this type were obtained in [11], where a minimum degree condition was added.

Theorem 9 *For a fixed integer $t \geq 2$, let G be a 2–connected graph of order p with $\delta(G) \geq t$. If $|N(u) \cup N(v)| \geq p - t$ for each pair u, v of nonadjacent vertices, then G is hamiltonian.*

Corollary *For a fixed integer $t \geq 2$, let G be a connected graph of order p with $\delta(G) \geq t - 1$. If $|N(u) \cup N(v)| \geq p - t$ for each pair u, v of nonadjacent vertices, then G is traceable.*

If G is the 2–connected graph $3K_{t-2} + K_2$, $t \geq 3$, of order $p = 3t - 4$, then G is 2–connected and satisfies the neighborhood condition of Theorem 9. However, $\delta(G) = t - 1$ and G is not hamiltonian.

Ore's degree sum condition guarantees the existence of *one* hamiltonian cycle in a graph. In [22] Ore's result was generalized to a condition giving multiple edge–disjoint hamiltonian cycles.

Theorem 10 *Let G be a graph of order p and minimum degree δ such that $p \geq 2\delta^2$. If $\deg u + \deg v \geq p$ for each pair u, v of nonadjacent vertices, then G contains $\lfloor (\delta - 1)/2 \rfloor$ edge–disjoint hamiltonian cycles.*

A neighborhood condition for multiple hamiltonian cycles was established in [19]. The *edge–connectivity* $\kappa_1(G)$ of a nontrivial graph G is the minimum number of edges whose removal from G results in a disconnected graph.

Theorem 11 *Let k be a fixed positive integer. Then there is a constant $c = c(k)$ such that if G is a graph of sufficiently large order p satisfying*
(1) $|N(u) \cup N(v)| \geq (2p + c) / 3$ for each pair u, v of nonadjacent vertices,
(2) $\delta(G) \geq 4k + 1$,
(3) $\kappa_1(G) \geq 2k$, and
(4) $\kappa_1(G - v) \geq k$ for every vertex v,
then G contains k edge–disjoint hamiltonian cycles.

Any theorem that gives a sufficient condition for a graph G to have k edge–disjoint hamiltonian cycles and is based on a neighborhood condition must have the types of restrictions listed in Theorem 11, as we shall see. However, only restrictions (3) and (4) are necessarily sharp.

(1) If G_1 is the graph $3K_n + K_2$, then G_1 is a graph of order $p = 3n + 2$ satisfying (2), (3), and (4) for p sufficiently large. Furthermore, $|N(u) \cup N(v)| = 2(p-2)/3$ for each pair u, v of nonadjacent vertices. However, G_1 is not hamiltonian.

(2) If G_2 is the graph $(K_{2k-1} \cup K_{p-2k-1}) + K_2$, then G_2 is a graph of order p satisfying (1), (3), and (4), and $\delta(G) = 2k$ for p sufficiently large. However, G_2 does not contain k edge–disjoint hamiltonian cycles; otherwise, each such cycle would necessarily contain a hamiltonian path in the subgraph K_{2k-1}, implying that K_{2k-1} has at least $k(2k-2)$ edges.

(3) If G_3 is the graph obtained from two copies of the complete graph K_n by adding $2k-1$ mutually nonadjacent edges between the copies, then G_3 is a graph of order $p = 2n$ satisfying (1), (2), and (4) and with $\kappa_1(G) = 2k-1$ for p sufficiently large. However, it is clear that G_3 contains at most $k-1$ edge–disjoint hamiltonian cycles.

(4) Let G_4 be the graph obtained from two copies of the complete graph K_n by adding edges from a fixed vertex v in the first copy of K_n to all of the vertices of the second copy of K_n, and edges from a second vertex in the first copy of K_n to $k-1$ vertices in the other copy of K_n. Then G_4 is a graph of order $p = 2n$ satisfying (1), (2), and (3) for p sufficiently large, and $\kappa_1(G - v) = k-1$. However, G contains at most $k-1$ edge–disjoint hamiltonian cycles.

A graph G of order $p \geq 3$ is *pancyclic* if contains a cycle of length r for $r = 3$, 4, ..., p. If each vertex of G lies on a cycle of every possible length, then G is *vertex–pancyclic*.

In [5] and [3] it was shown that the degree sum conditions in Theorems 2 and 3 are, in fact, "nearly sufficient" conditions for a graph to be pancyclic.

Theorem 12 *If G is a graph of order $p \geq 3$ such that $\deg u + \deg v \geq p$ for each pair u, v of nonadjacent vertices, then either G is pancyclic or else p is even and G is the graph $K_{p/2, p/2}$.*

Corollary *If G is a graph of order $p \geq 3$ such that $\deg u + \deg v \geq p + 1$ for each pair u, v of nonadjacent vertices, then G is pancyclic.*

Theorem 13 *If G is a 2–connected graph of order $p \geq 6$ such that $\deg u + \deg v + \deg w \geq 3p/2$ for each triple u, v, w of mutually nonadjacent vertices, then either G is pancyclic or else p is even and G is the graph $K_{p/2, p/2}$.*

Corollary *If G is a 2–connected graph of order $p \geq 6$ such that $\deg u + \deg v + \deg w > 3p/2$ for each triple u, v, w of mutually nonadjacent vertices, then G is pancyclic.*

Just as a slight increase in Ore's degree sum condition gives a sufficient condition for a graph to be pancyclic, so too with the neighborhood condition of Theorem 6 [11].

Theorem 14 *If G is a 2–connected graph of order p with $|N(u) \cup N(v)| \geq (2p + 5)/3$ for each pair u, v of nonadjacent vertices, then G is pancyclic.*

We have seen that, in the case of the properties of being hamiltonian and traceable, an imposed minimum degree condition allows the neighborhood condition to be relaxed. It is conjectured [11] that this is true for the property of being vertex pancyclic.

Conjecture *For a fixed integer $t \geq 2$, let G be a 2–connected graph of order p with $\delta(G) \geq t + 1$. If $|N(u) \cup N(v)| \geq p - t$ for each pair u, v of nonadjacent vertices of G, then G is vertex pancyclic.*

The conjecture has been verified for the case $t = 2$ [11].

Theorem 15 *If G is a 2–connected graph of order p with $\delta(G) \geq 3$ and such that $|N(u) \cup N(v)| \geq p - 2$ for each pair u, v of nonadjacent vertices, then G is vertex pancyclic.*

We note that if G is the 2–connected graph obtained from $K_{p-t-1} + \overline{K}_t$, $t \leq p/2$, by adding a new vertex which is adjacent to each vertex of the $\overline{K}_t$, then $\delta(G) = t$ and $|N(u) \cup N(v)| \geq p - t$ for each pair u, v of nonadjacent vertices. But G is not vertex pancyclic.

We next turn our attention to neighborhood conditions guaranteeing cycles and paths of at least a specified order [13].

Theorem 16 *Let G be a graph of order p satisfying $|N(u) \cup N(v)| \geq s$ for each pair u, v of nonadjacent vertices.*

> *(1)   If G is 2–connected, then G contains a path of order at least $3s/2 + 2$ or (if $p < 3s/2 + 2$) G is traceable.*
>
> *(2)   If G is 2–connected, then G contains a cycle of order at least $s + 2$ or (if $p < s + 2$) G is complete. If, in addition, s is odd and $p > s + 2$, then G contains a cycle of order at least $s + 3$.*

(3)   If G is connected, then G contains a path of order at least $s + 2$ or (if $p < s + 2$) G is complete. If, in addition, s is even and $p > s + 2$, then G contains a path of order at least $s + 3$.

(4)   If G is connected and $s \geq 3$, then G contains a cycle of order at least $(s + 2)/2$ or (if $p < s + 2$) G is complete.

(5)   If G is not connected, then at most one component of G has fewer than $(s + 2)/2$ vertices and if there is one, then it is complete. Every other component has at least $(s + 2)/2$ vertices, and is either complete or contains a path of order at least $s + 2$ and (if $s \geq 3$) a cycle of order at least $(s + 2)/2$..

The results of Theorem 16 are sharp, as the following examples indicate. In each case, the graph described has $|N(u) \cup N(v)| \geq s$ for each pair u, v of nonadjacent vertices.

(1) and (2): For s even, let G_1 be the graph $mK_{s/2} + K_2$, where $m \geq 4$. Then G_1 is 2–connected. Clearly, a longest path on G_1 has order $3s/2 + 2$ and a longest cycle has order $s + 2$.

(3) For s even, let G_2 be the graph $mK_{(s + 2)/2} + K_1$, where $m \geq 3$. Then G_2 is connected and a longest path in G_2 has order $s + 3$.

(4) For even $s \geq 4$, let G_3 be the graph obtained from $m \geq 1$ copies $H_1, H_2, ..., H_m$ of $K_{(s + 2)/2}$ by adding an edge joining a vertex of H_i and a vertex of H_{i+1} for i = 1, 2, ..., m − 1. Then G is connected, and a longest cycle in G_3 has order $(s + 2)/2$.

It is unknown if the results of Theorem 16 can be improved by imposing higher connectivity restrictions. For example, if G is a m–connected graph of sufficiently large order for which $|N(u) \cup N(v)| \geq s$, does G necessarily contain a cycle of order 2s ? The complete bipartite graph $K_{s, p-s}$ shows this would be best possible for $m \leq s \leq p/2$.

A *matching* in a graph G is a set of independent edges, i.e., a set of edges such that no two are adjacent. The *edge–independence number* $\beta_1(G)$ of G is the maximum number of edges in a matching in G. The next theorem [13] answers the questions: If $|N(u) \cup N(v)| \geq s$ for each pair u, v of nonadjacent vertices of a graph G, then how large is $\beta_1(G)$? Does the answer change if a connectivity condition is imposed?

Theorem 17 *Let G be a graph of order p such that $|N(u) \cup N(v)| \geq s$ for each pair u, v of nonadjacent vertices.*

(1)   If $s \leq (p - 2)/2$ then $\beta_1(G) \geq s$.

(2)   If $s \geq (p - 1)/2$ and p is odd, then $\beta_1(G) \geq (p - 3)/2$.

(3)   If $s > 2\lfloor p/3 \rfloor - 2$ and p is odd, then $\beta_1(G) = (p-1)/2$, unless $p = 5$ and $s = 1$.

(4)   If $s \geq (p-1)/2$ and p is odd and G is connected, then $\beta_1(G) = (p-1)/2$.

(5)   If $s \geq p/2$ and p is even, then $\beta_1(G) \geq (p-2)/2$.

(6)   If $s > 2(p-1)/3 - 1$ and p is even and G is connected, then $\beta_1(G) = p/2$.

(7)   If $s \geq p/2$ and p is even and G is 2–connected, then $\beta_1(G) = p/2$.

Certainly, $\beta_1(G) \leq \lfloor p/2 \rfloor$ for every graph G of order p. The following examples demonstrate the sharpness of the remaining parts of Theorem 17. Each graph satisfies the condition $|N(u) \cup N(v)| \geq s$ for all pairs u, v of nonadjacent vertices.

(1) The graph $K_{s,p-s}$, $s \leq p/2$, has edge–independence number s and indicates that the conclusion in (1) is best possible even for graphs with high connectivity.

(2) The graph $3K_n$ of order $p = 3n$ has edge–independence number $(p-3)/2$ for any odd integer n, and shows that (2) is best possible for values of s as large as $2\lfloor p/3 \rfloor - 2$.

(3) The graph $K_{1,4}$ is the excluded case in (3).

(5) The graph $K_1 \cup K_n$ (n odd) of order $p = n + 1$ has edge–independence number $(p-2)/2$, and shows that (5) is best possible for values of s as large as $p - 2$.

(6) The graph $3K_{n-1} + K_1$ (n even) shows that (6) is *not* true for $s = 2(p-1)/3 - 1$.

If G is a graph of even order p and M is a matching in G with $p/2$ edges, then M is called a *perfect matching* of G. We see from the previous theorem that if a connected graph G of even order p satisfies $|N(u) \cup N(v)| > 2(p-1)/3 - 1$, then G contains a perfect matching. We can, if fact, say more [19].

Theorem 18 *Let m be fixed positive integer. Then there is a constant $c = c(m)$ such that if G is a graph of sufficiently large even order p satisfying*

(1) $|N(u) \cup N(v)| \geq (2p + c)/3$ for each pair u, v of nonadjacent vertices,

(2) $\delta(G) \geq 2m$, and

(3) $\kappa_1(G) \geq m$,

then G contains m edge–disjoint perfect matchings.

Condition (3) is certainly sharp since any graph consisting of two odd order disjoint subgraphs with $m - 1$ edges between them contains at most $m - 1$ perfect matchings. Although condition (1) may not be sharp, it is of the correct order of magnitude as indicated by Theorem 17 (6). Similarly, if G has m perfect matchings, then $\delta(G) \geq m$ and so a condition like (2) is necessary.

The final result of this section involves the relationship between the neighborhood unions of pairs of nonadjacent vertices and "Menger path systems". For positive integers m and d, let $P_{d,m}$ denote the property that between each pair x, y of vertices of a graph G there are m vertex disjoint (except for x and y) paths, each of length at most d. For integers $d \geq 3$ and $m \geq 1$, let G_1 be the graph obtained from a path of order $d + 2$ by first replacing each of the first two and last two vertices with a copy of $K_{3t/2}$, t even, then replacing each of the remaining vertices with a copy of K_t, and finally adding all possible edges between "consecutive" complete graphs. If G is the graph $G_1 + K_{m-1}$, then G is an m–connected graph of order $p = (d + 4)t + m - 1$. Furthermore, $|N(u) \cup N(v)| \geq 5(p - m + 1)/(d + 4) + m - 3$ for each pair u, v of nonadjacent verices. However, if x is a vertex from the "first" copy of $K_{3t/2}$ and y is a vertex from the "last" copy of $K_{3t/2}$, then every $x - y$ path of length at most d contains a vertex of K_{m-1}. Consequently, G does not have property $P_{d,m}$. This example indicates that the neighborhood condition in the next theorem [18] is of correct order of magnitude.

Theorem 19 *Given integers $d \geq 2$ and $m \geq 1$, if G is an m-connected graph of sufficiently large order p with $|N(u) \cup N(v)| > 5p / (d + 2) + 2m$ for each pair u, v of nonadjacent vertices, then G satisfies $P_{d,m}$.*

3. All Sets of t Independent Vertices, $t \geq 2$

In Section 2 we examined sufficient conditions of the form $|N(u) \cup N(v)| \geq k$ for all sets S of two nonadjacent vertices, i.e., for all sets S of two independent vertices. An obvious question to ask is whether such conditions exist for sets S of t independent vertices, $t > 2$. It is not surprising that the number of known results here is small. In three instances of graphical properties, however, complete solutions have been obtained. The first of these generalizes Theorem 6 [20].

Theorem 20 *If G is a t–connected graph of order $p \geq 3$ and each set S of $t \geq 1$ independent vertices satisfies*

$$|N(S)| > t (p - 1)/(t + 1)$$

then G is hamiltonian.

This result is "nearly best possible" as is indicated by the graph G defined to be $(t + 1)K_q + K_t$. Then G is a t-connected graph of order $p = t + (t + 1)q$.

Furthermore, for each set S of t independent vertices, $|N(S)| = t + t(q - 1) = t(p-t)/(t + 1)$. However, G is not hamiltonian. We note that the case $t = 1$ in Theorem 20 is Dirac's theorem.

If $|N(S)|$ is "large enough" for sets S of t independent vertices of a graph G, then intuitively G contains a large number of well–distributed edges. Consequently, it is reasonable to expect the existence of a large complete subgraph in G. Such a result was established in [1].

Theorem 21 *Let t and m be fixed integers, $t \geq 1$ and $m \geq 3$. If G is a graph of sufficiently large order p such that each set S of t independent vertices satisfies*

$$|N(s)| > (m - 2)p / (m - 1)$$

then K_m is a subgraph of G.

If G is the complete $(m - 1)$–partite graph with $k \geq t$ vertices in each partite set (i.e., G is the Turan graph), then G has order $p = (m - 1)k$ and every set S of t independent vertices satisfies

$$|N(S)| = (m - 2)k = (m - 2)p / (m - 1).$$

However, K_m is not a subgraph of G. We note that the case $t = 1$ in Theorem 21 follows from the classic extremal result of Turan [27].

Theorem 21 was extended in [12], where a neighborhood union condition for independent t–sets was given that insures multiple copies of a complete graph of a fixed order.

Theorem 22 *Let t, m, n be fixed integers, $t \geq 1$, $m \geq 3$, $n \geq 1$. If G is a graph of sufficiently large order p such that each set S of t independent vertices satisfies*

$$|N(S)| \geq \frac{(m - 2)p + n}{m - 1}$$

then nK_m is a subgraph of G.

For fixed integers $t \geq 1$, $m \geq 3$, $n \geq 1$, let G be the graph $H + K_{n-1}$, where H is the complete $(m - 1)$–partite graph with $k \geq t$ vertices in each partite set. Then G has order $p = k(m - 1) + n - 1$. Furthermore, every set S of t independent vertices satisfies

$$|N(S)| = (m - 2)k + n - 1 = \frac{(m - 2)p + n - 1}{m - 1}.$$

But G does not contain nK_m as a subgraph. Thus the result in Theorem 22 is sharp.

In Section 2 we looked at neighborhood conditions based on all pairs of independent vertices and then, in this section, considered a generalization to all sets of t independent vertices. Very recently, a different direction from "all pairs of nonadjacent vertices" has been investigated by Lindquester [23]. For vertices u and v of a graph

G, let $d(u, v)$ denote the length of a shortest $u - v$ path in G. Then u and v are nonadjacent if and only if $d(u, v) \geq 2$.

Theorem 23 *If G is a 2–connected graph of order p such that* $|N(u) \cup N(v)| \geq (2p - 1)/3$ *for each pair u, v of vertices with* $d(u, v) = 2$, *then G is hamiltonian.*

Corollary *If G is a 2–connected graph of order p such that* $|N(u) \cup N(v)| > (2p - 4)/3$ *for each pair u, v of vertices with* $d(u, v) = 2$, *then G is traceable.*

Certainly Theorem 6 and its corollary follow from these two results. Moreover, Theorem 23 is an improvement of Theorem 6. For example, let I_r denote a graph with 2r vertices and r independent edges. Consider the graph G which consists of $I_r \cup K_{4r}$ plus a matching between the vertices of I_r and 2r of the vertices of K_{4r}. Then G has order $p = 6r$ and for vertices u, v with $d(u, v) = 2$, we have $|N(u) \cup N(v)| = 2p/3 > (2p - 1)/3$. Thus G is hamiltonian by Theorem 23. However, G does not satisfy the neighborhood condition of Theorem 6.

4. All Pairs of Vertices

The results of this section parallel, in some sense, those of Section 2. Here, however, we are considering neighborhood conditions of the type

$$|N(u) \cup N(v)| \geq k \text{ for } all \text{ pairs } u, v \text{ of vertices.}$$

The first results involve hamiltonian properties [9].

Theorem 24 *If G is a 2–connected graph of order p such that* $|N(u) \cup N(v)| \geq p/2$ *for each pair of u, v of vertices, then G is hamiltonian.*

Corollary *If G is a connected graph of order p such that* $|N(u) \cup N(v)| \geq (p - 1)/2$ *for each pair u, v of vertices, then G is traceable.*

Certainly $p/2$ cannot be lowered in Theorem 24 since, in fact, Dirac's minimum degree condition $\delta(G) \geq p/2$ is sharp. The graph $2K_{(p-1)/2} + K_1$ (p odd) shows that the connectivity condition cannot be relaxed.

In light of Theorem 24 one might hope that

$$|N(u) \cup N(v)| \geq (p + 1)/2 \text{ for each pair } u, v \, \varepsilon \, V(G) \rightarrow$$
$$G \text{ is hamiltonian–connected,}$$

since we know that

$$\delta(G) \geq (p + 1)/2 \rightarrow G \text{ is hamiltonian–connected}$$

and this minimum degree condition is best possible. At the present, the best result known is the following [15].

Theorem 25 *There is a constant c such that if G is a 3–connected graph of order p with $|N(u) \cup N(v)| \geq p/2 + c$ for each pair u, v of vertices, then G is hamiltonian–connected.*

Similarly, $\delta(G) \geq (p + 1)/2$ implies that G is pancyclic, and we have a corresponding neighborhood result [15].

Theorem 26 *There is a constant d such that if G is a 2–connected graph of order p with $|N(u) \cup N(v)| \geq p/2 + d$ for each pair u, v of vertices, then G is pancyclic.*

For shorter paths and cycles, the next two theorems [9] summarize the known results.

Theorem 27 *Let G be a connected graph of order p such that $|N(u) \cup N(v)| \geq s$ for each pair u, v of vertices and for some $3 \leq s < p/2$. Then G contains a path of order at least $2s - 1$ and a cycle of order at least s. Furthermore, $\beta_1(G) \geq s - 1$.*

Theorem 28 *Let G be a 2–connected graph of order p such that $|N(u) \cup N(v)| \geq s$ for each pair u, v of vertices and for some $3 \leq s \leq p/2$. Then G contains a path of order at least 2s and a cycle of order at least $2s - 2$. Furthermore, $\beta_1(G) \geq s$.*

For s, t ≥ 3, let G_1 be the graph $K_2 + tK_{s-2}$. Then G_1 is 2–connected, has order p = t(s − 2) + 2, satisfies $|N(u) \cup N(v)| \geq s$ for each pair u, v of vertices, and contains a cycle of order $2s - 2$ but no longer cycle. Thus the cycle result of Theorem 28 cannot be improved for $3 \leq s \leq (p + 4)/3$. For $3 \leq s \leq p/2$, let G_2 be the graph $K_s + \overline{K}_{p-s}$. Then G_2 is a 2–connected graph of order p and satisfies $|N(u) \cup N(v)| \geq s$ for each pair u, v of vertices. However, G_2 has no path of order $2s + 2$ and $\beta_1(G_2) = s$. Thus, only an improvement of 1 on the order of the path is possible in Theorem 28.

Theorem 28 gives a sufficient condition for a graph to contain a perfect matching. This was extended in [16].

Theorem 29 *For a fixed positive integer k, let G be a graph of even order p such that*
(1) $\delta(G) \geq k + 1$,
(2) $\kappa_1(G) \geq k$, and
(3) $|N(u) \cup N(v)| \geq p/2$ for each pair u, v of vertices.
If p is sufficiently large, then G contains k edge–disjoint perfect matchings.

Certainly condition (2) is necessary since any graph consisting of two odd order disjoint subgraphs with $k - 1$ edges between them contains at most $k - 1$ perfect matchings. For even p sufficiently large, the complete bipartite graph $K_{p/2 - 1, p/2 + 1}$ satisfies conditions (1) and (2), but not (3), and contains no perfect matchings. Finally, let G be any graph obtained by indentifying one vertex of a copy of $K_{p/2}$ with one vertex of another copy of $K_{p/2}$ and then adding a vertex x of degree k so that in the resulting graph, x is adjacent to the only vertex of degree $p - 1$. Then for $p \equiv 0 \pmod 4$ and p sufficiently large, G satisfies (2) and (3) but not (1), and the maximum number of edge–disjoint perfect matchings in G is $k - 1$.

The final all pairs neighborhood condition in this section is a sufficient condition for a graph to have property $P_{d,m}$ [17], i.e., for a graph to have the property that between each pair x, y of vertices there are m vertex disjoint (except for x and y) paths, each of length at most d.

Theorem 30 *Given integers $d \geq 2$ and $m \geq 1$ there is a constant $c = c(d, m)$ such that if G is an m-connected graph of order p with $|N(u) \cup N(v)| \geq 4p/(d + 2) + c$ for each pair u, v of vertices, then G satisfies $P_{d,m}$.*

Given integers $d \geq 2$ and $m \geq 2$ ($d \equiv 2 \pmod 4$), let H be the graph obtained from a cycle of order $d + 2$ by first replacing the first, second, fifth, sixth, ninth, tenth, etc. vertices with disjoint copies of K_n for some $n \geq 2$ and then adding all possible edges between "consecutive" complete graphs (including K_1) on the cycle. Define G to be the graph $H + K_{m-2}$. Then G is an m–connected graph of order

$$p = (n + 1)(d + 2)/2 + m - 2.$$

For all vertices u, v of G, $|N(u) \cup N(v)| \geq m - 2 + 2n + 1 =$

$$4p/(d + 2) + (md - 2m + 2 - 3d)/(d + 2).$$

However, G does not have property $P_{d,m}$. Thus, the order of magnitude of the previous theorem is correct.

There are relatively few known results of the type

$$|N(S)| \geq k \text{ for every set } S \text{ of } t \text{ vertices of } G \rightarrow$$

$$G \text{ has property } P$$

for $t > 2$. In the case of hamiltonian graphs we have the following extension of Theorem 24 [9].

Theorem 31 *Given a positive integer t there is a constant $c = c(t)$ such that if G is a 2–connected graph of order p with $\delta(G) \geq t$ and $|N(S)| \geq p/2 + c$ for each set S of t vertices, then G is hamiltonian.*

Some preliminary work has been done with neighborhood conditions based on all t–sets of vertices and the property $P_{d,m}$. Given integers m, t, and d with $2 \leq t \leq d - 1$, there are constants c_1 and c_2 depending only on m and d, and examples of m–connected graphs G of order p which satisfy

$$(1) \qquad |N(S)| \geq \max \begin{cases} 4p/(d + 4 - t) + c_1 \\ 5p(t - 1)/(d + 2)(t + 1) + c_2 \end{cases}$$

for each set S of t vertices. These graphs "just miss" satisfying property $P_{d,m}$. However, it is believed that (1) is of the correct order of magnitude for the desired sufficient condition for $P_{d,m}$.

5. Neighborhood Closures

In [6], a property P defined for all graphs of order p was called k–*degree stable* if, for all graphs G of order p, whenever G + uv has property P and deg u + deg v $\geq$ k, then G has property P. It was shown, for example, that the property of being hamiltonian is p–degree stable. The k–*degree closure* $D_k(G)$ of a graph G of order p is the graph obtained from G by recursively joining pairs of nonadjacent vertices whose degree sum is at least k (in the resulting graph at each stage) until no such pair remains. Since for $p \geq 3$ the complete graph of order p is hamiltonian, we have the following sufficient condition for a graph to be hamiltonian.

Theorem 32 *If G is a graph of order $p \geq 3$ such that $D_p(G)$ is complete, then G is hamiltonian.*

We note that Theorems 1 and 2 are immediate corollaries of this stronger degree result. However, the graph H described after the statement of Theorem 7 indicates that the neighborhood condition of Theorem 6 is distinct from the result in Theorem 32.

Analogously, let P be a property dfefined for all graphs G in a class Γ and let k be an integer. Then P is k-*neighborhood stable* in Γ if whenever G + uv has property P, where $G \varepsilon \Gamma$ and $|N(u) \cup N(v)| \geq k$, then G has property P [10]. The k–*neighborhood closure* $N_k(G)$ of a graph G is the defined to be the graph obtained from G by recursively joining pairs of nonadjacent vertices u, v for which the cardinality of the neighborhoods of u and v is at least k (in the resulting graph at each stage) until no such pair remains.

Theorem 33 *Let p, s, δ be integers satisfying $p \geq 2s \geq 2$. If Γ denotes the class of graphs of order p with $\delta(G) \geq \delta$ (connected graphs if p = 2s), then the property of containing sK_2 is $(2s - 1 - \delta)$–neighborhood stable in Γ.*

Corollary 1 *Let p, s, and δ be integers satisfying $p \geq 2s \geq 2$. If G is a graph of order p with $\delta(G) \geq \delta$ and $N_{2s-1-\delta}(G)$ is complete, then G contains sK_2 provided $n \geq 2s + 1$ or G is connected.*

Corollary 2 *Let p and s be positive integers satisfying $p \geq 2s$. If G is a graph of order p without isolated vertices for which $N_{2s-2}(G)$ is complete, then G contains sK_2 provided $p \geq 2s + 1$ or G is connected.*

The corresponding result for degree closure states that if G is a graph of order $p \geq 2s \geq 2$ and $D_{2s-1}(G)$ is complete, then G contains sK_2 (see [6]). If G is the graph obtained from two copies A and B of $K_s - e$, $s \geq 3$, where two vertices of degree $s - 2$, one from A and one from B, are joined by an edge and the remaining two vertices of degree $s - 2$ are joined by a path of length 2, $N_{2s-2}(G)$ is complete and consequently Theorem 33 gives that G contains sK_2. However, since $\Delta(G) = s - 1$, the $(2s - 1)$–degree closure $D_{2s-1}(G)$ is not complete and so the degree closure yields no information. By adding vertices to G, maintaining a maximum degree of at most $s - 1$, we obtain graphs G' of all orders $p \geq 2s + 1$ for which $N_{2s-2}(G')$ is complete but $D_{2s-1}(G')$ is not complete and is, in fact, just G'.

Theorem 34 *Let Γ denote the class of 2–connected graphs of order $p \geq 6$. Then the property of containing the cycle C_k, $6 \leq k \leq p$, is $(p - 2)$–neighborhood stable in Γ.*

Corollary *Let G be a 2–connected graph of order $p \geq 6$. If $N_{p-2}(G)$ is complete, then G contains the cycle C_k for $k = 6, 7, ..., p$.*

We note that the two previous results do not hold for values of $k < 6$.

The corresponding result for degree closure states that for integers p and k satisfying $p \geq s \geq 5$, if $D_{2p-s}(G)$ is complete, then G contains C_k (see [6]). For even $p \geq 8$, let G be the graph obtained from two copies A and B of $K_{p/2}$ by adding $p/2$ independent edges from the vertices of A to the vertices of B. Then $N_{p-2}(G)$ is complete and so, by Theorem 34, G contains cycles of lengths 6, 7, ..., p. However, since G is $p/2$–regular, $D_{2p-s}(G)$ is complete only if $s = p$.

REFERENCES

[1]   N. Alon, R.F. Faudree, Z. Furedi, A Turan–like condition and cliques in graphs. *Ann. NY Acas. Sci,* to appear.

[2]   I. Anderson, Sufficient conditions for matchings. *Proc. Edin. Math. Soc.,* 18 (1972) 129 – 136.

[3]   D. Bauer and E. Schmeichel, Pancyclic graphs and a degree condition of Bondy. Preprint.

[4]   J.A. Bondy, Longest paths and cycles in graphs of high degree. *Research report CORR 80 – 16,* University of Waterloo, Waterloo, Ontario.

[5]   J.A. Bondy, Pancyclic graphs I. *J. Combinatorial Theory* 11B (1971) 80 – 84.

[6]   J.A. Bondy and V. Chvatal, A method in graph theorey. *Discrete . Math.* 15 (1976) 111 – 135.

[7]   G. Chartrand and L. Lesniak, *Graphs & Digraphs,* Prindle, Weber, and Schmidt, Boston, 1986.

[8]   G.A. Dirac, Some theorems on abstract graphs, *Proc. London Math. Soc.* 2 (1952) 69 – 81.

[9]   R.J. Faudree, R.J. Gould, M.S. Jacobson, L.M. Lesniak, Neighborhood unions and a generalization of Dirac's theorem. Preprint.

[10] R.J. Faudree, R.J. Gould, M.S. Jacobson, L.M. Lesniak, Neighborhood closures for graphs. *Colloquia Mathematica Societatis Janos Bolyai* 52 (1987) 227 – 237.

[11] R.J. Faudree, R.J. Gould, M.S. Jacobson, L.M. Lesinak, Neighborhood unions and highly hamiltonian graphs. Preprint.

[12] R.J. Faudree, R.J. Gould, M.S. Jacobson, L.M. Lesniak, On a neighborhood condition implying the existence of disjoint complete graphs, *European J. Combinatorics,* to appear.

[13] R.J. Faudree, R.J. Gould, M.S. Jacobson, R.H. Schelp, Extremal problems involving neighborhood unions. *J. Graph Theory* 11 (1987) 555 – 564.

[14] R.J. Faudree, R.J. Gould, M.S. Jacobson, R.H. Schelp, Neighborhood unions and hamiltonian properties in graphs, *J. Combinatorial Theory B,* to appear.

[15] R.J. Faudree, R.J. Gould, L.M. Lesniak, Generalized degree conditions and hamiltonian properties in graphs. Preprint.

[16] R.J. Faudree, R.J. Gould, L.M. Lesniak, Neighborhood conditions and edge–disjoint perfect matchings. Preprint.

[17] R.J. Faudree, R.J. Gould, L.M. Lesniak, Neighborhood conditions and Menger path systems. Preprint.

[18] R.J. Faudree, R.J. Gould, R.H. Schelp, Menger path systems. Preprint.

[19] R.J. Faudree, R.J. Gould, R.H. Schelp, Neighborhood conditions and edge-disjoint hamiltonian cycles, *Congressus Numerantium* 59 (1987) 55 – 68.

[20] P. Fraisse, A new sufficient condition for hamiltonian graphs. *J Graph Theory* 10 (1986) 405 – 410.

[21] P. Hall, On representation of subsets. *J. London Math. Soc.* 10 (1935) 26 – 30.

[22] H. Li, Edge disjoint cycles in graphs. *J. Graph Theory,* to appear.

[23] T.E. Lindquester, The effects of distance and neighborhood union conditions on hamiltonian properties in graphs, *J. Graph Theory,* to appear.

[24] O. Ore, Hamiltonian connected graphs. *J. Math. Pures. Appl.* 42 (1963) 21 – 27.

[25] O. Ore, Note on hamilton circuits. *Amer. Math. Monthly* 67 (1960) 55.

[26] Posa, Hamiltonian circuits in random graphs. *Discrete Math.* 14 (1976) 359 – 364.

[27] P. Turan, Eine Extremalaufgabe aus der Graphtheorie. *Mat. Fiz. Lapok* 48 (1941) 436 – 452.

[28] D.R. Woodall, A sufficient condition for hamiltonian circuits. *J. Combinatorial Theory* B25 (1978) 184 – 186.

An Algorithm to Decide if a
Cayley Diagram is Planar

H. Levinson

Rutgers University

ABSTRACT

Given a presentation of a group on generators and defining relators, its associated Cayley diagram may or may not be planar. An algorithm is given which produces the set of all possible rotation sequences for embeddings, in the form of a tree. Only the defining relators are used, and a form of the word problem must be assumed solvable. The Cayley diagram is planar if and only if some "consistent" subset of these rotation sequences exists. Although Maschke in 1896 gave all finite Cayley diagrams, these results apply to any finite presentation with solvable word problems (most of which are infinite groups).

1. Introduction

The construction of the Cayley diagram of a presentation of a group is equivalent to solving the word problem for the presentation. Because of this, we must assume an effective solution of the word problem for a presentation in order to decide whether its Cayley diagram is planar. In particular, we shall confine out attention to those (finite) presentations for which the defining relators, considered as cyclic words, contain no proper segment equal to the group identity. We shall assume this condition throughout this article and not restate it again explicitly each time it is needed.

If we are confronted with a presentation P in which some defining relator R contains two or more segments $S_1,...,S_k$ which are themselves the identity, we may replace R by $S_1,...,S_k$ and obtain another presentation P' satisfying our condition and whose Cayley diagram is identical to that of P. For example, $P = \langle a,b;\ a^3 = a^4ba^{-1}b^{-1}a^3 = 1 \rangle$ has $aba^{-1}b^{-1}$ and a^3 both 1. $Q = \langle a,b;\ a^3 = a^3 = aba^{-1}b^{-1} = 1 \rangle$ has the same Cayley diagram as P.

2. Fundamental regions and effective construction.

We define the fundamental region F(P, g) at a point g of a Cayley diagram for P = $\langle x_1,...,x_n; R_1 = ... = R_k = 1 \rangle$ as follows.   Construct a colored directed cycle corresponding to each distinct cyclic permutation of the defining relators $R_1,...,R_k$. Each such cycle has a distinguished vertex from which the succession of edges, in some direction, is the sequence of generators, starting from the first, in the cyclic permutation of the relator it represents. Identify all these distinguished vertices together as one point g. If any two edges having the same direction and color are incident at the same point, coalesce them. When there are no more edges to coalesce, the remaining graph is our fundamental region. It is, in fact, a point g of the Cayley diagram C(P) of P with an incident cycle starting at g for each cyclic permutation of each defining relator of P. That F(P, g) is well defined results from the uniqueness of C(P) which, in turn, results from the group given by P being well defined.

The classic construction of C(P) starts with a tree T corresponding to the free group on $x_1,...,x_n$, and continues by identifying all points of T in equivalence classes of words which are "consequences" of the defining relators, $R_1,...,R_k$. The problem with this method is to discern which words are "consequences". A consequence is a juxtaposition of (a finite number of) conjugates of the defining relators and their inverses. After free cancellation (removing all adjacent pairs $x \cdot x^{-i}$), the remaining word may not be easily recognized. Moreover, there may be some very short words which only result from very large aggregates of conjugates of defining relators. "The word problem is not effectively decidable" means that there are presentations in which it is impossible to obtain an upper bound on the length of a string of conjugates of defining relators which collapses to a given word, should such a string exist.

We shall construct C(P) differently. Call a point p *saturated* if there is an entrant and a salient edge at p for each "color" $x_1,...,x_n$. In F(P, g), g is saturated. If a point p in F(P, g) is not saturated identify the g of a copy of F(P, g) with p, and coalesce adjacent edges having the same color and direction, as in the construction of F(P, g). We call this process *saturating p with F(P, g)*. Starting with F(P, g) and saturating each unsaturated point, including all newly introduced points, with F(P, g) results in C(P). We refer to this as the F(P, g) construction of C(P). Although this task is, in general, infinite, the condition we prescribe in the introduction gives us an advantage.

Theorem 1   *Each (finite) diagram obtained during the F(P, g) construction of C(P), in which all points are saturated, is a subgraph of C(P).*

Proof   Suppose there are two distinct points p and q of the saturated part of a diagram resulting during the F(P, g) construction of C(P) which will be identified in C(P). Then there is some finite sequence of saturations of points, starting with p, which results in p being identified with q, via edge identifications. However the identification of two saturated points will result in edge identifications leading to the collapse of some part of some F(P, g). This means that some cycle of F(P, g) has a proper segment equal to the group identity, which contradicts our introductory assumption on P.$\square$

Thus F(P, g) saturation, coupled with our assumption on the defining relators of P, gives us an effective method for constructing any finite portion of C(P). In addition, the identifications of edges during the process affects algebraic cancellation in a geometric manner.

3.   Embedding fundamental regions.

We shall show that is planar embeddings of F(P, g) exist and satisfy certain compatibility conditions, then any subgraph of C(P) is planar. This has already been exploited in the case where each generator appears exactly twice among all the symbols of all the defining relators of P. In this very special case, if C(P) is planar, it is point–symmetric planar or weakly point–symmetric planar and an algorithm to decide planarity has been given in [2]. This result made crucial use of the Edmonds–Heffter embedding technique [1].

The algorithm given in [2] has now been generalized using an idea due to A. Barnasconi of the Universidad Técnica Federico Santo Maria, Valparaiso, Chile. He suggested that the different occurrences of the generators in the defining relators be subscripted to distinguish them from each other. Figure 1 gives the algorithm which results from exploiting the adjacencies of generators at g prescribed by their adjacencies in the defining relators of P. If F(P, g) has a planar embedding, the Edmonds–Heffter embedding technique will produce it. Since the algorithm yields *all* embedding schemas (rotation systems), any particular embedding will appear. The algorithm produces a forest Φ of rooted trees of occurrences of generators within the defining relators. If there is some path, or some "feasible" collection of paths, in Φ which contains an instance of each occurrence of each generator in each defining relator of P, then F(P, g) may have a planar embedding. (The meaning of "feasible" will become clear later.) In such a planar embedding, should it be realizable, each relator cycle does *not* necessarily bound its own disk.

The sequence of generators is not necessarily the succession of edges appearing at a given vertex. Rather, a list of *successive inner corners of relator cycles* is what is

produced by the algorithm. These corners may be determined as follows from a
sequence, or collection of sequences, given by the forest. Number the entries of a
complete list of the defining relators and their distinct cyclic permutations. Next to each
subscripted generator with exponent +1 in the forest, write the number of (the cyclic
permutation of) the

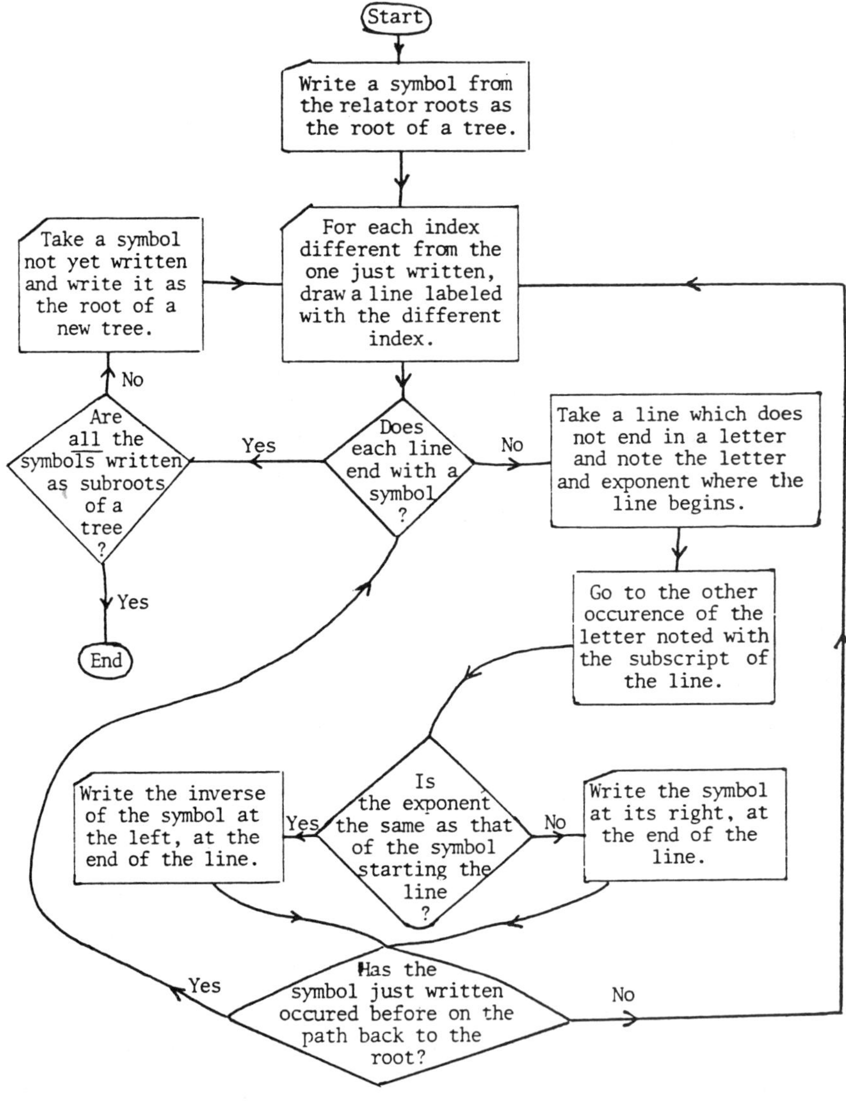

Figure 1

Take the set of all roots of the defining relators and subscript the distinct
occurrences of the generators in them. (i.e. ab is a root of $(ab)^{1\,3}$.) A symbol
will be considered distinct from its inverse.

defining relator in the numbered list, which starts with that numbered generator. Next to each generator with exponent -1, write the number of the relator which ends with that numbered generator. This induced list of relator numbers gives the relator corners as they appear around g in an embedding of $F(P, g)$. Some may be superimposed on others.

$$P = \langle a,\ b,\ c;\ abc = (ab)^3 = (ac)^3 = 1 \rangle$$

Subscripted roots:

$a_1 b_1 c_1$

$a_2 b_2$

$a_3 c_2$

Numbered cyclic permutations:

1 - abc

2 - bca

3 - cab

4 - $(ab)^3$

5 - $(ba)^3$

6 - $(ac)^3$

7 - $(ca)^3$

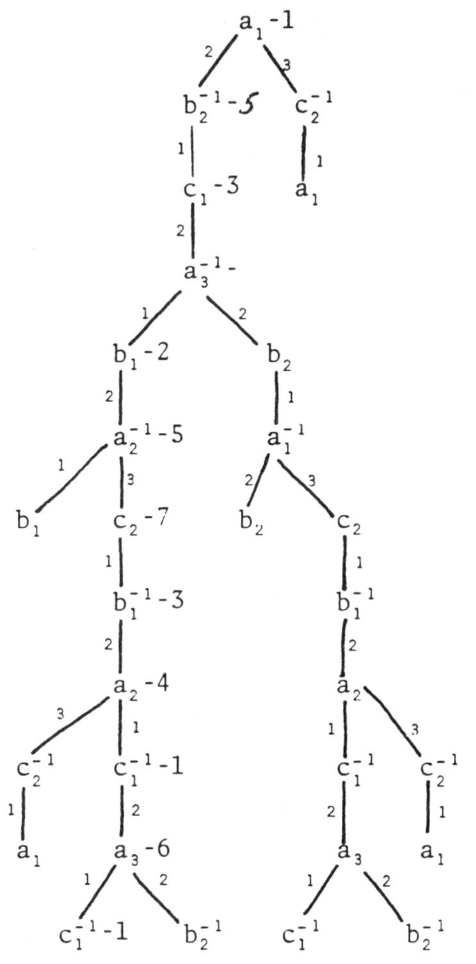

Figure 2

Figure 2 shows an example of a presentation for which the algorithm gives a single tree. The subscripted generators along a particular path have been numbered according to the directions given in the previous paragraph. Since numbers for all (7) distinct cyclic permutations of each defining relator appear on this list, there may be a planar embedding of F(P, g) which realizes this sequence of generators. A planar embedding of F(P, g) is given in Figure 3. Here the "sequence" involves jumps. From a_1, jump clockwise to b_2^{-1} to realize $4-(ab)^3$. From b_2^{-1}, go counterclockwise to c_1 to realize $3-cab$, etc. The list of relator cycles is also shown separately in Figure 3 since they are hard to discern individually within F(P, g).

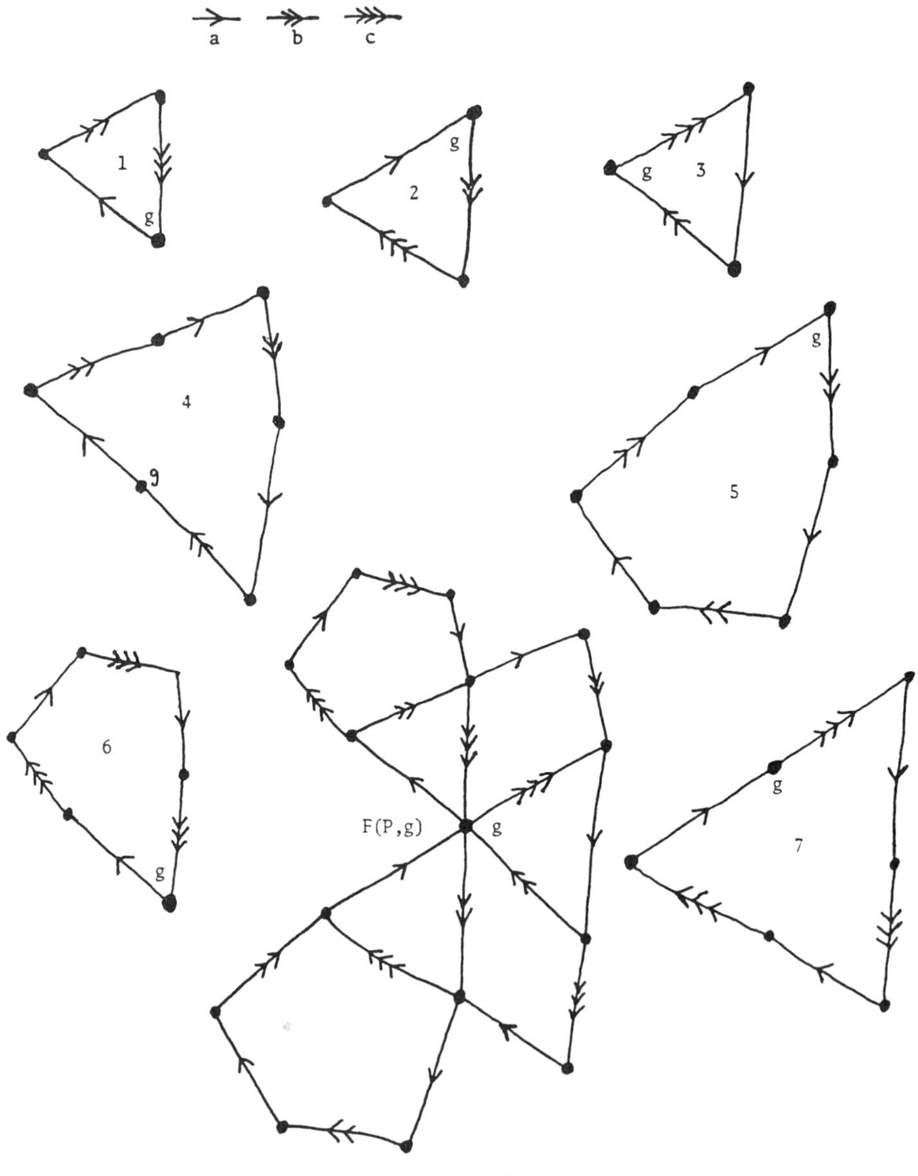

Figure 3

To produce a planar embedding, should one exist, construct cycles corresponding to the numbered defining relators, each with a distinguished vertex at the start of the leading generator of the relator it represents. Embed the first relator of the relator list induced by the path, or collection of paths, chosen in the forest. Next, superimpose the leading generator in the second relator from the list on the appropriate generator emanating from g in what has already been embedded. Unless there are adjacent edges of the same color and direction to be merged, there is now a choice of embedding the second relator clockwise or counterclockwise in various disks. Embed it whichever way does not give improper crossings with edges already embedded. (If several ways are possible, make separate embedding diagrams for all, each of which is to be continued, if possible.) Embed successive relators from the induced relator list similarly to the way in which the second was embedded. If it is impossible to continue at any step without introducing improper crossings, then that particular line of embedding, reflecting previous choices of cycle embeddings, is unfeasible. Should there be no way of continuing this embedding process to a complete embedding of F(P, g), then the path, or collection of paths, chosen from the forest does not result in a planar embedding, but some higher genus embedding. (Remember that if there is *some* planar embedding of F(P, g), the forest will yield it.)

If there is a collection of paths from the forest in which all the distinct cyclic permutations of the relators occur, it may be the case that the induced relator lists from these paths may be realized by a planar embedding. If so, we shall call the system of paths *feasible*. If the paths are relator–wise disjoint, the corresponding regions of F(P, g) they represent will be distinct blocks, and g will be a cut point. (all points in C(P) will be cut points.) In this case, any planar block may be embedded in any disk of any other, provided only that g be on the boundary of that disk (in order to identify the distinguished vertices of all the blocks.) Figure 4 shows an example of a presentation which yields a forest of paths. A collection of paths is indicated which gives the embedding of F(P, g) shown. (The presentation in Figure 4 is of the free group on 3 generators, *not* freely presented.)

It may seem that this algorithm is itself so complicated that one might just as well try all possible sequences of edges about a point by "brute force". However, reflection will show that $\frac{(2n-1)!}{2}$ sequences of edges must be examined for a presentation on n generators. In most cases, finding feasible sequences from the forest produced by the algorithm reduces the search for planar embeddings to a much smaller order of magnitude. Although the computational complexity of the algorithm presented is exponential, it is better than sheer brute force. Furthermore, it is reasonable to conjecture that the problem itself is only solvable in exponential time.

$$P = \langle a, b, c, d, e, f: \ adf^{-1} = bed^{-1} = cfe^{-1} = 1 \rangle$$

$$adf^{-1} \cdot fe^{-1}c = ade^{-1}c$$

$$ad \cdot d^{-1}be \cdot e^{-1}c = abc$$

Since abc is a consequence of the defining relators of P, adding it as a "new" relator won't change the Cayley diagram and each generator will appear at least twice. The subscripted roots are:

$$a_1 d_1 f_1^{-1} \qquad b_1 e_1 d_2^{-1} \qquad c_1 f_2 e_2^{-1} \qquad a_2 b_2 c_2$$

The forest is:

a_1	d_1	f_1^{-1}	b_1^{-1}	a_1^{-1}	a_2	b_2^{-1}	d_1^{-1}
$\mid$ 2	$\mid$ 2	$\mid$ 2	$\mid$ 2	$\mid$ 2	$\mid$ 1	$\mid$ 1	$\mid$ 2
c_2^{-1}	b_1	e_2^{-1}	c_2	b_2	f_1	e_1	e_1^{-1}
$\mid$ 1	$\mid$ 2	$\mid$ 1	$\mid$ 1	$\mid$ 1	$\mid$ 2	$\mid$ 2	$\mid$ 2
f_2	a_2^{-1}	d_2^{-1}	e_2	d_2	c_1^{-1}	c_1	f_2^{-1}
$\mid$ 1	$\mid$ 1	$\mid$ 1	$\mid$ 1	$\mid$ 1	$\mid$ 2	$\mid$ 2	$\mid$ 1
a_1	d_1	f_1^{-1}	b_1^{-1}	a_1^{-1}	a_2	b_2^{-1}	d_1^{-1}

The paths above the horizontal bracket form a compatible system yielding the planar embedding of $F(P, g)$ given below.

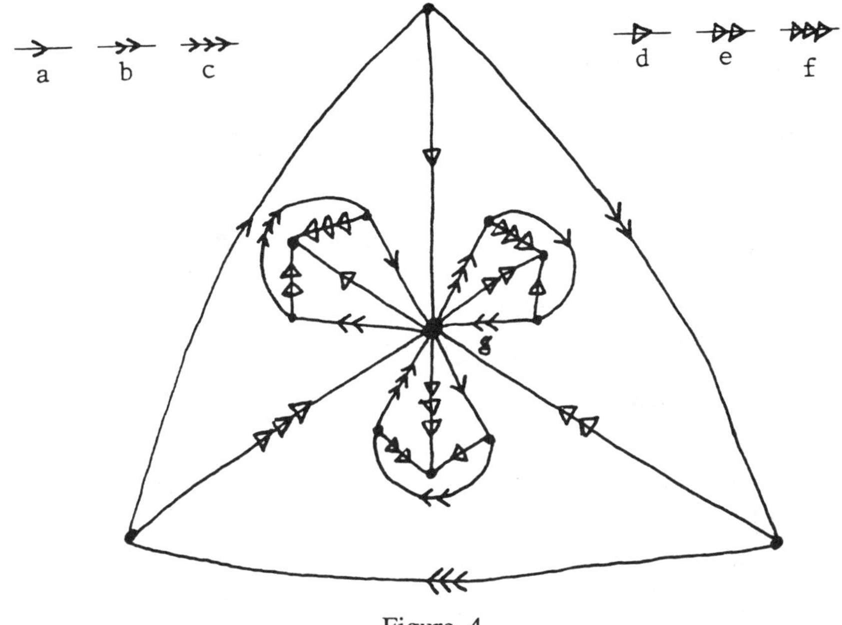

Figure 4

3. Compatible systems of embeddings.

We will now proceed from (local) planarity of F(P, g) to (global) planarity of C(P). We must devise a (finite) test to ensure that the planarity of an embedding of F(P, g) may be extended at each successive point saturation by F(P, g) to planar embeddings of the resulting regions of C(P).

Let us denote the sequence of successive edges about g, in a planar embedding of F(P, g) by s. We will not stipulate clockwise or counterclockwise, and we consider s as a cyclic word in the generators and inverses of generators of P, obviating the representation of cyclic permutations of s. We call the points in an embedding of F(P, g) whose sequences of edges are subsequences of s, *points of type s*. If there are points in our embedding of F(P, g) which are not of type s, we must find other new planar embeddings of F(P, g) such that their sequences are subsequences of the new sequences about g of these embeddings. Further, each "new" embedding must be inspected to discover if all points have sequences of edges about them which are of types already present. If not, we continue adding types (different embeddings of F(P, g)), if possible, until sequences exist about g to type all points present. Should there be no system of embeddings of F(P, g) for which each point in the embeddings is of one of the types of the embeddings, C(P) is not planar, although F(P, g) may be planar.

Should there exist a system of embedding for which each point is of the type of one of the embeddings, we must check each one to see if each point may be saturated using an embedding of F(P, g) of its type such that no types of embeddings conflict with each other on their intersections. If such is the case, we call the system of embeddings of F(P, g) *compatible*.

Theorem 2   *If a presentation P has a compatible system of planar embeddings of its fundamental region F(P, g), then its Cayley diagram may be embedded in the plane.*

Proof   We simply note that at each stage of the F(P, g) construction of C(P), each embedding of F(P, g) used to saturate an unsaturated point p, fits into the disks about p without improper crossings, as guaranteed by our compatibility assumption.◻

In the two examples shown, each point in the embedding of F(P, g) is of the same type as the sequence about g. It is easy to check that each respective embedding forms a self–compatible system, and in this case the Cayley diagrams for the presentations have point–symmetric embeddings.

REFERENCES

[1]   J. Edmonds, A combinatorial representation for polyhedral surfaces, *Notices Amer. Math. Soc.* 7 (1960) 646.

[2]   H. Levinson, Planarity of Cayley diagrams: weak point–symmetry, *Proc. 10th S-E Conf. Combinatorics, Graph Theory & Computing* (1979) 679–697.

CRITICALLY N-CONNECTED DIGRAPHS

W. Mader

Universität Hannover

Institut für Mathematik

Federal Republic of Germany

ABSTRACT

A graph or digraph G is critically n-connected, if it is n-connected, but for every vertex x of G, $G - x$ is not n-connected. In 1972, Chartrand, Kaugars, and Lick proved in [1] that every critically n-connected, finite (undirected) graph has a vertex of degree at most $\frac{3n}{2} - 1$. Generalizing this result it was shown in [5] that every critically n-connected, noncomplete, finite graph has a fragment of order at most $\frac{n}{2}$, where a fragment of a graph G of connectivity number n is an induced subgraph F of G of order less than $|G| - n$ such that F has exactly n neighbours in G. In this paper we will study the corresponding problems for directed graphs. We shall survey the results on critically n-connected digraphs and shall point out some open questions. We shall much simplify some proofs for results in [13] and shall prove some new results. Furthermore, we give counterexamples to a conjecture in [14].

1.   Introduction

First we need some definitions and notation. A digraph has neither loops nor multiple edges of the same direction. An edge (x, y) of a digraph D is called *symmetric*, if also the edge (y, x) from y to x belongs to D. A digraph without symmetric edges is called *antisymmetric*. For a digraph $D, \underline{D}$

811

denotes the multigraph arising from D by disregarding the orientation of the edges. The digraph $\overrightarrow{G}$ arises from the graph G by replacing every edge of G with a pair of oppositely directed edges; in particular, $\overrightarrow{K}_n$ denotes the complete digraph on n vertices. For a digraph, the term "connected" always means "strongly connected", and we use the terms "path", "circuit", and "component" always in the directed sense. A (simple) path P from vertex x_o to vertex x_n is written $x_o, x_1, ..., x_n$ and is called an x_o, x_n -path; for $0 \le i \le j \le n, P[x_i, x_j]$ denotes the subpath of P from x_i to x_j. For $A, B \subseteq V(D)$, an A, B - *path* P is an a,b-path for certain (not necessarily distinct) $a \in A$ and $b \in B$ such that $V(P-a) \cap A = \emptyset$ and $V(P-b) \cap B = \emptyset$ holds. An x, y-path P and an x', y'-path P' are *openly disjoint*, if $P \ne P'$ holds and $z \in P \cap P'$ implies $z = x = x'$ or $z = y = y'$. A (forward) *chord* of a path $x_0, x_1, ..., x_n$ in a digraph D is an edge $(x_i.x_j)$ of D such that $j \ge i + 2$ holds; if there is none, the path is called *chordless*. For $X \subseteq V(D)$, the subdigraph $D(X)$ induced by X is $D - (V(D) - X)$ and for a subdigraph $H \subseteq D$, set $D(H) := D(V(H))$. An *induced subdigraph* of D is a subdigraph of the form $D(X)$. A subdigraph H of a digraph D is called a *subcomponent of* D, if there is a component C of D so that $H = D(X)$ holds for an X maximal in $(\{Y \subsetneqq V(C) : Y \ne 0$ and $D(Y)$ connected$\}, \subseteq)$. Obviously, every vertex of a connected, finite digraph D of order at least 2 is contained in a subcomponent of D. For a subdigraph $F \subseteq D$ or $F \subseteq V(D)$, we define $N^+(F; D) := \{y \in V(D) : y \notin F$ and there is an $x \in F$ such that $(x, y) \in E(D)$ holds$\}$, and $N^-(F; D)$ is defined dually. For $x \in V(D)$, set $N^\epsilon(x; D) := N^\epsilon(\{x\}); D)$ for $\epsilon = +, -$ and $N(x; D) := N^+(x; D) \cup N^-(x; D)$, and $d^\epsilon(x; D) := |N^\epsilon(x; D)|$ denotes for $\epsilon = +$ the *outdegree* and for $\epsilon = -$ the *indegree of* x in D. For a digraph D, $D = \emptyset$ means $V(D) = \emptyset$ and $x \in D$ means $x \in V(D)$. For graphs or digraphs, $G_1 \cong G_2$ means that G_1 and G_2 are isomorphic. $\overline{K}_n$ denotes the graph of order n without edges and let L_m be an undirected circuit of length $m \ge 3$. For disjoint graphs G and H, the sum $G + H$ arises from

$G \cup H$ by adding all edges between G and H. If G and H are not disjoint, let $G + H$ be any graph isomorphic to $G' + H'$, where G' and H' are disjoint copies of G and H, respectively. Instead of $G_1 + G_2 + ... + G_k$ with $G_1 \cong G_2 \cong ... \cong G_k$, we write kG_1.

A digraph D is *n-connected* for a non-negative integer n, if it has more than n vertices, and for every $V' \subseteq V(D)$ with $|V'| < n, D - V'$ is connected. By Menger's Theorem, D is n-connected, iff for every pair of vertices x, y in D, there are n openly disjoint x, y-paths. The *connectivity number* $\kappa(D)$ of a digraph D is the largest n for which D is n-connected. (For the infinite digraphs we shall consider, such a maximal n will always exist.) In particular, $\kappa(\overleftrightarrow{K}_{n+1}) = n$ holds. A digraph D is *k-critically n-connected* for integers $0 \le k \le n$ (or *(n,k)-critical*), if $\kappa(D - V') = n - |V'|$ holds for every $V' \subseteq V(D)$ with $|V'| \le k$. In particular, for an (n, k)-critical digraph D, we have $\kappa(D) = n$. A digraph D is called *k-critical*, if it is $(\kappa(D), k)$-critical, it is called *critically n-connected*, if it is $(n, 1)$-critical, and it is called *critically connected*, if it is $(1, 1)$-critical. A *fragment of sign* $\epsilon \in \{+, -\}$ of a digraph D of (finite) connectivity number n is an induced subdigraph $F \ne \emptyset$ of D such that $|N^\epsilon(F; D)| = n$ and $V(F) \cup N^\epsilon(F; D) \subsetneq V(D)$ holds. A *positive fragment* is one of sign $+$ and a *negative fragment* is one of sign -, and a *fragment* is a positive or a negative fragment. If F is a positive fragment of D, then $\overline{F} := D - V(F) \cup N^+(F; D))$ is a negative fragment of D. Of course, every non-complete digraph of finite connectivity number has at least one positive fragment and at least one negative fragment.

Let $\mathbb{N}$ and $\mathbb{N}_n$ denote the set of positive integers and $\{m \in \mathbb{N} : m \le n\}$, respectively. $\mathbb{Z}_n$ denotes the integers modulo n. For a set M and $n \in \mathbb{N}$, define $(\binom{M}{n}) := \{' \subseteq M : |M'| = n\}$. Let us consider now some examples for critically n-connected digraphs.

2. Examples

(1) Let $m \ge n$ be positive integers and let $K_{m,m}$ be the complete bipartite graph with bipartition A, B into the independent vertex sets A

and B. Choose an n-factor F of $K_{m,m}$. The digraph $\vec{F}$ may arise from F by orienting all edges of F from A to B. Then the digraph $D :=$ $(A \cup B, E(\vec{F}) \cup (B \times A))$ is critically n-connected. Since $K_{m,m}$ is m-connected, we see that *for every $n \geq 1$, there are critically n-connected digraphs D so that the connectivity number of $\underline{D}$ is arbitrarily large.* In particular, there is no value $f(n)$ such that for all critically n-connected, finite digraphs D, $\min\limits_{x \in V(D)} (d^+(x; D) + d^-(x; D)) \leq f(n)$ holds.

(2) Let $n \geq 1$ and $m \geq 4$ be integers and choose $k \in \{3, ..., m-1\}$. Consider disjoint copies $D_1, ..., D_m$ of $\overleftrightarrow{K}_n$. The digraph D' arises from $\bigcup\limits_{i \in \mathbb{Z}_m} D_i$ by adding all edges $\bigcup\limits_{i \in \mathbb{Z}_m} V(D_i) \times V(D_{i+1})$. Then D' is critically n- connected. Every D_i is a positive and a negative fragment of D' is critically n- connected. Every D_i is a positive and a negative fragment of D' and every $x \in D$ has $d^+(x; D') = d^-(x; D') = 2n - 1$. If we add the further edges

$$\bigcup\limits_{\substack{i \in \mathbb{Z}_m \\ i \neq 1, k}} V(D_{i+1}) \times V(D_i) \text{ to } D',$$

getting D in this way, D remains critically n-connected. Now, D has exactly two positive fragments of order at most n (namely D_2 and D_{k+1}) and exactly two negative fragments of order at most n (namely D_1 and D_k), and these fragments have order n. In general, therefore, *in a critically n-connected, finite digraph we cannot find a vertex of outdegree or indegree less than $2n-1$ nor a fragment of order less than n.*

(3) Let $m \geq n \geq 2$ be integers. Consider a complete digraph $\overleftrightarrow{K}_m$ and subsets $S_1, ..., S_k \in \binom{V(\overleftrightarrow{K}_m)}{(}n$ such that $V(\overleftrightarrow{K}_m) = \bigcup\limits_{i=1}^{k} S_i$ holds. Let $C_1, ..., C_k$ be disjoint copies of $\overleftrightarrow{K}_n$ with $C_i \cap \overleftrightarrow{K}_m = \emptyset$ for $i \in \mathbb{N}_k$ and let f_i be a bijection of S_i to $V(C_i)$ for $i \in \mathbb{N}_k$. The digraph D may arise from $\overleftrightarrow{K}_m \cup \bigcup\limits_{i \in \mathbb{N}_k} C_i$ by adding all edges $\bigcup\limits_{i \in \mathbb{N}_k} \{(x, f_i(x)) : x \in S_i\}$

and $\bigcup\limits_{i\in\mathbb{N}_k} V(C_i) \times V(\overleftrightarrow{K}_m)$. The digraph D is critically n-connected, but $d^+(x; D) \geq m$ for all $x \in D$. Hence, *for every $n \geq 2$, there are critically n-connected, finite digraphs D with arbitrarily large $\min\limits_{x\in D} d^+(x; D)$.*

It is not an imperfection of our construction that we had to assume $n \geq 2$ in example 3, because it was shown in [13] that every critically connected, finite digraph has two vertices of outdegree 1. The proof given in [13] has become too complicated, because it took its rise as a by-product of the proof of the general case (see [14]). Se we shall give a simple proof here.

Theorem 1 Let C be a subcomponent of a critically connected digraph D of order at least 3. Then $m := |D - V(C)$ is finite and at least 2 and $D - V(C)$ has a chordless, hamiltonian path $P : x_1, x_2, ..., x_m$ such that $N^+(C; D) = \{x_1\}$ and $N^-(C; D) = \{x_m\}$ holds.

Proof Since D is critically connected, but $\kappa(C) \geq 1$ or $|C| = 1$ holds, we get $m \geq 2$. Of course, $N^\epsilon := N^\epsilon(C; D) \neq \emptyset$ holds for $\epsilon \in \{+, -\}$ and there is an N^+, N^- -path $P : x_1, \cdots, x_k$ in $D - V(C)$. Since $D(C \cup P)$ is connected and C is a subcomponent of D, P must be a hamiltonian path of $D - V(C)$, hence $m = k \in \mathbb{N}$. Since this is true for every N^+, N^- -path in $D - V(C)$, we see that $N^+ = \{x_1\}$ and $N^- = \{x_m\}$ holds and that P has no chord.

Remark Since, by Theorem 1, a subcomponent of a critically connected digraph D of order at least 3 is a positive (and a negative) connected fragment of D, Theorem 1 implies immediately Theorem 1 of [13].

Corollary 1 Every critically connected, finite digraph D of order at least 4 has disjoint, chordless paths P_1 and P_2, say, from x_i^+ to x_i^- for $i = 1, 2$ such that $x_i^+ \neq x_i^-$ and $N^\epsilon(D - V(P_i); D) = \{x_i^\epsilon\}$ holds for $i = 1, 2$ and $\epsilon = +, -$.

Proof We may assume that D is not a circuit. Then there is a subcomponent C of D with $|C| \geq 2$. By Theorem 1, there is an x_2^+, x_1^- -path P_1 with

$V(P_1) = V(D) - V(C)$ which has the properties described in Corollary 1. If there is a subcomponent of D containing P_1, then Theorem 1 provides an x_2^+, x_2^- -path P_2 disjoint from P_1 with the desired properties. So we may assume that every path P from $N^+(x_1^-; D) \cap C$ to $N^-(x_1^+; D) \cap C$ contains $V(C)$. There is such a path P_2 in C, say, an x_2^+, x_2^- -path. Then as above, P_2 has no chord and $N^\epsilon(P_1; D) = \{x_2^\epsilon\}$ holds for $\epsilon = +, -$, since $N^\epsilon(C; D) = \{x_1^\epsilon\}$ is true for $\epsilon = -, +$. Since $|C| \geq 2$, we have $x_2^+ \neq x_2^-$, and Corollary 1 is proved.

Since $d^+(x_i^+; D) = d^-(x_i^-; D) = 1$ holds for the endvertices of the x_i^+, x_i^- -path in Corollary 1, we get immediately as a slight improvement of Corollary 2 in [13].

Corollary 2   Every critically connected, finite digraph D of order at least 4 has 4 distinct vertices x_1, x_2, y_1, y_2 such that $d^+(x_i; D) = d^-(y_i; D) = 1$ holds for $i = 1, 2$.

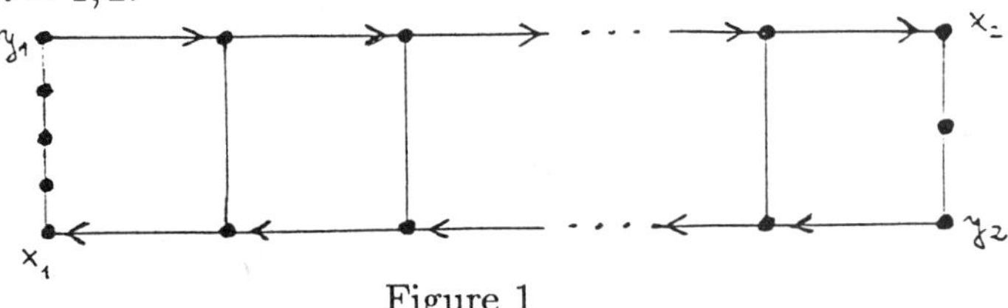

Figure 1

The second construction in example 2 for $n = 1$ shows that *for every integer $m \geq 4$, there are critically connected digraphs of order m containing exactly two vertices of outdegree 1 and exactly two vertices of indegree 1.* For $m = 3$ this is not true, because the circuit of length 3 is the only critically connected digraph of order 3.) Figure 1 displays a further critically connected digraph D with exactly two vertices x_1, x_2 of outdegree 1 and exactly two vertices y_1, y_2 of indegree 1 such that the vertex set $\{x_1, x_2, y_1 y_2\}$ is independent in D. (In figure 1, a pair of symmetric edges is drawn as an undirected edge.) It is not so by chance that D contains symmetric edges. For it is shown in [13] and easily derived from Corollary 1 that *every antisymmetric, critically connected, finite digraph D with exactly two vertices of outdegree 1* *must contain disjoint edges* (x_1, y_1) *and* (x_2, y_2) *such that* $d^+(x_i; D) = d^-(y_i; D) = 1$ *holds for* $i = 1, 2$. It is even possible to characterize all antisymmetric, critically connected, finite digraphs with exactly two vertices of outdegree 1. They are built up from almost 20 blocks in an analogous way as shown in Figure 2.

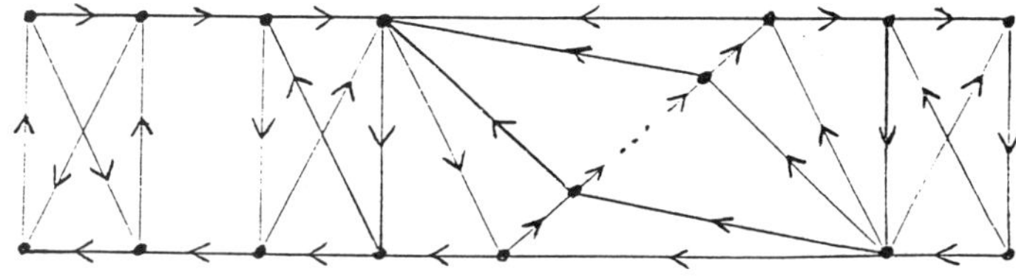

Figure 2

Let us turn now to critically n-connected digraphs for $n \geq 2$. Let us first have a glance at the undirected case. Chartrand et al. showed in [1] that every critically n-connected, non-complete, finite graph has a vertex of degree at most $\frac{3n}{2} - 1$. It was later proven in [5] that such a graph has a fragment F of order at most $\frac{n}{2}$. Of course, every vertex of such a fragment has degree at most $\frac{3n}{2} - 1$. As example 2 shows, in general, we cannot find a fragment of order less than n in a critically n-connected digraph. Example 3 shows that the minimal order of a positive fragment in a critically n-connected, finite graph is not bounded above by a function of n, because every vertex x of a positive fragment F of D satisfies $d^+(x; D) \leq |F| - 1 + \kappa(D)$. But there is a "small" positive or negative fragment in a critically n-connected, non- complete, finite digraph.

Theorem 2   [14]. Every critically n-connected, non-complete, finite digraph has a fragment of order at most n.

This had been conjectured by Y. O. Hamidoune in [2] and was proved for digraphs with vertex transitive automorphism group (which digraphs, of course, are 1-critical) in [4].

One of the main steps in the proof of Theorem 2 is the following lemma. For this we need further definitions. A set $\mathcal{F}$ of positive fragments of a digraph is called *non-crossing*, if for all $F, F' \in \mathcal{F}, F \cap F' \neq \emptyset$ implies $F \subseteq F'$ or $F' \subseteq F$. If F is a positive fragment and G is any fragment of D, then $F \leq G$ means $F \subseteq G$, if G is positive, and $V(F) \subseteq V(G) \cup N^-(G; D)$, if G is negative.

Lemma 1   [14]. Let G be a fragment of a finite digraph D of connectivity number N and let $\mathcal{F}$ be a non-empty, non-crossing set of positive fragments F' of D with $F' \leq G$. Define $\mathcal{F}_o := \{F \in \mathcal{F} : F \text{ minimal in } (\mathcal{F}, \subseteq)\}$ and assume

$$V(G) \cap \bigcup_{F \in \mathcal{F}_o} V(F) \subseteq \bigcup_{F \in \mathcal{F}} N^+(F; D).$$

Then $|\overline{G}| < n$ holds or there is an $F \in \mathcal{F}$ with $|F| \leq n$.

It is still more intricate to get such a system $\mathcal{F}$ we need for the application of lemma 1. Theorem 2 immediately implies

Corollary 3   [14. Every critically n-connected, finite digraph D contains a vertex x with $d^+(x; D) < 2n$ or $d^-(x; D) < 2n$.

This was also conjectured in [2]. Example 2 shows that this upper bound $2n - 1$ for $\min_{x \in D} \{d^+(x; D), d^-(x; D)\}$ is best possible, and example 3 shows that for $n \geq 2$, there is not always a vertex x in a critically n-connected, finite digraph D such that $d^+(x; D) < 2n$ holds.

For antisymmetric digraphs we get a better upper bound for the minimum degree, because an antisymmetric digraph F contains a vertex x with $d^+(x; F) \leq \frac{1}{|F|}\binom{|F|}{2}$ and a vertex y with $d^-(y; F) \leq \frac{|F|-1}{2}$. Hence Theorem 2 implies

Corollary 4   [14]. Every antisymmetric, critically n-connected, finite digraph D contains a vertex x with $d^+(x; D) \leq \lfloor \frac{3n-1}{2} \rfloor$ or $d^-(x; D) \leq \lfloor \frac{3n-1}{2} \rfloor$.

The following example shows that this bound is also best possible.

Example   (4). For $n \in \mathbb{N}$ and $m \geq 3$, define $H_n := (\mathbb{Z}'_n \{(x, x+i) : x \in \mathbb{Z}_n$ and $i = 1, ..., \lfloor \frac{n-1}{2} \rfloor\})$ and consider m disjoint copies $D_1, ..., D_m$ of H_n. The digraph $D := (\bigcup_{i \in \mathbb{Z}_m} V(D_i), \bigcup_{i \in \mathbb{Z}_m} E(D_i) \cup \bigcup_{i \in \mathbb{Z}_m} V(D_i) \times V(D_{i+1}))$ is critically n-connected and antisymmetric, and for every $x \in D, d^+(x; D) = d^-(x; D) = \lfloor \frac{3n-1}{2} \rfloor$ holds.

A digraph D is *n-edge-connected*, if $D - E'$ is connected for every $E' \subseteq E(D)$ such that $|E'| < n$, and it is called *minimally n-edge-connected*, if it is n-edge-connected, but for every $e \in E(D), D - e$ is not n-edge-connected, finite digraph D contains a vertex x such that $d^+(x; D) = d^-(x; D) = n$ holds. This result is now easily derived from Theorem 2 by the following construction.

Let D be a minimally n-edge-connected, finite digraph. We assign to every $x \in D$ a complete digraph $K(x)$ of order $d^+(x; D) + d^-(x; D)$ such that $K(x) \cap K(y) = \emptyset$ for $x \neq y$. We replace now every $x \in D$ with $K(x)$ and every $(x, y) \in E(D)$ with an edge from $K(x)$ to $K(y)$ in such a way that all edges corresponding to the edges of D are disjoint. The resulting digraph is critically n-connected, and an application of Corollary 3 easily provides

Corollary 5   [7, 14]. Every minimally n-edge-connected, finite digraph D has at least two vertices x such that $d^+(x; D) = d^-(x; D) = n$ holds.

It was conjectured in [11] that Corollary 5 is true also for minimally n-(vertex-) connected digraphs. This conjecture seems not derivable from Theorem 2 and remains open. - As in the undirected case, Theorem 2 and Corollaries 1 to 5 have no counterparts for infinite digraphs.

It is probable that in a critically n-connected, finite digraph, not only one outdegree or indegree is less than $2n$. But let us first consider the undirected case. It was proved in [3] that every critically n- connected, finite graph $(n \geq 2)$ has at least two vertices of degree at most $\frac{3n}{2} - 1$ and that for $n \geq 3$, there are critically n-connected, finite graphs of arbitrarily large order having exactly two such vertices. Generalizing this result, it was proved in [10] that every critically n-connected, non-complete, finite digraph has two disjoint fragments of order at most $\frac{n}{2}$.

Let us now turn to the directed case. The examples given in [14] suggested the conjecture that every critically n-connected, finite digraph D has at least four distinct pairs $(x, \epsilon) \in V(D) \times \{+, -\}$ with $d^\epsilon(x; D) < 2n$. This is true for $n = 1$, as stated in Corollary 2, but for $n \geq 2$, this conjecture is wrong, as the following examples show.

Examples (5) For integers $n \geq 2$ and $m \geq 3$, let $G_o, G_1, ..., G_m$ be disjoint graphs such that $G_o = (\{a\}, \emptyset)$ and $G_i \cong K_n$ for $i \in \mathbb{N}_m$ hold. Define $G := \bigcup_{i=1}^{m}(G_{i-1} + G_i)$ and choose any $b \in G_m$. The digraph D may arise from $\overleftrightarrow{G}$ by deleting the edges $\{(x, b) : x \in G_{m-1}\}$ and adding the edges $\{(a, x) : x \in \bigcup_{i=2}^{m} G_i\}$. The digraph D is n-connected. Take a copy $\overline{D}$ of D which is disjoint from D. The vertex $\overline{x} \in \overline{D}$ and the graph $\overline{G}_i$ may correspond to $x \in D$ and G_i, respectively. The digraph D_1 may arise from $D \cup \overline{D}$ by adding all edges $V(G_m) \times V(\overline{G}_m)$ and $V(\overline{G}_m) \times V(G_m)$. For $n = 2$ and $m = 3$, this construction is illustrated in Figure 3. *The digraph D_1 is critically n-connected. It has no vertex of outdegree less than $2n$ and exactly two vertices of indegree less than $2n$, namely the vertices a and $\overline{a}$.*

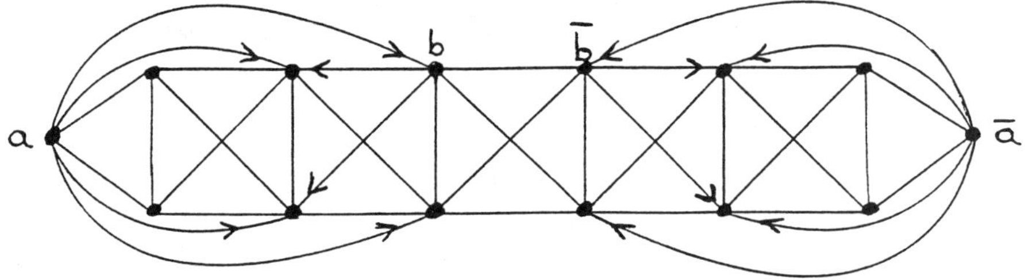

Figure 3

If we take in the above construction a disjoint copy of the dual D' of D instead of $\overline{D}$, we get a critically n-connected digraph D_2 with exactly two outdegrees or indegrees less than $2n$, one outdegree n and one indegree n.

(6) For integers $n \geq 2$ and $m \geq 2$, let $D_o, D_1, ..., D_m$ be disjoint digraphs such that $D_o = (\{a\}, \emptyset), D_1 \cong D_m \cong \overleftrightarrow{K}_{3n-1}$, and $D_i \cong \overleftrightarrow{K}_{4n-2}$ for $i = 2, .., m-1$ hold. For $i \in \mathbb{N}_m$, let A_i, A_i', B_i, B_i' be a partition of $V(D_i)$ such that for $i \in \mathbb{N}_m, |A_i| = |A_i'| = n$, for $i \in \mathbb{N}_{m-1}, |B_i| = n-1$, and $B_m = \emptyset$ hold. (Hence, $B_i' = \emptyset$ and for $i = 2, ..., m, |B_i'| = n-1$ holds.)

Define $A_o = A'_o = \{a\}$. The digraph D may arise from $\bigcup_{i=0}^{m} D_i$ by adding the edges $\bigcup_{i=1}^{m-1} B_i \times B'_{i+1}$ and $\bigcup_{i \in \mathbb{Z}_{m+1}} A_{i+1} \times A'_i$. For $n=2$ and $m=3$, this digraph D is displayed in Figure 4, where the edges of the complete digraphs D_1, D_2, D_3 are omitted. *The digraph D is critically n-connected, but every vertex of D has outdegree and indegree at least $2n$ with the only exception of the vertex a having $d^+(a;D) = d^-(a;D) = n$.*

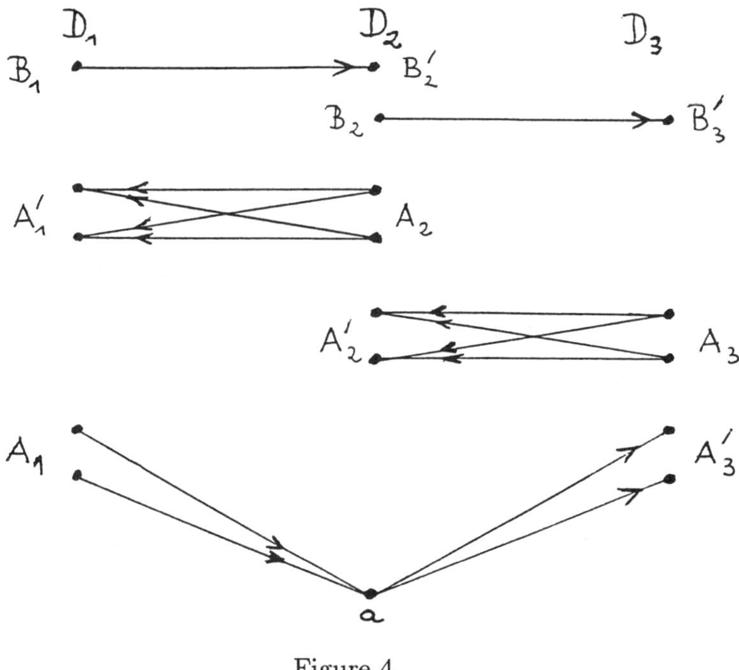

Figure 4

This example shows that in a critically n-connected, finite digraph D, there are not always two vertices x such that $\min\{d^+(x;D), d^-(x;D)\} < 2n$ holds. But the following conjecture should be true.

Conjecture 1 For every critically n-connected, finite digraph D, there are at least two pairs $(x, \epsilon) \in V(D) \times \{+, -\}$, such that $d^\epsilon(x;D) < 2n$ holds.

Example 6 also shows that in a critically n-connected, non-complete, finite digraph, one cannot always find two disjoint fragments of order at most n. The following conjecture, obviously, is stronger than conjecture 1 and, by example

5, in a sense, best possible.

Conjecture 2   If a critically n-connected, non-complete, finite digraph does not contain a negative fragment of order at most n, then it has two disjoint positive fragments of order at most n.

It was shown in [15] that there is no (n,n)-critical, finite graph G but K_{n+1}. This follows by considering a spanning tree T of G and deleting $\{x_1, ..., x_n\}$ from G, where x_i is an endvertex of $T - \{x_1, ..., x_{i-1}\}$. It was proved in [8] that every 2-critical graph is finite, but in [12], for every $n \in \mathbb{N}$, an example of an infinite, (n,n)-critical digraph was constructed. (In example 7 we shall give further constructions for infinite, (n,n)-critical digraphs.) Therefore, the question arises, if there are finite, (n,n)-critical digraphs for $n \geq 2$ besides $\overleftrightarrow{K}_{n+1}$. Applying Corollary 2, this question was answered in the negative in [13].

Theorem 3   [13] For $n \geq 2$, there is no finite, (n,n)-critical digraph besides $\overleftrightarrow{K}_{n+1}$.

The real reason for this fact had not become quite clear in the proof given in [13]. So we shall give here a proof more in the spirit of the proof in the undirected case, which makes this result intuitively clear. We need the following

Lemma 2   Let z be a vertex of a 2-connected digraph D and suppose there are disjoint subdigraphs H_1 and H_2 of $D - z$ such that for $i = 1, 2$, the digraph H_i contains vertices $x_i^+ \neq x_i^-$ with the following property: $N^\epsilon((D - z) - V(H_i); D - z) = \{x_i^\epsilon\}$ and $z \in N^\epsilon(x_i^\epsilon; D)$ for $\epsilon = +, -$.

Then $D - \{x_1^+, x_2^-\}$ is connected.

Proof   Consider any $x \in D - \{x_1^+, x_2^-, z\}$. We will show that there is an x, z-path in $D' := D - \{x_1^+, x_2^-\}$. Since D is 2-connected, there are openly disjoint x, z-paths P_1 and P_2 in D. We may assume that $x_1^+ \in P_1$ and $x_2^- \in P_2$ holds. If $x \notin H_2$, then $x_2^- \in P_2$ implies $x_2^+ \in P_2$, since $N^+(D - z - V(H_2); D - z) = \{x_2^+\}$. But then we get an x, z-path $P_2[x, x_2^+] \cup (x_2^+, z)$ in D', since $x_2^+ \neq x_2^-$ and $z \in N^+(x_2^+; D)$ holds. So we may assume $x \in H_2$. Then $x_1^+ \in P_1$ implies $x_2^- \in P_1$, since $x_1^+ \notin H_2$ and $N^-(D - z - N(H_2); D - z) = \{x_2^-\}$ holds. But $x_2 \in P_1$ contradicts $V(P_1) \cap V(P_2) = \{x, z\}$. In an analogous way, we see that there is a z, x-path in D'. Hence, D' is connected.

Proof of Theorem 3   It is enough to prove Theorem 3 for $n = 2$, since an (n,n)-critical digraph provides a $(2,3)$-critical one by deleting any $n - 2$ vertices. Let us suppose there is a $(2,2)$-critical, finite digraph D with $|D| \geq 4$. Since $D - x$ is critically connected for every $x \in D$ and since the only critically connected digraph of order 3 is a circuit of length 3, we must have $|D| \geq 5$. Choose $z \in D$. By Corollary 1, $D - z$ has two disjoint x_i^+, x_i^--paths P_i of length at least 1 such that $N^\epsilon(D - z - V(P_i); D - z) = \{x_i^\epsilon\}$ and $d^\epsilon(x_i^\epsilon; D - z) = 1$ holds for $\epsilon = +, -$ and $i = 1, 2$. Since D is 2-connected, $d_i^\epsilon; D - z) = 1$ holds for $\epsilon = +, -$ and $i = 1, 2$. Since D is 2-connected, $d^\epsilon(x_i^\epsilon; D - z) = 1$ implies $z \in N^\epsilon(x_i^\epsilon; D)$ for $\epsilon = +, -$ and $i = 1, 2$. So we can apply lemma 2 and get a contradiction.

Instead of deleting vertices, we may delete edges or vertices and edges. The case of deleting at least two edges is trivial: There are no graphs or digraphs besides L_m such that deleting any two edges, the connectivity number decreases by 2. (Consider any edge $e = [x, y]$ or $e = (x, y)$ of G and a system $P_1, ..., P_{\kappa(G)}$ of openly disjoint x, y-paths in G. Then for every $e' \in E(G) - \bigcup_{i=1}^{\kappa(G)} E(P_i), \kappa(G - \{e, e'\}) \geq \kappa(G) - 1$ holds). So only the case of mixed deletion remains. A graph or digraph G is called *vertex-edge-critical* or *v-e-critical*, if $\kappa(G) \geq 2$ and for all $v \in V(G)$ and $e \in E(G - v), \kappa(G - \{v, e\}) = \kappa(G) - 2$ holds. It is easy to see (in the same way as just before) that there are no infinite v-e-critical graphs or digraphs. The finite, v-e-critical graphs have been characterized in [6]: *The v-e-critical graphs are exactly the graphs* $G_{m,k,1} := k\overline{K}_m + 1L_{m+2}$, *where* $m \geq 1, k, 1$ *are non-negative integers such that* $\kappa(G_{m,k,1})^{\geq 2}$ *holds.* (The proof given in [6] was not so easy, but using Theorem 4 in [9], we get a straightforward proof.)

Of course, the digraphs $\overleftrightarrow{G}_{m,k,1}$ are v-e-critical. So the question arises, if there are still further ones. In an analogous way as in (a_2) on page 277 in [6], one can see that every vertex of a v-e-critical digraph of connectivity number n has outdegree and indegree equal to n. Using results of [11], one can show that every v-e-critical digraph must have a symmetric edge, but I do not know, if every edge must be symmetric. But this is the case for $n = 2$, as seen in the following way. Let D be a v-e- critical digraph of connectivity number 2. For $x \in D, D - x$ is minimally connected and, therefore, has two vertices of indegree and outdegree 1, by Corollary 5. Hence, the 4 edges incident to x are symmetric and $D = \overleftrightarrow{L}_{|D|}$ holds.

For every cardinal $\aleph_\alpha$ and every $n \in \mathbb{N}$, an (n,n)-critical digraph of order $\aleph_\alpha$ was constructed in [12]. There is not much hope to able to describe the structure of all (n,n)-critical digraphs for $n \geq 2$ precisely, but we can say something about the degrees in such digraphs. The examples of (n,n)-critical digraphs of order $\aleph_\alpha$ given in [12] have in every vertex outdegree equal to $\aleph_\alpha$ and in some vertices indegree equal to $\aleph_\alpha$. The next result shows that for $n \geq 3$, it was not so by chance that every vertex had outdegree equal to $\aleph_\alpha$.

Proposition 1   Let D be an (n,n)-critical digraph of order $\aleph_\alpha$ for an integer $n \geq 2$. Then for every $x \in V(D), d^+(x; D) + d^-(x; D = \aleph_\alpha$ holds. For $n \geq 3$, every vertex of D has outdegree $\aleph_\alpha$ or every vertex of D has indegree $\aleph_\alpha$.

Proof   For the first and for the second assertion, it is enough to consider the case $n = 2$ and $n = 3$, respectively. Choose $n \in \{2,3\}$ and let D be an n-connected digraph of order $\aleph_\alpha$ which contains vertices x,y such that $^-(x; D) < \aleph_\alpha$ and $d^+(y; D) < \aleph_\alpha$ holds, where x=y for $n = 2$ and $x \neq y$ for $n = 3$ is assumed. For every pair of vertices $a \in N^-(x; D)$ and $b \in N^+(y; D)$, there is an a,b-path $P_{a,b}$ in $D - \{x,y\}$, since $D - \{x,y\}$ is connected. Since $V' := \bigcup_- V(P_{a,b})$ has less than $\aleph_\alpha$ elements,

$$a \in N^-(x; D)$$

$$b \in N^+(y; D)$$

there is a $z \in V(D) - (V' \cup \{x, y\})$. If $D' := D - \{x, y, z\}$ were not connected, there were a partition $A \neq \emptyset, B \neq \emptyset$ of $V(D')$ such that $N^+(A; D') = \emptyset$. Then $N^+(A; D) = \{x, y, x\} = N^-(B; D)$ holds, since D is n-connected. But this implies $N^-(x; D) \cap A \neq \emptyset$, say $a \in N^-(x; D \cap A$, and $N^+(y; D) \cap B \neq \emptyset$, say $b \in N^+(y; D) \cap B$. But $P_{a,b} \subseteq D'$ holds, contradicting $N^+(A; D') = \emptyset$. Hence, D' is connected and D is not (n,n)-critical. From this, Proposition 1 follows.

Obviously, Proposition 1 is not true for $n = 1$. The examples given in [12] show that not at all every vertex of an infinite, (n,n)-critical digraph must have infinite out-degree and infinite indegree. *For every $n \geq 2$, we construct now even an example of an infinite, (n,n)-critical digraph where every vertex has indegree n.* (Of course, such a digraph cannot exist for $n = 1$.) Such a digraph must be countable and have infinite outdegree in every vertex by Proposition 1. Subsequently, we shall give an example which shows that there is no integer $f(n)$ such that $\min_{z \in D}\{d^+(z; D), d^-(z; D)\} \leq f(n)$ holds for every countable, (n,n)-critical digraph D.

Example 7   Let n be an integer at least 2 and set $C_o := \{z \in \mathbb{Z} : z \equiv 0 (\mathrm{mod}\ n + 1)\}$. First we consider the digraph $D_o := (\mathbb{Z}, \{(z, z + i) : z \in \mathbb{Z}$ and $i \in \mathbb{N}_{n-1}\} \cup \{(z, z + n + 1) : z \in \mathbb{Z} - C_o\})$. The digraph D_o has exactly one hamiltonian path and for every $T \subseteq \mathbb{Z}$ with $|T| \leq n - 2, D_o - T$ is still hamiltonian. If $D_o - T$ is not hamiltonian for a $T \in (\mathbb{Z}_{n-1})$, then T is an interval of $\mathbb{Z}$, say, $T = \{z_o + i : i \in \mathbb{N}_{n-1}\}$, and $D_o - (T \cup \{z_o\})$ or $D_o - (T \cup \{z_o + n\})$ is hamiltonian. Let $C_1, C_2, \dots$ be a partition of C_o such that for every $i \in \mathbb{N}, C_i$ is not bounded below, and let $z_o := \mathbb{Z}, z_1, z_2, \dots$ be disjoint countable sets. For $k \in \mathbb{N}$, let f_k be a bijection of $\left(\left(\bigcup_{i<k} z_i\right)_n\right)$ onto z_k. For $i \in \mathbb{N}$, we define now inductively digraphs

$$D_i := (V(D_{i-1}) \cup z_i, E(D_{i-1}) \cup \bigcup_{M \in \binom{V(D_{i-1})}{n}}$$

$$\{(x, f_i(M) : x \in M\}).$$

In $\tilde{D} := \bigcup_{i \in \mathbb{N}} D_i$, the vertices of C_o have indegree $n-1$, but all other vertices have indegree n. For every $x \in C_o$, we define exactly one further incoming edge in such a way that the arising digraph D becomes n-connected. For $x \in V' := \bigcup_{i \in \mathbb{N}} z_i$, define $B(x) := \{z \in \mathbb{Z} : $ there is a z,x-path in $\tilde{D} - (\mathbb{Z}^{i \in \mathbb{N}}\{z\})\}$. For all $x \in V', B(x)$ is finite and $|B(x)| \geq n$ holds. Choose $i \in \mathbb{N}$. We define now a bijection $h_i : Z_i \to C_i$. Write Z_i in a sequence $z_1, z_2, \ldots$. Since $B(z_1)$ is finite and C_i is not bounded below, $\{x \in C_i : x < \min B(z_1)\}$ is not empty and has a maximum which we denote by $h_i(x_1)$. Assume $h_i(z_i), \ldots, h_i(z_m)$ are defined. Then define $h_i(z_{m+1}) := \max\{x \in C_i - \{h_i(z_1), \ldots, h_i(z_m)\} : x < \min B(z_{m+1})\}$. Of course, $h_i : Z_i \to C_i$ is injective. But h_i is also subjective, because for every $x \in C_i$, $\min B(f_i(\{x+1, \ldots, x+n\})) = x+1$ holds. Define $E_i := \{(z, h_i(z)) : z \in Z_i\}$ and $D := \tilde{D} \bigcup_{i \in \mathbb{N}} E_i$. Then $d^-(z; D) = n$ for every $z \in D$, and D is an antisymmetric digraph.

We shall prove now that D is (n,n)-critical. Since for every $T \in \binom{V(D)}{n}$, $D - T$ has a vertex of indegree 0, it is enough to show that D is n-connected. Choose $T \in \binom{V(D)}{n-1}$. We will show that $D' := D - T$ is connected. First we prove

(α) All vertices of $D_o - T$ belong to the same component of D'.

Proof of (α)   Consider $x' < x$ in $D_o - T$. Since $N^+(x; D) \cap z_1$ is infinite and for all $y \in N^+(x; D) \cap Z_1, h_1(y) < x$ holds, there is a path (of length 2) in D' from x to a vertex $x'' < x'$ in $D_o - T$. This implies (α), if $D_o - T$ is hamiltonian. Suppose that $D_o - T$ is not hamiltonian. Then $T \subseteq \mathbb{Z}$ and T is an interval, say $T = \{z_o + i : i \in \mathbb{N}_{n-1}\}$, and $D_o - (T \cup \{z_o\})$ or $D - (T \cup \{z_o + n\})$ has a hamiltonian path P. Since P belongs again to the same component of D', it suffices to show that there is a $z_{o'}(z_o + n)$-path in D'. If z_o is not on P, then $(z_o + n) \in P$ and there is a path from z_o to an $x < z_o$ in D', as we have seen above. But $x \in P$ holds and, therefore, a $z_{o'}(z_o + n)$ - path exists in D'. Assume $z_o + n$ is not on P. Then z_o is on P. If $(z_o - 1, z_o + n) \in E(D_o)$ holds, then $D_o - (T \cup \{z_o\})$ is hamiltonian and we return to the case just considered. Hence we may assume $(z_o - 1, z_o + n) \notin E(D_o)$, i.e. $z_o + n \in C_o$. So there is an $i \in \mathbb{N}$ with $z_o + n \in C_i$ and a $z \in Z_i$ with $h_i(z) = z_o + n$. Then $z_o + n < \min B(z)$ and so $B(z) \subseteq V(P)$ holds. Since $(z, z_o + n) \in E(D')$, we get a $z_{o'}(z_o + n)$-path in D' via any $y \in B(z)$.

Since for every $x \in V'$, there are infinitely many disjoint edges from $N^+(x; D)$ to $\mathbb{Z}$, and n disjoint $\mathbb{Z}, N^-(x; D)$-paths in D, (α) immediately implies that D' is connected. We will show now that by addition of further edges we get an (n,n)-critical digraph from D which has also infinite indegree in every vertex. Consider any $i \in \mathbb{N}$ and any $x \in Z_i$. Defining $F(x) := \{f_{i+1}(M) : x \in M \in \binom{N^-(x;D) \cup \{x\}}{n}\}$, we can form $E(x) := \{(f_j(F(x)), x) : j \geq i + 2\}$, since $|F(x)| = n$ holds. The digraph $\overline{D} := (V(D), E(D) \cup \bigcup_{x \in V'} E(x) \cup \bigcup_{x \in \mathbb{Z}} \{(y, x) : y \in V' - N^+(x; D)\})$ is antisymmetric and has infinite indegree in every vertex. We will show that it is still (n,n)-critical. Choose any $T \in \binom{V(D)}{n}$. We define vertex sets $T_o := T \subseteq T_1 \subseteq T_2 \subseteq \cdots$ inductively by $T_i := T_{i-1} \cup \{x \in V' : N^-(x; D) \subseteq T_{i-1}\}$ for $i \in \mathbb{N}$. Then $\overline{T} := \bigcup_{i \in \mathbb{N}} T_i$ is an infinite, proper subset of $V(D)$. Suppose there is an edge $(y, x) \in E(\overline{D})$ such that $y \in V(D) - \overline{T}$ and $x \in \overline{T} - T$ holds. Then $x \in V'$, say $x \in Z_i$ for an $i \in \mathbb{N}$. Since $N^-(x; D) \subseteq \overline{T}$ for every $x \in \overline{T} - T$, we see $(y, x) \in E(\overline{D} - E(D))$, hence $(y, x) \in E(x)$. Hence, there is a $j \geq i + 2$ such that $f_j F(x)) = y$ holds. But $x \in \overline{T} - T$ implies $N^-(x; D) \subseteq \overline{T}$, therefore, also $f_{i+1}(M) \in \overline{T}$ for all $M \in \binom{N^-(x;D) \cup \{x\}}{n}$. So $F(x) \subseteq \overline{T}$ and hence $y = f_j(F(x)) \in \overline{T}$ follows. This contradiction shows $N^-(\overline{T} - T; \overline{D} - T) = \emptyset$. So $\overline{D}$ is (n,n)-critical.

Using this last idea, *for every cardinal* $\aleph_\alpha$ *and every* $n \in \mathbb{N}$, *it is possible to construct an (n,n)-critical digraph of order* $\aleph_\alpha$ *where every vertex has outdegree and indegree equal to* $\aleph_\alpha$. For this, one has to take for D a digraph built as is [12], but with the following modification: In every step, split z_i into $\aleph_\alpha$ many disjoint subsets $z_{i\beta}$ of cardinality $\aleph_\alpha$ and for every β, use a bijection $f_{i\beta}$ of $\left(\bigcup_{j \underset{\tilde{n}}{\leq} i} z_j\right)$ onto $z_{i\beta}$.

It remains open, if for any cardinal $\aleph_{\alpha'}$ there are (2,2)-critical digraphs of order $\aleph_\alpha$ which have vertices of outdegree less than $\aleph_\alpha$ and vertices of indegree less than $\aleph_\alpha$.

REFERENCES

[1] G. Chartrand, A. Kaugars, and D. R. Lick, "Critically n-connected graphs," *Proc. Amer. Math. Soc.* 32 (1972), 63-68.

[2] Y. O. Hamidoune (Mohamedou Yahya Ould Elmoctar), Contribution à l'étude de la connectivité d'un graphe, Thèse d'Etat, Paris 1980.

[3] Y. O. Hamidoune, "On critically h-connected simple graphs," *Discrete Math.* 32 (1980), 257-262.

[4] Y. O. Hamidoune, "Quelques problèmes de connexité dans les graphes orientés," *J. Combinatorial Theory Ser.* B 30 (1981), 1-10.

[5] W. Mader, "Eine Eigenschaft der Atome endlicher Graphen," Arch. Math. 22 (1971), 333- 336.

[6] W. Mader, "1-Faktoren von Graphen," Math. Ann. 201 (1973), 269-282.

[7] W. Mader, "Ecken vom Innen-und Aubengrad n in minimal n-fach kantenzusammenhängenden Digraphen," Arch. Math. 25 (1974), 107-112.

[8] W. Mader, "Endlichkeitssätze für k-kritische Graphen," Math. Ann. 229 (1977), 143-153.

[9] W. Mader, "Zur Struktur minimal n-fach zusammenhängender Graphen," Abh. Math. Sem. Univ. Hamburg 49 (1979), 49-69.

[10] W. Mader, "Disjunkte Fragmente in kritisch n-fach zusammenhängenden Graphen," *Europ. J. Combinatorics* 6 (1985), 353-359.

[11] W. Mader, "Minimal n-fach zusammenhängende Digraphen," *J. Combinatorial Theory Ser.* B 38 (1985), 102-117.

[12] W. Mader, "On infinite, n-connected graphs," to appear in the Proceedings of the Workshop on "Cycles and rays" in Montreal 1987.

[13] W. Mader, "On critically connected digraphs," to appear in the Journal of Graph Theory.

[14] W. Mader, "Ecken von Kleinem Grad in kritisch n-fach zusammenhängenden Digraphen," submitted to the Journal of Combinatorial Theory Series B.

[15] St. Maurer and P. Slater, "On k-critical, n-connected graphs," *Discrete Math.* 20 (1977), 255-262.

Walks in Covering Spaces

Bennet Manvel* and Richard Osborne
Colorado State University

ABSTRACT

Many topological problems on covering spaces can be restated as problems involving homomorphic images of graphs or directed graphs. In this paper we consider homomorphisms between directed graphs which are k–regular. The principal question is: which homomorphisms between such digraphs allow the lifting of all walks? We cannot answer that question, but we have found some surprising conditions on lifting homomorphisms. In the last section we briefly discuss some related questions, including: which digraphs have the same number of walks of every length?

1. Introduction

We begin with basic definitions. For this paper, a *digraph* may have loops and multiple arcs. A *homomorphism* between digraphs is any onto map sending vertices to vertices and arcs to arcs, preserving directions. (We do not require that the homomorphism be a local isomorphism, as is sometimes done when considering coverings of graphs, e.g. [2].) In the case of multiple arcs in the image, it is necessary to specify the image arc, rather that just the vertex map. We will be working with homomorphisms which are onto. If there is a homomorphism from D* to D then D* is said to *cover* D. A *walk* in a digraph is an alternating sequence of vertices and arcs $v_0, x_1, v_1, ..., x_n, v_n$ in which each arc x_i is directed from v_{i-1} to v_i. A digraph is *k–regular* if there are exactly k arcs into each vertex and k arcs out of each vertex. We call a digraph *regular* if it is k–regular for some k. The following definition is central to our discussion.

Definition Given a homomorphism h from D* to D, and a walk W then W *lifts* if there is some walk W* in D* such that h(W*) = W. We say that W* is a *lifting* of W.

Because homomorphisms allow identification of vertices, some walks in the digraph D, covered by D*, may not lift to walks in D*. If every walk in D lifts to a walk in D*, we say that the homomorphism h has the *lifting property*. (A topologist would say it has the path lifting property, but a topologist's path is a graph theorist's walk. We will not use the term path in this paper, to avoid confusion.)

When does a homomorphism h between digraphs D* and D have the lifting property? Why would we like to know? We cannot answer the first question, although the rest of this paper gives some rather surprising partial results. It is easier to answer the second question, by citing applications to topology and computer science.

Covering spaces and liftings are central to topology. It turns out that in the case of undirected walks, lifting occurs only when you have a local isomorphism, i.e. when the homomorphism is what is called a covering projection. The case of directed walks is more interesting, and has applications in the study of 3-manifolds. For more details, see [5].

In a more applied direction, finite state machines are just digraphs with certain functions (transition and output) attached. They are intended to model computers, while the functions model programs. Given two finite state machines D* and D, D* can be programmed to emulate anything that D can do if and only if there is a homomorphism h: D* → D with the lifting property. That is so because D* must be able to perform any sequence of operations which D can perform, in the same order.

So which digraph homomorphisms have the lifting property? The question for arbitrary triples (D*, D, h) is interesting, but seems extremely difficult. We focus our attention here on the case where both D* and D are regular. Two special classes of digraphs are particularly interesting.

The *deBruijn graph* $dB_{m,n}$ has m^n vertices, labelled with the m–ary n–tuples, and the vertex $e_1e_2...e_n$ has an arc going to the vertex $f_1f_2...f_n$ if and only if $e_i = f_{i-1}$ for $i = 2$ to n. Thus $dB_{m,n}$ is m–regular, for each n.

The *Kautz graph* $Kz_{m,n}$ is defined in a similar way, except that its vertices are labelled with m–ary n-tuples in which consecutive terms must be different. (We use Kz for Kautz in deference to the customary use of K for complete). Since there are $m(m-1)^{n-1}$ such n–tuples, the binary Kautz graphs have only two vertices, and are not interesting. Ternary Kautz graphs are 2–regular, and will turn up in our examples. In general, the m–ary Kautz graphs are (m–1)–regular. Of course the Kautz graph $Kz_{m,n}$ is a subgraph of the deBruijn graph $dB_{m,n}$.

In Figure 1 we offer some examples of digraphs and homomorphisms. In each case the digraph D has its vertices labelled, and the labelling of the digraph D* indicates the intended homomorphism h. In Figures 1a and 1b, D* is the Kautz graph $Kz_{3,3}$ and D is the deBruijn graph $dB_{2,2}$. In Figures 1c and 1d, D* is the deBruijn graph $dB_{2,3}$

and D is the deBruijn graph $dB_{2,1}$. The homomorphism of Figure 1d does not have the lifting property. For example, the walk 01001 does not lift. All of the other homomorphisms do have the lifting property, for reasons we will now explain.

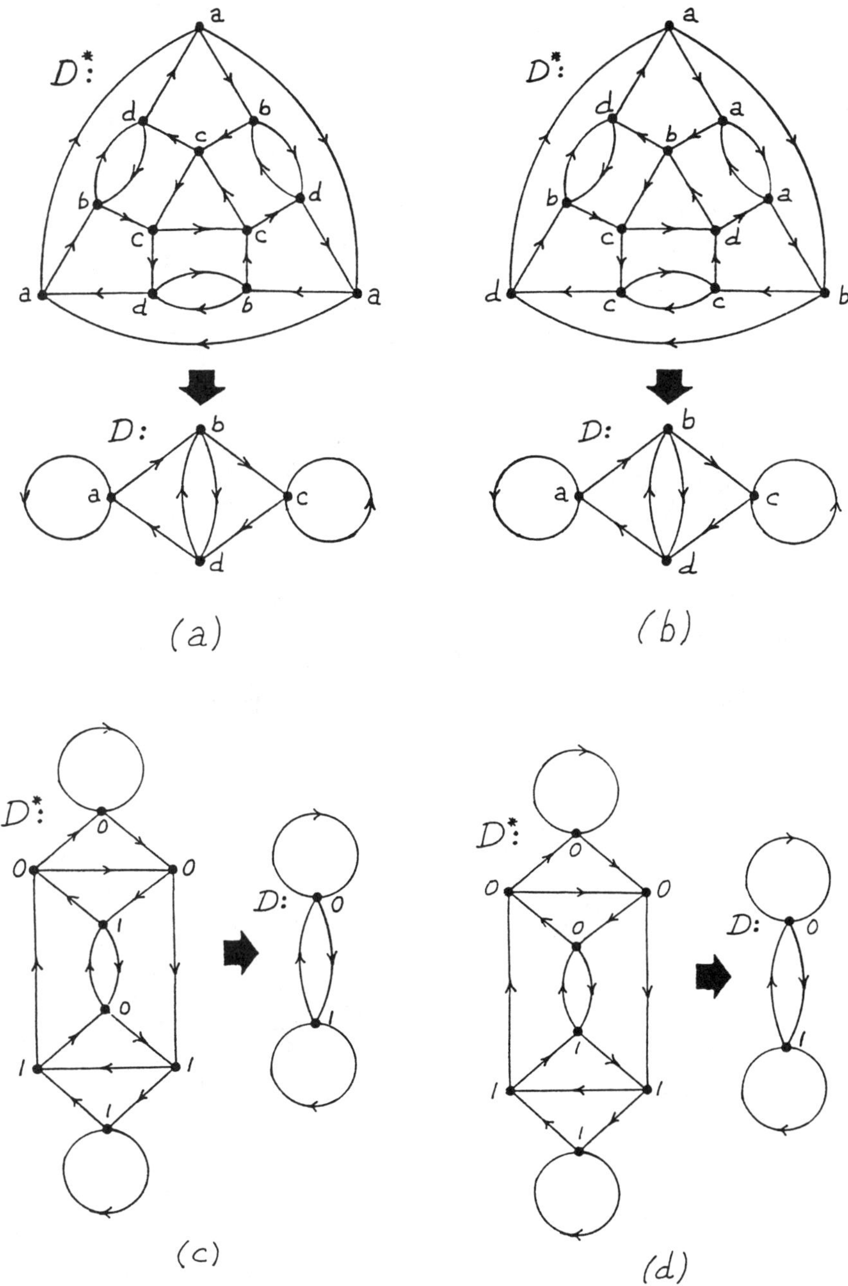

Figure 1

The *out–neighborhood* of a vertex v in a digraph is the set of vertices adjacent from v, and the *in–neighborhood* is the set of vertices adjacent to v. We say that a homomorphism h from D^* to D is *out–regular* if the out–neighborhood of each vertex v of D^* is mapped onto (not just into) the out–neighborhood of $h(v)$, and similarly for *in–regular*. If h is both in– and out–regular then it is *regular*.

Lemma 1 *If $h: D^* \rightarrow D$ is either out–regular or in–regular, then h has the lifting property.*

Proof Say h is out–regular, and we seek a lifting of the walk $v_0, x_1, v_1, ..., x_n, v_n$. Since h is onto, there is a vertex u_0 on D^* such that $h(u_0) = v_0$. By out–regularity, some vertex u_1 adjacent from u_0 is mapped to v_1, and so on. Thus the walk lifts. If h is in–regular, we use the same argument starting with v_n instead of v_0. $\square$

Since the homomorphism of Figure 1a is regular and that of Figure 1b is out–regular, they have the lifting property, by the Lemma. The homomorphism of Figure 1c is neither in– nor out–regular, so we need another argument to show lifting in that case. The following observation is obvious.

Lemma 2 *If homomorphisms $h^*: D^{**} \rightarrow D^*$ and $h: D^* \rightarrow D$ have the lifting property, then so does $hh^*: D^{**} \rightarrow D$.*

In view of this lemma, all we need to do to prove that the homomorphism h of Figure 1c has the lifting property is exhibit two homomorphisms with lifting whose composition is h. That is done in Figure 2, where the first homomorphism, indicated with the labels a, b, c, d, is in–regular and the second, indicated with the lables 0 and 1, is out–regular.

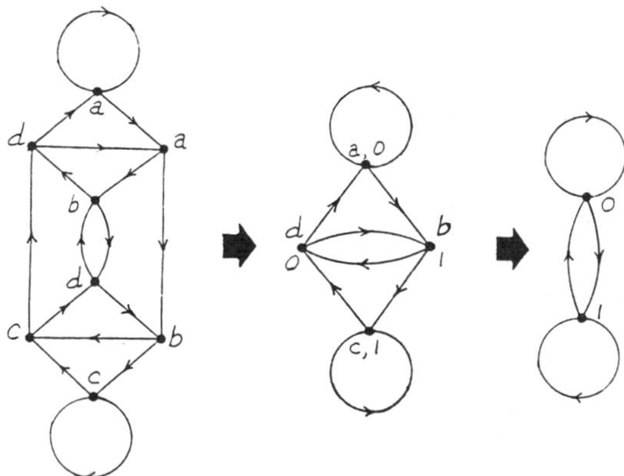

Figure 2

Since preparing this paper, we have been able to prove the existence of a function f(n) such that h has the lifting property if every walk of length at most f(n) lifts, where n the size of D*. We are still working on reducing the size of f(n), and will publish those results in a later paper. Here we present, in the next section, some observations about liftings, particularly in regular digraphs. In the final section we offer some ideas for further exploration.

2. Conditions for Lifting

For the remainder of this paper we will assume that the digraphs we are dealing with are k–regular (with k fixed within each homomorphism) and connected. (Because of the regularity, any sort of connectivity implies strong connectivity.) Several of our results depend on the following obvious lemma.

Lemma 3 *If W is any walk in a strongly connected digraph, almost all long walks contain W as a subwalk.*

Proof Say W is of length n and the diameter of the digraph is d. Given any vertex v of D, there is a walk starting at v of length at most d + n, containing W. Thus the number of walks of length d + n beginnning at v and not containing W is less than the total number of (d + n)–walks from v. Denote these numbers by N_v and M_v, respectively, and let R denote the maximum of the ratios N_v/M_v among all vertices v. Note that R < 1. To build a long walk, first build a (d + n)–walk. The number of ways to do that without including a copy of the walk W is at most R times the total number of ways to build a walk. Such a (d + n)–walk ends at some vertex. To continue and make a walk of length 2(d + n), go to an adjacent vertex and build another (d + n)–

walk. No matter what that adjacent vertex was, the number of ways to continue without including a copy of W is at most R times the total number of ways to continue without restriction. By proceeding in this way we see that the number of walks of length $k(d + n)$ including no copy of W is at most R^k times the total number of walks of that length. Since $R < 1$, almost every long walk contains a copy of the given walk W. ❏

For any homomorphism $h: D^* \to D$ there must be some walk W_m in D which has the minimum number of liftings, m. The homomorphism has the lifting property if and only if $m > 0$.

Lemma 4 *If h is a homomorphism from D^* to D, D and D^* are k regular, and W_m has the minimum number of liftings, m, then every walk containing W_m has exactly m liftings.*

Proof Clearly no walk has fewer than m liftings. Suppose, on the other hand, that some walk containing W_m has more than m liftings. Say that W is such a walk which is minimal, so that removing the vertex at either end of W results either in a walk with m liftings, or a walk not containing W_m. Remove an endvertex u which results in a walk W^- which has only m liftings, and contains W_m. Then the endvertex u is one of k vertices which could have been chosen to extend W^-. One of those k extensions, W, has more than m liftings, and the total number of liftings of those extensions is km. Thus, by the pigeonhole principle, some extension has fewer than m liftings, which contradicts the definition of m. ❏

Definition If two different walks W^* and W^{**} in D^* have the same beginning and ending vertices and both map to walk W under a homomorphism h, then (W^*, W^{**}) form a *split* in D^*.

Theorem 1 *A homomorphism between connected k–regular digraphs has the lifting property if and only if it has no splits.*

Proof Say that W does not lift. Then nothing containing W can lift. By Lemma 3, almost every long walk contains W, so almost no long walks lift. In D^*, there are k^r walks of length r, each a lifting of some walk in D. Hence, for r sufficiently large, if $k > 1$, then some walk in D must have more than n^2 liftings in D^*, where D^* has order n. But if D^* has more than n^2 liftings of any walk, two must have common starting and ending vertices, so there is a split. That leaves the case $k = 1$. But if $k = 1$, every homomorphism lifts and there can be no splits.

Suppose, on the other hand, that h has the lifting property, and suppose there is a split (W^*, W^{**}). Since D^* is connected, there is a closed walk W including a walk

W_m with the minimum number of liftings and also including the homomorphic image of W^*. Placing t copies of W end–to–end we obtain a walk W^t which includes the homomorphic image of the split t times, and includes W_m. Thus on the one hand W^t has at least 2^t liftings and on the other hand, by Lemma 4, it has only m liftings. Since t is arbitrary and m is fixed, this is a contradiction. ❑

Although this theorem gives a necessary and sufficient condition for lifting, it is not clear how useful that condition is, since there is no obvious way to check for the existence of a split. The next lemma and theorem may be more useful.

Lemma 5   *Under a homomorphism between connected k–regular digraphs, if the number of liftings of a walk is bounded, then every walk has the same number of liftings.*

Proof   The number of liftings is always an integer, so say that M is the least upper bound on the number of liftings, and W_M has exactly M liftings. Then every walk containing W_M has at most M liftings. By k–regularity, that implies in turn (as in the proof of Lemma 4) that every walk containing W_M has exactly M liftings. But D is strongly connected, so some walk contains both W_M and W_m, a walk with the minimum number of liftings. Hence, by Lemma 4, M = m, and every walk has the same number of liftings. ❑

Theorem 2   *A homomorphism between connected k–regular digraphs has the lifting property if and only if every walk has the same number of liftings.*

Proof   Certainly constant lifting implies lifting. On the other hand, Lemma 5 implies that every walk has the same number of liftings unless the number of liftings of some walk is arbitrarily large. But if D^* has n vertices and some walk has more than n^2 liftings, there must be a split. By Theorem 1, that implies we do not have lifting. ❑

In view of Lemmas 3 and 4, Theorem 2 is not very surprising. On the other hand, it has two immediate corollaries, obtained by applying it to 0–walks and 1–walks, which are not intuitively obvious.

Corollary 2.1   *A homomorphism between connected k–regular digraphs D^* and D can have the lifting property only if the number of vertices of D divides the number of vertices of D^*.*

If h is homomorphism between D^* and D, with lifting, we call an arc e of D^* *busy* if h: $(D^* - e) \to D$ does not have lifting.

Corollary 2.2 *In any homomorphism between connected k–regular graphs with lifting, every arc is busy.*

This corollary brings out just how strongly global, rather than local, the property of lifting is in k–regular digraphs.

3 . Related Problems

Work on liftings has suggested several directions for further investigation, which are interesting in their own right.

First, consider a question which arises from applying Theorem 2   to homomorphisms from 2–regular digraphs to the deBruijn graph $dB_{2,1}$. Since $dB_{2,1}$ is complete, with loops, any labelling of the vertices of a digraph D* with two labels, say 0 and 1, defines a homomorphism onto $dB_{2,1}$. If the homomorphism has lifting, then by Theorem 2 it has constant lifting. In particular, all walks which are strings of 0's or strings of 1's must lift exactly the same number of times. Thus the subgraph of D* induced by the vertices in D* labelled 0, the *0–subgraph*, is a digraph with exactly k vertices, k arcs, k directed 2–walks, and so on, for some k. It is somewhat surprising that such a digraph can exist, but a glance at the 0–subgraph in Figure 1c gives a non–trivial example.   Trivial examples include directed cycles and functional (or antifunctional) digraphs.

The general question of which digraphs have exactly k walks of length n, for every n, is somewhat complicated. The results will appear in full in [4], but we summarize them here. The first observation to make is that a digraph with constant walk numbers must have at least one directed cycle, but no two directed cycles are joined by a directed path. Any digraph with such a cycle structure will have walk numbers which are eventually periodic. (Note the similarity of this result to theorems on iterated line digraphs in Beineke [1], and Hemminger [3].) By reducing unilateral components off of the cycle to directed trees, numbers can be generated which determine the number of walks of every length. The conditions for eventually constant walk numbers are relatively neat, and are derived using eigenvalues of circulant matrices. Making the initial numbers constant as well is somewhat messier.

A second area for further study is to examine certain natural ways to define homomorphisms, particularly for the deBruijn and Kautz graphs, and see what can be said about the lifting property in each case.

As was pointed out before, any map onto a complete digraph with loops is a homomorphism -- every arc automatically maps to an arc. If we view the homomorphism as a labelling of a k–regular digraph D^*, the conditions for such a labelling to have lifting include: all vertex labels must appear the same number of times; the subgraphs induced by the vertices with a single label must have constant walk numbers; the bipartite subgraph with arcs joining vertices of two different labels must have constant walk numbers. Another image graph which gives interesting homomorphisms is the complete graph without loops. In that case each homomorphism corresponds to a labelling of D^* which is a proper vertex coloring. The same necessary conditions for lifting apply. A third interesting image digraph has a single vertex, and two directed loops. In this case a homomorphism is just a labelling of the arcs with two labels. In fact this digraph can be considered as the smallest deBruijn graph, and is of interest in studying factorings of homomorphisms, which we will discuss in a moment.

The connection between digraphs and their line digraphs gives other interesting homomorphisms. If $D^* = L(D)$, then the natural corresponcdence between vertices of D^* and arcs of D gives two natural homomorphisms. The *forward* homomorphism maps v, in D^*, to the head of the arc in D corresponding to v. The *backward* homomorphism maps v to the tail of the same arc. The deBruijn graphs and Kautz graphs are examples of sequences of iterated line digraphs, and so give examples of such line digraph homomorphisms. The homomorphisms in Figure 2, for example, are respectively backward and forward.

The labellings of the vertices of the deBruijn and Kautz graphs with n–tuples present many more opportunities for interesting homomorphisms. Each vertex is labelled $x_1x_2...x_n$, with x_i from Z_m. Every function from Z_m^n to Z_m defines a homomorphism from $dB_{m,n}$ to $dB_{m,1}$. Conversely, by using sums of products such as $(x_1)(x_2-1)(x_3-1)$ (which gives the value 1 at the vertex 100 and is 0 elsewhere), a function can be defined for any homomorphism. The question of lifting then becomes a question about the functions. For example, it is not hard to see which such functions will give in–regular or out–regular homomorphisms. Unfortunately, this formulation does not yield any obvious new theorems, although it does generate some nice examples. It is interesting to note that the natural homomorphism which maps each vertex in $dB_{2,n}$ with coordinates $x_1x_2...x_n$ to the vertex in $dB_{2,(n-1)}$ with coordinates $(x_1 + x_2)(x_2 + x_3) ... (x_{n-1} + x_n)$ is neither the foward nor the backward homomorphism obtained from the line digraph map.

Another approach which may offer answers to some questions about homomorphisms and liftings is factoring. It was already observed in Lemma 2 that the composition of maps which lift is a map which lifts. Given a homomorphism h from D^* to D, it may be possible to find another digraph E so that there is a homomorphism from D^* to E and another from E to D, whose composition is h. If there is no such homomorphism, h might be considered prime. Of course a similar definition might be made, with lifting demanded of each homomorphism. Then it is really only necessary to study prime homomorphisms with lifting, rather than all homomorphisms.

Clearly a great deal is still to be learned about homomorphisms and lifting of walks in digraphs.

REFERENCES

[1]   Beineke, L.W., On derived graphs and digraphs, in *Beitrage zur Graphentheorie* (ed. H Sachs et al),   Teubner–Verlag, Leipzig (1968) 17 – 23.

[2]   Fellows, Mike, On mutual and common covers and the complexity of covering, this Proceedings.

[3]   Hemminger, R.L., Digraphs with periodic line digraphs, *Studia Sci. Math. Hungar.* 9 (1974) 27 – 31.

[4]   Iyer, H., B. Manvel, and R. Osborne, Digraphs with constant walk numbers, in preparation.

[5]   Osborne, Richard, Lifting directed paths, in preparation.

Chordal and Interval Multigraphs

Terry A. McKee*

Wright State University

ABSTRACT

By allowing multiple edges to represent multiple intersection, much of the work done with intersection graphs can be upgraded to intersection multigraphs. In particular, starting from subtrees of trees and subpaths of paths leads to notions of chordal and interval multigraphs. We present efficient characterizations and applications of these concepts.

1. Introduction

Chordal (and interval) graphs may be characterized as intersection graphs of families of subtrees of a tree (respectively, subpaths of a path) . We propose their intersection multigraph analogs, where two subtrees or subpaths having k vertices in common correspond to two vertices being joined by k parallel edges in the intersection multigraph. Because such intersection multigraphs carry more information about the original tree or path representations, they can be more appropriate for applications. In [5] we use our characterizations of these multigraphs to determine which partially ordered sets have upper bound multigraphs (i.e., intersection multigraphs of the sets of upper bounds of elements) which are chordal or interval.

Intersection graphs are simply the underlying graphs of the analogous intersection multigraphs. But the characterization problem for such intersection multigraphs is quite different from (yet, as we shall see, closely related to) that for the corresponding intersection graphs. For instance, the complete tripartite graph $K_{1,1,2}$ (i.e., a rectangle with one diagonal) is a subpaths–of–a–path intersection graph, but not multigraph. A second parallel diagonal would have to be added before four subpaths could intersect in

---

* The author thanks the Western Michigan University Department of Mathematics & Statistics for its exuberant hospitality during his sabbatical stay there. The research was partially supported by Office of Naval Research contract N00014–88–K–0163.

precisely that fashion. (Yet adding parallel edges can sometimes destroy being such an intersection multigraph).

For any multigraph M with vertex set $V(M)$, let $M{\downarrow}$ be the underlying (simple) graph. For each pair $u, v \in V(M)$, let $\mu(u,v)$ be the number of parallel edges joining u with v; if u is adjacent to v, then $\mu(u,v)$ is called the *multiplicity* of the multiple edge. A *clique* of M is a simple subgraph of M whose vertex set induces an inclusion–maximal complete subgraph of $M{\downarrow}$ (i.e., a clique of $M{\downarrow}$). (Hence a clique of a multigraph contains exactly one edge from each bundle of parallel edges joining its vertices.) For $u, v \in V(M)$, let $\kappa(u,v)$ be the number of cliques of $M{\downarrow}$ which contain $\{u, v\}$. A *simplicial vertex* of M is a vertex in a unique clique of $M{\downarrow}$.

A multigraph M is said to satisfy the $\mu \geq \kappa$ *condition* if, for every pair $u, v \in V(M)$, $\mu(u,v) \geq \kappa(u,v)$. Thus M satisfies the $\mu \geq \kappa$ condition if and only if all the cliques of $M{\downarrow}$ can be simultaneously identified with edge–disjoint cliques of M. Any edges left over, i.e., $\mu(u,v) - \kappa(u,v)$ parallel edges joining each pair $u, v \in V(M)$, are called *residual edges* of M. The submultigraph of M consisting of residual edges is called the *residual submultigraph*, or *residuum*, of M, denoted rM. (We use r^2M for r(rM), etc.) The residuum plays a key role in relating the theories of intersection graphs and multigraphs.

2. Chordal Multigraphs

We define a *subtrees–of–a–tree* representation of a graph G to be a tree T whose vertices are subsets of $V(G)$ such that:
(a) For each $v \in V(G)$, $T_v = \{A \in V(T) : v \in A\}$ induces a subtree of T; and
(b) For each $u, v \in V(G)$, u and v are adjacent in G if and only if $T_u \cap T_v$ is nonempty.
A subtrees–of–a–tree representation of a multigraph M is defined similarly, except replacing (b) with
(b') For each $u, v \in V(M)$, $\mu(u,v) = |T_u \cap T_v|$.
Notice that, by (b) or (b'), each vertex of T induces a complete subgraph of M or $M{\downarrow}$.

Lemma *In any subtrees–of–a–tree representation of a multigraph M, each vertex set of a clique of M must equal one or more vertices of the tree.*

Proof We first show that whenever vertices v_1, v_2, v_3, form a triangle in $M{\downarrow}$, then some vertex of the representing tree T must contain them all. By (b), each two v_i, v_j are in a common vertex A_{ij} in T. If each such A_{ij} failed to contain the third vertex of the triple, then, by (a), there would be a unique vertex of G at the intersection of the three paths pairwise connecting A_{12}, A_{13}, and A_{23}, and this vertex would contain all three v_i's. Similarly, the vertex set of every complete subgraph of $M\downarrow$ must be contained in a common vertex of T. ❏

A *chord* of a circuit in a graph is an edge which joins vertices in, but not consecutive along the circuit. *Chordal graphs* are characterized by every circuit of length at least four having at least one *chord*. (See Chapter 4 of [2] for an extensive survey of chordal graphs, called there "triangulated graphs" and sometimes elsewhere "rigid circuit graphs".) Chordal graphs were characterized (independently) by Walter, Gavril, and Buneman [2, Theorem 4.8] as those graphs having subtrees–of–a–tree representations. This motivates our defining a multigraph to be a *chordal multigraph* if it has a subtrees–of–a–tree representation.

Theorem 1 *A multigraph M is a chordal multigraph if and only if:*
   (1) M satisfies the $\mu \geq \kappa$ condition; and
   (2) $M\downarrow$ is a chordal graph.

Proof Suppose first M has a subtrees–of–a–trees representation T. Since T is also a subtrees–of–a–tree representation of $M\downarrow$, condition (2) follows immediately. If vertices u and v of M are in κ common cliques of $M\downarrow$, then by the Lemma these are in at least κ common vertices of T, so $|T_u \cap T_v| \geq \kappa$, and so condition (1) follows from (b').

Conversely, suppose conditions (1) and (2) hold. Suppose M+ is obtained from M as follows: For each residual edge of M, add a new simplicial vertex (simply) adjacent to precisely the two endvertices of that edge. Repeat if necessary until M+ satisfies $\mu(u,v) = \kappa(u,v)$ for all $u, v \in V(M+)$. Let $G = M+\downarrow$. It is easy to check that (2) implies that G is also chordal.

Claim If T is the subtrees–of–a–tree representation for the graph G produced in Golumbic's proof of Theorem 4.8 of [2], then T is also a subtrees–of–a–tree representation for the multigraph M+. (Then removing the elements of $V(M+) - V(M)$ out of the vertices of T will leave the desired subtrees–of–a–tree representation for M itself.)

We show the Claim inductively, with the case when G is complete as basis. (T then has V(G) as its unique vertex.) For the inductive step, suppose G is not complete and that the Claim is true for all smaller graphs. Because G is chordal, there is [2, Lemma 4.2] a simplicial vertex a. (Our notation is chosen to be consistent with proof of Theorem 4.8 of [2].) Put A = N[a], the closed neighborhood of a. Thus A is a clique of G, and so must be a vertex of T by our Lemma. Letting N(u) be the open neighborhood of vertex u, put $U = \{u \in A : N(u) \subset A\}$ and $Y = A - U$. Put $G' = G - U$ and let B be any clique of G' which contains Y. The cases $B = Y$ and $B \neq Y$ are handled separately.

Suppose $B = Y$ and put $M' = M+ - U$. Note that M' satisfies (1) and (2) and that $G' = M'{\downarrow}$. Since U is nonempty, G' is smaller than G, and so by the inductive hypothesis, the subtrees–of–a–tree representation T' for G' produced as in Theorem 4.8 of [2] is also a subtrees–of–a–tree representation for M'. Define T to be the same as T' except with vertex B replaced by A, and verify that T is a subtrees–of–a–tree representation for M by comparing the multiplicities with the number of common vertices of T for each of the six possible types of vertex pairs from the partition U, Y, $V(M+) - A$ of $V(M+)$.

Now suppose $B \neq Y$ and put $M' = M+ - U$ except also remove one edge from between every two members of Y. Proceed as in the $B = Y$ case, except define T to be the same as T' except for adding the one new vertex A and one new edge between A and B. ❑

For comparison with Theorem 2 below, note that we have essentially proved an equivalent version of Theorem 1 with condition (2) replaced by:

(2') For $M+$ defined from M by adding a new simplicial vertex for each edge of rM (repeating if necessary until $\mu(u,v) = \kappa(u,v)$ for all u, v), $M+{\downarrow}$ is a chordal graph.

Corollary *If M is a chordal multigraph, then every circiut of length at least four must have at least two (possibly parallel) chords.*

Proof One chord exists by condition (2) of Theorem 1. The existence of the second follows by induction on the length of the circuit. The basis (length four) case follows from condition (1). For lengths greater than four, use a smaller circuit subtended by the first chord. ❑

3. Interval Multigraphs

We can define a *subpaths–of–a–path* representation of a graph or a multigraph exactly as we did a subtrees–of–a–tree representation, except that the tree T will be a path P such that each $P_v = \{A \in V(P) : v \in A\}$ is a subpath of P, with adjacency corresponding to the intersection of the P_v's in the expected manner.

Interval graphs are defined as those graphs having a subpaths–of–a–path representation and have been characterized in many ways; see Section 1.3 and Chapter 8 of [2] or Section 3.4 of [6]. The earliest characterization was that of Lekkerkerker and

Boland [2, Theorem 8.4], as being both chordal and having no 'asteroidal triples'. This latter statement can be viewed as a sort of generalized chordality: Given any closed walk passing through three distinguished "anchor" vertices, there must exist a chord joining one of the anchors to some vertex of the walk lying between (or at one of) the other two anchors.

We define a multigraph to be an *interval multigraph* if and only if it has a subpaths–of–a–path representation. It should be noted that, unlike chordal multigraphs, the family of interval multigraphs is not closed under duplicating edges; i.e., increasing an already positive multiplicity in an interval multigraph may produce a non interval multigraph.

Theorem 2   *A multigraph M is an interval multigraph if and only if:*
(1) M, rM, r^2M, … each satisfies the $\mu \geq \kappa$ condition; and
(2) For M+ defined from M by adding a new simplicial vertex for each clique of rM, r^2M, … ((simply) adjacent to precisely the vertices of that clique), M+$\downarrow$ is an interval graph.

Proof   Suppose first M has a subpaths–of–a–path representation P. Since this is also a subtrees–of–a–tree representation, M is a chordal multigraph and so, by Theorem 1, satisfies $\mu \geq \kappa$. By the Lemma, each clique of M corresponds to a vertex of P. Removing these vertices from P by contracting edges of P leaves a subtrees–of–a–tree representation for rM. So rM is a chordal multigraph and so, by Theorem 1, satisfies $\mu \geq \kappa$. Thus and similarly, condition (1) holds. Finally, adding the vertices of V(M+) − V(M) into the appropriate vertices of P produces a subpaths–of–a–path representation for M+$\downarrow$, thereby proving (2).

Conversely, suppose conditions (1) and (2) hold. By (2), M+$\downarrow$ has a subpaths–of–a–path representation P, where the vertices of P will be precisely the vertex sets of the cliques of M+$\downarrow$. By (1) and the construction of M+, each pair u, v $\in$ V(M) will occur in exactly μ(u,v) of the vertices of P. Hence, removing the elements of V(M+) − V(M) out of the vertices of P will leave a subpaths–of–a path representation for M itself. ❑

4. Algorithms, Applications, and Alterations

Known algorithms and results concerning chordal graphs (namely, Corollary 4.6 and Theorem 4.17 of [2]) can be combined with Theorem 1 to recognize chordal multigraphs efficiently. Theorem 1 also allows Theorem 2 to be equivalently stated with condition (1) replaced with:

(1') Each of M, rM, r^2M, … is a chordal multigraph.

This allows the recognition procedure for chordal multigraphs to be combined with known algorithms and results for interval graphs (namely, Theorem 8.3 of [2]) to recognize interval multigraphs efficiently.

Corollary *Chordal [and interval] multigraphs can be recognized in time, proportional to the order [times the maximum multiplicity occurring].*

Chordal and interval graphs serve as models in a wide variety of applications; Section 8.4 of [2] and Section 3.4 of [6] survey many of these. (When interval graphs are defined in terms of intervals of the real line, subpaths–of–a–path representa⁀⁀ ⁀an be obtained by taking left endpoints as vertices, or by using representatives of equivalence classes of points as vertices of P, where two points are equivalent whenever they are in precisely the same intervals.)

One of the leading applications in [1] was the construction of the "genealogy" of medieval manuscripts by tracing errors (or other distinctive features) which were made in handcopying and might be repeated (or possibly corrected) in later recopying. (Reconstructing evolutionary trees is a second example in [1], using the same ideas; both arose from real–life concerns.) The manuscripts can be viewed as the vertices of a rooted tree, with the errors corresponding to certain subtrees of the tree. Thus the intersection (multi)graph of errors must be a chordal (multi)graph.

As an oversimplified example, suppose there are four errors designated A, B, C, D and five manuscripts having the following combinations of errors: A; A & B; A & B & D; B & C; and C & D. The intersection graph of errors is the complete tripartite graph $K_{1,1,2}$ and is chordal. But the intersection multigraph ($K_{1,1,2}$ plus a residual edge between A and B) is not a chordal multigraph: By the Corollary to Theorem 1, a second chord is needed for the circuit A, B, C, D; the missing edge predicts a missing manuscript. (In fact, our Lemma implies there must be a manuscript with errors B, C and D.) The point is, if all one has is the intersection graph, all looks well; but by having the intersection multigraph, an inconsistency (or incompleteness) is seen.

Suppose we were to start with an actual chordal multigraph M which did represent errors in manuscripts. Suppose also that each manuscript has at least two errors (or other distinctive features). Each manuscript would then correspond to a clique of M+↓ (using the notation of the proof of Theorem 1). Forming the "clique (multi)graph" (i.e., the intersection (multi)graph of the family of cliques) of M+↓ would produce the intersection (multi)graph of the original family of manuscripts, each now viewed as a set of errors. (This "manuscript (multi)graph" would not have been similarly producible if we had started with a chordal graph rather than multigraph.) This then imparts structure to the manuscripts–as–sets–of–errors (multi)graph, as well as to the errors–as–sets–of–manuscripts (multi)graph.

A similar example concerns the detection of a linear structure underlying developmental psychology; see the bottom of page 120 in [6] for discussion and primary references. Children would correspond to vertices of what was hypothesized as a subpaths–of–a–path representation, with certain traits or characteristics as subpaths. If the hypothesis were correct, the intersection (multi)graph of traits would be an interval (multi)graph. Using the multigraph model, the appropriate multigraph criterion of Theorem 2 gives a more stringent test of the hypothesis. Missing edges now correspond to unrepresented types of children (i.e., to undiscovered combinations of traits).

There are also obvious alterations to our results made by defining "subthings–of–a–thing representations" and seeking corresponding characterization theorems having clauses involving (1) the $\mu \geq \kappa$ condition for, and the (2) underlying graph of a multigraph and its residuum. For instance, the "things" could easily be caterpillars or stars as well as paths or trees, or they could be mixed, as in subpaths–of–a–tree representations. But the problem would still remain of going from knowledge of such representations to suitable conditions for the intersection multigraphic characterizations. (The similar problem for intersection graphic characterizations is discussed in [3]; a somewhat unsatisfactory approach for intersection multigraphic characterizations is mentioned at the end of [4].)

As long as the "things" in a subthings–of–a–thing representation are varieties of trees, we can simultaneously characterize *edge intersection multigraphs*, where the subgraphs are considered as sets of edges instead of sets of vertices. Namely, the edge intersection multigraph is obtained from the (vertex) intersection multigraph by subtracting from it the edges of the (underlying) intersection graph. (This corresponds to each tree having exactly one more vertex than edge.) Thus of the four alternatives determined by the vertex/edge and simple/multi choices, the (vertex) intersection multigraph directly determines each of the others, while the standard (vertex) intersection (simple) graph determines none of the others.

In closing, we mention that our Lemma corresponds to saying that any family of subtrees–of–a–tree satisfies the *Helly Property*: Whenever all the subtrees of any subfamily have pairwise nonempty intersections, there is a vertex in common to all the subtrees of the subfamily; see Proposition 4.7 of [2]. If we define a *Helly multigraph* to be the intersection multigraph of a Helly family, then the arguments we made above can be rephrased as follows.

Theorem 3 *A multigraph* M *is a Helly multigraph if and only if* M *satisfies the* $\mu \geq \kappa$ *condition.*

As a corollary, every graph underlies a Helly multigraph, and so every graph would be a "Helly graph". As a corollary of Theorems 1, 2 and 3, we can sharpen the role of the $K_{1,1,2}$ example used at the beginning of this paper. A chordal (or interval or Helly) graph G is a chordal (or respectively, interval or Helly) multigraph if and only if each edge of G is in a unique clique, and so if and only if G contains no induced subgraph isomorphic to $K_{1,1,2}$.

REFERENCES

[1] P. Buneman, A characterization of rigid circuit graphs, *Discrete Math.* **9** (1974) 205 − 212.

[2] M.D. Golumbic, *Algorithmic Graph Theory and Perfect Graphs*, Academic Press (New York) 1980.

[3] T.A. McKee, Intersection properties of graphs, to appear.

[4] T.A. McKee, Foundations of intersection graph theory, to appear.

[5] T.A. McKee, Upper bound multigraphs for posets, to appear.

[6] F.R. Roberts, *Discrete Mathematical Models with Applications to Social, Biological and Environmental Problems*, Prentice–Hall (Englewood Cliffs NJ) 1976.

The Expected Number of Symmetries in Locally–Restricted Trees I

Kathleen A. McKeon

Connecticut College

ABSTRACT

Exact formulas are derived for the number of symmetries in several types of unlabeled trees with vertices of restricted degree. The trees are d–trees whose vertices have degree at most d and (1,d)–trees whose vertices have degree 1 or d. These results together with similar results for the number of such trees provide formulas for the expected number of symmetries in these trees.

These trees give rise to significant examples in polymer chemistry. For example, (1,4)–trees represent the alkanes and 4–trees represent the carbon skeletons of alkanes. The expected number of symmetries in such trees is important in the study of collections of molecular species formed during some chemical reaction process.

1. Introduction

The enumeration of trees is an important problem in graph theory with a distinguished history as well as applications to theoretical chemistry. The first major work in this area was performed by Cayley who determined exact formulas for the number of labeled trees [C89], the number of rooted trees [C57] and the number of free trees [C75, C81]. These results were extended and an asymptotic analysis of the numbers was provided by Pólya [P37] and Otter [O48].

Cayley's work [C75] was motivated by the problem of enumerating isomers of alkanes, compounds of carbon and hydrogen atoms which have valencies of 4 and 1 respectively. The alkanes have the general formula C_kH_{2k+2} and can be represented by (1,4)–trees. They are the best documented family of chemical compounds and provide a model for much of chemical theory [GoK73]. Generalizing (1,4)–trees, we have (1,d)–trees, which give rise to other meaningful examples in polymer chemistry. There is a correspondence between (1,d)–trees and d–trees that also has chemical significance.

While 4–trees correspond to the carbon skeletons of alkanes [GoK73], d–trees in general correspond to skeleton polymers, i.e., polymer molecules that have been stripped of their reactive end groups [GoT76].

The problem addressed in this paper, the enumeration of symmetries in (1,d)–trees and d–trees for d = 3, 4, is also motivated by chemistry. In the study of collections of molecular species, it is almost always the average of some property over an appropriate class of trees that is required. In computing such an average, it is necessary to assign weights to the various trees in the class so as to reflect the (not usually equal) proportions in which they are formed by the chemical process involved. The proper assignment of weights to the trees often involves the orders of their automorphism groups [GoL75]. Consequently, chemists are interested in the orders of the automorphism groups of large trees of various species such as (1,d)–trees and d–trees.

The tool used to do the counting is a two–variable generating function, an approach that seems to have originated in the work of Etherington [Et38]. For a given class C of trees, let $t(x,y)$ be the generating function in two variables x and y such that the coefficient of $y^m x^n$ is the number of trees T in C of order n in which m is the logarithm base 2 of the order of the automorphism group of T. In $t(x,2)$, the coefficient of x^n is the sum of the orders of the automorphism groups of all such trees.

The technique used to do the counting was developed by Pólya [P37], perfected by Otter [O48] and described as a twenty step algorithm for counting various types of trees by Harary, Robinson and Schwenk [HRS75]. The generating functions $t(x,y)$ and $t(x,2)$ satisfy functional equations from which recurrence relations for their coefficients are determined.

In this paper, the technique is illustrated and results given for (1,3)–trees. Exact formulas are determined for the number of symmetries in both planted and free unlabeled (1,3)–trees. These results together with similar results for the number of (1,3)–trees provide formulas for the expected number of symmetries in these trees. A study of the asymptotic behavior of the number of symmetries in such trees will appear in Part II of this series. A brief sketch of the general approach is provided in the short research announcement [KMPR].

2. Generating Functions

While equations are given for (1,3)–trees only, the method has been used to enumerate symmetries in four types of trees: d–trees and (1,d)–trees for d = 3, 4 [Mc87] and may be applied for higher values of d.

We begin by defining a generating function that counts symmetries in planted trees of the specified type. In general, the results for planted trees provide a means for obtaining the results for free trees. However, removing the root of a planted (1,3)–tree leaves a binary tree that has the same automorphism group as the original planted (1,3)–tree. Thus, when counting symmetries in planted (1,3)–trees we are also counting symmetries in binary trees.

For the planted trees of each type, a two–variable logarithmic generating function is defined as follows:

$$T(x,y) = \sum_{n=1}^{\infty} \sum_{m} T_{m,n} y^m x^n \qquad (2.1)$$

For d–trees, $T_{m,n}$ is the number of planted trees T of order $n+1$ in which $m = \log_2 |\Gamma(T)|$, where $\Gamma(T)$ denotes the automorphism group of T. Every (1,d)–tree, planted or free, has 2 modulo (d–1) vertices. This is taken into account in the definition of $T(x,y)$ for (1,d)–trees. In this example, since every planted (1,3)–tree has an even number of vertices, $T_{m,n}$ is defined to be the number of planted (1,3)–trees on 2n vertices (or binary trees on $2n - 1$ vertices) with 2^m symmetries .

The values which m may assume in the sum (2.1) depend on both d and the type of tree. Since an automorphism of a rooted tree must leave the root fixed, the order of the automorphism group of a planted (1,3)–tree is of the form 2^m where m is an integer ranging from 0 to $n - 1$.

Note that when $y = 1$ is substituted in (2.1), $T(x,1)$ counts the number of planted trees of the specified type. Substituting $y = 2$ in (2.1) results in a one–variable generating function which counts symmetries in planted trees of the specified type. Let S_n be the total number of symmetries in all planted (1,3)–trees on 2n vertices.

$$S_n = \sum_{m} T_{m,n} 2^n \qquad (2.2)$$

and

$$T(x,2) = \sum_{n=1}^{\infty} S_n x^n \qquad (2.3)$$

Similarly, $t(x,y)$ can be defined for free trees. However, we actually only work with $t(x,2)$. Thus, we define

$$t(x,2) = \sum_{n=1}^{\infty} s_n x^n \qquad (2.4)$$

which counts symmetries in free, i.e., unrooted trees. For (1,3)–trees, s_n is the total number of symmetries in all free (1,3)–trees on 2n vertices.

3. Functional Relations

To obtain the exact formulas for the number of symmetries in these trees, functional relations satisfied by $T(x,y)$, $T(x,2)$ and $t(x,2)$ are now derived.

First observe that rooted and planted trees of a specified type can be formed from planted trees of that type. A rooted tree in which the root has degree k is formed by taking a collection of k planted trees and identifying their roots to form the root of the new tree. Adding a new vertex adjacent to the root of this rooted tree results in a planted tree in which the degree of the vertex adjacent to the root is $k+1$. Based on this observation, relations expressing $T(x,y)$ in terms of $T(x,y)$, $T(x^2,y^2)$, and $T(x^3,y^3)$ are derived.

Theorem 3.1 *The generating functions $T(x,y)$ and $T(x,2)$ which count symmetries in planted $(1,3)$–trees satisfy*

$$T(x,y) = x + \frac{1}{2} T(x,y)^2 + \left(y - \frac{1}{2}\right) T(x^2,y^2) \tag{3.1}$$

and

$$T(x,2) = x + \frac{1}{2} T(x,2)^2 + \frac{3}{2} T(x^2,4) \tag{3.2}$$

Proof The vertex adjacent to the root of a planted $(1,3)$–tree has degree 1 or 3. The term x counts the symmetries in a planted K_2, the only planted $(1,3)$–tree in which the vertex adjacent to the root has degree 1.

To count symmetries in those trees in which the vertex adjacent to the root has degree 3, two cases must be considered. Suppose T is the planted $(1,3)$–tree formed from the planted $(1,3)$–trees T_1 and T_2 in the manner described above. If $T_1 \neq T_2$, then we have $|\Gamma(T)| = |\Gamma(T_1)|\,|\Gamma(T_2)|$. Then $\frac{1}{2}(T(x,y)^2 - T(x^2,y^2))$ counts symmetries in this case. If $T_1 = T_2$, then we have $|\Gamma(T)| = 2|\Gamma(T_1)|^2$ since the two branches T_1 and T_2 can be permuted. This case is handled by $yT(x^2,y^2)$ with the factor of y accounting for the additional factor of 2 in the group order. Now (3.2) is obtained by substituting $y = 2$ in (3.1). $\square$

Using the following lemma which relates the order of the automorphism group of a free tree to the orders of the automorphism groups of the vertex and edge–rooted versions of the tree, $t(x,2)$ is expressed in terms of $T(x,2)$, $T(x^2,4)$ and $T(x^3,8)$.

Lemma 3.2 *For any tree T,*

$$|\Gamma(T)| = \sum_{T_1} |\Gamma(T_1)| - \sum_{T_2} |\Gamma(T_2)| + |\Gamma(T_3)| \tag{3.3}$$

where the first sum is taken over all different vertex–rooted versions T_1 of T and the second sum is taken over all different edge–rooted versions T_2 of T. If T has a symmetry edge, an edge whose vertices are interchanged by some automorphism of T, then $T_3 = T$. If T does not have a symmetry edge, then T_3 is the empty graph and $|\Gamma(T_3)| = 0$.

Proof   This lemma is a variation of a lemma due to Otter [HP73]. As in the proof of Otter's lemma, the vertex and edge–rooted versions of T can be paired such that the paired vertex and edge–rooted versions of T have the same automorphism group. Recall that an automorphism of a graph must leave the root fixed while an automorphism of an edge–rooted graph must leave the vertices of the root edge fixed. For each vertex v that is not in the center of T, match the version of T that is rooted at v with the edge–rooted version that is rooted at the first edge on the path from v to the center of T.

If T has a symmetry edge, the center of T consists of two vertices, say u and v. Since the edge uv is a symmetry edge of T, rooting T at u is equivalent to rooting T at v. Hence if the version of T that is rooted at the vertex v is paired with the version of T that is rooted at the edge uv, then the difference of the two sums in (3.3) is 0 and $T_3 = T$. Thus, (3.3) holds in this case.

If T does not have a symmetry edge two cases must be considered. If the center of T consists of two vertices u and v, match the version of T that is rooted at v with the version of T that is rooted at the edge uv. In this case and the case that the center of T consists of just one vertex u, there is one vertex–rooted version of T that cannot be paired with an edge–rooted version. This is the tree that results from rooting T at the vertex u which is in the center of T. Since T does not have a symmetry edge, the vertices in the center of T are all fixed points of the automorphisms of the unrooted tree T. Hence this extra vertex–rooted version of T has the same automorphism group as T and (3.3) holds in this case also. $\square$

This lemma can be extended to a statement about the generating functions that count symmetries by multiplying (3.3) by x^n and summing over all trees of the appropriate order. Summing the result over all $n \geq 1$ gives $t(x,2)$ on the left side. The first sum on the right side gives the series that counts symmetries in rooted trees and the second sum gives the series that counts symmetries in edge–rooted trees while $|\Gamma(T_3)|x^n$ sums to the series that counts symmetries in trees with a symmetry edge. For $(1,3)$–trees we have the following functional relation.

Theorem 3.3 *The generating function $t(x,2)$ for symmetries in free (1,3)–trees is given by*

$$t(x,2) = \frac{1}{2x}T(x,2)^2 - \frac{1}{3x}T(x,2)^3 + \frac{3}{2x}T(x^2,4) + \frac{13}{3x}T(x^3,8).$$

Proof First we determine an expression for the series that counts symmetries in rooted (1,3)–trees. As previously described, this expression can be found by using planted (1,3)–trees to build rooted (1,3)–trees. The series for rooted (1,3)–trees is equal to

$$T(x,2) + \frac{3!}{x}T(x^3,8) + [\frac{2}{x}T(x^2,4)\,T(x,2) - T(x^3,8)]$$

$$+ [\frac{1}{3!\,x}(T(x,2)^3 - 3\,T(x^2,4)\,T(x,2) + 2\,T(x^3,8))]$$

Symmetries in rooted (1,3)–trees in which the root has degree 1, i.e., in planted (1,3)–trees are counted by $T(x,2)$. To count symmetries in rooted (1,3)–trees in which the root has degree 3, three cases must be considered. Suppose T is the rooted tree formed from the planted (1,3)–trees T_1, T_2 and T_3. The second term of (3.5) counts symmetries in the case that all three trees are the same. The case that exactly two of the three trees are the same and the case that all three are different are handled by the first and second bracketed terms of (3.5) respectively.

A tree rooted at an edge can be formed by identifying the edges incident to the roots of two planted trees. That edge is the root edge of the edge–rooted tree. When the two trees which are combined are the same, that edge is a symmetry edge. Thus,

$$\frac{1}{2x}(T(x,2)^2 + T(x^2,4))$$

counts symmetries in edge–rooted (1,3)–trees and

$$\frac{2}{x}T(x^2,4)$$

counts symmetries in (1,3)–trees that have a symmetry edge. Combining (3.5), (3.6) and (3.7) as in Lemma 3.2 and using the functional relation (3.2) to simplify gives equation (3.4). $\square$

4. Recurrence Relations

From the functional equation (3.1) recurrence relations for $T_{m,n}$, the coefficient of $y^m x^n$ in $T(x,y)$ can now be determined. Note that throughout this section the subscripts on the variables are always non–negative integers. Otherwise one can assume the value of the variable is zero.

The only planted $(1,3)$–tree with the identity group as its automorphism group occurs when $n = 1$. Otherwise, there is at least one pair of end vertices that can be permuted. Thus, $T_{0,1} = 1$ and $T_{0,n} = 0$ if $n \geq 2$. For $n \geq 2$ and $1 \leq m \leq n-1$, $T_{m,n}$, the number of planted $(1,3)$–trees on $2n$ vertices (binary trees on $2n-1$ vertices) with 2^m symmetries is as follows:

$$T_{m,n} = T_{(m-1)/2,\, n/2} - \frac{1}{2}T_{m/2,\, n/2} + \begin{cases} 0, \text{ if } m = n - 1 \\[2ex] \frac{1}{2}\sum_{k=1}^{n-1} \sum_{i} T_{i,k}T_{m-i,n-k}, \text{ if } m \neq n - 1 \end{cases}$$

Recall that S_n is the coefficient of x^n in $T(x,2)$ and let B_n and C_n be the coefficients of x^{2n} and x^{3n} in $T(x^2,4)$ and $T(x^3,8)$ respectively. Then as a consequence of equation (3.4), s_n, the coefficient of x^n in $t(x,2)$, can be expressed in terms of S_n, B_n and C_n. For $n \geq 2$,

$$s_n = \frac{1}{2}\sum_{k=1}^{n} S_k S_{n-k+1} - \frac{1}{3}\sum_{i=1}^{n-1}\sum_{j=1}^{n-i} S_i S_j S_{n-i-j+1} + \frac{3}{2}B_{(n+1)/2} + \frac{13}{3}C_{(n+1)/3}$$

Let T_n be the number of planted $(1,3)$–trees on $2n$ vertices (binary trees on $2n - 1$ vertices) and let t_n be the number of free $(1,3)$–trees on $2n$ vertices. Then equations for T_n and t_n [BaKP81] can be combined with (4.1) and (4.2) respectively to give formulas for the expected number of symmetries in these trees. That is, the expected number of symmetries in a planted $(1,3)$–tree on $2n$ vertices (binary tree on $2n - 1$ vertices) is S_n/T_n and the expected number of symmetries in a free $(1,3)$–tree on $2n$ vertices is s_n/t_n.

5. Numerical Results

Values of $T_{m,n}$, S_n and s_n were computed using the CDC Cyber 750 in the Computer Laboratory at Michigan State University. The computation of these numbers was limited by the available accuracy and storage restrictions. Another limiting factor was the time required to compute the values using the recurrence relations. In the case of $(1,3)$–trees, the Fortran programs used to compute S_n for $n \leq 50$ took 52 seconds while an additional 300 seconds were required to compute S_{68}. Table 1 contains values of $T_{m,n}$ for $n = 6$ to 13. Table 2 contains values of S_n and s_n for $n = 6$ to 25.

Table 1.   Coefficients of $T(x,y)$ for Planted $(1,3)$–trees

n	m	$T_{m,n}$	n	m	$T_{m,n}$
6	1	1	11	1	1
	2	2		2	16
	3	2		3	50
	4	1		4	58
				5	54
7	1	1		6	17
	2	4		7	8
	3	3		8	3
	4	3			
			12	1	1
8	1	1		2	20
	2	6		3	85
	3	7		4	119
	4	7		5	126
	5	1		6	61
	6	0		7	27
	7	1		8	9
				9	2
9	1	1		10	1
	2	9			
	3	14	13	1	1
	4	14		2	25
	5	6		3	135
	6	1		4	239
	7	1		5	273
				6	187
10	1	1		7	80
	2	12		8	32
	3	28		9	8
	4	28		10	3
	5	21			
	6	4			
	7	3			
	8	1			

Table 2.   Coefficients of T(x,2) and t(x,2) for (1,3)–trees

n	S_n	s_n
6	42	24
7	90	168
8	354	240
9	758	608
10	2290	920
11	6002	5680
12	18410	6104
13	51310	18416
14	154106	43008
15	449322	148152
16	1384962	325608
17	4089174	980840
18	12475362	2421096
19	37746786	7336488
20	116037642	19769312
21	355367310	58192608
22	1097869386	164776248
23	3393063162	502085760
24	10546081122	1427051544
25	32810171382	4261678656

Values of T_n and t_n and the ratios S_n/T_n and s_n/t_n were computed using the Macintosh II microcomputer. Table 3 contains values of S_n/T_n, the expected number of symmetries in a planted (1,3)–tree on $2n$ vertices (binary tree on $2n - 1$ vertices), and s_n/t_n, the expected number of symmetries in a free (1,3)–tree on $2n$ vertices, for $n = 6$ to 25.

Table 3.   Expected Number of Symmetries in Planted and Free (1,3)–trees

n	Planted	Free
6	7.000	12.000
7	8.182	42.000
8	15.391	40.000
9	16.478	55.273
10	23.367	51.111
11	28.995	153.514
12	40.820	92.485
13	52.197	136.415
14	70.723	162.294
15	92.644	268.391
16	127.002	287.640
17	166.017	406.988
18	222.731	474.911
19	295.100	665.743
20	395.295	829.041
21	525.569	1114.097
22	702.244	1435.383
23	935.721	1973.833
24	1250.072	2522.478
25	1667.160	3367.375

REFERENCES

[BaKP81]   C. K. Bailey, J. W. Kennedy and E. M. Palmer, Points by degree and orbit size in chemical trees I, *Theory and Applications of Graphs* (G. Chartrand et al, eds.) Wiley, New York (1981) 27–43.

[BaKP83]   C. K. Bailey, J. W. Kennedy and E. M. Palmer, Points by degree and orbit size in chemical trees II, *Discrete Appl. Math.*, **5** (1983) 157–164.

[C57]   A. Cayley, On the theory of the analytical forms called trees, *Philos. Mag.* (4) **13** (1857) 172–176 = *Math. Papers*, Vol. 3, 242–246.

[C75]   A. Cayley, On the analytical forms called trees, with application to the theory of chemical combinations, *Rep. Brit. Assoc. Advance. Sci.* **45** (1875) 257–305 = *Math. Papers,* Vol. 9, 427–460.

[C81]   A. Cayley, On the analytical forms called trees, *Amer. J. Math.* **4** (1881) 266–268 = *Math. Papers*, Vol. 11, 365–367.

[C89]   A. Cayley, A theorem on trees, *Quart. J. Pure Appl. Math.* **23** (1889) 376–378 = *Math. Papers*, Vol. 13, 26–28.

[Et38]   I. M. H. Etherington, On non–associative combinations, *Proc. Roy. Soc. Edinburgh* **59** (1938/39) 153–162.

[GoK73]   M. Gordon and J. W. Kennedy, The graph–like state of matter, Part 2, LGCI schemes for the thermodynamics of alkanes and the theory of inductive inference, *J. Chem. Soc.*, Farady II **69** (1973) 484–504.

[GoL75]   M. Gordon and C. G. Leonis, Combinatorial shortcuts to statistical weights and enumeration of chemical isomers, *Proc. 5th British Combinatorial Conf.* (1975) 231–238.

[GoT76]   M. Gordon and W. B. Temple, The graph–like state of matter and polymer science, Chapter 10 in *Chemical Applications of Graph Theory* (A. T. Balaban, ed.), Academic, London (1976).

[HP73]   F. Harary and E. M. Palmer, *Graphical Enumeration*, Academic, New York (1973).

[HRS75]   F. Harary, R. W. Robinson and A. J. Schwenk, Twenty–step algorithm for determining the asymptotic number of trees of various species, *J. Austral. Math. Soc. Ser. A* **20** (1975) 483–503.

[KMPR]   J. W. Kennedy, K. A. McKeon, E. M. Palmer and R. W. Robinson, Asymptotic number of symmetries in locally–restricted trees, to appear.

[Mc87]   K. A. McKeon, Enumeration of symmetries in locally–restricted trees, Dissertation, Michigan State University, (1987).

[O48]   R. Otter, The number of trees, *Ann. of Math.* **49** (1948) 583–599.

[P37]   G. Pólya, Kombinatorische Anzahlbestimmungen für Gruppen, Graphen und chemische Verbindungen, *Acta Math.*, **68** (1937) 145–254.

MULTIDIMENSIONAL BANDWIDTH IN RANDOM GRAPHS

Zevi Miller

Miami University

ABSTRACT

Let $[n]^k$ be the graph of the k-dimensional grid of side n. That is, $[n]^k$ has as its vertices the lattice points in k-dimensional space with coordinates between 0 and n, and as its edges those pairs of lattice points differing by exactly 1 in one coordinate. Let G be a random graphs with fixed edge probability and having $N = |[n]^k| = (n+1)^k$ vertices. Given a one to one map of vertices $f : G \to [n]^k$, let $[f] = max[dist_{[n]^k}(f(x), f(y)) : xy\epsilon E(G)]$ and $B(G) = min_f |f|$. Under the assumption $\Omega(k) = n \leq e^{e^{o(k)}}$, we show that $p[diam([n]^k) - C_1 \, log(log \, N) \leq B(G) \leq diam(diam([n]^k) - C_2 \frac{log(log \, N)}{log \, k}] \to 1$ as $k \to \infty$, where C_1 and C_2 are constants.

1.   Introduction

Let G and H be graphs with $|G| = |H|$, and $f : V(G) \to V(H)$ a one to one map of the vertex set of G to that of H. Let $|f| = max[dist_H(f(x), f(y)) : xy\epsilon E(G))]$ (where $dist_H$ denotes distance in H), and let $B(G, H) = min_f |f|$.

The investigation of $B(G, H)$ for certain classes of "host graphs" H, apart from its intrinsic interest as a graph theory problem is motivated by issues in VLSI and parallel computation. when H is a path the parameter $B(G, H)$, denoted $b(G)$, is known as the *bandwidth* of G.

The problem of determining $b(G)$ is NP-complete, even when G is a tree of maximum degree three $[GGJK]$. This has prompted research into the probabilistic analysis of $b(G)$. The probable behavior of $b(G)$ for random graphs with fixed edge probability p was investigated in [KMc], where it was shown that for almost all graph on n points we have $b(G) = n - (2 + 2^{\frac{1}{2}} + 0(1))\frac{\log(n)}{\log(\frac{1}{1-p})}$. A similar result is proved in [EHW]. In [Tu] the success of some well known heuristics for $b(G)$ is explained from a probabilistic point of view.

We now consider the case when H is a grid of arbitrary dimension. As notation, let $[n]^k$ denote the graph with vertex set $V = \{(x_1, x_2, ..., x_k) : 0 \leq x_i \leq n, x_1 \text{ integers}]$ and edges consisting of those unordered pairs $[x, y]$ from V for which $\sum_{i=1}^{d} |x_i - y_i| = 1$. Thus $V([n]^k)$ may be identified with the lattice points in k-dimensional space having coordinates between 0 and n, where two points are joined by an edge when they are "adjacent" as lattice points. Note that $[n]^k$ has $(n+1)^k$ points and diameter kn. The NP-completeness of finding $B(G, [n]^k)$ when $k = 2$ was shown independently in [Mil], [BCo], and [BSu], and the proofs extend readily to arbitrary dimension. Bounds for $B(G, [n]^k)$ may be derived from [Ro] and [RoSn], and other work by these and other authors.

Given certain upper and lower bounds for n in terms of k, our purpose in this paper is to give reasonably tight upper and lower bounds for $B(G, [n]^k)$ which are valid with probability approaching 1 as $k \to \infty$.

2. Grid Bandwidth in Random Graphs

We take as our probability model the set $G(N, p)$ of all labelled graphs on $N = (n+1)^k$ points, with labels from $\{1, 2, ..., N\}$, having edge probability p where p is fixed. We set $q = 1 - p$, and we denote an element of $G(N, p)$ by G. Under this model the probability that two points i and j are joined by an edge in G is p, and the two events $i_1 j_1 \epsilon E(G)$ and $i_2 j_2 \epsilon E(G)$ are independent for distinct pairs of points $i_1 j_1$ and $i_2 j_2$. For applications of this model to many different problems in graphs theory the reader is referred to [Bo] and [Pa].

In this section we will study the probable behavior of $B(G, H)$, where $G \in G(N, p)$ and $H = [n]^k$. As H will always have this meaning, we abbreviate

$B(G,H)$ by $B(G)$ from now own.

We use the standard notation $O(s(n)), o(s(n))$, and $\Omega(s(n))$ to denote any function $g(n)$ for which the absolute ration $|\frac{g(n)}{s(n)}|$ is bounded above by a constant, approaches 0, and is bounded below by a constant respectively as n approaches ∞.

We will need the following facts for later use. The first gives standard bounds on binomial coefficients, and the second follows readily from Taylor's theorem.

Lemma 1

a) $\left(\frac{n}{k}\right) \leq \binom{n}{k} \leq \left(\frac{ne}{k}\right)^k$ for any $k \leq n$.

b) Suppose $r(k)$ is a function satisfying $r(k) = o(k)$. Then $(1 + \frac{r(k)}{k})^k \sim \exp(O(r(k)))$ as $k \to \infty$.

We now introduce some notation. A point of $[n]^k$ will be called an *extreme point* if each of its coordinates is 0 or n. Note that there are 2^k such points in $[n]^k$. For each extreme point v let $D_i(v)$ be the set of points in $[n]^k$ at distance i from v and let $Nb(v,r) = \bigcup_{i=0}^{r} D_i(v)$. We call the sets $D_i(v)$ the *diagonals* at v and $Nb(v,r)$ the *r-neighborhood* about v, and note that by symmetry $|Nb(v,r)| = |Nb(w,r)|$ for any two extreme points v and w. For $i \leq n$ we can identify $D_i(v)$ with the set of nonnegative integer sequences $(x_1, x_2, ..., x_k)$ such that $\sum_{j=i}^{k} x_j = i$, so $|D_i(v)| = \binom{i+k-1}{k-1}$ for $i \leq n$. As $|D_i(v)|$ is independent of v, we denote it by D_i. Also for $r \leq n$ we then have $|Nb(v,r)| = \sum_{i=o}^{r} \binom{i+k-1}{k-1} = \binom{r+k}{k}$.

Let $r \geq 0$ be and consider a sequence of sets $(S_0, S_1, ..., S_r)$ where S_i is a subset of $G \epsilon G(n,p)$ of size D_i, and $S_i \cap S_j = \phi$ for $i \neq j$. The set $\bigcup_{i=0}^{r} S_i$ will be called an *r-corner* of G (or just corner when r is understood). Thus an r-corner in G has the same size as an r-neighborhood about an extreme point in $[n]^k$, and we will refer to the sets S_i as the *diagonals* of the corner. Now let $v(1), v(2), ..., v(2^k)$ be a fixed ordering of the extreme points of $[n]^k$. A sequence $(S^{(1)}, S^{(2)}, ..., S^{(2^k)})$ of r-corners of G will then be called an *r-system* if the set systems $\bigcup_{j=0}^{2^k} S_j$ and $\bigcup_{j=0}^{2^k} Nb(v(j), r)$ (with edge sets $[S^{(j)} : 1 \le j \le 2^k]$ and $[Nb(v(j), r) : 1 \le j \le 2^k]$ respectively) are isomorphic by an isomorphism under which the induced map on edges sends $S^{(j)}$ to $Nb(v(j), r)$ for each j. As motivation for these definitions note that if $f : G \to [n]^k$ is an embedding, then the inverse image of the collection $\{Nb(v, r) : v$ extreme $\}$ must be an r-system while the inverse image of each $Nb(v, r)$ must be an r-corner of that r-system.

We will assume now that the ordering $v(i)$ of the extreme points is such that for any odd i the pair $\{v(i), v(i + 1)\}$ are antipodal (i.e. at distance kn). Let S and T be two r-corners of G (with diagonal sets $\{S_i\}$ and $\{T_i\}$ respectively) such that each point $v \epsilon S_i$ is nonadjacent in G to all points in $\bigcup_{j=0}^{r-1} T_j$. (It follows symmetrically that any point $v \epsilon T_i$ is nonadjacent in G to all points in $\bigcup_{j=0}^{r-i} S_j$). We then call S and T *opposite corners*. An r-system will then be termed a *good r-system* if each pair $\{S^{(i)}, S^{(i+1)}\}$ with i odd are opposite corners. If S is a good r-system of G, then the required set of nonadjacencies among the 2^{k-1} opposite corner pairs of S will be referred to as the *nonadjacencies implied by S*.

The significance of a good r-system is seen in the following lemma.

Lemma 2 $B(G) \le \operatorname{diam}([n]^k) - r - 1 \leftrightarrow G$ has a good r-system.

Proof   Observe that points $s, t \epsilon [n]^k$ satisfy $\operatorname{dist}(s, t) \ge kn - r$ iff there are antipodal extreme points $v(i)$ and $v(i + 1)$ such that (without loss of generality) $s \epsilon D_j(v)$ and $t \epsilon \bigcup_{t=0}^{r-j} D_t(v(i + 1))$ for some $j \le r$.

Now let $f : G \to [n]^k$ be an embedding. It follows that $|f| \le kn - r - i$ if and only if for each extreme point $v(i) \epsilon [n]^k$ (say with i odd) the two sets $f^{-1}(Nb(v(i), r))$ and $f^{-1}(Nb(v(i + 1), r))$ are opposite corners in G. Thus $|f| \le kn - r - 1$ if and only if the sequence $(f^{-1}(Nb(v(1), r)),$

$f^{-1}(Nb(v(2),r)),...,f^{-1}(Nb(v(2^k),r)))$ is a good r-system in G.

We may now proceed to our theorem.

Theorem Let $N = (n+1)^k$. Assume $\Omega(k) = n \le e^{e^{o(k)}}$. Then there are constants C_1 and C_2 such that for p fixed and $G\epsilon G(N,p)$ we have
$$P[\operatorname{diam}([n]^k) - C_1 \log(\log N) \le B(G) \le \operatorname{diam}([n]^k) - C_2 \frac{\log(\log N)}{\log k}] \to \quad \text{as}$$
$k \to \infty$.

Toward proving this theorem we begin with a lemma.

Lemma 3 The number of nonadjacencies implies by an r-system for $r \le n$ is $2^{k-1}\binom{2k+r}{r}$.

Proof Consider an opposite corner pair S and T (with diagonal sets $\{S_i\}$ and $\{T_i\}$ respectively) belonging to an r-system. For a given $i \le r$ we have $|\bigcup_{m=0}^{r-1} T_m| = \sum_{m=0}^{r-i} \binom{m+k-1}{k-1} = \binom{r-i+k}{k}$. Thus the number of nonadjacencies between S_i and $\bigcup_{m=0}^{r-i} T_m$ is $\binom{i+k-1}{k-1}\binom{r-i+k}{k}$. Hence the number of implied nonadjacencies among S and T is

$$\sum_{i=0}^{r}\binom{i+k-1}{k-i}\binom{r-i+k}{k} = \sum_{i=0}^{r}\binom{i+k-1}{i}\binom{r-i+k}{r-i} = \binom{2k+r}{r}.$$

Since there are 2^{k-1} pairs of opposite corners in an r-system, the lemma follows.

As further notation, for $G\epsilon G(n,p)$ let $N_r(G)$ be the number of r-systems in G. Also if $n = rs$ then let $\binom{n}{rXs}$ denote the multinomial coefficient $\binom{n}{r,r,...r}$ where the r appears s times in the bottom.

Proof of Theorem First we show that $P[B(G) > \operatorname{diam}([n]^k) - C \log(\log N)] \to 1$ as $k \to \infty$.

First observe that with $r = C \log(\log N)$ and k sufficiently large we have

$$E(N_r) = \left(2^k\binom{r^N+k}{k}\right)\left(\frac{2^k\binom{r+k}{k}}{\binom{r+k}{k} \times 2^k}\right)\left(\frac{\binom{r+k}{k}}{\binom{k-1}{k-1}\binom{1+k-1}{k-1}...\binom{r+k-1}{k-1}}\right)^{2^k}(2^k)!q^{\binom{r+2k}{r}2^{k-1}}$$

To see this, note that since $r = \log(k) + \log\log(n) = o(k)$ and $n = \Omega(k)$ it follows that for k sufficiently large the r-corners making up an r-system are pairwise disjoint. Hence the number of points in an r-system for large k is $2^k\binom{r+k}{k}$. Thus the first factor on the right is the number of ways of choosing the ground set of $2^k\binom{r+k}{k}$ vertices for our r-system. The second factor is the number of ways to partition these vertices into a collection of 2^k corners. The third factor is the number of ways to partition each of the corners into the diagonals $S_0, S_1, S_2, ..., S_r$, these partitions being independent. The fourth factor is the number of different orderings of the 2^K corners. Thus the product of the first four factors is the number of r-systems. The exponent of q is the number of disallowed edges in a good r-system by lemma 3, so the fifth factor is the probability that an r-system is a good r-system. Thus the full product is the expectation of N_r.

We now estimate $E(N_r)$. First we will bound the multinomial coefficients by powers of binomial coefficients. The one with $2^k\binom{r+k}{k}$ as the top is clearly at most $\left(\frac{2^k\binom{r+k}{k}}{\binom{r+k}{k}}\right)^{2^k}$. The one with $\binom{r+k}{k}$ as the top may be written as the product $\prod_{j=0}^{r}\left(\frac{\binom{j+k}{k}}{\binom{j+k-1}{k-1}}\right)$ using the fact that for any integer $j \geq 0$ we have $\sum_{i=0}^{j}\binom{i+k-1}{k-1} = \binom{j+k}{k}$. Now some analysis of binomial coefficients shows that the biggest among the $r+1$ factors in this product is $\left(\frac{\binom{r+k}{k}}{\binom{r+k-1}{k-1}}\right)$. Hence the multinomial coefficient is bounded above by $\left(\frac{\binom{r+k}{k}}{\binom{r+k-1}{k-1}}\right)^{r+1}$.

Now substituting for the multinomial coefficients by the above bounds, and using lemma 1 to upper bound all binomial coefficients, one can show that if $r > C\log(\log N)$ for a suitable constant C, then $E(N_r) \to 0$ as $k \to \infty$. We omit details of this analysis here, as they will appear later in the full paper. Now using lemma 2 and Markov's inequality we have $P[B(G) \leq \text{diam}([n]^k) - r - 1] = P[N_r \geq 1] \leq E(N_r) \to 0$ as $k \to \infty$ for $r > C\log(\log N)$. Hence $P[B(G) > \text{diam}([n]^k) - C\log(\log N)] \to 1$ as $k \to \infty$, proving half of our theorem.

To prove that $\quad P[B(G) \leq \operatorname{diam}([n]^k) - C_2 \frac{\log(\log N)}{\log k}] \to 1 \quad$ we will apply the second moment method.

Let R be the set of r-systems. For each $\quad A \epsilon R \quad$ let

$$Z_A(G) = \begin{cases} 1 & \text{if } A \text{ is a good r-system} \\ 0 & \text{otherwise} \end{cases}$$

Note that $\quad N_r(G) = \sum_{A \epsilon R} Z_A(G).$

We begin by estimating the variance of $\quad N_r.$ Let F be a fixed member of $R.$ Then

$$\begin{aligned} \operatorname{Var}(N_r) &= E(N_r^2) - E(N_r)^2 \\ &= \sum_{A \epsilon R} \sum_{B \epsilon R} (E(Z_A Z_B) - E(Z_A)E(Z_B)) \\ &= |R| \sum_{A \epsilon R} (E(Z_F Z_A) - E(Z_F)E(Z_A)). \end{aligned}$$

To estimate $E(Z_F Z_A)$, note first that since Z_F and Z_A are 0-1 valued random variables we have $E(Z_F Z_A) = P(Z_F Z_A = 1) = P(Z_A = 1 | Z_F = 1).P(Z_F = 1).$

By lemma 2 we have $P(Z_F = 1) = q^{\binom{2k+r}{r} 2^{k-1}}.$ Now given that F is a good r-system the probability that A is a good r-system is just the probability that A has the required nonadjacencies not already implied by F. This probability is $q^{\binom{2k+r}{r} 2^{k-1} - c(A)}$, where $c(A)$ is the number of nonadjacencies implied by jointly by A and F. Thus $E(Z_F Z_A) = q^{\binom{2k+r}{r} 2^k - c(A)}.$

We can now develop the above expression a bit further. Let $h = 2^k \binom{r+k}{k}$, and let v_t be the set of elements of R having t points in common with F. Also let $c(t)$ be the maximum number of nonadjacencies implied jointly by F and any member of v_t. Then continuing from above we have

$$\mathrm{Var}(N_r) = |R|q^{\binom{2k+r}{r}2^k}\sum_{A\epsilon R}(q^{-c(A)}-1)$$

$$\leq |R|q^{\binom{2k+r}{r}2^k}\sum_{t=2}^{h}|\nu_t|q^{-c(t)}$$

$$= |R|^2 q^{\binom{2k+r}{r}2^k}\sum_{t=2}^{h}\frac{|\nu_t|}{|R|}q^{-c(t)}$$

$$= |E|(N_r)^2\sum_{t=2}^{h}\frac{|\nu_t|}{|R|}q^{-c(t)}$$

We therefore have

$$\frac{\mathrm{Var}(N_r)}{E(N_r)^2} \leq \sum_{t=2}^{h}\frac{|\nu_t|}{|R|}q^{-c(t)} \quad (A1)$$

It remains to estimate the right side of the last expression. First note that we may write $\nu_t = N_1 N_2$, where $N_1 =$ number of ways to choose h points from $\{1,2,...,N\}$ of which t are in the ground set of F, and $N_2 =$ number of ways to arrange any ground set counted in N_1 into a member of $|R|$.

We have $N_1 = \binom{h}{t}\binom{N-h}{h-t}$, while $N_2 = \frac{|R|}{\binom{N}{h}}$. Therefore from $(A1)$ we get

$$\frac{\mathrm{Var}(N_r)}{E(N_r)^2} \leq \sum_{t=2}^{h}\frac{\binom{h}{t}\binom{N-h}{h-t}}{\binom{N}{h}}q^{-c(t)} \quad (A2)$$

We recognize the coefficients of the $q^{-c(t)}$ in this sum as the terms of the hypergeometric distribution. In the course of approximating the hypergeometric by the Poisson distribution one can derive the following inequality [Bo pp. 7-8].

$$\frac{\binom{h}{t}\binom{N-h}{h-t}}{\binom{N}{h}} \leq \frac{(\frac{h^2}{N})^t \mathrm{ext}(-\frac{h^2}{N})}{t!}(1-\frac{t}{N})^{t-h}\mathrm{ext}(\frac{ht}{N})$$

(This is obtained by first approximating the hypergeometric by the binomial distribution b(t; h,p) with success probability $p = \frac{h}{N}$, and then the binomial by the Poisson distribution with mean $\lambda = ph = \frac{h^2}{N}$).

Next we need an estimate for $c(t)$. Consider then a set $Q\epsilon\nu_t$ at which the maximum defined by $c(t)$ is realized. Thus $c(t)$ is the number of nonadjacent pairs xy, x, y ϵ Q $\cap$ F, implied jointly by Q and F. Now $c(t) = \frac{1}{2}\sum_{x\epsilon Q\cap F} d(x)$, where $d(x)$ is the number of nonadjacencies implied jointly by Q and F in which x is a member. Since $Q\epsilon R$ we have $d(x) \le \binom{r+k}{k}$, with equality occurring precisely when there is a corner pair S, T in Q for which $x = S_0$ and the points counted in $d(x)$ comprise all of T. Thus $c(t) \le \frac{1}{2}t\binom{r+k}{k}$.

Now using Chebyshev's inequality and (A2) we therefore have $P(B > \mathrm{diam}([n]^k) - r - 1) = P(N_r = 0)$

$$\le \frac{\mathrm{Var}(N_r)}{E(N_r)^2}$$

$$\le \exp(-\frac{h^2}{N})\sum_{t=2}^{h}\frac{(\frac{1}{h^2}N)^t}{t!}(1-\frac{t}{N})^{t-h}\exp\frac{1}{ht}N)q^{-t\binom{r+k}{k}}$$

$$\le \exp(-\frac{h^2}{N})(1-\frac{h}{N})^{-h}[\exp\{\frac{h^2}{N}\exp\frac{1}{h}N)q^{-\binom{r+k}{k}}\} - 1 - \frac{h^2}{N}\exp\frac{h}{N}q^{-\binom{r+k}{k}}] \quad (A3)$$

Now using lemma 1 and the assumed upper bound on n one can show that if $r \le C\frac{\mathrm{loglog}(N)}{\log(k)}$ for an appropriate constant C, then the right side of (A3) approaches 0 as $k \to \infty$. Again the details will appear later in the full paper. It follows that almost every graph satisfies $B(G) \le \mathrm{diam}([n]^k) - C\frac{\mathrm{loglog}(N)}{\log(k)}$ and the theorem is proved.

Remark J. Spencer [Sp] points out the following. From random graph theory we know that almost every graph on N points has essentially $\frac{N}{\log N}$ disjoint independent sets of size $\log N$. We can use these sets as r-corners in constructing a good r-system. The value of r is determined by the requirement $\log N \sim \binom{r+k}{k} \sim r^k$ for fixed k as $n \to \infty$ (where $N = (n+1)^k$). Thus for fixed k we have $B(G) \le \mathrm{diam}([n]^k) - (\log N)^{1/k}$ for almost all graphs.

Now for $n \ge e^{e^{o(k^{1+\epsilon})}}$ this upper bound on $B(G)$ is an improvement on the one presented here. In fact it is less than the lower bound we have, which shows that the hypothesis $n \le e^{e^{o(k)}}$ in our theorem cannot be relaxed very much. But under this hypothesis the upper bound of our theorem in the range $n \ge e^{k^{1+\epsilon}}$ is an improvement on the one above.

REFERENCES

[1] [BSu] P. Bertolazzi, I. H. Sudborough, "The grid embedding problem is NP-complete even for edge length 2", *Technical Report* (1983), EE/CS Dept., Northwestern University.

[2] [BCo] S. Bhatt, S. Cosmadakis, "The complexity of minimizing wire lengths in VLSI layouts", manuscript (1983).

[3] [Bo] B. Bollobas, "Random Graphs", *Academic Press* (1985).

[4] [EHW] P. Erdös, P. Hell, P. Winkler, "Bandwidth versus bandsize", manuscript (1985).

[5] [GGJK] M. Garey, R. Graham, D. Johnson, D. Knuth, "Complexity results for bandwidth minimization", SIAM J. of Applied Mathematics 34 (1978) 477-495.

[6] [KMc] Y. Kuang, C. McDiarmid, "On the bandwidth of a random graph", manuscript (1985).

[7] [Mil] Z. Miller, J. Orlin, "NP-completeness for minimizing maximum edge length in grid embeddings", *J. of Algorithms* 6, (1985) 10-16.

[8] [Pa] E. M. Palmer, *Graphical Evolution: An Introduction to the Theory of Random Graphs*, John Wiley & Sons (1985).

[9] [Ro] A. Rosenberg, "On embedding graphs in grids", *IBM Research Report RC* 7559 (#2668) (1970).

[10] [RoSn] A. Rosenberg, L. Snyder, "Bounds on the costs of data encodings", *Mth. Systems Theory* 12 (1978) 9-39.

[11] [Sp] J. Spencer, private communication.

[12] [Tu] J. Turner, "Probabilistic analysis of bandwidth minimization algorithms", *Proc. 15th Annual ACM Symposium on Theory of Computing* (1983) 467-476.

THE LAPLACIAN SPECTRUM OF GRAPHS

Bojan Mohar

University of Ljubljana

Jadranska 19, 61111 Ljubljana, Yugoslavia

ABSTRACT

The paper is essentially a survey of known results about the spectrum of the Laplacian matrix of graphs with special emphasis on the second smallest Laplacian eigenvalue λ_2 and its relation to numerous graph invariants, including connectivity, expanding properties, isoperimetric number, maximum cut, independence number, genus, diameter, mean distance, and bandwidth-type parameters of a graph. Some new results and generalizations are added.

1. Introduction

The Laplacian matrix of a graph and its eigenvalues can be used in several areas of mathematical research and have a physical interpretation in various physical and chemical theories. The related matrix - the adjacency matrix of a graph and its eigenvalues were much more investigated in the past than the Laplacian matrix. The reader is referred to the monographs [CDS, CDGT]. However, in the author's opinion the Laplacian spectrum is much more natural and more important than the adjacency matrix spectrum. It is the aim of this survey paper to explain where this belief comes from.

The work supported in part by the Research Council of Slovenia, Yugoslavia. Part of the work was done while the author was a Fulbright Scholar at the Ohio State University, Columbus, Ohio.

We shall use the standard terminology of graph theory, as it is introduced in most textbooks on the theory of graphs (e.g., [Wi]). Our graphs are unoriented, but they may have loops and multiple edges. We also allow *weighted graphs* which are viewed as a graph which has for each pair u, v of vertices, assigned a certain *weight* $a_{u,v}$. The weights are usually real numbers and they must satisfy the following conditions:

(i) $a_{u,v} = a_{vu}, v, u \in V(G)$, and

(ii) $a_{vu} \neq 0$, if and only if v and u are adjacent in G.

Usually the additional condition on the non-negativity of weights is assumed:

(iii) $a_{uv} \geq 0, v, u \in V(G)$.

It will be clear from the context or otherwise explicitly specified if a graph is weighted. Unweighted graphs can be viewed as a special case of weighted graphs, by specifying, for each $u, v \in V(G)$, the weight a_{uv} be equal to the number of edges between u and v. The matrix $A = A(G) = [a_{uv}]_{u,v \in V(G)}$, is called the *adjacency matrix* of the graph G. We shall use the same name for the matrix of weights if the graph is weighted.

Let $d(v)$ denote the degree of $v \in V(G), d(v) = \sum_u a_{uv}$, and let $D = D(G)$ be the diagonal matrix indexed by $V(G)$ and with $d_{vv} = d(v)$. The matrix $Q = Q(G) = D(G) - A(G)$ is called the *Laplacian matrix* of G. It should be noted at once that loops have no influence on $Q(G)$. The matrix $Q(G)$ is sometimes called the *Kirchhoff matrix* of G due to its role in the well-known Matrix-Tree theorem (cf. §4)which is usually attributed to Kirchhoff. Another name, the *matrix of admittance*, comes from the theory of electrical networks (admittance = conductivity). It should be mentioned here that the rows and columns of graph matrices are indexed by the vertices of the graph, their order being unimportant. The matrix $Q(G)$ acts naturally on the vector space $\ell^2(V(G))$. For any vector $x \in \ell^2(V(G))$ we denote its coordinates by $x_v, v \in V(G)$.

Throughout the paper we shall denote by $\mu(G, x)$ the characteristic polynomial of $Q(G)$. Its roots will be called the *Laplacian eigenvalues* (or sometimes just *eigenvalues*) of G. They will be denoted by $\lambda_1 \le \lambda_2 \le \cdots \le \lambda_n (n = |V(G)|)$, always enumerated in increasing order and repeated according to their multiplicity. We shall use the notation $\lambda_k(G)$ to denote the k-th smallest eigenvalue of the graph G (counting multiplicities). The letter n will always stand for the order of G, so $\lambda_n(G)$ will be the maximal eigenvalue of $Q(G)$.

Let G be a given graph. Orient its edges arbitrarily, i.e. for each $e \in E(G)$ choose one of its ends as the *initial* vertex, and name the other end the *terminal* vertex. The *oriented incidence matrix* of G with respect to the given orientation is the $|V| \times |E|$ matrix $C = [c_{ve}]$ with entries

$$c_{ve} = \begin{cases} +1, & \text{if } v \text{ is the terminal vertex of } e, \\ -1, & \text{if } v \text{ is the initial vertex of } e, \\ 0, & \text{if } v \text{ and } e \text{ are not incident.} \end{cases}$$

It is well known that

$$Q(G) = CC^t \tag{1.1}$$

independent of the orientation given to the edges of G (cf., e.g., [Bi]). It should be noted that (1.1) immediately implies the formula (2.1) since the inner product $(Q(G)x, x)$ is equal to $(CC^t x, x) = (C^t x, C^t x)$.

The Laplace differential operator Δ is one of the basic differential operators in mathematical physics. One looks for non-trivial solutions of $\Delta \phi = \lambda \phi$ on a certain region Ω. By discretizing the Laplace equation one gets the Laplacian matrix Q of the discretisized space (usually a graph). We mention that, by this correspondence, the oriented incidence matrix C, as defined above, corresponds to the gradient operator, and so (1.1) has a clear physical interpretation.

In Section 2 we review the basic spectral properties of $Q(G)$. The next section presents the results on the spectra of graphs obtained by means of some operations on graphs, including the disjoint union, Cartesian product and the join of graphs, deleting or inserting an edge, the complement, the line graph, etc. Section 4 is devoted to the renowned application of $Q(G)$, the Matrix-Tree-Theorem, which expresses the number of spanning trees of a graph in terms of its non-zero eigenvalues.

There are many problems in physics and chemistry where the Laplacian matrices of graphs and their spectra play the central role. Some of the applications are mentioned in Section 5. It is worth noting that the physical background served as the idea of a well known algorithm of W. T. Tutte [T] for testing planarity and constructing "nice" planar drawings of 3-connected planar graphs.

The second smallest Laplacian eigenvalue λ_2 plays a special role. Recently its applications to several difficult problems in graph theory were discovered (e.g., the expanding properties of graphs, the isoperimetric number, and the maximum cut problem). Section 6 presents these applications, including the relation of λ_2 to the diameter and the mean distance of a graph. In addition, a relation of λ_2 to the independence number, genus, and bandwidth-type invariants is presented. The structure of the eigenvectors corresponding to λ_2 is discussed in the next section. The last section covers a few other results on $Q(G)$ and its applications.

There are some new results in this paper. Many of them are more or less trivial and have probably been known to researchers in the field, although not published before. The results surveyed in the paper are biased by the viewpoint of the author. We apologize to all who feel that their work is missing in the references, or has not been emphasized sufficiently in the text.

2. Basic Properties

The following properties were established by several authors [K3, V, AnM] for the case of unweighted graphs. The proofs carry over to the weighted case if all the weights are non-negative.

Theorem 2.1   Let G be a (weighted) graph with all weights non-negative.

Then:

(a) $Q(G)$ has only real eigenvalues,

(b) $Q(G)$ is positive semidefinite,

(c) its smallest eigenvalue if $\lambda_1 = 0$ and a corresponding eigenvector is $(1, 1, ..., 1)^t$. The multiplicity of 0 as an eigenvalue of $Q(G)$ is equal to the number of components of G.

We mention that the positive semidefiniteness of $Q(G)$ follows from the next useful expression for the inner product $(Q(G)x, x)$ which holds also in the weighted case:

$$(Q(G)x, x) = \sum_{vu \in E} a_{vu}(x_v - x_u)^2 \qquad (2.1)$$

Let $\lambda_1 \leq \lambda_2 \leq \cdots \leq \lambda_n$ be the eigenvalues of $Q(G)$ in increasing order and repeated according to their multiplicity. So, $\lambda_1 = 0$, and $\lambda_2 > 0$ if and only if G is connected. The following bounds for the eigenvalues are known.

Theorem 2.2　　Let G be a graph of order n. Then:

(a) [F1] $\lambda_2 \leq \frac{n}{n-1} \min\{d(v); v \in V(G)\}$.

(b) [AnM] $\lambda_n \leq \max\{d(u) + d(v); uv \in E(G)\}$. If G is connected then the equality holds if and only if G is bipartite semiregular.

(c) [K3] If G is a simple graph then $\lambda_n \leq n$ with equality if and only if the complement of G is not connected.

(d) $\sum_{i=1}^{n} \lambda_i = 2|E(G)| = \sum_{v} d(v)$.

(e) [F1] $\lambda_n \geq \frac{n}{n-1} \max\{d(v); v \in V(G)\}$.

(f) [MM, p. 168] $\lambda_n \geq \max\{\sqrt{(d(v) - d(u))^2 + 4a_{uv}^2}; v, u \in V(G), v \neq u\}$.

Let G be a (weighted) graph and $V_1 \cup V_2 \cup ... \cup V_k$ a partition of its vertex set. This partition is said to be *equitable* if for each $i, j = 1, 2, ..., k$ there is a number d_{ij} such that for each $v \in V_i$ there are exactly d_{ij} edges between v and vertices in $V_j (\sum_{u \in v_j} a_{vu} = d_{ij}, v \in V_i)$. The name equitable partition was introduced by Schwenk. There are several other terms used for the same thing (e.g., divisor [CDS], coloration degree refinement, etc.).

Theorem 2.3 Let $V_1 \cup V_2 \cup ... \cup V_k$ be an equitable partition of G with parameters $d_{ij}(i, m = 1, 2, ..., k)$, and let $B - [b_{ij}]_{i,j=1,...,k}$ be the matrix defined by

$$b_{ij} \begin{cases} -d_{ij}, & \text{if } i \neq j \\ (\sum_{s=1}^{k} d_{is}) - d_{ii}, & \text{if } i = j. \end{cases}$$

If λ is an eigenvalue of B then λ is also an eigenvalue of $Q(G)$.

Proof Let $Bx = \lambda x, x = (x_1, ..., x_k)^t$. Let $y = (y_v)_{v \in V(G)}$ be defined by: if $x \in V_i$ then let $y_v = x_i$. Now it is not too difficult to verify that $Q(G)y = \lambda y$. Let $v \in V_i$ be any vertex of G. Then

$$(Qy)_v = d(v)y_v - \sum_u a_{vu}y_u = (\sum_{j=1}^{k} d_{ij})x_i - \sum_{j=1}^{k} d_{ij}x_j =$$
$$= (Bx)_i = \lambda x_i = \lambda y_v.$$

Note We may view $D = [d_{ij}]$ as the matrix of a weighted *directed* graph with k vertices. Then B is just its Laplacian matrix.

Let us mention briefly that equitable partitions of vertices arise in many important situations. For example, if $p : \tilde{G} \to G$ is a graph covering projection (in the sense of topology) then the fibres $p^{-1}(v), v \in V(G)$, form an equitable partition of $\tilde{G}$. The corresponding matrix B is just $Q(G)$, and this shows that the Laplacian spectrum of $\tilde{G}$ contains the spectrum of G. Many examples of equitable partitions of a graph G are obtained by taking, as the classes of a partition, the orbits of some group of automorphisms of the graph G.

3. Operations on Graphs and the Resulting Spectra

Many published works relate the Laplacian eigenvalues of graphs with the eigenvalues of graphs obtained by means of some operations on the graphs we start with. The first result is obvious.

Theorem 3.1 Let G be the disjoint union of graphs $G_1, G_2, ..., G_k$. Then

$$\mu(G,x) \prod_{i=1}^{k} \mu(G_i,x).$$

Let G be a (weighted) graph and let $G' = G + 3$ be the graph obtained from G by inserting a new edge e into G (possibly increasing the multiplicity of an existing edge). Then $Q(G')$ and $Q(G)$ differ by a positive semidefinite matrix of rank 1. It follows by the Courant-Weyl inequalities (see, e.g., [CDS, Theorem 2.1]) that the following is true.

Theorem 3.2   The eigenvalues of G and $G' = G + e$ interlace:

$$0 = \lambda_1(G) = \lambda_1(G') \le \lambda_2(G) \le \lambda_2(G') \le \lambda_3(G) \le \ldots \le \lambda_n(G) \le \lambda_n(G').$$

We notice that $\sum_{i=1}^{n}(\lambda_i(G') - \lambda_i(G)) = 2$ by Theorem 2.2(d), so that at least one inequality $\lambda_i(G) \le \lambda_i(G')$ must be strict.

By inserting more than one edge we may loose interlacing of the eigenvalues. Nevertheless, there is an important result on λ_2.

Theorem 3.3   Let $G = G_1 \oplus G_2$ be a factorization of a graph G. Then
(a) [F1] $\lambda_2(G) \ge \lambda_2(G_1) + \lambda_2(G_2)$.
(b) $\max\{\lambda_n(G_1), \lambda_n(G_2)\} \le \lambda_n(G) \le \lambda_n(G_1) + \lambda_n(G_2)$.

Note   In Theorem 3.3, graphs may be weighted with non-negative weights. In that case the factorization means that the uv-weight in G is the sum of the uv-weights in G_1 and G_2.

Proof   (b) $G = G_1 \oplus G_2$ means that $Q(G) = Q(G_1) + Q(G_2)$. Then

$$\lambda_n(G) = \max_{||x||=1}(Q(G)x, x) = \max_{||x||=1}[(Q(G_1)x, x) + (Q(G_2)x, x)] \le$$
$$\le \max_{||x||=1}(Q(G_1)x, x) + \max_{||x||=1}(Q(G_2)x, x) = \lambda_n(G_1) + \lambda_n(G_2).$$

The other inequality follows from a similar consideration.

Corollary 3.4   [F1] If G_1 is a spanning subgraph of G_2 then $\lambda_2(G_1) \le \lambda_2(G_2)$.

Fiedler [F1] derived also a result about the Cartesian products of graphs.

Theorem 3.5   [F1] The Laplacian eigenvalues of the Cartesian product $G_1 \times G_2$ of graphs G_1 and G_2 are equal to all the possible sums of eigenvalues of the two factors:

$$\lambda_i(G_1) + \lambda_j(G_2), \ i = 1, ..., |V(G_1)|, j = 1, ..., |V(G_2)|.$$

By applying Theorem 3.5 we can easily determine the spectrum of "lattice" graphs. The $m \times n$ lattice graph is just the Cartesian product of paths, $P_m \times P_n$. The spectrum of P_k is [AnM]

$$\ell_i^{(k)} = 4\sin^2(\frac{\pi i}{2k}), \ i = 0, 1, ..., k - 1$$

so $P_m \times P_n$ has eigenvalues

$$\lambda_{i,j} = \ell_i^{(m)} + \ell_j^{(n)} = 4\sin^2(\frac{\pi i}{2m}) + 4\sin^2(\frac{\pi i}{2n}).$$

The next two results were first observed by Kel'mans [K1,K2].

Theorem 3.6   [K1,K2] If $\overline{G}$ denotes the complement of the graph G then

$$\mu(\overline{G}, x) = (-1)^{n-1}\frac{x}{n - x}\mu(G, n - x)$$

and so the eigenvalues of $\overline{G}$ are $\lambda_1(\overline{G}) = 0$, and

$$\lambda_{i+1}(\overline{G}) = n - \lambda_{n-i+1}(G), \ i = 1, 2, ..., n - 1.$$

Note   Theorem 3.6 has a generalization to the weighted case, see [MP], if we define the weights of $\overline{G}$ to be $a'_{uv} = 1 - a_{uv}(u \neq v)$.

Corollary 3.7   [K1,K2] Let $G_1 * G_2$ denote the join of G_1 and G_2, i.e. the graph obtained from the disjoint union of G_1 and G_2 by adding all possible edges $uv, u \in V(G_1), v \in V(G_2)$. Then

$$\mu(G_1 * G_2, x) = \frac{x(x - n_1 - n_2)}{(x - n_1)(x - n_2)}\mu(G_1, x - n_2)\mu(G_2, x - n_1).$$

where n_1 and n_2 are orders of G_1 and G_2, respectively.

Let G be a simple unweighted graph. The *line graph* $L(G)$ of G is the graph whose vertices correspond to the edges of G with two vertices of $L(G)$ being adjacent if and only if the corresponding edges in G have a vertex in common. The *subdivision graph* $S(G)$ of G is obtained from G by inserting, into each edge of G, a new vertex of degree 2. The *total graph* $T(G)$ of G has its vertex set equal to the union of vertices and edges of G, and two of them being adjacent if and only if they are incident or adjacent in G.

Theorem 3.8   [K3] Let G be a d-regular simple graph with m edges and n vertices. Then

(a) $\mu(L(G), x) = (x - 2d)^{m-n}\mu(G, x)$

(b) $\mu(S(G), x) = (-1)^m(2 - x)^{m-n}\mu(G, x(d + 2 - x))$

(c) $\mu(T(G), x) = (-1)^m(d + 1 - x)^n(2d + 2 - x)^{m-n}\mu(G, \frac{x(d+2-x)}{d+1-x})$.

The part (a) of Theorem 3.8 was also obtained by Vahovskii [V]. Theorem 3.8(a) can be proved also for bipartite semiregular graphs. Recall that a graph G is *r,s)-semiregular* if it is bipartite with a bipartition $V = U \cup W$ such that all vertices in U have degree r and all vertices in W have degree s.

Theorem 3.9   If G is a simple (r,s)-semiregular graph then

$$\mu(L(G), x) = (-1)^n(x - (r + s))^{m-n}\mu(G, r + s - x).$$

Proof   Orient the edges of G in the direction from U to W ($U \cup W$ is a semiregular bipartition) and let C be the oriented incidence matrix of G with respect to this orientation. Then

$$CC^t = Q(G) \quad \text{and} \quad C^tC = 2I + A(L(G)).$$

The line graph of an (r, s)-semiregular graph is $(r + s - 2)$-regular, hence $Q(L(G)) = (r+s-2)I - A(L(G))$. It is well-known that the matrices CC^t and $C^t C$ have the same eigenvalues with the exception of the possible eigenvalue 0. It follows that $\mu(G, x)$ and the characteristic polynomial of $(r + s)I - Q(L(G))$ have the same non-zero roots (including their multiplicities). The proof is finished by observing that the difference between the dimensions of $Q(L(G))$ and $Q(G)$ is $m - n$ and the fact that the leading coefficient of the characteristic polynomial is equal to 1.

Note a) If G is (r, s)-semiregular then $\lambda_n(G) = r + s$ and this eigenvalue corresponds, by the formula of Theorem 3.9, to the eigenvalue 0 of $\mu(L(G))$.

b) Let $\varphi(.,.)$ denote the characteristic polynomial of the adjacency matrix of the graph. It is clear from the proof of Theorem 3.9 that $\varphi(L(G), x), x) = (x + 2)^{m-n} \mu(G, x + 2)$ for any bipartite graph G.

Subdivision graphs, with many vertices subdividing each edge of the original graph, and their spectra are particularly important in the study of thermodynamic properties of crystalline solids (cf. §5). This practical problem led B. E. Eichinger and J. E. Martin [EM] to devise an algorithm for computing the Laplacian eigenvalues of a subdivided graph by applying numerical linear algebraic methods only to the matrix of the unsubdivided graph.

4. The Matrix-Tree-Theorem The most renowned application of the Laplacian matrix of a graph is in the well-known Matrix-Tree- Theorem. This result is usually attributed to Kirchhoff [Ki].

Theorem 4.1 (**Matrix-Tree-Theorem**). Let u, v be vertices of a graph G, and let $Q_{(uv)}$ be the matrix obtained from $Q(G)$ by deleting the row u and the column v. The absolute value of the determinant of $Q_{(uv)}$ is equal to the number of spanning trees $\kappa(G)$ of the graph G.

Corollary 4.2 The number $\kappa(G)$ of spanning trees of the graph G of order n is equal to

$$\frac{1}{n}(-1)^{n-1}\mu'(G,0) = \frac{1}{n}\lambda_2(G)\lambda_3(G)\cdots\lambda_n(G).$$

A generalization of the Matrix-Tree Theorem was obtained by Kel'mans [K3] who gave a combinatorial interpretation to all the coefficients of $\mu(G,x)$ in terms of the numbers of the certain subforests of the graph. This result has been obtained even in greater generality (for weighted graphs) by fiedler and Sedlácek [FS].

Theorem 4.3   [FS,K3] If $\mu(G,x) = x^n + c_1 x^{n-1} + \cdots + c_{n-1}x$ then

$$c_i = (-1)^i \sum_{\substack{s \subset V \\ |S|=n-i}} \kappa(G_S)$$

where $\kappa(H)$ is the number of spanning trees of H, and G_S is obtained from G by identifying all points of S to a single point.

In [K4] graphs are compared by their polynomial $\mu(G,x)$ with application to the number of spanning trees. Kel'mans and Chelnokov [KC] consider the problem of determining the graphs with extreme number of spanning trees (minimal, or maximal) in the family of graphs with a given number of vertices and edges. They make use of Corollary 4.2 Constantine [Co] further generalizes the results of [KC].

5. Physical and Chemical Applications

The Laplace differential operator Δ is one of the basic differential operators in mathematical physics. There are two boundary problems connected with this operator. In each of them one has to look for non-trivial solutions of $\Delta\phi = \lambda\phi$ on Ω. If we add the boundary condition $\phi_{|\partial\Omega} = 0$ we get the *Dirichlet problem*. The same equation with the *Neumann condition* at the boundary describes the vibration of a membrane which does not have its boundary fixed. The same two problems have been studied on Riemannian manifolds. It also makes sense to consider Riemannian manifolds without boundary, in which case there is no distinction between both problems (see, e.g.

[Cha]). The discretization of these problems gives rise to the Laplacian matrix of a graph (possibly infinite) and the eigenvalue problem for this matrix. The first problem we mention is the vibration of a membrane. It is described by the Laplace equation

$$\Delta z = -\lambda z, \ z = 0 \ \text{ on } \ \Gamma \tag{5.1}$$

where Γ is a simple closed curve in the z-plane. Discrete analogue of Δ is the Laplacian matrix of a graph which discretizes the region where the equation (5.1) is studied, cf. [CDS, p. 257].

Fisher [Fi] discusses a discrete model of a vibrating membrane where interaction occurs only between neighbouring atoms (vertices of a graph). The discretization of the vibration of a membrane in this model leads to the Laplacian matrix of the graph with its eigenvalues corresponding to the characteristic frequencies of the membrane. (It seems that the author of [Fi] realizes, because of the regularity of his "lattice" graphs, the connection with adjacency matrix eigenvalues, and addresses the general problem to $A(G)$.) The Laplacian problem on graphs in this interpretation determines the so-called *combinatorial drum* [CDS, p. 256]. It means vibrating of a drum membrane without boundary.

Viewing a graph as a system of vertices joined by elastic springs representing its edges, and observing a kinematic system which vibrates in the xy-plane tends to its equilibrium (stationary) state, led W. T. Tutte [T] to a very interesting algorithm for convex straight-line embeddings of 3-connected planar graphs in the plane. First, one has to select a non-separating induced cycle C of the given 3-connected graph G. Let $(x_1, y_1), (x_2, y_2), \cdots, (x_k, y_k)$ be the vertices of a convex k gon in the plane where k is the length of C. These points will be the coordinates of vertices of C in the constructed embedding. There are unique solutions x, y of

$$Q(G)x = x^o \ \text{ and } \ Q(G)y = y^o \tag{5.2}$$

where the components of $x = (x_v)_{\in V}$ corresponding to vertices on C are equal to their coordinates x_i, and the other components of x are unknowns. The vector x^o on the right side of (5.2) has all coordinates corresponding to vertices not on C equal to 0 and the others unknown. Similar holds for y and y^p. It turns out [T] that the solution x, y of (5.2) determines the coordinates of a planar convex, straight-line embedding of G if and only if G is planar.

The Laplacian matrix appears also in the theory of electrical currents and flows - the incidence matrices C and $Q = CC^t$ can be found in the famous Kirchhoff laws. As a reference we give the classical Kirchhoff's paper [Ki].

C. Maas showed in [Ma] that the Laplacian eigenvalues of the underlying graph determine the kinematic behaviour of a liquid flowing through a system of communicating pipes. It turns out that the second smallest eigenvalue λ_2 (see also §6) determines the basic behaviour of the flow (e.g., whether the flow is of periodic, or aperiodic type). Let G be a graph representing a system of beads as vertices and edges representing the mutual interactions between these beads. Then the potential, or the kinematic energy of such a system is a quadratic form which can be expressed (cf. (2.1)) by the use of the Laplacian matrix of G. Many related physical quantities have the same relation to $Q(G)$. Eichinger, *et al.* [E1, E2, E3, E4, EM] for example showed that the eigenvalues of the Laplacian matrix of a molecular graph determine the distribution function of the so-called radius of gyration of the molecule, and that the non-zero eigenvalues and their eigenvectors can be used efficiently to compute the scattering functions for Gaussian molecules. See [GS] for some additional references. It is worth mentioning that the asymptotic behaviour of the distribution function of the radius of gyration of a molecule depends mostly upon the magnitude and multiplicity of λ_2 [E1].

6. λ_2 - The Algebraic Connectivity of Graphs

The second smallest Laplacian eigenvalue λ_2 of graphs is probably the most important information contained in the spectrum of a graph. This eigenvalue is related to several important graph invariants, and it has been extensively investigated. Most of the results are consequences of the well- known Courant-Fischer principle which states that

$$\lambda_2(G) = \min_{\substack{x \perp 1 \\ x \neq 0}} \frac{(Q(G)x, x)}{x, x} \tag{6.1}$$

where $0 = (0, 0, ..., 0)^t$, and $1 = (1, 1, ..., 1)^t$ is an eigenvector of $\lambda_1 = 0$. Fiedler [F2] obtained another expression for λ_2.

6.1 Proposition [F2] Let G be a weighted graph with non-negative weights a_{uv}. Then

$$\lambda_2(G) = 2n \min_{x \in \Phi} \frac{\displaystyle\sum_{uv \in E(G)} a_{uv}(x_u - x_v)^2}{\displaystyle\sum_{u \in V}\sum_{u \in V}(x_u - x_v)^2} \tag{6.2}$$

where Φ is the set of all non-constant vectors $x \in \ell^2(V)$.

It can be shown easily, using the fact that $\lambda_n(G) = |V(G)| - \lambda_2(\overline{G})$, that a similar formula holds for the maximal eigenvalue of a graph:

$$\lambda_n(G) = 2n \max_{x \in \Phi} \frac{\displaystyle\sum_{uv \in E(G)} a_{uv}(x_u - x_v)^2}{\displaystyle\sum_{u \in V}\sum_{u \in V}(x_u - x_v)^2} \tag{6.2'}$$

Fiedler [F1, F3] calls the number $\lambda_2(G)$ the *algebraic connectivity* of the graph G. This is influenced by its relation to the classical connectivity parameters of the graph - the *vertex connectivity* $\nu(G)$ and the *edge connectivity* $\eta(G)$.

Theorem 6.2 [F1] Let G be a graph of order n and with maximal valency $\Delta(G)$, and denote by $\omega = \frac{\pi}{n}$. Then

(a) $\lambda_2(G) \leq \nu(G) \leq \eta(G)$,

(b) $\lambda_2(G) \geq 2\eta(G)(1 - \cos \omega)$, and

(c) $\lambda_2(G) \geq 2(\cos \omega - \cos 2\omega)\eta(G) - 2\cos \omega(1 - \cos \omega)\Delta(G)$.

It was discovered recently that graphs with large λ_2 (with respect to the maximal degree) have some properties which make them to be very useful objects in several applications. It is important that λ_2 imposes reasonably good bounds on several properties of graphs which are, for an explicit graph, very hard to compute. We shall mention three such applications. It should be noted that, in all of them, the graph invariants, on which λ_2 imposes non-trivial bounds, can be viewed as measures of connectivity.

Concentrators and expanders are graphs with certain high connectivity properties. They are used in the construction of switching networks that exhibit high connectivity, in the recent parallel sorting of Ajtai, Komlós, and Szermerédi [AKS], in the construction of linear-sized tolerant networks which arise in the study of fault tolerant linear arrays [AC], for the construction of the so-called superconcentrators which are extensively used in the theoretical computer science (e.g., the study of lower bounds in the algorithmic complexity (cf. [Va]), in the establishment of time space tradeoffs for computing various functions [Ab, JJ, To], the construction of graphs that are hard to pebble [LT, Pi, PTC], the construction of low complexity error-correcting codes, etc.), etc. Tanner [Ta] was probably the first who realized that the concentration and expanding properties of a graph can be analyzed by its (adjacency) eigenvalues. He observed that a small ratio of the subdominant adjacency eigenvalue to the dominant eigenvalue implies good expansion properties. Alon [A2] and Alon and Milman [AM1, AM2] followed Tanner's approach, but later they realized [A1, AM3, AGM] that the Laplacian spectrum of a graph (in particular the second smallest eigenvalue) appears more naturally in the study of expanding properties of graphs. [Ro] is an overview article about superconcentrators, and it includes also an exposition of some of the eigenvalue methods which we are trying to summarize here. In [AM3] the authors present several inequalities of the isoperimetric nature relating λ_2 and several other quantities in graphs. These results have analytic analogues [GM] in the theory of Riemannian manifolds where the role of λ_2 is played by the smallest positive eigenvalue of the

Laplacian differential operator on the Riemannian manifold (cf. also [Cha]).

The basic lemma of [AM3] is the following equality. Let A and B be subsets of $V(G)$ at distance ρ (this is the minimal distance between a vertex in A and a vertex in B), and let F be the set of edges which do not have both ends in A and do not have both ends in B. Then

$$|F| \geq \rho^2 \lambda_2(G) \frac{|A||B|}{|A| + |B|} \tag{6.3}$$

In particular, when $B = V \backslash A$, then $F = \delta A = \delta B$ (the *coboundary* of A, or of B) is the set of edges with one end in A and the other end outside A. In this case $\rho = 1$ and (6.3) implies

$$|\delta A| \geq \lambda_2(G) \frac{|A|(n - |A|)}{n} \tag{6.4}$$

A refinement of (6.3) is derived in [M3]. If A and B are subsets of $V(G)$ at distance $\rho > 1$ then

$$(\rho - 1)^2 < \frac{\lambda_n(G)}{4\lambda_2(G)} \frac{(n - |A| - |B|)(|A| + |B|)}{|A||B|} \tag{6.5}$$

This paper [AGM] considers the expansion properties of graphs and their applications. The main eigenvalue based lemma of [AGM] gives a lower bound on the number of neighbours of a set $X \subseteq V$. If $N(X)$ is the set of those neighbours of vertices of X which do not lie in X, then

$$|N(X)|^2 - 2(n - 2|X| - \alpha n)|N(X)| - 4|X|(n - |X|) \geq 0 \tag{6.6}$$

where $\alpha = \frac{1}{2}(1 + \frac{\Delta}{\lambda_2})$, and Δ is the maximum valency in G.

In [A1] expanders and graphs with large λ_2 are related. Expanders can be constructed from graphs which are *c-magnifiers* $(c \in \mathbb{R}^+)$. These are graphs which are highly connected according to the following property. For every set X of vertices of G with $|X| \leq \frac{n}{2}$ the neighbourhood $N(X)$ of X contains at least $c|X|$ vertices. In [A1] it is shown that a graph G is $\frac{2\lambda_2}{\Delta + 2\lambda_2}$-magnifier and, conversely, if G is a c-magnifier then $\lambda_2(G) \geq \frac{c^2}{4+2c^2}$. The first result is based on (6.4), while the second one is a discrete version of the Cheeger inequality [Che] from the theory of Riemannian manifolds.

A strong improvement over the Alon's discrete version of the Cheeger inequality was obtained by the author [M2] in connection with another problem. The *isoperimetric number* $i(G)$ of a graph G is equal to

$$i(G) = \min\{\frac{\delta X}{|X|}; X \subset V, 0 < |X| \leq \frac{|V|}{2}\}.$$

This graph invariant is very hard to compute, and even obtaining any lower bounds on $i(G)$ seems to be a difficult problem. It is shown in [M2] that

$$i(G) \geq \frac{\lambda_2(G)}{2} \tag{6.7}$$

and, moreover, a strong discrete version of the Cheeger inequality holds [M2]:

$$i(G) \leq \sqrt{\lambda_2(2\Delta - \lambda_2)} \tag{6.8}$$

where Δ is, as usual, the maximal degree in G. The reader is also referred to [M1]. We mention that these results are very important since they yield efficient checking procedures of several graph properties. For example, if $\lambda_2(G) = 2$ then we know by (6.7) that $i(G) \geq 1$. If, moreover, we find a cut X with $|\delta X| = |X|$ then we can conclude that $i(G) = 1$.

Besides the expansion properties and the isoperimetric numbers of graphs, an eigenvalue based inequality can be used of the max-cut problem [MP] (also the weighted case) which is known to be NP-hard. It is shown in [MP] that the number of edges MC(G) in a maximal cut in a graph G is bounded above by

$$MC(G) \leq \frac{n\lambda_n(G)}{4}. \tag{6.9}$$

Notice that $\lambda_n(G)$ is related to the second smallest eigenvalue of the complement of G (cf. Theorem 3.6).

The second eigenvalue is also related to some other graph invariants. One of the most interesting connections is its relation to the diameter and the mean distance of graphs. There is a lower bound

$$\text{diam}\,(G) \geq \frac{4}{n\lambda_2(G)} \tag{6.10}$$

This bound was obtained by Brendan McKay [McK] but its proof appeared for the first time in [M3].

To get an upper bound one may use the inequality (6.5) which gives rise to an eigenvalue-based upper bound on the diameter of a graph [M3]:

$$\text{diam}\,(G) \leq 2\left\lceil\sqrt{\frac{\lambda_n(G)}{\lambda_2(G)}}\sqrt{\frac{a^2-1}{4a}}+1\right\rceil\left\lceil\log_a\frac{n}{2}\right\rceil \tag{6.11}$$

where γ is any real number which is > 1. For any particular choice of n, λ_n, and λ_2 one can find the value of γ which imposes the lowest upper bound on the diameter of the graph. See [M3] for details. A good general choice is $a = 7$.

In [M3] another upper bound on the diameter of a graph is obtained

$$\text{diam}\,(G) \leq 2\left\lceil\frac{\Delta+\lambda_2(G)}{4\lambda_2(G)}\ln(n-1)\right\rceil. \tag{6.12}$$

This improves a bound of Alon and Milman [AM3]. It should be noted that Chung, Faber and Manteuffel [CFM] found another bound:

$$\text{diam}(G) \leq \left\lceil\frac{\log(n-1)}{\log((\lambda_n+\lambda_2)/(\lambda_n-\lambda_2))}\ln(n-1)\right\rceil.$$

In [M3], some bounds on the mean distance $\bar{\rho}(G)$ are derived. Recall that the mean distance is equal to the average of all distances between distinct vertices of the graph. A lower bound is

$$(n-1)\bar{\rho}(G) \geq \frac{2}{\lambda_2(G)} + \frac{n-2}{2} \tag{6.13}$$

and an upper bound, similar to (6.12), is

$$\bar{\rho}(G) \leq \frac{n}{n-1}\left\lceil\frac{\Delta+\lambda_2(G)}{4\lambda_2(G)}\ln(n-1)\right\rceil. \tag{6.14}$$

There is also an upper bound on $\overline{\rho}(G)$ related to the inequality (6.11). Cf. [M3].

Some inequalities relating graph invariants to the spectrum of the adjacency matrix of a graph can as well be formulated in terms of the Laplacian spectrum - usually obtaining even stronger results this way. As an example we extend a Hoffman-Lovász' bound [CDS, Lo] on the *independence number* $\alpha(G)$ of a graph. They proved that a d-regular graph G has $\alpha(G) \le n(1 - d/\lambda_n)$.

Let G be a graph of order n with vertices of degrees $d_1 \le d_2 \le \cdots \le d_n$. Set

$$e_r = \frac{1}{r}(d_1 + d_2 + \cdots + d_r), \quad 1 \le r \le n.$$

Assume we have an independent set R of vertices of size r. Define a vector $x \in \ell^2(V(G))$ by setting

$$x_v = \begin{cases} 0, & v \in R \\ 1, & v \notin R \end{cases}$$

By (6.2'),

$$\lambda_n(G) \sum_{u \in V} \sum_{v \in V} (x_u - x_v)^2 \ge 2n \sum_{uv \in E} (x_u - x_v)^2$$

which reduces to $\lambda_n r(n - r) \ge n|\delta R| \ge nre_r$. It follows that

$$r \le \frac{n(\lambda_n - e_r)}{\lambda_n} \tag{6.15}$$

Theorem 6.3 If r_0 is the smallest number r for which (6.15) fails then

$$\alpha(G) \le r_0 - 1. \tag{6.16}$$

It is interesting that large graphs of bounded genus and with bounded maximal degree have small λ_2.

Theorem 6.4 Let G be a graph of order n, with maximal vertex degree Δ and genus g. In $n > 18(g + 2)^2$ then

$$\lambda_2(G) \le \frac{6(g + 2)\Delta}{\sqrt{n/2} - 3(g + 2)}.$$

Proof   Boshier [Bo] proved that under the hypothesis of the theorem

$$i(G) \leq \frac{3(g+2)\Delta}{\sqrt{\frac{n}{2}} - 3(g+2)}$$

where $i(G)$ is the isoperimetric number of G. The inequality (6.7) now completes the proof.

It would be interesting to have a "direct" proof of Theorem 6.4. We believe that such a proof would yield even better inequality,, between λ_2 and the genus of a graph, than outlined above.

C. Maas [Ma] studied extensively how can $\lambda_2(G)$ change if we delete or insert an edge into G. He derived several upper and lower bounds on this change, some of them being quite non- obvious. In [Ma] there is also a result which was obtained independently by R. Merris [Me2]. See also [F4].

Theorem 6.5   For a tree T, $\lambda_2(T) \leq 1$, with equality if and only if T is a star.

Three interesting notions are introduced in [Ma], depending on λ_2 and the corresponding eigenspace: *permeability* of a graph, the *well-connectedness* of pairs of vertices, and a measure to relate how *good position* have particular vertices. It is shown on some examples that the notions introduced behave in correspondence with our intuitive notion of permeability, well connected pairs of vertices, and a good (strategic, central) position in a network. Merris [Me2] also found that a kind of a central position in a tree can be defined by using λ_2 and the corresponding eigenvectors.

Let us finish this section with a few words about infinite graphs. Let G be a locally finite countable graph with bounded vertex degrees. Its Laplacian matrix $Q(G)$ gives rise to a self-adjoint linear operator on the Hilbert space $\ell^2(V)$. Its spectrum $\sigma(G)$ is called the Laplacian spectrum of G. It is worth mentioning that the role of λ_2 is replaced by $\lambda = \inf \sigma(G)$. It might happen that $\lambda = 0$ for a connected graph G. Indeed, this is true for any graph of polynomial growth. More details can be found in [BMS, M1, MW], especially about the relation between λ, the growth, and the isoperimetric number of G.

7. Characteristic Valuations

An eigenvector of $\lambda_2(G)$ is called a *characteristic valuation* of G. The characteristic valuations, especially their sign structure, have been studied by Fiedler [F2], and Merris and Grone [Me2, GM1, GM2]. We shall collect here only the most interesting results of these papers.

Theorem 7.1 [F2, F4] Let G be a connected weighted graph with non-negative weights, and $y = (y_v)_{v \in V(G)}$ a characteristic valuation of G. For $r \geq 0$ let

$$S(r) = \{v \in V(G) | y_v \geq -r\}.$$

Then the subgraph induced on $S(r)$ is connected.

A similar result holds for $r \leq 0$ and $S'(r) = \{v | y_v \leq -r\}$. It is an interesting corollary to Theorem 7.1 that if $c \geq 0$ is such a constant that $y_v \neq c$ for all vertices v of G, and $S_1 = \{v | y_v > c\}, S_2 = V(G) \backslash S_1$, then the subgraphs of G induced on S_1 and S_2 are both connected.

Theorem 7.2 [F2] Let G be a connected graph, $y = (y_v)_{v \in V(G)}$ a characteristic valuation of G, and v is a cut-vertex of G. Denote by $G_1, G_2, ..., G_r$ the components of $G - v$. Then:

(a) If $y_v > 0$ then exactly one G_i contains a vertex u with $y_u < 0$. All vertices in other components G_j have positive y-value.

(b) If $y_v = 0$ and some G_i contains positively and negatively valuated vertices, then all the remaining components are 0-valuated.

Theorem 7.3 [F2] Let G be a connected graph with a characteristic valuation y. Two possibilities arise:

(a) There exists a block B_0 of G with both, positive and negative y-values. All other blocks have either all positive, or all negative, or only zero values.

(b) No block of G contains positive and negative values simultaneously. In this case there is a *unique* cut-vertex v with $y_v = 0$ which has a neighbour u with $y_u \neq 0$. In the case when G is a tree, the above results are strengthened in [F2] and discussed in greater details in [Me2, GM1, GM2]. In [GM1] the *trees of type I* are investigated. These are trees with a characteristic valuation y such that $y_v = 0$ for some vertex v of the tree. It is shown that every tree T of type I contains a unique vertex w such that $y_w = 0$ for every characteristic valuation y of T. This vertex is called a *characteristic vertex* of T. *Branches*, i.e. the components of $T - w$ are characterized as *active* or *passive* (every characteristic valuation is 0 on a passive branch). Their properties with respect to $\lambda_2(T)$ are investigated. In [GM2] the authors consider a more general family of matrices, $Q^{\alpha,\beta} = \alpha D(G) + \beta A(G) (\alpha, \beta \in R)$.

Characteristic valuations can be efficiently used to obtain well-behaved heuristic algorithms for various problems. Let y be a characteristic valuation of G and let $A = \{v \in V(G) | y_v \geq 0\}, B = V \backslash A$. It follows from the proof of Theorem 4.2 in [M2] that the partition $V = A \cup B$ is not too far from an optimal partition $V = A^o \cup B^o$ where the optimum means that A^o and B^o have minimal possible average outdegree. This fact can be applied to devise several *divide-and-conquer* algorithms: One solves a problem separately on A and on B and then tries to combine solutions to get an acceptable solution for the graph G.

Another type of problem where characteristic valuations can be used are *optimal labeling problems*, e.g., the bandwidth, or the min-sum problem. It is requested to arrange the vertices of a graph in a linear order $v_1, v_2, \cdots, v_n$ in such a way that the edges will not do too long jumps (if $v_i v_j$ is an edge then $|i - j|$ should be small). A reasonably good ordering is obtained by ordering the vertices of G with respect to the values of a characteristic valuation y, i.e., $v \leq u$ if $y_v \leq y_u$. The eigenvalue $\lambda_2(G)$ also gives a lower bound on the average square of jumps for *any* linear ordering $v_1, \cdots, v_n$ of $V(G)$. If $e = v_i v_j$ then $jump(e) := i - j|$. Then

$$\sum_{e \in E(G)} [\text{jump}(e)]^2 \geq \lambda_2(G) \frac{n(n^2-1)}{12}. \tag{7.1}$$

The details about (7.1) with many different extensions will appear elsewhere [JM].

8. Miscellaneous

In case of regular graphs, all results about the adjacency spectrum of graphs carry over to results about the Laplacian spectrum since for a d-regular graph G

$$\mu(G, x) = (-1)^n \varphi(G, d - x) \tag{8.1}$$

where φ is the characteristic polynomial of $A(G)$. But even in the general, non- regular case, the Laplacian spectrum of G is related to the adjacency spectrum of some graph G'. Let Δ be the maximal valency of G, and let G' be the graph obtained from G by adding, at each vertex $v \in V(G) = V(G'), \Delta - d(v)$ loops. Thus G' is Δ-regular and $Q(G) = Q(G')$. Consequently,

$$\mu(G, x) = \mu(G', x) - (-1)^n \varphi(G', \Delta - x) \tag{8.2}$$

D. L. Powers has assembled a catalogue containing the eigenvalues and eigenvectors of the adjacency and the Laplacian matrices of all connected graphs with up to 6 vertices [P2] and all trees with up to 9 vertices [P1].

Kel'mans [K3] found a class of graphs which is characterized by the Laplacian spectrum. In general there are many cospectral non-isomorphic graphs. For example, it can be shown that almost all tee are cospectral.

R. Merris [Me2] derived some inequalities between the coefficients of the Laplacian polynomial $\mu(G, x)$ and the coefficients of the chromatic polynomial of G (G is a connected simple graph).

Let G be a simple graph. For a vertex $v \in V(G)$, let the *star degree* of v be defined as

$$\text{stardeg}(v) = \begin{cases} 0, & \text{if no neighbour of } v \text{ is a pendant vertex} \\ k - 1, & \text{if } v \text{ has } k \geq 1 \text{ pendant neighbours} \end{cases}$$

The *star degree* of the graph G is then equal to the sum of star degrees of all its vertices. I. Faria [Fa] proved that the star degree of G is equal to the multiplicity of $x = 1$ as the root of the permanental polynomial per $(xI - B(G)), B(G) = D(G) + A(G)$. If G is bipartite then the permanental polynomial of $B(G)$ is equal to the permanental polynomial of the Laplacian matrix $Q(G)$. The author mentions (without an explicit proof) that in the case of characteristic polynomials of $B(G)$ and $Q(G)$, we can only conclude that the star degree of G is at most equal to the multiplicity of $x = 1$ as the zero of these polynomials. Some other authors studied the permanental polynomial of $Q(G)$ [BG, Me1].

C. D. Godsil [Go] proved that the polynomial $\psi(G, S; x) = \sum f_k x^k$, where $S \subseteq E(G)$ and f_k is the number of spanning trees of G with exactly k edges in S, has only real zeros. A corresponding result holds for a generalization of $\psi(G, S; x)$ to unimodular matroids and numbers of bases with specified number of elements in given subsets of the matroid. Is there a corresponding generalization of $\mu(G, x)$ to matroids where the coefficients of the polynomial relate to bases of the matroid in the same way as the coefficients of $\mu(G, x)$ related to spanning trees of G (cf. Theorem 4.3)?

REFERENCES

[1] [Ab] H. Abelson, "A note on time space tradeoffs for computing continuous functions," *Infor. Proc. Letters 8* (1979) 215-217.

[2] [AKS] M. Ajtai, J. Komlós, and E. Szemerédi, "Sorting in $c \log n$ parallel steps," *Combinatorica 3* (1983) 1-9.

[3] [A1] N. Alon, "Eigenvalues and expanders," *Combinatorica 6* (1986) 83-96.

[4] [A2] N. Alon, "Eigenvalues, geometric expanders, sorting in rounds, and Ramsey theory," *Combinatorica 6* (1986) 207-219.

[5] [AC] N. Alon, F. R. K. Chung, "Explicit construction of linear sized tolerant networks, *Proc. 1st Japan Conf. on Graph Theory and Applications*.

[6] [AGM] N. Alon, Z. Galil, and V. D. Milman, "Better expanders and superconcentrators, *J. Algorithms 8* (1987) 337-347.

[7] [AM1] N. Alon, V. D. Milman, "Concentration of measure phenomena in the discrete case and the Laplace operator of a graph," Semin. analyse fonct., Paris 1983-84, Publ. Math. Univ. Paris VII 20 (1984) 55-68.

[8] [AM2] N. Alon, V. D. Milman, "Eigenvalues, expanders and superconcentrators," *Proc. 25th FOCS*, Florida, 1984, pp. 320-322 (Extended abstract).

[9] [AM3] N. Alon, V. D. Milman, "λ_1, isoperimetric inequalities for graphs and superconcentrators," *J. Combin. Theory*, Ser. B 38 (1985) 73-88.

[10] [AnM] W. N. Anderson, T. D. Morley, "Eigenvalues of the Laplacian of a graph," *Lin. Multilin. Algebra* 18 (1985) 141-145.

[11] [Bi] N. L. Biggs, "Algebraic graph theory," Cambridge Univ. Press, Cambridge, 1974.

[12] [BMS] N. L. Biggs, B. Mohar, and J. Shawe-Taylor, "The spectral radius of infinite graphs," *Bull. London Math. Soc.* 20 (1988) 116-120.

[13] [Bo] A. G. Boshier, "Enlarging properties of graphs," Ph.D. Thesis, Royal Holloway and Bedford New College, University of London, 1987.

[14] [BG] R. A. Brualdi, J. L. Goldwasser, "Permanent of the Laplacian matrix of trees and bipartite graphs," *Discrete Math.* 48 (1984) 1-21.

[15] [Cha] I. Chavel, "Eigenvalues in Riemannian geometry," *Academic Press*, New York, 1984.

[16] [Che] J. Cheeger, "A lower bound for the smallest eigenvalue of the Laplacian, in *Problems in analysis*," (R. C. Gunnig, ed.), Princeton Univ. Press, 1970, pp. 195-199.

[17] [CFM] F. R. K. Chung, V. Faber, and T. Manteuffel, "An estimate of the diameter of a graph from the singular values and eigenvalues of its adjacency matrix," preprint, 1988.

[18] [Co] G. M. Constantine, "Schur convex functions on the spectra of graphs," *Discr. Math.* 45 (1983) 181-188.

[19] [CDGT] D. M. Cvetković, M. Doob, I. Gutman, and A. Torgasev, "Recent results in the theory of graph spectra," *Ann. Discr. Math.* 36, North-Holland, 1988.

[20] [CDS] D. M. Cvetković, M. Doob, and H. Sachs, "Spectra of graphs - Theory and applications," VEB Deutscher Verlag d. Wiss., Berlin, 1979; Acad. Press, New York, 1979.

[21] [E1] B. E. Eichinger, "An approach to distribution functions for Gaussian molecules," *Macromolecules* 10 (1977) 671-675.

[22] [E2] B. E. Eichinger, "Scattering functions for Gaussian molecules," *Macromolecules* 11 (1978) 432-433.

[23] [E3] B. E. Eichinger, "Scattering functions for Gaussian molecules. 2. Intermolecular correlation," *Macromolecules* 11 (1978) 1056-1057.

[24] [E4] B. E. Eichinger, "Configuration statistics of Gaussian molecules," *Macromolecules* 13 (1980) 1-11.

[25] [EM] B. E. Eichinger, J. E. Martin, "Distribution functions for Gaussian molecules. II. Reproduction of the Kirchhoff matrix for large molecules," *J. Chem. Phys.* 69 (10) (1978) 4595- 4599.

[26] [Fa] I. Faria, "Permanental roots and the star degree of a graph," Linear Algebra Appl. 64 (1985) 255-265.

[27] [F1] M. Fiedler, "Algebraic connectivity of graphs," *Czech. Math. J* 23 (98) (1973) 298-305.

[28] [F2] M. Fiedler, "A property of eigenvectors of nonnegative symmetric matrices and its application to graph theory," Czech. Math. J.25 (100) (1975) 619-633.

[29] [F3] M. Fiedler, "An algebraic approach to connectivity of graphs, in *Recent advances in graph theory*", Academia, Prague, 1975, pp. 193-196.

[30] [F4] M. Fiedler, "Algebraische Zusammenhangszahl der Graphen und ihre numerische Bedeutung, in *Numerische Methoden bei graphentheoretischen und kombinatorischen Problemen*," Birkhäuser, Basel, 1975, pp. 69-85.

[31] [FS] M. Fiedler, J. Sedlácek, "O w-basich orientovaných grafu," Cas. Pest. Mat. 83 (1958) 214-225 (Czech.).

[32] [Fi] M. E. Fisher, "On hearing the shape of a drum," *J. Combin. Theory 1* (1966) 105 125.

[33] [GS] H. Galina, M. Syslo, "Some applications of graph theory to the study of polymer configuration," *Discr. Appl. Math.* 19 (1988) 167-176.

[34] [Go] C. D. Godsil, "Real graph polynomials, in *Progress in graph theory*," Academic Press, 1984, pp. 281-293.

[35] [GM] M. Gromov, V. D. Milman, "A topological application of the isoperimetric inequality," American J. Math. 105 (1983) 843-854.

[36] [GM1] R. Grone, R. Merris, "Algebraic connectivity of trees," *Czech. Math. J. 37* (112) (1987) 660-670.

[37] [GM2] R. Grone, R. Merris, "Cutpoints, lobes and the spectra of graphs, preprint, 1987.

[38] [H] K. M. Hall, "r-dimension quadratic placement algorithm" *Management Sci. 17* (1970) 219-229.

[39] [JJ] J. Ja'Ja, "Time space tradeoffs for some algebraic problems," *Proc. 12th Ann. ACM Symp. on Theory of Computing*, 1980, pp. 339-350.

[40] [JM] M. Juvan, B. Mohar, "Optimal linear labelings and algebraic properties of graphs, submitted.

[41] [K1] A. K. Kel'mans, "The number of trees in a graph. I." *Automat. i Telemeh. 26* (1965) 2194-2204 (in Russian); transl. Automat. Remote Control 26 (1965) 2118-2129.

[42] [K2] A. K. Kel'mans, "The number of trees in a graph. II.," *Automat. i Telemeh. 27* (1966) 56-65 (in Russian); transl. Automat. Remote Control 27 (1966) 233-241.

[43] [K3] A. K. Kel'mans, "Properties of the characteristic polynomial of a graph," Kibernetiky - na sluzbu kommunizmu," Vol. 4, Energija, Moskva - Leningrad, 1967, pp. 27-41 (in Russian).

[44] [K4] A. K. Kel'mans, "Comparison of graphs by their number of spanning trees," *Discrete Math. 16* (1976) 241-261.

[45] [KC] A. K. Kel'mans, V. M. Chelnokov, "A certain polynomial of a graph and graphs with an extremal number of trees," *J. Combin. Theory*, Ser. B 16 (1974) 197-214.

[46] [Ki] G. Kirchhoff, "Über die Auflösung der Gleichungen, auf welche man bei der Untersuchung der linearen Verteilung galvanischer Ströme geführt wird," *Ann. Phys. Chem* 72 (1847) 497-508. Translated by J. B. O'Toole in I.R.E. Trans. Circuit Theory, CT-5 (1958) 4.

[47] [LT] T. Lengauer, R. E. Tarjan, "Asymptotically tight bounds on time space trade-offs in a pebble game," J. ACM 29 (1982) 1087-1130.

[48] [Lo] L. Lovász, "On the Shannon capacity of a graph," *IEEE Trans. Inform. Theory*, IT-25 (1979) 1-7.

[49] [Ma] C. Maas, "Transportation in graphs and the admittance spectrum," *Discr. Appl. Math.* 16 (1987) 31-49.

[50] [MM] M. Marcus, H. Minc, "A survey of matrix theory and matrix inequalities," Allyn and Bacon, Boston, Mass., 1964.

[51] [McK] B. D. McKay, private communication.

[52] [Me1] R. Merris, "The Laplacian permanental polynomial for trees," *Czech. Math. J.* 32 (107 (1982) 391-403.

[53] [Me2] R. Merris, "Characteristic vertices of trees," *Lin. Multilin. Alg.* 22 (1987) 115-131.

[54] [M1] B. Mohar, "Isoperimetric inequalities, growth, and the spectrum of graphs," *Linear Algebra Appl.* 103 (1988) 119-131.

[55] [M2] B. Mohar, "Isoperimetric numbers of graphs," *J. Combin. Theory*, Ser. B 47 (1989).

[56] [M3] B. Mohar, "Eigenvalues, diameter, and mean distance in graphs," submitted.

[57] [MP] B. Mohar, S. Poljak, "Eigenvalues and the max-cut problem," submitted.

[58] [MW] B. Mohar, W. Woess, "A survey on spectra of infinite graphs," *Bull. London Math. Soc.* 21 (1989) 209-234.

[59] [PTC] W. J. Paul, R. E. Tarjan, and J. R. Celoni, "Space bounds for a game on graphs," Math. Sys. Theory 20 (1977) 239-251.

[60] [Pi] N. Pippenger, "Advances in pebbling," *Internat. Colloq. on Autom. Lang. and Prog.* 9 (1982) 407-417.

[61] [P1] D. L. Powers, "Tree eigenvectors, preprint, 1986.

[62] [P2] D. L. Powers, "Graph eigenvectors," preprint, 1986.

[63] [Ro] A. Rouault, "Superconcentrateurs," Publ. Math. d'Orsay, No. 87-01.

[64] [Ta] R. M. Tanner, "Explicit concentrators from generalized n-gons," *SIAM J. Alg. Discr. Meth.* 5 (1984) 287-293.

[65] [To] M. Tompa, "Time space tradeoffs for computing functions, using connectivity properties of their circuits," *J. Comp. and Sys. Sci.* 20 (1980) 118-132.

[66] [T] W. T. Tutte, "How to draw a graph," *Proc. London Math. Soc.* 52 (1963) 743-767.

[67] [V] E. B. Vahovskii, "On the characteristic numbers of incidence matrices for non- singular graphs," Sibirsk. Mat. Zh. 6 (1965) 44-49 (in Russian).

[68] [Va] L. G. Valiant, "Graph theoretic properties in computational complexity," *J. Comp. and Sys. Sci.* 13 (1976) 278-285.

[69] [Wi] R. J. Wilson, "Introduction to graph theory," Longman, New York, 1972.

Partitioning Points and Graphs to Minimize the Maximum or the Sum of Diameters

Clyde Monma and Subhash Suri
BellCore
Morristown, New Jersey

ABSTRACT

*Let V be a set of n points in the Euclidean plane. We consider the problem of partitioning V into two sets X and Y such that the sum of the diameters of X and Y is minimized. Our main result is an $O(n^2)$–time and $O(n)$–space algorithm for this problem. More generally, if V is the vertex–set of a weighted graph having m edges, then we can partition V into two sets with a minimum sum of diameters in time O(nmlogn). The previously known best algorithm for these problems required $O(n^3 logn)$ time. We also consider a related problem where one seeks a bipartition of V into two sets so that the maximum of the two diameters is minimized. We solve this problem in time O(nlogn) for the Euclidean case, and in time O(mlog*n) for an arbitrary weighted graph. Our algorithm for the Euclidean case is simpler than the one recently proposed by Asano et al., while for arbitrary graphs the best algorithm previously known required time O(mlogn). The general problem of partitioning a set into k ≥ 3 subsets under either criterion of minimization (the maximum or the sum) is known to be NP–complete.*

1. Introduction

The problem of partitioning a set of entities into clusters arises frequently in many disciplines. Due to the wide range of applications, many different objective functions have been considered, thus, giving rise to many variations of this problem. Generally speaking, the basic problem of cluster analysis is to partition a set of entities into homogeneous and well-separated classes, called clusters. Separation and homogeneity are often expressed by describing dissimilarities between pairs of entities.

For optimization criteria, the notions of split and diameter have been used widely (see Delattre and Hansen [DH]).

This paper is concerned with the problem of partitioning a set of points in the Euclidean plane or the vertices of a weighted graph so as to minimize certain functions of diameters. To be precise, let V be a set of n points in the Euclidean plane. We want to partition V into two sets X and Y such that the sum of the diameters of X and Y is minimized. We present an $O(n^2)$–time and $O(n)$–space algorithm for this problem. More generally, if V is the vertex set of an arbitrary weighted graph, then we can partition V into two subsets having a minimum sum of diameters in time $O(nm\log n)$, where m is the number of edges in the graph. These results improve over the previous best algorithm, which is due to Hansen and Jaumard and requires $O(n^3\log n)$ time [HJ]. We also give an optimal solution for the following decision problem: Given an arbitrary graph G with real–valued edge weights and two real numbers r_1 and r_2, can the vertex set of G be partitioned into two subsets X and Y such that diameters of X and Y are at most r_1 and r_2, respectively? An $O(n^2)$ algorithm for this problem is presented in Hansen and Jaumard [HJ]; the Euclidean version of this problem is analogous to the "specified diameter" problem considered by Avis, also solved in $O(n^2)$ time [Av]. We give an $O(n + m)$ algorithm for this problem, where G has n vertices and m edges, which is an improvement over the previous algorithms for sparse graphs.

A related problem is to partition V into subsets X and Y such that the *larger* of the two diameters is minimized. Our results for this problem are as follows. If V is a set of points in the Euclidean plane, then we can find a minimum bipartition in time $O(n\log n)$, which is opitmal in the algebraic computation–tree model. Although an $O(n\log n)$ algorithm for this problem already exists (due to Asano et al. [ABKY]), the merit of our algorithm lies in its simplicity. If V is the vertex set of an arbitrary graph having m edges, then our algorithm finds a minimum bipartition in time $O(m\log^* n)$, where $\log^* n$ is the iterated logarithm. The latter time bound improves upon the previous best result, which runs in $O(m\log n)$ time [Av].

Partitions with either a minimum diameter or a minimum sum of diameters are of interest in many situations where an analyst's main concern is the homogeneity of the clusters. For instance, as mentioned by Hansen and Jaumard [HJ], this is the case when patients suffering from a multiform disease need to be split into clusters for treatment, or when counties within a state need to be grouped for the efficient enactment of an economic policy. Partitions that minimize the maximum diameter, however, suffer from the *dissection* effect, i.e., similar entities may be assigned to different clusters, see Cormack [Co]. This happens because clusters tend to have roughly equal diameters, which may entail the dissection of a natural cluster. Dissection effect is less damaging if the sum of the diameters is minimized, see Hansen and Jaumard [HJ]. Finally, while two clusters often are not enough for a detailed classification, one may apply our bipartitioning algorithm recursively to obtain an approximate partition into $k \geq 3$ clusters.

The paper is organized in five sections. Section 2 contains a basic lemma concerning the maximum diameter of a bipartition. In Section 3, we prove our results for minimizing the maximum diameter in a bipartition. Section 4 contains our main results, namely, the algorithms for minimizing the sum of diameters. Finally, in Section 5, we close with a few remarks concerning various other functions of diameters to which our algorithms are applicable.

2. Preliminaries

Let $G = \{V, E, d: E \to R^+\}$ be a weighted graph on the set of vertices $V = \{v_1, v_2, \ldots, v_n\}$. If two vertices v_i and v_j are joined by an edge in G, then we denote this fact by $v_i v_j \in E$. The weight of $v_i v_j$ is denoted $d(v_i, v_j)$. In the special case where V is a set of points in the plane, the Euclidean distance function $d(\cdot)$ induces a complete weighted graph, which we call the *Euclidean graph* of V. That is, in the Euclidean graph $G = \{V, E, d: E \to R^+\}$, $v_i v_j \in E$, and $d(v_i, v_j)$ equals the Euclidean distance between the points v_i and v_j, for all $1 \leq i < j \leq n$. Throughout, we assume $|V| = n$ and $|E| = m$. For any subset $V' \subseteq V$, the *diameter* of V', denoted $\mathrm{diam}(V')$, is the maximum weight of an edge in the subgraph of G induced by V', i.e.,

$$\mathrm{diam}(V') = \max \{d(v_i, v_j) \mid v_i, v_j \in V', v_i v_j \in E\}.$$

A *maximum spanning tree* of G is a spanning tree of G having a maximum total edge weight; in general, a maximum spanning tree is not unique. In the following, we let $\mathrm{MXST}(G)$ denote an arbitrary but fixed maximum spanning tree of G. Consider an edge $v_i v_j \in E$ that does not belong to $\mathrm{MXST}(G)$. There exists an unique cycle formed by the edge $v_i v_j$ and the path joining v_i and v_j in $\mathrm{MXST}(G)$. We denote this cycle by $cycle(v_i, v_j)$. If $cycle(v_i, v_j)$ has an odd (resp. even) number of edges, we call it an odd (resp. even) cycle. Th following lemma is established in Hansen and Jaumard [HJ]; for the sake of completeness, we include its proof.

Lemma 1 *Let $G = \{V, E, d: E \to R^+\}$ be an arbitrary graph and let $\{V_1, V_2\}$ be a bipartition of V, where $\mathrm{diam}(V_1) \geq \mathrm{diam}(V_2)$. Let $v_p v_q$ be a maximum–weight edge in E that forms an odd cycle with $\mathrm{MXST}(G)$. Then the following holds.*
(1) $\mathrm{diam}(V_1) \geq d(v_p, v_q)$.
(2) If $\mathrm{diam}(V_1) > d(v_p, v_q)$, then there exists an edges $v_k v_l \in \mathrm{MXST}(G)$ such that $\mathrm{diam}(V_1) = d(v_k, v_l)$.

Proof of (1)   Suppose, to the contrary, that $\text{diam}(V_1) < d(v_p, v_q)$. Clearly, v_p and v_q must lie in different components of the partition. Assume, without loss of generality, that $v_p \in V_1$ and $v_q \in V_2$. Let $v_p = x_1, x_2, \ldots, x_m = v_q$ be the ordered list of vertices on $\text{cycle}(v_p, v_q)$, where m is odd and $x_i x_{i+1} \in \text{MXST}(G)$, for $i = 1, 2, \ldots, m-1$. By the maximality of the spanning tree $d(x_i, x_{i+1}) \geq d(v_p, v_q)$, for $i = 1, 2, \ldots, m-1$. Hence, x_i and x_{i+1} lie in different components of the partition, for $1 \leq i \leq m-1$. That is, starting with $x_1 \in V_1$, x_i must alternate between V_1 and V_2, ending in $x_m \in V_2$. That, however, is impossible since m is odd, which contradicts the assumption that $\text{diam}(V_1) < d(v_p, v_q)$.

Proof of (2)   Suppose that $\text{diam}(V_1) = d(v_i, v_j) > d(v_p, v_q)$ but the edge $v_i v_j$ does not belong to $\text{MXST}(G)$. It follows that $v_i v_j$ forms an even cycle with $\text{MXST}(G)$. We, thus, have $v_i, v_j \in V_1$ and $d(x_i, x_{i+1}) \geq d(v_i, v_j)$, for all edges $x_i x_{i+1} \in \text{cycle}(v_i, v_j)$. Since $\text{cycle}(v_i, v_j)$ is even, at least two other adjacent vertices on the cycle, say v_k and v_l are assigned to the same component of the partition. Since having $d(v_k, v_l) > d(v_i, v_j)$ would contradict our assumption that $\text{diam}(V_1) = d(v_i, v_j)$, the equality $d(v_i, v_j) = d(v_k, v_l)$ holds and, thus, the edge $v_k v_l$ satisfies the claim. This completes the proof. $\square$

3. Algorithms for Minimizing the Maximum Diameter

Let $G = \{V, E, d{:}E{\rightarrow}R^+\}$ be an arbitrary weighted graph, and let $\{V_1, V_2\}$ be a bipartition of V. We say that $\{V_1, V_2\}$ is a *minimax* bipartition if $\max\{\text{diam}(V_1),$ $\text{diam}(V_2)\}$ is a minimum over all bipartitions of V. Our first theorem states our results for minimax bipartitions.

Theorem 1   *Let $G = \{V, E, d{:}E{\rightarrow}R^+\}$ be an arbitrary weighted graph. There is an $O(m\log^* n)$–time and $O(m)$–space algorithm for computing a minimiax bipartition of V. If G is an Euclidean graph, then the algorithm runs in $O(n\log n)$ time and uses $O(n)$ space. The bounds for the Euclidean case are optimal in the algebraic computation–tree model.*

Proof   We first describe the algorithm and then discuss its time and space complexity. Compute a maximum spanning tree of G, $\text{MXST}(G)$. Let $\chi\colon V{\rightarrow}\{1, 2\}$ be a 2–coloring of $\text{MXST}(G)$ and let $\{V_1, V_2\}$ be the bipartition of V induced by χ, i.e., $V_k = \{v \in V | \chi(v) = k\}$, for $k = 1, 2$. We claim that $\{V_1, V_2\}$ is a minimax bipartition. To prove the claim, assume without loss of generality that $\text{diam}(V_1) \geq \text{diam}(V_2)$. If v_i and v_j are two vertices in V_1, then $v_i v_j$ forms an odd cycle with $\text{MXST}(G)$. Let $v_p v_q$ be a maximum–weight edge in E that forms an odd–cycle with $\text{MXST}(G)$. Then

$\mathrm{diam}(V_1) \le d(v_p, v_2)$. The claim now follows easily from Lemma 1 since, for every bipartition $\{V', V''\}$ of V,

$$\max\{\mathrm{diam}(V'), \mathrm{diam}(V'')\} \ge d(v_p, v_q).$$

Next, we analyze the time complexity of our algorithm. Given a maximum spanning tree of G, a 2–coloring χ can be obtained in time $O(n)$. To compute a maximum spanning tree of G, we first negate all the edge weights and then use the minimum spanning tree algorithm of Fredman and Tarjan [FT], which runs in time $O(m\log^* n)$. In the Euclidean case, a maximum spanning tree can be computed in time $O(n\log n)$, using an algorithm of Monma, Paterson, Suri and Yao [MPSY]. The time complexity in the latter case is optimal since the problem of computing the diameter of a planar point set (which is known to require $\Omega(n\log n)$ time in the algebraic computation–tree model) is $O(n)$–time reducible to the minimax bipartition problem. This completes the proof of the theorem. $\square$

Asano et al. [ABKY] have shown that, for the Euclidean case, there always exists a *linearly separable* minimax bipartition $\{V_1, V_2\}$, i.e., there exists a line that separates the points of V_1 and V_2. The solution obtained by using Theorem 1 may not have this property. However, Asano et al. show that, any bipartition can be transformed into a linearly separable bipartition without increasing the maximum diameter (see [ABKY]). Consequently, if desired, the solution reparted by Theorem 1 can be converted into a linearly separable minimax bipartition in additional $O(n\log n)$ time.

4. Algorithms for Minimizing the Sum of Diameters

A bipartition $\{V_1, V_2\}$ of V is called a *minimum–sum* bipartition if $\mathrm{diam}(V_1) + \mathrm{diam}(V_2)$ is minimum over all bipartitions of V. In this section, we develop algorithms for computing minimum–sum bipartitions. Our algorithm for arbitrary graphs runs in time $O(nm\log n)$ and the algorithm for Euclidean graphs runs in time $O(n^2)$. The two algorithms are quite similar in spirit but certain steps can be speeded up for Euclidean graphs. We first consider the problem for an arbitrary graph $G = \{V, E, d: E \to R^+\}$.

4.1 General Graphs

We start with the following decision problem: Given two real numbers r_1 and r_2, where $r_1 \ge r_2$, is there a bipartition $\{V_1, V_2\}$ of V satisfying $\mathrm{diam}(V_1) \le r_1$ and $\mathrm{diam}(V_2) \le r_2$? If such a bipartition exists, we call it a (r_1, r_2)–partition. We show that the problem of deciding whether a (r_1, r_2)–partition exists can be solved in time $O(m)$.

Our method is based on labeling the vertices of G with labels 1 and 2. Initially, all vertices are unlabeled. Our procedure then either labels all the vertices of G successfully, in which case the vertices labeled i are assigned to V_i, $i = 1, 2$, or it halts having discovered that no (r_1, r_2)–partition exists.

Let $v_i v_j$ be an edge of E such that v_i is labeled but v_j is not. Our procedure uses the following two propagation rules to decide what label to use for v_j. Rule (R1) applies if $d(v_i, v_j) > r_1$, and rule (R2) applies if $r_1 \geq d(v_i, v_j) > r_2$.

(R1) [*Applies when* $d(v_i, v_j) > r_1$.] If v_i is labeled with 1 (resp. 2), then label v_j with 2 (resp. 1).

(R2) [*Applies when* $r_1 \geq d(v_i, v_j) > r_2$.] If v_i is labeled with 2, then label v_j with 1. Otherwise, leave v_j unlabeled.

We say that a *label–violation* occurs at an edge $v_i v_j$ if either v_i and v_j both are labeled 1 and $d(v_i, v_j) > r_1$, or v_i and v_j are both labeled 2 and $d(v_i, v_j) > r_2$. Clearly, a (r_1, r_2)–partition of V exists if and only if there is a way to label all the vertices without a label–violation.

Our procedure starts by labeling an arbitrary vertex v^* with label 1. The two propagation rules (R1) and (R2) are then used to label as many vertices as possible. One of the following two events must occur:

(i) A label–violation occurs at some edge $v_i v_j$.

(ii) A connected subgraph of G gets labeled without any label–violations and no more vertices can be labeled using (R1) and (R2).

If event (i) occurs, we erase all the labels that propagated from v^* and restart the procedure by labeling v^* with 2. If a label–violation occurs again, we halt the procedure and report that a (r_1, r_2)–partition of V is impossible.

If event (ii) occurs and all the vertices of V have been labeled, the procedure stops and reports the (r_1, r_2)–partition given by the labels. Otherwise, we assign to V_i all the vertices labeled i, for $i = 1, 2$, and delete from G these vertices. Let $G' \subseteq G$ be the subgraph induced by the remaining (unlabeled) vertices. Start the procedure afresh on G'.

The time complexity of the procedure is clearly bounded by $O(m)$. To prove the correctness of the procedure, we reason as follows. Any (r_1, r_2)–partition of V must induce a violation–free labeling of V. All the labels propagated via rules (R1) and (R2) therefore are correct, up to the choice of the label of the start vertex v^*. Therefore, if a label–violation occurs for both the choices of labels for v^*, our procedure correctly determines that (r_1, r_2)–partition of V is infeasible. Otherwise, a (connected) subgraph of G receives a violation–free labeling. Let $V_\ell \subseteq V$ be the set of vertices that received labels and let $V_m = V - v_\ell$ be the set of remaining (unlabeled) vertices. Consider a pair of vertices $v_\ell \in V_\ell$ and $v_m \in V_m$ that are joined by an edge in G. Since both rules (R1) and (R2) fail to apply for the edge $v_\ell v_m$, one of the following two conditions must hold:

(*) $r_1 \geq d(v_\ell, v_m) > r_2$ and v_ℓ is labeled 1, or

(**) $r_2 \geq d(v_\ell, v_m)$.

In either case, v_m can be labeled either 1 or 2 without causing a label–violation at $v_\ell v_m$. In other words, the vertices of V_m can be labeled independently of labels of V_ℓ, which shows that event (ii) is handled correctly by the procedure. This completes our discussion of the decision problem, and we summarize the result as follows.

Theorem 2 *Let $G = \{V, E, d: E \to R^+\}$ be an arbitrary weighted graph, where $|V| = n$ and $|E| = m$, and let r_1 and r_2 be two real numbers. Then, there is an $O(m)$ time and space algorithm that either finds a (r_1, r_2)–partition of V, or determines that no such bipartition is possible.*

Now, we use the result of Theorem 2 to design an algorithm for finding a minimum–sum bipartition of V.

Let $MXST(G)$ be a maximum spanning tree of G. Let $v_p v_q$ be a maximum–weight edge in E that forms an odd cycle with $MXST(G)$. Let K be the set of edges consisting of $v_p v_q$ and the edges in $MXST(G)$ whose weight exceeds that of $v_p v_q$, i.e.,

$$K = \{v_p v_q\} \cup \{v_i v_j \in MXST(G) \mid d(v_i, v_j) > d(v_p, v_q)\}$$

Let $\{V_1, V_2\}$ be any bipartition of V. Without loss of generality, assume $\text{diam}(V_1) \geq \text{diam}(V_2)$. We know from Lemma 1 that there always exists an edge $v_k v_l \in K$ such that $\text{diam}(V_1) = d(v_k, v_l)$. Therefore, in our search for a minimum–sum bipartition, we always can choose an edge of K to define the larger of the two diameters. Our algorithm is described as follows.

Algorithm Minimum–Sum–Partition–1

(1) Sort edges of E in non–increasing order of weight. Let E_s be the resulting list of edges.

(2)   Compute a maximum spanning tree of G, MXST(G), and determine the set of
       edges K.

(3)   For each edge–weight r_1 in K, perform a binary search on E_s to determine the
       smallest edge–weight r_2 for which there exists a (r_1, r_2)–partition. Report that
       (r_1, r_2)–partition for which $r_1 + r_2$ is minimum.

Step 1 takes $O(m\log m)$ time, by a standard sorting algorithm. Computing MXST(G)
takes time $O(m\log^* n)$; use the minimum spanning tree algorithm of Fredman and Tarjan
[FT] after negating all the edge weights. Given MXST(G), the set of edges K can be
found in time $O(m)$, as follows. Let χ be a 2–coloring of MXST(G). Then, edges of
E that form an odd cycle with MXST(G) are those edges whose both endpoints have
the same color under χ. Thus, in a single scan of E, we can find the maximum–weight
edge that forms an odd cycle with MXST(G). Let $v_p v_q$ be this edge. Given $v_p v_q$, the
remaining edges of K can be determined from MXST(G) in $O(n)$ time. The most
expensive step of the algorithm Minimum–Sum–Partition–1 is Step 3. By Theorem 2,
feasibility of a (r_1, r_2)–partition can be determined in time $O(m)$ for any fixed pair of
values r_1 and r_2. Since K has at most n edges and, for any fixed r_1, at most
$(\log m +1)$ pairs of r_1 and r_2 are tested, Step 3 requires $O(nm\log n)$ time. We,
thus, have the following result.

Theorem 3   *Let $G = \{V, E, d:E \to R^+\}$ be an arbitrary weighted graph, where
$|V| = n$ and $|E| = m$. There is an $O(nm\log n)$–time and $O(m)$–space algorithm for
determining a minimum–sum bipartition of V.*

4.2. Euclidean Graphs

For Euclidean graphs, Theorem 3 gives $O(n^3 \log n)$ time bound. In this section,
we show that the geometric structure of the problem can be exploited to improve the time
complexity to $O(n^2)$. We start with a few preliminaries.

Let V be a set of points in the plane in general position, i.e., no three points lie on a
straight line. Following [MPSY], we adopt a general "tie–breaking" convention that
guarantees distinctness of distances between all pairs of points. The convention briefly is
as follows. First, assume that the points $\{v_1, v_2, ..., v_n\}$ are lexicographically ordered
by their coordinates. Then, for any pair of points v_i and v_j, with $i < j$, define the
augmented distance between them as the triple $< d(v_i, v_j), j, -i >$. The lexicographic
order on this augmented distance gives the required total ordering on distances. This
ordering is consistent in the following sense: for any $\varepsilon > 0$, there is another point set V'
in the plane such that (1) distances between points of V' are all distinct, (2) each point

of V' is within distance ε of its corresponding point in V, and (3) ordering between points of V' is exacly the same as the augmented ordering between points of V. (The reader is referred to Monma, Paterson, Suri and Yao [MPSY] for further details regarding this augmented ordering.) Therefore, in the following, we may assume that the distances between all pairs of points of V are distinct. $CH(V)$ denotes the convex hull of V.

Lemma 2 *Let V be a set of points in the plane in general position and let $\{V_1, V_2\}$ be a bipartition of V such that $CH(V_1)$ and $CH(V_2)$ intersect. Then $diam(V_1) + diam(V_2) \geq diam(V_1 \cup V_2)$.*

Proof   If either $CH(V_1) \subseteq CH(V_2)$ or vice versa, then the claim trivially holds. Otherwise, let v_p and v_q be two points such that $diam(V_1 \cup V_2) = d(v_p, v_q)$. If either $v_p, v_q \in V_1$ or $v_p, v_q \in V_2$, then the claim holds again, since $diam(V_1), diam(V_2) > 0$. So, assume without loss of generality that $v_p \in V_1$ and $v_q \in V_2$. Let x be a point where the boundaries of $CH(V_1)$ and $CH(V_2)$ intersect. The triangle inequality in the plane implies that

$$diam(V_1 \cup V_2) = d(v_p, v_q) \leq d(v_p, x) + d(x, v_q) \leq diam(V_1) + diam(V_2).$$

This finishes the proof. $\square$

Recall that a bipartition $\{V_1, V_2\}$ is called linearly separable if there exists a straight line that separates $CH(V_1)$ and $CH(V_2)$. An easy consequence of Lemma 2 is that, for every set of points V in the plane, there exists a linearly separable minimum–sum bipartition.

Let v_a and v_b be two points. The common intersection of the two closed discs, each of radius $d(v_a, v_b)$, centered at v_a and v_b, is called the *lune of $v_a v_b$*, and is denoted $lune(v_a, v_b)$. The following fact is proved easily.

Lemma 3 *Let v_a, v_b, v_c and v_d be four points in the plane in general position such that $d(v_c, v_d) > d(v_a, v_b)$. If v_c and v_d both lie in $lune(v_a, v_b)$, then the line segments $v_a v_b$ and $v_c v_d$ intersect.*

Let $G = \{V, E, d: E \to R^+\}$ be the Euclidean graph of the point set V. Let $MXST(G)$ be a maximum spanning tree of G; distinctness of all the distances among points of V ensures that $MXST(G)$ is unique. Let v_p and v_q be the two points such that, among all edges of E that form an odd cycle with $MXST(G)$, $d(v_p, v_q)$ is maximum. Let K be the set consisting of $v_p v_q$ and the edges of $MXST(G)$ whose weight exceeds $d(v_p, v_q)$, i.e.,

$$K = \{v_p v_q\} \cup \{v_i v_j \in MXST(G) \mid d(v_i, v_j) > d(v_p, v_q)\}$$

By Lemma 1, for any bipartition of V, there exists an edge in K that defines the larger of the two diameters. Therefore, our problem remains to find, for each edge weight r_1 in K, a bipartition $\{V_1, V_2\}$, with $\mathrm{diam}(V_1) = r_1$ and $\mathrm{diam}(V_2) = r_2$, such that r_2 is as small as possible. Obviously, we are only interested in those bipartitions $\{V_1, V_2\}$ that satisfy $\mathrm{diam}(V_1) + \mathrm{diam}(V_2) < \mathrm{diam}(V_1 \cup V_2)$. Our procedure, again, is based on labeling the vertices of V either 1 or 2 and is described below in more detail.

Let $v_a v_b$ be an edge in K and let r_1 be its weight, i.e., $r_1 = d(v_a, v_b)$. A *label–violation* is said to occur in a labeling of V if two vertices at distance greater than r_1 are labeled 1. The procedure starts by deleting from $\mathrm{MXST}(G)$ all the edges whose weights are less than or equal to r_1. Let $F(r_1)$ denote the resulting forest and let T_1, T_2, ..., T_k be the subtrees of $F(r_1)$.

Lemma 4 *Let v_i and v_j be two vertices that belong to the subtrees T_i and T_j, respectively, where $1 \leq i \neq j \leq k$. Then $d(v_i, v_j) \leq r_1$.*

Proof   Let V' be the vertex set of T_i and let $V'' = V - V'$. Let $v'v''$ be the maximum–weight edge that goes between V' and V''. It is well known that $v'v''$ belongs to $\mathrm{MXST}(G)$. (This is the familiar theorem of Prim stated in the context of maximum spanning trees, see Graham and Hell [GH].) However, V' and V'' are disconnected in $F(r_1)$, implying that $d(v', v'') \leq r_1$. Since $v_i \in V'$ and $v_j \in V''$, it follows that $d(v_i, v_j) \leq d(v', v'') \leq r_1$. $\square$

It follows from Lemma 4 that if T_i consists of a single vertex v_i, then v_i can be labeled safely with 1; this cannot cause a label–violation since all other points of V are at distance no more than r_1 from v_i. We, therefore, assume from now on that all subtrees T_i have at least two vertices. Let $v_i v_j$ be an edge of some subtree T_i. Then v_i and v_j cannot be given the same label, because $d(v_i, v_j) > r_1$. Therefore, labeling any vertex of T_i, $1 \leq i \leq k$, forces labels of all the vertices of T_i (recall the propagation rule (R1) from the previous section). Since each subtree in the forest can be labeled in two different ways, there still are exponentially many labelings that do not have a label–violation. We are interested in a labeling that minimizes the diameter of the points labeled 2. In the following, we show that it suffices to check only two different labelings of the points.

Without loss of generality, assume that $v_a v_b$ is placed vertically in the plane such that the endpoint v_a is above v_b, where $v_a v_b$ is the unique edge in K having the weight equal to r_1. Let T_a and T_b be the subtrees in $F(r_1)$ that contain v_a and v_b, respectively; $T_a = T_b$ is possible. Clearly, v_a and v_b both must be labeled with 1. These labels force a unique labeling of all other vertices of $T_a \cup T_b$. It remains to label the vertices of other subtrees. Consider a subtree T_i, where $i \neq a, b$, and a vertex v_i in it. By Lemma 4, $d(v_a, v_i)$ and $d(v_b, v_i)$ are both less than $d(v_a, v_b) = r_1$ and, hence, $v_i \in$ *lune*(v_a, v_b). Furthermore, if $v_i v_j$ is an edge of T_i, then by Lemma 3, $v_i v_j$ and $v_a v_b$ intersect; recall that $d(v_i, v_j) > r_1$. Thus, all the vertices of T_i, $i \neq a, b$, lie in *lune*(v_a, v_b) and all edges of T_i intersect $v_a v_b$.

Let V_L (resp. V_R) be the vertices of $\bigcup_{i \neq a,b} T_i$ that lie to the left (resp. right) of $v_a v_b$; since points of V are in general position, no vertex lies on the segment $v_a v_b$. Note that, due to the earlier exclusion of single–vertex subtrees, each T_i has vertices both in V_L and V_R.

Lemma 5 *Let $\{V_1, V_2\}$ be a bipartition such that $diam(V_1) = r_1$ and $diam(V_1) + diam(V_2) < diam(V_1 \cup V_2)$. Then, either $V_L \subseteq V_1$ and $V_R \subseteq V_2$ or vice versa.*

Proof   Suppose the lemma were false. Then, there are two edges in $F(r_1)$, $v_i v_i'$ and $v_j v_j'$, belonging to subtrees T_i and T_j, respectively, such that (1) v_i, v_j are in V_L and v_i', v_j' are in V_R, and (2) v_i, v_j' are labeled 1 and v_i', v_j are labeled 2. Elementary geometry shows that $CH(V_1) \cap CH(V_2) \neq \varnothing$, which together with Lemma 2 contradicts the hypothesis that $diam(V_1) + diam(V_2) < diam(V_1 \cup V_2)$. Therefore, either $V_L \subseteq V_1$ and $V_R \subseteq V_2$ or vice versa. $\square$

We are now ready to describe the algorithm.

Algorithm Minimum–Sum–Partition–2

(1)   Compute a maximum spanning tree of G, MXST(G), and determine the set of edges K.

(2)   For each edge $v_a v_b \in K$, perform steps (2.1)–(2.4).

    (2.1) Compute $F(r_1) = \{T_1, T_2, ..., T_k\}$, where $r_1 = d(v_a, v_b)$.

    (2.2) Label v_a, v_b with 1 and propagate labels to all other vertices of the subtrees containing v_a and v_b. Label all single–vertex subtrees with 1.

    (2.3) Partition the remaining vertices into V_L and V_R, depending on whether they lie on the left or right of $v_a v_b$.

    (2.4) Label all vertices of V_L with 1 and all vertices of V_R with 2. Compute the diameter of the points of V labeled 2. Next, reverse the labels of V_L and

V_R and recomputed the diameter of points labeled 2. Keep the bipartition that gives the smaller r_2.

(3)   Report the bipartition for which $r_1 + r_2$ is minimum.

Next, we analyze the time complexity of Minimum–Sum–Partition–2. In Step 1, the maximum spanning tree of G can be computed in time $O(n\log n)$ using the algorithm of Monma, Paterson, Suri and Yao [MPSY]. Computing the set of edges K takes additional $O(n\log n)$ time, as follows. Let χ be a 2–coloring of MXST(G). The edges of E that form an odd cycle with MXST(G) are exactly the ones whose both endpoints have the same color. Therefore, the maximum–weight edge in E that forms an odd cycle with MXST(G) is the diameter of either the points labeled 1 or the points labeled 2. Let $v_p v_q$ be this edge. We can find $v_p v_q$ in time $O(n\log n)$ because there is an $O(n\log n)$–time algorithm for computing the diameter of n planar points (see Preparata and Shamos, pp. 176 [PS]). Given $v_p v_q$, the remaining edges of K can be determined in $O(n)$ time from the edge set of MXST(G).

Steps (2.1), (2.2), and (2.3) are straightforward and require $O(n)$ time. Step (2.4) consists of computing diameters of certain point sets, each of which has at most n points. Since we need to perform this step repeatedly, we can improve upon the obvious $O(n\log n)$ time bound by preprocessing the points of V. In particular, after $O(n\log n)$–time and $O(n)$–space preprocessing, the diameters in Step (2.4) can be computed in time $O(n)$, as is described below. Let v' and v" be the two points that define the diameter of V, i.e., $d(v', v") = \text{diam}(V)$. Sort the points of V angularly around v' and v". Let S'(V) and S"(V), respectively, be the two lists thus obtained. Now, in Step (2.4), let V_1 and V_2 be the sets of points labeled 1 and 2, respectively. Note that v' and v" both cannot be labeled 1 (or 2); that would imply that $\text{diam}(V_1) + \text{diam}(V_2) > \text{diam}(V_1 \cup V_2)$. Assume without loss of generality that $v' \in V_1$ and $v" \in V_2$. Use S' and S" to arrange V_1 and V_2 in sorted order around v' and v", respectively. Connect points of V_1 and V_2 in the sorted order to obtain two (star–shaped) simple polygons, say, P_1 and P_2. Clearly, the diameter of P_i equals the diameter of V_i, $i = 1, 2$. However, the diameter of a simple polygon with n vertices can be computed in time $O(n)$ (see Preparata and Shamos, pp. 176 [PS]). Thus, each iteration of Step (2) can be performed in time $O(n)$ and there are at most n iterations. The $O(n\log n)$ cost of preprocessing is incurred once. The total time required by the algorithm Minimum–Sum–Partition–2 is, therefore, $O(n^2)$. We summarize this result.

Theorem 4 *Let V be a set of n points in the Euclidean plane. There is an $O(n^2)$– time and $O(n)$–space algorithm for finding a minimum–sum bipartition of V.*

5. Closing Remarks

We considered the problem of partitioning a set of entities V subject to minimizing the maximum or the sum of the diameters. The entities are either points in the plane with the Euclidean measure of interdistance, or they are the vertices of an arbitrary graph with real–valued weights on edges. Most of our work, however, is applicable to a wider class of functions of the diameters. We briefly describe some of the generalizations in the following.

Let $\{V_1, V_2\}$ be a bipartition of V, where V is either a set of points or the vertices of an arbitrary graph. Let $d_1 = \mathrm{diam}(V_1)$ and $d_2 = \mathrm{diam}(V_2)$ be the diameters of V_1 and V_2, respectively, where we assume that $d_1 \geq d_2$. Let $f(d_1, d_2)$ be a function of the two diameters such that, for any choice of parameters, f is computable in $O(1)$ time. Consider the problem of determining a bipartition of V subject to minimizing f.

All the results of Section 4 extend to the functions f that are *monotone* in d_2, i.e., if $f(d_1, d_2) \leq f(d_1, d_2')$ whenever $d_2 \leq d_2'$. If, in addition, f also is *unimodal* with respect to d_1, then a bipartition minimizing f can be determined in $O(m\log^2 n)$ time for an arbitrary n–node m–edge graph. The time complexity can be improved to $O(n\log n)$ for the Euclidean graphs. This follows because, instead of iterating Step (3) of the algorithm Minimum–Sum–Partition–1 or Step (2) of the algorithm Minimum–Sum–Partition–2 for each edge–weight in K, we now can perform a binary search on the sorted lists of weights in K. This reduces the number of iterations from n to at most $(\log n + 1)$. Observe that the function $\max\{d_1, d_2\}$ is trivially unimodal with respect to d_1. However, $f(d_1, d_2) = d_1 + d_2$ is not unimodal.

For Euclidean graphs, it is easy to realize an $O(n^3\log n)$–time algorithms under any function f that permits an optimal linearly–separable bipartition; simply compute f for all $O(n^2)$ linearly–separable bipartitions. We remind the reader that such bipartitions exist for minimizing the maximum diameter as well as for minimizing the sum of the diameters. Finally, as observed by Hansen and Jaumard [HJ], *any* function f of the diameters can be minimized in time $O(n^5)$.

The general problem of partitioning a graph into an arbitrary number of subsets is NP–complete under both the minimax or the minimum–sum diameter measure. Similarly, it is NP–complete to decide if a set of n points in the Euclidean plane can be partitioned into k subsets so that the diameter of each subset is less than some given bound d^* (see Johnson [Jo]). Finally, for the minimum–sum diameter problem, one can easily obtain a polynomial–time algorithm for any fixed number of subsets, since Lemma 2 guarantees the existence of an optimal partition where every pair of subsets is linearly separable.

REFERENCES

[Av] D. Avis, Diameter Partitioning, *Discrete and Computational Geometry*, 1 (1986) 265–276.

[ABKY] T. Asano, B.K. Bhattacharya, J.M. Keil and F.F. Yao, Clustering algorithms based on minimum and maximum spanning trees, *Proc. of the Fourth Annual Symposium on Computational Geometry* (1988) 252–257.

[Co] R.M. Cormack, A review of classification, *Journal of Royal Statistical Society*, A(134), (1971) 321–367.

[DH] M. Delattre and P. Hansen, Bicriterion cluster analysis, *IEEE Transactions on Pattern Analysis and Machine Intelligence,* (2), (1980) 277–291.

[FT] M. Fredman and R.E. Tarjan, Fibonacci heaps and their uses in improved network optimization algorithms, *Journal of the Association for Computing Machinery*, 34, (1987) 596–615.

[GH] R. Graham and P. Hell, On the history of minimum spanning tree problem, *Annals of History of Computing*, 7, (1985)

[HD] P. Hansen and M. Delattre, Complete link cluster analysis by graph coloring, *Journal of the American Statistical Association*, 73, (1978) 397–403.

[HJ] P. Hansen and B. Jaumard, Minimum sum of diameters clustering, *Journal of Classification*, (1987) 215–226.

[Jo] D.S. Johnson, NP–completeness column, *Journal of Algorithms* (3) (1982) 103–195.

[MPSY] C. Monma, M. Paterson, S. Suri and F.F. Yao, Computing Euclidean Maximum Spanning Trees, *Proc. of the Fourth Annual Symposium on Computational Geometry* (1988) 241–251.

[PS] F.P. Preparata and M.I. Shamos, *Computational Geometry*, Springer Verlag, New York, NY, (1985).

On the Number of Well–Covered Trees

J. W. Moon

University of Alberta

ABSTRACT

Results are obtained on the number of well–covered trees in various families of trees.

1. Introduction

A subset I of nodes of a graph is an *independent set* if no two nodes of I are joined to each other. If, in addition, every node not in I is joined to at least one node in I, then I is a *maximal independent* set. A graph is said to be *well–covered* (cf. [15] or [19]) if all maximal independent sets have the same size. A non–trivial tree is well–covered if and only if every node that is not an end–node is joined to one and only one end–node. This result apparently first appeared implicitly in [18; p. 13]; see also [17], [2] and [3]. Our object here is to enumerate the well–covered trees in various families of trees.

We introduce some terminology and preliminary results in §2. Then we consider certain simply–generated families in §3 and some non–simply–generated families in §4; it follows from our results for these families that the probability that a tree T_{2m} with 2m nodes is well–covered is asymptotic to $\alpha \cdot \beta^m$ where α and β are constants that depend on the family being considered. We conclude in §5 with some remarks on the average number of maximal independent sets in well–covered trees in certain families.

2. Preliminaries

We recall that *plane* trees – or *ordered* trees, as they are called by some authors [7; p. 306] – are rooted trees with a specified ordering for the branches incident with each note. Let $\mathcal{F}$ denote a give family of weighted plane trees in which the tree T_n has weight $\omega(T_n)$. The *out–degree* of a node u in a rooted tree is the number of edges incident with u that lead away from the root; let $D_i(T_n)$ denote the number of nodes of out–degree i in the rooted tree T_n. We say that the family $\mathcal{F}$ is a *simple–generated* family if there exists a sequence of non–negative constants $c_0 \ (=1)$, c_1, c_2, ..., such that

(2.1) $$\omega(T_n) = \Pi c_i^{D_i(T_n)}$$

for every tree T_n in $\mathcal{F}$. Let Y_n denoted the number of trees T_n in the family $\mathcal{F}$ where (here and elsewhere) the weights are taken into account, i. e.,

$$Y_n = \Sigma \omega(T_n)$$

where the sum is over all plane trees with n nodes. It is not difficult to see (cf. [8; p. 999] or [20; p. 24]) that if $\mathcal{F}$ is a simply–generated family, then its generating functions $Y = \sum_1^\infty Y_n x^n$ satisfies the relation

(2.2) $$Y = x\Phi(Y)$$

where

$$\Phi(Y) = 1 + c_1 Y + c_2 Y^2 + \dots .$$

Different sets of coefficients c_i give rise to different simply–generated families. Thus, for example, if $c_i = 1$ for $i \geq 0$ we obtain the ordinary family of plane trees; and if $c_i = 1/i!$ for $i \geq 0$ we obtain (in effect) the family of rooted labelled trees.

In the next section we shall use the following result which is equivalent to a result provided in [8; Theorem 3.1]; a closely related result was proved earlier in [14].

Lemma 1 *Suppose* $\theta(t) = a_0 + a_1 t + a_2 t^2 + \dots$ *is a regular function of* t *when* $|t| < R < +\infty$, *and let* $F = F(z) = F_1 z + F_2 z^2 + \dots$ *denote the solution of* $F(z) = z\theta(F(z))$ *in the neighborhood of* $z = 0$. *If*

   (i) $a_i \geq 0$ *for* $i \geq 0$,

   (ii) $a_0 > 0$ *and* $a_i > 0$ *and* $a_j > 0$ *for some distinct integers* i *and* j *such that* $\gcd(i, j) = 1$, *and*

   (iii) $v\theta'(v) = \theta(v)$ *for some* v, *where* $0 < v < R$, *then*

$$F_m \sim a\delta^{-m} m^{-3/2}$$

as $m \to \infty$, *where* $\delta = v/\theta(v)$ *and* $a = (\theta(v)/2\pi\theta''(v))^{1/2}$.

Suppose T is a non–trivial well–covered tree that is rooted at node r. We shall call T a *P–tree* if r has out–degree one and we shall call T a *Q–tree* if r is joined to a node of out–degree zero. Notices that every non–trivial well–covered tree T is either a P–tree or a Q–tree and that the rooted tree with two nodes is the only such tree that is both a P–tree and a Q–tree. The *principal branches* of a rooted tree T are the maximal subtrees not containing the root–node; these branches are considered as being rooted at the nodes joined to the root of T. The *secondary branches* of T are the principal branches of the principal branches of T. The following results, which we shall use in enumerating well–covered trees, are straightforward consequences of the definitions and the characterization of well–covered trees stated in the introduction.

Lemma 2   *A rooted tree T is a P–tree if and only if the root–node has out–degree one and all the secondary branches are Q–trees.*

Lemma 3   *A rooted tree T is a Q–tree if and only if one principal branch consists of a single node and all the remaining principal branches are Q–trees.*

3.   Results for simply–generated families

Let $\mathcal{F}$ denote some given simply–generated family of trees. We assume that the function Φ that appears in relation (2.2) satisfies the hypothesis of Lemma 1 with $\Phi(0) = 1$ and that, in particular, $\tau\Phi'(\tau) = \Phi(\tau)$ for some τ, where $0 < \tau < R$, so that

$$(3.1) \qquad\qquad Yn \sim c\rho^{-n}n^{-3/2}$$

as $n \to \infty$, where $\rho = \tau/\Phi(\tau)$ and $c = (\Phi(\tau)/2\pi\Phi''(\tau))^{1/2}$.

There are no non–trivial well–covered trees with an odd number of nodes; this follows immediately from the characterization of well–covered trees. Hence to enumerate the well–covered trees in $\mathcal{F}$ it will suffice to consider the generating functions $P(z) = \sum_{1}^{\infty} P_m z^m$ and $Q(z) = \sum_{1}^{\infty} Q_m z^m$ where P_m denotes the number of P–trees T_{2m} in $\mathcal{F}$ and Q_m denotes the number of Q–trees T_{2m} in $\mathcal{F}$.

Theorem 1 $P(z) = c_1 z \Phi(Q(z))$.

Theorem 2 $Q(z) = z\Phi'(Q(z))$.

Proof   It follows from Lemma 2 and relation (2.1) that the number of P–trees T_{2m} with k secondary branches (all of which are Q–trees) equal the coefficient of z^m in $c_1 c_k z Q^k(z)$ for $k \geq 0$. For $Q^k(z)$ enumerates the ordered k–tuples of Q–trees, the factor z takes into account the root–node of T_{2m} and the (unique) node u joined to the root, and the factors c_1 and c_k record the weights associated with the nodes r and u. Consequently

$$P(z) = c_1 z \sum_0^\infty c_k Q^k(z) = c_1 z \Phi'(Q(z)).$$

Similarly, it follows from Lemma 2 that the number of Q–trees T_{2m} with k principal branches (one of which consists of a single node v and $k-1$ of which are Q–trees) equals the coefficient of z^m in $zkc_k Q^{k-1}(z)$ for $k \geq 1$. For, $Q^{k-1}(z)$ enumerates the ordered $(k-1)$–tuples of Q–trees, the factor k counts the number of choices for the position of the trivial branch consisting of a single node v, the factor z takes into account the root–node and the node v, and the factor ck records the weight associated with the root–node. Consequently,

$$Q(z) = z \sum_1^\infty kc_k Q^{k-1}(z) = z \Phi'(Q(z))$$

We now determine the asymptotic behavior of P_m and Q_m subject to some additional assumptions on the function Φ.

Theorem 3 *Suppose that $c_1 > 0$ and that $c_i > 0$ and $c_j > 0$ for some distinct integers i and j such that $i > 1$, $j > 1$, and $\gcd(i-1, j-1) = 1$. Suppose further that $v\Phi''(v) = v\Phi'(v)$ for some v where $0 < v < R$. Then*

(3.2) $$Q_m \sim a\delta^{-m} m^{-3/2}$$

and

(3.3) $$P_m \sim vc_1 Q_m$$

as $m \to \infty$, where $\delta = v/\Phi'(v)$ and $a = (\Phi'(v)/2\pi\Phi''(v))^{1/2}$.

Proof   Relation (3.2) follows upon applying Lemma 1 to the relation $F(z) = z\theta(F(z))$ with $F(z) = Q(z)$ and $\theta(t) = \Phi'(t)$. And it follows from Theorem 1 and a result in [9; Lemma 4] that

$$P_m \sim c_1 \Phi'(Q(\delta)) Q_{m-1}$$

as $m \to \infty$. Hence

$$P_m \sim c_1 \delta \Phi'(\upsilon) Q_m = \upsilon c_1 Q_m$$

by (3.2) and the definitions of δ and υ. (We remark that it is possible to give examples of functions $\Phi(t)$ such that the equation $\tau\Phi'(\tau) = \Phi(\tau)$ has a solution τ where $0 < \tau < R$ but the equation $\upsilon\Phi''(\upsilon) = \Phi'(\upsilon)$ does not have a solution where $0 < \upsilon < R$; so the hypothesis that υ exists cannot simply be dropped.)

We conclude this section by illustrating the preceding results for two particular simply–generated families of trees. Let $\mathcal{F}$ denoted the family of rooted labelled trees. Then (see, e. g., [11; p. 26] or [4; p. 23]) $Y = xe^Y$ and

$$Y_n = n^{n-1}/n! \sim (2\pi)^{-1/2}e^n n^{-3/2}.$$

In this case $\Phi'(t) = \Phi(t) = e^t$ so $P = Q = ze^Q$ and

$$P_m = Q_m = m^{m-1}/m! \sim (2\pi)^{-1/2}e^m m^{-3/2}.$$

These results agree with relations (3.2) and (3.3) with $\upsilon = 1$ and $\delta = e^{-1}$. (We remark that in this particular case the expression for P_m and Q_m can readily be derived from first principals upon appealing to the characterization stated in the introduction.)

Now let $\mathcal{F}$ denote the family of plane trees. Then (see,.e. g., [16; p. 197] or [4; p. 67]) $Y = x(1 - Y)^{-1}$ and

$$Y_n = \frac{1}{n}\binom{2n-2}{n-1} \sim \pi^{-1/2}4^{n-1}n^{-3/2}$$

In this case $\Phi(t) = (1 - t)^{-1}$ and $\Phi'(t) = (1 - t)^{-2}$ so $Q = z(1 - 1)^{-2}$. When we apply Lagrange's inversion formula [1;p. 148] to this relation for Q we find that

$$Q_m = \frac{1}{m}\binom{3m-2}{m-1} \sim (27\pi)^{-1/2}(27/4)^{m-1}m^{-3/2}.$$

Furthermore, it follows from Theorems 1 and 2 that

$$P = z(1 - Q)^{-1} = Q(1 - Q).$$

Appealing again to Lagrange's formula, we find that

$$P_m = \frac{1}{3m-2}\binom{3m-2}{m-1} = \frac{m}{3m-2}Q_m$$

These results agree with relations (3.2) and (3.3) with $\upsilon = 1/3$ and $\delta = 4/27$.

4. Result for some non–simply–generated families

We now consider some non–simply–generated families for which relation (2.1) does not hold. A tree T_n with n labelled nodes rooted at node 1 is a *recursive tree* if

$n = 1$ or if $n > 1$ and T_n can be constructed by joining node n to one of the $n - 1$ nodes of a recursive tree T_{n-1}; or, equivalently, T_n is a recursive tree if the labels of the nodes on any path leading away from the root form an increasing sequence (see, e. g., [12] or [8]). Let y_n denote the number of recursive trees T_n. These definitions readily imply that $y_n = (n - 1)!$ and that the generating function $y = \sum_1^\infty y_n x^n/n!$ satisfies the differential equation $y' = e^y$. Let p_m denote the number of recursive P–trees T_{2m} and let q_m denote the number of recursive Q–trees T_{2m}. Lemmas 1 and 2 can be applied to obtain differential equations for the (exponential) generating functions of the numbers p_m and q_m, and these equations can then be solved to yield formulas for p_m and q_m. In this case, however, it is possible to derive formulas for p_m and q_m by appealing directly to the characterization of well–covered trees stated in the introduction.

Theorem 4 *Let $\mathcal{F}$ denote the family of recursive trees. Then*

(4.1) $$q_m = (2m - 1)!/2^{m-1}$$

and

(4.2) $$p_m = (2m - 2)!/2^{m-1}.$$

Proof There are $(2m)!/m!2^m$ ways of partitioning $2m$ labelled nodes $(1, 2, ..., 2m)$ into m pairs of nodes (x_i, y_i) where $x_i < y_i$ for $1 \le i \le m$; we may suppose that $x_1 = 1$. Now construct a recursive tree T_m with the nodes $x_1, ..., x_m$; this can be done in $(m - 1)!$ ways. If we then join node y_i to node x_i for $1 \le i \le m$, it is not difficult to see that the resulting tree T_{2m} is a recursive Q–tree and that all such trees can be constructed in this way. It follows, therefore, that

$$q_m = \frac{(2m)!}{m!2^m} \cdot (m - 1)! = \frac{(2m - 1)!}{2^{m-1}}.$$

Before enumerating the recursive P–trees we observe that if T_{2m} is a recursive P–tree, then node 2 is the (unique) node joined to the root–node 1. There are $(2m - 2)!/(m - 1)!2^{m-1}$ ways of partitioning $2m$ labelled nodes $(1, 2, ..., 2m)$ into m pairs of nodes (x_i, y_i) where $x_1 = 1$, $y_1 = 2$ and $x_i < y_i$ for $2 \le i \le m$. Now construct one of the $(m - 1)!$ recursive trees T_m with the nodes $2, x_2, ..., x_m$. Finally, join node 1 to node 2 and node y_i to node x_i for $2 \le i \le m$. As before, it is not difficult to see the the resulting tree T_{2m} is a recursive P–tree and that all such trees can be constructed in this way. Hence,

$$P_m = \frac{(2m-2)!}{(m-1)!2^{m-1}} \cdot (m-1)! = \frac{(2m-2)!}{2^{m-1}}.$$

Now let $\mathcal{F}$ denote the family of nonisomorphic rooted unlabeled trees. It is well–known ([4]; p. 52) that the generating function $Y(x) = \sum_1^{\infty} Y_n x^n$ for this family satisfies the relation

$$Y(x) = x \exp \sum_1^{\infty} T(x)^k/k.$$

Otter [13] (see also [4; p. 213]) showed that $Y_n \sim c\rho^{-n}n^{-3/2}$ where $c = 4.399\ldots$ and $\rho = .3383\ldots$. It is very easy to enumerate the Q–trees and the P–trees in this family. suppose we join each node of a tree T_m in $\mathcal{F}$ to a different new node. If the resulting tree T_{2m} is regarded as being rooted at the root–node of the original tree T_m then T_{2m} is clearly a Q–tree; and if T_{2m} is regarded as being rooted at the new node joined to the root–node of T_m then T_{2m} is a P–tree. It is not difficult to see that this implies that

$$P_m = Q_m = Y_m$$

for this particular family.

Finally, let $y(x) = \sum_1^{\infty} y_n x^n$ where y_n denote the number of trees T_n in the family $\mathcal{G}$ of nonisomorphic unrooted unlabeled trees. Otter [13] (see also [4; pp. 57 and 214]) showed that

$$y(x) = Y(x) + \frac{1}{2}(Y^2(x) - Y(x^2)),$$

where $Y(x)$ denotes the generating function for the nonisomorphic unrooted unlabeled trees, from which he deduced that $y_n \sim b\rho^{-n}n^{-5/2}$ where $b = .5349\ldots$ and, as in the rooted case, $\rho = .3833\ldots$. Since the trees in $\mathcal{G}$ are not rooted we cannot talk about P–trees and Q–trees now. But if we let w_m denoted the number of well–covered trees T_{2m} in $\mathcal{G}$, then it follows essentially the same argument as was used in the rooted case that $w_m = y_m$.

5. Maximal Independent Sets

Let $M(T_{2m})$ denote the number of maximal independent set in the well–covered tree T_{2m} and let $N(T'_m)$ denote the number of (not necessarily maximal) independent sets in the "stripped" tree T'_m remaining after removing all the end–nodes of T_{2m}; the empty set is to be counted in $N(T'_m)$ and we assume that $m > 1$. Since every maximal independent set S of T_{2m} has a unique decomposition of the form $S = I \cup E$ where I is an independent set of T'_m and E consists of the end–nodes of T_{2m} that are not joined to any node of I, it follows that $M(T_{2m}) = N(T'_m)$. Thus to determine the total number of maximal independent sets in all well–covered trees T_{2m} in a given simply–generated family $\mathcal{F}$, it suffices to determine the total number of independent sets in the corresponding "stripped" trees T'_m. The problem of determining the total number of independent sets in certain simply–generated families of trees has been considered in [5], [6] and [10]. The approach used in theses papers can, in principle, be adapted to the present problem for simply–generated families by considering the "stripped" trees obtained from Q–trees and from P–trees separately and making use of Lemmas 2 and 3. We shall not pursue this further here other than to remark that if $e(2m)$ denotes the average number of maximal independent sets in well–covered trees T_{2m} belonging to any given family $\mathcal{F}$, then it can be shown that $(e(2m))^{1/2m}$ tends to $1.286...$ and to $1.293...$ for the labelled trees and the plane trees, respectively; by contrast, if $\mu(n)$ denotes the average number of maximal independent sets in all trees T_n in $\mathcal{F}$, then (see [10]) $(\mu(n))^{1/n}$ tends to $1.273...$ and to $1.239...$ for these same two families respectively.

6.   Acknowledgements

I am indebted to Professor B. Hartnell for introducing me to well–covered trees. I am also indebted to W. Aiello for performing some calculations for me. The preparation of this paper was assisted by a grant from the Natural Sciences and Engineering Research Council of Canada.

REFERENCES

[1]   L. Comtet, *Advanced Combinatorics*, Reidel, Dordrecht, 1974.

[2]   O. Favaron, Very well covered graphs, *Discrete Math.* **42** (1982) 177–187.

[3]   A. S. Finbow and B. L. Hartnell, A game related to covering by stars, *Ars Combinatoria* **16A** (1983) 189–198.

[4]   F. Harary and E. M. Palmer, *Graphical Enumeration*, Academic Press, New York, 1973.

[5]   P. Kirschenhofer, H. Prodinger and R. F. Tichy, Fibonacci numbers of graphs: II, *Fibonacci Quarterly* **21** (1983) 219–229.

[6]   P. Kirschenhofer, H. Prodinger and R. F. Tichy, Fibonacci numbers of graphs: III, *Fibonacci Numbers and their Applications,* Reidel, Dordrecht, (1986) 105–120.

[7] D. E. Knuth, *The Art of Computer Programming*, vol. 1, Addison–Wesley, Reading, 1973.

[8] A. Meir and J. W. Moon, On the altitude of nodes in random trees, *Can. J. Math.* **30** (1978) 997–1015.

[9] A. Meir and J. W. Moon, Games on random trees, *Congressus Numerantium* **44** (1984) 293–303.

[10] A. Meir and J. W. Moon, On maximal independent sets of nodes in trees, *J. Graph Th.* **12** (1988) 265–283.

[11] J. W. Moon, *Counting Labelled Trees*, Canadian Mathematical Congress, Montreal, 1970.

[12] B. R. Myers and M. A. Tapia, Generation of concave node–weighted trees, *IEEE Trans. Cir Th.* CT–14 (1967) 229–330.

[13] R. Otter, The number of trees, *Ann. of Math.* **49** (1984) 583–399.

[14] R. Otter, The multiplicative process, *Ann. of Math. Statis.* **20** (1949) 206–224.

[15] M. D. Plummer, Some covering concepts in graphs, *J. Comb. Th.* **8** (1970) 91–98.

[16] G. Pólya, Kombinatorische Anzahlbestimmungen für Gruppen, Graphen und chemische Verbindungen, *Acta Math.* **68** (1937) 145–254.

[17] G. Ravindra, Well covered graphs, *J. Combin Inform System. Sci.* **2** (1977) 20–21.

[18] J. A. W. Staples, *On some subclasses of well–covered graphs*, Ph.D. Thesis, Vanderbilt University, 1975.

[19] J. A. W. Staples, On some subclasses of well–covered graphs, *J. Graph Th.* **3** (1979) 197–204.

[20] J.–M. Steyaert and P. Flajolet, Patterns and pattern–matchings in trees: an analysis, *Informations and Control* **58** (1983) 19–58.

THE UNIVERSAL GRAPHS OF FIXED FINITE DIAMETER

Lawrence S. Moss

University of Michigan

at Ann Arbor

ABSTRACT

In [3], we showed that for each natural number N there is a graph U_N with the following two properties: (1) U_N is a countable graph of diameter N, and (2) for every countable graph H of diameter N, there is an isometric embedding of H into U_N. Our proof there was from first principles. In this paper, we present a new proof of the existence of these universal graphs of finite diameter. Our proof here is a little simpler than in [3], especially for even N, but it relies on some related results from [2]. In that paper, we constructed a countable connected graph U_∞ with the property that every countable connected graph is isometrically embedded in U_∞. These results are reviewed below, so this paper may be read on its own. Then we prove the existence of U_N. We also show that each U_N is distance homogeneous, and in fact it is the unique distance homogeneous graph with properties (1) and (2).

The first results concerning universal graphs are due to Rado [4]. He showed that there is a countable graph which isomorphically embeds every countable graph. In our notation, Rado's graph is U_2, the universal graph of diameter 2. (It is obvious that U_0 is a single point, and

This research was supported in part by NSF grant DMS-85-01752.

U_1 is the complete graph on infinitely many vertices.) Our graphs U_N are thus higher diameter analogs of the Rado universal graph.

Our last section contains results relating the different U_N. We investigate first-order theories of the universal graphs, and our main result there is that a first order sentence is satisfied by U_∞ if and only if it is satisfied by U_N for almost all N. So the theory of U_∞ is the limit of the theories of the U_N.

1.   Preliminaries

If G is any graph, then we write $d_G(x, y) = N$ to mean that the shortest path in G from x to y has length N. If N is "∞", then this means that x and y belong to different connected components of G. A map between graphs is an *isometric embedding* if it preserves all distances. All of the maps between graphs in this paper are isometric embeddings. An induced subgraph H of G is an *isometric subgraph* of G if the inclusion map is an isometric embedding. We write $H \hookrightarrow G$ in this situation. G is *distance finite* if every finite set X of vertices is contained in a finite isometric subgraph of G.

For two tuples $\overline{x}$ and $\overline{y}$ of the same length k taken from two graphs G and H, respectively, we will write $\overline{x} =_N \overline{y}$ to mean that whenever $1 \leq i, j \leq k$ and either $d_G(x_i, x_j) \leq N$ or $d_H(y_i, y_j) \leq N$, then $d_G(x_i, x_j) = d_H(y_i, y_j)$. A graph G is *distance homogeneous* if for every pair of tuples $\overline{x}$ and $\overline{y}$ of the same length such that $\overline{x} =_\infty \overline{y}$ there is some automorphism if G taking $\overline{x}$ to $\overline{y}$ pointwise.

G is *distance injective* if for all finite connected F and H, if $i : F \to G$ and $J : F \to H$, then there is an isometric embedding $k : H \to G$ such that $k \deg j = i$.

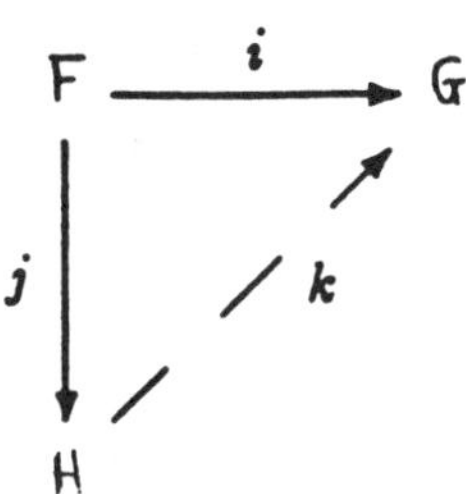

In this paper, we will need a refinement of this concept. G is *N-distance injective* if for all $i : F \to G$ and $j : F \to H$ where H is a finite graph of *diameter N*, there is some $k : H \to G$ such that $k o j = i$.

Clearly, G is distance injective iff it is N-distance injective for all N, and if G is N-distance injective, then it is also M-distance injective for all $M < N$.

Finally, G is *distance full* if G is distance injective and distance finite.

In [2] we studied these concepts, and the reader is referred there for a more thorough discussion. Here is a summary of some of the results: Using the Amalgamation Lemma for distances in graphs, it is routine to show that there is a countable distance full graph. One constructs such a graph by repeatedly amalgamating all of the finite connected graphs, in all possible ways. In addition, every countable distance full graph isometrically embeds every countable connected graph. (So the disjoint union of countably many such graphs embeds isometrically every countable graph.) Every countable distance full graph is distance homogeneous. There is exactly one countable connected distance full graph, and we let U_∞ denote this graph. Our last preliminary is a general construction concerning graphs of finite diameter. Let G be any graph, and $N \in \omega$. Then we build a graph G^+, the *canonical graph of diameter N determined by G* as follows: If N is even, then we take a new point z and add chains of length $N/2$ from the z the elements of G. The chains consist of new elements, and the only neighbors that these elements have are other elements on the chains. If N is odd, then we take a new clique $C = \{c_g : g \in G\}$ in one-to-one correspondence with the elements of G. We then add a new chain of length $(N-1)/2$ from each c_g to the corresponding g.

If G has diameter N, then G is an isometric subgraph of G^+. If G did not have diameter N, then the inclusion will not preserve all distances, but it will preserve all distances of length $\leq N$. This is because all of the paths in G^+ which contain new points have length $\geq N$.

2. Saturated Cliques

The study of universal graphs of odd diameter calls for an investigation of cliques in U_∞. The idea is that to get the universal graph of diameter $2N+1$, we take a clique C, and then consider the set of all points whose distance to some element of C is N. However, it turns out that this will not work unless C has a special saturation property which we isolate below.

The main results in this section are Lemmas 1, 2 and 5. At first the reader may wish to omit all of the details of this section except the two definition paragraphs and the statements of these results.

Definitions   Let C be a clique in a graph G, and let $S \subseteq G$ be finite. A *type of S* is a function $t : S \to \omega$. The type t is *consistent with* C if there is an isometric extension H of G which contains a point y such that $C \cup \{y\}$ is a clique in H, and for all $x \in S, d_H(x,y) = t(x)$. A type t is *realized in* C if there is some $y \in C$ such that for all $x \in S, d_H(x,y) = t(x)$. C is *saturated* if every type consistent with C is realized in C.

Before proving the existence of saturated cliques in U_∞, we should make a few remarks which will be used below. If t is consistent with C, then t is also consistent with every subset $D \subseteq C$. If t is realized in C, then it is consistent with C - take H to be G with a new point z whose neighbors are y and the neighbors of y in G. Also, if C is saturated, then every type consistent with it is realized by infinitely many elements of C.

Every saturated clique $C \subseteq U_\infty$ has the property that for all $x \in U_\infty$ there are exactly two distances from x to the points of C. (If not, add a new point z to U_∞, making it a neighbor only to the points of C.) We call such a clique *two-valued*, and we let $m = m_c : U_\infty \to \omega$ be such that $m(x)$ is the least distance from x to an element of C.

It is important to reformulate the notion of consistency as follows in terms of the function m: Let C be a two-valued clique in a graph G, and let S be finite. A type t of S is consistent iff (1) for all $x \in S, t(x) = m(x)$ or $t(x) = m(x) + 1$; and (2) if for all $x, y \in S, d_G(x, y) \le t(x) + t(y)$ and $t(x) \le t(y) + d_G(x, y)$. (1) is necessary because a point in a graph can have at most two distances to any clique, and (2) is necessary by the triangle inequality. To see that these two are sufficient, assume that t has these properties. Let $H = H(G, C, t)$ arise by adding a new point y to G with edges to all of the points in C and with a chain of length $t(x)$ to each $x \in S$. Then triangle inequalities of (2) imply that $G \hookrightarrow H$ and that for all $x \in S, d_H(x, y) = t(x)$.

Recapitulating, the consistency of a type t of S may be decided solely on the basis of the distance relations among the points of S and the action of m_C on S.

Lemma 1 There exist saturated cliques in U_∞. In fact, every finite clique D is included in a saturated clique C.

Proof Let D be any finite clique. Let $< t_i : i \in \omega >$ enumerate the types of U_∞. We construct finite cliques C_i for $i \in \omega$ as follows: Let $C_0 = D$. Given C_i, if t_i is not consistent with C_i, set $C_{i+1} = C_i$. If it is consistent, let F be a finite isometric subgraph of U_∞ containing $C_i \cup dom(t_i)$. In the notation above, let $H = H(F, C_i, t)$. So $F \hookrightarrow H$, and by distance injectivity, we can assume that $H \hookrightarrow U_\infty$. So let $C_{i+1} = C_i \cup \{y\}$; this insures that if t_i is consistent with an extension D of C_{i+1}, it is realized in D.

Now the union $\bigcup_{o \in \omega} C_i$ is saturated.

Before going on, we should state a reformulation of the notion of a saturated clique. Consider the category of cliqued graphs; i.e., the objects are structures $< G, C >$ such that G is a graph and C is a clique in G. The morphisms are the isometric embeddings which preserve membership in the clique in both directions. This category has the amalgamation property, and

$C \subseteq U_\infty$ is saturated iff $< U_\infty, C >$ has the injectivity property in the category of cliqued graphs.

Lemma 2   Let G be any countable connected graph and let D be a clique in G. Then there exists an isometric embedding $i : G \to U_\infty$ and a saturated clique $C \subseteq U_\infty$ such that $i(D) \subseteq C$.

   Proof   We construct C and i together, alternating the steps. The steps in constructing C are as in Lemma 1. And the steps in constructing i are as in the proof that U_∞ embeds G.

   The following lemma shows that saturation is a strong form of maximality in the poset of all cliques. It will not be used in the sequel.

Lemma 3   Every saturated clique is maximal. There are maximal cliques of U_∞ which are not saturated.

   Proof   Let C be saturated. If $x \in U_\infty - C$, then there are points in C of different distances to x. In particular, x is not a neighbor of every element of C. So C is maximal.

   For the other direction, let x and y be neighbors in U_∞. Let $< a_i : i \in \omega >$ enumerate the neighbors of x, and let $< b_i : i \in \omega >$ enumerate the neighbors of y. We construct finite cliques C_i by recursion on i. Let $C_0 = \{y\}$. Given C_{2i}, let C_{2i+1} be $C_{2i} \cup \{b_i\}$ if this is a clique and b_i is not a neighbor of x (and let it be C_{2i} otherwise). Given C_{2i+1}, let z be a neighbor of every element of C_{2i+2} but not a neighbor of a_i or x. Such a point exists by injectivity. Let $C_{2i+2} = C_{2i+1} \cup \{z\}$.

   This defines the sequence, and let C be the union. Now C is a clique, and the only neighbor of x in C is y. So C is not saturated. We claim that C is maximal. If not, then for some $i, b_i \notin C$, but $C \cup \{b_i\}$ is a clique. Now in defining C_{2i+1}, we did not put b_i into C. Thus b_i is a neighbor of x. Suppose $b_i = a_j$. Then at step C_{2j+2} we added to C to some z which is not a neighbor of b_i. Thus $C \cup \{b_i\}$ is not a clique, and this is a contradiction.

Definition   Let C and D be saturated cliques in U_∞. Let $\overline{x}$ and $\overline{y}$ be two n-tuples from U_∞. We write $\overline{x} \equiv_{C,D} \overline{y}$ if $\overline{x} = \infty \overline{y}$ and if $m_C(x_i) = m_D(y_i)$ for $1 \leq i \leq n$.

Lemma 4   Suppose $\overline{x} \equiv_{C,D} \overline{y}$. Then for every $x^* \in U_\infty$ there is some $y^* \in U_\infty$ such that $\overline{x}, x^* \equiv_{C,D} \overline{y}, y$.

Proof   We first consider the special case when $x^* \in C$. Define t on $\overline{x}$ by $t(x_i) = d(x_i, x^*)$. This type is consistent since it is realized. So it satisfies the triangle inequalities. Now define a type u on $\overline{y}$ by $u(y_i) = t(x_i)$. Then u satisfies the same triangle inequalities, so it is consistent with D. By saturation, we may just take y^* to be a point in D realizing u.

The general case uses this observation. Consider $\overline{x}, x^*$, and let a_1 and a_2 be points from C such that $d(x^*, a_1) = m_C(x^*)$ and $d(x^*, a_2) = m_C(x^*) + 1$. By the first paragraph we can find b_1 and b_2 from D such that $\overline{x}, a_1, a_2 \equiv_{C,D} \overline{y}, b_1, b_2$. By distance homogeneity, we can find some y^* such that $\overline{x}, a_1, a_2, x^* = \infty \overline{y}, b_1, b_2, y^*$. It follows from this that $m_D(y^*) = m_C(x^*)$. Therefore $\overline{x}, x^* \equiv_{C,D} \overline{y}, y^*$.

Lemma 5   Suppose that C and D are saturated cliques in U_∞. Then there is an automorphism $\imath$ of U_∞ such that $\imath(C) = D$. In fact, given any tuples $\overline{x}$ and $\overline{y}$, such that $\overline{x} \subset_{C,D} \overline{y}$, there is an automorphism $\imath$ taking the first tuple to the second pointwise.

Proof   We build the automorphism one step at a time using the last lemma.

3. Existence of Universal Graphs

Theorem 6   For each N, there is a countable graph U_N of diameter N such that for all countable G of diameter N there is an isometric embedding $i : G \to U_N$.

Proof   First we consider the case of even N. Let x be any point of U_∞, and let U_{2N} be the subgraph of U_∞ induced by the set

$$S_x = \{y \in U_\infty : d_{U\infty}(x,y) = n\}.$$

Clearly U_{2N} has diameter $2N$. To show that U_{2N} is universal, let F be any countable graph of this diameter. Let z be a new point, and consider a graph H obtained by adding a new chain of length N to each point of F. This H has diameter $2N$, and F is isometrically embedded in it. Since it is countable, there is an isometric embedding $j : G \to U_\infty$. But also, there is an automorphism $\imath$ of U_∞ taking $j(z)$ to x. Let k be the restriction of $j \circ \imath$ to H. So k is an isometric embedding of H into U_∞ whose range is contained in U_{2N}.

We claim k is an isometric embedding when considered as a map into U_{2N}. Let $x, y \in F$. Since U_{2N} is an induced subgraph of U_∞, $d_{U_\infty}(k(x), k(y)) \leq d_{U_N}(k(x), k(y))$. But there is a path in U_{2N} between $k(x)$ and $k(y)$ whose length is $d_F(x,y)$. Therefore, $d_{U_{2N}}(k(x), k(y)) \leq d_F(x,y) = d_{U\infty}(k(x), k(y))$.

Second, we get universal graphs for the odd diameters. Let C be a saturated clique, and let U_{2N+1} be the subgraph induced by

$$S_C = \{y \in U_\infty : (\exists x \in C)d_{U_\infty}(x,y) = N \,\&\, (\exists x \in C)d_{U_\infty}(x,y) = N+1\}.$$

We first need to see that S_C has diameter $2N+1$. Clearly the diameter is at most $2N+1$. Let x and y be any two points in C, let F be $\{x,y\}$ considered as a subgraph of U_∞, and let H be a chain $z_1, ..., z_{2N+2}$ of length $2N+1$. Then $F \hookrightarrow U_\infty$, and $F \hookrightarrow W$ via $x \mapsto z_{N+1}, y \mapsto z_{N+2}$. By distance injectivity, we can assume H is a subgraph of $U\infty$. Then z_1 and z_{2N+2} are in U_{2N+1} and their distance is $2N+1$.

Now we prove universality. Let F be a countable graph of diameter $2N+1$, and let F^+ be the canonical graph of diameter $2N+1$ determined by F. By Lemma 2, there is an isometric embedding $i : H \to U_\infty$ with the property that $i(D)$ is contained in a saturated clique. The rest of the proof is the same as in the even case, using Lemma 5.

Henceforth we will let U_N denote the graph built in the preceding proof. It is easy to see that U_{2N} does not depend on the choice of x. This is because if x and y are any two points in U_∞, then there is an automorphism taking x to y and hence S_x to S_y. Similarly, Lemma 5 implies that U_{2N+1} does not depend on the choice of C.

4. Distance Homogeneity and Uniqueness of the Universal Graphs

In this section, we characterize the universal graph $_N$ as the unique countable distance homogenous graphs of diameter N which isometrically embeds every countable (or even every finite) graph of diameter N. In so doing, we need to prove a few other properties of these graphs. These are all consequences of the following technical lemma which is a refinement of Theorem 6.

Lemma 7 Let F be a finite isometric subgraph of U_∞ of diameter $\leq N$. There is an automorphism $\imath$ of U_∞ such that $\imath(F) \subseteq U_N$ and $\imath$ is the identity on $F \cap U_N$.

Proof We need to consider the cases of even and odd diameter separately. We omit the even case, because it is easier and because it is similar to the proof of Theorem 6.

For the odd case, let C be such that in the notation from the proof of Theorem 6, $U_{2N+1} = S_C$. Given F, let F^+ be the canonical graph of diameter N determined by F, and let D be the new clique in F^+. Now F^+ is isometrically embedded in U_∞, and by distance homogeneity, we may assume that the embedding is the identity on F. Let E be a saturated clique extending D by Lemma 1. Consider the set $F \cap U_{2N+1}$ and recall the function m which gives the minimum distance to a clique. By the definition of U_{2N+1} and the construction of D we see that $m_E(x) = m_E(x) = N$ for $x \in F \cap U_{2N+1}$. Thus $\overline{x} \equiv_{C,E} \overline{x}$. Now by Lemma 5, let $\imath$ be an automorphism of U_∞ taking E to C and fixing $\{x_1, ..., x_n\}$ pointwise. This $\imath$ takes F to a subset of U_{2N+1}.

Corollary 8 For all N, U_N is an isometric subgraph of U_∞.

Proof   Given x and y, let F be a path in U_∞ between them.

Henceforth we shall use Corollary 8 without mentioning it. For example to show that a subgraph F of U_N is an isometric subgraph, it is sufficient to show that $F \hookrightarrow U_\infty$.

Corollary 9 U_N is distance finite and distance homogeneous for all N.

Proof   Let S be a finite subset of U_∞. Let F be S together with paths between the elements of S, and let F^+ be the canonical graph of diameter N determined by F. The distances between the points in S are the same when measured in F^+ as in F, and F^+ is isometrically embedded in U_∞. By Lemma 7, we can assume that the embedding maps into U_N and that it is the identity on S. Thus U_∞ is distance finite.

For distance homogeneity, suppose $x_1, \cdots, x_n =_\infty y_1, \cdots, y_n$ are tuples from U_N, and let x^* also be a point in U_N. We know that there is some $y^* \in U_\infty$ such that $x_1, \cdots, x_n x^* =_\infty y_1, \cdots, y_n, y^*$. Let F be a finite isometric subgraph of U_∞ containing $y_1, ..., y_n$ and y^*. The diameter of F may be taken to be at most N by the argument of the last paragraph, since the distances between the points in the tuples are at most N. Apply Lemma 7. The point we want is $\imath(y^*)$.

For the next result, recall the definition of N-distance injectivity from Section 1.

Corollary 10 U_N is N-distance injective for all N.

Proof   Any $i : F \to U_N$. may be regarded as an isometric embedding into U_∞. So there is some k such that $koj = i$. This k is a map into U_∞, but since its image $k(H)$ has diameter N, we may assume by Lemma 7 that it maps into U_N.

Lemma 11   Fix N. U_N is (up to isomorphism) the only countable distance finite N-distance injective graph of diameter N.

Proof   This is a routine back and forth argument.

The following is our main result in this section:

Theorem 12   For all N, U_N is (up to isomorphism) the only countable distance homogeneous graph of diameter N which isometrically embeds each finite graph of diameter N.

Proof   We know by Corollaries 9 and 10 that U_N has the properties mentioned above. For uniqueness, suppose that G has them. We show that G is distance finite and N-distance injective. The arguments are virtually the same as those for Corollaries 9 and 10, and we omit them. Then by Lemma 11, $G = U_N$.

5. Additional Characterizations

One of the basic characterizations of the Rado universal graph U_2 is that it is the only countable graph G with the property P_N for all N. P_N is the following condition:

(P_N) For every two disjoint sets U and V of vertices of size at most N, there is a point $x \in G$ which is a neighbor of every element of U and of no element of V.

In this section we present two properties of a similar flavor that capture the other universal graphs. Consider first the following property of a graph G:

(Q_N) Let $V_1, V_2, ... V_{N-1}$ and W be finite subsets of G. Suppose that whenever $1 \le i \le j \le N-1, x \in V_i$, and $y \in V_j$, we have $j - i \le d_G(x, y) \le j + i$. Finally, suppose that whenever $1 \le i \le N - 1, x \in V_i$ and $y \in W$, we have $N - i \le d_G(x, y)$. Then there is some $z \in G$ such that (a) for all $i \le N$ and all $x \in V_i, d_G(x, z) = i$; and (b) for all $y \in W, d_G(y, z) \ge N$. The motivation for the conditions is as follows: Let z be a point in any graph, and for $1 \le i \le N - 1$, let V_i be the set of points whose distance from z is i. Let W be the set of points whose distance is at least N. Then $V_1, \cdots V_{N-1}, W$ have the properties above. The condition Q_N says that whenever such a point z could possibly exist, it must exist.

Lemma 13   For all N, if $N \le M \le \infty$, then U_M has Q_N.

Proof Let F be a finite isometric subgraph of U_N containing $V_1 \cup \cdots \cup V_{N-1} \cup W$. Let H_0 be obtained by adding a new point w together with chains of length i to each element of V_i. It should be checked that $F \hookrightarrow H_0$, and that w satisfies the conclusion of Q_N. If M is finite, we cannot apply injectivity unless we know that the diameter of H_0 is at most M. This is not in general true, but we can consider instead the canonical graph of diameter M determined by H_0. Call this graph H. H_0 need not be isometric subgraph of H, but F still is. Also, the distances between w and the points in F are not shortened as we pass to H. Then by M-distance injectivity, we get a map $k : H \to U_N$, and we let $z = k(w)$.

It follows from results below that U_N is the only countable graph of diameter N with Q_N. Our proof involves another property, R_N given as follows:

(R_N) Let H be any graph, and let $m \in \omega$ be arbitrary. Suppose $\overline{x}$ and $\overline{y}$ are tuples of length m from G and H respectively, and suppose that $\overline{x} =_{2(N-1)} \overline{y}$. Then for any $y^* \in H$ there is some $x^* \in G$ such that $\overline{x}, x^* =_{N-1} \overline{y}, y^*$.

Lemma 14 For all G, Q_N iff R_N.

Proof Let G be any graph with Q_N. Fix $\overline{x}, \overline{y}$, and y^*. For $1 \leq i \leq N-1$, let $V_i = \{x_a : 1 \leq a \leq n \, \& \, d_H(y^*, y_a) = i\}$. Let $W = \{x_a : d_H(y^*, y_a) \geq N\}$. We verify the hypothesis of Q_N for these sets. Suppose $i \leq j \leq N-1, x_a \in V_i$, and $x_b \in V_j$. Then the triangle inequality tells us that $d_H(y_a, y_b) \leq 2(N-1)$. Hence $d_H(y_a, gy_b) = d_G(x_a, x_b)$. And this distance is between $j - i$ and $j + i$.

Next, suppose that $x_a \in V_i$ and $x_b \in W$. We need to see that $d_G(x_a, x_b) \geq N - i$. Suppose not. Then $d_H(y_a, y_b)$ is the same, and so

$$d_H(y^*, y_b) \leq d_H(y^*, y_a) + d_H(y_a, y_b) < i + (N - i) = N.$$

This contradicts the assumption that $x_b \in W$.

Thus all of the hypotheses for Q_N are satisfied. Let x^* be the point z given by Q_N. This proves R_N. The argument that R_N implies Q_N is similar to the proof of Lemma 13.

Theorem 15   For all N, U_N is the only countable graph of diameter N with property R_N.

Proof   By Lemmas 13 and 14, U_N has R_N. To prove uniqueness, we use a back and forth argument to show that any two graphs G and H of diameter N with R_N must be isomorphic. Suppose that $\overline{x}$ and $\overline{y}$ are two n-tuples from these graphs, and suppose that $\overline{x} =_\infty \overline{y}$. A fortiori, $\overline{x} =_{2(N-1)} \overline{y}$. Let $x^* \in G$. By R_N, we get some y^* such that $\overline{x}, x^* =_{N-1} \overline{y}, y^*$. The important point is that since both G and H have diameter N, we actually have " $=_\infty$ " here.

We do not know whether there is a uniqueness result for the graphs of diameter N which have R_M for $M < N$. (We suspect that there are many such graphs.) We also do not know the relation of R_N to the other properties studied in this paper, such as distance finiteness and distance homogeneity.

6.   The First Order Theories of the Universal Graphs

The goal of this section is to prove a result which illustrates a sense in which U_∞ is a kind of limit of the graphs U_N. Since the U_N do not sit inside each other in any obvious way, we cannot simply take a union. Our result is based on the first order logic. We have in mind the very simple language containing exactly one nonlogical symbol, a binary relation symbol R. All of the results in this paper concern this language. We show that the first order theory of U_∞ is the limit of the first order theories of its approximations U_N.

Consider a game $\mathcal{G}_{m,r}(G, H)$ played by two players called I and II. This game has m rounds, and in the course of these rounds, the players pick m-tuples $\overline{x}$ from G and $\overline{y}$ from H. More concretely, a play of the game is as follows: player I picks a point in either G or H, and then player II responds with a point in the other graph. They repeat this m times, and at in each of the m rounds, I may pick in either structure. In this way, they select tuples $\overline{x}$ from G and $\overline{y}$ from H. Note that $\overline{x}$ consists of the points chosen from G by either player; the order of points in this tuple is the order in which the points were selected. Given such a round of the game, we say that II wins iff $\overline{x} =_r \overline{y}$. We are interested in the question of which player has a winning strategy in this game.

Theorem 16   Let $m \geq 2$, and suppose that $r \cdot 2^{m-2} < N, M \leq \infty$. Then player II has a winning strategy in $\mathcal{G}_{m,r}(U_N, U_M)$.

Proof   By induction on m. When $m = 2$ this follows from the $(r+1)$-distance injectivity of U_N and U_M. Assume it true for m. Fix r, and suppose $r \cdot 2^{m-1} < N, M$. Therefore $(2r) \cdot 2^{m-2} < N, M$, so we have a winning strategy σ for II in $\mathcal{G}_{m,2r}(U_N, U_M)$. We get a winning strategy for II in $\mathcal{G}_{m+1,r}(U_N, U_M)$ by playing the first m rounds according to σ. This gives m-tuples $\overline{x} =_{2r} \overline{y}$. For the last move, note that since $r \cdot 2^{m-1} \geq r+1, U_N$ and U_M both have property R_{r+1}. So II can always play to insure that the $(m+1)$-tuples will be r-equivalent.

The well-known Ehrenfeucht-Fraïssé game of length m is $\mathcal{G}_{m,1}(G, H)$. It has an important connection to first order logic by a theorem which we will need. To state it, we need to recall the inductive definition of the *quantifier rank* of a formula ϕ of first order logic, $qr(\phi)$:

$$qr(\phi) := 0 \text{ if } \phi \text{ is atomic}$$

$$qr(\neg\phi) := qr(\phi)$$

$$qr(\phi \wedge \psi) = qr(\phi \vee \psi) := \max(qr(\phi), qr(\psi))$$

$$qr((\exists x)\phi) = qr((\forall x)\phi) := 1 + qr(\phi).$$

For example, consider the formulas $\psi_m(x, y)$ defined by recursion on m: $\psi_0(x, y)$ is $(x = y) \vee R(x, y)$, and given $\psi_m, \psi_{m+1}(x, y)$ is $(\exists z)(\psi_m(x, z) \wedge \psi_m(z, y))$. Then each $\psi_m(x, y)$ has quantifier rank m. And for all graphs

$G, G \models \psi_m[x, y]$ iff $d_G(x, y) \leq 2^m$.

Recall the Ehrenfeucht-Fraïssé Theorem (cf. [1]): Suppose that II has a winning strategy in $\mathcal{G}_{m,1}(G, H)$, and let ϕ be a sentence of quantifier rank m. Then $G \models \phi$ iff $H \models \phi$.

From this and Theorem 16, we see that if ϕ is a sentence of quantifier rank $m \geq 2$ and $N > 2^{m-2}$, then $U_\infty \models \phi$ iff $U_N \models \phi$. (This lower bound on N is sharp. Consider the sentence $(\exists x)(\exists y)\neg\psi_{m-2}(x, y)$. This sentence has quantifier rank m and it is true in U_∞ but false in $U_{2^{m-2}}$.)

Thus with the natural meaning of limit,

$$\text{Theory}(U_\infty) = \lim_{N \to \infty} \text{Theory}(U_N).$$

We might also note that the results of this section show that the first order theories of the U_N are all complete. (A set of axioms for Theory (U_N) may be obtained from Theorem 15.) If we extend the language of graphs to include additional predicates for finite distances, then Theory(U_N) is the model completion of the theory of graphs of diameter N. We conjecture that every finite subtheory of it has a finite model. One way to express this conjecture is as an injectivity principle: For all N, there is a finite G such that for all F and H of size at most N and all $i : F \to G$ and $j : F \to H$ there is some $k : H \to G$ such that $k \circ j = i$.

REFERENCES

[1] Ehrenfeucht, A., "An Application of Games to the Completeness Problem for Formalized Theories," *Fundamenta Mathematicae* 49 (1961), 129-141.

[2] Moss, L S., "Distanced Graphs," to appear.

[3] Moss, L. S. "Existence and Nonexistence of Universal Graphs," *Fundamenta Mathematicae*, to appear.

[4] Rado, R., "Universal Graphs and Universal Functions," *Acta Arithmetica* 9 (1964), 331-340.

On the Difference Between the Domination and Independent Domination Numbers of Cubic Graphs

Christine Mynhardt

University of South Africa

ABSTRACT

Barefoot, Harary and Jones [2] conjectured that $K(3,3)$ and $C_5 \times K_2$ are the only 3–connected cubic graphs for which the domination and independent domination numbers, γ and i respectively, differ. They also conjectured the existence of an infinite class of cubic graphs with connectivity one for which $i - \gamma$ becomes unbounded. We disprove the former conjecture by describing an infinite class of 3–connected cubic graphs for which $\gamma \neq i$, and prove the latter conjecture by constructing a class of graphs satisfying the given requirements.

1. Introduction

Unless stated otherwise we use the notation and terminology of [3]. A set D of vertices of a graph $G = (V, E)$ is a *dominating set* if each vertex of G not in D is adjacent to at least one vertex in D. A set I of vertices is *independent* in G if no two vertices in I are adjacent; if, in addition I is also a dominating set then I is called an *independent dominating set* of G. Note that the independent dominating sets of G are exactly the maximal independent sets of G (see [4, p. 309]). A vertex x of a subset X of V is *redundant* in X if its closed neighbourhood is contained in the union of the closed neighbourhoods of the vertices in $X - \{x\}$, and X is *irredundant* if it contains no redundant vertices.

Two parameters are associated with each of the above concepts in the following way. The *domination number* $\gamma(G)$ *(upper domination number* $\Gamma(G)$ *)* of G is the smallest (largest) number of vertices in a minimal dominating set of G. (In [3], $\sigma(G)$ is used instead of $\gamma(G)$.) The *independent domination number* $i(G)$ *(independence number* $\beta(G)$ *)* of G is the smallest (largest) number of vertices in a maximal independent set of G, and finally, the *irredundance number* $ir(G)$ *(upper irredundance number* $IR(G)$ *)* of G is the smallest (largest) number of vertices in a maximal irredundant set in G. Since every maximal independent set is a minimal dominating set (see [4, p. 309]), and every minimal dominating set is a maximal irredundant set (see [7]), it follows that for any graph G,

$$\mathrm{ir}(G) \le \gamma(G) \le i(G) \le \beta(G) \le \Gamma(G) \le \mathrm{IR}(G).$$

Various authors have found sufficient conditions for two or more of these parameters to be equal. These include [1], [5], [8] and [13] for the lower parameters and [6], [8], [11] and [12] for the upper parameters. Of specific importance to the present paper are conditions under which equality of the domination and independent domination numbers occurs. In this respect Allan and Laskar proved in [1] that if G is K(1, 3)–free, then $i(G) = \gamma(G)$. This extended an earlier result by Mitchell and Hedetniemi [13] that if G is the line graph of a tree, then $i(G) = \gamma(G)$. Harary and Livingston characterised the trees and the caterpillars for which these two parameters are equal in [9] and [10] respectively, and in [2], Barefoot, Harary and Jones conjectured that the only 3–connected cubic graphs for which $i \ne \gamma$ are the graphs K(3, 3) and $C_5 \times K_2$. We disprove this conjecture by describing a known infinite class of 3–connected cubic graphs containing $C_5 \times K_2$ for which $i = \gamma + 1$.

Barefoot, Harary and Jones [2] further constructed an infinite class of cubic graphs of connectivity 2 for which the difference $i - \gamma$ becomes unbounded, and conjectured that an infinite class of cubic graphs of connectivity 1 with the same property existed. We prove this conjecture by constructing a class of graphs with the desired properties.

2. A Class of 3–Connected Cubic Graphs with $i \ne \gamma$

For any positive integer k, let $G_k = C_k \times K_2$ – note that G_k is 3–connected and 3–regular. Suppose that $V(G_k) = U \cup W$ with $U \cap W = \phi$ and $G_k \langle U \rangle = G_k \langle W \rangle = C_k$. Say $U = \{u_0, ..., u_{k-1}\}$, $W = \{w_0, ..., w_{k-1}\}$ and assume that u_j and w_j are adjacent for each $j = 0, ..., k - 1$.

Theorem 1　*If $k \equiv 5(mod\ 12)$, then*
$$\gamma(G_k) = \lceil k/2 \rceil \ and \ i(G_k) = \lceil k/2 \rceil + 1.$$

Proof　For any graph G,
$$\gamma(G) \ge \lceil |V(G)| / (\Delta(G) + 1) \rceil.$$
Hence to show that $\gamma(G_k) = \lceil k/2 \rceil$ we only need to find a dominating set of G_k of cardinality $\lceil k/2 \rceil$. Let $k = 12m + 5$ for some integer m and let $D = \{ u_{4j} \mid j = 0, ..., 3m \} \cup \{w_{4j+2} \mid j = 0, ..., 3m\} \cup \{w_{12m+3}\}$. Then D dominates G_k (the routine details are omitted) and $|D| = 6m + 3 = \lceil k/2 \rceil$. Since w_{12m+2} and w_{12m+3} are adjacent, D is not independent.

Let $I = (D - \{w_{12m+2}\}) \cup \{w_{12m+1}, u_{12m+2}\}$. Then I is an independent dominating set of G_k of cardinality $\lceil k/2 \rceil + 1$. Suppose that S is an independent dominating set of G_k of cardinality $\lceil k/2 \rceil = 6m + 3$. Since G_k is 3–regular and S is independent, each vertex of S dominates exactly three vertices of $G_k - S$. But $G - S$ has $18m + 7$ vertices whereas $3|S| = 18m + 9$, which by the pigeon hole principle implies that $G_k - S$ either contains two vertices, each of which is dominated by two vertices in S, or one vertex which is dominated by three vertices in S, while all remaining vertices of $G_k - S$ are dominated by exactly one vertex in S.

Suppose firstly that u_0 (say) is dominated by the vertices w_0, u_1 and u_{k-1} in S. Then w_{k-1} and w_1 are dominated by u_{k-1}, w_0 and u_1, w_0 respectively, which is impossible. Therefore $G - S$ has two vertices which are each dominated by two vertices in S.

Say u_0 is dominated by two vertices in S – without losing generality we may assume that u_1 is one such vertex. If $u_{k-1} \in S$, then in order for w_0 to be dominated, w_0 is in S since S is independent. But we have already shown this to be impossible, hence $u_{k-1} \notin S$ and $w_0 \in S$, so that w_1 is the remaining vertex of $G_k - S$ dominated by two vertices in S. Since w_2 is dominated and S is independent, it follows that $w_3 \in S$. (Note that $w_2 \notin S$ for otherwise w_1 is dominated by three vertices in S.) Similarly, u_4 is dominated and $u_4 \notin S$, consequently $u_5 \in S$. Continuing in this way we deduce that $S = \{w_0\} \cup \{w_{4\ell+3} \mid \ell = 0, ..., 3m\} \cup \{u_{4\ell+1} \mid \ell = 0, ..., 3m\}$ – however, this is impossible since u_{k-1} is not dominated while w_{k-1} is dominated by w_0 and w_{k-2}.

This proves that $i(G_k) = \lceil k/2 \rceil + 1$. $\square$

The graphs $C_k \times C_2$ for $k \equiv 5 \pmod{12}$ therefore form an infinite class of 3–connected cubic graphs for which $i = \gamma + 1$.

3. A Class of Cubic Graphs with Connectivity One for which $i - \gamma$ becomes Unbounded

Let F, G and J with $V(F) = \{v_1, ..., v_5\}$ and $V(G) = \{w_1, ..., w_6\}$ be the graphs depicted in Figure 1. We construct the graph H_k as follows: Let $V(H_k) = U \cup V \cup W$ (disjoint union), where

$$U = \bigcup_{i=1}^{2k-1} U_i \qquad \text{for } U_i = \{u_{i1}, u_{i2}\},$$

$$V = \bigcup_{i=0}^{4k-1} V_i \qquad \text{for} \quad V_i = \{v_{i1}, ..., v_{i5}\} \quad \text{and}$$

$$W = \bigcup_{i=1}^{2k-2} W_i \qquad \text{for} \quad W_i = \{w_{i1}, ..., w_{i6}\}.$$

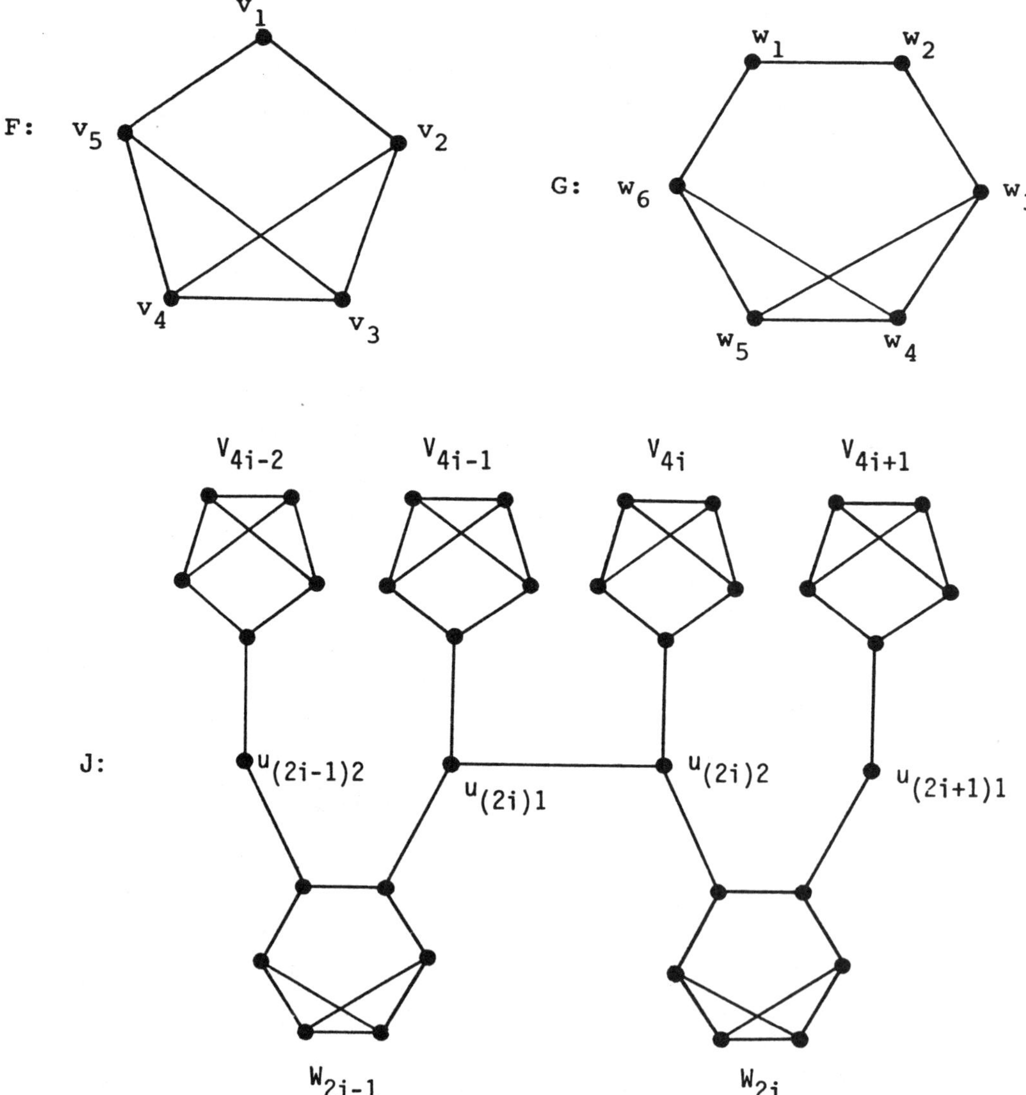

Figure 1. The graphs F, G and J used in the construction of H_k.

Add edges in such a way that

$$H_k\langle U_i\rangle = K_2 \quad \text{for each } i = 1, \ldots, 2k-1,$$

$$H_k\langle V_i\rangle = F \quad \text{such that } v_{ij} \text{ corresponds to the vertex } v_j \text{ of } F$$

for every $j = 1, \ldots, 5$ and $i = 0, \ldots, 4k-1$, and

$$H_k\langle W_i\rangle = G \quad \text{such that } w_{ij} \text{ corresponds to the vertex } w_j \text{ of } G$$

for every $j = 1, \ldots, 6$ and $i = 1, \ldots, 2k-2$. Add further edges such that for

$$X_i = (V_{4i-2} \cup V_{4i-1} \cup V_{4i} \cup V_{4i+1}) \cup (W_{2i-1} \cup W_{2i}) \cup$$

$$\{u_{(2i-1)2}, u_{(2i)1}, u_{(2i)2}, u_{(2i+1)1}\},$$

where $i = 1, \ldots, k-1$,

$$H_k\langle X_i\rangle = J_i = J.$$

Finally add the edges $\{v_{01}u_{11}, v_{11}u_{11}, v_{(4k-2)1}u_{(2k-1)2}, v_{(4k-1)1}u_{(2k-1)2}\}$. The sresulting graph is H_k. Note that H_k contains a string of $k-1$ copies $J_1, \ldots, J_{k-1}$ of J, where for each $i = 1, \ldots, k-2$, the copy J_i is joined to J_{i+1} by the edge $u_{(2i+1)1}u_{(2i+1)2}$. We illustrate H_1 and H_2 in Figures 2 and 3 respectively. Note that H_k is 3–regular and has connectivity one.

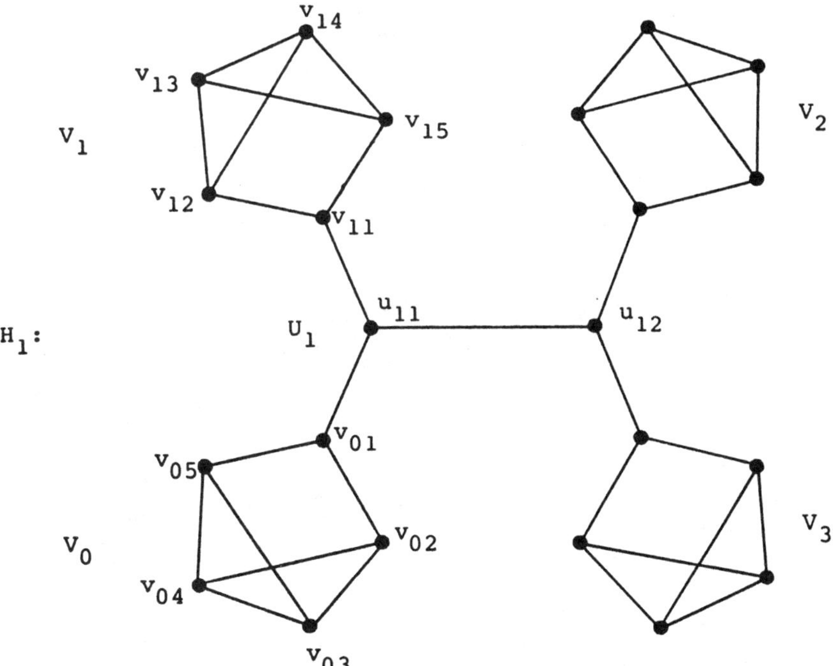

Figure 2. The graph H_1

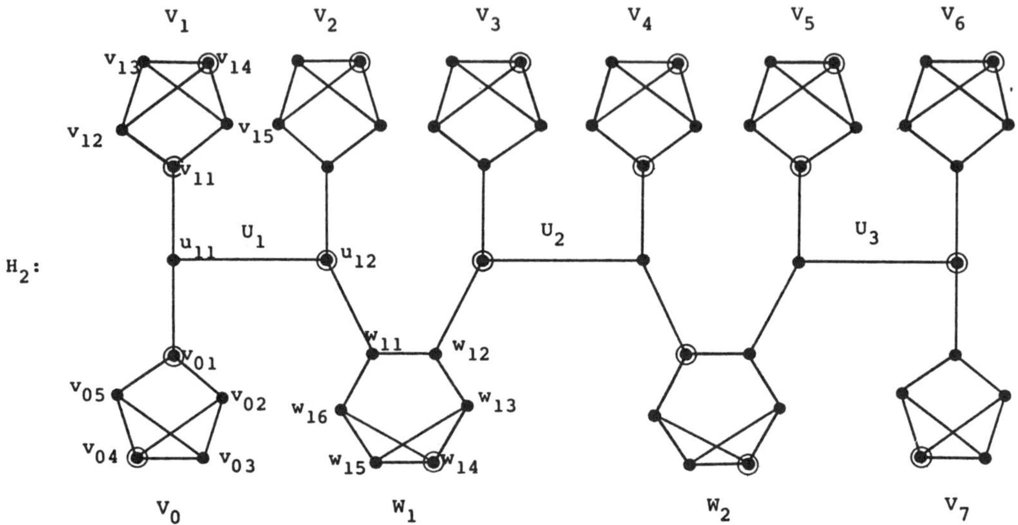

Figure 3. The graph H_2

We claim that $\gamma(H_k) = 10k - 4$ while $i(H_k) = 11k - 4$. In order to prove this, we first prove that it holds for $k = 1$.

Lemma 1 $\gamma(H_1) = 6$ *and* $i(H_1) = 7$.

Proof   A dominating set of H_1 of cardinality six is given by $\{u_{11}, u_{12}, v_{04}, v_{14}, v_{24}, v_{34}\}$ and since

$$\lceil |V(H_1)| / (\Delta(H_1) + 1) \rceil = 6,$$

we have that $\gamma(H_1) = 6$. An independent dominating set of H_1 of cardinality seven is given by $\{u_{11}, v_{04}, v_{14}, v_{22}, v_{25}, v_{32}, v_{35}\}$. Further, if D is any independent dominating set of H_1, then at most one of u_{11} and u_{12} is in D. If (say) $u_{12} \notin D$, then at least two vertices in each of V_2 and V_3 are in D in order to dominate $V_2 \cup V_3$. No vertex in $V_2 \cup V_3$ dominates any vertex in $V_0 \cup V_1 \cup \{u_{11}\}$ and no subset of $V_0 \cup V_1 \cup \{u_{11}\}$ containing fewer than three vertices dominates $V_0 \cup V_1 \cup \{u_{11}\}$; hence $|D| \geq 7$. We have thus proved that $i(H_1) = 7$. $\square$

Now let $J = J_1$, $L = J_1 - \{u_{12}\}$ and $M = J_1 - \{u_{12}, u_{31}\}$. In our next result we determine the domination and independent domination numbers of J, L and M.

Lemma 2 $\gamma(J) = 10,$

$\gamma(L) = i(J) = i(L) = 11$ *and*

$\gamma(M) = i(M) = 12.$

Proof   Throughout the proof we use the fact that if D is a set which dominates $\langle V_i \rangle$, and v_{i1} is not dominated by $D - V_i$, then $|D \cap V_i| \geq 2$, and if D dominates $\langle W_i \rangle$ and at least one of w_{i1} and w_{i2} is not dominated by $D - W_i$, then $|D \cap W_i| \geq 2$.

   Let $N = H_k \langle V_3 \cup V_4 \cup U_2 \rangle$. By the above observation it is clear that $\gamma(N) = i(N) = 4$. Also by the above observation, if D is any dominating set of M, then D contains at least two vertices of each of V_2, V_5, W_1 and W_2, and it follows that $|D| \geq 12$. The set

$$D = \left(\bigcup_{i=2}^{5} \{v_{i2}, v_{i5}\} \right) \cup \{w_{12}, w_{14}, w_{21}, w_{24}\}$$

is an independent dominating set of M of cardinality 12 and hence $\gamma(M) = i(M) = 12$. Similarly, $\gamma(L) = i(L) = 11$ and $\gamma(J) = 10$.

   To prove that $i(J) \geq 11$, note that if S is an independent dominating set of J, then at most one of u_{21} and u_{22} is in S. If (say) $u_{22} \notin S$, then S contains at least two vertices of each of V_4 and W_2 and the result follows. An independent dominating set of J of cardinality 11 is given by

$$S = \{u_{12}, v_{24}, w_{14}, u_{21}, v_{34}, v_{42}, v_{45}, w_{21}, w_{24}, u_{31}, v_{54}\},$$

consequently $i(J) = 11$. $\square$

   Note that D dominates H_k, then for any $i = 1, ..., k-1$, $D \cap V(J_i)$ corresponds to a dominating set of J, L or M, depending on whether two, one or none of the vertices $u_{(2i-1)2}$ and $u_{(2i+1)1}$ of J_i are contained in D.

Theorem 2   *For any positive integer k,*

$$\gamma(H_k) = 10k - 4 \quad and \quad i(H_k) = 11k - 4.$$

Proof   A dominating set of H_k of cardinality $10k - 4$ is given by

$$U \cup \left(\bigcup_{i=0}^{4k-1} \{v_{i4}\} \right) \cup \left(\bigcup_{i=1}^{2k-2} \{w_{i4}\} \right).$$

We also claim that

$$\left(\bigcup_{i=0}^{4k-1} \{v_{i4}\} \right) \cup \left(\bigcup_{i=1}^{2k-2} \{w_{i4}\} \right) \cup \left(\bigcup_{i=1}^{k-1} \{u_{(2i-1)2}, u_{(2i)1}, w_{(2i)1}\} \right) \cup \left(\bigcup_{i=0}^{k-1} \{v_{(4i)1}, v_{(4i+1)1}\} \right) \cup \{u_{(2k-1)2}\}$$

is an independent dominating set of H_k of cardinality $11k - 4$. That this is indeed so, can most easily be seen by noticing that the relevant vertices are the circled vertices of H_2 in Figure 3; the genaral case is similar.

If, for $k \geq 2$, H_k has a dominating set D with $|D| < 10k - 4 = 6 + 10(k - 1)$, then either D contains at most five vertices of $Y = V_0 \cup V_1 \cup V_{4k-2} \cup V_{4k-1} \cup \{u_{11}, u_{(2k-1)2}\}$, or D contains at most nine vertices of a subgraph J_i of H_k. Since neither case is possible, $\gamma(H_k) = 10k - 4$.

If H_k has an independent dominating set S with $|S| < 11k - 4 = 7 + 11(k - 1)$, then either $|S \cap Y| \leq 6$ (in which case $|S \cap Y| = 6$), or S contains at most ten vertices of J_i for some i. By Lemma 2 the latter possibility does not occur, hence suppose $|S \cap Y| = 6$. Then $\{u_{11}, u_{(2k-1)2}\} \subseteq S$ and $\{u_{12}, u_{(2k-1)1}\} \cap S = \phi$. But then for at least one i, say $i = 1$, $S \cap V(J_1)$ corresponds to an independent dominating set of M and hence $|S \cap V(J_1)| \geq 12$. Since Lemma 2 implies that $|S \cap V(J_i)| \geq 11$, it now follows that

$$|S| \geq 6 + 12 + 11(k - 2) = 11k - 4,$$

a contradiction. Therefore $i(H_k) = 11k - 4$. $\square$

REFERENCES

[1]   R.B. Allan and R. Laskar, On domination and independent domination numbers of a graph, *Discrete Math.* **23** (1978), 73 – 76.

[2]   C. Barefoot, F. Harary and K.F. Jones, What is the difference between α and α' of a cubic graph?   Preprint.

[3]   M. Behzad, G. Chartrand and L.Lesniak–Foster, *Graphs and Digraphs*, Prindle, Weber and Schmidt,   Boston, 1979.

[4]   C. Berge, *Graphs and Hypergraphs*, North–Holland, Amsterdam, 1973.

[5]   B. Bollobás and E.J. Cockayne, Graph–theoretic parameters concerning domination,   independence and irredundance, *J. Graph Theory* **3** (1979), 241 – 249.

[6]   E.J. Cockayne, O. Favaron, C. Payan and A.G. Thomason, Contributions to the theory of domination,   independence and irredundance in graphs, *Discrete Math.* **33** (1981), 249 – 258.

[7]   E.J. Cockayne, S.T. Hedetniemi and D.J. Miller, Properties of hereditary hypergraphs and middle   graphs, *Canad. Math. Bull.* **21** (4) (1978), 461 – 468.

[8]   O. Favaron, Stability, domination and irredundance in graphs, *J. Graph Theory* **10** (1986), 429 – 438.

[9]   F. Harary and M. Livingston, Characterisation of trees with equal domination numbers, *Congressus   Numerantium* (to appear).

[10] F. Harary and M. Livingston, The caterpillars with equal domination and independent domination   numbers, *Annals of Discrete Math* (to appear).

[11] M.S. Jacobson and K. Peters, Chordal graphs and upper irredundance, upper domination and   independence. Preprint.

[12] M.S. Jacobson and K. Peters, Upper domination, independence and irredundance numbers for   peripheral graphs.Private communication.

[13] S. Mitchell and S.T. Hedetniemi, Independent domination in trees, *Proc. of Eight S.E. Conference on   Combinatorics, Graph Theory and Computing* (1977), 489 – 509.

A REPORT ON THE ALLY RECONSTRUCTION PROBLEM

Wendy Myrvold

University of Victoria

British Columbia

ABSTRACT

The ally-reconstruction number of a graph G, ally-rn(G), is the minimum number of vertex-deleted subgraphs required in order to identify G up to isomorphism. In this paper, we report on some recent results for this parameter.

1.   Introduction

In this paper, we report on some new results for the reconstruction problem. Readers unfamiliar with this problem should refer to Bondy and Hemminger [4] which is a good survey of papers up to 1976 with a comprehensive bibliography. All graphs considered in this paper are simple, finite and undirected. The *order* of a graph is the number of vertices. Any further graph theoretic terminology can be found in [3].

The ally-reconstruction number of a graph G, *ally-rn(G)*, is the minimum number of vertex- deleted subgraphs of G required in order to identify G up to isomorphism. This parameter was introduced as the reconstruction number of a graph by Harary and Plantholt in 1985 [5]. In this paper, we survey what is known about the ally-reconstruction number of a graph and explain how this problem relates to the reconstruction problem.

2. Graphs with Small Ally-Reconstruction Number

One simple observation that was made by Harary and Plantholt [5, p. 451, Observation 1] is that the ally- reconstruction number of any graph is always at least three. We have included a proof here since it was omitted in [5].

Observation 2.1   For any graph G, ally-rn$G \geq 3$.

Proof   Suppose ally-rn$(G) = 2$. Then there are two cards, $G - u$ and $G - v$, which uniquely identify G. Assume without loss of generality that G does not have the edge (u, v). Then $H = G + (u, v)$ also has these two cards. The graph H is not isomorphic to G because it has a different number of edges.

We now show that almost all graphs can be reconstructed from any three cards of their deck and as a result, almost all graphs have ally-reconstruction number three. By 'almost all', we mean that the percentage of graphs on n vertices for which the statement is true approaches one hundred percent as n approaches infinity. This result was conjectured by Harary and Plantholt [5, p. 454, Conjecture 4] and proved by Bollobás [2]. We give an alternative proof which is a simple extension of the proof by Müller [13] that almost all graphs are reconstructible.

Lemma 2.2   Almost all graphs are reconstructible from any three cards.

Proof   In [13], Müller proves that given $\epsilon > 0$, the k-vertex- deleted subgraphs of almost all graphs are pairwise nonisomorphic if $k \leq \frac{n}{2}(1 - \epsilon)$. All that is needed for this proof is that almost all graphs have mutually nonisomorphic one, two, and three-vertex-deleted subgraphs. Because this is true, the following reconstruction algorithm will reconstruct almost all graphs from any three cards.

Reconstruction Algorithm for a Random Graph   **Input**: Any three cards, say $G - a, G - b,$ and $G - c$ from a graph G.

Output: If the output of this algorithm is a graph H, then we are guaranteed that any graph with the three cards input is isomorphic to H. Otherwise, the output is the message "Algorithm failed to reconstruct G". In this case, there still may only be one reconstruction from the cards given or there may be several.

1. Find all vertices b' of $G - a$ so that $G - a - b'$ is a card of $G - b$. If there is more than one such vertex b', report "Algorithm failed to reconstruct G".

2. Find all vertices a' of $G - b$ so that $G - b - a'$ is a card of $G - a$. If there is more than one such vertex a', report "Algorithm failed to reconstruct G".

3. If there is more than one isomorphism f from the vertices of $G-a-b'$ to those of $G - b - a'$, then report "Algorithm failed to reconstruct G". Otherwise, add a vertex b to $G - b$ which is incident to vertex w on $G - b$ if and only if vertex b' is adjacent to w' on $G - a$, and $f(w') = w$.

4. At this point, the structure of G has almost been determined. The only thing not known is whether or not G has edge (a, b). This is determined from $G - c$ by labeling vertices a and b on this card as in step one.

Since for almost all graphs, the 2-vertex-deleted subgraphs are mutually nonisomorphic, steps one, two and four are almost always carried out successfully. Since almost all graphs have mutually nonisomorphic 3-vertex-deleted subgraphs, there is almost always only one isomorphism from $G - a - b'$ to $G - b - a'$ in step three. Hence, given any three cards from a random graph, the preceding algorithm almost always succeeds in creating a unique reconstruction.

3. Graphs With Large Ally-Reconstruction Number

What is the largest ally-reconstruction number that a graph on n vertices can have? It has been conjectured by Harary and Plantholt [5, p. 453, Conjecture 1] that the answer is $\frac{n}{2} + 2$. A proof would be remarkable since this conjecture implies the reconstruction conjecture. This conjecture is true for the families of graphs considered in the following section. We present here the only known family of graphs with ally-reconstruction number equal to this value.

Lemma 3.1 [5]   There exists an infinite family of r-regular graphs which have ally-reconstruction number equal to

$$\lfloor \frac{n}{2} \rfloor + 2.$$

Proof   Let G be the r-regular graph consisting of two copies of K_{r+1}. Let H consist of K_{r+2} and K_r. This graph pair is shown in Figure 1 for $r = 3$. All cards of G are isomorphic to the cards created from H be deleting any of the $r + 2$ vertices in the K_{r+2}. Hence, any subset of the deck of G which the ally chooses has at least $r + 3$ cards.

Finally, we would like to quickly mention a counterexample to a remark made by Harary and Plantholt [5, p. 454, Conjecture 4]. They suggested that if G is of odd order, then perhaps $ally\text{-}rn(G) = 3$. The graph family consisting of $G_p = 3K_p$ is a family of graphs with $ally\text{-}rn(G_p) = p + 1$ (the proof of this is similar to that of Lemma 3.1). These graphs have odd order when p is odd (the number of vertices is $3\,p$).

(a) G.   (b) H.

Figure 1. Regular graph G with ally-rn 6

4.   Results for Specific Families of Graphs

As is customary for various reconstruction problems, the ally-reconstruction problem has been studied for various classes of graphs. The *class-reconstruction number* of a graph with respect to some particular class C of graphs is the number of cards needed to reconstruct a graph G assuming that we already know that G is a member of the class C. The class-reconstruction number has been studied for trees [1], unicyclic graphs [6], and maximal planar graphs [7]. Since these results are surveyed fairly thoroughly by Lauri [8], we will not cover them here. Instead, we will concern ourselves with what is known about the ally-reconstruction number of graphs in particular classes. The difference between these two problems is that in the latter case, we must also be able to determine that a graph is a member of the class being studied.

Harary and Plantholt [5, p. 454, Conjecture 3] conjectured that trees on five or more vertices have ally- reconstruction number three. Their conjecture is restricted to trees on five or more vertices since the ally- reconstruction number of a path on four vertices is four. This conjecture is proved in [11].

Theorem 4.1   For any tree T with $n \geq 5$ vertices, $ally\text{-}rn(T) = 3$.

In the proof of this result, we first describe how to select three cards for any tree T from which T is reconstructible. Next, we show that any graph having these cards must be a tree. Finally, we prove that any tree with the three cards selected is isomorphic to T. The main idea used in this proof is different from that used by Harary and Lauri for the class-reconstruction number of a tree in [1].

The ally-reconstruction problem has also been solved for disconnected graphs. In [10], it is proved that unless a disconnected graph G has all its components isomorphic, then the ally-reconstruction number of G is three. If G has isomorphic components of order c, then it has ally-reconstruction number at most $c + 2$. The family of graphs described in Lemma 3.1 can be used to construct families of graphs which show that this result is tight.

Theorem 4.2   [10, pp. 59-65]   Let G be a disconnected graphs. If all components of G are isomorphic with order c, then $ally\text{-}rn(G) \leq c + 2$. Otherwise, $ally\text{-}rn(G) = 3$.

If a graph G is reconstructible from some set C of its deck the complement $\overline{G}$ of G is reconstructible from the corresponding cards of $\overline{G}$. This is true because the cards of $\overline{G}$ are the complements of the cards of G. One result of this observation is that Theorem 4.2 also applies to graphs which are the complements of disconnected graphs (and less interesting perhaps, Theorem 4.1 applies to graphs whose complements are trees).

The only other class of graphs for which this problem has been solved is for regular graphs. To understand where the following theorem comes from, observe that the complement of an r-regular graph is an $(n-1-r)$-regular graph.

Theorem 4.3   [12, pp. 18-24]   An r-regular graph G has

$$\text{ally-rn}(G) \leq \min \{r+3, [(n-1)-r]+3\}.$$

This result is proved by first showing that the degree sequence of a regular graph is reconstructible from the number of cards mentioned above. The regular graph can then easily be reconstructed from any one card by adding a new vertex to the card and placing edges from the new vertex to all the vertices on the card of degree $r-1$. With little extra work, the following corollary can be obtained.

Corollary 4.4   [10, pp. 22-23]   If G is an r-regular graph on n vertices, then $\text{ally-rn}(G) \leq \lfloor \frac{n}{2} + 2 \rfloor$.

For all the classes mentioned here, we have been able to prove that the ally-reconstruction number is at most $\lfloor \frac{n}{2} \rfloor + 2$.

5.   Open problems

In this final section, we make some remarks concerning why we feel the ally-reconstruction problem is an important new area for reconstruction fanatics. No attempt has been made to screen out statements that may be controversial.

Nobody has been able to find a family of graphs or even a single graph having ally-reconstruction number greater than $\frac{n}{2} + 2$. If we define the ally-reconstruction number of a non-reconstructible graph on n vertices to be $n+1$, even the graphs in the non-reconstructible pair of graphs on two vertices satisfy the condition that the ally-reconstruction number is at most $\frac{n}{2} + 2$. If there does exist a graph with ally-reconstruction number $n + 1$ (i.e. a non-reconstructible graph), then one might also expect to find reconstructible graphs with reconstruction number greater than $\frac{n}{2} + 2$. The apparent absence of such graphs suggests strongly that perhaps the reconstruction conjecture is true.

The ally-reconstruction number could be thought of as a measure of how hard it is to reconstruct a graph. Regular graphs are one of the easiest cases for the general reconstruction problem, in spite of their incredible symmetry. Using the ally-reconstruction number as a measure, they appear to be one of the hardest cases. This is perhaps more appropriate.

The results mentioned in the previous section were generally proved by first explaining how to select the ally's cards from the deck of the graph. The next step was to explain how to reconstruct the graph from these cards together with a proof that there was only one way to do this (i.e. a proof that all reconstructions of the cards selected are isomorphic). This type of argument gives much more information concerning why a graph is reconstructible than previous results which were proved by contradiction (see for example [9]). Hopefully, someone will be able to use this additional insight to make further advances on the reconstruction conjecture.

At a time when most work on the reconstruction conjecture has come to a standstill, ally-reconstruction is flourishing. It is hoped that the ideas presented here will spark new interest in the reconstruction problem and that someday this (in)famous conjecture will be resolved.

REFERENCES

[1] F. Harary and J. Lauri, "On the class-reconstruction number of trees," *Oxford Quarterly Journal of Mathematics*, 1987. To appear.

[2] B. Bollobás, "Almost every graph has reconstruction number three," *Journal of Graph Theory*, 1987. To appear.

[3] J. A. Bondy and U. S. R. Murty, *Graph Theory with Applications*, Elsevier North Holland, Inc., New York, 1976.

[4] J. A. Bondy and R. L. Hemminger, "Graph reconstruction - A survey," *Journal of Graph Theory*, vol. 1, no. 3, pp. 227-268, 1977.

[5] F. Harary and M. Plantholt, "The graph reconstruction number," *Journal of Graph Theory*, vol. 9, pp. 451-454, 1985.

[6] F. Harary and M. Plantholt, "The class-reconstruction numbers of unicyclic graphs," 1987. Submitted.

[7] F. Harary and J. Lauri, "The class-reconstruction number of maximal planar graphs," *Graphs and Combinatorics*, vol. 3, no. 1, pp. 45-53, 1987.

[8] J. Lauri, "Graph reconstruction - Some techniques and new problems," *Ars Combinatoria*, 1987. Submitted.

[9] W. J. Myrvold, M. N. Ellingham, and D. G. Hoffman, "Bidegreed graphs are Edge Reconstructible," *Journal of Graph Theory*, vol. 11, pp. 281-302, 1987.

[10] W. J. Myrvold, *"Ally and Adversary Reconstruction Problems*, Ph.D. thesis, University of Waterloo, 1988.

[11] W. J. Myrvold, "The ally-reconstruction number of a tree with five or more vertices is three, " *Journal of Graph Theory*, 1989. To appear.

[12] W. J. Myrvold, "The ally-reconstruction number of a disconnected graph," *Ars Combinatoria*, 1989. To appear.

[13] V. Müller, "Probabilistic reconstruction from subgraphs," *Commentationes Mathematicae Universitatis Carolinae*, vol. 17, pp. 709-719, 1976.

Random Superposition: Multigraphs

E.M. Palmer

Michigan State University

Dedicated to J. Howard Redfield, 1879 – 1944

ABSTRACT

A new probability model for graphs based on the superposition principle of J.H. Redfield is introduced, thereby revealing a great array of problems involving certain types of multigraphs. We focus on the simplest possible special case in which each superposition is the union of a collection of isolated edges. Some fundamental results are derived for these random multigraphs. In particular, the distribution of multiple edges is sketched and a sharp threshold for connectivity is determined. Numerical evidence to support the theory is provided for small multigraphs with less than 7000 vertices.

1. Introduction

Let H be any graph of order n and let $s(H)$ denote the order of the automorphism group of H. Hence, the number of ways to label the vertices of H is $n! / s(H)$. Now select a random list of r labeled copies

$$H_1, H_2, ..., H_r$$

of H, with replacement. So we have an r–permutation, with repetition, of a k–set, where $k = n! / s(H)$. Now form the random superposition graph, denoted $G(H)$, from the list by identifying vertices with the same label. Note that $G(H)$ is a multigraph of order n in which the r isomorphic factors retain their identity. Thus we have a probability model in shich the sample space has

$$(1.1) \qquad\qquad (n! / S(H))^r$$

equally likely elements.

Suppose we are given a particular superposition graph $G(H)$ of order n and the identities of the r copies of H are revealed by attaching the label 1 to each edge of the

957

first copy of H, 2 to the second copy and so on. Then the contribution to (1.1) of all superpositions isomorphic to G(H) is

$$(1.2) \qquad\qquad\qquad\qquad (r! \,/\, a)\,(n! \,/\, b)$$

where a is the number of permutations of the edge labels that leave G invariant, and b is the number of permutations of the vertex labels that leave G invariant, i.e. preserve edge labels. Otherwise put, (1.2) is the number of ways to label a superposition G(H).

Redfield introduced the notion of superposition in his famous 1927 paper [R27], in which he developed a very broad method for counting unlabeled superpositions. For background material and the development of this topic there are the important articles [H60], [HR84], [Re59] and [Re60] by Harary, Read and Robinson mentioned in the references. See especially the 1984 issue vol.8(2) of the *Journal of Graph Theory* that is dedicated to the memory of J.H. Redfield and contains a beautiful biographical sketch by Keith Lloyd [L84].

Many questions could be raised about the nature of the random superposition graph G(H) that results from various choices of the basic factor H and the number of copies in the superposition. In this article we focus on the simple special case in which H is the graph of order n with exactly one edge, i.e. the disjoint union $H = K_2 \cup \overline{K}_{n-2}$. There are $\binom{n}{2}$ ways to label H and we form the sample space of superpositions with q copies of H. Hence there are $(\,n(n-1)\,/\,2\,)^q$ superpositions G(H) and each one is a multigraph with q edges. We call q the *size* of G(H). Recall that the copies of H retain their identity in the superposition so each of the edges may be regarded as having a different label from 1 to q.

We use C(n, k) to denote the binomial coefficient $\binom{n}{k}$ and illustrate the sample space with n = 4 and q = 3. There are $C(4, 2)^3 = 216$ superpositions in the sample space. There are 6 basic types described as follows:

G_1 has one edge of multiplicity 3 and two isolated vertices

G_2 has one edge of multiplicity 2 and one isolated vertex

G_3 has one edge of multiplicity 2 and no isolated vertices

G_4 is a path

G_5 is the bipartite tree $K_{1,3}$

G_6 consists of a triangle and an isolated vertex.

The next equation illustrates the application of formula (1.2) to the graphs in the order listed above, thereby counting the number of elements in the sample space in another way:

$$(1.3) \qquad (3!/3!)(4!/4) + (3!/2)(4!/4) + (3!/2)(4!/1) + (3!/2)(4!/1)$$
$$+ (3!/3!)(4!/1) + (3!/3!)(4!/1) = 216.$$

2. Distribution of Multiple Edges

It is convenient to denote the binomial coefficient $C(n, 2)$ by the single letter N. Then our sample space of superpositions of q copies of H has N^q elements. Before any investigation of the structure of a random superposition, some basic facts about multiple edges must be determined. Let $X_m(G)$ denote the number of edges of multiplicity m in the superposition graph $G = G(H)$. Then the expected value of this random variable X_m for $m \geq 0$ is

$$(2.1) \qquad E(X_m) = N\, C(q, m)\, (1/N)^m\, (1 - 1/N)^{q-m}.$$

The justification of this formula proceeds as follows. The first factor N is the number of ways to select two vertices, say u and v. The second factor $C(q, m)$ counts the number of ways to select labels (or different copies of H) for the m edges joining u and v. Then $(1/N)^m$ is the probability that the m selected edges actually join u and v, and finally $(1 - 1/N)^{q-m}$ is the probability that the other $q - m$ edges do not join u and v. We want to know about the likelihood of edges of fixed multiplicity m in a random superposition graph when n and q are large. For example, suppose $m = 3$ and n is very large. If q is too small, it is quite likely that the random superposition will not have any edges of multiplicity 3. But if we increase q suffciently, edges of multiplicity 3 will appear with high probability. But if q is increased by too great an amount, every pair of vertices may be adjacent and even with multiplicity greater then 3. The critical values of q can be determined from the expectation $E(X_m)$ as can be seen in the next theorem. All limits in this article are taken with respect to $n \to \infty$.

Theorem 2.1 *With fixed $m \geq 1$, let*
$$(2.2) \qquad q = n^{2-2/m}\, \omega_n \text{ where } \omega_n \to \infty$$
but
$$(2.3) \qquad \omega_n = o(\, n^{2/(m(m+1))}\,).$$
Then $E(X_m) \to \infty$ but $E(X_{m+1}) \to 0$.

Proof First observe that the hypothesis (2.2) implies that
$$(1 - 1/N)^q \sim 1.$$
Then we observe that since m is fixed,
$$(2.4) \qquad E(X_m) \sim (2^{m-1}/m!)\, (q/(n^{2-2/m}))^m,$$
and so $E(X_m) \to \infty$.
On the other hand,
$$(2.5) \qquad E(X_{m+1}) \sim (2^m/(m+1)!)\, (q/(n^{2-2/(m+1)}))^{m+1},$$

and now it follows from (2.3) that $E(X_{m+1}) \to 0$. $\square$

Let $\mathcal{A}_m$ denote all the superpositions of order n that have edges of multiplicity exactly m. As usual $P(\mathcal{A}_m)$ denotes the probability of this event. The second moment method (see [Pa85] section 3.1 or [Sp87] lecture 3) can be used to get the first part of the next corollary. The second part follows from the fact that $E(X_{m+1}) \to 0$.

Corollary 2.1 *Under the same hypothesis as the theorem above, $P(\mathcal{A}_m) \to 1$ but $P(\mathcal{A}_m) \to 0$, i.e. almost all superpositions have edges of multiplicity m but have none of multiplicity $m + 1$.*

If $q = n^{2 - 2/m} c$, with constant $c > 0$, then
$$(2.6) \qquad E(X_m) \sim (1/2)\,(2c)^m / m! = \mu$$
and with a bit of work it can be shown that the edges of multiplicity m are distributed according to Poisson's Law with mean μ. Here is an application. Let $\langle x \rangle$ denote the nearest integer to x and suppose
$$(2.7) \qquad q = \langle c\, n \log n \rangle$$
Then it is easy to see from Theroem 2.1 that
$$E(X_2) \to \infty \quad \text{but} \quad E(X_3) \to 0,$$
so almost all superpositions have edges of multiplicity 1 and 2 but none of multiplicity 3. In fact, with q defined by (2.7), it follows fron (2.1) that
$$E(X_1) \sim c\, n \log n \qquad \text{and} \qquad E(X_2) \sim (c \log n)^2$$
so the edges of multiplicity 1 still dominate. Recall that in the Erdös–Rényi probability model for graphs [ErR59], almost all are connected if q has the form (2.7) and $c > 1/2$. The fact that the edges of multiplicity 1 dominate in our model indicates that the threshold for connectivity may be the same in both models.

If the number of edges is a bit larger than N, then edges of every (fixed) multiplicity appear. The next theorem describes the range of values of q for which this phenomenon occurs.

Theorem 2.2 *For constant $c > 0$ and $\omega_n \to \infty$ with $\omega_n = o(N)$, if q is in the interval*
$$(2.8) \qquad cN < q < N \log (N/\omega_n),$$
then $E(X_m) \to \infty$ for every fixed $m \geq 0$.

Proof Since $q = o(N^2)$, we have
$$(1 - 1/N)^q \sim e^{-q/N}$$
and hence from (2.1)
$$(2.9) \qquad E(X_m) \sim (1/m!)\, q^m\, e^{-q/N} / N^{m-1}.$$

It follows from (2.8) that the right side of (2.9) is bounded below by $(c^m/m!)\omega_n$. ❏

An application of the second moment method yields the next corollary.

Corollary 2.2   *Under the same hypothesis as the theorem above*
$$P(\mathcal{A}_m) \to 1,$$
i.e. almost every superposition has edges of every (fixed) multiplicity.

As the edges accumulate in the random superpostion graph the edge multiplicities increase until every pair of vertices is joined by at least one edge. That is, the minimum multiplicity is at least 1 and we say that the graph is *complete* or *full*. Then the edges of multiplicity 1 diminish in number and eventually disappear, and so on. Next we determine when these events take place.

Theorem 2.3   *With fixed $m \geq 0$, let*
$$(2.10) \qquad q = N(\log N + m \log \log N + \log \omega_n)$$
where $\omega_n \to \infty$ but
$$(2.11) \qquad \omega_n = o(\log N).$$
Then $E(X_m) \to 0$ but $E(X_{m+1}) \to \infty$.

Proof   Since $q = o(N^2)$, we have (2.9). On substitution of q from (2.10) in (2.9), we find
$$(2.12) \qquad E(X_m) \sim (1/m!)/\omega_n \to 0$$
and similarly
$$(2.13) \qquad E(X_{m+1}) \sim (1/(m+1)!)(\log n)/\omega_n \to \infty. \quad ❏$$

As before, there is an accompanying result that can be established with a bit more effort.

Corollary 2.2   *Under the same hypothesis as the theorem above*
$$P(\mathcal{A}_m) \to 0 \quad but \quad P(\mathcal{A}_{m+1}) \to 1,$$
i.e. almost all superpostion graphs have no edges of multiplicity m but edges of multiplicity m + 1 persist.

Note in particular that if $m = 0$ in (2.10) then for
$$(2.14) \qquad q = N(\log N + \log \omega_n)$$
almost all superpositions are complete but there are still edges of multiplicity 1. This observation should be contrasted with the Erdös–Rényi Model where X_0 also stands for the number of pairs of vertices of multiplicity 0. In this model the sample space

consists of $C(N, q)$ equiprobable labeled graphs of order n and size q. Then obviously $E(X_0) = N - q$ and so

$$(2.15) \qquad E(X_0) \to 0 \quad \text{if } q = N + o(1)$$

but

$$(2.16) \qquad E(X_0) \to \infty \quad \text{if } q = N - \omega_n \text{ where } \omega_n \to \infty.$$

Thus the duplication of edges in the superposition model causes a substantial delay in achieving a complete graph. This delay is also apparent in the variation of the Erdös–Rényi Model in which the q edges are selected with repetition allowed. Then the sample space consists of $C(N + q - 1, q)$ equally likely labeled multigraphs of order n with q (unlabeled) edges. And we have

$$(2.17) \qquad E(X_0) = N\, C(N + q - 2, q) / C(N + q - 1, q),$$

from which the following facts emerge. If $\omega_n \to \infty$, then

$$(2.18) \qquad E(X_0) \to 0 \quad \text{if } q = N^2 \omega_n$$

but

$$(2.19) \qquad E(X_0) \to \infty \quad \text{if } q = N^2 / \omega_n.$$

The effect of the duplication of edges is really apparent here where far more edges are necessary to gaurantee a complete multigraph. For edges of any multiplicity $m \geq 0$ in this model we have the formula

$$(2.20) \qquad E(X_m) = N\, C(N + q - m - 2, q - m) / C(N + q - 1, q).$$

A little algebra shows that if $q = o(N)$, the behavior of $E(X_m)$ is exactly the same as in Theorem 2.1. But if $N = o(q)$, we find the following:

$$(2.21) \qquad E(X_m) \to 0 \quad \text{if } q = N^2 \omega_n, \ \omega_n \to \infty$$

and

$$(2.22) \qquad E(X_m) \to \infty \quad \text{if } q = N^2 / \omega_n, \ \omega_n = o(N).$$

Note that this result is independent of m. Evidently for q roughly between N and N^2, edges of all finite multiplicities appear but when q passes N^2, they vanish. Hence in this model they last longer and their termination is sudden compared to the random superposition model in which their endurance is brief and extinction comes in stages.

We conclude this section with some remarks about the comparison of the number q of edges in each superposition G and the number of edges in the underlying graph, namely $N - X_0(G)$. From formula (2.1) it follows that $E(X_0)^2 \sim E(X_0^2)$ whenever $E(X_0) \to \infty$. So almost all superpositions will have X_0 fairly close to the mean $E(X_0)$. We also find that if $q = N/\omega_n$, where $\omega_n \to \infty$ but $\omega_n = o(N)$, then

$$(2.22) \qquad N - E(X_0) \sim q,$$

i.e., the expected number of edges in the underlying graph of the random superposition and the total number of edges in the superposition are asymptotically equivalent. This result covers the range of values of q in Theorem 2.1. Hence some properties of random superpositions may be similar to those of random graphs in the Erdös–Rényi model. Of course, the arithmetic difference between q and $N - E(X_0)$ is $o(1)$ only if almost all superpositions have no edges of multiplicity 2. But this requires $q/n = o(1)$.

When q lies in the gap specified by (2.8), the expected number of edges in the underlying graph is substantially less then q. For example if $q = N$,

$$(2.23) \qquad N - E(X_0) \sim (1 - 1/e)q.$$

And if $q = N \log n$,

$$(2.24) \qquad N - E(X_0) \sim (1 / \log n)q.$$

There are many other comments that could be made about the distribution of multiple edges in these models. But the theorems above are sufficient for our investigation which concentrates on the range of q for which there are still edges of multiplicity zero. Of course the results of this section are of a fundamental combinatorial nature. They are just straightforward solutions of standard occupancy problems for several probability models. They have been recast in the terminology of random graphs for the convenience of the reader. The corresponding results for superposition of graphs with more than one edge will be substantially more difficult.

For another probability model for random multigraphs see [Go79] which is briefly mentioned in [Bo85].

3. Connectivity

How many edges should a random superposition have so that it is almost surely connected? Erdös and Rényi [ErR59] found that about $n \log n$ edges are needed when multiple edges are forbidden. In the superposition model when q has the form of (2.7), we find there are sure to be edges of multiplicity 2. But the next theorem shows that despite these redundancies, the threshold for connectivity is exactly the same as that found by Erdös and Rényi. In fact, the theorem also holds for the Erdös–Rényi model with repetition of edges allowed. The proof generally follows the pattern set by the founding fathers [ErR59] but there are interesting differences here and there. For example, we need the following lemmas because the edge probability for one pair of vertices is not independent of the edge probability of another pair. Let $\{(u_i, w_i) \mid i = 1$ to $k\}$ be a set of k different pairs of vertices and let $\mathcal{A}_i$ be the event that u_i and w_i are joined by an edge of multiplicity at least 1.

Lemma 1 *If $q < N$, then the probability that k specified pairs of vertices are adjacent has the bound:*

$$(3.1) \qquad P\left(\bigcap_{i=1}^{k} \mathcal{A}_i \right) < (q/N)^k.$$

Proof From the definition of conditional probability

$$(3.2) \qquad P(\bigcap_{i=1}^{k} \mathcal{A}_i) = P(\bigcap_{i=1}^{k-1} \mathcal{A}_i \mid \mathcal{A}_k) \, P(\mathcal{A}_k),$$

and from the definition of the superposition model,

$$(3.3) \qquad P(\bigcap_{i=1}^{k-1} \mathcal{A}_i \mid \mathcal{A}_k) < P(\bigcap_{i=1}^{k-1} \mathcal{A}_i).$$

Hence we have

$$(3.4) \qquad P(\bigcap_{i=1}^{k} \mathcal{A}_i) < \prod_{i=1}^{k} P(\mathcal{A}_i).$$

The rules of complementary probabilities tell us that

$$(3.5) \qquad P(\mathcal{A}_i) = 1 - (1 - 1/N)^q$$

$$= \sum_{j=1}^{q} (-1)^{j+1} \, C(q,j) \, (1/N)^j$$

Now an application of the ratio test shows that the terms of the sum are decreasing, i.e.

$$(3.6) \qquad C(q, j) \, (1/N)^j > C(q, j + 1) \, (1/N)^{j+1}$$

provided that $q < N$. Therfore for each $i = 1$ to k

$$(3.7) \qquad P(\mathcal{A}_i) < q/N,$$

and combining this inequality with (3.4) finishes the proof. $\square$

We also need to estimate thr probability of the event that some pairs of vertices are adjacent and others are not. Suppose

$$(3.8) \qquad \mathcal{A} = \bigcap_{i=1}^{r} \mathcal{A}_i \quad \text{and} \quad \mathcal{B} = \bigcap_{i=r+1}^{r+s} \bar{\mathcal{A}}_i.$$

Note that Lemma 1 can be applied when the number N of pairs of vertices that are allowed to have edges is reduced to $N - s$. Then we have the conditional probability of the next lemma.

Lemma 2 *If $q < N - s$, the probability that r pairs of vertices are adjacent given that s other pairs are not adjacent is bounded by*

$$(3.9) \qquad P(\mathcal{A} \mid \mathcal{B}) < (q/(N - s))^r.$$

But we have the simple exact formula for $P(\mathcal{B})$ of the next lemma.

Lemma 3 *The probability that s specified pairs of vertices are not adjacent is*

$$(3.10) \qquad P(\mathcal{B}) = (1 - s/N)^q.$$

Hence Lemmas 2 and 3 can be combined to obtain an upper bound for the probability of the intersection of $\mathcal{A}$ and $\mathcal{B}$.

Lemma 4 *If $q < N - s$, the probability that r specified pairs of vertices are adjacent and s others are not is bounded by*

(3.11) $P(\mathcal{A} \cap \mathcal{B}) < (q/(N - s))^r (1 - s/N)^q.$

It is interesting to observe that an exact formula for the probability $P(\mathcal{A} \cap \mathcal{B})$ can be worked out from formula (3.10) and the following recurrenece relation in which we have used $P(r, s)$ to denote $P(\mathcal{A} \cap \mathcal{B})$:

(3.12) $P(r, s) = P(r - 1, s) - P(r - 1, s + 1).$

On solving the recurrence relation we find

(3.13) $P(\mathcal{A} \cap \mathcal{B}) = \sum_{k=0}^{r} C(r, k) \, (-1)^k \, (1 - (s + k) / N)^q.$

On dividing the right side of (3.13) by the right side of (3.10), after some simplification we arrive at an exact formula for the conditional probability:

(3.14) $P(\mathcal{A} \mid \mathcal{B}) = \sum_{i=0}^{q} C(q, i) \, (-1 / (N - s))^i \, \sum_{k=0}^{r} C(r, k) \, (-1)^k \, k^i.$

Next we find the neat identity (see formula 1.13 in Gould's Tables [Go72]) for the second sum in (3.14):

(3.15) $\displaystyle \sum_{k=0}^{r} C(r, k) \, (-1)^k \, k^i = \begin{cases} 0 & \text{if } 0 \le i < r \\ (-1)^r \, r! & \text{if } i = r \end{cases}.$

So the first sum in (3.14) can begin at $i = r$ and the contribution to $P(\mathcal{A} \mid \mathcal{B})$ of this term is $C(q, r) / (N - s)^r$ from which (3.9) can be derived after more work on the remaining terms of the sum. Hence it seems that the exact formulas are harder to deal with than the more conbinatorial approach that was taken in the lemmas. At any rate, we are now ready for the main result of this section.

Theorem 3.1 *Let x be any (fixed) real number and set*

(3.16) $q = \langle (n/2) (\log n + x) \rangle$

and denote by C_n the set of connected superposition graphs of order n and size q. Then

(3.17) $P(C_n) \to e^{-e^{-x}}.$

Proof For $i = 1, 2, ..., \lfloor n/2 \rfloor$, let $Y_i(G)$ be the number of components of order i in the superposition graph G. Now define

$\qquad\qquad X = Y_i$

and

$\qquad\qquad Y = Y_2 + ... + Y_{\lfloor n/2 \rfloor}.$

Then we have the upper and lower bounds for $P(C_n)$:

(3.18) $P(X = 0) - P(Y \ge 1) \le P(C_n) \le P(X = 0).$

Since $P(Y \geq 1) \leq E(Y)$, we have

$$(3.19) \qquad P(X = 0) - E(Y) \leq P(C_n) \leq P(X = 0).$$

Now we estimate the upper bound $P(X = 0)$ by using the binomial moments. First observe that

$$(3.20) \qquad E(X) = n(C(n-1, 2) / N)^q$$
$$= n(1 - 2/n)^q$$

and a few simple steps show that

$$(3.21) \qquad \log E(X) = -x + o(1).$$

Hence the first binomial moment $S_1 = E(X)$ is asymptotic to e^{-x} and just as observed by Erdös and Rényi [ErR59] in the case of random graphs, it can be shown that the binomial moments for $k = 0, 1, 2, \ldots$ are given by

$$(3.22) \qquad S_k \sim (e^{-x})^k / k!.$$

And so the isolated vertices are distributed according to Poisson's law with mean $\mu = e^{-x}$ and we have

$$(3.23) \qquad P(X = 0) \to e^{-e^{-x}}.$$

It remains to show that $E(Y) \to 0$, i.e. almost all random graphs have one large component and perhaps some isolated vertices. We begin with the next equality which holds for $k \geq 1$ and follows from the definition of expectation:

$$(3.24) \qquad E(Y_k) = C(n, k) P(\mathcal{D} \cap \mathcal{B}),$$

where the binomial coefficient $C(n, k)$ counts the number of ways to select k labels for the vertices of the component of order k, $\mathcal{D}$ is the event that the superposition graph G has a connected subgraph spanning the specified vertices with labels 1 to k and $\mathcal{B}$ is the event that it has no edges joining these k vertices with the other $n - k$ vertices. Recall that Cayley's theorem says that there are k^{k-2} labeled trees of order k. Since every connected graph of order k contains a spanning tree of order k, we have

$$(3.25) \qquad P(\mathcal{D} \cap \mathcal{B}) \leq k^{k-2} P(\mathcal{A} \cap \mathcal{B}),$$

where $\mathcal{A}$ is the event that $k - 1$ specified edges are present in the subgraph on the vertices with labels 1 to k. Now we apply lemma 4 to (3.25), combine (3.24) and arrive at the next inequality, which holds for $k \geq 1$:

$$(3.26) \qquad E(Y_k) \leq C(n, k) k^{k-2} (q / (N - k(n-k)))^{k-1} (1 - k(n-k) / N)^q.$$

To estimate the right side of (3.26) we begin by applying Stirling's formula for $k!$ and find that

$$(3.27) \qquad C(n, k) k^{k-2} \leq o(1) n^k e^k.$$

Since $k(n-k)$ is largest when $k = n/2$, we have for $n \geq 3$:

$$(3.28) \qquad 1 - k(n-k) / N \geq 1/4,$$

and this can be used to show that

$$(3.29) \qquad (q / (N - k(n-k)))^{k-1} \leq (4q/N)^{k-1}.$$

Since $1 \le k \le n/2$,

$$(3.30) \qquad k/(n-1) \le k(n-k)/N,$$

and so

$$(3.31) \qquad (1 - k(n-k)/N)^q \le (1 - k/(n-1))^q < e^{-qk/n}.$$

Now we can use the formula (3.16) for q together with (3.27), (3.29) and (3.31) to obtain the bound

$$(3.32) \qquad E(Y_k) = \mathcal{O}(n/\log n)\,(\,\mathcal{O}(1)\log n/\sqrt{n}\,)^k.$$

Hence if $k \ge 3$, then $E(Y_k) \to 0$ but if $k = 2$, formula (3.32) tells us that $E(Y_2) = \mathcal{O}(\log n)$. So when $k = 2$, we can improve the bound from (3.31) by using

$$(3.33) \qquad (1 - 2(n-2)/N)^q \le \mathcal{O}(n^{-2}).$$

Then $E(Y_2) \to 0$ and for $k \ge 3$

$$(3.34) \qquad \sum_{k=3}^{\infty} E(Y_k) = \mathcal{O}(1)\,(\log n)^2/\sqrt{n} \to 0.$$

Hence (3.17) is established. $\square$

The theorem also holds for the Erdös–Rényi model with replacement. In this case, an explicit formula for $P(\mathcal{A} \cap \mathcal{B})$ is immediate:

$$(3.35) \qquad P(\mathcal{A} \cap \mathcal{B}) = C(N - s + q - r - 1, q - r)/C(N + q - 1, q),$$

and from this follows the approproate inequality corresponding to formula (3.11) of Lemma 4:

$$(3.36) \qquad P(\mathcal{A} \cap \mathcal{B}) < (q/(N - s - 1))^r\,(1 + q/(N-1))^{-s}.$$

Now (3.36) can be used used to show that $E(Y) \to 0$ just as in the superposition model.

4. Numerical Results

For small values of n and q, we can calculate the exact number $F_{n,q}$ of connected superpositions of order n and size q from the following recurrence relation. For $n \ge 2$ and $q \ge n - 1$:

$$(4.1) \qquad N^q = F_{n,q} + (n-1)F_{n-1,q} + C(n-1,2)^q$$
$$+ \sum_{k=2}^{n-2} C(n-1, k-1) \sum_{i=k-1}^{q} C(q, i)\, F_{k,i}\, C(n-k, 2)^{q-i}.$$

This relation is derived by expressing the total number N^q of superpositions as the sum for $k = 1$ to n of the number for which the component that contains the vertex with label number 1 has order k. The first three terms on the right side of (4.1) arise when $k = n, n - 1$ and 1 respectively. In the double sum $C(n-1, k-1)$ is the number of ways to select the other $k - 1$ labels for the component containing vertex #1. Next $C(q, i)$ is the number of ways to choose i edges for this component. Then $F_{k,i}$ is the number of components with these specified vertex and edge labels. Finally $C(n - k, 2)^{q-i}$ is the number of ways to fill in the remaining $n - k$ vertices with the other $q - i$ edges.

Now we introduce notation for the probability of connectivity for $n \geq 1$ and $q \geq n - 1$:

$$(4.2) \qquad P_{n,q} = F_{n,q} / N^q.$$

Then (4.1) can be transformed into a recurrence relation for probabilities by dividing both sides by N^q and then using (4.2) to eliminate each occurence of F. The result for $n \geq 2$ and $q \geq n - 1$ is a recurrence relation for the probability $P_{n,q}$ that a random superposition of order n and size q is connected:

$$(4.3) \qquad P_{n,q} = 1 - (n - 1)((n - 2)/n)^q P_{n-1,q} - ((n - 2)/n)^q$$
$$+ \sum_{k=2}^{n-2} C(n - 1, k - 1) \sum_{i=k-1}^{q} C(q, i) P_{k,i} C(k, 2)^i C(n - k, 2)^{q-i} / N^q.$$

A double precision Fortran program used this relation on a Zenith–158 PC to calculate the numbers in the Table. If $x = 0$ in formula (3.16) of Theorem 3.1, then $q = \langle (n \log n) / 2 \rangle$ as in the table and according to the theorem, $P_{n,q}$ should approach $1/e = .367879...$. For $n > 40$ the computations required by the recurrence relation were beyond the range of the program. But $P_{n,q}$ can then be calculated using the bounds in equation (3.19). However, as in the case of the Erdös–Rényi model (see [BoT85] for extensive data and relevant equations for computing the probability of connectivity for ordinary random graphs), to achieve better accuracy it was found necessary to redefine X and Y as follows:

$$X = Y_1 + Y_2$$

and

$$Y = Y_3 + ... + Y_{\lfloor n/2 \rfloor}.$$

The formula for the binomial moments is rather more complicated now. We need S_r, the expected number of r-sets of components of order one or two. And now we must take into consideration the multiplicity of the edges in the components of order two. If $n = 100$ and $q = 230$, then the expected number of components of order two with edges of multiplicity 3 is less than .0000085 and this is also an upper bound on the probability of such a component. Hence in the next formula we only include edges of multiplicity 2. For $r \geq 1$, we find

$$(4.4) \qquad S_r = (1/r!) \sum_{s=0}^{r} (n)_{r+s} C(r,s) (1/2)^s (1/N)^s (C(n - r - s, 2) / N)^{q-s}$$
$$\sum_{i=0}^{s} C(s, i) (q)_{s+1} (1/2)^i (1/N)^i.$$

In this formula, r is the number of components of order one or two, s is the number of order two. So $r - s$ is the number of isolated vertices. Then in the second sum, i is the number of components of order two with edges of multiplicity 2, while $s - i$ is the number of multiplicity 1. In our interactive program, that could only handle n up to about 7000, only about ten binomial moments (i.e. $r \leq 10$) were needed to obtain the first 5 digits of $P(X = 0)$. Estimates of $E(Y_k)$ were made from (3.26) but these declined so rapidly that the contributions for $k \leq 10$ were always sufficient to obtain good results. It was found that for $n = 500, 1000, 2000, 3000, 5000$, and 7000, the upper and lower bounds for $P_{n,q}$ when rounded to three digits, all resulted in the same number, namely 0.368.

n	q	Pn,q
5	4	.30000
10	12	.37487
15	20	.32261
20	30	.35469
25	40	.34851
30	51	.35874
35	62	.35553
40	74	.36649

Table 1 Probability of connectivity with $q = \langle (n \log n) / 2 \rangle$.

REFERENCES

[Bo85] B. Bollobás, *Random Graphs*, Academic, New York (1985).

[BoT85] B. Bollobás and A. Thomason, Random graphs of small order, *Annals of Discrete Math.* **28** (1985) 47 – 97.

[ErR59] P. Erdös and A. Rényi, On random graphs I, *Publ. Math. Debrecen* **6** (1959) 290 – 297.

[Go79] E. Godehart, An extension of the theorems of Erdös–Rényi–type to random multigraphs, *Proc. Sixth Conf. Probability Theory,* (B. Bereanu *et al.*, eds.) Ed. Acad. R.S. România, Burcharest (1981) 417 – 425.

[Go72] H.W. Gould, *Combinatorial Identies,* H.W. Gould, Morgantown (1972).

[H60] F. Harary, Unsolved problems in the enumeration of graphs, *Magyar Tud. Akad. Mat. Kutató Int. Közl* **5** (1960) 63 – 95.

[HR84] F. Harary and R.W. Robinson, The rediscovery of Redfield's papers, *J. Graph Theory* **8** (1984) 191 – 193.

[L84] E.K. Lloyd, J. Howard Redfield 1879 – 1944, *J. Graph Theory* **8** (1984) 195 – 203.

[Pa85]   E.M. Palmer, *Graphical Evolution: an introduction to the theory of random graphs,* Wiley–Interscience Series in Discrete Mathematics, New York (1985).

[Re59]   R.C. Read, The enumeration of locally restricted graphs I, *J. London Math. Soc.* **34** (1959) 417 – 436.

[Re60]   R.C. Read, The enumeration of locally restricted graphs II, *J. London Math. Soc.* **35** (1960) 334 – 351.

[R27]   J.H. Redfield, The theory of group–reduced distributions, *Amer. J. Math.* **49** (1927) 433 – 455.

[Sp87]   J. Spencer, *Ten lectures on the probabilistic method,* SIAM, Philadelphia (1987).

Graphical Designs[1]

T.D. Parsons[2]
the late of the California State University at Chico, CA

Tomaž Pisanski
University of Ljubljana, Yugoslavia

ABSTRACT

Let G be a graph on n vertices $1, 2, ..., n$. Let
$$G = G_1 \oplus G_2 \oplus ... \oplus G_p$$
be an arbitrary factorization [= edge decomposition] of G. Then there exist an integer $d > 0$, and n vectors $x_1, x_2, ..., x_n$ from $\{-1, +1\}^d$, and a set of distinct integers $\{a_0, a_1, ..., a_p\}$ such that $a_0 = 0$, $a_k < 0$, for $k \neq 0$ and that vertex i is adjacent to vertex j in G_k if and only if $\langle x_i, x_j \rangle = a_k$, and $\langle x_i, x_j \rangle = a_0$ if i is not adjacent to j in G, where $\langle x_i, x_j \rangle$ is the inner product of x_i and x_j.

In previous work [PP1, PP2, PP3] we have investigated vector representation of graphs. Vertices are represented as vectors in some vector space with a given bilinear form. Two vertices are adjacent if and only if the form evaluated for the corresponding two vectors attains a value from a specified set of values. The purpose of this work is to extend this notion to an edge–disjoint decomposition of a graph.

Motivation for our work can be found, for instance, in [PP3]. For the sake of self–sufficiency we rephrase here the definition from [PP3] of a good Hadamard representation. For an arbitrary graph G on the vertex set $\{1, 2, ..., n\}$ let $M(G)$ be an n by d matrix with entries $+1$ or -1 such that $M(G)M(G)^t = dI_n + t A(G)$, where t is a negative integer and $A(G)$ is the adjacency matrix of G. If such a matrix exists we call it a *representation matrix* of a *good Hadamard representation* of the graph G. Moreover, t is called the *symbol* and d is the *dimension* of the representation. As usal $M(G)^t$ denotes the transpose of $M(G)$. If the rows of $M(G)$ are considered as vectors $x_1, x_2, ..., x_n$, from a real d–space with cordinates $+1, -1$, then any two

1. Work supported in part by a grant from the California State University at Chico and by the NSF (Grant DMS – 8717441)
2. Tory suddenly died on April 2, 1987, a few days after we obtained this result.

971

vectors x_i and x_j are orthogonal if and only if the corresponding vertices i and j are not adjacent in G. Let $d_1{}^*(G)$ denote the minimum dimension of any Hadamard representation of graph G.

By Theorem 12 of [PP3] we have $d_1{}^*(G) \leq 2n^2 - 2n$; that is $2n^2 - 2n$ is an upper bound for the minimum dimension of a good Hadamard representation. If M is the matrix of a representation of G and t is the symbol of the representation we may easily produce infinitely many representations with infinitely many different symbols. Let M_k denote the matrix $M_k = [M\,|\,M\,|\,...\,|\,M]$ which is obtained by putting k copies of M next to each other. Clearly M_k is matrix of a good (kd)–dimensional Hadamard representation of G with the symbol kt.

Lemma   *Given an arbitrary graph H and an arbitrary finite set X of nonpositive integers there is a negative integer c such that c does not belong to X and that H has a representation with the symbol c.*

Proof   By Theorem 12 of [PP3] H admits a good Hadamard representation M with symbol t, for some negative integer t. By the above argument $M_k = [M\,|\,M\,|\,...\,|\,M]$ (k-times) is a Hadamard representation with the symbol kt. For $k = 1, 2, 3, ...$, we get infinitely many distinct values for c where $c = kt$. Therefore there exists a positive integer k such that $c \notin X$ and that H admits a good Hadamard representation with the symbol c. $\square$

Theorem   *Let G be a graph on n vertices $1, 2, ..., n$. Let*
$$G = G_1 \oplus G_2 \oplus ... \oplus G_p$$
be an arbitrary factorization [= edge decomposition] of G. Then there exists an integer $d > 0$, and n vectors $x_1, x_2, ..., x_n$ from $\{-1, +1\}^d$, and a set of distinct integers $\{a_0, a_1, ..., a_p\}$ such that $a_0 = 0, a_k < 0,$ for $k \neq 0$ and that vertex i is adjacent to vertex j in G_k if and only if $\langle x_i, x_j \rangle = a_k$, and $\langle x_i, x_j \rangle = a_0$ if i is not adjacent to j in G, where $\langle x_i, x_j \rangle$ is the inner product of x_i and x_j.

Proof   By induction on p. For $p = 1$ we simply use Theorem 12 of [PP3]. Let $G = G' \oplus G_{p+1}$ and $G' = G_1 \oplus G_2 \oplus ... \oplus G_p$. By the induction hypothesis the Theorem holds true for any factorization of G' with p factors. Let us denote by d', a_0', a_1', ..., a_p', x_1', x_2', ..., x_n', the parameters corresponding to the above factorization of G' with representation M'. Let $X = \{a_0', a_1', ..., a_p'\}$ and let $H = G_{p+1}$. If we apply our Lemma to these values we get a representation, say with vectors y_i, a representation matrix M_0 and with the symbol c for G_{p+1}. By combining $M = [M'\,|\,M_0]$ we get the required representation with the parameters $d = d' + d_0$, $x_i = [x_i'\,|\,y_i]$ and $a_k = a_k'$ for $k = 0, 1, ..., p$ and $a_{p+1} = a_0' + c$. $\square$

There are at least two other interpretations of our Theorem.

Corollary A   *Let us select an arbitrary (not necessarily proper) edge–coloring of the complete graph K_n. Then there exists a d–cube graph Q_d, and a set of vertices $v_1, v_2, ..., v_n$ of Q_d, such that two edges ab and cd of K_n have the same color if and only if the distance from v_a to v_b in Q_d is the same as the distance from v_c to v_d.*

Proof   By replacing each occurrence of -1 by 0 in x_i in our Theorem we transform x_i into v_i. Thus v_i is a d–dimensional 0–1 vector which naturally represents a vertex of Q_d. If $\langle x_i, x_j \rangle = s$, then the distance between v_i and v_j is $(d - s) / 2$. $\square$

Corollary B   *Let us select an arbitrary (not necessarily proper) edge coloring of the complete graph K_n. Then there exists a family of sets $\{V_1, V_2, ..., V_n\}$ such that two edges ab and cd of K_n have the same color if and only if the symmetric difference of V_a and V_b has the same number of elements as the symmetric difference of V_c and V_d.*

Proof   If 0–1 vectors are considered as characteristic vectors of subsets of some d–set then the Corollary A translates directly into Corollary B. $\square$

During the conference in Kalamazoo Tom Tucker pointed out that Corollary B remains true if we replace the symmetric difference by intersection. A simple modification of the well–known proof that each graph is an intersection graph applies.

Proposition   *Let us select an arbitrary (not necessarily proper) edge–coloring of the complete graph K_n. Then there exists a family of sets $\{U_1, U_2, ..., U_n\}$ such that any two edges ab and cd of K_n have the same color if and only if the intersection of U_a and U_b has the same number of elements as the intersection of U_c and U_d.*

Proof   Assume that the edges of K_n are being colored by positive integers. For each vertex i we construct the set U_i as follows. If an edge $ij = ji = e$ is colored C then let U_i contain the C pairs $(e, 1), (e, 2), ..., (e, C)$. If ab and cd are two edges colored with the same color, say C, then the intersection of U_a with U_b has cardinality C and so does the intersection of U_c and U_d. $\square$

REFERENCES

[PP1] T.D. Parsons and T. Pisanski, Inner product representation of graphs,
       Proceedings of the Sixth Yugoslav Seminar on Graph Theory, Dubrovnik 1985,
       151 – 157.

[PP2] T.D. Parsons and T. Pisanski, Exotic n–universal graphs, J. Graph Theory, 12
       (1988) 155–158.

[PP3] T.D. Parsons and T. Pisanski, Vector representations of graphs, Discrete Math.,
       to appear.

Properties of Non–Minimum Crossings
for Some Classes of Graphs

B. L. Piazza

The University of Southeren Mississippi

R. D. Ringeisen*

Clemson University

S. K. Stueckle

The University of Idaho

ABSTRACT

In this paper we investigate drawings of some graphs which are not minimum crossing drawings. Interest in the study of non-minimum drawings has arisen in many papers and has been of growing interest in recent years. Our goal here will be to present an upper bound for the number of crossings in a good drawing of a graph, modify this bound somewhat and display graphs which meet this new bound. In some cases we examine the possibility of interpolation results for numbers of crossings which lie between the usual and the maximum crossing numbers. After looking at graphs which do meet this new upper bound, we briefly look at the unsolved problem of determining the maximum crossing number of the n–cube.

1. Introduction

In this paper we investigate drawings of some graphs which are not minimum crossing drawings. Interest in the study of non-minimum drawings has arisen in many papers and has been of growing interest in recent years. Among those papers which have examined this phenomenon have been those by Harborth [4,5], Kleitman [6], Eggleton [2], Woodall [10], Ringel [9], and several authors in [3]. Our goal here will be to present an upper bound for the number of crossings in a good drawing of a graph,

---

* Partial support from the U. S. Office of Naval Research.

modify this bound somewhat and display graphs which meet this new bound. In some cases we examine the possibility of interpolation results for numbers of crossings which lie between the usual and the maximum crossing numbers. After looking at graphs which do meet this new upper bound, we briefly look at the unsolved problem of determining the maximum crossing number of the n–cube.

Throughout the paper we will only consider the so called "good" drawing of graphs in the plane. Informally, a drawing is a *good drawing* if no edge crosses itself, no two edges cross more than once, no two edges incident with the same vertex cross, and no more than two edges cross at a point. The *crossing number* of a graph, which is well studied but known for very few classes of graphs (see Kleitman [7], Beineke and Ringeisen [1], and Ringeisen and Beineke [8]), is the minimum number of crossings among all such drawings. The maximum crossing number is the maximum number of crossings among all good drawings in the plane. We will not study the crossing number here and will denote the maximum crossing number as $cr^M(G)$.

The first graph for which the maximum crossing number was studied was the complete graph. Ringel [9] showed that $cr^M(K_n)$ is n choose four with the key to such a determination being that the maximum crossing number of the complete four graph is one. An immediate corollary to this result is the following, which we single out as a remark because of its importance later in this paper.

Remark 1   *The maximum crossing number of the four cycle is one.*

Of course the result on complete graphs gives an upper bound for the maximum crossing number of any graph. In this paper we look at some other upper bounds and distinguish some graphs which meet them.

2. Upper Bounds and Some Previous Results

Given any edge in a good drawing of a graph, it is clear that it many not cross itself or any edge incident with it. The following lemma states this fact formally.

Lemma 1   *For an edge $e = uv$ in a graph G the number of edges which it may cross is bounded above by $E - (degree\ u + degree\ v - 1)$, where E is the number of edges of G.*

If we then tally up this bound over all such edges of the graph we obtain an upper bound for the maximum crossing number, which we give as the following theorem because of the significance of the result, although the proof is immediate.

Theorem 1 *For any graph G with E edges, we have*

$cr^M(G) \leq (\Sigma(E - \deg u - \deg v + 1))/2,$ *where the sum is taken over all edges $e = uv$*
  $= (E^2 + E - \Sigma \deg^2 u)/2,$ *where the sum is all over vertices u.*

For convenience, we will denote the number on the right side of the inequality in Theorem 1 as $\mathcal{D}(G)$. We note that for the complete graphs this number is larger than the actual maximum crossing number. However, there are graphs which achieve the bound. The following theorem is due to Harborth [5] in a upcoming paper.

Theorem 2 *For a cycle C_n, $n \neq 4$, we have that $cr^M(C_n) = n(n - 3)/2 = \mathcal{D}(Cn)$*

In order that the reader may get a feeling for this work we display in figures 1 and 2 the drawings which realize the upper bound. Throughout the remainder of the paper, if a drawing has more than two edges crossing at a point then it is presumed that the edges may be slightly perturbed to give a proper drawing.

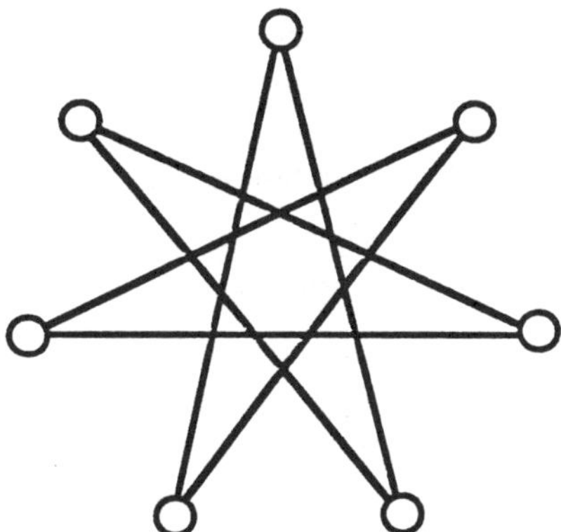

Figure 1. Drawing the odd cycles

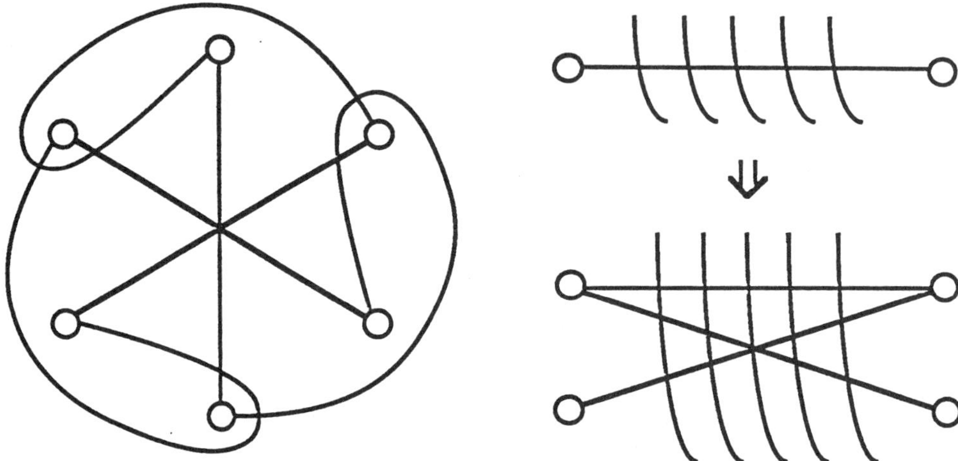

Figure 2. Drawing the even cycles

Note that the odd cycles meet the upper bound rather "naturally," while a bit more sophistication is needed in the even case. What is illustrated in figure 2 is a drawing for C_6 and then a method to go from one even number to the next by deleting an edge in the given drawing and replacing it with three edges as shown. The curved lines show that each edge would still cross all edges it may "legally" cross according to Theorem 1.

The drawings for cycles also illustrate another property of drawings which a given class of graphs may or may not possess. We will use the words *interpolation property* to mean that the graph can attain any number of crossing between its maximum crossing number and its crossing number. Notice that by modifying the drawings for cycles, one can drop the numbers of crossings by exactly one, thereby showing that cycles have the interpolation property. This property for the complete graphs is studied by Eggleton, Guy, Harborth, and Ringeisen in [3]. Kleitman [5] had earlier show that the parity of the number of crossings for any two drawings of odd complete graphs must be the same, thereby showing that these graphs do not have the interpolation property. In the four author paper [3], it is shown that there are "gaps" considerably larger than one when trying to obtain all possible number of crossings for complete graphs. These graphs then are graphs which are greatly lacking relative to the interpolation property.

Main Results

It seems natural to examine drawings of other simple planar graphs. We look next at trees and then at unicyclic graphs. Throughout this section we will use the phrase "all eligible edges" to mean those edges which a given edge may cross relative to Theorem 1 in a good drawing.

The fact that trees and connected unicyclic graphs meet the upper bound is proved using the language of "thrackles" in [10]. However, we give different proofs here which allow us to also conclude that the graphs have interpolation property.

Theorem 3 *Let T be a tree with p vertices. Then T has a drawing on the plane with r crossings, for all r, $0 \le r \le \mathcal{D}(T) = (p(p-1) - \Sigma \, deg^2 \, u)/2$, where the sum is taken over all vertices u.*

Proof We first prove that any tree can be drawn with $\mathcal{D}(T)$ crossings and then examine those drawings in order to show the interpolation property. We proceed by induction on the number of vertices p. When $p \le 2$, it is clear that T can only be drawn with zero crossings. Assume that any tree on $p \ge 2$ vertices can be drawn with the appropriate number of crossings. Let T be a tree with $p + 1$ vertices with an endvertex v adjacent to a vertex u and let $T' = T - v$. By the induction hypothesis, we find a drawing D of T' with $\mathcal{D}(T')$ crossings.

Let uw, $w \ne v$, be an edge of T and thus an edge of T'. Then uw crosses all eligible edges of T' in D. We create a drawing of T by adding vertex v of T' in such a way that the edge uv is drawn "parallel" to uw, crossing all edges which uw crosses and then looping around vertex w so that it crosses all edges incident with w except uw. (See figure 3 for an illustration of the path with 6 vertices.) Note that none of the edges which uv crosses are incident with u and that uv crosses no edge more than once. In the newly created drawing uv rosses all of its eligible edges and such remains true of all of the other edges of T. Hence we have created a drawing of T in which every edge crosses all of its eligible edges and hence a drawing which has $\mathcal{D}(T)$ crossings.

To show that interpolation holds for any tree we only need note that a similar induction argument can be made. In this case we move from the drawings for T' to the drawings for T in a similar manner. However, when placing the "parallel" edge uv into a drawing we can create the various drawings necessary for T by placing the vertex v so as to cross one of its eligible edges at a time. The proof is then complete.

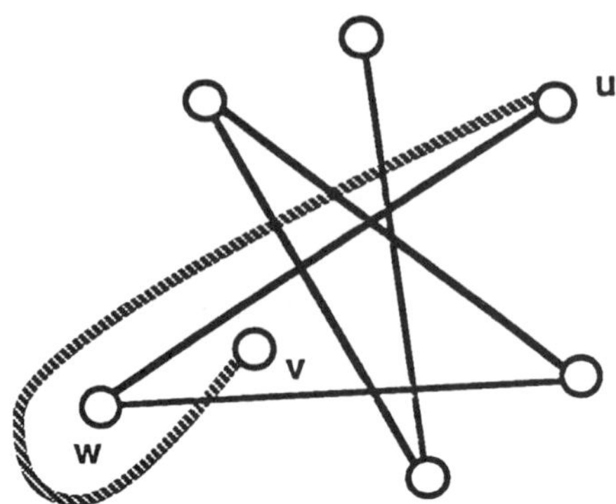

Figure 3. A tree (path) with max crossings.

The above proof can be modified so as to give a nice corollary.

Corollary 1   *If G is a connected unicyclic graph, where the cycle does not have four edges, then $cr^M(G) = \mathcal{D}(G)$, and G has the interpolation property.*

Proof   If G is a cycle, then the aforementioned result of Harborth applies. Otherwise, G must have a vertex of degree one and again such may be deleted and the induction hypothesis invoked on the remaining graph. The deleted edge may then be replaced by using the same argument as in the previous proof. The interpolation property follows in exactly the same manner and the proof is complete. ❑

Note that if the cycle has 4 edges then a similar induction may be used giving the $cr^M(G) = \mathcal{D}(G) - 1$.

When one begins to further examine maximum crossing number, it looks as if no other graphs G are likely to reach $\mathcal{D}(G)$. In fact the following conjecture which we make here is similar to the "thrackel conjecture," although the terminology is different. (See Woodall [10]).

Conjecture   *A connected graph G has $cr^M(G) = \mathcal{D}(G)$ if and only if G has at most one cycle C and $C \neq C_4$.*

An area of current interest, albeit of little progress, is in determining the maximum crossing number of the n–cubes. Such is not known even for n = 3. One notes that the n–cube has many four cycles, and that such are likely to play an important role in an upper found for any graph since a four cycle itself does not meet the previous upper bound. Hence, one needs to subtract the number of nonidentical C_4's from the upper bound. Also, since $\mathcal{D}(K_4)$ is three while at the same time K_4 has exactly three four cycles, the number of nonidentical K_4's must be added back in. Consequently, the following modified upper bound seems appropriate.

Remark 2 *For any graph G, $cr^M(G) \le \mathcal{D}(G) - \mathcal{N} + \mathcal{M}$, where $\mathcal{N}$ and $\mathcal{M}$ are the number of nonidentical C_4's and K_4's in G, respectively.*

In this sequel we will denote $\mathcal{D}(G) - \mathcal{N} + \mathcal{M}$ by $\mathcal{D}'(G)$. Notice that if G is a tree or a unicyclic graph, where the cycle is not a four cycle, then $\mathcal{D}'(G) = \mathcal{D}(G)$ and if G is unicyclic where the cycle is a four cycle, then $\mathcal{D}'(G) = \mathcal{D}(G) - 1$, the bound which was met earlier. Furthermore, if G is a complete graph, then a simple counting argument shows that $\mathcal{D}(G) = \mathcal{N}$, giving the result that $\mathcal{D}'(G) = \mathcal{M}$, which is the result in [9].

An interesting question is whether any collection of graphs other than these immediate examples meet the modified bound $\mathcal{D}'$. The following results came about as a direct consequence of attempts to solve this problem for the n–cubes. Of course, the so called "ladder" graphs and "prism" graphs are important classes of graphs in their own right.

Theorem 4 *For the ladder graphs $L_{2n} = P_n \times K_2$, where P_n i s the path with n vertices, we have $cr^M(L_{2n}) = \mathcal{D}'(L_n)$.*

Proof We need to display drawings of the L_{2n} which have the proper number of crossings. The drawings are illustrated in figure 4 for L_{12}, where 4a is a modification of the drawing with the proper number of crossings, while 4b shows the labeling of vertices when the graph is drawn in what we refer to as "standard form".

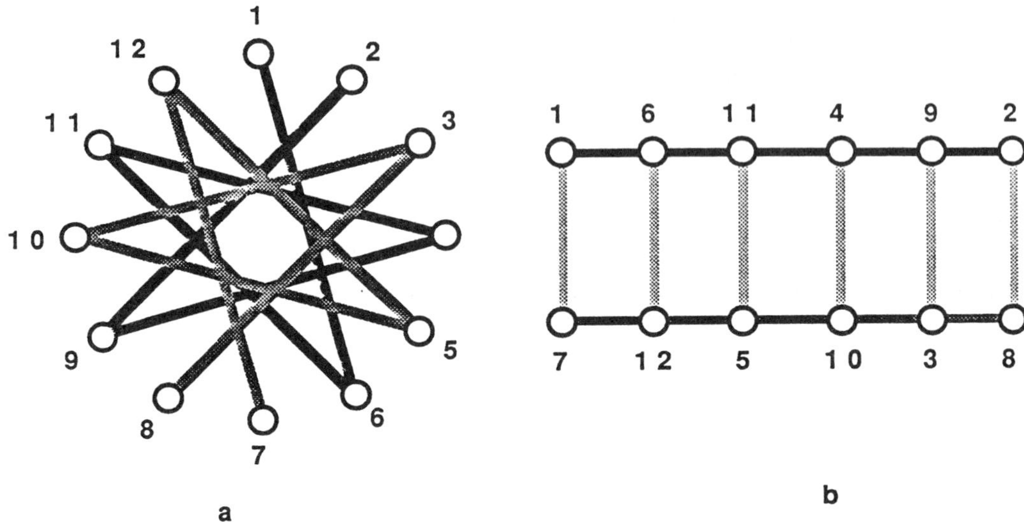

Figure 4. The ladder graph

The only four cycles in the ladder graph are the obvious ones and their are no K_4 subgraphs. To see that the desired drawing may be created for any ladder graph notice that the path P_n can be drawn with the maximum number of crossings by placing the vertices labeled 1 through $2n$ in a circular fashion, then joining vertex 1 to vertex $1 + (n - 1)$, then this vertex n to $n + (n - 1)$, etc. Now referring to figure 4a, notice that another copy of P_n can be drawn by beginning at vertex $n + 1$ and again adding $(n - 1)$ to reach vertex $2n$. Continue to create this second path by adding $(n - 1)$ to each successive vertex until the path is completed. One can use modular arithmetic to show that such a drawing always results in any "upper horizontal" edge in the standard drawing of L_{2n} (figure 4b) parallel to the corresponding "lower horizontal" edge in its four cycle. Notice that the "vertical" edges in the standard drawing each have endpoints which are diametrically opposite in figure 4a. When these edges are added directly across the interior of the drawing one obtains a drawing of L_{2n} which has $\mathcal{D}(L_{2n})$ crossings. The proof is then complete. ❏

Although the prism graphs are only a slight modification of the ladder graphs, the fact that they also meet the $\mathcal{D}$ bound is much more difficult to see. Nevertheless, we do have the following result.

Theorem 5 *The prism graphs $R_{2n} = C_n \times K_2$, for $n \neq 4$ and C_n is the cycle graph on n vertices, have $cr^M(R_{2n}) = D'(R_{2n})$.*

Proof   Again, drawings need to be displayed which meet the proper criteria. Since drawings which maximize the number of crossings for the cycle graphs are more complicated for even cycles than for the odd, one might expect that the prism formed when the base cycle is even might be more complicated than that for odd cycles. Such is certainly the case. We first display a drawing for the odd cycle case and then use a procedure similar to that used by Harborth [5] to obtain the even cycle case. We note that the only four cycles in R_{2n} are the obvious ones and that there are no K_4 subgraphs.

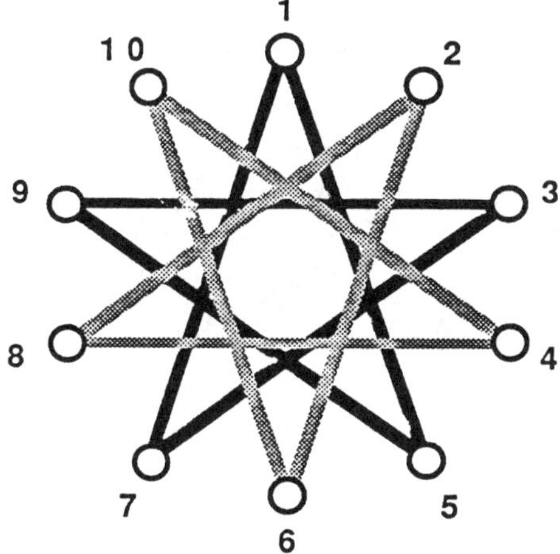

Figure 5. Drawing the Odd Cycle Prisms

Suppose n is odd. Then a drawing of R_{2n} with D' crossings can be obtained in a manner similar to that used for the ladder graphs. (Refer to figure 5.) The cycle C_n can be drawn with the maximum number of crossings by placing vertices labeled 1 through 2n in a circular fashion, then joining vertex 1 to vertex $1 + (n - 1)$, then this vertex n to $n + (n - 1)$ and so on until vertex 1 is reached again. Notice that another copy of C_n can be drawn beginning at vertex 2 and again adding $(n - 1)$ to each new vertex until vertex 2 is reach. (In what follows we consider the "standard drawing" of the prism to be that illustrated for R_{12} i n figure 6b.) As in the proof to Theorem 4, such a drawing results in any "upper horizontal " edge in the standard drawing parallel to the "lower horizontal" edge in its four cycle. Again, the "vertical" edges in the standard drawing each have endpoints which are diametrically opposite in figure 5. Adding these edges directly across the interior results in a drawing of R_{2n} with the proper number of crossings.

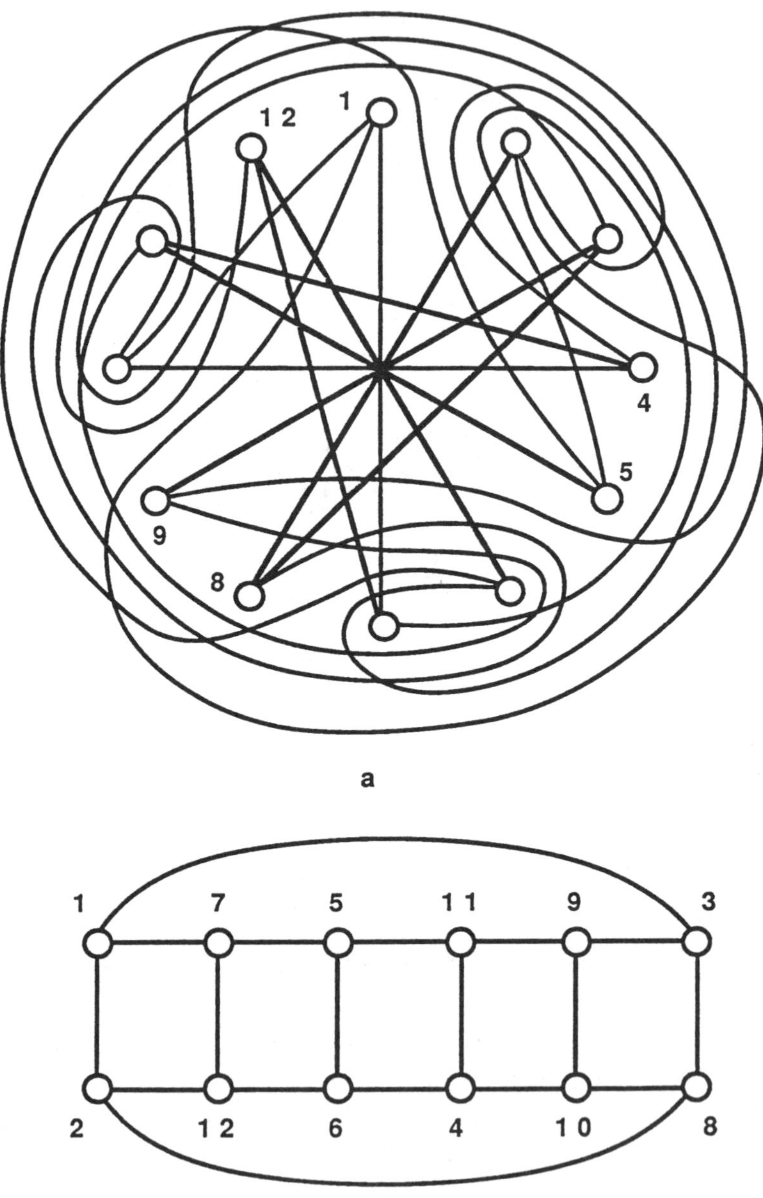

a

b

Figure 6. Drawing R_{12}

The drawing for R_{12} with $\mathcal{D}'$ crossings is given in figure 6, where the vertices are assumed to be labeled clockwise. Notice that the drawing induced by the odd numbered vertices and that induced by even number vertices are both essentially the even cycle drawing of figure 2. So we now consider R_{2n} where n i s even and $n \geq 8$.

Henceforth we consider the two n cycles which comprise the vertices of R_{2n} to be labeled $1, 2, \ldots, n$ and $1^*, 2^*, \ldots, n^*$, where we assume that the edges are i, i*, i (i + 1), and i* (i+1)*, for $i = 1, \ldots, n$ and all arithmetic is modulo n. $\mathcal{D}'(R_{16}) = 220$ and a drawing with 220 crossings is obtained from figure 7 by taking the union of the drawings in 7a and 7b, where the vertices in part a are superimposed on the vertices in part b. The proof now proceeds by induction, where the just illustrated case $n = 8$ is the basis and the induction is in increments of two.

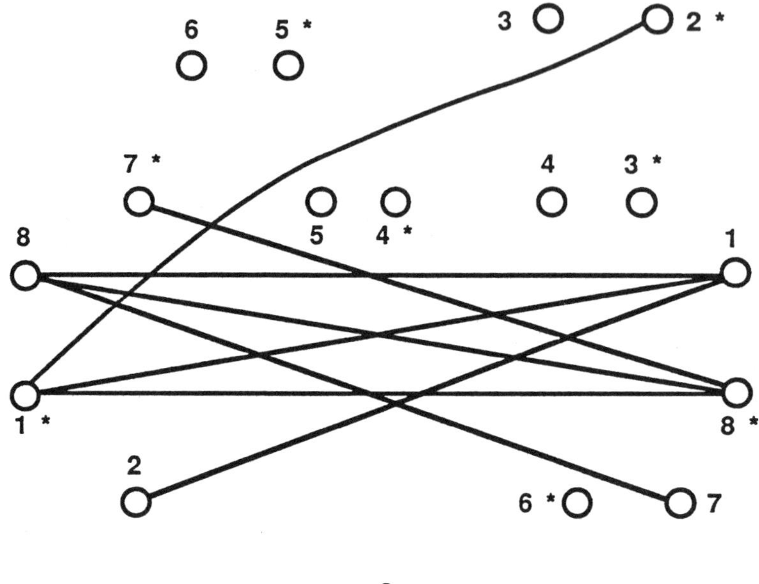

a

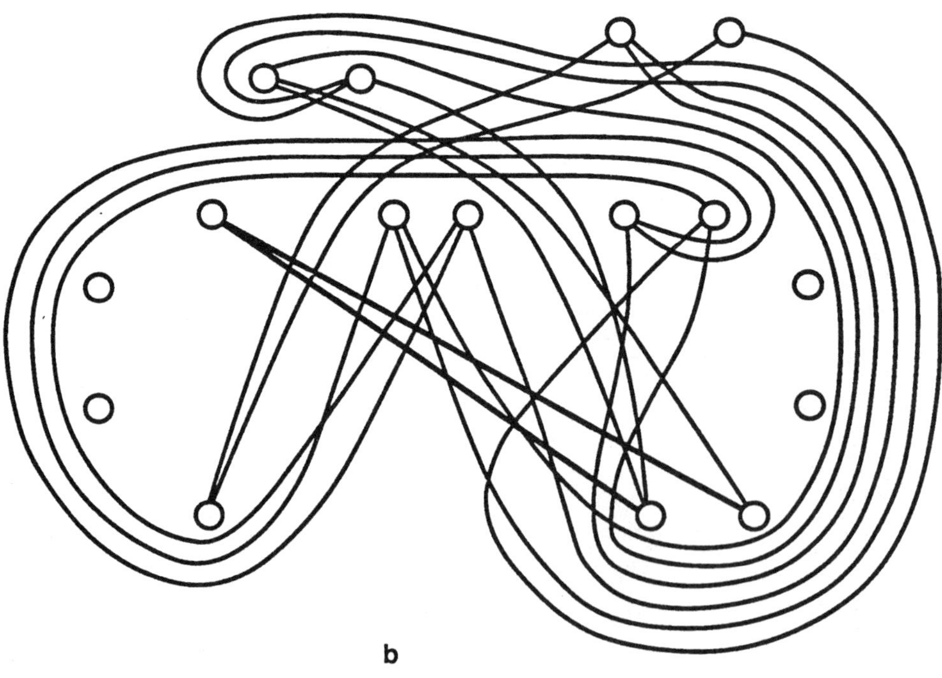

b

Figure 7. Drawing R_{16}

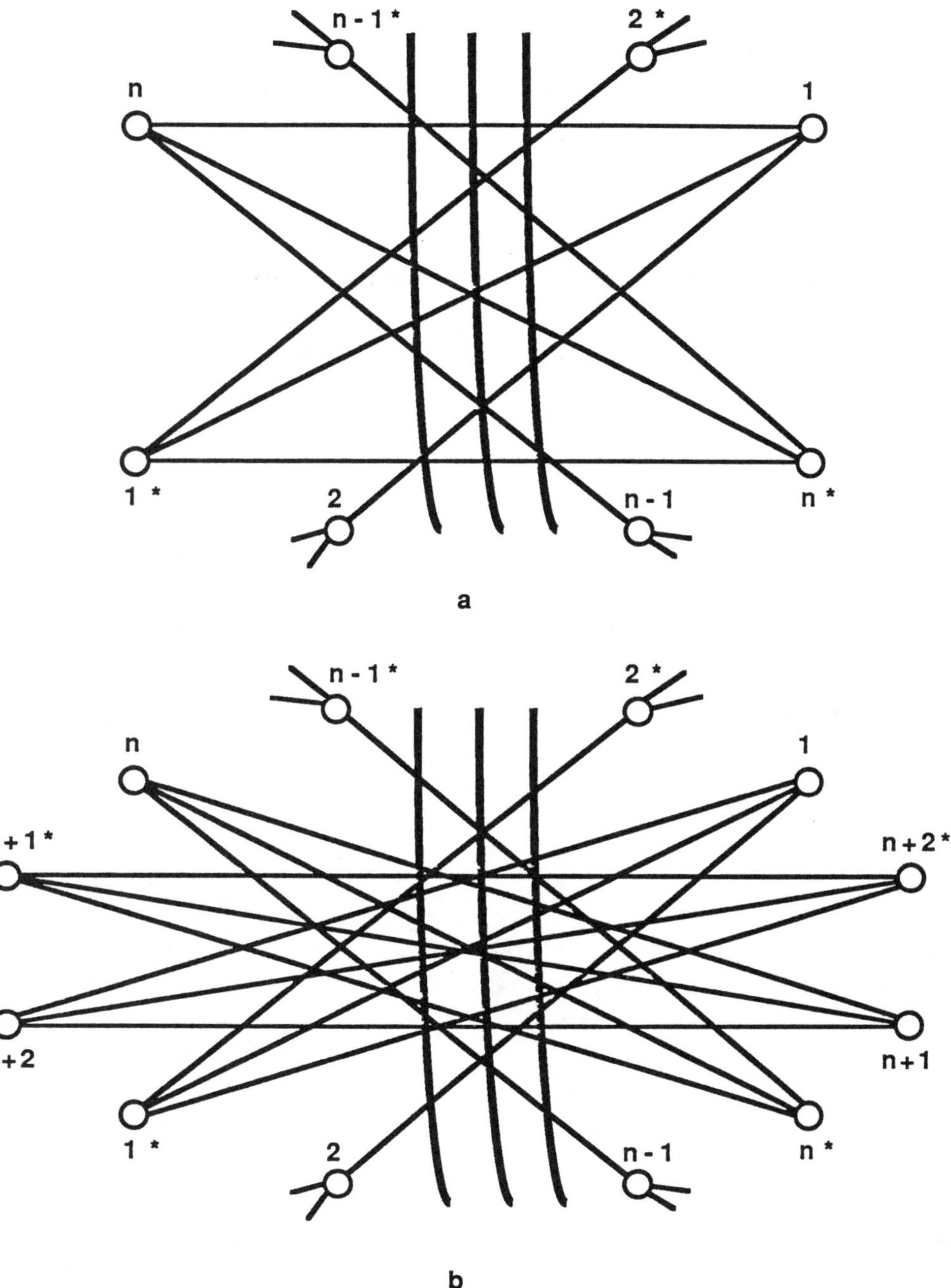

Figure 8. Induction step for even cycle prisms

The induction step is illustrated in figure 8. Figure 8a shows part of a special drawing of R_{2n} where the vertices labeled 1, 2, n − 1, n, 1*, 2*, (n − 1)*, and n* are drawn as indicated. Notice that all the legal crossings for edges among these vertices do occur. The curved lines indicate that all the other edges of the graph pass through the drawing in that way. Notice that the drawing for R_{16} has the appropriate vertices drawn in this manner. Hence, we may assume that R_{2n} has such a drawing.

To obtain the desired drawing of $R_{2(n+1)}$, delete the edges 1 n and 1* n*. Then place the vertices (n + 1), (n + 2), (n + 1)*, and (n + 2)* along with the edges n (n + 1), (n + 1) (n + 2), (n + 2) 1, n* (n + 1)*, (n + 1)* (n + 2)*, (n + 2)* 1* (n+1)(n+1)*, and (n+2)(n+2)* on the drawing as in figure 8b. Notice that all the other edges of $R_{2(n+1)}$ still cross through the structure as desired. Furthermore, in the new drawing, the vertices 1, n, n + 1, n + 2, 1*, n*, (n + 1)*, and (n + 2)* along with their appropriate edges form the structure determined by figure 8a. Hence the induction is complete and the theorem follows.

We finish this paper with a look at a drawing of the three cube. Figure 9 displays a drawing of Q_3, which is isomorphic to the excluded case R_8 from the last theorem, with 34 crossings. Of course $\mathcal{D}(Q_3) = 36$ and thus we do not know if this drawing gives the maximum crossing number for the graph. In fact, we are at this point uncertains as to whether to conjecture that the cube graphs will meet the upper bound $\mathcal{D}$ or whether they will not. The problem is certainly and interesting one. So we close with the

Question   Do the cube graphs meet the upper bound $\mathcal{D}$?

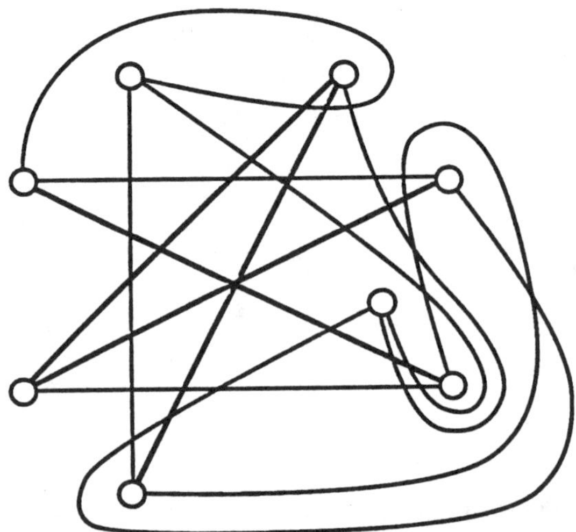

Figure 9. Drawing Q_3

REFERENCES

[1]   L. W. Beineke and R.D. Ringeisen, On the Crossing Number of Products of Cycles and Graphs of Order of Four, *J. of Graph Theory* **4** (1980), 145–155.

[2]   R. B. Eggleton, *Crossing Number of Graphs*, Ph.D. Dissertation, University of Calgary 1973.

[3]   R. B. Eggleton, R. Guy, H. Harborth, and R. D. Ringeisen, Non-Minimal Crossings for the Complete Graphs, in progress.

[4]   H. Harborth, Drawings of the Cycle Graph, *Congressus Numerantium*, to appear.

[5]   H. Harborth, Parity of Number of Crossings for Complete n-partite Graphs, *Math.Slov.* **26** (1976), 77-95.

[6]   D. J. Kleitman, A note on the Parity of Numbers of Crossings of a Graph, *Journal of Combinatorial Theory (B)* **21** (1976), 88-89.

[7]   D. J. Kleitman, The Crossing Number of $K_{5,n}$, *J. Combinatorial Theory* **9** (1970), 315-323.

[8]   R. D. Ringeisen and L. W. Beineke, The Crossing Number of $C_3 \times C_n$, *J. Combinatorial Theory* **24** (1978), 134-136.

[9]   G. Ringel, Extremal Problems in the Theory of Graphs, in *Theory of Graphs and its Applications*, Proceedings of the Symposium in Smolenice, June 1963 (M. Fiedler, Ed.), 85-90.

[10] R. D. Woodall, Thrackles and Deadlock, in *Combinatorial Mathematics and its Applications*, Proceedings of a Conference held at the Mathematical Institute, Oxford, 1969 (D. A. Welsh, ed), 335-348.

Partially Distance–Regular Graphs

David L. Powers
Clarkson University

ABSTRACT

A graph G is distance(s)–regular if, for each vertex u, there is a partition of the vertices into cells $V_0(u)$, $V_1(u)$, ..., $V_t(u)$, such that: $V_k(u) = \{v : dist(v, u) = k\}$ for k = 0, 1, ..., s; and for all j, k and u, there are b_{jk} edges from each vertex of $V_j(u)$ to vertices of $V_k(u)$ (independent of u). The eigenvalues and eigenvectors of the adjacency matrix A of G are determined by those of the matrix $B = [b_{jk}]$. If the columns of a matrix Z are orthogonal basis for an eigenspace of A, then the mapping that takes a vertex to the corresponding row of Z allows a geometric interpretation of G. For a distance(1)–regular graph, there are conditions that guarantee that the mapping is 1–1, adjacency in G becomes proximity in euclidean space, and the automorphism group of G is isomorphic to the orthogonal symmetry group of the rows of Z.

1.   Introduction and Definitions

The objective of this paper is to show that certain results that are known for distance–regular graphs can be generalized to less regular graphs, including vertex– or arc–transitive graphs. The book of Biggs [1] contains theorems showing that the eigenvalues of a distance–regular graph (i.e., of the adjacency matrix), their multiplicities, and the associated eigenvectors are determined by the eigenvalues and eigenvectors of a related, but usually much smaller, matrix (intersection array) that summarizes some graph properties. Terwilliger [8] and Godsil [4] used these ideas to relate eigenvalue multiplicites to structure of distance–regular graphs. In [7] connections were found between eigenvectors and the automorphism group of a distance–regular graph.

This work was supported by the Office of Naval Research under grant N00014–85–04097.

Throughout, G is a connected graph of diameter d, with vertex set $V = \{1, 2, ..., n\}$ and adjacency matrix A; e is a column matrix of 1's, and e_i is column i of the identity matrix (dimensions determined by context).

A *coloration* or *equitable partition* of G is a partition of the vertex set into cells $V_0, V_1, ..., V_t$, such that each vertex in V_i is adjacent to b_{ij} vertices in V_j. Let X be the *indicator* of the partition: the v, i–entry of X is 1 if vertex v is in V_i or 0 if not. A partition with indicator matrix X is a coloration if and only if $AX = XB$, where $B = [b_{ij}]$ is called the *coloration matrix*.

We begin by defining two parametrized classes of graphs, suggested by the work of Terwilliger [8]. We will refer to these sets of vertices: $D_k(u) = \{v : \text{dist}(u, v) = k\}$, $k = 0, 1, ..., d$.

Definition   A graph G is *distance(s)–transitive* if G is vertex–transitive and, for each vertex u, $D_k(u)$ is an orbit of the stabilizer of u for $k = 0, 1, ..., s$. A graph G is *distance(s)–regular* if, for each vertex u there is a coloration with cells $V_0(u)$, $V_1(u)$, ..., $V_t(u)$, such that $V_k(u) = D_k(u)$ for $k = 0, 1, ..., s$, and the coloration matrix is independent of u. We say that such a coloration with $V_0(u) = \{u\}$ is *centered at* u.

It is easy to see that distance(s)–transitive is just vertex–transitive if $s = 0$, arc–transitive if $s = 1$ and distance–transitive if $s = d$; similarly, distance(d)–regular is distance–regular. Distance(s)–regular implies that the leading principal submatrix of B of order $s + 1$ is tridiagonal. A distance(0)–regular graph is degree–regular; we denote its degree by r.

Lemma 1   *If G is distance(s)–transitive, then G is distance(s)–regular.*

Proof   Let u be a fixed vertex, and partition the vertices of G into the orbits of the stabilizer of u, $V_0(u)$, $V_1(u)$, ..., $V_t(u)$. We may assume that the cells of the partition are labeled so that $V_k(u) = D_k(u)$ for $k = 0, 1, ..., s$. It is well known that the orbits of any subgroup of the automorphism group of G is a coloration [2, p. 119]; thus we have a partition centered at u that is a coloration. If v is any vertex, let g be an automorphism of G that takes u to v. Then $V_k(v) = g(V_k(u))$ for $k = 0, 1, ..., t$ defines a partition cenetered at v that is a coloration. It is easy to show that the partitions thus constructed are independent of the initial vertex u and of the automorphisms used to transfer the initial coloration to other vertices, and that all have the same coloration matrix. ❑

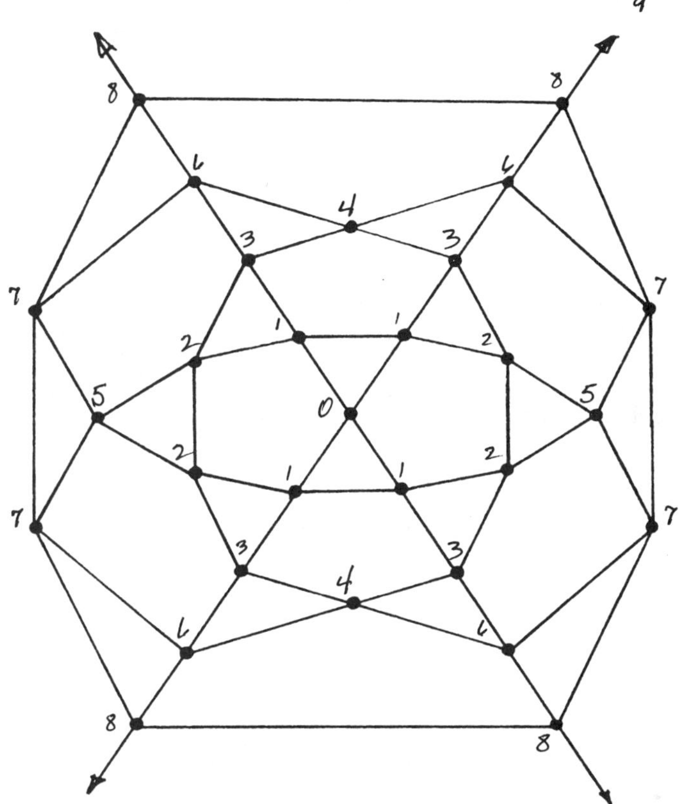

Figure 1. Icosidodecahedron

Example   Figure 1 shows the skeleton of an icosidodecahedron, which is an arc–transitive graph. The number next to a vertex shows the orbit it is in.

The connection of the distance(s)–regular graph to linear algebra comes through the coloration. For each vertex u we define the indicator matrix X_u for the partition centered at u. It is easy to see that $X_u^T X_u = \text{diag}\{r_0, r_1, ..., r_t\} = R$, where $r_k = |V_k(u)|$. The fact that the partition is a coloration is expressed by the equation $AX_u = X_u B$, where B is the coloration matrix of order $t + 1$. We also define the *relation matrices* $A_0, A_1, ..., A_t$ by requiring the v, u–entry of A_k to be 1 if vertex v is in $V_k(u)$ or 0 if not. The following lemma, although fundamental, is trivial to prove.

Lemma 2   *For any vertex u and any integer $k = 0, 1, ..., t$, column u of A_k is column k of X_u : $A_k e_u = X_u e_k$.*

Thus the information contained in the indicator matrices is the same as that in the relation matrices, in a different arrangement. Let us also note that in a distance(0)–regular graph, $A_0 = I$, the sum of all the relation matrices is ee^T, and the adjacency matrix A of G is a sum of some relation matrices. (If G is distance(1)–regular then A is the relation matrix A_1.) In general, the relation matrices are not symmetric. For an example, note that in Figure 1, if u is the central vertex and v is in $V_5(u)$, then u is in $V_4(v)$.

2. Matrices

A key result in the theory of distance–regular graphs concerns the algebra of polynomials in A. Here we consider instead the subspace $\mathcal{V}$ of $n \times n$ matrices that is spanned by the relation matrices.

Theorem 1 *Let G be a distance(0)–regular graph with adjacency matrix A, relation matrices $A_0, A_1, ..., A_t$, and coloration matrix B.*

(1) $A_0, A_1, ..., A_t$ form a basis of $\mathcal{V}$.

(2) $\mathcal{V}$ is invariant under left multiplication by A.

(3) For $k = 0, 1, ..., t$,

$$A_k = \sum_{j=0}^{t} b_{jk} A_j.$$

That is, left multiplication by A is represented by B, relative to the basis $A_0, A_1, ..., A_t$.

(4) The space $\mathcal{V}$ contains the algebra of polynomials in A.

Proof (1) Since each of the A's has 0's where another has 1's, they are linearly independent. (2) & (3) For any vertex u and any $k = 0, 1, ..., t$,

$$AA_k e_u = AX_u e_k = X_u B e_k$$
$$= X_u \sum_{j=0}^{t} b_{jk} e_j = \sum_{j=0}^{t} b_{jk} A_j e_u$$

This equality holds for each $u = 1, 2, ..., n$; thus both statments are proved. (4) Since $A_0 = I$, it follows from (3) that every power of A, and hence every polynomial in A, is in $\mathcal{V}$. $\square$

Corollary 1.1 *Let G be a vertex–transitive graph. The number of orbits of a vertex stabilizer is greater than or equal to the number of distinct eigenvalues of A.*

Proof The number of distinct eigenvalues of A is the number of independent powers of A, which is not more than the dimension of $\mathcal{V}$, which is $t + 1$, the number of orbits of a vertex stabilizer. $\square$

In the case of distance−regular graphs, the number of distinct eigenvalues of A is exactly $1 + t = 1 +$ the diameter of G. For the graph of Figure 1, the number of orbits of a vertex stabilizer is 10, while A has only 7 distinct eigenvalues. The difference can be partly explained by the fact that the powers of A span a space of symmetric matrices, which is a proper subspace of $\mathcal{V}$ if any of the A_k are not symmetric.

Theorem 2 *Let G be a distance(0)−regular graph with adjacency matrix A, relation matrices $A_0, A_1, ..., A_t$, and coloration matrix B. Let L be the projector of A associated with an eigenvalue α. Then there is a column $y = [y_0, y_1, ..., y_t]^T$, such that $L = y_0A_0 + y_1A_1 + ... + y_tA_t$, and $By = \alpha y$.*

Proof The existence of y follows directly from Theorem 1, since L is a polynomial in A. Now

$$Le_u = y_0A_0e_u + ... + y_tA_te_u$$
$$= y_0X_ue_0 + ... + y_tX_ue_t = X_uy.$$

With two applications of this equality, the equation $AL = \alpha L$ gives

$$ALe_u = AX_uy = X_uBy = \alpha Le_u = \alpha X_uy.$$

Since X_u has independent columns, it follows that $By = \alpha y$. Clearly $y \neq 0$. $\square$

Corollary 2.1 *Every eigenvalue of A is an eigenvalue of B.*

This is also a corollary of a theorem of Godsil and McKay [5] that applies to the more general class of walk−regular graphs: those for which each power A^k has uniform diagonal.

It remains to identify the eigenvector y, especially in the case where α is a multiple eigenvalue of B. Suppose that p is the Lagrange interpolating polynomial that takes value 1 at α and value 0 at the other eigenvalues of A. Then the projector for A associated with α is $L = p(A)$. Moreover, since A and B have the same spectrum, $M = p(B)$ is the projector for B associated with α, and $LX_u = X_uM$. We have seen that $Le_u = X_uy$, but also $Le_u = LX_ue_0 = X_uMe_0$. Since X_u has independent columns, we may conclude that $y = Me_0$.

The multiplicity m of the eigenvalue α in the spectrum of A can be found very simply from y, for $m = \text{tr}(L) = ny_0$, since $A_0 = I$ is the only relation matrix with nonzero diagonal. However, if y can be found by means other that the equation $y = Me_0$ (e.g. if α is a simple eigenvalue of B), it will be useful to have another relation for the multiplicity. We start from an elementary property of projectors.

$$m = \text{tr}(L) = \text{tr}(L^2) = \text{tr}(L^TL)$$
$$= \sum_{u=1}^{n} \| Le_u \|^2 = n \| Le_u \|^2 = n \| X_uy \|^2 = n \sum_{i=0}^{t} r_iy_i^2.$$

Now let x be defined by $y = y_0 x$. By comparing the expression above for m with $m = n y_0$, we determine that the multiplicity of is [1, p. 143]

$$m = n \Big/ \sum_{i=0}^{t} r_i x_i^2.$$

and $y_0 = m/n$.

For future convenience, define x to be a *leading eigenvector* of B if x is any positive multiple of $M e_0$, where M is the projector of B associated with the same eigenvalue. The x above is a leading eigenvector with $x_0 = 1$.

3. Geometry

Now we investigate properties of eigenvectors of distance(0)–regular graph G. If α is an eigenvalue of A with multiplicity m, a *complete eigenmatrix* assocaited with α is an $n \times m$ matrix Z satisfying $AZ = \alpha Z$ and $Z^T Z = I$. Row u of Z is denoted by $w_u = e_u^T Z$. Thus we have a mapping from the vertices of the graph to a set of points in m–dimensional space. The algebraic investigation of this mapping depends for the most part on viewing the projector L of A as the Gram matrix of the rows of $Z : L = Z Z^T = [(w_u, w_v)]$, where $(w_u, w_v) = w_u w_v^T$.

Lemma 3 *Let G be a distance(0)–regular graph, Z a complete eigenmatrix associated with eigenvalue α of A, x a leading eigenvector of B associated with α.*

(1) All rows of Z have the same euclidean length, and there is a positive constant b such that $(w_u, w_u) = b x_0$;

(2) the inner product of two rows of Z is given by $(w_u, w_v) = b x_k$ if vertex u is in $V_k(v)$.

Proof By Theorem 2, the projector has the representation

$$L = b(x_0 A_0 + x_1 A_1 + \dots + x_t A_t) \tag{1}$$

where x is a leading eigenvector and b is a positive constant. For any u, the product $(w_u, w_u) = (e_u, L e_u) = b x_0$, since $A_0 = I$ is the only relation matrix with a nonzero diagonal. (In fact, it is clear that $b x_0 = m/n$.) Similarly, (w_u, w_v) is the u, v–entry of L and hence equals $b x_k$ if A_k is the relation matrix that has a 1 in the u, v–position. $\square$

Of course, it is desirable that the mapping from vertices of G to the points whose coordinates are the rows of Z be one–to–one; that is, that the rows of Z be distinct. This will be true if a simple condition is met by a leading eigenvector of B.

Theorem 3 *Let G be a distance(0)–regular graph, Z a complete eigenmatrix associated with an eigenvalue α of A, x a leading eigenvector of B associated with α. The rows of Z are distinct if and only if $x_0 > x_k$ for all $k > 0$.*

Proof Since the rows of Z all have the same euclidean length, rows u and v ($v \neq u$) are distinct if and only if their inner product, (w_u, w_v), is strictly less than (w_u, w_u). That is, if and only if the u, u–entry of L is larger than any other entry in row u. Form Eq. (1) this is equivalent to the condition that $x_0 > x_k$ for all $k > 0$ in any leading eigenvector x. ❏

To see that the theorem is not trivial, consider the arc–transitive graph G with $2m$ vertices ($m \geq 5$) in which vertex u is adjacent to vertices $u + 1, u - 1, u + m - 1, u - m + 1$ (addition modulo $2m$). If Z is a complete eigenmatrix associated with any eigenvalue $\alpha \neq 0$, then rows u and $u + m$ of Z are identical.

From here on, we consider distance(1)–regular graphs. In order to avoid tedious repetitions, we will assume that the following notational hypotheses apply: G is a connected, distance(1)–regular graph with adjacency matrix A, α is an eigenvalue of A with multiplicity m, Z is a complete eigenmatrix associated with α, $L = ZZ^T$, and x is a leading eigenvector of B associated with α. Recall that the relation matrix A_1 is the adjacency matrix A of G. The first row of the coloration matrix B is

$$[0 \quad r \quad 0...0]$$

Hence, it is easy to see that in a leading eignevector x, $x_1/x_0 = \alpha/r$.

We will refer to the group of permutations that commute with a given matrix, defined by

$$\text{perm}(M) = \{P : P \text{ a permutation and } PM = MP\}.$$

Theorem 4 *Let G be distance(1)–regular. If $x_0 > x_k$ for all $k \neq 0$ and $x_1 \neq x_k$ for all $k \neq 1$, then the group*

$$\text{orth}(Z) = \{S : ZS = PZ \text{ for some permutation } P\}$$

is isomorphic to the group of automorphisms of G.

Proof It is easy to show that orth(Z) is a group of orthogonal matrices, and that the mapping of perm(ZZ^T) onto orth(Z) given by $R = Z^TPZ$ is a group homomorphism with kernel $\{P : PZ = Z\}$ [3], [7]. However, since the rows of Z are distinct by Theorem 3, the kernel is trivial. Thus, the isomorphism of orth(Z) and perm(L) is established.

Next consider the representation of L in the proof of Lemma 3. Since $x_1 \neq x_k$ for all $k \neq 1$, the number bx_1 appears as an entry of L exactly where 1 appears as an entry of $A_1 = A$. Thus, a permutation matrix P that commutes with L must also commute with A. On the other hand, L is a polynomial in A, so any matrix that commutes with A commutes with L. Therefore, $\text{perm}(L) = \text{perm}(A)$, and the latter is well known to be isomorphic to the group of automorphisms of G. ❑

Theorem 5 *Let G be distance(1)–regular. If $x_0 > x_1$ and $x_1 > x_k$ for $k > 1$, then u and v are adjacent vertices of G if and only if w_u and w_v are nearest neighbors among the rows of Z.*

Proof Note that the rows of Z are distinct by Theorem 3, since $x_0 > x_k$ for $k > 0$. Next, $x_1 > x_k$ for all $k > 1$ if and only if the inner product (w_u, w_v) for v adjacent to u (v in $V_1(u)$) is greater than the inner product (w_u, w_v) for v not adjacent to and not equal to u (v in $V_k(u), k > 1$). ❑

Note that under the hypotheses of this theorem, the mapping from vertices to points of m–space is one–to–one, preserves the automorphism group, and carries adjacency into proximity. The conditions of the theorem are always fulfilled for the second eigenvalue of a distance–regular graph [4], [7].

Godsil [3] introduced the idea of the polytope associated with a graph eigenvalue: if Z is a complete eigenmatrix associated with α, then the convex hull of the points whose coordinates are the rows of Z is the polytope $C(\alpha)$. If the hypotheses of Theorem 3 are satisfied, $C(\alpha)$ has n extreme points; the group orth(Z) referred to in Theorem 4 is the symmetry group of $C(\alpha)$, including reflections. This polytope inherits some properties of the graph.

Theorem 6 *Let G be distance(1)–regular graph, let $0 \le \alpha < r$, and let $x_0 > x_1$ and $x_1 > x_k$ for $k > 1$. If U is a set of vertices of G that induce a complete graph K_q in G, then the corresponding points, w_u, are extreme points of a face F of $C(\alpha)$, and F is a simplex.*

Proof Let K be the $q \times q$ submatrix of L formed from entries whose indices are in U. (See Eq. 1.) Since all the vertices of U are mutually adjacent, $K = b(x_1 ee^T + (x_0 - x_1)I)$. The eigenvalues of K are $b(x_0 - x_1) = bx_0(r - \alpha)$ and $b(x_0 + (q - 1)x_1) = bx_0(r + (q - 1)\alpha)$, both positive. Thus the rank of K is q, and the rows of Z whose indices are in U are independent.

Next, let g be the sum of the rows w_u for which u is in U. then $(g, w_v) = b(x_0 + (q - 1)x_1) = \gamma$, say, if v is in U, but $(g, w_v) \le qbx_1 < \gamma$ if not. Then the hyperplane $H = \{w : (g, w) = \gamma\}$ supports $C(\alpha)$ and contains only those extreme points of $C(\alpha)$ corresponding to vertices in U. These must be extreme points of a $(q - 1)$–face F of $C(\alpha)$, and, since they are equidistant, must form a simplex. ❑

Corollary 6.1 *If u and v are adjacent vertices in G, then w_u and w_v are adjacent in the skeleton of $C(\alpha)$.*

Proof   Vertices u and v induce a K_2. By the theorem, the corresponding points form a 1–face of $C(\alpha)$. ❑

Corollary 6.2 *If q is the clique number of G, then the multiplicity of α is at least q.*

Corollary 6.3 *If the multiplicity of α in the spectrum of A is 3, then G is planar. If the multiplicity is 2, then G is an n–cycle.*

Proof   By Corollary 6.1, a copy of G is an edge subgraph of the skeleton of $C(\alpha)$. If the multiplicity of α is 3, then $C(\alpha)$ is a polyhedron. If the multiplicity is 2, the same argument applies, but $C(\alpha)$ must be a regular n–gon. The copy of G that is an edge subgraph of skel$(C(\alpha))$ is connected and vertex–transitive, so it must be an n-cycle. ❑

In [3] Godsil mentions that in an edge– and vertex–transitive graph G, the value of (w_u, w_v) must be the same for all pairs u, v of adjacent vertices. Thus, in such a graph, assume that the cells of the coloration centered at u are numbered so that $V_1(u), ..., V_q(u)$ contain all the vertices adjacent to u. Let ξ be the common value of $x_1, ..., x_q$ in a leading eignevector of B. The results if Theorems 4, 5, and 6 may be extended to vertex– and edge–transitive graphs by substituting the following hypotheses. (4') Let $x_0 > \xi$ and $\xi \neq x_k$ for $k > q$. (5') and (6') Let $x_0 > \xi$ and $\xi > x_k$ for $k > q$. D.F. Holt [6] gives an example of a vertex– and edge–transitive graph that is not distance(1)–regular, and for which these special hypotheses are satisfied.

REFERENCES

[1]   N. Biggs, *Algebraic Graph Theory*. Cambridge University Press, London (1974).

[2]   D.M. Cvetković, M. Doob and H. Sachs, *Spectra of Graphs*. VEB, Berlin/Academic Press, New York (1980).

[3]   C.D. Godsil, Graphs, groups and polytopes. In *Combinatorial Mathematics VI (Canberra 1977)*, D.A. Holton and J. Seberry, eds. Springer, Berlin/New York (1978), 157 – 164.

[4]   C.D. Godsil, Bounding the diameter of distance–regular graphs. Submitted (1987).

[5]   C. Godsil and B.D. McKay, Feasibility conditions for the existence of walk–regular graphs, Linear Algebra and Applications 30 (1980), 51 – 61.

[6]   D.F. Holt, A graph which is edge transitive but not arc transitive. Journal of Graph Theory, 5 (1981) 201 – 204.

[7]   D.L. Powers, Eignevectors of distance–regular graphs. SIAM Journal on Matrix Analysis and Appl., 9 (1988), 399 – 407.

[8]   P. Terwilliger, Eigenvalue multiplicities of highly symmetric graphs, Discrete Math., 41 (1982), 295 – 302.

On Enumeration of Complete Matchings in Hexagonal Lattices

M. Randić
Drake University

ABSTRACT

We consider a class of polyhex graphs which are lattices, i.e., graph which when suitably oriented allow identifiaction of the "master" and the "slave" vertices. An elegant algorithm for enumeration of the perfect matchings in such lattices is outlined. The algorithm combines elements of an earlier algorithm valid for chain-like fused polyhex graphs with an addition rule which counts "interior" paths from the "master" to the slave "vertex".

1. Introduction

An important part of organic chemistry is concerned with the properties of the so called polycyclic conjugated hydrocarbons, the essential structural element of which is a six-membered ring of carbon atoms. If we ignore the peripheral hydrogen atoms such molecules can be viewed as fused polyhex units. Of interest in mathematical chemistry is to see to what extent the properties of these compounds are reflected in the combinatorial and the topological variations associated with different mode of fusion of polyhex units.

The simplest polyhex corresponds to molecule C_6H_6, benzene, which is customarily taken as a representative compound of the so called aromatic hydrocarbons to which other compounds are compared. Benzene is very stable, the "extra" stability is attributed to the fact that for this molecule one can write two so called valence structures:

August Kekulé in 1876 proposed the above valence structures of benzene. He thus indicated that there are molecules for which theoretical models require two or more molecular structural formula. Such structural formulas are refered to in chemical litereature as Kekulé valence structures. It soon became apparent that molecules showing

1001"

a greater "extra" stability are those having a larger number of Kekulé valence forms. Hence, it has been of a considerable interest in theoretical organic chemistry to find for a given molecule the number of possible valence structures, or in the language of mathematics, to find the number of perfect matchings. As has been customary in chemistry for long time, we will refer to the number of perfect matchings as K, to honor August Kekulé.

2. Catacondensed polyhexes – Algorithm of Gordon and Davison

If each polyhex ring in a structure, with the exception of the terminal rings, has at most two neighbors we obtain so called cata-condensed system. In Fig. 1 we illustrate one such case. Gordon and Davison (1) already in 1952 arrived at an unusually elegant algorithm for enumeration of perfect matchings in such polyhex cata-condensed systems. The algorithm is illustrated in the upper part of Fig. 1, where numbers are inserted in hexagons successively, from the left to the right. The numbers represent the values of K for the structures obtained by truncation at any particular ring. According to Gordon and Davison one inscribes 2 in the first ring, which signifies the fact that benzene (a single hexagon system) has two Kekulé structures, or two perfect matchings. For each linearly added ring one increases the number by one, thus obtaining 3, 4. However, at the site of "change of direction" of fusion, instead of increasing the number by "one" we add the number assigned to the previous ring. Hence $3 + 4$ gives 7. One continues adding this "previous" number (3 in our example) until one again comes to a ring with a "change" of direction of fusion. Hence we have, 7, 10, 13. The process of assignment of numbers to rings continues until all rings are assigned. In fully linear systems the number of perfect matching is therefore simply given by the number of fused rings + 1. If at each ring there is a "change of direction" of fusion we obtain as result Fibonacci numbers 2, 3, 5, 8, 13, 21, ... Recent revived interest in simple and elegant algorithms for finding K in general polyhex systems has been for the most part stimulated by the pioneering work of Gordon and Davison.

Although the algorithm of Gordon and Davison is remarkably simple, it has been recently pointed out that it could be even further simplified (2). As illustrated in the lower part of Fig. 1, instead of constructing successive terms from the left to the right we selected a single "kink" ring in the center of the structure, and applied the algorithm of Gordon and Davison to the two disjoint fragments obtained by erasing the selected ring (shown as shaded). The K for the molecule is obtained by multiplying the results for the two fragments to which one adds a similar result obtained from the even smaller fragments derived by erasing also all "linearly" fused rings to the ring previously selected.

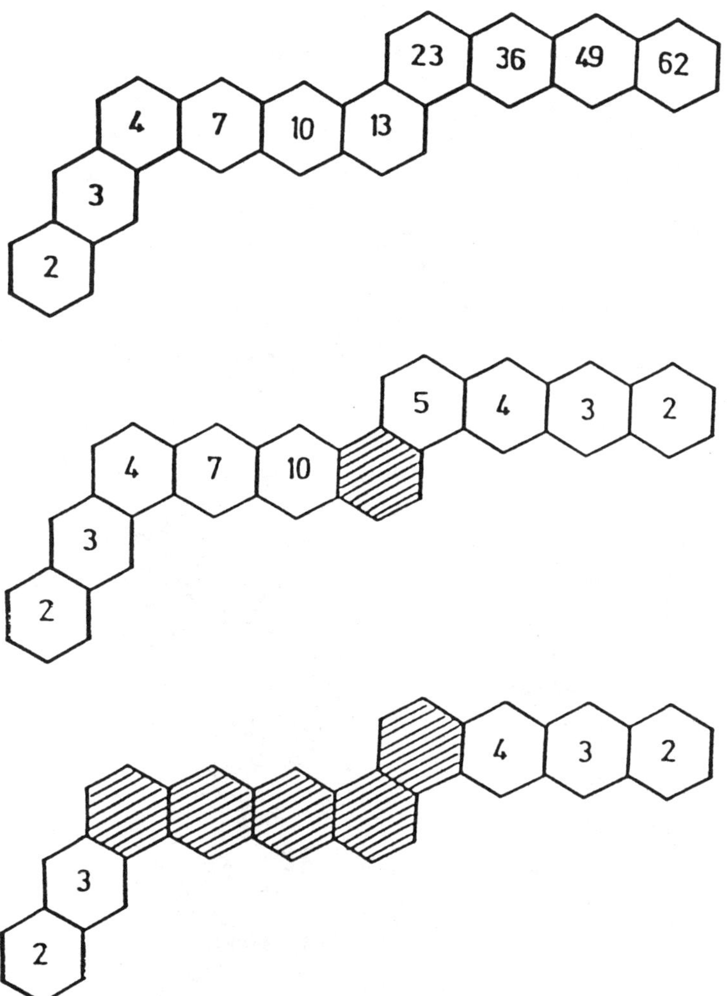

$$K = 10 \bullet 5 + 3 \bullet 4 = 62$$

Figure 1

As discussed by Gordon and Davison their algorithm can be extended to branched cata-condensed systems. Their particular generalization, however, involves some non-unique "surgery" at the branching site in the structure and thus apparently much of its original elegance is lost. However, this is not the case with extension of the above outlined "erasure" procedure. As illustrated in Fig. 2, one obtains K in branched systems immediately by applying the same "erasure" algorithm to the ring with three neighbors. The only difference is that now we have to combine the results from three disjoint fragments, rather than two, as was the case before.

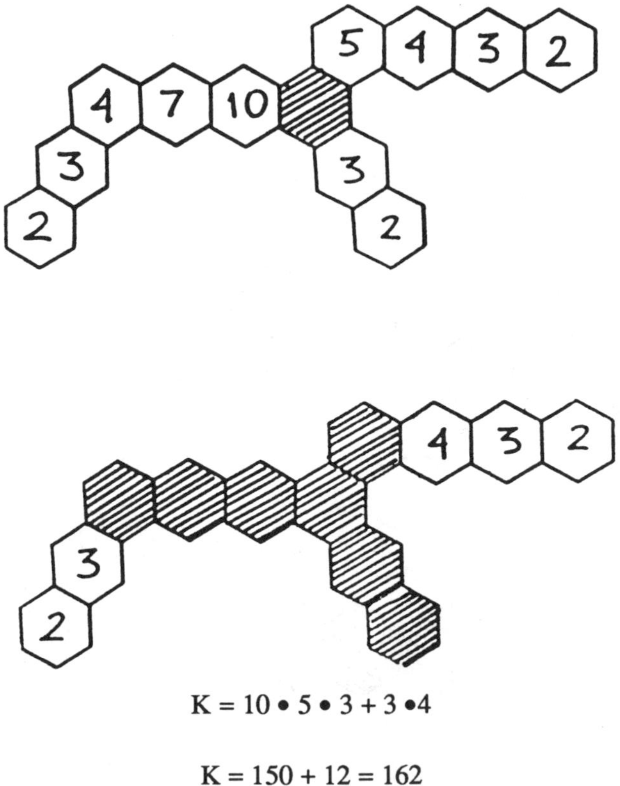

$$K = 10 \cdot 5 \cdot 3 + 3 \cdot 4$$

$$K = 150 + 12 = 162$$

Figure 2

3. Pericondensed Systems

There have been some attempts to extend the Gordon and Davison type of approach to pericondensed benzenoid systems, that is, to polyhex structures obtain by unrestricted fusion of hexagonal rings. There are no analogous simple algorithms that allow one to determine K in these more general cases. Some progress was reported for construction of polyhex systems for which $K = 0$ (3), but these cases, in view of a postulate of E. Clar, a doyen of chemistry of benzenoid hydrocarbons, that such systems cannot exist (4), are of limited chemical relevance. More pertinent is the recently outlined algorithm which allows one to construct all "augmented" polyhex systems of prescribed K, if the number of Kekulé structures of the smaller components and their fragments is known (5).

Gutman and Cyvin in several papers (6) considered selected subclasses of polyhex molecules for which they were apparently able to extend Gordon and Davison type algorithm. Their work has shown that simple algorithms are possible in some special cases of pericondensed polyhexes. It is unclear, however, from their work if there exist additional structures for which application of a simple algorithm will work, and how one finds such structures.

4. Polyhex Lattices Algorithm for K

We would like to report here that in the case of polyhex (benzenoid) lattices we can derive K, the number of Kekulé structures or perfect matchings, by a very simple and elegant manner, which can be viewed as an extension of the Gordon and Davison algorithm. Polihex lattice is a lattice composed of fused hexagonal rings, i.e., the vertices of fused polyhex rings form a lattice. Such graphs can be embedded in a plane and oriented so that one vertex (ring) is at the top (refered to as "master" vertex) and one vertex (ring) is at the bottom (refered to as "slave" vertex). The shape of the lattice is not important, and it may be "triangular", convex or not. In such lattices we can differentiate the "peripheral" rings and "inside" rings. However, for the algorithm to be outlined it is more useful to differentiate between rings with two predecessor rings and the rings with only one predecessor. The former class allows immediately the assignment of the partial K-values by simply adding the known predecessor values. The assignment of the value for the other class follows Gordon and Davison algorithm. The algorithm for finding the number of perfect matchngs is as follows:

Start with the "slave ring" at the bottom and assign to it value 2 (same as in benzene and in the initialization of Gordon and Davison algorithm). Consider now separately each of the two sides of peripheral rings and assign to them the corresponding numbers as dictated by the Gordon and Davison procedure. These are the rings with one "predecessor". Consider the assignment of partial K–values for rings which have both lower rings assignment completed, by adding the values of the two predecessor rings. By exhausting the assignment in this step one continue to apply Gordon and Davison algorithm to additional peripheral rings for which this was not possible previously. In

Fig. 3 we have illustrated the process, step by step, on a relatively large polyhex lattice. Observe that at each step we enlarge the sublattice of labeled rings of the lattice initially considered.

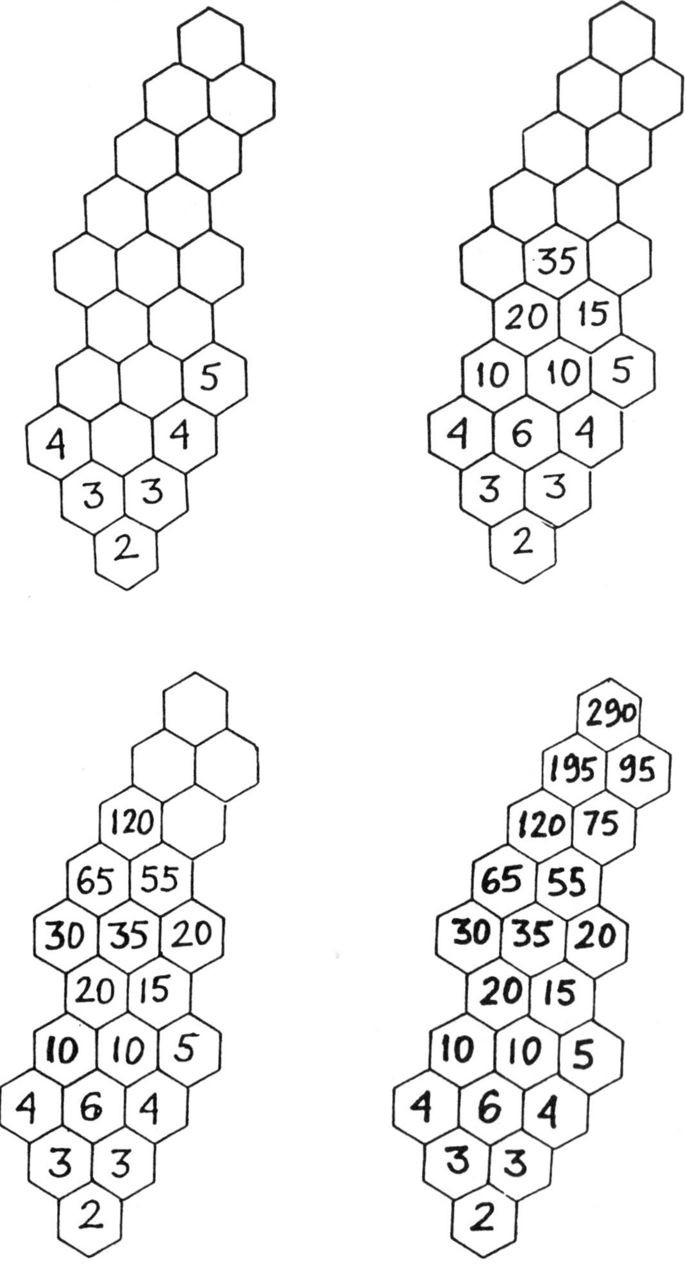

Figure 3

The basis for the outlined algorithm follows from the fact that counting paths from the master to the slave vertex, as has been discussed in the literature (7), is equivalent to counting perfect matching. The "seed" of this observation is already visible in work of Gordon and Davison, but use of count of paths has been fully recognized by a number of more recent investigators.

In concluding the brief outline of the algorithm we would like to point out that the assigned partial K-values hold for any proper sublattice of the original lattice as is illustrated in Fig. 4. By "proper" we mean any sublattice which defines a "catchment" domain for the "slave" ring.

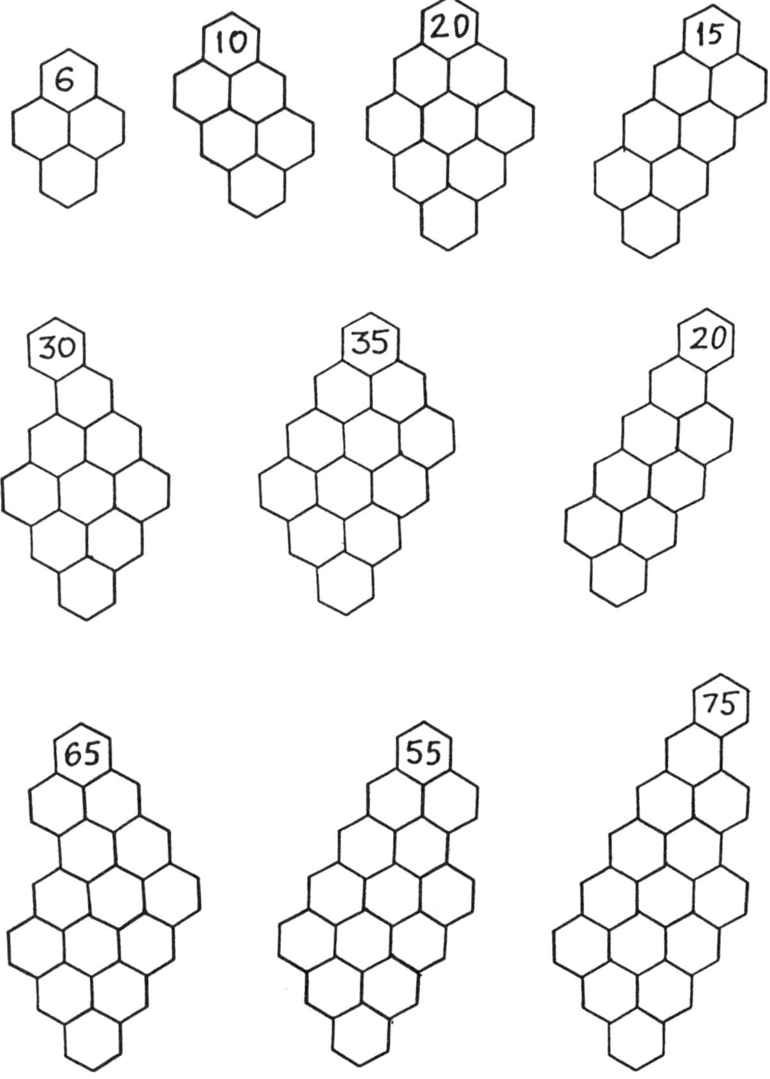

Figure 4

REFERENCES

[1]   M. Gordon and W. H. T. Davison, Theory of Resonance Topology of Fully Aromatic Hydrocarbons. I., *J. Chem. Phys.* 20 (1952), 428-435.

[2]   M. Randić, On the Enumeration of Kekulé Structures in Cata-Condensed Benzenoids, *Chem. Phys. Lett.* (submitted)

[3]   H. Hosoya, How to Design Non-Kekulé Polyhex Graphs ?, *Croat. Chem. Acta,* 59 (1986), 583-590.

[4]   E. Clar, *Aromatic Sextet,* J. Wiley & Sons, New York (1972)

[5]   M. Randić, On Construction of Benzenoids Having a Same Number of Kekulé Valence Structures, in *Valence bond Theory and Chemical Structure* (D. J. Klein and N. Trinajstić, editors), Elsevier Sci. Publ., Amsterdam (in press)

[6]   S. J. Cyvin and I. Gutman, Topological Properties of Benzenoid Systems. Part XXXVI.   Algortihm for the Number of Kekulé Structures in Some Peri-Condensed Benzenoids, *MATCH* (Comm. Math. Chem.), 19 (1986), 229 – 242.

S.J. Cyvin, Extended Application of and Algorithm for the Number of Kekulé Structures, *MATCH* (Comm. Math. Chem.), 22 (1987), 101-103.

[7]   N. Trinajstić and P. Krivka, On the Reduced Graph Model, in *Matheamtics and Computational Concepts in Chemistry*, (N. Trinajstić, ed.)

W. He and W. He, On Kekulé Structure and P-V Path Method, in *Graph Theory and Topology in Chemistry*, (R.B. King and D.H. Rouvray, eds.), Elsevier Sci. Publ., Amsterdam (1987), 476-483.

Drake University, Des Moines, Iowa 50311 and Ames Laboratory, DOE, Iowa State University, Ames, Iowa 50011

Chromatic Roots of Families of Graphs

Ronald C. Read
University of Waterloo

Gordon F. Royle
University of Waterloo

ABSTRACT

It has been known for a long time that when the zeros of the chromatic polynomials of certain families of graphs are plotted in the complex plane, the points display a clearly non–random pattern, seeming to lie on or near to certain curves in the plane.

A recent theorem of Beraha, Kahane and Weiss enables some of this behaviour to be explained. Specifically, the limit points of the chromatic roots for recursive families of graphs can now be recognized, and the curves on which they lie can be identified. We present a number of examples for which the appropriate computations have been carried out.

This is not the whole story, however. In addition to the roots lying near the limit points of zeros, some recursive families of polynomials have isolated roots, which themselves exhibit regular patterns, as yet unexplained. Moreover, the chromatic roots of other, non–recursive, families of graphs, such as cubic graphs, show further kinds of regularity which also call for explanation.

1. Introduction

We shall assume that the reader is familiar with the basic facts and theorems about chromatic polynomials as given in, for example, [6]. In the usual definition of a chromatic polynomial as the number of ways of colouring a graph in a given number of colours, the variable is, of course, an integer; once the polynomial is defined however, the variable can be regarded as a complex number, and this we shall do in the present paper. For this reason we shall use the letter z in place of the traditional λ. We denote the chromatic polynomial of G by $P_G(z)$.

* The research for this paper was supported by grant A8142 from the Natural Sciences and Engineering Research Council, Canada.

The chromatic roots of a graph G are the roots of the polynomial equation

$$P_G(z) = 0.$$

Consider a family of graphs defined by some common property. If the chromatic roots of the graphs in this family are plotted in the complex plane it is usually found that the points representing these roots display a markedly nonrandom pattern. The problem that we address in this paper is that of explaining some of these patterns.

2. Recursive families of graphs

A recursive family of graphs is a sequence of graphs – one for each integer n – such that the Tutte polynomials satisfy a linear recurrence. See [2] or [10] for further details. It follows that the chromatic polynomials for such a family also satisfy a linear recurrence, which will be of the form

$$P_n(z) = \sum_{r=1}^{k} a_r(z)\, P_{n-r}(z).$$

In general, such a recurrence will have a solution of the form

$$P_n(z) = \sum_{i=1}^{k} \alpha_i(z)\, [\lambda_i(z)]^n$$

where the functions $\lambda_i(z)$ are the roots of the auxiliary equation

$$t^k - a_1(z)\, t^{k-1} - a_2(z)\, t^{k-2} - \ldots - a_k(z) = 0.$$

As an example consider the family of graphs called ladders or prisms as illustrated in Figure 1. Let $L_n(z)$ be the chromatic polynomial of the ladder on $2n$ vertices. It was shown in [2] that $L_n(z)$ satisfies a linear recurrence, the solution to which is

$$L_n(z) = (z^2 - 3z + 3)^n + (z-1)(3-z)^n + (z-1)(1-z)^n + (z^2 - 3z + 1).$$

Thus these chromatic polynomials can be easily found and their complex roots determined, provided n is not too large. If the complex roots are plotted in the complex plane we get a distribution something like that of Figure 2, which shows the points for $n \leq 15$. It will be seen that the points appear to lie either on a line perpendicular to the real axis or on a C–shaped curve. Biggs, Damerell and Sands [2] give a similar plot, but at that time there was no proof that the roots did, in fact, approximate to the line or the curve, and no indication of what the C–shaped curve might be.

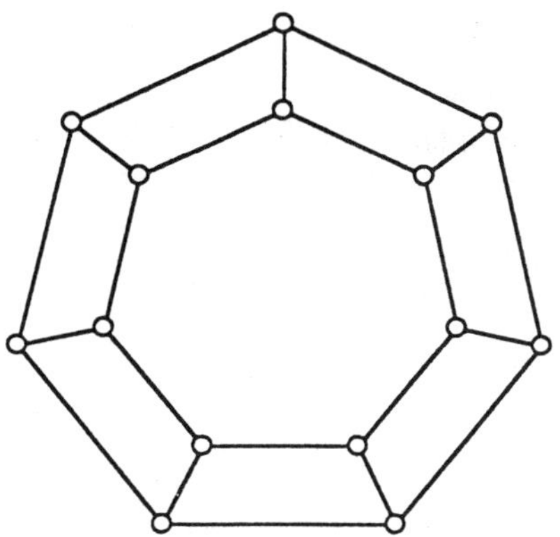

Figure 1

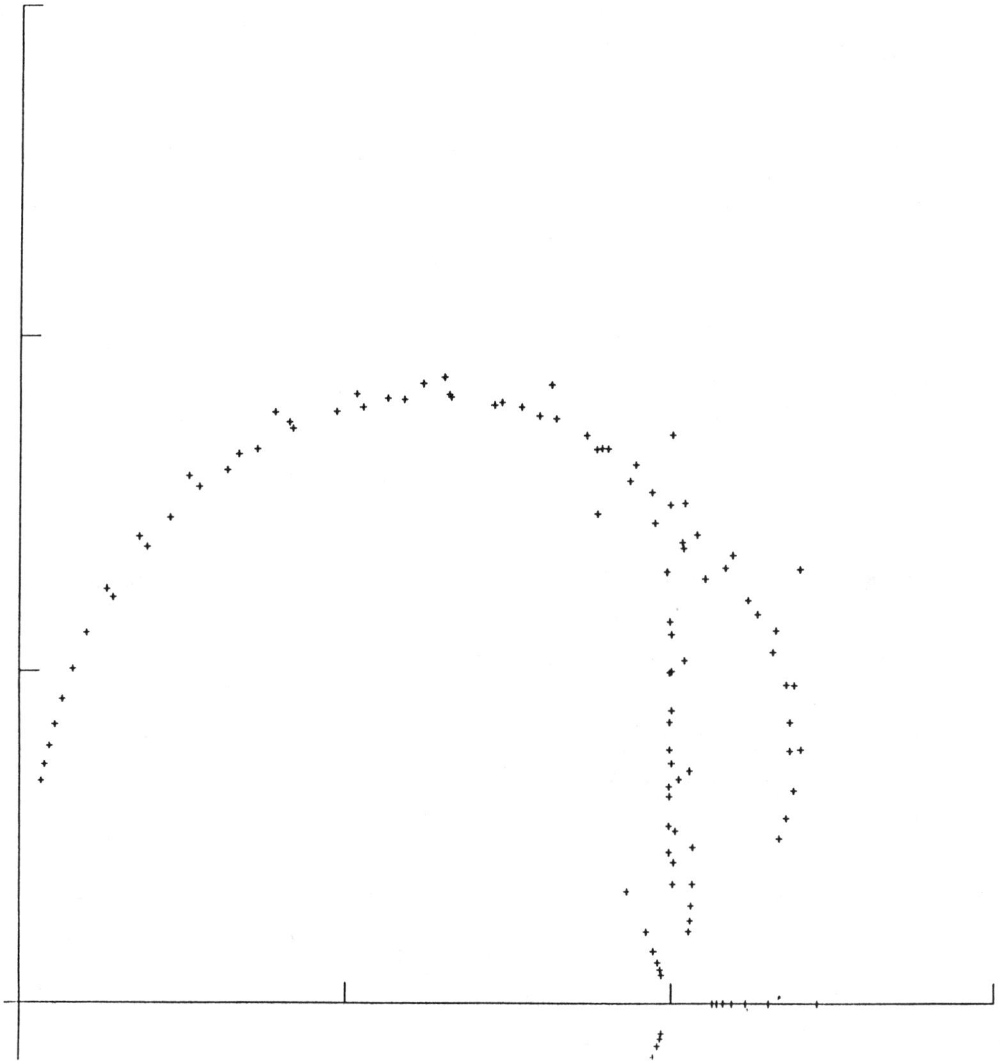

Figure 2

3. Limit points of zeros

Much of the behaviour of the chromatic roots for recursive families of graphs can be explained by a recent result of Beraha, Kahane and Weiss. Suppose we have a family of polynomials $\{P_n(z)\}$ given by

$$P_n(z) = \sum_{i=1}^{k} \alpha_i(z)\,[\lambda_i(z)]^n .$$

A complex number ζ is defined to be a limit point of zeros of this family if there exists a sequence $\{z_i\}$ tending to ζ, such that for each i, z_i is a root of $P_i(z)$. We then have the following theorem.

Theorem   *(Beraha – Kahane – Weiss 1980). Under the non–degeneracy conditions that $\{P_n(z)\}$ does not satisfy a lower order recurrence, and $\lambda_i(z)/\lambda_j(z)$ is not identically a constant of unit modulus for any $i \neq j$, the complex number ζ is a limit point of zeros of $\{P_n(z)\}$ if and only if, at $z = \zeta$, one of the following two conditions holds:*

1.   *One of the roots $\lambda_i(z)$ dominates all the other roots (that is, has larger modulus), and the corresponding $\alpha_i(z) = 0$.*

or

2.   *Two or more of the roots $\lambda_i(z)$ are of equal modulus, and dominate the others.*

For the proof of this theorem, and more information, see [1].

Note that 1. will give isolated points, and that these points will depend on the functions $\alpha_i(z)$, while 2. will give curves of points, and that these curves will be independent of the coefficients $\alpha_i(z)$. Thus these curves will depend only on the recurrence for the polynomials $P_n(z)$ and not on the initial conditions; this means that they depend only on the coefficients in the auxiliary equation. In this paper we shall be interested only in the curves of limit points of zeros and therefore confine our attentions to condition 2.

4.   Applications to ladders

For the case of ladders we have the following α_i and λ_i.

$$\alpha_1(z) = 1 \qquad\qquad \lambda_1(z) = z^2 - 3z + 3$$
$$\alpha_2(z) = z - 1 \qquad\qquad \lambda_2(z) = 3 - z$$

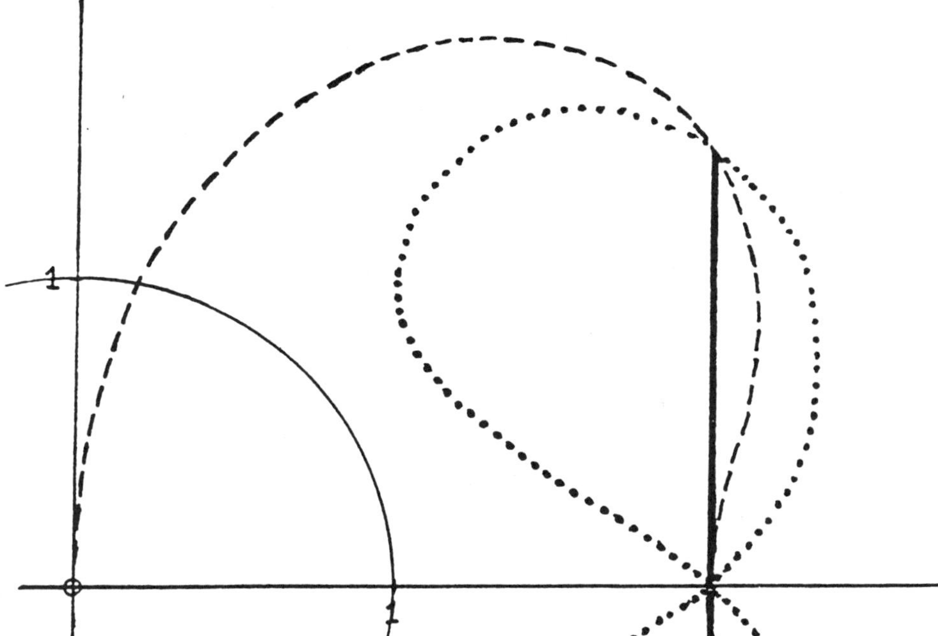

Figure 3

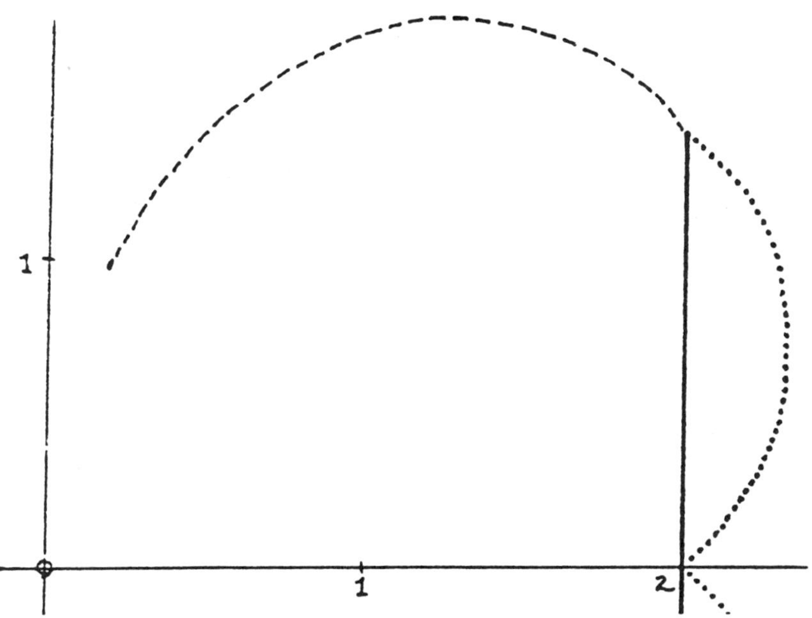

Figure 4

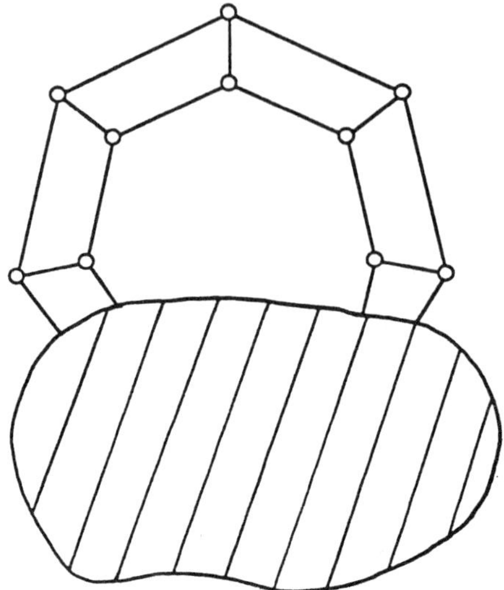

Figure 5

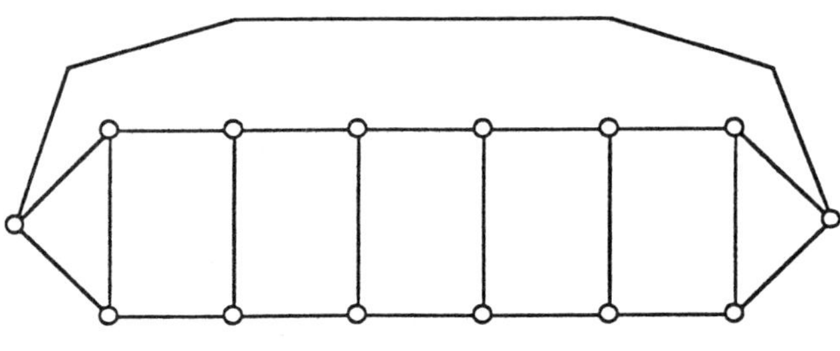

Figure 6

$$\alpha_3(z) = z - 1 \qquad\qquad \lambda_3(z) = 1 - z$$
$$\alpha_4(z) = z^2 - 3z + 1 \qquad\qquad \lambda_4(z) = 1.$$

It is easy to verify that, applying the Beraha–Kahane–Weiss theorem we get the following three portions of curves:

(i) $|\lambda_2| = |\lambda_3|$ gives the line $\Re(z) = 2$ between the points $2 - i\sqrt{2}$ and $2 + i\sqrt{2}$.

(ii) $|\lambda_1| = |\lambda_2|$, that is, $|z^2 - 3z + 3| = |3 - z|$ gives the quartic curve whose equation is

$$y^4 + 2(x^2 - 3x + 1)y^2 + (x^4 - 6x^3 + 14x^2 - 12x) = 0$$

where in order to satisfy the requirement for domination the points must be outside the unit circle and to the left of the line $\Re(z) = 2$

(iii) $|\lambda_1| = |\lambda_3|$ gives the quartic curve with equation

$$y^4 + 2(x^2 - 3x + 1)y^2 + (x^4 - 6x^3 + 14x^2 - 16x + 8) = 0$$

to the right of $\Re(z) = 2$.

These curves are shown in full in Figure 3, together with the portion of the straight line. Figure 4 shows only those parts of the curves which satisfy the requirements of the theorem, and it will be seen that there is excellent agreement between these portions of curves and the positions of the roots shown in Figure 2.

This problem of explaining the behaviour of the chromatic roots of ladders was briefly discussed by one of us as part of a talk given at the combinatorial conference in Barbados in January of this year [8]. We now report on more recent developments.

5. Some observations

We have already noted that the curves of limit points of zeros depend only on the recurrence satisfied by the family of graphs, and not on the coefficients $\alpha_i(z)$ in the solution. Now the recurrence for ladders can be obtained by means of operations on the graph – deletion of edges and identification of vertices – applied to a small portion of the graph involving just two or three "rungs". It follows from these remarks that if we consider a family of graphs such as is shown in Figure 5, where the shaded portion denotes an arbitrary graph and the successive members of the family are obtained by increasing the number of rungs in the ladder–like portion of the graph, then the chromatic polynomials of the graphs in this family will satisfy the same recurrence as do ladders.

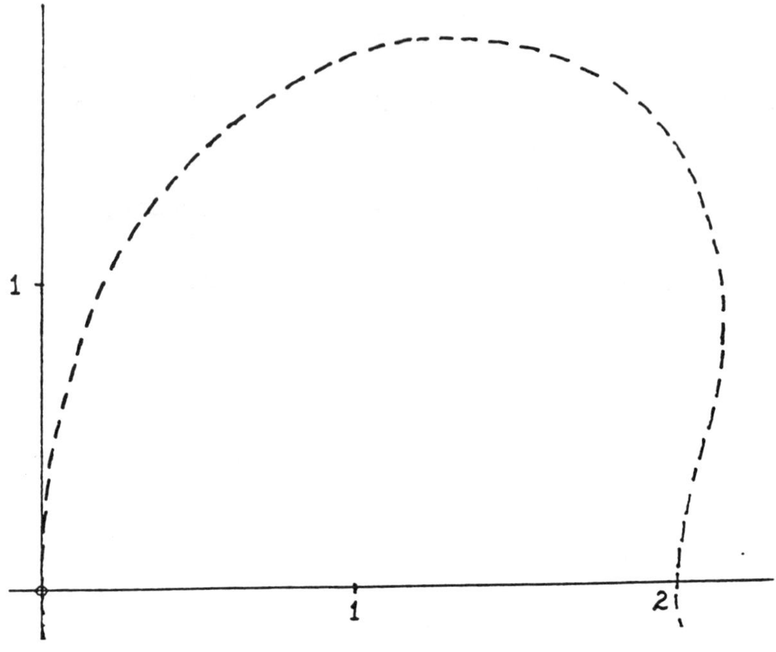

Figure 7

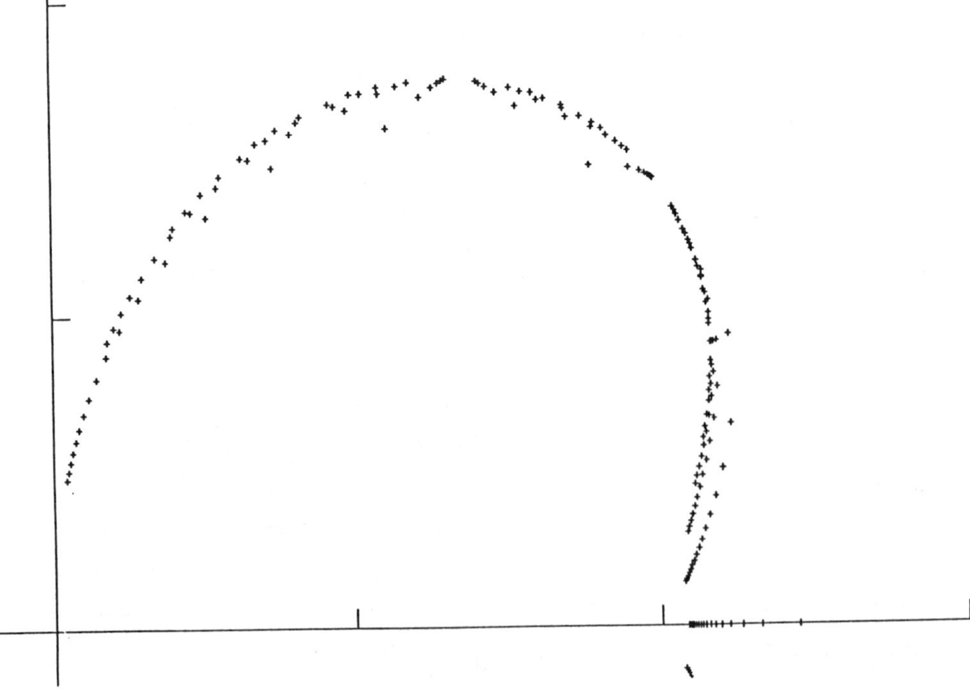

Figure 8

In particular, the Möbius ladders obtained from an ordinary ladder by crossing the two sides in between a pair of rungs, satisfy the same recurrence, and hence have the same curves of the limit points of zeros.

Note however that it is possible for such a family to satisfy a *simpler* recurrence. As an example of this consider the family of graphs of which a typical member is that shown in Figure 6, graphs which we call "rope ladders". It can be shown that these satisfy a second order recurrence and that the polynomial for the rope ladder on 2n vertices is given by

$$R_{2n}(z) = (z-1)^2(z-2)^2(z^2 - 3z + 3)^{n-2} - 2(z-1)(z-2)(3-z)^{n-2}$$

In this case we have only two roots to the auxiliary equation. They are $\lambda_1(z)$ and $\lambda_2(z)$ in the notation used above. Thus there is only one curve, given by $|\lambda_1| = |\lambda_2|$, and since there are no other roots to dominate, every point of this curve is a limit point of zeros. This curve is shown in Figure 7, and a plot of the chromatic roots of the first few graphs of this family is shown in Figure 8.

6.   Problems

It might seem that the Beraha–Kahane–Weiss theorem tells us the whole story of what happens to the chromatic roots of a family of graphs as the number of vertices increases. Unfortunately, this is far from being the case.

Consider the family of bipyramids – the graphs obtained by joining each of two non–adjacent vertices to all the vertices of a cycle, as shown in Figure 9. It is easy to show that the chromatic polynomial of the bipyramid on $n + 2$ vertices is given by

$$z(z-1)(z-3)^n + z(z-2)^n + (-1)^n z(z^2 - 3z + 3)$$

thus we have $\lambda_1 = z - 3$; $\lambda_2 = z - 2$ and $\lambda_3 = -1$. Applying the Beraha–Kahane–Weiss theorem we see that

$$|\lambda_1| = |\lambda_2|$$

gives a portion of the line $\Re(z) = 2.5$.

$$|\lambda_1| = |\lambda_3|$$

gives an arc of the circle centre 2 radius 1.

$$|\lambda_2| = |\lambda_3|$$

gives an arc of the circle centre 3 radius 1.

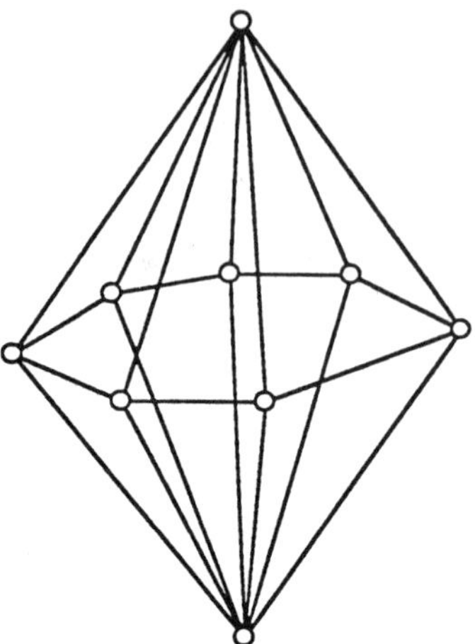

Figure 9

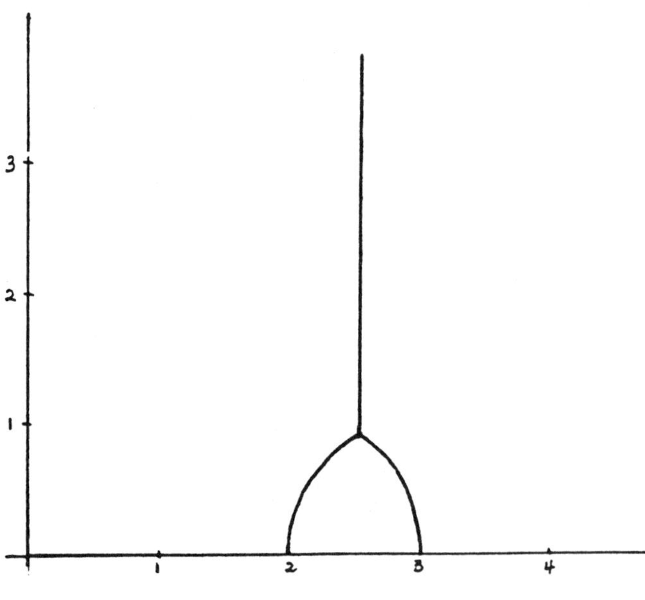

Figure 10

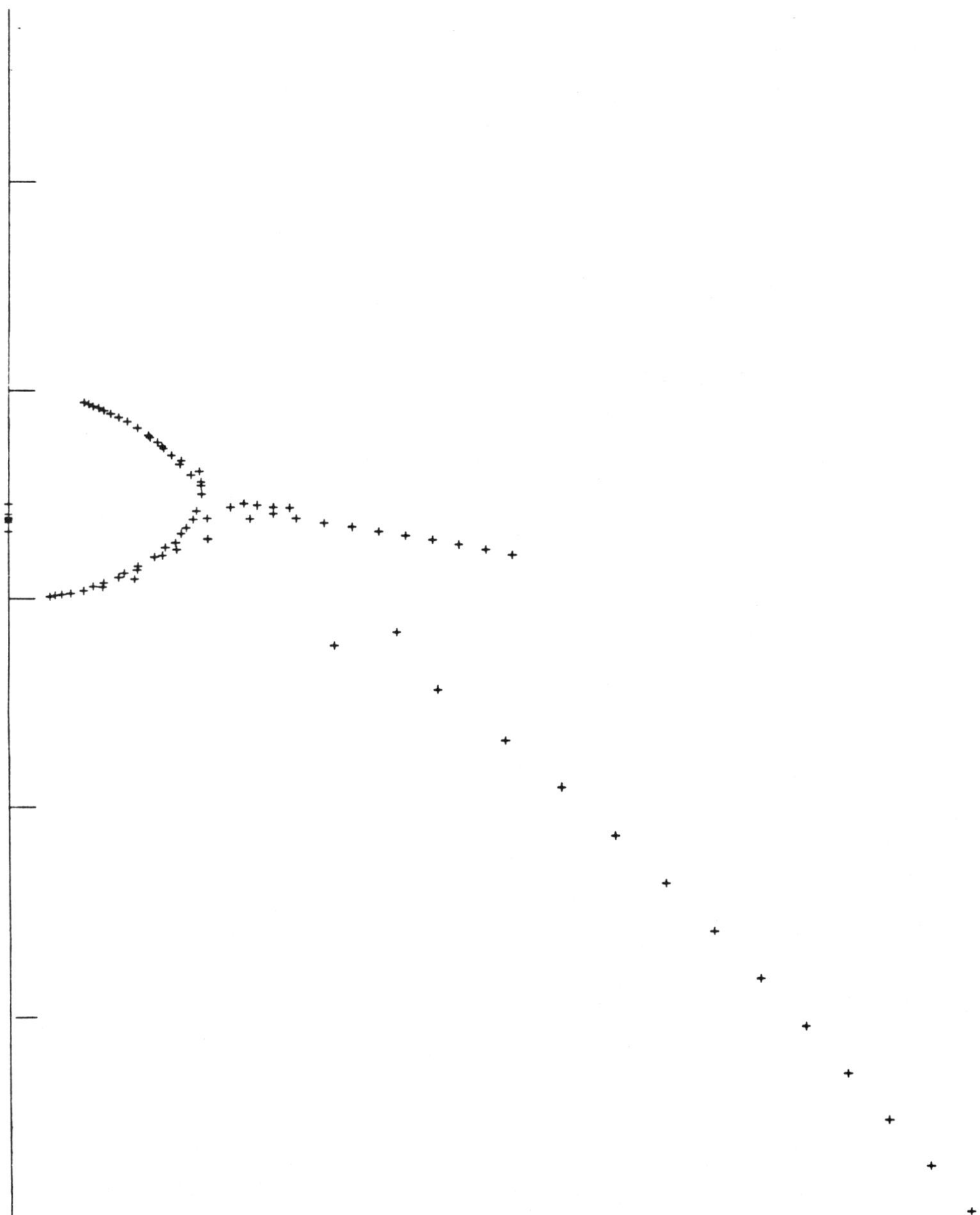

Figure 11

The relevant parts of this line and the circles are shown in Figure 10. Figure 11 shows the actual locations of the chromatic roots of the first few members of the family. It will be seen that there is excellent agreement with the two arcs of the circles, to which a large number of the roots approximate very closely. However, the remaining roots seem to arrange themselves in a number of "sprays", three of which can be seen in Figure 11. Presumably, if we could extend the calculation to very large graphs of this family, there would be further sprays getting closer and closer to the theoretical line of limit points of zero. Thus the theoretical result is not contradicted by the plot of the roots, but a new problem is introduced, namely that of explaining the behaviour of these sprays of roots. They appear to define smooth curves, and it would be of interest to know what these curves are; but up to now we have no answers to questions of this kind.

Thus even for recursive families of graphs there are still some open problems.

7.   Non–recursive families

In this section we consider a somewhat different aspect of the problem of the location of chromatic roots. We concentrate on complete classes of graphs on a fixed number of vertices, rather than infinite families of directly related graphs. Unfortunately, when it comes to chromatic roots of non–recursive families of graphs there is, as yet, no theory at all to guide us. We therefore merely draw attention to some interesting cases that we have observed.

A natural class of graphs to examine is that of regular graphs. As the problem is trivial for graphs of valency 1 and 2, we shall start by considering cubic graphs. A complete catalogue of all cubic graphs up to 20 vertices has been prepared [4], and the following table gives their numbers.

Number of vertices	4	6	8	10	12	14	16	18	20
Number of vertices	1	2	5	19	85	509	4060	41301	510489

The algorithm described in [7] was used to compute the chromatic polynomials (in the tree basis) of all the cubic graphs on up to 16 vertices, and a standard library routine used to find the roots of these polynomials. The results obtained when these roots were plotted on the complex plane are shown in Figures 12, 13 and 14 for 12, 14 and 16 vertices respectively. Once again the roots clearly display a non–random pattern, tending to cluster about $(n-4)/2$ points in the complex plane arranged roughly equidistantly on an ellipse–like curve in the upper right quadrant (here we are ignoring the complex conjugates of the roots shown and the real roots). Moreover, the clusters for the 14 vertex cubic graphs start slightly to the left of those for the 12 vertex cubic graphs and interleave with them; similarly the 16 vertex clusters interleave with the 14 vertex clusters.

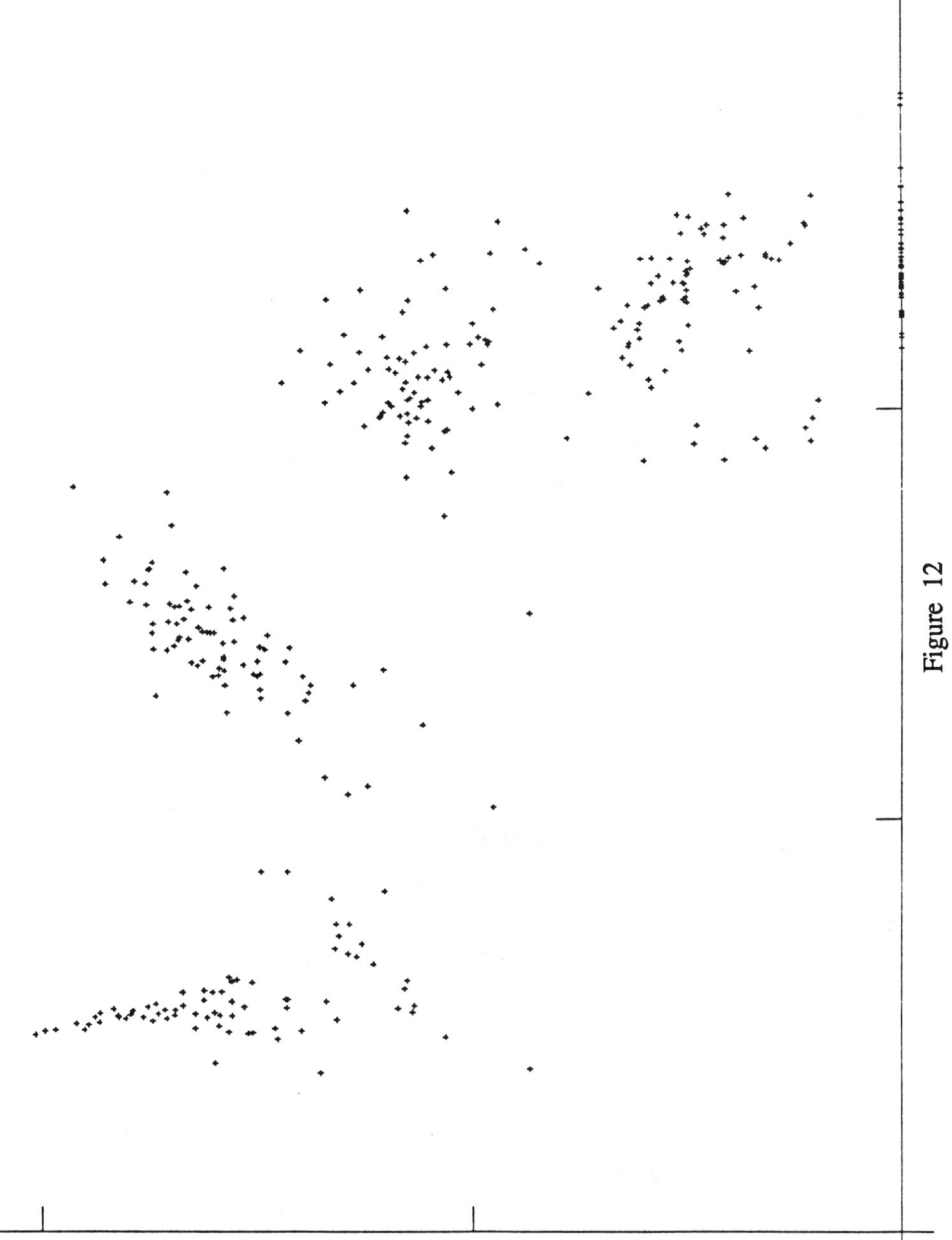

Figure 12

1021

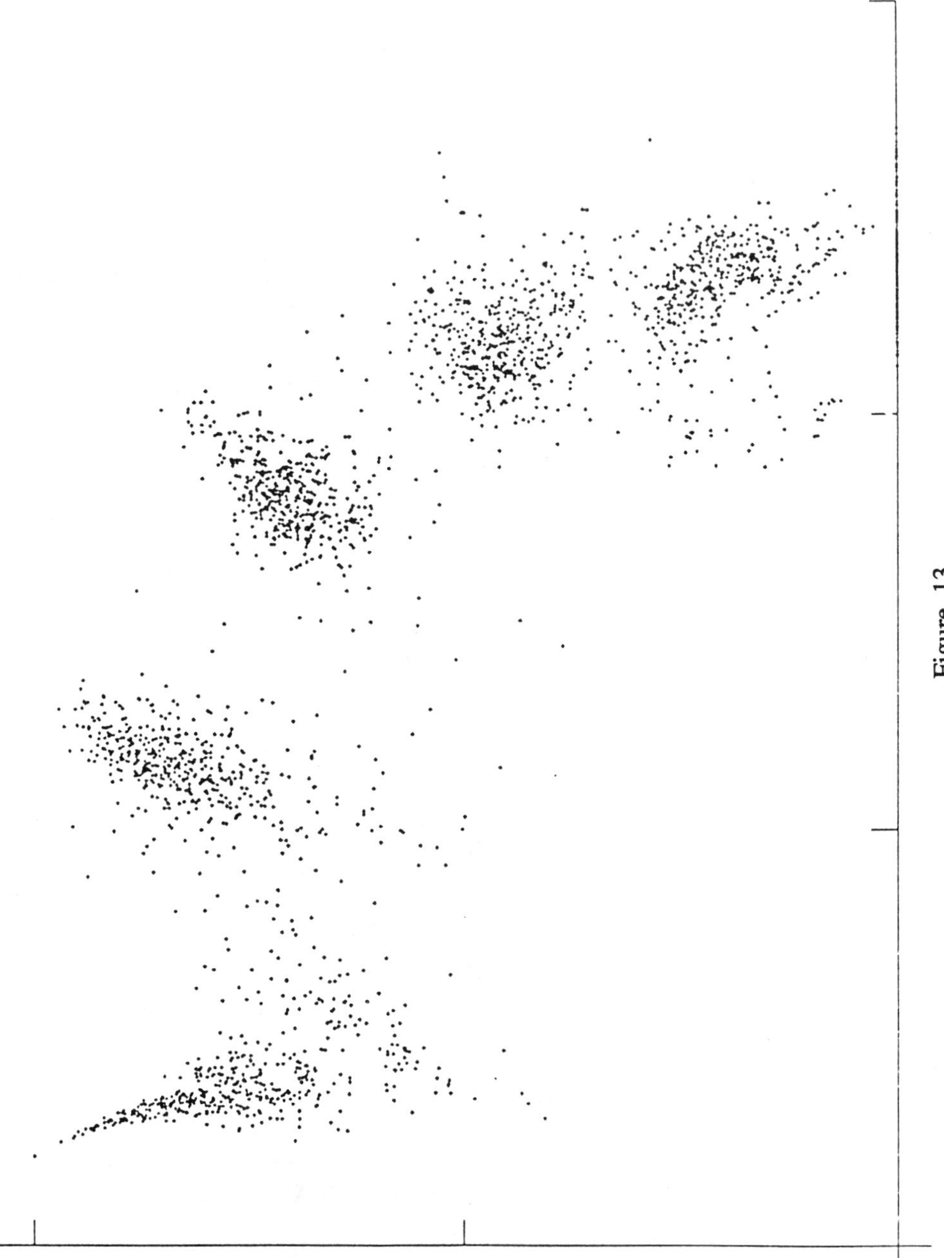

Figure 13

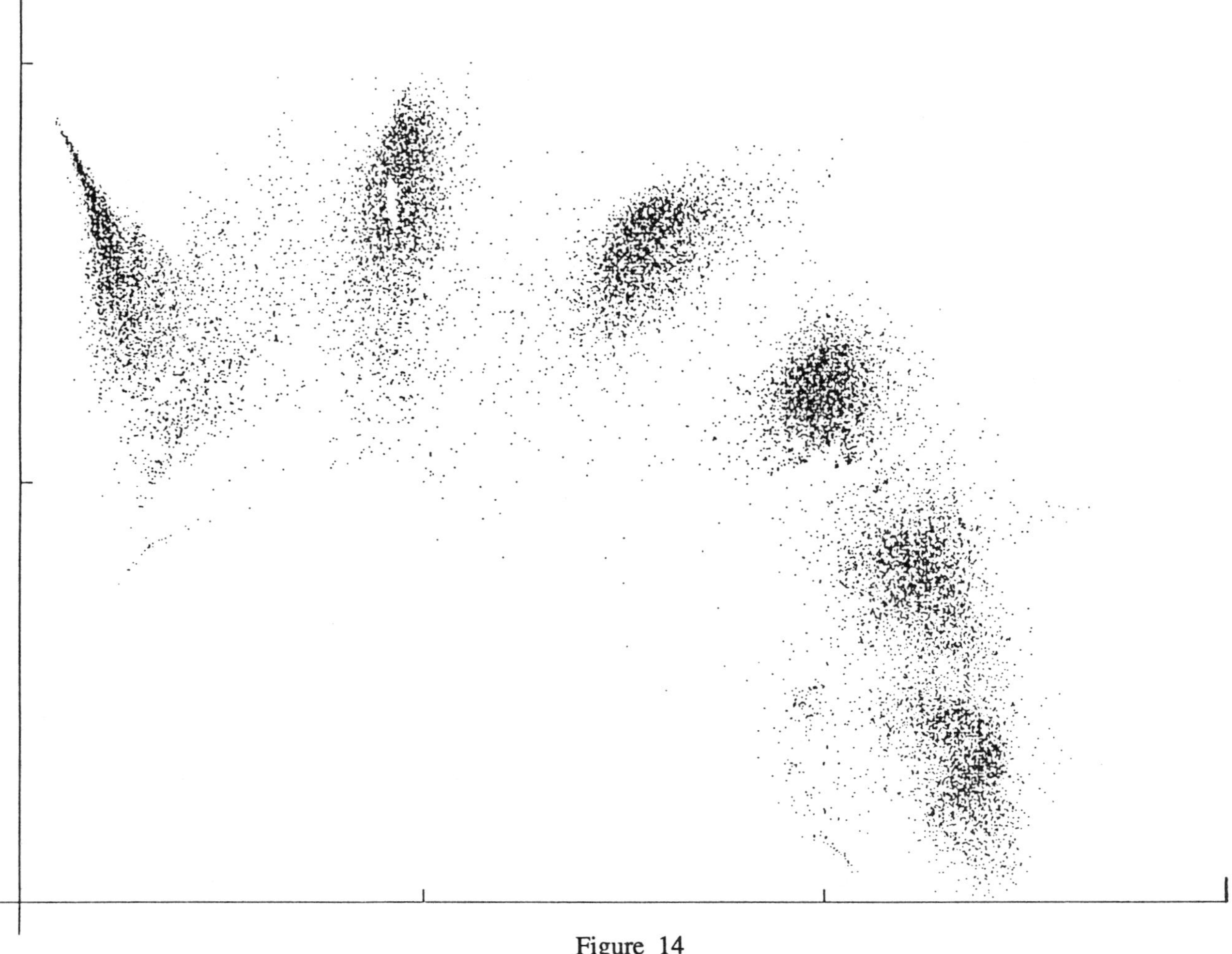

Figure 14

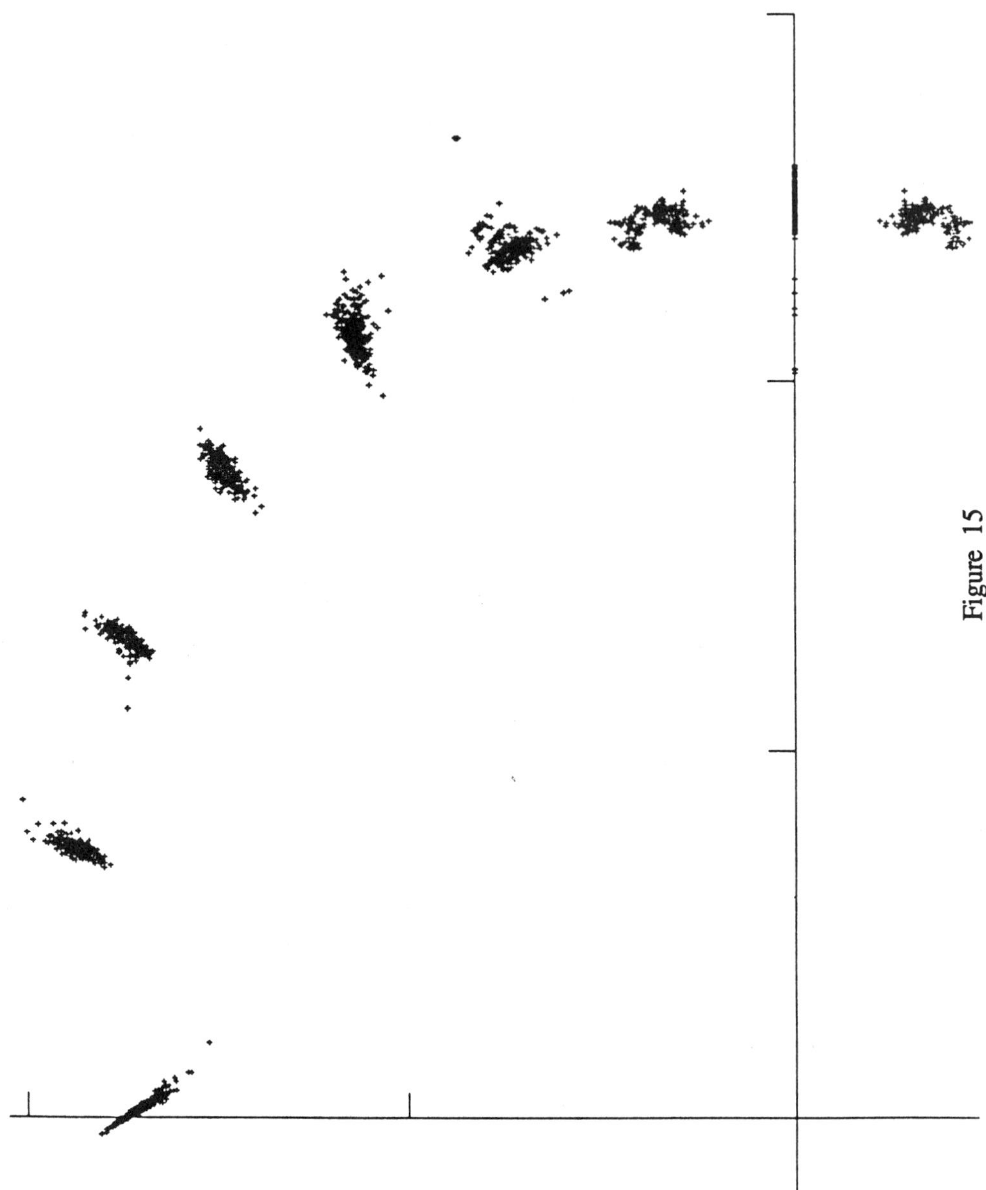

Figure 15

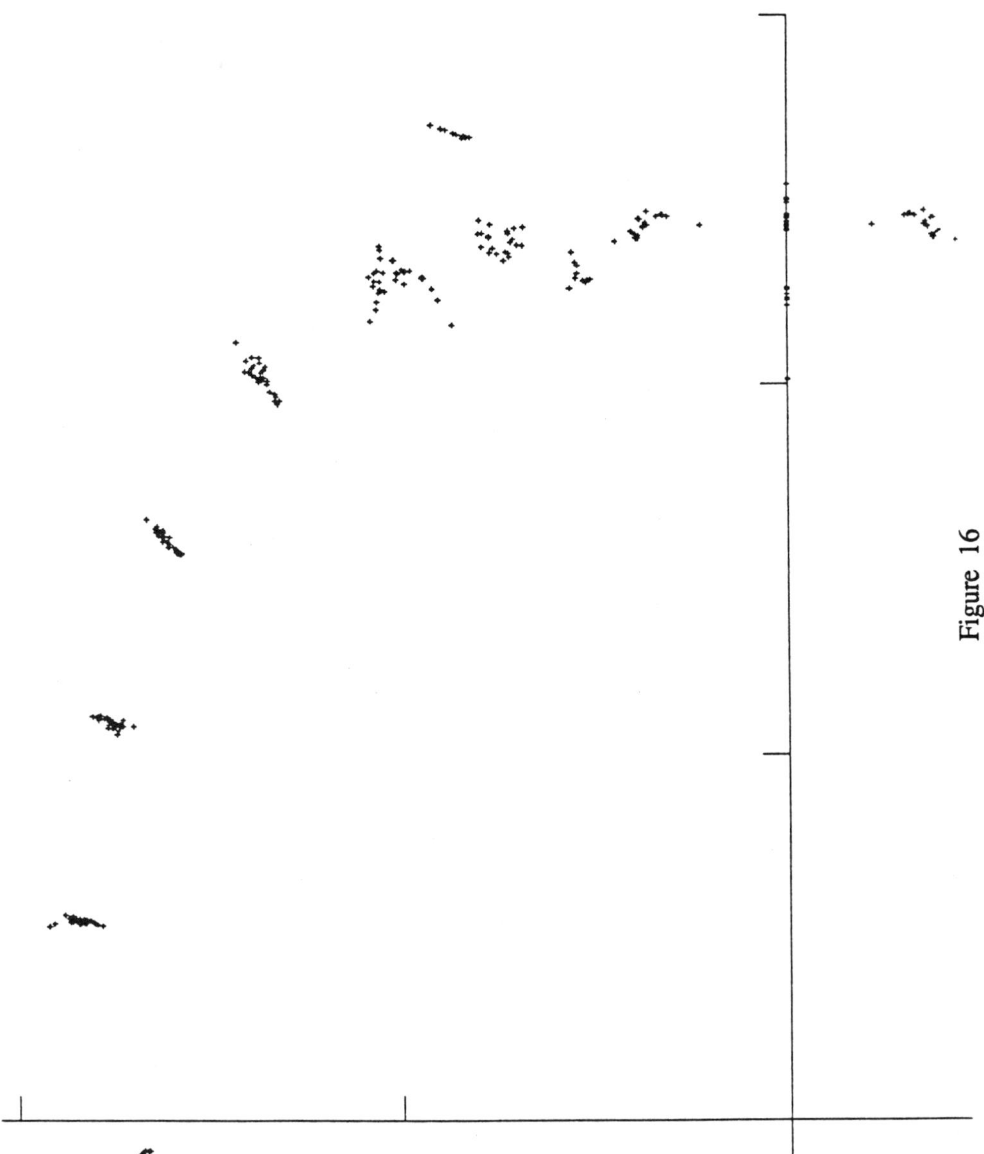

Figure 16

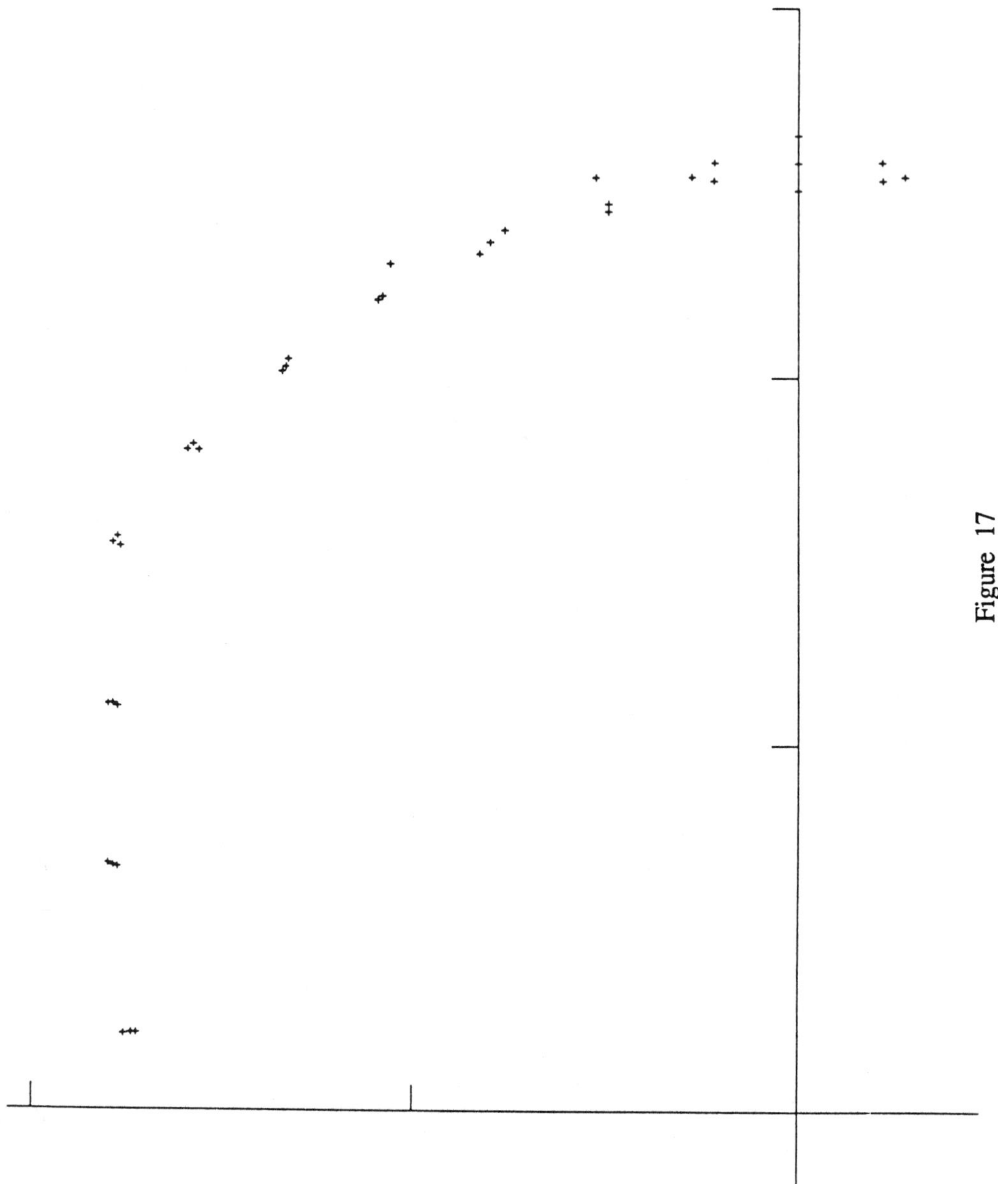

Figure 17

8. Two conjectures

We now describe two conjectures that have apparently been widely believed to hold; see [2], [3], [10].

Conjecture 1   For cubic graphs, the chromatic roots z all have modulus $|z| \leq 3$.

In fact this is a special case of the more general conjecture

Conjecture 1'   There is a function $B: R \to R$, such that the chromatic roots z of k–regular graphs all have modulus $|z| \leq B(k)$.

Thus Conjecture 1 states that $B(3) = 3$.

Our results clearly support this conjecture, and as we have tested a large number of graphs, and (more importantly) have checked *every* graph of the classes in question, we regard this as strong evidence in favour of this conjecture. The second conjecture stems from the fact that all previously known chromatic roots have positive real parts.

Conjecture 2   For all chromatic roots z, $\Re(z) \geq 0$.

Below we demonstrate that this conjecture is false. In [10] the dodecahedron is presented as a counterexample to Conjecture 2. However the chromatic polynomial is given incorrectly there, and although the conjecture *is* false, the dodecahedron is not a counterexample.

As remarked above, the leftmost cluster of roots is closer to the imaginary axis for the 16 vertex cubic graphs than the 14 vertex cubic graphs. We wish to know the behaviour of this cluster as the number of vertices is increased still further, to determine whether it actually crosses the imaginary axis or, as Conjecture 2 suggests, whether it 'squeezes–up' against the imaginary axis.

However, it is not possible to test all the cubic graphs on 18 and 20 vertices because computing chromatic polynomials is time–consuming even for graphs of this size. Therefore we wish to determine which graphs are contributing the roots of smallest real part, and what property distinguishes them. Examination of the extreme graphs on 14 and 16 vertices leads us to the conclusion that the important factor is the girth of the graph; in particular graphs with high girth contribute the roots of smallest real part. It is not surprising that girth is a significant factor because the leading coefficients of the chromatic polynomial take their maximum absolute values when the girth is high. In fact if

$$P_G(z) = \sum_{i=0}^{n} (-1)^i a_{n-i} \lambda_{n-i}$$

and q is the number of edges of G, then $a_{n-s} \leq \binom{q}{s}$. Furthermore, $a_{n-s} = \binom{q}{s}$ if the girth of the graph is greater than $s + 1$, and $a_{n-s} = \binom{q}{s} - p$, if the girth is precisely $s + 1$, where p is the number of circuits of length $s + 1$ [5]. Thus one would expect such graphs to be extreme in some sense.

This is useful because it is simple to extract the cubic graphs of high girth from the catalogues on 18 and 20 vertices, and in fact further catalogues of all the maximal girth cubic graphs on up to 30 vertices have been produced [9].

Figure 15 shows the chromatic roots of the high girth (≥ 5) cubic graphs on 18 vertices demonstrating the smallest known counterexample to Conjecture 2, Figure 16 shows the chromatic roots of the cubic graphs of girth ≥ 6 on 20 vertices and Figure 17 shows the chromatic roots of the cubic graphs of girth ≥ 7 on 26 vertices. Despite the small number of graphs the pattern is clear, and there are roots with quite large negative real parts.

Once again, further examination is halted by the prohibitive time required to directly compute the chromatic polynomials of even a small number of graphs with a large number of vertices. One immediate question raised by this is determining just how large a negative real part a chromatic root can have. To this end it would be particularly interesting to examine the behaviour of the *cages* – cubic graphs with a minimum number of vertices for a given girth – if their chromatic polynomials could be found.

REFERENCES

[1]   S. Beraha, J. Kahane and N. J. Weiss, Limits of chromatic zeros of some families of maps. *J.Comb. Theory Ser. B.* **28** (1980) 52–65.

[2]   N. L. Biggs, R. M. Damerell and D. A. Sands, Recursive families of graphs. *J.Comb. Theory Ser. B.* **12** (1972) 123-131.

[3]   E. J. Farrell, Chromatic roots – some observations and conjectures. *Discrete Math.* **29** (1980) 161–167.

[4]   B. D. McKay and G. F. Royle, Constructing the cubic graphs on up to 20 vertices. *Ars Combinatoria* **21–A** 129–140.

[5]   G. H. J. Meredith, Coefficients of chromatic polynomials. *J. Comb. Theory Ser. B,* **13** (1972) 14-17.

[6]   R. C. Read, Introduction to chromatic polynomials. *J. Comb. Theory* **4** (1968) 52-71.

[7] R. C. Read, An improved method for computing the chromatic polynomials of sparse graphs. *Research Report* CORR 87-20, C&O Dept., U.of Waterloo (1987).

[8] R. C. Read, Recent Advances in Chromatic Polynomial Theory. *Proceedings of the 5th Caribbean Conference on Combinatorics and Computing.* Barbados, 1988/ To appear.

[9] G. F. Royle, Constructive enumeration of graphs. PhD Thesis. University of Western Australia (1987).

[10] D. A. Sands, Dichromatic polynomials of linear graphs. PhD Thesis. Royal Holloway College, London (1972).

From Garbage to Rainbows: Generalizations of Graph Coloring and Their Applications

Fred S. Roberts

Rutgers University

ABSTRACT

We describe a variety of applications of ordinary graph coloring and some generalizations and variations of ordinary graph coloring which are motivated by these applications. The generalizations and variations discussed are called T–colorings, H–colorings, I–colorings, J–colorings, n–tuple colorings, consecutive colorings, set colorings in general, and list colorings. We then discuss combinations of some of these variants, describing set T–colorings, n–tuple T–colorings, list T–colorings, and set list T–colorings.

1. Introduction

The basic message of this paper is that graph coloring has many applications and that these applications have often led to very interesting generalizations of ordinary graph coloring. The paper starts with a grab–bag of applications of ordinary graph coloring. It then considers a variety of generalizations of ordinary graph coloring which are motivated by these applications. For each generalization, some recent results are mentioned and usually an open problem.

The generalizations to be discussed in this paper are called T–colorings, H-colorings, I–colorings, J–colorings, n–tuple colorings, consecutive colorings, set colorings in general, and list colorings. Some of these ideas are then combined with the first one, to obtain set T–colorings, n–tuple T–colorings, list T–colorings, and set list T–colorings.

Recall that if $G = (V, E)$ is a graph, a *coloring* of G is an assignment of a color $F(x)$ to each vertex x so that for all $x \neq y$ in V,

$$\{x, y\} \in E \rightarrow F(x) \neq F(y).$$

The *chromatic number* of G, $\chi(G)$, is the smallest k so that G has a coloring using k colors. We begin in the next section with a collection of applications of graph coloring.

2. Applications of Graph Coloring

Graph coloring has a long history, and from the beginning it has been closely tied to one well-known application, namely map coloring. Modern applications of graph coloring are made to many important applied problems. We mention some of these problems here.

Scheduling Meetings of Committees in a State Legislature

Graph coloring has been used in the New York State Assembly (see Bodin and Friedman [1971]). In this application, we wish to assign a meeting time to each legislative committee. If two committees have a member in common, they must get a different meeting time. We build a schedule graph whose vertices are the committees and which has an edge between x and y if and only if x and y have a member in common. Then the colors are the meeting times. The chromatic number gives the smallest number of meeting times required.

Channel Assignment Problem

In radio and television applications, we sometimes wish to assign a channel to each t.v. or radio transmitter. Some transmitters interfere. In the simplest version of this problem, they interfere if and only if they are within a certain distance of each other. However, interference can take place for reasons other than proximity. Transmitters which interfere must get different channels. We build an interference graph whose vertices are the transmitters and which has an edge between x and y if and only if x and y interfere. Then the colors are the channels. We often wish to use as little of the spectrum as possible. Thus, we are interested in the chromatic number of the interference graph. This problem, with various complications, has been studied at the Federal Communications Commission, AT&T Bell Laboratories, the National Telecommunications and Information Administration, the Department of Defense, and elsewhere. The idea of studying channel assignment graph–theoretically goes back to Metzger [1970] and Zoellner and Beall [1977], and the formulation of the channel assignment problem we shall study below is due to Hale [1980].

Garbage Collection

A *tour* of a garbage truck is a schedule of sites it visits on a given day. If two tours visit the same site, they must be run on different days. We wish to assign to each tour a day of the week on which it will be scheduled so that if two tours visit a given site, they get a different day. We build a tour graph with vertices the set of tours and with an edge between x and y if and only if tours x and y visit a common site. Then the colors are the days. If we wish to use no more than the 6 days of the week (excluding Sunday) to schedule garbage collection, then we are interested in determining whether or not the tour graph is 6–colorable. This problem arises as part of a larger garbage truck scheduling problem posed by the New York City Department of Sanitation and solved heuristically by Beltrami and Bodin [1973] and by Tucker [1973]. In the heuristic solution, it is necessary to solve the tour graph coloring problem over and over again.

Mobile Radio Frequency Assignment

We wish to assign frequencies to mobile radio telephones in vehicles. Each vehicle in a given region is to receive a frequency set over which it can transmit. Some regions interfere because of reasons of proximity, meteorological reasons, etc. Interfering regions must get different frequency sets. We build an interference graph by taking as vertices the regions and including an edge between x and y if and only if x and y interfere. Then the colors are the frequency sets. If we wish to minimize the number of frequency sets, we seek the chromatic number of the interference graph. This problem was introduced by Gilbert [1972] at Bell Labs and has been studied by a wide variety of others.

Fleet Maintenance

A maintenance facility awaits vehicles (trucks, planes, ships, etc.). Each vehicle has a time period during which it will be in the facility for regularly scheduled maintenance. We want to be sure that we can assign a space in the facility to each vehicle. If two vehicles are in the facility in overlapping time periods, they must get different spaces. We build an overlap graph whose vertices are the vehicles and which has an edge between x and y if and only if x and y are in the facility in overlapping time periods. Then the colors are the spaces. If we wish to minimize the total space in the facility, we want the chromatic number of the overlap graph. This problem has been studied for ship maintenance (with linear facilities or docks) by Alan Hoffman and Ellis Johnson at IBM (see Golumbic [1980] and Opsut and Roberts [1981]).

Traffic Phasing

There is a stream of requests for use of a facility (classroom, computer, traffic intersection). Each use is to be assigned a period of time, a green or "up" time, during which it may use the facility. Certain uses are incompatible with each other. Then we wish to assign green times so that incompatible uses receive different green times. We build an incompatibility graph whose vertices are the traffic streams and which has an edge between x and y if and only if x and y are incompatible. Then the colors are the green times. If we wish to minimize the total amount of time used, we seek the chromatic number of the incompatibility graph. This problem was introduced by Stoffers [1968] in the study of traffic lights and in the more general interpretation of traffic phasing by Opsut and Roberts [1981].

Task Assignment

A large task has been divided into subtasks. Some subtasks are incompatible because they require the same resources, tools, space, workers, etc. The problem is to schedule the subtasks so that incompatible subtasks get different time periods. We build an incompatibility graph whose vertices are the subtasks and which has an edge between x and y if and only if x and y are incompatible. Then the colors are the time periods. If we seek to minimize the total amount of time taken in all of the subtasks, we seek the chromatic number of the incompatibility graph. (This problem is a common one in the scheduling literature. See for instance Opsut and Roberts [1981].)

There are many other applications of graph coloring. For a discussion of some of these, and an elaboration of some of those above, see Golumbic [1980], Roberts [1976, 1978, 1984], or Bondy and Murty [1976].

Most of the problems posed above have various complications which imply that ordinary graph coloring is a simplification. These complications have given rise to interesting generalizations of graph coloring, which we describe in this paper. Before discussing these generalizations, we need two graph–theoretic preliminaries, which we introduce in the next section.

3. Intersection Graphs

We say that a graph G is an *intersection graph* of certain kinds of sets if there is an assignment of a set $S(x)$ of the given kind to each $x \in V(G)$ so that for all $x \neq y$,

$$\{x, y\} \in E(G) \leftrightarrow S(x) \cap S(y) \neq \phi.$$

The function S is called an *intersection assignment* for G. G is an *interval graph* if each set $S(x)$ is a real interval. For a general introduction to the theory of intersection graphs, and in particular of interval graphs, the reader is referred to Roberts [1976, 1978] or Golumbic [1980] or Fishburn [1985]. Intersection graphs in general, and interval graphs in particular, play an important role in what follows.

4. T–Colorings

The first generalization of graph coloring which we study arises from the channel assignment problem. Let us think of channels as positive integers. One complication in channel assignment is that certain separations between interfering transmitters are disallowed. Let T be a set of nonnegative integers representing disallowed separations between interfering transmitters. Assume $0 \in T$. We want to find a function $F: V(G) \to Z^+$ such that for all $x \neq y$ in $V(G)$,

$$\{x, y\} \in E \to |F(x) - F(y)| \notin T.$$

Such an F is called a *T–coloring* of G. For instance, suppose $T = \{0\}$. Then a T-coloring is an ordinary coloring. To give another example, suppose $T = \{0, 1\}$. Then, interference implies that channels are different and nonadjacent. The sets T can get rather complicated. For instance, in UHF television, sets such as $T = \{0, 7, 14, 15\}$ arise.

The formulation of the channel assignment problem as a T–coloring problem is due to Hale [1980]. Hale points out that there are several criteria for efficiency of a T-coloring. Let us define the *order* of a T–coloring F as the number of different colors $F(x)$ and the *span* of F as the maximum of $|F(x) - F(y)|$ over all x, y. Then we might wish to minimize the order or the span. The *T–chromatic number* of G, denoted $\chi_T(G)$, is defined to be the minimum order of a T–coloring of G, while the *T–span* of G, denoted $\mathrm{sp}_T(G)$, is defined to be the minimum span of a T–coloring of G. The difference between these two concepts can be illustrated by considering the complete graph $G = K_3$ and the set $T = \{0, 1, 4, 5\}$. Suppose we color one vertex of this graph with color (channel) 1. If we try to color a second vertex with the lowest possible channel, we are forced to use channel 3. But then the lowest possible channel available for the third vertex is channel 9. We have a T–coloring of G with order 3 and span 8. Using the colors 1, 4, and 7 gives us the same order, but a smaller span, 6. There are examples of graphs where it is not possible to optimize on both criteria at once. For instance, consider the 5–cycle C_5 with the set $T = \{0, 1, 4, 5\}$. In this graph, a minimum span assignment uses successively around the cycle the colors 1, 4, 2, 5, and 3. However, a minimum order assignment uses successively the colors 1, 4, 1, 4, 7.

Roberts [1986] is a recent survey paper on the theoretical work on T–colorings. Here, we summarize a number of the important results.

Note that if $T = \{0\}$, then $\chi_T = \chi$ and $sp_T = \chi - 1$. Therefore, the problems of computing χ_T and sp_T are NP–complete. In general, as Cozzens and Roberts [1982] observe, $\chi_T = \chi$, so sp_T is the interesting new number.

Theorem 1 *(Cozzens and Roberts 1982) If $\omega(G)$ is the size of the largest clique of G, then*

$$sp_T(K_{\omega(G)}) \leq sp_T(G) \leq sp_T(K_{\chi(G)}).$$

Corollary *If G is weakly γ–perfect, i.e., if $\omega(G) = \chi(G)$, then*

$$sp_T(G) = sp_T(K_{\chi(G)}).$$

Many of the known results about T–colorings are obtained either using special assumptions about T or special assumptions about G. We begin with some results under special assumptions about T. A set $T = \{0, 1, ..., r\} \cup S$ is called *r–initial* if S contains no multiple of $r + 1$. $T = \{0, 1, 2, 5\}$ is an example of a 2–initial set.

Theorem 2 *(Cozzens and Roberts 1982) If T is r–initial, then for all graphs G,*
$$sp_T(G) = (r + 1)[\chi(G) - 1] = sp_T(K_{\chi(G)}).$$

A set T is a *k–multiple of s set* if $T = \{0, s, 2s, ..., ks\} \cup S$, where $s \geq 1, k \geq 1$, and S is contained in $\{s + 1, s + 2, ..., ks - 1\}$.

Theorem 3 *(Raychaudhuri 1985, 1988) If T is a k–multiple of s set, then for all graphs G,*

$$sp_T(G) = sp_T(K\chi(G))$$

$$= \begin{cases} st + skt - sk - 1 & \text{if } \chi(G) = st \\ st + skt + p - 1 & \text{if } \chi(G) = st + p \end{cases}.$$

The following *greedy algorithm* sometimes gives us $\chi_T(G)$ and $sp_T(G)$. Order $V(G)$ as $x_1, ..., x_v$. Let $F(x_1) = 1$. Having assigned $F(x_1), ..., F(x_k)$, let $F(x_{k+1})$ be the smallest channel so that $F(x_1), ..., F(x_{k+1})$ do not violate the requirements. We used this algorithm earlier in obtaining the coloring of K_3 with colors 1, 3, 9. It is a basic open question to determine for what graphs and what sets T there is an ordering for which the greedy algorithm correctly computes χ_T, and similarly for sp_T.

We turn next to results under special assumptions about the graph G. Even the case where G is a complete graph is nontrivial. Note that $sp_T(K_q)$ is fundamental because for many graphs, $sp_T(G) = sp_T(K_{\chi(G)})$.

Wang [1985], Cozzens and Wang [1984] and Tesman [1988] obtain many results about this problem for different sets T and integers q. However, $sp_T(K_q)$ remains unknown for such simple sets T as $\{0, 1, 4, 5\}$, $\{0, 1, 4, 6\}$, and $\{0, 1, 4, 7\}$. It also remains open to characterize sets T and integers q such that the greedy algorithm gives $sp_T(K_q)$. The answer to this question would have wide applicability, as the next theorem shows:

Theorem 4 *(Roberts 1986)* *Suppose G is weakly γ-perfect. Then there is an ordering of G such that the greedy algorithm gives $sp_T(G)$ if and only if the greedy algorithm gives $sp_T(K_{\chi(G)})$.*

Probably the most important class of graphs for which to study T-colorings is the class of R-unit sphere graphs. An *R-unit sphere graph* is the intersection graph of closed spheres in R-space of unit diameter. The channel assignment problem is of particular interest for these graphs, because when transmitters are located in R-space ($R = 1, 2,$ or 3), interference sometimes takes place if and only if two transmitters are within m miles. In this case, the interference graph is an R-unit sphere graph. (These graphs are studied by Maehara [1984] and by Havel [1982], Havel, Kuntz, and Crippen [1983], and Havel, et al. [1983] in connection with biochemistry.) Unfortunately, we do not know how to characterize R-unit sphere graphs except when $R = 1$. Moreover, even for 2-unit sphere graphs, computation of the ordinary chromatic number $\chi(G)$ is already an NP-complete problem (James Orlin, unpublished). In general, we do not even know, when $R > 1$, good heuristic or approximate methods for computing $\chi_T(G)$ or $sp_T(G)$ for G an R-unit sphere graph.

The case $R = 1$ is an interesting special case. The 1-unit sphere graphs are sometimes called unit interval graphs or indifference graphs. They have been characterized by Roberts [1969] and others. The interference graph is a 1-unit sphere graph if the transmitters are located in a linear corridor and interference corresponds to being within m miles.

Theorem 5 *(Cozzens and Roberts 1982)* *Suppose G is a 1-unit sphere graph. Then for a special vertex ordering called compatible (which always exists), the greedy algorithm computes $\chi_T(G)$. Moreover, if T is an r-initial set, the algorithm computes $sp_T(G)$ for such an ordering. Even if the ordering is not given, the algorithm has complexity $O(v^2 t)$ where v is the number of vertices of G and $t = |T|$. If T is r-initial, the complexity is $O(v^2)$.*

The T-coloring problem has also been studied for chordal graphs and for perfectly orderable graphs, which are defined in Golumbic [1980] and in Chvátal [1984], respectively. The reader can also refer to these papers for the definitions of the orderings in the following theorems.

Theorem 6 *(Raychaudhuri 1985, 1988) Suppose G is a chordal graph. The same results as in Theorem 5 hold for the reverse of a perfect elimination ordering and for either r–initial sets or k–multiple of s sets.*

Theorem 7 *(Raychaudhuri 1985, 1988)   Suppose G is a perfectly orderable graph. The same results as in Theorem 5 hold for an admissible (perfect) ordering of G and for either r–initial sets or k–multiple of s sets, except that the complexity results only hold if the admissible ordering is given.*

5. H–Colorings

Recall that a *homomorphism* from a graph G to a graph H is a mapping F from $V(G)$ to $V(H)$ so that for all $x \neq y$ in $V(G)$,

$$\{x, y\} \in E(G) \ \rightarrow \ \{F(x), F(y)\} \in E(H).$$

An *H–coloring* of G is defined to be a homomorphism from G into H. Note that an ordinary graph coloring with q colors is a K_q–coloring, i.e., a homomorphism from G into K_q.

Here we note that a T–coloring is just an H–coloring where H has a set of integers as its vertices and for all $x \neq y$ in $V(H)$,

$$\{x, y\} \in E(H) \ \leftrightarrow \ |x - y| \notin T.$$

The H–coloring problem is the problem of deciding if a graph is H–colorable. If $H = K_q$, the H–coloring problem is of course NP–complete. If H is bipartite, it is easy to see that the problem is polynomial. However, Hell and Nesetril [1986] have proved that for all other graphs H, the H–coloring problem is NP–complete.

6. Set Colorings

In many of the problems we have talked about, it is more appropriate to think of assigning more than one color to a vertex. In channel assignment or mobile radio frequency assignment, a transmitter might be allowed to transmit over more than one frequency. In traffic phasing or task assignment, a use of a facility or a subtask of a task might be given more than one time period during which it can take place. In fleet maintenance, a vehicle in the maintenance facility might be given more than one space.

Suppose S is a function which assigns to each vertex x of graph G a set $S(x)$ of colors. We call S a *set assignment* for G. We call S a *set coloring* for G if for all $x \neq y$ in $V(G)$,

$$\{x, y\} \in E \ \rightarrow \ S(x) \cap S(y) = \phi.$$

A *(set) intersection assignment* is a set assignment $S(x)$ so that for all $x \neq y$ in $V(G)$,

$$\{x, y\} \in E \ \leftrightarrow \ S(x) \cap S(y) \neq \phi.$$

The terminology set assignment and set coloring is due to Roberts [1979].

There are different parameters we might try to optimize for a set assignment, in particular for a set coloring. The *order* of a set assignment S is the size of $\cup S(x)$ over all x in V. We might wish to minimize the order. The *score* of a set assignment S is the sum of the sizes of the sets $S(x)$ over all x. We might wish to maximize the score, subject to the order not being too large. For instance, in channel assignment, the order of a set coloring is the total number of channels used altogether, and the score is a count of the total number of channels assigned counting multiplicity. The latter is some measure of the efficiency of the assignment. In traffic phasing, the order of a set coloring is the total amount of time used to complete all the activities, while the score is the sum of the green times, a measure of how well the facility is utilized. To give an example, consider the graph C_4. If we label the vertices in order around the cycle as x_1, x_2, x_3, x_4, then one set coloring takes $S(x_1) = \{a, b, c\}$, $S(x_2) = \{d, e\}$, $S(x_3) = \{a, b, c\}$, and $S(x_4) = \{f\}$. The order of this set coloring is $|\cup S(x)| = |\{a, b, c, d, e, f\}| = 6$. The score of this set coloring is $\Sigma |S(x)| = 3 + 2 + 3 + 1 = 9$.

With no restrictions on the sizes of the sets $S(x)$ other than that they be nonempty, the minimum order of a set coloring of C_4 is 2: Simply let $S(x_1) = S(x_3) = \{a\}$, $S(x_2) = S(x_4) = \{b\}$. Moreover, the maximum score of a set coloring if the order is at most 5 is now 10. To see why, note that $|S(x_1) \cup S(x_2)| \le 5$ and $|S(x_3) \cup S(x_4)| \le 5$, which implies that the score of any set coloring is at most 10. However, the score can be as high as 10: Take $S(x_1) = S(x_3) = \{a, b, c\}$, $S(x_2) = S(x_4) = \{d, e\}$.

A basic observation relates the minimum order or maximum score of a set coloring of a graph to the minimum order or maximum score of a set intersection assignment of spanning subgraphs of the complement. Let $\chi_\lambda(G)$ be the minimum order of a set coloring of G where all the sets $S(x)$ satisfy some property λ. Let $i_\lambda(G)$ be defined analogously for set intersection assignments. These numbers are undefined if no assignments of the specified types exist. The following result relates these numbers.

Theorem 8 *(Roberts 1979)   For all conditions λ of interest in this paper, $\chi_\lambda(G)$ is the minimum $i_\lambda(H)$ for all spanning subgraphs H of G^c for which there is a set intersection assignment satisfying condition λ.*

Let $\chi^{\lambda,N}(G)$ be the maximum score over all set colorings of G which use sets satisfying condition λ and which have order at most N. Let $i^{\lambda,N}(G)$ be defined analogously for set intersection assignments. Again, these numbers are undefined if no assignment of the specified type exists. Also as above, these two numbers are related:

Theorem 9   *(Roberts 1979) $\chi^{\lambda,N}(G)$ is the maximum of $i^{\lambda,N}(H)$ for all spanning subgraphs H of G^c for which there is a set intersection assignment satisfying condition λ and having order at most N.*

6.1. n–tuple Colorings

The simplest example of a set coloring arises when condition λ says that each set $S(x)$ is a set of n colors. We call a set coloring satisfying this condition λ an *n–tuple coloring*. Such colorings were introduced by Gilbert [1972] in connection with the mobile radio frequency assignment problem. If we use the notation for the vertices introduced above, a 2–tuple coloring of C_4 is given by taking $S(x_1) = S(x_3) = \{a, b\}$ and $S(x_2) = S(x_4) = \{c, d\}$.

In the mobile radio frequency assignment problem, we are often interested in minimizing the total number of frequencies used, i.e., minimizing the order, i.e., minimizing $|\cup S(x)|$. This minimum is $\chi_\lambda(G)$, where λ is the condition that each set $S(x)$ is a set of n elements. Here, $\chi_\lambda(G)$ is called the *n–tuple chromatic number of G* and is denoted $\chi_n(G)$.

Obviously, $\chi_2(C_4) = 4$. Thus, $\chi_n(G) = n\chi(G)$. If G is bipartite (and has an edge), $\chi_n(G) = 2n$, i.e., $\chi_n(G) = n\chi(G)$. Is $\chi_n(G)$ always equal to $n\chi(G)$? The answer is: lots of the time, but not always.

Theorem 10   *(Roberts 1978)   If G is weakly γ–perfect, then $\chi_n(G) = n\chi(G)$.*

However, the theorem is false in general. It is easy to show that $\chi_2(C_5) = 5$.

A useful early result about n–tuple colorings, which is used to prove Theorem 10, is the following:

Theorem 11   *(Stahl 1976)   For every graph G,*
$$\chi_n(G) = \chi(G[K_n]),$$
where $G[K_n]$ is the lexicographic product of G with K_n.

There was a flurry of activity about the computation of χ_n in the 1970's. Very little has been done since. Two sample results follow.

Theorem 12   *(Irving 1983)   It is NP–complete to determine if $\chi_n(G) \leq 2n + 1$.*

Theorem 13   *(Raychaudhuri 1985): $\chi_n(G)$ can be calculated as the solution to an integer programming problem in which the variables correspond to the maximal cliques of G^c.*

The maximum score of an n–tuple coloring is trivial to compute since $\Sigma |S(x)| = vn$ for all n–tuple colorings.

6.2. Consecutive Colorings

Many variations of n–tuple colorings have been studied. One variation is to allow the sets $S(x)$ to have different sizes. Here, $i_\lambda(G)$, the minimum order of a set intersection assignment, is called the *intersection number* of the graph. This concept was introduced by Erdös, Goodman, and Pósa [1966] and has been widely studied. The minimum order and maximum score of a set coloring in this situation have been studied by Roberts [1979], Opsut and Roberts [1981], and Opsut [1984].

Another variation of n–tuple colorings is to let the sets have different sizes, but for each to have a certain minimum size.

If the sets $S(x)$ are finite, there is no loss of generality in thinking of them as being sets of integers. Another variation on n–tuple colorings is to let $S(x)$ be a set of consecutive integers. This is a natural variation in connection with the various applications, especially the scheduling applications. This variation has been studied by Roberts [1979], Opsut and Roberts [1981], and Opsut [1984].

If we ask for sets of consecutive integers, it is also natural to ask that each set $S(x)$ have a certain minimum size, r_x. We call a set coloring which satisfies these conditions a *consecutive coloring*. Let $\chi_{c,r_x}(G)$ denote the minimum order of a consecutive coloring. An upper bound on this number is given by the following theorem:

Theorem 14   *(de Werra and Hertz 1988)*
$$\chi_{c,r_x} \leq \max_x [r_x + \{\Sigma r_y : x, y \text{ adjacent}\}].$$

As Gilbert [1972] points out in another context, it is easy to see that a lower bound on the minimum order of a consecutive coloring is given by

$$\chi_{c,r_x} \geq \omega(r_x) = \max \{ \sum_{x \in K} r_x : K \text{ is a clique}\}.$$

Graphs for which this inequality is an equality for all sets of nonnegative integers r_x are called *superperfect*.[1] Alan Hoffman has shown that all comparability graphs are

---

[1]Technically, superperfection was originally defined using the notion of I–coloring defined in the next section. Then G is defined to be superperfect if for all sets of nonnegative real numbers r_x, $\chi'_{I \geq r_x} = \omega(r_x)$, where $\chi'_{I \geq r_x}(G)$ is like $\chi_{I \geq r_x}(G)$ as defined in the next section except that all intervals are of length exactly r_x. It is easy to show that $\chi'_{I \geq r_x} = \chi_{I \geq r_x}$ and that if all r_x are integers, these two numbers equal χ_{c,r_x}. Moreover, Golumbic [1980] observes that $\chi'_{I \geq r_x} = \omega(r_x)$ for all sets r_x of nonnegative reals if and only if this is true for all sets r_x of nonnegative integers. Hence, the two notions of superperfection are equivalent.

superperfect. Characterization of the superperfect graphs is still an open problem. For more on superperfection, see Golumbic [1980].

To illustrate some of these ideas, consider first the graph C_4 and take $r_x = 2$ for all x. Then if the vertices around the cycle are in order x_1, x_2, x_3, x_4, we find that the following is a consecutive coloring: $S(x_1) = S(x_3) = \{1, 2\}$, $S(x_2) = S(x_4) = \{3, 4\}$. It follows that for C_4, $\chi_{c,r_x} = \omega(r_x) = 4$. However, for C_5 with $r_x = 2$ for all x, $\omega(r_x)$ still equals 4, while it is easy to see that $\chi_{c,r_x} > 4$.

6.3. I–Colorings

For some problems, it makes sense to take the set $S(x)$ to be a real interval. For instance, in traffic phasing or task assignment, $S(x)$ is an interval of time, in channel assignment and mobile radio frequency assignment, $S(x)$ is a frequency band, and in fleet maintenance, $S(x)$ is a space along a linear dock (as in the shipbuilding problem). A set coloring in which each $S(x)$ is a real interval is called an *I–coloring*. A set intersection assignment in which each $S(x)$ is a real interval is called an *I–intersection assignment*. Figure 1 shows a graph and an I–coloring.

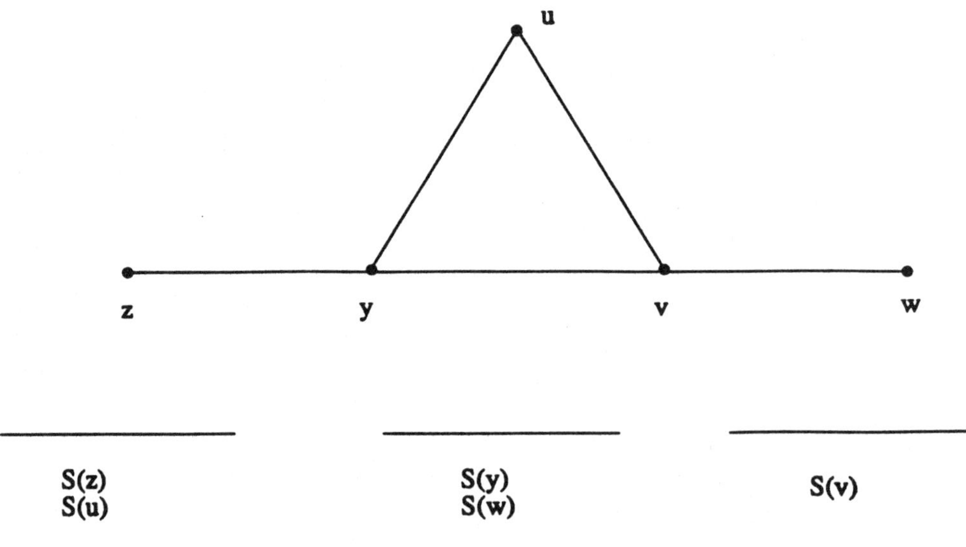

Figure 1.

In dealing with set assignments using real intervals, it makes sense to interpret the size of a set as its measure or length rather than its cardinality and to compute the infimum order rather than the minimum order. For set colorings, this infimum is always 0: we take all the intervals to be very small. The problem is more interesting, and more realistic, if we require (as in Section 6.2) that the interval assigned to x have a length which is at least as large as a specified minimum r_x. The requirement of specified minimum length makes sense in applications such as traffic phasing or fleet maintenance.

We define $\chi_{I \geq r_x}(G)$ to be $\chi_\lambda(G)$ if λ is the property that each set $S(x)$ is a real interval of length at least r_x. We define $i_{I \geq r_x}(G)$ analogously in terms of $i_\lambda(G)$. Thus, $\chi_{I \geq r_x}(G)$ is the infimum of the measures of $\cup S(x)$ over all set colorings S such that each $S(x)$ is a real interval of length $\geq r_x$.

To give an example, consider the graph of Figure 1 and suppose $r_u = 1, r_z = 1$, $r_y = 2, r_v = 1, r_w = 1$. By using the I–coloring shown in the figure and taking the lengths of the intervals assigned to z, u, and v as 1 and the lengths of the intervals assigned to y and w as 2, we see that $\chi_{I \geq r_x}(G) \leq 4$. It is clear that $\chi_{I \geq r_x}(G) = 4$.

In general, computation of $\chi_{I \geq r_x}(G)$ is an NP–complete problem. By a result of Larry Stockmeyer (see Golumbic [1980]), it is NP–complete even if all r_x are 1 or 2 and G is an interval graph.

However, by the following theorem, $i_{I \geq r_x}(G)$ can be computed in polynomial time for those graphs for which it is defined, namely the interval graphs.

Theorem 15   *(Opsut and Roberts 1983b)   The number $i_{I \geq r_x}(G)$ can be computed by solving a linear program whose variables depend on the maximal cliques of G. Hence, $i_{I \geq r_x}(G)$ can be computed in polynomial time for all interval graphs.*

The second part of the theorem follows because for an interval graph, the maximal cliques can be found in polynomial time (see Golumbic [1980]). Since linear programs can be solved in polynomial time, the result follows.

For the reader who is interested, the linear program mentioned in Theorem 15 is defined as follows. If $K_1, ..., K_p$ are the maximal cliques of G, $i_{I \geq r_x}(G)$ is the solution to the following linear programming problem:

$$\text{Minimize} \quad \sum_{i=1}^{p} d_i$$

$$\text{Subject to} \quad \sum_{i:x \in K_i} d_i \geq r_x \qquad (\text{for all } x \in V(G))$$

$$d_i \geq 0.$$

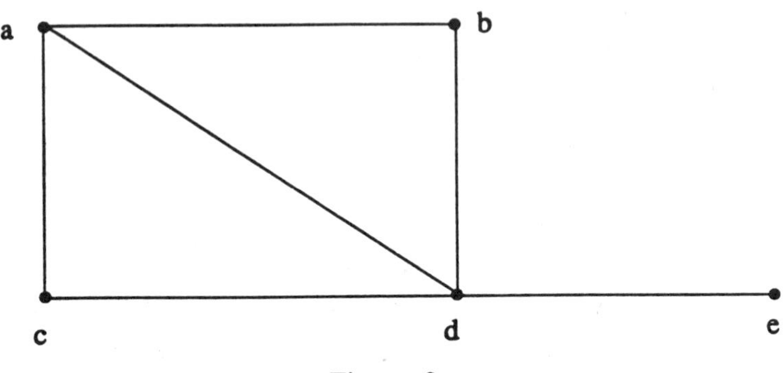

Figure 2.

Consider for example the graph of Figure 2 where $r_a = 2$, $r_b = 1$, $r_c = 3$, $r_d = 1$, and $r_e = 2$. The maximal cliques are $K_1 = \{a, b, d\}$, $K_2 = \{a, c, d\}$, $K_3 = \{d, e\}$. Then we have the linear programming problem

$$
\begin{array}{lll}
\text{Minimize} & d_1 + d_2 + d_3 & \\
\text{Subject to} & d_1 + d_2 \geq 2 & (x = a) \\
& d_1 \geq 1 & (x = b) \\
& d_2 \geq 3 & (x = c) \\
& d_1 + d_2 + d_3 \geq 1 & (x = d) \\
& d_3 \geq 2 & (x = e).
\end{array}
$$

The solution to this problem is: $d_1 = 1$, $d_2 = 3$, $d_3 = 2$, $i_{I \geq r_x}(G) = 6$.

Recall the observation of Theorem 8 that $\chi_{I \geq r_x}(G)$ can be computed as the minimum $i_{I \geq r_x}(H)$ over spanning subgraphs H of G^c which have a set intersection of the kind described. Thus, $\chi_{I \geq r_x}(G)$ can be computed by solving these linear programs for each such H. Unfortunately, there might be lots of these linear programs. However, this observation leads to some useful conclusions, such as:

Theorem 16 *(Opsut and Roberts 1983a)   If G^c is an interval graph, then $\chi_{I \geq r_x}(G)$ is $i_{I \geq r_x}(G^c)$ and hence can be computed in polynomial time.*

There are analogous results for scores. For scores with assignments of intervals, it again makes sense to interpret size as measure or length and it now makes sense to take the supremum rather than the maximum. Thus, let $\chi^{I \geq r_x, N}(G)$ be the supremum score of any set coloring where each set $S(x)$ is a real interval of length at least r_x and where the total order is at most N. Let $i^{I \geq r_x, N}(G)$ be defined analogously for set intersection assignments.

Theorem 17 *(Opsut and Roberts 1983b)* *The numbers $i^{I \geq r_x, N}(G)$ can be computed by solving a linear program whose variables depend on the maximal cliques of G. Hence, $i^{I \geq r_x, N}(G)$ can be computed in polynomial time for all interval graphs.*

It follows from the observation of Theorem 9 that $\chi^{I \geq r_x, N}(G)$ can be computed by solving these linear programs for each spanning subgraph H of G^c which has a set intersection assignment of the kind described. One corollary of this observation is that $\chi^{I \geq r_x, N}(G)$ can be computed in polynomial time (indeed $O(v^2)$ time) for G the complement of an interval graph. (See Opsut and Roberts [1983a]).

It still remains an open problem to determine if $\chi_{I \geq r_x}(G)$ and $\chi^{I \geq r_x, N}(G)$ are computable in polynomial time for G the complement of a chordal graph.

6.4. J–Colorings

Sometimes it makes more sense to think of each set $S(x)$ as a union of intervals. This occurs, for example, in traffic phasing, task assignment, channel assignment, or mobile radio frequency assignment. We call a set coloring in which each $S(x)$ is a union of real intervals a *J–coloring*. A *J–intersection assignment* is defined analogously.

It also makes sense to study J–colorings and J–intersection assignments in which $\cup S(x)$ has total length at least a specified number r_x. We denote this condition by $J \geq r_x$. Then the numbers $\chi_{J \geq r_x}$, $i_{J \geq r_x}$, $\chi^{J \geq r_x, N}$, and $i^{J \geq r_x, N}$ are defined in the usual manner as χ_λ, i_λ, $\chi^{\lambda, N}$, and $i^{\lambda, N}$, respectively. Note that the second number is always defined because every graph is the intersection graph of sets each of which is a union of real intervals (Griggs and West [1980], Trotter and Harary [1979]).

Theorem 18 *(Raychaudhuri 1985)* *For any graph G,*
$$\chi_{J \geq r_x}(G) = i_{J \geq r_x}(G^c)$$
and both numbers can be calculated by a linear program whose variables are defined in terms of the maximal cliques of G^c.

Note that this theorem does not say that $\chi_{J \geq r_x}(G)$ is computable in a polynomial number of steps for every G. It is necessary first to show that the maximal cliques of G^c can be indentified in a polynomial number of steps. The latter is true for G^c a chordal graph and hence for G^c an interval graph (see Golumbic [1980]).

The next theorem relates I–colorings to J–colorings.

Theorem 19 *(Raychaudhuri 1985)   If G is the complement of an interval graph, then*

$$\chi_{J \geq r_x}(G) = \chi_{I \geq r_x}(G).$$

Thus, we have the surprising result that for complements of interval graphs, going to unions of intervals does not allow us to improve the total order of a coloring.

Raychaudhuri obtains analogous results for scores.

We close this subsection by mentioning a number of open problems. An important one is to find ways to compute the minimum order and maximum score for *D–colorings*, colorings in which each set $S(x)$ is a union of at most two real intervals (the union being of at least a certain minimum size r_x). One difficulty here is that the characterization of the intersection graphs of families of sets each of which is a union of at most two intervals, the so–called *double interval graphs*, remains an open question. (See Trotter and Harary [1979] for partial results.)

Another open problem is to investigate the minimum order or maximum score when the sets $S(x)$ are rectangles in the plane (of certain minimum measure). This question arises in the fleet maintenance problem if each vehicle is assigned a rectangular space in the floor of a maintenance facility.

7. List Colorings

In many practical coloring problems a choice of color to assign to x is restricted. A set or *list* $R(x)$ of possible colors to be assigned to x is specified, and we seek a graph coloring $F(x)$ so that $F(x)$ always belongs to $R(x)$. We say that G can be *list–colored* if there is such a coloring for G . List colorings arise for instance in the channel assignment problem (when we specify possible acceptable channels) or the traffic phasing problem (when we specify possible green times say for use of a classroom). Erdös, Rubin, and Taylor [1979] introduced the idea of considering when a graph G can be list–colored for every assignment of lists $R(x)$ of k colors each to the vertices. If G can always be list–colored for every such assignment, we say that G is *k-choosable*. We define the *choice number* of G , ch(G), to be the smallest k so that G is k–choosable. (Bollobas and Harris [1985] consider similar concepts for edge colorings.) To illustrate this concept, we note that C_5 is not 2–choosable: Let $R(x) = \{1, 2\}$ for every x . However, C_4 is 2–choosable.

Among the known results about graph list colorings is a characterization of 2-choosability. To present the characterization, let us say that a graph is called a *θ-graph* if it consists of two distinguished vertices u and v and three chains joining u to v which are otherwise vertex–disjoint. The graph $\theta_{i,j,k}$ is the θ–graph in which the chains between u and v have lengths i, j, and k. The *core* of a graph is obtained by successively removing vertices of degree 1 until it is impossible to do so.

Theorem 20   *(Erdös, Rubin, and Taylor 1979)   A connected graph is 2–choosable if and only if its core is either a single vertex, an even cycle, or the graph $\theta_{2,2,2m}$ for some $m \geq 1$.*

A more recent result about list colorings is the following:

Theorem 21   *(Tesman 1988, 1989)   If G is chordal, the choice number of G is $\chi(G)$.*

By way of open problems, we mention two conjectures which still seem to be unsettled:

Conjecture   *(Erdös, Rubin, and Taylor 1979)   Every planar graph is 5–choosable.*

Conjecture   *(Erdös, Rubin, and Taylor 1979)   There is a planar graph which is not 4–choosable.*

8. Set T–Colorings

We have described so far a variety of generalizations or variants of ordinary graph coloring. It seems reasonable to consider as well different combinations of these. In what follows, we describe some results about combinations of set coloring and T-coloring and about combinations of list coloring and T–coloring, and we suggest the possibility of combining all three ideas. A *set T–coloring* of G is an assignment of a set $S(x)$ of positive integers to each vertex x of G so that for all $x \neq y$ in $V(G)$,

$$\{x, y\} \in E \;\rightarrow\; |F_x - F_y| \notin T$$
$$\text{for all } F_x \in S(x), F_y \in S(y).$$

We shall concentrate on *n–tuple T–colorings*, set T–colorings in which each set is a set of n positive integers.

The *order* of an n–tuple T–coloring is the number of distinct integers used in all of the sets $S(x)$. The *span* of such a coloring is the difference between the largest and smallest integers used in any of the sets $S(x)$. The minimum order of an n–tuple T-coloring of G is denoted $\chi_T^n(G)$. The minimum span over all n–tuple T-colorings of

G is denoted $\text{sp}_T^n(G)$. Of course, if $T = \{0\}$, $\chi_T^n = \chi_n$.

Many of the early Cozzens–Roberts [1982] results about ordinary T–colorings go over for n–tuple T–colorings.

Theorem 22   *(Tesman 1988)* $\chi_T^n(G) = \chi_n(G)$.

Theorem 23   *(Tesman 1988)* $sp_T^n(K_{\omega(G)}) \le sp_T^n(G) \le sp_T^n(K_{\chi(G)})$.

Theorem 24   *(Tesman 1988)*   *The greedy algorithm on a compatible ordering and the reverse of a perfect elimination ordering computes* χ_T^n *and* sp_T^n *for 1-unit sphere graphs and chordal graphs, respectively, if* T *is* $\{0, 1, ..., r\}$.

However, this greedy algorithm fails for general r–initial sets, even for complete graphs. It also fails for simple k–multiple of s sets $\{0, s, 2s, ..., ks\}$, even for complete graphs.

One case where the Cozzens–Roberts results do not go over easily is captured in the following theorem.

Theorem 25   *(Tesman 1988, Füredi, Griggs, and Kleitman 1988)*   *Suppose* $T = \{0, 1, ..., r\}$. *Then if* G *has an edge,*
$$(r + 1)[\chi(G) - 1] + 2(n - 1) \le sp_T^n(G) \le (n + r)\chi(G) - (r + 1).$$

The following conjecture presents an open problem in this area:

Conjecture   *(Füredi, Griggs, and Kleitman 1988)*   *All values in the interval in the above theorem are attained by suitable graphs* G.

Theorem 26   *(Füredi, Griggs, and Kleitman 1988)*   *The conjecture is true when* $n = 2$.

9. List T–Colorings

In this section, we combine the idea of a list coloring with the idea of a T–coloring. Given lists $R(x)$ for G and a set T of nonnegative integers, a *list T–coloring* of G is a T–coloring F in which each $F(x)$ belongs to the set $R(x)$. This idea is mentioned in Hale [1980] and Roberts [1986]. We say that G is *T–k–choosable* if there is a T–coloring for every assignment of lists $R(x)$ in which each list has k elements. The *T–choice number* of G, T–ch(G), is the smallest k so that G is T–k–choosable.

For instance, suppose $T = \{0, 1\}$. Then K_3 is not T–4–choosable. For, suppose the vertices of K_3 are x, y, and z and we take $R(x) = R(y) = R(z) = \{1, 2, 3, 4\}$. Then it is easy to show that we cannot find $F(x), F(y), F(z)$ from $R(x), R(y), R(z)$, respectively, with
$$|F(x) - F(y)| \ne 0, 1$$

$$|F(x) - F(z)| \neq 0, 1$$
$$|F(y) - F(z)| \neq 0, 1.$$

On the other hand, suppose again that $T = \{0, 1\}$. Then K_3 is T–7–choosable. To see why, pick $F(x) = a$. Thus, $F(y) \neq a, a - 1, a + 1$. Since $|R(y)| = 7$, we can pick $F(y) = b \neq a, a - 1, a + 1$. Now we need $F(z) \neq a, a - 1, a + 1, b, b - 1, b + 1$. Hence, since $|R(z)| = 7$, we can pick $F(z) = c$ for some $c \in R(z)$.

The following is a recent result about list T–colorings.

Theorem 27   *(Tesman 1988, 1989)   For G chordal and $T = \{0, 1, ..., r\}$,*
$$(r + 1)[\chi(G) - 1] + 1 \leq T\text{–}ch(G) \leq (2r + 1)[\chi(G) - 1] + 1.$$

It is easy to show that both bounds in Theorem 27 are tight, the lower bound for arbitrary χ and the upper bound at least for $\chi = 2$; moreover, the lower bound is tight for 1–unit sphere graphs. However, it remains an open question to determine if the upper bound is tight for 1–unit sphere graphs.

10. Set List T–Colorings

If we combine three ideas, we can think of set list T–colorings, at least n–tuple list T–colorings. There are two possible variants of this idea, one where the lists $R(x)$ consist of acceptable single elements and the other where the lists $R(x)$ consist of acceptable sets (n–element sets). To the best of the author's knowledge, no work has been done on either of these variants, and they appear to be attractive variants for consideration.

Acknowledgement: The author thanks Pey–chun Chen, Garth Isaak, Suh–ryung Kim, and Barry Tesman for their helpful comments and the Air Force Office of Scientific Research for its support under grant number AFOSR–85–0271 to Rutgers University.

REFERENCES

Beltrami, E., and Bodin, L., "Networks and Vehicle Routing for Municipal Waste Collection," *Networks*, **4** (1973), 65 – 94.

Bodin, L.D., and Friedman, A.J., "Scheduling of Committees for the New York State Assembly," Tech. Report USE No. 71–9, Urban Science and Engineering, State University of New York, Stony Brook, 1971.

Bollobas, B., and Harris, A.J., "List Colourings of Graphs," *Graphs and Combinatorics*, **1** (1985), 115 – 127.

Bondy, J.A., and Murty, U.S.R., *Graph Theory with Applications*, Elsevier, New York/MacMillan, London, 1976.

Chvátal, V., "Perfectly Ordered Graphs," in C. Berge and V. Chvátal (eds.), *Topics on Perfect Graphs,* North–Holland, Amsterdam, 1984, 63 – 65.

Cozzens, M.B., and Roberts, F.S., "T–Colorings of Graphs and the Channel Assignment Problem," *Congressus Numerantium,* **35** (1982), 191 – 208.

Cozzens, M.B., and Wang, D–I., "The General Channel Assignment Problem," *Congressus Numerantium,* **41** (1984), 115 – 129.

de Werra, D., and Hertz, A., "Consecutive Colorings of Graphs," *Zeitschrift für Operations Research,* **32** (1988), 1 – 8.

Erdös, P., Goodman, A., and Pósa, L., "The Representation of a Graph by Set Intersections," *Can. J. Math.,* **18** (1966), 106 – 112.

Erdös, P., Rubin, A.L., and Taylor, H., "Choosability in Graphs," *Congressus Numerantium,* **26** (1979), 125 – 157.

Fishburn, P.C., *Interval Orders and Interval Graphs,* Wiley, New York, 1985.

Füredi, Z., Griggs, J.R., and Kleitman, D.J., "Pair Labellings with Given Distance," mimeographed, Institute for Mathematics and its Applications, University of Minnesota, Minneapolis, 1988. To appear in *SIAM J. on Discr. Math.*

Gilbert, E.N., Unpublished Technical Memorandum, Bell Telephone Laboratories, Murray Hill, New Jersey, 1972.

Golumbic, M.C., *Algorithmic Graph Theory and Perfect Graphs,* Academic Press, New York, 1980.

Griggs, J.R., and West, D.B., "Extremal Values of the Interval Number of a Graph," *SIAM J. Alg. Disc. Math.,* **1** (1980), 1 – 7.

Hale, W.K., "Frequency Assignment: Theory and Applications," *Proc. IEEE,* **68** (1980), 1497 – 1514.

Havel, T., "The Combinatorial Distance Geometry Approach to the Calculation of Molecular Conformation," *Congressus Numerantium,* **35** (1982), 361 – 371.

Havel, T., Kuntz, I.D., and Crippen, G.M., "The Combinatorial Distance Geometry Approach to the Calculation of Molecular Conformation I: A New Approach to an Old Problem," *J. Theor. Biol.,* **104** (1983), 359 – 381.

Havel, T., Kuntz, I.D., Crippen, G.M., and Blaney, J.M., "The Combinatorial Distance Geometry Approach to the Calculation of Molecular Conformation II: Sample Problems and Computational Statistics," *J. Theor. Biol.,* **104** (1983), 383 – 400.

Hell, P., and Nesetril, J., "On the Complexity of H–Coloring," Tech. Rep. TR–86–4, Department of Computing Science, Simon Fraser University, Burnaby, British Columbia, 1986.

Irving, R.W., "NP–Completeness of a Family of Graph–Colouring Problems," *Discr. Appl. Math.,* **5** (1983), 111 – 117.

Maehara, H., "Space Graphs and Sphericity," *Discr. Appl. Math.,* **7** (1984), 55 – 64.

Metzger, B.H., "Spectrum Management Technique," paper presented at 38th National ORSA Meeting, Detroit, MI, 1970.

Opsut, R.J., *Optimization of Set Assignments for Graphs*, Ph.D. Thesis, Department of Mathematics, Rutgers University, New Brunswick, N.J., May 1984.

Opsut, R.J., and Roberts, F.S., "On the Fleet Maintenance, Mobile Radio Frequency, Task Assignment, and Traffic Phasing Problems," in G. Chartrand, Y. Alavi, D.L. Goldsmith, L. Lesniak–Foster, and D.R. Lick (eds.), *The Theory and Applications of Graphs,* Wiley, New York, 1981, 479 – 492.

Opsut, R.J., and Roberts, F.S., "I–colorings, I–phasings, and I–intersection Assignments for Graphs, and their Applications," *Networks*, **13** (1983), 327 – 345. (a)

Opsut, R.J., and Roberts, F.S., "Optimal I–intersection Assignments for Graphs: A Linear Programming Approach," *Networks,* **13** (1983), 317 – 326. (b)

Raychaudhuri, A., *Intersection Assignments, T Coloring, and Powers of Graphs,* Ph.D. thesis, Department of Mathematics, Rutgers University, New Brunswick, N.J., 1985.

Raychaudhuri, A., "Further Results on T–Colorings and Frequency Assignment Problems," mimeographed, Department of Mathematics, College of Staten Island (CUNY), Staten Island, New York, 1988.

Roberts, F.S., "Indifference Graphs," in F. Harary (ed.), *Proof Techniques in Graph Theory*, Academic Press, New York, 1969, 139 – 146.

Roberts, F.S., *Discrete Mathematical Models, with Applications to Social, Biological, and Environmental Problems,* Prentice–Hall, Englewood Cliffs, N.J., 1976.

Roberts, F.S., *Graph Theory and its Applications to Problems of Society,* CBMS–NSF Monograph No. 29, SIAM, Philadelphia, 1978.

Roberts, F.S., "On the Mobile Radio Frequency Assignment Problem and the Traffic Light Phasing Problem," *Annals New York Acad. Sci.,* **319** (1979), 466 – 483.

Roberts, F.S., *Applied Combinatorics*, Prentice–Hall, Englewood Cliffs, N.J., 1984.

Roberts, F.S., "T–Colorings of Graphs: Recent Results and Open Problems," Research Rep. RRR 7–86, Rutgers Center for Operations Research, Rutgers University, New Brunswick, N.J., May 1986. To appear in *Annals of Discrete Mathematics.*

Stoffers, K.E., "Scheduling of Traffic Lights – A New Approach," *Transportation Res.,* **2** (1968), 199 – 234.

Stahl, S., "n–tuple Colorings and Associated Graphs," *J. Comb. Theory,* **20** (1976), 185 – 203.

Tesman, B., Ph.D. Thesis, Department of Mathematics, Rutgers University, New Brunswick, New Jersey, in preparation, 1988.

Tesman, B., "Vertex List T–Colorings of Graphs," mimeographed, Department of Mathematics, Rutgers University, New Brunswick, NJ 1989.

Trotter, W.T., and Harary, F., "On Double and Multiple Interval Graphs," *J. Graph Theory,* **3** (1979), 205 – 211.

Tucker, A.C., "Perfect Graphs and an Application to Optimizing Municipal Services," *SIAM Rev.,* **15** (1973), 585 – 590.

Wang, D–I., "The Channel Assignment Problem and Closed Neighborhood Containment Graphs," Ph.D. Thesis, Northeastern University, Boston, MA, 1985.

Zoellner, J.A., and Beall, C.L., "A Breakthrough in Spectrum Conserving Frequency Assignment Technology," *IEEE Trans. on Electromag. Comput.,* **EMC–19** (1977), 313 – 319.

EDGE DOMINATING NUMBERS OF COMPLEMENTARY GRAPHS

Seymour Schuster

ABSTRACT

An edge dominating set in a graph G is a set S of edges of G such that every edge of G belongs to S or is adjacent to an edge of S. The edge dominating number (also called the edge-edge-covering number) $\alpha_{11}(G)$ of a graph G is the cardinality of a smallest dominating set of edges in G. Inequalities of the Nordhaus-Gaddum type are established, providing best possible upper and lower bounds for $\alpha_{11}(G) + \alpha_{11}(\overline{G})$ and $\alpha_{11}(G) \cdot \alpha_{11}(\overline{G})$, where $\overline{G}$ is the complement of G.

1.   Introduction

In 1956, Nordhaus and Gaddum [9] established the following inequalities involving the chromatic number $\chi(G)$ of a graph G of order p and the chromatic number $\chi(\overline{G})$ of its complement $(\overline{G})$:

$$2\sqrt{p} \leq \chi(G) + \chi(\overline{G}) \leq p + 1$$
$$p \leq \chi(G) \cdot \chi(\overline{G} \leq \lfloor \frac{p+1}{2} \rfloor^2.$$

Inequalities of this "Nordhaus-Gaddum type" have also been found for a host of other graphical parameters: achromatic number by Gupta [7]; edge chromatic number by Alavi and Behzad [2] and by Vizing [10]; domination number by Jaeger and Payan [8]; connectivity and edge connectivity by Alavi and Mitchem [3]; diameter, girth, circumference, covering number and edge covering number by Xu [11]; independence and edge independence number by Chartrand and Schuster [5] and by Erdös and Schuster [6].

In the current paper, we study the edge domination number (which has also been called the edge-edge covering number (with the aim of obtaining Nordhaus-Gaddum inequalities for this graphical parameter. We begin by presenting some necessary definitions and accompanying notation. A set of vertices is called *independent* if no two vertices in the set are adjacent. The *vertex independence number* $\beta_0(G)$ of a graph G is the cardinality of a largest independent set of vertices of G. A *vertex dominating set* S for a graph G is the set of vertices such that every vertex of G belongs to S or is adjacent to a vertex of S. The cardinality of a smallest vertex dominating set is called the *vertex dominating number* $\alpha_{00}(G)$ *of* G. For euphonic consideration, the word "dominating" is occasionally replaced by "domination". Also, the vertex dominating number has been called the *vertex-vertex covering number* since the dominating set S covers the vertex set $V(G)$ of G. The *edge independence number* $\beta_1(G)$, and *edge dominating set* and the *edge dominating number* $\alpha_{11}(G)$ are defined analogously.

2. Relations Between Edge Domination and Edge Independence

Since the principal focus of our study is on the edge dominating number α_{11}, we begin with some immediate observations concerning edge domination.

Observation 1 If S is a smallest edge dominating set in G and $uv \in E(G)$, then at least one of u and v belong to an edge of S.

This is so, for otherwise uv would not be covered by S.

Observation 2 A smallest edge dominating set cannot possess three consecutive edges. This is true, for if v_1v_2, v_2v_3, v_3v_4 are in a dominating set, then v_1v_2 and v_3v_4 already cover the edges covered by v_2v_3. So edge v_2v_3 is superfluous in the dominating set.

Observation 3 For any graph G of order $p, \alpha_{11}(G) \leq \beta_1(G) \leq \lfloor \frac{p}{2} \rfloor$.

Proof Let T be a maximal independent set of edges of G. Then T covers $E(G)$; otherwise, there would exist $e \in (E(G) - T)$ that would not be adjacent to any edge of T, and this would contradict the maximality of T. Hence, $\alpha_{11}(G) \le |T| \le \beta_1(G) \le \lfloor \frac{p}{2} \rfloor$. See [1].

Observation 4 If K_p is the complete graph of order p, then $\alpha_{11}(K_p) = \lfloor \frac{p}{2} \rfloor$.

Observation 5 For the bipartite graph $K_{m,n}$ with $m \le n, \alpha_{11})(K_{m,n}) = m$.

Theorem 1 For any graph G, there exists a smallest edge dominating set in G that is also a maximal independent set of edges.

Proof Let S be a smallest edge dominating set that is not an independent set. Then there are two adjacent edges $v_1 v_2, v_2 v_3 \in S$. The vertex v_3 is not an endvertex (i.e., $\deg v_3 > 1$); otherwise $v_2 v_3$ would be superfluous in S. Also, we assert that none of the other edges of G that are incident with v_3 can belong to S, for otherwise Observation 2 would be contradicted. We observe, still further, that if all of these other edges incident with v_3 were adjacent to an edge of S other than $v_2 v_3$, then again $v_2 v_3$ would be superfluous in S. Hence, there is an edge $v_3 v_4$ that is not adjacent to any edge of S other than $v_2 v_3$. We therefore form a new edge-cover

$$S' = (S - \{v_2 v_3\}) \cup \{v_3 v_4\}$$

Clearly, S' is also a smallest dominating set, and S has one fewer pair of adjacent edges than S. Repeating this process of constructing smallest dominating sets while reducing the number of adjacencies among the edges, we obtain a smallest dominating set that is simultaneously an independent set of edges. Moreover, this cover is a maximal independent set of edges.

In consequence of this theorem, we obtain the following result of Chartrand and Lesniak [4, p. 249]:

Corollary 1 If $\beta_1^*(G)$ is the minimum cardinality among the maximal independent sets of edges of any graph G, then $\alpha_{11}(G) = \beta_1^*(G)$.

Another result - the analogue of a theorem which Allan and Laskar proved for vertex domination in [1] - also follows immediately.

Corollary 2 For any graph G, the independence edge dominating number $\alpha'_{11}(G)$, which is the cardinality of a smallest independent dominating set of edges, equals the edge dominating number of G. That is, $\alpha'_{11}(G) = \alpha_{11}(G)$.

3. The Nordhaus-Gaddum Inequalities

We first establish inequalities of the Nordhaus-Gaddum type that hold for graphs of any order $p \geq 2$; later, we shall see that these may be sharpened in accordance with the number-theoretic character of p.

Theorem 2 For any graph G of order $p \geq 3$,

$$\lfloor \tfrac{p}{2} \rfloor \leq \alpha_{11}(G) + \alpha_{11}(\overline{G}) \leq 2\lfloor \tfrac{p}{2} \rfloor. \tag{1}$$

Proof Theorem 1 guarantees the existence of a set S of independent dominating set of edges. If $|S| = k$, then the edges of S are incident with exactly $2k$ vertices. If $u, v \in V(G)$ are not among these $2k$ vertices, then $uv \notin E(G)$ for uv is not adjacent to any edge of S. Hence, the remaining $p - 2k$ vertices induce a subgraph H of G which must be empty. Therefore, $\overline{G} \supset K_{p-2k}$. Since $\alpha_{11}(Kp - 2k) = \lfloor \tfrac{p-2k}{2} \rfloor$ (by Observation 4), we have

$$\alpha_{11}(G) + \alpha_{11}(\overline{G}) \geq k + \lfloor \frac{p - 2k}{2} \rfloor = \lfloor \tfrac{p}{2} \rfloor,$$

which establishes the lower bound.

Applying Observation 3 to G and $\overline{G}$, we get

$$\alpha_{11}(G) + \alpha_{11}(\overline{G}) \leq \lfloor \tfrac{p}{2} \rfloor + \lfloor \tfrac{p}{2} \rfloor = 2\lfloor \tfrac{p}{2} \rfloor,$$

which produces the upper bound.

Theorem 3 For every graph G of order $p \geq 2$

$$0 \leq \alpha_{11}(G) \cdot \alpha_{11}(\overline{G}) \leq \lfloor p2 \rfloor^2.$$

Proof Applying the arithmetic-geometric mean inequality to $\alpha_{11}(G)$ and $\alpha_{11}(\overline{G})$ together with Theorem 2 gives

$$\sqrt{\alpha_{11}(G) \cdot \alpha_{11}(\overline{G})} \le \lfloor p2 \rfloor. \tag{2}$$

The question of whether Theorems 2 and 3 are "best possible" will be discussed in our next section.

4. Realizability

It is easy to find graphs for which the upper and lower bounds in Theorems 2 and 3 are attained provided $p \equiv 0, 1,$ or $3 \pmod 4$: the lower bounds of the theorems are attained if $G = K_p$; if $p \equiv 0 \pmod 4$, the upper bounds in (1) and (2) are attained when $G = K_{\frac{p}{2},\frac{p}{2}}$. If $p \equiv 1$ or 3, the upper bounds in (1) and (2) are attained when $G = K_{\lceil \frac{p}{2} \rceil, \lfloor \frac{p}{2} \rfloor}$.

Attaining the upper bound if $p \equiv 2 \pmod 4$ is a different matter, for the likely candidate $G = K_{\frac{p}{2},\frac{p}{2}}$ fails to yield the upper bound of (1)-hence, of (2) as well. This is so, for $\overline{G} = 2K_{\frac{p}{2}}$ with $\frac{p}{2}$ odd; hence $\alpha_{11}(\overline{G}) = 2\lfloor \frac{p}{4} \rfloor = 2\lfloor \frac{p}{2} \rfloor - 1 < \frac{p}{2}$. Therefore, if $p \equiv 2 \pmod 4$ then $\alpha_{11}(K_{\frac{p}{2},\frac{p}{2}}) + a_{11}(\overline{K}_{\frac{p}{2},\frac{p}{2}}) = \frac{p}{2} + (\frac{p}{2} - 1) = p - 1$, which leads us to our next result.

Theorem 4 For every graph G of order p with $p \equiv 2 \pmod 4$,

$$\frac{p}{2} \le \alpha_{11}(G) + \alpha_{11}(\overline{G}) \le p - 1 \tag{3}$$

$$0 \le \alpha_{11}(G) \cdot \alpha_{11}(\overline{G}) \le \lfloor (\frac{p-1}{2}) \rfloor^2. \tag{4}$$

Proof Let us write $p = 4k + 2$, where k is a non-negative integer. Then $|E(G)| + |E(\overline{G})| = \frac{p(p-1)}{2} = \frac{(4k+2)(4k+1)}{2} = (2k + 1)(4k + 1)$, which is odd. Thus, $|E(G)|$ and $|E(\overline{G})|$ are of different parity.

Since $\alpha_{11}(G) \le \frac{p}{2}$, we establish (3) by proving that $\alpha_{11}(G)$ and $\alpha_{11}(\overline{G})$ cannot both equal $\frac{p}{2}$ simultaneously. We shall do this by showing that $\alpha_{11}(G) = \frac{p}{2}$ only if $|E(G)|$ is odd (in which case $\alpha_{11}(\overline{G}) < \frac{p}{2}$ since $|E(\overline{G})|$ must be even).

We suppose that $\alpha_{11}(G) = \frac{p}{2}$ and that a smallest edge dominating set S consists of the $\frac{p}{2}$ independent edges $v_1 v_1', v_2 v_2', ..., v_{\frac{p}{2}} v_{\frac{p}{2}}'$. With $\equiv 2 \pmod 4$, we know that $\frac{p}{2}$ is odd.

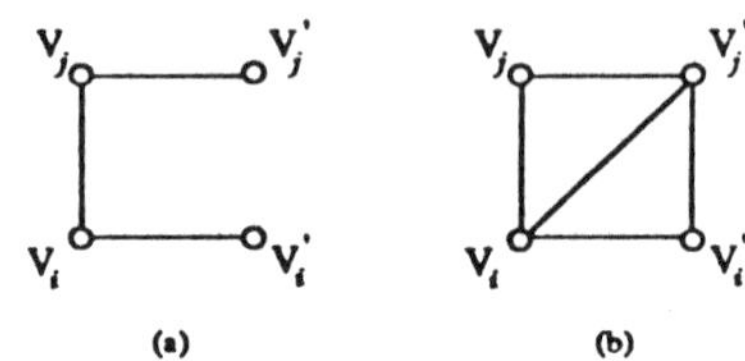

Figure 1

In order to determine the parity of $|E(G)|$, we examine the graphs induced by pairs of edges of S, namely the graph $< v_i, v_i', v_j, v_j' >$ where $1 \le i < j \le \frac{p}{2}$. If any of these graphs have an odd number of edges, then it must be isomorphic to one of the graph in Figure 1. In both of these cases, we would have a contradiction of the fact that S is a smallest dominating set, for the two edges $v_i v_i'$ and $v_j v_j'$ could be replaced by a single edge ($v_i v_j$ in (a) or $v_i v_j'$ in (b)) to obtain a smaller dominating set. Thus the induced graphs have 0, 2 or 4 edges in addition to the two in S, namely $v_i v_i'$ and $v_j v_j'$. Since the number of edges in S is odd and the number not in S is even, we have that $|E(G)|$ is odd. Of course, this implies that $|E(\overline{G})|$ is even so $\alpha_{11}(\overline{G}) < \frac{p}{2}$. This establishes the upper bound of (3).

The remaining bounds follow immediately from Theorems 2 and 3.

REFERENCES

[1] R. B. Allen and R. Laskar, "On domination and independent domination numbers of a graph," *Discrete Math.* 23 (1978), 73-76.

[2] Y. Alavi and M. Behzad, "Complementary graphs and edge chromatic numbers," *SIAM J. Appl. Math.* 20 (1971) 161-163.

[3] Y. Alavi and J. Mitchem, "Connectivity and line-connectivity of complementary graphs," in *Recent Trends in Graph Theory*, M. Capobianco, J. B. Frechen and M. Krolik, eds., Springer, New York, 1971.

[4] G. Chartrand and L. Lesniak, *Graphs and Digraphs*, 2nd ed., Wadsworth, Inc., Belmont, CA, 1986.

[5] G. Chartrand and S. Schuster, "On the independence number of complementary graphs," Trans. New York. Acad. Sci. Ser. II, 36 (1974) 247-251.

[6] P. Erdös and S. Schuster, "Existence of complementary graphs with specified independence numbers," on *Theory and Applications of Graphs*, G. Chartrand, et al., eds., John Wiley and Sons, Inc., 1981, 343-349.

[7] R. P. Gupta, "Bounds on the chromatic and achromatic numbers of complementary graphs," in *Recent Progress in Combinatorics*, W. T. Tutte, ed., Academic Press, New York, 1969, 229.

[8] F. Jaeger and C. Payan, "Relations du type Nordhaus-Gaddum pour le Nombre d'absorption d'un graphe simple," C. R. Acad. Sci. Paris, Ser. A, t. 274 (1972) 728-730.

[9] E. Nordhaus and J. W. Gaddum, "On complementary graphs," *Amer. Math. Monthly* 63 (1956) 175-177.

[10] V. G. Vizing, "The chromatic class of a multigraph," *Cybernetics* 1 (3) (1965) 32-41.

[11] S.-j. Xu, "Some parameters of graph and its complement," *Discrete Math.* 65 (1987) 197-207.

MINIMUM DOMINATING, OPTIMALLY INDEPENDENT
VERTEX SETS IN GRAPHS

W. J. Selig

NASA/Marshall Space Flight Center

Huntsville, AL

P. J. Slater

The University of Alabama in Huntsville

ABSTRACT

As one example of a new class of problems (namely, evaluating a "single set, prioritized multiproperty" parameter) we introduce the following problem. Find the minimum possible number of edges in the subgraph induced by a minimum dominating set for a graph G. We show this to be an NP-hard problem in general, and we present a linear time algorithm to compute this parameter for a tree.

1.　Introduction

Given the graph $G = (V, E)$, vertex set $S \subseteq V$ is independent if no two vertices in S are adjacent, and S is a dominating set if each vertex in $V - S$ is adjacent to at least one vertex in S. Let $\beta_0(G)$ denote the maximum cardinality of an independent vertex set in G, and let $\gamma(G)$ be the minimum cardinality of a dominating set. Determining if $\beta_0(G) \geq K$ is an NP-complete problem (Garey,

---

*Research supported in part by the U.S. Office of Naval Research Grant N00014-86-K-0745.

Johnson and Stockmeyer [3]), as is deciding if $\gamma(G) < k$ (Garey and Johnson [2]). Finding a maximum independent set (a β_0-set) and finding a minimum dominating set (a γ-set) are examples of finding a single set that is extremal relative to a single property.

Independence and domination have been widely studied, and extensive bibliographies on the subjects have been compiled by P. L. Hammer (RUTCOR) and S. T. Hedetniemi and R. Laskar (Clemson University), respectively.

On the other hand we can form a partition of the vertex set $V = S_1 \bigcup S_2 \bigcup \cdots \bigcup S_k$ where each S_i has some property P. The chromatic number $\chi(G)$ is the minimum number of vertex sets in a partition of V into independent sets, and the domatic number $do(G)$ is the maximum number of vertex sets in a partition of V into dominating sets. Deciding if $\chi(G) < K$ is NP-complete (Karp [11]), as is determining if $do(G) \geq K$ (Garey, Johnson and Tarjan [4]).

An intermediate type of problem is to optimize a parameter involving a fixed number (such as two) of sets. In Grinstead and Slater [6] we introduced this type of problem, specifically asking for two minimum dominating sets with minimum possible intersection. It was that simply determining if there exist two disjoint minimum dominating sets is an NP-hard problem for arbitrary bipartite graphs. Further results on this particular problem appear in these proceedings [7], and further treatment of the "multiset single parameter" problem type will appear in [5,8].

Here we introduce another problem type, the "single set, prioritized multiproperty" problem type. As an example, we consider the two properties of domination and independence with priority on the domination parameter. Thus we seek a minimum dominating set (a γ-set) which will be "as independent as possible."

We first note that the fairly well studied parameter $i(G)$, the independent domination number of G, is not really a multiproperty parameter. We defined $\gamma(G)$ as the minimum number of vertices in a (minimal) dominating set, and $\Gamma(G)$ has been defined as the maximum number of vertices in a minimal dominating set. Similarly, $\beta_0(G)$ is the maximum number of vertices in a (maximal) independent set, and $i(G)$ is the minimum number of vertices in a maximal independent set (see Beyer, et al [1]).

Because each independent set is maximal if and only if it is a dominating set, $i(G)$ equals the minimum number of vertices in a dominating set that is independent. But the single property of independence suffices to define $i(G)$.

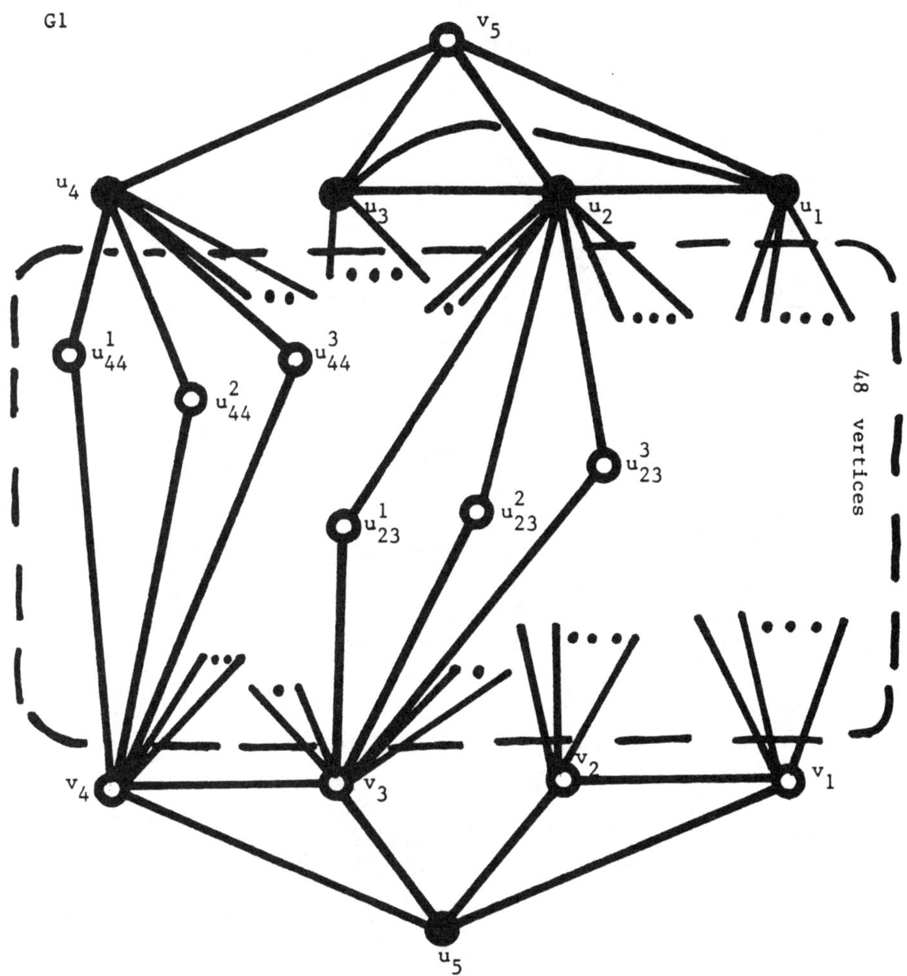

Figure 1. Graph $G1$ with $\gamma(G1) = 5$ and $\gamma(G1)$ sets
$\{u_1, u_2, u_3, u_4, u_5\}$ and $\{v_1, v_2, v_3, v_4, v_5\}$

Several measures of what "as independent as possible" means could be used. We illustrate two of these in Figure 1. The graph $G1$ has 58 vertices with $V(G1) = \{u_1, u_2, u_3, u_4, u_5, v_1, v_2, v_3, v_4, v_5\} \bigcup \{u^k_{ij} : 1 \leq i, j \leq 4, 1 \leq k \leq 3\}$. The edge set

contains $\{u_1u_2, u_2u_3, u_1u_3, u_1v_5, u_2v_5, u_3v_5, u_4v_5, v_1v_2, v_3v_4, v_1u_5, v_2u_5, v_3u_5, v_4u_5\}$.
Further, each u_{ij}^k is a vertex of degree two adjacent to u_i and v_j. That is, each
of the sixteen pairs $\{u_i, v_j\}$ share three common neighbors, each of degree two. It is
fairly obvious that $\gamma(G1) = 5$, and the only γ-sets are $D1 = \{u_1, u_2, u_3, u_4, u_5\}$ and
$D2 = \{v_1, v_2, v_3, v_4, v_5\}$. Vertex set $D1$ can be considered to be more independent
than $D2$ because $< D1 >$, the subgraph generated by $D1$, has two isolated vertices
while $< D2 >$ only has one. On the other hand, $D2$ can be considered to be more
independent than $D1$ because $< D2 >$ has only two edges while $< D1 >$ has
three. It is this latter measure we will use here.

2.   The NP-Hard Dipset Problem

Given a graph $G = (V, E)$ we are interested in finding a minimum Dominating
set D which is as Independent as Possible under the criterion that $< D >$ has as
few edges as possible. We define $\text{DIP}(G)$ to be the minimum number of edges in a
subgraph induced by a minimum dominating set of G. In particular, $\text{DIP}(G) = 0$ if
and only if $\gamma(G) = i(G)$. For example consider the tree $T2$ in Figure 2. We have
$i(G) = 9$ and $\{2, 8, 10, 13, 15, 17, 20, 24, 27\}$ is an $i(G)$-set. Also, $\gamma(G) = 8$ and
$\{2, 8, 10, 14, 17, 20, 24, 27\} = D1$ is a $\gamma(G)$-set with $< D1 >$ containing four
edges. $D1$ is the minimum dominating set obtained by the "anti-greedy" algorithm
applied to $T2$ with vertex 17 as the root, and $D1$ is "locally optimal" in the sense
that any $\gamma(T2)$-set $D2$ satisfying $D2 = D1 - v + u$ where $v \epsilon D1$ has at least four
edges in $< D2 >$. Note however that $D = D1 - 17 - 14 + 16 + 13$ is a $\gamma(T2)$-set
with only two edges in $< D >$. In fact, $\text{DIP}(T2) = 2$. We call a $\gamma(G)$-set, D, a
DIP-set if $< D >$ has DIP(G) edges.

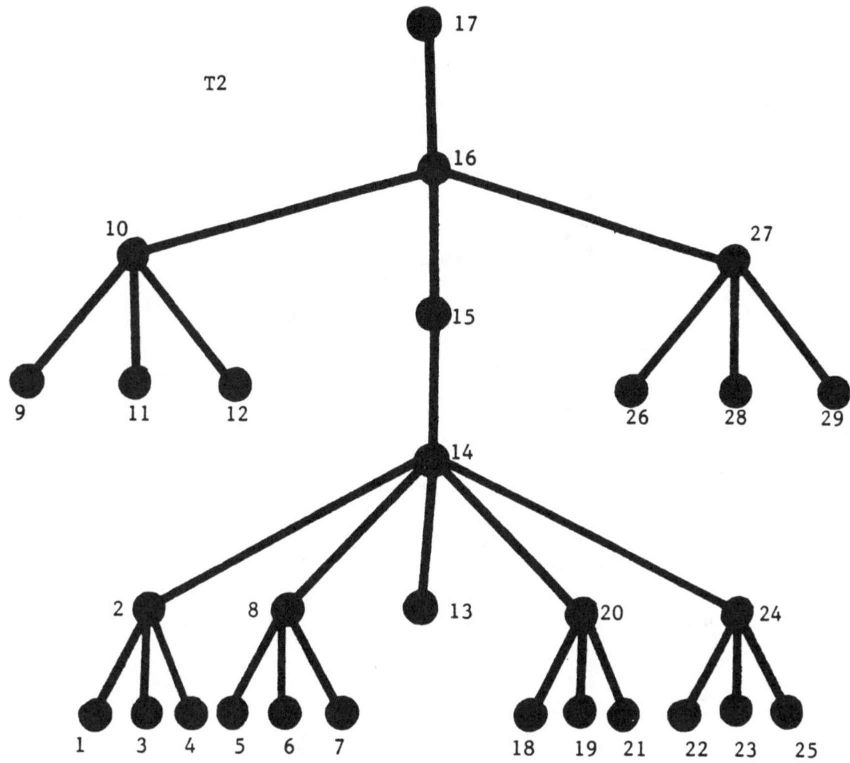

Figure 2. Tree T2 with $\gamma(\text{T2}) = 8$ and $\text{DIP}(\text{T2}) = 2$

We next show that determining whether or not $\text{DIP}(\text{G}) < \text{K}$ for a non-negative inter K is NP-hard. We describe a polynomial time reduction from 3-Satisfiability, one of the six basic NP-complete problems in Garey and Johnson [2], implying that our problem is NP-hard. We note that DIPSET is in NP^{NP}, but it appears not to be in NP because verification of a proposed solution set D includes verifying that D includes verifying that D is a *minimum* dominating set.

3-Satisfiability

Instance: Collection $C = \{c_1, c_2, \cdots, c_m\}$ of clauses on a finite set $U = \{u_1, u_2, \cdots, u_n\}$ of variables such that $|c_i| = 3$ for $1 \leq i \leq m$.

Question: Is there a truth assignment for U that satisfies all the clauses in C?

Dipset

Instance: Graph $G = (V, E)$, positive integer K.

Question: Is DIP(G) $\leq$ K? Given an instance of 3-satisfiability we construct a graph G as follows. First, for each variable $u_i \epsilon U$ we construct the 7-vertex, 9-edge graph H_i illustrated in Figure 3.

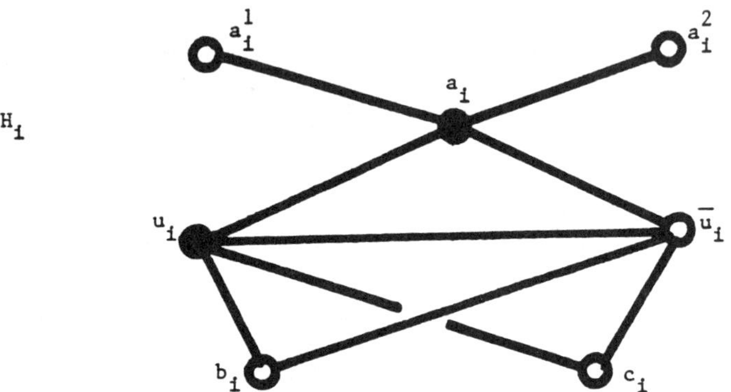

Figure 3. Graph H_i corresponding to variable v_i.

Second, for each clause $c_j \epsilon C$ construct the 21-vertex, 24-edge graph M_j illustrated in Figure 4.

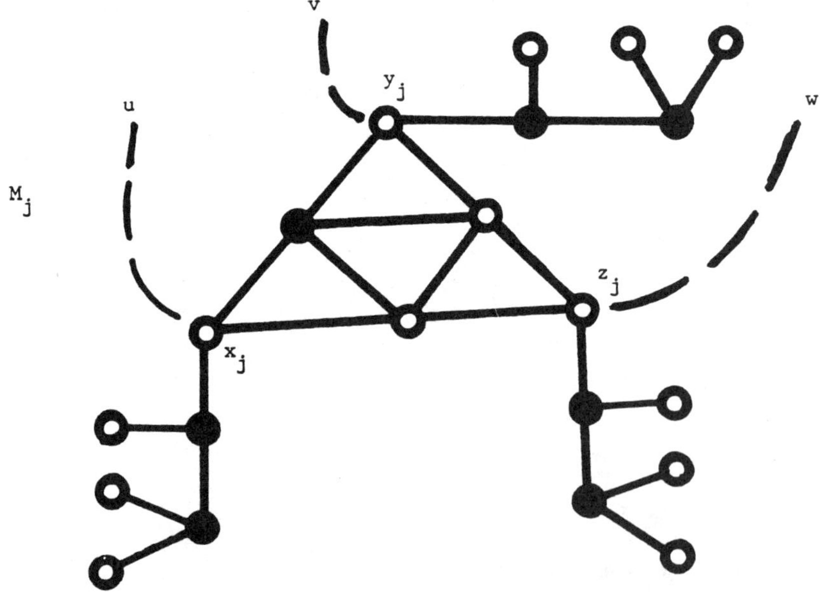

Figure 4. Graph M_j corresponding to clause c_j.

Finally, if $c_j = (u \lor v \lor w)$ where each of u, v, w is a literal u_k or $\overline{u}_k$ then add edges $x_j u, y_j v,$ and $z_j w$ as indicated in Figure 4. The graph G so constructed has $7n + 21m$ vertices and can be constructed from C in polynomial time. We have $\gamma(G) = 2n + 7m$. (One can use each a_i and u_i and each set of seven darkened vertices from each M_j to form a $\gamma(G)$-set). Because each $\gamma(G)$-set must contain each a_i and either u_i or $\overline{u}_i$, we always have DIP(G) $\geq$ n. It remains only to see, as can be easily verified, that there exists a truth assignment for U if and only if DIP(G) $\leq$ n.

Indeed, a referee has improved our result and shown the stronger result that *zero-dipset* is NP-hard.

Zero-Dipset

Instance: Graph $G = (V, E)$.

Question: Is DIP(G) $= 0$? (That is, does $\gamma(G) = i(G)$?)

3. A Linear Algorithm for Determining DIP(T) and a Corresponding DIP-set for a Tree T

In this section we will describe a linear algorithm for determining DIP(T), the minimum cardinality of the edge set of the induced subgraph of $d \subseteq V(T)$, where D is a minimum dominating set of T. We note that this algorithm can be extended to allow us to find such a set D, also in linear time. Throughout this section we will use $\delta(T)$ to denote DIP(T).

Without loss of generality, it will be assumed that all trees are rooted at some vertex that may be chosen arbitrarily. This will enable us to use recursive representations of trees. Given a rooted tree T, we will represent T by the number of nodes in T, say n, an endnode list $EL = (v_1, v_2, \cdots, v_n)$, and an associated parent list $PA = (u_1, u_2, \cdots, u_{n-1})$. The endnode list is any enumeration of the nodes of T in which each node precedes its parent. In the associated parent list, each u_i is the parent of v_i in T. Note that PA has length $n-1$ and not n, since the root v_n has no parent. For example, the tree of Figure 2 may be represented by n = 29, EL = (1, 5, 18, 22, 23, 19, 6, 3, 4, 7, 21, 25, 13, 2, 9, 8, 26, 20, 11, 24, 28, 14, 12, 15, 29,

10, 27, 16, 17) and PA = (2, 8, 20, 24, 24, 20, 8, 2, 2, 8, 20, 24, 14, 14, 10, 14, 27, 14, 10, 14, 27, 15, 10, 16, 27, 16, 16, 17).

These lists can be constructed for a tree of n nodes in time $O(n)$ (see [12] and [14]), so requiring them does not increase the order of execution time for our algorithm.

We will also make use of the following notation. As the vertex v is reached in left-to-right processing of the endnote list, let Tv be the subtree induced by v and all of its descendants. (Note that all of the descendants of v have already been processed since they precede v in EL.) Let u be the parent of v (which is determined using PA), and let Tu' be the subtree induced by u, the children of u that precede v in EL, and the descendants of all such children. Finally, let tu be composed of Tu', Tv and the edge (u, v). Note that Tu does not necessarily contain all of the descendants of u since there may be children of u which appear after v in EL. See Figure 5.

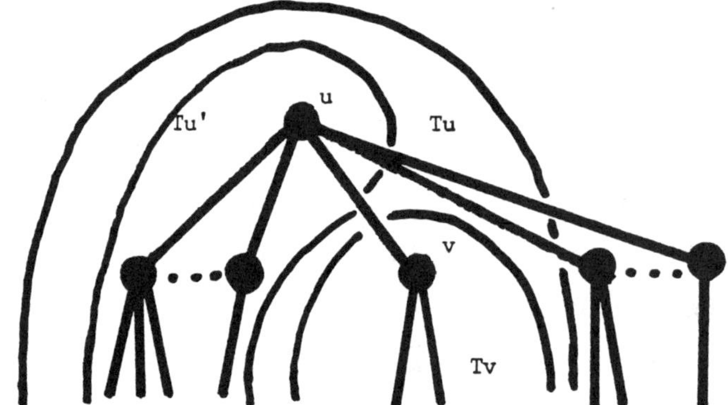

Figure 5. Illustration of Tv, Tu and Tu' notation

The first three parameters we introduce are used to insure that we have a minimum dominating set. Note that $\gamma(T)$ and the corresponding minimum dominating set D can be found without the use of these parameters by using an "anti-greedy" algorithm, but we employ them here because they will be used later in this section in determining $\delta(T)$. For a vertex $u\epsilon V(T)$, define

$$\gamma_y(Tu) = \text{MIN }\{|D| : u \in D, D \text{ dominates } Tu\}$$

$$\gamma_y(Tu) = \text{MIN }\{|D| : u \notin D, D \text{ dominates } Tu\}$$

$$\gamma_{\overline{n}}(Tu) = \text{MIN }\{|D| : u \notin D, D \text{ dominates } Tu - u,$$

$$D \text{ does not dominate } u\}$$

where $D \subseteq V(tu)$. (Note that we need to require $D \subseteq V(Tu)$ only for the case $\gamma_n(Tu)$ where u is an endnode. In this case we will define $\gamma_n(Tu) = \infty$.) That is, let $\gamma_y(Tu)$ be the minimum cardinality of a set containing u that dominates Tu; let $\gamma_n(Tu)$ be the minimum cardinality of a set not containing u that dominates Tu; and let $\gamma_{\overline{n}}(Tu)$ be the minimum cardinality of a set not containing u that dominates $Tu - u$ but specifically does not dominate u itself. This third parameter will be useful when u is to be dominated by its parent or by an as yet unprocessed child.

Similar parameters have been used in other work dealing with minimum dominating sets. See Grinstead and Slater [6]. Note that our parameters differ slightly from theirs in that they used $\gamma_{\overline{n}}(Tu)$ to mean it does not matter whether u is dominated or not. We, however, use it to mean that u specifically *does not* get dominated.

If D is a minimum dominating set of Tu and if $u \epsilon D$, then we may write $D = U \cup V$ where U is a minimum dominating set of Tu', and V is a minimum set of vertices in $V(Tv)$ which dominated $Tv - v$ (since v is already dominated by u). The set V may contain v, not contain v but dominate it, or not contain v and not dominate it.

$$\gamma_y(Tu) = \gamma_y(Tu') + \text{MIN }\{\gamma_y(Tv), \gamma_n(Tv), \gamma_{\overline{n}}(Tv)\} \tag{1}$$

Similarly, it is straightforward to derive the following:

$$\gamma_n(Tu) = \text{MIN }\{\gamma_n(Tu') + \gamma_n(Tv), \gamma_n(Tu') + \gamma_y(Tv),$$
$$\gamma_{\overline{n}}(Tu') + \gamma_y(Tv)\} \tag{2}$$

$$\gamma_{\overline{n}}(Tu) = \gamma_{\overline{n}}(Tu') + \gamma_n(Tv) \tag{3}$$

To see how these parameters should be initialized, consider a subtree consisting of a single vertex v. The minimum number of vertices needed to dominate the subtree using v is one; it is not possible to dominate the subtree without sign v; and zero vertices are required to dominate the subtree minus v. Thus for any endnote v_i of a tree T we may initialize $\gamma_y(Tv_i) = 1, \gamma_n(Tv_i) = \infty$, and $\gamma_{\overline{n}}(Tv_i) = 0$. For any internal node u_i of T we initialize its parameters similarly. After this initialization, we may proceed left-to-right through the endnode list evaluating equations (1) through (3) for u_i, where u_i is the parent of the current endnode list entry v_i. Once v_n is reached, we can determine $\gamma(T) = \text{MIN } \{\gamma_y(Tv_n), \gamma_n(Tv_n)\}$.

As previously stated, similar parameters have been used before. We now present an additional three parameters which are used to determine $\delta(T)$, the minimum cardinality of the edge set of the induced subgraph of $D \subseteq V(T)$, where D is a minimum dominating set (MDS) of T. While the domination parameters are used to maintain information about possible minimum dominating sets, these new parameters are used to maintain information about the "non-independence" of those sets. For a vertex $u \epsilon V(T)$, define

$$\delta_y(Tu) = \text{MIN } \{|E(< D >)| : u \in D, D \text{ is an MDS of } Tu\}$$

$$\delta_n(Tu) = \text{MIN } \{|E(< D >)| : u \notin D, D \text{ is an MDS of } Tu\}$$

$$\delta_{\overline{n}}(Tu) = \text{MIN } \{|E(< D >)| : u \notin D, D \text{ is an MDS of } Tu - u$$

$$D \text{ does not dominate } u\}$$

That is, let $\delta_y(Tu)$ be the minimum cardinality of the edge set of the subgraph induced by a minimum dominating set of Tu that contains vertex u; let $\delta_n(Tu)$ be the minimum cardinality of the edge set of the subgraph induced by a minimum dominating set of Tu that does not contain vertex u; and let $\delta_{\overline{n}}(Tu)$ be the minimum cardinality of the edge set of the subgraph induced by a minimum dominating set of $Tu - u$ that does not contain vertex u and specifically does not dominate u.

These new parameters are used to keep track of the order of the edge set induced by a dominating set, but in this case we only consider minimum dominating sets as determined by the γ-parameters. Thus, the calculation of the δ-parameters closely parallels the calculation of the δ- parameters but only involves those terms corresponding to the terms in the γ-parameters that lead to a minimum for that γ-parameter, as follows.

For a parent vertex $u \epsilon V(T)$, let $M1 = \text{MIN } \{\gamma_y(Tv), \gamma_n(Tv), \gamma_{\overline{n}}(Tv)\}$ from equation (1). We calculate $\delta_y(Tu)$ as follows.

$$\text{IF } \gamma_y(Tv) = M1 \text{ THEN } A = \delta_y(Tu') + 1 + \delta_y(Tv)$$

$$\text{ELSE } A = \infty$$

$$\text{IF } \gamma_n(Tv) = M1 \text{ THEN } B = \delta_y(Tu') + \delta_n(Tv)$$

$$\text{ELSE } B = \infty$$

$$\text{IF } \gamma_{\overline{n}}(Tv) = M1 \text{ THEN } C = \delta_y(Tu') + \delta_{\overline{n}}(Tv)$$

$$\text{ELSE } C = \infty$$

$$\delta_y(Tu) = \text{ MIN}\{A, B, C\}$$

Similarly, let $M2 = \gamma_n(Tu)$ from equation (2) and calculate $\delta_n(Tu)$ as follows.

$$\text{IF } \gamma_n(Tu') + \gamma_n(Tv) = M2 \text{ THEN } A = \delta_n(Tu') + \delta_n(Tv)$$

$$\text{ELSE } A = \infty$$

$$\text{IF } \gamma_n(Tu') + \gamma_y(Tv) = M2 \text{ THEN } B = \delta_n(Tu') + \delta_y(Tv)$$

$$\text{ELSE } B = \infty \tag{5}$$

$$\text{IF } \gamma_{\overline{n}}(Tu') + \gamma_y(Tv) = M2 \text{ THEN } C = \delta_{\overline{n}}(Tu') + \delta_y(Tv)$$

$$\text{ELSE } C = \infty$$

$$\delta_n(Tu) = \text{ MIN}\{A, B, C\}$$

Finally, to calculate $\delta_{\overline{n}}(Tu)$

$$\delta_{\overline{n}}(Tu') + \delta_n(Tv) \tag{6}$$

In a manner similar to that for the domination parameters, the initializations for these new parameters for a vertex v_i of T are $\delta_y(Tv_i) = 0, \delta_n(Tv_i) = \infty$ and $\delta_{\overline{n}}(Tv_i) = 0$. After initializing all six parameters, we may proceed through the endnode list as before evaluating equations (1) through (6) for u_i, where u_i is the parent of the current endnode list entry v_i. Once v_n is reached, we can determine $\gamma(T)$ as before and $\delta(T)$ as follows.

$$M = \gamma(T)$$

$$\text{IF} \quad \gamma_y(Tv_n) = M \quad \text{THEN} \quad A = \delta_y(Tv_n)$$

$$\text{ELSE} \quad A = \infty$$

$$\text{IF} \quad \gamma_n(Tv_n) = M \quad \text{THEN} \quad B = \delta_n(Tv_n)$$

$$\text{ELSE} \quad B = \infty$$

$$\delta(T) = \text{MIN} \ \{A, B\}$$

These six parameters let us determine the parameters $\gamma(T)$ and $\delta(T)$. We note than one can find a minimum dominating set, D, corresponding to $\gamma(T)$ and $\delta(T)$ in one additional right-to-left scan of the endnode list if one calculates some additional information during the initial left-to-right scan. Details are available in [13].

REFERENCES

[1] T. Beyer, A. Proskurowski, S. Hedetniemi, and S. Mitchell, "Independent domination in trees," Proc. 8th S. E. Conference on Combinatorics, *Graph Theory and Computing*, Utilitas Mathematica, Winnipeg, 1977, 321-328.

[2] M. R. Garey and D. S. Johnson, *Computers and Intractability*, W. H. Freeman and Company, San Francisco, 1979.

[3] M. R. Garey, D. S. Johnson, and L. Stockmeyer, "Some simplified NP-complete graph problems," *Theor. Comp. Sci.*, 1, 1976, 237-267.

[4] M. R. Garey, D. S. Johnson, and R. E. Tarjan, 1976, unpublished.

[5] D. L. Grinstead, Ph.D. thesis, in preparation.

[6] D. L. Grinstead and P. J. Slater, "On minimum dominating sets with minimum intersection," *Annals of Discrete Math*, to appear.

[7] D. L. Grinstead and P. J. Slater, "On minimum dominating sets with minimum intersection in series-parallel graphs," these proceedings.

[8] D. L. Grinstead and P. J. Slater, "A multiset, single property problem: maximum independent sets with minimum intersection," in preparation.

[9] P. L. Hammer, "A bibliography on domination in graphs," personal communication.

[10] S. T. Hedetniemi and R. Laskar, "A bibliography on domination in graphs," personal communication.

[11] R. M. Karp, "Reducibility among combinatorial problems," in R. E. Miller and J. W. Thatcher (eds.), *Complexity of Computer Computations,* Plenum Press, New York, 1972, 85-103.

[12] D. E. Knuth, "The Art of Computer Programming, Vol. 1: Fundamental Algorithms," Addison-Wesley, Reading (1968) 334-338.

[13] W. J. Selig and P. J. Slater, "Minimum dominating, optimally independent vertex sets in graphs," The University of Alabama in Huntsville Mathematical Sciences Technical Report, 1988-81.

[14] R. E. Tarjan, "Depth-first search and linear graph algorithms," *SIAM J. Computing* 1 (1972) 146-160.

Duality Theorems for the Maximum Concurrent Flow Problem

Farhad Shahrokhi

Department of Computer Science

New Mexico Institute of Mining and Technology

ABSTRACT

The maximum concurrent flow problem (MCFP) is a multicommodity flow problem. The objective in solving the MCFP is to maximize the portion of the demands which could be realized in the network, subject to the capacity constraints. In this paper, we develop duality theorems similar to the max–flow min cut–theorem for the MCFP.

1. Introduction

The single commodity flow problem [FF62] is a well known problem with many applications to the other combinatorial problems [FF62, PS82]. The elegant maximum–flow minimum–cut theorem has two important features: (i) it relates the maximum flow problem to an important dual combinatorial problem, the minimum cut problem, and (ii) it provides for the efficient algorithms which simultaneously solve both the maximum flow problem and the minimum cut problem [PS82]. Unfortunately, not much of the elegant structure of the single commodity flow problem has been discovered for the multicommodity flow problems. That is, the max–flow min–cut theorem, in general, does not hold for the multicommodity flow problems. (For very restricted multicommodity flow problems the max–flow min–cut theorem holds [Hu69, OS81, Se81].)

The maximum concurrent flow problem (MCFP) [Ma85] is the optimization version of the feasibility problem in multicommondity flows [Hu69], [On70], [Ir71]. The objective is to maximize the ratio of the flow supplied for every commodity to the demand for that commodity, subject to the capacity constraints. The ratio of the utilized flow to the demand for a commodity is termed the throughput and is the same for all the commodities. In [SM88] we developed an efficient approximation algorithm and constructed combinatorial duals for the MCFP. One of these duality theorems combined the distance dual [Ir71], [On70] to the concept of a *k–partite* cut and provided for a more natural combinatorial path–cut dual. The duality results in [MS88] have been verified only for the MCFP with uniform capacity, that is, when all the edge capacities are the same. The main contribution of this paper is to extend the duality results for the MCFP with uniform capacity in [SM88] to the general MCFP. Our first duality theorem (which extends the results in [On70], [Ir71] for feasibility problem to the optimization problem) relates the concurrent flow problem to the problem of associating costs (distances) to the edges so as to maximize the minimum cost of routing the concurrent flows. Our effort to obtain a natural combinatorial "cut dual" for the MCFP is, then, embodied in a second duality theorem which is our main result. Our duailty proofs are short and employ the dulaity theory of the linear programming. Our work emphasizes on utilizing the duality theory of linear programming as a tool to explore and analyze the combinatorial structure of the problem. The work in this paper is non-algorithmic and has focussed on exploring the combinatorial structure of the MCFP.

It is worth mentioning that although the linear programming had been initially devised for solving the continuous optimization problems, it has been proven to be fundamental to the study of many strictly combinatorial problems. The idea of applying the LP duality to a network problem and explore the structure of the problem has been useful in the desing of specialized algorithms for many combinatorial optimization problems such as, the single source shortest paths problem, the single commodity flow problem, the minimum cost flow problem, and the weighted matching proble [PS82].

Our paper is organized as follows: Section 2 formally defines the MCFP and the related concepts. In Section 3 tow duality theorems are presented. Finally, in Section 4 we give our concluding remarks regarding the computational aspects of our duality results.

2. The Maximum Concurrent Flow Problem (MCFP)

We assume the reader is familiar with basic graph terminology as in [CL86]. Throughout this paper we assume that graph $G = \langle V, E \rangle$ has the vertex set $V = \{1, 2, ..., n\}$. Let $G = \langle V, E \rangle$ be a graph. We denote by P_e the set of all paths containing the edge e, for each edge $e \in E$, and by P_{ij} the set of all paths between end vertices i, j (termed ij paths), for each distinct pair of vertices $i, j \in V$. Also let P denote the set of all paths in G. A capacity function on G is a function $C : E \rightarrow R^+$. A demand fucntion D on G is a function $D : V \times V \rightarrow R^+$ such that $D(i, j) = D(j, i)$ and $D(i, i) = 0$ for all $i, j \in V$. To eliminate the trivial cases we assume, $D(i, j) > 0$ for some $i, j \in V$.

A *concurrent flow* of throughput Z in G is a function $f : P \to R^+$ such that

i) $\sum_{p \in P_{ij}} f(p) = Z\,D(i,j)$ for all distinct pairs of vertices $i, j \in V$ and

ii) $\sum_{p \in P_e} f(p) \le C(e)$ for every edge $e \in E$,

where D and C are the given demand and capacity function on G. We extend the domain of a concurrent flow f to $P \cup E$ using

$$f(e) = \sum_{p \in P_e} f(p) \text{ for all } e \in E.$$

$P' = \{p \mid f(p) > 0\}$ denotes the set of active paths for the concurrent flow f. The *maximum concurrent flow problem* (MCFP) is to find the largest throughput $\hat{Z}$ achieved by a concurrent flow function f which is explicitly defined on the set of active paths P'. The MCFP can be expressed as the linear program,

$$\left|\begin{array}{ll} \text{Maximize } Z, \text{ subject to} & \\[4pt] \sum_{p \in P_{i,j}} f(p) - Z\,D(i,j) = 0 & \text{for all } i, j \in V \text{ with } i < j, \\[4pt] \sum_{p \in P_{i,j}} f(p) \le C(e) & \text{for all } e \in E, \\[4pt] f(p) \ge 0 & \text{for all } p \in P, \end{array}\right. \qquad (2.1)$$

where D, C are the demand and capacity function on $G = \langle V, E \rangle$. The set of *saturated edges* for f is $\{e \mid f(e) = C(e)\}$.

A distance (toll) function on $G = \langle V, E \rangle$ is a function $d : E \to R^+$ such that $d(e) > 0$ for at least one edge e. Let d be a distance function; we set,

$d(p) = \sum_{e \in p} d(e)$ for all $p \in P$.

$d(i, j) = \min_{p \in P_{ij}} \{d(p)\}$ for all $i, j \in V$.

A shortest distance ij path $p \in P_{ij}$ has $d(p) = d(i, j)$ and the expression

$$\left\{ \frac{\sum_{e \in E} d(e)C(e)}{\sum_{i<j} d(i, j)D(i, j)} \right\}$$

which is denoted by Z_d is termed the *distance ratio* of d. Let $(A, \overline{A})$ be a cut in G.

Let $C(A, \overline{A}) = \sum_{e \in (A, \overline{A})} C(e)$ and $D(A, \overline{A}) = \sum_{i \in A, j \in \overline{A}} D(i, j)$. The expression $\dfrac{C(A, \overline{A})}{D(A, \overline{A})}$

is termed density of $(A, \overline{A})$, and is denoted by *den* $(A, \overline{A})$. A cut of minimum density is termed a *sparsest cut*. The following lemma is easy to verify.

Lemma 2.1   *[Ma85], [SM87] Let $G, D,$ and C be an instance of the MCFP. Then any maximum concurrent flow $\hat{f}$ of throughput $\hat{Z}$ satisfies*

(i) *[Cut Upper Bound] $\hat{Z} \le den(A, \overline{A})$, where $(A, \overline{A})$ is any cut in G with $D(A, \overline{A}) > 0$, and*

(ii) *[Distance Upper Bound] $\hat{Z} \le Z_d$, where Z_d is the distance reatio correpsonding to any distance function d on G with $\sum_{i<j} d(i, j)D(i, j) > 0$.*

3. Duality for the MCFP

It is interesting to observe that the cut upper bound in Lemma 2.1 will not provide a dual to the MCFP, that is, the largest throughput does not necessarily equal to the density of a sparest cut. To see this, consider the MCFP for the graph $K_{2,3}$ with unit capacities on the edges and unit demands between all vertex pairs. This instance of the MCFP has an optimal throughput of $3/7$ with a sparest cut of density $1/2$ [Se81], [Ma85]. Therefore, in general, the cut upper bound is subject to an inherent gap. Our first duality result establishes the more gneral distance upper bound of Lemma 2.1 as the true "dual", meaning that for some distance fucntion the distance upper bound inequality becomes equality.

It is sometimes more convenient to treat maximzing the throughput of the concurrent flow as minimizing the utilized capacity to realize the demands. For a given G, D and C, the minimum capacity ultilization problem (MCUP) is

$$\left| \begin{array}{ll} \text{Minimize } u, \text{ subject to} & \\[4pt] \sum_{p \in P_{i,j}} f(p) = D(i, j) & \text{for all } i, j \in V \text{ with } i < j, \\[4pt] uC(e) - \sum_{p \in P_e} f(p) \geq 0 & \text{for all } e \in E, \\[4pt] f(p) \geq 0 & \text{for all } p \in P, \end{array} \right. \qquad (3.1)$$

Note that in the MCUP we are seeking smallest percentage of (uniform) increase on the edge capacities such that all demands are satisfied. Observe that if the optimal values of objective functions for (3.1) and (2.1) are denoted by u^*, $\hat{Z}$ respectively, then $\hat{Z} = 1/u^*$. In this paper we refer to the maximum concurrent flow problem to include both MCFP and MCUP, with the application or algorithm dictating the preferred formulation.

Let G, D, C be an instance of the MCFP. The *Min Distance Ratio Problem* (MDRP) is to determine a distance function $\hat{d}$ such that

$$Z_{\hat{d}} = \min \{Z_d \mid d \text{ is a distance function on } G\} \qquad (3.2)$$

Our first duality theorem is the following.

Theorem 3.1 *[Max–Throughput Min–distance Ratio Theorem.] Let G, D, C be an instance of the MCFP with the optimal throughput $\hat{Z}$. Then*

$$\hat{Z} = Z_{\hat{d}}.$$

Proof We will show this, using the dulaity theory of linear programming. Consider the dual of (3.1) which is the LP,

$$\left\{ \begin{array}{ll} \text{maximize } \Sigma_{i<j\in V} \, Q(i,j)D(i,j) \text{ subject to,} \\[4pt] \Sigma_{e\in p} \, Q(e)C(e) \le 1, \\[4pt] Q(i,j) - \Sigma_{e\in p} \, Q(e) \le 0, & \text{for all } ij \text{ paths } p, \ i,j \in V, i<j \\[4pt] Q(e) \ge 0 & \text{for all } e \in E \\[4pt] Q(i,j) <> 0 & \text{for all } i,j \in V, i<j \end{array} \right. \tag{3.3}$$

($Q(i,j) <> 0$ indicates that $Q(i,j)$ is unconstrained in sign.) Let $\hat{Q}$ be an optimal solution to (3.3) and assume $\overline{d}$ is the distance function identified by $\overline{d}(e) = \hat{Q}(e)$ for all $e \in E$. Let p be any ij path with $\overline{d}(p) = \overline{d}(i,j)$ and $D(i,j) > 0$. We have, $\hat{Q}(i,j) \le \Sigma_{e\in p} \overline{d}(e) = \overline{d}(i,j)$, implying that $\overline{d}$ is also an optimal solution to (3.3). Thus

$$\frac{1}{\hat{Z}} = u^{*} = \sum_{i<j} D(i,j)\overline{d}(i,j) \quad \text{or} \quad \hat{Z} = \frac{1}{\Sigma_{i<j}\overline{d}(i,j)D(i,j)} = Z_{\overline{d}}.$$

(Note that $\Sigma_{e\in E} \overline{d}(e)c(e) = 1$, since $\overline{d}$ is optimal.) However, by Lemma 2.1, $\hat{Z} \le Z_{\overline{d}}$. Thus, $Z_{\overline{d}}$ is the smallest distance ratio which implies, $\hat{Z} = Z_{\overline{d}} = Z_{\hat{d}}$. $\square$

The following simple observation which derives from Theorem 3.1 is important; let f, d be a concurrent flow and a distance function feasible in (3.1) and (3.3) respectively, then by the complementary slackness of the linear programming f, d are jointly optimal if and only if

$$f(p)(d(p) - d(i,j)) = 0 \text{ for all } p \in p_{ij},$$

$$[\textit{every active ij path is a shortest ij path}] \tag{3.4}$$

and

$$d(e)(uC(e) - f(e)) = 0 \text{ for all } e \in E,$$

$$[\textit{non–saturated edges are assigned zero distances}] \tag{3.5}$$

In order to present our second duality theorem, additional terms must be defined.

Let $G = <V, E>$ be a graph; denote by $(A_1, A_2, ..., A_k)$, for some fixed $k \ge 2$, the set of all edges in G whose end vertices are in distinct elements of the partition $\{A_1, A_2, ..., A_k\}$ of V. We term $(A_1, A_2, ..., A_k)$ a *k–partitie cut* in G. Let $(A_1, A_2, ..., A_k)$ denote any k–partite cut of G. For $1 \le i, j \le k$, let $C(A_i, A_j)$ denote the total capacity of all edges between A_i and A_j, and let $D(A_i, A_j)$ denote the total demand between vertices of A_i and A_j. Also, define a weight function to be any function

$$w : \{1, 2, ..., k\} \times \{1, 2, ..., k\} \to R^{+}$$

satisfying $w(i,j) + w(j,l) \ge w(i,l)$ and $w(i,j) = w(j,i)$ for any $1 \le i, j, l \le k$, with $\Sigma_{1\le i<j\le k} \, w(i,j)D(A_i, A_j) \ne 0$. Let $\text{den}(A_1, A_2, ..., A_k)$ denote the *density* of the k-partite cut $(A_1, A_2, ..., A_k)$ defined by

$$\text{den}(A_1, A_2, ..., A_k) = \min_{w} \frac{\displaystyle\sum_{1 \le i < j \le k} w(i,j)C(A_i,A_j)}{\displaystyle\sum_{1 \le i < j \le k} w(i,j)D(A_i,A_j)} \tag{3.6}$$

where the minimum is over all weight functions. Note that for any 2–partite cut $(A, \overline{A})$ of G, the density is simply

$$\text{den}(A,\overline{A}) = \frac{C(A,\overline{A})}{D(A,\overline{A})} = \frac{\sum_{e \in (A,\overline{A})} C(e)}{\sum_{i \in A,\, j \in \overline{A}} D(i,j)} ,$$

which is the density of $(A, \overline{A})$.

The set of *critical edges* in G consists precisely of all those edges that are saturated under any maximum concurrent flow function in G. The following theorem explores the structure of the critical edges and will be used in our main result.

Theorem 3.2   *[Ma85], [SM88] The set of critical edges is a k–partite cut.*

We now present our main result which exhibits the "cut dual" of the MCFP.

Theorem 3.3   *[Max concurrent flow Min k–partite cur Theorem] For any G, D and C, the maximum throughput of any concurrent flow equals the minimum density of any k–partite cut, and in particular, equals the density of the k-partite cut of critical edges.*

Proof   Assume that, $(A_1, A_2, ..., A_j)$, $2 \le j \le n$ is any j–partite cut. Let w' be a weight function minimizing (3.6). We can construct a distance function d', such that $d'(e) = w'(i,j)$ whenever the end vertices of e are in different parts A_i, A_j, and $d'(e) = 0$ otherwise. The shortest distance function extension of d' in G to $V \times V$ must have $d'(u,v) \ge w'(i,j)$ for $u \in A_i$ and $v \in A_j$, $i \ne j$, since w' satifies the triangle inequality. Furthermore, $d'(u,v) = w'(i,j)$ if some edge $e \in E$ joins a vertex in A_i to a vertex in A_j. Note that, if there are no edges between A_i and A_j, then $C(A_i, A_j) = 0$, hence $\sum_{e \in e} d'(e)C(e) = \sum_{1 \le i < j \le k} w'(i,j)C(A_i, A_j)$. But $d'(u,v) \ge w'(i,j)$ for $u \in A_i$, $v \in A_j$, thus we have $w'(i,j) = d'(u,v)$ since w' minimizes the right hand side of (3.6). Thus $\text{den}(A_1, A_2, ..., A_j) = Z_{d'} \ge \hat{Z}$.

Conversely, observe that by (3.5) any optimal distance function d must assign non–zero distances to the critical edges, which compose the k–partite cut $E_c = (A_1, A_2, ..., A_k)$. Furthermore, for any $e_1 = u_1v_1$ and $e_2 = u_2v_2$, with $u_t \in A_i$ and $v_t \in A_j$, $t = 1, 2$, it is not difficult to verify that $d(e_1) = d(e_2) = d(u_1,v_1) = d(u_2,v_2)$. Now let $u \in A_i$ and $v \in A_j$ and set $w(i,j) = d(u,v)$, for all $1 \le i, j \le k$, $i \ne j$, and obtain $\text{den}(A_1, A_2, ..., A_k) = \hat{Z}$, which implies the result. $\square$

4. Computational Aspects of the MCFP

The general instances of MCFP can be solved in polynomially bounded time [Ta86], [SM88], however, the corresponding algorithms are not efficient. In [SM88, Sh89] we have developed an efficient fully polynomial time approximation algorithm [PS82] for the MFCP. It is interesting to note that the correctness of the algorithm in [SM88] can be verified using the duality results presented in this paper (this fact was not obvserved in [SM88]). In particular, the algorithm in [SM88] computes a concurrent f and a distance d such that (3.4) and (3.5) are almost satisfied (here, almost means with small error). To (almost) satisfy (3.5), the algorithm assigns the saturated edges much larger distances thant the distances assigned to those edges which are not saturated. To (almost) satisfy (3.4), the algorithm reroutes the flow from the long ij paths to the short ij paths and recomputes the distances on the edges. It can be verified, then, that after polynomially many iteration, both (3.4) and (3.5) will be (almost) satisfied. This, however, implies that the near optimal solutions to the MCFP and its dual are obtained [SM88].

Finally, the feasibility version of the MCFP problem [Hu69, On70] in planar networks has been solved using efficient combinatorial algorithms [MNS85, MNS86]. These algorithms are based on the elegant duality theorems in [Se81] and [Os81]. The duality results in [Se81, Os81] indicate that for the MCFP in certain planar networks, the density of a sparsest cut equals to the largest throughput.

REFERENCES

[CL86] Chartrand, G., and Lesniak, L., Graphs and Digraphs: Wadsworth and Books/Cole Mathematics series 1986.

[FF62] Ford, L.R., and Fulerson, D.R., Flows in Networks, Princeton: Princeton University Press, 1962.

[Hu69] Hu, T.C., Integer Programming and Network Flows. Reading, Mass.: Addison–Wesley.

[Ir71] Iri, M., On an Extension of the Maximum–Flow Minimum–Cut Theorem to the Multicommodity Flows, J. of Operation Research Society of Japan, vol. 5, no. 4, Dec. 1967, pp. 691 – 703.

[Ma85] Mastula, D.W., Concurrent Flow and Concurrent Connectivity in Graphs, in Graph Theory and its Applications to Algorithms and Computer Science, ed. Alavi, Y., et al., Wiley, New York, 1985, pp. 543 – 559.

[MNS85] Matsumoto, K., Nishizeki, T., and Saito, N., An Efficient Algorithm for Finding Multicommodity Flows in Planar Networks, SIAM J. on Compt., 14, 1985, pp. 289 – 301.

[MNS86] Matsumoto, K., Nishizeki, T. and Saito, N., Planar Multicommodity Flows, Maximum Matchings and Negative Cycles, SIAM J. on Compt., 15, 1986, pp. 495 – 510.

[OS81] Okamura, H., Seymour, P.D., Multicommodity Flows in Planar graphs, J. Combin. Theory B, 31 (1981), pp. 75 – 81.

[On70] Onaga, K., A Multicommodity Flow Theorem, Trans. on IECE, Japan, vol. 53–A, No. 7, 1970, pp. 350 – 356.

[PS82] Papadimitriou, H., and Stieglitz, K., Combinatorial Optimization: Algorithms and Complexity, Englewood Cliffs: Prentice Hall, 1982.

[Se81] Seymour, P.D., On Odd Cuts and Planar Multicommodity Flows, Proc. London, Math. Soc., (3) 42 (1981), pp. 178 – 192.

[SM87] Shahrokhi, F., and Matula, D.W., The Maximum Concurrent Flow Problem, to appear in JACM.

[Sh89] Shahrokhi, F., Approximation Algorithms for the Maximum Concurrent Flow Problem, to appear in ORSA J. on Computing.

[Ta86] Tardos, E., A Strongly Polynomial Algorithm to Solve Combinatorial Linear Programs, J. Operations Research, Vol. 34, No. 2, 1986.

Who is the Best Doubles Tennis Player?
An Introduction to k–tournaments

Paul K. Stockmeyer
College of William and Mary

ABSTRACT

The graph–theoretic concept of a (round–robin) tournament, *which models a competition in which each of n participants plays a match with each of the others, can be generalized to team competitions. Thus a k–tournament of order n can be defined in the obvious way to model a competition in which each of the $\binom{n}{k}$ k–subsets of a set of n participants plays a match with each of the $\binom{n-k}{k}$ k-subsets disjoint from it.*

In this paper we begin to study the structure of k–tournaments, examining questions such as the one implied by the title: under what circumstances is it possible to rank a set of individual participants based upon team performances?

1. Introduction

Suppose that four people with the rather unimaginative names of A, B, C, and D wish to determine their relative talents in doubles tennis. They first play a match with A and B competing against C and D, which we may assume without loss of generality is won by A and B. Next A and C team up against B and D, and by symmetry we may as well suppose that A and C win. Finally, A and D compete against B and C. Here the situation is not symmetric. Prior to this third match, A and D have win–loss records of 2–0 and 0–2, respectively, while B and C are both 1–1.

If the team of A and D wins this third match, then A has a final record of 3–0, while B, C, and D all end up with records of 1–2. On the other hand, if B and C defeat A and D, then A, B, and C tie for first place with records of 2–1 while D comes in last at 0–3. Moreover, except for possible renaming, these two outcomes are the only ones possible. Either one player will dominate, winning all three of his matches, with the other three players tie for last place, or else one player will lose all three of his matches while the others tie for first place. It is not possible to linearly order the players based on team competitions; depending on the relative strengths of the

"

players, it is possible to determine either the best player or the worst player, but not both. These sets of matches illustrate the concept of a k–tournament, which we now define.

Definition A *k–tournament of order* n is a triple $\langle P, T, M \rangle$ consisting of

1). a set P of n players,

2). the set T of all k–element subsets (teams) of P, and

3). a set M of ordered pairs (matches) of disjoint teams such that exactly one of the pairs $\langle t_i, t_j \rangle$ and $\langle t_j, t_i \rangle$ is in M for each disjoint pair of teams t_i and t_j. If $\langle t_i, t_j \rangle \in$ M, we say that t_i *dominates* t_j; this will occasionally be denoted $t_i > t_j$.

There are two comments that should be made here. First, a traditional graph–theoretic (round–robin) tournament is the same as a 1–tournament if we identify players with 1–person teams. Thus k–tournaments are true generalizations of standard tournaments. Second, we allow the possibility that $n < 2k$. When $n < k$, a k–tournament of order n consists of players only, with no teams. For $k \le n \le 2k$, a k–tournament has teams but no matches. Such k–tournaments are called *trivial*; a k–tournament with $n \ge 2k$ is called *non–trivial*.

Although the concept of a k–tournament seems quite natural, it does not appear to have been studied previously. In this paper we begin an exploration of these structures, presenting some elementary results and suggesting areas for further investigation.

2. Enumeration

There are several elementary counting results for k–tournaments that we list here for future reference. The proofs are immediate.

Theorem 1 *In any k–tournament of order n,*

a). there are exactly $\binom{n}{k}$ teams;

b). each team participates in $\binom{n-k}{k}$ matches;

c). there are $\dfrac{\binom{n}{k} \binom{n-k}{k}}{2}$ total matches;

d). each player participates in $\binom{n-1}{k-1} \binom{n-k}{k}$ matches. Moreover,

e). there are exactly $2^{\binom{n}{k} \binom{n-k}{k}/2}$ total (labeled) k–tournaments of order n.

In the case of 2–tournaments of order 4, we have already seen that there are 6 teams; each team and each player participates in all 3 matches. There are $2^3 = 8$ possible outcomes for this set of 3 matches, and hence 8 (labeled) 2–tournaments. They partition into 2 equivalence classes under the relation of isomorphism, which we now formally define.

Definition　　Two k–tournaments $\mathbf{T_1} = \langle P_1, T_1, M_1 \rangle$ and $\mathbf{T_2} = \langle P_2, T_2, M_2 \rangle$ are *isomorphic* if there exists a bijection σ from P_1 to P_2 such that

$$\langle \{p_{i_1}, p_{i_2}, ..., p_{i_k}\}, \{p_{j_1}, p_{j_2}, ..., p_{j_k}\} \rangle \in M_1$$

if and only if

$$\langle \{\sigma(p_{i_1}), \sigma(p_{i_2}), ..., \sigma(p_{i_k})\}, \{\sigma(p_{j_1}), \sigma(p_{j_2}), ..., \sigma(p_{j_k})\} \rangle \in M_2.$$

Such a mapping σ is an *isomorphism* from $\mathbf{T_1}$ to $\mathbf{T_2}$.

The number of unlabeled k–tournaments, or more precisely, the number of isomorphism classes of k–tournaments, can be determined by making fairly standard use of Pólya–type techniques, as described, for example in [1]. Specifically, the number of nonisomorphic k–tournaments of order n is computed by modifying the cycle index of the symmetric group S_n of degree n. Each term in the cycle index is replaced by the number of (labeled) k–tournaments that are left fixed by the permutation of the players indicated by that term; evaluating the resulting numerical expression yields the desired results. Relatively simple formulas exist for carrying this out in the case of the standard 1–tournaments (see, for example, [2, p. 85 – 89], but the details get increasingly more complex as k gets larger. Table 1 gives the number of non–isomorphic k–tournaments of all orders $n \le 6$.

				n		
k	1	2	3	4	5	6
1	1	1	2	4	12	56
2			2	324	48, 869, 983, 488	
3						13

Table 1. The number of k–tournaments of order n.

As with most graphical enumeration problems, the first term dominates all of the others. This leads to the following asymptotic result.

Theorem 2　*For each positive integer k, the number $f_k(n)$ of non–isomorphic k-tournaments of order n satisfies*

$$f_k(n) \sim \frac{1}{n!}\, 2^{\binom{n}{k}} \binom{n-k}{k}/2$$

as n tends to infinity.

3. Automorphisms

As with any graphical structure, a k–tournament posses automorphisms. In this case, they form a permutation group acting on the set P of players.

Definition An *automorphism* of a k–tournament **T** is an isomorphism from **T** to itself. The *automorphism group* of **T** is the group of all automorphisms of **T** under the operation of composition.

Each of the two non–ismorphic 2–tournaments of order 4 has the group $E_1 + S_3$ of order 6 as its automorphism group. Amoug the non–isomorphic 1–tournaments of order 4, two of them have group $E_1 + C_3$, while the other two have group E_4.

The number of k–tournaments with each possible automorphism group can in theory be determined by the methods in [3], although the required knowledge of the lattice of subgroups of S_n makes it pratical only for fairly small orders. In Table 2 we give the number of non–isomorphic k–tournaments of order 5 with each possible auto–morphism group, for $k = 1$ and 2.

		k
Group	1	2
E_5	7	232
$E_3 + S_2$		76
$E_2 + C_3$	4	2
C_5	1	2
$E_2 + S_3$		8
$C_2 + C_3$		4
Total	12	324

Table 2. The number of k–tournaments of order 5
with each possible automorphism group.

An intriguing problem is to determine which groups can serve as the automorphism group of a k–tournament, either isomorphically as an abstract group or identically as a concrete permutation group. The first interpretation of the problem has been solved for the case of 1–tournaments in the following theorem of Moon [2, p. 74 – 78].

Theorem 3 *A finite group G is abstractly isomorphic to the automorphism group of a 1–tournament if and only if the order of G is odd.*

As with most graphical structures, the group that most often occurs as the automorphism group of a k–tournament is the identity group, E_n. Let $g_k(n)$ be the number of non–isomorphic k–tournaments of order n with group E_n. The next theorem is proved in essentially the same way as Theorem 2.

Theorem 4   *Almost every k–tournament has the identity group as its automorphism group, in the sense that for each positive integer k, $f_k(n) \sim g_k(n)$ as n tends to infinity.*

4. Self–Converse k–tournaments

The concept of the converse of a standard tournament generalizes in an obvious fashion.

Definition   The *converse* of the k–tournament $\langle P, T, M \rangle$ is the k–tournament $\langle P, T, M' \rangle$, where

$$\langle t_1, t_2 \rangle \in M' \text{ if and only if } \langle t_2, t_1 \rangle \in M.$$

A k–tournament is *self–converse* if it is isomorphic to its converse.

Intuitively, the converse of a k–tournament is obtained by reversing the outcome of all of the matches. Each of the two non–isomorphic 2–tournaments of order 4 is the converse of the other, so there are no self–converse tournaments with these parameters. Self–converse k–tournaments can be enumerated by using techniques similar to those used for self–converse digraphs, as illustrated in [1]. The trick in each case is to determine the number of structures that are mapped onto their converse by each permutation in S_n. The next theorem summarizes what is known about the number of self–converse k–tournaments.

Theorem 5

   a). *There exist self–converse 1–tournaments of all orders. The number of such tournaments of order n for $n = 1, 2, ..., 8$ is 1, 1, 2, 2, 8, 12, 88, and 176, respectively.*

   b). *There exist no non–trivial self–converse 2–tournaments of any order.*

   c). *There exist exactly 7 self–converse 3–tournaments of order 6.*

Proof   Parts a) and c) are obtained through fairly simple application of standard counting results. For part b), suppose there exists some permuation σ that maps a 2-tournament **T** onto its converse. It follows that if $\langle t_1, t_2 \rangle \in M$, then $\langle \sigma^j(t_1), \sigma^j(t_2) \rangle \in M$ when j is even, and $\langle \sigma^j(t_2), \sigma^j(t_1) \rangle \in M$ when j is odd. We consider 4 cases.

Case 1.   The disjoint cycle decomposition of σ contains a cycle π of the form $\pi = (p_1, p_2, ..., p_{2j+1})$ for some $j \geq 2$. Let team $t_1 = \{p_1, p_2\}$, and $t_2 = \{p_3, p_4\}$. Then $\sigma^{2j+1}(t_i) = t_i$ for $i = 1$ and 2, which is impossible.

Case 2. The disjoint cycle decomposition of σ contains a cycle π of the form $\pi = (p_1, p_2, ..., p_{4j})$ for some $j \geq 1$. Let team $t_i = \{p_1, p_{j+1}\}$, and $t_2 = \{p_{2j+1}, p_{3j+1}\}$. Then $\sigma^{2j}(t_1) = t_2$, and $\sigma^{2j}(t_2) = t_1$, which is impossible.

Case 3. The disjoint cycle decompostion of σ contains cycle π of the form $\pi = (p_1, p_2, ..., p_{4j+2})$ for some $j \geq 1$. Let team $t_1 = \{p_1, p_{2j+2}\}$, and $t_2 = \{p_2, p_{2j+3}\}$. Then $\sigma^{2j+1}(t_i) = t_i$ for $i = 1$ and 2, which is impossible.

Case 4. All of the cycles in the disjoint cycle decomposition of σ have length ≤ 3. The various subcases can all easily be shown to be impossible.

Thus there is no possible permutation σ that can map a 2–tournament onto its converse. ❑

Similar investigations have confirmed that there do not exist any self–converse 4-tournaments of order 8, suggesting that the following conjecture may be true.

Conjecture *There exist no non–trivial self–converse k–tournaments when k is even.*

5. Score Sequences

The score sequence is another concept from standard tournament theory that we want to extend to k-tournaments. Although we could assign scores to either the teams or the players, the second alternative seems more natural.

Definition The *score* of player p_i in the k–tournament **T** of order n is the number s_i of matches in which p_i is a member of the dominant team. The *score sequence* of **T** is the n–tuple $\langle s_1, s_2, ..., s_n \rangle$, with the players numbered so that $s_1 \leq s_2 \leq ... \leq s_n$.

The two 2–tournaments of order 4 have score sequences $\langle 0, 2, 2, 2 \rangle$ and $\langle 1, 1, 1, 3 \rangle$. Two obvious questions arise: which n–tuples are in fact the score sequences of k–tournaments, and how many distinct score sequences exist for a given n and k? The second question is quite difficult to answer even in the case $k = 1$, and we will not pursue it here. The first question, though, has a very nice answer for $k = 1$, as presented for example in [2, p. 61 – 62].

Theorem 6   *The sequence $\langle s_1, s_2, ..., s_n\rangle$, with $s_1 \le s_2 \le ... \le s_n$ is the score sequence for some 1–tournament **T** if and only if*

$$\sum_{i=1}^{j} s_i \ge \binom{j}{2}$$

for $j = 1, 2, ..., n$, with equality when $j = n$.

The necessity of these conditions follows from the fact that the right hand side of the inequality expresses the number of matches that take place totally within a set of j players and hence the minimum total score of any such set. Similar necessary conditions can be derived for other values of k by selecting a set S of j players and imagining a k–tournament in which each match is won by the team with fewer players from S. The sum of the scores of the players in S is then a sharp lower bound for the sum of the scores of the j worst players. For $k = 2$ and $n = 4$, for instance, we have $s_1 \ge 0$, $s_1 + s_2 \ge 2$, $s_1 + s_2 + s_3 \ge 3$, and $s_1 + s_2 + s_3 + s_4 = 6$.

Unfortunately, these necessary conditions are not sufficient to insure that a sequence is in fact the score sequence of some 2–tournament **T**. The sequence $\langle 1, 1, 2, 2\rangle$ satisfies the conditions but is not the score sequence of either 2–tournament. Furthermore, these inequalities can not be sharpened. We are forced to conclude that no set of conditions based on lower bounds for the sum of the j smallest scores can be both necessary and sufficient in this case. Similar complications arise for larger values of n and k. Thus the problem of characterizing legitimate score sequences of k–tournaments appears to be quite difficult for $k \ge 2$.

A *regular* k–tournament is one in which the scores of all players are equal. In this case each player must win exactly half of his matches, so an obvious necessary condition for the existence of a regular k–tournament is that each player competes in an even number of matches. For $k = 1$, this means that n must be odd, and it is well–known that there exist regular 1–tournaments of all odd orders. The following generalization is an easy exercise.

Theorem 7   *The number*

$$\binom{n-1}{k-1}\binom{n-k}{k} = \frac{(n-1)!}{k!(k-1)!(n-2k)!}$$

of matches in which each player in a k–tournament of order n participates is even except when k is a power of 2 and n is a multiple of 2k.

It appears likely that this necessary condition is also sufficient.

Conjecture   *There exist regular k–tournaments of order n in all cases except those excluded by Theorem 7.*

6. Other Topics

We conclude with a short list of other question about k–tournaments that we consider worthy of study. Almost nothing is known about any of them.

A 1–tournament is called *transitive* if $t_1 > t_2$ and $t_2 > t_3$ imply $t_1 > t_3$ for any three (one person) teams in the tournament. There is exactly one transitive 1–tournament of each order, up to isomorphism. How should this concept be extended to more general k–tournaments? Using exactly the same definition does not work, as t_1 and t_3 will not always be disjoint and hence $t_1 > t_3$ will not be possible. There are many equivalent characterizations of transitivity in 1–tournaments, (see [2, p. 15], for example), but none of them seem to properly capture the concept for higher values of k. Being free of directed cycles among the teams, for example, does not seem to possess the property we desire to capture. On the other hand, demanding that the scores be all distinct seems overly restricitve. What is the most appropriate way to extend the concept of transitivity?

At the other extreme are the strong touranments. Here again, there are many equivalent characterizations of this concept in the case of 1–tournaments, and their extensions are not all equivalent for larger values of k. What is the best way to extend the concept of strength? Can the set of players in a k–tournament be partitioned into strong components that can be contracted to form the set of players for a new, transitive k–tournament?

What do random k–tournament look like? Suppose the outcome of each match is determined by an independent flip of a fair cion. We know from Theorem 4 that such a k–tournament will almost certainly have the identity group E_n as its automorphism group. But how transitive or strong is it likely to be? What is the expected value of the largest score?

Are k–tournaments of order n ever reconstructable (uniquely determined up to isomorphism) from their sub–k–tournaments of order $n - 1$? It is known (see [4], for example), that the 1–tournaments of order 7 are so reconstructable, but that there are non–reconstructable 1–tournaments of all other orders from 2 to 10. Moreover, there are non–reconstructable 1–tournaments for infinite number of orders. What can be said for other values of k? Since there are only two distinct 2–tournaments of order 4, there are only 6 possible decks that could serve as the collection of subtournaments of a 2-tournament of order 5. It is easy to verify that each of these 6 possible decks is in fact the deck derived from at least two of the 324 2–tournaments of order 5, so that none of these is reconstructable.

REFERENCES

[1] F. Harary and E.M. Palmer, *Graphical Enumeration,* Academic Press, New York, 1973.

[2] J.W. Moon, *Topics on Tournaments*, Holt, Rinehart and Winston, New York, 1968.

[3] P.K. Stockmeyer, *Enumeration of Graphs with Prescribed Automorphism Group*, Ph.D. Dissertation, University of Michigan, 1971.

[4] P.K. Stockmeyer, "A census of non–reconstructable digraphs, I: Six related families," *J. Combinatorial Theory, Series B* **31** (1981), 232 – 239.

Recent Results on Graph Embeddings

Carsten Thomassen

ABSTRACT

We survey some recent results on embeddings of graphs on surfaces. We indicate short graph theoretic proofs of the Jordan–Schönflies theorem and of the classification of the surfaces. We describe a polynomially bounded algorithm which determines the genus of a large class of graphs. We also explain why the graph genus problem for all graphs is NP–complete. Details can be found elsewhere.

1. Introduction

A *curve* in a topological space X is the image of a continuous 1–1 map $f: [0,1] \to X$. We say that a graph G can be *embedded* into X if G can be represented in X such that the vertices of G are distinct elements (points) in X, and each edge xy in G is a curve in X joining the points corresponding to x and y. Moreover, two edges in X do not intersect except possibly at an end. There is an immense literature on graph embeddings. The main source of motivation was the following problem: Given a surface S, determine the smallest natural number $h(S)$ such that every map on S can be colored in $h(S)$ colors in such a way that no two neighboring countries receive the same color. When S is the sphere this is the *4–color–problem* and for other surfaces it is the *Heawood conjecture*. More recently, graph embeddings are also studied because of the applications to real world problems. In such problems there are often complicated additional constraints imposed which can make the embedding problems complicated even for "simple" topological spaces X. We shall here treat embeddings with no additional constraints.

If X is the Euclidean space $\mathbf{R}^3$, then the embedding problem is uninteresting because every graph can be embedded into X: Simply put every vertex on the moment curve $\{(t, t^2, t^3) \mid t \in [0,1]\}$ and join every pair of adjacent vertices by a straight line segment. If X is $\mathbf{R}^2$ (or the sphere in $\mathbf{R}^3$), then the embedding problem is solved by

Kuratowski's theorem: A graph G can be embedded into $\mathbf{R}^2$ if and only if G contains no subdivision of any of the Kuratowski graphs K_5 or $K_{3,3}$. In view of this it seems natural to focus on topological spaces which are "smaller" than $\mathbf{R}^3$ but "larger" than $\mathbf{R}^2$. One such example is the *k-book* which is obtained from k disjoint squares of side length 1 (the pages) by choosing one side (the spine) on each page and identifying these. Clearly, a graph can be embedded into a k–book ($k \leq 2$) if and only if the graph is planar. A surprising result of G. Atneosen (see [2]) says that every (other) graph can be embedded in the 3–book. This was a corollary of a more general topological result. A trivial proof is given in [2]. So, also the k–books are uninteresting from the general embedding point of view.

The most interesting spaces from a graph embedding point of view are probably the surfaces. (A *surface* is a compact connected space which is locally homeomorphic to $\mathbf{R}^2$). When dealing with graph embeddings one makes often use of the *classification theorem*: Every surface is homeomorphic to the surface S_g obtained from the sphere S_0 by adding g handles, or the the surface N_k obtained from the sphere by adding k crosscaps. Another fundamental result needed for embeddings is the *Jordan curve theorem* or the stronger *Jordan–Schönflies theorem*: If f is a homeomorphism of a closed curve C_1 in $\mathbf{R}^2$ onto a closed curve C_2 in $\mathbf{R}^2$, then f can be extended to a homeomorphism of the whole plane $\mathbf{R}^2$. We shall here indicate the ideas behind simple graph theoretic proofs of the Jordan–Schönflies theorem and the classification theorem. Detailed proofs are given in [5].

2 . The Jordan–Schönflies Curve Theorem

We recall that a *curve* in the plane $\mathbf{R}^2$ is the image of a continuous 1–1 map f: $[0,1] \rightarrow \mathbf{R}^2$. We say that the curve *joins* $f(0)$ and $f(1)$. A *closed curve* is defined analogously except that $f(0) = f(1)$. A (closed) polygonal curve is a (closed) curve which is the union of a finite number of straight line segments. If Ω is an open set in $\mathbf{R}^2$, then a *region* Ω' of Ω is a maximal subset of Ω such that any two points of Ω' are joined by a curve (in Ω). We say that Ω is *connected* if it has only one region. A *planar graph* is a graph that can be embedded in $\mathbf{R}^2$. The following statements are easy exercises (see [1, 5]).

(1) If $\Omega \subseteq \mathbf{R}^2$ is open and connected, then any two points of Ω can be connected by a polygonal curve in Ω.

(2) If G is a planar graph, then G can be drawn (embedded) in the plane such that all edges are polygonal curves.

(3) If C is a closed polygonal curve in $\mathbf{R}^2$, then $\mathbf{R}^2 \setminus C$ has two regions each of which has C as boundary.

Using (1), (2), (3) one easily proves

    (4) K_5 and $K_{3,3}$ are nonplanar.

Also (5) below has a short graph theoretic proof [5].

    (5) If P is a curve in $\mathbf{R}^2$, then $\mathbf{R}^2 \setminus P$ is connected.

We now indicate how (1) – (5) imply

The Jordan Curve Theorem   *If C is a closed curve in the plane* $\mathbf{R}^2$, *then* $\mathbf{R}^2 \setminus C$ *has precisely two regions each of which has C as boundary.*

Proof   Let L_1 and L_2 be the vertical straight lines both intersecting C such that C is in the right (respectively left) closed half plane of L_1 (respectively L_2). Let P_1, P_2 be two curves on C joining q_1 and q_2 where q_i is the top point of $C \cap L_i$ for i = 1, 2. Consider a vertical straight line between L_1 and L_2. That line contains a segment P_3 joining P_1 and P_2 and having only its ends in common with C. Let P_4 be a horizontal straight line segment joining L_1 and L_2 above C. Now P_3 and P_4 must be in distinct regions of $\mathbf{R}^2 \setminus C$. Otherwise, a curve in $\mathbf{R}^2 \setminus C$ from P_3 to P_4 together with $C \cup P_3 \cup P_4$ and segments of L_1, L_2 would form an embedding of $K_{3,3}$ in $\mathbf{R}^2$ contradicting (4). Hence $\mathbf{R}^2 \setminus C$ has at least two regions.

To see that $\mathbf{R}^2 \setminus C$ has at most two regions, let us assume that q_1, q_2, q_3 are points in distinct regions of $\mathbf{R}^2 \setminus C$. Let p_1, p_2, p_3 be any three points on C. Let D_1, D_2, D_3 be pairwise disjoint discs around p_1, p_2, p_3, respectively, such that none of them contains any of q_1, q_2, q_3. Let P_i be a segment of C inside D_i (i = 1, ,2 ,3). By (5), $R_i = \mathbf{R}^2 \setminus (C \setminus P_i)$ is connected for i = 1, 2, 3. Any curve from q_1 to q_2 in R_i interesects P_i. Hence $\mathbf{R}^2 \setminus C$ contains a polygonal curve from q_i to D_j for i = 1, 2, 3 and j = 1, 2, 3. We can assume that these nine curves do not intersect except at q_1, q_2, q_3. Now we obtain a plane representation of $K_{3,3}$ from these nine curves by adding straight line segments in $D_1 \cup D_2 \cup D_3$. This contradicts (5) and shows that $\mathbf{R}^2 \setminus C$ has precisely two regions. $\square$

The unbounded region of $\mathbf{R}^2 \setminus C$ is called the *exterior* of C and is denoted ext(C). The other region of $\mathbf{R}^2 \setminus C$ is the *interior* int(C). A point p on C is *accessible* from int(C) if, for some (and hence each) point q in int(C), there is a polygonal curve from q to p having only p in common with C. The proof of the Jordan Curve Theorem shows that the points on C which are accessible from int(C) form a dense set on C.

We shall now sketch a simple graph theoretic proof of the Jordan–Schönflies Theorem which extends the Jordan Curve Theorem and which is normally regarded as a difficult result. A detailed proof can be found in [5].

The Jordan–Schönflies Theorem *If f is a homeomorphism of a closed curved C_1 onto a closed curve C_2 (both in the plane), then f can be extended to a homeomorphism of the whole plane.*

Sketch of Proof   Without loss of generality we can assume that C_2 is a closed polygonal curve. We shall extend f to $\text{int}(C_1)$. (The extension to $\text{ext}(C_1)$ is done in a similar way). Let A be a countable set of accessible points on C_1 (from $\text{int}(C_1)$) such that A is dense on C_1. We shall consider a sequence of plane 2–connected graphs Γ_0, $\Gamma_1, \Gamma_2, \ldots$ such that $C_1 \subseteq \Gamma_0 \subseteq \Gamma_1 \subseteq \ldots$ and such that each Γ_k consists of C_1 (which is a cycle in Γ_k) together with polygonal curves in $C_1 \cup \text{int}(C_1)$. Furthermore, $\Gamma_k \setminus C_1$ is connected for each k. Finally we shall construct the sequence $\Gamma_0, \Gamma_1, \ldots$ such that $V(\Gamma_0) \cup V(\Gamma_1) \cup \ldots$ contains A and is dense in $\text{int}(C_1)$. At the same time we construct a sequence of graphs $\Gamma'_0, \Gamma'_1, \ldots$ such that Γ'_k consists of C_2 and polygonal curves in $\text{int}(C_2) \cup C_2$ and such that there exists a graph isomorphism $g_k : \Gamma_k \rightarrow \Gamma'_k$. Moreover, $\mathbf{R}^2 \setminus \Gamma_k$ and $\mathbf{R}^2 \setminus \Gamma'_k$ have a finite number of regions each of which has a cycle of Γ_k (or Γ'_k) as boundary. Finally, a cycle S in Γ_k is such a boundary if and only if also $g_k(S)$ is the boundary of a region in $\mathbf{R}^2 \setminus \Gamma'_k$.

We construct Γ_0 as follows: We pick two points p, q in A and join them by a polygonal curve P such that $P \setminus \{p,q\} \subseteq \text{int}(C_1)$. Then we put $\Gamma_0 = C \cup P$ and consider Γ_0 as a graph with vertices p, q joined by three edges. An easy extension of the Jordan Curve Theorem shows that $\mathbf{R}^2 \setminus \Gamma_0$ has three regions whose boundaries are the three cycles of Γ_0. We let Γ'_0 be obtained from C_2 by adding a polygonal curve joining $f(p)$ and $f(q)$. Suppose we have already defined $\Gamma_0, \Gamma_1, \ldots, \Gamma_k$ and Γ'_0, $\Gamma'_1, \ldots, \Gamma'_k$ and the isomorphisms $g_0, g_1, \ldots, g_k$. Then we select a region of $\mathbf{R}^2 \setminus \Gamma_k$ bounded by the cycle S, say. In $\text{int}(S)$ we add a polygonal curve joining two points on S on edges a and b, say. In $\text{int}(g_k(S))$ we add a corresponding curve joining two points on $g_k(a)$ and $g_k(b)$ This results in Γ_{k+1} and Γ'_{k+1}, respectively. The isomorphism g_{k+1} is defined in the obvious way. We extend successively f such that it is defined on C_1 and the vertex set of Γ_k and agrees with g_k.

Since A is countable we can construct $\Gamma_0, \Gamma_1, \ldots$ such that each point of A is a vertex of some Γ_k. We can also assume that each point of A is a vertex of some Γ_k. We can also assume that each point in $\text{int}(C_1)$ which has rational corrdinates is in some Γ_k. Hence f will be defined on a dense set in C_1 and a dense set in $\text{int}(C_1)$. Furthermore, with a little additional care we can make sure that the following holds: For every point p in $\text{int}(C_1)$ (respectively $\text{int}(C_2)$) there is a cycle S in some Γ_k (respectively Γ'_k) such that $p \in \text{int}(S)$ and S has arbitrarily small diameter. This will guarantee that f when considered as a function defined on $C_1 \cup V(\Gamma_0) \cup V(\Gamma_1) \cup \ldots$ can be extended to a function (which is also denoted f) on $\text{int}(C_1)$ which is 1–1 and maps $\text{int}(C_1)$ onto $\text{int}(C_2)$ and which is continuous on $\text{int}(C_1)$ and has a continuous inverse on $\text{int}(C_2)$. (Note that the construction of f implies that f maps $\text{int}(S)$ into $\text{int}(g_k(S))$ and $\text{ext}(S)$ into $\text{ext}(g_k(S))$ and hence S into $g_k(S)$ when S is the above cycle in Γ_k. This also holds if S intersects C_1). It only remains to show that f is continuous on C_1. For this we consider a point p on C_1 and a sequence $q_1, q_2, \dots$ of points in $C_1 \cup \text{int}(C_1)$ converging towards p. We shall prove that $f(q_n) \to f(p)$ as $n \to \infty$. Suppose therefore that this is not the case. Then we can assume (by considering a subsquence if necessary) that $f(q_n) \to f(q) \neq f(p)$ as $n \to \infty$. Since f is continuous in $\text{int}(C_1)$ and also continuous on C_1 (when considered as a function defined on C_1) we can assume that $q_n \in \text{int}(C_1)$ and $q \in C_1$. In some Γ_k there is a polygonal curve P joining two points p_1, p_2 on C_1 such that one of the segments of C_1 from p_1 to p_2 contains p and the other contains q. Let S be the cycle in $C_1 \cup P$ containing p. Then $q_n \in \text{int}(S)$ for n sufficiently large. Then $f(q_n) \in \text{int}(f(S))$, by the above parenthetical remark. But this contradicts the assumption that $f(q_n) \to f(q)$ as $n \to \infty$. $\square$

3. The Classification of Surfaces

In Section 2 it is shown how graph theory can be an efficient tool in topology. Conversely, the classification theorem is a topological result of great importance to graph embeddings. If we wish to draw (embed) a given graph G on a given surface S, then instead of starting with S and trying to draw G on it we start out with G and use it as the "skeleton" of a surface obtained by pasting discs together on G. It may not be immediately obvious whether or not the resulting surface we get is homeomorphic to S. But that can be decided using the classification theorem (combined with Euler's formula). We shall here indicate a simple graph theoretic proof of the classification theorem. The advantage of the present proof is that it depends only on local considerations and it is very easy to make it rigorous. Here only the ideas are described. Again, the details are in [5].

A surface is defined as a connected compact topological space which is locally homeomorphic to a disc. The sphere S_0 is a surface. Adding a crosscap to the sphere means that we cut out a disc and identify diametrically opposite points on its boundary. Adding a *handle* to a disc may be described as cutting out two disjoint discs and identifying the boundary of one with the boundary in the other such that the orientations of these boundaries are the same. If the orientations do not agree, we say that we add a *twisted handle*. Let S_g and N_k be obtained from the sphere S_0 by adding g handles, respectively k crosscaps. It is an easy exercise to show that adding a twisted handle amounts to the same as adding two crosscaps and that adding a handle after a crosscap has been added amounts to the same as adding a twisted handle (i.e. two crosscaps). So, if we add handles, twisted handles and crosscaps to the sphere, we obtain one of the surfaces S_g, N_k.

There is another natural way of forming a surface. Let us consider a collection of disjoint triangles of side length 1 in the plane. We identify each side of a triangle with precisely one side in another triangle. This results in a compact topological space S and a graph G whose vertices and edges are the corners and sides, respectively, of the triangles. If G (and hence also S) is connected and S is locally homeomorphic to a disc (i.e., G is locally isomorphic to a wheel), then S is a surface. If G has no vertices of degree 2, then we will call G a *triangulation of S*. Thus the smallest triangulation is obtained by pasting four triangles together such that G is K_4 and S is the tetrahedron (or, equivalently, the sphere). It is well–known that every surface S can be triangulated. Intuitively, S can be triangulated as follows: For every point p on S we consider a small disc containing p. In that disc we consider a closed curve having p in the interior. Since S is compact there is a finite collection of such curves such that the union of their interior is S. In [5] it is shown, using the Jordan–Schöflies theorem combined with elementary graph theory, that the above curves can be chosen such that only finitely many points on S are on two or more of the curves. Then we triangulate the interior of each of them. Using this we may proceed to:

The Classification Theorem *Every surface S is homeomorphic to one of the surfaces S_g $(g \geq 0)$ or N_k $(k \geq 1)$.*

Sketch of proof Let G be a triangulation of S with n vertices, e edges, and f regions which are all bounded by triangles of G. We shall prove that S is homeomorphic to S_g or N_k. We shall do this by contradiction assuming that g or k is minimal. Euler's formula (which we haven't proved yet) says that

$$n - e + f = 2 - 2g \text{ or } 2 - k.$$

Therefore, we assume that the counter example S, G is chosen such that

(i) $2 - n + e - f$ is minimum. (An easy exercise shows that $2 - n + e - f$ is nonnegative, see [5]).

Among those counter examples we assume that

(ii) n is minimal

and that

(iii)   the smallest vertex degree of G is minimal.

If G has a vertex v of degree 3, then also $G - v$ triangulates S and we have a contradiction to (ii).

If G has a vertex v of minimal degree ≥ 4 and with neighbors $v_1, v_2, ..., v_n$ in that cyclic order, then we can assume that G contains the edges $v_1 v_3, v_2 v_4, ...$. For if $v_1 v_3$ is missing then $G - v v_2 + v_1 v_3$ is a triangulation of S contradicting (iii).

Now we consider the triangle T: $v v_1 v_3$ in G. Note that T is not the boundary of a region of $S \setminus G$ and that T does not separate G or S since $v_2 v_4$ is an edge in G. Then we form a new surface S_1 as follows: We cut along T such that each vertex and edge of T is cut into two vertices (or edges). This results in a "near–surface" S' and a graph G' which has $n' = n + 3$ vertices and $e' = e + 3$ edges. The triangle T in G corresponds in G' either to two disjoint triangles T_1, T_2 or a cycle C of length 6. Note that S' is not locally homeomorphic to a disc on $T_1 \cup T_2$ or C. We extend S' into a surface S_1 by pasting discs on each of T_1, T_2 or on C. Then S_1 is a surface with $f' = f + 2$ or $f + 1$ regions. In case G' contains C, we extend G' into a triangulation G" of S_1 by adding a vertex inside C and joining it to the six vertices of C. Now (S_1, G') or $(S_1, G")$ cannot be a counterexample to the theorem because that would contradict (i). So, S_1 is one of the surfaces S_g, N_k. Since S is obtained from S_1 by adding a handle, or a twisted handle or a crosscap, the proof is complete. ❏

Note that the above proof also implies Euler's formula for triangulations. With a little additional reasoning, it can be extended to include Euler's formula in its full generality. It also implies the invariance of the genus (i.e. the fact that all the surfaces $S_0, S_1, ..., N_1, N_2, ...$ are pairwise nonhomeomorphic). For details, see [5].

4. Finding the Genus of a Graph

The *genus* g(G) of a graph G is the smallest number g such that G can be embedded into S_g. The *crosscap number* of G is defined anaolgously. We shall here concentrate on the genus and we shall only consider connected graphs G. Let V(G) = $\{v_1, v_2, ..., v_n\}$ and let π_i be a cyclic permutation (which we shall refer to as a *clockwise orientation*) of the edges inicident with v_i, i = 1, 2, ..., n. Suppose we walk towards vertex v_i along the edge a. If we continue walking away from v_i along $\pi_i(a)$ we say that we turned *sharp left* at v_i. If we continue walking sharp left at every vertex we encounter, we obtain a *facial walk*. Let $e = |E(G)|$ and f the number of facial walks. For each of the f facial walks we select a disc and identify the boundary of the disc with the facial walk. This results in a surface S. One can show, for example using methods in the previous section that $S = S_g$ where g satisfies Euler's formula

$$n - e + f = 2 - 2g.$$

Furthermore, one can show that every embedding of G on a surface of minimum genus is of this form. So, determining g(G) amounts to specifying clockwise orientations (called a *rotation system*) such that f is maximized. This is a purely combinatorial problem and it makes sense to speak of the computational complexity of the problem.

The genus has been determined for some graphs of a high degree of symmetry, such as complete graphs, complete bipartite graphs, and also of graphs obtained by using special graph operations. For g fixed, a polynomially bounded algorithm for deciding if a graph can be embedded into S_g was found by Filotti, Miller and Reif and is also a special case of a more general algorithm in the Robertson–Seymour theory on graph minors. These results are surveyed in various books or papers, for example [2].

We shall here confine ourselves to the description of an algorithm which determines the genus of a large class of graphs without any symmetry properties and with unbounded genus. This algorithm and the theory behind it is described purely combinatorially in [3]. Here we shall describe it partly in topological terms in order to emphasize the ideas behind the algorithm. If G is a graph embedded in S_g then a cycle C in G is *contractible* if $S_g \setminus G$ has two connected components, one of which is homeomorphic to a disc. (That disc is called the *interior* of C). Otherwise, C is *noncontractible*. It is very easy to decide if a given cycle is noncontractible (see [3]). Moreover, it is shown in [3] that the shortest noncontractible cycle can be found as follows: Pick a vertex v in G and grow a breadth–first tree (distance tree) T from v. Look at each fundamental cycle with respect to T and let C_v be a shortest noncontractible one. (It is possible that C_v does not exist). Repeat this for every vertex. The shortest cycle found in this way is a shortest noncontractible cycle.

Now define the *edge–width* ew(G) of the embedded graph G as the length of a shortest noncontractible cycle. We say that the embedding has *large edge–width* if all facial walks have length $\leq$ ew(G) and we say that G is *LEW–embedded*. This concept, which was introduced by J. Hutchinson, is very useful for studying embeddings since LEW–embeddings share many properties with planar embedding as demonstrated in [3]. Let G be a connected graph and C a cycle in G. Following W.T. Tutte we define a *C–bridge* as a chord of C (together with its ends) or a connected component of $G - V(C)$ together with all edges joining it to C and the ends of these edges. If H is a C–bridge, then $H \cap C$ are the *vertices of attachment* of C. Two bridges H_1, H_2 *avoid one another* if C has two edge–disjoint paths P_1, P_2 such that H_i has all its vertices of attachment on P_i for $i = 1, 2$. Otherwise, H_1 and H_2 *overlap*. If G is drawn in the plane and two C–bridges overlap, then one of them is in ext(C) and the other is in int(C). The *overlap graph* O(G,C) has the C–bridges as vertices such that two vertices are adjacent iff the corresponding bridges overlap. So if G is planar, then O(G,C) is bipartite. We say that C is *induced* and *nonseparating* if G has only one C–bridge. Tutte proved that the facial cycles in a 3–connected planar graph are precisely the induced nonseparating cycles. This extends Whitney's result that a 3–connected planar graph has only one embedding in the plane. These results were generalized in [3].

Theorem 1 *If G is a 3–connected LEW–embedded graph, then the embedding is the unique minimum genus embedding of G. The facial walks are precisely the induced nonseparating cycles of length < ew(G).*

Theorem 2 *If G is a cycle of length < ew(G) in a 3–connected LEW–embedded graph, then O(G,C) is connected and bipartite. Moreover, if G is nonplanar, then G has precisely one C–bridge which together with C forms a nonplanar graph.*

Theorems 1 and 2 are structural results which imply the following algorithmic result.

Theorem 3 *There exists a polynomially bounded algorithm for describing an LEW–embedding of an arbitrary 3–connected graph G or deciding that G has no such embedding.*

By Theorem 1, the algorithm determines the genus of a large class of graphs, namely those that are 3–connected and have LEW–embeddings. In order to describe the embedding it is sufficient to describe the (unique) rotation system or the induced nonseparating cycles of length < ew(G). The problem is that we do not know ew(G) in advance. We can overcome this problem by using Theorem 2. We consider any edge e = vu and, by Theorem 1, all we have to do is to find the shortest and second shortest induced nonseparating cycle through e. Also, this is an unsolved problem in general, (see Problem 1 below). So, let C be any shortest cycle through e. If G has an LEW–embedding, then O(G,C) is bipartite and G has only one C-bridge H such that H ∪ C is nonplanar. We consider all the C–bridges in the partite class of O(G,C) not containing H. These can be drawn (in polynomial time) as a planar graph inside C. In this way we find one of the two shortest induced nonseparating cycles containing e (if G has an LEW–embedding). In order to find the other one we select any edge e' incident with v (and going out from the above drawing of C) and we let C' be a shortest cycle through e and e'. We repeat the above procedure with C' instead of C. We have already found the successor or predecessor (say the latter) of e around v. When e' is the successor of e the following will happen in the above procedure: The interior of C' contains either all or none of the edges incident with v and distinct from e, e'. When this happens we have found the successor and predecessor of e around v. If something goes wrong (for example if O(G,C) is not bipartite or has more than one nonplanar bridge or some edge has no successor), then G has no LEW–embedding and the algorithm stops.

If a graph G is embedded on S_g where $g = g(G)$, then it is easy to subdivide G and add vertices and edges such that the resulting graph is 3–connected and LEW–embedded. This suggests that Theorem 3 might be extended to a larger class of graphs although it is not clear how to add the new vertices and edges. The following result of [4], which answers one of the basic questions on NP–completeness raised by Garey and Johnson, indicates that Theorem 3 can probably not be extended to all graphs.

Theorem 4 *The following problem is NP–complete: Given a connected graph G and a natural number m, is $g(G) \leq m$?*

We shall here indicate how the well–known NP–complete problem of finding the independence number $\alpha(G)$ of a graph G can be reduced, in polynomial time, to that in Theorem 4. If $|V(G)| = n$ and $|E(G)| = e$, then it is easy to see that

$$g(G) \leq e - n + 1.$$

First draw a spanning tree of G on the sphere. Then add successively the remianing $e - n + 1$ edges by adding a new handle at each step. The above upper bound is in general a poor one because G does not really use the handles fully. Let G_k by the graph obtained from G by replacing each edge xy by a cycle of length k, i.e., we delete the edge xy and add instead a cycle of length k joined completely to x and y. The above argument shows that also G_k has genus at most $e - n + 1$ (let the new cycles of length k go around the handles). In [4] it is shown that, if $k > 3n^2$, then G_k has genus precisely $e - n + 1$. We now add to G_k a new vertex v and join v to precisely one vertex in each of the new cycles of length k. It is shown in [1] that the resulting graph G'_k has genus $e - \alpha(G)$. Now the problem: Is $\alpha(G) \geq m$? (which is NP–complete) reduces to the problem: Is $g(G'_k) \leq e - m$?

5. Unsolved Problems

Theorem 3 would follow quickly from Theorem 1 (without the use of Theorem 2) if we could find, in polynomial time, the shortest and second shortest induced nonseparating cycle through a given edge in a 3-connected graph. W.T. Tutte has described an algorithm for finding two distinct induced nonseparating cycles through a given edge but they need not be short. So, we suggest the following.

Problem 1 Does there exist a polynomially bounded algorithm for finding a shortest induced nonseparating cycle through a given edge in a 3–connected graph?

In [3] is described a general algorithm for finding shortest cycles of many different types, for example a shortest noncontractible cycle in an embedded graph. J. Hutchinson raised the following question which was not covered in [3].

Problem 2   Does there exist a polynomially bounded algorithm for finding a shortest contractible cycle in an embedded graph?

We shall here point out that the answer is affirmative in the important special case where all vertices have degree at least 3. We first find all facial walks. Among those which are cycles we select a shortest one. Then we consider all cycles of length 3, 4, 5. If one of them is contractible we select a shortest contractible one. This algorithm is clearly polynomially bounded. We claim that it produces a shortest contractible cycle if there is one. For suppose that C is a shortest contractible cycle and that C is not a facial walk and that C has length ≥ 6. Then C and its interior is a planar graph and all vertices in int(C) have degree ≥ 3. It is now an easy exercies to use Euler's formula and find a facial cycle in int(C) which is shorter than C.

REFERENCES

[1]   C. Thomassen, *Kuratowski's theorem*, J. Graph Theory 5 (1981), 225 – 241.

[2]   C. Thomassen, *Embeddings and minors*, in "Handbook of Combinatorics (eds. R.L. Graham, M.   Grötschel and L. Lovász)," North–Holland (to appear).

[3]   C. Thomassen, *Embeddings with no short noncontractible cycles*, J. Combinatorial Theory, Ser. B (to   appear).

[4]   C. Thomassen, *The graph genus problem is NP–complete*, J. Algorithms (to appear).

[5]   C. Thomassen, *The Jordan–Schönflies theorem and the classification of surfaces* (to appear).

Symmetric Embeddings of Cayley Graphs in Nonorientable Surfaces

Thomas W. Tucker[*]
Colgate University

ABSTRACT

An embedding of a Cayley graph for a group A in an orientable or nonorientable surface S is symmetric if the natural action of the group A on the Cayley graph extends to the surface S. It is shown that every irredundant Cayley graph for the group A has a symmetric nonorientable embedding if and only if A is not a 2-group. Parameters analogous to the genus and symmetric genus of a group are introduced for the nonorientable case and are computed for a few examples. The projective plane and klein bottle groups are classified. Some inequalities relating the orientable and nonorientable parameters are summarized.

1. Introduction

Given a group A and a generating set X for A, the *Cayley graph* $C(A, X)$ has the elements of A as vertices, and edges from a to ax for each $a \in A$ and $x \in X$. If x has order two, the pair of edges from a to ax and from ax to $axx = a$ is sometimes identified to a single edge, sometimes not; we allow both possibilities, even in the same Cayley graph, with one restriction -- all pairs of edges corresponding to the same generator must be treated the same. Left multiplication by an element of A induces an automorphism of any Cayley graph for A. This *natural action* of a group on any of its Cayley graphs is transitive and fixed–point free on the vertex set. In fact, if a group A acts transitively and fixed–point freely on the vertex set of a graph, the graph must be a Cayley graph for A (see [S] or [GT]). If the edge from a to ax is directed and labeled x for each a and x, then the natural action preserves edge directions and labels.

An embedding of a Cayley graph $C(A, X)$ in a closed surface S is *symmetric* if the natural action of A on the Cayley graph $C(A, X)$ extends to an action by a group of homeomorphisms of the surface S (all embeddings in this paper are 2–cell: the interior

---

[*] Supported by NSF Contract DMS–8601760.

of each face is an open disk). If, in addition, the surface S is orientable and the extended action preserves the orientation of S, then the embedding is *strongly symmetric*. For orientable surfaces, the symmetry of an embedding simply means that the embedding looks locally the same at every vertex: that is, the cyclic ordering of edge labels and directions encountered in a small "clockwise" trip around any vertex is the same (strongly symmetric) or sometimes the same and sometimes the reverse (symmetric but not strongly symmetric). For nonorientable surfaces, the symmetry of an embedding is more technical and less intuitive. It is explained in the next section.

The *genus* of a graph is the minimum genus of an orientable surface containing an embedding of the graph. If the graph is a Cayley graph and embeddings are restricted to being symmetric (respectively, strongly symmetric), we have the *symmetric* (respectively, *strong symmetric*) *genus* of a Cayley graph. For nonorientable surfaces, one can likewise define a *crosscap number* for any graph and a *symmetric crosscap number* for a Cayley graph. White [Wl] has defined the *genus of a group* A, denoted $\gamma(A)$, to be the minimum genus of any of its Cayley graphs. We have defined similarly [T2] the *symmetric* (respectively, *strong symmetric*) *genus* of a group A, denoted $\sigma(A)$ (respectively, $\sigma^{\circ}(A)$) to be the minimum symmetric (respectively, strong symmetric) genus of any Cayley graph for A. For nonorientable surfaces, there are corresponding definitions of the *crosscap number of a group*, denoted $\overline{\gamma}(A)$, and *symmetric crosscap number of a group,* denoted $\overline{\sigma}(A)$. The last has not appeared explicitly in the literature.

Generally speaking, the most convenient parameter for surfaces is not the genus g or number of crosscaps c, but rather its Euler characteristic χ (for orientable surfaces $\chi = 2 - 2g$ and for nonorientable surfaces $\chi = 2 - c$). For that reason, let us define some Euler characteristic parameters for a group. Let $\chi(A)$ (respectively, $\Sigma(A)$, $\Sigma^{\circ}(A)$) denote the maximal Euler characteristic of any orientable surface containing an embedding (respectively, symmetric, strong symmetric) of a Cayley graph for A. Let $\overline{\chi}(A)$ (respectively, $\overline{\Sigma}(A)$) be the maximal Euler characteristic of any nonorientable surface containing an embedding(respectively, symmetric) of a Cayley graph for A. Thus $\chi = 2 - 2\gamma$, $\Sigma = 2 - 2\sigma$, and $\Sigma^{\circ} = 2 - 2\sigma^{\circ}$. Similarly, $\overline{\chi} = 2 - \overline{\gamma}$ and $\overline{\Sigma} = 2 - \overline{\sigma}$, where we agree that $\overline{\sigma} \geq 1$ (for graphs one usually allows the crosscap number to be 0, even though a surface with no crosscaps is orientable).

In [T2], it is proved that a group A acts on a surface S if and only if there is a Cayley graph for A symmetrically embedded in S. Thus symmetric genus etc. can be defined without any reference to Cayley graphs or generating sets. In this guise, the strong symmetric genus has a long history, which goes back to Burnside [B] and which has strong connections to groups of conformal automorphisms of Riemann surfaces (see for example [M] or [LM]). For more information about the genus and symmetric genus of a group, see [W2], [T2], or [GT].

The groups in this paper are all finite. The genus of an infinite group [L] is either 0 or ∞, so in that sense the genus of an infinite group is less interesting. For a partial classification of infinite groups of genus 0, see [MPTW].

The remainder of this paper is organized as follows. Section 2 considers nonorientable symmetric embeddings in general. Section 3 relates such embedding to the orientable double covering of a nonorientable surface and classifies groups with $\bar{\Sigma} = 1$. Section 4 recalls the Riemann–Hurwitz equation and classifies groups with $\bar{\Sigma} = 0$. Section 5 computes $\bar{\Sigma}$ for some examples. Section 6 summarizes the known inequalities relating the various χ and Σ parameters. Section 7 presents some problems for further study.

2. Nonorientable Symmetric Embeddings

To understand nonorientable symmetry, it is best to take the "band decomposition" viewpoint of a graph embedding, as given in [GT]. An embedding of a graph G in the surface S is pictured as a thickened neighborhood of the graph in the surface S: a small disk around each vertex, a long narrow band for each edge, and a hole in the surface for each face (which can be capped off with a disk to give the closed surface S). For an orientable surface, one can choose a local orientation for each vertex–disk and edge–band so that the orientation of each edge–band is consistent with the orientation of the vertex–disks at its endpoints. Thus to describe an orientable embedding it suffices to give the cyclic ordering or *rotation* of the edges at each vertex given by a single fixed orientation of the surface. The rotation at each vertex tells how to attach the incident edge–bands to the vertex–disk, automatically creating the thickened neighborhood of the graph in the embedding surface.

For nonorientable surfaces, each vertex–disk can be assigned some local orientation, but these orientations may not be consistent with that of the edge–bands. Given a choice of orientation for the vertex–disks, call an edge–band *type–0* if the directions induced on the boundary of the edge–band by the orientation of its endpoint vertex disks agree; call the edge–band *type–1* otherwise. Figure 1 illustrates a type–0 and a type–1 edge–band. Notice that if the vertex–disks at the end of an edge–band are both placed flat in the plane so that both vertex–disk local orientations are, say, clockwise, then a type–1 edge–band has a half–twist, while a type–0 edge–band lies flat in the plane. Thus, one often thinks of a type–1 band as "twisted", but please note that this twisting is only an artifact of how one chooses to view the vertex–disks. If the two vertex–disks are laid in the plane so that one orientation is clockwise and the other is counterclockwise, then a type–0 edge–band is twisted while a type–1 band is not! The type of an edge–band is thus not an intrinsic property of the embedding, but rather depends on the arbitrary choice of local orientations of the vertex–disks.

Figure 1. A type–0 edge–band and a type–1 edge–band.

The presence of a type–1 edge does not even guarantee that the embedding surface is nonorientable. Although vertex–disk orientations of an orientable embedding can be chosen so that all edge–bands are type–0, that does not mean that vertex–disk orientations must be chosen that way. For example, if one chooses the vertex–disk orientations for a 4–cycle embedded in the sphere so that they alternate clockwise, counterclockwise around the 4–cycle, then every edge–band is type–1 even though the embedding surface, the sphere, is orientable. The criterion for deciding when an embedding surface, given in terms of rotations and edge types, is orientable is not hard: the surface is nonorientable if and only if there is a cycle in the graph containing an odd number of type–1 edges.

Suppose that a Cayley graph $C(A, X)$ is symmetrically embedded in a possibly nonorientable surface S. Since the natural action of A preserves edge labels and directions, in order to map one vertex–disk onto another the rotation at each vertex (in terms of the X–label and direction of each edge) must be the same or reversed. Choose any vertex and fix one of the two orientations for that vertex–disk and now choose the orientation of every other vertex–disk to be the same. For this choice of vertex–disk orientations, some edge–bands may be type–0 and some type–1, but since the action of A preserves edge labels, in order for the action of A to extend to the embedding surface, all edges corresponding to the same generator $x \in X$ must have the same type: a surface homeomorphism cannot preserve vertex–disk orientations and take a type–0 edge–band to a type–1 edge–band or vice versa.

Conversely, if vertex–disk orientations of an embedding of the Cayley graph $C(A, X)$ in the surface S can be chosen so that every rotation is the same and all edges corresponding to the same generator have the same type, then the natural action of A on the Cayley graph $C(A, X)$ extends to the surface S. Just define the extended action first for the vertex–disks, then for edge–bands, then the face–holes. The consistency of rotations and edge types guarantees that no obstructions arise. To summarize this discussion we state the following:

Proposition 2.1 *An embedding of a Cayley graph $C(A, X)$ in the surface S is symmetric if and only if vertex–disk orientations can be chosen so that every rotation is the same and every edge corresponding to the same generator $x \in X$ has the same type. Given such a choice of vertex–disk orientations, the given surface S is nonorientable only if there is a relator in the generating set X in which some generator appears an odd number of times.*

Proof The first statement follows from the previous discussion. The surface S is nonorientable if and only if there is a cycle in the Cayley graph $C(A, X)$ containing an odd number of type–1 edges. Every cycle in $C(A, X)$ corresponds to a relator in the generating set X, that is, a finite sequence of elements of X or inverses of elements of X, whose product is the identity in the group A. Since every edge corresponding to a given generator (or its inverse) has the same edge type, in order for a cycle to contain an odd number of type–1 edges, the corresponding relator must have an odd number of occurrences of some generator (the inverse of a generator counts as an occurrence). ❑

Proposition 2.1 has the surprising consequence that some Cayley graphs have no symmetric nonorientable embeddings at all. For example, let $A = Z_4$, the cyclic group of order 4, and let x be a generator for A. Then $C(A, \{x\})$ has no symmetric nonorientable embedding since the only relator is x^4. Is there a group A having no symmetric nonorientable embeddings for any of its Cayley graphs? No, because one can always include the identity element in the generating set X and declare all edge–bands for that element to have type–1. If one doesn't like multiple edges or loops, one can always introduce a redundant generator $z = x^2$ or $z = xy$, where $x, y \in X$, and declare all edge types for z to be type–1 while all other edge types are type–0 (the resulting graph has no multiple edges unless A has order 2 or 3). For example, the Cayley graph $C(Z_4, \{x, x^2\})$ has a symmetric nonorientable embedding obtained by making all the x^2 edges type–1; the resulting surface is the projective plane, if multiple edges corresponding to the generator x^2 are identified, and the nonorientable surface of Euler characteristic -2, otherwise.

But what if we do not allow redundant generators? Call a generating set X for the group A *irredundant* if no proper subset of X generates A. Call the Cayley graph $C(A, X)$ irredundant if X is irredundant. Then no irredundant Cayley graph for Z_4 has a symmetric nonorientable embedding. In general, we have the following somewhat mysterious characterization of 2–groups:

Theorem 2.2   *The group A has a symmetric nonorientable embedding for at least one of its irredundant Cayley graphs if and only if the order of A is not a power of 2.*

Proof   Suppose that no irredundant Cayley graph for A has a symmetric nonorientable embedding. Then no irredundant generating set for A has an element of odd order by Proposition 2.1, since a symmetric nonorientable embedding can be obtained by declaring all edges for an odd order generator to be type–1. For each prime p dividing the order of A, let B_p be a p–Sylow subgroup of A. Let X be the union of the subgroups B_p. Then the subgroup B of A generated by X has order divisible by the highest power of p dividing the order of A for each prime p, since B contains B_p. Thus X generates A. The generating set X contains an irredundant generating set X'. Since no irredundant generating set for A contains any element of odd order, X' must be contained in B_2. But then $A = B_2$ and hence the order of A is a power of 2.

Conversely, suppose the order of A is a power of 2. Let F be the Frattini subgroup of A, namely the intersection of all maximal subgroups of A. The following facts about F are well–known (see [G] for example):

(1) F is normal in A with quotient group A/F a direct product of Z_2's.

(2) if $X \cup F$ generates A, then so does X.

Let X be an irredundant generating set for A and suppose C(A, X) has a symmetric nonorientable embedding. Then some generator x must appear in some relator an odd number of times. The image of x in the quotient group $A/F = Z_2 \times ... \times Z_2$ therefore appears exactly once in a relator in A/F involving the images of the other elements of X. Thus x can be expressed in terms of the other elements of X and the elements of F. By fact 2, this means $X - \{x\}$ generates A, contradicting the irredundancy of X. We conclude that every symmetric embedding of C(A, X) is orientable.   ❑

The second half of the proof of this theorem is essentially contained in [PTW]. Part of the proof also applies to groups other than 2–groups. If A has a homomorphism onto the direct product of n Z_2's and X is a generating set for A of n elements, then C(A, X) has no symmetric nonorientable embedding.

3. Orientable Double Coverings

Let $\tilde{S}$ be an orientable surface. Call a homeomorphism t of $\tilde{S}$ to be an *antipodal map* if the order of t is two, t has no fixed points, and t reverses the orientation of $\tilde{S}$. If $\tilde{S}$ is embedded in R^3 symmetrically with respect to the origin, then the map of R^3 taking (x, y, z) to $(-x, -y, -z)$ is an antipodal map of $\tilde{S}$ when restricted to $\tilde{S}$. In fact, all antipodal maps of $\tilde{S}$ are conjugate to this one in the group of homeomorphism of $\tilde{S}$. Given an antipodal map t of the orientable surface $\tilde{S}$, one can form the quotient space $\tilde{S}/t = S$ by identifying t and $t(u)$ for each point $u \in \tilde{S}$. It follows from standard topology [T2] that the natural projection of $\tilde{S}$ onto S is a 2–sheeted, unbranched covering and that S is a closed, nonorientable surface of half the Euler characteristic of $\tilde{S}$. Every nonorientable surface can be obtained in this way. The surface $\tilde{S}$ is called the *orientable double covering of* S *with antipodal map* t. The following is the fundamental theorem about orientable double coverings and group actions:

Theorem 3.1 *The group A acts on the nonorientable surface S if and only if A acts on the orientable double covering $\tilde{S}$ so that the action preserves orientation and commutes with the antipodal map of $\tilde{S}$ (that is, $Z_2 \times A$ acts on $\tilde{S}$ with the A factor preserving orientation and the Z_2 factor generated by an antipodal map).*

Proof The "only if" direction is Theorem 15 of [T2]. Conversely, suppose such an action of A on $\tilde{S}$ exists. Thus for each $a \in A$ there is an orientation preserving homeomorphism $\tilde{h}_a : \tilde{S} \to \tilde{S}$ such that $\tilde{h}_a t = t \tilde{h}_a$, where t is the antipodal map of $\tilde{S}$. Define a map $h_a : S \to S$ as follows. Let $p : \tilde{S} \to S$ be the natural projection. For each $u \in S$, let $\tilde{u}$ be one of the two points of $p^{-1}(u)$. Then $h_a(u) = p\tilde{h}_a(u)$. We must check that h_a is well defined: $p\tilde{h}_a(t(\tilde{u})) = pt\tilde{h}_a(\tilde{u}) = p\tilde{h}_a(\tilde{u})$. It follows that h_a is a homeomorphism because $\tilde{h}_a$ is, and that $h_{ab} = h_a h_b$ because $\tilde{h}_{ab} = \tilde{h}_a \tilde{h}_b$. Thus we have an action of A on the surface S. $\square$

Determining when an orientation preserving action commutes with an antipodal map is not easy. This is related to symmetries of Riemann surfaces [Sn] and reflexible maps [CM]. One case at least is well understood: finite group actions on the sphere. These groups are just the automorphism groups (or subgroups) of the regular prisms and platonic solids. They are generated by standard rotations of the unit sphere in R^3. Thus the following classification of groups with $\bar{\Sigma} = 1$ is probably a folk theorem to many topologists. We denote the cyclic group of order m by Z_m and the dihedral group of order $2m$ by D_m.

Theorem 3.2 *(Classification of projective plane groups). $\bar{\Sigma}(A) = 1$ if and only if A is Z_m, D_m, the symmetric group on 4 symbols, or the alternating groups on 4 or 5 symbols.*

Proof By Theorem 3.1 and the discussion of Section 2, the group A has a Cayley graph symmetrically embedded in the projective plane if and only if A acts on the sphere (the orientable double covering of the projective plane) preserving orientation and commuting with an antipodal map. The groups given in the statement of this theorem are precisely the groups which act on the sphere preserving orientation (see Theorem 6.3.1 of [GT]). Each of these actions can be given as a group of orthogonal 3×3 matrices acting on the unit sphere in R^3. The antipodal map of the unit sphere is represented by the negative of the identity matrix and hence commutes with all these actions. $\square$

4. The Classification of Klein Bottle Groups

We wish to classify as well groups A such that $\overline{\Sigma}(A) = 0$. To do so we need first the Riemann–Hurwitz equation. An action of a group A on the closed surface S is *pseudo–free* if the number of points left fixed by some element of A is finite. If A acts pseudo–freely on the closed surface S then the quotient surface S/A is a closed surface and the projection $p: S \to S/A$ is a regular branched covering. Each branch point u in S/A is the image of some fixed point of some element of A acting on S. The elements of A leaving a point in $p^{-1}(u)$ fixed form a subgroup of A. Subgroups leaving different points in $p^{-1}(u)$ fixed are conjugate, and by standard facts about surfaces the subgroup is cyclic. Its order is called the *order* of the branch point u and is denoted here $r(u)$. By simply counting vertices, edges, and faces for a graph embedded in S/A whose vertices include all branch points, one obtains the following *Riemann–Hurwitz equation* relating the Euler characteristics $\chi(S)$ and $\chi(S/A)$:

$$\chi(S) = |A| \left(\chi(S/A) - \sum_u (1 - 1/r(u)) \right).$$

The sum in this equation is over all branch points.

Not all actions on surfaces are pseudo–free. If h is a homeomorphism of finite order of the surface S, there are basically two possibilities. Either h has a finite number of fixed points or h has order two and leaves some embedded simple closed curve fixed. In the latter case for orientable surfaces, h also reverses orientation (see Proposition 1 of [T2]), and the curves left invariant by h separate S into "halves" which are switched by h. In [T2], such an h is called a *reflection* and we retain that terminology for nonorientable surfaces, but if should be noted that for nonorientable surfaces reflections do not have to interchange halves of a surface. For more about reflections of surfaces see [B, Sn]. The important observation here is that if h has odd order it cannot be a reflection, and if A has odd order, then no action of A on any surface can contain any reflections.

With this brief discussion of pseudo–free actions, we are ready to classify groups with $\overline{\Sigma} = 0$. The klein bottle has a simple closed curve dividing the surface into two projective planes with a disk removed. Since the cyclic group Z_n and dihedral group

D_m act on the projective plane it is not hard to construct actions $Z_2 \times Z_n$ and $Z_2 \times D_m$ on the klein bottle. Surprisingly enough, these are the only possibilities.

Theorem 4.1   *(Classification of klein bottle groups)* $\overline{\Sigma}(A) = 0$ *if and only if* A *is* $Z_2 \times Z_{2m}$ *or* $Z_2 \times D_{2m}$.

Proof   By Theorem 3.1, if $\overline{\Sigma}(A) = 0$ then $Z_2 \times A$ acts on the double cover of the klein bottle, namely the torus, with the Z_2 factor generated by an antipodal map t and A preserving orientation. The groups which act on the torus preserving orientation are the finite quotients of the five Euclidean space groups that are orientation preserving. The corresponding partial presentations from [GT] are:

   a) $\langle x, y : x\,y\,x^{-1}\,y^{-1} = 1, \ldots\rangle$
   b) $\langle x, y, z : y^2 = z^2 = 1, (xy)^2 = (xz)^2 = 1, \ldots\rangle$
   c) $\langle x, y : x^3 = y^3 = (xy)^3 = 1, \ldots\rangle$
   d) $\langle x, y : x^4 = y^2 = (xy)^4 = 1, \ldots\rangle$
   e) $\langle x, y : x^3 = y^2 = (xy)^6 = 1, \ldots\rangle$

Each of the group actions on the torus given by presentations a) – e) can be obtained by taking the corresponding Euclidean space group C and projecting it onto the torus by "rolling up" the Euclidean plane in two independent directions. The space group C in this case preserves orientation and is of the type p1, p2, p3, p4, p6 (see [L] or [CM]). It is generated by translations h_1 and h_2 by independent vectors V_1 and V_2 together, in cases b) – e), with a rotation of order 2, 3, 4, 6, respectively. The set T of all translations, which is generated by h_1 and h_2, is an abelian normal subgroup of C and its quotient is the cyclic group generated by the added–on rotation (or in fact by any rotation in C of maximal order). The finite group A is then the quotient A/N where N is an normal subgroup of C. The intersection of N with the translation subgroup T must have finite index in T and hence is generated by translations in independent directions. Since C is generated by rotations in cases b) – e), if N contains a rotation, C/N is trivial, or cyclic of low order, or dihedral, none of which are of interest here. Thus N contained in T.

The antipodal map t, which commutes with the action of A on the torus, also is the projection of some isometry g of the Euclidean plane. The only orientation reversing Euclidean isometries are reflection in a line or a glide along a line (a translation in the direction of the line composed with a reflection in the line). Since a reflection leaves a line fixed and t has no fixed points, g must be a glide. Since t has order 2, g^2 is in N. Let f be the reflection part of the glide g. Then if h is a translation in C, $g\,h\,g^{-1}$ $= f\,h\,f^{-1}$ because the translation part of g commutes with h. Elementary geometry shows that if h is translation by the vector U, then $f\,h\,f^{-1}$ is translation by the reflected vector $f(U)$.

We can immediately eliminate presentations c), d), and e). It follows from elementary geometry that if r is a rotation by θ about the point p, then $g\,r\,g^{-1}$ is rotation by the angle $-\theta$ about the point $g(p)$. Since the projection of g on the torus, namely the antipodal map t, commutes with the action of A, it follows that only rotations by $180°$ are allowed in the group C. Since presentation c) has $120°$ rotations, d) has $90°$ rotations, and e) has $60°$ rotations, none of these are possibilities for the group C.

For the remaining cases a) and b), we first claim that the quotient group T/N is $Z_2 \times Z_{2m}$ or Z_n. Again assume that T is generated by translations h_1 and h_2 by vectors V_1 and V_2, respectively. Suppose the glide g is in the direction of $aV_1 + bV_2$ where a and b are relatively prime integers. If c and d satisfy $ad - bc = 1$, then translations by $aV_1 + bV_2$ and $cV_1 + dV_2$ also generate T. Thus we can and will assume that the glide g is in the direction of V_1. As before, let f denote the reflection part of g. Then $h_2\,g\,h_2^{-1}\,g^{-1}$ is a translation by the vector $U = V_2 - f(V_2)$ and is in N since the projections of h_2 and g in the torus commute. By elementary geometry, $V_2 + f(V_2) = cV_1$ for some integer c. Suppose that c is even. Since $U = 2V_2 - cV_1$, the translation h'_2 by $(1/2)U$ is in T. Since $V_2 = (1/2)U + (c/2)V_1$, it follows that h_1 and h'_2 generate T. Moreover, translation by U is in N and hence $(h'_2)^2$ is in N. Therefore T/N is $Z_2 \times Z_k$ for some integer k. If k is odd, then T/N is cyclic. This establishes our claim when c is even. Suppose instead that c is odd. Then the translation h''_2 by the vector $(1/2)(U + V_1)$ is in T. Again since $V_2 = (1/2)(U + V_1) + (c - 1)/2\ V$, if follows that h_1 and h''_2 generate T. Moreover $h_1(h''_2)^{-2}$ is translation by $-U$ and hence is in N. Therefore in T/N the images of h_1 and $(h''_2)^2$ are equal, so h''_2 generates T/N and hence T/N is cyclic, as claimed.

The cases a) and b) are now easily analyzed. In case a), there are no rotations and $C = T$. Thus $A = C/N = T/N$ has the desired form $Z_2 \times Z_{2m}$ or is cyclic, in which case $\overline{\Sigma}(A) = 1$. In case b) the space group C is obtained by adding to T a rotation by $180°$. Conjugation by such a rotation takes any translation to its inverse. Thus if $T/N = Z_2 \times Z_{2m}$, the group $A = C/N$ has a presentation

$$\langle x, y, z : x^{2m} = y^2 = z^2 = 1,\ y\,x\,y = x,\ z\,x\,z = x^{-1},\ z\,y\,z = y \rangle.$$

The subgroup generated by x and z is just the dihedral group D_{2m}, and since y commutes with x and z, the group A is just $Z_2 \times D_{2m}$. If $T/N = Z_n$ instead, then A has the presentation $\langle x, z : x^n = 1,\ z\,x\,z = x^{-1} \rangle$, which is just the dihedral group D_n, in which case $\overline{\Sigma}(A) = 1$. $\square$

5. Some Examples

In order to understand $\bar{\Sigma}$ and its relationship to the orientable parameters Σ, Σ°, and χ, we present the following examples. We restrict our attention mostly to groups of odd order so that group actions are pseudo–free and the Riemann–Hurwitz equation can be applied. The idea is as follows. If A has odd order, then no generator of A has order 2 and any action of A on a surface is pseudo–free. Thus if a Cayley graph for A is symmetrically embedded in a surface S, that Cayley graph projects to a bouquet of circles embedded in the quotient surface S/A. Each face of that quotient embedding contains a branch point of order equal to the order of the group element of A given by taking the product of the generators encountered in tracing the boundary of the face. By considering possible quotient embeddings, we can use the Riemann–Hurwitz equation to compute the possible symmetric embeddings of $C(A, X)$. The voltage graph construction [GT] actually constructs the given symmetric embedding.

Example 1 Let $A = Z_n \times Z_n$, n odd and $n > 3$. Clearly $\Sigma^\circ = \Sigma = \chi = 0$. Thus $\bar{\chi} = 0$ or -1; it is not known which. Consider $\bar{\Sigma}$. Any Cayley graph for A involves at least two generators, so we need to consider bouquets of at least two circles in various quotient surfaces. A bouquet of two circles in the projective plane has two faces. If x and y are the generators, those face boundaries correspond to group elements $x^{\pm 1}$, $y^{\pm 1}$, $x^{\pm 1}y^{\pm 2}$ or $x^{\pm 2}y^{\pm 1}$ all of which have order n. By the Riemann–Hurwitz equation, the embedding surface has Euler characteristic
$$\chi(S) = |A| \,(1 - 2(1 - 1/n)) = |A| \,(-1 + 2/n).$$
A similar computation when S/A is the klein bottle, gives $\chi(S) = |A| \,(-1 + 1/n)$. If $\chi(S/A) \leq -1$, then $\chi(S) \leq |A| \,(-1)$. Therefore, the best we can do is when S/A is the projective plane, and hence $\bar{\Sigma}(A) = |A| \,(-1 + 2/n)$. $\square$

Observe that in this example, $\bar{\Sigma}$ can be made to differ by an arbitrary amount from Σ°, Σ, χ, and $\bar{\chi}$. Also, because a one face embedding for a degree 4 Cayley graph for A has Euler characteristic $|A| - 2|A| + 1$, the given value for $\bar{\Sigma}(A)$ is not far from being the "worst" it can be. As n approaches infinity, the ratio $\bar{\Sigma}(A)/|A|$ approaches -1, which means that the ratio of the number of faces to the order of A for the best nonorientable symmetric embedding approaches 0.

When n is even in Example 1, the situation becomes more complicated. If the generating set X has only two elements, then no relator uses either generator an odd number of times so by Proposition 2.1 all symmetric embeddings are orientable. Thus Cayley graphs of degree 5 or greater must be used, driving $\bar{\Sigma}$ down to be even more negative. On the other hand, group actions may now involve reflections, which generally drives $\overline{\Sigma}$ back up again.

It is probably feasible to compute $\overline{\Sigma}$ as well for odd order abelian groups of rank greater than two. The parameters χ and $\overline{\chi}$ have been computer for most abelian groups [JW, PW], and Σ° has been computed for all abelian groups [M] (and for odd groups $\Sigma^\circ = \Sigma$).

Example 2   Let A be the metacyclic group of order 27 with the presentation $<x, y: y^3 = x^9 = 1, y^{-1} x y = x^4>$. Then $\overline{\chi}(A) = -3$ and $\chi(A) = -6$ by [BRS]. It can be shown that A is not generated by elements of order 3. By analyzing bouquets of two or more circles, it follows that if A acts on the surface S, then the quotient surface S/A must have 3 or more branch points at least two of which have order 9 when S/A is the sphere, 2 or more branch points at least one of which has order 9 when S/A is the projective plane, and at least one branch point when S/A is the torus or klein bottle. We therefore conclude that

$$\Sigma(A) = |A| (2 - (1 - 1/3) - 2(1 - 1/9)) = -12$$

and $$\overline{\Sigma}(A) = |A| (1 - (1 - 1/3) - (1 - 1/9)) = -15. \quad \square$$

This group is the first known one for which χ and $\overline{\chi}$ differ by more than 1. We suspect also that the ratio $\Sigma / \overline{\chi}$ is the largest of any group with $\overline{\chi} < 0$. Cayley graphs for which χ and $\overline{\chi}$ differ by an arbitrary amount are given in [PTW].

The other, more famous group of order 27, namely $Z_3 \times Z_3 \times Z_3$, can also be analyzed. By [MPSW] and [BS], we know that $\chi = -12$ and $\overline{\chi} = -11$ or -12. The Riemann–Hurwitz equation can be used once again to show that $\chi = -18$ and $\overline{\Sigma} = -27$.

Example 3   Let A be the alternating group on n symbols, $n > 167$. Conder [C] has shown that A has a presentation of the form

$$<x, y, z : x^2 = y^2 = z^2 = 1, (xy)^2 = (yz)^3 = (xz)^7 = 1, ...>.$$

If we choose the same rotation at every vertex of $C(A, \{x, y, z\})$ and every edge is type–1, we obtain an embedding in which every face corresponds to one of the relators $(xy)^2$, $(yz)^3$, or $(xz)^7$. Counting faces, we find the embedding surface has Euler characteristic $-|A| / 84$. Hurwitz's theorem [GT] states the $\Sigma^\circ(B) \leq -|B| / 42$ for any group B. Since the alternating group A is simple, it has no index 2 subgroup and hence any action of A on an orientable surface preserves orientation. Thus the given embedding surface of Euler characteristic $-|A|/84$ must be nonorientable. In addition, the action of A on the induced double covering achieves the Hurwitz bound. Thus $\overline{\Sigma}(A) = -|A| / 84$ and $\Sigma^\circ(A) = \Sigma(A) = -|A| / 42$. The Hurwitz theorem for embedded Cayley graphs [T3] implies that $\overline{\chi}(A) = -|A| / 84$ as well. On the other hand, $\chi(A)$ is still unknown. $\quad \square$

By the double covering Theorem 3.1, we know that $\overline{\Sigma} \leq \Sigma^\circ/2$, but in most of the examples we have given $\overline{\Sigma}$ is less than Σ°, sometimes a lot less. In addition, $\overline{\Sigma}$ has also been much less than $\Sigma, \chi,$ and $\overline{\chi}$. This example shows that symmetric nonorientable embeddings can be large, in fact as large as the Hurwitz bound allows. In particular, it is possible to have $\overline{\Sigma} > \chi, \overline{\Sigma} = \Sigma^\circ/2,$ and $\overline{\Sigma} = \overline{\chi}.$

6. Inequalities

There are a number of inequalities relating the various parameters $\chi, \Sigma, \Sigma^\circ, \overline{\chi}, \overline{\Sigma}$ to each other and to the order of the given group. Some of these are trivial. Some are classical and deep. Some are new, difficult and not well understood. We begin with the easy ones.

Easy Inequalities $\Sigma^\circ \leq \Sigma \leq \chi, \overline{\Sigma} \leq \overline{\chi}, \overline{\chi} \geq \chi - 1, \overline{\Sigma} \leq \Sigma^\circ/2.$ $\square$

The next to last inequality comes from the observation that one can always twist a single edge in any orientable embedding of any graph (other than a tree) to obtain a nonorientable embedding at the expense of decreasing the Euler characteristic by at most 1. The last inequality follows of course from Theorem 3.1.

The next collection of inequalities are the Hurwitz inequalities relating our various parameters to the order of the group. Whenever the order of a group is sufficiently large compared to the Euler characteristic of the embedding surface, the structure of the group becomes restricted by the presence of the short relators necessary to construct the many small faces of the embedding. This leads to some complicated refinements of the basic Hurwitz inequality. The following notation and terminology is helpful. Call A a $(p, q, r)^\circ$ *group* if A has a presentation of the form

$$\langle x, y, z : x^2 = y^2 = z^2 = 1, (xy)^p = (yz)^q = (xz)^r = 1, ...\rangle.$$

Call a (p, q, r) group A *proper* if the subgroup generated by xy and yz is a proper subgroup (it is easily shown to have index at most 2); call it *improper* otherwise. Call A a $(p, q, r)^\circ$ *group* if it has a presentation of the form

$$\langle u, v : u^p = v^q = (uv)^r = 1, ...\rangle.$$

Hurwitz Inequalities   Let A be a group with $\chi(A) < 0$. Then $\chi, \Sigma, \overline{\chi},$ and $\overline{\Sigma}$ are all bounded above by $-|A|/84$. This bound is achieved if and only if A is a proper $(2, 3, 7)$ group (for χ and Σ) or an improper $(2, 3, 7)$ group (for $\overline{\chi}$ and $\overline{\Sigma}$). If A is not a $(2, 3, 7)$ group, then $\chi, \Sigma, \overline{\chi}, \overline{\Sigma}$ are all bounded above by $-|A|/48$ and this bound is achieved only for proper and improper $(2, 3, 8)$ groups. Similarly, Σ° is bounded above by $-|A|/42$, with equality only when A is a $(2, 3, 7)^\circ$ group. $\square$

Hurwitz's original inequality [H] is the one for Σ° and appears in the context of the conformal automorphisms of Riemann surfaces. The version for embedded Cayley graphs is first given in [T1]. There are many further and important refinements [T4], which we have not presented here.

The inequalities we give in this last collection are new. Proofs will appear elsewhere [T5]. These inequalities also will have a number of refinements, including applications to the genus of a quotient group. We state here only the simplest versions.

New Inequalities   Let A be any group with $\chi(A) < 0$. Then the following inequalities hold:

     a) $\chi > 6\bar{\chi},\ \Sigma > 6\bar{\chi},\ \Sigma > 6\chi$

     b) $\chi \geq 4\bar{\chi},\ \Sigma \geq 4\bar{\chi},\ \Sigma \geq 4\chi$ if A has no Cayley graph of degree 5 or less

     c) $\Sigma^\circ > 6\bar{\chi},\ \Sigma^\circ > 6\chi,\ \Sigma^\circ > 6\Sigma$ unless A is a $(2, 3, 7)$ or $(2, 3, 8)$ group. $\Box$

It is expected that the numbers 6 and 4 can be lowered in many cases in some of the inequalities. Notice however for the b) inequalities that $\Sigma = 4\bar{\chi}$ in Example 2, and for the c) inequalities, it can be shown that $\Sigma^\circ = 8\Sigma$ for the group of genus two [T5].

7. Problems

We conclude with some problems for further study. Some are purely computational, others are wider ranging.

Problem 1   Compute $\bar{\Sigma}(Z_n \times Z_n)$ for n even.

Problem 2   Compute $\bar{\Sigma}$ for various 2–groups.

Problem 3   Determine all groups A with $\bar{\Sigma} = -1$. By work similar but easier to that in [T4], it can be shown that if $\bar{\chi} \geq -1$, then $\chi \geq 0$. Also since $\Sigma(Z_2 \times A) = -2$, the computations in [T7] can be applied to $Z_2 \times A$, and it seems that $|A| \leq 24$. The Riemann–Hurwitz equation can be used to show that there are no pseudo–free group actions of order greater than 6 on the surface of Euler characteristic -1. We conjecture that there are no groups with $\bar{\Sigma} = -1$.

Problem 4   Find a lower bound for $\bar{\Sigma}$ in terms of $\chi, \bar{\chi}, \Sigma$, or Σ°. The New Inequalities say nothing about $\bar{\Sigma}$. None of the methods used in [T5] apply directly to $\bar{\Sigma}$. Notice in Example 2 that $\bar{\Sigma} = 5\bar{\chi}$ and for the group of genus two we believe that $\bar{\Sigma} = 9\Sigma = 9\chi$.

Problem 5   Determine all the surfaces containing a symmetric Cayley graph embedding for a given group. Graph embeddings satisfy an interpolation theorem: if a graph has a 2–cell embedding in orientable surfaces of genus g_1 and g_2, then it has embeddings in all orientable surfaces of genus between g_1 and g_2. Symmetric embeddings do not satisfy this result, mostly because there are so few symmetric embeddings of a Cayley graph. For example, a Cayley graph of degree three only has 4 or 8 possible symmetric embeddings, depending on whether the generating set has 2 or 3 elements (there are only two possible rotations and one is the reverse of the other, so the only choice is edge types).

Problem 6   Classify irredundant Cayley graphs embeddable in the klein bottle. Theorem 4.1 only classifies symmetric embeddings. The full classification of groups with $\chi = 0$ given in [T4] implies that any irredundant Cayley graph embeddable in the klein bottle embeds symmetrically in the torus, but the original klein bottle embedding of that Cayley graph could be nonsymmetric. We conjecture that $\overline{\chi} = 0$ if and only if $\overline{\Sigma} = 0$. A first step would be to show that $\overline{\chi}(Z_m \times Z_n) = -1$, where m divides n and $m > 2$. This problem is solved for the projective plane: the only group with $\overline{\chi} = 1$ but $\overline{\Sigma} < 1$ is $Z_3 \times Z_3$ (see Exercises in [GT]).

REFERENCES

[BRS]   M.G. Brin, D.E. Rauschenberg, and C.C. Squier, On the genus of the semidirect product of Z_9 by Z_3, J. Graph Theory, to appear.

[BS]   M.G. Brin and C.C. Squier, On the genus of $Z_3 \times Z_3$, Europ. J. Combin., to appear.

[BSn]   E. Bujalance and D. Singerman, the symmetry type of a Riemann surface, Proc. London Math Soc. 51 (1985), 501 – 509.

[B]   W. Burnside, *Theory of Groups of Finite Order*, Cambridge Univ. Press, Cambridge, 1897.

[CM]   H.S.M. Coxeter and W.O.J. Moser, *Generators and Relations for Discrete Groups*, 4th Ed., Springer–Verlag, New York, 1980.

[G]   D. Gorenstein, *Finite Groups,* Chelsea, New York, 1980.

[GT]   J.L. Gross and T.W. Tucker, *Topological Graph Theory*, Wiley–Interscience, New York, 1987.

[H]   A. Hurwitz, Uber algebraische Gebilde mit eindeutigen Transformationen in sich, Math. Ann. 41 (1892), 403 – 442.

[JW]   M. Jungerman and A.T. White, On the genus of finite abelian groups, Europ. J. Combin. 1 (1978), 243 – 251.

[L]   H.M. Levinson, On the genera of graphs of group presentations, Ann. New York Acad. Sci. 175 (1970), 277 – 284.

[LM]   H.M. Levinson and B. Maskit, Special embeddings of Cayley diagrams, J. Combin. Theory Ser. B 18 (1975), 12 – 17.

[Ly]   R.C. Lyndon, *Groups and Geometry*, Lon. Math. Soc. Notes 101, Cambridge Univ. Press, Cambridge, 1985.

[M]   C. Maclachlin, Abelian groups of automorphisms of compact Riemann surfaces, Proc. Lond. Math Soc. 15 (1965), 699 – 712.

[MPSW]   B. Mohar, T. Pisanski, M. Skoviera, and A.T. White, The cartesian product of three triangles can be embedded into a surface of genus 7, Discrete Math. 56 (1985), 87 – 89.

[MPTW]   B. Mohar, T. Pisanski, T.W. Tucker, and M.E. Watkins, On the classification of infinite planar groups, in preparation.

[PTW]   T. Pisanski, T.W. Tucker, D. Witte, The nonorientable genus of some metacyclic groups, preprint.

[PW]   T. Pisanski and A.T. White, Nonorientable embeddings of groups, Europ. J. Combin. 9 (1988), 445 – 461.

[S]   G. Sabidussi, On a class of fixed–point free graphs, Proc. Amer. Math. Soc. 9 (1958), 800 – 804.

[Sn]   D. Singerman, Symmetries of Riemann surfaces with large automorphism group, Math Ann. 210 (1974), 17 – 32.

[T1]   T.W. Tucker, The number of groups of a given genus, Trans. Amer. Math Soc. 258 (1980), 167 – 179.

[T2]   T.W. Tucker, Finite groups acting on surfaces and the genus of a group, J. Combin. Theory Ser. B 34 (1983), 82 – 98.

[T3]   T.W. Tucker, A refined Hurwitz theorem for imbeddings of irredundant Cayley graphs, J. Combin. Theory Ser. B (1984), 244 – 268.

[T4]   T.W. Tucker, There is one group of genus two, J. Combin. Theory Ser. B (1984), 269 – 275.

[T5]   T.W. Tucker, Bounds on the genus parameters of a group and its quotient groups, in preparation.

[W1]   A.T. White, *Graphs, Groups, and Surfaces*, North–Holland, Amsterdam, 1973 (rev. ed. 1984).

[W2]   A.T. White, The genus parameter for groups, Scientia, to appear.

The Number of Isomorphism Classes of Spanning Unicyclic Subgraphs of a Graph

Preben Dahl Vestergaard

Aalborg University

Denmark

ABSTRACT

In [3], B. Zelinka proved that the spanning trees of a graph containing n disjoint circuits can be partitioned into at least n + 1 isomorphism classes.

We generalize this result in several directions: if a graph contains n disjoint circuits, then it contains at least n isomorphism classes of spanning unicyclic graphs (Corolllary 1), and we give a lower bound depending on the girth of G (Theorem 1), also we determine the extremal graphs (Theorem 3).

Definitions

A *tree* is a connected graph with no circuit. A *rooted graph* (G, v) is a graph G together with a distinguished vertex v, called the root, from V(G). If it is clear which root $v \in V(G)$ is intended, we may just write the rooted graph G. A *unicyclic graph* G is a connected graph with precisely one circuit C, a component of $G - E(C)$ is called a *pendant tree,* and it is said to be *attached* to C at the unique vertex which it has in common with C. In a graph a *block* is defined to be a maximal subgraph spanned by a set of edges, any two of which belong to a common circuit. An edge which belongs to no circuit is a block. A *cactus* is a connected graph in which each block is either a circuit or an edge. An *n–cactus* is a cactus with precisely n circuits. A *spanning subgraph* of G is G itself, or a subgraph of G obtained by deleting edges from G. The *distance* between two subsets of V(G) is the length of a shortest path joining a vertex from one subset to a vertex from the other subset. Two graphs are said to be *isomorphic,* if there exists a bijection from one graph onto the other which preserves incidence. Two rooted graphs are said to be *root–isomorphic,* if there exists an isomorphism between the graphs which maps one root onto the other root. $(v_1, v_2, \ldots, v_r)$ denotes a circuit with the vertices in that order. $\lceil t \rceil$ denotes the least integer not less than t, and $\lfloor t \rfloor$ denotes the largest integer not greater than t.

A circuit C in a cactus G is called an *internal* circuit if $G - E(C)$ has at least two connected components each containing a circuit, if $G - E(C)$ has at most one connected component containing a circuit, then C is said to be *external*. For a unicyclic graph H with circuit C, we define the weight of H to be

$$w(H) = \sum_{v \in V(C)} \sum_{x \in V(G)} d(x,v)$$

Theorem 1   *If the graph G contains a spanning n–cactus, $n \geq 2$, with an external circuit of length r, $r \geq 3$, then G contains at least $n - 2 + \frac{r}{2}$ pairwise non–isomorphic spanning unicyclic subgraphs.*

Corollary 1   *If a connected graph G contains n pairwise disjoint circuits, $n \geq 1$, then G contains at least n pairwise non–isomophic spanning unicyclic subgraphs.*

Proof of Corollary 1   Any connected graph G which contains a circuit C also contains a spanning unicyclic subgraph: delete if necessary edges from $G - E(C)$ until all other circuits are destroyed and such that the resulting graph remains connected.

This proves the Corollary for n = 1. For $n \geq 2$ we see that $r \geq 3$ implies that an integer $\geq n - 2 + \frac{r}{2}$ must be at least n. Since a connected graph with n disjoint circuits contains a spanning n–cactus, we can apply Theorem 1 to obtain the desired reult that G contains at least n pairwise non-isomorphic spanning unicyclic subgraphs. This proves Corollary 1. ❏

Proof of Theorem 1   Theorem 1 is only claimed to hold for $n \geq 2$, and in fact the statement of Theorem 1 does not hold for n = 1, $r \geq 5$, because then $\lceil n - 2 + \frac{r}{2} \rceil \geq 2$, but if G is unicyclic, then by definition G contains only one spanning unicyclic subgraph, namely G itself.

It is enough to prove Theorem 1 for an n–cactus, so we shall suppose that G is an n–cactus and we shall do the proof by induction on n.

Let n = 2, and let G be a 2–cactus with the two circuits C_1 and C_2. Let the vertices of C_1 be $v_1, v_2, v_3, \ldots, v_r$ in that order, and let v_1 be the uniquely determined vertex on C_1 having minimum distance in G to C_2. Suppose r is odd. We shall now list $\frac{r+1}{2}$ spanning unicyclic subgraphs of G, no two of which are isomorphic to each other, namely:

$$G - (v_1, v_2), \; G - (v_2, v_3), \; \ldots, \; G - (v_{\frac{r-1}{2}}, v_{\frac{r+1}{2}}), \; G - (v_{\frac{r+1}{2}}, v_{\frac{r+3}{2}})$$

Each of the $\frac{r+1}{2}$ graphs above contains all vertices of G, contains C_2 as its only circuit and is connected, so each one is a unicyclic spanning subgraph of G. No two of the graphs are isomorphic to each other because no two of them have the same weight.

It is easy to see that the weights of the graphs above form a strictly decreasing sequence, because as i increases from 1 to $\frac{r+1}{2}$ access from C_2 through v_1 to vertices round $C_1 - (v_i, v_{i+1})$ happens through paths some of whose lengths are shorter and some of whose lengths are unchanged. E. g., the distance from v_1, are hence from vertices on C_2, to v_2 is $r - 1$ in $G - (v_1, v_2)$ but only 1 in $G - (v_2, v_3)$.

If r is even we can analogously enumerate $\frac{r}{2}$ pairwise non–isomorphic spanning unicyclic subgraphs for G. This proves Theorem 1 for $n = 2$ and for all $r \geq 3$.

Let $n \geq 3$. Suppose that Theorem 1 holds for $n - 1$ and all $r \geq 3$, we shall then prove that it also holds for n and all $r \geq 3$.

Let G be an n–cactus and let $C_1 = (v_1, v_2, ..., v_r)$ be an external circuit in G. Since C_1 is external, it contains a uniquely determined vertex, say v_1, through which any path joining the other vertices of C_1 to another cirucit of G must pass. Suppose r is even, the case r odd can be treated analogously. The graph $G - (v_1, v_2)$ by induction hypothesis contains at least $n - 1$ pairwise non–isomorphic spanning unicyclic subgraphs. Among them all let H be one with minimum weight. Denote the graph $H + (v_1, v_2)$ by H'. We shall now list $\frac{r}{2} - 1$ spanning unicyclic subgraphs of G, all of which have weights which are pairwise distinct and also less than the weight of H. This will give us at least $(n-1) + (\frac{r}{2}-1)$ pairwise non–isomorphic spanning unicyclic subgraphs for G as desired. The $\frac{r}{2} - 1$ graphs are:

$$H' - (v_2, v_3), \; H' - (v_3, v_4), \; ..., \; H' - (v_{\frac{r}{2}}, v_{\frac{r}{2}+1}).$$

Their weights form a stricly decreasing sequence of numbers less than $w(H)$ because for each i, $i = 2, 3, ..., \frac{r}{2}$, the distance from v_1 to v_i is shorter in $H' - (v_i, v_{i+1})$ than in $H' - (v_{i-1}, v_i)$. This proves Theorem 1. ❑

The following example shows that Theorem 1 is best possible:

Example Let G be the n–cactus consisting of n triangles v, v_{2i-1}, v_{2i}, $i = 1, 2,$ ..., n, with the common vertex v. Each triangle is an external circuit of length $r = 3$ and G must by Theorem 1 contain $\geq n - 2 + 1\frac{1}{2}$, i. e., at least n, pairwise non–isomorphic spanning unicyclic subgraphs, and we can verify that G in fact contains exacly n

pairwise non–isomorphic spanning unicyclic subgraphs: a spanning unicyclic subgraph for G consists of one triangle with a vertex v from which emanates $2i$ paths of length 1 and $n - 1 - i$ paths of length 2, where i is one of the numbers $0, 1, 2, ..., n - 1$.

We shall show in Theorem 3 that the graphs which are extremal for Theorem 1 are all quite similar to G in the Example above.

For the proof of Theorem 3 we shall need the following Theorem 2 and Lemma 1.

Theorem 2 *If a graph contains a spanning 2–cactus and has exactly two isomorphism classes of spanning unicyclic subgraphs then*

either (1) G can be constructed thus:

> *1a. Let v_1, v_2, v_3 and w_1, w_2, w_3 be two 3–circuits which are disjoint except that possibly $v_1 = w_1$. If v_1 and w_1 are distinct, then they are joined by a v_1w_1–path which has nothing but v_1 and w_1 in common with the two 3–circuits.*

> *1b. Attach to each of v_2, v_3, w_2, w_3 a copy of a rooted tree A.*

> *1c. If $v_1 = w_1$, then attach to v_1 a rooted tree B. Otherwise, let the v_1w_1–path be $t_1, t_2, ..., t_n$ with $t_1 = v_1$ and $t_n = w_1$. For each $i = 1, 2, ..., \lfloor \frac{n+1}{2} \rfloor$ attach to each of t_i, t_{n-i+1} a copy of a rooted tree B_i.*

or (2) G can be constructed thus:

> *2a. Let v_1, v_2, v_3, v_4 and w_1, w_2, w_3, w_4 be two 4–circuits which are disjoint except that possibly $v_1 = w_1$. If v_1 and w_1 are distinct, then they are joined by a v_1w_1–path which has nothing but v_1 and w_1 in common with the two 4–circuits.*

> *2b. Attach to each of v_2, v_4, w_2, w_4 a copy of a rooted tree A. Attach to each of w_3, w_3 a copy of a rooted tree B.*

> *1c. If $v_1 = w_1$, then attach to v_1 a rooted tree C. Otherwise, let the v_1w_1–path be $t_1, t_2, ..., t_n$ with $t_1 = v_1$ and $t_n = w_1$. For each $i = 1, 2, ..., \lfloor \frac{n+1}{2} \rfloor$ attach to each of t_i, t_{n-i+1} a copy of a rooted tree C_i.*

Theorem 2 can be derived from results in [2] and it states that if a graph G with a spanning 2–cactus has exactly two isomorphism classes of spanning unicyclic subgraphs then G is a symmetric 2–cactus with circuit length 3 or 4. ❏

Lemma 1   *If G contains a spanning n–cactus, $n \geq 3$, and has exactly n isomorphism classes of spanning unicyclic subgraphs and if e is an edge of an external circuit of the cactus then $G - e$ has exactly $n - 1$ isomorphism classes of spanning unicyclic subgraphs.*

Proof of Lemma 1 $G - e$ contains a spanning $(n - 1)$–cactus and by Theorem 1 has at least $n - 1$ isomorphism classes of spanning unicyclic subgraphs. Suppose $G - e$ has n or more isomorphism classes of spanning unicyclic subgraphs then as in the proof of Theorem 1 we take H, a spanning unicyclic subgraph of $G - e$ with maximum/minimum weight and on the external circuit containing e we can find another edge f such that $H + e - f$ is another spanning unicyclic subgraph of G with larger/smaller weight than H and thus G will have at least $n + 1$ isomorphism classes of spanning unicyclic subgraphs. This contradicts the hypothesis about G and Lemma 1 is proven. $\square$

We are now ready to state Theorem 3:

Theorem 3   *If the graph G contains a spanning n–cactus, $n \geq 3$, and if G contains exactly n isomorphism classes of spanning unicyclic subgraphs, then G can be obtained by the following construction:*

1. *Let (M, v) denote a unicyclic graph M with root v and with the circuit having length 3 or 4. In both cases, the circuit has exactly two vertices x an y which are equidistant from v. Further, let the pendant trees in M at x and y, respectively, be root–isomorphic to each other.*

2. *Form the disjoint union of n copies of (M, v) with v identified, and*

3. *Attach to v a rooted tree (T, v), possibly $T = v$.*

Theorem 3 obviously does not hold for $n = 1$ because Theorem 3 prescribes a circuit length of 3 or 4, but also a unicyclic graph circuit with length 5 or more has exactly one spanning unicyclic subgraph. Neither is Theorem 3 quite true for $n = 2$, but that case is completely described by Theorem 2.

Examples of G obtained by the construction of Theorem 3:

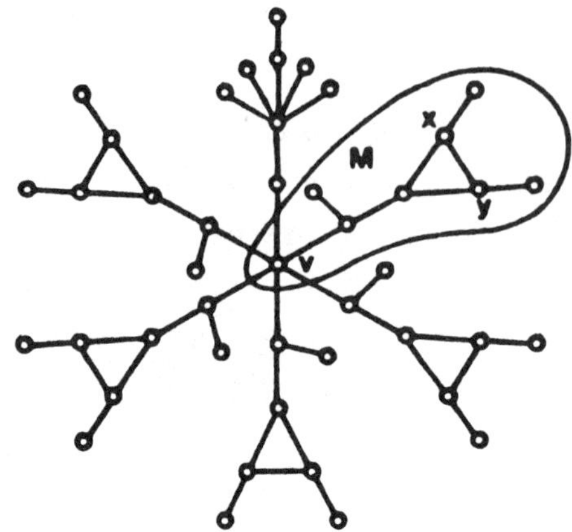

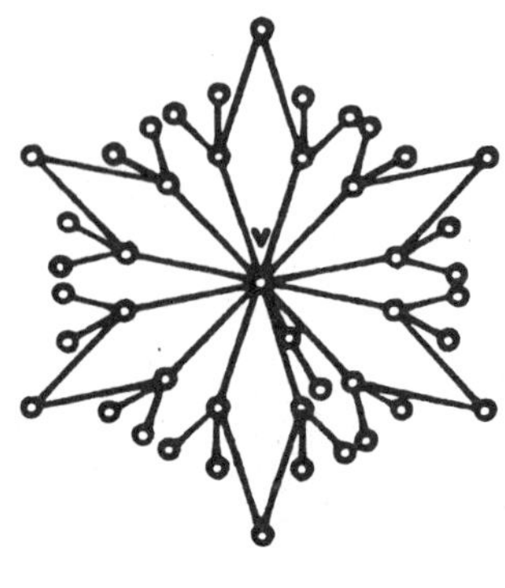

The graph G with $n = 5$.   The graph G with $n = 6$.

M has a circuit of length 3.   M has a circuit of length 4.

Proof of Theorem 3   First we shall prove Theorem 3 under the assumption that G is an n–cactus. We shall use induction on n. Let $n = 3$. Any 3–cactus has at least two external circuits, so G has at most one internal circuit and one of the following two cases occur:

Case 1. G has one internal circuit.

Case 2. G has no internal circuit.

Suppose Case 1 occurs, then G contains one internal circuit, C_2, and two external circuits, C_1, C_3. Let $v_1 \in V(C_1)$ have minimum distance from C_2, and let $C_1 = (v_1, v_2, ..., v_r)$. For r odd, let $e = (v_{\frac{r+1}{2}}, v_{\frac{r+3}{2}})$, and for r even, let $e = (v_{\frac{r}{2}}, v_{\frac{r}{2}+1})$. $G - e$ is a 2–cactus, and, by Lemma 1, $G - e$ has exactly two isomorhpism classes of spanning unicyclic subgraphs, and, by Theorem 2, $G - e$ is a symmetric 2–cactus, i. e., the pendant trees to C_3 are the same as those to C_2.

Let $f = (v_1, v_2)$, and apply now Lemma 1 and Theorem 2 to the graph $G - f$, then we also obtain that the pendant trees to C_3 are the same as those to C_2. but for $G - e$ and $G - f$ the pendant trees of C_3 remain unchanged while those of C_2 do change. This is a contradiction which proves that Case 1 cannot occur.

Thus Case 2 must occur, i. e. all 3 circuits C_1, C_2, C_3 in G are external. By Lemma 1, $G - e$ has for any $e \in E(C_1)$ exactly two isomorphism classes of spanning unicyclic subgraphs. Hence $G - e$, by Theorem 2, is a symmetric 2–cactus.

Since this holds for all edges of C_1, we can conclude that the tree containing $C_1 - e$ necessarily is attached to the central vertex of the unique path joining C_2 to C_3. considering $G - f$ for $f \in E(C_i)$, $i = 2, 3$, we find that G has the desired structure and Theorem 3 is proven for $n = 3$.

Let $n \geq 4$. Suppose Theorem 3 holds for $n - 1$, we shall then prove that it also holds for n.

Let C_1 be an external circuit for G. For any $e \in E(C_1)$, $G - e$ is an $(n - 1)$–cactus which by Lemma 1 has exactly $n - 1$ isomorphism classes of spanning unicyclic subgraphs. By induction hypothesis $G - e$ is a symmetric $(n - 1)$–cactus. By considering $G - f$, where f belongs to a circuit of $G - E(C_1)$ we find that G has the desired structure and Theorem 3 is proven under the assumption that G is a cactus.

To finish the proof we only have to demonstrate that if a graph G satisfies the hypothesis of Theorem 3 and hence contains a spanning cactus G' then $G = G'$.

Suppose $E(G') \subset E(G)$ and let $e \in E(G) \setminus E(G')$, then it is easy to see that $G' + e$ and hence G contains at least $n + 1$ isomorphism classes of spanning unicyclic subgraphs. This completes the proof of Theorem 3. ❏

Problem 1   *How does the number of spanning unicyclic subgraphs depend on circuit length?*

It is probably not true that the number of spanning unicyclic subgraphs in some sense increases monotonically with the girth or with, say, the sum of lengths of all circuits in the graph, but Theorem 1 shows that at least for special graphs as the cacti circuit length plays a role.

Problem 2   *What can be said about the number of spanning 2–cacti, the number of spanning subgraphs with cyclomatic number $r = 0, 1, 2, 3, ... ?*

REFERENCES

[1]   B. L. Hartnell, Some problems on spanning trees, *Proc. Fifth Manitoba Conf. Numer. Math.* (1975) 375–384.

[2]   P. D. Vestergaard, On graphs with two isomorphism classes of spanning unicyclic subgraphs, in preparation.

[3]   Bohdan Zelinka, The number of isomorphism classes of spanning trees of a graph, *Math. Slovaca.* **28** (1978) no. 4, 385–388.

A Survey of Snarks

John J. Watkins
The Colorado College

Robin J. Wilson
The Open University, England

ABSTRACT

A snark *is a connected bridgeless cubic graph with edge–chromatic number 4 —
that is, it cannot be 3–edge–colored. In this paper we survey the work that has
been done on snarks, including their history, various constructions, and possible
generalizations. This paper updates the survey of snarks given in [10].*

1. Introduction

The story of snarks began in 1880. P.G. Tait, a mathematical physicist at the
University of Edinburgh, was attempting to find a short proof of the four–color theorem.
An accepted 'proof' had been given by A.B. Kempe in 1879, but P.J. Heawood found a
flaw in it in 1890. Tait translated the problem of coloring maps to one of coloring the
edges of cubic graphs [29]. In particular, he showed that *the four–color theorem is
equivalent to the statement that every bridgeless cubic planar graph is 3–edge–colorable*
(see [14], pp. 26–27).

Tait believed that all cubic graphs are 3–edge–colorable, and was thus convinced that
he had proved the four–color theorem. However, he overlooked some cubic graphs that
cannot be 3–edge–colored: those with bridges, such as the graph in Figure 1, and certain
non–planar graphs, such as the Petersen graph (Figure 2).

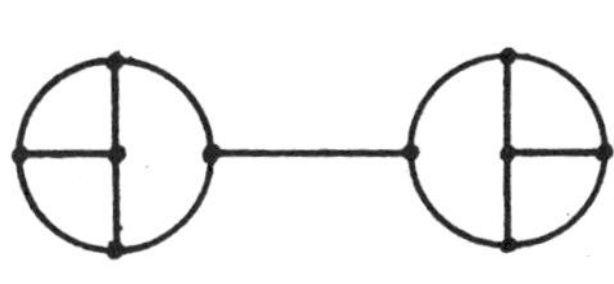

Figure 1

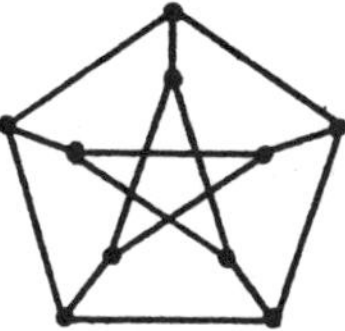

Figure 2

Non–trivial cubic graphs that cannot be 3–edge–colored, such as the Petersen graph, captured the fancy of Martin Gardner and led him to introduce the term *snark* in a Scientific American column in 1976 [17].

2. Examples of Snarks

The earliest snark to appear was the Petersen graph, discussed by Petersen [25] in 1898; Figure 3 shows an alternative drawing, due to Kempe [22] some years earlier.

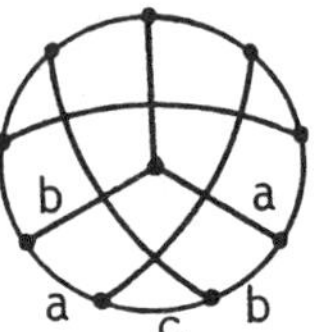

Figure 3

To see that this graph cannot be 3–edge–colored, we assume the contrary, and begin (without loss of generality) by 3–coloring any 5–cycle a, b, a, b, c as shown. This immediately determines the colors of five additional edges, and leads to an impasse since we are then forced to color two adjacent edges with the same color.

The next snark appeared in 1946, in a chemical paper of D. Blanuša [2]. We now know that this type of snark comes in two varieties (Figure 4), each of which is composed from two 'copies' of the Petersen graph, in a sense to be explained later.

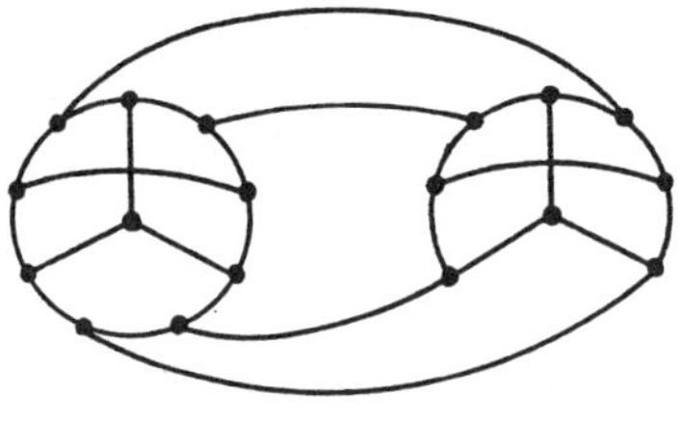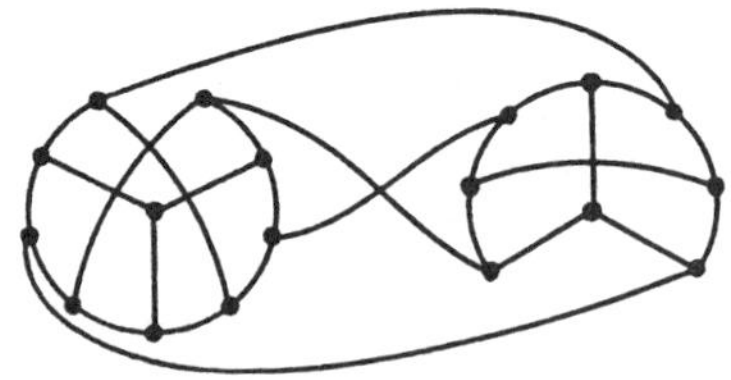

Type 1   Type 2

Figure 4

Another snark was found by Blanche Descartes [12] in 1948; it has 210 vertices, and was constructed by adding a 'star configuration' to each edge of the Petersen graph (see [14]). A further snark was found in 1973 by G. Szekeres [28]. As shown in Figure 5, it is composed of five 'copies' of the Petersen graph.

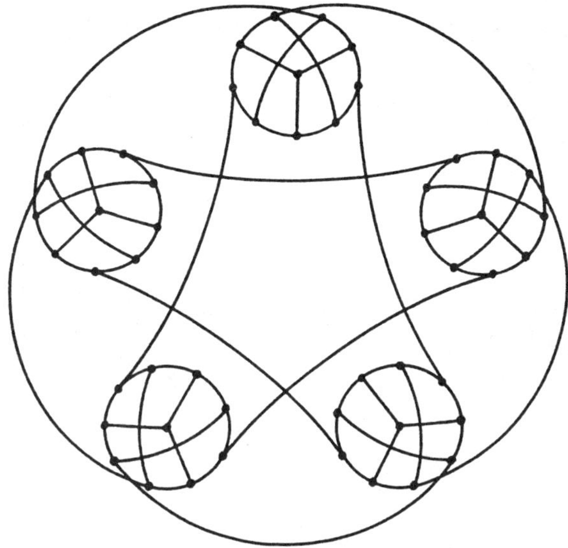

Figure 5

Such painstakingly slow progress in the discovery of snarks underwent a dramatic change when an infinite family of snarks was discovered, by E. Grinberg in 1972, and independently by R. Isaacs in 1974 [19]; they are known as the *flower snarks* $J_3, J_5, J_7, \ldots$ of orders 12, 20, 28, ... The first flower snark J_3 is really just the Petersen graph, with the central vertex replaced by a triangle. We show the first two flower snarks in Figure 6.

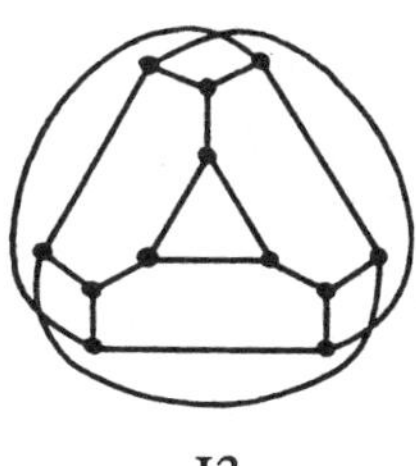

J3

J5

Figure 6

Isaacs also presented a method for joining two snarks to get a new snark; for example, two copies of the Petersen graph combine in this way to give the two types of

Blanuša snark. In this process, two snarks G_1 and G_2 are joined to form a new snark $G_1 \cdot G_2$, called a *dot product*, as follows:

(1) remove any two non–adjacent edges ab and cd from G_1;

(2) remove any two adjacent vertices x and y from G_2;

(3) join the vertices, as shown in Figure 7.

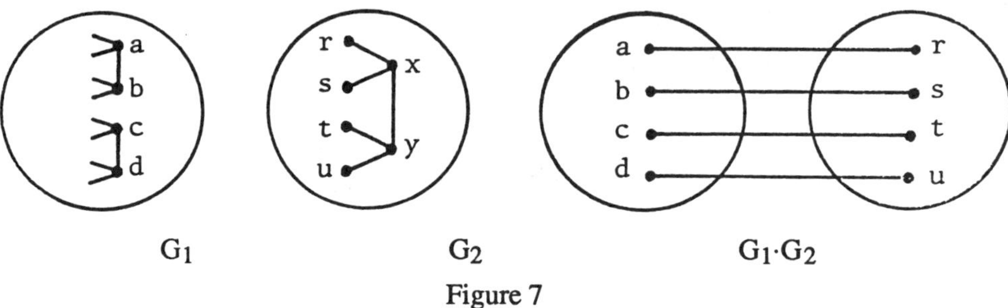

Figure 7

A dot product depends both on G_1 and G_2, and on the choices made in (1) and (2). For example, the two types of Blanuša snark arise from the Petersen graph because there are essentially two different choices in (1).

With his remarkable paper, Isaacs transformed snark hunting. Where before there had been essentially four snarks, there were now infinitely many snarks: the flower snarks and those formed by dot products. Furthermore, these two families are distinct and contain all of the previously known snarks.

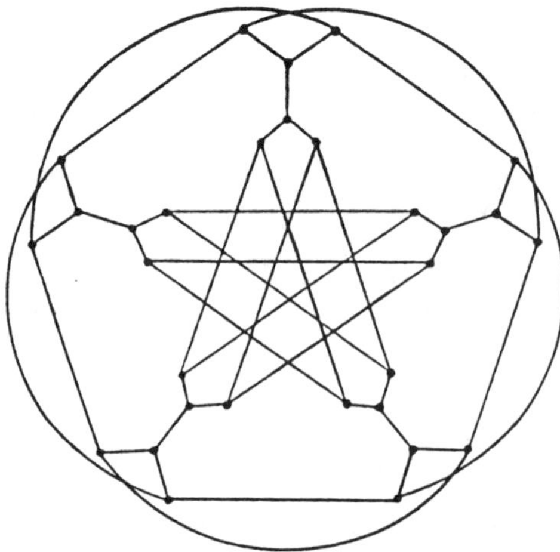

Figure 8

Isaacs [19] also discovered another snark which is not a member of either of these infinite families. This is the *double–star snark* D(5,2), shown in Figure 8; it has five pieces in each of the inner and outer 'rims', and the connections from the outer rim are made to every second piece of the inner rim. The double–star graph D(n,k) is defined similarly. A.G. Chetwynd [8] has shown that D(n,k) is a snark if and only if it is D(5,2) or D(n,k), where $\frac{1}{3}$n is an integer and gcd(n,k) = $\frac{1}{3}$n. (However, the latter snarks are often considered trivial, a notion we discuss in the next section.) Surprisingly, a similar generalization of the Petersen graph yields no additional snarks [5].

3.　Trivial Snarks

Before describing some further constructions of snarks, we must deal with questions of triviality. Some snarks are trivial modifications of others, and we should eliminate such examples. This topic has attracted considerable discussion in the literature (see, for example, [10], [17], [19], [27], and [32]). But exactly which modifications of snarks are we to consider trivial?

If we take any snark and replace any vertex by a triangle, then the resulting cubic graph still cannot be 3–edge–colored. Similarly, a digon (a pair of multiple edges) can be inserted into any edge, as in Figure 9. Conversely, contracting a triangle to a vertex does not affect 3–edge–colorability, nor does contracting a digon to an edge. Thus we may assume that snarks have girth 4 or more.

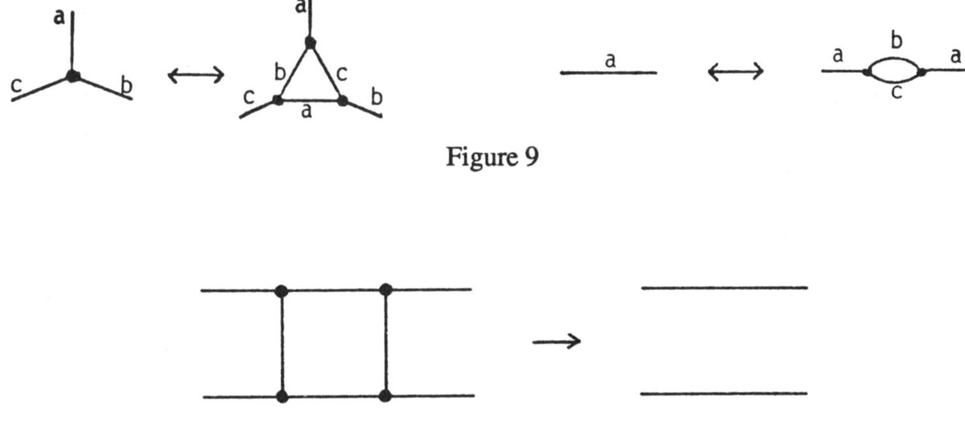

Figure 9

Figure 10

We can go further, since a quadrilateral in a snark may be replaced by two edges, as shown in Figure 10. This produces a smaller snark, since any 3–coloring of the smaller graph would immediately yield a 3–coloring of the original snark, which is impossible. On the other hand, inserting a quadrilateral along two edges of a snark *can* affect the colorability. For example, a quadrilateral inserted along any two non–adjacent edges of the Petersen graph yields a 3–edge–colorable graph. The status of quadrilaterals is thus somewhat different from that of digons and triangles. Nonetheless, it has been customary to exclude quadrilaterals and require that snarks have girth 5 or more.

It has also been customary to require a snark to be cyclically 4–edge–connected; that is, deleting fewer than four edges does not disconnect the graph into two components each containing a cycle. For if a cubic graph has a bridge, then it is not 3–edge–colorable; if it has a cutset with two edges, then in any 3–edge–coloring of the graph these two edges must be colored the same, and thus we can take a snark and a cubic graph and construct a new snark by cutting any edge in each graph and joining the resulting semi–edges; similarly, if it has a cyclically–disconnecting cutset with three edges, then in any 3–edge–coloring of the graph these three edges are colored differently, and thus we can take a snark and a cubic graph and construct a new snark by deleting a vertex from each graph and joining the three pairs of semi–edges.

It has recently been shown that similar results hold for snarks with cutsets of size 4 and 5 (see [4], [18] and [27]). Thus these snarks can also be considered to be trivial, because they can easily be constructed from smaller snarks. However, it is *not* usual to make this restriction.

It seems, therefore, that the elusive notion of triviality has escaped us. We could impose the restriction that snarks be cyclically 6–edge–connected, but there may also be a decomposition result for cutsets of size 6. It therefore seems premature to exclude such snarks at present.

An alternative approach is to focus on the structure of snarks, leaving the definition of a snark as broad as possible — simply, a connected bridgeless cubic graph with chromatic index 4. The goal is then to specify a collection of 'prime' snarks, and a list of basic constructions with which every snark can be built up from prime snarks. We cannot yet offer a definition of a prime snark. However, a prime snark should contain no multiple edges, triangles or quadrilaterals and (apart from the Petersen graph) should be cyclically 6–edge–connected.

4. Some Specific Constructions

The dot product is a very general and powerful method for generating new snarks from known snarks. In this section we narrow our focus to some *specific* constructions for snarks; when these constructions yield cyclically 4–edge–connected snarks, there is automatically an alternative construction using the dot product.

Loupekine Snarks

If we remove a path of length 3 from a snark, we obtain a graph with five semi–edges. In any coloring of this graph, one pair of semi–edges has the same color (a match) and another pair has different colors (a mismatch). The color of the fifth edge does not concern us.

This idea was used by F. Loupekine to construct an infinite family of snarks (see [10], [20] and [32]). For example, a snark results if we join an *odd* number of these graphs, as in Figure 11; here we have removed a path of length 3 from the Petersen graph.

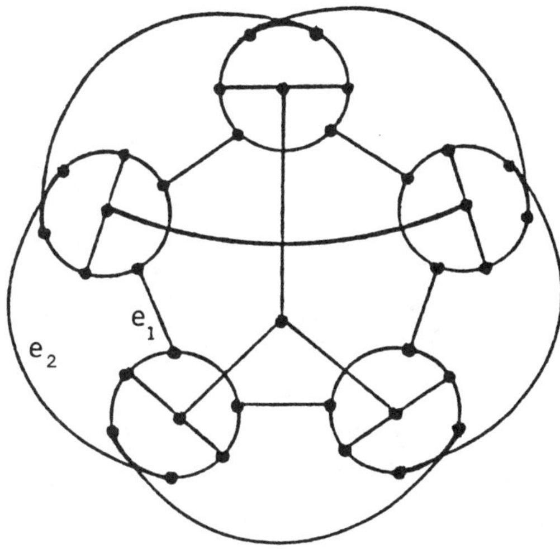

Figure 11

In this construction, any snark may be used, we may cross some edges (such as e_1 and e_2 in Figure 11), or we may insert arbitrary graphs into a pair of connecting edges. We may also treat the central edges in any fashion we like, since they are not involved in considerations of colorability.

Loupekine also obtained snarks by joining an *even* number of these graphs together. Here, the matched and mismatched pairs work out, but he created snarks by attaching the central semi–edges to appropriate graphs (see [32]). It turns out that successive pairs of central semi–edges match and can easily be joined to something to make a snark. For example, if we join three successive central edges to a single vertex, and join the remaining edges in any appropriate way, then we obtain a snark. Figure 12 shows such an example. Note that this snark is cyclically 4–edge–connected and could also be derived using the dot product.

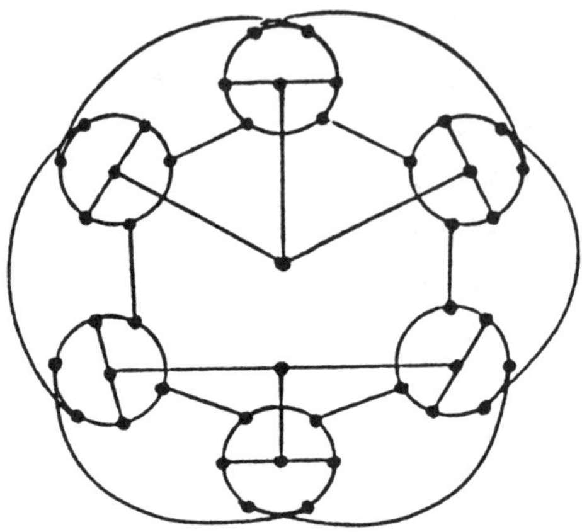

Figure 12

Another construction of Loupekine uses the notion of a 'critical' cycle. Let C be a cycle in a snark G, and construct $G - C$, keeping all the semi–edges. There are two possible cases:

(1) if $G - C$ is not 3–edge–colorable, then we can construct a variety of snarks by adding whatever we like to $G - C$;

(2) if $G - C$ is 3–edge–colorable, then we call the cycle C *critical*. An example of a critical cycle is any 5–cycle of the Petersen graph.

Loupekine [32] has found a general method for constructing snarks, which he calls *cartwheel snarks*, using an odd number of copies of $G - C$, when C is critical.

Goldberg Snarks

M. Goldberg [18] has constructed an infinite family of snarks, which can be used to give infinitely many counter–examples to the critical graph conjecture [11]. We illustrate two of these snarks in Figure 13. The Goldberg snarks are all cartwheel snarks.

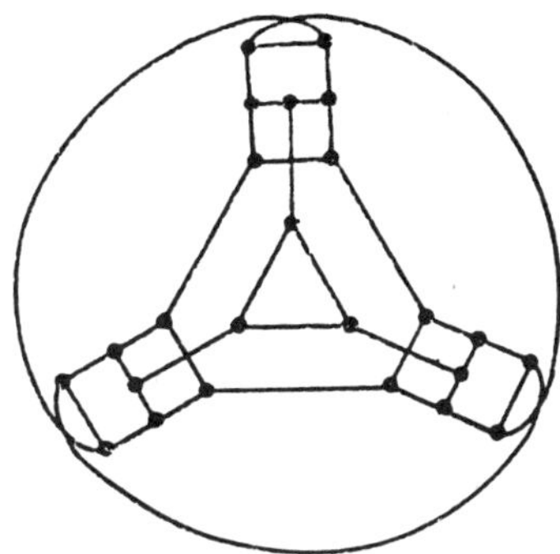 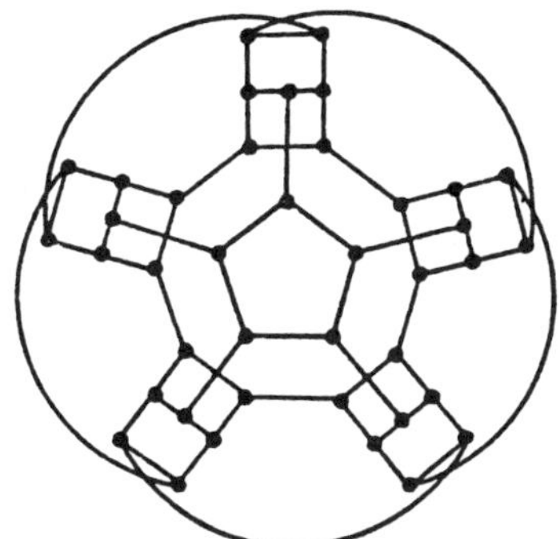

Figure 13

Celmins and Swart Snarks

Several additional snarks have been constructed by Celmins and Swart [7] from planar configurations which are not D–reducible, in the sense of the four–color theorem. Two of these snarks are shown in Figure 14.

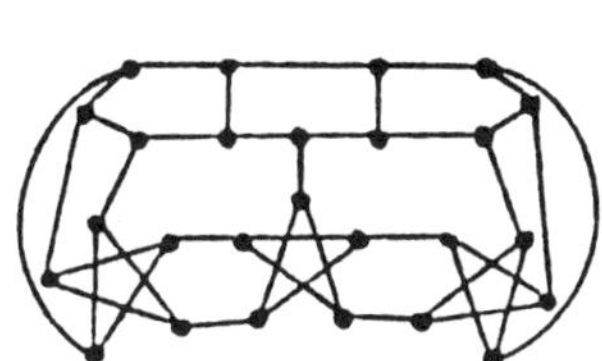 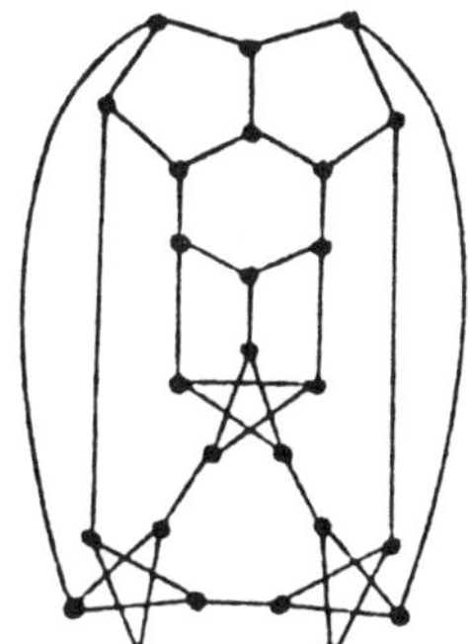

Figure 14

Watkins Snarks

In 1983, J. Watkins [31] constructed two specific families of snarks of orders 10, 18, 26, ..., containing the two Blanuša snarks. The idea is to join graphs with mutually incompatible coloring properties. We illustrate the member of order 42 from each family in Figure 15. These snarks can also be derived using the dot product.

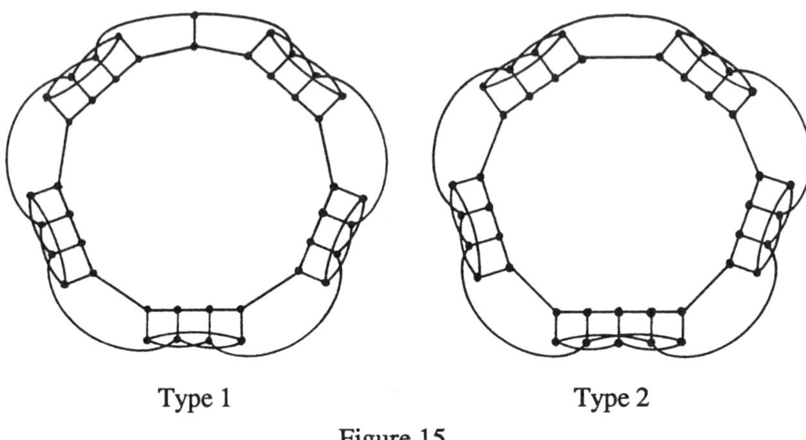

Type 1   Type 2

Figure 15

Two specific families of snarks of orders 50, 90, 130, ... were also constructed by Watkins. One family includes the Szekeres snark; the other gives a different snark of order 50, shown in Figure 16. Moreover, these families can be mixed to yield additional snarks; for example, there are six snarks of order 50 that can be formed in this way. Again, the dot product could also be used.

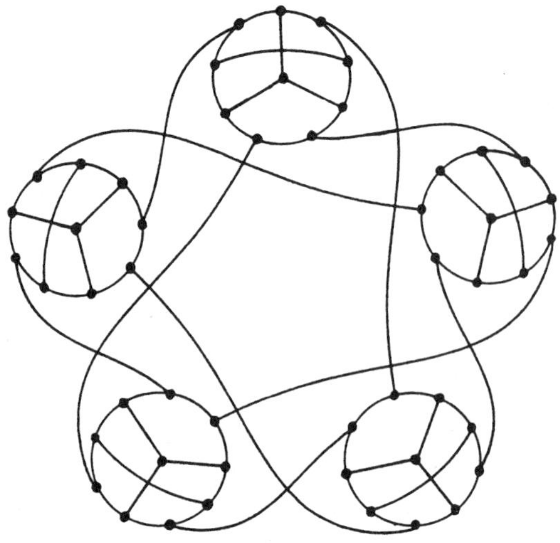

Figure 16

There are two ways of removing two non–adjacent edges from the Petersen graph. Watkins observed that replacing an edge by either of these two graphs does not affect 3–edge–colorability. This process is a special case of the dot product, but provides an easy and specific way to construct snarks. For example, if each 'spoke' of the Petersen graph (Figure 2) is replaced by one of these graphs, then the result is the Szekeres snark. If each 'spoke' is replaced by the other graph, then the result is the snark of Figure 16.

5. Small Snarks

In their 1981 paper [10], Chetwynd and Wilson listed ten snarks of order not exceeding 30. Many more have now been constructed (see, for example, Fouquet et al. [16]). The Petersen graph is the only snark of order 10, and there are no snarks of orders 12, 14 or 16 (see, for example, [6], [13], [14] and [15]). It has been shown independently by a number of people that there are just two snarks of order 18 (see, for example, M. Preissmann [26]); they were incorrectly depicted in [10]. For other small orders the situation has changed dramatically. For example, Chetwynd and Wilson presented only one snark of order 20 (the flower snark J_5) and two snarks of order 22 (Loupekine snarks). There are precisely six snarks of order 20 (see Figure 17 for the five additional ones), and precisely twenty snarks of order 22 (see [1] and [31]). A recent paper of Leizhen Cai [3] presents a snark of order 24; also included in this paper is a proof of a conjecture of Isaacs that snarks exist for all even orders greater than or equal to 18.

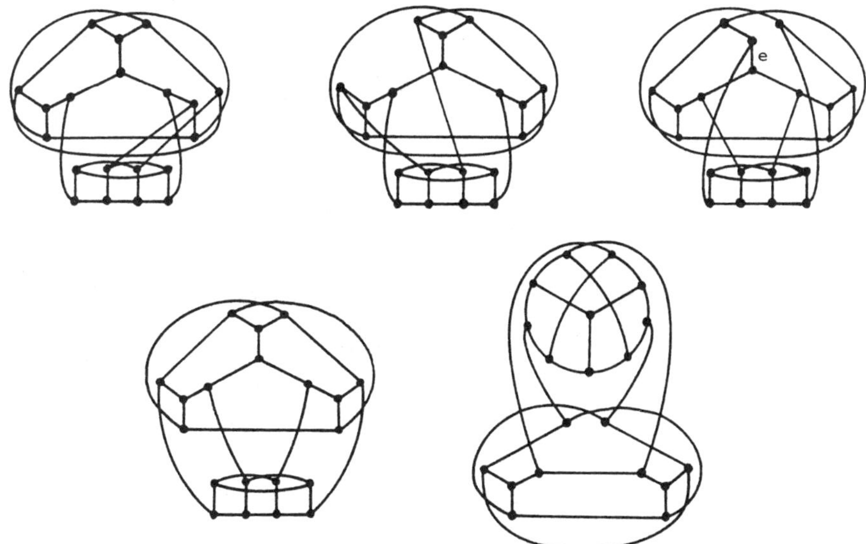

Figure 17

Computational work on small snarks has recently been carried out by F. Holroyd (unpublished). He defined an *antipodal snark* as a snark formed by identifying antipodally opposite ends of a plane cubic tree. A computer search showed that, up to order 20, the Petersen graph is the only antipodal snark. He also defined a *snarkmaker* to be an even cycle with a semi–edge incident with each vertex, and a pairing defined on the semi–edges such that, for every 3–edge–coloring of the graph, at least one of the defined pairs is colored alike; it is so–called, because if each semi–edge is joined to a new vertex, and if the new vertices are joined in *any* way to produce a cubic graph with girth not exceeding 5, then a snark is produced. A computer search showed that, up to order 14, the only snarkmaker is as shown in Figure 18 with antipodal pairing, giving rise to the Petersen graph. It will be interesting to see which other snarks are produced by this approach.

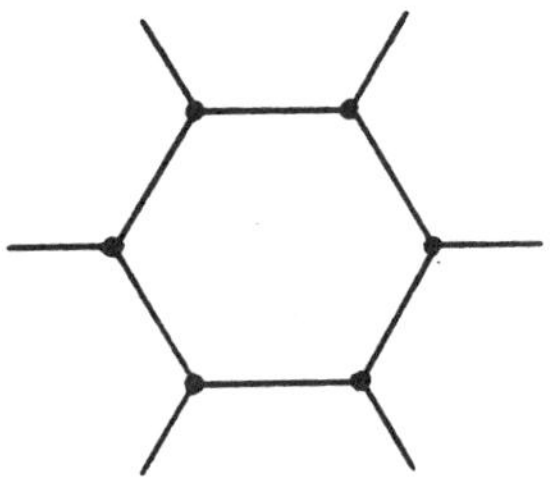

Figure 18

Holroyd has also defined a *bicycle snark* of order n; this is a snark consisting of two n–cycles and n 'spokes', each joining a vertex of one n–cycle to a vertex of the other. A computer search correctly yielded the Petersen and Blanuša snarks, confirmed the non–existence of bicycle snarks of order 14, showed the non–existence of bicycle snarks of order 22, and produced over 50 bicycle snarks of order 26.

6. k–Snarks

We have seen that a snark is essentially a 3–regular graph with edge–chromatic number 4. It is natural to generalize this idea, and define a *k–snark* (for $k \geq 3$) to be a k–regular graph with edge–chromatic number k + 1. Thus a 3–snark is simply a snark.

As with snarks, there are a number of 'trivial' graphs which we exclude from our consideration. In addition, since every k–regular graph of odd order automatically has edge–chromatic number k + 1, we exclude them and concentrate only on regular graphs of even order.

As was shown by Kotzig (unpublished), an important class of 4–snarks is obtained by taking 3–snarks and forming their line graphs. For example, the line graphs of the flower snarks of orders 12, 20, 28, ... are 4–snarks of orders 18, 30, 42, ... As shown in [10] and [13], these 4–snarks can be used to obtain further counter-examples of the critical graph conjecture. A 4–snark of order 18 is shown in Figure 19.

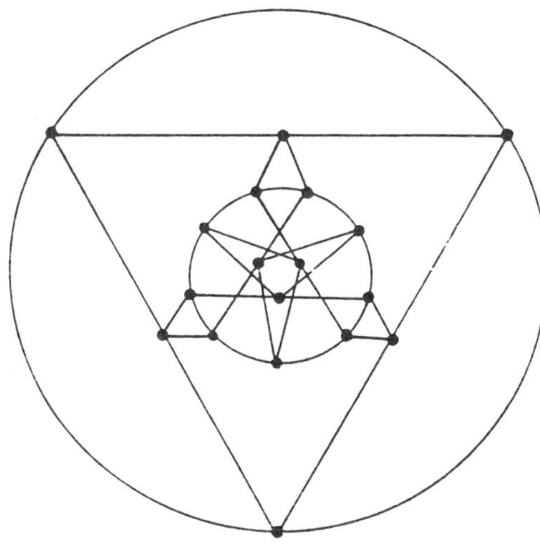

Figure 19

Meredith [24] has given a construction which yields a k–snark for each $k \geq 3$. To construct such a k–snark, we take ten copies of the complete bipartite graph $K_{k,k-1}$, and join them as in Figure 20.

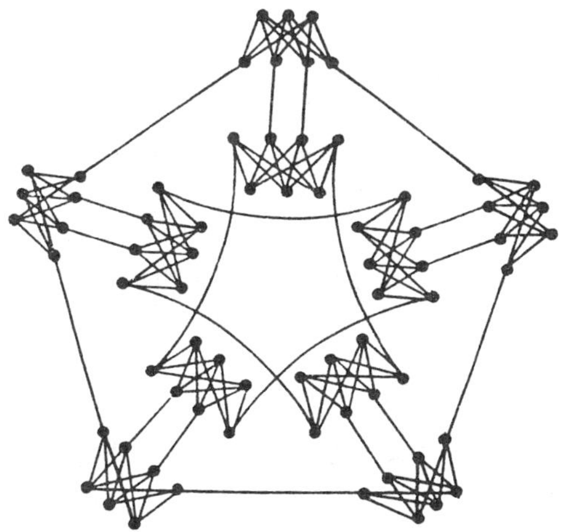

Figure 20

Some new constructions for k–snarks and multisnarks have recently been given by Chetwynd and Hilton [9]. Further information about k–snarks in general can be found in [10].

7. Open Problems

We conclude this survey by mentioning some unsolved conjectures. These conjectures have been around for some time, and appeared in [10].

The first conjecture, due to Tutte, relates to the apparent omnipresence of the Petersen graph in the study of snarks:

Tutte's conjecture. Every snark contains

    (a) a subgraph homeomorphic to the Petersen graph;
    (b) a subgraph contractible to the Petersen graph.

Another celebrated conjecture relates to the fact that all snarks so far constructed are of girth 5 or 6. For example, the Petersen graph has girth 5, whereas the flower snarks have girth 6:

The girth conjecture. Every snark has a 5–cycle or a 6–cycle.

Much of the interest and importance of snarks arises from the fact that the following general conjectures can (without loss of generality) be restricted to snarks — more precisely, any minimal counter–example must necessarily be a snark. Celmins [6] has shown that all three conjectures are true for the flower snarks and the double–star snark D(5,2):

The 5–flow conjecture. Every bridgeless graph has a 5–flow.

The double cover conjecture. Every bridgeless graph can be twice–covered with cycles.

Fulkerson's conjecture. Every cubic bridgeless graph can be twice–covered with 1–factors.

The relationships between these conjectures and their connections with snarks are discussed in the recent survey of Jaeger [21].

REFERENCES

[1]   R. E. L. Aldred, B. Sheng, D. A. Holton and G. F. Royle, All the small snarks, to appear.

[2]   D. Blanuša, Problem ceteriju boja (The problem of four colors), *Hrvatsko Prirodoslovno Društvo Glasnik Mat.–Fiz. Astr. Ser. II*, **1** (1946), 31–42.

[3]   L. Cai, A snark of order 24, to appear.

[4]   P. J. Cameron, A. G. Chetwynd and J. J. Watkins, Decomposition of snarks, *J. Graph Theory* **11** (1987), 13–19.

[5]   F. Castagna and G. Prins, Every generalized Petersen graph has a Tait coloring, *Pacific J. Math.* **40** (1972), 53–58.

[6]   U. A. Celmins, A study of three conjectures on an infinite family of snarks, *Department of Combinatorics and Optimization, University of Waterloo, research report* CORR 79–19.

[7]   U. A. Celmins, and E. R. Swart, The construction of snarks, *Department of Combinatorics and Optimization, University of Waterloo, research report* CORR 79–18.

[8]   A. G. Chetwynd, Ph.D. thesis, The Open University, England.

[9]   A. G. Chetwynd and A. J. W. Hilton, Snarks and k–snarks, *Ars Combinatoria* **25C** (1988), 39–54.

[10]   A. G. Chetwynd and R. J. Wilson, Snarks and supersnarks, *The Theory and Applications of Graphs,* John Wiley & Sons, New York, (1981), 215–241.

[11]   A. G. Chetwynd and R. J. Wilson, The rise and fall of the critical graph conjecture, *J. Graph Theory*, **7** (1983), 154–157.

[12]   B. Descartes, Network colourings, *Math Gazette*, **32** (1948), 67–69.

[13]   M. A. Fiol, Contribución a la teoria de grafos regulares, Universidad Politécnica de Barcelona, 1979.

[14]   S. Fiorini and R. J. Wilson, Edge–Colourings of Graphs, *Research Notes in Math.* **16**, Pitman, London, (1977).

[15]   J.–L. Fouquet, Note sur la non existence d'un snark d'ordre 16, *Discrete Math.* **38** (1982), 163–171

[16]   J.–L. Fouquet, J.–L. Jolivet and M. Rivière, Graphes cubiques d'indice trios, graphes cubiques isochromatiques, graphes cubiques d'indice quatre, *J. Combinatorial Theory (B)*, **31** (1981), 262–281.

[17] M. Gardner, Mathematical Games: Snarks, Boojums and other conjectures related to the four–color–map theorem, *Scientific American*, **234(:4)** (April 1976), 126–130.

[18] M. K. Goldberg, Construction of class 2 graphs with maximum vertex degree 3, *J. Combinatorial Theory (B)*, **31** (1981), 282–291.

[19] R. Isaacs, Infinite families of non–trivial trivalent graphs which are not Tait colorable, *Amer. Math. Monthly*, **82** (1975), 221–239.

[20] R. Isaacs, Loupekine's snarks: A bifamily of non–Tait–colorable graphs, unpublished.

[21] F. Jaeger, Nowhere–zero flow problems, *Selected Topics in Graph Theory*, **3** (ed. L. W. Beineke and R. J. Wilson), Academic Press, London, (1988).

[22] A. B. Kempe, A memoir on the theory of mathematical form, *Phil. Trans. Roy. Soc. London*, **177** (1886), 1–70.

[23] F. Loupekine and J. J. Watkins, Cubic graphs and the four–color theorem, *Graph Theory and its Applications to Algorithms and Computer Science*, John Wiley & Sons, New York, (1985), 519–530.

[24] G. H. J. Meredith, Regular n–valent, n–connected, non–Hamiltonian, non–n–edge–colorable graphs, *J. Combinatorial Theory (B)*, **14** (1973), 55–60.

[25] J. Petersen, Sur le théoreme de Tait, *Intermed. Math.*, **15** (1898), 225–227.

[26] M. Preissmann, Snarks of order 18, *Discrete Math.*, **42** (1982), 125–126.

[27] M. Preissmann, C–minimal snarks, *Annals of Discrete Math.*, **17** (1983), 559–565.

[28] G. Szekeres, Polyhedral decompositions of cubic graphs, *Bull. Austral. Math. Soc.*, **8** (1973), 367–387.

[29] P. G. Tait, Remarks on the colouring of maps, *Proc. Roy. Soc. Edinburgh*, **10** (1880), 729.

[30] F. C. Tinsley and J. J. Watkins, A study of snark embeddings, *Graphs and Applications*, John Wiley & Sons, New York (1985), 317–332.

[31] J. J. Watkins, On the construction of snarks, *Ars Combinatoria*, **16B** (1983), 111–123.

[32] J. J. Watkins, Snarks, *Proc. First China–U.S.A. Graph Theory Conf.*, to appear.

Graphical Combinatorial Families and Unique Representations of Integers

Herbert S. Wilf*

University of Pennsylvania

In 1978 I published ([1], [2]) two papers on a unified approach to several different algorithmic problems in combinatorics, which grew out of joint work [3] (particularly chapter 13) with A. Nijenhuis. Aside from these algorithmic uses, however, the method has purely mathematical consequences, and these have not been sufficiently brought out. So I would like to take this opportunity to review the unified approach here, this time discussing the graph theoretical side of the story, and its implications, more fully.

Let G be a strongly connected acyclic digraph. By a *terminal vertex* of G we mean a vertex of outdegree 0. We assume that G has exactly one terminal vertex, τ. We assume further that the edges of G are numbered, computer–science style. That is to say, if v is a fixed vertex, with ρ outgoing edges, then these edges are numbered 0, 1, ..., $\rho - 1$.

Such a digraph will be called a graphical combinatorial family.

Let $v \in V(G)$. By a *combinatorial object of order* v we mean a walk from v to τ. Associated with each combinatorial object of order v is its edge code word, which is the sequence of edge numbers that occur in the walk. Evidently the order v of the walk, and the code word determine the walk uniquely.

Many garden–variety combinatorial families are special cases of these structures. Take $V(G)$ to the set of lattice points (n, k) $(n \geq k \geq 0)$ in the plane. If $v = (n, k)$ with $n > k \geq 1$ then out of v there go two edges, one, numbered 0, goes to $(n - 1, k)$, and the other, numbered 1, goes to $(n - 1, k - 1)$. If $n = k \geq 1$ there is just one edge, numbered 0, and it goes to $(n - 1, k - 1)$. If $n > 0$ and $k = 0$ there is also just one edge, numbered 0, and it goes to $(n - 1, k)$. The unique terminal vertex is $\tau = (0, 0)$.

Now, for a fixed $v = (n, k)$, there is a 1–1 correspondence between the combinatorial objects of order v and the subsets of k elements chosen from n: given a

* Research supported by the United States Office of Naval Research

walk from v to (0, 0), insert into the set the x–coordinate of the initial vertex of every edge numbered 1 that occurs in the walk.

That was one example. There is an extensive list of combinatorial objects that can be brought into the fold, including set partitions, number partitions, permutations by cycles, permutations by runs, Young tableaux, etc. etc. Hence it is a worthwhile matter to study the general properties of these systems.

The original reason for studying them was to carry out four algorithms: listing, ranking, unranking, and random selection. It turns out that we can list the walks from v to τ in lex order in any such digraph. Hence we can list the objects in any such family. For example, we can make a list of all of the permutations of 19 letters that have 8 cycles, or of all of the partitions of [1..15] that have 8 classes, etc., by the same algorithm.

The ranking problem is this: given an object in a family of objects. Determine where it is on the list. More precisely, determine its position in the lex ordered list of all such objects. For instance, in the list of all 225 permutations of 6 letters that have 3 cycles, where is (14) (523) (6)?

Ranking problems can also be done by a general method, in the graph–theoretic setting of combinatorial families. The problem is the, given a walk from v to τ. Where is it in the lex ordered list of all walks from v to τ?

In order to present the new applications of the method, it will be helpful to review the ranking procedure. Suppose we are given a path from v to τ in a graphical combinatorial family, as shown in Fig. 1.

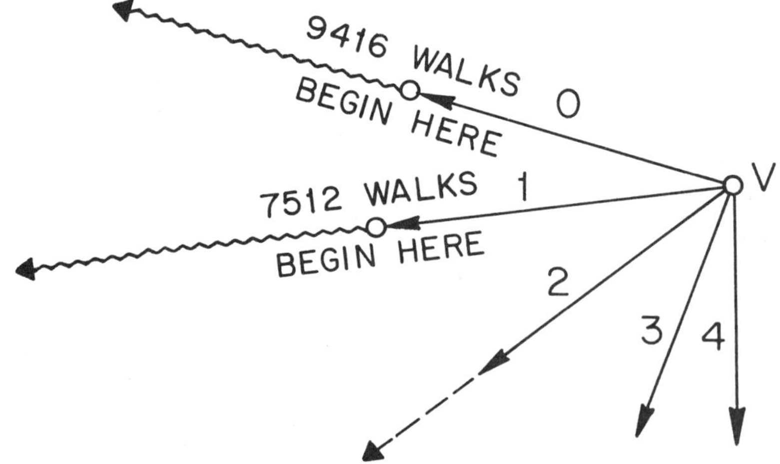

Fig. 1: This step costs $16,928.

Suppose the walk has arrived at v, and that it next uses edge number 2 outbound from v. It will then ignore all of the walks that would have used edges 0 or 1 next, instead. Furthermore, all of those walks *precede* the walk that uses edge 2 since the letters 0, 1 precede 2. Therefore, if a total of 9416 walks begin at the terminal vertex of edge 0, and if 7512 walks begin at the terminal vertex of edge 1, we will have jumped to a place in the alphabet that is below all 16, 928 of those walks. Hence, if edge 2 is in the walk that we are considering, its rank is augmented by 16, 928 on that single step.

Furthermore, the rank of a complete walk from v to τ is the sum of all of the numbers, like 16,928, that occur on every edge of the walk. We call 16,928 the *cost* of the edge. The costs of all the edges can be precomputed, without reference to some particular walk. The cost of an edge e, whose initial vertex is v and whose terminal vertex is w, is the sum of the numbers of walks from v' to τ, over all vertices v' such that $(v, v') \in E(G)$, and edge–number of edge (v, v') is < e.

If these costs have been precomputed, the *the rank of a walk is the sum of the costs of the edges in the walk.*

Corollary *In a graphical combinatorial family G, let v be a vertex, and suppose that there are exactly b(v) walks from v to τ. Then every integer r, $0 \leq r \leq b(v) - 1$ is uniquely representable as the sum of the costs on the edges of one of these walks.*

Proof   Every such integer is the rank of exactly one such walk. ❑

This observation contains as special cases a number of well known, and some new, unique representation theorems for integers.

Example 1. Fibonacci Numbers

Consider the graph G whose vertices are the nonnegative integers. There is an edge from m to $m - 1$, numbered 0, and from m to $m - 2$, numbered 1, for each $m \geq 2$, and edge (1, 0), numbered 0. The number of walks from m to 0 is F_m, and consequently *every integer $r, 0 \leq r \leq F_m - 1$ is uniquely representable as $F_{m_1} + F_{m_2} +$ ..., where all $m_{i+1} \leq m_i - 2$.* ❑

Example 2. Partitions of Integers

Define the graph G as follows. Its vertices are the plane lattice points (n, k) with $n \geq k \geq 1$ and the terminal vertex $\tau = (0, 0)$. There is an edge from (n, k) to (n - k, k), for $n \geq k \geq 1$, numbered 0, and from (n, k) to $(n - 1, k - 1)$, numbered 1. The number of walks from (n, k) to (0, 0) is then p(n, k), the number of partitions of n into parts $\leq k$, because both p(n, k) and the number of walks satisfy the same recurrence relation with the same initial conditions, viz.

$$p(n, k) = p(n - 1, k - 1) + p(n - k, k).$$

It follows that *every integer* r *such that* $0 \leq r \leq p(n, k) - 1$ *is uniquely expressible as*

$$r = p(n_0, k) + p(n_1, k - 1) + \dots$$

where, for all $j \geq 0$, $n_{j+1} = n_j - t_j(k - j) - 1$, $n_0 = n - (t + 1)k$, *and* $t, t_0, t_1, \dots$ *are nonnegative integers.* $\square$

Example 3. Binomial Coefficients

Here the vertices of G are the lattice points (n, k) with $n \geq k \geq 0$. There is an edge from each (n, k), $n > k \geq 1$, to $(n - 1, k)$, numbered 0, and an edge from (n, k) to $(n - 1, k - 1)$, numbered 1. If $n > 0$ and $k = 0$ there is one edge, numbered 0, from (n, k) to $(n - 1, k)$. The terminal vertex is $(0, 0)$. The number of walks from (n, k) to τ is $\binom{n}{k}$.

We obtain at once the fact that *every integer* r, $0 \leq r \leq \binom{n}{k} - 1$, *is uniquely of the form*

$$r = \binom{n_1}{k} + \binom{n_2}{k - 1} + \dots + \binom{n_k}{1}$$

where $n > n_1 > \dots \geq 0$. $\square$

Example 4. Stirling Numbers

Let $S(n, k)$ be the number of partitions of $[n]$ into k classes (the Stirling numbers of the second kind.) The same method yields the result that *every integer* r, $0 \leq r \leq S(n, k) - 1$ *is uniquely expressible as*

$$r = j_1 S(n - 1, k_1) + j_2 S(n - 2, k_2) + \dots$$

where for all i, $0 \leq j_i \leq k_i$, $2 \leq k_i \leq n - i$, *and*

$$k_{i+1} = \begin{cases} k_i & \text{if } j_i < k_i - 1 \\ k_i & \text{if } j_i = k_i, \end{cases}$$

and $k_1 = k$. $\square$

Similar resluts hold for Stirling numbers of the first kind, q–binomial coefficients, hook–formula numbers, etc. All of these formulas simply express the fact that in a suitable lex ordered list, every element has a unique rank.

One other area of application of the method of combinatorial families deserves mention, because of interesting results that have been obtained by Bruce Sagan [4].

Let a sequence $\{c_j\}$ count certain sets $S_1, S_2, \dots$ Suppose we want to prove that the sequence is log concave, i.e., that $c_{j+1} c_{j-1} < c_j^2$, for all j. One way to do that would be to exhibit an injection $S_{j+1} \times S_{j-1} \to S_j \times S_j$, for all j.

If the sets in question are representable as walks in a graphical combinatorial family, then we want an injection from a pair of walks with nearby initial vertices to a pair of walks with the same initial vertex. The geometric picture of walks on the plane lattice, which occurs in several important examples, opens the possibility of cutting and pasting pieces of the given pair of walks to obtain the final pair of walks, and yields injections that are natural to discuss in that context. Refer to [4] for the details, but here is one of the results that he obtained. It relates the concavity of the solution of a recurrence to the concavity of the coefficients of the recurrence.

Theorem   *(Sagan [4]) Let $\{t_{n,k}\}$ satisfy a recurrence of the form*
$$t_{n,k} = c_{n,k}t_{n-1,k-1} + d_{n,k}t_{n-1,k}   (t, c, d, \geq 0)$$
where the $c_{n,k}$ and the $d_{n,k}$ are themselves log concave in k, and further satisfy
$$c_{n,k-1}d_{n,k+1} + c_{n,k+1}d_{n,k-1} \leq 2c_{n,k}d_{n,k}.$$
Then $\{t_{n,k}\}$ is log concave in k.

As a problem for future research, it would seem that the method has the capability of producing unimodality results, where log–concavity doesn't exist. This would require injections from some S_j to an S_{j+1}, that might easily be facilitated by cutting and pasting operations suggested by the lattice pictures.

REFERENCES

[1]   Herbert S. Wilf, A unified setting for sequencing, ranking and random selection of combinatorial objects, Adv. Math. **24** (1977), 281 – 291.

[2]   Herbert S. Wilf, —, II, Ann. Discr. Math. **2** (1978), 281 – 291.

[3]   A. Nijenhuis and H.S. Wilf, Combinatorial Algorithms (2nd ed.), Academic Press, New York, 1978.

[4]   Bruce Sagan, Inductive and injective proofs of log concavity results, Discr. Math. **68** (1988), 281 – 292.

Number Theory for Graphs

Robin J. Wilson

The Open University, England

and

The Colorado College

ABSTRACT

In this paper we show how the definitions of several standard number-theoretic functions, such as the Möbius function μ *and the Euler phi-function* ϕ, *can be extended to connected graphs, using the operation of Cartesian product.*

1. The Cartesian product of graphs

Let G_1 and G_2 be finite graphs with vertex–sets $V(G_1)$ and $V(G_2)$, respectively. The *Cartesian product* $G_1 \times G_2$ is the graph with vertex–set $V(G_1) \times V(G_2)$ in which the vertices (v_1, w_1) and (v_2, w_2) are joined by an edge if and only if

either v_1 is adjacent to v_2 in G_1, and $w_1 = w_2$

or $\quad v_1 = v_2$, and w_1 is adjacent to w_2 in G_2.

For example:

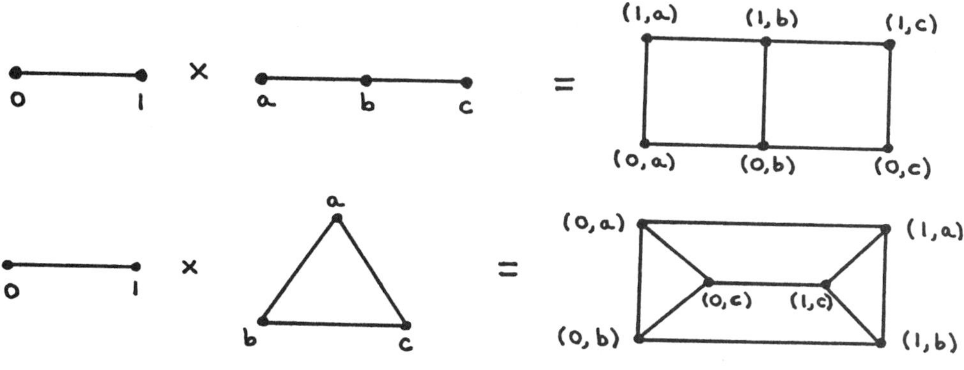

Figure 1

1151

It is easy to check that the Cartesian product $\times$ is associative and commutative, and that the one–vertex graph K_1 is an identity element for $\times$.

2. Prime graphs

A connected graph is *prime* if it cannot be decomposed into non–trivial factors with respect to $\times$. (A factor is non–trivial if it is connected and has at least two vertices.) For example, the following graphs are all prime:

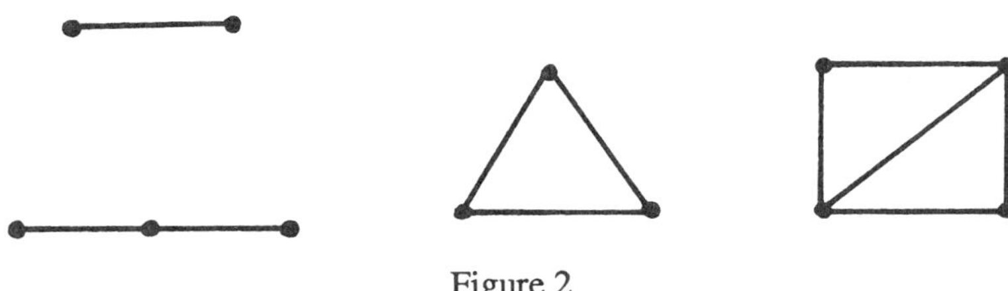

Figure 2

Note that, if G has a prime number of vertices, then G must be a prime graph, but the converse result is false. It is known (see, for example, Feigenbaum [2] or Winkler [10]) that almost all connected graphs are prime; in particular, Feigenbaum has proved that the proportion of graphs of order n that are non–trivially factorizable vanishes as 2^{-cn^2} as $n \to \infty$, where c is a positive constant.

The following table gives the numbers of connected graphs and prime graphs of order n, for $2 \leq n \leq 8$:

n	2	3	4	5	6	7	8
connected graphs	1	2	6	21	112	853	11117
prime graphs	1	2	5	21	110	853	11111

The main result on prime graphs is the following fundamental theorem, proved independently, and in various versions, by Sabidussi [8], Vizing [9], Miller [7], and Imrich [5], [6]:

Unique Factorization Theorem *Any connected graph G can be expressed uniquely as the Cartesian product of prime graphs:*

$$G = P_1^{k_1} \times P_2^{k_2} \times \ldots \times P_r^{k_r}.$$

Feigenbaum, Hershberger and Schäffer [3] and Winkler [10] have shown that there exists a polynomial algorithm for deciding whether a given connected graph G is a non–trivial Cartesian product, and for finding its prime decomposition if so. The corresponding statement for disconnected graphs is false; indeed, Feigenbaum [2] and Winkler [10] have shown that the factorizing problem for disconnected graphs is equivalent to the graph isomorphism testing problem.

3. Divisibility

If G and H are connected graphs, then H *divides* G (written $H \mid G$) if $G = H \times K$ for some graph K. The following results follow immediately from this definition; K_1 denotes, as usual, the one–vertex graph:

 (i) for all connected graphs G, $G \mid G$ and $K_1 \mid G$;

 (ii) if $H_1 \mid H_2$ and $H_2 \mid G$, then $H_1 \mid G$;

 (iii) if $H_1 \mid G_1$ and $H_2 \mid G_2$, then $H_1 \times H_2 \mid G_1 \times G_2$.

If G and H are connected graphs, we define their greatest common divisor and least common multiple in the obvious way – namely:

 $\gcd (G, H) = K$ if (i) $K \mid G$, $K \mid H$, and (ii) $L \mid G$, $L \mid H \Rightarrow L \mid K$;

 $\operatorname{lcm} (G, H) = L$ if (i) $G \mid L$, $H \mid L$, and (ii) $G \mid K$, $H \mid K \Rightarrow L \mid K$.

We can then derive such results as

 $\gcd (G, H) \times \operatorname{lcm} (G, H) = G \times H.$

If $\gcd (G, H) = K_1$, then G and H are *relatively prime*.

Just as for the set of positive integers, the set of connected graphs can be represented as a distributive lattice under divisibility, as shown below. It is the direct product of chains of the form $K_1 - P - P^2 - P^3 - \ldots$, where P is a prime graph.

Note that, if P is a prime graph, and if $P \mid G \times H$, then $P \mid G$ or $P \mid H$. Also, if $G = P_1^{k_1} \times \ldots \times P_r^{k_r}$, and if $P \mid G$, then $P = P_i$ for some $i = 1, \ldots, r$.

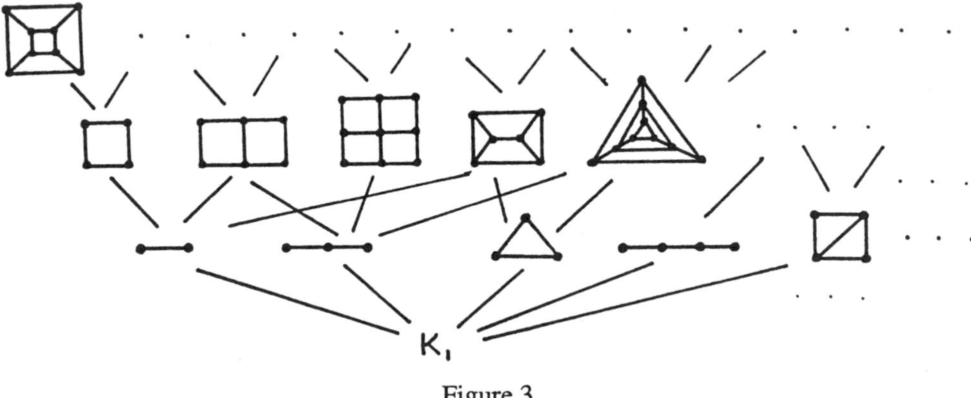

Figure 3

4. Multiplicative functions

An *arithmetical function* is a function f that associates with each connected graph G a real number f(G). Such a function f is *multiplicative* if $f(G \times H) = f(G)\, f(H)$, whenever gcd $(G, H) = K_1$.

Examples of multiplicative functions are the following:

(1) *The Möbius function*

We define the *Möbius function* μ by

$$\mu(G) = \begin{cases} 1, \text{ if } G = K_1 \\ 0, \text{ if } P^2 \,|\, G, \text{ for some prime graph } P \\ (-1)^r, \text{ if } G = P_1 \times ... \times P_r \text{ (distinct primes).} \end{cases}$$

It is then a simple matter to derive the following basic properties of the Möbius function; we prove only (ii):

(i) μ *is multiplicative;*

(ii) $\displaystyle\sum_{H \,|\, G} \mu(H) = \begin{cases} 1 & \text{if } G = K_1 \\ 0 & \text{if } G \neq K_1. \end{cases}$

Proof Since μ is multiplicative, it is sufficient to prove the result when G is a prime power, $G = P^k$. In this case,

$$\sum_{H|G} \mu(H) = \mu(1) + \mu(P) + ... + \mu(P^k)$$

$$= 1 - 1 + 0 + ... + 0$$

$$= 0, \text{ as required.}$$

(iii) *(The Möbius inversion formula) If f is an arithmetical function, and if*

$$F(G) = \sum_{H|G} f(H), \text{ then } f(G) = \sum_{H|G} \mu(G/H) \, F(H),$$

where G/H is the graph K defined by $G = H \times K$.

(2) *The divisor function*

We define the *divisor function* $\tau(G)$ to be the number of graphs which divide G; thus, $\tau(G) = \sum_{H|G} 1$. It follows that τ is multiplicative, and that

$$\text{if } G = P_1^{k_1} \times ... \times P_r^{k_r}, \text{ then } \tau(G) = (k_1 + 1) \, ... \, (k_r + 1).$$

We can also use the multiplicativity to prove results such as the following:

(i) $\sum_{H|G} \mu(H) \, \tau(H) = (-1)^r$, and $\sum_{H|G} \mu(G/H) \, \tau(H) = 1$;

(ii) $\sum_{H|G} \tau(H)^3 = (\sum_{H|G} \tau(H))^2.$

Proof of (ii) Since τ is multiplicative, it is sufficient to prove the result when G is a prime power, $G = P^k$. In this case,

$$\sum_{H|G} \tau(H)^3 = \tau(1)^3 + \tau(P)^3 + ... + \tau(P^k)^3$$

$$= 1^3 + 2^3 + ... + (k + 1)^3$$

$$= (1 + 2 + ... + (k + 1))^2$$

$$= (\tau(1) + \tau(P) + ... + \tau(P^k))^2$$

$$= (\sum_{H|G} \tau(H))^2.$$

(3) *The Liouville function*

If $G = P_1^{k_1} \times ... \times P_r^{k_r}$, we define the *Liouville function* λ by $\lambda(G) = (-1)^{k_1 + ... + k_r}$. It follows that λ is multiplicative, and that

(i) $\sum\limits_{H|G} \lambda(H) = \begin{cases} 1 & \text{if } G = K^2 \text{ for some graph } K \\ 0 & \text{if not.} \end{cases}$

(ii) $\sum\limits_{H|G} \mu(H)\, \lambda(H) = 2^r.$

There are, however, no obvious analogues of the Euler ϕ–function and the sum–of–divisors function σ. In order to define these, we introduce the idea of a norm function.

5. Norm functions

A *norm function* is a function N that associates with each connected graph G a positive real number $N(G)$, satisfying

$$N(K_1) = 1, \text{ and } N(G_1 \times G_2) = N(G_1)\, N(G_2) \text{ for all graphs } G_1, G_2.$$

It follows that $N(P_1^{k1} \times \ldots \times P_r^{kr}) = N(P_1)^{k1} \ldots N(P_r)^{kr}$, and hence that a norm function is specified completely by its effect on prime graphs. We normally assume that $N(G) > 1$ if $G \neq K_1$.

Examples of possible norm functions (some of which were suggested by M. Petkovsek and P. Winkler) are:
  (i) $N(G) =$ the number of vertices of G;
  (ii) $N(G) = 2^d$, where d is the average vertex–degree in G;
  (iii) $N(G) = 2^\Delta$, where Δ is the maximum vertex–degree in G;
  (iv) list all the prime graphs in some order, and assign 2 to the first prime graph P_1, 3 to the second prime graph P_2, and so on; then define $N(G) = 2^{k1}\, 3^{k2} \ldots$ for $G = P_1^{k1} \times P_2^{k2} \times \ldots$.

Once we have found a norm function for graphs, we can easily define the Euler phi–function and the sum–of–divisors function as follows:

(1) The Euler phi–function ϕ

Rather than defining $\phi(G)$ to be the number of graphs H such that $N(H) < N(G)$ and $\gcd(G, H) = K_1$, generalizing the usual definition of the Euler phi–function $\phi(n)$, we choose to generalize the results

$$\sum\limits_{d|n} \phi(d) = n \text{ and } \phi(n) = \sum\limits_{d|n} \mu(n/d)\, d \text{ (obtained by Möbius inversion).}$$

We therefore define

$$\phi(G) = \sum\limits_{H|G} \mu(G/H)\, N(H).$$

It follows that ϕ is multiplicative, and that

(i) $\sum\limits_{H|G} \phi(H) = N(G)$;

(ii) $\phi(P^k) = N(P^k) - N(P^{k-1})$;

(iii) $\phi(P_1{}^{k_1} \times \ldots \times P_r{}^{k_r}) = N(G) \prod\limits_{i} \{1 - (1/N(P_i))\}$.

(2) *The sum–of–divisors function* σ

We define the *sum–of–divisors function* $\sigma(G)$ by

$$\sigma(G) = \sum\limits_{H|G} N(H).$$

It follows that σ is multiplicative, and that

$$\sigma(P_1{}^{k_1} \times \ldots \times P_r{}^{k_r}) = \prod\limits_{i} \{(N(P_i)^{k_i+1} - 1)/(N(P_i) - 1)\}.$$

6. Dirichlet series

If f is an arithmetical function, we define its formal *Dirichlet series* F by

$$F(s) = \sum\limits_{G} f(G) N(G)^{-s}.$$

(We are not concerned here with questions of convergence.) For example, taking $f(G) = 1$ for all G gives the *Riemann zeta function*

$$\zeta(s) = \sum\limits_{G} N(G)^{-s}.$$

When f is completely multiplicative, we also obtain the *Euler product formula*

$$F(s) = \prod\limits_{P} (1 - f(P) N(P)^{-s})^{-1},$$

where the product extends over all prime graphs P; for example,

$$\zeta(s) = \prod\limits_{P} (1 - N(P)^{-s})^{-1}.$$

The *product* of two Dirichlet series

$$F(s) = \sum\limits_{G} f(G) N(G)^{-s} \quad \text{and} \quad G(s) = \sum\limits_{G} g(G) N(G)^{-s}$$

is the Dirichlet series H defined by

$$H(s) = \sum_G h(G)\, N(G)^{-s}, \text{ where } h(G) = \sum_{H|G} f(H)\, g(G/H).$$

We then have the following results:

(i) $1/\zeta(s) = \sum_G \mu(G)\, N(G)^{-s};$

(ii) $\zeta(s-1)/\zeta(s) = \sum_G \phi(G)\, N(G)^{-s};$

(iii) $\zeta^2(s) = \sum_G \tau(G)\, N(G)^{-s};$

(iv) $\zeta(2s)/\zeta(s) = \sum_G \lambda(G)\, N(G)^{-s};$

(v) $\zeta(s)\, \zeta(s-1) = \sum_G \sigma(G)\, N(G)^{-s}.$

The proofs of these results are elementary, and depend on the results of Sections 4 and 5; as an example, we prove (ii):

Proof of (ii) $\quad \zeta(s) \sum_G \varphi(G)\, N(G)^{-s} = \sum_G \varphi(G)\, N(G)^{-s} \cdot \sum_G N(G)^{-s}$

$$= \sum h(G)\, N(G)^{-s}, \text{ where } h(G) = \sum_{H|G} \phi(H)$$

$$= \sum N(G)\, N(G)^{-s}, \text{ by Section 5(1)(i)}$$

$$= \sum N(G)^{1-s} = \zeta(1-s).$$

7. Conclusion

In this paper we have shown how many of the standard definitions and results in number theory can be extended to connected graphs with a norm function defined on them. Using these ideas, we can obtain graph–theoretic analogues of many results in number theory, such as the prime number theorem. We hope to develop these ideas in future papers.

REFERENCES

All of the number–theoretic results in this paper can be found in [1] or [4].

[1]   D.M. Burton, *Elementary Number Theory*, Allyn and Bacon, Boston, 1980.

[2]   J. Feigenbaum, Product Graphs: Some Algorithmic and Combinatorial Results, Stanford University Technical Report STAN–CS–86–1121, Ph.D. Thesis, 1986.

[3]   J. Feigenbaum, J. Hershberger and A. Schäffer, A polynomial time algorithm for finding the prime factors of Cartesian–product graphs, *Discrete Appl. Math.* 12 (1985), 123–130.

[4]   G.H. Hardy and E.M. Wright, *An Introduction to the Theory of Numbers* (fifth edition), Oxford University Press, Oxford, 1979.

[5]   W. Imrich, Über das schwache Kartesiche Produkt von Graphen, *J. Combinatorial Theory* 11 (1971), 1–16.

[6]   W. Imrich, Embedding graphs into Cartesian products, to appear.

[7]   D.J. Miller, Weak Cartesian product of graphs, *Colloq. Math.* 21 (1970), 55–74.

[8]   G. Sabidussi, Graph multiplication, *Math Z.* 72 (1960), 446–457.

[9]   V.G. Vizing, The Cartesian product of graphs (Russian), *Vyčisl. Sistemy* 9 (1963), 30–43; English translation in *Comp. Elec. Syst.* 2 (1966), 352–365.

[10] P. Winkler, Factoring a graph in polynomial time, *Europ. J. Combinatorics* 8 (1987), 209–212.

THE DESTRUCTIBILITY NUMBER OF A VERTEX

P. A. Winter

University of Natal

Durban, South Africa

ABSTRACT

*A connected graph G of order $p = |V|$ and size $q = |E|$ is said to be a_i, b_i)-destructible (with respect to E_i and V_i say) if a_i, b_i are integral factors of p and an a_i-set of edges E_i exists whose removal from G results in exactly b_i components isomorphic to K_1, i.e. whose removal from G isolates the vertices in a b_i-set V_i. Let G posses exactly m distinct (a_i, b_i)-destructions, each with respect to E_i and V_i and let $\overline{V} = V \backslash \bigcup_{i=1}^{m} V_i$. If $v \in V$, then the number of such distinct (a_i, b_i)-destructions with $v \in V_i$ is called the destructibility number of v and denoted by $dest_G(v)$. We define $dest_G(v) = 0$ if and only if $v \in \overline{V}$. The graph G is said to be destruction-regular of degree k if $dest_G(v) = k$ for each $v \in V$. In this paper we consider destruction-regular graphs of degree 0 and 1. We show that the only **regular** graphs which are destruction-regular of degree 1 are cycles of prime order and K_2 which is equivalent to a characterization of regular graphs reconstructible from their collection of uniquely destruction-labeled vertices.*

1. Introduction

The terminology and notation of [2] will be used throughout this paper. We will let G denote a simple, connected graph of order $p = |V|$ and size $q = |E|$. A cycle of length n will be denoted by C_n and, for $n \geq 2, P_n$ will denote a path of length n. The length of the longest path in G will be denoted by $\ell(G)$. Finally $\Delta(G) = max\{d_G(v)|v \in V\}, \delta(G) = min\{d_G(v)|v \in V\}$.

The end, or absence of a proof, will be denoted by $\square$.

2. Graphs which are (a,b)-destructible

The definition and subsequent investigation of the concept of (a, b)-destructibility of simple graphs was partially motivated by consideration of the reconstruction problem.

The Reconstruction Conjecture, first formulated by S.M. Ulam in collaboration with P. J. Kelly in 1941, may be stated as follows:

Every simple (finite) graph G on at least three vertices is determined, up to isomorphism, by its collection of vertex-deleted subgraphs, i.e. by the collection of all subgraphs $G_i = G - v_i, v_i \in V(G)$.

If G is uniquely determined by its collection of vertex-deleted subgraphs we say that G is *reconstructible* from this collection of subgraphs.

Advances made towards the solution of the problem and variations thereof have been extensively surveyed (cf. [1], [2], [4]). One variation involves the reconstruction of G from its n-vertex-deleted subgraphs, $n \geq 2$. Another variation involves the reconstruction of G from its m-edge- deleted subgraphs).

If we remove edges *and* vertices from K_2 we obtain $\emptyset$ (which is not a graph) and since K_2 is the only (simple) graph on one edge and two vertices, K_2 is uniquely reconstructible from $\emptyset$ together with the labels 1(edge) and 2(vertices) (i.e. $\emptyset$ labeled 1,2). Similarly K_3 is uniquely determined by $\emptyset$ labeled 3,3. The graph $K_n + \{u\} + \{uv\}(v \in V(K_n))$, is uniquely reconstructible from K_n labeled 1,1. The graph $K_4 - e(e \in E(K_4))$ is uniquely reconstructible from K_2 labeled 4,2 and K_3 labeled 2.1. The labels 1,2; 3,3; 1,1; 4,2; 2,1 involve factors of the order of each graph involved so that in [6] a further variation of the problem was introduced by considering the reconstruction of a graph G of order $p \geq 2$ from the subgraphs obtained by deleting a edges and b vertices where we imposed upon a, b the condition that they be factors of p - thus also providing a necessary bound on the number of edges and vertices removed which enabled us to make advances towards the solution of Ulam's problem. This concept generalizes the deletion of an end vertex and gave rise to a structural characterization of a particular class of prime-order graphs (cf. [7]), (m,n)- matroids (cf. [5]) and a new vertex chromatic number (cf. [8]). In this paper we consider a reconstruction problem involving labeled vertices.

Formally, if a, b are integral factors of p, G is defined to be (a, b)-*destructible* (with respect to E' and V') if an a-set E' of edges exists such that $G - E'$ contains exactly b isolated vertices (vertices incident with no edge of G) in a b-set V'. We denote $G - E' - V'$ (to $\oslash$) by $D_{a,b}(G)$ if $b \neq p$ (or $b = p$, respectively). The operation $D_{a,b}$ is called an *(a,b)-destruction* of G. The graph G is *stable* if it is no *(a,b)*- destructible for any a, b which divide p (cf. [6], [7]).

We say that the graph G is *immediately annihilable* if it is (q, p)-destructible. Clearly the only immediately annihilable graphs are those simple, connected graphs obtained by the insertion of a single edge into a tree of order $p \geq 3$ (graphs of *nullity one*), and K_2.

3. A Reconstruction Problem

Since each vertex v of a graph G can either be isolated by some (a, b)-destruction (and then removed) or not isolated by a destruction at all, we can associate each vertex with either the label a,b or the number 0, respectively. By assigning either this label a,b or 0 or v we say that v has been *non-zero destruction -labeled* or *zero destruction -labeled*, respectively. A vertex v is non-zero destruction -labeled for each distinct destruction which isolates v. Note that a vertex can have many (not necessarily different) non-zero destruction-labels associated with it and not vertex and have both a zero and a non-zero destruction-label. We now have the following reconstruction problem:

Determine those graphs which are reconstructible from their collection of destruction-labeled vertices.

In this paper we shall determine those regular graphs which are reconstructible from their collection of uniquely non-zero destruction-labeled vertices. We formally discuss the above concept of destruction- labeled vertices in the next section.

4. The Destructibility Number of a Vertex

Let G possess exactly m distinct (a_i, b_i)-destructions each with respect to E_i and V_i and let $\overline{V} = V \setminus \cup_{i=1}^m V_i$. For a vertex v of G, the number of such distinct (a_i, b_i)-destructions with $v \in V_i$ is called the *destructibility number* of v and is denoted by $dest_G(v)$ (this number is the number of non-zero destruction-labels associated with v). We define $dest_G(v) = 0$ if and only if $v \in \overline{V}$. The graph G is said to be *destruction-regular* of degree k if $dest(v) = k$ for each $v \in V$. We say that G is destruction-regular if G is destruction-regular of degree k for some k i.e., for $k \neq 0$ each vertex of G has the same number of non- zero destruction-labels associated with it and for $k = 0$ each vertex has a zero destruction-label.

The following results are easily verified:

(1) A graph G is destruction-regular of degree 0 if and only if G is stable.

(2) The graph K_2 is destruction-regular of degree 1.

(3) All cycles are destruction-regular.

(4) Cycles of prime order are destruction-regular of degree 1.

(5) Complete graph K_n are destruction-regular of degree $\frac{(n-1)(n-2)}{2}$ for $n \geq 3$.

Conjecture 4.1 The only graphs which are destruction-regular of degree 1 are cycles of prime order and K_2.

or

Conjecture 4.2 The only graphs which are reconstructible from their collection of uniquely non-zero destruction labeled vertices are cycles of prime order and K_2.

We shall provide a partial solution to these conjectures.

Lemma 4.1   No graph on $p \geq 3$ vertices containing an end vertex can be destruction-regular.

Proof   Let G have an end vertex v adjacent to a vertex w. Clearly $d(w) \geq 2$. Since every (a_i, b_i)-destruction of G which isolates w isolates v as well, we have that $dest(v) \geq dest(w)$. However v can be isolated by a $(1,1)$-destruction which does not isolate w so that $dest(v) > dest(w)$. Hence G cannot be destruction-regular.

Corollary 4.1   No tree on $p \geq 3$ vertices can be destruction-regular.

Lemma 4.2   The only graphs of nullity one which are destruction-regular of degree 1 are cycles of prime order.

Proof   Let G be a graph of nullity one which is destruction-regular of degree 1. Clearly G must be a cycle since by lemma 4.1 we have that G cannot have an end vertex. If G is a cycle of non-prime order than $dest(v) \geq 2$ for each $v \in V$ since G is immediately annihilable and from [3] each vertex of G is $(a, 1)$-destructible for some a which divides p. Thus G must be a cycle of prime order.

Lemma 4.3   Let G be a graph on $p \geq 4$ vertices which possesses an $(a, 1)$-destruction $D_{(a,1)}$ associated with E' and $V' = \{v\}$ satisfying $d(v) < a$ and let F be the set of edges covered by v. If $dest(v) = 1$, then $D_{(a,1)}(G)$ is the disjoint union of stars $S_i, i = 1, 2..., h; h \geq 2$, such that

(i) If $p(S_j) > 2$, no edge of $E' \backslash F$ is incident with an end vertex of S_j, and

(ii) The graph G cannot contain a cycle whose edges are alternately edges are alternately edges from $E' \backslash F$ and the S_j.

Proof   Let G be as above and let $H = D_{(a,1)}(G) = G - E' - v$ with $dest(v) = 1$. Note that $d(v) \geq 2$, for if $d(v) = 1$ then G will possess two distinct destructions which isolate v. Let F be the set of edges incident with v in G.

Suppose that H contains a path $v_1, v_2, \cdots, v_n, n \geq 4$. Then G is $(a,1)$- destructible with respect to E'' and V' where $E'' = (E' \setminus \{e\}) \cup \{v_2 v_3\}$, where $e \in (E' \setminus F) \neq \emptyset$. This means that $dest(v) \geq 2$, which is a contradiction. Thus $\ell(H) \leq 2$ so that H is the disjoint union of star graphs $S_i, i, 2, ..., h$.

(i) Let $p(S_j) > 2$ from some j. Suppose there is an edge $e \in E' \setminus F$ incident with an end vertex, w say, of S_j, and let z be the center of S_j. Then G is $(a,1)$-destructible with respect to $(E' \setminus \{e\}) \cup \{wz\}$ and V', a contradiction.

(ii) Suppose G contains a cycle C_n whose edges are alternately in $E' \setminus F$ and the S_j. Then n must be even, say $n = 2t$. Let $E(C_{2t}) = \{f_1, f_2, \cdots, f_t, e_1, e_2, \cdots e_t\}$, where $e_i \in E' \setminus F$ and $f_i \in H$. Then G is $(a,1)$-destructible with respect to $E' \setminus \{e_1, \cdots, e_t\}) \cup \{f_1, \cdots, f_t\}$, and V', which is a contradiction.

Let $h = 1$. Then $p(S_1) > 2$, for if $p(S_1) = 2$ we have that $p(G) = 3$, a contradiction. Thus we must have at least one edge of $E' \setminus F$ incident with an end vertex of S_1 which is impossible by case (i) above. Thus $h \geq 2$.

5. Regular Graphs Which Are Destruction-Regular of Degree 1

From [3] we know that every graph on a non-prime number p of vertices is $(a,1)$-destructible for some a which divides p. Thus cycles of non-prime order are destruction-regular of degree k where $k \geq 2$ and cycles of prime order are destruction-regular of degree 1. Complete graphs of order $p \geq 4$ cannot be destruction-regular of degree 1. We shall show that the only regular graph of even order which is destruction-regular of degree 1 is K_2.

Lemma 5.1   Let G be a regular graph of even order $p \geq 4$. If G has an $(a,1)$-destruction $D_{a,1}$ associated with E' and $V' = \{v\}$ with $d(v) = a$, then $dest(v) \geq 2$.

Proof   Let G be a regular graph of even order $p \geq 4$ and which possess an $(a,1)$-destruction $D_{a,1}$ with respect to E' and $V' = \{v\}$.

Let $D_{a,1}(G) = H_1$ and suppose that $d(v) = a$. Clearly $a < p(G)$, so that, because a is a factor of $p, a \leq p/2$. If $a = 2$ then G is a cycle and $des(v) \geq 2$. Thus $a \geq 3$.

Case (A). Let $a = p/2$. There exists at least two nonadjacent vertices v, w of G such that $G - v - w$ contains no isolated vertices. Thus G is also $(2a, 2)$-destructible and $dest(v) \geq 2$.

Case (B). suppose $a < p/2$. Let $s = a + t$ be the next divisor of p after a.

The removal of an edge e_1 from $H_1 = G - v$ does not isolate any vertices of H_1 so that if $(a + 1)$ is a factor of p then $dest(v) \geq 2$.

We shall show that there exists a set of t edges of H_1 whose removal from H_1 does not isolate any vertices of H_1 and hence of G.

If $p = 4$, then $a < 2$ which is impossible. If $p = 6$, then $a \leq 2$ which is a contradiction. Clearly the cases $p = 8$ and $p = 10$ cannot occur. Thus $p \geq 12$. Let us remove an edge e_1 from H_1 to obtain H_2. Let

$$A_2 = \{w | d_{H2}(w) = 1\},$$

$$B_2 = \{w | d_{H2}(w, A_2) \leq 1\}.$$

The set B_2 consists of vertices of H_2 which are either end vertices of H_2 or adjacent to end vertices of H_2. Let $\overline{B_2} = V_{(2)} \backslash B_2$. Since $|A_2| \leq 2$ and $|B_2| \leq 4, \overline{B_2} = p - 1 - |B_2| \geq p - 1 - 4 \geq 7$. Thus there exists a non-end-vertex y_2 of H_2 adjacent to a vertex x_2 of H_2 such that $x_2 \notin A_2$. If we let $e_2 = x_2 y_2$, then the removal of e_2 from H_2 does not isolate any vertices of H_2. Form $H_i, A_i, B_i, \overline{B_i}$ and $e_i = x_i y_i$. If $i < t$, form $H_{i+1} = H_i - e_i, A_{i+1}$ etc. and stop when $i + 1 = t$. Clearly $|A_i| \leq (i - 1)2, |B_i| \leq (i - 1)4$ so that

$$|\overline{B_t}| \geq p - 1 - 4(t - 1)$$

$$= p - 1 - 4t + 4$$

$$= p - 4t + 3.$$

If $p \geq \frac{4a-2}{3}$, then, since $a + t \leq p/2$ implies that $-t \geq \frac{2a-p}{2}$ we have that

$$|\overline{B_t}| \geq p - 4(\frac{2a - p}{2}) + 3$$

$$= 3p - 4a + 3 \geq 1.$$

If $p < \frac{4a-2}{3}$ then we must have $a + t \leq 2a$ otherwise

$$4a \leq p \leq \frac{4a - 2}{3}$$

which is impossible. Hence $t \leq a$ so that $p \geq 4t$. It follows that $|\overline{B_t}| \geq p - 4t + 3 \geq 3$.

Thus there exists a t-set of edges $E_t = \{e_1, e_2, \cdots, e_t\}$ whose removal from H_1 does not isolate any vertices of H_1 so that G is $(a + t, 1)$-destructible where $t \geq 1$. Hence $dest(v) \geq 2$.

Lemma 5.2   Let G be a regular graph of even order which possesses an $(a, 1)$-destruction with respect to E' and $V' = \{v\}$ such that $d(v) < a$. Then $dest(v) \geq 2$.

Proof   Let G be a regular graph of even order $p \geq 4$ as stated above with $dest(v) = 1$. Then by lemma 4.3 we have that $H = G - E' - v$ is the disjoint union of star graphs. Let F be the set of edges of G covered by the vertex v. Then there exists at least one edge e of $E'\backslash F$ whose one end u is adjacent to an end vertex w of H where uw is an edge of H. The graph G is now $(a, 2)$-destructible with respect to $(E'\backslash\{e\}) \cup \{uw\}$ and $\{v, w\}$ so that $dest(v) \geq 2$.

Corollary 5.1   The only regular graph of even order which is destruction-regular of degree 1 is K_2.

We shall now consider regular graphs of odd order.

Lemma 5.3   Let G be a regular graph of odd order. Then G cannot possess an $(a, 1)$-destruction with respect to E' and $V' = \{v\}$ such that $d(v) = a$.

Proof   Let G be a regular graph of odd order which possess an $(a, 1)$-destruction with respect to E' and $\{v\}$ such that $d(v) = a$. Then, since

a must be a factor of p, a must be odd. But this means that G contains an odd number of vertices of odd degree which is a contradiction.

Lemma 5.4 Let G be a regular graph of odd order which possesses an $(a,1)$- destruction with respect to E' and $V' = \{v\}$ such that $d(v) < a$. Then $dest(v) \geq 2$.

Proof Let G be a regular graph defined above and suppose $dest(v) = 1$. Let F be the set of edges covered by v. From lemma 4.3 we have that $H - G - E' - v$ must be the union of disjoint stars $S_1, S_2, \cdots, S_h$ where $h \geq 2$. If one of the star graphs has two edges in H then the end vertices of this star graph must be adjacent to v in G and no edge of $E' \backslash F$ can be incident in G with any of these end vertices so that G is a cycle which is a contradiction. Thus G is the disjoint union of K_2 components. If we let $H' = G - v$, then $\delta(H') \geq 2$ since $\delta(G) \geq 3$. Let Q be any component of H'. Every vertex of Q must be an end of exactly one K_2 component of H so that Q, and hence G, must contain an even cycle whose edges are alternately in $E' \backslash F$ and the K_2 components of H which contradicts lemma 4.3. Hence $dest(v) \geq 2$.

Corollary 5.2 The only regular graphs of prime order $p \geq 3$ which are destruction-regular of degree 1 are cycles of prime order.

Proof Let G be a regular graph of prime order $p \geq 3$. If G is a cycle then G is destruction-regular of degree 1. Let $\delta(G) \geq 3$. Then G is either stable, in which case G is destruction-regular of degree 0, or G can only be $(p,1)$- destructible with respect to E' and $\{v\}$ say, in which case $dest(v) \geq 2$ by the above lemma.

Theorem 5.1 The only regular graphs which are destruction-regular of degree 1 are cycles of prime order and K_2, i.e. the only regular graphs which are reconstructible from their collection of uniquely non-zero destruction-labeled vertices are cycles of prime order and K_2.

[1] Capobianco, M. and Moluzzo, J. C. *Examples and Counterexamples in graph theory*, Elsevier North-Holland, New York (1978).

[2] Chartrand, G. and Lesniak-Foster L., *Graphs and Digraphs*, Wadsworth, Monterey (1986).

[3] Goddard W. and Winter P. A. *All graphs on a non-prime number of vertices are destructible*, Quaestiones Math. 8 (1986), 381-385.

[4] Harary, F., *On the reconstruction of a graph from a collection of subgraphs*, Theory of Graphs and its Applications (Proceedings of the Symposium held in Prague, 1964), edited by M. Fiedler, Czechoslovak Academy of Sciences, Prague, 1964, 47-52, reprinted, Academic Press. New York.

[5] Swart, Henda C. and Winter, P. A. *The reconstruction of graphs from (m,n)- matroids*, Journal of Combinatorics, Information and System Sciences, 11 (1986), 115-123.

[6] Winter P. A. and Swart, Henda C. *On (α, β)-destructible graphs*, Quaestiones Math. 7 (1984), 161-178.

[7] Winter, P. A. and Swart, Henda C. *On stable graphs*, Quaestiones Math. 7 (1984), 397-405.

[8] Winter, P. A. *On the (a_i, b_i)-chromatic number of a graph*, ARS Combinatoria, Vol. 23A (1987), 337-348.